Organophosphorus Chemistry 2018

Organophosphorus Chemistry 2018

Special Issue Editor

György Keglevich

MDPI • Basel • Beijing • Wuhan • Barcelona • Belgrade • Manchester • Tokyo • Cluj • Tianjin

Special Issue Editor
György Keglevich
Department of Organic
Chemistry and Technology,
Budapest University of
Technology and Economics
Hungary

Editorial Office
MDPI
St. Alban-Anlage 66
4052 Basel, Switzerland

This is a reprint of articles from the Special Issue published online in the open access journal *Molecules* (ISSN 1420-3049) (available at: https://www.mdpi.com/si/molecules/Organophosphorus_Chemistry).

For citation purposes, cite each article independently as indicated on the article page online and as indicated below:

LastName, A.A.; LastName, B.B.; LastName, C.C. Article Title. *Journal Name* **Year**, *Article Number*, Page Range.

ISBN 978-3-03928-236-4 (Pbk)
ISBN 978-3-03928-237-1 (PDF)

Contents

About the Special Issue Editor

György Keglevich graduated from the Technical University of Budapest in 1981 as a chemical engineer. He received his "Doctor of Chemical Science" degree in 1994, in the subject of organophosphorus-chemistry. He has been the Head of the Department of Organic Chemistry and Technology since 1999. Within organophosphorus chemistry, his major field embraces P-heterocycles, involving selective syntheses as well as bioactive and industrial aspects. He also deals with environmentally friendly chemistry involving MW chemistry, its theoretical aspects, phase transfer catalysis, the development of new chiral catalysts, and the use of ionic liquids. He is the author or co-author of about 550 papers (the majority of which appeared in international journals) including around 70 review articles and 40 book chapters. He is a member of the Editorial Boards of *Molecules*, *Heteroatom Chemistry* and *Phosphorus, Sulfur and Silicon, and the Related Elements*, and *Current Microwave Chemistry*. He is the Editor-in-Chief of *Current Organic Chemistry and Current Green Chemistry*, the co-Editor-in-Chief of *Current Catalysis*, an Associate Editor of *Current Organic Synthesis* and *Letters in Drug Design and Discovery*, and a Regional Editor of *Letters in Organic Chemistry*.

Preface to "Organophosphorus Chemistry 2018"

These days, organophosphorus (OP) chemistry forms an integrant part of synthetic organic chemistry. OP compounds are used as starting materials, intermediates, reagents, catalysts (phase transfer catalysts or P(III)-transition metal complexes) and solvents (ionic liquids (IL)) in research laboratories and in the industry. There are many frequently applied reactions, such as reductions, the Wittig reaction and its variations, especially the catalytic versions, the Arbuzov reaction, the Pudovik reaction, the Mitsunobu reaction, or the Kabachnik–Fields reaction, that apply P-containing reagents. The simple deoxygenation of phosphine oxides is also a challenging field. Other reactions e.g. homogeneous catalytic transformations or C-C coupling reactions involve P-ligands in the catalysts. There have been major developments in the field of chiral OP compounds. Methods have been elaborated for the resolution of tertiary phosphine oxides and for stereoselective OP transformations. The optically active P(III) species may be used as ligands in transition metal (Pt, Pd, etc.) complex catalysts, making possible enantioselective transformations. The heterocyclic discipline may include P-heterocycles and classical O- and N-heterocycles with P-functions. A special field comprises P-containing or P-functionalized macrocycles and other macromolecules, like dendrimers. An up-to-date approach is to perform syntheses in the OP discipline in an environmentally-friendly manner. This may include the use of microwave or ultrasound. Solvent-free accomplishments are also interesting. At the other end, OP species (e.g., catalysts and ILs) may be tools in organic chemistry. Monitoring the reactions in order to optimize the conditions, or to observe reactive species, is a challenging field. Theoretical calculation within OP chemistry is a developing field; these days, however, stereostructures and mechanisms may be easily evaluated. A very important segment of OP chemistry, the driving force for development, is the pool of biologically active OP compounds (e.g., bisphosphonic derivatives and aminophosphonic species) that are searched for and used as drugs or plant protecting agents. The natural (e.g., peptide and amino acid) analogue P-compounds should also be mentioned. Many new phosphine oxides, phosphinates, phosphonates and phosphoric esters that may achieve different kinds of application have been described.

The OP Special Issue of *Molecules* will welcome submissions that fit in any way in the above outline.

György Keglevich
Special Issue Editor

molecules

MDPI

Editorial

Editorial to the Organophosphorus Chemistry Special Issue of Molecules (2012–2014)

György Keglevich

Department of Organic Chemistry and Technology, Budapest University of Technology and Economics, 1521 Budapest, Hungary; gkeglevich@mail.bme.hu; Tel.: +36-1-463-1111 (ext. 5883); Fax: +36-1-463-3648

Received: 19 September 2014; Accepted: 22 September 2014; Published: 26 September 2014

check for updates

The review entitled "Organophosphorus Chemistry for the Synthesis of Dendrimers" gives an overview of the methods of synthesis of phosphorus-containing dendrimers, with emphasis on the various roles played by the chemistry of phosphorus [1]. It is demonstrated that the presence of phosphorus atom(s) at each branching point of the dendrimeric structure is particularly important and highly valuable.

The review "Stoichiometric and Catalytic Synthesis of Alkynylphosphines summarizes the possibilities for the preparation of alkynylphosphines that, or their borane complexes, are available either through C–P bond forming reactions, or through modification of the phosphorus or the alkynyl function of various alkynyl phosphorus derivatives [2]. The latter strategy involving phosphoryl reduction is the method of choice for preparing primary and secondary alkynylphosphines, while the former strategy is employed for the synthesis of tertiary alkynylphosphines, or their borane complexes. Recently efficient catalytic procedures, involving copper complexes and either an electrophilic or a nucleophilic phosphorus reagent have emerged.

New developments of the Kabachnik–Fields (K-F) reaction were surveyed in the review "The Kabachnik-Fields Reaction: Mechanism and Synthetic Use" [3]. The monitoring of a few K–F reactions by *in situ* Fourier transform IR spectroscopy has indicated the involvement of the imine intermediate that was also justified by theoretical calculations. The K–F reaction was extended to >P(O)H species, comprising cyclic phosphites, acyclic and cyclic *H*-phosphinates, as well as secondary phosphine oxides. The synthesis under solvent-free microwave conditions is a good method of choice, as sophisticated and environmentally unfriendly catalysts suggested by literature methods are completely unnecessary under microwave conditions. The double K–F reaction made available bis(phosphonomethyl)amines.

The paper "Hydrophosphonylation of Nanoparticle Schiff Bases as a Mean for Preparation of Aminophosphonate-Functionalized Nanoparticles" describes results on magnetic nanoparticles with a modified surface that are attractive alternatives to deliver therapeutic agents [4]. The surface of the iron oxide nanoparticles was modified with aminophosphonic acids by applying the classical hydrophosphonylation protocol.

Recent achievements in the field of phosphinic dipeptide derivatives bearing appropriate side-chain substituents are summarized in the review entitled "Synthesis and Modifications of Phosphinic Dipeptide Analogues" [5]. Improved methods for the formation of the phosphinic peptide backbone, including stereoselective and multicomponent reactions, are presented. Parallel modifications leading to the structurally diversified substituents are also described, and selected examples of the biomedical applications of the title compounds are given.

In the paper "Chemistry of Phosphorylated Formaldehyde Derivatives. Part I." a few structurally related compounds, such as thioacetals, aminonitriles, aminomethylphosphinoyl compounds, are discussed [6]. The halogen aminals and acetals of phosphorylated formaldehyde, and a phosphorylated gem-diol of formaldehyde were discussed separately.

The chemical synthesis of DNA and RNA is usually carried out using nucleoside phosphoramidites or *H*-phosphonates as synthons. The review "Synthesis of DNA/RNA and their Analogs via Phosphoramidite and *H*-Phosphonate Chemistries" focuses on the phosphorus chemistry behind these synthons, and how it has been developed [7]. Additionally, the synthesis and properties of certain DNA and RNA analogues that are modified at the phosphorus atom were also discussed. These analogues include boranephosphonates, metallophosphonates, and alkylboranephosphines.

The research "Cesium Ion Removal from Aqueous Solutions through Adsorption onto Florisil® Impregnated with Trihexyl(tetradecyl)phosphonium Chloride" describes the adsorption performance of Florisil® impregnated with trihexyl(tetradecyl)phosphonium chloride in the process of cesium ion removal from aqueous solutions [8]. The adsorption process has been investigated as a function of pH, solid:liquid ratio, adsorbate concentration, contact time, and temperature.

References

1. Caminade, A.-M.; Laurent, R.; Zablocka, M.; Majoral, J.-P. Organophosphorus Chemistry for the Synthesis of Dendrimers. *Molecules* **2012**, *17*, 13605–13621. [CrossRef]
2. Bernoud, E.; Veillard, R.; Alayrac, C.; Gaumont, A.-C. Stoichiometric and Catalytic Synthesis of Alkynylphosphines. *Molecules* **2012**, *17*, 14573–14587. [CrossRef]
3. Keglevich, G.; Bálint, E. The Kabachnik–Fields Reaction: Mechanism and Synthetic Use. *Molecules* **2012**, *17*, 12821–12835. [CrossRef] [PubMed]
4. Siemieniec, J.; Kafarski, P.; Plucinski, P. Hydrophosphonylation of Nanoparticle Schiff Bases as a Mean for Preparation of Aminophosphonate-Functionalized Nanoparticles. *Molecules* **2013**, *18*, 8473–8484. [CrossRef] [PubMed]
5. Mucha, A. Synthesis and Modifications of Phosphinic Dipeptide Analogues. *Molecules* **2012**, *17*, 13530–13568. [CrossRef] [PubMed]
6. Morgalyuk, V.P. Chemistry of Phosphorylated Formaldehyde Derivatives. Part I. *Molecules* **2014**, *19*, 12949–13009. [CrossRef] [PubMed]
7. Roy, S.; Caruthers, M. Synthesis of DNA/RNA and Their Analogs via Phosphoramidite and *H*-Phosphonate Chemistries. *Molecules* **2013**, *18*, 14268–14284. [CrossRef] [PubMed]
8. Lupa, L.; Negrea, A.; Ciopec, M.; Negrea, P. Cs^+ Removal from Aqueous Solutions through Adsorption onto Florisil® Impregnated with Trihexyl(tetradecyl)phosphonium Chloride. *Molecules* **2013**, *18*, 12845–12856. [CrossRef] [PubMed]

molecules

MDPI

Review

The Kabachnik–Fields Reaction: Mechanism and Synthetic Use

György Keglevich [1,*] and Erika Bálint [2]

[1] Department of Organic Chemistry and Technology, Budapest University of Technology and Economics, 1521 Budapest, Hungary

[2] Research Group of the Hungarian Academy of Sciences, Department of Organic Chemistry and Technology, Budapest University of Technology and Economics, 1521 Budapest, Hungary

* Author to whom correspondence should be addressed; gkeglevich@mail.bme.hu; Tel.: +36-1-463-1111 (ext. 5883); Fax: +36-1-463-3648.

Received: 18 October 2012; Accepted: 26 October 2012; Published: 1 November 2012

check for updates

Abstract: The Kabachnik–Fields (phospha-Mannich) reaction involving the condensation of primary or secondary amines, oxo compounds (aldehydes and ketones) and >P(O)H species, especially dialkyl phosphites, represents a good choice for the synthesis of α-aminophosphonates that are of significant importance due to their biological activity. In general, these three-component reactions may take place via an imine or an α-hydroxy-phosphonate intermediate. The monitoring of a few Kabachnik–Fields reactions by *in situ* Fourier transform IR spectroscopy has indicated the involvement of the imine intermediate that was also justified by theoretical calculations. The Kabachnik–Fields reaction was extended to >P(O)H species, comprising cyclic phosphites, acyclic and cyclic *H*-phosphinates, as well as secondary phosphine oxides. On the other hand, heterocyclic amines were also used to prepare new α-amino phosphonic, phosphinic and phosphine oxide derivatives. In most cases, the synthesis under solvent-free microwave (MW) conditions is the method of choice. It was proved that, in the cases studied by us, there was no need for the use of any catalyst. Moreover, it can be said that sophisticated and environmentally unfriendly catalysts suggested are completely unnecessary under MW conditions. Finally, the double Kabachnik–Fields reaction has made available bis(phosphonomethyl)amines, bis(phosphinoxidomethyl)amines and related species. The bis(phosphinoxidomethyl)amines serve as precursors for bisphosphines that furnish ring platinum complexes on reaction with dichlorodibenzonitriloplatinum.

Keywords: Kabachnik–Fields reaction; α-aminophosphonates; reaction pathway; environmentally friendly; microwave; solventless

1. Introduction

The basic method for the preparation of α-aminophosphonates, valuable synthetic equivalents and biologically active substrates, involves the condensation of a primary or secondary amine, a carbonyl compound (aldehyde or ketone) and dialkyl phosphite (Scheme 1) [1,2].

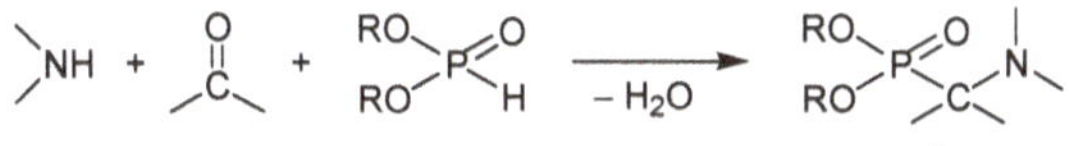

Scheme 1. General scheme for the Kabachnik–Fields reaction.

α-Aminophosphonic acids, considered as phosphorus analogues of α-amino acids, have attracted much attention in drug research due to their low mammalian toxicity. They are important targets in the development of antibiotics, antiviral species, antihypertensives, and antitumour agents based on their effect as inhibitors of GABA-receptors, enzyme inhibitors and anti-metabolites [3–9]. Diaryl α-amino-phosphonate derivatives are selective and highly potent inhibitors of serine proteases, and thus can mediate the patho-physical processes of cancer growth, metastasis, osteoarthritis or heart failure [10]. Dialkylglycine decarboxylase [9] and leucine aminopeptidase [11] are also inhibited by α-amino-phosphonates. Cyanoacrylate [12] and amide derivatives [13] of α-aminophosphonates are active antiviral compounds and inactivators of the tobacco mosaic virus. Certain α-aminophosphonates were proved to be suitable for the design of continuous drug release devices due to their ability to increase the membrane permeability of a hydrophilic probe molecule [14].

2. Possible Pathways for the Kabachnik–Fields Reaction

Cherkasov *et al.* studied the mechanism of the Kabachnik–Fields reaction in detail. One possibility is that an imine (a Schiff base) is formed from the carbonyl compound and the (primary) amine, and then the dialkyl phosphite is added on the C=N unit of the intermediate. The other route assumes the formation of an α-hydroxyphosphonate by the addition of the dialkylphosphite to the carbonyl group of the oxo component, then the hydroxyphosphonate undergoes substitution by the amine to furnish the α-aminophosphonate. On the basis of kinetic studies, it was concluded that the mechanism is dependent on the nature of the reactants. For example, the condensation of aniline, benzaldehyde and a dialkyl phosphite was assumed to follow the "imine" mechanism. Interestingly it was found that before the condensation of the aniline and the benzaldehyde, an H-bond is formed between the P=*O* function of the phosphite and the *H*N unit of the amine (Scheme 2) [15,16].

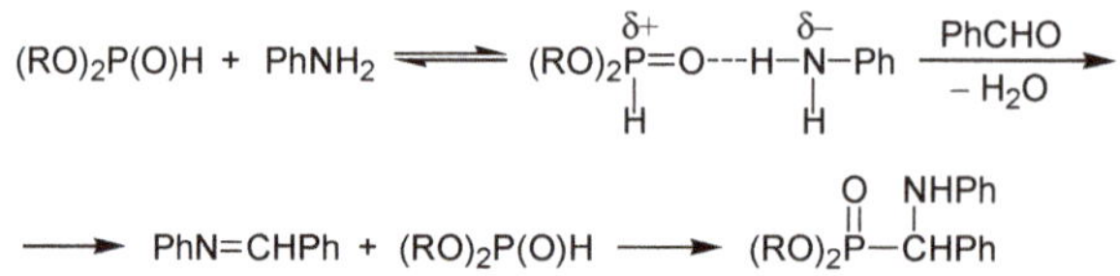

Scheme 2. The "imine" mechanism proposed for a Kabachnik–Fields reaction [15,16].

In another case, Cherkasov *et al.* suggested that the reaction of the more nucleophilic cyclohexyl-amine, benzaldehyde and a dialkyl phosphite takes place via the "hydroxyphosphonate" route. Here again an interaction was substantiated to precede the addition of the dialkylphosphite on the C=O unit of the oxo-compound. According to this, an H-bond is formed between the P(O)*H* moiety of the phosphite and the nitrogen atom of the amine (Scheme 3) [15,17].

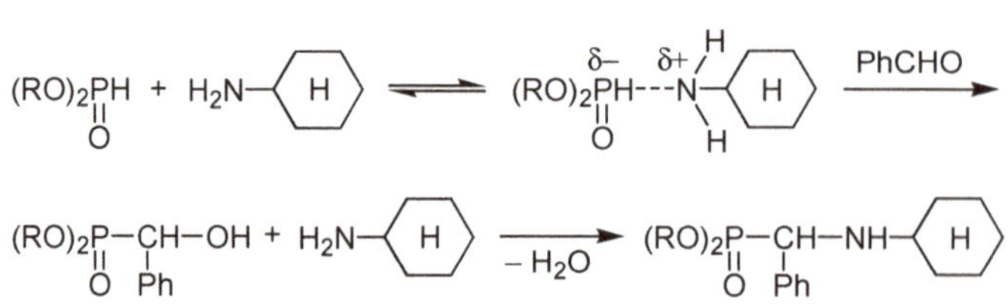

Scheme 3. The "α-hydroxyphosphonate" mechanism proposed for a Kabachnik–Fields reaction [15,17].

Later, however, Zefirov and Matveeva proved that the condensation of cyclohexylamine, benzaldehyde and dialkyl phosphite follows the "imine route", and concluded that there is no real experimental evidence for the hydroxyphosphonate route [18]. It is also worth mentioning that the reaction of cyclohexylamine, benzaldehyde and dibutylphosphine oxide, that may be regarded as an extended Kabachnik–Fields condensation, was shown to proceed according to the "imine" mechanism [15,19]. It seems probable that the actual mechanism is dependent on the components of the reaction, although the "imine" route seems to be more general than the route involving an "α-hydroxy-phosphonate" intermediate [3]. R. Gancarz and I. Gancarz substantiated that a reversible formation of the α-hydroxyphosphonate may also occur, and if it is rearranged to the corresponding phosphate, this becomes a "dead-end" route [20]. It can be said that in the Kabachnik–Fields reaction, a soft nucleophile (the dialkyl phosphite) and a hard nucleophile (the amine) compete for the electrophilic carbonyl compound. The softer the carbonyl compound is, the faster it reacts with the softer P-nucleophile and the slower it reacts with the harder amine nucleophile [21].

We wished to investigate the phospha-Mannich condensation of *n*-propylamine, benzaldehyde and diethyl phosphite (Scheme 4) by following the reaction utilizing *in situ* Fourier transform (FT) Infra Red (IR) spectroscopy [22].

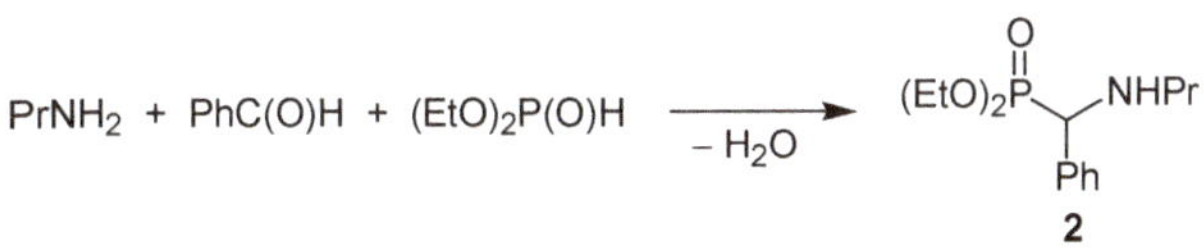

Scheme 4. The Kabachnik–Fields reaction studied by us.

The possible reaction paths are shown in Scheme 5. The question was whether the imine **3** or the α-hydroxyphosphonate **4** is the intermediate during the formation of the corresponding α-aminophosphonate **2**.

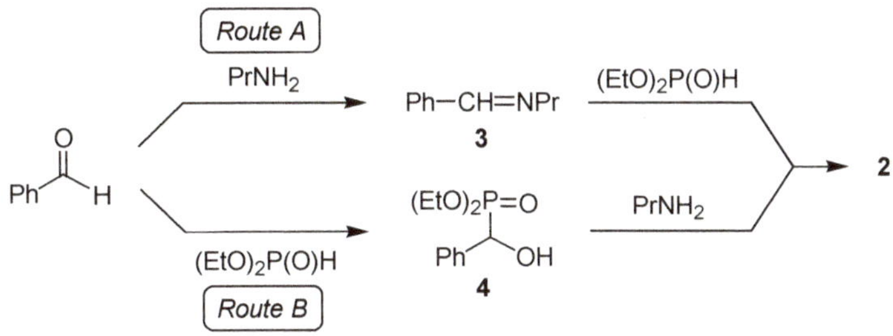

Scheme 5. Possible routes for the Kabachnik–Fields reaction studied by us.

The reaction carried out at 80 °C in acetonitrile was monitored by registering a 3D IR diagram. On the basis of the characteristic $\nu_{C=N}$ stretching vibration at 1,648 cm^{-1}, the imine **3** could be observed as a transient species. It was possible to obtain a relative concentration—time diagram for the components (Figure 1) by deconvolution of the 3D IR diagram. It can be seen that the imine intermediate **3** reaches its maximum concentration after a 10 min reaction time [22].

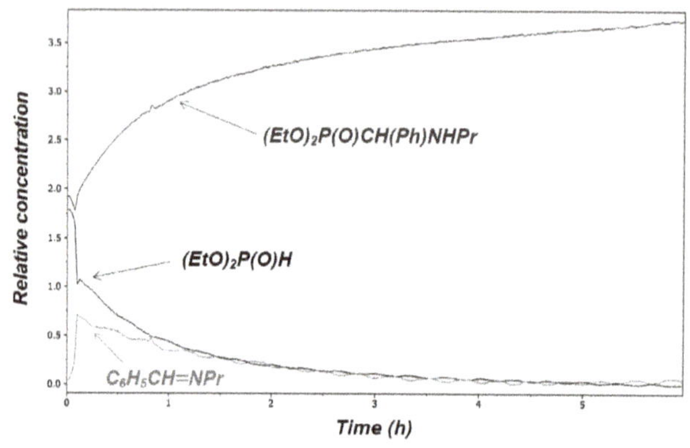

Figure 1. Concentration profile for the Kabachnik–Fields reaction studied at 80 °C in acetonitrile.

It was shown above that there was also controversy over the mechanism of the Kabachnik–Fields condensation of cyclohexylamine, benzaldehyde and dialkyl phosphites (Scheme 6) [15,17,18]. We sought to clarify the situation by *in situ* FT IR spectroscopy [23].

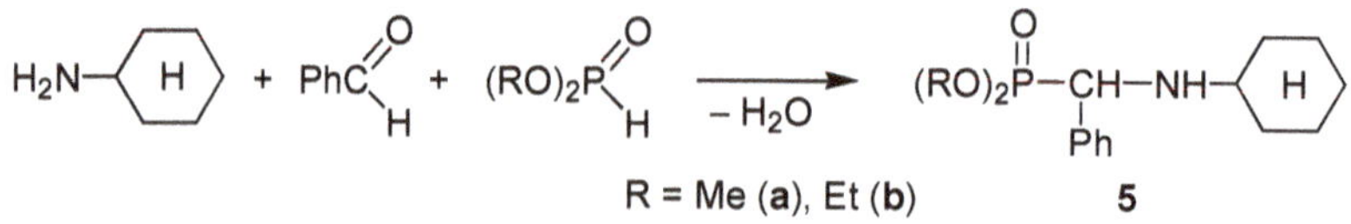

Scheme 6. Another Kabachnik–Fields reaction investigated by us.

From among the two possible intermediates **6** and **7**, again the imine **6a** could be detected on the basis of the $\nu_{C=N}$ = 1,644 cm^{-1} absorption as the transient species for α-aminophosphonate **5a** (Scheme 7). The intermediacy of imine **6** can be seen in Figure 2.

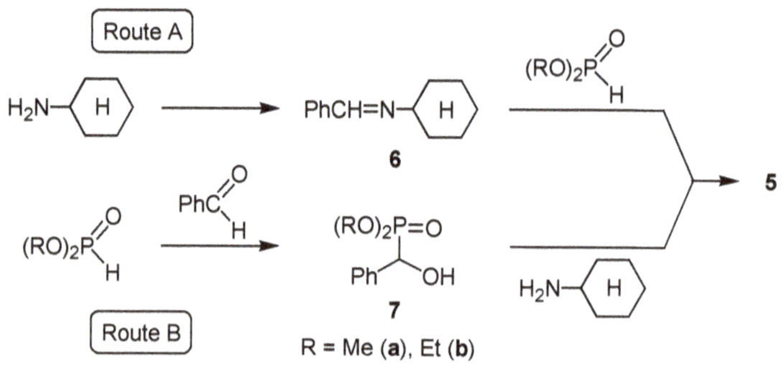

Scheme 7. Possible pathways for the second model investigated.

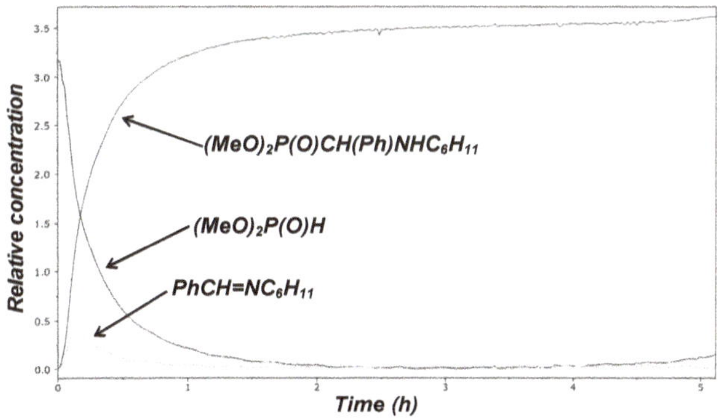

Figure 2. Concentration profile for the Kabachnik–Fields reaction studied at 80 °C in acetonitrile.

Relative energies for the possible intermediates **6** and **7** and for α-aminophosphonate **5** were calculated by the B3LYP/6-31G** method and then refined by the the B3LYP/6-311G**++ method provided that dimethyl phosphite is the reactant. It can be seen from Table 1 that the formation of the imine **6** goes with significantly lower energy gain than that of the α-hydroxyphosphonate **7**. On the one hand, the imine **6** would like to be stabilized further by reaction with the dimethyl phosphite on way to the α-aminophosphonate **5**. On the other hand, the hydroxyphosphonate **7** is too stable to react further to the aminophosphonate **5**. The conversion of **7** to **5** represents only a slight energy gain of 2.4 kJ/mol. In other words, there is no significant driving force for the substitution [23].

Table 1. Relative energies for the four states calculated.

Species	Relative energy (kJ/mol)
Reactants (benzaldehyde, cyclohexylamine and dimethyl phosphite)	0.0
Imine intermediate **6**	−18.6
α-Hydroxyphosphonate intermediate **7**	−40.5
Product **5**	−42.9

3. Microwave-Assisted Solvent- and Catalyst-Free Approach for the Synthesis of α-Amino-phosphonates and Related Derivatives

Although a lot of catalytic variations to carry out three-component Kabachnik–Fields condensations have been described, we found that the most straightforward synthesis is when the reactants are irradiated with microwave (MW) in the absence of any catalyst or solvent. The solventless and MW-assisted approach was useful in the synthesis of a few α-aminomethylphosphonates [24]. We used aniline or benzylamine as the amine, formaldehyde, benzaldehyde, acetophenone and cyclohexanone as the oxo-component and dialkyl phosphites and diphenylphosphine oxide as the >P(O)H reactant. The α-aminophosphonates and α-aminophosphine oxide products are represented by structure **8** in general Scheme 8 [25]. The detailed results are listed in Table 2. The comparative results of the catalytic versions were also included. A detailed account on the conditions of the catalytic reactions is provided in Table 3.Table InTable aTable partTable ofTable theTable casesTable suchTable asTable inTable theTable exampleTable coveredTable byTable reference[26], the catalytic versions could already be carried out at room temperature.

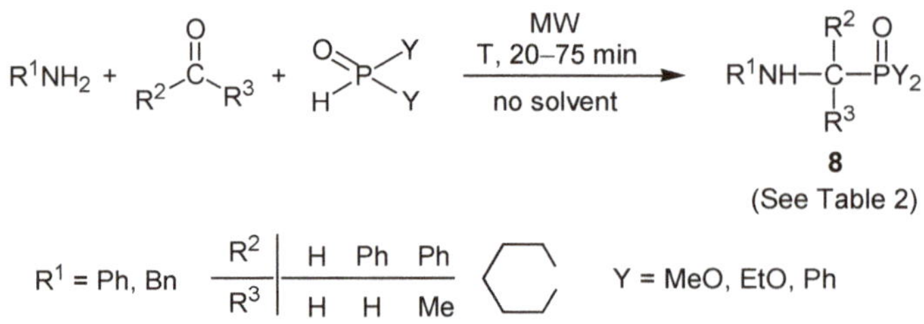

Scheme 8. General scheme for the solventless, catalyst-free MW-assisted Kabachnik–Fields reactions studied.

Table 2. Kabachnik–Fields reactions carried out without the use of a solvent and a catalyst under MW irradiation [25].

Entry	R^1	R^2	R^3	Y	Product	T (°C)	Yield (%)	Yield (%) of catalytic methods [ref.] †
1	Ph	H	H	EtO	**8a**	80 [a] 100 [b]	91	
2	Ph	H	H	MeO	**8b**	80 [a] 80 [b]	80	
3	Ph	H	H	Ph	**8c**	80	94	
4	Bn	H	H	EtO	**8d**	100	81	
5	Bn	H	H	Ph	**8e**	80	88	
6	Ph	H	Ph	EtO	**8f**	100	93	98 [27], 85 [28], ~95 [29], 88 [30], 79 [31], 93 [32], 92 [33], ~90 [34], 96 [35], 60 [36], 86 [37], 92 [38]
7	Ph	H	Ph	MeO	**8g**	100	86	98 [24], 98 [27], 92 [33]
8	Ph	H	Ph	Ph	**8h**	80	87	
9	Bn	H	Ph	EtO	**8i**	100	83	99 [28], 84 [29], 92 [30], 85 [31], 91 [32], 91 [33], 92 [35], 92 [36], 93 [38]
10	Bn	H	Ph	MeO	**8j**	100	87	95 [27], 82 [33]
11	Ph	Me	Ph	EtO	**8k**	120	80	75 [27], 74 [30], 63 [35], 18 [36]
12	Bn	Me	Ph	EtO	**8l**	120	84	92 [26], 80 [27], 81 [38]
13	Bn	Me	Ph	Ph	**8m**	100 [a] 120 [b]	80	
14	Ph			EtO	**8n**	120	81	92 [27], ~72 [29], 86 [30], 47 [31], 87 [37]
15	Bn			EtO	**8o**	120	91	85 [26], 90 [27], 83 [31], 80 [33], 85 [38]
16	Bn			MeO	**8p**	120	85	92 [27]
17	Bn			Ph	**8q**	100 [a] 120 [b]	80	

† for details see Table 3; [a] condensation of the oxo-component and the amine; [b] addition of the >P(O)H species to the Schiff-base.

Table 3. Kabachnik–Fields reactions carried out in the presence of catalysts.

Catalyst	Solvent	MW/Δ	T [°C]	t	Yield (Product) [%]	Ref.
Phthalocyanine-AlCl	CH_2Cl_2	–	26 [a]	12 h	92 (**8b**), 85 (**8p**)	[26]
$Mg(ClO_4)_2$	–	–	26	2 min/8 h	90–98 (**8f**, **8g**, **8j**, **8n-p**)	[27]
$Mg(ClO_4)_2$	–	Δ	50–80	45 min–12 h	80–99 (**8f**, **8i**, **8l**, **8n-p**)	[27,28]
$Mg(ClO_4)_2$	EtOH	Δ	50	5 h/12 h	85 (**8f**), 99 (**8i**)	[28]
$M(OTf)_n$ M = Li, Mg, Al, Cu, Ce	–	Δ	80	20 min–3.5 h	72–95 (**8f**, **8i**, **8n**)	[29]
GaI_3	CH_2Cl_2	–	26	3–6 h	74–92 (**8f**, **8i**, **8k**, **8n**)	[30]
$In(OTf)_3$	THF	Δ	66	21–35 h	47–85 (**8f**, **8i**, **8n**, **8o**)	[31]
$BiNO_3$	–	– [b]	26	10 h	93 (**8f**), 91 (**8i**)	[32]
$BiCl_3$	MeCN	Δ	80	6–15 h	80–92 (**8f**, **8g**, **8i**, **8j**, **8o**)	[33]
$FeCl_3$	EtOH (or solvent free)	–	26		~90 (**8f**)	[34]
$YbCl_3$	MeCN	–	26	24 h	63–96 (**8f**, **8i**, **8k**)	[35]
SmI_2 (+ 4 Å mol sieves)	MeCN	Δ	80	24 h	18–92 (**8f**, **8i**, **8k**)	[36]
ceric ammonium nitrate	MeCN	–	26	3 h	86 (8f), 87 (8n)	[37]
$InCl_3$	THF	Δ	66	9–12 h	81–93 (**8f**, **8i**, **8l**, **8o**)	[38]
$InCl_3$	DMF	MW	no data	2 min	82 (**8f**) [c]	[39]
$InCl_3$	[bmim][PF_6]	MW	no data	2 min	91 (**8f**) [c]	[39]
$Ln(OTf)_3$ Ln = Yb, Sc, Dy, Sm	DMF	MW	no data	2 min	72 (**8f**) [c]	[39]
$Ln(OTf)_3$ Ln = Yb, Sc, Dy, Sm	[bmim][PF_6]	–	26	27 h	92 (**8f**) [c]	[40]
$Ln(OTf)_3$ Ln = Yb, Sc, Dy, Gd	[bmim][PF_6]	MW	no data	2 min	89 (**8f**) [c]	[39]
the solvent acts as catalyst	[bmim][BF_4]	–	26	5 h/8 h	90 (**8f**), 84 (**8f**)	[41]

[a] Diethyl phosphite was added to preformed imines; [b] Was also performed under MW; [c] The product was extracted with benzene.

On the basis of our experimental data, there is no need to use exotic (expensive and environmentally unfriendly) catalysts. In most cases, the solvent- and catalyst-free MW-assisted reactions give excellent results. Further exploration of catalysts does not seem to be justified in this field. Next, our method was extended to phospha-Mannich condensations involving heterocyclic amines (Scheme 9) [42]. Diphenylphosphine oxide was also used as the P-component.

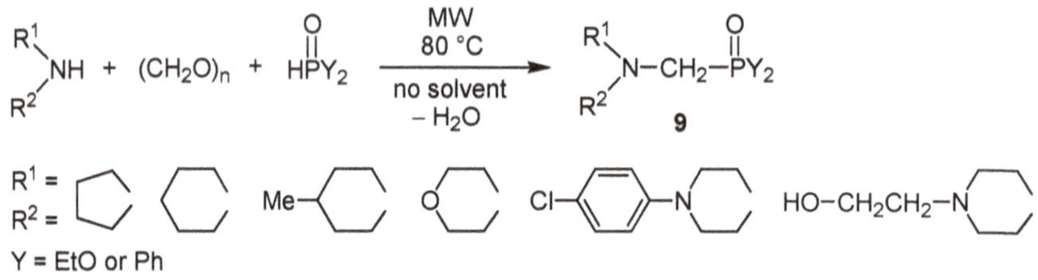

Scheme 9. Kabachnik–Fields reactions applying *N*-heterocycles as the amine component.

In another series of reactions, 1,3,2-dioxaphosphorine oxide (**10**) was used as the phosphite (Scheme 10) [43]. In this way "double" heterocyclic derivatives were prepared. These reactions were more efficient in the presence of a solvent.

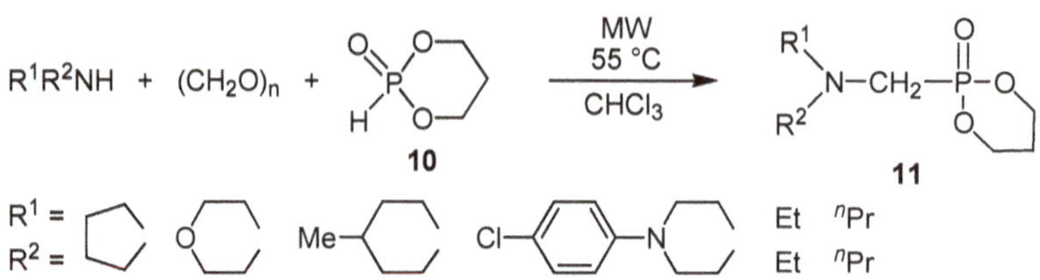

Scheme 10. Kabachnik–Fields reactions applying 1,3,2-dioxaphosphorine oxide as the P-reactant.

Applying 1,3,2-dioxaphosphorine oxide (**10**) together with benzaldehyde, the steric hindrance prevented the efficient condensation. It was better to prepare the imine **12** first and to react it separately with the cyclic phosphite **10** (Scheme 11) [43].

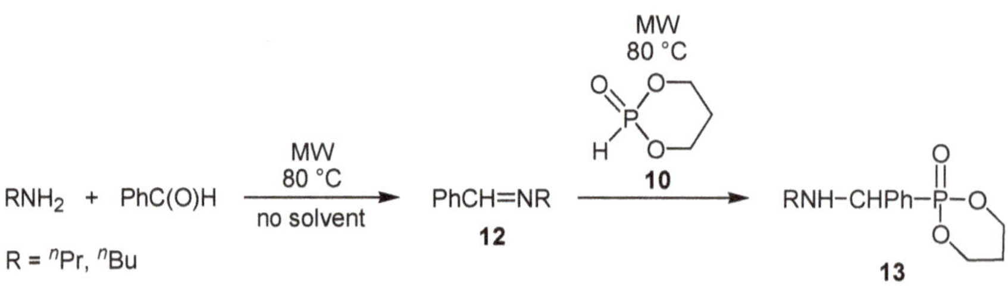

Scheme 11. Synthesis of α-aminophosphonates via the imine intermediate.

Dibenzo[*c.e*][1,2]oxaphosphorine (**14**) was also utilized in the synthesis of P-heterocyclic derivatives. In this case, the primarily formed product **15** underwent opening of the hetero ring by reaction with the water formed (Scheme 12) [43].

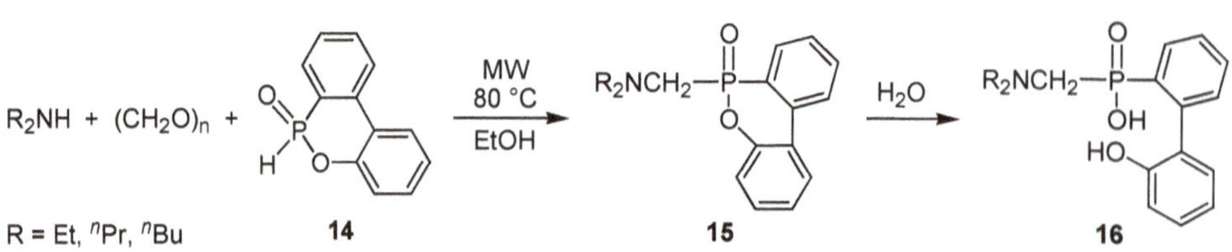

Scheme 12. Kabachnik–Fields reaction applying a dibenzooxaphosphorine oxide as the P-reactant.

For the preparation of diethyl α-diethylaminophenylmethylphosphonate (**17**), the two-step approach led to better results. The aldehyde–amine adduct formed primarily was reacted with diethyl phosphite to afford product **17** (Scheme 13) [43].

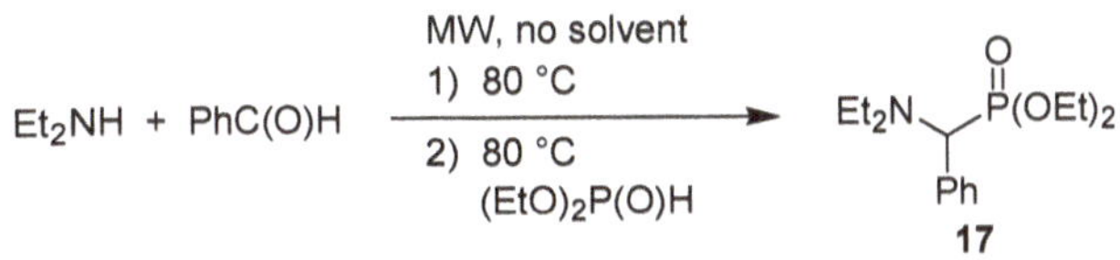

Scheme 13. Synthesis of an α-aminophosphonates in two steps.

The MW-assisted solventless procedure was useful in the synthesis of a series of bis(phosphonomethyl)amines and related derivatives marked as **18** (Scheme 14) [44–46]. Product **18** could be obtained in yields, mostly above 80%. The double Kabachnik–Fields reaction was then extended to the synthesis of bis(phosphinoxidomethyl)amines **19**. In these cases, heterogenity of the reaction mixture requested the use of a solvent that was acetonitrile (Scheme 15) [44–46]. The use of aniline as the amine component led to by-product **20** besides the expected product **19** (Y = Ph) [46].

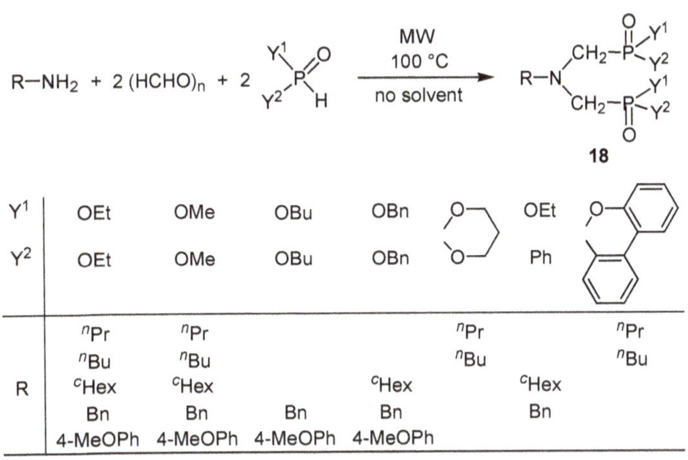

Scheme 14. Synthesis of bis(phosphonomethyl)amines and related derivatives by the double Kabachnik–Fields reaction.

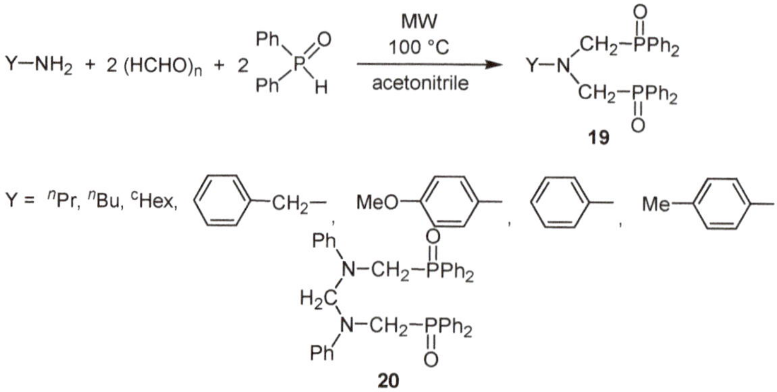

Scheme 15. Bis(phosphinoxidomethyl)amines by the double phospha-Mannich reaction.

The bis(phosphinoxidomethyl)amines **19** served as precursors for bis(phosphinomethyl)amine bidentate P-ligands **21** by double deoxygenation. The bisphosphines **21** so formed were reacted with half an equivalent of dichlorodibenzonitriloplatinum to furnish ring platinum complexes **22** (Scheme 16) [45,46].

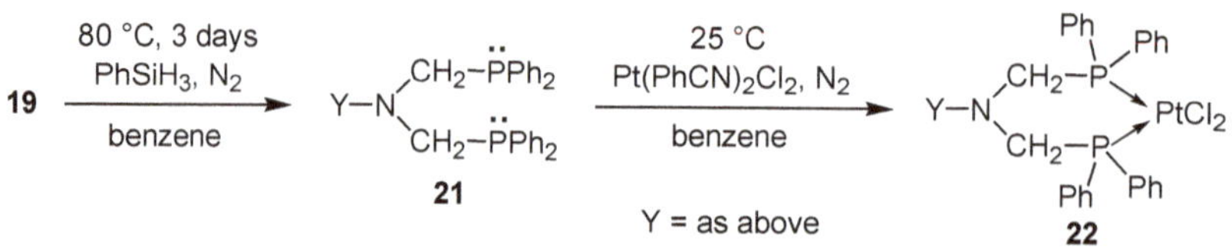

Scheme 16. Synthesis of ring platinum complexes from bis(phosphinoxidomethyl)amines.

The bidentate P-ligands may be stored as their phosphine-borane complexes. This is shown in the example of the **23** → **24** conversion (Scheme 17). In general, the phosphine can be regenerated from the phosphine-borane by heating with a secondary amine, such as diethylamine, in an aromatic solvent [47].

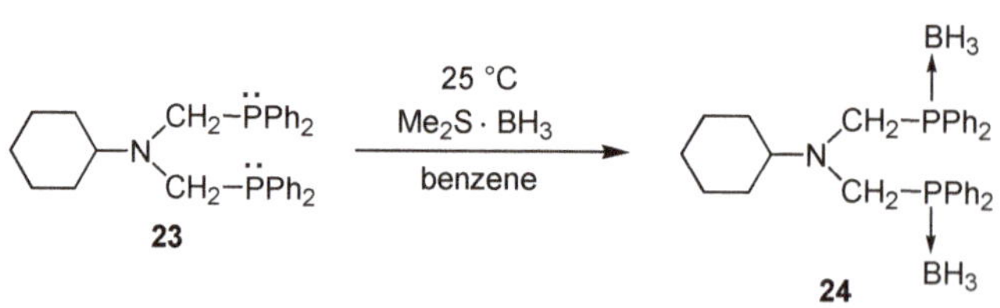

Scheme 17. Stabilization of a bis(phosphinomethyl)amine as a bis(borane complex).

The MW-assisted catalytic addition of dialkyl phosphites on the carbonyl group of a series of benzaldehyde derivatives was also elaborated (Scheme 18) [48]. The α-hydroxyphosphonates (**25**, Y = RO) are potential intermediates of the Kabachnik–Fields reaction. The use of diphenylphosphine oxide in the addition led to the formation of α-hydroxyphosphine oxides (**25**, Y = Ph).

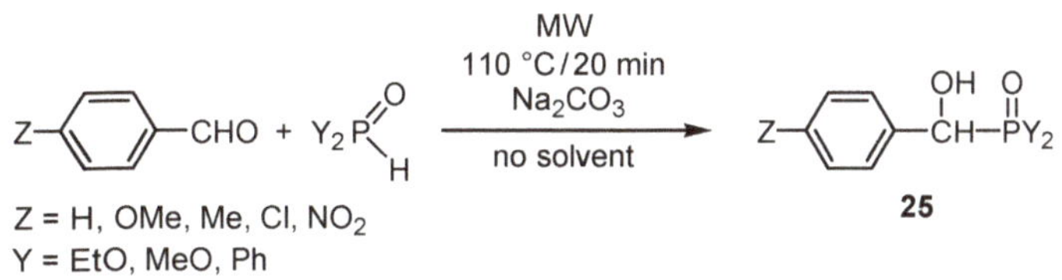

Scheme 18. MW-assisted synthesis of α-hydroxyphosphonates and α-hydroxyphosphine oxides.

The addition of dialkyl phosphites to α-ketophosphonates led to 1-hydroxymethylene-bisphosphonates [49,50]. It was interesting to find that, as a consequence of the neighboring group effect of the P=O moiety, α-hydroxyphosphonate **25** (Y = EtO) could be readily converted to the corresponding α-aminophosphonates (**26**) (Scheme 19) [51]. Quantum chemical calculations justified the beneficial neighboring group effect of the P=O moiety [51].

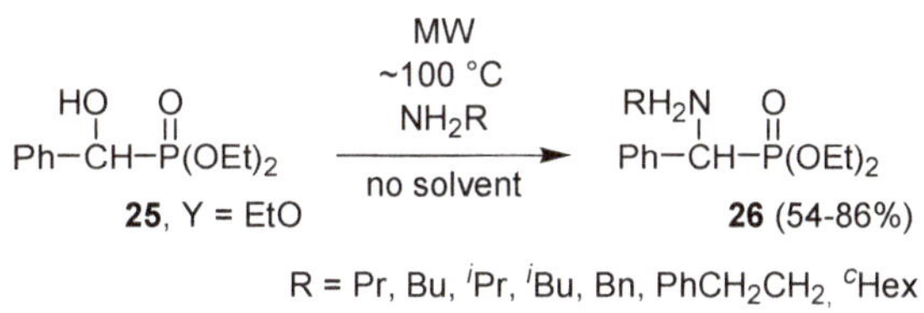

Scheme 19. Preparation of α-aminophosphonates by substitution of α-hydroxyphosphonates.

4. Conclusions

In conclusion, recent results obtained in the study of the Kabachnik–Fields reaction have been summarized. This mini-review sheds light on the new developments regarding mechanistic and synthetic aspects showing that the phospha-Mannich reaction remains an evergreen topic for organic chemists. On the one hand, the mechanism of the Kabachnik–Fields reaction still reserves some surprises, on the other hand, the 3-component condensation is an ideal subject for green chemical reactions. In addition, the α-aminophosphonate and α-aminophosphine oxide products are biologically active substrates.

Acknowledgments: The authors are grateful for Hungarian Scientific Research Fund (No: OTKA K 83118).

References

1. Kabachnik, M.I.; Medved, T.Y. New synthesis of aminophosphonic acids. *Dokl. Akad. Nauk SSSR* **1952**, *83*, 689–692.
2. Fields, E.K. The synthesis of esters of substituted amino phosphonic acids. *J. Am. Chem. Soc.* **1952**, *74*, 1528–1531. [CrossRef]
3. Zefirov, N.S.; Matveeva, E.D. Catalytic Kabachnik-Fields reaction: New horizons for old reaction. *ARKIVOC* **2008**, 1–17.
4. Kukhar, V.P.; Hudson, H.R. (Eds.) *Aminophosphonic and Aminophosphinic Acids: Chemistry and Biological Activity*; Wiley: Chichester, UK, 2000.
5. Fields, S.C. Synthesis of natural products containing a C-P bond. *Tetrahedron* **1999**, *55*, 12237–12272. [CrossRef]
6. Kafarski, P.; Lejczak, B. Aminophosphonic acids of potential medical importance. *Curr. Med. Chem. Anticancer Agents* **2001**, *1*, 301–312. [CrossRef] [PubMed]
7. Bird, J.; De Mello, R.C.; Harper, G.P.; Hunter, D.J.; Karran, E.H.; Markwell, R.E.; Miles-Williams, A.J.; Rahman, S.S.; Ward, R.W. Synthesis of novel *N*-phosphonoalkyl dipeptide inhibitors of human collagenase. *J. Med. Chem.* **1994**, *37*, 158–169. [CrossRef] [PubMed]
8. Liu, W.S.; Rogers, C.J.; Fisher, A.J.; Toney, M.D. Aminophosphonate inhibitors of dialkylglycine decarboxylase: Structural basis for slow binding inhibition. *Biochemistry* **2002**, *41*, 12320–12328. [CrossRef]

9. Mucha, A.; Kafarski, P.; Berlicki, L. Remarkable potential of the α-aminophosphonate/phosphinate structural motif in medicinal chemistry. *J. Med. Chem.* **2011**, *54*, 5955–5980. [CrossRef]
10. Sienczyk, M.; Oleksyszyn, J. Irreversible inhibition of serine proteases—Design and *in vivo* activity of diaryl alpha-aminophosphonate derivatives. *Curr. Med. Chem.* **2009**, *16*, 1673–1687. [CrossRef]
11. Grembecka, J.; Mucha, A.; Cierpicki, T.; Kafarski, P. The most potent organophosphorus inhibitors of leucine aminopeptidase. Structure-based design, chemistry, and activity. *J. Med. Chem.* **2003**, *46*, 2641–2655. [CrossRef]
12. Long, N.; Cai, X.J.; Song, B.A.; Yang, S.; Chen, Z.; Bhadury, P.S.; Hu, D.Y.; Jin, L.H.; Xue, W. Synthesis and antiviral activities of cyanoacrylate derivatives containing an alpha-aminophosphonate moiety. *J. Agric. Food Chem.* **2008**, *56*, 5242–5246. [CrossRef]
13. Hu, D.Y.; Wan, Q.Q.; Yang, S.; Song, B.A.; Bhadury, P.S.; Jin, L.H.; Yan, K.; Liu, F.; Chen, Z.; Xue, W. Synthesis and antiviral activities of amide derivatives containing the alpha-aminophosphonate moiety. *J. Agric. Food Chem.* **2008**, *56*, 998–1001. [CrossRef] [PubMed]
14. Danila, D.C.; Wang, X.Y.; Hubble, H.; Antipin, I.S.; Pinkhassik, E. Increasing permeability of phospholipid bilayer membranes to alanine with synthetic alpha-aminophosphonate carriers. *Bioorg. Med. Chem. Lett.* **2008**, *18*, 2320–2323. [CrossRef] [PubMed]
15. Cherkasov, R.A.; Galkin, V.I. The Kabachnik–Fields reaction: synthetic potential and the problem of the mechanism. *Russ. Chem. Rev.* **1998**, *67*, 857–882. [CrossRef]
16. Galkin, V.I.; Zvereva, E.R.; Sobanov, A.A.; Galkina, I.V.; Cherkasov, R.A. Kinetics and mechanism of Kabachnik–Fields reaction in dialkylphosphite-benzaldehyde-aniline system. *Zhur. Obsch. Khim.* **1993**, *63*, 2224–2227.
17. Galkina, I.V.; Sobanov, A.A.; Galkin, V.I.; Cherkasov, R.A. Kinetics and mechanism of the Kabachnik–Fields reaction: IV. Salicyaldehyde in the Kabachnik–Fields reaction. *Russ. J. Gen. Chem.* **1998**, *68*, 1398–1401.
18. Matveeva, E.D.; Zefirov, N.S. On the mechanism of the Kabachnik–Fields reaction: Does a mechanism of nucleophilic amination of alpha-hydroxyphosphonates exist? *Doklady Chem.* **2008**, *420*, 137–140. [CrossRef]
19. Galkina, I.V.; Galkin, V.I.; Cherkasov, R.A. Kinetics and the mechanism of Kabachnik–Fields reaction—V. Effect of the nature of hydrophosphoryl compound on the mechanism of Kabachnik–Fields reaction. *Zhur. Obsch. Khim.* **1998**, *68*, 1469–1475.
20. Gancarz, R.; Gancarz, I. Failure of aminophosphonate synthesis due to facile hydroxyphosphonate-phosphate rearrangement. *Tetrahedron Lett.* **1993**, *34*, 145–148. [CrossRef]
21. Gancarz, R. Nucleophilic addition to carbonyl compounds. competition between hard (amine) and soft (phosphite) nucleophile. *Tetrahedron* **1995**, *51*, 10627–10632. [CrossRef]
22. Keglevich, G.; Fehérvári, A.; Csontos, I. A study on the Kabachnik–Fields reaction of benzaldehyde, propylamine and diethyl phosphite by *in situ* Fourier transform (FT) IR spectroscopy. *Heteroatom Chem.* **2011**, *22*, 599–604. [CrossRef]
23. Keglevich, G.; Kiss, N.Z.; Menyhárd, D.; Fehérvári, A.; Csontos, I. A study on the Kabachnik–Fields reaction of benzaldehyde, cyclohexylamine and dialkyl phosphites. *Heteroatom Chem.* **2012**, *23*, 171–178. [CrossRef]
24. Mu, X.-J.; Lei, M.-Y.; Zou, J.-P.; Zhang, W. Microwave-assisted solvent-free and catalyst-free Kabachnik–Fields reactions for α-amino phosphonates. *Tetrahedron Lett.* **2006**, *47*, 1125–1127. [CrossRef]
25. Keglevich, G.; Szekrényi, A. Eco-friendly accomplishment of the extended Kabachnik–Fields reaction; a solvent- and catalyst-free microwave-assisted synthesis of α-aminophosphonates and α-aminophosphine oxides. *Lett. Org. Chem.* **2008**, *5*, 616–622. [CrossRef]
26. Matveeva, E.D.; Podrugina, T.A.; Tishkovskaya, E.V.; Tomilova, L.G.; Zefirov, N.S. A novel catalytic three-component synthesis (Kabachnick-Fields reaction) of α-aminophosphonates from ketones. *Synlett* **2003**, 2321–2324. [CrossRef]
27. Bhagat, S.; Chakraborti, A.K. An extremely efficient three-component reaction of aldehydes/ketones, amines, and phosphites (Kabachnik–Fields reaction) for the synthesis of α-aminophosphonates catalyzed by magnesium perchlorate. *J. Org. Chem.* **2007**, *72*, 1263–1270. [CrossRef]
28. Wu, J.; Sun, W.; Xia, H.-G.; Sun, X. A facile and highly efficient route to α-amino phosphonates via three-component reactions catalyzed by $Mg(ClO_4)_2$ or molecular iodine. *Org. Biomol. Chem.* **2006**, *4*, 1663–1666. [CrossRef]
29. Firouzabadi, H.; Iranpoor, N.; Sobhani, S. Metal triflate-catalyzed one-pot synthesis of α-aminophosphonates from carbonyl compounds in the absence of solvent. *Synthesis* **2004**, 2692–2696. [CrossRef]

30. Sun, P.; Hu, Z.; Huang, Z. Gallium triiodide catalyzed organic reaction: a convenient synthesis of α-amino phosphonates. *Synth. Commun.* **2004**, *34*, 4293–4299. [CrossRef]
31. Ghosh, R.; Maiti, S.; Chakraborty, A.; Maiti, D.K. In(OTf)(3) catalysed simple one-pot synthesis of α-amino phosphonates. *J. Mol. Catal. A* **2004**, *210*, 53–57. [CrossRef]
32. Bhattacharya, A.K.; Kaur, T. An efficient one-pot synthesis of alpha-amino phosphonates catalyzed by bismuth nitrate pentahydrate. *Synlett* **2007**, 745–748. [CrossRef]
33. Zhan, Z.-P.; Li, J.-P. Bismuth(III) chloride-catalyzed three-component coupling: Synthesis of alpha-amino phosphonates. *Synth. Commun.* **2005**, *35*, 2501–2508. [CrossRef]
34. Wu, J.; Sun, W.; Wang, W.-Z.; Xiu, H.-G. A highly efficient catalyst $FeCl_3$ in the synthesis of α-amino phosphonates via three-component reactions. *Chin. J. Chem.* **2006**, *24*, 1054–1057. [CrossRef]
35. Xu, F.; Luo, Y.; Wu, J.; Shen, Q.; Chen, H. Facile one-pot synthesis of α-amino phosphonates using lanthanide chloride as catalyst. *Heteroatom Chem.* **2006**, *17*, 389–392. [CrossRef]
36. Xu, F.; Luo, Y.; Deng, M.; Shen, Q. One-pot synthesis of α-amino phosphonates using samarium diiodide as a catalyst precursor. *Eur. J. Org. Chem.* **2003**, 4728–4730. [CrossRef]
37. Ravinder, K.; Vijender Reddy, A.; Krishnaiah, P.; Venkataramana, G.; Niranjan Reddy, V.L.; Venkateswarlu, Y. CAN catalyzed one-pot synthesis of α-amino phosphonates from carbonyl compounds. *Synth. Commun.* **2004**, *34*, 1677–1683. [CrossRef]
38. Ranu, B.C.; Hajra, A.; Jana, U. General procedure for the synthesis of α-amino phosphonates from aldehydes and ketones using indium(III) chloride as a catalyst. *Org. Lett.* **1999**, *1*, 1141–1143. [CrossRef]
39. Lee, S.; Lee, J.K.; Song, C.E.; Kim, D.-C. Microwave-assisted Kabachnik–Fields reaction in ionic liquid. *Bull. Korean Chem. Soc.* **2002**, *23*, 667–668.
40. Lee, S.; Park, J.H.; Kang, J.; Lee, J.K. Lanthanide triflate-catalyzed three component synthesis of α-amino phosphonates in ionic liquids. A catalyst reactivity and reusability study. *Chem. Commun.* **2001**, 1698–1699. [CrossRef]
41. Yadav, J.S.; Reddy, B.V.S.; Sreedhar, P. An eco-friendly approach for the synthesis of α-aminophosphonates using ionic liquids. *Green Chem.* **2002**, *4*, 436–438. [CrossRef]
42. Prauda, I.; Greiner, I.; Ludányi, K.; Keglevich, G. Efficient synthesis of phosphono- and phosphinoxidomethylated N-heterocycles under solvent-free microwave conditions. *Synth. Commun.* **2007**, *37*, 317–322. [CrossRef]
43. Keglevich, G.; Szekrényi, A.; Sipos, M.; Ludányi, K.; Greiner, I. Synthesis of cyclic aminomethylphosphonates and aminomethyl-arylphosphinic acids by an efficient microwave-mediated phospha-Mannich approach. *Heteroatom Chem.* **2008**, *19*, 207–210. [CrossRef]
44. Keglevich, G.; Szekrényi, A.; Szöllősy, Á.; Drahos, L. Synthesis of bisp(hosphonatomethyl)-, bis(phosphinatomethyl)-, and bis(phosphinoxidomethyl)amines, as well as related ring bis(phosphine) platinum complexes. *Synth. Commun.* **2011**, *41*, 2265–2272. [CrossRef]
45. Bálint, E.; Fazekas, E.; Pintér, G.; Szöllősy, Á.; Holczbauer, T.; Czugler, M.; Drahos, L.; Körtvélyesi, T.; Keglevich, G. Synthesis and utilization of the bis(>P(O)CH_2)amine derivatives obtained by the double Kabachnik–Fields reaction with cyclohexylamine; Quantum chemical and X-ray study of the related bidentate chelate platinum complexes. *Curr. Org. Chem.* **2012**, *16*, 547–554. [CrossRef]
46. Bálint, E.; Fazekas, E.; Pongrácz, P.; Kollár, L.; Drahos, L.; Holczbauer, T.; Czugler, M.; Keglevich, G. *N*-benzyl and *N*-aryl bis(phospha-Mannich adducts): Synthesis and catalytic activity of the related bidentate chelate platinum complexes in hydroformylation. *J. Organomet. Chem.* **2012**, *717*, 75–82. [CrossRef]
47. Gourdel, Y.; Pelon, P.; Toupet, L.; Le Corre, M. Stereoselective synthesis of new functionalized bisphosphines. *Tetrahedron Lett.* **1994**, *35*, 1197–1200. [CrossRef]
48. Keglevich, G.; Tóth, V. R.; Drahos, L. Microwave-assisted synthesis of α-hydroxy-benzylphosphonates and -benzylphosphine oxides. *Heteroatom Chem.* **2011**, *22*, 15–17. [CrossRef]
49. Grün, A.; Molnár, I.G.; Bertók, B.; Greiner, I.; Keglevich, G. Synthesis of α-hydroxy-methylenebisphosphonates by the microwave-assisted reaction of α-oxophosphonates and dialkyl phosphites under solventless conditions. *Heteroatom Chem.* **2009**, *20*, 350–354. [CrossRef]
50. Keglevich, G.; Grün, A.; Molnár, I.G.; Greiner, I. Phenyl-, benzyl- and unsymmetrical hydroxy-methylenebisphosphonates as dronic acid ester analogues from α-oxophosphonates by microwave-assisted synthesis. *Heteroatom Chem.* **2011**, *22*, 640–648. [CrossRef]

51. Kiss, N.Z.; Kaszás, A.; Drahos, L.; Mucsi, Z.; Keglevich, G. A neighbouring group effect leading to enhanced nucleophilic substitution of amines at the hindered α-carbon atom of an α-hydroxyphosphonate. *Tetrahedron Lett.* **2012**, *53*, 207–209. [CrossRef]

molecules

MDPI

Review

Synthesis and Modifications of Phosphinic Dipeptide Analogues

Artur Mucha

Department of Bioorganic Chemistry, Faculty of Chemistry, Wrocław University of Technology, Wybrzeże Wyspiańskiego 27, 50-370 Wrocław, Poland; artur.mucha@pwr.wroc.pl; Tel.: +48-71-320-3446; Fax: +48-71-320-2427

Received: 17 October 2012; Accepted: 12 November 2012; Published: 15 November 2012

Abstract: Pseudopeptides containing the phosphinate moiety (-P(O)(OH)CH_2-) have been studied extensively, mainly as transition state analogue inhibitors of metalloproteases. The key synthetic aspect of their chemistry is construction of phosphinic dipeptide derivatives bearing appropriate side-chain substituents. Typically, this synthesis involves a multistep preparation of two individual building blocks, which are combined in the final step. As this methodology does not allow simple variation of the side-chain structure, many efforts have been dedicated to the development of alternative approaches. Recent achievements in this field are summarized in this review. Improved methods for the formation of the phosphinic peptide backbone, including stereoselective and multicomponent reactions, are presented. Parallel modifications leading to the structurally diversified substituents are also described. Finally, selected examples of the biomedical applications of the title compounds are given.

Keywords: phosphinic peptides; P-C bond formation; multicomponent reactions; side-chain diversification; enzyme inhibition

---

Abbreviations: $Aa_1\psi$[P(O)(OH)X]-Aa_2: phosphorus-containing dipeptide analogues (X = NH, phosphonamidates; X = O, phosphonates; X = CH_2, phosphinates); AaPH: α-amino-*H*-phosphinic (phosphonous) acid; Ac: acetyl; Ad: 1-adamantyl; Alk: alkyl; Alloc, allyloxycarbonyl; APN: alanyl aminopeptidase; Ar: aryl; ATP: adenosine triphosphate; Boc: *t*-butyloxycarbonyl; BOP: (benzotriazol-1-yloxy)tris(dimethylamino)phosphonium hexafluorophosphate; BSA: *N,O*-bis(trimethylsilyl)-acetamide; *i*-Bu: isobutyl; *n*-Bu: *n*-butyl; *sec*-Bu: *sec*-butyl; *t*-Bu: *tert*-butyl; Bz: benzoyl; Bzl: benzyl; Cbz: benzyloxy-carbonyl; CPB: carboxypeptidase B; dba: dibenzylideneacetone; DEAD: diethyl azodicarboxylate; DCC: *N,N'*-dicyclohexylcarbodiimide; *de*: diastereomeric excess; DIPEA: diisopropylethylamine; DMAP: 4-dimethylaminopyridine; DME: 1,2-dimethoxy-ethane; DMF: dimethylformamide; EDC: 1-ethyl-3-(3'-dimethylaminopropyl)carbo-diimide; Et: ethyl; EWG: electron withdrawing group; Fmoc: 9-fluorenylmethoxy-carbonyl; HMDS: 1,1,1,3,3,3-hexamethyldisilazane; HOBt: 1-hydroxybenzotriazole; HPLC: high performance liquid chromatography; LAP: leucine aminopeptidase; Me: methyl; MurD: UDP-*N*-acetylmuramoyl-L-alanyl:D-glutamate ligase; Nu: nucleophile; Pac: phenacyl; PG: protecting group; Ph: phenyl; Phth: phthaloyl; *i*-Pr: isopropyl; PyBOP: (benzotriazol-1-yloxy)tris(pyrrolidino)phosphonium hexafluorophosphate; SPPS: solid-phase peptide synthesis; Su: succinimide; TAFIa: thrombin-activatable fibrinolysis inhibitor; TFA: trifluoroacetic acid; THF: tetrahydrofuran; Tl: triflate (trifluoromethane- sulfonate); TLC: thin layer chromatography; TMSCl: chlorotrimethyl-silane; Tr: trityl (triphenylmethyl); TS: transition state; Ts: tosyl; Trs: 2,4,6-triisopropylphenylsulfonyl.

1. Introduction

Phosphinic pseudodipeptides (Figure 1) are typically defined as dipeptide analogues that replace the amide bond with the phosphinate moiety. The rationale behind the replacement is not to mimic the peptide linkage alone, explicitly in its ground state, but the assumed analogy concerns isosteric and isoelectronic resemblances to the high-energy tetrahedral transition state (TS) of the amide hydrolysis.

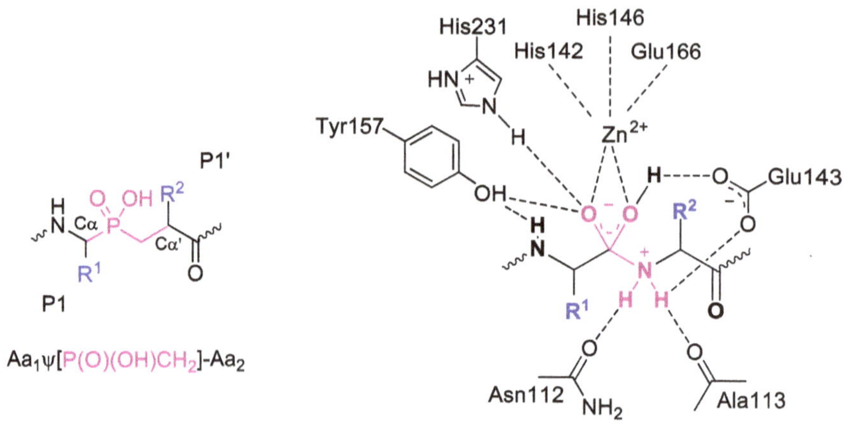

Figure 1. A general structure for phosphinic dipeptide analogues and the resemblance of this structure to the tetrahedral transition state of peptide hydrolysis postulated for thermolysin [1]. Side chains R^1 and R^2 are specific to the S1 and S1′ binding pocket and are also referred to as the P1 and P1′ substituents, according to the nomenclature proposed by Schechter and Berger [2].

Translation of the TS arrangement into the inhibitor structure is one of the fundamental concepts in drug design, originating directly from the early suggestion of evolutionary orientation of the enzyme active sites to bind substrates optimally in this state, which diminishes the energy of activation of the catalyzed reaction [3]. Formally, the binding of TS inhibitors should be tighter than the binding of the substrates by the factor of the enzymatic rate enhancement [4].

Three different types of phosphorus-containing peptide analogues, which comprise the structural fragment $Aa_1\psi[P(O)(OH)X]$-Aa_2, have been constructed to follow this idea. The phosphorus-containing peptide analogues include: phosphonamidates (X = NH, the closest TS analogues), phosphonates (X = O, pseudodepsipeptides) and phosphinates (X = CH_2). All these compounds appear particularly effective in regulating the activity of metalloproteases. Nevertheless, investigation of other proteases (e.g., aspartyl) and, in general, other classes of enzymes (e.g., ligases) also brought inhibition to an impressive level [5–7]. In the early period of these studies (during the 80s of the 20th century), phosphonamidate and phosphonate pseudopeptides attracted major attention. These pseudopeptides emerged as invaluable tools in fundamental structural and mechanistic studies carried out on prototypical metalloproteases, thermolysin and carboxypeptidase A [1,8–14]. Certain drawbacks, such as hydrolytic instability of the P-N bond and frequent limited activity of the P-O derivatives excluded them however from later practical applications.

Phosphinic compounds avoid these inconveniences. Phosphinates are stable over the whole pH range and equipotent with or only slightly less potent than the corresponding phosphonamidate species [15]. The utility potential of phosphinic peptides in drug design was greatly increased by development of a synthetic procedure leading to the Fmoc-$Aa_1\psi[P(O)(OAd)CH_2]$-Aa_2 building block [16]. This synthon is compatible with standard methodologies of solid-phase peptide synthesis (SPPS) and combinatorial synthesis. Since then, numerous active sequences have been obtained, and this progress is relevantly and comprehensively reviewed elsewhere [5–7,17–20]. The phosphinate chemistry was also historically systematized in an excellent paper by Yiotakis *et al.* in 2004 [21]. Despite considerable achievements, construction of a fundamental phosphinic α,α′-dipeptide, which comprises appropriate side-chains remains the main challenge in the field. The current review is dedicated entirely to this synthetic aspect with particular attention to diversification of the P1 and P1′ substituents. The contents are limited to recent achievements. In principle, the literature data published after 2000 are discussed, and earlier papers are only selectively highlighted to give the proper background.

2. Synthesis of the Phosphinic α,α′-Dipeptide Backbone

2.1. Retrosynthetic Analysis

Approaches based on a phospha-Michael addition to obtain the C-P-C pseudodipeptide skeleton predominate. The most traditional approach involves the addition of a suitably *N*-protected α-amino-*H*-phosphinic (phosphonous, AaPH) acid or its ester (alkyl *H*-phosphinate) to an acrylate (P-C disconnection, synthetic direction from the N to C terminus, Scheme 1). The *H*-phosphinic acid component requires activation to the more nucleophilic tervalent ester form that is typically achieved in the presence of a silylation agent. On the other hand, the reaction of alkyl *H*-phosphinate esters can also be catalyzed with a strong base. α-Amino-*H*-phosphinic substrates are conveniently available from appropriate aldehydes, amino components and hypophosphorous acid. Acrylates (α-substituted α,β-unsaturated esters) also need to be synthesized separately, typically by a Knoevenagel reaction of α-substituted malonate monoesters and formaldehyde.

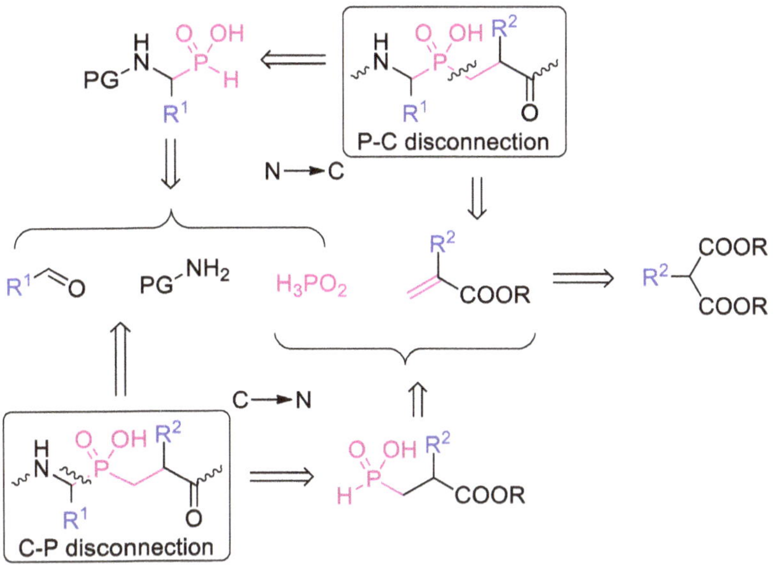

Scheme 1. Retrosynthetic analysis of the phosphinic dipeptide scaffold.

Alternatively, a Michael-type addition can precede incorporation of the *N*-terminal fragment. The essential substrate, β-substituted β-alkoxycarbonylethyl-*H*-phosphinic acid, is readily obtained in the addition of bis(trimethylsilyl) phosphonite to an acrylate, similarly to the reactions previously mentioned. The product is then subjected to a three-component condensation with an aldehyde and amino/amido component (C-P disconnection, C → N, Scheme 1). Depending on the exact structures of the reactants, this reaction is a variant of the Kabachnik-Fields, phospha-Mannich or amidoalkylation reaction.

2.2. *N* → *C Strategy*

Historically, addition of α-amino-*H*-phosphinates (acids and esters) to acrylates has been the most commonly recognized approach to construct phosphinic dipeptide analogues (Scheme 2). This popularity has been stimulated by an easy access to suitably protected phosphorus substrates. Different variations for obtaining these P-H substrates, suggested originally by Baylis *et al.*, are based mainly on a three-component condensation of an aldehyde, hypophosphorous acid and diphenylmethylamine as the nitrogen source, or addition of H_3PO_2 to diphenylmethylimines [22]. The diphenylmethyl group can be hydrolyzed from the adduct under acidic conditions to obtain free α-aminophosphonous acids. The main advantages of the approach are simplicity and low cost, even though the preparative yield is not high. The amino group can subsequently be protected by the benzyloxycarbonyl group (Cbz) in a water solution [22]. The procedure is particularly convenient for alkyl and aryl substituents with a single and unpredictable exception of GlyPH. An additional side-chain functionality needs attention with respect to its orthogonal protection during the synthesis of α-aminophosphonous acids and their further transformations.

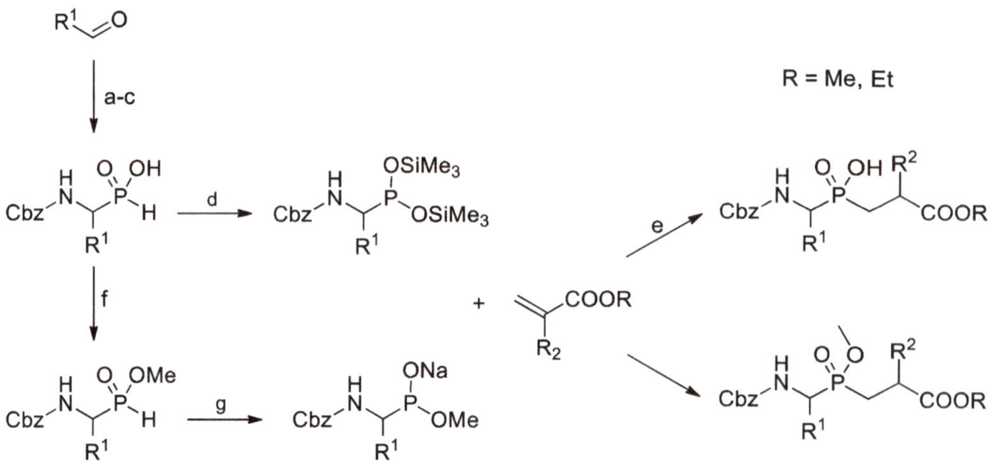

Typical reaction conditions: (**a**) $Ph_2CHNH_3^+$ $H_2PO_2^{\cdot -}$, EtOH or dioxane, reflux; (**b**) HCl_{aq}, reflux or HBr_{aq}, 100 °C; (**c**) CbzCl, NaOH (pH = 9), H_2O/dioxane, 0 °C → rt; (**d**) HMDS, heating or TMSCl, Et_3N, CH_2Cl_2, 0 °C → rt, then acrylate; (**e**) MeOH or EtOH; (**f**) CH_2N_2, Et_2O; (**g**) MeONa, MeOH.

Scheme 2. A general overview of the synthetic strategies leading to phosphinic dipeptide analogues by addition of α-amino-*H*-phosphinates (acids or esters) to acrylates.

Benzyloxycarbonyl is a standard group used for N-protection, but other carbamates are compatible with the phospha-Michael addition conditions, particularly with the mild TMSCl/tertiary amine version. These other carbomates commonly include *t*-butyloxycarbonyl (Boc) [23] and 9-fluorenylmethoxycarbonyl (Fmoc) [24]. Other specific protecting groups will be also mentioned in the text.

The *H*-phosphinic component requires activation to the tervalent nucleophilic ester form, typically accomplished by silylation (Scheme 2). Alternatively, the effect can be achieved by heating with hexamethyldisilazane (HMDS), by action of *N*,*O*-bis(trimethylsilyl)acetamide (BSA), or in the presence of chlorotrimethylsilane (TMSCl) and a tertiary amine [21,25–30]. After addition to an acrylate, the silylated adduct is decomposed with an alcohol or water.

Transformation of *N*-protected α-amino-*H*-phosphinic acid into its ester provides another opportunity for activation. This reaction involves the action of a strong base, e.g., sodium methoxide or potassium *t*-butoxide, to shift the equilibrium from the P(V) tautomer to a more nucleophilic P(III) species (Scheme 2) [21,31–34].

As mentioned above, this method for obtaining a phosphinic dipeptide backbone is not problematic for alkyl, aryl or alkylaryl P1 residues that do not contain an additional functionality, and accordingly, numerous structures have become available in this manner [16,21,35,36]. However, more complex Cα side-chains that incorporate a certain heteroatom moiety can greatly influence the chemical behavior of the phosphorus substrate. For example, the neighboring assistance of a carboxylate group (free or esterified) in the phosphonous analogue of aspartic acid did not permit the Michael addition to occur [23]. Even though the target phosphinic compound is obtained, further transformations of this phosphinic compound are frequently altered by the complexity of the protection/deprotection strategy. Cbz-Lys(Phth)ψ[P(O)(OH)CH_2]-Ala-OMe can serve as a representative example [37]. The attempts to prepare its free carboxylate form (with all other three functionalities blocked), suitable for peptide coupling, failed. In this case, the issue had to be addressed by the use of alternative synthetic approaches.

Selected rare examples of phosphinic dipeptide structures that have been prepared by incorporation of an additionally functionalized α-amino-*H*-phosphinic acid to an acrylate are listed in Scheme 3. The first illustration addresses the preparation of P1 methionine derivatives as inhibitors of metalloproteases [36,38]. To present a less popular version of obtaining the P-H substrate, Liboska *et al.* added methanethiol to acrolein (**1**) and converted the product into the corresponding oxime, which, in turn, reacted with anhydrous hypophosphorous acid (Scheme 3A) [38]. Boc-protected *H*-phosphinic acid was added to acrylates to yield **2** under typical conditions. The second example (Scheme 3B) shows a molecule of C_2 symmetry obtained by Kabudin and Saadati from benzaldehyde (**3**). The final modification to aminobisphosphinic acid **4** was performed in a standard way, however, without any protection of the phosphorus component [39]. Scheme 3C presents the construction of a pseudodipeptide containing a 5-membered ring of proline (**7**) starting from hydroxyaldehyde **5** [40]. To perform this transformation, the Michael addition followed cyclative alkylation of α-amino-ω-bromobutylphosphonous acid (**6**). The overall yield of the five-step procedure was only 6%, whereas the strategy proceeding in the opposite direction (C → N, compare Scheme 1 and chapter 2.3.1) provided the target molecule in 52% yield in three steps [40].

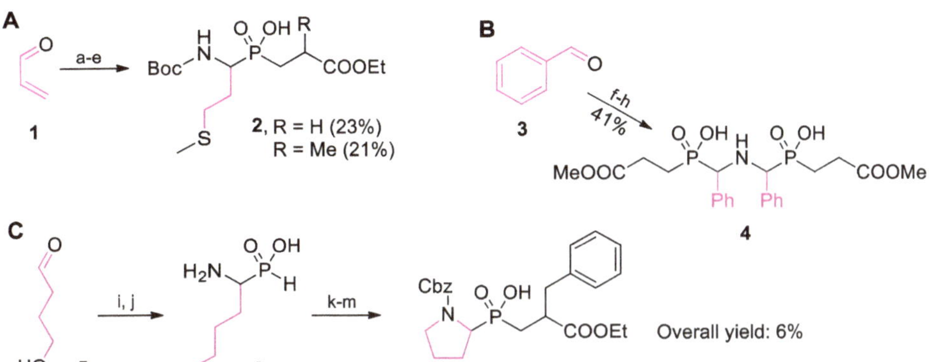

Reaction conditions: (**a**) MeSH, 0 °C; (**b**) $NH_2OH \times HCl$, pyridine, rt; (**c**) H_3PO_2, MeOH, reflux, (**d**) Boc_2O, Na_2CO_3, H_2O/dioxane, 0 °C → rt; (**e**) TMSCl, DIPEA, then ethyl acrylate or methacrylate, rt; (**f**) NH_4OH, reflux; (**g**) H_3PO_2, EtOH; (h) HMDS, 110 °C, then methyl acrylate, heating, then EtOH; (**i**) $Ph_2CHNH_3^+$ $H_2PO_2\cdot^-$, dioxane, reflux; (**j**) 47% HBr_{aq}, reflux; (**k**) *N*,*N*-dimethyldodecylamine, *n*-Bu_4NOH, CH_2Cl_2, H_2O; (**l**) CbzCl, Na_2CO_3, H_2O, 0 °C → rt; (**m**) HMDS, 110 °C, then ethyl α-benzylacrylate, heating, then EtOH.

Scheme 3. Selected examples of preparation of additionally functionalized α-amino-*H*-phosphinic acids and their application in addition reactions to acrylates.

In contrast to the limited tolerance of a P1 functionality, the addition of the P-H moiety to the activated double bond leads to an extensive choice of structurally variable acrylates (Figure 2). Cyclic compounds [41], as well as compounds containing a carboxyl group [42–44], a phosphonate [31], a hydroxyl group [35,45], a thiol [31,35] or an amino function [46] in the side chain (typically, suitably protected) and residues of an unsaturated character [47,48] allow an effective reaction. Some of the unsaturated ester substrates are obvious precursors of proteinogenic P1′ fragments that ensure specificity of the products in biological studies. Other acrylates are designed to introduce a structural fragment that can be subjected to a further diversification to the phosphinic scaffold.

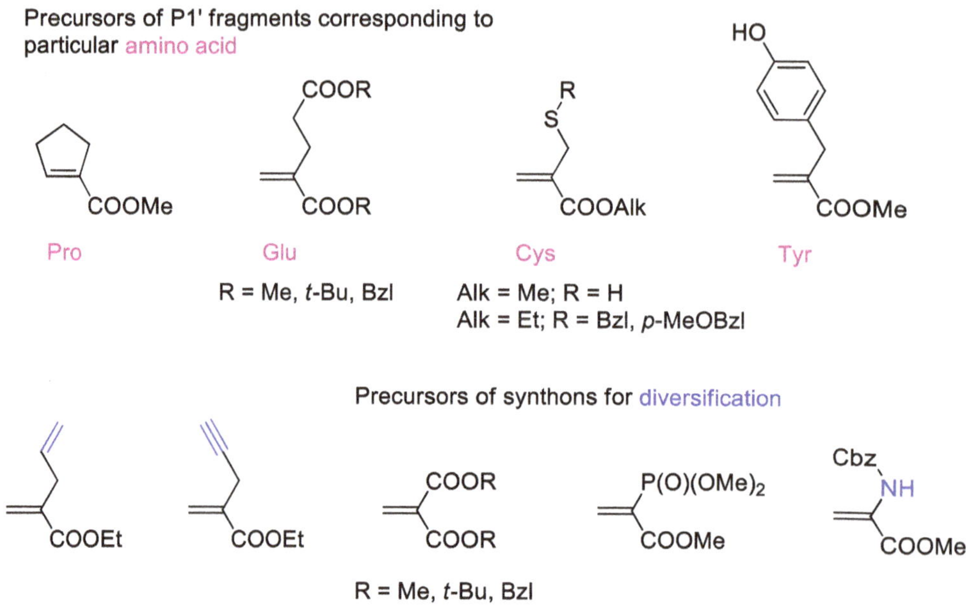

Figure 2. A selection of additionally functionalized α-substituted acrylate derivatives that were found to be compatible with phospha-Michael addition chemistry.

Obtaining the acrylates may present individual synthetic challenges (selected examples are shown in Scheme 4). Preparing the appropriate acrylate usually involves a Knoevenagel condensation of appropriately substituted malonate monoester with formaldehyde [49]. The key malonate substrate (e.g., **10**, Scheme 4A) can be prepared by condensation of an aldehyde (such as **8**) with subsequent reduction of the product **9** [45] or by a standard alkylation (e.g., **15**, Scheme 4B) with alkyl halides (such as imidazole derivative **13**) [50]. The total yield of the product (here, **11** and **16**) varies significantly and depends on the structure of the substituent. Other methods have been developed exclusively for particular target derivatives [21,35,36]. For example, β-alkoxy or β-alkylthio derivatives (**18**, Scheme 4C) are prepared efficiently by alkylation of sodium alkoxides or thiolates with α-bromomethyl acrylate [35].

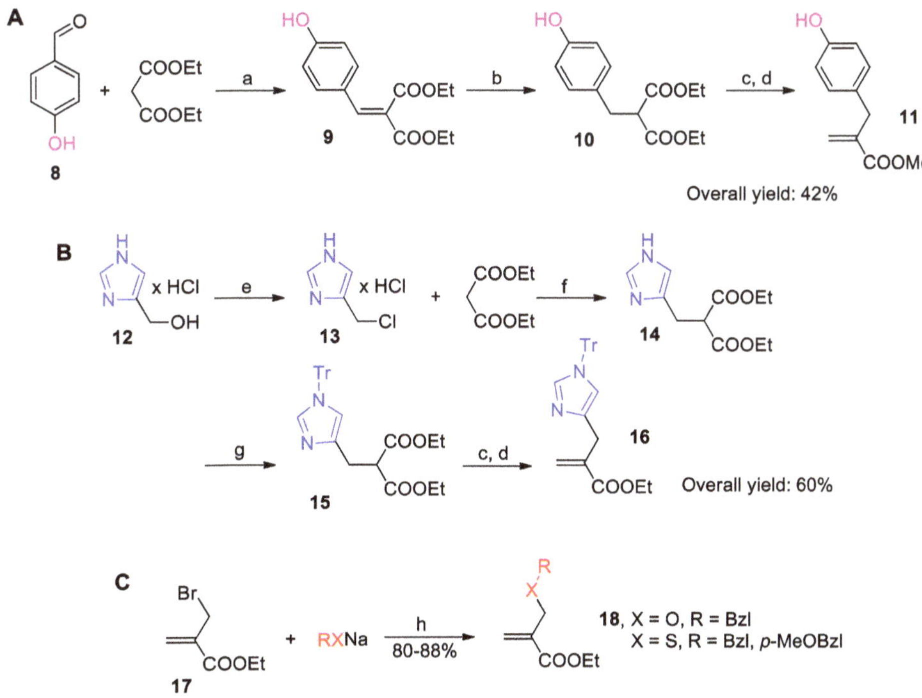

Reaction conditions: (**a**) piperidine, benzoic acid, benzene (-H_2O); (**b**) $NaBH_4$, MeOH, 0 °C; (**c**) KOH, MeOH or EtOH, then HCl_{aq}; (**d**) $(CHO)_n$, piperidine, pyridine; (**e**) $SOCl_2$, benzene, reflux; (**f**) EtONa, EtOH, 50 °C; (**g**) TrCl, DIPEA, DMF, rt; (**h**) Et_2O, 15 °C → rt.

Scheme 4. Selected synthetic procedures leading to additionally functionalized α-substituted acrylate derivatives.

Multistep preparations of the phosphinic dipeptide from acrylates and α-aminophosphonous acids that converge the substrates in the phospha-Michael addition are reproducible and show reasonable yields. Consequently, these preparations have generally been employed for the construction of pseudodipeptides corresponding to both natural and non-natural amino acid sequences. Certain limitations, mostly concerning the P1 substituent structure, were proposed for the multicomponent approaches (see below). Several attempts to modify the original N → C Michael addition have also been undertaken. Most of the attempts at modification involved simplification of the protection/deprotection procedure and merging both P-C bond formation processes in a one-pot manner that reduced the number of the synthetic steps [33,34,51,52]. Typically, an appropriate ester of hypophosphorous acid is treated consecutively with an imine (or triazine) and then by an acrylate. Because these modifications were scarcely noticed in the scientific community, a single and arbitrarily selected case is presented here.

In the course of studies on the reactivity of trimethylsilylated compounds, including $(Me_3SiO)_2P$-H (**20**), Prishchenko *et al.* discovered several reaction variants that led to phosphinic compounds [53–55]. Among others, *N*-(methoxymethyl)bis(trimethylsilyl)amine (**19**) reacted with bis(trimethylsilyl) phosphonite in the presence of $ZnCl_2$ to yield the GlyPH analogue **21**, which was subsequently added to trimethylsilyl acrylate (**22**) [53]. Free Glyψ[P(O)(OH)CH_2]-Gly (**23**) was obtained after methanolysis with a total yield of 44% (Scheme 5).

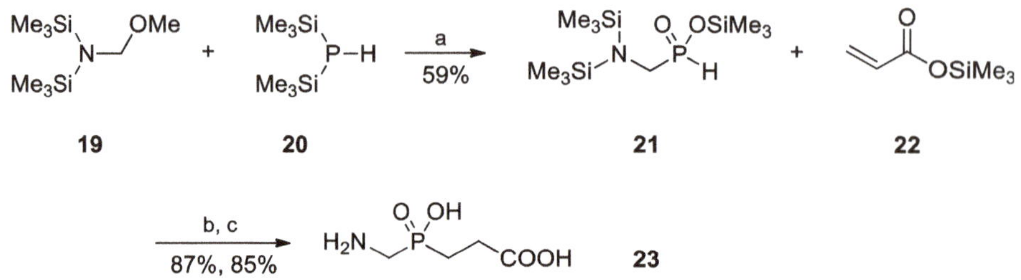

Reaction conditions: (**a**) $ZnCl_2$, TMSCl, 120–140 °C; (**b**) pyridine, CH_2Cl_2, rt → 100 °C; (**c**) MeOH, Et_2O, rt → reflux.

Scheme 5. Glyψ[P(O)(OH)CH_2]-Gly preparation from silylated substrates.

An interesting possibility of tandem P-C and C-C bond formation on a phosphinic dipeptide scaffold was recognized in a preliminary way in the Yiotakis group [56]. The tandem bond formation involved phospha-Michael addition of a P-H substrate to allyl acrylate (**25**) followed by a Claisen-type rearrangement (Scheme 6). Unexpectedly, the product of the addition of optically active α-*N*-benzyloxycarbonylaminoalkylphosphonous acid [(***R***)**-24**] rearranged to a lesser extent than structurally less complex phenylphosphinic acid. The ratio of the final product **26** to un-rearranged allyl ester **27** was 55:45, compared to 91:9 for phenylphosphinic acid. An incompletely understood role for the amide N-H in intramolecular proton transfer in the silylated intermediate was suggested as a possible reason. Indeed, bis-N-protection of the α-aminoalkyl-*H*-phosphinic acid substrate increased the preparative yield of the target product from 52% to 84%, comparable to the yield obtained for phenylphosphinic acid. The reaction can be carried out conveniently in a three-component variant using acryloyl chloride (or other α-substituted analogues) and various allyl and propargyl alcohols.

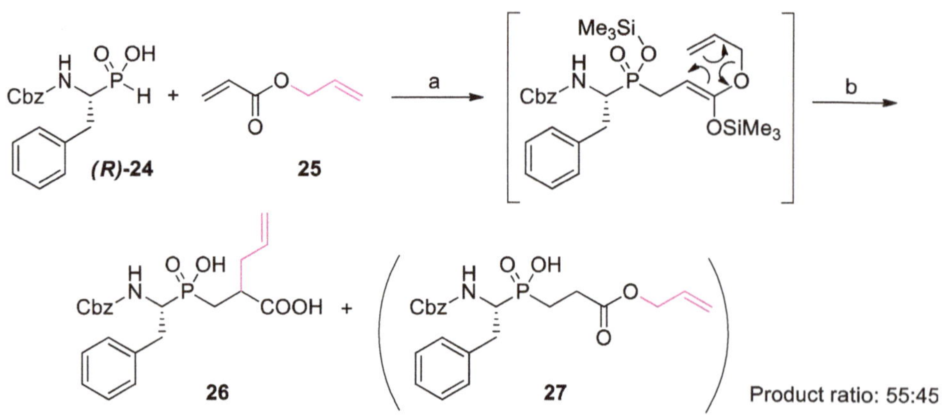

Reaction conditions: (**a**) DIPEA, CH_2Cl_2, then TMSCl, −78 °C→ rt; (**b**) H^+.

Scheme 6. Phospha-Michael addition of α-*N*-benzyloxycarbonylaminoalkyl-*H*-phosphinic acid to allyl acrylate followed by the Ireland-Claisen rearrangement of the product.

Phospha-Michael Addition on Solid Phase

Formation of the phosphinic bond could also occur on a solid support. Dorff *et al.* modified Wang resin with acrylic or metacrylic acid (**28**, R^2 = H or Me) under Mitsunobu conditions (Scheme 7A) [57]. Subsequently, FmocGlyPH (**29**, R^1 = H, PG = Fmoc), activated by BSA, was added to the immobilized double bond of **30** to give product **31**. After esterification of the phosphinate with diazomethane and Fmoc removal, the N-terminus could easily be modified, e.g., by acyl chlorides or isocyanates [57].

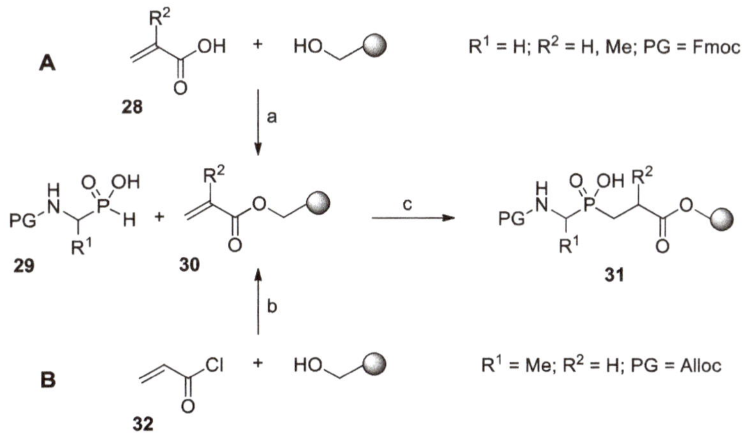

Reaction conditions: (**a**) Ph_3P, DEAD, THF; (**b**) Et_3N, DMAP; CH_2Cl_2; (**c**) BSA, CH_2Cl_2 or CH_2ClCH_2Cl, then MeOH.

Scheme 7. Phospha-Michael addition of N-protected α-aminoalkyl-*H*-phosphinic acid to acrylates immobilized on a solid phase.

In a similar approach, Buchardt and Meldal acylated a polyethylene glycol/polystyrene resin with acryloyl chloride (**32**), followed by addition of AllocAlaPH (**29**, R^1 = Me, PG = Alloc, Scheme 7B) [58]. The procedure was then validated in a solid-phase synthesis of an undecapeptide containing an internal Alaψ[P(O)(OH)CH_2]-Gly unit. Although these methodologies could provide an interesting alternative for the preparation of longer sequences in an automatic manner, the methodologies failed to attract further attention. Introduction of a pre-synthesized pseudodipeptidic building block to the classical SPPS appeared to be appreciated, for the most part.

2.3. *C → N Strategy*

Acrylates are also indispensable starting materials in an alternative strategy for phosphinic α,α′-dipeptide synthesis. Similarly to N → C strategy, this approach involves incorporation of structural fragments to the central phosphorus core, but an acrylate, which determines the P1′ portion, is added first to the activated silyl ester of hypophosphorous acid (Scheme 8) [29,30]. The reaction, starting from ammonium hypophosphate under the same conditions as described above for α-aminophosphonous acids (see Scheme 2), is carried out and yields a β-substituted β-alkoxycarboxyl-*H*-phosphinic acid. Subsequently, the product is used in a condensation reaction with carbonyl and amine/amide components to build the N-terminal fragment of the molecule.

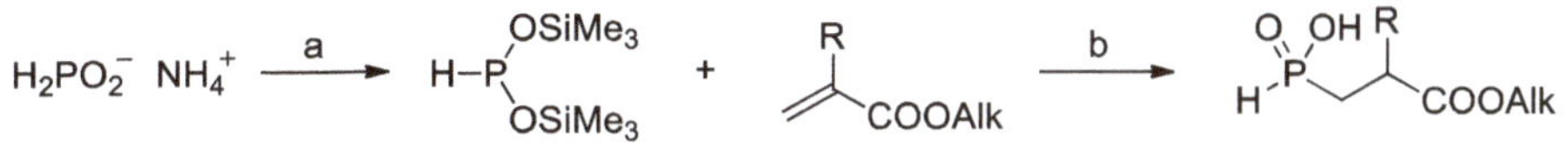

Typical reaction conditions: (**a**) HMDS, heating or TMSCl, Et_3N, CH_2Cl_2, 0 °C → rt, then acrylate; (**b**) MeOH or EtOH (for HMDS), or HCl_{aq} (for TMSCl).

Scheme 8. A general scheme for the synthesis of β-substituted β-alkoxycarboxyl-phosphonous acids, precursors of the P1′ fragment of phosphinic dipeptides.

2.3.1. Amidoalkylation

Obtaining phosphinic pseudodipeptides in a three-component condensation of carbamate, an aldehyde and a β-alkoxycarboxylphosphonous acid was initially suggested by Chen and Coward (Scheme 9) [59].

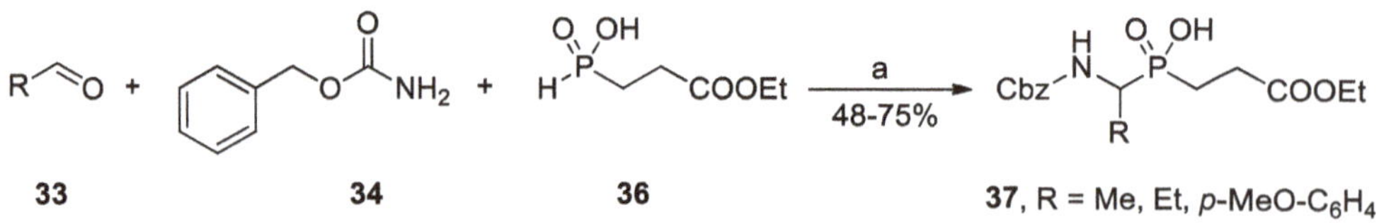

Reaction conditions: (**a**) AcCl, 0 °C → rt.

Scheme 9. Synthesis of phosphinic dipeptides in a three-component condensation of an aldehyde, benzyl carbamate and 2-ethoxycarbonylethylphosphinic acid.

This reaction was an analogy to the previous applications of less complex phosphorus components (phosphites) as a method to synthesize α-aminophosphonic acid derivatives [60]. Similarly, aliphatic and aromatic aldehydes (**33**, R = Me, Et, *p*-MeO-C_6H_4) reacted with benzyl carbamate (**34**) and 2-ethoxycarbonylethylphosphonous acid (**35**) in the presence of acetyl chloride. Conversion proceeded with a good yield (48–75%) to produce the final products **37** after an aqueous workup. The precise role of acetyl chloride as a dehydration (acetylation) solvent/reagent remained ambiguous.

Matziari and Yiotakis used the corresponding approach to obtain pseudodipeptides that were directly suitable for solid-phase peptide synthesis [61]. The modification demanded the use of $FmocNH_2$ (**38**) instead of $CbzNH_2$ and a carboxylic phosphonous acid **39** (R^2 = H, Me, *i*-Bu, Bzl), instead of an ester to react with different aldehydes **33** in AcCl/AcOH (Scheme 10). *N*-Fmoc protected C-terminal free compounds **40** were isolated with a yield comparable to the yield obtained in the original report of Chen and Coward. The products incorporated P1′ residues of glycine, alanine, valine, leucine, isoleucine, phenylglycine and histidine, but also the appropriately protected glutamic acid and serine. Their utility in solid phase preparation of elongated peptide sequences was also confirmed. An EDC/HOBt coupling ensured lack of byproducts despite the presence of a non-protected phosphinate moiety.

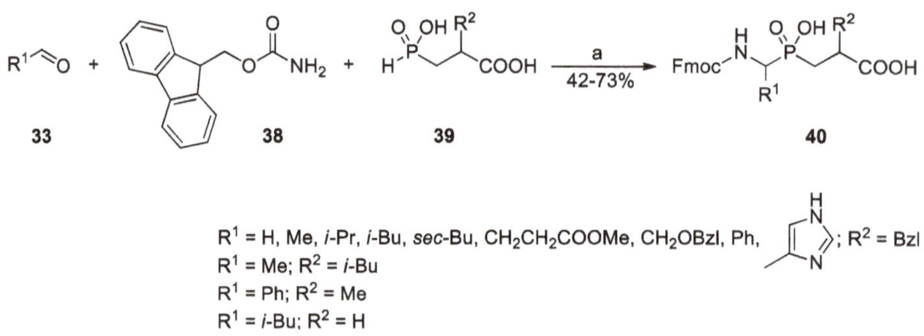

Reaction conditions: (**a**) AcCl/AcOH (5:1), 0 °C → rt.

Scheme 10. Synthesis of phosphinic dipeptide building blocks suitable for solid phase peptide synthesis in a three-component condensation of an aldehyde, 9-fluorenylmethyl carbamate and 2-substituted 2-carboxyethyl-*H*-phosphinic acid.

A similar amidoalkylation reaction performed in an intramolecular manner was employed to obtain P1 constrained peptidomimetics **42** (Scheme 11) [62]. Azacyclic compounds of a 5-, 6- and 7-membered ring were readily synthesized from α,ω-carbamoylaldehydes **41** and phosphinic acid **39** (R = Bzl). Formation of rings of these specific sizes was strongly favored. Attempts to obtain azetidine derivatives were not successful. The cyclization was apparently the driving force for the condensation. No concurrent open-chain products of intermolecular reactions were indicated in competitive experiments.

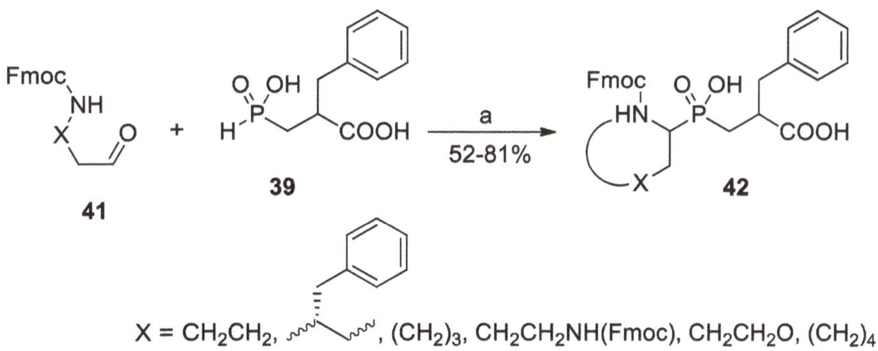

Reaction conditions: (**a**) AcCl, rt, dilution of substrates to concentration of 1.4 M.

Scheme 11. Synthesis of constrained phosphinic dipeptides in an intramolecular three-functional two-component condensation of a 9-fluorenylmethyl carbamate-derived aldehyde and 2-carboxy-3-phenylpropyl-*H*-phosphinic acid.

Based on the same general methodology, Rozhko and Ragulin synthesized fully deprotected analogues in a two-step procedure [63]. The condensation of benzaldehyde (**3**), acetamide (**43**), and a phosphonous acid **44** (R = H, *i*-Bu, CH_2COOEt) in acetic anhydride was simply followed by acid hydrolysis (Scheme 12). Phosphinic compounds **45** were isolated with a low yield (29–37%) using ion exchange chromatography.

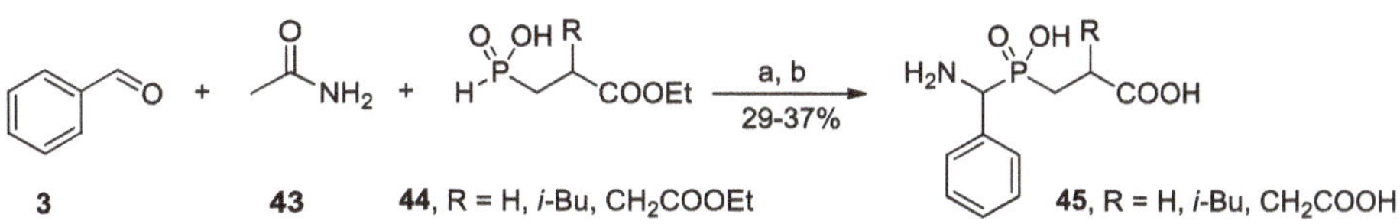

Reaction conditions: (**a**) Ac_2O, reflux; (**b**) 6–8 M HCl_{aq}, reflux, then column chromatography on Dowex 50 W (H^+).

Scheme 12. Synthesis of deprotected phosphinic dipeptides in a three-component condensation of benzaldehyde, acetamide, and 2-substituted 2-ethoxycarbonylethyl-*H*-phosphinic acid.

As mentioned, the precise role of active carboxyl compounds (in the majority of cases, AcCl) as a driving force for the condensation remained incompletely clear. Recently, this aspect has been studied in detail by Dmitriev and Ragulin [64–66]. Although acetyl chloride was proven to give the highest yield (approximately 70% yield in the reaction of benzaldehyde, an alkyl carbamate and an alkyl acrylate or methacrylate-derived phosphorus component), other systems were almost equally efficient. These systems included acetic anhydride alone or with the addition of trifluoroacetic or *p*-toluenesulfonic acid, which gave rise to 55-65% effectiveness of the condensation [64,65]. More importantly, careful inspection of the byproducts allowed the isolation of a biscarbamate **46** (Scheme 13) and suggest its structure as the key intermediate of the process [65].

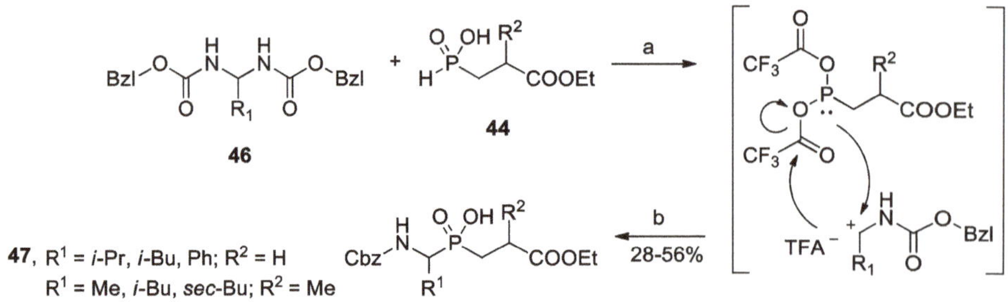

Reaction conditions: (**a**) TFAA (1 eq.), rt, ambient solvent; (**b**) H_2O.

Scheme 13. Synthesis of phosphinic dipeptides in the reaction of benzyl biscarbamate and a 2-ethoxycarbonylethyl-*H*-phosphinic acid in the presence of trifluoroacetic anhydride (TFAA) and a suggested mechanism for substrate activation.

Indeed, pre-synthesized biscarbamates reacted with phosphorous species **44** to the same extent as both substrates separately under analogous conditions [66]. To this end, an Arbuzov-type reaction was postulated as an amidoalkylation mechanism responsible for the novel P-C bond formation in **47** (Scheme 13). According to the suggestion, formation of a reactive nucleophilic phosphite and electrophilic acyliminium ion species from the corresponding substrates is mediated by carboxylic acid chloride or anhydride (here, trifluoroacetic anhydride) [66].

Currently, the amidoalkylation approach to phosphinic pseudodipeptides seems to be considered advantageous over the phospha-Michael reaction. First of all, the scope of the amidoalkylation is broader, in particular with respect to the structural complexity of the P1 substituent (Scheme 14).

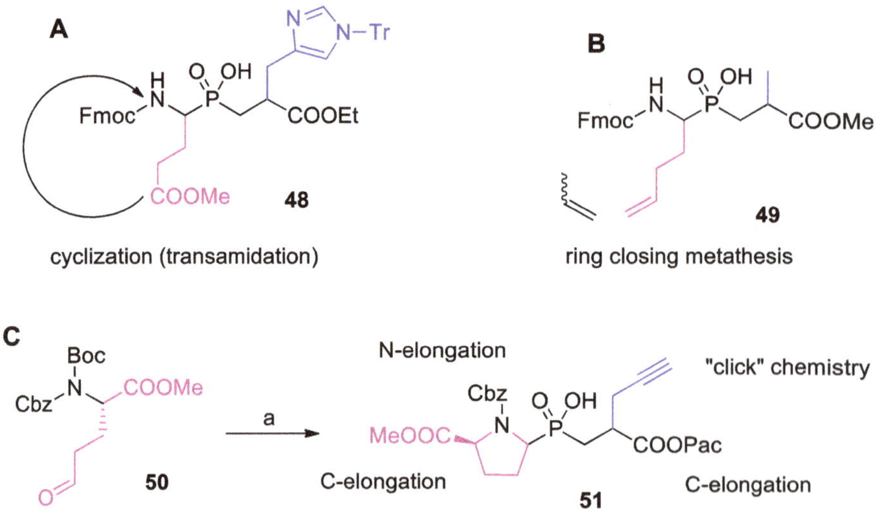

Typical reaction conditions: (**a**) an aldehyde, $FmocNH_2$ (or **50** in panel C), β-substituted β-alkoxycarbonyl-ethyl-*H*-phosphinic acid, AcCl (AcOH), rt.

Scheme 14. Examples of preparation of multifunctional phosphinic dipeptide analogues in the amidalkylation approach. Further modification points are highlighted.

P1 functionalities are well tolerated, as can be illustrated by effective preparation of a phosphinic compound bearing a glutamyl residue (**48**, Scheme 14A). Previous attempts to use phosphorus substrates incorporating additional carboxylate in an N → C silylation-activated process failed [23]. Alternative acetyl chloride-mediated amidoalkylation allowed the target molecule **48** to be obtained starting from complex substrates in one step and with a reasonable yield (48%) [50]. After incorporation into the target peptide structure (the analogue of thyrotropin-releasing hormone, cycloGlu-His-Pro-NH_2), cyclization to the pseudoglutamyl residue occurred easily. Similarly, Huber *et al.* constructed a dipeptide that holds an unsaturated P1 side-chain (**49**, Scheme 14B, yield 64%) [67]. The properly protected building block was subsequently introduced into an octapeptide sequence. Ring closing metathesis involving another homoallylglycine residue produced a cyclic constrained inhibitor of β-secretase (BACE) with improved serum stability. Finally, a highly functionalized building block **51**, the precursor of P1 constrained peptidomimetics, is also shown. This building block was obtained in a cyclative manner from a suitably protected amino acid derivative bearing a distal side-chain aldehyde function **50** (Scheme 14C) [62]. Three points of reactivity of the product are designated for elongation of the structure in standard amide chemistry. The terminal alkyne group in the P1′ position is a dipolarophilic site that offers further expansion to the heteroaromatic fragments.

2.3.2. Phospha-Mannich and Kabachnik-Fields Reaction

A related three-component condensation (phospha-Mannich reaction) was used by Hermann and co-workers to produce multifunctional complexing agents that hold a phosphinic glycyl-glycine unit. Starting from 2-carboxyethylphosphonous acid **39** (R = H) and a secondary polyamine, 1,4,7,10-tetraazacyclododecane-1,4,7-triacetate (**52**) or 1,4,7-trazacyclononane (**53**), phosphinate-functionalized products (**54** and **55**) were obtained (Scheme 15) [68,69]. As the condensation mechanism involves addition of the P-H function to the imminium ion formed *in situ* from the secondary amine and the carbonyl component, the process can be also considered as a variant of the Kabachnik-Fields reaction.

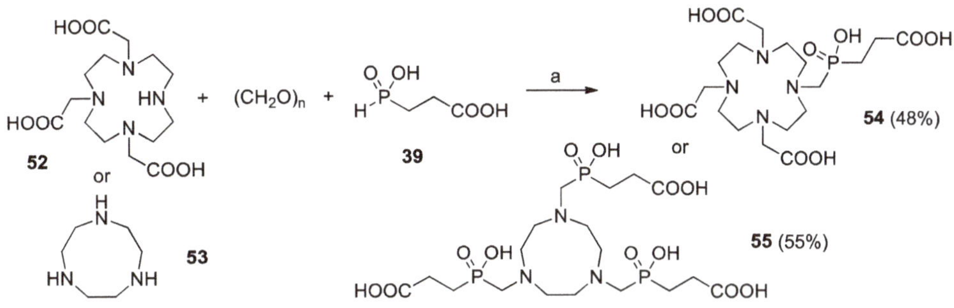

Reaction conditions: (**a**) HCl_{aq}, heating.

Scheme 15. Synthesis of Glyψ[P(O)(OH)CH_2]-Gly-substituted polyamines in a phospha-Mannich condensation with *H*-phosphinic acid and formaldehyde.

The reaction proceeded with acidic catalysis and, although the reaction gave a moderate yield, this reaction was found to be simple and inexpensive. New ligands **54** and **55** are suitable for labeling biomolecules with transition metal radioisotopes in nuclear medicine [68–72]. For example, 1,4,7-trazacyclononane-based phosphinate has recently been described as an exceptionally convenient platform to develop gadolinium ^{68}Ga(III) complexes as radiopharmaceuticals for positron emission tomography [72].

A similar idea was earlier employed by Ragulin to obtain phosphinic dipeptides bearing glycine as the P1 fragment (**58**, Scheme 16) [63,73]. The whole synthesis was performed in a convenient manner involving a one-pot two-step sequential reaction of unsaturated compounds, first an acrylate **56**, then *N*-trityl formimine (**57**), with the silyl ester of hypophosphorous acid (**20**). The methodology for subsequent modification of the phosphorus component is complementary with one-pot modifications of the C → N approach (compare chapter 2.2), but the order of P-C bond formation is just reversed.

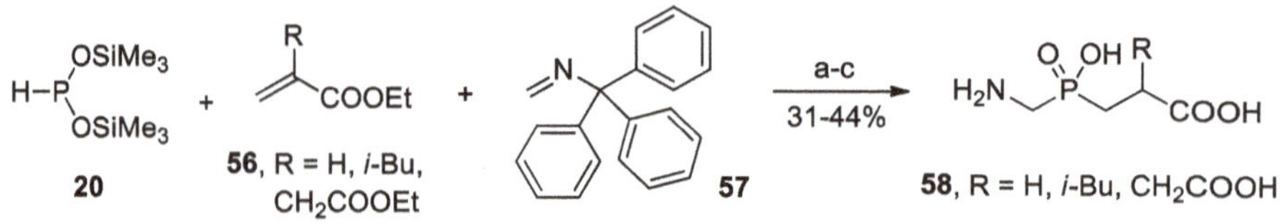

Reaction conditions: (**a**) acrylate, 40 °C; (**b**) imine, toluene, reflux; (**c**) 6–8 M HCl_{aq}, reflux, then column chromatography on Dowex 50W(H^+).

Scheme 16. Synthesis of Glyψ[P(O)(OH)CH_2]-Aa derivatives in subsequent addition reactions of an acrylate and *N*-trityl formimine to bis(trimethylsilyl) phosphonite.

The overall yield of the reaction was diminished by 10–15% because of bis(2-carboxyethyl)-phosphinic acid formation. Appearance of the symmetrical byproduct can be suppressed entirely when at least a 5-fold excess of **20** to acrylate is used [29,30]. However, in such a case, an additional workup after the first stage would be demanded.

3. Modifications Following the C-P-C Formation

Formation of the Cbz-Aa_1ψ[P(O)(OH)CH_2]-Aa_2-OAlk scaffold assembles a fundamental phosphinic structure that can rarely be used directly for further studies. At a minimum, the products of the synthesis need to be deprotected, which is achieved under standard conditions (for carbamates and esters). Alternatively, subsequent multidirectional transformations can be envisaged. These transformations typically concern two categories of changes: sequence propagation by amide bond formation at the N- and/or C-termini and/or structural modification of the R^1/R^2 substituent. Diverse chemistry has been applied for the latter purpose, including redox processes and synthesis of novel carbon-carbon and carbon-heteroatom bonds. All these reactions need careful orthogonal protection for at least three functionalities present in the molecule. Providing this protection is not an obvious task because these groups can behave unusually in this particular arrangement. The neighboring participation in the reactivity is best recognized for the pair phosphinate-β-carboxylate [74,75]. In the presence of the free carboxyl group, phosphinate protection is particularly labile, both under basic and acidic conditions. Such a hydrolysis is mediated by presumptive formation of a five-member cyclic phosphorus species (Scheme 17).

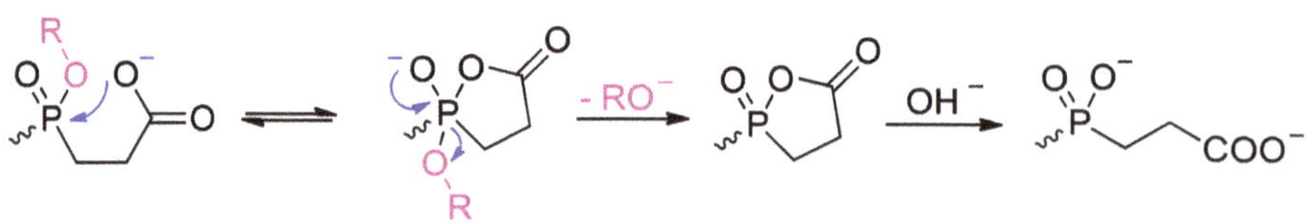

Scheme 17. Suggested mechanism for a facilitated hydrolysis of a phosphinate ester under basic conditions. The reaction proceeds *via* a five-membered cyclic intermediate with neighboring-group assistance of β-carboxylate.

Other examples of cooperative effects were also reported [45,56,76]. One such example is cyclization of terminal groups. A free amino moiety and a methyl ester form the phosphinic analogue of diketopiperazine under basic conditions [45]. In this context, proper manipulations on these groups can be quite demanding. Achievement in this field is adequately illustrated by preparation of phosphinic dipeptides for solid-phase peptide synthesis.

3.1. *Building Blocks for Solid-Phase Peptide Synthesis*

Application of phosphinic dipeptides as building blocks for peptide synthesis on a solid phase demanded the development of a convenient orthogonal protective group for the phosphinic moiety. After extensive studies in the Yiotakis group, the 1-adamantyl (Ad) group was found to be fully compatible with Fmoc methodology [16]. The blocking group was conveniently introduced in a silver oxide-mediated reaction (Scheme 18).

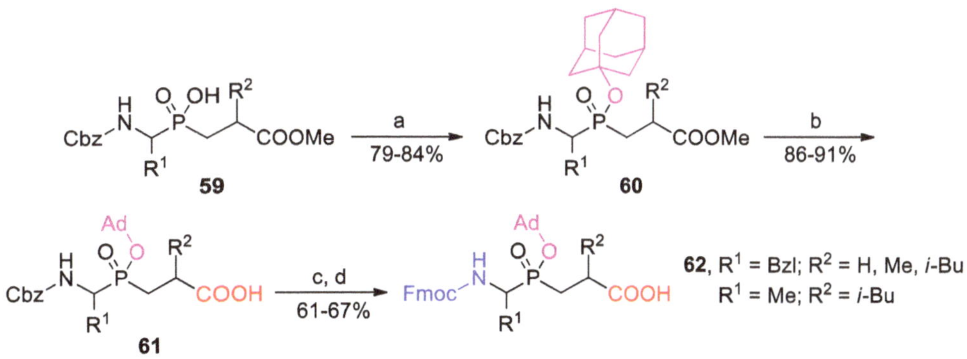

Reaction conditions: (**a**) AdBr, Ag_2O, CH_3Cl, reflux; (**b**) NaOH, H_2O/EtOH; (**c**) $HCOONH_4$, 10% Pd-C, MeOH; (**d**) FmocCl, Na_2CO_3, H_2O/dioxane.

Scheme 18. Preparation of phosphinic dipeptides as building blocks suitable for SPPS—the original procedure.

Typically, the whole procedure of Fmoc-$Aa_1\psi$[P(O)(OAd)CH_2]-Aa_2-OH (**62**) preparation is a four-step protection/deprotection reaction sequence starting from Cbz-$Aa_1\psi$[P(O)(OH)CH_2]-Aa_2-OR (**59**). The sequence consists of: (a) phosphinate adamantylation to obtain **60**; (b) C-terminal hydrolysis leading to acid **61**; (c and d) replacement of Cbz by Fmoc at the N-terminus. Cbz hydrogenolysis had to be carried out under particularly mild conditions to avoid cleaving the adamantyl group, so ammonium formate was applied as a hydrogen donor for a short-term catalytic process [16].

Since then, some modifications of the original approach have been developed. For example, the Fmoc protecting group is not sensitive to mild conditions of phosphinic acid activation (TMSCl/tertiary amine). Accordingly, Fmoc-AaPH could be applied in the Michael addition to synthesize **63** [77]. This reaction saved two manipulations at the N-terminus in the last stages of the procedure outlined in Scheme 18. However, to avoid basic conditions of alkyl carboxylate hydrolysis (also causing Fmoc cleavage), acrylate benzyl esters had to be applied (Scheme 19). Target compounds **62** were obtained with an excellent yield after adamantylation and hydrogenation.

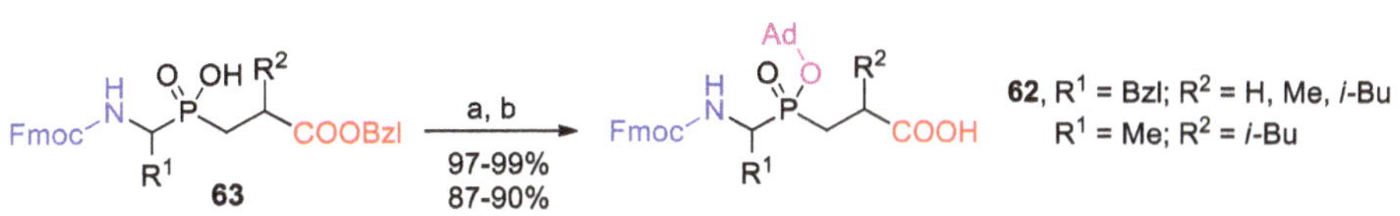

Reaction conditions: (**a**) AdBr, Ag_2O, CH_3Cl, reflux; (**b**) H_2, 10% Pd-C.

Scheme 19. Final stages of phosphinic dipeptide preparation starting from N-Fmoc-α-amino-*H*-phosphinic substrates.

To avoid problems in a large scale and expensive silver-mediated adamantylation, Buchardt *et al.* suggested a two-step esterification that proceeded *via* the appropriate phosphinochloridate, and Buchardt *et al.* validated this approach for the Gly-Leu analogue (Scheme 20) [78]. Phosphinochloridate **65**, prepared quantitatively by oxalyl chloride reaction with substrate **64**, spontaneously reacted with sodium 1-adamantyl oxide to yield the P-protected compound **66**. *N*-Terminus protection and deprotection leading from **61** to **62** (R^1 = H, R^2 = *i*-Bu) were coalesced into one step by hydrogenolysis performed in the presence of the FmocOSu acylation agent.

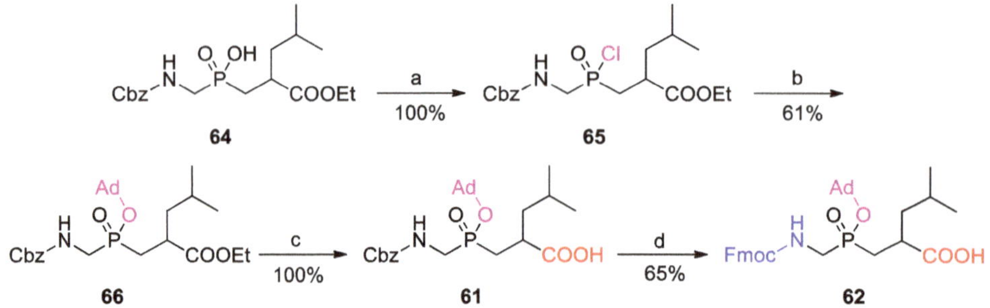

Reaction conditions: (**a**) $(COCl)_2$, DMF, THF, 0 °C; (**b**) AdONa; (**c**) NaOH, H_2O/EtOH; (**d**) H_2, 5% Pd-C, FmocOSu, $NaHCO_3$, AcOEt/MeOH/H_2O.

Scheme 20. Modification of Fmoc-$Aa_1\psi[P(O)(OAd)CH_2]$-Aa_2-OH preparation, performed for Gly-Leu analogue.

Finally, all specific improvements were combined. Following the preparation of adamantylated products **67** (*via* the corresponding phosphinochloridates), all three remaining manipulations (tandem C-benzyl and N-Cbz removal together with simultaneous *N*-Fmoc introduction) proceeded in a single reduction step to give **62** (R^1 = H, R^2 = *i*-Pr or *i*-Bu, Scheme 21) [79].

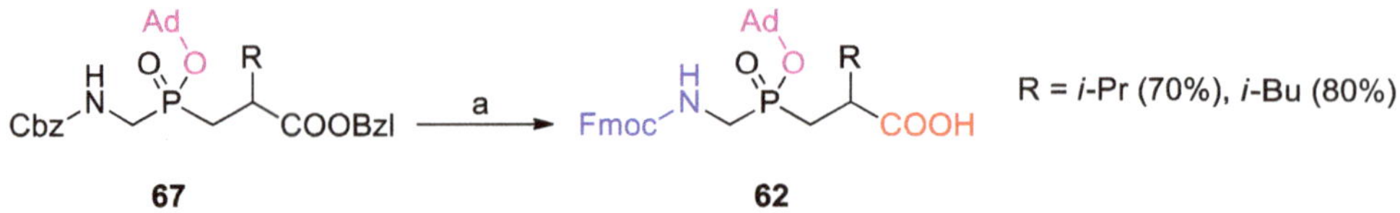

Reaction conditions: (**a**) H_2, 5% Pd-C, FmocOSu, MeOH.

Scheme 21. Transformation of Cbz-Gly$\psi[P(O)(OAd)CH_2]$-Aa_2-OBzl into Fmoc-Gly$\psi[P(O)(OAd)CH_2]$-Aa_2-OH in a single synthetic step.

N-Fmoc-protected phosphinic pseudodipeptides **40** with both acidic functionalities free are easily accessible using a three-component condensation (compare Scheme 10) [61]. Having these pseudopeptides in hand, a unique chemoselective protection strategy was envisaged by Yiotakis and co-workers (Scheme 22) [76]. The phenacyl (Pac) group was installed on both the phosphinate and carboxylate of **68** under common reaction conditions. However, the action of a mild acid caused fully selective removal only from the hydroxyphosphinyl moiety (**69**). Such susceptibility to hydrolysis was tentatively explained by the assistance of the carbamate neighboring group, which might also participate in the formation of a five-membered reactive intermediate [76]. Two subsequent steps consisting of a typical adamantyl ester formation and Pac cleavage from the C-terminus completed the reaction sequence.

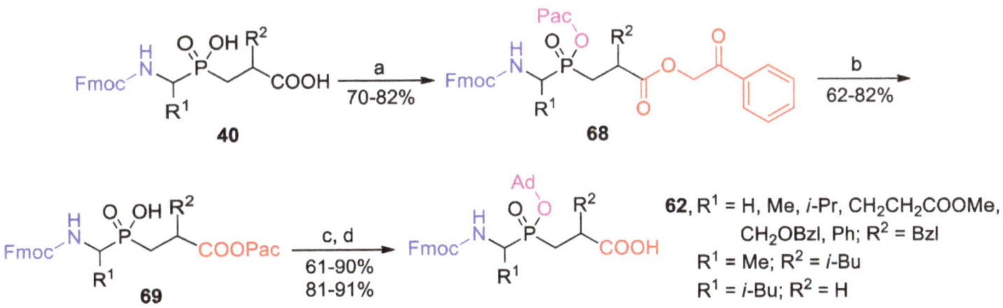

Reaction conditions: (**a**) PacBr, Et_3N, AcOEt, 0 °C → rt; (**b**) TFA/CH_2Cl_2 (4:1), 50 °C; (**c**) AdBr, Ag_2O, CH_3Cl, reflux; (**d**) Mg, AcOH, MeOH/DMF (4:1), rt.

Scheme 22. Chemoselective protection of a phosphinic dipeptide.

Fmoc-$Aa_1\psi$[P(O)(OAd)CH_2]-Aa_2-OH building blocks (**62**) are commonly used in standard, automated Fmoc/benzotriazole solid-phase and combinatorial peptide synthesis [16,21,80–83]. Nevertheless, several reports consider P-protection unnecessary when particular short sequences are being targeted. The phosphinate functionality is poorly activated by standard coupling agents. The carboxylate can therefore be chemoselectively converted into the corresponding amide in the presence of the free phosphinic acid moiety, both in solution and on the solid phase by means of DCC (EDC)/HOBt, BOP, PyBOP, *i*-butyl chloroformate or other activators [32,36,61,84,85].

3.2. Side-Chain Substituents Modifications and Parallel Diversification

Post-synthetic modifications of the phosphinate scaffold substituents concern mainly the C-terminal fragment of the molecule. One of the rare examples of the P1 transformation is an oxidative conversion of the phenyl ring of **70** into a carboxylate group. Mild conditions of oxidation allowed Georgiadis *et al.* to obtained Asp and Glu acidic residues in the $Aa_1\psi$[P(O)(O)CH_2]-Ala sequence (**71**) in moderate yield (Scheme 23) [23]. This problematic sequence is difficult to obtain in a typical phospha-Michael approach.

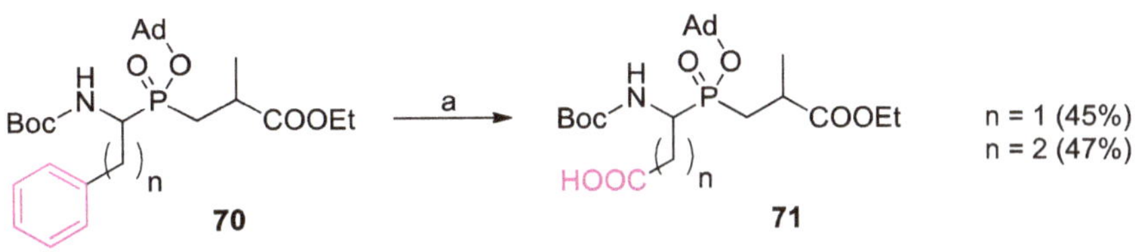

Reaction conditions: (**a**) $RuCl_3$, $NaIO_4$, MeCN, H_2O.

Scheme 23. An oxidative approach to Aspψ[P(O)(O)CH_2]-Ala and Gluψ[P(O)(O)CH_2]-Ala phosphinic dipeptides.

Among chemical entities implemented in multidirectional and parallel P1′ diversification chemistry, active methylene phosphinic synthons have been found to be one of the most interesting options (Scheme 24).

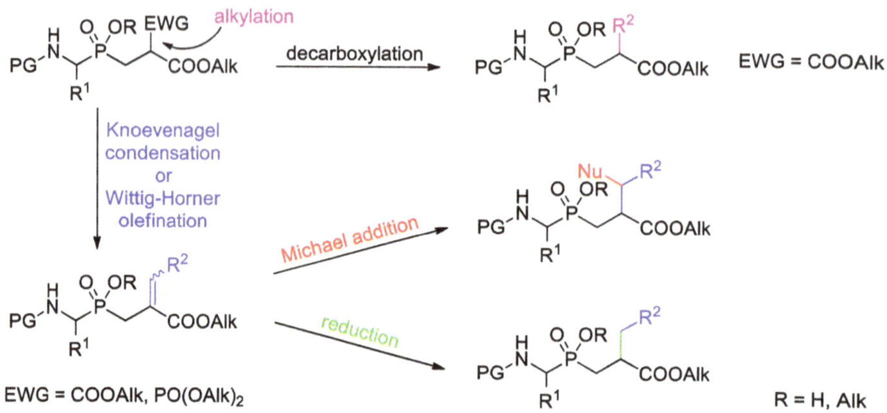

Scheme 24. A general scheme of multidirectional modifications that have been envisaged starting from P1′ active methylene phosphinic compounds.

Such synthons are easily prepared by addition of an alkyl methylidenemalonate or α-phosphonoacrylate to an α-amino-*H*-phosphinate. The Cα′ proton acidity of the adduct can subsequently be utilized in several different ways. Two main routes involve alkylation of the malonate followed by decarboxylation of the product and a Wittig-Horner type of olefination (or, alternatively, Knoevenagel condensation) to introduce a double bond that can optionally be reduced or added with a nucleophile. Both methods have served to introduce a specific side-chain substituent to P1′ portion of the molecule.

The alkylation strategy was used, for example, in the Ebetino group to prepare a precursor for the phosphinic Phe-Arg dipeptide isostere (**74**, Scheme 25) [44]. Di-*t*-butyl malonate-substituted pseudodipeptide **73**, obtained from **24** and malonate **72**, allowed effective alkylation with 1-azido-3-iodopropan (yield 88%). The building block was incorporated into the target peptide sequence and then subjected to one-pot modifications. The azido group was reduced to the amino group that finally was guanidinated *in situ*.

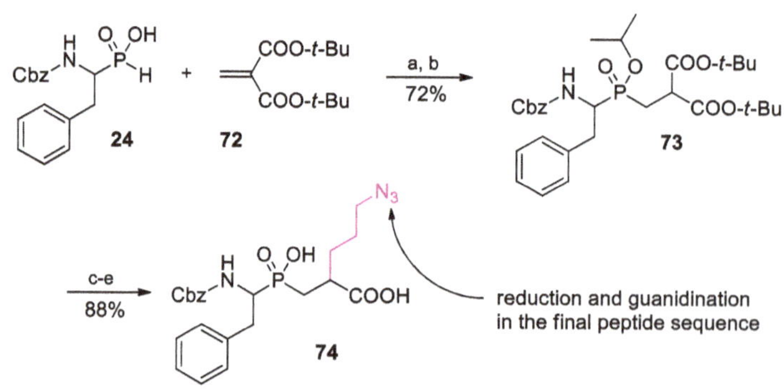

Reaction conditions: (**a**) TMSCl, DIPEA, CH_2Cl_2, 0 °C; (**b**) *i*-PrOH, EDC, CH_2Cl_2; (**c**) *t*-BuOK, $I(CH_2)_3N_3$, DME; (**d**) TFA, CH_2Cl_2; (**e**) toluene, reflux.

Scheme 25. Construction of phosphinic dipeptide precursor of Pheψ[P(O)(OH)CH_2]-Arg.

Similarly, Matziari *et al.* introduced bulky arylmethyl P1′ substituents to **75** using appropriate bromides and an alkylation approach (Scheme 26) [86]. The overall yield of products **76** was roughly 50%, outscoring the standard Michael-type addition that had previously been applied to provide the same 2-naphthyl compound (6.5%).

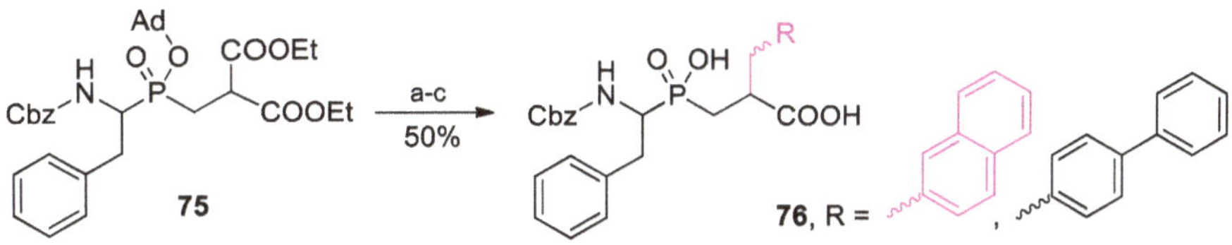

Reaction conditions: (**a**) NaH, RCH_2Br, THF, 0 °C → rt; (**b**) TFA, CH_2Cl_2; (**c**) LiOH, MeOH.

Scheme 26. Alkylation of a malonate-derived phosphinic dipeptide.

As an early example of the second variant (double bond formation and subsequent reduction, compare Scheme 24), Parsons *et al.* performed a stereoselective synthesis of D-Alaψ[P(O)(OH)CH_2]-D-Ala inhibitor (**80**) of D-Ala-D-Ala ligase [31], a bacterial enzyme involved in cell wall biosynthesis. After addition of trimethyl phosphonoacrylate (**78**) to the *H*-phosphinic ester **77** and reaction with formaldehyde, stereoselective reduction of the pseudodehydroalanine compound **79** was employed (Scheme 27).

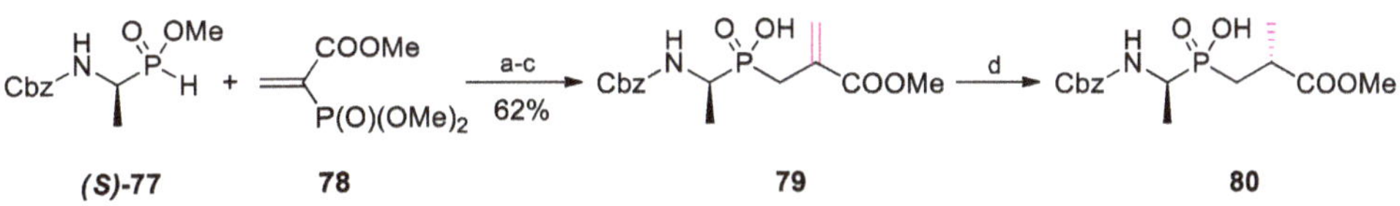

Reaction conditions: (**a**) MeONa, methanol, 0 °C; (**b**) CH_2O, 0 °C → rt; (**c**) LiI, THF; (**d**) H_2, $[(COD)RhCl]_2$, (−)-DIOP ((−)-*O*-isopropylidene-2,3-dihydroxy-1,4-bis(diphenylphosphino)butane) then recrystallization from acetic acid.

Scheme 27. Synthesis of D-Alaψ[P(O)(OH)CH_2]-D-Ala by a stereoselective reduction of dehydroalanine.

More recently, an analogous approach was utilized to synthesize phosphinic dehydro analogues of cyclohexylalanine **81**-based inhibitors of renal dipeptidase (Scheme 28) [87,88]. After olefination of the phosphonoacetate with cyclohexylaldehyde or structurally diversified benzaldehydes, *Z* and *E* diastereoisomeric products **82** and **83** were separated chromatographically. *Z* isomers slightly predominated, typically in a 40%:30% ratio. Deprotected compounds, tested against the target peptidase, revealed the enzyme preference for the *Z* configuration (IC_{50} of a low nanomolar value).

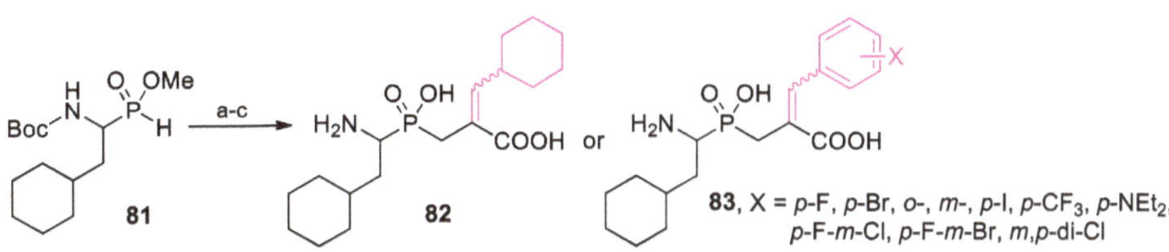

Reaction conditions: (**a**) trimethyl-2-phosphonoacrylate, MeONa, MeOH, 0 °C; (**b**) cyclohexylaldehyde or ArCHO, rt, then chromatographic separation; (**c**) TFA, CH_2Cl_2, rt, then HCl_{aq}, 50 °C.

Scheme 28. Synthesis of P1′ dehydro phosphinic dipeptide inhibitors of renal dipeptidase.

To obtain P1′ dehydroamino acid residues **85**, Matziari *et al.* employed a Knoevenagel-type condensation (Scheme 29) [86]. Different aldehydes reacted with phosphinic dipeptide-derived malonic acid **(*S*)-84** with a good yield and a preference for the *E* diastereoisomer up to 100%. The attempts at asymmetric reduction with rhodium catalysts and chiral ligands gave poor *de*.

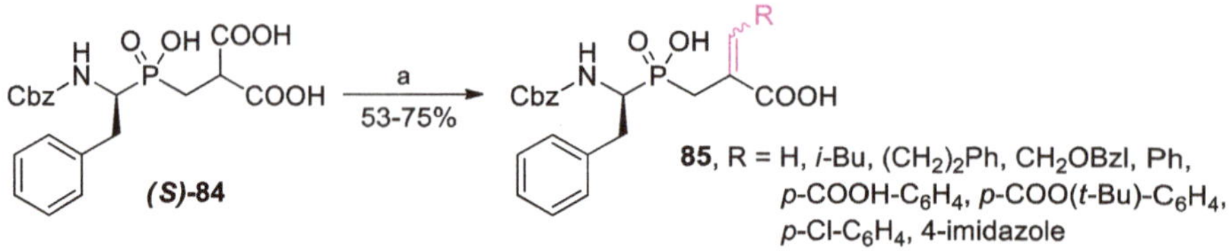

Reaction conditions: (**a**) RCHO, piperidine, pyridine, reflux.

Scheme 29. Synthesis of phosphinic pseudodipeptides containing P1′ dehydroamino acid fragments by a Knoevenagel condensation.

A mild, one-step and high-yielding conversion of P-H substrates, such as **24**, to P1′ dehydroalanine phosphinic analogues **87** was also proposed in the Yiotakis group [89]. This conversion involved reaction of α-(bromomethyl)acrylic ester (**86**) with the trimethylsilylated phosphinate *via* a tandem Arbuzov reaction followed by allylic rearrangement (Scheme 30). The same results were obtained in the reaction of the unsaturated substrate with the acetoxy (instead of bromo) leaving group. A diversification potential for the dehydroalanine fragment was used to develop novel inhibitors of matrix metalloproteases (see below).

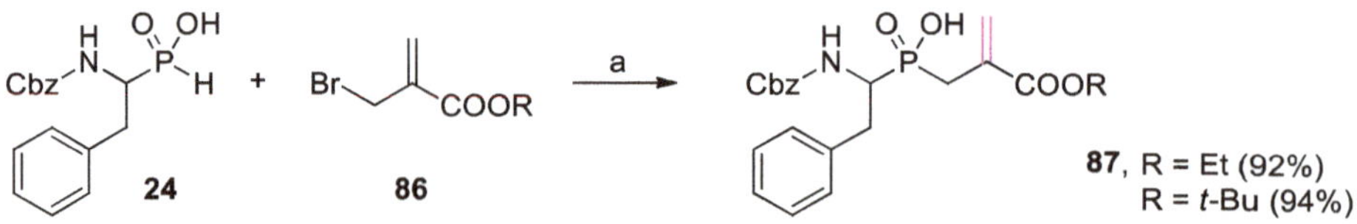

Reaction conditions: (**a**) TMSCl, DIPEA, CH_2Cl_2, 0 °C → rt or HMDS, 110 °C; then acrylate, 0 °C → rt (for TMSCl variant) or 110 °C (for HMDS variant), then EtOH.

Scheme 30. Reaction of α-amino-*H*-phosphinic acid with α-(bromomethyl)acrylates leading to dehydroalanine phosphinic dipeptides.

The tandem Arbuzov reaction (employing the acetoxy derivative **88**), combined with double bond reduction, was used to introduce the constrained P1′ fragment of proline to the phosphinic dipeptide scaffold **89** (Scheme 31) [90]. Incorporated as a building block into a tripeptide sequence, it allowed to establish structural determinants of highly potent and selective inhibitors of the C-terminal domain of angiotensin-converting enzyme.

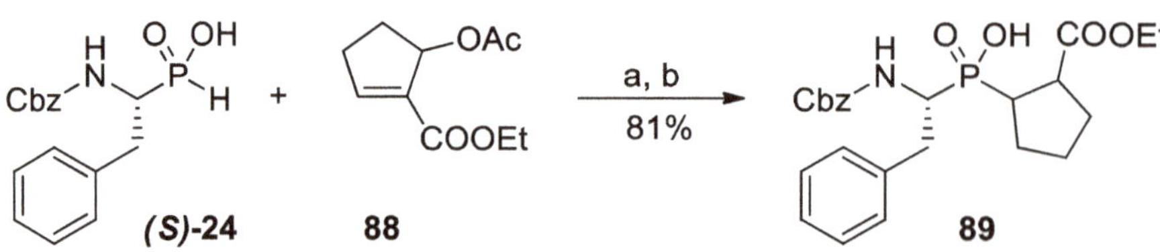

Reaction conditions: (**a**) HMDS, 110 °C, then acrylate, 90 °C, then EtOH; (**b**) $NaBH_4$, $NiCl_2$, MeOH, −30 °C.

Scheme 31. Preparation of a phosphinic dipeptide bearing the pseudoproline P1′ fragment.

Dehydroalanine synthon **87** can undergo a variety of subsequent transformations. As indicated, nitrogen, carbon and sulfur nucleophiles can easily be added to the activated double bond to yield products **90–93** (Scheme 32) [89]. The reactions were carried out under mild conditions and without the necessity of phosphinate group protection. The potential of P1′ diversification in this manner was best recognized in the addition of thiols to acrylate implemented in a tripeptide sequence. The products were studied as inhibitors of matrix metalloproteases with particular attention to MMP-11 [91]. As a result, a diversity of tripeptide analogues of a general structure Cbz-(*L*)-Pheψ[P(O)(OH)CH_2]-Cys(*R*)-(*L*)-Trp-NH_2 were obtained and tested for complementarity to the S1′ pocket of MMP-11. Remarkable selectivity of two orders of magnitude *versus* other members of MMP family was found for particular R substituents, for example, *o*-bromo or *o*-methoxyphenyl [91].

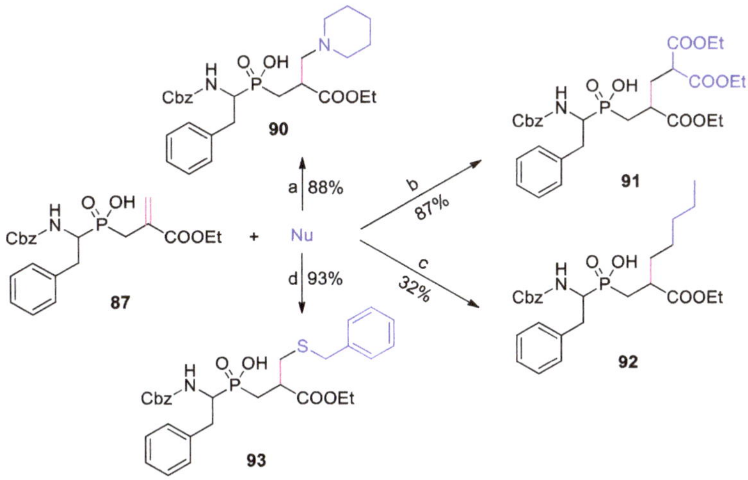

Reaction conditions: (**a**) piperidine, CH_2Cl_2, rt; (**b**) sodium diethylmalonate, EtOH, rt; (**c**) *n*-BuLi, CuI, TMSCl, Et_2O/THF, −20 °C; (**d**) BzlSH, Et_3N, CH_2Cl_2, rt.

Scheme 32. Addition of various nucleophiles (Nu) to P1′ dehydroalanine phosphinic dipeptide.

Other unsaturated groups present in the Cα′ side-chain substituent of the phosphinic dipeptides can also be used for specific transformations. In particular, terminal alkenes and alkynes have been recognized in "click" chemistry to produce heterocyclic systems (Scheme 33) [40,47,48]. Both unsaturated systems (**94** and **96**) were successfully subjected to 1,3-dipolar cycloaddition as dipolarophilic substrates. The unsaturated systems reacted with nitrile oxides generated *in situ* from aryl oximes by oxidative chlorination. As a result, aryl-substituted isoxazoles or isoxazolines (**95** and **97**) were obtained with a good yield. The transformations were found to be equally effective in a dipeptidic building block and in an elongated peptide structure. Similarly to the previous example, the products were utilized to explore the specificity of the S1′ binding pocket of selected metalloproteases, e.g., matrix metalloproteases, angiotensin- and endothelin-converting enzymes [40,47,48].

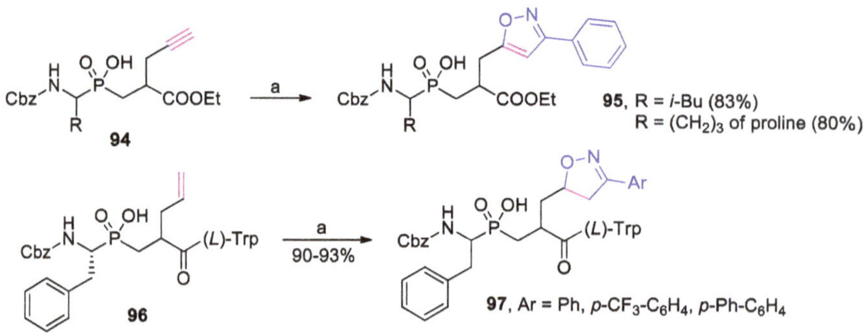

Reaction conditions: (**a**) ArCH=NOH, NCS, pyridine, $CHCl_3$, 45 °C, then an alkene or alkyne, Et_3N, 45 °C.

Scheme 33. Synthesis of P1′ isoxazole and isoxazoline derivatives of phosphinic peptide in 1,3-dipolar cycloaddition.

An expected anti-aminopeptidase inhibitory activity inspired elaboration of a phosphinic dipeptide building block containing a β′ amino group that was suitable for parallel substitution [46]. According to the molecular modeling results, this group should be close to the position of the nitrogen atom in the transition state of the cleaved amide bond and, thus, favorably bound at the enzyme active site. The appropriate synthons were obtained by addition of the *H*-phosphinic analogue of homophenylalanine (**98**) to a dehydroalanine derivative **99** (Scheme 34).

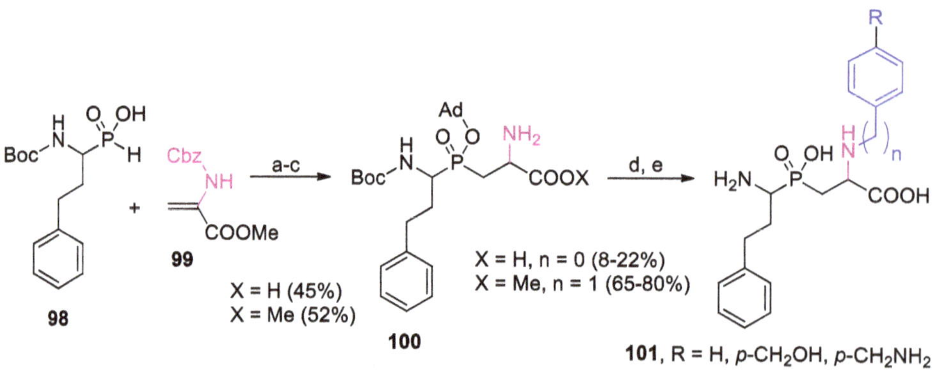

Reaction conditions: (**a**) HMDS, 110 °C, then dehydroalanine, 90 °C; (**b**) AdBr, Ag_2O, $CHCl_3$, reflux; (**c**) NaOH, MeOH, rt, then H_2, 10% Pd/$BaSO_4$, MeOH (for X = H) or H_2, 10% Pd/$BaSO_4$, MeOH (for X = Me); (**d**) ArI, CuI, K_2CO_3, 65 °C (for X = H, n = 0) or $ArCH_2Br$, Cs_2CO_3, DMF, 45 °C (for X = Me, n = 1), (**e**) TFA, CH_2Cl_2, rt, then HPLC (for X = H) or NaOH, MeOH, rt, then TFA, CH_2Cl_2, rt, then HPLC (for X = Me). When necessary, the additional substituents R of the substrate halides were suitably protected.

Scheme 34. Parallel modification of phosphinic peptides containing a β′ amino group in cross-coupling and alkylation reactions.

After typical protection/deprotection manipulations of the protecting group, two building blocks (**100**, X = H, n = 0 or X = Me, n = 1) were constructed and subjected to an cross-coupling with corresponding aryl iodides or an alkylation reaction with benzyl bromides [46]. Because of the structural complexity (namely, the presence of multiple functionality groups in both substrates), the yield of the cross-coupling reaction was rather low. The final products **101** needed to be purified on HPLC, which, in turn, allowed separation of diastereoisomers in the majority of cases.

4. Stereoselective Approaches

Enantiomerically pure α-amino-*H*-phosphinic acids are relatively readily available. *N*-Benzyloxycarbonyl-protected compounds form diastereomeric salts with commercially available chiral amines, e.g., 1-phenylethylamine. The salts crystallize smoothly to yield the required enantiomer [22]. The resolved phosphonous acids are not susceptible to racemization and tolerate diverse reaction conditions. Accordingly, optically pure α-amino-*H*-phosphinic acids are commonly used as the substrates in the phospha-Michael addition to produce P1 stereo-defined phosphinic pseudodipeptides. Even though the enantiomeric phosphonous acids rarely cause a significant induction of the newly appearing asymmetric center (Cα′ atom), diastereomeric products can be separated simply by crystallization or chromatography [21,47]. The resolution is particularly effective when pseudodipeptides are included in a stereo-defined sequence of elongated peptides. In those cases, differentiation of even four diastereoisomers (epimeric on both Cα and Cα′ atoms) on HPLC is not problematic [35,36]. Similarly, an astonishing difference in solubility in common solvents allowed separation of the four diastereoisomers of a phosphinic tripeptide by simple recrystallization [92]. With the aid of quantum mechanical calculations and molecular dynamics simulation these distinguishing physicochemical properties were attributed to the conformationally-specific pattern of inter- and intramolecular interactions among solute and solvent molecules. In accordance with this argument, stereoselective synthesis of phosphinic dipeptide analogues is not popular in the literature and concerns basically the P1′ induction. Evans oxazolidinone-type auxiliaries were applied by the Ebetino group to induce the stereochemistry of the P1′ position of Phe-Phe phosphinic dipeptide **104** (Scheme 35) [93].

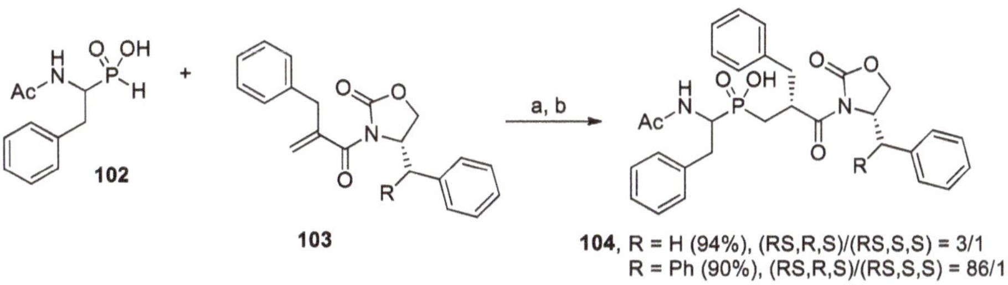

Reaction conditions: (**a**) TMSCl, DIPEA, CH_2Cl_2, 0 °C → rt, then acrylate addition; (**b**) EtOH, −10 °C.

Scheme 35. Diastereoselective addition of chiral oxazolidinone-derived acrylamides to α-amino-β-phenylethyl-*H*-phosphinic acid.

To prepare chiral substrates **103**, α-benzylacrylic acid was coupled with the lithium salt of an oxazolidinone. Low-temperature trimethylsilyl-mediated addition of **102** to the acrylamides was found to produce an optimal induction. Quite naturally, top results of the *de* were achieved for the most crowded 4-diphenylmethyl-substituded oxazolidinone. Semiempirical AM1 energy calculation for the presumed enol ether intermediates pointed out that the *R* isomer is the favored product.

A double diastereoselective approach to phosphinic dipeptide formation in a Michael addition followed by a Cα′ alkylation was presented by Yamagishi *et al.* (Scheme 36) [94]. Using enantiomerically (R^1 = H) or diastereomerically (R^1 = *i*-Bu, Bn) pure P-chiral α-aminoalkyl-*H*-phosphinic acid esters (**105**) as substrates, the authors carried out addition of *t*-butyl acrylate (**106**) without a loss of the phosphorus chirality. The reaction, catalyzed efficiently by magnesium alkoxide, produced a single isomer **107** with an excellent chemical yield (Scheme 36A). The stereochemistry was controlled by the phosphorus atom configuration to give the complete retention for various P1 side chain substituents and N-protection groups.

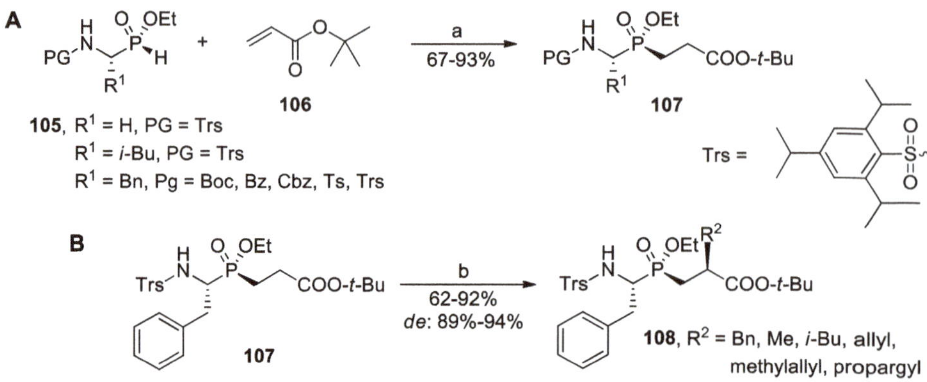

Reaction conditions: (**a**) *t*-BuOMgBr, 0 °C; (**b**) LiHMDS, R^2X, −78 °C.

Scheme 36. Diastereoselective phospha-Michael addition and Cα′ alkylation of the resulting phosphinate dipeptide controlled by the defined phosphorus atom configuration.

Contrarily, the subsequent lithium-mediated alkylation step depended strongly on the steric parameters of the N-terminus. Excellent diastereoselectivity was observed for the most hindered Trs sulfonate protection (**107**, PG = Trs, R^1 = Bzl), whereas induction was markedly poorer for other blocking groups [94]. Bulkiness of the side chain on the Cα position was not critical. Several different residues (alkyl, benzyl, alkenyl and alkynyl) were introduced in this way to the phosphinate dipeptide backbone as novel P1′ substituents with a good yield (compounds **108**, Scheme 36B). Difficult N-deprotection can be considered as the single disadvantage of the method. The deprotection is a moderately effective three-step procedure including additional Cbz carbamoylation, Trs removal by SmI_2 treatment and final Cbz hydrogenolysis [94].

The addition of cyclopentenyl carboxylate to TrsLeuPH ethyl ester (**105**, R = *i*-Bu, PG = Trs) proceeded with difficulty. The methodology failed to prepare a Leuψ[P(O)(OH)CH_2]-Pro derivative in the same manner. To overcome this drawback, a cross-coupling of the phosphorus substrate with the appropriate triflate **109** was envisaged (Scheme 37) [95]. After careful optimization of conditions, a palladium-catalyzed reaction carried out in the presence of K_2CO_3 as a base and DPEphos ((oxydi-2,1-phenylene)bis(diphenylphosphine)) as a chiral ligand yielded the unsaturated dipeptide **110** with a very good yield and an excellent *de*. Quite naturally, the subsequent reduction gave predominantly the *cis*-cyclopentane diastereoisomer **111** that could be isolated in a pure form after recrystallization or preparative TLC. Epimerization, occurring spontaneously to some extent during the hydrogenolysis, was preparatively induced by increase the temperature of reduction to yield the *trans*-cyclopentane configuration (**112**) [95].

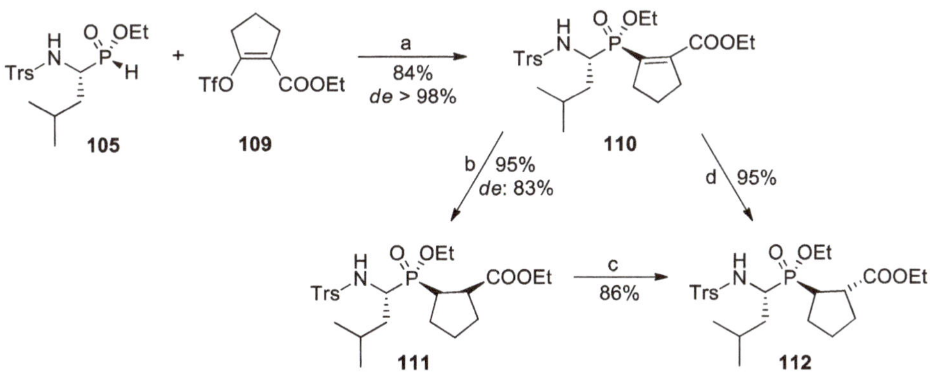

Reaction conditions: (**a**) K_2CO_3, $Pd_2(dba)_3$, DPEphos, triflate (2 eq.), toluene, 80 °C; (**b**) H_2 (3.5 atm), PtO_2, EtOH, rt; (**c**) H_2, PtO_2, EtOH, 80 °C; (**d**) H_2, PtO_2, EtOH, rt then 80 °C.

Scheme 37. Diastereoselective cross-coupling and subsequent reduction leading to Leuψ[P(O)(OH)CH_2]-Pro derivatives of the controlled pseudoproline configuration.

Diastereoselective Michael-type addition of 2*H*-2-oxo-1,4,2-oxazaphosphinane **113** to olefins, including α,β-unsaturated esters (**114**, R = Me or Bzl, Scheme 38), was investigated by Monbrun *et al.* [96]. The reactions catalyzed by potassium *t*-butoxide proceeded with an excellent chemical yield and with complete retention of configuration of the phosphorus atom.

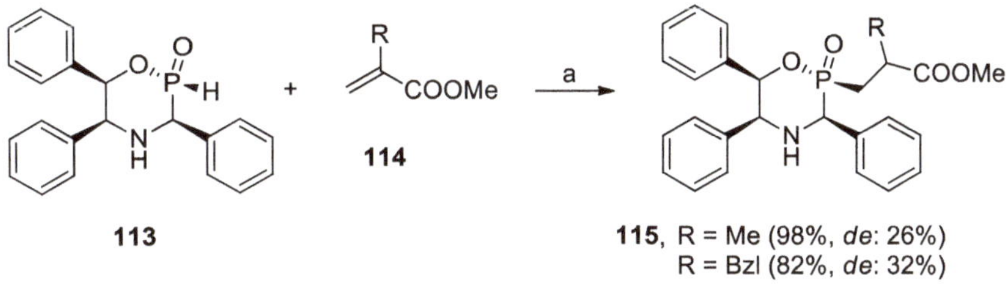

Reaction conditions: (**a**) CH_2Cl_2, *t*-BuOK, −5 °C → rt.

Scheme 38. Diastereoselective addition of 2*H*-2-oxo-1,4,2-oxazaphosphinanes to acrylates.

The chiral induction of the newly formed stereogenic center in **115** depended strongly on the structure of the unsaturated substrates. For acrylates, the results were rather poor (~30%). More impressive data were indicated for substituents located closer to the sterically crowded heteroatom system.

In addition to the synthetic methods, stereoselective instrumental techniques were also applied to differentiate/resolve enantiomeric phosphinic dipeptide analogues. For example, an anion exchange chiral stationary phase based on cinchona alkaloids allowed resolution of dipeptide analogues in the reversed-phase HPLC mode [97]. The most effective *O*-9-(*t*-butylcarbamoyl)quinidine selector immobilized on a solid support was successfully utilized in chiral chromatography or electrochromatography separation of four stereoisomers of hPheψ[P(O)(OH)CH_2]-Phe, a nanomolar inhibitor of leucine aminopeptidase [98,99].

5. Applications and Conclusions

The main field of application of phosphinic pseudopeptides concerns fundamental and practical aspects of inhibition of selected enzymes. This field of application mostly means representatives of two classes of catalytic proteins: ligases and hydrolases, in particular metallo-dependent proteases.

Among others, these enzymes can catalyze, respectively, formation or cleavage of an amide bond in a metal ion-mediated process that proceeds *via* a tetrahedral transition state. C-P-C dipeptides comprise fundamental structural features of advantageous transition state analogue inhibitors. First, the dipeptides contain P1 and P1′ fragments dedicated to explore the specificity of the corresponding S1 and S1′ enzyme binding pockets. Second, the central phosphinate group is tetrahedrally-shaped and mimics the geometry and electron distribution of a diolate intermediate in the hydrolysis process. Third, phosphinic acid can be complexed to the central metal ion and block its catalytic function. Accordingly, phosphinic dipeptide isosters are broadly recognized as reversible, competitive inhibitors of many enzymatic targets [5–7,21].

Regulation of D-Ala-D-Ala ligase activity by D-Alaψ[P(O)(OH)CH_2]-D-Ala (**115**) is one of the most prominent examples of ligase inhibition. The enzyme catalyzes condensation of two alanine molecules with ATP participation to form the terminal peptide of a peptidoglycan monomer. As the process is specific for bacteria, D-Ala-D-Ala ligase is an attractive antimicrobial target. The phosphinic analogue of the product of the reaction was found to be a micromolar inhibitor of the enzyme [31,100]. Interestingly, careful inspection of the mechanism of its action revealed that the phosphinate is not a typical transition state analogue. The phosphinate binds to the ligase in the phosphorylated form (**116**, Scheme 39A). Thus, the pseudodipeptide can be considered as a suicide substrate [100,101]. Identical phosphorylation of D-Alaψ[P(O)(OH)CH_2]-D-Ala was discovered upon binding with VanA, a ligase responsible for natural resistance of vancomycin [102,103]. Other bacterial ligases involved in peptidoglycan biosynthesis have also been exploited as targets for antibacterial drug design using phosphinate dipeptide inhibitors [42,104–106]. Representative examples of UDP-*N*-acetylmuramoyl-L-alanyl:D-glutamate ligase (MurD) inhibitors (**117** and **118**) are shown in Scheme 39B. Recently, the L-Alaψ[P(O)(OH)CH_2]-L-Phe ligand was also used for structural characterization of an L-amino acid ligase from *Bacillus subtilis* [107].

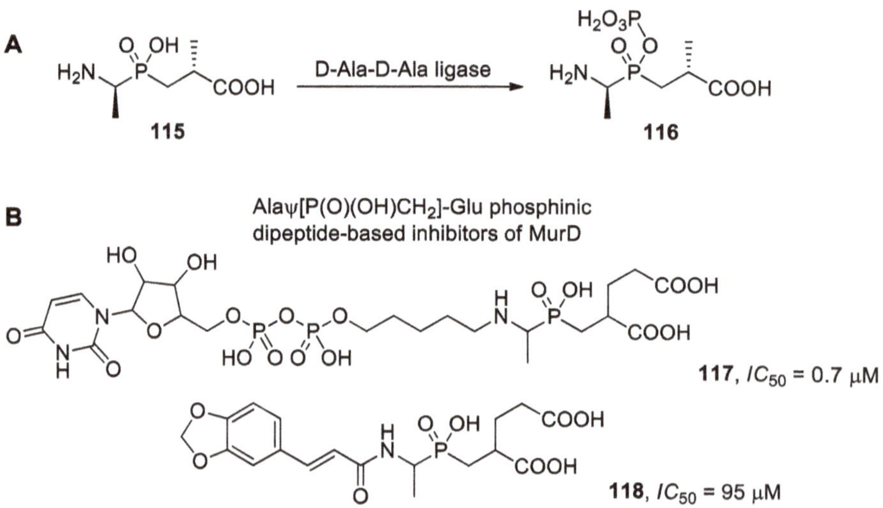

Scheme 39. (**A**) The mode of action of D-Alaψ[P(O)(OH)CH_2]-D-Ala, a phosphinic dipeptide inhibitor of D-Ala-D-Ala ligase; (**B**) Phosphinic dipeptide-derived inhibitors of MurD, another ligase involved in peptidoglycan biosynthesis.

As far as metalloprotease targets are concerned, inhibition of leucine and alanyl aminopeptidases (LAP and APN) seems to be the most recognized example of biological activity of phosphinic dipeptide analogues [45,46,99,108–111]. LAP and APN are multifunctional broad-band specificity aminopeptidases involved in biological functions in both eukaryotic and prokaryotic cells.

hPheψ[P(O)(OH)CH_2]-Phe (**119**) and hPheψ[P(O)(OH)CH_2]-Tyr (**120**), phosphinic compounds of an optimized structure, exhibited nanomolar activity towards mammalian (porcine kidney) [45], protozoan (recombinant *Plasmodium falciparum*) [112,113] and bacterial (recombinant *Neisseria meningitides*) [114] enzymes (Figure 3).

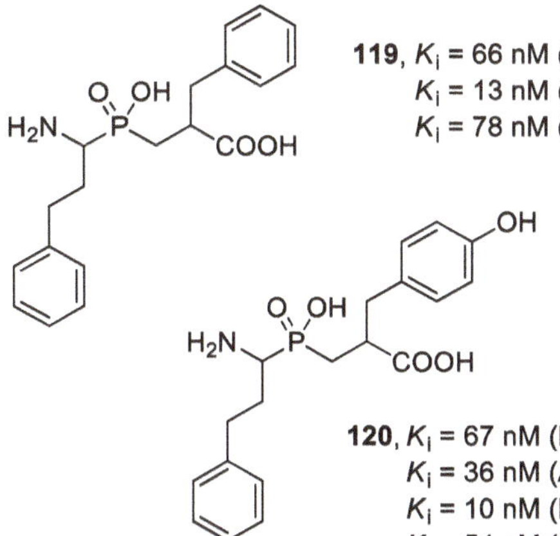

119, K_i = 66 nM (LAP, porcine kidney)
K_i = 13 nM (LAP, rec. *P. falciparum*)
K_i = 78 nM (APN, rec. *P. falciparum*)

120, K_i = 67 nM (LAP, porcine kidney)
K_i = 36 nM (APN, porcine kidney)
K_i = 10 nM (LAP, rec. *P. falciparum*)
K_i = 54 nM (APN, rec. *N. meningitides*)

Figure 3. The structure and activity of hPheψ[P(O)(OH)CH_2]-Phe and hPheψ[P(O)(OH)CH_2]-Tyr phosphinic dipeptide inhibitors of leucine and alanyl aminopeptidases.

In the most advanced studies, the phosphinates efficiently controlled the growth of *P. falciparum* cell lines and infection of a malaria model *in vivo* [112,113]. The structural basis of their activity was established by resolution of the crystal structures of the enzyme-ligand for both protozoan aminopeptidases complexed with the hPhe-Phe analogue [115,116]. The data altogether validated *P. falciparum* aminopeptidases as promising targets to treat malaria [113].

Plasma procarboxypeptidase B (thrombin-activatable fibrinolysis inhibitor, TAFIa) is a zinc-based exopeptidase proteolytically activated by thrombin into the active enzyme (carboxypeptidase B, CPB) that down-regulates fibrinolysis by removing the C-terminal lysine from fibrin fiber. Altering the action of CPB is a new way to target thrombosis-related diseases [117]. Phosphinic dipeptide analogues that comprise a lysine mimetic in the P1′ position (**121** and **122**) were found to be excellent tools for this purpose. Compounds known in the literature as EF6265 and BX 528 (Figure 4) inhibited CPB with high affinity and excellent selectivity *versus* other carboxypeptidases and are prospective drugs for fibrinolytic therapy [118–121].

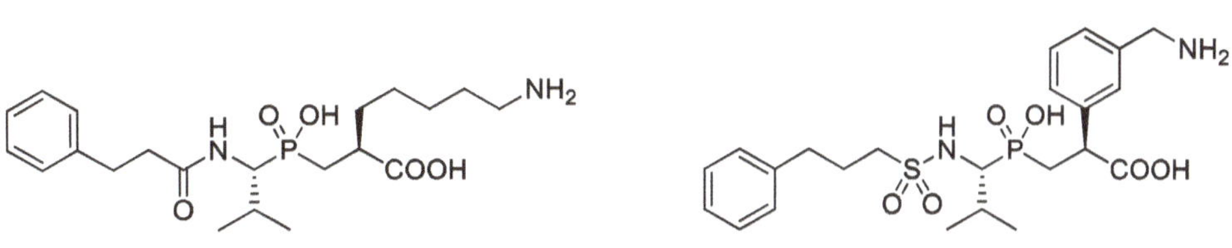

121, EF6265, IC_{50} = 8 nM (CPB, human plasma) **122**, BX 528, IC_{50} = 2 nM (TAFIa)

Figure 4. The structure and activity of phosphinic dipeptide inhibitors of TAFIa/carboxypeptidase B.

Other hydrolases, mostly zinc-containing, probed with phosphinic pseudodipeptide tools are listed below:

- angiotensin-converting enzyme 2, a mono-zinc carboxypeptidase, the first known human homologue of ACE [40];
- aminopeptidase A (glutamyl aminopeptidase) [122];
- NAALADase (*N*-acetylated-α-linked acidic dipeptidase), a metallo-dependent neuropeptidase [43];
- human renal dipeptidase and its bacterial homologue [87,88,123];
- M18 aspartyl aminopeptidase of *Plasmodium falciparum* [124];
- dinuclear zinc aminopeptidase PepV from *Lactobacillus delbrueckii* [125];
- D-Ala-D-Ala dipeptidase VanX, required for vancomycin resistance [126];
- chymotrypsin, cathepsin G and neutrophil elastase, serine proteases [127].

These examples do not complete the wide scope of phosphinic dipeptide utility. The majority of the currently-studied phosphinic compounds have used the fundamental building blocks for subsequent evolution into more complex systems. With respect to biological activity, the developed optimized sequences possess a refined potency/affinity to their targets that is also associated with an improved selectivity. Phosphinic fragments included in extended structures also acquire interesting metal complexing properties that have only occasionally been mentioned along the text. To conclude, phosphinic dipeptide preparation and diversification seem to be fundamental elements in a much broader and multifaceted area of phosphorus chemistry and biology.

Acknowledgments: The work was financed by a statutory activity subsidy from the Polish Ministry of Science and Higher Education for the Faculty of Chemistry of Wrocław University of Technology. The work is dedicated to Vincent Dive and Athanasios Yiotakis who introduced the author to phosphinic peptide chemistry.

References

1. Matthews, B.W. Structural basis of the action of thermolysin and related zinc peptidases. *Acc. Chem. Res.* **1988**, *21*, 333–340. [CrossRef]
2. Schechter, I.; Berger, A. On the size of the active site in proteases. I. Papain. *Biochem. Biophys. Res. Commun.* **1967**, *27*, 157–162. [CrossRef]
3. Pauling, L. Chemical achievement and hope for the future. *Am. Sci.* **1948**, *36*, 51–58. [PubMed]
4. Wolfenden, R. Transition state analogues for enzyme catalysis. *Nature* **1969**, *223*, 704–705. [CrossRef] [PubMed]
5. Kafarski, L.; Lejczak, B. The biological activity of phosphono- and phosphinopeptides. In *Aminophosphonic and Aminophosphinic Acids. Chemistry and Biological Activity*; Kukhar, V.P., Hudson, H.R., Eds.; John Wiley & Sons: Chichester, UK, 2000; pp. 407–442.
6. Collinsová, M.; Jiráček, J. Phosphinic acid compounds in biochemistry, Biology and medicine. *Curr. Med. Chem.* **2000**, *7*, 629–647. [CrossRef] [PubMed]
7. Mucha, A.; Kafarski, P.; Berlicki, Ł. Remarkable potential of the α-aminophosphonate/ phosphinate structural motif in medicinal chemistry. *J. Med. Chem.* **2011**, *54*, 5955–5980. [CrossRef]
8. Jacobsen, N.E.; Bartlett, P.A. A phosphorus-containing dipeptide analog as an inhibitor of carboxypeptidase A. *J. Am. Chem. Soc.* **1981**, *103*, 654–657. [CrossRef]
9. Bartlett, P.A.; Marlowe, C.K. Phosphonamidates as transition state analog inhibitors of thermolysin. *Biochemistry* **1983**, *22*, 4618–4624. [CrossRef]
10. Holden, H.M.; Tronrud, D.E.; Monzingo, A.F.; Weaver, L.H.; Matthews, B.W. Slow- and fast-binding inhibitors of thermolysin display different modes of binding: Crystallographic analysis of extended phosphonamidate transition-state analogues. *Biochemistry* **1987**, *26*, 8542–8553. [CrossRef]
11. Bartlett, P.A.; Marlowe, C.K. A possible role for water dissociation in the slow binding of phosphorus-containing transition state analog inhibitors of thermolysin. *Biochemistry* **1987**, *26*, 8553–8561. [CrossRef]
12. Christianson, D.W.; Lipscomb, W.N. Comparison of carboxypeptidase A and thermolysin: Inhibition by phosphonamidates. *J. Am. Chem. Soc.* **1988**, *110*, 5560–5565. [CrossRef]

13. Hanson, J.E.; Kaplan, A.P.; Bartlett, P.A. Phosphonate analogs of carboxypeptidase A are potent transition state analog inhibitors. *Biochemistry* **1989**, *28*, 6294–6305. [CrossRef] [PubMed]
14. Kaplan, A.P.; Bartlett, P.A. An inhibitor of carboxypeptidase A with a K_i value in the femtomolar range. *Biochemistry* **1991**, *30*, 8165–8170. [CrossRef] [PubMed]
15. Morgan, B.P.; Scholtz, J.M.; Ballinger, M.D.; Zipkin, I.D.; Bartlett, P.A. Differential binding energy: A detailed evaluation of the influence of hydrogen-bonding and hydrophobic groups on the inhibition of thermolysin by phosphorus-containing inhibitors. *J. Am. Chem. Soc.* **1991**, *113*, 297–307. [CrossRef]
16. Yiotakis, A.; Vassiliou, S.; Jiráček, J.; Dive, V. Protection of the hydroxyphosphinyl function of phosphinic dipeptides by adamantyl. Application to the solid-phase synthesis of phosphinic peptides. *J. Org. Chem.* **1996**, *61*, 6601–6605. [CrossRef]
17. Dive, V.; Lucet-Levannier, K.; Georgiadis, D.; Cotton, J.; Vassiliou, S.; Cuniasse, P.; Yiotakis, A. Phosphinic peptide inhibitors as tools in the study of the function of zinc metallopeptidases. *Biochem. Soc. Trans.* **2000**, *28*, 455–460. [CrossRef]
18. Dive, V.; Georgiadis, D.; Matziari, M.; Makaritis, A.; Beau, F.; Cuniasse, P.; Yiotakis, A. Phosphinic peptides as zinc metalloproteinase inhibitors. *Cell. Mol. Life Sci.* **2004**, *61*, 2010–2019. [CrossRef]
19. Cuniasse, P.; Devel, L.; Makaritis, A.; Beau, F.; Georgiadis, D.; Matziari, M.; Yiotakis, A.; Dive, V. Future challenges facing the development of specific active-site-directed synthetic inhibitors of MMPs. *Biochimie* **2005**, *87*, 393–402. [CrossRef]
20. Matziari, M.; Dive, V.; Yiotakis, A. Matrix metalloproteinase 11 (MMP-11; stromelysin-3) and synthetic inhibitors. *Med. Res. Rev.* **2007**, *27*, 528–552. [CrossRef]
21. Yiotakis, A.; Georgiadis, D.; Matziari, M.; Makaritis, A.; Dive, V. Phosphinic peptides: Synthetic approaches and biochemical evaluation as Zn-metalloprotease inhibitors. *Curr. Org. Chem.* **2004**, *8*, 1135–1158. [CrossRef]
22. Baylis, E.K.; Campbell, C.D.; Dingwall, J.G. 1-Aminoalkylphosphonous acids. Part 1. Isosteres of the protein amino acids. *J. Chem. Soc. Perkin Trans. 1* **1984**, 2845–2853. [CrossRef]
23. Georgiadis, D.; Matziari, M.; Vassiliou, S.; Dive, V.; Yiotakis, A. A convenient method to synthesize phosphinic peptides containing an aspartyl or glutamyl aminophosphinic acid. Use of the phenyl group as the carboxyl synthon. *Tetrahedron* **1999**, *55*, 14635–14648. [CrossRef]
24. Dumy, P.; Escale, R.; Girard, J.-P.; Parello, J.; Vidal, J.-P. A convenient synthetic approach to new α-(9-fluorenylmethoxycarbonylamino)alkylphosphonic acid derivatives. *Synthesis* **1992**, 1226–1228. [CrossRef]
25. Thottathil, J.K.; Przybyla, C.A.; Moniot, J.L. Mild Arbuzov reactions of phosphonous acids. *Tetrahedron Lett.* **1984**, *25*, 4737–4740. [CrossRef]
26. Thottathil, J.K.; Ryono, D.E.; Przybyla, C.A.; Moniot, J.L.; Neubeck, R. Preparation of phosphinic acids: Michael additions of phosphonous acids/esters to conjugated systems. *Tetrahedron Lett.* **1984**, *25*, 4741–4744. [CrossRef]
27. Giannousis, P.P.; Bartlett, P.A. Phosphorus amino acid analogs as inhibitors of leucine aminopeptidase. *J. Med. Chem.* **1987**, *30*, 1603–1609. [CrossRef]
28. Sampson, N.S.; Bartlett, P.A. Synthesis of phosphonic acid derivatives by oxidative activation of phosphinate esters. *J. Org. Chem.* **1988**, *53*, 4500–4503. [CrossRef]
29. Boyd, E.A.; Corless, M.; James, K.; Regan, A.C. A versatile route to substituted phosphinic acids. *Tetrahedron Lett.* **1990**, *31*, 2933–2936. [CrossRef]
30. Boyd, E.A.; Reagan, A.C.; James, K. Synthesis of alkyl phosphinic acids from silyl phosphonites and alkyl halides. *Tetrahedron Lett.* **1994**, *35*, 4223–4226. [CrossRef]
31. Parsons, W.H.; Patchett, A.A.; Bull, H.G.; Schoen, W.R.; Taub, D.; Davidson, J.; Combs, P.L.; Springer, J.P.; Gadebusch, H.; Weissberger, B.; *et al.* Phosphinic acid inhibitors of D-alanyl-D-alanine ligase. *J. Med. Chem.* **1988**, *31*, 1772–1778. [CrossRef]
32. Allen, M.C.; Fuhrer, W.; Tuck, B.; Wade, R.; Wood, J.M. Renin inhibitors. Synthesis of transition-state analogue inhibitors containing phosphorus acid derivatives at the scissile bond. *J. Med. Chem.* **1989**, *32*, 1652–1661. [CrossRef] [PubMed]
33. Cristau, H.-J.; Coulombeau, A.; Genevois-Borella, A.; Pirat, J.-L. A convenient one-pot synthesis of phosphino-dipeptide analogs. *Tetrahedron Lett.* **2001**, *42*, 4491–4494. [CrossRef]

34. Cristau, H.-J.; Coulombeau, A.; Genevois-Borella, A.; Sanchez, F.; Pirat, J.-L. Preparation of phosphinodipeptide analogs as building blocks for pseudopeptides synthesis. *J. Organomet. Chem.* **2002**, *643/644*, 381–391. [CrossRef]
35. Vassiliou, S.; Mucha, A.; Cuniasse, P.; Georgiadis, D.; Lucet-Levannier, K.; Beau, F.; Kannan, R.; Murphy, G.; Knäuper, V.; Rio, M.-C.; *et al.* Phosphinic pseudo-tripeptides as potent inhibitors of matrix metalloproteinases: a structure-activity study. *J. Med. Chem.* **1999**, *42*, 2610–2620. [CrossRef] [PubMed]
36. Chen, H.; Noble, F.; Mothé, A.; Meudal, H.; Coric, P.; Danascimento, S.; Roques, B.P.; George, P.; Fournié-Zaluski, M.-C. Phosphinic derivatives as new dual enkephalin-degrading enzyme inhibitors: Synthesis, biological properties, and antinociceptive activities. *J. Med. Chem.* **2000**, *43*, 1398–1408. [CrossRef] [PubMed]
37. Miller, D.J.; Hammond, S.M.; Anderluzzi, D.; Bugg, T.D.H. Aminoalkylphosphinate inhibitors of D-Ala-D-Ala adding enzyme. *J. Chem. Soc. Perkin Trans. 1* **1998**, 131–142. [CrossRef]
38. Liboska, R.; Pícha, J.; Hančlová, I.; Buděšínský, M.; Šanda, M.; Jiráček, J. Synthesis of methionine- and norleucine-derived phosphinopeptides. *Tetrahedron Lett.* **2008**, *49*, 5629–5631. [CrossRef]
39. Kaboudin, B.; Saadati, F. A simple, novel and convenient method for the synthesis of 1-aminophosphinic acids: synthesis of a novel C2-symmetric phosphinic acid pseudodipeptide. *Tetrahedron Lett.* **2009**, *50*, 1450–1452. [CrossRef]
40. Mores, A.; Matziari, M.; Beau, F.; Cuniasse, P.; Yiotakis, A.; Dive, V. Development of potent and selective phosphinic peptide inhibitors of angiotensin-converting enzyme 2. *J. Med. Chem.* **2008**, *51*, 2216–2226. [CrossRef]
41. Demange, L.; Dugave, C. Synthesis of phosphinic alanyl-proline surrogates Alaψ(PO_2R-CH)Pro as potential inhibitors of the human cyclophilin hCyp-18. *Tetrahedron Lett.* **2001**, *42*, 6295–6297. [CrossRef]
42. Tanner, M.E.; Vaganay, S.; van Heijenoort, J.; Blanot, D. Phosphinate inhibitors of the D-glutamic acid-adding enzyme of peptidoglycan biosynthesis. *J. Org. Chem.* **1996**, *61*, 1756–1760. [CrossRef] [PubMed]
43. Jackson, P.F.; Tays, K.L.; Maclin, K.M.; Ko, Y.-S.; Li, W.; Vitharana, D.; Tsukamoto, T.; Stoermer, D.; Lu, X.-C.M.; Wozniak, K.; Slusher, B.S. Design and pharmacological activity of phosphinic acid based NAALADase inhibitors. *J. Med. Chem.* **2001**, *44*, 4170–4175. [CrossRef] [PubMed]
44. Kende, A.S.; Dong, H.-Q.; Liu, X.; Ebetino, F.H. A useful synthesis of the Phe-Arg phosphinic dipeptide isostere. *Tetrahedron Lett.* **2002**, *43*, 4973–4976. [CrossRef]
45. Grembecka, J.; Mucha, A.; Cierpicki, T.; Kafarski, P. The most potent organophosphorus inhibitors of leucine aminopeptidase. Structure-based design, chemistry, and activity. *J. Med. Chem.* **2003**, *46*, 2641–2655. [CrossRef] [PubMed]
46. Vassiliou, S.; Xeilari, M.; Yiotakis, A.; Grembecka, J.; Pawełczak, M.; Kafarski, P.; Mucha, A. A synthetic method for diversification of the P1′ substituent in phosphinic dipeptides as a tool for exploration of the specificity of the S1′ binding pockets of leucine aminopeptidases. *Bioorg. Med. Chem.* **2007**, *15*, 3187–3200. [CrossRef]
47. Makaritis, A.; Georgiadis, D.; Dive, V.; Yiotakis, A. Diastereoselective solution and multipin-based combinatorial array synthesis of a novel class of potent phosphinic metalloprotease inhibitors. *Chem. Eur. J.* **2003**, *9*, 2079–2094. [CrossRef]
48. Jullien, N.; Makritis, A.; Georgiadis, D.; Beau, F.; Yiotakis, A.; Dive, V. Phosphinic tripeptides as dual angiotensin-converting enzyme C-domain and endothelin-converting enzyme-1 inhibitors. *J. Med. Chem.* **2010**, *53*, 208–220. [CrossRef]
49. Stetter, H.; Kuhlmann, H. Eine einfache Herstellung von α-Alkylacrylsäure-estern. *Synthesis* **1979**, 29–30. [CrossRef]
50. Matziari, M.; Bauer, K.; Dive, V.; Yiotakis, A. Synthesis of the phosphinic analogue of thyrotropin releasing hormone. *J. Org. Chem.* **2008**, *73*, 8591–8593. [CrossRef]
51. Borloo, M.; Jiao, X.-Y.; Wójtowicz, H.; Rajan, P.; Verbruggen, C.; Augustyns, K.; Haemers, A. A convenient one-pot preparation of disubstituted phosphinic acids derived from simple amino acids and proline. *Synthesis* **1995**, 1074–1076. [CrossRef]
52. Fougère, C.; Guénin, E.; Hardouin, J.; Lecouvey, M. Rapid and efficient synthesis of unsymmetrical phosphinic acids R′P(O)OHR″. *Eur. J. Org. Chem.* **2009**, 6048–6054. [CrossRef]

53. Prishchenko, A.A.; Livantsov, M.V.; Novikova, O.P.; Livantsova, L.I.; Petrosyan, V.S. Synthesis of new organophosphorus-substituted mono- and bis(trimethylsilyl)amines with PCH_2N fragments and their derivatives. *Heteroatom Chem.* **2010**, *21*, 71–77. [CrossRef]
54. Prishchenko, A.A.; Livantsov, M.V.; Novikova, O.P.; Livantsova, L.I.; Petrosyan, V.S. Synthesis of new organophosphorus-substituted derivatives of functionalized propionates and their analogues. *Heteroatom Chem.* **2008**, *19*, 418–428. [CrossRef]
55. Livantsov, M.V.; Prishchenko, A.A.; Novikova, O.P.; Livantsova, L.I.; Grigor'ev, E.V. Synthesis of phosphorus-substituted derivatives of methylsuccinic acid. *Russ. J. Gen. Chem.* **2003**, *73*, 659–660. [CrossRef]
56. Rogakos, V.; Georgiadis, D.; Dive, V.; Yiotakis, A. A modular rearrangement approach toward medicinally relevant phosphinic structures. *Org. Lett.* **2009**, *11*, 4696–4699. [CrossRef]
57. Dorff, P.H.; Chiu, G.; Goldstein, S.W.; Morgan, B.P. Solid phase synthesis of phosphinopeptoids as transition state analog inhibitors. *Tetrahedron Lett.* **1998**, *39*, 3375–3378. [CrossRef]
58. Buchardt, J.; Meldal, M. Novel methodology for the solid-phase synthesis of phosphinic peptides. *J. Chem. Soc. Perkin Trans. 1* **2000**, 3306–3310. [CrossRef]
59. Chen, S.; Coward, J.K. A general method for the synthesis of N-protected α-aminophosphinic acids. *Tetrahedron Lett.* **1996**, *37*, 4335–4338. [CrossRef]
60. Yuan, C.; Wang, G.; Chen, S. Studies on organophosphorus compounds; XLVI. A facile and direct route to dialkyl 1-(benzyloxycarbonylamino)alkylphosphonates and dialkyl or diphenyl α-(benzyloxycarbonylamino)benzylphosphonates. *Synthesis* **1990**, 522–524. [CrossRef]
61. Matziari, M.; Yiotakis, A. Shortcut to Fmoc-protected phosphinic pseudodipeptidic blocks. *Org. Lett.* **2005**, *7*, 4049–4052. [CrossRef]
62. Nasopoulou, M.; Georgiadis, D.; Matziari, M.; Dive, V.; Yiotakis, A. A versatile annulation protocol toward novel constrained phosphinic peptidomimetics. *J. Org. Chem.* **2007**, *72*, 7222–7228. [CrossRef]
63. Rozhko, L.F.; Ragulin, V.V. Method for the synthesis of phosphinic acids from hypophosphites V. The synthesis of pseudo-α,α-dipeptides. *Amino Acids* **2005**, *29*, 139–143. [CrossRef]
64. Dmitriev, M.E.; Ragulin, V.V. New opinions on the amidoalkylation of hydrophosphorylic compounds. *Tetrahedron Lett.* **2010**, *51*, 2613–2616. [CrossRef]
65. Dmitriev, M.E.; Rossinets, E.A.; Ragulin, V.V. Amidoalkylation of hydrophosphoryl compounds. *Russ. J. Gen. Chem.* **2011**, *81*, 1092–1104. [CrossRef]
66. Dmitriev, M.E.; Ragulin, V.V. Arbuzov-type reaction of acylphosphonites and N-alkoxycarbonylimine cations generated *in situ* with trifluoroacetic anhydride. *Tetrahedron Lett.* **2012**, *53*, 1634–1636. [CrossRef]
67. Huber, T.; Manzenrieder, F.; Kuttruff, C.A.; Dorner-Ciossek, C.; Kessler, H. Prolonged stability by cyclization: Macrocyclic phosphino dipeptide isostere inhibitors of β-secretase (BACE1). *Bioorg. Med. Chem. Lett.* **2009**, *19*, 4427–4431. [CrossRef]
68. Foersterova, M.; Svobodova, I.; Lubal, P.; Taborsky, P.; Kotek, J.; Hermann, P.; Lukes, I. Thermodynamic study of lanthanide(III) complexes with bifunctional monophosphinic acid analogues of H4dota and comparative kinetic study of yttrium(III) complexes. *J. Chem. Soc. Dalton Trans.* **2007**, 535–549. [CrossRef]
69. Notni, J.; Hermann, P.; Havlickova, J.; Kotek, J.; Kubicek, V.; Plutnar, J.; Loktionova, N.; Riss, P.J.; Roesch, F.; Lukes, I. A triazacyclononane-based bifunctional phosphinate ligand for the preparation of multimeric ^{68}Ga tracers for positron emission tomography. *Chem. Eur. J.* **2010**, *16*, 7174–7185. [CrossRef]
70. Foersterova, M.; Petrik, M.; Laznickova, A.; Laznicek, M.; Hermann, P.; Lukes, I.; Melichar, F. Complexation and biodistribution study of ^{111}In and ^{90}Y complexes of bifunctional phosphinic acid analogs of H4dota. *Appl. Radiat. Isotopes* **2009**, *67*, 21–29. [CrossRef]
71. Lacerda, S.; Marques, F.; Campello, P.; Gano, L.; Kubicek, V.; Hermann, P.; Santos, I. Chemical, radiochemical and biological studies of Sm and Ho complexes of H4dota analogues containing one methylphosphonic/phosphinic acid pendant arm. *J. Labelled Comp. Radiopharm.* **2010**, *53*, 36–43. [CrossRef]
72. Notni, J.; Simecek, J.; Hermann, P.; Wester, H.-J. TRAP, a powerful and versatile framework for gallium-68 radiopharmaceuticals. *Chem. Eur. J.* **2011**, *17*, 14718–14722. [CrossRef]
73. Ragulin, V.V. One-pot synthesis of α-amino phosphinic acids. *Russ. J. Gen. Chem.* **2004**, *74*, 142–143. [CrossRef]

74. Georgiadis, D.; Dive, V.; Yiotakis, A. Synthesis and comparative study on the reactivity of peptidyl-type phosphinic esters. Intramolecular effects in the alkaline and acidic cleavage of methyl β-carboxyphosphinates. *J. Org. Chem.* **2001**, *66*, 6604–6610. [CrossRef] [PubMed]
75. Reiter, L.A.; Jones, B.P. Amide-assisted hydrolysis of β-carboxamido-substituted phosphinic acid esters. *J. Org. Chem.* **1997**, *62*, 2808–2812. [CrossRef]
76. Nasopoulou, M.; Matziari, M.; Dive, V.; Yiotakis, A. Chemoselective protection of solid-phase compatible Fmoc-phosphinic building blocks. *J. Org. Chem.* **2006**, *71*, 9525–9527. [CrossRef]
77. Georgiadis, D.; Matziari, M.; Yiotakis, A. A highly efficient method for the preparation of phosphinic pseudodipeptidic blocks suitably protected for solid-phase peptide synthesis. *Tetrahedron* **2001**, *57*, 3471–3478. [CrossRef]
78. Buchardt, J.; Ferreras, M.; Krog-Jensen, C.; Delaissé, J.-M.; Foged, N.T.; Meldal, M. Phosphinic peptide matrix metalloproteinase-9 inhibitors by solid-phase synthesis using a building block approach. *Chem. Eur. J.* **1999**, *5*, 2877–2884. [CrossRef]
79. Bhowmick, M.; Sappidi, R.R.; Fields, G.B.; Lepore, S.D. Efficient synthesis of Fmoc-protected phosphinic pseudodipeptides: Building blocks for the synthesis of matrix metalloproteinase inhibitors. *Biopolymers* **2011**, *96*, 1–3. [CrossRef]
80. Jiráček, J.; Yiotakis, A.; Vincent, B.; Lecoq, A.; Nicolau, A.; Checler, F.; Dive, V. Development of highly potent and selective phosphinic peptide inhibitors of zinc endopeptidase 24–15 using combinatorial chemistry. *J. Biol. Chem.* **1995**, *270*, 21701–21706. [CrossRef] [PubMed]
81. Jiráček, J.; Yiotakis, A.; Vincent, B.; Checler, F.; Dive, V. Development of the first potent and selective inhibitor of the zinc endopeptidase neurolysin using a systematic approach based on combinatorial chemistry of phosphinic peptides. *J. Biol. Chem.* **1996**, *271*, 19606–19611. [CrossRef]
82. Dive, V.; Cotton, J.; Yiotakis, A.; Michaud, A.; Vassiliou, S.; Jiráček, J.; Vazeux, G.; Chauvet, M.-T.; Cuniasse, P.; Corvol, P. RXP 407, a phosphinic peptide, is a potent inhibitor of angiotensin I converting enzyme able to differentiate between its two active sites. *Proc. Natl. Acad. Sci. USA* **1999**, *96*, 4330–4335. [CrossRef] [PubMed]
83. Buchardt, J.; Schiødt, C.B.; Krog-Jensen, C.; Delaissé, J.-M.; Foged, N.T.; Meldal, M. Solid phase combinatorial library of phosphinic peptides for discovery of matrix metalloproteinase inhibitors. *J. Comb. Chem.* **2000**, *2*, 624–638. [CrossRef] [PubMed]
84. Campagne, J.-M.; Coste, J.; Guillou, L.; Heitz, A.; Jouin, P. Solid phase synthesis of phosphinic peptides. *Tetrahedron Lett.* **1993**, *34*, 4181–4184. [CrossRef]
85. Mucha, A.; Pawełczak, M.; Hurek, J.; Kafarski, P. Synthesis and activity of phosphinic tripeptide inhibitors of cathepsin C. *Bioorg. Med. Chem. Lett.* **2004**, *14*, 3113–3116. [CrossRef] [PubMed]
86. Matziari, M.; Nasopoulou, M.; Yiotakis, A. Active methylene phosphinic peptides: A new diversification approach. *Org. Lett.* **2006**, *8*, 2317–2319. [CrossRef]
87. Gurulingappa, H.; Buckhaults, P.; Kumar, S.K.; Kinzler, K.W.; Vogelstein, B.; Khan, S.R. Design, synthesis and evaluation of new RDP inhibitors. *Tetrahedron Lett.* **2003**, *44*, 1871–1873. [CrossRef]
88. Gurulingappa, H.; Buckhalts, P.; Kinzler, K.W.; Vogelstein, B.; Khan, S.R. Synthesis and evaluation of aminophosphinic acid derivatives as inhibitors of renal dipeptidase. *Bioorg. Med. Chem. Lett.* **2004**, *14*, 3531–3533. [CrossRef]
89. Matziari, M.; Georgiadis, D.; Dive, V.; Yiotakis, A. Convenient synthesis and diversification of dehydroalaninyl phosphinic peptide analogues. *Org. Lett.* **2001**, *3*, 659–662. [CrossRef]
90. Georgiadis, D.; Cuniasse, P.; Cotton, J.; Yiotakis, A.; Dive, V. Structural determinants of RXPA380, a potent and highly selective inhibitor of the angiotensin-converting enzyme C-domain. *Biochemistry* **2004**, *43*, 8048–8054. [CrossRef]
91. Matziari, M.; Beau, F.; Cuniasse, P.; Dive, V.; Yiotakis, A. Evaluation of P1′-diversified phosphinic peptides leads to the development of highly selective inhibitors of MMP-11. *J. Med. Chem.* **2004**, *47*, 325–336. [CrossRef]
92. Matziari, M.; Dellis, D.; Dive, V.; Yiotakis, A.; Samios, J. Conformational and solvation studies *via* computer simulation of the novel large scale diastereoselectively synthesized phosphinic MMP Inhibitor RXP03 diluted in selected solvents. *J. Phys. Chem. B* **2010**, *114*, 421–428. [CrossRef] [PubMed]

93. Liu, X.; Hu, X.E.; Tian, X.; Mazur, A.; Ebetino, F.H. Enantioselective synthesis of phosphinyl peptidomimetics *via* an asymmetric Michael reaction of phosphinic acids with acrylate derivatives. *J. Organomet. Chem.* **2002**, *646*, 212–222. [CrossRef]
94. Yamagishi, T.; Ichikawa, H.; Haruki, T.; Yokomatsu, T. Diastereoselective synthesis of α,β′-disubstituted aminomethyl(2-carboxyethyl)phosphinates as phosphinyl dipeptide isosteres. *Org. Lett.* **2008**, *10*, 4347–4350. [CrossRef] [PubMed]
95. Yamagishi, T.; Tashiro, N.; Yokomatsu, T. Diastereoselective synthesis of the Leu-Pro type phosphinyl dipeptide isostere. *J. Org. Chem.* **2011**, *76*, 5472–5476. [CrossRef]
96. Monbrun, J.; Dayde, B.; Cristau, H.-J.; Volle, J.-N.; Virieux, D.; Pirat, J.-L. Diastereoselective Michael addition of 2*H*-2-oxo-1,4,2-oxaza phosphinanes to olefins. *Tetrahedron* **2011**, *67*, 540–545. [CrossRef]
97. Lämmerhofer, M.; Hebenstreit, D.; Gavioli, E.; Lindner, W.; Mucha, A.; Kafarski, P.; Wieczorek, P. High-performance liquid chromatographic enantiomer separation and determination of absolute configurations of phosphinic acid analogues of dipeptides and their α-aminophosphinic acid precursors. *Tetrahedron: Asymmetry* **2003**, *14*, 2557–2565. [CrossRef]
98. Preinerstorfer, B.; Lubda, D.; Mucha, A.; Kafarski, P.; Lindner, W.; Lämmerhofer, M. Stereoselective separations of chiral phosphinic acid pseudodipeptides by CEC using silica monoliths modified with an anion-exchange-type chiral selector. *Electrophoresis* **2006**, *27*, 4312–4320. [CrossRef]
99. Mucha, A.; Lämmerhofer, M.; Lindner, W.; Pawełczak, M.; Kafarski, P. Individual stereoisomers of phosphinic dipeptide inhibitor of leucine aminopeptidase. *Bioorg. Med. Chem. Lett.* **2008**, *18*, 1550–1554. [CrossRef]
100. Ellsworth, B.A.; Tom, N.J.; Bartlett, P.A. Synthesis and evaluation of inhibitors of bacterial D-alanine: D-alanine ligases. *Chem. Biol.* **1996**, *3*, 37–44. [CrossRef]
101. Fan, C.; Park, I.-S.; Walsh, C.T.; Knox, J.R. D-Alanine:D-alanine ligase: Phosphonate and phosphinate intermediates with wild type and the Y216F mutant. *Biochemistry* **1997**, *36*, 2531–2538. [CrossRef]
102. Kuzin, A.P.; Sun, T.; Jorczak-Baillass, J.; Healy, V.L.; Walsh, C.T.; Knox, J.R. Enzymes of vancomycin resistance: The structure of D-alanine-D-lactate ligase of naturally resistant *Leuconostoc mesenteroides*. *Structure* **2000**, *8*, 463–470. [CrossRef]
103. Roper, D.I.; Huyton, T.; Vagin, A.; Dodson, G. The molecular basis of vancomycin resistance in clinically relevant *Enterococci*: Crystal structure of D-alanyl-D-lactate ligase (VanA). *Proc. Natl. Acad. Sci. USA* **2000**, *97*, 8921–8925. [CrossRef] [PubMed]
104. Gegnas, L.D.; Waddell, S.T.; Chabin, R.M.; Reddy, S.; Wong, K.K. Inhibitors of the bacterial cell wall biosynthesis enzyme Mur D. *Bioorg. Med. Chem. Lett.* **1998**, *8*, 1643–1648. [CrossRef]
105. Štrancar, K.; Blanot, D.; Gobec, S. Design, synthesis and structure-activity relationships of new phosphinate inhibitors of MurD. *Bioorg. Med. Chem. Lett.* **2006**, *16*, 343–348. [CrossRef]
106. Štrancar, K.; Boniface, A.; Blanot, D.; Gobec, S. Phosphinate inhibitors of UDP-*N*-acetylmuramoyl-L-alanyl-D-glutamate: L-Lysine ligase (MurE). *Arch. Pharm. Chem. Life Sci.* **2007**, *340*, 127–134. [CrossRef]
107. Shomura, Y.; Hinokuchi, E.; Ikeda, H.; Senoo, A.; Takahashi, Y.; Saito, J.; Komori, H.; Shibata, N.; Yonetani, Y.; Higuchi, Y. Structural and enzymatic characterization of BacD, an L-amino acid dipeptide ligase from *Bacillus subtilis*. *Protein Sci.* **2012**, *21*, 707–716. [CrossRef]
108. Chen, H.; Roques, B.P.; Fournié-Zaluski, M.-C. Design of the first highly potent and selective aminopeptidase N (EC 3.4.11.2) inhibitor. *Bioorg. Med. Chem. Lett.* **1999**, *9*, 1511–1516. [CrossRef]
109. Yang, K.-W.; Golich, F.C.; Sigdel, T.K.; Crowder, M.W. Phosphinate, sulfonate, and sulfonamidate dipeptides as potential inhibitors of *Escherichia coli* aminopeptidase N. *Bioorg. Med. Chem. Lett.* **2005**, *15*, 5150–5153. [CrossRef]
110. Pícha, J.; Liboska, R.; Buděšínský, M.; Jiráček, J.; Pawełczak, M.; Mucha, A. Unusual activity pattern of leucine aminopeptidase inhibitors based on phosphorus containing derivatives of methionine and norleucine. *J. Enz. Inhib. Med. Chem.* **2011**, *26*, 155–161. [CrossRef]
111. Mucha, A.; Drąg, M.; Dalton, J.P.; Kafarski, P. Metallo-aminopeptidase inhibitors. *Biochimie* **2010**, *92*, 1509–1529. [CrossRef] [PubMed]
112. Skinner-Adams, T.S.; Lowther, J.; Teuscher, F.; Stack, C.M.; Grembecka, J.; Mucha, A.; Kafarski, P.; Trenholme, K.R.; Dalton, J.P.; Gardiner, D.L. Identification of phosphinate dipeptide analog inhibitors directed against the *Plasmodium falciparum* M17 leucine aminopeptidase as lead antimalarial compounds. *J. Med. Chem.* **2007**, *50*, 6024–6031. [CrossRef] [PubMed]

113. Skinner-Adams, T.S.; Stack, C.M.; Trenholme, K.R.; Brown, C.L.; Grembecka, J.; Lowther, J.; Mucha, A.; Drąg, M.; Kafarski, P.; McGowan, S.; *et al. Plasmodium falciparum* neutral aminopeptidases: New targets for anti-malarials. *Trends Biochem. Sci.* **2010**, *35*, 53–61. [CrossRef] [PubMed]
114. Węglarz-Tomczak, E.; Poręba, M.; Byzia, A.; Berlicki, Ł.; Nocek, B.; Mulligan, R.; Joachimiak, A.; Drąg, M.; Mucha, A. An integrated approach to the ligand binding specificity of *Neisseria meningitidis* M1 alanine aminopeptidase by fluorogenic substrate profiling, inhibitory studies and molecular modeling. *Biochimie* **2012**. [CrossRef] [PubMed]
115. McGowan, S.; Porter, C.J.; Lowther., J.; Stack, C.M.; Golding, S.J.; Skinner-Adams, T.S.; Trenholme, K.R.; Teuscher, F.; Donnelly, S.M.; Grembecka, J.; *et al.* Structural basis for the inhibition of the essential *Plasmodium falciparum* M1 neutral aminopeptidase. *Proc. Natl. Acad. Sci. USA* **2009**, *106*, 2537–2542. [CrossRef]
116. McGowan, S.; Oellig, C.A.; Birru, W.A.; Caradoc-Davies, T.T.; Stack, C.M.; Lowther, J.; Skinner-Adams, T.; Mucha, A.; Kafarski, P.; Grembecka, J.; *et al.* Structure of the *Plasmodium falciparum* M17 aminopeptidase and significance for the design of drugs targeting the neutral exopeptidases. *Proc. Natl. Acad. Sci. USA* **2010**, *107*, 2449–2454. [CrossRef] [PubMed]
117. Willemse, J.L.; Heylen, E.; Nesheim, M.E.; Hendriks, D.F. Carboxypeptidase U (TAFIa): A new drug target for fibrinolytic therapy? *J. Thromb. Haemost.* **2009**, *7*, 1962–1971. [CrossRef]
118. Suzuki, K.; Muto, Y.; Fushihara, K.; Kanemoto, K.; Iida, H.; Sato, E.; Kikuchi, C.; Matsushima, T.; Kato, E.; Nomoto, M.; *et al.* Enhancement of fibrinolysis by EF6265 [(*S*)-7-amino-2-[[[(*R*)-2-methyl-1-(3-phenylpropanoylamino)propyl]hydroxyphosphinoyl]methyl] hepta-noic acid], a specific inhibitor of plasma carboxypeptidase B. *J. Pharmacol. Exp. Ther.* **2004**, *309*, 607–615. [CrossRef]
119. Wang, Y.X.; Zhao, L.; Nagashima, M.; Vincelette, J.; Sukovich, D.; Li, W.; Subramanyam, B.; Yuan, S.; Emayan, K.; Islam, I.; *et al.* A novel inhibitor of activated thrombin-activatable fibrinolysis inhibitor (TAFIa)—Part I: Pharmacological characterization. *Thromb. Haemost.* **2007**, *97*, 45–53.
120. Wang, Y.X.; da Cunha, V.; Vincelette, J.; Zhao, L.; Nagashima, M.; Kawai, K.; Yuan, S.; Emayan, K.; Islam, I.; Hosoya, J.; *et al.* A novel inhibitor of activated thrombin activatable fibrinolysis inhibitor (TAFIa)—part II: Enhancement of both exogenous and endogenous fibrinolysis in animal models of thrombosis. *Thromb. Haemost.* **2007**, *97*, 54–61.
121. Adler, M.; Buckman, B.; Bryant, J.; Chang, Z.; Chu, K.; Emayan, K.; Hrvatin, P.; Islam, I.; Morser, J.; Sukovich, D.; *et al.* Structures of potent selective peptide mimetics bound to carboxypeptidase B. *Acta Crystallogr. D Biol. Crystallogr.* **2008**, *64*, 149–157. [CrossRef]
122. Georgiadis, D.; Vazeux, G.; Llorens-Cortes, C.; Yiotakis, A.; Dive, V. Potent and selective inhibition of zinc aminopeptidase A (EC 3.4.11.7, APA) by glutamyl aminophosphinic peptides: Importance of glutamyl aminophosphinic residue in the P1 position. *Biochemistry* **2000**, *39*, 1152–1155. [CrossRef] [PubMed]
123. Cummings, J.A.; Nguyen, T.T.; Fedorov, A.A.; Kolb, P.; Xu, C.; Fedorov, E.V.; Shoichet, B.K.; Barondeau, D.P.; Almo, S.C.; Raushel, F.M. Structure, Mechanism, And substrate profile for Sco3058: The closest bacterial homologue to human renal dipeptidase. *Biochemistry* **2010**, *49*, 611–622. [CrossRef] [PubMed]
124. Teuscher, F.; Lowther, J.; Skinner-Adams, T.S.; Spielmann, T.; Dixon, M.W.A.; Stack, C.M.; Donnelly, S.; Mucha, A.; Kafarski, P.; Vassiliou, S.; *et al.* The M18 aspartyl aminopeptidase of the human malaria parasite *Plasmodium falciparum*. *J. Biol. Chem.* **2007**, *282*, 30817–30826. [CrossRef] [PubMed]
125. Jozic, D.; Bourenkow, G.; Bartunik, H.; Scholze, H.; Dive, V.; Henrich, B.; Huber, R.; Bode, W.; Maskos, K. Crystal structure of the dinuclear zinc aminopeptidase PepV from *Lactobacillus delbrueckii* unravels its preference for dipeptides. *Structure* **2002**, *10*, 1097–1106. [CrossRef]
126. Wu, Z.; Walsh, C.T. Phosphinate analogs of D-, D-dipeptides: Slow-binding inhibition and proteolysis protection of VanX, a D-, D-dipeptidase required for vancomycin resistance in *Enterococcus faecium*. *Proc. Natl. Acad. Sci. USA* **1995**, *92*, 11603–11607. [CrossRef]
127. Walker, B.; Wharry, S.; Hamilton, R.J.; Martin, S.L.; Healy, A.; Walker, B.J. Asymmetric preference of serine proteases toward phosphonate and phosphinate esters. *Biochem. Biophys. Res. Commun.* **2000**, *276*, 1235–1239. [CrossRef]

molecules

MDPI

Review

Organophosphorus Chemistry for the Synthesis of Dendrimers

Anne-Marie Caminade [1,2,*], Régis Laurent [1,2], Maria Zablocka [1,2,3] and Jean-Pierre Majoral [1,2]

1 CNRS, LCC (Laboratoire de Chimie de Coordination), 205 route de Narbonne, BP44099, F-31077 Toulouse CEDEX 4, France; rlaurent@lcc-toulouse.fr (R.L.); majoral@lcc-toulouse.fr (J.-P.M.)

2 Université de Toulouse, UPS, INPT, F-31077 Toulouse CEDEX 4, France

3 Centre of Molecular and Macromolecular Studies, The Polish Academy of Sciences, Sienkiewicza 112, 90363 Lodz, Poland; zabloc@cbmm.lodz.pl

* Correspondence: caminade@lcc-toulouse.fr; Fax: +33-561-553-003

Received: 31 October 2012; Accepted: 12 November 2012; Published: 16 November 2012

check for updates

Abstract: Dendrimers are multifunctional, hyperbranched and perfectly defined macromolecules, synthesized layer after layer in an iterative manner. Besides the nature of the terminal groups responsible for most of the properties, the nature of the internal structure, and more precisely of the branching points, is also of crucial importance. For more than 15 years, we have demonstrated that the presence of phosphorus atom(s) at each branching point of the dendrimeric structure is particularly important and highly valuable for three main reasons: (i) the versatility of phosphorus chemistry that allows diversified organochemistry for the synthesis of dendrimers; (ii) the use of ^{31}P-NMR, which is a highly valuable tool for the characterization of dendrimers; (iii) some properties (in the fields of catalysis, materials, and especially biology), that are directly connected to the nature of the internal structure and of the branching points. This review will give an overview of the methods of synthesis of phosphorus-containing dendrimers, as well on the ways to graft phosphorus derivatives as terminal groups, with emphasis on the various roles played by the chemistry of phosphorus.

Keywords: dendrimers; phosphorus; hydrazone; phosphine; phosphonate

1. Introduction

Dendrimers [1] have an aesthetic structure constituted of branching units emanating radially from a central core. They are synthesized step-by-step in an iterative fashion. Each time the number of terminal groups is multiplied, a new generation is created. Due to this highly controlled synthesis, dendrimers offer a perfect modularity of size (a few nanometers), functionality, and solubility, mainly depending on the type of their terminal groups. Scheme 1 displays the principles of the divergent process, most generally used for the synthesis of dendrimers.

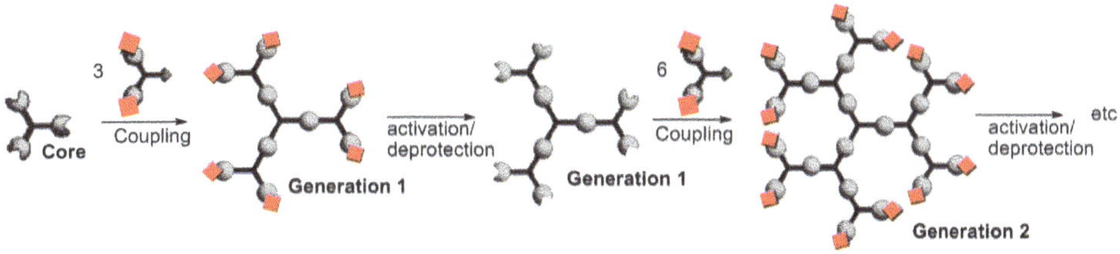

Scheme 1. The principle of the divergent synthesis of dendrimers.

Among all types of dendrimers [2], phosphorus-containing dendrimers [3] that have one phosphorus atom at each branching point, play an important role, with applications ranging from catalysis [4], materials [5], and even biology/nanomedicine [6]. This review will focus on our work, emphasizing the role of phosphorus [7]. It will be organized depending on the type of reactions that will occur on phosphorus atoms, whatever their location. All the other reactions of phosphorus-containing dendrimers, but not occurring on the P atoms will not be displayed, except if they are a necessary pathway towards the chemistry of phosphorus, or for the grafting of phosphorus entities.

2. Substitution Reactions on P-Cl Functions for the Synthesis and Functionalization of Dendrimers

Our first and main method of synthesis of phosphorus dendrimers [8] consists in the repetition of two quantitative reactions, the first step being the nucleophilic substitution of Cl by 4-hydroxybenzaldehyde in basic conditions. The second step is the condensation of the aldehydes with the dichlorophosphothiohydrazide. This compound is also issued from the organic chemistry of phosphorus (substitution of one Cl of $P(S)Cl_3$ with methylhydrazine, at low temperature). This second step generates $P\text{-}Cl_2$ functions suitable to perform again substitutions with HOC_6H_4CHO (Scheme 2).

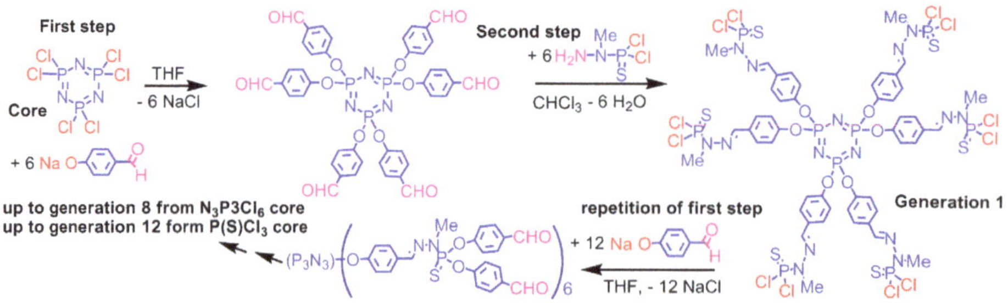

Scheme 2. The most important method of synthesis of phosphorus dendrimers.

This method is very powerful and has been carried out up to generation 8 starting from $N_3P_3Cl_6$ [9], and to generation 12 (the highest generation ever synthesized for any type of dendrimers) from $P(S)Cl_3$ [10]. The substitution reaction of P-Cl by phenols is quantitative in most cases, using <5% excess of reagents. It is particularly powerful for the functionalization of the surface of dendrimers, with variously functionalized phenols depending on the desired properties. Several examples are displayed in Figure 1. One can cite in particular the aldehyde (for the elaboration of DNA chips) [11], various ferrocenes (for studying electrochemical properties [12], evolution of chirality [13], and for catalysis [14]), various ligands suitable for catalysis such as derivatives of triphenylphosphine [15] (also precursors of phosphoniums [16]), thiazolylphosphines [17], iminophosphines [18], diphosphines [19], diketones [20], or azabis-oxazolines [21], dithioesters for thioacylation reactions (R = Me) [22] or as precursors of star polymers ($R = CH_2Ph$) [23]. Several fluorophores such as maleimide derivatives [24], dansyl derivatives [25] and also dabsyl dyes and protected tyramine [26], or fluorophores having two-photon absorption (TPA) properties [27], with eventually interchromophoric activities [28], or third order non-linear properties [29], have been grafted thanks to the reactivity of phenols, as well as D-xylose derivatives [30], phosphonates as precursors of anti-HIV derivatives [31], and azabisphosphonates (and isosteric carboxylic esters analogues [32]) precursors of symmetrical [33] or non-symmetrical [34] azabisphosphonic salts having important biological properties. In all cases, ^{31}P-NMR is a precious tool for characterizing these dendrimers and the achievement of the reactions [35].

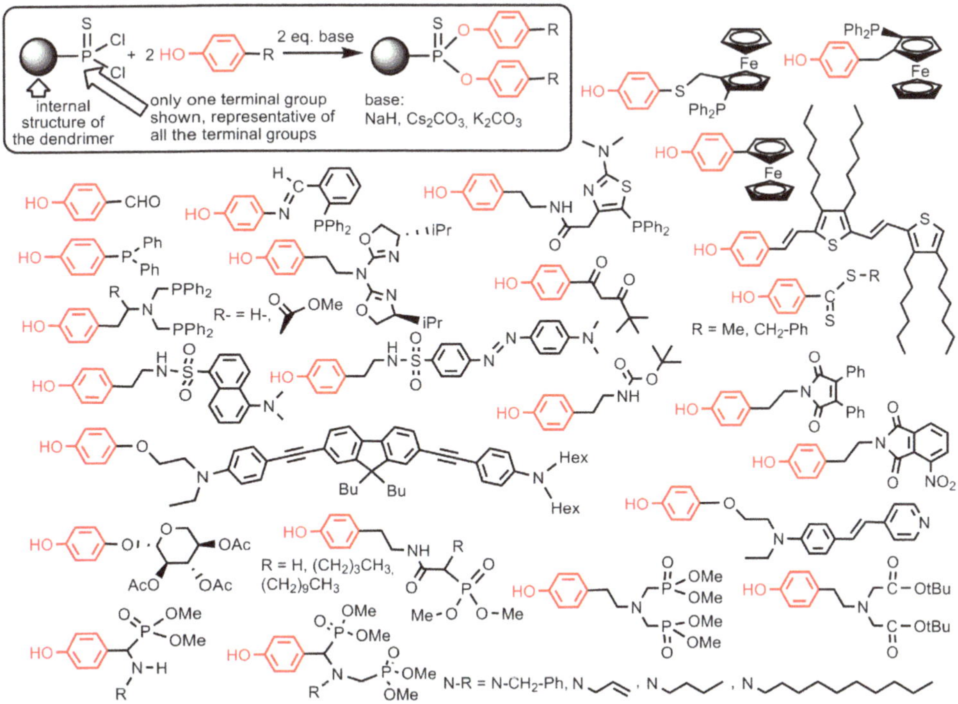

Figure 1. Functionalized phenols that have been grafted to dendrimers ended by $P(S)Cl_2$ functions.

The substitution reactions with phenols are also usable for varying the internal structure of dendrimers, by replacing 4-hydroxybenzaldehyde by other phenol aldehydes (Figure 2). Among them, one can cite the possibility to have various ferrocenes [12,36,37], longer branches [38], fluorophores [39], azobenzenes [40], or dialdehydes [41] for multiplying rapidly the number of terminal functions.

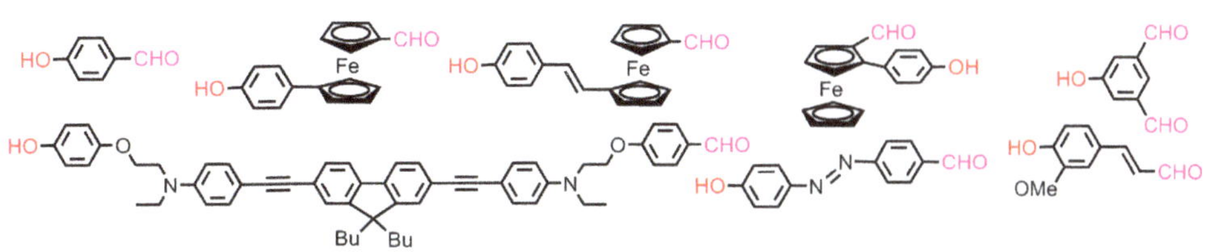

Figure 2. Some phenol aldehydes used instead of HOC_6H_4CHO for the synthesis of dendrimers.

We have also attempted to use the substitution reactions of hydrazines for the synthesis of dendrimers [42,43]. The most recent example is shown on Scheme 3 [44]. However, none of them give perfectly quantitative yields, and they have been carried out only up to the first generation.

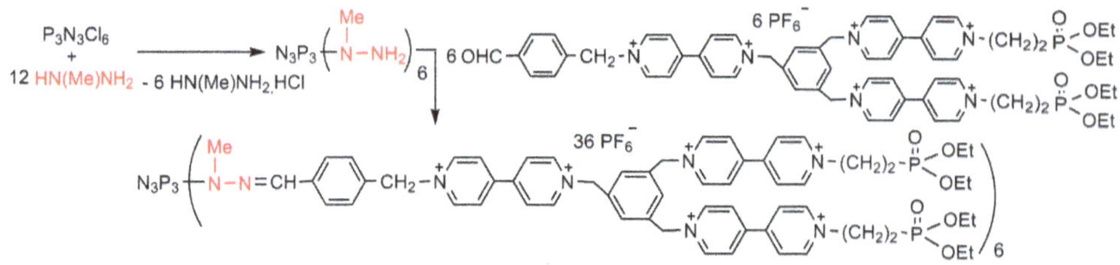

Scheme 3. Synthesis of a small dendrimer via substitutions with methylhydrazine.

On the contrary, the substitution reactions with amines are very powerful for functionalizing the surface of dendrimers, starting from $P(S)Cl_2$ end groups (Figure 3). Among them, one can cite in particular allyl and propargyl amines [45], and also diethylethylenediamine, which affords in a single step water-soluble dendrimers; HCl generated by the substitution reaction is trapped by the tertiary amine [46]. Water-soluble dendrimers [47] have important biological properties [48], but those ones possess interesting properties both in the fields of materials (for the elaboration of nano-tubes [49] and micro-capsules [50], of highly sensitive DNA chips [51]), and biology (transfection agents [46], anti-prion agents [52], anti-aggregation agent of Alzheimer peptides [53]). Other types of diamines, such as morpholine or piperidine derivatives have also been used [54]. In another example, both Cl linked to the same P react with a single diamine, creating a diazaphospholane cycle. This cycle can be obtained from various macrocycles [55], or can be linked to a macrocycle that is able to complex Pd^0 [56] or Pt^0 [57], to create nanoparticles of these metals, and even to organize them in dendritic networks [58].

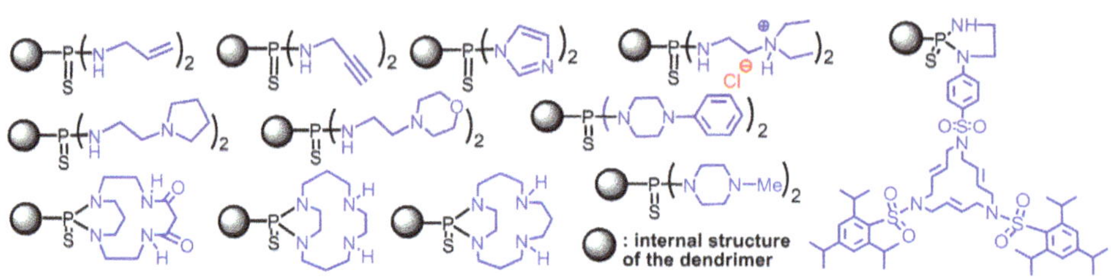

Figure 3. Functionalization of the surface of dendrimers by amino derivatives (only one function is shown, representative of all the terminal groups).

The reactions with amines are also suitable to perform clean monosubstitutions on each $P(X)Cl_2$ (X = S, O) end group. The reaction is regiospecific, but not enantioselective. The monosubstitution with amines leads to dendrimers with two, three, and even four unique functional groups on each chain end [59]. The second substitution can be performed with another amine, but also with phenols, in particular HOC_6H_4CHO, leading to dendrimers having functions in the internal structure [60] (Figure 4).

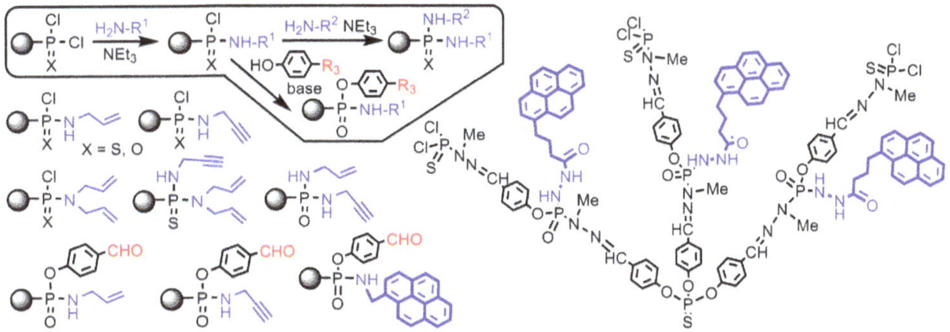

Figure 4. Multifunctionalization of the surface of dendrimers, and of the internal structure.

3. Diverse Ways for Grafting Phosphorus Entities as Terminal Groups

Diverse reactions have been performed to graft phosphorus derivatives as terminal groups of dendrimers. They can be divided into two types: those occurring on phosphorus, and those occurring on a function linked to phosphorus. In the first case, two different types of reactions have been performed: the addition of P-H onto unsaturated bonds such as aldehydes and imines [61], and the substitution reactions of P-Cl with N-H functions, from hydrazones [62] or amines [63] (Scheme 4).

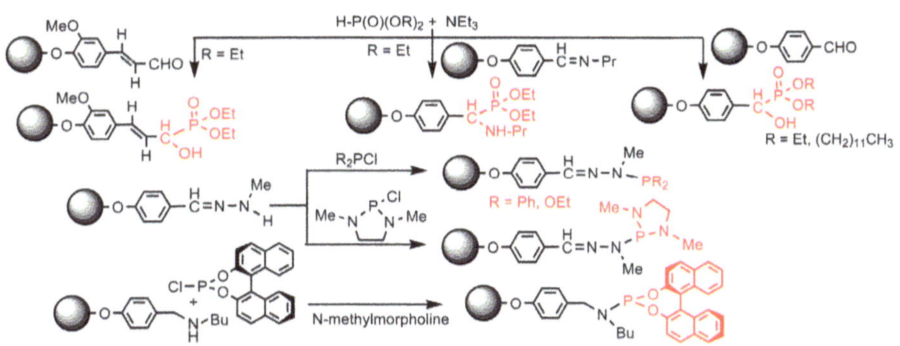

Scheme 4. Functionalization of terminal groups by direct reaction of phosphorus derivatives.

Various phosphorus derivatives, in particular phosphines and phosphonates or phosphates, have been grafted to the terminal groups of dendrimers essentially through condensation reactions, addition reactions and "click" reactions. The condensation reaction of hydrazones with aldehydes has afforded phosphites, phosphates or aminophosphates as terminal groups [61], whereas the condensation with Ph_2PCH_2OH on chiral amines (or hydrazone [64]) has led to chiral phosphines [65]. Addition reactions of amino groups onto unsaturated bonds have led to the grafting of ylides [61], or gem-bisphosphonates [66]. Alkylation of one nitrogen of PTA (phosphatriazaadamantane) has led to the grafting of one [67] or two [68] PTA per terminal function. Finally, the "click" reaction (azides with alkynes) has led to the grafting of azabisphosphonate groups [69] (Scheme 5).

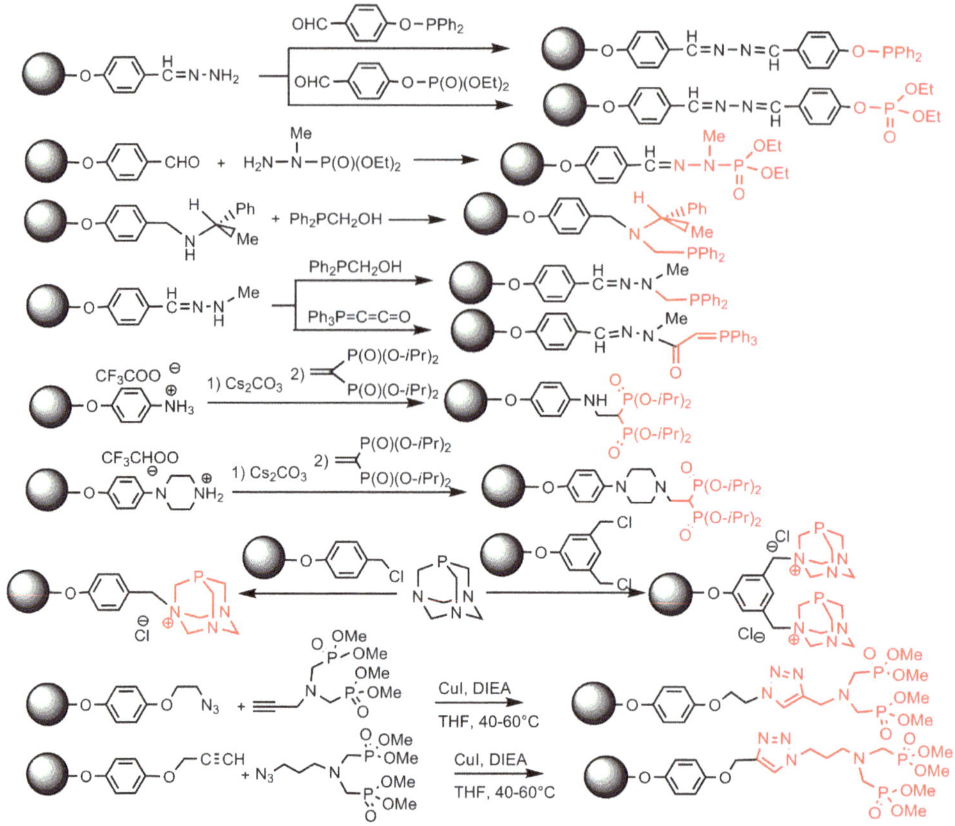

Scheme 5. Other types of reactions for the grafting of phosphorus derivatives as terminal groups.

4. Staudinger Reactions and Subsequent Reactions

The Staudinger reaction of phosphines with azides creates P=N functions; which are generally sensitive to hydrolysis. However, if the P=N function is conjugated, its stability is largely increased. Thus, instead of using organic azides, we have used thiophosphoryl azides, to generate P=N-P=S functions (or eventually P=N-P=N functions when using azides linked to the cyclotriphosphazene, as shown in the following scheme). We have synthesized several types of monomers to use alternatively the condensation reaction (aldehyde with hydrazine) and the Staudinger reaction. These monomers comprise either phosphines and hydrazine, or aldehydes and azide, in a 2/1 [70] or 5/1 ratio [71], eventually in combination [72]. Using these monomers allows a rapid multiplication of the number of terminal groups, and creates a new generation at each step and not every two steps as usual. This method of synthesis is also compatible with the first one mentioned in Scheme 2 (Scheme 6) [73].

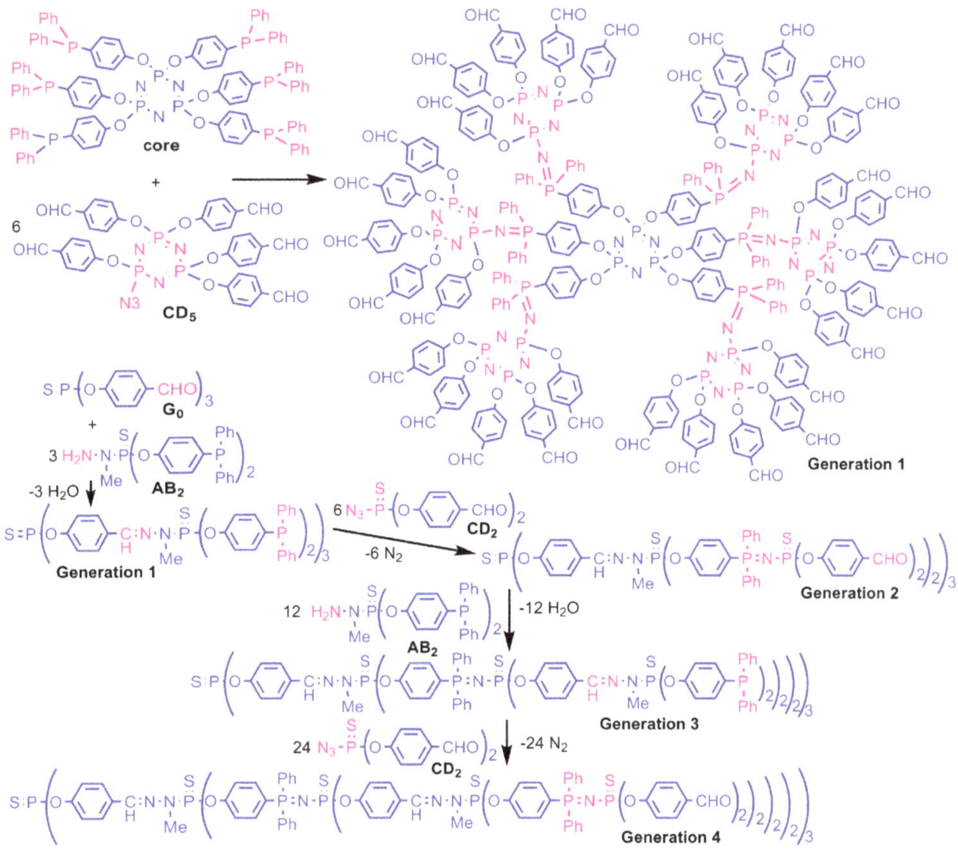

Scheme 6. Two methods of synthesis of dendrimers by Staudinger reactions.

The compatibility of the Staudinger reaction with the condensation reaction allows having P=N-P=S linkages selectively at one or two layers. The P=S groups linked to a P=N group have distinguishable properties compared to the other P=S groups, due to a delocalized form $^{+}$P-N=P-S^{-}, with a negative charge on S, which renders it sensitive to alkylation reactions [74] using various triflates [75] whereas the other P=S groups do not react. It is also suitable for the complexation of gold [76]. The alkylation induces a weakening of the PS bond, which can be cleaved using a nucleophilic phosphine such as $P(NMe_2)_3$. This reaction generates tricoordinated phosphorus atoms (P^{III}) at specific layers of the internal structure, that can be used for alkylation reactions, and can undergo Staudinger reactions creating P=N-P=N-P=S linkages [77]. The presence of aldehydes inside the dendrimers allows either the step-by-step growing of new branches [78] (Scheme 7), or the grafting of dendrons, leading to highly sophisticated dendrimeric structures [79], still unprecedented for any type of dendrimers, but also the grafting of new functions such as fluorescent groups [80], or zwitter-ions [81].

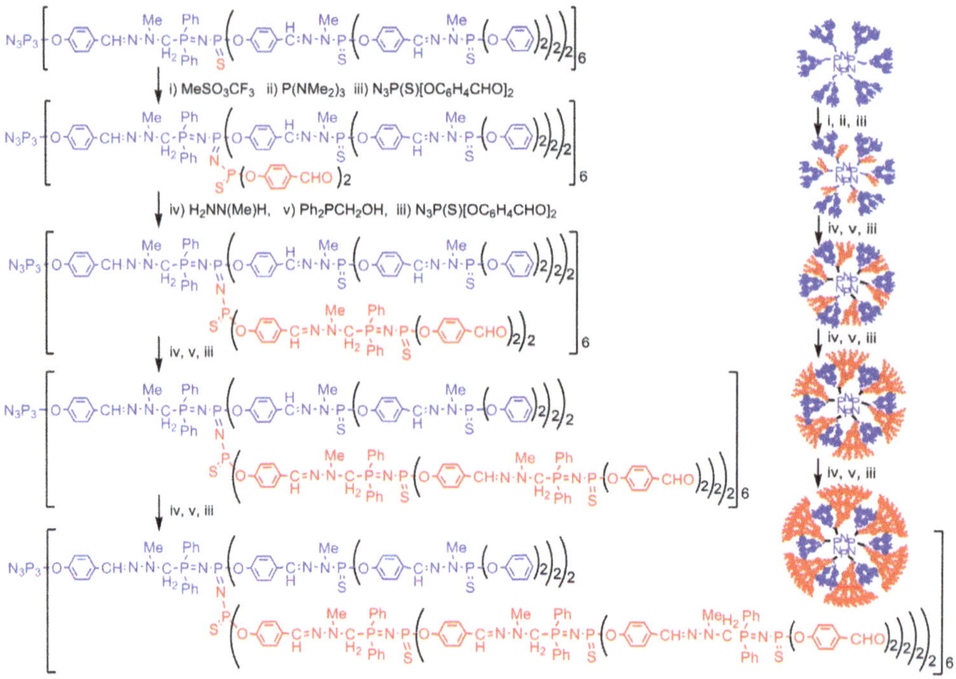

Scheme 7. Reactivity of the P=N-P=S linkages and growing of new branches inside the dendrimer.

The P=N-P=S linkage is also able to activate vinyl groups linked to the phosphazene, and used as core of dendrons (dendritic wedges). Different types of amines were used for Michael-type additions, suitable for grafting together by their core two dendrons which differ by their terminal functions such as nitrile, amine or phosphine [82] but also amine and carboxylate [83], leading to "Janus" dendrimers [84] (Scheme 8).

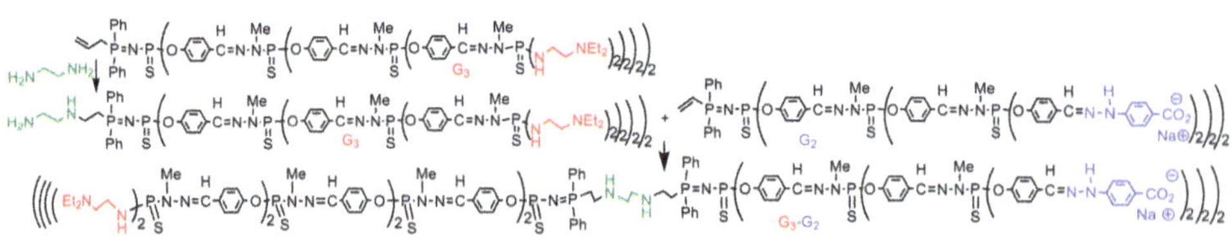

Scheme 8. Example of synthesis of a Janus dendrimer, thanks to the presence of P=N-P=S linkages.

5. Wittig and Horner-Wadsworth-Emmons Reactions

We have used this classical phosphorus reaction for the functionalization of the terminal groups of dendrimers, starting from the aldehyde functions. The Wittig reaction was used in particular for the grafting of ketone and nitrile [45], or tetrathiafulvalene (TTF) derivatives, including one with a macrocyclic substituent, suitable for the electrochemical sensing of Ba^{2+} [85]. The Wittig reaction was also applied when only half of the terminal groups were aldehydes [59], or ylides [61] (Scheme 9).

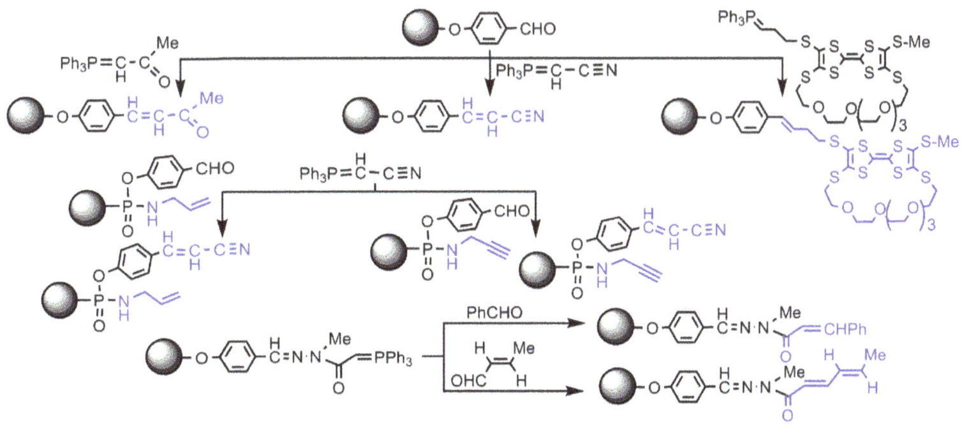

Scheme 9. Use of the Wittig reaction for the functionalization of the surface of dendrimers.

The Horner-Wadsworth-Emmons reaction has been applied to the aldehyde terminal functions, affording predominantly the E isomers, in particular for the grafting of aminoacids [86] (Scheme 10).

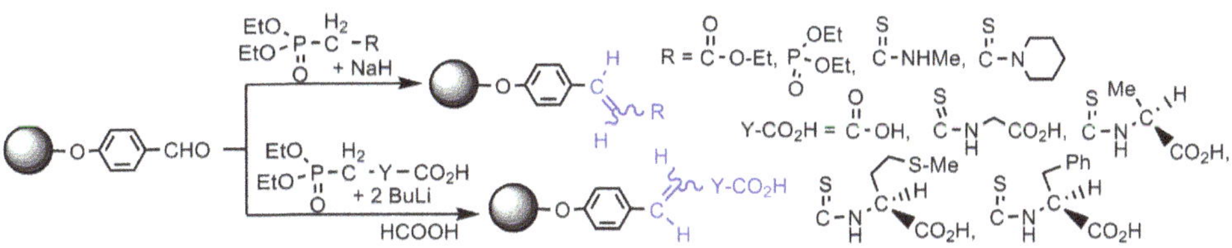

Scheme 10. Horner-Wadsworth-Emmons reaction for the grafting of aminoacids.

6. Cleavage of P-OR Bonds

In the course of our studies about the physico-chemical properties of phosphorus dendrimers, we have studied their thermal stability, and discovered that in many cases the first mass loss corresponds to the peeling of the surface, thus to the cleavage of the P-OR terminal groups at high temperature (above 200 °C for the least stable, but generally above 350 °C) [87]. Such cleavage has been also observed in the case of diketone terminal groups used for the complexation of copper, then for catalyzing diarylether formation at 120 °C. The efficiency of the catalysis was found independent of the generation of the dendrimer, and no reuse was possible, contrarily to what we had observed in all previous examples of catalysis [88]. Studying in details the reaction media after catalysis, we found a large amount of the monomer, resulting from the cleavage of the surface of the dendrimers. It must be noted that the cleavage is due to the catalysis, since the dendrimer is recovered intact in the same conditions, but in the absence of metal (Scheme 11) [17].

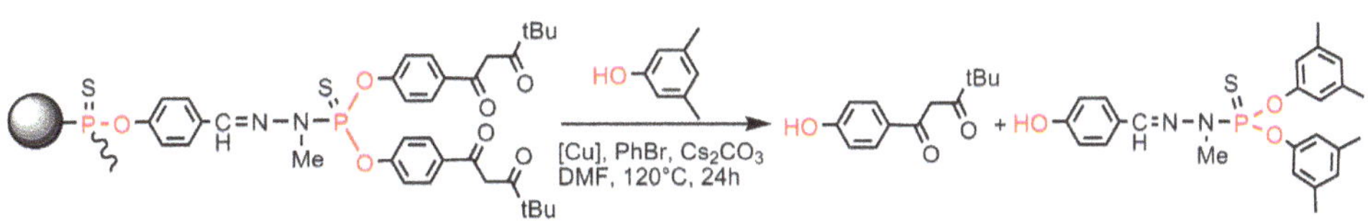

Scheme 11. Cleavage of P-OR bonds in catalysis conditions.

The dendrimers ended by azabisphosphonate groups are not easily soluble in water, thus we tried to obtain phosphonic acid instead of phosphonate terminal groups. For this purpose, the first step is the reaction with bromotrimethylsilane, which generates P-O-$SiMe_3$ groups, subsequently hydrolyzed. The last step is the reaction with NaOH (Scheme 12), affording water-soluble dendrimers [89], which possess very important biological properties [90], in particular towards the human immune system [91], as anti-inflammatory drug [92], and against rheumatoid arthritis [93].

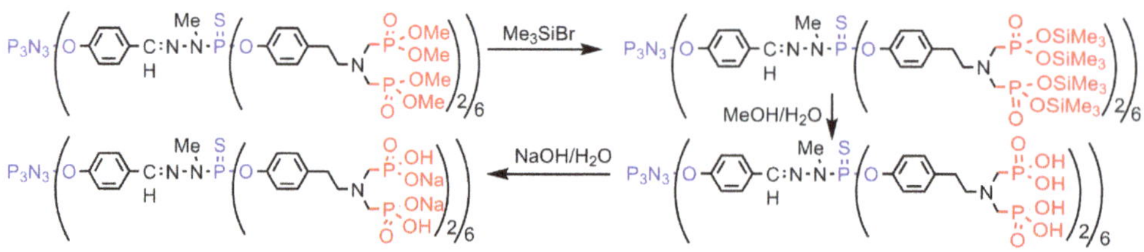

Scheme 12. Cleavage of P-OMe bonds while preserving P-OAr bonds.

7. Conclusions

A large panel of organophosphorus reactions has been used for the synthesis of phosphorus-containing dendrimers. Besides the efficiency of these reactions, the simplicity of characterization of these large compounds by ^{31}P-NMR has to be emphasized. Indeed, even highly sophisticated structures can be totally analyzed by ^{31}P-NMR [78]. It must be emphasized also that the presence of phosphorus leads to unprecedented properties, particularly in the fields of catalysis, materials, and biology.

Besides our work, which has been largely displayed in this review, a few other groups have reported the synthesis of phosphorus-containing dendrimers. We have to mention in particular the pioneering work made by R. Engel (polyphosphonium dendrimers) [94], M. J. Damha (nucleic acid dendrimers) [95], and D. L. DuBois (small polyphosphines) [96]. Later on, large polyphosphine dendrimers have been proposed by A. K. Kakkar [97], and also thiophosphate dendrimers by G. M. Salamonczyk [98] based on the use of phosphoramidite reagents. Taken all together, these researches demonstrate the rich diversity of the chemistry of phosphorus, even when applied to nano-objects such as dendrimers.

Acknowledgments: Thanks are due to the CNRS for financial support, and to the ANR DENDSWITCH.

References

1. Caminade, A.M.; Turrin, C.O.; Laurent, R.; Ouali, A.; Delavaux-Nicot, B. (Eds.) *Dendrimers. Towards Catalytic, Material and Biomedical Uses*; John Wiley & Sons: Chichester, UK, 2011; pp. 1–528.
2. Astruc, D.; Boisselier, E.; Ornelas, C. Dendrimers Designed for Functions: From Physical, Photophysical, and Supramolecular Properties to Applications in Sensing, Catalysis, Molecular Electronics, Photonics, and Nanomedicine. *Chem. Rev.* **2010**, *110*, 1857–1959. [CrossRef]
3. Majoral, J.P.; Caminade, A.M. Dendrimers containing heteroatoms (Si, P, B, Ge, or Bi). *Chem. Rev.* **1999**, *99*, 845–880. [CrossRef]
4. Caminade, A.M.; Servin, P.; Laurent, R.; Majoral, J.P. Dendrimeric phosphines in asymmetric catalysis. *Chem. Soc. Rev.* **2008**, *37*, 56–67. [CrossRef]
5. Knoll, W.; Caminade, A.M.; Char, K.; Duran, H.; Feng, C.L.; Gitsas, A.; Kim, D.H.; Lau, A.; Lazzara, T.D.; Majoral, J.P.; *et al.* Nanostructuring Polymeric Materials by Templating Strategies. *Small* **2011**, *7*, 1384–1391. [CrossRef]
6. Caminade, A.M.; Turrin, C.O.; Majoral, J.P. Biological properties of phosphorus dendrimers. *New J. Chem.* **2010**, *34*, 1512–1524. [CrossRef]

7. Majoral, J.P.; Caminade, A.M.; Maraval, V. The specific contribution of phosphorus in dendrimer chemistry. *Chem. Commun.* **2002**, 2929–2942. [CrossRef]

8. Launay, N.; Caminade, A.M.; Lahana, R.; Majoral, J.P. A General Synthetic Strategy for Neutral Phosphorus-Containing Dendrimers. *Angew. Chem. Int. Ed. Engl.* **1994**, 1589–1592. [CrossRef]

9. Launay, N.; Caminade, A.M.; Majoral, J.P. Synthesis of bowl-shaped dendrimers from generation 1 to generation 8. *J. Organomet. Chem.* **1997**, *529*, 51–58. [CrossRef]

10. Lartigue, M.L.; Donnadieu, B.; Galliot, C.; Caminade, A.M.; Majoral, J.P.; Fayet, J.P. Large dipole moments of phosphorus-containing dendrimers. *Macromolecules* **1997**, *30*, 7335–7337. [CrossRef]

11. Trevisiol, E.; Le Berre-Anton, V.; Leclaire, J.; Pratviel, G.; Caminade, A.M.; Majoral, J.P.; Francois, J.M.; Meunier, B. Dendrislides, dendrichips: A simple chemical functionalization of glass slides with phosphorus dendrimers as an effective means for the preparation of biochips. *New J. Chem.* **2003**, *27*, 1713–1719. [CrossRef]

12. Turrin, C.O.; Chiffre, J.; de Montauzon, D.; Daran, J.C.; Caminade, A.M.; Manoury, E.; Balavoine, G.; Majoral, J.P. Phosphorus-containing dendrimers with ferrocenyl units at the core, within the branches, and on the periphery. *Macromolecules* **2000**, *33*, 7328–7336. [CrossRef]

13. Turrin, C.O.; Chiffre, J.; Daran, J.C.; de Montauzon, D.; Caminade, A.M.; Manoury, E.; Balavoine, G.; Majoral, J.P. New chiral phosphorus-containing dendrimers with ferrocenes on the periphery. *Tetrahedron* **2001**, *57*, 2521–2536. [CrossRef]

14. Routaboul, L.; Vincendeau, S.; Turrin, C.O.; Caminade, A.M.; Majoral, J.P.; Daran, J.C.; Manoury, E. New phosphorus dendrimers with chiral ferrocenyl phosphine-thioether ligands on the periphery for asymmetric catalysis. *J. Organomet. Chem.* **2007**, *692*, 1064–1073. [CrossRef]

15. Merino, S.; Brauge, L.; Caminade, A.M.; Majoral, J.P.; Taton, D.; Gnanou, Y. Synthesis and characterization of linear, hyperbranched, and dendrimer-like polymers constituted of the same repeating unit. *Chem. Eur. J.* **2001**, *7*, 3095–3105. [CrossRef]

16. Lacour, M.A.; Zablocka, M.; Caminade, A.M.; Taillefer, M.; Majoral, J.P. Design of phosphonium ended dendrimers bearing functionalized amines. *Tetrahedron Lett.* **2009**, *50*, 4870–4873. [CrossRef]

17. Keller, M.; Hameau, A.; Spataro, G.; Ladeira, S.; Caminade, A.M.; Majoral, J.P.; Ouali, A. An efficient and recyclable dendritic catalyst able to dramatically reduce palladium leaching in Suzuki couplings. *Green Chem.* **2012**, *14*, 2807–2815. [CrossRef]

18. Koprowski, M.; Sebastian, R.M.; Maraval, V.; Zablocka, M.; Cadierno, V.; Donnadieu, B.; Igau, A.; Caminade, A.M.; Majoral, J.P. Iminophosphine palladium complexes in catalytic Stille coupling reactions: From monomers to dendrimers. *Organometallics* **2002**, *21*, 4680–4687. [CrossRef]

19. Servin, P.; Laurent, R.; Romerosa, A.; Peruzzini, M.; Majoral, J.P.; Caminade, A.M. Synthesis of dendrimers terminated by bis(diphenylphosphinomethyl)amino ligands and use of their palladium complexes for catalyzing C-C cross-coupling reactions. *Organometallics* **2008**, *27*, 2066–2073. [CrossRef]

20. Keller, M.; Ianchuk, M.; Ladeira, S.; Taillefer, M.; Caminade, A.M.; Majoral, J.P.; Ouali, A. Synthesis of Dendritic beta-Diketones and Their Application in Copper-Catalyzed Diaryl Ether Formation. *Eur. J. Org. Chem.* **2012**, 1056–1062. [CrossRef]

21. Gissibl, A.; Padie, C.; Hager, M.; Jaroschik, F.; Rasappan, R.; Cuevas-Yanez, E.; Turrin, C.O.; Caminade, A.M.; Majoral, J.P.; Reiser, O. Synthesis and application of phosphorus dendrimer immobilized azabis(oxazolines). *Org. Lett.* **2007**, *9*, 2895–2898. [CrossRef]

22. Marchand, P.; Griffe, L.; Caminade, A.M.; Majoral, J.P.; Destarac, M.; Leising, F. Thioacylation reactions for the surface functionalization of phosphorus-containing dendrimers. *Org. Lett.* **2004**, *6*, 1309–1312. [CrossRef]

23. Darcos, V.; Dureault, A.; Taton, D.; Gnanou, Y.; Marchand, P.; Caminade, A.M.; Majoral, J.P.; Destarac, M.; Leising, F. Synthesis of hybrid dendrimer-star polymers by the RAFT process. *Chem. Commun.* **2004**, 2110–2111. [CrossRef] [PubMed]

24. Franc, G.; Mazeres, S.; Turrin, C.O.; Vendier, L.; Duhayon, C.; Caminade, A.M.; Majoral, J.P. Synthesis and properties of dendrimers possessing the same fluorophore(s) located either peripherally or off-center. *J. Org. Chem.* **2007**, *72*, 8707–8715. [CrossRef] [PubMed]

25. Fuchs, S.; Pla-Quintana, A.; Mazeres, S.; Caminade, A.M.; Majoral, J.P. Cationic and Fluorescent "Janus" Dendrimers. *Org. Lett.* **2008**, *10*, 4751–4754. [CrossRef] [PubMed]

26. Hameau, A.; Fuchs, S.; Laurent, R.; Majoral, J.P.; Caminade, A.M. Synthesis of dye/fluorescent functionalized dendrons based on cyclotriphosphazene. *Beilstein J. Org. Chem.* **2011**, *7*, 1577–1583. [CrossRef] [PubMed]

27. Mongin, O.; Krishna, T.R.; Werts, M.H.V.; Caminade, A.M.; Majoral, J.P.; Blanchard-Desce, M. A modular approach to two-photon absorbing organic nanodots: Brilliant dendrimers as an alternative to semiconductor quantum dots? *Chem. Commun.* **2006**, 915–917. [CrossRef] [PubMed]
28. Terenziani, F.; Parthasarathy, V.; Pla-Quintana, A.; Maishal, T.; Caminade, A.M.; Majoral, J.P.; Blanchard-Desce, M. Cooperative Two-Photon Absorption Enhancement by Through-Space Interactions in Multichromophoric Compounds. *Angew. Chem. Int. Ed.* **2009**, *48*, 8691–8694. [CrossRef]
29. Fuks-Janczarek, I.; Nunzi, J.M.; Sahraoui, B.; Kityk, I.V.; Berdowski, J.; Caminade, A.M.; Majoral, J.P.; Martineau, A.C.; Frere, P.; Roncali, J. Third-order nonlinear optical properties and two-photon absorption in branched oligothienylenevinylenes. *Opt. Commun.* **2002**, *209*, 461–466. [CrossRef]
30. Hadad, C.; Majoral, J.P.; Muzart, J.; Caminade, A.M.; Bouquillon, S. First phosphorous D-xylose-derived glycodendrimers. *Tetrahedron Lett.* **2009**, *50*, 1902–1905. [CrossRef]
31. Perez-Anes, A.; Spataro, G.; Coppel, Y.; Moog, C.; Blanzat, M.; Turrin, C.O.; Caminade, A.M.; Rico-Lattes, I.; Majoral, J.P. Phosphonate terminated PPH dendrimers: Influence of pendant alkyl chains on the in vitro anti-HIV-1 properties. *Org. Biomol. Chem.* **2009**, *7*, 3491–3498. [CrossRef]
32. Rolland, O.; Turrin, C.O.; Bacquet, G.; Poupot, R.; Poupot, M.; Caminade, A.M.; Majoral, J.P. Efficient synthesis of phosphorus-containing dendrimers capped with isosteric functions of amino-bismethylene phosphonic acids. *Tetrahedron Lett.* **2009**, *50*, 2078–2082. [CrossRef]
33. Poupot, M.; Griffe, L.; Marchand, P.; Maraval, A.; Rolland, O.; Martinet, L.; L'Faqihi-Olive, F.E.; Turrin, C.O.; Caminade, A.M.; Fournié, J.J.; *et al.* Design of phosphorylated dendritic architectures to promote human monocyte activation. *FASEB J.* **2006**, *20*, 2339–2351. [CrossRef] [PubMed]
34. Marchand, P.; Griffe, L.; Poupot, M.; Turrin, C.O.; Bacquet, G.; Fournie, J.J.; Majoral, J.P.; Poupot, R.; Caminade, A.M. Dendrimers ended by non-symmetrical azadiphosphonate groups: Synthesis and immunological properties. *Bioorg. Med. Chem. Lett.* **2009**, *19*, 3963–3966. [CrossRef]
35. Caminade, A.M.; Laurent, R.; Turrin, C.O.; Rebout, C.; Delavaux-Nicot, B.; Ouali, A.; Zablocka, M.; Majoral, J.P. Phosphorus dendrimers as viewed by P-31 NMR spectroscopy; synthesis and characterization. *C. R. Chim.* **2010**, *13*, 1006–1027. [CrossRef]
36. Turrin, C.O.; Chiffre, J.; Daran, J.C.; de Montauzon, D.; Balavoine, G.; Manoury, E.; Caminade, A.M.; Majoral, J.P. New phosphorus-containing dendrimers with ferrocenyl units in each layer. *C. R. Chim.* **2002**, *5*, 309–318. [CrossRef]
37. Turrin, C.O.; Chiffre, J.; de Montauzon, D.; Balavoine, G.; Manoury, E.; Caminade, A.M.; Majoral, J.P. Behavior of an optically active ferrocene chiral shell located within phosphorus-containing dendrimers. *Organometallics* **2002**, *21*, 1891–1897. [CrossRef]
38. Magro, G.; Marchand, P.; Sebastian, R.M.; Guyard-Duhayon, C.; Caminade, A.M.; Majoral, J.P. Synthesis and characterization of phosphorus dendrimers containing long, conjugated branches. *Eur. J. Org. Chem.* **2005**, 1340–1347. [CrossRef]
39. Mongin, O.; Rouxel, C.; Vabre, J.M.; Mir, Y.; Pla-Quintana, A.; Wei, Y.Q.; Caminade, A.M.; Majoral, J.P.; Blanchard-Desce, M. Customized multiphotonics nanotools for bioapplications: Soft organic nanodots as an eco-friendly alternative to quantum dots. In *Nanobiosystems: Processing, Characterization, and Applications II Book Series, Proceedings of SPIE-International Society Optical Engineering: San Diego, CA, USA, 2009*; Kobayashi, N., Ouchen, F., Rau, I., Eds.; SPIE- International Society Optical Engineering: Bellingham, WA, USA, 2009; Volume 7403, pp. 740303:1–740303:12.
40. Sebastian, R.M.; Blais, J.C.; Caminade, A.M.; Majoral, J.P. Synthesis and photochemical behavior of phosphorus dendrimers containing azobenzene units within the branches and/or on the surface. *Chem. Eur. J.* **2002**, *8*, 2172–2183. [CrossRef]
41. Servin, P.; Rebout, C.; Laurent, R.; Peruzzini, M.; Caminade, A.M.; Majoral, J.P. Reduced number of steps for the synthesis of dense and highly functionalized dendrimers. *Tetrahedron Lett.* **2007**, *48*, 579–583. [CrossRef]
42. Launay, N.; Galliot, C.; Caminade, A.M.; Majoral, J.P. Synthesis of small phosphorus dendrimers from (S)P[N(Me)-NH_2]$_3$. *Bull. Soc. Chim. Fr.* **1995**, *132*, 1149–1155.
43. Kraemer, R.; Galliot, C.; Mitjaville, J.; Caminade, A.M.; Majoral, J.P. Hexamethylhydrazino-cyclotriphosphazene $N_3P_3(NMeNH_2)_6$: Starting reagent for the synthesis of multifunctionalized species, macrocycles and small dendrimers. *Heteroat. Chem.* **1996**, *7*, 149–154. [CrossRef]

44. Katir, N.; Majoral, J.P.; El Kadib, A.; Caminade, A.M.; Bousmina, M. Molecular and Macromolecular Engineering with Viologens as Building Blocks: Rational Design of Phosphorus-Viologen Dendritic Structures. *Eur. J. Org. Chem.* **2012**, 269–273. [CrossRef]
45. Launay, N.; Slany, M.; Caminade, A.M.; Majoral, J.P. Phosphorus-containing dendrimers. Easy access to new multi-difunctionalized macromolecules. *J. Org. Chem.* **1996**, *61*, 3799–3805. [CrossRef] [PubMed]
46. Loup, C.; Zanta, M.A.; Caminade, A.M.; Majoral, J.P.; Meunier, B. Preparation of water-soluble cationic phosphorus-containing dendrimers as DNA transfecting agents. *Chem. Eur. J.* **1999**, *5*, 3644–3650. [CrossRef]
47. Caminade, A.M.; Majoral, J.P. Water-soluble phosphorus-containing dendrimers. *Prog. Polym. Sci.* **2005**, *30*, 491–505. [CrossRef]
48. Rolland, O.; Turrin, C.O.; Caminade, A.M.; Majoral, J.P. Dendrimers and nanomedicine: Multivalency in action. *New J. Chem.* **2009**, *33*, 1809–1824. [CrossRef]
49. Feng, C.L.; Zhong, X.H.; Steinhart, M.; Caminade, A.M.; Majoral, J.P.; Knoll, W. Functional quantum-dot/dendrimer nanotubes for sensitive detection of DNA hybridization. *Small* **2008**, *4*, 566–571. [CrossRef]
50. Kim, B.S.; Lebedeva, O.V.; Kim, D.H.; Caminade, A.M.; Majoral, J.P.; Knoll, W.; Vinogradova, O.I. Assembly and mechanical properties of phosphorus dendrimer/polyelectrolyte multilayer microcapsules. *Langmuir* **2005**, *21*, 7200–7206. [CrossRef]
51. Feng, C.L.; Yin, M.Z.; Zhang, D.; Zhu, S.M.; Caminade, A.M.; Majoral, J.P.; Mullen, K. Fluorescent Core-Shell Star Polymers Based Bioassays for Ultrasensitive DNA Detection by Surface Plasmon Fluorescence Spectroscopy. *Macromol. Rapid Commun.* **2011**, *32*, 679–683. [CrossRef]
52. Solassol, J.; Crozet, C.; Perrier, V.; Leclaire, J.; Beranger, F.; Caminade, A.M.; Meunier, B.; Dormont, D.; Majoral, J.P.; Lehmann, S. Cationic phosphorus-containing dendrimers reduce prion replication both in cell culture and in mice infected with scrapie. *J. Gen. Virol.* **2004**, *85*, 1791–1799. [CrossRef]
53. Wasiak, T.; Ionov, M.; Nieznanski, K.; Nieznanska, H.; Klementieva, O.; Granell, M.; Cladera, J.; Majoral, J.P.; Caminade, A.M.; Klajnert, B. Phosphorus Dendrimers Affect Alzheimer's (A beta(1-28)) Peptide and MAP-Tau Protein Aggregation. *Mol. Pharm.* **2012**, *9*, 458–469. [CrossRef] [PubMed]
54. Padie, C.; Maszewska, M.; Majchrzak, K.; Nawrot, B.; Caminade, A.M.; Majoral, J.P. Polycationic phosphorus dendrimers: Synthesis, characterization, study of cytotoxicity, complexation of DNA, and transfection experiments. *New J. Chem.* **2009**, *33*, 318–326. [CrossRef]
55. Prevote, D.; Donnadieu, B.; Moreno-Manas, M.; Caminade, A.M.; Majoral, J.P. Grafting of tetraazamacrocycles on the surface of phosphorus-containing dendrimers. *Eur. J. Org. Chem.* **1999**, 1701–1708. [CrossRef]
56. Badetti, E.; Caminade, A.M.; Majoral, J.P.; Moreno-Manas, M.; Sebastian, R.M. Palladium(0) nanoparticles stabilized by phosphorus dendrimers containing coordinating 15-membered triolefinic macrocycles in periphery. *Langmuir* **2008**, *24*, 2090–2101. [CrossRef] [PubMed]
57. Franc, G.; Badetti, E.; Duhayon, C.; Coppel, Y.; Turrin, C.O.; Majoral, J.P.; Sebastian, R.M.; Caminade, A.M. An efficient synthesis combining phosphorus dendrimers and 15-membered triolefinic azamacrocycles: Towards the stabilization of platinum nanoparticles. *New J. Chem.* **2010**, *34*, 547–555. [CrossRef]
58. Franc, G.; Badetti, E.; Colliere, V.; Majoral, J.P.; Sebastian, R.M.; Caminade, A.M. Dendritic structures within dendritic structures: Dendrimer-induced formation and self-assembly of nanoparticle networks. *Nanoscale* **2009**, *1*, 233–237. [CrossRef]
59. Lartigue, M.L.; Slany, M.; Caminade, A.M.; Majoral, J.P. Phosphorus-containing dendrimers: Synthesis of macromolecules with multiple tri- and tetrafunctionalization. *Chem. Eur. J.* **1996**, *2*, 1417–1426. [CrossRef]
60. Severac, M.; Leclaire, J.; Sutra, P.; Caminade, A.M.; Majoral, J.P. A new way for the internal functionalization of dendrimers. *Tetrahedron Lett.* **2004**, *45*, 3019–3022. [CrossRef]
61. Prevote, D.; Caminade, A.M.; Majoral, J.P. Phosphate-, phosphite-, ylide-, and phosphonate-terminated dendrimers. *J. Org. Chem.* **1997**, *62*, 4834–4841. [CrossRef]
62. Galliot, C.; Prevote, D.; Caminade, A.M.; Majoral, J.P. Polyaminophosphines Containing Dendrimers - Syntheses and Characterizations. *J. Am. Chem. Soc.* **1995**, *117*, 5470–5476. [CrossRef]
63. Garcia, L.; Roglans, A.; Laurent, R.; Majoral, J.P.; Pla-Quintana, A.; Caminade, A.M. Dendritic phosphoramidite ligands for Rh-catalyzed [2+2+2] cycloaddition reactions: Unprecedented enhancement of enantiodiscrimination. *Chem. Commun.* **2012**, *48*, 9248–9250. [CrossRef] [PubMed]
64. Slany, M.; Bardaji, M.; Caminade, A.M.; Chaudret, B.; Majoral, J.P. Versatile complexation ability of very large phosphino-terminated dendrimers. *Inorg. Chem.* **1997**, *36*, 1939–1945. [CrossRef] [PubMed]

65. Lartigue, M.L.; Caminade, A.M.; Majoral, J.P. Chiroptical properties of dendrimers with stereogenic end groups. *Tetrahedron: Asymmetry* **1997**, *8*, 2697–2708. [CrossRef]
66. Franc, G.; Turrin, C.O.; Cavero, E.; Costes, J.P.; Duhayon, C.; Caminade, A.M.; Majoral, J.P. Gem-Bisphosphonate-Ended Group Dendrimers: Design and Gadolinium Complexing Properties. *Eur. J. Org. Chem.* **2009**, 4290–4299. [CrossRef]
67. Servin, P.; Laurent, R.; Gonsalvi, L.; Tristany, M.; Peruzzini, M.; Majoral, J.P.; Caminade, A.M. Grafting of water-soluble phosphines to dendrimers and their use in catalysis: Positive dendritic effects in aqueous media. *Dalton Trans.* **2009**, 4432–4434. [CrossRef]
68. Servin, P.; Laurent, R.; Dib, H.; Gonsalvi, L.; Peruzzini, M.; Majoral, J.P.; Caminade, A.M. Number of terminal groups versus generation of the dendrimer, which criteria influence the catalytic properties? *Tetrahedron Lett.* **2012**, *53*, 3876–3879. [CrossRef]
69. Cavero, E.; Zablocka, M.; Caminade, A.M.; Majoral, J.P. Design of Bisphosphonate-Terminated Dendrimers. *Eur. J. Org. Chem.* **2010**, 2759–2767. [CrossRef]
70. Brauge, L.; Magro, G.; Caminade, A.M.; Majoral, J.P. First divergent strategy using two AB_2 unprotected monomers for the rapid synthesis of dendrimers. *J. Am. Chem. Soc.* **2001**, *123*, 6698–6699. [CrossRef]
71. Maraval, V.; Caminade, A.M.; Majoral, J.P.; Blais, J.C. Dendrimer design: How to circumvent the dilemma of a reduction of steps or an increase of function multiplicity? *Angew. Chem. Int. Ed.* **2003**, *42*, 1822–1826. [CrossRef]
72. Maraval, V.; Pyzowski, J.; Caminade, A.M.; Majoral, J.P. "Lego" chemistry for the straightforward synthesis of dendrimers. *J. Org. Chem.* **2003**, *68*, 6043–6046. [CrossRef]
73. Sebastian, R.M.; Magro, G.; Caminade, A.M.; Majoral, J.P. Dendrimers *with N,N*-disubstituted hydrazines as end groups, useful precursors for the synthesis of water-soluble dendrimers capped with carbohydrate, carboxylic or boronic acid derivatives. *Tetrahedron* **2000**, *56*, 6269–6277. [CrossRef]
74. Larre, C.; Caminade, A.M.; Majoral, J.P. Chemoselective polyalkylations of phosphorus-containing dendrimers. *Angew. Chem. Int. Ed. Engl.* **1997**, *36*, 596–599. [CrossRef]
75. Larre, C.; Donnadieu, B.; Caminade, A.M.; Majoral, J.P. Phosphorus-containing dendrimers: Chemoselective functionalization of internal layers. *J. Am. Chem. Soc.* **1998**, *120*, 4029–4030. [CrossRef]
76. Larre, C.; Donnadieu, B.; Caminade, A.M.; Majoral, J.P. Regioselective gold complexation within the cascade structure of phosphorus-containing dendrimers. *Chem. Eur. J.* **1998**, *4*, 2031–2036. [CrossRef]
77. Larre, C.; Bressolles, D.; Turrin, C.; Donnadieu, B.; Caminade, A.M.; Majoral, J.P. Chemistry within megamolecules: Regiospecific functionalization after construction of phosphorus dendrimers. *J. Am. Chem. Soc.* **1998**, *120*, 13070–13082. [CrossRef]
78. Galliot, C.; Larre, C.; Caminade, A.M.; Majoral, J.P. Regioselective stepwise growth of dendrimer units in the internal voids of a main dendrimer. *Science* **1997**, *277*, 1981–1984. [CrossRef]
79. Maraval, V.; Laurent, R.; Donnadieu, B.; Mauzac, M.; Caminade, A.M.; Majoral, J.P. Rapid synthesis of phosphorus-containing dendrimers with controlled molecular architectures: First example of surface-block, layer-block, and segment-block dendrimers issued from the same dendron. *J. Am. Chem. Soc.* **2000**, *122*, 2499–2511. [CrossRef]
80. Brauge, L.; Caminade, A.M.; Majoral, J.P.; Slomkowski, S.; Wolszczak, M. Segmental mobility in phosphorus-containing dendrimers. Studies by fluorescent spectroscopy. *Macromolecules* **2001**, *34*, 5599–5606. [CrossRef]
81. Cadierno, V.; Igau, A.; Donnadieu, B.; Caminade, A.M.; Majoral, J.P. Dendrimers containing zwitterionic [phosphonium anionic zirconocene(IV)] complexes. *Organometallics* **1999**, *18*, 1580–1582. [CrossRef]
82. Maraval, V.; Laurent, R.; Merino, S.; Caminade, A.M.; Majoral, J.P. Michael-type addition of amines to the vinyl core of dendrons - Application to the synthesis of multidendritic systems. *Eur. J. Org. Chem.* **2000**, 3555–3568. [CrossRef]
83. Maraval, V.; Maraval, A.; Spataro, G.; Caminade, A.M.; Majoral, J.P.; Kim, D.H.; Knoll, W. Design of tailored multi-charged phosphorus surface-block dendrimers. *New J. Chem.* **2006**, *30*, 1731–1736. [CrossRef]
84. Caminade, A.M.; Laurent, R.; Delavaux-Nicot, B.; Majoral, J.P. "Janus" dendrimers: Syntheses and properties. *New J. Chem.* **2012**, *36*, 217–226. [CrossRef]
85. Le Derf, F.; Levillain, E.; Trippe, G.; Gorgues, A.; Salle, M.; Sebastian, R.M.; Caminade, A.M.; Majoral, J.P. Immobilization of redox-active ligands on an electrode: The dendrimer route. *Angew. Chem. Int. Ed.* **2001**, *40*, 224–227. [CrossRef]

86. Prevote, D.; LeRoyGourvennec, S.; Caminade, A.M.; Masson, S.; Majoral, J.P. Application of the Horner-Wadsworth-Emmons reaction to the functionalization of dendrimers: Synthesis of amino acid terminated dendrimers. *Synthesis-Stuttgart* **1997**, 1199–1207. [CrossRef]
87. Turrin, C.O.; Maraval, V.; Leclaire, J.; Dantras, E.; Lacabanne, C.; Caminade, A.M.; Majoral, J.P. Surface, core, and structure modifications of phosphorus-containing dendrimers. Influence on the thermal stability. *Tetrahedron* **2003**, *59*, 3965–3973. [CrossRef]
88. Majoral, J.P.; Caminade, A.M.; Laurent, R. Metallo groups linked to the surface of phosphorus-containing dendrimers. In *Metal-Containing and Metallosupramolecular Polymers and Materials*; Schubert, U.S., Newkome, G.R., Manners, I., Eds.; American Chemical Society: Washington, DC, USA, 2006; Volume 928, pp. 230–243.
89. Rolland, O.; Griffe, L.; Poupot, M.; Maraval, A.; Ouali, A.; Coppel, Y.; Fournie, J.J.; Bacquet, G.; Turrin, C.O.; Caminade, A.M.; *et al.* Tailored control and optimisation of the number of phosphonic acid termini on phosphorus-containing dendrimers for the *ex-vivo* activation of human monocytes. *Chem. Eur. J.* **2008**, *14*, 4836–4850. [CrossRef]
90. Griffe, L.; Poupot, M.; Marchand, P.; Maraval, A.; Turrin, C.O.; Rolland, O.; Metivier, P.; Bacquet, G.; Fournie, J.J.; Caminade, A.M.; *et al.* Multiplication of human natural killer cells by nanosized phosphonate-capped dendrimers. *Angew. Chem. Int. Ed.* **2007**, *46*, 2523–2526. [CrossRef]
91. Portevin, D.; Poupot, M.; Rolland, O.; Turrin, C.O.; Fournie, J.J.; Majoral, J.P.; Caminade, A.M.; Poupot, R. Regulatory activity of azabisphosphonate-capped dendrimers on human CD4(+) T cell proliferation enhances *ex-vivo* expansion of NK cells from PBMCs for immunotherapy. *J. Transl. Med.* **2009**, *7*, 13. [CrossRef]
92. Fruchon, S.; Poupot, M.; Martinet, L.; Turrin, C.O.; Majoral, J.P.; Fournie, J.J.; Caminade, A.M.; Poupot, R. Anti-inflammatory and immunosuppressive activation of human monocytes by a bioactive dendrimer. *J. Leuk. Biol.* **2009**, *85*, 553–562. [CrossRef]
93. Hayder, M.; Poupot, M.; Baron, M.; Nigon, D.; Turrin, C.O.; Caminade, A.M.; Majoral, J.P.; Eisenberg, R.A.; Fournie, J.J.; Cantagrel, A.; *et al.* A Phosphorus-Based Dendrimer Targets Inflammation and Osteoclastogenesis in Experimental Arthritis. *Sci. Transl. Med.* **2011**, *3*, 11. [CrossRef]
94. Rengan, K.; Engel, R. Phosphonium Cascade Molecules. *J. Chem. Soc. Chem. Commun.* **1990**, 1084–1085. [CrossRef]
95. Hudson, R.H.E.; Damha, M.J. Nucleic-Acid Dendrimers—Novel Biopolymer Structures. *J. Am. Chem. Soc.* **1993**, *115*, 2119–2124. [CrossRef]
96. Miedaner, A.; Curtis, C.J.; Barkley, R.M.; Dubois, D.L. Electrochemical Reduction of Co2 Catalyzed by Small Organophosphine Dendrimers Containing Palladium. *Inorg. Chem.* **1994**, *33*, 5482–5490. [CrossRef]
97. Petrucci-Samija, M.; Guillemette, V.; Dasgupta, M.; Kakkar, A.K. A new divergent route to the synthesis of organophosphine and metallodendrimers via simple acid-base hydrolytic chemistry. *J. Am. Chem. Soc.* **1999**, *121*, 3248–3248. [CrossRef]
98. Salamonczyk, G.M.; Kuznikowski, M.; Skowronska, A. A divergent synthesis of thiophosphate-based dendrimers. *Tetrahedron Lett.* **2000**, *41*, 1643–1645. [CrossRef]

molecules

MDPI

Review

Stoichiometric and Catalytic Synthesis of Alkynylphosphines

Elise Bernoud, Romain Veillard, Carole Alayrac * and Annie-Claude Gaumont *

Laboratoire de Chimie Moléculaire et Thioorganique, UMR CNRS 6507, INC3M FR3038, ENSICAEN & Université de Caen Basse-Normandie, 14050 Caen, France; e.bernoud@gmail.com (E.B.); romain.veillard@ensicaen.fr (R.V.)

* Correspondence: carole.witulski-alayrac@ensicaen.fr (C.A.); annie-claude.gaumont@ensicaen.fr (A.-C.G.); Tel.: +33-2-31-45-28-86 (C.A.); +33-2-31-45-28-73 (A.-C.G.); Fax: +33-2-31-45-28-77

Received: 23 November 2012; Accepted: 5 December 2012; Published: 7 December 2012

check for updates

Abstract: Alkynylphosphines or their borane complexes are available either through C–P bond forming reactions or through modification of the phosphorus or the alkynyl function of various alkynyl phosphorus derivatives. The latter strategy, and in particular the one involving phosphoryl reduction by alanes or silanes, is the method of choice for preparing primary and secondary alkynylphosphines, while the former strategy is usually employed for the synthesis of tertiary alkynylphosphines or their borane complexes. The classical C–P bond forming methods rely on the reaction between halophosphines or their borane complexes with terminal acetylenes in the presence of a stoichiometric amount of organometallic bases, which precludes the access to alkynylphosphines bearing sensitive functional groups. In less than a decade, efficient catalytic procedures, mostly involving copper complexes and either an electrophilic or a nucleophilic phosphorus reagent, have emerged. By proceeding under mild conditions, these new methods have allowed a significant broadening of the substituent scope and structure complexity.

Keywords: organophosphorus chemistry; phosphines; phosphine-boranes; homogeneous catalysis

---

1. Introduction

Alkynylphosphines are relevant versatile compounds due to the presence of two important functions, a phosphorus atom—a key-atom in metal coordination [1]—and an alkynyl moiety, which can also coordinate to metals and is known to display a rich and diverse chemistry [2].

In the past, these phosphines were mainly used in coordination chemistry for the assembly of various mono- or bi-metallic complexes and clusters [3], by taking advantage of their diverse coordination modes to transition metals, via the phosphorus atom and/or the triple bond. It is noteworthy that the range of alkynylphosphines involved in those complexes and clusters is almost exclusively covered by diphenylphosphino-phenylacetylene and bis(diphenylphosphino)acetylene.

It is only recently that the high potential of alkynylphosphines as ligands for homogeneous catalysis has emerged through a few relevant catalytic applications [4]. Notably P-stereogenic alkynylphosphines were shown to induce excellent enantioselectivity in several asymmetric transition metal-catalyzed reactions [5]. Moreover the easy transformation of alkynylphosphines into valuable functionalized alkenyl mono- or di-phosphines through addition reactions on the triple bond was recently demonstrated [6]. These new applications as ligands for homogeneous catalysis and building blocks for organophosphorus chemistry have aroused the need for new general synthetic methods allowing the access to a broad substitution pattern.

The classical method for preparing alkynylphosphines is based on the nucleophilic substitution reaction at the phosphorus atom of a halophosphine by a metal (lithium, magnesium, titanium or zinc)

acetylide at low temperature (Scheme 1, path A). This methodology was recently successfully applied to the synthesis of a few enantioenriched P-stereogenic alkynylphosphine-boranes. The reaction however suffers from severe limitations due to the weak configurational stability of halophosphines.

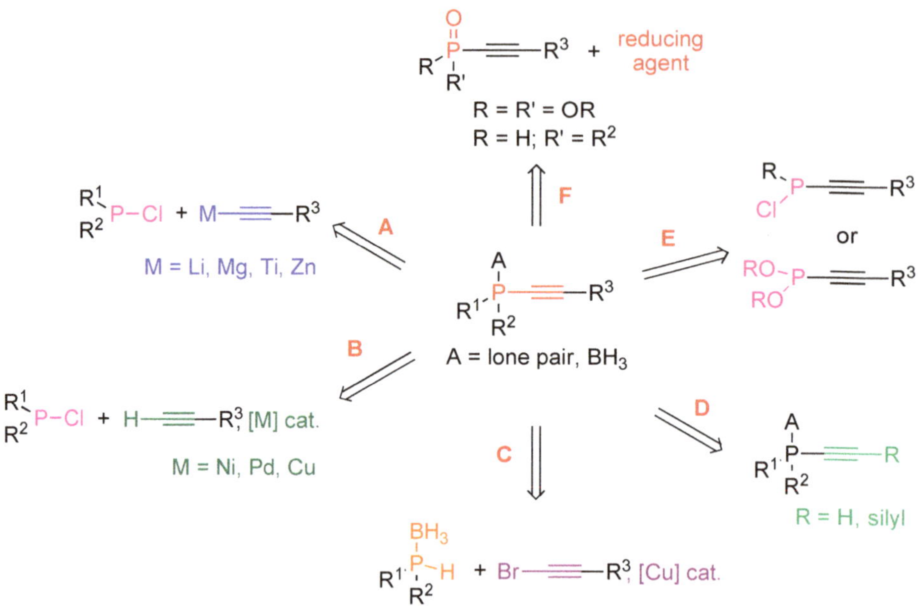

Scheme 1. Strategies A–F for the synthesis of alkynylphosphines and/or their borane complexes.

Catalytic variants of this reaction have been recently described as useful alternative approaches when sensitive functional groups are present (Scheme 1, path B). They rely on the cross-coupling reactions of terminal alkynes with chlorophosphines in the presence of a catalytic amount of nickel, palladium or copper complexes. Both stoichiometric and catalytic methods are mainly based on an electrophilic phosphorus derivative—a halophosphine—as starting material. The use of a nucleophilic phosphorus derivative has been only little explored so far. In the stoichiometric series, attempts involving treatment of alkynylphosphonium chlorides with a base failed to afford the expected tertiary alkynylphosphines because their oxidation could not be avoided under the reaction conditions [7]. On the other hand the catalytic synthesis of tertiary alkynylphosphine-boranes based on the copper(I)-catalyzed cross-coupling reaction between secondary phosphine-boranes and bromoalkynes was shown to proceed readily (Scheme 1, path C). A further strategy towards alkynylphosphines relies on the transformation of alkynylphosphorus derivatives through modification of the alkyne (Scheme 1, path D) or the phosphorus function (Scheme 1, paths E and F). Notably the reduction of the phosphoryl moiety of alkynylphosphorus derivatives (Scheme 1, path F) is highly relevant for the synthesis of primary and secondary alkynylphosphines, which are air- and heat-sensitive compounds.

The scope of this review encompasses the synthesis of alkynylphosphines and their borane complexes [8]. The stoichiometric methods for C–P bond formation will be first surveyed followed by the catalytic procedures. Finally the methods based on the transformation of alkynylphosphorus derivatives will be detailed.

2. Nucleophilic Substitution at P-atom of Halophosphines

Tertiary alkynylphosphines are classically prepared through the reaction of chlorophosphines and metal acetylides. For secondary alkynylphosphines, only those bearing sterically encumbered substituents are available through this procedure.

2.1. Synthesis of Secondary Alkynylphosphines

Märkl and Reitinger reported the synthesis of a set of secondary alkynylphosphines based on the reaction of the bulky secondary chlorophosphine **1** with various alkyne Grignard reagents in diethyl ether at low temperature [9] (Scheme 2).

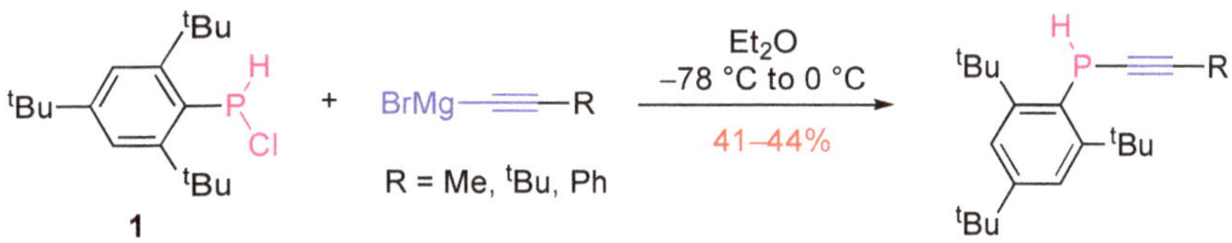

Scheme 2. Synthesis of secondary alkynylphosphines via nucleophilic substitution.

The analogous reactions of chlorophosphine **1** with lithium acetylides in THF did not afford alkynylphosphines but their 3*H*-phosphaallene tautomers, because of the stronger basicity of lithium acetylides in comparison to the related Grignard reagents. One exception was found with lithium trimethylsilylacetylide, which led to the alkynyl product under the same conditions. The reaction of secondary chlorophosphine **1** with metal acetylides proved to be not only metal- and substrate- but also solvent-dependent, probably because of the importance of the acetylide solvation. For example the reaction of **1** with lithium *tert*-butylacetylide led to the alkynyl product in diethyl ether and to the allenyl product in THF (Scheme 3).

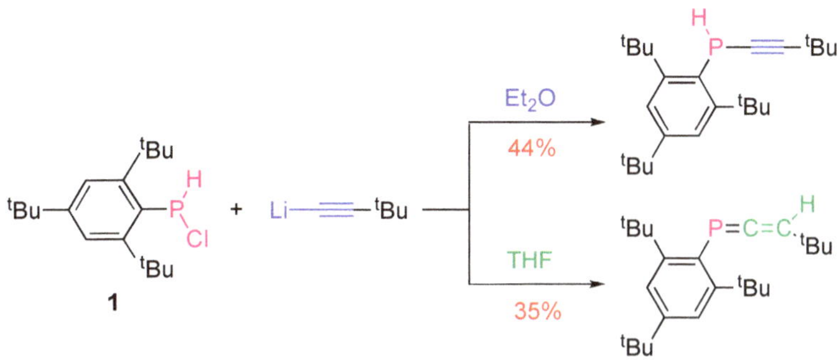

Scheme 3. Solvent-effect on the outcome of the reaction of lithium acetylides with a secondary chlorophosphine.

2.2. Synthesis of Tertiary Alkynylphosphines

2.2.1. Version with Achiral Reagents

The oldest methods for the preparation of tertiary alkynylphosphines are based on the nucleophilic substitution reaction at the phosphorus atom of halophosphines by lithium [10] or magnesium [11] acetylides. For example, Cadiot and Chodkiewicz described the synthesis of a set of alkynylphosphines in good to excellent yields (50%–93%) through the reaction of Grignard reagents and dialkyl- or diarylchlorophosphines in THF at −30 °C to 40 °C (Scheme 4).

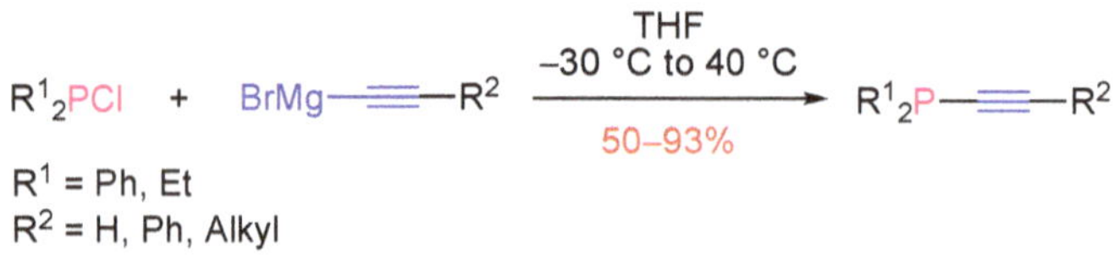

Scheme 4. Reaction of magnesium acetylides with chlorophosphines towards alkynylphosphines.

These procedures involving magnesium or lithium acetylides are still commonly used nowadays. For example Montchamp [12] described the synthesis of several alkynylphosphines using *sec*-butyllithium to generate *in situ* the lithium acetylide reagent. Subsequent treatment with BH_3•SMe_2 afforded the corresponding borane adducts in good yields (Scheme 5).

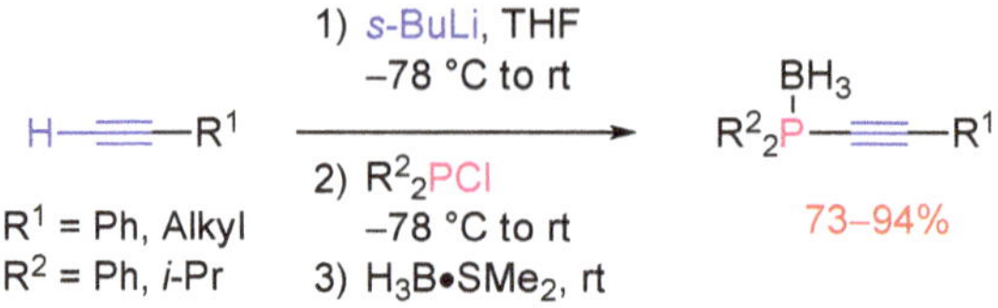

Scheme 5. Synthesis of alkynylphosphine-borane complexes from lithium acetylides.

Besides magnesium and lithium reagents, copper- [13], tin- [14], or silver- [15] acetylides have been involved in a few examples. In view of their high reactivity, magnesium and lithium acetylides are suitable for the formation of alkynylphosphines which do not bear functional groups sensitive to nucleophilic attack such as esters. To prevent the occurrence of side reactions, the use of more selective organometallic reagents like zinc [16] and titanium [17] acetylides was explored. Knochel demonstrated that zinc organometallics were effective for the synthesis of polyfunctionalized phosphines and described one example in the alkynyl series, but on a simple substrate. The alkynylzinc bromide reagent resulting from the transmetalation of 1-hexynyllithium with zinc bromide reacted with chlorodiphenylphosphine to provide the corresponding alkynylphosphine-borane in 80% yield after complexation with borane (Scheme 6).

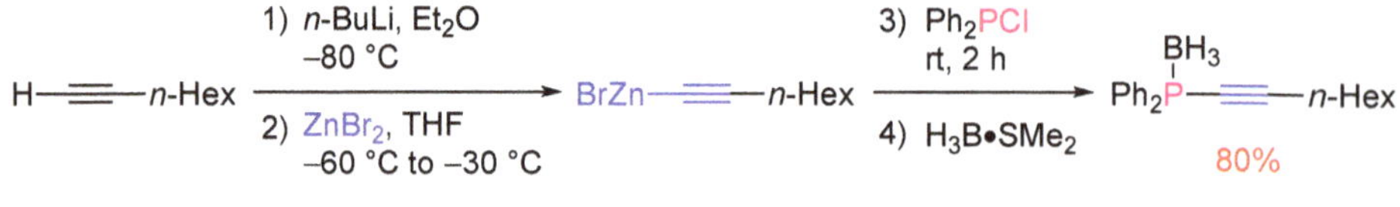

Scheme 6. Synthesis of alkynylphosphines using a zinc acetylide.

Periasamy [17] reported the synthesis of alkynylphosphines using alkynyl titanium reagents. The latter were not obtained by transmetalation of alkynyllithium reagents [18], but were generated *in situ* by direct metalation of 1-alkynes with the combination of titanium tetrachloride and triethylamine in dichloromethane (Scheme 7, Equation 1). Alkyne dimerisation may occur as a side reaction. For example, the reaction of phenyl acetylene with diphenylchlorophosphine afforded the corresponding alkynylphosphine in 54% yield, together with 2% of the diyne. Not only diphenylchlorophosphine but also triethylphosphite could serve as an electrophile. The substitution of all ethoxy groups was observed leading to tris(phenylethynyl)phosphine in 59% yield (Scheme 7, Equation 2).

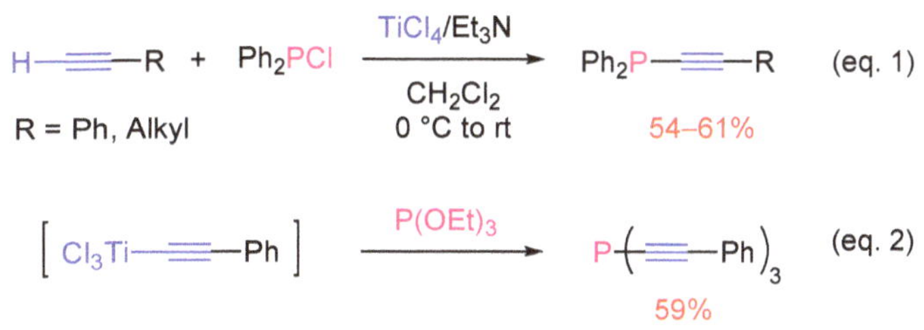

Scheme 7. Synthesis of alkynylphosphines using titanium acetylides.

2.2.2. Asymmetric Version

The single example relating the synthesis of a set of enantiopure P-stereogenic alkynylphosphines was described by Imamoto [5] (Scheme 8). The procedure relies on the *in situ* generation of a P-stereogenic bromophosphine-borane from an enantiopure secondary phosphine-borane via deprotonation by *n*-butyllithium in diethyl ether at −78 °C and subsequent halogenation of the resulting lithium phosphido-borane by treatment with dibromoethane. The bromophosphine-borane was not isolated because of its tendency to racemize at room temperature, but was treated *in situ* with a set of alkynyl lithium reagents. The nucleophilic substitution reaction was shown to occur with clean inversion of configuration at the phosphorus atom, delivering the alkynylphosphines in high yields and excellent stereospecificity.

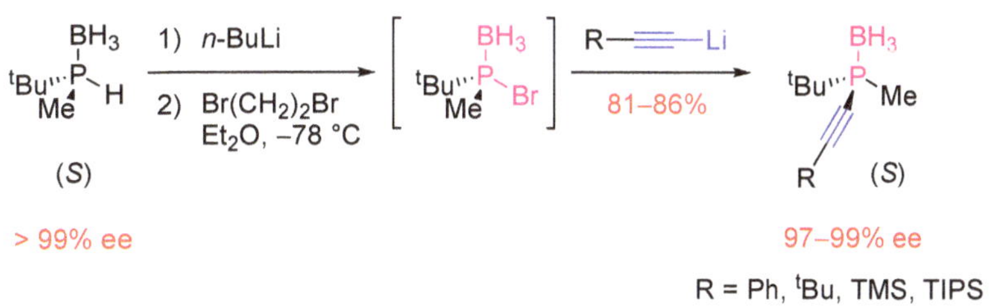

Scheme 8. Synthesis of enantioenriched P-stereogenic alkynylphosphine-boranes.

The procedure is up to now only applicable to enantiopure *tert*-butylmethylbromophosphine-borane, which displays a sufficient configurational stability at low temperature. This successful example is probably difficult to extend to a wide variety of phosphines due to the weak configurational stability of halophosphine-boranes.

3. Catalytic C-P bond Forming Reactions

Within the last ten years a few catalytic approaches towards alkynylphosphines have emerged. They rely on the use of either (i) electrophilic phosphorus derivatives, using nickel, palladium or copper complexes as catalytic system, or (ii) nucleophilic phosphorus derivatives using a copper(I) complex.

3.1. Electrophilic Phosphorus Reagent

As previously mentioned, alkynylphosphines are classically available through the reaction of chlorophosphines with terminal alkynes in the presence of stoichiometric amounts of organometallic bases. The Beletskaya group reported in 2003 [19] a catalytic variant, which was based on the use of nickel or palladium complexes in the presence of an excess of triethylamine. Nickel complexes were shown to display a higher activity than the palladium ones, the best catalysts being Ni(cod)$_2$ and Ni(acac)$_2$. The reaction with aryl or alkyl acetylenes readily proceeded in toluene at 80 °C for

10 to 30 min or at room temperature for 6 to 10 h (Scheme 9). It is noteworthy that the selective monosubstitution of one chloride atom from di- or tri-chlorosubstrates is not possible.

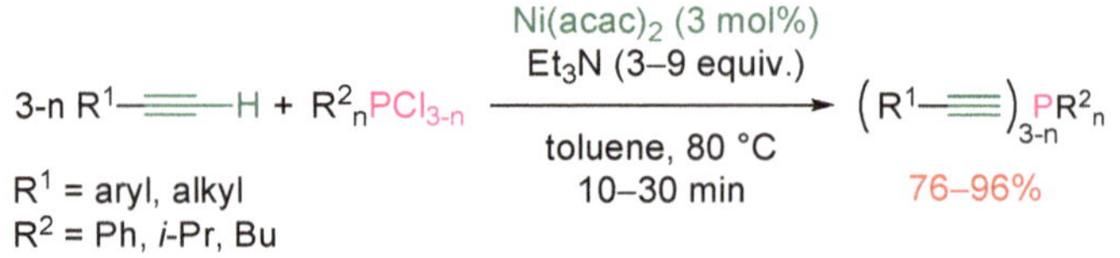

Scheme 9. Nickel-catalyzed cross-coupling of terminal alkynes with chlorophosphines.

The simplified postulated mechanism first involves the insertion of the metal into the P–Cl bond of the chlorophosphine. The resulting metal phosphido complex undergoes transmetalation with the *in situ* formed acetylide ion, followed by reductive elimination (Scheme 10).

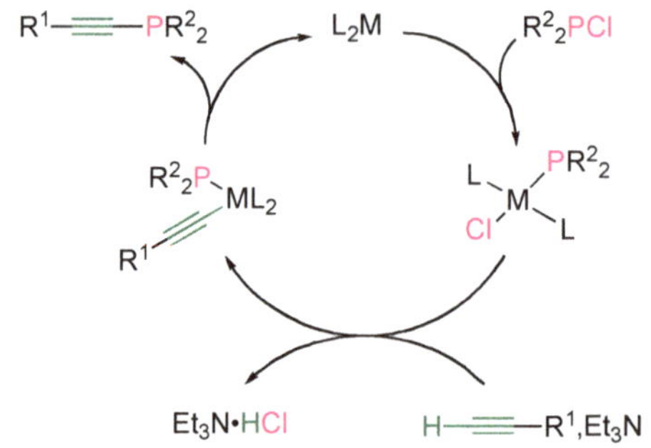

Scheme 10. Simplified postulated mechanism of nickel- or palladium-catalyzed cross-coupling between terminal alkynes and chlorophosphines.

The procedure was applied to the synthesis of a set of functionalized alkynylphosphines bearing carbonyl groups, which would have been difficult to access via the stoichiometric method (Figure 1). However in the case of aliphatic alkynes, a full conversion could not be reached and poor yields (around 30%) were obtained [20].

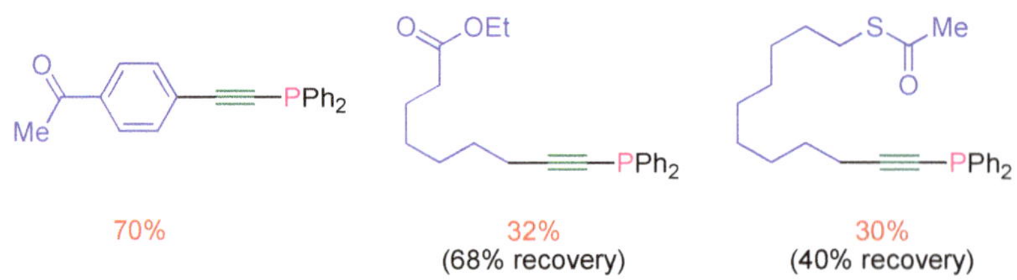

Figure 1. Functionalized alkynylphosphines synthesized via the nickel-catalyzed procedure.

Another limitation of this nickel-catalyzed procedure was its lack of efficiency with substrates bearing an alkoxy or amino group. To overcome these limitations a more versatile catalytic method was developed by the same authors. The method is based on the use of copper salts [21] and proceeds under mild conditions. Copper(I) iodide was used as the copper source in the presence of an excess of triethylamine in toluene at room temperature (Scheme 11). The reaction scope is broad and encompasses alkyl and (het)aryl acetylenes. Regarding the phosphorus coupling partner, aryl and alkyl chlorophosphines are suitable substrates.

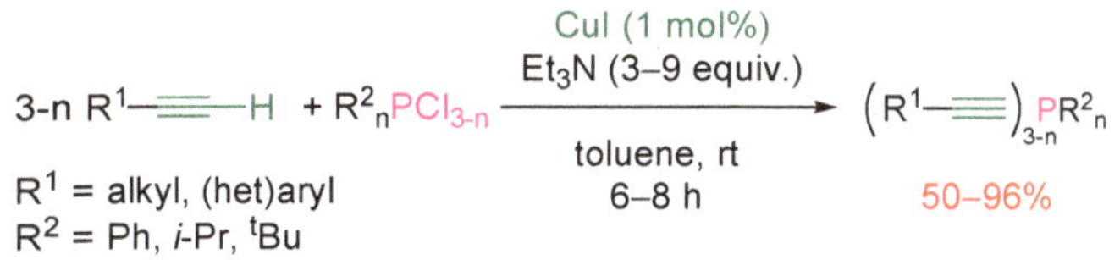

Scheme 11. Copper-catalyzed cross-coupling of terminal alkynes with chlorophosphines.

The postulated mechanism does not rely on P–Cl bond activation by the metal, as proposed in the case of Ni or Pd catalysis, but copper π-complexation of the terminal alkyne, followed by nucleophilic substitution at the phosphorus atom of the chlorophosphine in the presence of triethylamine, as outlined on Scheme 12. This hypothesis was supported by the isolation of the tetrahedral complex $(Ph_2PCl)_3CuI$, which is the first active species involved in the suggested catalytic cycle.

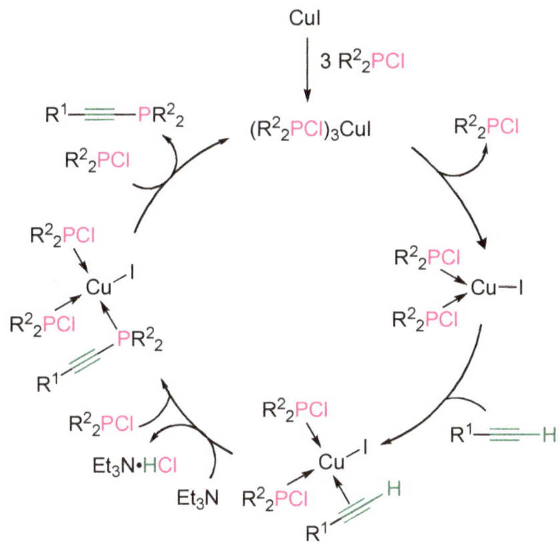

Scheme 12. Postulated mechanism of copper-catalyzed cross-coupling between terminal alkynes and chlorophosphines.

This general and mild procedure was successfully applied to the synthesis of bulky triethynylphosphine ligands [22,23] (Scheme 13).

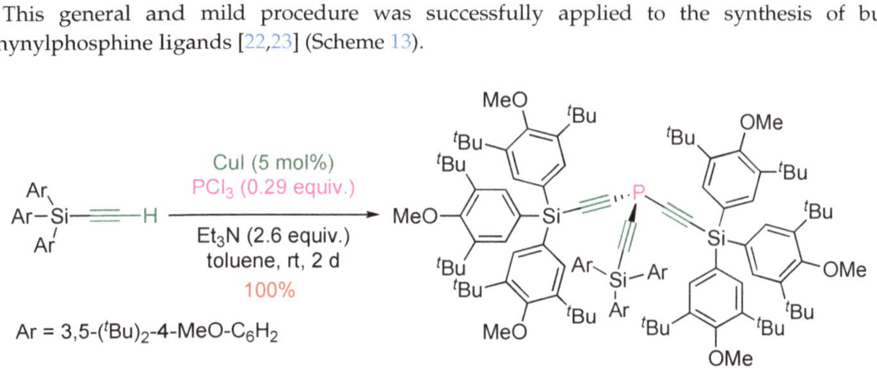

Scheme 13. Copper(I)-catalyzed synthesis of functionalized triethynylphosphines.

3.2. Nucleophilic Phosphorus Reagent

The Gaumont group reported the only example of a catalytic synthesis of alkynylphosphines, relying on nucleophilic phosphorus reagents [24]. The alkynylphosphines were obtained as their borane complexes. The methodology is based on the copper-catalyzed cross-coupling of secondary phosphine-boranes with alkyl or aryl 1-bromoalkynes using CuI/1,10-phenanthroline as catalytic source (Scheme 14). The reaction proceeds in toluene under mild conditions (20 to 60 °C, weak base).

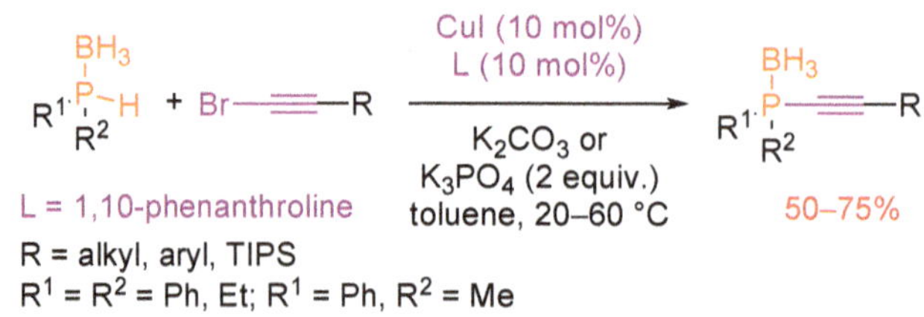

Scheme 14. Copper(I)-catalyzed reaction of secondary phosphine-boranes with 1-bromoalkynes towards alkynylphosphine-boranes.

Preliminary mechanistic studies are consistent with the involvement of a copper(I) phosphido-borane complex as the reactive species. Indeed neutral copper phosphido-boranes [$R_2PHCuPPh_2BH_3$phen], which could be isolated for the first time and fully characterized via experimental and computational methods [25], were shown to serve as pre-catalysts in the P-alkynylation of secondary phosphine-boranes with 1-bromoalkynes (Scheme 15).

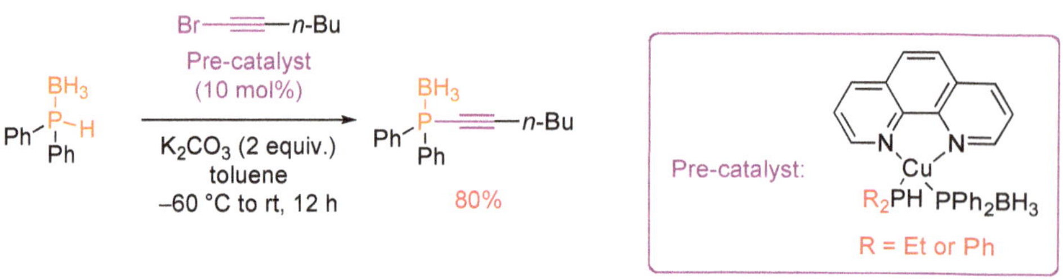

Scheme 15. P-alkynylation of secondary phosphine-boranes catalyzed by well-defined copper(I) phosphido-borane complexes.

The putative mechanism of the reaction involves oxidative addition of the copper(I) phosphido-borane complex to the 1-bromoalkyne, followed by reductive elimination, leading to the cross-coupling product and allowing the release of the copper(I) catalyst (Scheme 16).

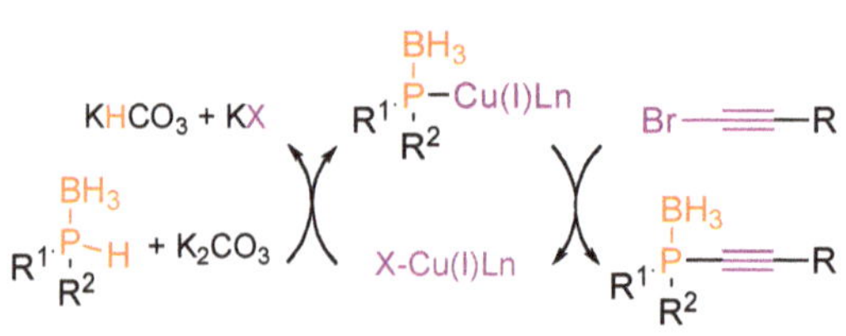

Scheme 16. Putative mechanistic cycle for the copper(I)-catalyzed P-alkynylation of secondary phosphine-boranes.

4. Transformation of Alkynylphosphorus Derivatives

Regarding the synthesis of tertiary alkynylphosphines, modification of the alkynyl or the phosphorus function of alkynyl phosphorus derivatives is a less commonly used method in comparison to C–P bond formation. On the other hand, in the case of secondary and primary alkynylphosphines, the reduction of the phosphoryl moiety of phosphine oxides and phosphonates respectively, is the method of choice for their preparation.

4.1. Functionalization of the Triple bond Moiety

Classical alkyne chemistry has been applied in the functionalization of alkynylphosphines through their metalation and subsequent reaction with an electrophile. With the aim of preparing dendrimers, Caminade and Majoral [26] generated the Grignard reagent of ethynylphosphine as building block for the synthesis of mono-, di-, tri-, and tetra-phosphines (Scheme 17). It is noteworthy that the lithium acetylide of ethynylphosphine was not suitable for these transformations, complex mixtures being formed in this case.

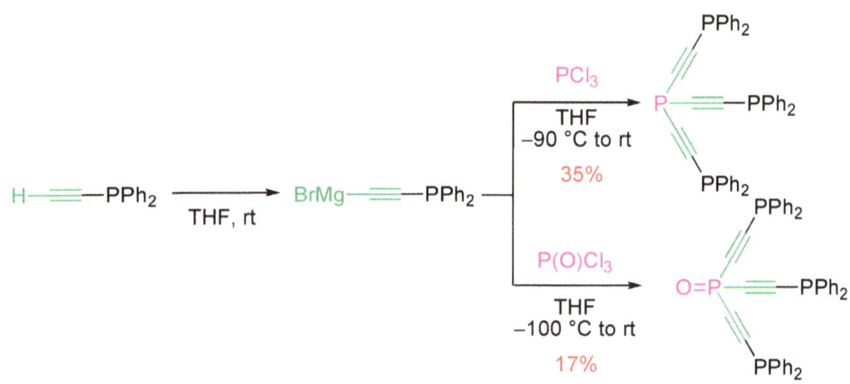

Scheme 17. Metalation and functionalization of ethynylphosphine.

The Imamoto group [5] described a 4-step reaction sequence allowing the synthesis of an alkynyl variant of BisP* ligands, analogous to DIPAMP [27], from an enantiopure silyl alkynylphosphine-borane (Scheme 18). The first step was a dimerization reaction through copper(II)-catalyzed oxidative coupling leading to the bis-alkynylphosphine-borane **2**. The next step was the silyl deprotection by treatment with TBAF. It was followed by deprotonation with *sec*-BuLi and methylation of the resulting lithium derivative with methyl iodide. Finally borane removal by treatment with DABCO (1,4-diazabicyclo[2.2.2]octane) afforded the bis-alkynylphosphine.

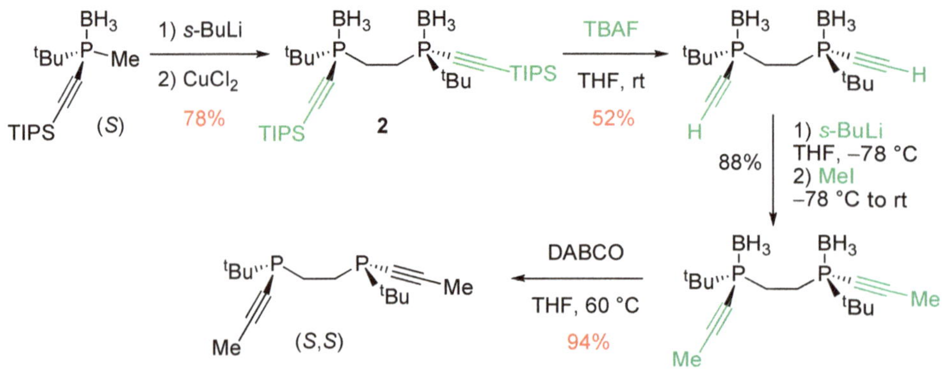

Scheme 18. Synthesis of an alkynyl analogue of DIPAMP.

4.2. Nucleophilic Substitution at Phosphorus Atom

The nucleophilic substitution at the phosphorus atom of alkynylchlorophosphines by Grignard reagents has been used for the building of complex structures. As an example, the Yoshifuji group [28] described the synthesis of the dialkynyldiphosphine **4** from alkynyldichlorophosphine **3** (Scheme 19).

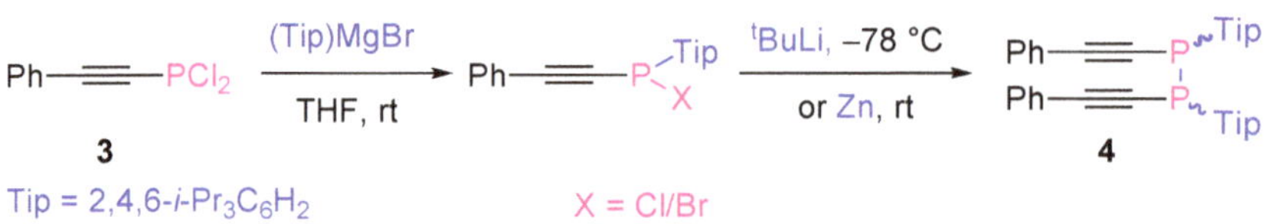

Scheme 19. Synthesis of alkynylphosphines from alkynyl-chlorophosphines.

Besides chlorophosphines, phosphonites can serve as precursors of alkynylphosphines. The Aguiar group [29] reported the synthesis of a set of alkynylphosphines through the reaction of the *in situ* generated diethyl alkynyl-1-phosphonite **5** and Grignard reagents (Scheme 20).

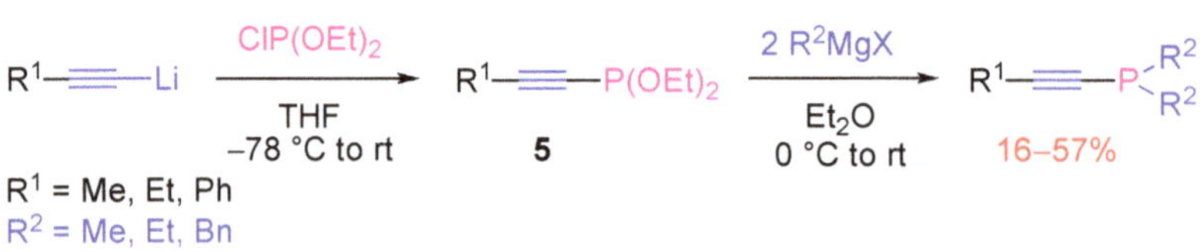

Scheme 20. Synthesis of alkynylphosphines from *in situ* generated alkynylphosphonites.

4.3. Reduction of the Phosphoryl Moiety

As described in the Section 2.1, a few secondary alkynylphosphines were prepared via the reaction of chlorophosphines and metal acetylides. However it concerned specific substrates stabilized by the presence of bulky substituents on the phosphorus atom.

A more straightforward and general access to unstabilized primary and secondary alkynylphosphines relies on the reduction of phosphoryl derivatives by alanes or silanes [30]. Guillemin and Denis [31] developed the chemoselective reduction of alkynylphosphonates using $AlHCl_2$ as reducing agent. The best yields were obtained when the reaction was performed at low temperature in diethyl ether or THF (Scheme 21, Equation 1). The same alane reducing agent was less suitable for the preparation of secondary alkynylphosphines from alkynylphosphinates. Indeed C–P bond cleavage occurred as a side-reaction [32], thus lowering the yields (Scheme 21, Equation 2).

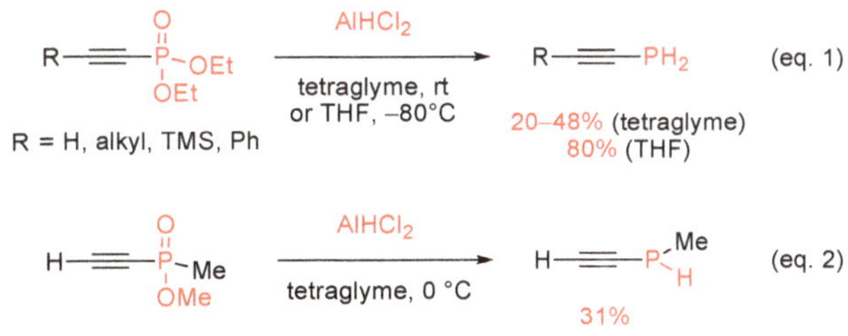

Scheme 21. Synthesis of primary and secondary alkynylphosphines via phosphoryl reduction by alanes.

A mild and practical synthesis of secondary alkynylphosphines was reported by Gaumont and Denis [33]. It is based on the reduction of secondary alkynylphosphine oxides in the presence of silane derivatives. The starting phosphine oxides were available through acidic cleavage of aminophosphine precursors. The appropriate silane reducing agent depended on the substituents at the phosphorus atom. Phenylsilane was used as reducing agent in the case of P-alkyl derivatives (Scheme 22, Equation 1) while the reduction of P-phenyl derivatives was only partial under these conditions. It was finally achieved with the combination of phenylsilane and trichlorophenylsilane in a 3:1 ratio (Scheme 22, Equation 2). The active reducing agent is believed to be the *in situ* generated $PhSiHCl_2$. The presence of chlorine atoms on silicon, by increasing the oxophilic properties of the silane reagent, would explain its higher reactivity.

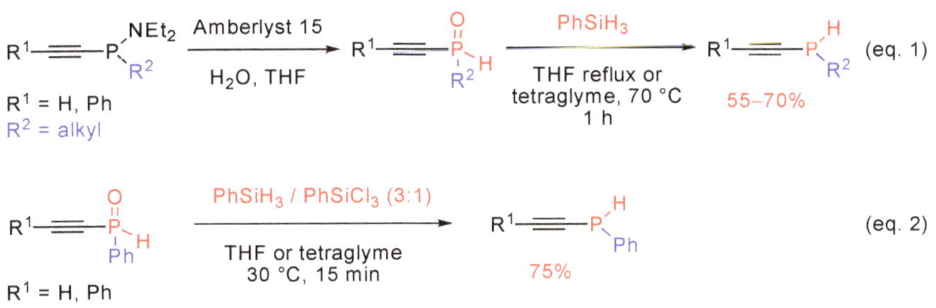

Scheme 22. Synthesis of secondary alkynylphosphines via phosphoryl reduction by silanes.

5. Conclusions

The synthetic pathways towards alkynylphosphines used to be restricted to stoichiometric procedures involving the use of organometallic bases, which limited the scope of the substitution pattern to robust functional groups. Since less than a decade, benefiting from the renaissance of copper catalysis for C–C and C–Heteroatom bond forming reactions, catalytic procedures have emerged, which involve either an electrophilic or a nucleophilic phosphorus derivative. By proceeding under mild conditions, these new catalytic methods have broadened the substituent scope and have been successfully applied to the synthesis of functionalized complex phosphine structures.

Future developments comprise catalytic procedures avoiding the use of halogen-containing derivatives, such as oxidative cross-couplings, which have proven their efficiency for the synthesis of alkynylphosphonates [34], but whose application to the synthesis of alkynylphosphines is challenging because of the high sensitivity of phosphines towards oxidation. Another challenge is the extension to the asymmetric series through the use of chiral ligands associated with the metal catalyst.

Acknowledgments: We gratefully acknowledge financial support from CNRS and Crunch (interregional organic chemistry network), the "Ministère de la Recherche et des Nouvelles Technologies", the "Région Basse-Normandie" and ERDF funding (ISCE-Chem & INTERREG IVa program). COST ACTION CM0802 "PHOSCINET" is also gratefully thanked for support.

References

1. Börner, A. *Phosphorus Ligands in Asymmetric Catalysis*; Wiley-VCH: Weinheim, Germany, 2008.
2. Diederich, F.; Stang, P.; Tykwinski, R.R. *Acetylene Chemistry*; Wiley-VCH: Weinheim, Germany, 2004.
3. Weymiens, W.; Slootweg, J.C.; Lammertsma, K. Phosphine Acetylenic Macrocycles and Cages: Synthesis and Reactivity. In *Phosphorus Compounds Advanced Tools in Catalysis and Material Science*, 1st ed.; Peruzzini, M., Gonsalvi, L., Eds.; Springer: New York, NY, USA, 2011; Volume 37, pp. 21–55.
4. Ito, H.; Ohmiya, H.; Sawamura, M. Construction of methylenecycloheptane frameworks through *7-exo-dig* cyclisation of acetylenic silyl enol ethers catalyzed by triethynylphosphine-gold complex. *Org. Lett.* **2010**, *12*, 4380–4383. [CrossRef] [PubMed]
5. Imamoto, T.; Saitoh, Y.; Koide, A.; Ogura, T.; Yoshida, K. Synthesis and enantioselectivity of P-chiral phosphine ligands with alkynyl groups. *Angew. Chem. Int. Ed.* **2007**, *46*, 8636–8639. [CrossRef] [PubMed]
6. Kondoh, A.; Yorimitsu, H.; Oshima, K. 1-Alkynylphosphines and their derivatives as key starting materials in creating new phosphines. *Chem. Asian J.* **2010**, *5*, 398–409. [CrossRef] [PubMed]
7. D'yachkova, S.G.; Nikitin, M.V.; Beskrylaya, E.A.; Arbuzova, S.N.; Kashik, T.V.; Gusarova, N.K.; Trofimov, B.A. Reaction of alkylthio(chloro)acetylenes with bis(2-phenylethyl)phosphine. *Russ. J. Gen. Chem.* **1999**, *69*, 767–770.
8. Pietrusiewicz, K.M.; Stankevic, M. 1-Phosphorus-functionalized alk-1-ynes. In *Science of Synthesis: Houben-Weyl Methods of Molecular Transformations*; de Meijere, A., Ed.; Thieme: Stuttgart, Germany, 2005; Volume 24, pp. 1073–1085.
9. Märkl, G.; Reitinger, S. 3H-Phosphaallen—Alkinyl-1H-Phosphan—Tautomere. *Tetrahedron Lett.* **1988**, *29*, 463–466. [CrossRef]
10. Voskuil, W.; Arens, J.F. Chemistry of acetylenic ethers. LXII. Tertiary phosphines with an acetylene-phosphorus bond. *Recl. Trav. Chim. Pays-Bas* **1962**, *81*, 993–1008. [CrossRef]
11. Charrier, C.; Chodkiewicz, W.; Cadiot, P. Préparations, propriétés chimiques et spectrographiques de phosphines α-acétyléniques. *Bull. Soc. Chim. Fr.* **1966**, 1002–1011.
12. Ortial, S.; Montchamp, J.-L. Synthesis of Z-alkenyl phosphorus compounds through hydroalumination and carbocupration of alkynyl precursors. *Org. Lett.* **2011**, *13*, 3134–3137. [CrossRef]
13. Sladkov, A.M.; Ukhin, L.Y.; Korshak, V.V. Reaction of copper acetylides with halogen compounds. *Izv. Akad. Nauk SSR Ser. Khim.* **1963**, 2213–2215.
14. Hartmann, H. Exchange reactions of some ethynyl-, diethynyl-, and butadiynyltin compounds. *Justus Liebigs Ann. Chem.* **1968**, *714*, 1–7. [CrossRef]
15. Gol'ding, I.R.; Sladkov, A.M. Silver acetylenides in ethynylation reactions. *Izv. Akad. Nauk SSR Ser. Khim.* **1972**, 529–530.
16. Langer, F.; Püntener, K.; Stürmer, R.; Knochel, P. Preparation of polyfunctional phosphines using zinc organometallics. *Tetrahedron: Asymmetry* **1997**, *8*, 715–738. [CrossRef]
17. Bharathi, P.; Periasamy, M. Direct metalation of 1-alkynes using $TiCl_4/Et_3N$ and the reactions of the organotitanium intermediates with electrophiles. *Organometallics* **2000**, *19*, 5511–5513. [CrossRef]
18. Krause, N.; Seebach, D. Chemistry of acetylenic titanium compounds. *Chem. Ber.* **1987**, *120*, 1845–1851. [CrossRef]
19. Beletskaya, I.P.; Afanasiev, V.V.; Kazankova, M.A.; Efimova, I.V. New approach to phosphinoalkynes based on Pd- and Ni-catalyzed cross-coupling of terminal alkynes with chlorophosphanes. *Org. Lett.* **2003**, *5*, 4309–4311. [CrossRef]
20. Kondoh, A.; Yorimitsu, H.; Oshima, K. Copper-catalyzed anti-hydrophosphination reaction of 1-alkynylphosphines with diphenylphosphine providing (Z)-1,2-diphosphino-1-alkenes. *J. Am. Chem. Soc.* **2007**, *129*, 4099–4104. [CrossRef]
21. Afanasiev, V.V.; Beletskaya, I.P.; Kazankova, M.A.; Efimova, I.V.; Antipin, M.U. A convenient and direct route to phosphinoalkynes via copper-catalyzed cross-coupling of terminal alkynes with chlorophosphanes. *Synthesis* **2003**, 2835–2838. [CrossRef]
22. Ochida, A.; Ito, H.; Sawamura, M. Using triethynylphosphine ligands bearing bulky end caps to create a holey catalytic environment: Application to gold(I)-catalyzed alkyne cyclizations. *J. Am. Chem. Soc.* **2006**, *128*, 16486–16487. [CrossRef]

23. Ochida, A.; Sawamura, M. Phosphorus ligands with a large cavity: synthesis of triethynylphosphines with bulky end caps and application to the rhodium-catalyzed hydrosilylation of ketones. *Chem. Asian J.* **2007**, *2*, 609–618. [CrossRef]
24. Bernoud, E.; Alayrac, C.; Delacroix, O.; Gaumont, A.-C. Copper-catalyzed synthesis of alkynylphosphine derivatives: unprecedented use of nucleophilic phosphorus compounds. *Chem. Commun.* **2011**, *47*, 3239–3241. [CrossRef]
25. Abdellah, I.; Bernoud, E.; Lohier, J.-F.; Alayrac, C.; Toupet, L.; Lepetit, C.; Gaumont, A.-C. Neutral copper-phosphido-borane complexes: Synthesis, characterization, and use as precatalysts in C_{sp}-P bond formation. *Chem. Commun.* **2012**, *48*, 4088–4090. [CrossRef] [PubMed]
26. Huc, V.; Balueva, A.; Sebastian, R.-M.; Caminade, A.-M.; Majoral, J.-P. Synthesis of functionalized mono-, di-, tri- and tetraphosphines: attempted application to prepare hyperbranched polymers and dendrimers built with phosphines at each branching point. *Synthesis* **2000**, 726–730. [CrossRef]
27. Knowles, W.S.; Sabacky, M.J.; Vineyard, B.D.; Weinkauff, D.J. Asymmetric hydrogenation with a complex of rhodium and a chiral bisphosphine. *J. Am. Chem. Soc.* **1975**, *97*, 2567–2568. [CrossRef]
28. Yoshifuji, M.; Ichikawa, Y.; Yamada, N.; Toyota, K. Preparation and X-ray analysis of novel carbonyltungsten(0) complexes of diphosphinidenecyclobutenes. *Chem. Commun.* **1998**, 27–28. [CrossRef]
29. Aguiar, A.M.; Irelan, J.R.S.; Morrow, C.J.; John, J.P.; Prejean, G.W. A convenient, synthetic pathway to dialkyl-1-alkynylphosphines. *J. Org. Chem.* **1969**, *34*, 2684–2686. [CrossRef]
30. Keglevich, G.; Gaumont, A.-C.; Denis, J.-M. Selective reductions in the sphere of organophosphorus compounds. *Heteroat. Chem.* **2001**, *12*, 161–167. [CrossRef]
31. Guillemin, J.-C.; Savignac, P.; Denis, J.-M. Primary alkynylphosphines and allenylphosphines. *Inorg. Chem.* **1991**, *30*, 2170–2173. [CrossRef]
32. Guillemin, J.-C.; Janati, T.; Denis, J.-M.; Guenot, P.; Savignac, P. Unstabilized 1-phosphaallenes: synthesis and characterization. *Tetrahedron Lett.* **1994**, *35*, 245–248. [CrossRef]
33. Pilard, J.-F.; Baba, G.; Gaumont, A.-C.; Denis, J.-M. Reduction of phosphine oxides, a mild and practical synthesis of secondary P-ethynylphosphines. *Synlett* **1995**, 1168–1170. [CrossRef]
34. Gao, Y.; Wang, G.; Chen, L.; Xu, P.; Zhao, Y.; Zhou, Y.; Han, L.-B. Copper-catalyzed aerobic oxidative coupling of terminal alkynes with *H*-phosphonates leading to alkynylphosphonates. *J. Am. Chem. Soc.* **2009**, *131*, 7956–7957. [CrossRef]

molecules

MDPI

Article

Hydrophosphonylation of Nanoparticle Schiff Bases as a Mean for Preparation of Aminophosphonate-Functionalized Nanoparticles

Justyna Siemieniec [1], Pawel Kafarski [1] and Pawel Plucinski [2,*]

1 Department of Bioorganic Chemistry, Faculty of Chemistry, Wrocław University of Technology, Wybrzeże Wyspiańskiego 27, 50-370 Wroclaw, Poland

2 Department of Chemical Engineering, University of Bath, Bath BA2 7AY, UK

* Correspondence: p.plucinski@bath.ac.uk; Tel.: +44-0-1225-386-961

Received: 9 June 2013; Accepted: 15 July 2013; Published: 18 July 2013

Abstract: The development of nanotechnology is responsible for an increase in the achievements in medical diagnostics and in the preparation of new therapeutic vehicles. In particular, magnetic nanoparticles with a modified surface are a very attractive alternative to deliver therapeutic agents. We describe the modification of the surface of the iron oxide nanoparticles with aminophosphonic acids by applying the classic hydrophosphonylation approach.

Keywords: magnetic nanoparticles; aminophosphonates; surface modification

1. Introduction

The dynamic development of nanotechnology (including nanoparticle technology) and molecular biology is the cause of a rapid increase in the achievements in medical diagnostics and in the preparation of new therapeutic substances [1–3]. Nanoparticles with a suitably chemically modified surface may be used for many applications, such as magnetic resonance imaging contrast enhancement, immunoassay, detoxification of biological fluids, drug delivery devices, and in cell separation [4,5]. Magnetic nanovehicles are especially very attractive for the delivery of therapeutic agents. They can be targeted to specific locations in the body through the application of an external magnetic field gradient, as has been documented by a number of promising animal and clinical trials [6–8].

In this paper we describe the usefulness of hydro-phosphonylation of the C=N bond for the modification of the surface of the iron oxide nanoparticles with aminophosphonic acids. Such aminophosphonate nanoparticles should act as interesting, surfactant-like species because depending on the pH of the solution, they may appear as negatively (because of the presence of phosphonate groups in their structure), positively (because of the presence of ammonium groups) or zwitterionic species, thus being able to exhibit non-typical properties, similar to those of recently obtained gold-supported negatively charged surfactant-like nanoparticles [9]. This property should facilitate their ability to bind organic compounds, proteins and metal ions.

The combination of the biological properties of aminophosphonates [10,11] and the potential of the magnetic nanoparticles for effective targeted delivery using an external magnetic force, should lead to magnetic bio-conjugates for selective targeting of physiologically active aminophosphonates to organs which require their activity. The nanoparticles covered with aminophosphonates may also be used as the supports for the immobilization of other drugs (especially anticancer ones) or for the immobilization of enzymes. Additionally, they can be considered as a system for the removal of metal ions from environmental and body fluids [12]. This could be achieved because aminophosphonic acids are bifunctional molecules and possess two reactive moieties—amino and phosphonate groups. Thus, when present at the surface of nanoparticles they could be used to bind other substances (including proteins) either covalently or by adsorption. This property should facilitate their ability to bind organic compounds, proteins and metal ions. In this paper we describe our preliminary studies on the modification of superparamagnetic iron oxide nanoparticles with aminophosphonic acids.

2. Results and Discussion

2.1. Preparation of Modified Nanoparticles

Iron oxide nanoparticles obtained by the standard co-precipitation method from iron(II) and iron(III) chlorides in the presence of ammonium hydroxide solution [13], were coated with tetraethyl orthosilicate and their surface was further functionalized with (3-aminopropyl)triethoxysilane (APTMS) [14].

Magnetic iron oxide nanoparticles coated with structurally variable aminophosphonates have been obtained by hydrophosphonylation of Schiff bases. We have applied its reacted version commonly used in aminophosphonate synthesis [15]. The direct application of this procedure when (3-aminopropyl)-triethoxysilane functionalized nanoparticles were used gave unsatisfactory results, because of strong aggregation of the resulting nanoparticles. This might be due to the high density of amino groups on the nanoparticles' surface and their simultaneous low flexibility. Therefore, we decided to apply the double-functionalization approach using aminopyridine or diaminopyrimidine as linkers. The synthetic procedure, schematically shown in Figure 1, provided two series of nanoparticles **1a–j** and **2d** and **2g**. Finally, one phosphonate moiety was hydrolyzed with sodium iodide in acetone followed by acidification with 1M hydrochloric acid, a reaction which provided nanoparticles functionalized with aminophosphonate esters. Finally, one phosphonate moiety was desterilized with sodium iodide in acetone followed by acidification with 1M hydrochloric acid, a reaction which provided nanoparticles functionalized with aminophosphonate esters. Hydrolysis to monoester is of strategic importance—in further studies, the resulting compound library will be used to further immobilization, including enzymes.

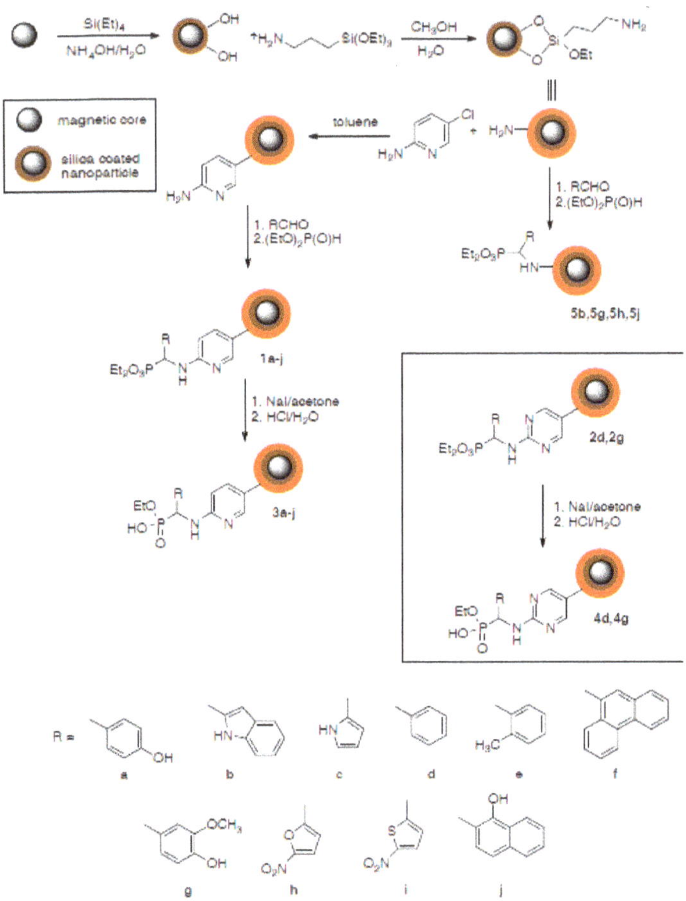

Figure 1. Synthetic scheme leading to aminophosphonate-functionalized magnetic nanoparticles.

2.2. Characterization of Nanoparticles

The formation of functionalized nanoparticles is accompanied by the sharpening of the IR spectra, which is in good accordance with the literature [16,17]. The infrared spectra of unmodified and aminophosphonate-modified iron oxide nanoparticles are shown in Figure 2. The IR spectrum of pure iron oxide nanoparticle shows a strong and wide band around 580 cm^{-1}, characteristic for the Fe–O stretching vibrations related to the magnetic core [Figure 2a]. A broad peak at around 1078 cm^{-1} corresponds to the Si-O and Si-C antisymmetric out of plane stretching bands, which are overlapping [Figure 2b]. Binding of aminophosphonate results in significant changes in the spectra [Figure 2c]. They are similar to those which are expected for aminophosphonates. The presence of P=O results in the stretching band at 1,230 cm^{-1}, whereas the peak at 1,030 cm^{-1} corresponds to both P-O-C and Ar-O-C vibrations (the latter one in the particular cases of particles **1g** and **1j**). The peak at 1,600 cm^{-1} most likely corresponds to the N-H bending scissoring vibrations, while C-N bond stretching is seen at 1,480 cm^{-1}.

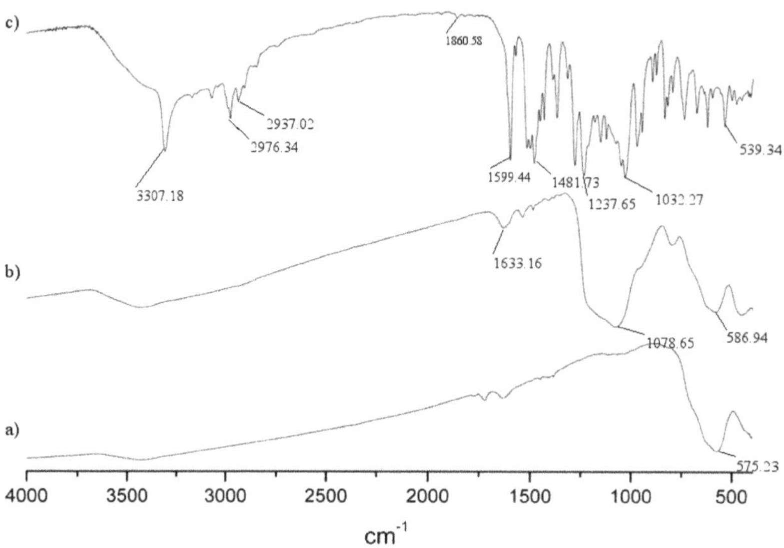

Figure 2. IR spectra of (**a**) iron oxide nanoparticle; (**b**) nanoparticle coated with silica and functionalized with APTMS; (**c**) aminophosphonate-functionalized nanoparticle **1g**.

In order to determine if there is a phosphorus atom present at the surface of the obtained nanoparticles we have applied energy-dispersive X-ray spectroscopy (SEM-EDX), which gave expected results (Figure 3). During sample scanning the obtained bands clearly showed the presence of the following atoms: Fe (6.5 and 0.8 KeV), Si (1.9 KeV), P (2.0KeV), O (0.5KeV) and C (0.1 KeV). Intensified signals at 0.1, 0.3, 2.5 are derived from elements contained in the matrix: C (0.1 KeV) and Cl (0.3 KeV, 2.5 KeV).

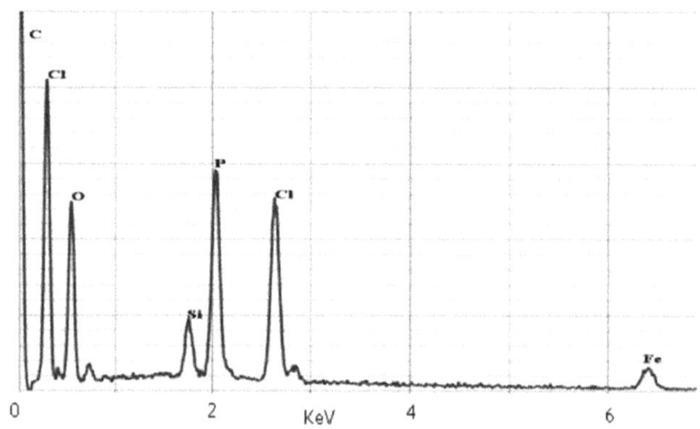

Figure 3. EDX spectrum of **1g** nanoparticles.

Due to the magnetic properties of iron oxide nanoparticles we could not apply ^{31}P-NMR, a technique typically used for the determination of the presence of carbon-to-phosphorus bonds. Therefore we have decided to obtain pure silica oxide nanoparticles and functionalize their surface with representative aminophosphonates in a manner identical to the magnetic ones. We expected that the infrared spectra of these nanoparticles were going to be similar to those obtained for functionalized magnetic nanoparticles showing only the lack of the band corresponding to the Fe-O vibrations.

On the other hand, the fact that this system is not magnetic allows us to perform ^{31}P-NMR measurements. The silica nanoparticles, lacking the iron oxide magnetic core, were synthesized from tetraethylortosilicate following the procedure described in the literature. In the next step we modified their surfaces with (3-aminopropyl)-trimethoxysilane and then with aminophosphonates following the procedure elaborated in this work. The IR spectra received for these particles, lacking the magnetic core, were identical to those obtained for magnetic ones. The ^{31}P-NMR spectra obtained for the nanoparticles clearly indicated the presence of a phosphonate moiety appearing at about 20 ppm, which corresponded to the expected products (Figure 4). This finding unequivocally confirms that the procedure described in this paper results in iron oxide nanoparticles surface covered with aminophosphonates. Between 5 and 10 ppm additional peaks corresponding to diethyl and monoethyl phosphites could be observed. Their appearance results from the fact that we had not purified the functionalized silica.

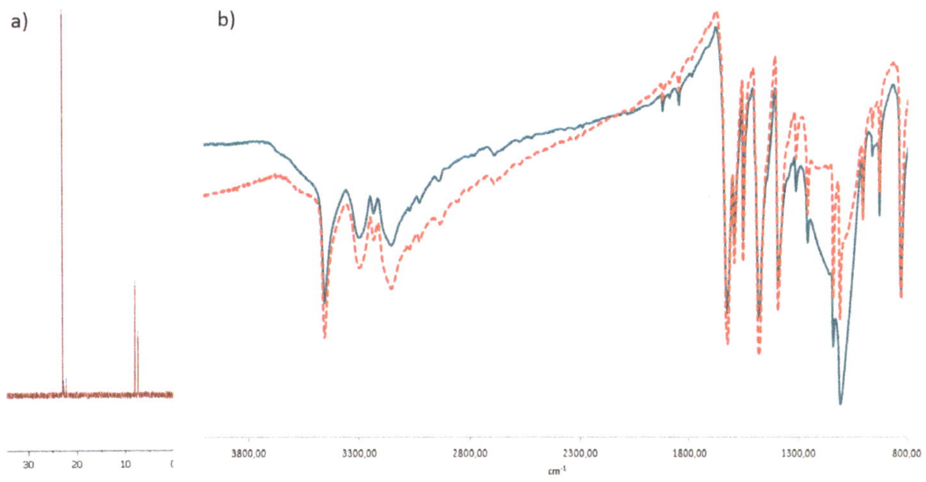

Figure 4. (**a**) ^{31}P-NMR spectrum of nanoparticle **1g** lacking magnetic core (**b**) Comparison of IR spectrum with magnetic core (blue line) and without (red line) of nanoparticles modified with 2,6-diamino-4-chloropyrimidine.

2.3. Nanoparticles Size and Dispersion

The size and distribution of oxide nanoparticles have been measured using dynamic light scattering. The size distribution of non-functionalized iron oxide nanoparticles, is between 15–50 nm, with a maximum value at 32 nm (Figure 5). The size distribution of functionalized particles clearly shows that they have a strong tendency to agglomerate. This is clearly seen from the transmission electron microscopy measurements (TEM), which support both the nearly uniformly sized uncoated iron oxide and the wide distribution of the size of the functionalized ones results from aggregation. The degree of aggregation is dependent on the structure of nanoparticle coating (presence or absence of linker), the structure of the aminophosphonate side chain and on the solvent used (measurements were done in water, methanol, toluene and acetone). The measurements indicated that the average diameter of the obtained nanoparticles appeared to be the smallest in methanol, with compounds **1a** (hydrodynamic size 144 nm), **1c** (72 nm), **1f** (70 nm), **1i** (65 nm) and **1j** (139nm) being representative examples. The respective acids **1a**, **1c**, **1d**, **4d** and **4g** also exhibit low size of dispersion of 50–200 nm.

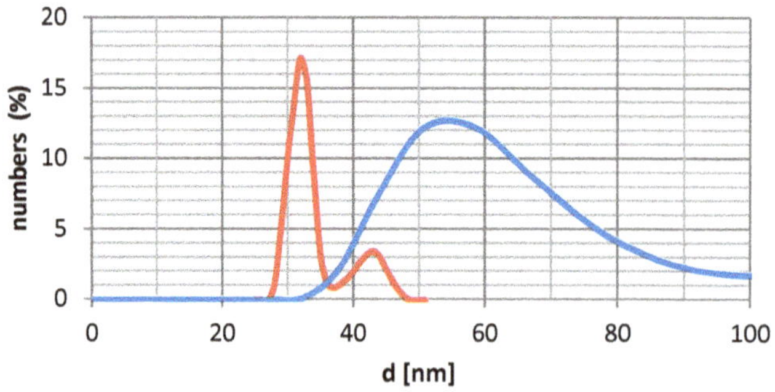

Figure 5. DLS measurements of iron oxide nanoparticles (red line) and with silica core (blue line).

The measurements in acetone also gave satisfactory results—some nanoparticles (**1f**, **1c** and **5b**) exhibit a low range of size dispersion, whereas others yielded quite large aggregates, while in toluene mostly highly aggregated particles were observed. It is also worth noting that nanoparticles after hydrolysis **3a–j** gave the best results in acetone. It is not possible to draw any meaningful size-dispersion—structure relationship for these species from these results. In Table 1 all DLS results in four solvents are summarized.

Table 1. DLS results in four solvents.

No	Hydrodynamic Size [nm] (Fraction %)			
	Water	Methanol	Acetone	Toluene
1a	168 (79%), 954 (21%)	144 (100%)	85 (100%)	Above 1 μm
1b	710 (100%)	184 (100%)	399 (100%)	507 (10%)
1c	485 (100%)	**72 (100%)**	39 (49%), **81 (51%)**	194 (36%), 808 (38%)
1d	**74 (47%)**, 259 (53%)	180 (100%)	**96 (100%)**	280 (23%), 870 (77%)
1e	180 (91%), 840 (8%)	170 (100%)	121 (100%)	697 (100%)
1f	820 (100%)	**70 (40%)**, 337 (60%)	**73 (40%)**, 280 (60%)	560 (40%)
1g	402 (25%), 930 (75%)	159 (100%)	294 (100%)	Above 1 μm
1i	296 (100%)	**65 (100%)**	114 (100%)	495 (100%)
1j	519 (100%)	139 (100%)	294 (100%)	Above 1 μm
2d	177 (28%)	202 (100%)	Powyżej 1 μm	Above 1 μm
2g	36 (15%), 116 (85%)	178 (100%)	**95 (100%)**	Above 1 μm
3a	105 (100%)	120 (78%)	**86 (100%)**	Above 1 μm
3c	**100 (39%)**, 319 (61%)	130 (100%)	**82 (100%)**	349 (100%)
3d	**93 (38%)**, 277 (62%)	109 (100%)	**77 (100%)**	Above 1 μm
3e	158 (15%), 406 (85%)	**82 (22%)**, 125 (78%)	**96 (100%)**	377 (58%), 840 (42%)
3j	630 (100%)	209 (100%)	**86 (100%)**	635 (100%)
4d	602 (100%)	196 (100%)	**80 (100%)**	249 (46%), 748(54%)
4g	128 (52%), 380 (48%)	150 (100%)	198 (100%)	175 (100%)
5b	490 (30%), 770 (70%)	**71 (100%)**	**86 (100%)**	215 (25%), 743 (75%)
5g	123 (100%)	130 (18%), 346 (82%)	**93 (100%)**	Above 1 μm
5h	105 (100%)	118 (100%)	160 (100%)	222 (100%)

The particles obtained without the application of linker are characterized by an enormous dispersion of size in water and therefore we decided to modify their structure by the application of linkers. The latter structures (a representative one is shown in Figure 6) contain multiple protonation sites, which means that they also tend to aggregate in a pH-dependent manner. They may be considered as polyionic surfactant-like species.

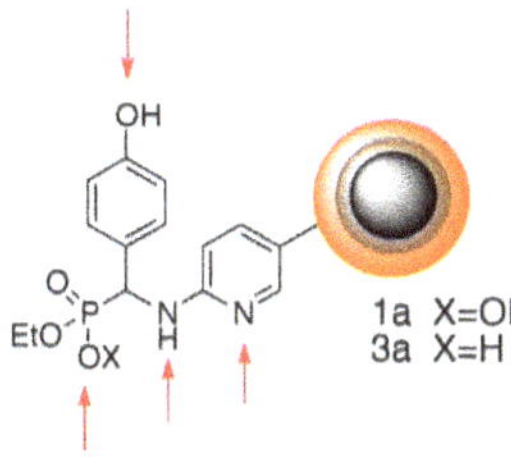

No	Hydrodynamic Size [nm] (Fraction %)			
	Water	Methanol	Acetone	Toluene
1a	168 (79%), 954 (21%)	144 (100%)	85 (100%)	Above 1 μm
1b	710 (100%)	184 (100%)	399 (100%)	507 (10%)
1c	485 (100%)	**72 (100%)**	39 (49%), **81 (51%)**	194 (36%), 808 (38%)
1d	**74 (47%)**, 259 (53%)	180 (100%)	**96 (100%)**	280 (23%), 870 (77%)
1e	180 (91%), 840 (8%)	170 (100%)	121 (100%)	697 (100%)
1f	820 (100%)	**70 (40%)**, 337 (60%)	**73 (40%)**, 280 (60%)	560 (40%)
1g	402 (25%), 930 (75%)	159 (100%)	294 (100%)	Above 1 μm
1i	296 (100%)	**65 (100%)**	114 (100%)	495 (100%)
1j	519 (100%)	139 (100%)	294 (100%)	Above 1 μm
2d	177 (28%)	202 (100%)	Powyżej 1 μm	Above 1 μm
2g	36 (15%), 116 (85%)	178 (100%)	**95 (100%)**	Above 1 μm
3a	105 (100%)	120 (78%)	**86 (100%)**	Above 1 μm
3c	**100 (39%)**, 319 (61%)	130 (100%)	**82 (100%)**	349 (100%)
3d	**93 (38%)**, 277 (62%)	109 (100%)	**77 (100%)**	Above 1 μm
3e	158 (15%), 406 (85%)	**82 (22%)**, 125 (78%)	**96 (100%)**	377 (58%), 840 (42%)
3j	630 (100%)	209 (100%)	**86 (100%)**	635 (100%)
4d	602 (100%)	196 (100%)	**80 (100%)**	249 (46%), 748(54%)
4g	128 (52%), 380 (48%)	150 (100%)	198 (100%)	175 (100%)
5b	490 (30%), 770 (70%)	**71 (100%)**	**86 (100%)**	215 (25%), 743 (75%)
5g	123 (100%)	130 (18%), 346 (82%)	**93 (100%)**	Above 1 μm
5h	105 (100%)	118 (100%)	160 (100%)	222 (100%)

Figure 6. Representative structure of nanoparticle—arrows indicate protonation sites.

Nanoparticles (both esters and acids) gave satisfactory size-dispersion when directly suspended in water, being usually in the range of 100 nm. The average sizes of the obtained particles were, however, usually higher that those obtained in methanol. Therefore we had measured their dependence on the pH of the aqueous solution by applying pH of 2, 6.8 and 12. At neutral pH, a strong tendency to aggregation was observed. A representative example obtained is shown in Figure 7. As expected, a significant reduction of the average size and size dispersion was obtained in strongly basic (pH 12) solutions, where all the residues are unprotonated. Also no aggregation was observed at this pH. More complex, and dependent on the chemical structure of aminophosphonate side chain, is the influence of acidic pH. In the cases when the aminophosphonate side chain is hydrophobic it usually has a positive influence on nanoparticle dispersion. Also in this case it is difficult to draw any reasonable structure-behavior relationship.

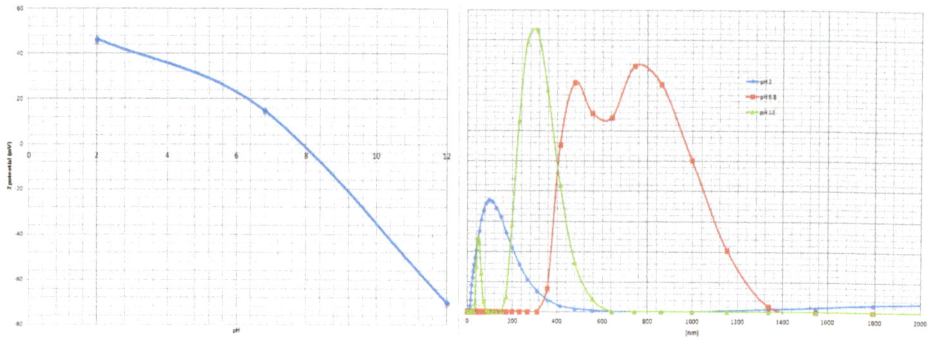

Figure 7. Size distribution of nanoparticles **2g** in different pH values.

The stability of nanoparticles dispersions correlates well with the ζ—potential measurements. It is generally accepted that the potential higher than 30 meV or lower than −30 meV characterizes stable systems. This is also the case in these studies.

3. Experimental

3.1. General

All solvents and reagents were purchased from commercial suppliers (Aldrich, Sigma, Merck, POCh), were of analytical grade and were used without further purification. Infrared spectra were measured on a 1600 FT-IR Perkin-Elmer spectrometer. The size of the nanoparticles was determined by dynamic light scattering using a Malvern Zetasizer Nano ZS in Laboratory of Biomedical Chemistry, Department of Experimental Oncology, Institute of Immunology and Experimental Therapy, PAS, Wroclaw. NMR experiments were performed on a Bruker DRX AVANCE TM 300MHz spectrometer. SEM-EDX and TEM analyses were performed in the Microscopy and Analysis Suite, University of Bath, on a JEOL JSM6480LV with an Oxford INCA X-ray analyzer and an OEL JEM 1200EXII, respectively.

3.2. The Synthesis of Iron Oxide Nanoparticles Coated with Silica

To a one-litre three neck flask of water (188 mL) and 23% aqueous solution of ammonia (10 mL) was added. The solution was stirred under a nitrogen atmosphere. At the same time a mixture of $FeCl_3{\cdot}6H_2O$ and $FeCl_2{\cdot}4H_2O$ (in a molar ratio of 1:2) was dissolved in water (10 mL) containing 36% HCl solution (0.125 mL). This solution of chloride salts was very slowly added dropwise to the solution of NH_4OH with vigorous stirring. A black precipitate of iron oxide nanoparticles was formed and isolated through magnetic decantation. Magnetic nanoparticles were washed several times with distilled water and finally with 0.1 M solution of tetramethylammoniun hydroxide (TMAOH, 100 mL). This mixture was sonicated for 1 h, and the nanoparticles were isolated through magnetic decantation and dried.

Iron oxide nanoparticles (0.1 g) were placed in distilled water (100 mL) and sonicated for 1 h. Then 2-propanol (100 mL) was added and the mixture sonicated for an additional 3 h. After that 23% aqueous solution of ammonia (1 mL) was added. The mixture was vigorously stirred and tetraethylortosilicate (0.2 mL) was added. After 2 h of stirring the nanoparticles were separated through magnetic decantation, washed with water and dried.

3.3. The Functionalization of Silica Coated Nanoparticles with (3-aminopropyl)trimethoxysilane

Freshly coated iron oxide nanoparticles (0.5 g), water (25 mL) and methanol (25 mL) were placed in a flask and sonicated for 1 h. After this time (3-aminepropyl)trimethoxylsilane (0.75 mL) was added in one portion and the mixture stirred at 80 °C for 5 h. The brown precipitate was isolated through magnetic decantation, washed with water and dried.

3.4. The General Procedure Followed for the Synthesis of Aminophosphonate-Functionalized Magnetic Nanoparticles

Nanoparticles functionalized surfacially with amino-groups (0.5 g) were mixed with 2-amino-5-chloropyridine (0.5 g) or 2,6-diamino-4-chloropyrimidine (0.5 g) in toluene and refluxed for 8h. After this time, the solvent was evaporated, the precipitate washed with fresh toluene and magnetically decanted. The synthesised nanoparticles were then used as amine reagent for the synthesis of aminophosphonates. Thus, the appropriate amount of nanoparticles (0.5 g) and aldehyde (1 g) were mixed in toluene and refluxed until the amount of collected water remained unchanged in the Dean-Stark trap. The solvent was evaporated and the crude nanoparticle Schiff base was used in the next step without purification. It was suspended in a new portion of toluene and the equivalent of diethyl phosphite (in 1:1 molar ratio to aldehyde) was added followed by refluxing for 4 h. The solvent was then evaporated, the residue washed with toluene, and water and isolated through magnetic decantation.

1a: IR (KBr, cm^{-1}): 570 (m, Fe-O), 1049 (s, P-O-C), 1093 (s, Si-O), 1176 (s, Si-C), 1218 (s, P=O), 1479 (s, N-H), 1596 (s, C_{Ar}-N_{Ar}), 2919, 2990 (m, C-H), 3326 (m, Ar-O-H); SEM-EDX (KeV): 0.1 (C), 0.5 (O), 1.9 (Si), 2.0 (P), 6.7 (Fe).

1b: IR (KBr, cm^{-1}): 569.8 (m, Fe-O), 1037 (s, P-O-C), 1060 (s, Si-O), 1172 (s, Si-C), 1284 (m, P=O), 1479 (s, N-H), 1596 (s, C_{Ar}-N_{Ar}), 2919 (m, C-H); SEM-EDX (KeV): 0.1 (C), 0.5 (O), 1.9 (Si), 2.0 (P), 6.7 (Fe).

1c: IR (KBr, cm^{-1}): 541 (m, Fe-O), 1020 (s, P-O-C), 1048 (s, Si-O), 1230 (s, P=O), 1478 (s, N-H), 1598 (s, C_{Ar}-N_{Ar}), 2899, (m, C-H); SEM-EDX (KeV): 0.1 (C), 0.5 (O), 1.9 (Si), 2.0 (P), 6.7 (Fe).

1d: IR (KBr, cm^{-1}): 567 (m, Fe-O), 1029 (s, P-O-C), 1054 (s, Si-O), 1230 (s, P=O), 1247 (s, Si-C), 1484 (s, N-H), 1598 (s, C_{Ar}-N_{Ar}), 2865 (m, C-H), 3285,14 (s, N-H); SEM-EDX (KeV): 0.1 (C), 0.5 (O), 1.9 (Si), 2.0 (P), 6.7 (Fe).

1e: IR: (KBr, cm^{-1}): 577 (m, Fe-O), 1022 (s, P-O-C), 1053 (s, Si-O), 1475 (s, N-H), 1598 (s, C_{Ar}-N_{Ar}), 2906 (m, C-H), 3294 (m, N-H); SEM-EDX (KeV): 0.1 (C), 0.5 (O), 1.9 (Si), 2.0 (P), 6.7 (Fe).

1f: IR: (KBr, cm^{-1}): 533 (m, Fe-O), 1036 (s, P-O-C), 1224 (s, P-O), 1288 (m, Si-C), 1481 (s, N-H), 1596 (s, C_{Ar}-N_{Ar}), 2901, 2979 (m, C-H), 3283 (s, N-H); SEM-EDX (KeV): 0.1 (C), 0.5 (O), 1.9 (Si), 2.0 (P), 6.7 (Fe).

1g: IR (KBr, cm^{-1}): 583 (m, Fe-O), 1032 (s, P-O-C), 1052 (m, Si-C), 1238 (s, P=O), 1369 (m, N-O), 1481 (m, N-H), 1533 (m, C_{Ar}-N_{Ar}), 2976 (m, C-H), 3307 (s, N-H); SEM-EDX (KeV): 0.1 (C), 0.5 (O), 1.9 (Si), 2.0 (P), 6.7 (Fe).

1h: IR (KBr, cm^{-1}): 583 (m, Fe-O), 1028 (s, P-O-C), 1028 (m, Si-C), 1238 (s, P=O), 1354 (m, N-O), 1499 (m, N-H), 1533 (m, C_{Ar}-N_{Ar}), 2986 (m, C-H), 3419 (s, N-H); SEM-EDX (KeV): 0.1 (C), 0.5 (O), 1.9 (Si), 2.0 (P), 6.7 (Fe).

1i: IR (KBr, cm^{-1}): 568 (m, Fe-O), 1045 (s, P-O-C), 1162 (m, Si-O), 1224 (m, P=O), 1336 (m, N-O), 1439 (m, N-H), 1508 (m, C_{Ar}-N_{Ar}), 2982 (m, C-H), 3413 (s, N-H); SEM-EDX (KeV): 0.1 (C), 0.5 (O), 1.9 (Si), 2.0 (P), 6.7 (Fe).

1j: IR (KBr, cm^{-1}): 571 (m, Fe-O), 1096 (s, P-O-C), 1148 (s, Si-O), 1213 (s, P-O), 1246 (s, Si-C), 1459 (m, N-H), 2832, 2938 (m, C-H), 3187 (m, Ar-O-H), 3240 (s, N-H); SEM-EDX (KeV): 0.1 (C), 0.5 (O), 1.9 (Si), 2.0 (P), 6.7 (Fe).

2d: IR (KBr, cm^{-1}): 560 (m, Fe-O), 1051 (s, P-O-C), 1162 (s, Si-O), 1495 (m, N-H), 1579 (s, C_{Ar}-N_{Ar}), 2931, 2982, (m, C-H), 3399 (s, N-H); SEM-EDX (KeV): 0.1 (C), 0.5 (O), 1.9 (Si), 2.0 (P), 6.7 (Fe).

2g: IR (KBr, cm^{-1}): 546 (m, Fe-O), 1049 (s, P-O-C), 1162 (m, Si-O, Si-C), 1296 (m, Ar-O-C), 1467 (s, N-H), 1579 (s, C_{Ar}-N_{Ar}), 2980 (m, C-H), 3100 (m, Ar-O-H), 3303 (s, N-H); SEM-EDX (KeV): 0.1 (C), 0.5 (O), 1.9 (Si), 2.0 (P), 6.7 (Fe).

3.5. The General Synthesis of Monoesters

The nanoparticle diester (0.5 g) and the sodium iodide (0.5 g) were placed in acetone and refluxed for 3 h. The sodium salt of the monoester was isolated through magnetic decantation, washed with acetone and converted into acid form by washing with 1 M HCl. The obtained nanoparticles were washed with water and dried.

3a: R (KBr, cm^{-1}): 577 (m, Fe-O), 1050 (s, P-O-C), 1068 (s, Si-O), 1189 (s, Si-C), 1223 (m, P=O), 1607 (s, N-H), 1656 (s, C_{Ar}-N_{Ar}), 2737 (m, C-H), 3257 (m, AO-H), 3417 (s, N-H).

3b: IR (KBr, cm^{-1}): 570 (m, Fe-O), 1037 (s, P-O-C), 1060 (s, Si-O), 1173 (s, Si-C), 1284 (m, P=O), 1479 (s, N-H), 1596 (s, C_{Ar}-N_{Ar}), 2919, 3228 (m, C-H).

3c: IR (KBr, cm^{-1}): 541 (m, Fe-O), 1012 (s, P-O-C), 1044 (s, Si-O), 1212 (s, P=O), 1473 (s, N-H), 2849, 2917 (s, C-H), 3429 (m, O-H).

3d: IR (KBr, cm^{-1}): 569 (m, Fe-O), 1095 (s, P-O-C), 1228 (s, P=O), 1482 (s, N-H), 1598 (s, C_{Ar}-N_{Ar}), 3469 (s, O-H).

3e: IR (KBr, cm^{-1}): 611 (m, Fe-O), 1027 (m, P-O-C), 1054 (m, Si-O), 1479 (m, N-H), 1598 (m, C_{Ar}-N_{Ar}), 3416 (s, O-H).

3f: IR: (KBr, cm^{-1}): 536 (m, Fe-O), 1026 (s, P-O-C), 1052 (s, Si-O), 1228 (m, P=O), 1478 (s, N-H), 1597 (s, C_{Ar}-N_{Ar}), 2984, (m, C-H), 3468 (s, O-H).

3g: IR (KBr, cm^{-1}): 546 (m, Fe-O), 1024 (m, P-O-C), 1431 (w, N-H), 1605 (m, C_{Ar}-N_{Ar}), 3398 (s, O-H).

3h: IR (KBr, cm^{-1}): 611 (m, Fe-O), 1044 (m, P-O-C), 1162 (M, Si-O, Si-C), 1239 (m, P=O), 1355 (m, N-O), 1499 (w, N-H), 3416 (s, N-H), 3468 (s, O-H).

3i: R (KBr, cm^{-1}): 570 (m, Fe-O), 1047 (m, P-O-C), 1162 (m, Si-O), 1227 (m, P=O), 1336 (m, N-O), 1440 (w, N-H), 1510 (m, C_{Ar}-N_{Ar}), 3415 (s, N-H), 3471 (s, O-H).

3j: IR (KBr, cm^{-1}): 571 (m, Fe-O), 1096 (s, P-O-C), 1148 (s, Si-O), 1213 (s, P=O), 1460 (m, N-H), 1625 (s, C_{Ar}-N_{Ar}), 2832, 2938 (m, C-H), 3187 (m, Ar-O-H), 3240 (m, N-H), 3412 (m, O-H).

4. Conclusions

Hydrophosphonylation surfacial Schiff bases obtained from iron oxide magnetic nanoparticles functionalized with amino groups appeared to be an effective method for preparation of aminophosphonate-functionalized nanoparticles. By being surfacially charged (both negative and positive) these particles behave as micelles.

Acknowledgments: This work is dedicated to Henri-Jean Cristau on the occasion of his 70th birthday. The work was financed by a statutory activity subsidy from the Polish Ministry of Science and Higher Education for the Faculty of Chemistry of Wroclaw University of Technology and co-financed by the European Union from the European Social Fund. Authors thank Tomasz Goszczyński for helpful discussion of the DLS measurements. JS visiting postgraduate scholar stay at the University of Bath was funded by an EPSRC project EP/G028141/1.

Conflicts of Interest: The authors declare no conflict of interest.

References

1. Allen, T.M.; Cullis, P.R. Drug delivery systems: Entering the mainstream. *Science* **2004**, *303*, 1818–1822. [CrossRef] [PubMed]
2. Boyer, C.; Whittaker, M.R.; Bulmus, V.; Liu, J.; Davis, T.P. The design and utility of polymer-stabilized iron-oxide nanoparticles for nanomedicine applications. *NPG Asia Mater.* **2010**, *2*, 23–30. [CrossRef]

3. Morteza, M.; Sant, S.; Wang, B.; Laurent, S.; Sen, T. Superparamagnetic iron oxide nanoparticles (SPIONs): Development, Surface modification and applications in chemotherapy. *Adv. Drug Deliv. Rev.* **2011**, *63*, 24–46.
4. Coti, K.K.; Belowich, M.E.; Liong, M.; Ambrogio, M.W.; Lau, Y.A.; Khatib, H.A.; Zink, J.I.; Khashab, N.M.; Stoddart, J.F. Mechanised nanoparticles for drug delivery. *Nanoscale* **2009**, *1*, 16–39. [CrossRef] [PubMed]
5. Mout, R.; Moyano, D.R.; Rana, S.; Rotello, V.M. Surface functionalization of nanoparticles for nanomedicine. *Chem. Rev.* **2012**, *41*, 2539–2544. [CrossRef] [PubMed]
6. Cho, K.; Wang, X.; Nie, S.; Chen, Z.; Shin, D.M. Therapeutic nanoparticles for drug delivery in cancer. *Clin. Cancer Res.* **2008**, *14*, 1310–1316. [CrossRef] [PubMed]
7. Davis, M.E.; Chen, Z.G.; Shin, D.M. Nanoparticle therapeutics: An emerging treatment modality for cancer. *Nat. Rev. Drug Discov.* **2012**, *7*, 771–782. [CrossRef] [PubMed]
8. Wang, A.Z.; Langer, L.; Farokhzad, O.C. Nanoparticle delivery of cancer drugs. *Ann. Rev. Med.* **2012**, *63*, 185–198. [CrossRef] [PubMed]
9. Wang, D.; Tejerina, B.; Lagzi, I.; Kowalczyk, B.; Grzybowski, B. Bridging interactions and selective nanoparticle aggregation mediated by monovalent cations. *ACS Nano* **2011**, *5*, 530–536. [CrossRef] [PubMed]
10. Kafarski, P.; Lejczak, B. Aminophosphonic acids of potential medical importance. *Curr. Med. Chem.–Anti-Cancer Agents* **2001**, *1*, 301–312. [CrossRef] [PubMed]
11. Mucha, A.; Kafarski, P.; Berlicki, Ł. Remarkable potential of the -aminophosphonate/phosphinate structural motif in medicinal chemistry. *J. Med. Chem.* **2011**, *54*, 5955–5980. [CrossRef] [PubMed]
12. Kiss, T.; Lazar, I.; Kafarski, P. Chelating tendencies of bioactive aminophosphonates. *Metal-Based Drugs* **1994**, *1*, 247–264. [CrossRef] [PubMed]
13. Massart, R. Preparation of aqueous magnetic liquids in alkaline and acidic media. *IEEE Trans. Magn.* **1981**, *MAG-17*, 1247–1248. [CrossRef]
14. Liu, X.; Ma, Z.; Xing, J.; Liu, H. Preparation and characterization of amino–silane modified superparamagnetic silica nanospheres. *J. Magn. Mag. Mat.* **2004**, *270*, 1–6. [CrossRef]
15. Tyka, R. Novel Synthesis of α-Aminophosphonic Acids. *Tetrahedron Lett.* **1970**, 677–680. [CrossRef]
16. Netto, C.G.C.M.; Andrade, L.H.; Toma, E. H Enantioselective transesterification catalysis by *Candida antarctica* lipase immobilized on superparamagnetic nanoparticles. *Tetrahedron: Asymmetry* **2009**, *20*, 2299–2304. [CrossRef]
17. Benyettou, F.; Lalatonne, Y.; Chebbi, I.; Di Benedetto, M.; Serfay, J.-M.; Lecouvey, M.; Motte, L. A multimodal magnetic resonance imaging nanoplatform for cancer theranostics. *PhysChemChemPhys* **2011**, *13*, 10020–10027. [CrossRef] [PubMed]

Sample Availability: Samples of the nanoparticles are available from the authors.

molecules

MDPI

Article

Cs^+ Removal from Aqueous Solutions through Adsorption onto Florisil® Impregnated with Trihexyl(tetradecyl)phosphonium Chloride

Lavinia Lupa, Adina Negrea, Mihaela Ciopec * and Petru Negrea

Faculty of Industrial Chemistry and Environmental Engineering Blvd., "Politehnica" University of Timisoara, Vasile Parvan no. 6, Timisoara 300223, Romania; lavinia.lupa@chim.upt.ro (L.L.); adina.negrea@chim.upt.ro (A.N.); petru.negrea@chim.upt.ro (P.N.)

* Correspondence: mihaela.ciopec@chim.upt.ro; Tel.: +40-256-404-192

Received: 1 August 2013; Accepted: 10 October 2013; Published: 16 October 2013

Abstract: This research determined the adsorption performance of Florisil® impregnated with trihexyl(tetradecyl)phosphonium chloride (Cyphos IL-101) in the process of Cs^+ removal from aqueous solutions. The obtained Florisil® impregnated with the studied ionic liquid was characterized through energy dispersive X-ray analysis and Fourier transform infrared spectroscopy in order to verify that the impregnation with the ionic liquid had occurred. The adsorption process has been investigated as a function of pH, solid:liquid ratio, adsorbate concentration, contact time and temperature. The isotherm data was well described by a Langmuir isotherm model. The maximum adsorption capacities of the Florisil® impregnated with the studied ionic liquid was found to be 3.086 mg Cs^+/g of adsorbent. The results indicated that the adsorption fitted well with the pseudo-second order kinetic model.

Keywords: Cs^+ adsorption; trihexyl(tetradecyl)phosphonium chloride; Florisil®

1. Introduction

In the last decades the attention of researchers was focused on the problem of environmental pollution with radioactive contaminants. Among all radioactive components Cs^+ is an important radionuclide from several reasons: relatively long-live, high solubility/mobility and strong γ emitting radiation. Furthermore, it can be easily incorporated into terrestrial and aquatic organisms because of its chemical similarity to potassium [1–3]. Therefore it should be removed from waste radioactive solutions. One of the most studied methods of radionuclide removal from waste waters is the liquid-liquid extraction using various solvents [4–8]. In recent years ionic liquids (ILs) have gained considerable attention and were investigated as substitutes for the organic solvents normally used in this application due to their advantages: high-thermal stability, non-inflamability and non-volatility [8–11]. Based on literature survey was observed that the phosphonium ionic liquids, compared to their imidazolium- or pyridinium-based counterparts, present some advantages: improved thermal and chemical stability, unique miscibility behaviour and solvating properties [10–12]. Even so they have not been studied so intensely, and were not used for Cs^+ removal. Trihexyl(tetradecyl)phosphonium bis-2,4,4-trimethylpentylphosphinate has been used in lactic acid separation [13] and metal ions separation [11,14–17]. Trihexyl(tetradecyl)phosphonium chloride (Cyphos IL-101) has been used as effective extractant for the removal of bismuth, zinc or platinum [18–21]. We have also focused on the phosphonium-based ionic liquid, because was observed that from the studied ion exchanges those based on phosphonium (like ammonium molybdophosphate) have been intensively investigated and found to be effective for the removal of caesium, but from acidic liquids [22,23]. Therefore in this paper a fairly cheap IL—trihexyl(tetradecyl)phosphonium chloride (Cyphos IL-101)—was used as the separation medium for Cs^+ removal from aqueous solutions. Due to its commercial availability and low price, Cyphos IL-101 is a favourable extractant for Cs^+ removal. In the liquid-liquid extraction some drawbacks may arise like the large amount of ionic liquid used, high viscosity leading unfavourably to dissolution and diffusion, difficulties of separation and recovery and low interface area [24–26]. In order to overcome these shortcomings and to enhance the separation capacities of the studied IL this was impregnated onto a solid support. Doing this is combined the advantages of the IL with those of the heterogeneous support materials [11,14,17–20,24–31]. As a support material an inorganic one (Florisil®) was used because of its crystalline and well-ordered periodic pore structure which will develop higher resistance to chemical, thermal and radiation degradation and also its well known remarkable selectivity [14,28,31]. In order to determine the adsorption performance of the Florisil® impregnated with Cyphos IL-101 in the process of Cs^+ removal from aqueous solutions, equilibrium and kinetic studies were used.

2. Results and Discussion

2.1. *Characterization of the Florisil® Impregnated with IL*

In order to obtain an impregnated solid support with a higher adsorption capacity in the process of Cs^+ removal from aqueous solutions, the optimum quantity of the studied IL which can be impregnated onto the Florisil® support was determined in the first step. The experimental data regarding the dependence of the Cs^+ uptake *versus* ratio ionic liquid: solid support (IL:SS) are presented in Figure 1.

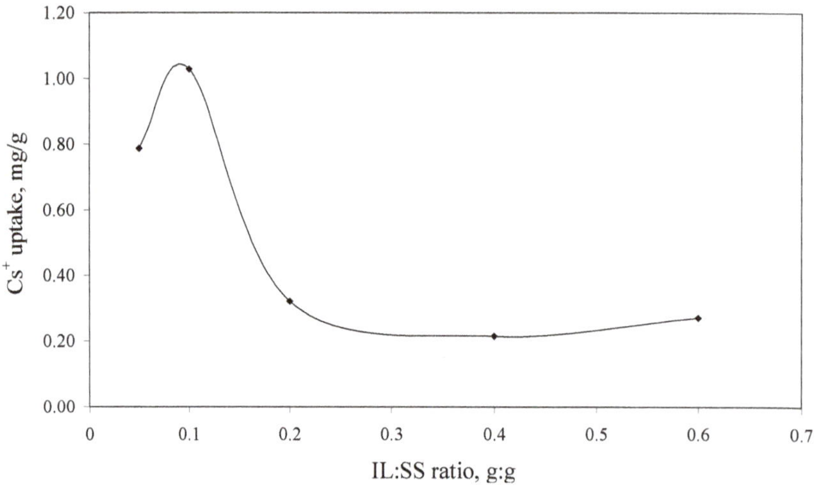

Figure 1. The dependence of the Cs^+ uptake *versus* of IL:SS ratio (C_i = 10 mg/L, V = 25 mL, t = 2 h, pH = 8, m = 0.1 g).

Without impregnation with the studied IL the Florisil® support didn't present any adsorption capacity in the process of Cs^+ removal from aqueous solutions. It can be observed that by increasing the quantity of the IL impregnated onto the Florisil® support the Cs^+ ions uptake increases, but for a quantity of IL impregnated onto the solid support higher than 0.1 g the trend is reversed. This happened because at higher quantities of IL, instead of the impregnation of the Florisil® the surface becomes clogged, which leads to the agglomeration of the Florisil® particles, decreasing in this way the surface contact between the adsorbent and adsorbate. The conglomeration of the Florisil® particles impregnated with a higher quantity of IL is visible with the naked eye. The optimum IL:SS ratio used in the further studies is 0.1:1.

The Florisil® impregnated with the optimum quantity of IL was characterized through energy dispersive X-ray analysis (EDX) in order to demonstrate that the impregnation with IL had occurred. The SEM images of the Florisil® support before and after impregnation with the studied IL are presented elsewhere [32]. The EDX spectrum of the impregnated Florisil® is presented in Figure 2 and in it the characteristic peaks of P and Cl which are components of the ionic liquid can be observed. This analysis thus proved that the impregnation of the Florisil® solid support with the studied IL occurred.

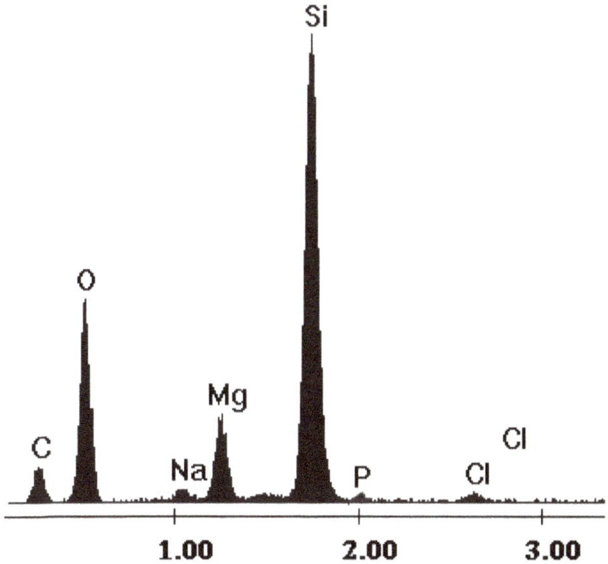

Figure 2. The EDX spectrum of the Florisil® impregnated with IL.

The Florisil® impregnated with IL and used in the process of Cs^+ removal from aqueous solutions was also characterized by Fourier transform infrared spectroscopy (FTIR). The FTIR spectrum of the Florisil® impregnated with IL after adsorption of Cs^+ from aqueous solutions is presented in Figure 3 and can be observed that the spectrum is in good agreement with those from the literature [32–35]. Table 1 shows the characteristic IR bands in [cm^{-1}] for the Florisil® impregnated with IL after adsorption of Cs^+ and the corresponding assignments. The impregnation of the Florisil® with the studied IL is confirmed by the IR peaks around 2,957 cm^{-1}, 2,933 cm^{-1}, 2,858 cm^{-1} and 1,395 cm^{-1} corresponding to the CH_3 and CH_2 stretching vibrations of Cyphos IL 101. The bands at 1,100 cm^{-1}, 1,050 cm^{-1} and 800 cm^{-1} are attributed to the group of ν_a (C-C) + ν (M-O) + P-C stretching vibrations [33–36]. The IR spectrum shows an intense and broad band with a maximum at 3,455 cm^{-1}, due to the ν(Cs-OH, Cs-OH_2) vibration, also confirming the retention of Cs^+.

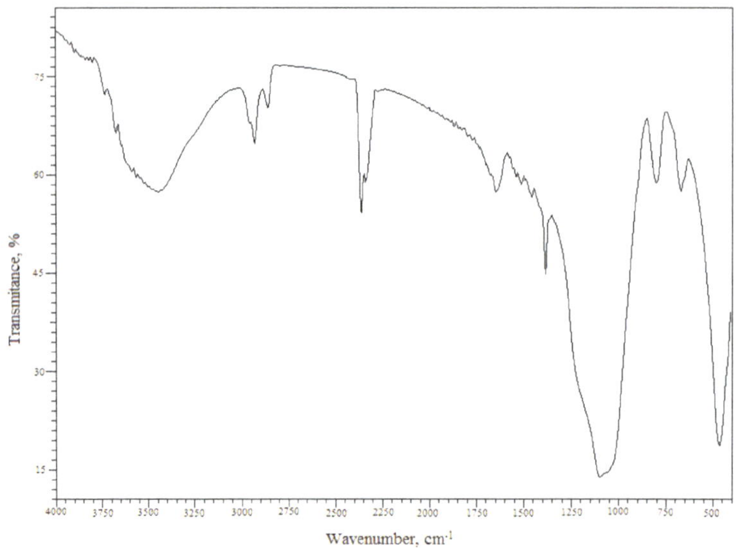

Figure 3. FTIR spectrum of the studied material.

Table 1. Assignments of the IR vibrations of the studied material [33–35].

Wavenumber [cm^{-1}]	Assignments
3455 (s,b)	νOH bonded ν(Cs-OH, Cs-OH_2)
2957 (w)	ν_a (CH_3)
2933 (w)	ν_a (CH_2)
2858 (w)	ν_s (CH_3)
1395 (m)	P-C stretching ν_s (CH_2)
1100 (vs)	P-C stretching ν_a (C-C)
1050 (vs)	P-C stretching ν (M-O)
800 (w)	P-C deformation (out of the plane) ν (M-O)
670 (w)	ν (Si-O)
465 (s)	ν (M-O)

s-strong; v-very; b-broad; m-medium; w-weak; (M = Si, Mg).

2.2. pH Influence on the Cs^+ Adsorption Process

The experimental data regarding the dependence of the Cs^+ uptake *versus* the pH of the solution, using an aqueous solution with a Cs^+ concentration of 10 mg/L and a S:L ratio 0.1 g:0.025 L, under a stirring time of 2 h, was studied. This behaviour is in agreement with other research from the literature [3,4,36]. This is explained by the fact that Cs^+ is stabilized with the increasing pH (6–8) with the formation of hydroxo complexes ($[Cs(OH)_2(OH_2)_{2....4}]^-$). This is also in agreement with the results obtained at the FTIR analysis. In accord with the experimental data we supposed that Cs^+ was retained from aqueous solutions through the attraction of the formed negative hydroxo complex of Cs^+ by the positive phosphonium ions from the IL impregnated onto the Florisil®. Therefore for the further experiments we use Cs^+ solution with an initial pH around value 8.

2.3. Influence of the Amount of Impregnated Florisil® on the Cs^+ Adsorption Process

The experimental data regarding the dependence of the Cs^+ uptake *versus* the amount of the impregnated Florisil® used in 0.025 L of Cs^+ solution, are presented in Figure 4.

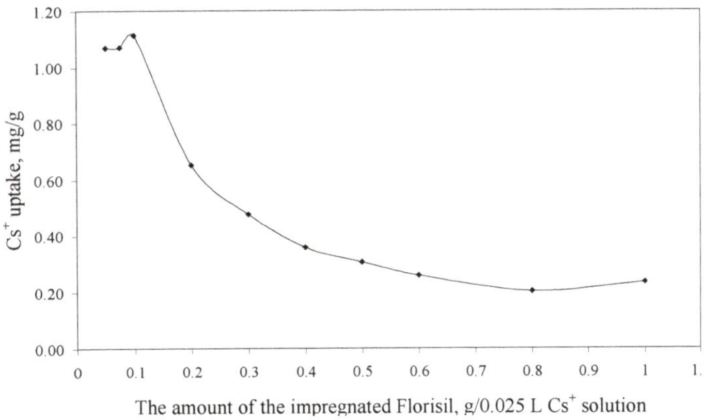

Figure 4. The dependence of the Cs^+ uptake *versus* the amount of the impregnated Florisil® (C_i = 10 mg/L, V = 25 mL, t = 2 h, pH = 8).

From Figure 4 can be observed that an increase in the amount of the impregnated Florisil® used in the process of Cs^+ removal from aqueous solutions lead to a decrease of the Cs^+ uptake, because the adsorption capacity is related to the amount of the adsorbent used in the removal process. The optimum S:L ratio used for the further experiments is 0.1 g of impregnated Florisil® for 0.025 L of aqueous Cs^+ solutions. This parameter is important for the scale up of the studies from the laboratory scale.

2.4. Adsorption Isotherm

The aim of the adsorption isotherm are: (1) to express the affinity of the impregnated Florisil® with the studied IL as a function of its surface properties; (2) to obtain the maximum adsorption capacity of the Florisil® impregnated with the IL in the process of Cs^+ removal from aqueous solutions and to compare it with other sorbents used in the literature; (3) the obtained results can be used to upscale laboratory batch experiments to a pilot or industrial scale. The equilibrium data were fitted with the Langmuir and Freundlich equations. The isotherms are presented in Figure 5 and the obtained parameters and the correlation coefficients (R^2) are presented in Table 2.

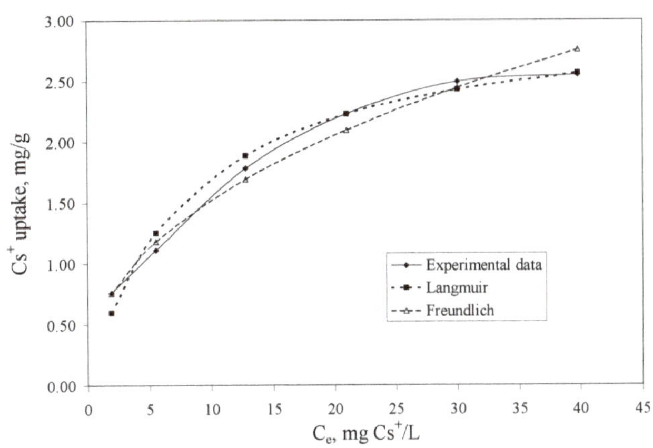

Figure 5. The adsorption isotherms of Cs^+ onto the impregnated Florisil® with the studied IL (t = 2 h, V = 25 mL, pH = 8, m = 0.1 g, C_i = 5–50 mg/L).

Table 2. Parameters of Langmuir and Freundlich isotherms for Cs^+ ion adsorption onto Florisil® impregnated with the studied IL.

Langmuir isotherm				Freundlich isotherm		
$q_{m,\,\mathrm{exp}}$ mg/g	K_L L/mg	$q_{m,\,\mathrm{calc}}$ mg/g	R^2	K_F mg/g	$1/n$	R^2
2.6	0.123	3.086	0.9907	0.5658	0.4299	0.9863

The constant K_F can be defined as an adsorption coefficient which represents the quantity of adsorbed metal ions for a unit equilibrium concentration. The slope *1/n* is a measure of the adsorption intensity or surface heterogeneity. For $1/n = 1$, the partition between the two phases is independent on the concentration. The situation $1/n < 1$ is the most common and correspond to a normal L-type Langmuir isotherm, whilst $1/n > 1$ is indicative of a cooperative adsorption which involves strong interactions between the molecules of adsorbate. Values of $1/n < 1$ show favourable adsorption of Cs^+ ions onto the Florisil® impregnated with IL. Both models showed good correlation coefficients, but those obtained in case of the Langmuir model is more closed to unity, and also the maximum adsorption capacity of the impregnated Florisil® with IL obtained from the Langmuir plot is very close to that obtained experimentally, therefore the experimental data are best described by the Langmuir isotherm suggesting that the adsorption of Cs^+ from aqueous solution take place onto the homogenous surfaces of the Florisil® impregnated with IL, forming a monolayer on the adsorption surface [1,19]. The adsorption capacities (q_m) of different adsorbents for adsorption of Cs^+ ions are compared in Table 3. It may be seen that the Florisil® impregnated with Cyphos IL 101 exhibits a good adsorption capacity in the process of Cs^+ removal from aqueous solutions, compared with other studied materials.

Table 3. Adsorption capacity of various adsorbents in the process of Cs^+ removal from aqueous solutions.

Adsorbent	q_m, mg/g	Reference
Nano-zirconium vanadate	9.1	[3]
Resorcinol-formaldehyde RF	5.56	[37]
Ceiling tiles	0.2128	[38]
Vermiculite	0.646	[39]
Florisil® impregnated with Cyphos IL-101	3.086	Present work

2.5. *The Influence of Contact Time and Temperature on the Cs^+ Adsorption Process and Adsorption Kinetics*

The experimental data regarding the dependence of the Cs^+ uptake *versus* the contact time, at three temperatures (298, 308 and 318 K) using a Cs^+ solution having a concentration of 10 mg/L, are presented in Figure 6. From the experimental data can be observed that the equilibrium between Cs^+ and Florisil® impregnated with IL is achieved after 120 min of contact for all the studied temperatures. This is a very good time comparing with other results from the literature where the equilibrium is achieved in up to 10 days [38]. The increase of the temperature leads to an increase of the quantity of Cs^+ uptake by the studied adsorbent, but this increase is not significant after the equilibrium is achieved, therefore from the economic point of view is not recommended to increase the temperature.

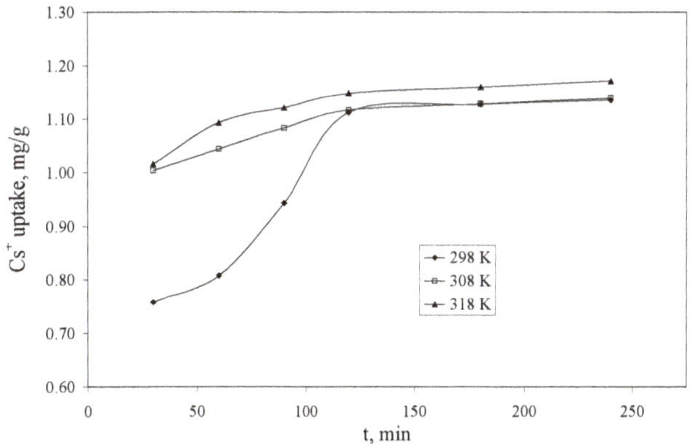

Figure 6. The dependence of the Cs^+ uptake *versus* the contact time (C_i = 10 mg/L, V = 25 mL, pH = 8, m = 0.1 g).

The kinetics of the adsorption describing the rate of the removal of Cs^+ is one of the important characteristics that defines the efficiency of adsorption. In order to evaluate the kinetic mechanism that controls the adsorption process, the pseudo-first-order and pseudo-second-order models were tested to interpret the experimental data. Because of the poor regression coefficient (R^2) values of the Lagergren pseudo-first-order model the results are not included in the text.

The pseudo second-order rate constant k_2 and q_e were calculated from the slope and intercept of the plots of t/q_t *versus* t (Figure 7) [1]. The experimental and calculated q_e values, pseudo-second-order rate constants, R^2 values are presented in Table 4. The calculated q_e values are in agreement with the experimental q_e values and the plots show good linearity, with R^2 higher than 0.99 for all three studied temperatures. Hence, this study suggested that the pseudo-second-order kinetic model better represent the adsorption kinetics, and the adsorption process has the profile of chemisorptions.

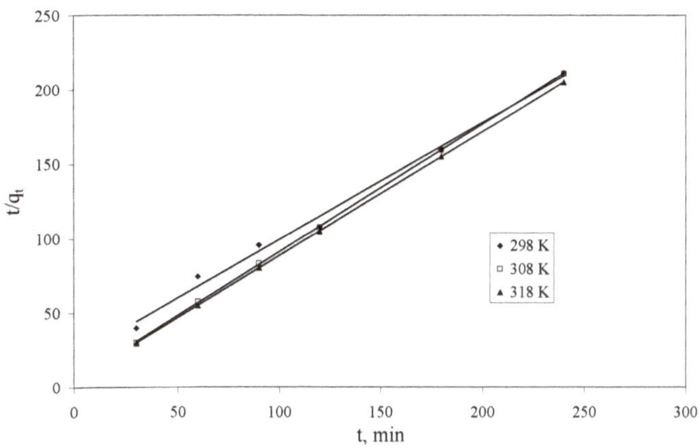

Figure 7. Pseudo-second-order kinetics plots for Cs^+ adsorption onto Florisil® impregnated with IL.

Table 4. Pseudo-second-order parameters for Cs^+ adsorption onto Florisil® impregnated with IL.

Temperature, K	Parameter			
	$q_{e,\,exp}$, mg/g	$q_{e,\,calc}$, mg/g	k_2, g/(min·mg)	R^2
298	1.15	1.27	0.0304	0.9926
308	1.15	1.16	0.144	0.9999
318	1.18	1.20	0.146	1

3. Experimental

3.1. *Materials*

The trihexyl(tetradecyl)phosphonium chloride (Cyphos IL-101) was purchased from Sigma-Aldrich (Oakville, ON, Canada), and the Florisil® support was supplied by Merck (Darmstadt, Germany). The Cs^+ aqueous solutions were prepared by adequate dilution of stock solution of 1000 mg/L (Merck standard solution). In all experiments distilled water was used.

3.2. *Preparation and Characterization of the Florisil® Impregnated with IL*

The IL was dissolved in acetone (analytical reagent provided by SC Chemical Company SA (Bucuresti, Romania) in a ratio of 0.1 g of IL in 5 mL of acetone. The Florisil® was impregnated with the desired quantity of IL via the dry impregnation method [40]. The EDX analysis of the Florisil® impregnated with IL was investigated by Scanning Electron Microscopy (SEM) using a Quanta FEG 250 instrument (EDAX Inc., Mahwah, NJ, USA), equipped with an Energy Dispersive X-ray quantifier (EDAX ZAF) (EDAX Inc., Mahwah, NJ, USA). The FTIR spectra (KBr pellets) of the studied materials were recorded on a Shimadzu Prestige-21 FTIR spectrophotometer (SHIMADEZU, Tokio, Japan) in the range 4000–400 cm^{-1}.

3.3. *Adsorption Experiments*

The Cs^+ adsorption from aqueous solutions onto Florisil® impregnated with IL was realised using batch studies. For the batch studies a Julabo SW23 shaker (JULABO Labortechnik GmbH, Sellbach, Germany) with a constant shaking rate was used. To determine the optimum pH the initial pH of the Cs^+ solutions was adjusted in the 2–12 range using 0.1 N NaOH or 0.1 N HCl, using a S:L ratio of 0.1 g in 0.025 L Cs^+ solution, keeping the samples under stirring for 2 h. Adsorption isotherms were studied by mixing a known amount of Florisil® impregnated with IL with various intial Cs^+ concentrations ranging from 5 to 50 mg/L at 298 K and pH = 8. The experimental data obtained were tested with Langmuir and Freundlich isotherms. The effect of contact time was studied varying the time from 30 to 240 min at three temperatures (298 K, 308 K and 318 K). Cs^+ concentration was determined through atomic emission spectrometry using a Varian SpectrAA 280 type atomic absorption spectrometer (Varian medical systems, Inc., Melbourne, Australia) using an air/acetylene flame. The Cs^+ uptake was determined using the corresponding mass balance.

4. Conclusions

In this work the adsorption properties of Florisil® impregnated with Cyphos Il-101 in the process of Cs^+ removal from aqueous solutions have been studied. The experiments showed that the adsorption depended on the pH, the maximum desorption capacity being achieved at an initial solution pH of 8. This conclusion is in accordance with the results obtained from the FTIR analysis where was observed that the Cs^+ ions was retained as hydroxo complexes, which are stable at neutral pH values. The equilibrium data were fitted with the Langmuir and Freundlich isotherm, the best correlaion being obtained by the Langmuir one. The maximum sorption capacity obtained with the Florisil®impregnated with Cyphos IL-101 is 3.086 mg Cs^+/g of adsorbent. The kinetic studies revealed that the process of Cs^+ adsorption onto Florisil® impregnated with the studied IL followed the pseudo-second order kinetic model, suggesting the chemisorptions profile of the adsorption process. The proposed method for Cs^+ removal from aqueous solutions, by using as adsorbent the Florisil® impregnated with Cyphos IL-101, is an efficient one, because it combines the advantages of the ionic liquids with those of the solid support. In this way a smaller amount of the ionic liquid is used for the removal of Cs^+ from aqueous solutions than in the case of liquid-liquid extraction techniques and there is no risk of loss of the extractant in the aqueous phase. The combined results showed that the obtained new impregnated material represents an efficient adsorbent in the process of Cs^+ ion removal from aqueous solutions compared with other materials studied in the literature. The experimental results can be easily used to scale up the studies from the laboratory scale.

Acknowledgments: This work was supported by a grant of the Romanian National Authority for Scientific Research, CNCS—UEFISCDI, project number PN-II-RU-TE-2012-3-0198.

Conflicts of Interest: The authors declare no conflict of interest.

References

1. Nilchi, A.; Saberi, R.; Moradi, M.; Azizpour, H.; Zarghami, R. Adsorption of caesium on copper hexacyanoferrate-PAN composite ion exchanger from aqueous solution. *Chem. Eng. J.* **2011**, *172*, 572–580. [CrossRef]
2. El-Kamash, A.M. Evaluation of zeolite A for the sorptive removal of Cs^+ and Sr^{2+} ions from aqueous solutions using batch and fixed bed columns operations. *J. Hazard. Mater.* **2008**, *151*, 432–445. [CrossRef] [PubMed]
3. Abd El-Latif, M.-M.; Elkady, M.-M. Kinetics study and thermodynamic behaviour for removing cesium, cobalt and nickel ions from aqueous solution using nano-zirconiuim vanadate ion exchange. *Desalination* **2011**, *271*, 41–54. [CrossRef]
4. Vendilo, A.G.; Djigailo, D.I.; Smirnova, S.V.; Torocheshnikova, I.I.; Popov, K.I.; Krasovsky, V.G.; Pletnev, I.V. 18-Crown-6 and dibenzo-18-crown-6 assisted extraction of cesium from water into room temperature ionic liquids and its correlation with stability constants for cesium complexes. *Molecules* **2009**, *14*, 5001–5016. [CrossRef]
5. Xu, J.Z.; Rajapkse, N.; Finston, H.L. Homogeneous liquid-liquid extraction of Uranium from fission-products. *Radiochim. Acta* **1990**, *49*, 135–140. [CrossRef]
6. Murata, K.; Ikeda, S.; Yokoyama, Y. Homogeneous liquid-liquid extraction method-Extraction of Iron(III) thenoyltifluoroacetonate by propylene carbonate. *Anal. Chem.* **1972**, *44*, 805–810. [CrossRef]
7. Preston, J.S. Solvent extraction of metals by carboxylic-acids. *Hydrometallurgy* **1985**, *14*, 805–810. [CrossRef]
8. Hoogerstraete, T.V.; Onghena, B.; Binnemans, K. Homogeneous liquid-liquid extraction of metal ions with a functionalized ionic liquid. *J. Phys. Chem. Lett.* **2013**, *4*, 1659–1663. [CrossRef]
9. Ha, S.H.; Menchavez, R.N.; Koo, Y.M. Reprocessing of spent nuclear waste using ionic liquids. *Korean J. Chem. Eng.* **2010**, *27*, 1360–1365. [CrossRef]
10. Fraser, K.J.; MacFarlane, D.R. Phosphonium-based ionic liquids: An overview. *Aust. J. Chem.* **2009**, *62*, 309–321. [CrossRef]
11. Yang, X.Y.; Zhang, J.P.; Guo, L.; Zhao, H.; Zhang, Y.; Chen, J. Solvent impregnated resin prepared using ionic liquid Cyphos IL 104 for Cr(VI) removal. *Trans. Nonferr. Met. Soc. China* **2012**, *22*, 3126–3130. [CrossRef]
12. Stojanovic, A.; Morgenbesser, C.; Kogelnig, D.; Krachler, R.; Keppler, B.D. *Quaternary Ammonium and Phosphonium Ionic Liquids in Chemical and Environmental Engineering, Ionic Liquids: Theory, Properties, New Approaches*; Kokorin, A., Ed.; Publisher: Rijeka, Croatia, 2011; pp. 657–680.
13. Martak, J.; Schlosser, S. Extraction of lactic acid by phosphonium ionic liquid. *J. Sep. Purif. Technol.* **2007**, *57*, 483–494. [CrossRef]
14. Liu, Y.; Zhu, L.; Sun, X.; Chen, J.; Luo, F. Silica materials doped with bifunctional ionic liquid extractants for Yttrium extraction. *Ind. Eng. Chem. Res.* **2009**, *48*, 7308–7313. [CrossRef]
15. Liu, Y.; Zhu, L.; Sun, X.; Chen, J. Toward greener separations of rare earths: Bifunctional ionic liquid extractants in biodiesel. *AlChE J.* **2010**, *56*, 2338–2346. [CrossRef]
16. Guo, L.; Liu, Y.; Zhang, C.; Chen, J. Preparation of PVDF-based polymer inclusion memebrane using ionic liquid plasticizer and Cyphos IL 104 carrier for Cr(VI) transport. *J. Membr. Sci.* **2011**, *372*, 314–321. [CrossRef]

17. Guibal, E.; Pinol, A.-F.; Ruz, M.; Vincent, T.; Jounanin, C.; Sastre, A. Immobilization of Cyphos ionic liquids in alginate capsules for Cd(II) sorption. *Sep. Sci. Technol.* **2010**, *45*, 1935–1949. [CrossRef]
18. Campos, K.; Domingo, R.; Vincent, T.; Ruiz, M.; Sastre, A.M.; Guibal, E. Bismuth recovery from acidic solutions using Cyphos IL-101 immobilized in a composite biopolymer matrix. *Water Res.* **2008**, *42*, 4019–4031. [CrossRef]
19. Gallardo, V.; Navarro, R.; Saucedo, I.; Avila, M.; Guibal, E. Zinc(II) extraction from hydrochloric acid solutions using Amberlite XAD7 impregnated with Chyphos IL101 (tetradecyl(trihexyl) phosphonium chloride). *Sep. Sci. Technol.* **2008**, *43*, 2434–2459. [CrossRef]
20. Vincent, T.; Parodi, A.; Guibal, E. Pt recovery using Cypos IL-101 immobilized in biopolymer capsule. *Sep. Purif. Technol.* **2008**, *62*, 470–479. [CrossRef]
21. Regel-Rosocka, M.; Wisniewski, M. Selective removal of znic(II) from spent pickling solutions in the presence of iron ions with phosphonium ionic liquid Cyphos IL101. *Hydrometallurgy* **2011**, *110*, 85–90. [CrossRef]
22. Todd, T.A.; Mann, N.R.; Tranter, T.J.; Sebesta, F.; John, J.; Motl, A. Cesium sorption from concentrated acidic tank wastes using ammonium molybdophosphate-polyacrylonitrile composite sorbents. *J. Radioanal. Nucl. Chem.* **2002**, *254*, 47–52. [CrossRef]
23. Tranter, T.J.; Herbst, R.S.; Todd, T.A.; Olson, A.L.; Eldredge, H.B. Evaluation and testing of ammonium molybdophosphate-polyacrylonitrile (AMP-PAN) as a cesium selective sorbent for the removal of cesium-137 from INTEC acidic waste. *Adv. Environ. Res.* **2002**, *6*, 107–121. [CrossRef]
24. Zhu, L.; Guo, L.; Zhang, Z.; Chen, J.; Zhang, S. The preparation of supported ionic liquids (SILs) and their application in rare metals separation. *Sci. China Chem.* **2012**, *55*, 1479–1487. [CrossRef]
25. Sun, X.; Preng, B.; Ji, Y.; Chen, J.; Li, D. The solid-liquid extraction of yttrium from rare earths by solvent (ionic liquid) impregnated resin coupled with complexing method. *Sep. Purif. Technol.* **2008**, *63*, 61–68. [CrossRef]
26. Sun, X.Q.; Xu, A.M.; Chen, J.; Li, D.Q. Application of room temperature ionic liquid-based extraction for metal ions. *Chin. J. Anal. Chem.* **2007**, *35*, 597–604.
27. Sun, X.; Li, Y.; Chen, J.; Ma, J. Solvent impregnated resin prepared using task-specific ionic liquids for rare earth separation. *J. Rare Earth* **2009**, *27*, 932–936. [CrossRef]
28. Mahmoud, M.E. Surface loaded 1-methyl-3-ethylimidazolium bis(trifluoromethylsulfonyl)imide [$EMIM^+Tf_2N^-$] hydrophobic ionic liquid on nano-Silica sorbents for removal of lead from water samples. *Desalination* **2011**, *266*, 119–127. [CrossRef]
29. Kalidhasan, S.; Kumar, A.-S.-K.; Rajesh, V.; Rajesh, N. An efficient ultrasound assisted approach for the impregnation of room temperature ionic liquid onto Dowex 1 × 8 resin matrix and its application toward the enhanced adsorption of Cr(VI). *J. Hazard. Mater.* **2012**, *213–214*, 249–257. [CrossRef]
30. Lemus, J.; Palomar, J.; Gilarranz, M.; Rodriquez, J. Characterization of supported ionic liquid phase (SILP) materials prepared from different supports. *Adsorption* **2011**, *17*, 561–571. [CrossRef]
31. Zhang, A.; Wang, W.; Chai, Z.; Kuraoka, E. Modification of a novel macroporous Silica-based Crown ether impregnated polymeric composite with 1-dodecanol and its adsorption for some fission and non-fission product contained in high level liquid waste. *Eur. Polym. J.* **2008**, *44*, 3899–3907. [CrossRef]
32. Negrea, A.; Ciopec, M.; Lupa, L.; Negrea, P.; Gabor, A. Influence of the solid support base impregnated with IL on the sorption of various radionuclides from aqueous solutions. *AWERProcedia Adv. Appl. Sci.* **2013**, *1*, 241–250.
33. Cholico-Gonzalez, D.; Avila-Rodriguez, M.; Cote, G.; Chagnes, A. Chemical properties of trihexyl(tetradecyl)phosphonium chloride and bis(2,4,4-trimethylpentyl)phosphinic acid mixtures: Interaction study by FT-IR and NMR spectroscopies. *J. Mol. Liq.* **2013**, *187*, 165–170. [CrossRef]
34. Plinio, I. Infrared spectroscopy of sol-gel derived silica-based films: A spectra-microstructure overview. *J. Non-Cryst. Solids* **2003**, *316*, 309–319.
35. Nakamoto, K. *Infrared Spectra of Inorganic and Coordination Compounds*, 4th ed.; A Wiley-Interscience Publishers John Wiley & Sons: New York, NY, USA, 1968.
36. Sinha, P.K.; Panicker, P.K.; Amalraj, R.V. Treatment of radioactive liquid waste containing caesium by indigenously available synthetic zeolites: A comparative study. *Waste Manag.* **1995**, *15*, 149–157. [CrossRef]
37. Hassan, N.M.; Adu-Wusu, K. Cesium removal from Hanford tank waste solution using resorcinol-formaldehyde resin. *Solvent Extr. Ion Exch.* **2005**, *23*, 375–389. [CrossRef]
38. Miah, M.Y.; Volchek, K.; Kuang, W.; Tezel, F.H. Kinetic and equilibrium studies of cesium adsorption on ceiling tiles from aqueous solutions. *J. Hazard. Mater.* **2010**, *183*, 712–717. [CrossRef]

39. Haladi, N.; Kananpanah, S.; Abolghasemi, H. Equilibrium and thermodynamic studies of cesium adsorption on natural Vermiculite and optimization of the operation conditions. *Iran. J. Chem. Chem. Eng.* **2009**, *28*, 29–36.
40. Negrea, A.; Ciopec, M.; Lupa, L.; Davidescu, C.M.; Popa, A.; Ilia, G.; Negrea, P. Removal of As^{V} by Fe^{III}—Loaded XAD7 impregnated resin containing di(2-ethylhexyl) phosphoric acid (DEHPA): Equilibrium, kinetic, and thermodynamic modeling studies. *J. Chem. Eng. Data* **2011**, *56*, 3830–3838. [CrossRef]

Sample Availability: *Sample Availability*: Samples of the Florisil®impregnated with Cyphos IL 101 are available from the authors.

molecules

MDPI

Review

Synthesis of DNA/RNA and Their Analogs via Phosphoramidite and H-Phosphonate Chemistries

Subhadeep Roy * and Marvin Caruthers

Department of Chemistry and Biochemistry, University of Colorado, Boulder, CO 80309, USA; Marvin.Caruthers@Colorado.edu

* Correspondence: subhadeep.roy@colorado.edu; Tel.: +1-303-492-7723; Fax: +1-303-735-2135

Received: 8 October 2013; Accepted: 8 November 2013; Published: 18 November 2013

Abstract: The chemical synthesis of DNA and RNA is universally carried out using nucleoside phosphoramidites or H-phosphonates as synthons. This review focuses on the phosphorus chemistry behind these synthons and how it has been developed to generate procedures whereby yields per condensation approach 100% with very few side products. Additionally the synthesis and properties of certain DNA and RNA analogs that are modified at phosphorus will also be discussed. These analogs include boranephosphonates, metallophosphonates, and alkylboranephosphines.

Keywords: DNA; RNA; phosphoramidites; H-phosphonates; boranephosphonates; metallophosphonates; boranealkylphosphines

1. Introduction

The ability to chemically synthesize 2′-deoxyoligonucleotides and oligoribonucleotides (oligonucleotides) in high yield and purity has revolutionized biochemical research as well as enabled a large number of other technologies such as DNA/RNA based therapeutics, DNA forensics, and high-throughput sequencing. The chemical structure of DNA/RNA consists of purines and pyrimidines attached to 2′-deoxyribose or ribose sugars. These 2′-deoxyribonucleosides or ribonucleosides are then joined via phosphodiester linkages. Thus phosphorus chemistry is central to the development of synthetic methods for preparing these important biological molecules and for achieving the high yield, high-throughput synthesis of oligonucleotides.

In this review we will focus on one aspect of organophosphorus chemistry, namely the reactivity of P (III) centers that enables the synthesis of DNA/RNA. We will describe these developments by first reviewing the work of Robert Letsinger and co-workers as their research led to the phosphoramidite approach in our laboratory. We will also discuss the H-phosphonate methodology that is an alternative P (III) chemistry for the synthesis of oligonucleotides. Both approaches have been used to prepare a plethora of nucleic acid analogs modified at phosphorus. This is a vast field of research in itself and a comprehensive description of all the different derivatives is outside the scope of this review. We will instead discuss a recent and promising class of analogs in which a P (III) center is bonded to a Lewis acid such as borane and forms stable internucleotide linkages.

2. Synthesis of DNA Using Chlorophosphites and Tetrazoylphosphites

Development of the phosphoramidite method of DNA synthesis was based upon two observations. One was that chloro- and dichlorophosphites react rapidly with the 3′-hydroxyl group of a 2′-deoxynucleoside, whereas the corresponding phosphorochloridates require several hours at room temperature [1]. Yields were also much higher with chloro and dichlorophosphites. Thus the possibility existed that DNA could be prepared rapidly and in high yield from activated phosphites. The other observation was that Robert Letsinger's laboratory had used dichlorophosphites to synthesize a

2′-deoxythymidine pentanucleotide. Stepwise yields for addition of each mononucleotide were 69–82%. Most encouraging was the observation that reaction times were only 5 min in comparison to hours for earlier approaches.

These observations led us to explore the use of 2′-deoxynucleoside P (III) derivatives for synthesizing 2′-deoxyoligonucleotides on polymeric supports [2,3]. For several reasons, our research was predicated on the use of HPLC-grade silica (also known as controlled pore glass (CPG)) as a support. One was that all previous research utilizing organic polymers, such as polystyrene, to synthesize DNA [4] had failed in part because organic supports readily adsorbed the synthons and other reagents. Thus removal of these materials after each condensation step was difficult. This problem, when coupled with the knowledge that HPLC-grade silica had been designed for efficient mass transfer, dictated its investigation. Our expectation was that both solvents and reagents could be removed rapidly and completely from the matrix. This was observed in our early work. Silica was also expected to be chemically inert towards all the reagents we contemplated using and was a rigid, non-swelling matrix in common organic solvents. Therefore, it could be packed into a column and reactants merely pumped through the column.

The initial approach we developed is outlined in Scheme 1. The first step was activating the silica matrix by attaching (3-aminopropyl)triethoxysilane to silica and then treating the product of this reaction with succinic anhydride in order to generate **1**. The next step was condensing 5′-dimethoxytrityl-2′-deoxythymidine to the support using dicyclohexylcarbodiimide (DCC) to activate the carboxylic acid. After removal of the 5′-dimethoxytrityl group with $ZnBr_2$ to yield **2**, condensation with a 2′-deoxynucleoside 3′-phosphite **3** was carried out. Of the synthons we examined, the most reactive while generating the fewest side-products, were the 5′-dimethoxytrityl 3′-methyltetrazoyl phosphites. Using these synthons, 95% yields per condensation were observed during synthesis. After oxidation of the phosphite to phosphate with aqueous iodine, capping with diethoxytriazolephosphine to remove any unreacted intermediates, and removal of the dimethoxytrityl group with $ZnBr_2$, the product **4** of this reaction sequence could be extended by repetitive use of this cycle in order to generate a 2′-deoxyoligonucleotide. Finally after completion of oligonucleotide synthesis, methyl groups were removed from phosphates using thiophenol. This oligomer was then cleaved from the support and base protecting groups (*N*-isobutyrylguanine, *N*-benzoylcytosine and *N*-benzoyladenine) removed with ammonium hydroxide. By measuring the amount of dimethoxytrityl cation released following synthesis of a 12 mer, the overall yield was determined to be 55%, which was unprecedented at that time in the nucleic acid field.

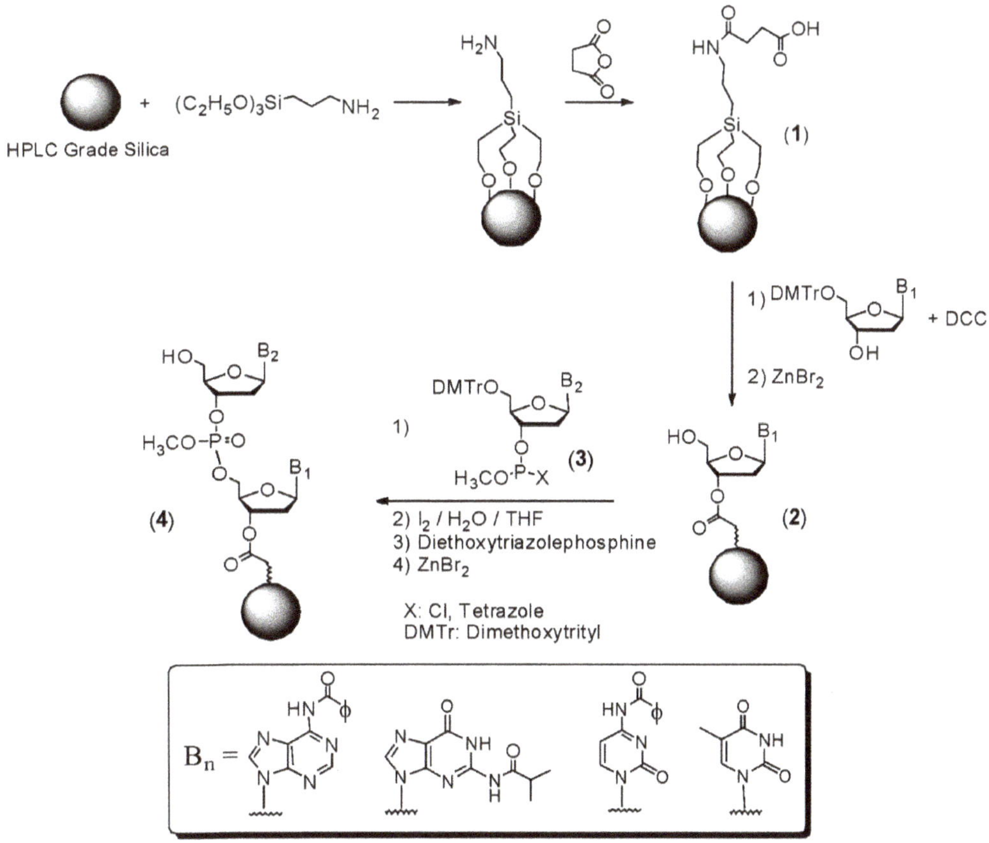

Scheme 1. DNA Synthesis on HPLC Silica.

During the course of this work, we also developed a semiautomatic machine where one cycle of synthesis on a silica column, including the use of all reagents and solvents, could be programmed and completed automatically. The operator then just added the next appropriately protected 2′-deoxy-nucleoside-3′-phosphite to the column and initiated another synthesis cycle. Although more successful than any previous method, it was still far from acceptable. The main problem was that the 2′-deoxy-nucleoside 3′-tetrazoylphosphites had to be prepared at −78 °C and preferably used the same day. The most significant outcome from this research that continues to the present, was the successful development of silica as a matrix for DNA synthesis.

3. The Phosphoramidite Approach for Synthesis of DNA and RNA

Our discovery of phosphoramidites as synthons was initiated with a study aimed at activating aminophosphines. The strategy was to activate 5′-dimethoxytrityl 2′-deoxynucleoside 3′-amino-phosphines via insertion of CO_2, CS_2 or COS to form mixed anhydrides [5,6] that we posited would condense with a 2′-deoxynucleoside. In our hands, this strategy was not successful. However during the attempted synthesis of the aminophosphines by reacting 5′-*O*-dimethoxytrityl-2′-deoxythymidine with N,N-dimethylaminomethoxychlorophosphine in pyridine, we observed the appearance of a new dimethoxytrityl positive spot on TLC (thin-layer chromatography) that moved similar to a dinucleotide. Based upon the previous literature on the acid activation of aminophosphines [7], we reasoned that the initial product, 5′-dimethoxytrityl

2′-deoxythymidine 3′-*N,N*-dimethylamino-methoxyphosphine (a phosphoramidite), was activated by pyridine hydrochloride that was produced in situ during the reaction. The protonated 5′-dimethoxytrityl 2′-deoxythymidine 3′-*O*-methoxy-*N,N*-dimethylammonium phosphine would then react with the free 3′-hydroxyl of excess 5′-dimethoxytrityl 2′-deoxythymidine present in the reaction mixture to form *bis*(5′-dimethoxytrityl 2′-deoxythymidine 3′-*O*) methylphosphite. Subsequently by repeating this reaction in the presence of a stronger base (Hunig's base) in the reaction mixture in order to remove the acid produced, we were able to isolate the desired phosphoramidites. This allowed us to confirm that phosphoramidites could indeed be activated with ease towards nucleophilic substitution using a number of weak acids. We then proceeded to combine this discovery with the use of CPG as a solid support to develop a high yielding method for oligonucleotide synthesis that has remained essentially unmodified since our original report [8].

The synthesis of DNA and RNA using the phosphoramidite approach is depicted in Scheme 2. The synthetic cycle begins with the removal of 5′-DMTr from **5** by treatment with a solution of 3% trichloroacetic acid in dichloromethane to yield **2**. Condensation of a 5′-dimethoxytrityl-2′-deoxynucleoside 3′-phosphoramidite **6** with **2** is then completed using tetrazole as an activator to produce the phosphite triester **7**. The most commonly used phosphoramidites contain a diisopropyl-amino group. The steric bulk of this group was found to provide an ideal balance between stability during preparation of these synthons and ease of activation [9]. Although many activators have been proposed, tetrazole remains a popular choice as it can be obtained as a solid by sublimation and kept stably as an anhydrous solution on the DNA synthesis machine for long periods of time.

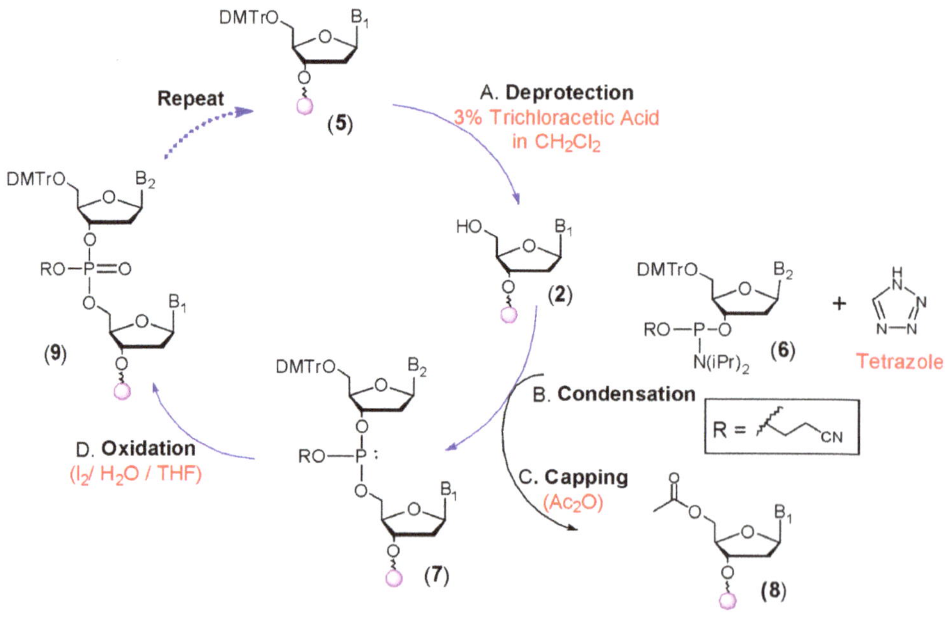

Scheme 2. The Phosphoramidite Approach for Oligonucleotide Synthesis.

The next step is capping of unreacted 5′-hydroxyl groups using acetic anhydride activated by N-methylimidazole in pyridine (Step C). This is important as it prevents any unreacted oligonucleotides from continuing to grow. Failing to do this leads to a final product mixture whereby the major oligonucleotide impurities are hard to remove because they are shorter by only a few nucleotides. For the synthesis of DNA (≥150 mers) on glass chips using inkjet printers, the capping step was found to be unnecessary [10]. This likely resulted from surface effects that allow even better condensation yields than possible with traditional CPG supports. The final step is oxidation of the phosphite triester to phosphate **9** using aqueous iodine. The entire sequence can then be repeated iteratively to obtain the desired oligomer. For the synthesis of various analogs such as phosphorothioates and phosphoramidates, oxidation can alternatively be carried out using a sulfurizing agent or iodine and amines respectively.

At this stage, the oligomer can be hydrolyzed from the support using an aqueous solution of ammonia which also removes the amide groups used to protect the exocylic amines of cytosine, guanine and adenine as well as the cyanoethyl protection [11] on phosphate. The oligomers can then be purified by reverse phase HPLC or gel electrophoresis.

During an exploration of various phosphorus protecting groups, we observed that *o*-methylbenzyl was removed during the iodine oxidation step [12]. Moreover a deoxyoligonucleotide 20 in length, as prepared with this protecting group, gave a very high yield of a homogeneous product. At the same time, van Boom and co-workers published similar results with the 2-cyano-1,1-dimethylethyl phosphorus protecting group [13]. These observations led us to propose that alkyl deoxynucleoside phosphites (**10a–c**, Scheme 3), having alkyl groups that stabilize the S_N1 character of the second stage of an Arbuzov reaction, upon reaction with most electrophiles, would lead to elimination of the alkyl protecting group with concomitant formation of the phosphoryl bond. Thus upon oxidation with iodine (the electrophile), the tertiary alkyl protecting group would be eliminated and generate the phosphoroiodate. Nucleophilic substitution of iodine would yield various desired compounds. For example if the reaction were performed in aqueous iodine, the product would be the natural, internucleotide phosphate diester linkage (compound **13**, R^6 = H). If however oxidation were to take place under anhydrous conditions in the presence of amines, azides or alcohols then the appropriate phosphoramidates **11**, phosphoroazoamidates **12**, or phosphate triesters **13** (R^6 = alkyl or aryl) would be the final products. In contrast if these phosphites were oxidized with *tert*-butylhydroperoxide or sulfur, the product was the dinucleotide phosphate triester or thiophosphate triester respectively. Removal of the 2-cyano-1,1-dimethylethyl protecting group with base would then generate the phosphate diester or phosphorothioate diester. Thus this approach, depending upon oxidation conditions for each synthesis cycle, can be used to introduce different analogs selectively throughout an oligonucleotide.

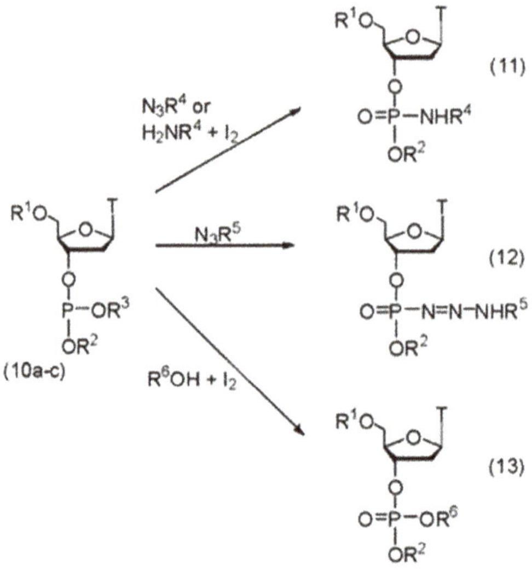

Scheme 3. (**a–c**): R^1 = 4,4′-dimethoxytrityl, R^2 = 3′-acetylthymidylyl; a: R^3 = 2-cyano-1,1-dimethylethyl, R^4 = *n*-butyl; b: R^2 = 3′-acetylthymidylyl, R^5 = 3-(*N*-ethylcarbazoyl), R^3 = *o*-methylbenzyl; c: R^2 = methyl, R^3 = 2-cyano-1,1-dimethylethyl, R^6 = H, alkyl, or aryl.

4. The H-Phosphonate Approach for Synthesis of DNA and RNA

H-Phosphonates are a distinct category of tervalent P (III) compounds that contain a phosphoryl group (P=O) and a hydrogen atom bonded to the phosphorous center (e.g., **18**). H-Phosphonates have a tetrahedral geometry that is similar to P (V) compounds. The phosphorus atom in H-phosphonates is electrophillic and lacks a lone pair of electrons. Therefore it is much more resistant towards oxidation under ambient conditions than most P (III) compounds. At the same time, upon activation, H-Phosphonate monoesters can be induced to undergo nucleophillic substitutions efficiently. This combination of properties has made them attractive synthons for preparing oligonucleotides.

H-Phosphonates were first used in nucleotide chemistry by Sir Alexander Todd [14,15]. They demonstrated that treatment of benzyl H-phosphonate monoester **16** with diphenyl phosphorochloridate (Scheme 4) led to the putative formation of the corresponding activated mixed anhydride **17**. This compound was reacted in situ with the 5′-hydroxyl of a 2′,3′-isopropylidine nucleoside to produce the H-phosphonate diester **18** that was in turn oxidized by *N*-chlorosuccinimide to the phosphorochloridate diester **19**. They later adapted this methodology for the first chemical synthesis of a natural 5′-3′ linked dinucleotide [16]. Ogilvie and Nemer in 1980 [17] as well as Kume et al. in 1984 [18] next demonstrated the synthesis of dinucleotides containing H-phosphonate internucleotide linkages. However the use of H-phosphonates as synthons for oligonucleotide synthesis was not developed till three decades after Todd's initial reports, when Garegg et al. [19] and later Froehler and Matteucci [20] described the synthesis of 2′-deoxynucleoside H-phosphonate diesters by activating 2′-deoxynucleoside H-phosphonate monoesters with several reagents such as 2,4,6-tri- isopropylbenzenesulfonylchloride, *N,N-bis*(2-oxo-3-oxazolidinyl)-phosphorodiamide chloride, diphenyl-chlorophosphate and pivaloyl chloride.

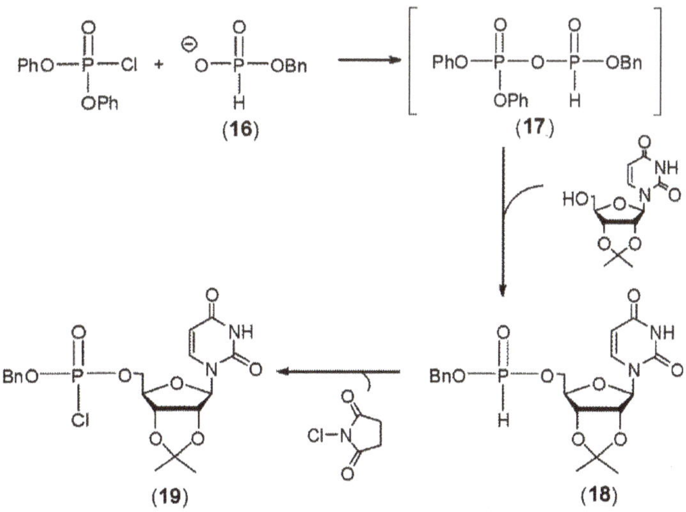

Scheme 4. Mixed Anhydride Activation of H-Phosphonates.

Froehler et al. [21] and Garegg et al. [22,23] next demonstrated the use of this method for the solid phase synthesis of oligonucleotides using 2′-deoxy and ribonucleoside H-phosphonate monoesters as synthons. The general method is shown in Scheme 5A. Following acid-mediated removal of the 5′-dimethoxytrityl protecting group, condensation is carried out by addition of an appropriately protected 2′-deoxy or ribonucleoside H-phosphonate monoester **20** and an activating agent—typically pivaloyl chloride or adamantoyl chloride. Reaction of the acid chloride with the H-phosphonate monoester produces the corresponding mixed phosphonocarboxylic anhydride. This mixed anhydride undergoes a nucleophilic attack at the phosphorous by the 5′- hydroxyl of the nucleoside joined to a polymer support. The product of this reaction is the H-phosphonate diester **21**. Unlike the phosphite triester **7**, this linkage is stable to the acidic conditions (3% trichloroacetic acid in dichloromethane) required for removal of the 5′-dimethoxytrityl group. Thus it is unnecessary to carry out oxidation at every cycle. Chain elongation therefore consists of two steps. Oxidation to the phosphodiester with aqueous iodine is completed following oligonucleotide synthesis. Alternatively oxidation can be carried out under non-aqueous conditions using various reagents which has enabled the synthesis of a number of modified oligonucleotides such as phosphoramidates [24], phosphorothioates [25], and phosphoro-selenoates [26]. After the oxidation step, aqueous ammonia is used to remove nucleobase amide protecting groups and to cleave the oligonucleotide from the support.

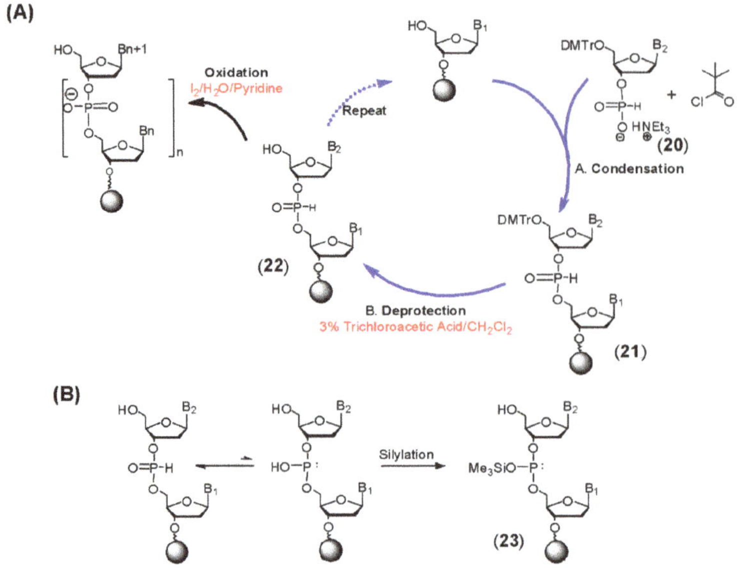

Scheme 5. The H-Phosphonate Approach to DNA/RNA Synthesis.

The H-phosphonate method is particularly well suited for RNA synthesis as some of the problems associated with this approach, such as double activation and phosphorus acylation, are ameliorated when using ribonucleosides having a 2′-protecting group [27]. Coupling of ribonucleoside H-phosphonates on a solid support when compared to coupling ribonucleoside phosphoramidites is less sensitive to steric effects that arise due to the 2′-protecting group. RNA molecules up to 50 to 60 nucleotides in length can be made in high yields by this method.

An additional attraction of this chemistry for synthesis of certain DNA or RNA derivatives stems from the fact that the H-phosphonate diester exists in equilibrium with its phosphite form (Scheme 5B). Ordinarily the H-phosphonate is more stable and the predominant species. However silylation can be used to trap the phosphite triester form **23**, which creates a nucleophilic phosphorus center having a lone pair of electrons. This phosphite triester can then be treated with electrophiles to allow synthesis of various phosphorus analogs. As the silylation procedure can be performed at the end of the synthetic cycles, repetitive exposure of the desired derivative to the acidic deprotection solution is avoided. This is an important advantage when preparing acid sensitive analogs. However a limitation of this protocol is that only uniformly modified oligomers can be produced. H-Phosphonate diesters also have stable stereochemical configurations and undergo oxidation or conversion to phosphite triesters (as in Scheme 5B) in a stereospecific manner [28]. This has been exploited as a method for the synthesis of stereodefined DNA derivatives.

H-Phosphonates offer a useful alternative to the phosphoramidite method particularly for the synthesis of RNA and acid labile oligonucleotide analogs. Although phosphoramidites continue to be much more widely adopted because of higher stepwise yields and fewer side-products, it should be noted that at present H-phosphonates are the only other commercially available DNA/RNA synthons. It is expected that H-phosphonate chemistry will continue to be useful for the preparation of certain novel DNA derivatives.

5. DNA Containing Stable Complexes of P (III) with Lewis Acids

The presence of a lone pair of electrons on the phosphorous atom in P (III) compounds allows it to form stable complexes with Lewis acids. Such complexation stabilizes the phosphorous towards oxidation or attack by electrophiles. Barbara Ramsay Shaw and co-workers were the first to exploit this bonding scheme in their synthesis of 2′-deoxythymidylyl-(3′-5′)-2′-deoxythymidine which contained a hydrolytically stable boranephosphonate (also referred to as boranophosphate in the literature) linkage **24** (Figure 1) [29,30]. These compounds are conventionally described as having one of the non-bridging phosphate oxygens replaced by borane (compound **24a**). This depiction emphasizes that the stability of these compounds is similar to phosphate diesters. On the other hand this linkage can also be described as a P (III) compound complexed with a borane group (compound **24b**). This representation ties these analogs to a wider category of similar compounds and helps to illustrate the fact that bonding of this type can be a general approach to creating varied backbone structures with engineered properties. For example we have reported that methyphosphine diesters complexed with borane (compound **25**) [31] yield a stable structure that is neutral as opposed to the negatively charged boranephosphonate. Oligomers containing these neutral linkages demonstrate enhanced cellular uptake. This approach can be extended to include a number of substituents including amines (compound **26**) and acetate (*vide infra*) with each having unique and tailored properties. In addition to borane, other Lewis acids can also be used as demonstrated by the synthesis of dinucleotides containing a P (III) center complexed to pentacarbonyltungstate(-1) and pentacarbonyl-molybdate(-1) (**27**) [32,33]. These metallophosphonate derivatives should prove to be useful molecules for the construction of DNA based circuits and metallic nanostructures. Because extensive research on the chemical and biochemical properties of boranephosphonate DNA and RNA (bpDNA or bpRNA) has already been completed, the primary emphasis in the following section is on bpDNA with only brief mention of the synthetic methods used to prepare the other analogs shown in Figure 1.

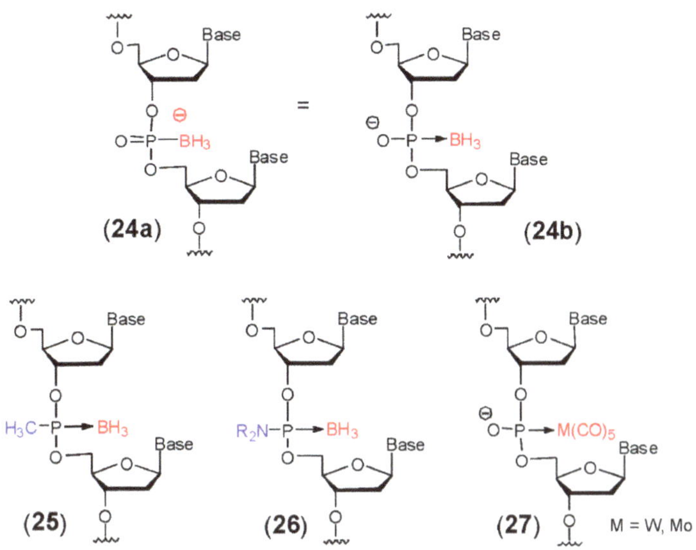

Figure 1. Lewis Acid—Phosphorous Analogs of DNA.

Although the first report of dithymidine having a boranephosphonate diester linkage was published more than twenty years ago, the lack of high yielding chemical methods till recently has severely limited the applicability of these molecules. However in the interim, the Ramsay-Shaw laboratory developed an elegant enzymatic method for synthesizing bpDNA and bpRNA. Their procedure used 2′-deoxynucleoside or ribonucleoside triphosphates that contained a borane moiety at the α phosphorous. When these triphosphates were used with certain polymerases and an appropriate primer/template, bpDNA or bpRNA could readily be prepared [34–37]. Enzymatically synthesized bpDNA (or bpRNA) enabled early studies that were important in demonstrating the biochemical potential of boranephosphonate oligonucleotides. Unfortunately enzymatic synthesis was not only expensive and low yielding but, because only one of the dNTPs could be replaced by its corresponding borane derivative during any one synthesis, the flexibility of positioning and density of boranephosphonate linkages within an oligomer was severely restricted. However it should be noted that an advantage of enzymatic synthesis was that polymerases recognize only one of the two enantiomers of α-P-borano dNTPs. Thus diastereomerically pure oligomers were produced by enzymatic synthesis whereas chemical methods generate mixtures of diastereomers.

Conceptually, the chemical synthesis of boranephosphonates simply requires substituting oxidation in the synthesis cycle with boronation. Chemically this means that a BH_3 group is exchanged between a labile borane complex (such as THF-borane) and the phosphite triester (*e.g.*, **7**, Scheme 2). However in practice two problems must be addressed. First, BH_3 reduces the amide protecting groups that are conventionally used with the exocylic amines of cytosine, adenine and guanine. The resulting alkyl amines cannot be removed and therefore represent an undesirable DNA modification. Additionally the trityl cation, which is generated during deprotection of the 5′-hydroxyl, reacts with the phosphite-borane linkage and causes degradation.

The earliest chemical synthesis of boranephosphonate oligomers yielded 2′-deoxyoligothymidines via the H-phosphonate approach (Scheme 6) [38–40]. Following synthesis of an oligonucleotide, the H-phosphonate linkages were silylated and the resulting phosphite triesters were boronated to obtain the phosphite triester-borane linkage. The silyl groups were then removed with NH_3 which generated boranephosphonate-linked 2′-deoxyoligothymidines **28**. Using this approach, 10–14 mers were successfully prepared in high yields. As this approach has the advantage of not exposing the P (III)-Lewis Acid complex to multiple rounds of strong acids, it was found to be well suited for preparing metallo phosphonate dinucleotides as well (compound **29**) [33]. In place of borane the silyltriester was treated with pentacarbonyltungstate or pentacarbonylmolybdate.

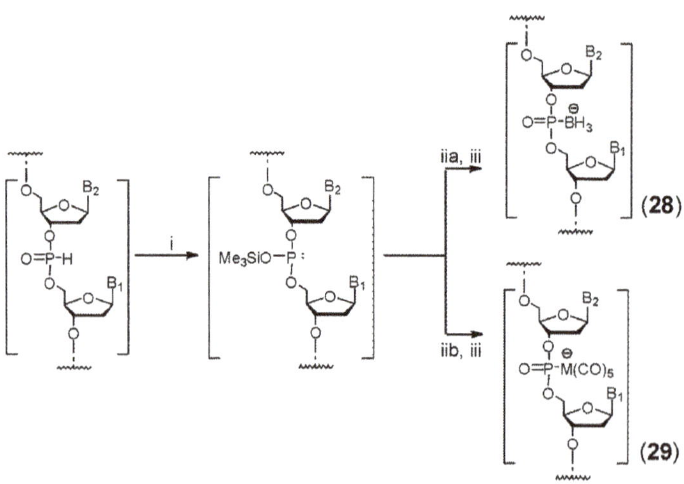

Scheme 6. H-Phosphonate method for synthesis of boranephosphonate DNA and metallophosphonate DNA. (*Reagents*: (i). *N,O*-bistrimethylsilylacetamide/DMF, (iia). BH_3.THF or $BH_3.SMe_2$, (iib). $M(CO)_5$.THF, (iii). aq. NH_3.)

In order to prepare boranephosphonate linked oligomers containing all four nucleobases, Wada has developed the phosphotriester approach [41,42]. In this method a pre-boronated phosphitylating agent (triethylammonium *bis*(2-cyanoethyl) boranophosphate) was coupled to a 5′-dimethoxytrityl-2′-deoxynucleoside to yield

5′-dimethoxytrityl-3′-boranephosphonate-2′-deoxynucleoside diester. Coupling of this synthon to a 2′-deoxynucleoside attached to a support through the 3′-hydroxyl was carried out using 3-nitro-1, 2, 4-triazol-1-yl-*tris*(pyrrolidin-1-yl)phosphonium hexafluorophosphate as an activator. Triethylsilane was added during acid deprotection in order to scavenge trityl cations. The low step-wise coupling yields however limited the utility of this approach.

We have reported an alternative method, adapted from RNA synthesis, that uses 5′-*O*-silyl 2′-deoxynucleosides 3′-phosphoramidites containing *N*-trimethoxytrityl (TMTr)-protected nucleobases [43]. These protecting groups were found to be stable toward reduction by borane. Thus boronation could be performed after each coupling step to obtain the phosphite triester-borane linkage. Removal of the TMTr groups at the end of the synthesis using aqueous acetic acid did not lead to trityl cation induced degradation of the boranephosphite triester linkages. Mixed base 14mer oligonucleotides containing phosphodiester and boranephosphonate diester linkages could be synthesized in good yields. The same strategy was also useful for the preparation of oligomers containing up to six methylphosphineborane modifications within a 16-mer 2′-deoxyoligonucleotide [31]. For borane-phosphonates as well as methylphosphineboranes, attempts to synthesize longer oligomers however led to many degradation products. These results were perhaps due to the instability of these linkages towards iterative treatment with basic fluoride solutions that were required for removal of 5′-silyl protection. In addition DNA synthesizers had to be equipped to handle basic fluoride. This component of the synthesis procedure therefore limited the broad adoption of the methodology.

To circumvent these obstacles we have recently developed *N*-silyl protected 5′-dimethoxytrityl-2′-deoxynucleoside 3′-phosphoramidites (**30a–c**; Scheme 7) [44]. This silyl group (di-*tert*-butylisobutylsilyl) is stable towards boronation as well as all other reagents used in DNA synthesis. These synthons in conjunction with the use of a more efficient trityl scavenger (trimethylphosphite borane; TMPB) allowed us to prepare 24–30 mer bpDNA oligomers in yields similar to those obtained for standard phosphate linked DNA. Moreover these conditions allow synthesis of bpDNA essentially using the same cycle as described in Figure 2 for standard DNA synthesis. The only modification involves replacement of the iodine oxidation reaction with boronation (Step 3). This leads to conversion of the phosphite triester **31** to a trialkylphosphite-borane derivative **33**. If desired, oxidation of **31** with *tert*-butylhydroperoxide can be performed in order to produce the phosphate triester **32**. Thus oligonucleotides having any desired combination of phosphate and boranephosphonate linkages can be synthesized by this procedure.

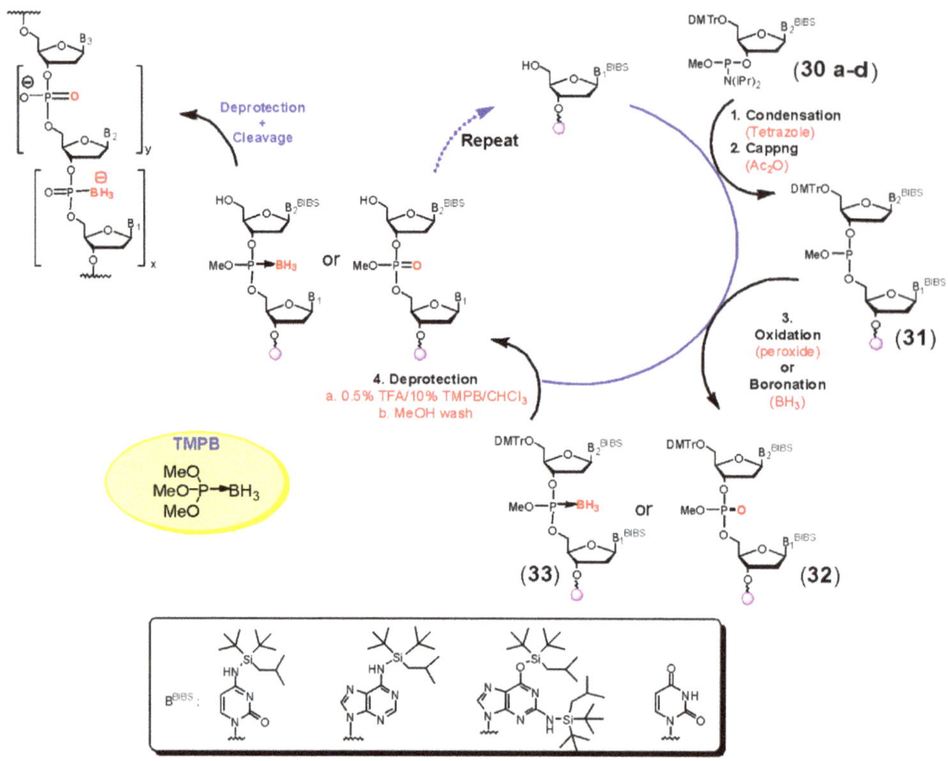

Scheme 7. N-Silyl Protected Phosphoramidites as Synthons for bpDNA.

A future goal is to prepare stereoregular bpDNA and bpRNA oligomers using chemical methods. Such oligomers as demonstrated by the Ramsay-Shaw lab using enzymatically synthesized bpDNA or bpRNA have superior biochemical properties when compared to the diastereomeric mixtures that are currently obtained through chemical synthesis. In this respect a noteworthy report by Iwamoto et al. [45] achieved the stereoregular synthesis of boranephosphonate 2′-deoxyoligothymidines using 3′-*O*-oxazaphospholidine derivatives of 2′-deoxythymidine. A pure stereoisomer of this synthon (compound **34**), when activated, generates stereodefined phosphite triesters **35** that can be converted to boranephosphonates **37** or **38** in a stereospecific manner via an H-phosphonate intermediate **36** (Scheme 8). Taken together these advances raise the possibility of mixed based, stereoregular synthesis of boranephosphonate linked DNA.

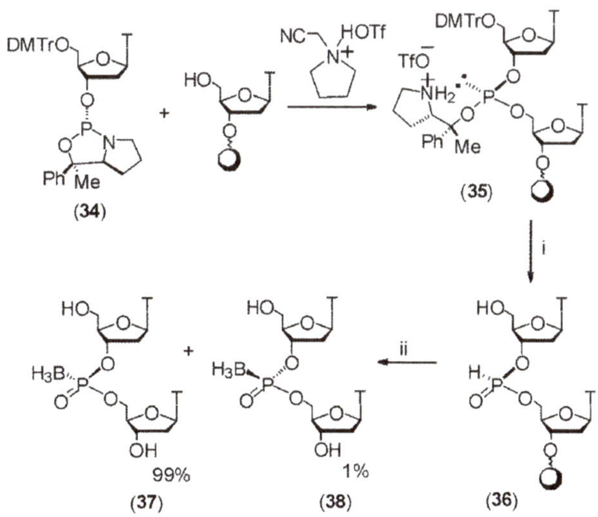

Scheme 8. Synthesis of stereodefined boranephosphonate linked DNA. (*Reagents and Conditions*: (i). 1% Trifluoroacetic acid in CH_2Cl_2. (ii). $BH_3.SMe_2$/*N,O*-bistrimethylsilylacetamide/DMF (1:1:8, v/v/v), RT, 15 min; then sat. NH_3/CH_3OH, RT, 1 h.).

Boranephosphonate diesters posses many properties that are intermediate between P (III) and P (V) compounds. For example they have a ^{31}P-NMR chemical shift of 94 ppm, which lies between 140 ppm for phosphite triesters and 0 ppm for phosphates. These compounds are stable to mild acids (pH > 2) and oxidizing agents (e.g., *tert*-butylhydroperoxide) but react with stronger acids or oxidizers such as iodine [46]. The significantly greater stability of borane-P (III) complexes, relative to analogous nitrogen complexes, stems from the fact that, in addition to donation of the lone pair from the P atom, back donation from the σ obital of the BH_3 group to empty orbitals on the P atom also occurs. The reduced hydridic character of BH_3 is indirect evidence for this explanation. Substituents on phosphorous also affect the strength of the backbonding interaction [47] and thus the overall stability of the P-B bond. Unpublished experiments from our laboratory demonstrate that replacement of the non-bridging oxygen in boranephosphonate diesters by the less electronegative methyl group significantly increases the hydridic properties of the BH_3 group. Systematic evaluation of such factors in the future has the potential to create new DNA derivatives with desired chemical and biochemical properties.

The biochemical properties of boranephosphonate linked nucleic acid oligomers have been investigated. They are stable towards hydrolysis by nucleases [36], are effective substrates for RNaseH and siRNA [48], and are more lipophilic than unmodified DNA. Current work in our laboratory has also revealed that these oligomers can be transfected into HeLa cells in the absence of lipid.

We have recently reported that bpDNA reacts with metal ions that possess a high reduction potential (Ag^+, Au^{3+} and Pt^{2+}) [49]. The products of these reductions are metallic nanoparticles. Investigations of this reaction further revealed that the boranephosphonate linkage is converted into a phosphate diester when carried out in water or to a phosphate triester in simple alcohols. We have taken advantage of this combination of reducing properties and an ability to undergo Watson-Crick base pairing to incorporate bpDNA at specific locations within DNA assemblies. This in turn allowed us to carry out deposition of metal nanoparticles onto these assemblies with high spatial resolution [44].

6. Conclusions

Chemical methods that allow synthesis of DNA/RNA and various derivatives have had tremendous effects on a large number of disciplines. At the heart of all these developments are fundamental discoveries in the chemistry of phosphorous. In particular the reactivity of P (III) compounds led to phosphoramidites as extremely stable, but easily activated synthons for condensation with various nucleophiles to generate DNA or RNA in very high yields. This discovery was a significant milestone, initially in the nucleic acids field, but more recently in other research areas as well. Chemists have also taken advantage of the versatility of phosphorous chemistry in order to create a large number of oligonucleotide derivatives. The majority of these analogs have been developed for use as therapeutics and several such candidates are in advanced clinical trials. In addition DNA derivatives that can combine the binding properties of oligonucleotides with chemically useful functionality are beginning to find applications in new research areas such as nanotechnology and data storage.

The ability to chemically synthesize DNA that is several hundred nucleotides in length is another challenge that awaits chemists. While current methods lead to the synthesis of DNA 20–30 nucleotides in length and more recently to oligomers 150 in length [10] in very high yields and purity, extension to ultra long regimes raises many issues. Thus the field is ripe once again for fresh developments in the chemistry of phosphorous that will allow realization of these goals.

Acknowledgments: Over the years, the Caruthers' laboratory received support from The National Institutes of Health, The National Science Foundation, The American Cancer Society, and The University of Colorado. Acknowledgment for the research contributions of various colleagues can be found in the references.

Conflicts of Interest: The authors declare no conflict of interest.

References

1. Letsinger, R.L.; Lunsford, W.B. Synthesis of thymidine oligonucleotides by phosphite triester intermediates. *J. Am. Chem. Soc.* **1976**, *98*, 3655–3661. [CrossRef] [PubMed]
2. Matteucci, M.D.; Caruthers, M.H. The synthesis of oligodeoxyprimidines on a polymer support. *Tetrahedron Lett.* **1980**, *21*, 719–722. [CrossRef]
3. Matteucci, M.D.; Caruthers, M.H. Studies on nucleotide chemistry iv: Synthesis of deoxyoligonucleotides on a polymer support. *J. Am. Chem. Soc.* **1981**, *103*, 3185–3191. [CrossRef]
4. Letsinger, R.L.; Mahadevan, V. Oligonucleotide synthesis on a polymer support 1,2. *J. Am. Chem. Soc.* **1965**, *87*, 3526–3527. [CrossRef] [PubMed]
5. Vetter, H.J.; Noeth, H. Dialkylaminophosphines. IV. Additions and substitutions on *tris*(Dimethylamino)phosphine. *Chem. Ber.* **1963**, *96*, 1308–1315. [CrossRef]
6. Oertel, G.; Malz, H.; Holtschmidt, H. Addition reactions to Amides of Silicon, Phosphorus, Arsenic, and Sulfur. *Chem. Ber.* **1964**, *97*, 891–902. [CrossRef]
7. Batyeva, E.S.; Al'fonsov, V.A.; Pudovik, A.N. Acid catalysis of derivatives of trivalent phosphorus acids. In *Chemistry of Organophosphorus Compounds*; Mir Publishers: Moscow, Russia, 1989; pp. 68–92.
8. Beaucage, S.L.; Caruthers, M.H. Studies on Nucleotide Chemistry V: Deoxynucleoside Phosphoramidites—A New Class of Key Intermediates for Deoxypolynucleotide Synthesis. *Tetrahedron Lett.* **1981**, *22*, 1859–1862. [CrossRef]
9. McBride, L.J.; Caruthers, M.H. An investigation of several deoxynucleoside phosphoramidites useful for synthesizing deoxyoligonucleotides. *Tetrahedron Lett.* **1983**, *24*, 245–248. [CrossRef]
10. Leproust, E.M.; Peck, B.J.; Spirin, K.; McCuen, H.B.; Moore, B.; Nomsaraev, E.; Caruthers, M.H. Synthesis of high-quality libraries of long (150mer) oligonucleotides by a novel depurination controlled process. *Nucleic Acids Res.* **2010**, *38*, 2522–2540. [CrossRef]
11. Sinha, N.D.; Biernat, J.; Köster, H. β-Cyanoethyl N,N-dialkylamino/N-morpholinomonochloro phosphoamidites, new phosphitylating agents facilitating ease of deprotection and work-up of synthesized oligonucleotides. *Tetrahedron Lett.* **1983**, *24*, 5843–5846. [CrossRef]
12. Nielsen, J.; Caruthers, M.H. Directed Arbuzov-type reactions of 2-cyano-1,1-dimethylethyl deoxynucleoside phosphites. *J. Am. Chem. Soc.* **1988**, *110*, 6275–6276. [CrossRef] [PubMed]
13. Marugg, J.E.; Burik, A.; van der Marel, G.A.; van Boom, J.H.; Nielsen, J.; Dahl, O. (2-Cyano-1,1-dimethylethoxy)bis(diethylamino)phosphine: A convenient reagent for the synthesis of DNA fragments. *Recl. Trav. Chim. Pays-Bas* **1987**, *106*, 72–76. [CrossRef]

14. Corby, N.S.; Kenner, G.W.; Todd, A.R. 704. Nucleotides. Part XVI. Ribonucleoside-5[prime or minute] phosphites. A new method for the preparation of mixed secondary phosphites. *J. Chem Soc.* **1952**, 3669–3675. [CrossRef]
15. Hall, R.H.; Todd, A.; Webb, R.F. 644. Nucleotides. Part XLI. Mixed anhydrides as intermediates in the synthesis of dinucleoside phosphates. *J. Chem. Soc.* **1957**, 3291–3296. [CrossRef]
16. Michelson, A.M.; Todd, A.R. Synthesis of a Dithymidine Dinucleotide Containing a 3′:5′-Internucleotide Linkage. *J. Chem. Soc.* **1955**, 2632–2638. [CrossRef]
17. Ogilvie, K.K.; Nemer, M.J. The synthesis of phosphite analogues of ribonucleotides. *Tetrahedron Lett.* **1980**, *21*, 4145–4148. [CrossRef]
18. Kume, A.; Fujii, M.; Sekine, M.; Hata, T. Acylphosphonates. 4. Synthesis of dithymidine phosphonate. A new method for generation of phosphonate function via aroylphosphonate intermediates. *J. Org. Chem.* **1984**, *49*, 2139–2143. [CrossRef]
19. Garegg, P.J.; Regberg, T.; Stawinski, J.; Stromberg, R. Formation of internucleotidic bonds via phosphonate intermediates. *Chem. Scr.* **1985**, *25*, 280–282.
20. Froehler, B.C.; Matteucci, M.D. Nucleoside H-phosphonates: Valuable intermediates in the synthesis of deoxyoligonucleotides. *Tetrahedron Lett.* **1986**, *27*, 469–472. [CrossRef]
21. Froehler, B.C.; Ng, P.G.; Matteucci, M.D. Synthesis of DNA via deoxynucleoside H-phosphonate intermediates. *Nucleic Acids Res.* **1986**, *14*, 5399–5407. [CrossRef]
22. Garegg, P.J.; Lindh, I.; Regberg, T.; Stawinski, J.; Strömberg, R.; Henrichson, C. Nucleoside H-phosphonates. III. Chemical synthesis of oligodeoxyribonucleotides by the hydrogenphosphonate approach. *Tetrahedron Lett.* **1986**, *27*, 4051–4054. [CrossRef]
23. Garegg, P.J.; Lindh, I.; Regberg, T.; Stawinski, J.; Strömberg, R.; Henrichson, C. Nucleoside H-phosphonates. IV. Automated solid phase synthesis of oligoribonucleotides by the hydrogenphosphonate approach. *Tetrahedron Lett.* **1986**, *27*, 4055–4058. [CrossRef]
24. Froehler, B.; Ng, P.; Matteucci, M. Phosphoramidate analogues of DNA: Synthesis and thermal stability of heteroduplexes. *Nucleic Acids Res.* **1988**, *16*, 4831–4839. [CrossRef]
25. Agrawal, S.; Tang, J.Y. Efficient synthesis of oligoribonucleotide and its phosphorothioate analogue using H-phosphonate approach. *Tetrahedron Lett.* **1990**, *31*, 7541–7544. [CrossRef]
26. Stawinski, J.; Thelin, M. Nucleoside H-phosphonates. 14. Synthesis of nucleoside phosphoroselenoates and phosphorothioselenoates via stereospecific selenization of the corresponding H-phosphonate and H-phosphonothioate diesters with the aid of new selenium-transfer reagent, 3H-1,2. *J. Org. Chem.* **1994**, *59*, 130–136. [CrossRef]
27. Kraszewski, A.; Stawinski, J. *H*-Phosphonates: Versatile synthetic precursors to biologically active phosphorus compounds. *Pure Appl. Chem.* **2007**, *79*, 2217–2227. [CrossRef]
28. Stawinski, J.; Stromberg, R.; Westman, E. Studies on reaction conditions for ribonucleotide synthesis via the H-phosphonate approach. *Nucleic Acids Symp Ser.* **1991**, 228.
29. Sood, A.; Shaw, B.R.; Spielvogel, B.F. Boron-containing nucleic acids. 2. Synthesis of oligodeoxynucleoside boranophosphates. *J. Am. Chem. Soc.* **1990**, *112*, 9000–9001. [CrossRef]
30. Li, P.; Sergueeva, Z.A.; Dobrikov, M.; Shaw, B.R. Nucleoside and oligonucleoside boranophosphates: Chemistry and properties. *Chem. Rev.* **2007**, *107*, 4746–4796. [CrossRef]
31. Krishna, H.; Caruthers, M.H. Solid-phase synthesis, Thermal denaturation studies, Nuclease resistance, and Cellular uptake of (Oligodeoxyribonucleoside)methylborane Phosphine-DNA chimeras. *J. Am. Chem. Soc.* **2011**, *133*, 9844–9854. [CrossRef]
32. Toma, J.M.D.R.; Bergstrom, D.E. Transition Metal Labeling of Oligodeoxyribonucleotides: Synthesis and Characterization of (Pentacarbonyl)tungsten(0) Nucleoside Phosphites. *J. Org. Chem.* **1994**, *59*, 2418–2422. [CrossRef]
33. Pav, O.; Caruthers, M.H. Synthesis of (pentacarbonyl)tungstate(-1) and (pentacarbonyl)molybdate(-1) dinucleotides. *Tetrahedron Lett.* **2009**, *50*, 5015–5017. [CrossRef]
34. Wan, J.; Shaw, B.R. Incorporation of ribonucleoside 5′-(alpha-P-borano)triphosphates into a 20-mer RNA by T7 RNA polymerase. *Nucleosides Nucleotides Nucleic Acids* **2005**, *24*, 943–946. [CrossRef] [PubMed]
35. Li, H.; Porter, K.; Huang, F.; Shaw, B.R. Boron-containing oligodeoxyribonucleotide 14mer duplexes: enzymatic synthesis and melting studies. *Nucleic Acids Res.* **1995**, *23*, 4495–4501. [CrossRef] [PubMed]

36. Porter, K.W.; Briley, J.D.; Shaw, B.R. Direct PCR sequencing with boronated nucleotides. *Nucleic Acids Res.* **1997**, *25*, 1611–1617. [CrossRef] [PubMed]
37. He, K.; Shaw, B.R. Diastereomers of 5′-O-adenosyl 3′-O-uridyl Boranophosphate [Up(BH3)A]: Synthesis and Nuclease Resistant Property. In *Nucleic Acids Symposium Series No. 41, Proceedings of Symposium on RNA Biology III: RNA, Tool and Target, Research Triangle Park, NC, USA, 1999*; Oxford University Press: Oxford, UK, 1999; pp. 99–100.
38. Higson, A.P.; Sierzchala, A.B.; Brummel, H.; Zhao, Z.; Caruthers, M.H. Synthesis of an oligothymidylate containing boranophosphate linkages. *Tetrahedron Lett.* **1998**, *39*, 3899–3902. [CrossRef]
39. Zhang, J.; Terhorst, T.; Matteucci, M.D. Synthesis and hybridization study of a boranophosphate-linked oligothymidine deoxynucleotide. *Tetrahedron Lett.* **1997**, *38*, 4957–4960. [CrossRef]
40. Sergueev, D.S.; Shaw, B.R. H-Phosphonate approach for solid-phase synthesis of oligodeoxyribonucleoside boranophosphates and their characterization. *J. Am. Chem. Soc.* **1998**, *120*, 9417–9427. [CrossRef]
41. Shimizu, M.; Wada, T.; Oka, N.; Saigo, K. A novel method for the synthesis of dinucleoside boranophosphates by a boranophosphotriester method. *J. Org. Chem.* **2004**, *69*, 5261–5268. [CrossRef]
42. Shimizu, M.; Saigo, K.; Wada, T. Solid-phase synthesis of oligodeoxyribonucleoside boranophosphates by the boranophosphotriester method. *J. Org. Chem.* **2006**, *71*, 4262–4269. [CrossRef]
43. McCuen, H.B.; Noe, M.S.; Sierzchala, A.B.; Higson, A.P.; Caruthers, M.H. Synthesis of mixed sequence borane phosphonate DNA. *J. Am. Chem. Soc.* **2006**, *128*, 8138–8139. [CrossRef] [PubMed]
44. Roy, S.; Olesiak, M.; Shang, S.; Caruthers, M.H. Silver nanoassemblies constructed from boranephosphonate DNA. *J. Am. Chem. Soc.* **2013**, *135*, 6234–6241. [CrossRef] [PubMed]
45. Iwamoto, N.; Oka, N.; Sato, T.; Wada, T. Stereocontrolled solid-phase synthesis of oligonucleoside H-phosphonates by an oxazaphospholidine approach. *Angew. Chem. Int. Ed.* **2009**, *48*, 496–499. [CrossRef] [PubMed]
46. Nahum, V.; Fischer, B. Boranophosphate salts as an excellent mimic of phosphate salts: preparation, characterization, and properties. *Eur. J. Inorg. Chem.* **2004**, *2004*, 4124–4131. [CrossRef]
47. Gilheany, D.G. Ylides, phosphoniumno d Orbitals but Walsh Diagrams and Maybe Banana Bonds: Chemical bonding in phosphines, Phosphine oxides, and phosphonium ylides. *Chem. Rev.* **1994**, *94*, 1339–1374. [CrossRef] [PubMed]
48. Hall, A.H.; Wan, J.; Spesock, A.; Sergueeva, Z.; Shaw, B.R.; Alexander, K.A. High potency silencing by single-stranded boranophosphate siRNA. *Nucleic Acids Res.* **2006**, *34*, 2773–2781. [CrossRef]
49. Roy, S.; Olesiak, M.; Padar, P.; McCuen, H.; Caruthers, M.H. Reduction of metal ions by boranephosphonate DNA. *Org. Biomol. Chem.* **2012**, *10*, 9130–9133. [CrossRef]

molecules

Review

Chemistry of Phosphorylated Formaldehyde Derivatives. Part I

Vasily P. Morgalyuk

Laboratory of Organophosphorus Compounds, Nesmeyanov Institute of Organoelement Compounds, Russian Academy of Sciences, Vavilova str., 28, Moscow 119991, Russia; morgaliuk@mail.ru

Received: 8 July 2014; Accepted: 15 August 2014; Published: 25 August 2014

Abstract: The underinvestigated derivatives of unstable phosphorylated formaldehyde acetals and some of the structurally related compounds, such as thioacetals, aminonitriles, aminomethylphosphinoyl compounds, are considered. Separately considered are halogen aminals of phosphorylated formaldehyde, acetals of phosphorylated formaldehyde of H-phosphinate-type and a phosphorylated gem-diol of formaldehyde. Synthetic methods, chemical properties and examples of practical applications are given.

Keywords: phosphorylated acetals; -thioacetals; -aminonitriles; -aminomethylphosphinoyl compounds; -chloro(or bromo)aminals; -gem-dioles; -ketene acetale; H-phosphinate

1. Introduction

Among organophosphorus compounds, α-phosphorylated carbonyl compounds stand out by the capacity of cleavage of phosphorus–carbon bond under mild conditions when reacted with nucleophiles [1–5]. The cleavage of phosphorus–carbon bond may proceed spontaneously as well. At the same time, α-oxoalkylphosphinoyl compounds also retain properties inherent in carbonyl compounds, for example, they undergo cross aldol condensation.

The least stable compounds among them are phosphorylated formaldehyde derivatives, dialkyl formylphosphonates (**1**) [6–11] and *N,N,N′,N′*-tetraalkyl formylphosphondiamides (**2**) [12] (Figure 1), whose existence was even disputed in the first half of the 1970s [13,14].

1 2

Figure 1. Structures of dialkyl formylphosphonates (**1**) and *N,N,N′,N′*-tetraalkyl formylphosphondiamides (**2**).

However, the first synthesis of **1** by reaction of sodium derivatives of dialkyl phosphites with acetic formic anhydride was described in 1974 [15]. Compounds **1** were shown to be unstable and prone to spontaneous degradation [9,10,16,17]; thus, diethyl formylphosphonate (**3**) prepared by the reaction of triethyl phosphite (**4**) with acetic formic anhydride (**5**) at low temperature begins to undergo decarbonylation even at −10 °C [9,10] to give diethyl phosphite (**6**) (Scheme 1).

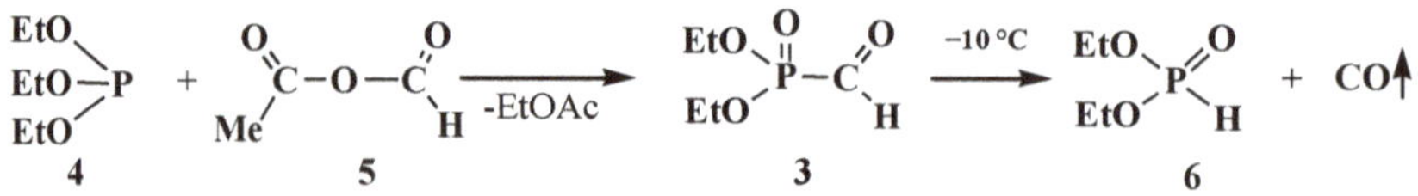

Scheme 1. Syntheses and destruction of diethyl formylphosphonate (**3**).

N,N,N′,N′-Tetraisopropylformylphosphondiamide (**7**) obtained by saponification of *N,N,N′,N′*-tetraisopropyl[(*N″,N″*-diisopropylamino)methylydeniminium]phosphondiamide dichlorophosphate (**8**) with potassium hydroxide in tetrahydrofuran at 20 °C proved to be slightly more stable; it undergoes decarbonylation only above 40 °C [12] to give *N,N,N′,N′*-tetraisopropylphosphonicdiamide (**9**) (Scheme 2).

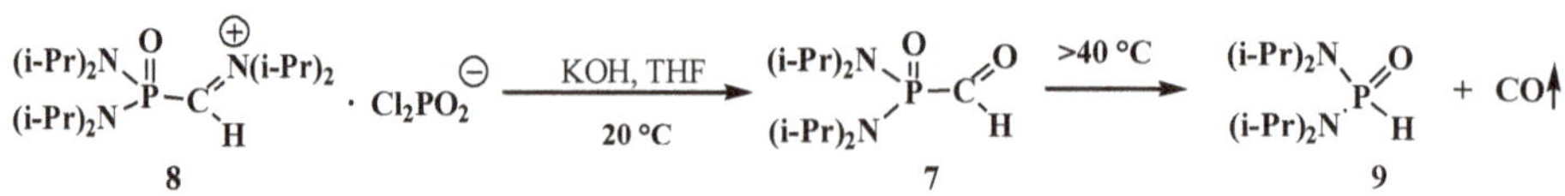

Scheme 2. Syntheses and destruction of *N,N,N′,N′*-tetraisopropyl formylphosphondiamide (**7**).

The stability of the formylphosphondiamide **7** allowed the study some of its chemical properties. It was shown that in its reactions with methylene(triphenyl)phosphorane ($Ph_3P{=}CH_2$) and 2,4-dinitrophenylhydrazine (NH_2NH-DNP) **7** behaves as a typical aldehyde, yielding the corresponding α-phosphorylated olefin **10** and hydrazone **11** [12] (Scheme 3).

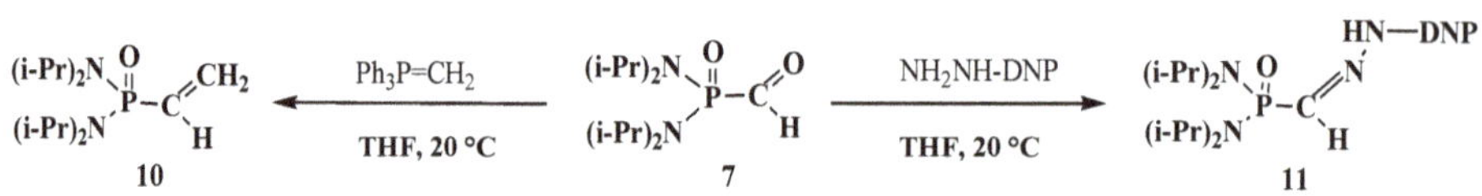

Scheme 3. Reactions *N,N,N′,N′*-tetraisopropylformylphosphondiamide (**7**) as an aldehyde (DNP means a 2,4-dinitrophenyl moiety).

However, many authors showed that formylphosphinoyl compounds are also stable in the form of formylphosphonic acid (**12**) [12,14,15], its disodium salt [15], aldehyde group derivatives such as acetals **13** [18–20] and structurally related compounds. These compounds are thioacetals **14** [21–23], aminonitriles **15** [24–26], diphosphinoyl (*N,N*-dialkylaminomethyl)methanes **16** [27–29], chloro(or bromo)aminals **17** [30–34], mixed S,O-thioacetals **18** [35–37], aminals **19** [30,38], aminoacetals **20** [31,32,38,39], aminothioacetals **21** [32], chloroacetals **22** [40,41], chloro- (or bromo-) thioacetals **23** [40,42,43], α-chlorosulfinyl derivatives **24** [44,45], α-alkoxynitriles **25** [46–48], α-thionitriles **26** [46], α-dihalo derivatives **27** [49–51], α-alkoxydiphosphoryl compounds **28** [41], α-mercaptodiphosphoryl compounds **29** [40,52,53], α-alkoxysilyl derivatives **30** [54], α-aminosilyl derivatives **31** [29,55], or α-mercaptosilyl derivatives **32** [56] (Figure 2):

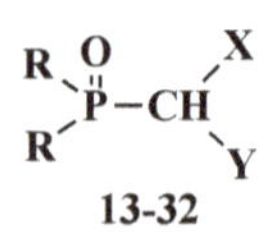

Figure 2. Molecular structure of phosphorylated formaldehyde acetals (**13**) and related compounds (**14–32**), where R = OAlk, Alk_2N, Ph; X, Y are the combinations of OAlk, Alk_2N, CN, $(AlkO)_2P(O)$, S(Alk, Ar), S(O)(Alk, Ar), N(H)C(O)Alk, C(O)Ar, $N(H)S(O)_2Alk$, $S(O)_2Ar$, Hal, Me_3Si [18–56].

Among these compounds, phosphorylated acetals **13**, thioacetals **14**, α-dimethylaminonitriles **15**, aminodiphosphinoyl compounds **16** and chloro- (or bromo-) aminals **17** are used in contemporary organic synthesis. Nonetheless, the phosphorylated acetals of formaldehyde and structurally related compounds remain poorly studied types of organophosphorus compounds until now. The chemical properties of this type of compounds were most studied on an example of dialkyl(dialkoxymethyl)phosphonates **33**. The properties of other compounds **18–32** have been studied in much less detail, and the majority of them are examined in one-sided manner, only as precursors for the synthesis of ketene acetals and similar compounds by the Horner reaction.

To date, it is known that—except for formylphosphonic acid (**12**), phosphorylated acetals **13**, and similar compounds **14–32**—the phosphorylated formaldehyde derivatives are stable in the form of H-phosphinate acetals: alkyl (diethoxymethyl)phosphinate **34** [57–59], and geminal dioles, phosphorylated formaldehyde hydrates (hydrates of phosphorylated formaldehyde) **35** [60–62] (Figure 3).

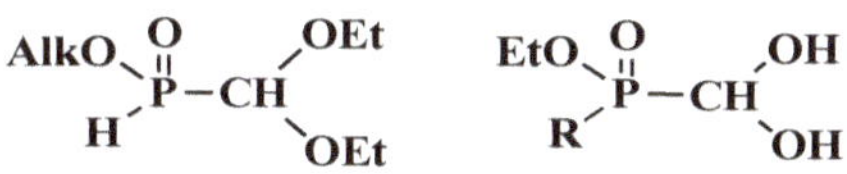

34

35
R = OEt, Ph

Figure 3. Molecular structure of alkyl (diethoxymethyl)phosphinate **34** and phosphorylated formaldehyde hydrates **35**.

Noteworthy are nitrogen-containing analogs of phosphorylated formaldehyde: *N*-substituted imines (**36**) [16,63,64], *N*-alkylnitrones **37** [8,65,66], phosphorylated diazomethane (**38** [67–69] and O-alkylated oximes of diethyl formylphosphonates **39** [7,70], used in contemporary organic synthesis (Figure 4).

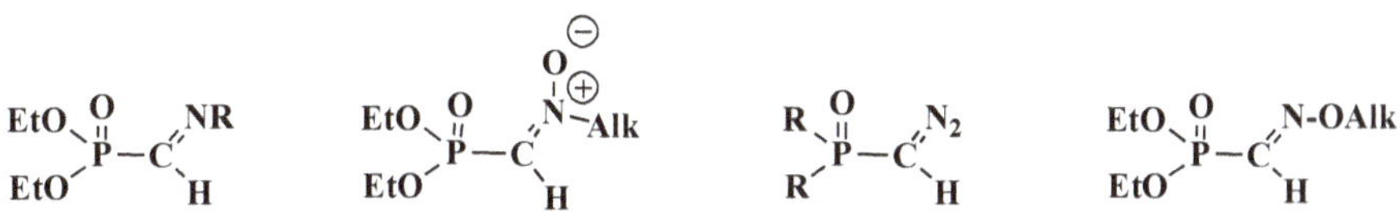

36
R=Alk, AlkC(O), Ar, ArC(O)

37

38
R = OAlk, Ph

39

Figure 4. Molecular structures of formylphosphonates of *N*-substituted imines **36**, N-alkylnitrones **37**, phosphorylated diazomethane **38** and alkylated oximes **39**.

However, the properties of compounds **36–39** differ significantly from those of compounds **13–32**. Therefore, compounds **36–39** will be considered in a separate publication.

2. Chemistry of Phosphorylated Formaldehyde Derivatives

2.1. *Syntheses and Chemical Properties of Formylphosphonic Acid* (**12**)

Formylphosphonic acid (**12**), which shows distinct antiviral activity [71] was obtained for the first time in 1974 [14] as a byproduct of the electrolysis of an aqueous solution of nitrilotrimethylphosphonic acid (**40**) (Scheme 4). Later it was shown that **12** might be also obtained in high yields by the catalytic oxidation of nitrilotrimethylphosphonic acid (**40**) with hydrogen peroxide in the presence of vanadyl sulfate or potassium tungstate (88–93% yield) [17] or activated carbon (up to 82% yield) [72]. Compound **12** can be also prepared by the oxidation of hydroxymethylphosphonic acid (**41**) in aqueous solution with air oxygen or hydrogen peroxide in the presence of Raney copper in a yield up to 65% [73] (Scheme 4).

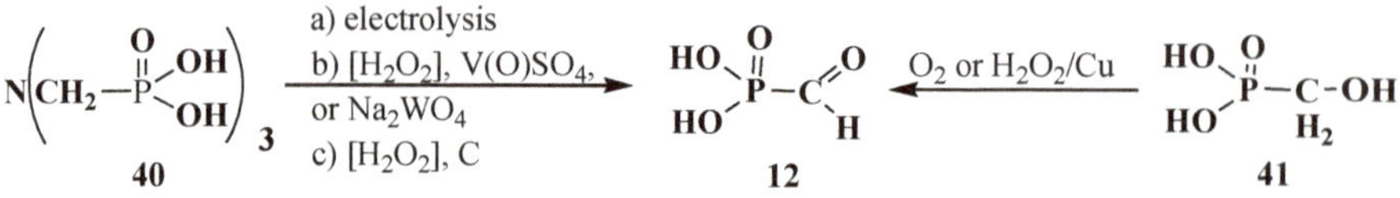

Scheme 4. Possible routes of syntheses of formylphosphonic acid (**12**).

In a third method of synthesis, **12** was obtained via a two-stage synthesis starting from dimethyl (dimethoxymethyl)phosphonate (**42**), which was converted by treatment with bromotrimethylsilane (**43**) into bis(trimethylsilyl)(dimethoxymethyl)phosphonate (**44**). Hydrolysis of the latter resulted in formylphosphonic acid (**12**) [74] (Scheme 5). See also Section 3.2.—Synthesis of formacetalphosphonic acids.

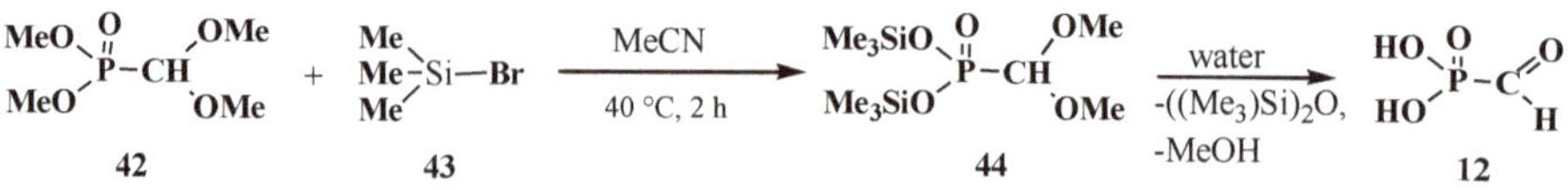

Scheme 5. Synthesis of formylphosphonic acid (**12**) from dimethyl (dimethoxymethyl)phosphonate (**42**).

However, the chemical properties of formylphosphonic acid (**12**) have been poorly studied until now. It is known that heating **12** in acidic medium leads to the cleavage of the P–C bond to form phosphoric and formic acids **45** and **46**. In the presence of bases **12** undergoes disproportionation (Cannizzaro reaction) to give hydroxymethylphosphonic **41** and carboxyphosphonic acids **47**. Acid **47** undergoes rapid decarboxylation on acidification of the reaction medium [14] (Scheme 6).

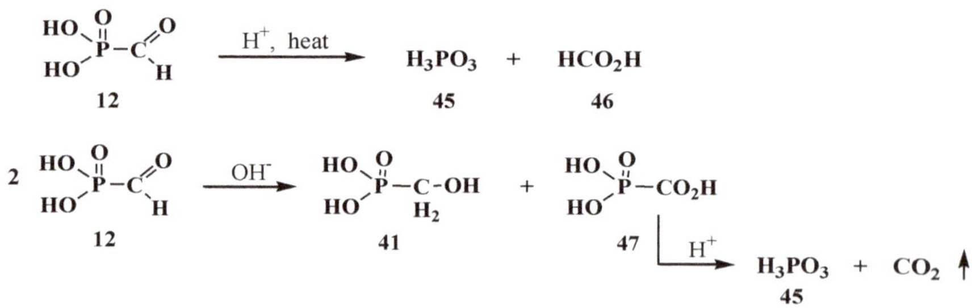

Scheme 6. Degradation of formylphosphonic acid (**12**) to form phosphoric **45** and formic acids **46** and disproportionation of **12** in aqueous solutions.

2.2. *Chemistry of Phosphorylated Formaldehyde Acetals* 13

Chemistry of phosphorylated formaldehyde acetals **13** started as a chemistry of dialkyl (dialkoxymethyl)phosphonates **33** due to their more ready availability as compared with the analogs—*N,N,N′,N′*-tetraalkyl(dialkoxymethyl)phosphondiamides **48**, dialkyl (dialkoxymethyl)-phosphine oxides **49** or diphenyl(dialkoxymethyl)phosphine oxide **50**. Many types of acetals are known to date, but their chemical properties are still insufficiently studied, although they are more studied than the other derivatives of phosphorylated formaldehyde.

2.2.1. Methods of Synthesis of Phosphorylated Formaldehyde Acetals 13

First phosphorylated formaldehyde acetals **13** were obtained by the reaction of hydrophosphinoyl compounds **51–53** with orthoformate esters **54** on heating. The reaction with dialkyl phosphites **51** [19,75] is conducted by heating to 182 °C [76,77] or at 60 °C in the presence of $BF_3{\cdot}Et_2O$ as catalyst (without the catalyst the yields decrease from 69–90% to 25% [75]). In the case of *sec*-phosphine oxides—dialkylphosphine oxides **52** [77] and diphenylphosphine oxide **53** [77,78], the reaction proceeds at 100 °C [19,77]. This method provides high yields of compounds **33, 49** and **50** (Scheme 7).

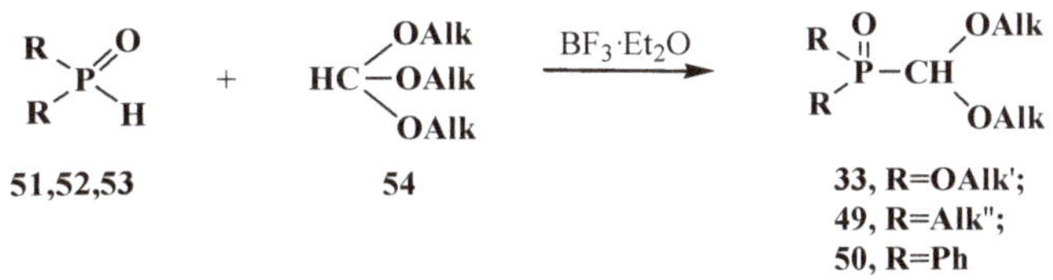

Scheme 7. Synthesis of phosphorylated formaldehyde acetals **33, 49, 50** from dialkyl phosphites **51** or *sec*-phosphine oxides **52, 53** by means of orthoformate esters **54**.

A variant of this method consists in the reaction of phosphoric acid (H_3PO_3, **45**) or phosphoric anhydride (**55**) with triethyl orthoformate (**56**) in 1:3 ratio that results in diethyl (diethoxymethyl)phosphonate (**57**) with minimal effort [79] (Scheme 8).

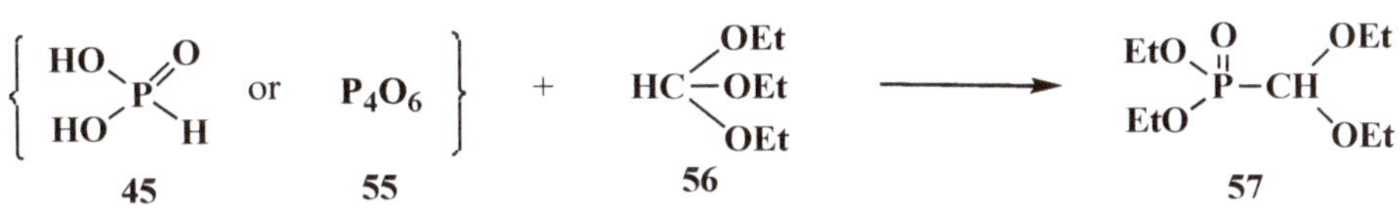

Scheme 8. Synthesis of diethyl (diethoxymethyl)phosphonate (**57**) from phosphoric acid (**45**) or phosphorous anhydride (**55**).

Compounds **33** can be also prepared by the reaction of trialkyl phosphites **58** with orthoformate esters **54**. It was shown that phosphites **58** do not react directly with **54** even on heating [19]. However, as in the case of dialkyl phosphites **51**, orthoformates **54** react with trialkyl phosphites **58** and their analogs $(AlkO)_2POX$ **59**, where X = Me_3Si, $(AlkO)_2P$, $(AlkO_2)P(O)$ on heating in the presence of catalytic amounts of boron trifluoride etherate $BF_3{\cdot}Et_2O$ [19]. The reaction of **58** with orthoformate esters **54** is also possible in the presence of phosphorus trichloride (PCl_3, **60**) or diethoxychloromethane $(EtO)_2C(H)Cl$ [80,81] (Scheme 9).

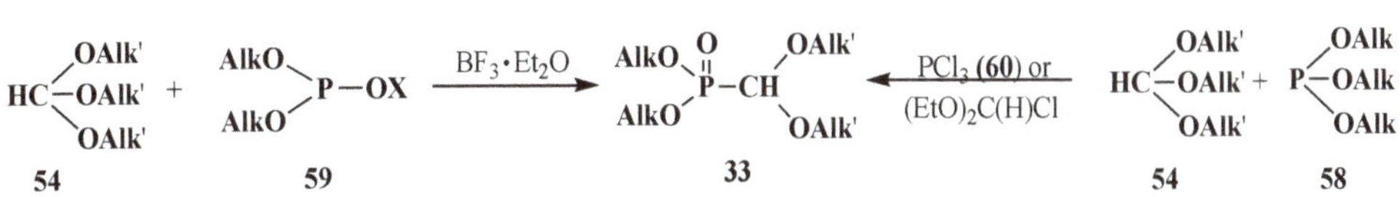

Scheme 9. Synthesis of dialkyl (dialkoxymethyl)phosphonates **33** from orthoformate esters **54**, trialkyl phosphites **58** and their analogs $(AlkO)_2POX$ **59**.

It was reported that the reaction of trialkyl phosphites **58** with acetoxy(diethoxy)methane (**61**), which is more reactive derivative than orthoformate esters **54** [19,82], results in dialkyl (diethoxymethyl)phosphonates **62** (Scheme 10).

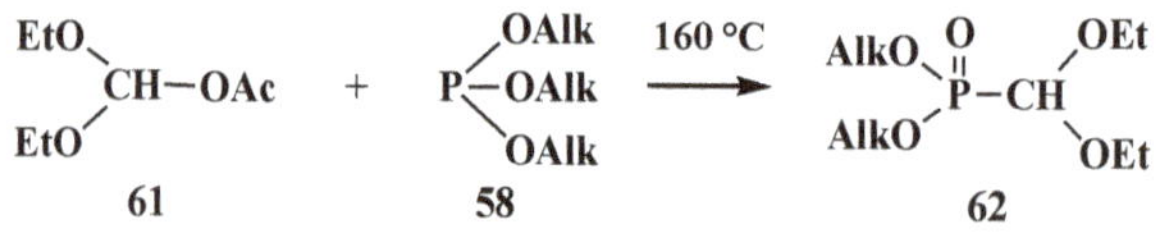

Scheme 10. Synthesis of dialkyl (diethoxymethyl)phosphonates **62** from orthoformate esters **54** and acetoxy(diethoxy)methane (**61**).

The most general method of synthesis of phosphorylated formaldehyde acetals **13** is the reaction of phosphorus trichloride derivatives **60** with orthoformate esters **54**. Excess of **54** reacts on heating with **60** [79,81,83–85] and mono- and dialkyl $AlkPCl_2$ **63**, Alk_2PCl **64**, alkoxy $(AlkO)PCl_2$ **65**, $(AlkO)_2PCl$ **66**, and phenyl $PhPCl_2$ **67**, Ph_2PCl **68** [80,84,86–89] derivatives of **60** (Scheme 11), to form symmetrical and unsymmetrical acetals **33**, **49**, **50** and **69**, **70**, respectively. It was noted that the reaction of 2-chloro-1,2,3-dioxaphospholanes **71** with **54** leads to the opening of the dioxaphospholane ring to form ethyl (β-chloroethyl) ([1,3]-dioxolan-2-yl)phosphonates **72** [90].

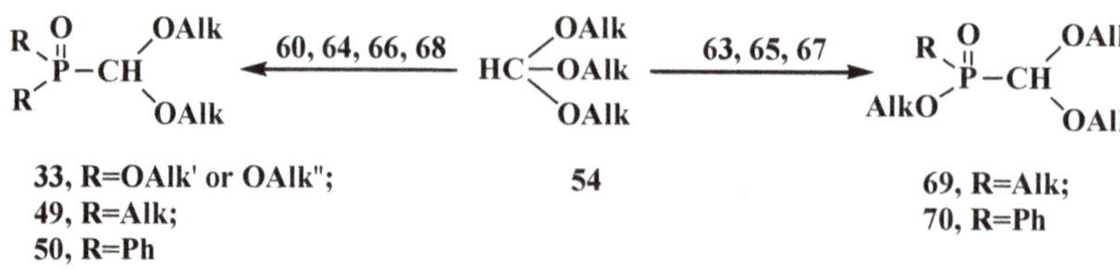

Scheme 11. Syntheses of phosphorylated formaldehyde acetals **33**, **49**, **50**, **69**, **70** from phosphorus trichloride derivatives **63–68** with orthoformate esters **54**. See the text above.

The reaction of *N,N,N′,N′*-tetraalkyl(chloro)phosphindiamides **73** with orthoformates **54** proceeds with a partial exchange of dialkylamino groups at phosphorus atom of **73** to give both *N,N,N′,N′*-tetraalkyl (dialkoxymethyl)phosphondiamides **48** and mixed phosphonamidates **74** [19,91,92]. In the case of *N,N*-dialkyl(dichloro)phosphinamides **75**, a total exchange of dialkylamino groups at the phosphorus atom for alkoxy groups takes place to yield dialkyl (dialkoxymethyl)phosphonates **33** [85] (Scheme 12).

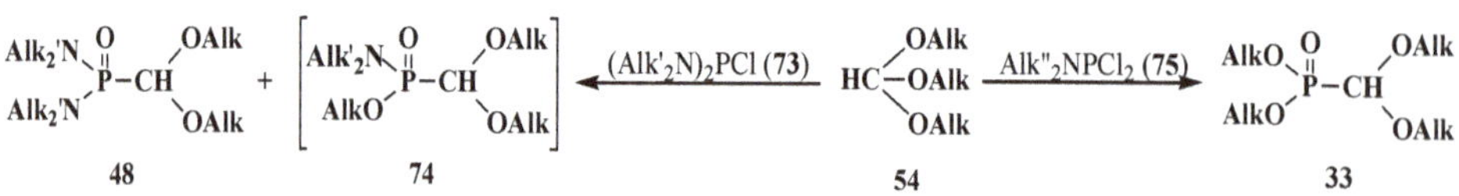

Scheme 12. Interaction of *N,N,N′,N′*-tetraalkyl(chloro)phosphindiamides **73** and *N,N*-dialkyl(dichloro)phosphinamides **75** with orthoformates **54**.

Quaternary ammonium salts of dimethylformamide acetals can be used instead of orthoformates **54** [18,93]. The method only allows preparation of cyclic phosphorylated formaldehyde dialkyl acetals **76** since the quaternary ammonium salts of dialkylformamide acetals of linear structure are unstable and undergo fast degradation [93] (Scheme 13).

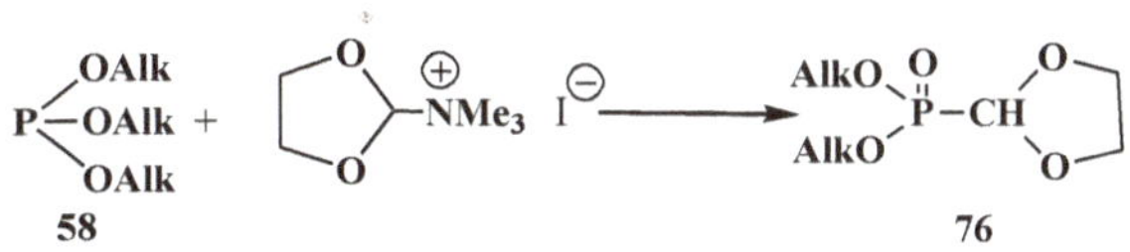

Scheme 13. Synthesis of cyclic acetals **76** by interaction of trialkyl phosphites **58** with the quaternary ammonium salts of dimethylformamide acetals.

Nonetheless, linear compounds **33** can be obtained by this method in low yield (15–25%) using a one-pot method from methyl iodide, dialkylformamide acetals **77**, and trialkyl phosphites **58** [18].

The reaction of phosphorus trichloride (**60**) with dimethylformamide dimethylacetal (**78**) also leads to the formation of dimethyl (dimethoxymethyl)phosphonate (**42**) (along with a certain amount of tetramethyl (*N,N*-dimethylaminomethyl)diphosphonate (**79**)) [83] (Scheme 14).

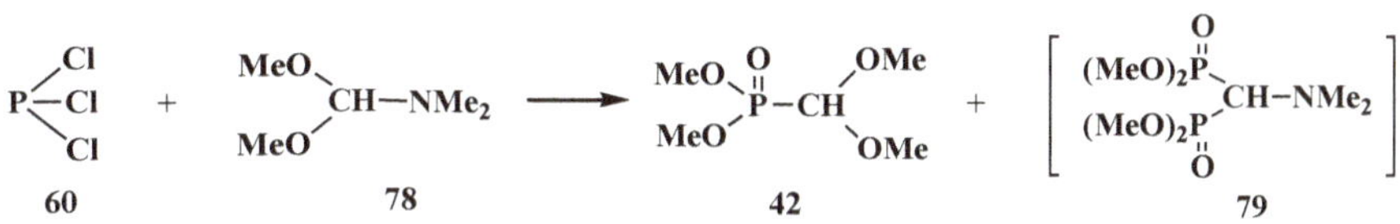

Scheme 14. Syntheses of dimethyl (dimethoxymethyl)phosphonate (**42**) from phosphorus trichloride (**60**) and dimethylformamide dimethylacetal (**78**).

Dialkyl acetals **33** can be also prepared by the reaction of alcohols with tetraalkyl bis(dialkoxymethyl)pyrophosphonates **80**. The method provides a possibility to obtain formaldehyde dialkyl acetals with different alkoxy substituents at phosphorus atom **81** [89] that are difficult to prepare from trivalent phosphorus derivatives and orthoformates **54** or their derivatives because of competitive exchange of substituents at phosphorus atom [19,89] (Scheme 15). Alkyl (dialkoxymethyl)phosphonic acid **82** is also formed at the same time. See also Schemes 22 and 23.

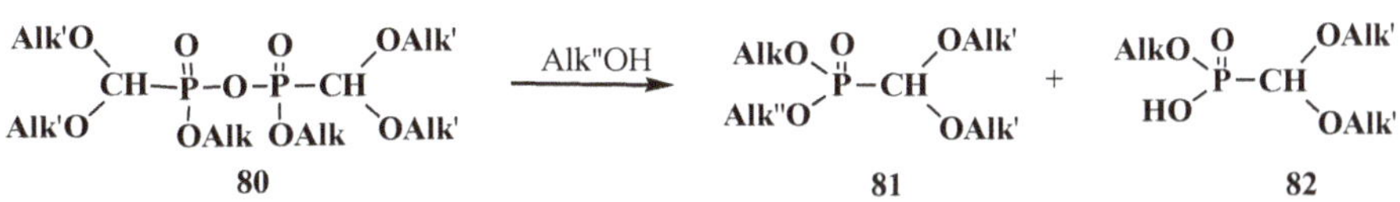

Scheme 15. Preparation of acetals **81** with different alkoxy substituents at phosphorus atom by means of alcoholysis of tetraalkyl bis(dialkoxymethyl)pyrophosphonates **80**.

Alkyl alkyl(dialkoxymethyl)phosphinates **83** unsymmetrically substituted at the phosphorus atom were synthesized by the Arbuzov reaction of dialkyl (dialkoxymethyl)phosphonites **84** with alkyl iodides [82,84], for example see Scheme 16.

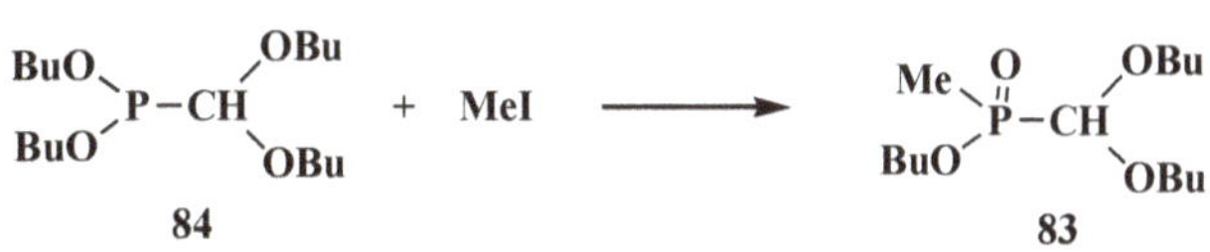

Scheme 16. Example synthesis of acetals with different substitutients at phosphorus atom by the Arbuzov reaction. See also Scheme 11, Scheme 15, Scheme 17, Schemes 21–23 and 111–114.

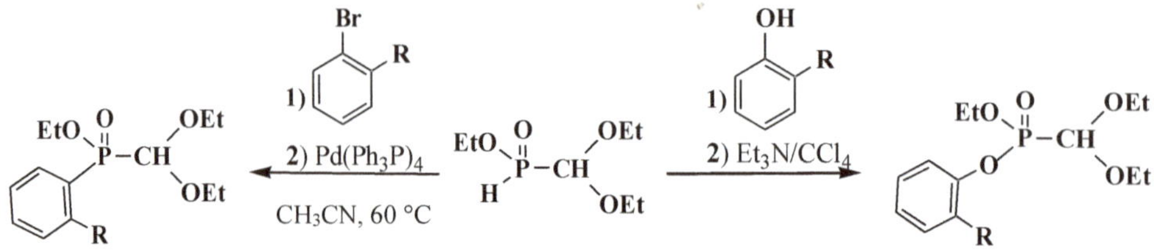

85 **86** **87**

Scheme 17. Synthesis of ethyl aryl(diethoxymethyl)phosphinates **85** and ethyl aryl (diethoxymethyl)phosphonates **87** from ethyl (diethoxymethyl)phosphinate (**86**), R are Alk or Hal.

The catalytic synthesis of ethyl aryl(diethoxymethyl)phosphinates **85** by the reaction of ethyl (diethoxymethyl)phosphinate (**86**) with *ortho*-substituted aryl bromides in the presence of tetrakis(triphenylphosphine)palladium(0) $Pd(Ph_3P)_4$ in 75–90% yields was reported in 1995 [94]. The Todd-Atherton reaction of **86** with *ortho*-substituted phenols in the presence of triethylamine at 0–23 °C leads to ethyl aryl (diethoxymethyl)phosphonates **87** [94] (Scheme 17).

The attempted preparations of novel acetales **33** by the transesterification of phosphorus ester groups failed because they gave rise to intractable mixtures of compounds [95]. Ethoxyphosphoryl groups in acetal **57** were replaced by butoxyphosphoryl groups only under cathode electrolysis conditions. As a result, butyl ethyl (diethoxymethyl)phosphonate (**88**) and dibutyl (diethoxymethyl)phosphonate (**89**) were obtained in low yields—11% and 5%, respectively [96] (Scheme 18).

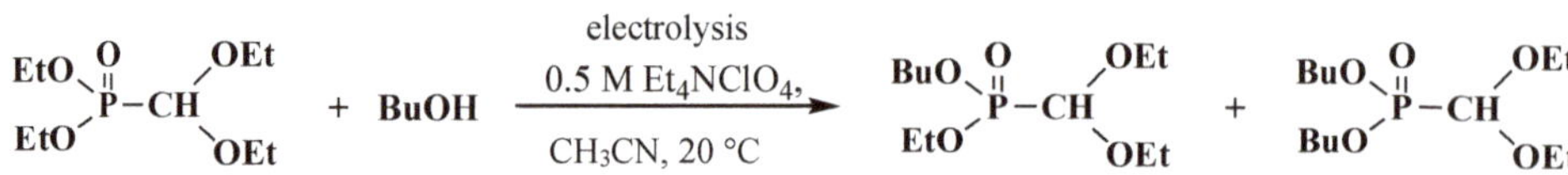

57 **88** **89**

Scheme 18. Synthesis of acetales **88** and **89** by transesterification under the conditions of electrolysis.

However, the heating of **57** with 1,3-dimethylpropanediols in the presence of a catalytic amount of benzenesulfonic acid ($PhSO_3H$) results in the replacement of ethoxy groups of the acetal fragment to give cyclic diethyl (5-dimethyl-[1,3]-dioxan-2-yl)phosphonates **90** [97], for example, see Scheme 19.

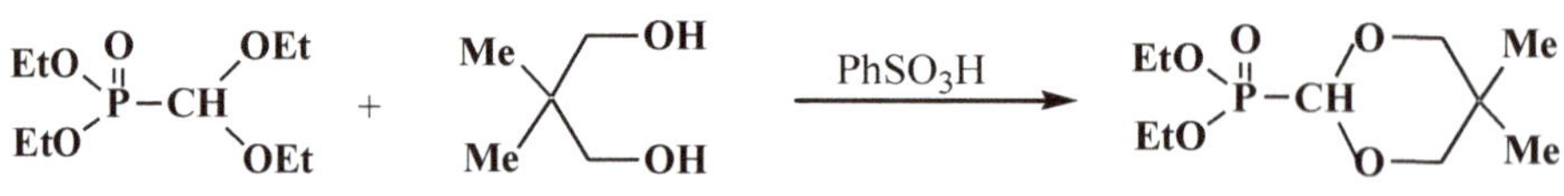

57 **90**

Scheme 19. Preparation of cyclic diethyl (5-dimethyl-[1,3]-dioxan-2-yl)phosphonates **90** by the transesterification under acid catalysis.

2.2.2. Chemical Properties of Phosphorylated Formaldehyde Acetals 13

The chemistry of phosphorylated formaldehyde acetals **13** was initially developed for the most part as a chemistry of available dialkyl (dialkoxymethyl)phosphonates **33**. Therefore the properties of acetals as a separate type of organophosphorus compounds were studied mainly by the examples of compounds **33** whose reactivity is affected by the presence of both phosphorus ester and acetal groups.

Hydrolysis of acetals **33** was studied by the example of compound **57** and a compound with aromatic substituents in the acetal group, diethyl (5,6-dichloro-1,3-benzodioxomethyl)phosphonate

(**91**). However, the attempted acid hydrolysis of acetal **57** on heating lead to the cleavage of phosphorus–carbon bond [77,95,98]. Compound **91** underwent acid hydrolysis on heating to give (5,6-dichloro-1,3-benzodioxomethyl)phosphonic acid (**92**) (Scheme 20). See also the section "Cleavage of Phosphorus–Carbon Bond under the Action of Acids and Acidic Reagents", Scheme 54.

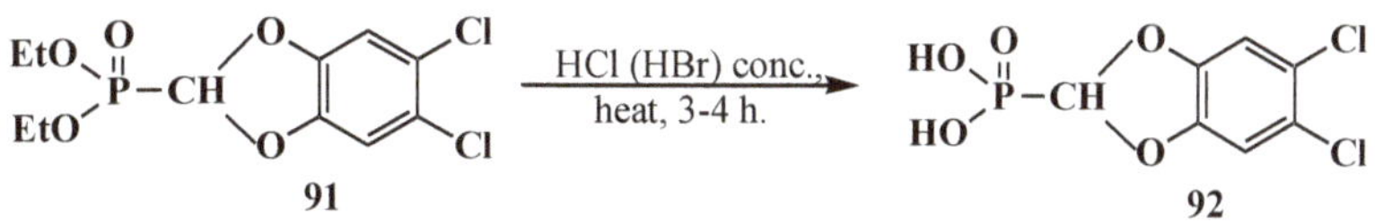

Scheme 20. Acid hydrolysis of diethyl (5,6-dichloro-1,3-benzodioxomethyl)phosphonate (**91**).

The heating of a solution of diethyl (dialkoxymethyl)phosphonate **93** in absolute ethanol with sodium ethoxide (NaOEt) leads to dealkylation of one of the ethoxy groups by phosphorus atom to form ethyl sodium (dialkoxymethyl)phosphonate **94**, which produces the free acid **95** on acidification [95]. Heating of **93** with sodium iodide NaI leads to the same result [93]. The reaction of ethyl sodium (diethoxymethyl)phosphonate (**96**) with electrophilic reagents brings about the formation of phosphonates **97** with different substituents at the phosphorus atom [95] (Scheme 21).

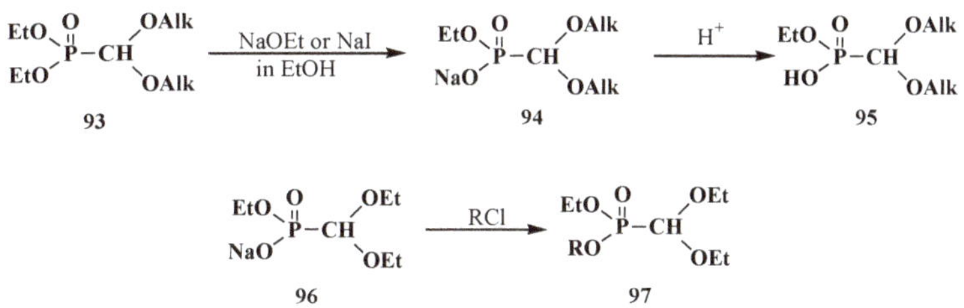

Scheme 21. Acetals **93** dealkylation at their interaction with sodium ethoxide or sodium iodide. Synthesis of acetals **97** with different substituents at phosphorus atom, where R = Alk, Ac, Me_3Si, $MeOCH_2$, $ArOCH_2$.

Acids **95** show typical properties of hydroxy compounds. Using as example ethyl (diethoxymethyl)phosphonic acid (**98**) it is shown that they react with diazomethane and thionyl chloride. The reaction products are ethyl methyl (diethoxymethyl)phosphonate (**99**) and ethyl (diethoxymethyl)phosphonic chloride (**100**) [95], which reacts with phenylmagnesium bromide in tetrahydrofuran to produce ethyl phenyl(diethoxymethyl)phosphinate (**101**) (Scheme 22).

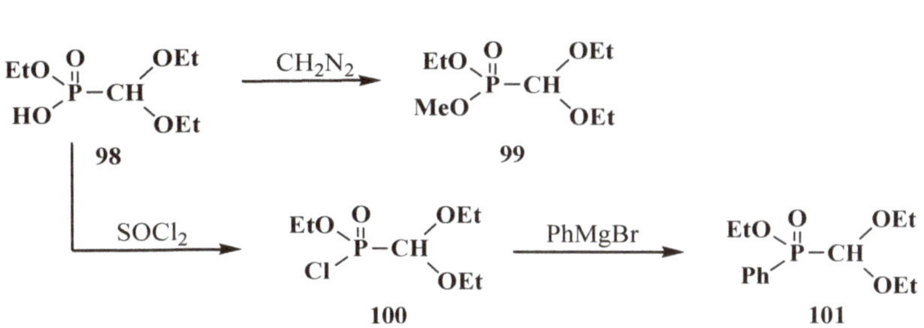

Scheme 22. Ethyl (diethoxymethyl)phosphonic acid (**98**) transformations.

Unsymmetrical ethyl aryl (diethoxymethyl)phosphonates **87** containing aryloxy substituents at the phosphorus atom [94] undergo rearrangement in the presence of equimolar amount of lithium diisopropylamide LDA in tetrahydrofuran at −70 °C to yield ethyl (2-hydroxyaryl)-(diethoxymethyl)phosphinates **102** (Scheme 23).

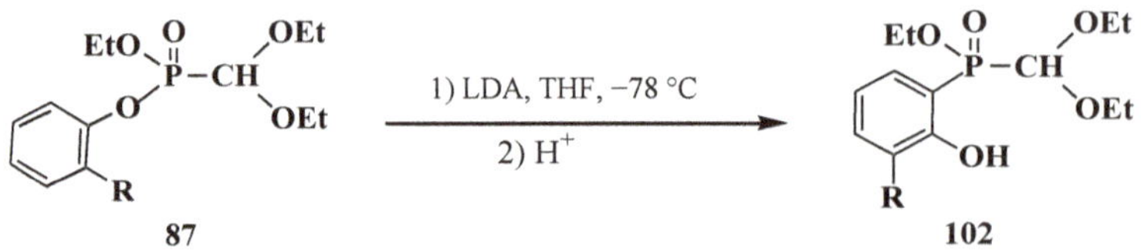

Scheme 23. Rearrangement of compounds **87**, R = Alk, Hal.

Heating of diethyl (diethoxymethyl)phosphonate (**57**) with a catalytic amount of $BF_3{\cdot}Et_2O$ gives rise to formation of tetraethyl (ethoxymethyl)diphosphonate (**103**) in a 14% yield [19] (Scheme 24).

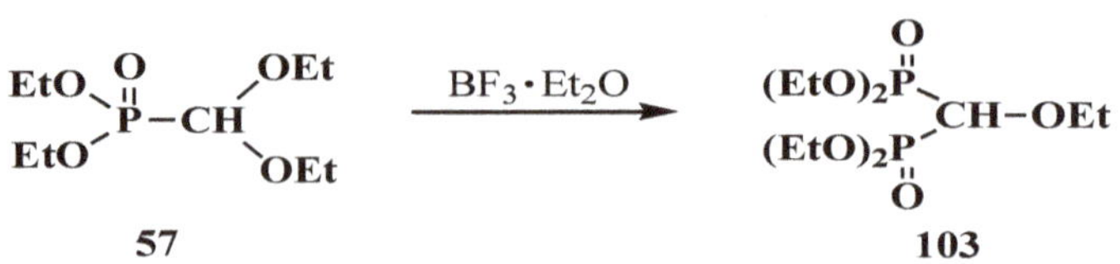

Scheme 24. Formation of tetraethyl (ethoxymethyl)diphosphonate (**103**) from acetal **57** under a catalysis by $BF_3{\cdot}Et_2O$.

The acetal group of compounds **13** is rather stable to the action of co-reactants. Nonetheless, a series of transformations of phosphorylated formaldehyde acetals (**33**) that involves dialkoxyacetal group is described.

The halogenation of compounds **33** with *N*-bromosuccinimide leads to alcoxycarbonylphosphonates **104**. Five-membered cyclic acetals **76** by halogenation with *N*-chlorosuccinimide and azodiisobutyro-nitrile mix produce β-haloethoxycarbonylphosphonates **105** [99]. Reactions proceed via a radical mechanism (Scheme 25).

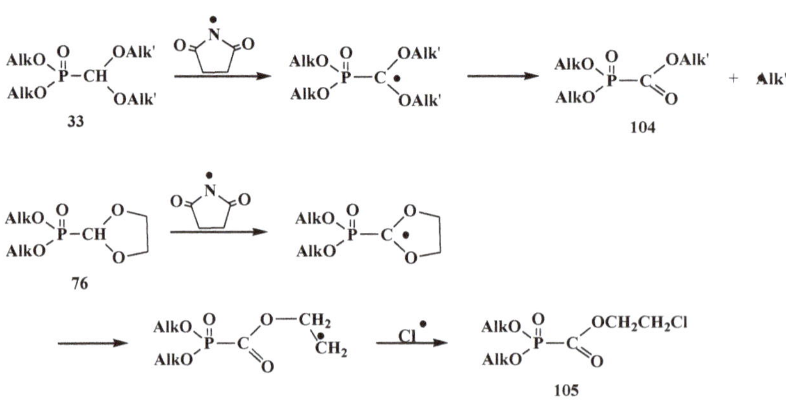

Scheme 25. Interaction of compounds **33** and **76** with N-bromosuccinimide or N-chlorosuccinimide.

The reaction of **57** with o-aminophenol in oxygen flow at 160 °C results in diethyl (2-benzoxazolyl)phosphonate (**106**) [100] (Scheme 26).

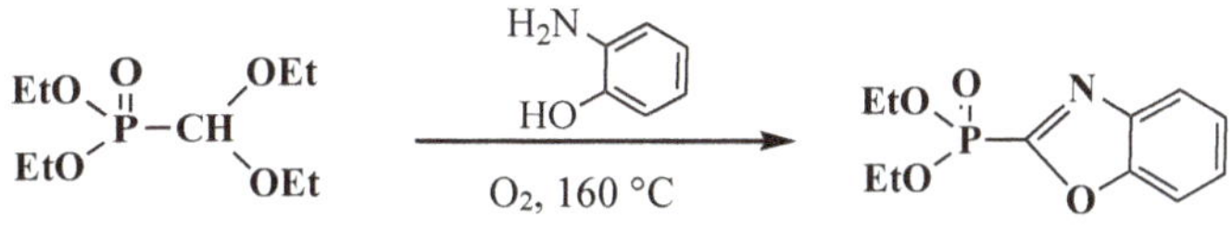

57 106

Scheme 26. The reaction of diethyl (diethoxymethyl)phosphonate (57) with o-aminophenol.

When compound 33 reacted with thiols in a 1:1 mixture of acetic and hydrochloric acid at 0 °C, dialkyl (dialkylthiomethyl)phosphonates 107 were obtained in 61–64% yield [21] (Scheme 27).

33 107

Scheme 27. Transformation of compounds 33 into dialkyl (dialkylthiomethyl)phosphonates 107.

Diethyl (diethoxymethyl)phosphonate (57) reacts with titanium tetrachloride $TiCl_4$ or tetrabromide $TiBr_4$ in diethyl ether to give diethyl [(ethoxy)chloromethyl]phosphonate (108) or diethyl [(ethoxy)bromomethyl]phosphonate (109) [61] (Scheme 28).

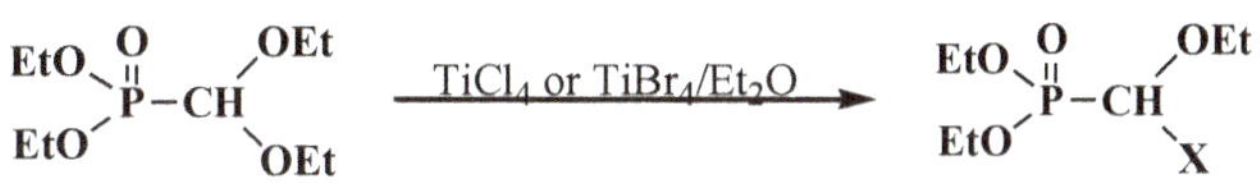

57 108, X=Cl 109, X=Br

Scheme 28. Syntheses of phosphorylated halogenacetals 108, 109 from diethyl (diethoxymethyl)phosphonate (57).

Cyclic diisopropyl [(4-bromomethyl-[1,3]-dioxolan)-2-yl]phosphonates 110 were used in the synthesis of analogs of natural purine and pyrimidine nucleotides [101]. Guanine analog 111 was prepared by the scheme (Scheme 29).

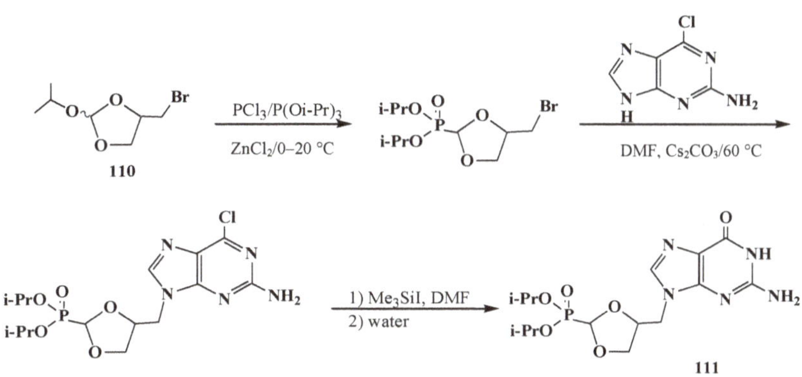

Scheme 29. Synthesis of [1,3]-dioxolane analog guanine nucleotide 111.

The corresponding uracil analog was obtained in a similar manner. Both compounds showed moderate activity in vitro toward human cytomegalovirus [101]. See also Scheme 16, Scheme 18, Scheme 19, Schemes 54, 57, 58, 60, 62, 86–88 and 90–92.

2.3. *Phosphorylated Formaldehyde Thioacetals* **14**

The first syntheses of compounds **14**, linear dialkyl (dialkylthiomethyl)phosphonates **107** or dialkyl (diphenylthiomethyl)phosphonates **112** [21,22], cyclic dialkyl ([1,3]-dithiolan-2-yl)phosphonates **113**, dialkyl ([1,3]-dithian-2-yl)phosphonates **114** [102] and dialkyl (1,3-benzodithiolylmethyl)phosphonates **115** [103] (Figure 5) were reported almost simultaneously in 1976–1977.

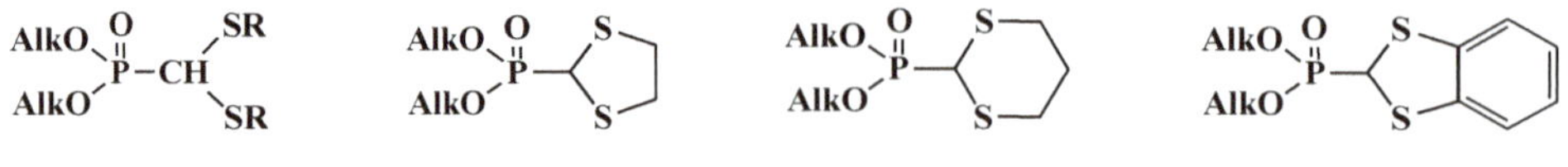

107 (R=Alk), **112** (R=Ph) **113** **114** **115**

Figure 5. First representatives of the phosphorylated formaldehyde thioacetals **107, 112–115**.

2.3.1. Methods of Synthesis of Phosphorylated Formaldehyde Thioacetals **14**

The first syntheses of linear and cyclic dialkyl (dialkylthiomethyl)phosphonates **107, 112–114** were performed by the analogy with the syntheses of dialkyl (dialkoxymethyl)phosphonates **33** (Schemes 9 and 13), namely, by the Arbuzov reaction of trialkyl phosphites **58** with linear (dialkylthio)chloromethanes **116** or (diphenylthio)chloromethane (**117**) [21,22], and cyclic 2-chloro-1,3-dithiolane [22,102] (**118**) or 2-chloro-1,3-dithiane (**119**) with 93–95% yields (or quaternary ammonium salts of dimethylformamide thioacetal (only for **113**), 50–85%) [21,22] (Scheme 30).

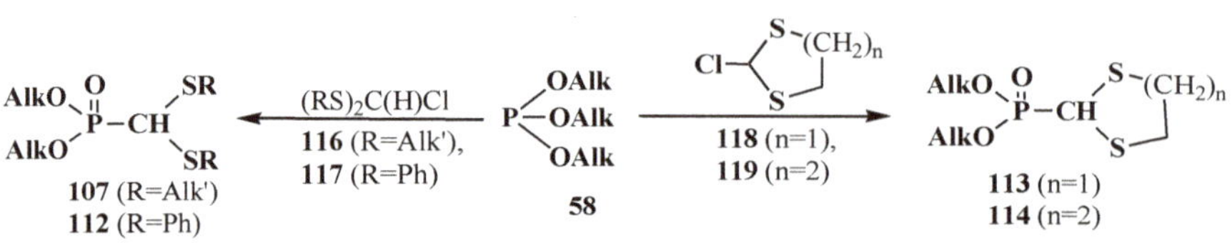

Scheme 30. The first syntheses of linear **107, 112** and cyclic **113, 114** phosphorylated formaldehyde thioacetals.

At present, the method is used for the synthesis of cyclic dialkyl (dialkylthiomethyl)phosphonates **113, 114**, because initial 2-chloro-1,3-dithiolane (**118**) and 2-chloro-1,3-dithiane (**119**) are readily prepared by the reaction of 1,3-dithiolane and 1,3-dithiane with *N*-chlorosuccinimide [102,104–106]. The method allows one to obtain in good yields both cyclic dialkyl (dialkylthiomethyl)phosphonates **113, 114** (55–96%), [102,104,106]) and diphenyl(dialkylthiomethyl)phosphine oxides **121, 122** (57–85%), [56,104, 105]) (Scheme 31).

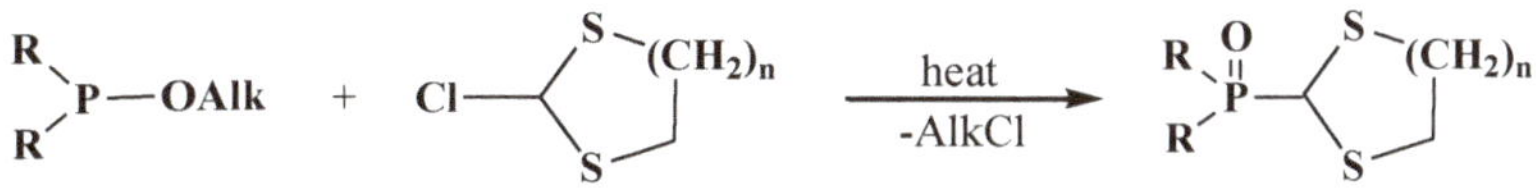

58 (R=OAlk)
121 (R=Ph)

118 (n=1),
119 (n=2)

113, n=1; R=OAlk;
114, n=2; R=OAlk;
121, n=1; R=Ph;
122, n=2; R=Ph

Scheme 31. Syntheses of cyclic dialkyl (dialkylthiomethyl)phosphonates **113**, **114** and diphenyl(dialkylthiomethyl)phosphine oxides **121**, **122**.

However, the poor availability of linear dialkylthiochloromethanes [21] confined the use of this method of synthesis of phosphorylated thioacetals **14** and stimulated the search for alternative methods for their synthesis.

In 1979, a new two-step method was proposed for the synthesis of unsymmetrical diethyl (dialkylthiomethyl)phosphonates **123** starting from diethyl (methylthiomethyl)phosphonate (**124**). After one-pot treatment with butyllithium and elemental sulfur (sulfenylation) followed by aqueous treatment, **124** was converted into diethyl [(methylthio)mercaptomethyl]phosphonate (**125**), which under phase transfer catalysis conditions was further alkylated to give final unsymmetrical thioacetals **123** [107] (Scheme 32). In further work, alkyl halides (AlkHal) were introduced in the reaction medium immediately after sulfenylation, which enabled the preparation of thioacetals **123** by a one-pot method [108].

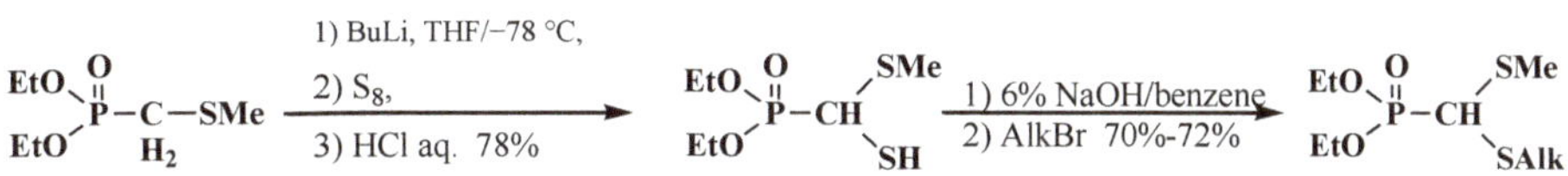

124 **125** **123**

Scheme 32. The synthesis of unsymmetrical diethyl (dialkylthiomethyl)phosphonates (**123**).

Diphenyl(diphenylthiomethyl)phosphine oxide (**126**) [109] was previously obtained in similar manner in 75% yield from diphenyl(phenylthiomethyl)phosphine oxide (**127**) by its interaction with diphenyldisulfide (Scheme 33).

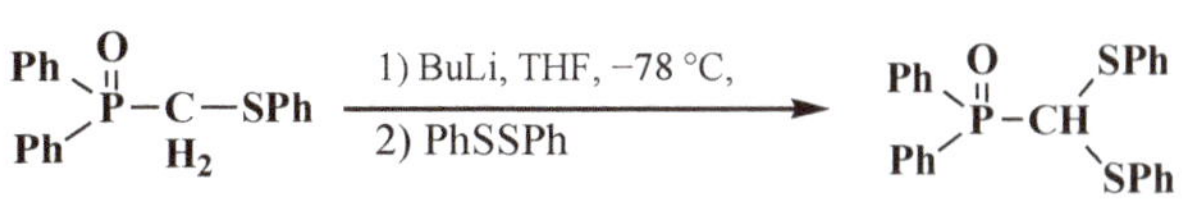

127 **126**

Scheme 33. Syntheses of diphenyl(diphenylthiomethyl)phosphine oxide (**126**) from diphenyl(phenylthiomethyl)phosphine oxide (**127**).

It was shown later that diethyl (diphenylthiomethyl)phosphonate (**128**) can be prepared from dialkyl methylphosphonate (**129**) [108] by a one-pot technique in 84% yield (Scheme 34).

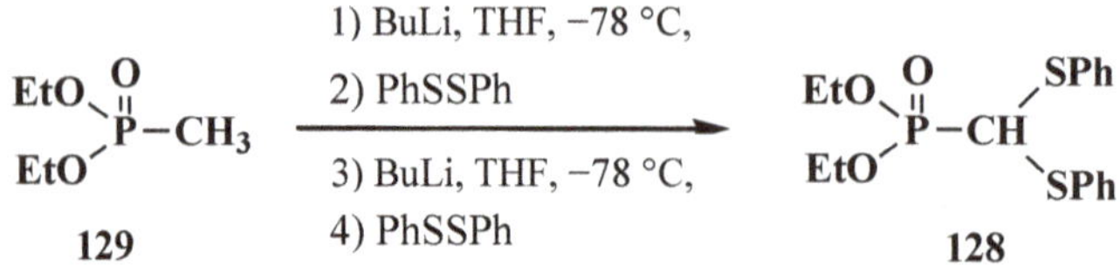

Scheme 34. Syntheses of diethyl (diphenylthiomethyl)phosphonate (**128**) from dialkyl methylphosphonate (**129**) by a one-pot technique.

Diethyl [(methylthio)(trimethylsilyl)methyl]phosphonate (**130**) undergoes a similar transformation with diethyldisulfide to form diethyl (diphenylthiomethyl)phosphonate (**131**) (yield 62–75%) [110] (Scheme 35).

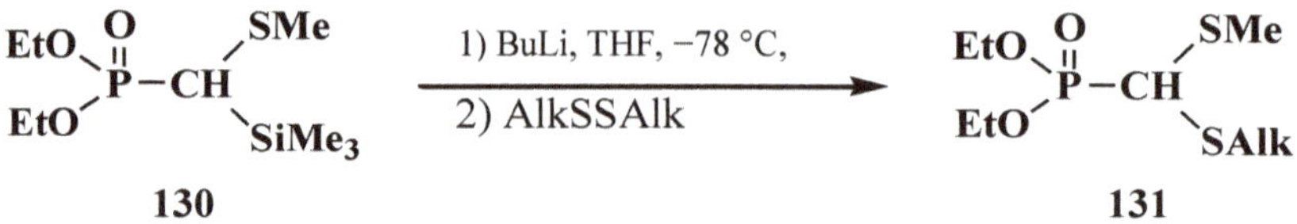

Scheme 35. Interaction of diethyl [(methylthio)(trimethylsilyl)methyl]phosphonate (**130**) with diethyldisulfide.

This method [107] and its variations [108–110] provide a possibility to synthesize linear dialkyl (dialkylthiomethyl)phosphonates **107** containing different substituents in the thioacetal group.

The reduction of phosphonodithioformates **132** with sodium borohydride ($NaBH_4$) or borane–dimethyl sulfide adduct $BH_3{\cdot}SMe_2$ [111] followed by alkylation of the resulted dialkyl [(methylthio)mercaptomethyl]phosphonates **133** finally results in unsymmetrical dialkyl [(methylthio)(alkylthio)methyl]phosphonates **134** (Scheme 36).

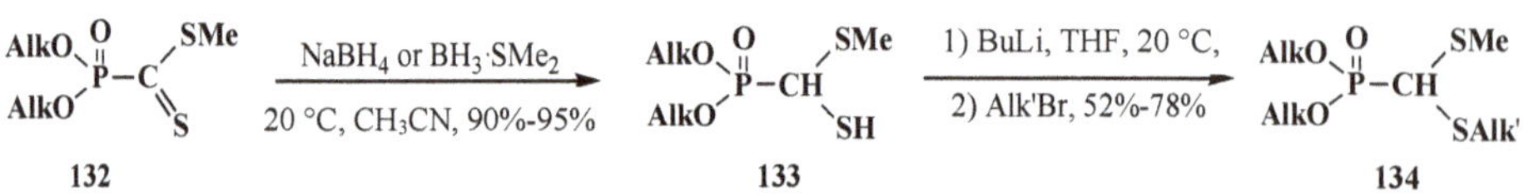

Scheme 36. Transformation of phosphonodithioformates **132** into dialkyl [(methylthio)(alkylthio)methyl]phosphonates **134**.

It was shown later that dialkyl and diaryl disulfides in the presence of catalysts (Cat.) like $BF_3{\cdot}Et_2O$, rhodium(II) tetraacetate $Rh_2(OAc)_4$, or copper(II) sulfate $CuSO_4$, can react with diazomethanephosphonates **38**, where R=OAlk. Phosphorylated carbenes produced in the reaction undergo insertion into S–S bond to afford dialkyl (dialkylthiomethyl)phosphonates **107** or dialkyl (diarylthiomethyl)phosphonates **135** in 42–93% yields [56,112]. Reactions proceed by the carbene mechanism (Scheme 37) [112].

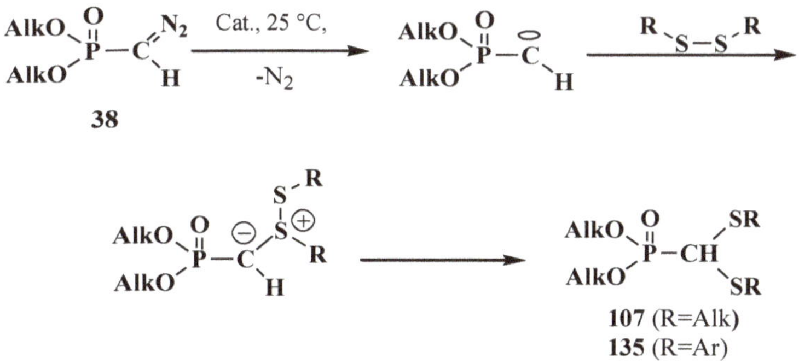

Scheme 37. Diazomethanephosphonates **38** reaction with organic disulfides with formation of dialkyl (dialkylthiomethyl)phosphonates **107** or dialkyl (diarylthiomethyl)-phosphonates **135**.

A method of synthesis of substituted diphenyl([1,3]-dithian-2-yl)phosphine oxides **136** (yields 34–85%) starting from substituted 1,3-dithianes and chlorodiphenylphosphine (**68**) was also suggested [104,113,114]. Initially formed diphenyl([1,3]-dithian-2-yl)phosphines **137** undergo further oxidation with molecular oxygen to final **136** (Scheme 38). Diphenyl([1,3]-dithian-2-yl)phosphine sulfides **138** or selenides **139** can be obtained by this method when elemental sulfur or selenium are used as oxidants for **137** [104,113].

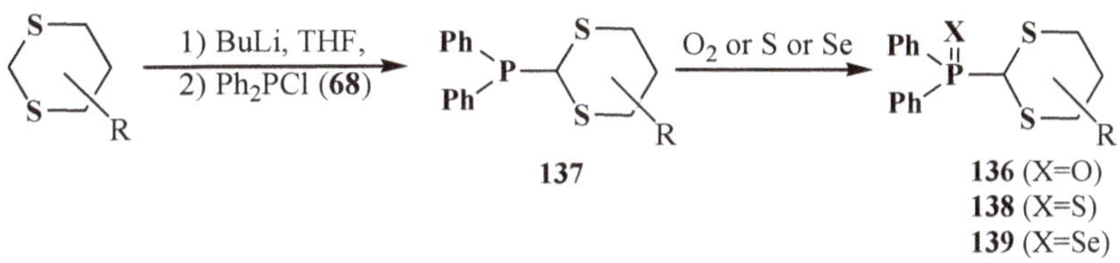

Scheme 38. Obtaining of substituted diphenyl([1,3]-dithian-2-yl)phosphine oxides **136** and corresponding phosphine sulfides **138** and phosphine selenides **139** from dithianes.

The attempted preparation of **107** and **114** in one step from dialkylthiomethanes or 1,3-dithianes and dialkyl chlorophosphate in the presence of strong bases failed, as evidenced by the negligible yield of final products [21,115]. The syntheses of **107** by the interaction of thioles with dialkyl [(*N*,*N*-dimethylamino)alkoxymethyl]phosphonates **140** or dialkyl [(alkylthio)chloromethyl]phosphonates **141** also failed [21].

Hovewer a method of synthesis of linear phosphorylated formaldehyde thioacetals **14** by the reaction of dialkyl [(arylthio)chloromethyl]phosphonates **142** with thiols at 0 °C in the presence of equimolar amount of tin(IV) tetrachloride $SnCl_4$ was succesful. This route provided a preparation of unsymmetrical dialkyl [(alkylthio)(arylthio)methyl]phosphonates **143** in 73–84% yields [43] (Scheme 39).

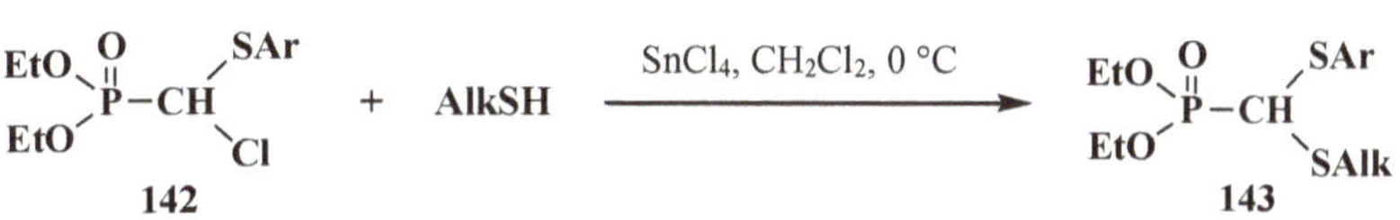

Scheme 39. Obtaining of dialkyl [(alkylthio)(arylthio)methyl]phosphonates **143** by the reaction of dialkyl [(arylthio)chloromethyl]phosphonates **142** with thiols.

Dialkyl (dialkylthiomethyl)phosphonates **107** were also obtained by the reaction of dialkyl (dialkoxymethyl)phosphonates **33** with thiols in the presence of acids [21] (see Scheme 27).

Dialkyl (1,3-benzodithiolylmethyl)phosphonates **115** were prepared for the first time by the reaction of 1,3-benzodithiolyl tetrafluoroborate with **58** in the presence of NaI [103]. The method is used at present without changes [116] (Scheme 40).

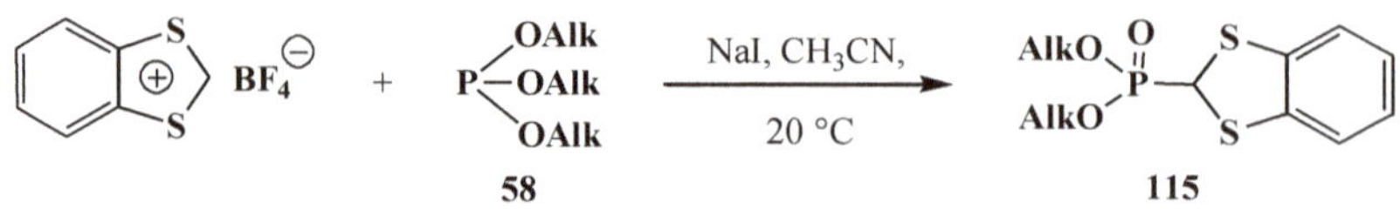

Scheme 40. The synthesis of dialkyl (1,3-benzodithiolylmethyl)phosphonates **115** from 1,3-benzodithiolyl tetrafluoroborate.

2.3.2. Chemical Properties of Phosphorylated Formaldehyde Thioacetals 14

The chemical properties of phosphorylated formaldehyde thioacetals **14** have been studied in much less detail compared with the corresponding acetals **13**. Their properties are determined by the presence of sulfur atoms and a disubstituted phosphoryl group. Both sulfur atoms are oxidized when diethyl ([1,3]-dithian-2-yl)phosphonate (**144**) is treated with sodium periodate $NaIO_4$ in aqueous methanol at 20 °C [106] to yield the dioxo form **145** (Scheme 41).

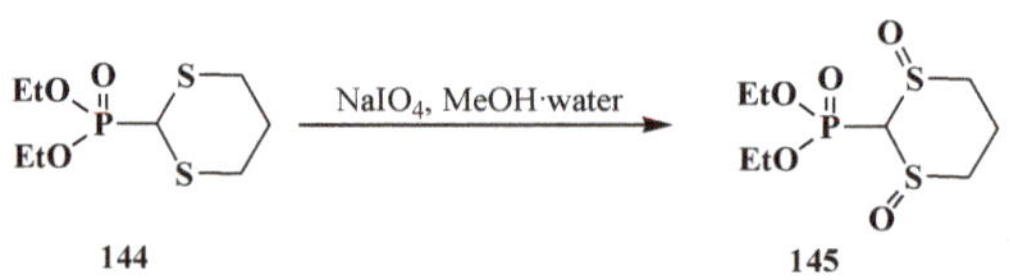

Scheme 41. Diethyl ([1,3]-dithian-2-yl)phosphonate (**144**) oxidation by periodate.

The heating of diphenyl([1,3]-dithiolan-2-yl)phosphine oxides **121** [115] and diphenyl([1,3]-dithian-2-yl)phosphine oxides **122** [56,104] with phosphorus pentasulfide P_2S_5 in benzene or *N,N*-diethylaniline $PhNMe_2$ leads to the replacement of phosphinoyl oxygen by sulfur to yield diphenyl([1,3]-dithiolan-2-yl)phosphine sulfide (**146**) and diphenyl([1,3]-dithian-2-yl)phosphine sulfide, respectively (**147**) (Scheme 42). See also Scheme 38.

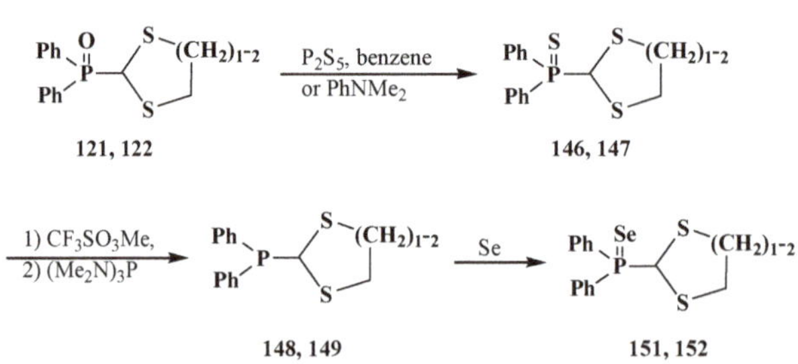

Scheme 42. Transformation of phosphorylated cyclic thioacetals **122, 123** into corresponding phosphines **148, 149**, phosphine sulfides **146, 147** and phosphine selenides **151, 152**.

The sequential treatment of the prepared phosphine sulfides **146** and **147** with trifluoromethylsulfonate CF_3SO_3Me and tris(dimethylamino)phosphite $(Me_2N)_3P$ finally affords diphenyl([1,3]-dithiolan-2-yl)phosphine (**148**) [105] and diphenyl([1,3]-dithian-2-yl)phosphines **149** [56,104]. The reaction of diphenyl(5-*tert*-butyl-[1,3]-dithian-2-yl)phosphine sulfide (**150**) with trichlorosilane leads to the same result [56]. The obtained phosphines **146, 147** combine with elemental selenium to yield diphenyl([1,3]-dithiolan-2-yl)phosphine selenide (**151**) [105] and diphenyl([1,3]-dithian-2-yl)phosphine selenide (**152**) [56,104]. See also Schemes 59, 60, 64, 65, 74–81.

2.4. *(N,N-dialkylamino)cyanomethyl Derivatives of Phosphorylated Formaldehyde (α-dialkylamino-nitriles)* **15**

For 1982 till now, only two derivatives of this type of organophosphorus compounds of phosphonate series were obtained, diethyl [(*N*,*N*-dimethylamino)cyanomethyl]phosphonate (**153**) [24] and diethyl [(*N*-morpholino)cyanomethyl]phosphonate (**154**) [26].

2.4.1. Methods of Synthesis of Diethyl [(N,N-Dialkylamino)cyanomethyl]phosphonates **15**

Diethyl [(*N*,*N*-dimethylamino)cyanomethyl]phosphonate (**153**) was obtained for the first time by a reaction similar to the preparation of phosphorylated acetals **13** (Scheme 7) from diethyl phosphite (**6**) and (*N*,*N*-dimethylamino)(methoxy)cyanomethane [24,25] (Scheme 43).

Scheme 43. Syntheses of diethyl [(*N*,*N*-dimethylamino)cyanomethyl]phosphonate (**153**).

Diethyl [(*N*-morpholino)cyanomethyl]phosphonate (**154**) was prepared by the second method from diethyl chlorophosphate (**155**) and (*N*-cyanomethyl)morpholine [26] (Scheme 44) and was used in a subsequent reaction without isolation.

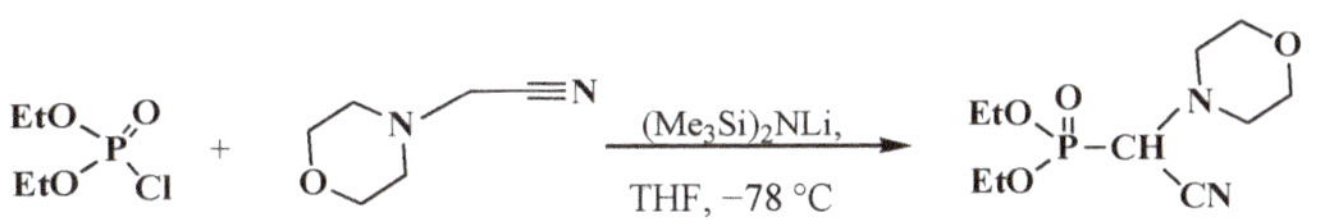

Scheme 44. Syntheses of diethyl [(*N*-morpholino)cyanomethyl]phosphonate (**154**) from diethyl chlorophosphate (**155**).

2.4.2. Chemical Properties of Diethyl [(N,N-Dialkylamino)cyanomethyl]phosphonates **15**

The chemical properties of this type of compounds have been studied almost completely using the example of diethyl [(*N*,*N*-dimethylamino)cyanomethyl]phosphonate (**153**). The treatment of **153** with methyl iodide (MeI) or dimethyl sulfate Me_2SO_4 [117,118] gives rise to methylation of nitrogen atom (quaternization) of the dimethylamino group and formation of diethyl [(*N*,*N*,*N*-trimethylammonio)cyanomethyl]phosphonate cation (**156**), whose α proton shows enhanced acidity and breaks off under the action of aqueous potassium carbonate solution K_2CO_3. The resultant diethoxy [(*N*,*N*,*N*-trimethylammonio)cyanomethylidium]phosphonate (**157**) can react with electrophiles—MeI and molecular chlorine Cl_2 [117], to give diethyl [(1-(*N*,*N*,*N*-trimethylammonio))(1-cyano)ethan-1-yl]phosphonate cation (**158**) and diethyl [chloro(1-*N*,*N*,*N*-trimethylammonio)cyanomethyl]phosphonate cation (**159**), respectively (Scheme 45).

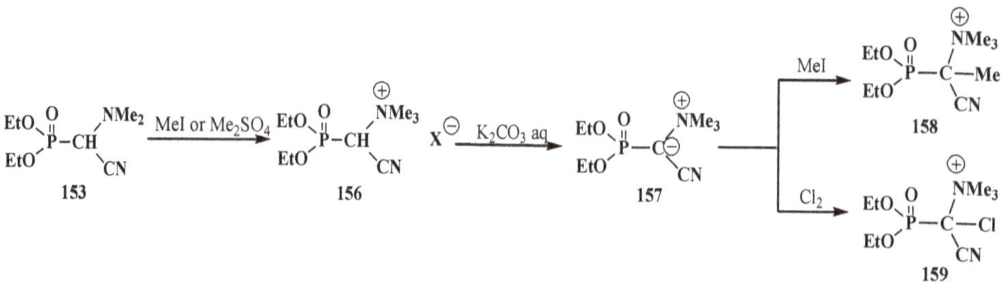

Scheme 45. Transformation of diethyl [(*N,N*-dimethylamino)cyanomethyl]phosphonate (**153**).

Under phase transfer catalysis conditions—50% KOH solution, $[Et_3NCH_2Ph]^+Cl^-$ (TEBAC) [24] phosphonate **153** combines with nitrosobenzene PhN=O similarly to the Horner reaction to afford [(*N,N*-dimethyl)(*N'*-phenyl)amidinoyl]oxalnitrile (**160**) in 58% yield (Scheme 46). See also Schemes 55, 60, 66, 67 and 82–85.

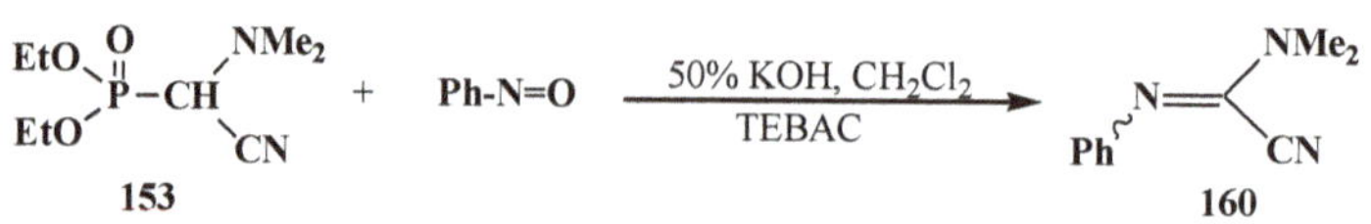

Scheme 46. Horner—analog reaction of phosphonate **153** with nitrosobenzene.

2.5. *Diphosphinoyl N,N-Dialkylaminomethanes* 16

2.5.1. Methods of Synthesis of Diphosphinoyl N,N-Dialkylaminomethanes 16

The first report on the synthesis of tetraethyl (*N,N*-dimethylaminomethyl)diphosphonate (**161**) was published in 1968. Compound **161** was obtained in 62% yield by heating a 2:1 mixture of diethyl phosphite (**6**) and dimethylformamide dimethylacetal (**78**) [27]. Dialkyl phosphites **51**, dialkyl- and diphenylphosphine oxides **52**, **53** can be also involved in this reaction [31,32,119]. Intermediate dialkyl [(*N,N*-dimethylamino)methoxymethyl]phosphonates (**162**) or dialkyl[(*N,N*-dimethylamino)-methoxymethyl]phosphine oxides **163** may be isolated in many cases [32,119]. This allows preparation of symmetrical **164** and unsymmetrical diphosphinoyl *N,N*-dimethylaminomethanes **165** [31,32,119,120] (Scheme 47).

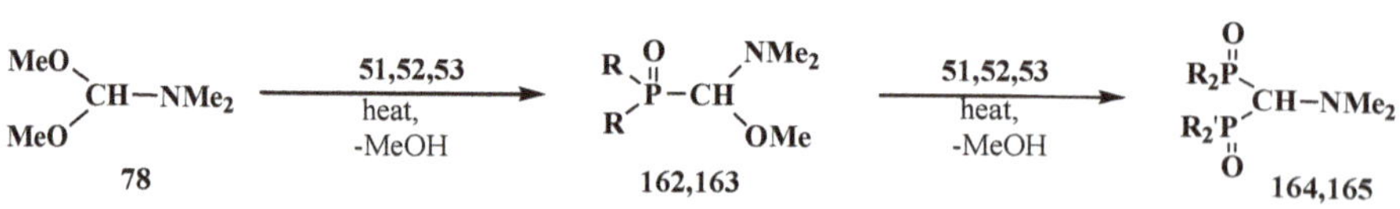

Scheme 47. Synthesis of diphosphinoyl *N,N*-dimethylaminomethanes **164** and **165** from dimethylformamide dimethylacetal (**78**), where R, R' = OAlk, Alk, Ph. See text above.

Other methods of synthesis of symmetrical tetraalkyl (*N,N*-dimethylaminomethyl)diphosphonates **166** were proposed later. Since trialkyl phosphites **58** do not react with dimethylformamide dimethylacetal (**78**) [18], mixed dialkyl trimethylsilyl phosphites $(EtO)_2PSiMe_3$ were successfully employed in the reaction. The reaction proceeds spontaneously at 20 °C [121] in 36–66% yield or upon heating in the presence of zinc chloride in 72–77% yield [122,123] (Scheme 48).

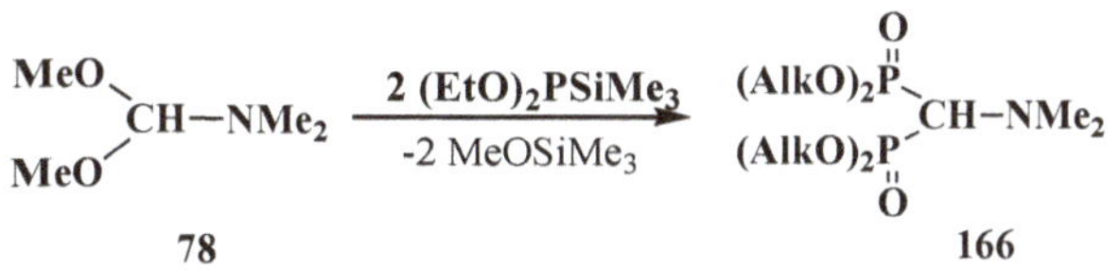

Scheme 48. Use mixed dialkyl trimethylsilyl phosphites for the syntheses of tetraalkyl diphosphonates **166** from dimethylformamide dimethylacetal (**78**).

A convenient method of synthesis of unsymmetrical tetraalkyl (*N*,*N*-dialkylaminomethyl)-diphosphonates **167** by the reaction of trialkyl phosphites (**58**) with *N*,*N*-dialkylhalo-methylideneiminium halides $[HalC(H)=NAlk_2]^{+}Hal^{-}$, where Hal = Cl (compounds **168**) [28,31,32,120] or Br (compounds **169**) [29] in 2:1 ratio was proposed in 1969.

It was shown that the reaction of trialkyl phosphites **58** with *N*,*N*-dimethylchloro-methylideneiminium chloride (**170**) proceeds via intermediate formation of dialkyl [(*N*,*N*-dimethylamino)chloromethyl]phosphonate **171** [31,32], specially prepared dimethyl [(*N*,*N*-dimethylamino)chloromethyl]phosphonate (**172**) and diethyl [(*N*,*N*-dimethylamino)chloromethyl]- phosphonate (**173**) may be also involved in the reaction [30,31,124]. This method enables preparation of symmetrical and unsymmetrical tetraalkyl phosphonates **174** as well (Scheme 49).

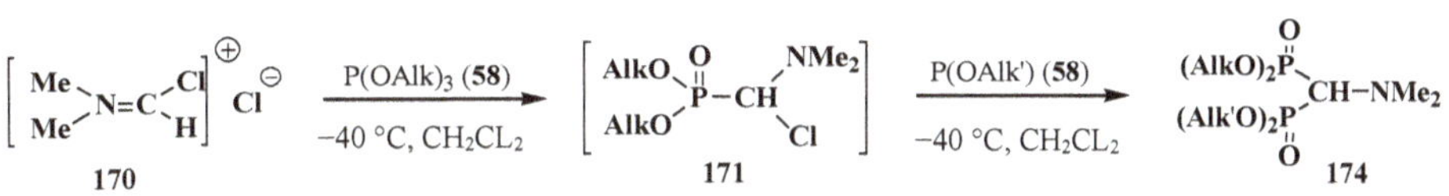

Scheme 49. Method of synthesis of symmetrical and unsymmetrical dialkyl (*N*,*N*-dimethylaminochloromethyl)phosphonates **174** from *N*,*N*-dimethylchloromethylideneiminium chloride (**170**) and trialkyl phosphites (**58**).

Dialkyl phosphites **51**, dialkylphosphine oxides **52** and diphenylphosphine oxide (**53**) also react with *N*,*N*-dialkylchloromethylideneiminium chlorides **168** [31,120] and their analogs **175** obtained from *N*,*N*-dialkylformamides and phosphorus oxychloride [28] (Scheme 50). The reaction also proceeds via intermediate compounds **176**, where X=Cl or Cl_2PO_2. This allows also the preparation of symmetrical or unsymmetrical diphosphinoyl compounds **177** and **178**, respectively.

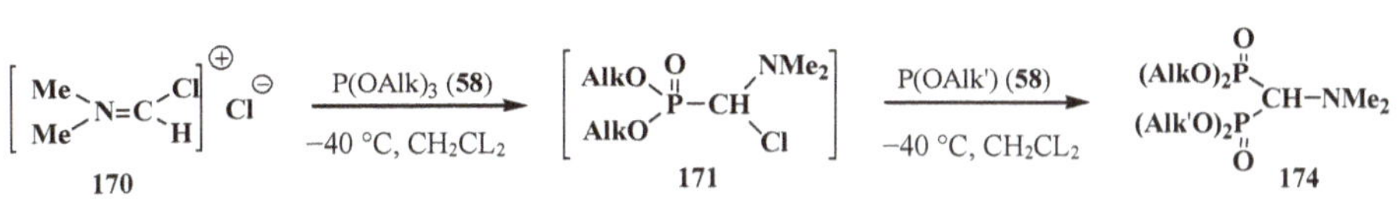

Scheme 50. Synthesis of diphosphinoyl *N*,*N*-dialkylaminomethanes **177** and **178** from *N*,*N*-dimethylchloromethylideneiminium chloride (**170**) and hydrophosphorylic compounds **51**, **52**, **53**. See text above.

See also Scheme 14.

2.5.2. Chemical Properties of Diphosphinoyl N,N-Dialkylaminomethanes 16

The chemical properties of this type of organophosphorus compounds are studied insufficiently and almost exclusively by the example of tetraethyl (*N*,*N*-dimethylaminomethyl)diphosphonate (**161**). Their properties are attributable to the presence of both an amino group and disubstituted phosphoryl groups.

Hydrolysis of tetraalkyl (*N,N*-dimethylaminomethyl)diphosphonates **174** was studied using the example of compound **161**. Boiling diphosphonate **161** with concentrated hydrochloric acid leads to (*N,N*-dimethylaminomethyl)diphosphonic acid (**179**) [98,120] in almost quantitative yield (Scheme 51).

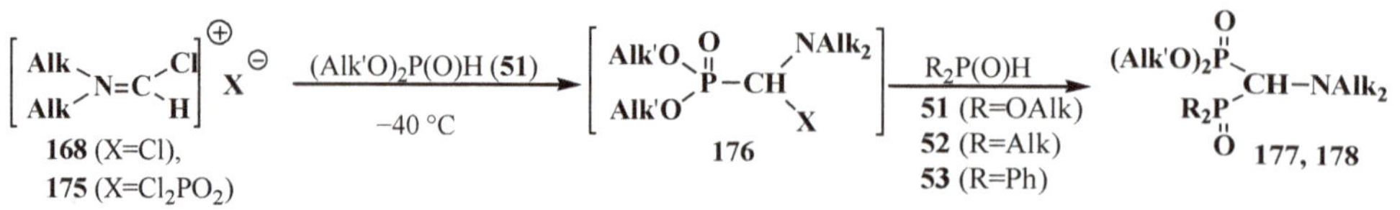

Scheme 51. Acid hydrolysis of tetraethyl (*N,N*-dimethylaminomethyl)diphosphonate (**161**).

The amino group of diphosphonate **161** undergoes methylation (quaternization) when reacted with methyl iodide (MeI) or dimethyl sulfate (Me_2SO_4) to give tetraethyl (*N,N,N*-trimethylammoniomethyl)diphosphonate cation (**180**) in 76 and 85% yields, respectively [118, 125]. The α-proton of the latter shows enhanced acidity and undergoes elimination under the action of aqueous solution of potassium carbonate (K_2CO_3) to afford tetraethyl [(*N,N,N*-trimethylammonio)methylidium]diphosphonate (**181**) [120,125] (Scheme 52).

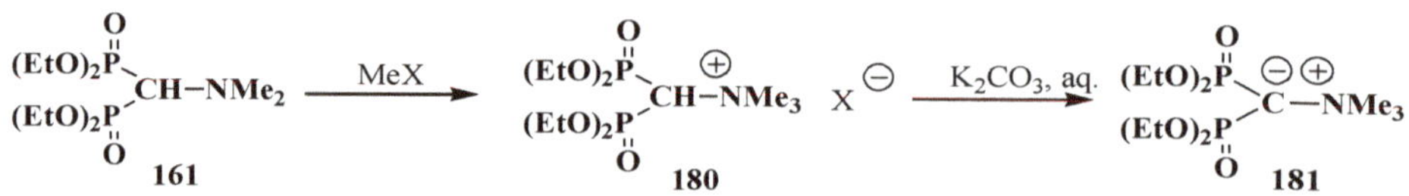

Scheme 52. Transformation of tetraethyl (*N,N*-dimethylaminomethyl)diphosphonate (**161**) into methylidiumdiphosphonate **181**, where X = I, $MeSO_3$.

Ylide **181** exhibits an enhanced stability: it can be stored for a long time in air, it is thermally stable and does not react with methyl iodide [126]. However, ylide **181**, like cyanomethylide **154**, reacts with molecular chlorine to yield tetraethyl [(*N,N,N*-trimethylammonio)chloromethyl)diphosphonate chloride (**182**) that undergoes fast dealkylation on storage to afford the corresponding betaine **183** in 85% yield [120] (Scheme 53).

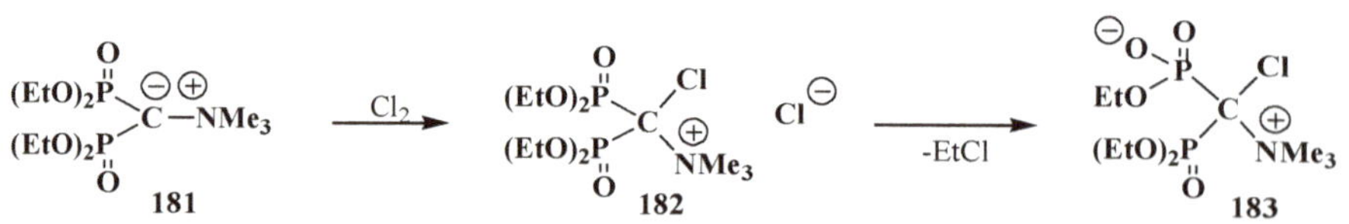

Scheme 53. Interaction of methylidiumdiphosphonate **181** with chlorine with the subsequent formation of betaine **183**.

See also Schemes 60, 68 and 72–74.

3. General Chemical Properties of Phosphorylated Formaldehyde Acetals 13 and Structurally Related Compounds 14–16

3.1. Phosphorus–Carbon Bond Cleavage

3.1.1. Cleavage of Phosphorus–Carbon Bond under the Action of Acids and Acidic Reagents

The general property of phosphorylated formaldehyde acetals **13** and structurally related compounds **14–16** is the possibility of phosphorus–carbon bond cleavage under the action of acids and acidic reagents [24,77,95,98], organic reagents [19,119], and bases [37,127,128].

The cleavage of phosphorus–carbon bond under acidic conditions compounds **33** is studied on examples of diethyl (diethoxymethyl)phosphonate (**57**). Phosphonate **57** was shown to undergo cleavage of the phosphorus–carbon bond on heating with 5% hydrochloric acid [77,98] to form diethyl phosphites (**6**), ethyl formate (**184**) and ethanol. Similar degradation of compound **57** occurs in a flow of hydrogen chloride at 20 °C [95] (ethyl chloride is formed as a byproduct) (Scheme 54). Unlike **57**, interaction of diethyl (diphenoxymethyl)phosphonate (**185**) with dry hydrogen chloride does not lead to the cleavage of phosphorus–carbon bond, but leads to the dealkylation of one ethoxy substituent at the phosphorus atom with obtaining ethyl (diphenoxymethyl)phosphonic acid (**186**) (Scheme 54).

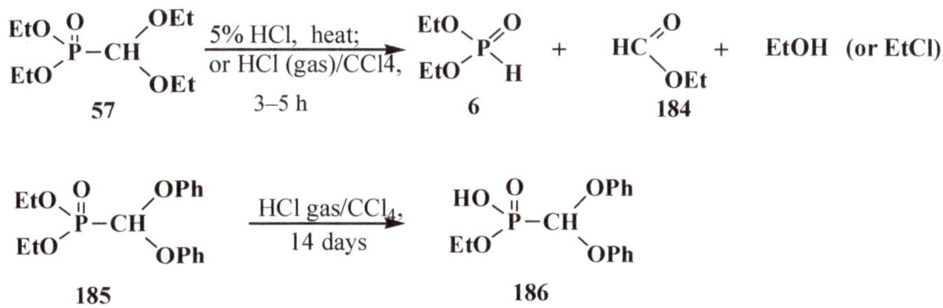

Scheme 54. Interaction of acetals **57** and **185** with hydrochloric acid (only for **57**) and hydrogen chloride.

It was shown that acid hydrolysis of diethyl (diethylthiomethyl)phosphonates (**187**) is accompanied by a partial cleavage of the phosphorus–carbon bond [98]. Prolonged refluxing of diethyl [(*N*,*N*-dimethylamino)cyanomethyl]phosphonate (**153**) in concentrated hydrochloric acid leads to *N*,*N*-dimethylaminoacetic acid (**188**) in 87% yield [24] (Scheme 55).

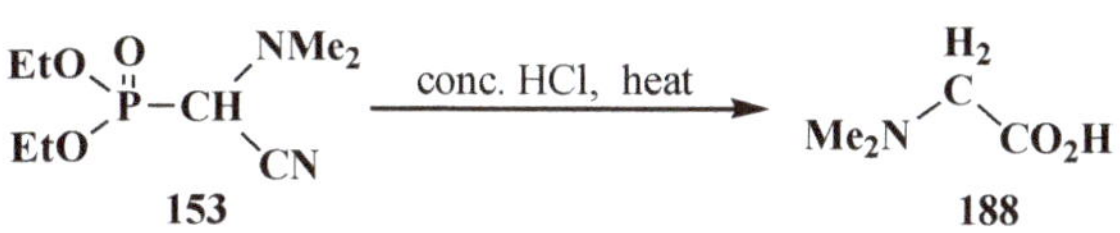

Scheme 55. Obtaining *N*,*N*-dimethylaminoacetic acid (**188**) by acid hydrolysis of cyanomethylphosphonate **153**.

It was shown in [95] that the attempted transesterification of ethoxy substituents at the phosphorus atom of phosphonate **57** by catechol residue also leads to the cleavage of the phosphorus–carbon bond (Scheme 56).

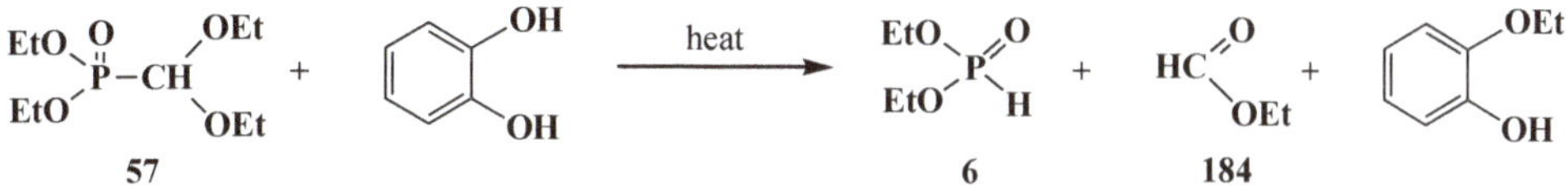

Scheme 56. Destruction of acetales **57** at its interaction with catechol.

The cleavage of phosphorus–carbon bond may also occur when diethyl (dialkoxymethyl)phosphonates **93** are exposed to phosphorus pentachloride [95].

3.1.2. The Cleavage of Phosphorus–Carbon Bond in Reactions with Organic Coreactants

The reactions of **57** with dichloromethoxymethane $Cl_2C(H)OMe$ or acetyl bromide MeC(O)Br was shown to be accompanied by the cleavage of phosphorus–carbon bond [19] to form diethoxychlorophosphine (**189**) and diethyl acethylphosphonate (**190**) (Scheme 57).

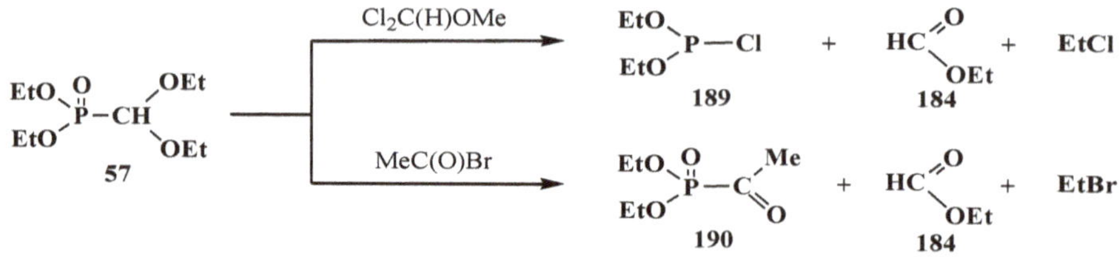

Scheme 57. Acetales **57** interaction with $Cl_2C(H)OMe$ and MeC(O)Br leading to the cleavage of phosphorus–carbon bond.

The reaction of diethyl phosphite (**6**) with dipropyl [(*N*,*N*-dimethylamino)ethoxymethyl]phosphonate (**191**) is also accompanied by the partial cleavage of the phosphorus–carbon bond [119].

3.1.3. Phosphorus–Carbon Bond Cleavage under the Action of Bases

The reaction of the lithium derivatives of diphenyl(dialkoxymethyl)phosphine oxides **50** with *n*-octanal and *p*-isopropylbenzaldehyde leads to α-phosphorylated alcohols **192**, the products of addition of diphenylphosphinite anion (**193**) to the carbonyl group (Abramov reaction) [127,128], rather than ketene *O*,*O*-acetals **194** as expected products of the Horner reaction [9,10] (Scheme 58). The reason of this course is the instability of diphenyl (dialkoxymethyl)phosphine oxide anion (**195**) that compound **50** decomposes after long exposition under the reaction conditions to form anion **193**, which further undergoes addition to the aldehyde carbonyl group [127,128].

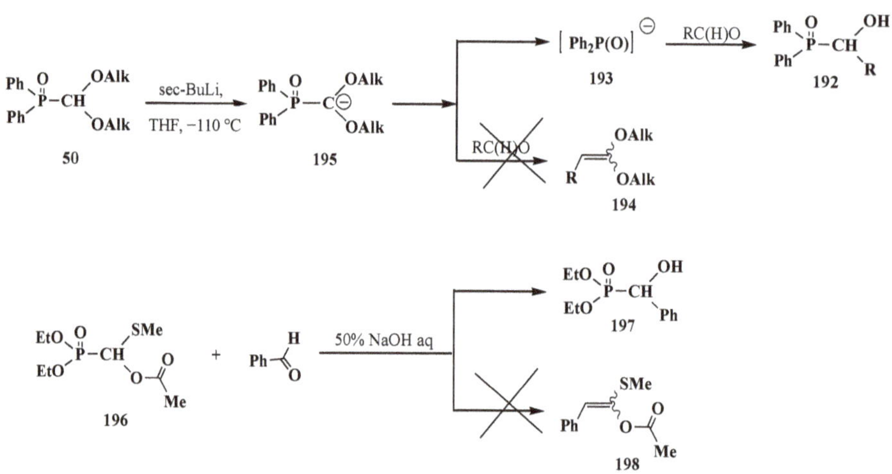

Scheme 58. Cleavage of phosphorus–carbon bond of acetal **57** and O,S-acetal **196** leading to the formation of α-phosphorylated alcohols **192** and **198**.

Like phosphine oxides **50**, the reaction of diethyl [(acetoxy)methylthiomethyl]phosphonate (**196**) with benzaldehyde PhC(H)O under phase transfer catalysis conditions gives rise to diethyl [(hydroxy)phenylmethyl]phosphonate (**197**), the Abramov reaction product, instead of the expected ketene O,S-acetal **198** resulting from the Horner reaction [37] (Scheme 58).

Lithiated anions **199** of diethyl [(methylthio)(alkylthio)methyl]phosphonates **123** react quickly with molecular oxygen with the cleavage of phosphorus–carbon bond and formation of dialkyl dithiocarbonates **200** (yields 71–72%) [129] (Scheme 59).

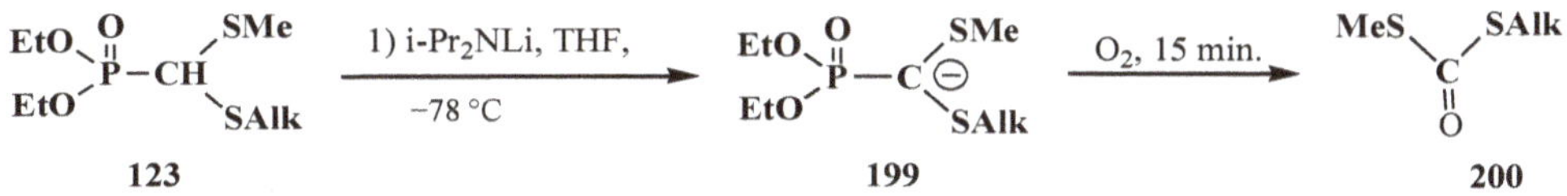

Scheme 59. Destruction of thioacetal **123** in the presence of oxygen, leading to formation of dithiocarbonates **200**.

See also Schemes 63, 96, 103, 104, 113–117 and 120.

3.2. *Synthesis of Formacetalphosphonic Acids*

Attempted preparations of formacetalphosphonic acids **201**, **202** and **203** from the corresponding phosphorylated acetal **57**, thioacetal **187** and aminonitrile **153** by acid hydrolysis lead to cleavage of the phosphorus–carbon bond, but in the case of compound **91** with a aromatic substituent in the acetal group, it undergoes acid hydrolysis to give the expected phosphonic acid **92** (Scheme 20). See also Section 3.1.1. "Cleavage of phosphorus–carbon bond under the action of acids and acidic reagents".

However, the target formacetalphosphonic acids **201–203** may be successfully prepared by the reaction of acetals **33**, diethyl (diethylthiomethyl)phosphonate (**187**), diethyl [(*N*,*N*-dimethyl-amino)cyanomethyl]phosphonate (**153**), and tetraethyl(*N*,*N*-dimethylaminomethyl)diphosphonate (**161**) with trimethylsilyl bromide (Me_3SiBr) in acetonitrile. Alkyl groups at the phosphorus atom are eliminated as alkyl halides to afford the corresponding intermediate bis(trimethylsilyl) phosphonates **204** (or tetrakis(trimethylsilyl) phosphonate (**205**) from diphosphonate **161**), which are further readily hydrolyzed by water treatment to give:

- (dialkoxymethyl)phosphonic acids **201** in 93–100% yield from **33** [19,98]. It was shown that the reaction **33** with an equimolar mixture of trimethylsilyl chloride Me_3SiCl and NaBr or LiBr [130] in acetonitrile of trimethylsilyl chloride and NaI in methylene chloride [131] leads to the same result,
- (diethylthiomethyl)phosphonic acid (**202**) from **187** in 58% yield [98],
- (1-dimethylamino-1-cyanomethyl)phosphonic acid (**203**) from **153** in 55% yield [24],
- (*N*,*N*-dimethylaminomethyl)diphosphonic acid (**179**) from **161** in 65% yield [98] (Scheme 60).

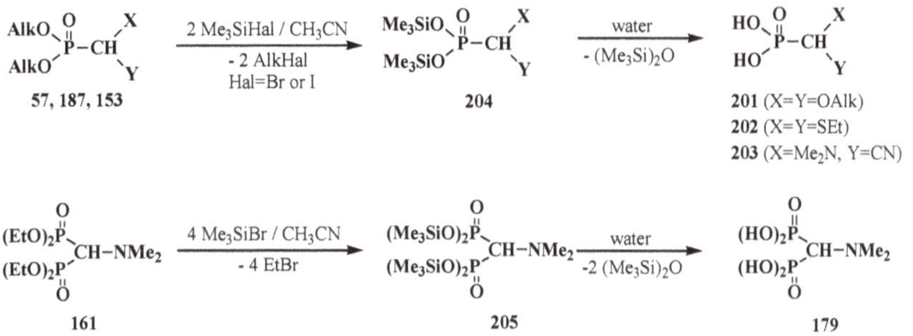

Scheme 60. Syntheses of formacetalphosphonic acids **201–203** and **179** from compounds **57**, **187**, **153**, **161** by means of Me_3SiBr or Me_3SiI. See text above.

See also Schemes 15, 20–22, 29 and 54.

3.3. Alkylation (Acylation) of the Formacetal Carbon Atom

Dialkyl (dialkoxymethyl)phosphonates **33** produce no stable phosphorylated carbanion **206** when reacted with bases (no metallation occurs, even under the action of *tert*-butyllithium (*t*-BuLi), which provides no possibility for further alkylation and acylation of the formacetal group [23,132] (Scheme 61).

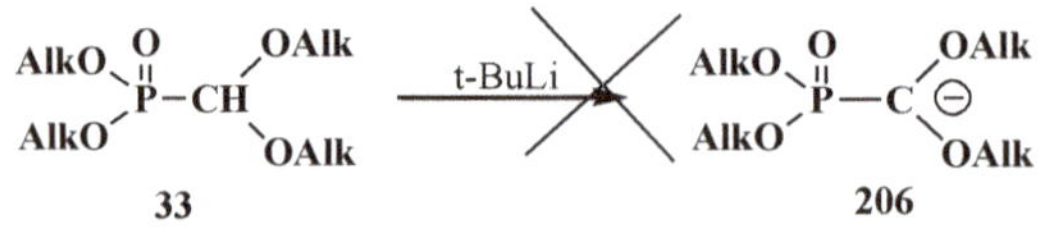

Scheme 61. Phosphorylated acetals **33** do not produce carbanions **206**.

This fact was explained by insufficient stabilization of the negative charge of carbanion on the two oxygen atoms in the α-position [23]. However, it was shown in 1983 [133] that, in contrast to phosphonates **33**, diphenyl(dialkoxymethyl)phosphine oxides **50** produce phosphorylated anions **195** at −110 °C that undergo metallation. The reason for the stability of the lithium derivatives of phosphine oxides **50** is the ability of diphenylphosphinoyl group to delocalize the negative charge of carbanion **195** [20,134] (Scheme 58). By the example of anion **207** of diphenyl(dimethoxy-methyl)phosphine oxide (**208**), it was shown that it is rather stable to subsequent alkylation with alkyl halides and acylation with benzoyl chloride [20]. The reactions afford diphenyl[(dimethoxy)alkylmethyl]phosphine oxides **209**, in 30–94% yields, and diphenyl-[(dimethoxy)benzoylmethyl]phosphine oxide (**210**) (60%) (Scheme 62).

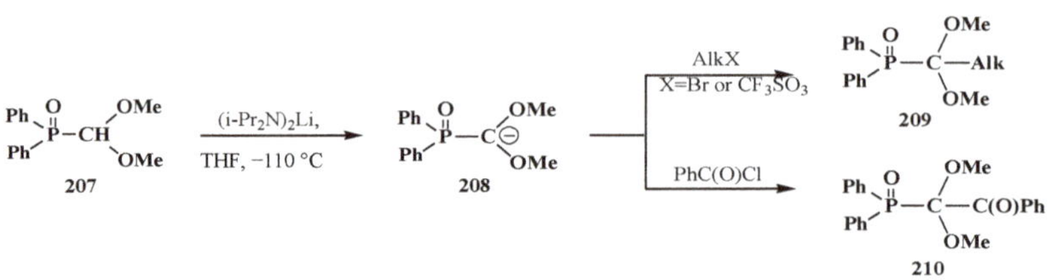

Scheme 62. Alkylation and acylation of carbanion **207**.

In acidic medium at 20 °C, phosphine oxides **209** are readily decomposed with cleavage of the phosphorus–carbon bond. The resultant methyl carboxylates **211** are homologous to the initial alkyl halides—carbon chain elongation by one atom (Scheme 63).

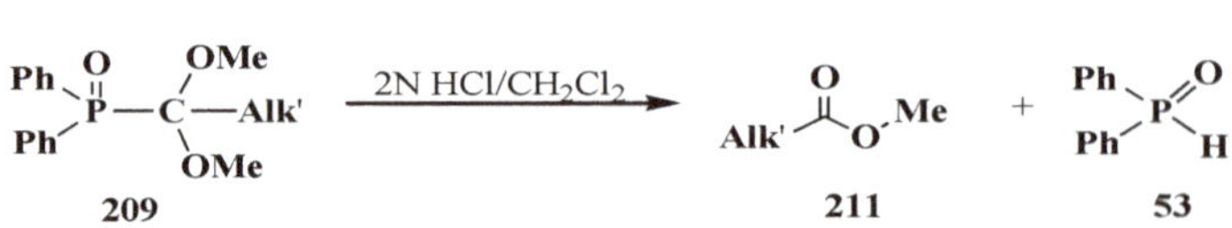

Scheme 63. Hydrolysis of phosphine oxide **209** leads to phosphorus–carbon bond cleavage.

Similarly, according to ^{1}H-NMR spectroscopy, the methanolysis of diphenyl[(1,1-dialkoxy)nonan-1-yl)phosphine oxide (**212**) in the presence of trifluoroacetic acid leads to 1,1,1-trimethoxynonane (**213**) in 65% yield [20].

Nonetheless, the storage of a solution of lithiated anion **207** for two hours even at −110 °C causes the cleavage of phosphorus–carbon bond (see Scheme 58).

Distinct from acetals **33** [23,132], the two sulfur atoms of thioacetals **113**, **114** stabilize well the neighboring carbanion [23], therefore the α-hydrogen atom in the thioacetal group is readily removed under the action of strong bases [23,132] in both dialkyl (dialkythiomethyl)phosphonates **113**, **114** [135] and diphenyl(dialkythiomethyl)phosphine oxides **121**, **122** [113]. Further, the carbanions are readily alkylated with alkyl halides [113,135]. For example, dialkyl ([1,3]-dithian-2-yl)phosphonates **114** in this reaction produce dialkyl [(2-alkyl-[1,3]-dithian)-2-yl]phosphonates **214**, and their oxidative decomposition may result in α-phosphorylated carbonyl compounds **215** (Scheme 64).

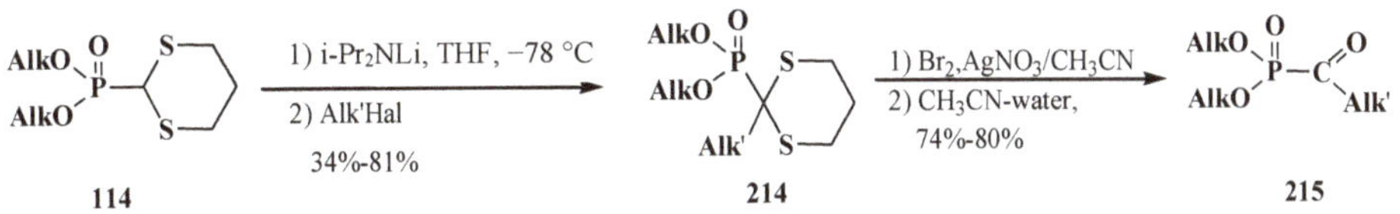

Scheme 64. Thioacetals **114** alkylation with the subsequent transformation of thioketals **214** into α-phosphorylated carbonyl compounds **215**.

The possibility to alkylate diethyl ([1,3]-dithian-2-yl)phosphonate (**144**) was used for the elongation of the hydrocarbon chain in the synthesis of 3-deoxy-D-manno-octulosonic acid (**216**), through compound **217** as alkylated form **144** [136] (Scheme 65).

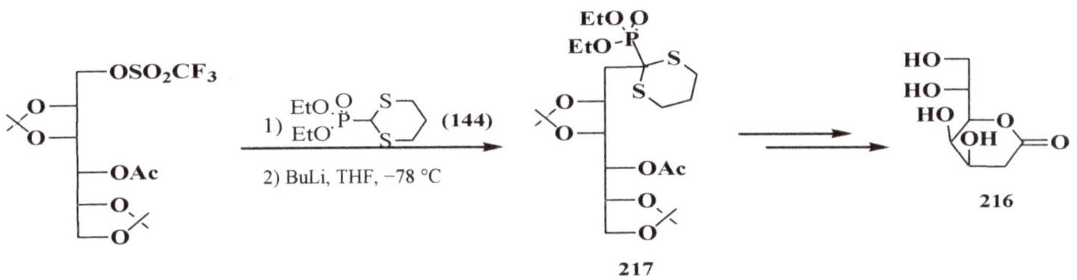

Scheme 65. Synthesis of 3-deoxy-D-manno-octulosonic acid (**216**) by the alkylation of thioacetal **144**.

In contrast to dialkyl (dialkoxymethyl)phosphonates **33** and similarly to phosphorylated formaldehyde thioacetals **14**, diethyl [(*N*,*N*-dimethylamino)cyanomethyl]phosphonate (**153**) is readily deprotonated under the action of sodium hydride in dioxane or dimethyl sulfoxide or 50% KOH solution under phase transfer catalysis conditions [24] as well as with butyllithium BuLi in THF [25]. Lithium derivative **218** of diethyl [(*N*,*N*-dimethylamino)cyanomethyl]phosphonate (**153**) proved to be so stable that it could be stored without decomposition for several months [25] (Scheme 66).

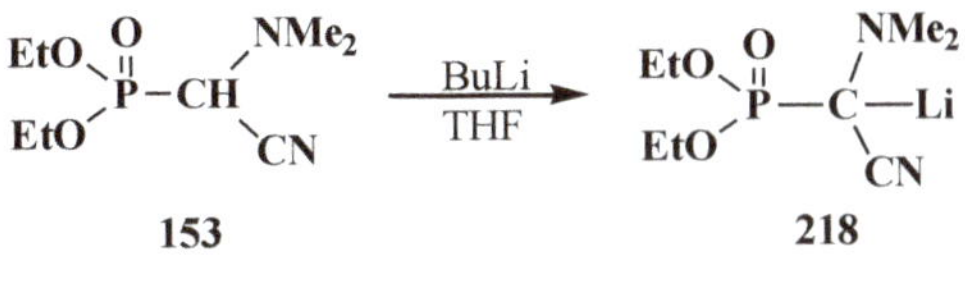

Scheme 66. Synthesis of stable lithium derivative **218** of diethyl [(*N*,*N*-dimethylamino)-cyanomethyl]phosphonate (**153**).

The anion of **218** undergoes alkylation at the carbon atom when treated with methyl iodide MeI [24,25,117], dimethyl sulfate Me_2SO_4 [117] or benzyl chloride $PhCH_2Cl$ [24] to give α-alkylated derivatives **219** and **220**. Benzyl derivative **220** prepared by this method eliminates hydrogen cyanide on heating to give α-phosphorylated enamine **221** [24] (Scheme 67).

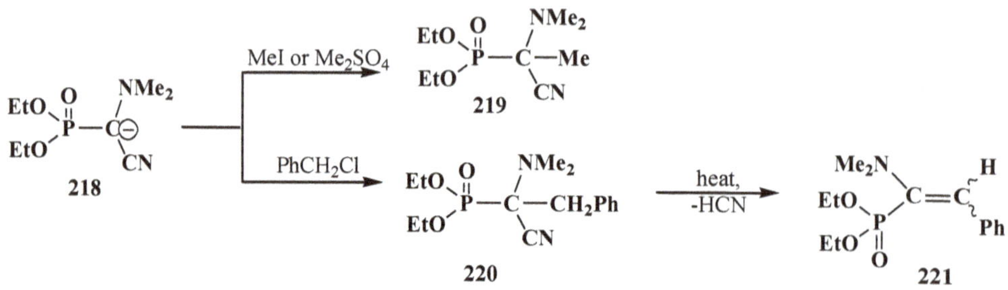

Scheme 67. Anion **218** alkylation.

Similarly to diphenyl(dialkoxymethyl)phosphine oxides (**50**), cyclic dialkyl (dialkylthiomethyl)-phosphonates **113**, **114**, cyclic diphenyl(dialkylthiomethyl)phosphine oxides **121**, **122**, and diethyl [(*N,N*-dimethylamino)cyanomethyl]phosphonate (**153**), and tetraethyl (*N,N*-dimethylaminomethyl)-diphosphonate (**161**) under the action of strong bases readily eliminate a proton from the acetal carbon atom to give anion **222**. This provides an opportunity for its further alkylation (compounds **223**) [28], which has been used in the synthesis of pesticides (Scheme 68).

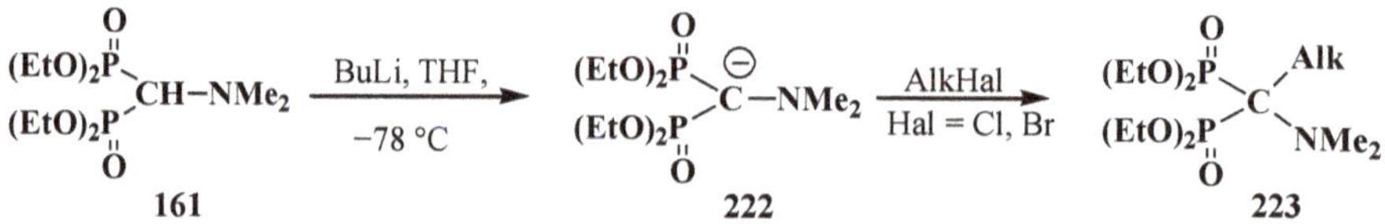

Scheme 68. Alkylation of methyldiphosphonate **161**, (Hal = Cl, Br).

3.4. Horner Reaction

In 1958 and 1959 L. Horner and co-authors reported their discovery of a new reaction [137, 138] that they named as "P=O-activated olefination". The authors showed that the reaction of alkyl(diphenyl)phosphine oxides **224** and dialkyl alkylphosphonates **225** with aldehydes and ketones in the presence of strong bases produce olefins **226** (Scheme 69). Key reaction intermediates—lithium derivatives of carbanion of the initial phosphinoyl compounds **227** and β-phosphorylated hydroxy derivatives **228** were identified on the example reaction of benzyl(diphenyl)phosphine oxide (**229**) with benzaldehyde in the presence of phenyllithium [138].

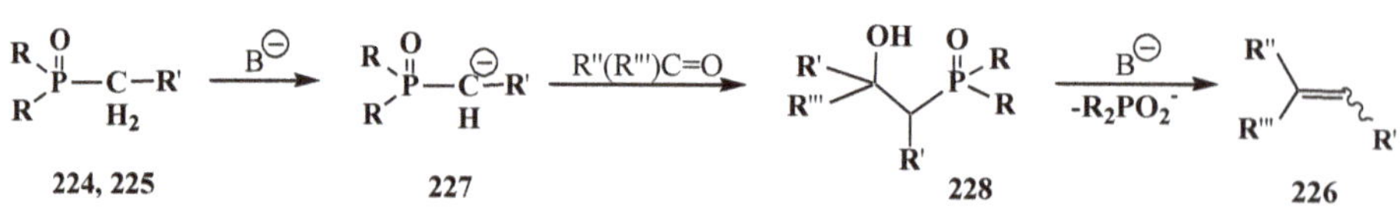

Scheme 69. Syntheses of olefins **226** from aldehydes and ketones by means of Horner's reaction, R = OAlk, Ph; R′ = Alk; R″, R‴ = H, Alk, Ar.

The reaction has a number of advantages in comparison with similar reaction of phosphorus ylides previously described by L. Wittig [139] where ketones are difficult to react, whereas both aldehydes and ketones undergo the Horner reaction. It was further shown that Horner reaction has a larger synthetic potential and is applicable for the synthesis of other types of organic compounds, for example, allenes, cyclopropanes, terminal [140] and disubstituted alkynes [132]. The involvement of phosphonates functionalized at the α-position with dialkylamino, alkoxy or alkylthio groups in the reaction leads to enamines, vinyl ethers [132,141–143] and vinyl thioethers [141,143]. Their subsequent hydrolysis affords aldehydes and ketones with elongated hydrocarbon chain in high yields (homologation).

The Horner reaction also provides the possibility to prepare carboxylic acids homologized by one carbon atom via the shortest route starting from phosphinoyl compounds functionalized at the α-position with two heteroatoms, namely, phosphorylated formaldehyde acetals and structurally related compounds [132,141]. In this case, carbanions **230** of phosphorylated formaldehyde acetals and structurally related compounds **13–16**, **18–32** behave as a masked form of triply functionalized carbanions **231** that may be considered as a synthetic equivalent or carrier of reversed-polarity formate carbanion $[O{=}C{-}OH]^-$ **232** [20,141] (Scheme 70).

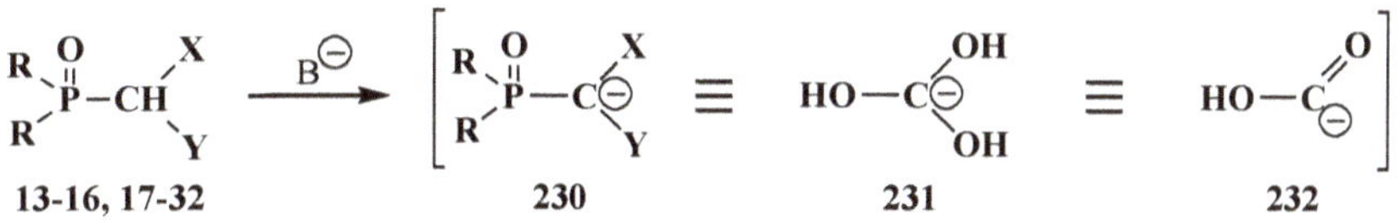

Scheme 70. Compounds **13–16**, **18–32** as hidden form of reversed-polarity formate carbanion **232**, where R = OAlk, Ph; X, Y = AlkO, AlkS, Alk_2N, CN, $R_2P(O)$, Hal, AlkS(O).

Carbanions **230**, prepared by the deprotonation of the initial phosphoryl compound, react with carbonyl compounds to afford β-phosphorylated alcohols **233**, which can be isolated. The subsequent treatment of alcohols **233** with strong bases, usually potassium *tert*-butoxide, leads to ketene acetals and structurally related compounds **234** that are valuable precursors in the synthesis of organic compounds of different kinds [23,24,132,133,143]. Further acid hydrolysis of compounds **234** produces carboxylic acids **235** (Scheme 71) or their derivatives, for example esters **236**, or thioesters **237**, depending on the conditions.

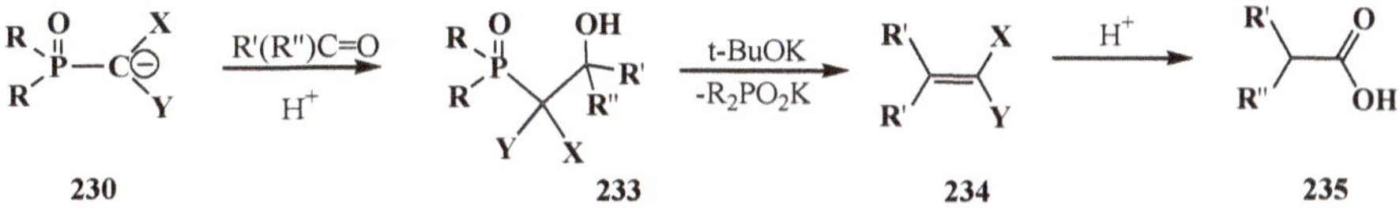

Scheme 71. Syntheses of carboxylic acids **235** by means of Horner's reaction, R′,R″ = H, Alk, Ar.

However, the simplest and most available phosphorylated formaldehyde acetals, dialkyl (dialkoxymethyl)phosphonates **33**, do not form stable carbanions [23,132], therefore the attempted synthesis of carboxylic acids and their derivatives by Horner reaction failed for a long time. Among acetals of phosphonate type compounds, only diethyl (5,6-dichloro-1,3-benzodioxomethyl)phosphonate (**91**) participated in the reaction with ketones at 90 °C in dioxane in the presence of sodium hydride NaH to give ketene acetals in 19–32% yields [23]. See also Section 3.3 "Alkylation of formacetal carbon atom".

Carboxylic acids were obtained for the first time by Horner reaction in 75–90% yields in 1968 by reacting tetraethyl (*N*,*N*-dimethylaminomethyl)diphosphonate (**161**) with aliphatic and aromatic aldehydes [27]. After formation of the carbanion **222**, the reaction proceeds through the sequential formation of 1-dimethylaminoalkenylphosphonates—α-phosphorylated enamines **236**, then α-phosphinoylacyl derivatives **237**, and finally yields free linear acids **238** (Scheme 72).

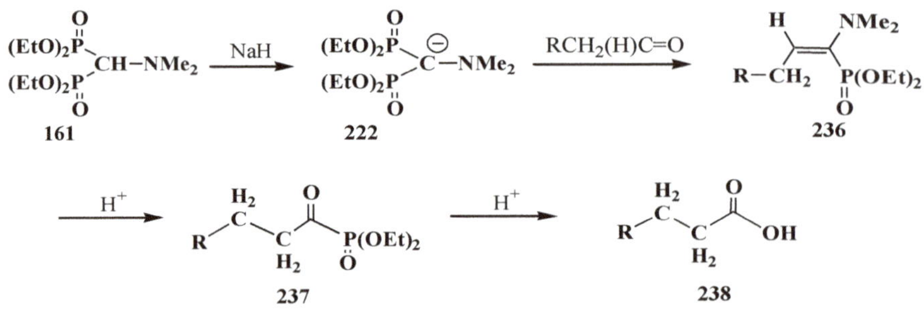

Scheme 72. Synthesis of carboxylic acids **238** by reacting tetraethyl (*N*,*N*-dimethylaminomethyl)diphosphonate (**161**) with aliphatic and aromatic aldehydes.

Since phosphorylated enamines **236** (synthesized from aliphatic aldehydes only) contain an anion-stabilizing diethoxyphosphinoyl group in the α-position, the methylene group in the γ-position is readily deprotonated under the action of strong bases. The resulting phosphorylated aminoallyl anions **239** react with alkyl halides and finally form carboxylic acids **240** branched at the β-position via the phosphorylated enamines **241**. The reaction of anions **239** with aldehydes gives rise to hydroxy compounds (**242**) and then to β,γ-disubstituted γ-butyrolactones **243** over unisolated γ-hydroxy carboxylic acids that undergo fast cyclization under the reaction conditions [144] (Scheme 73).

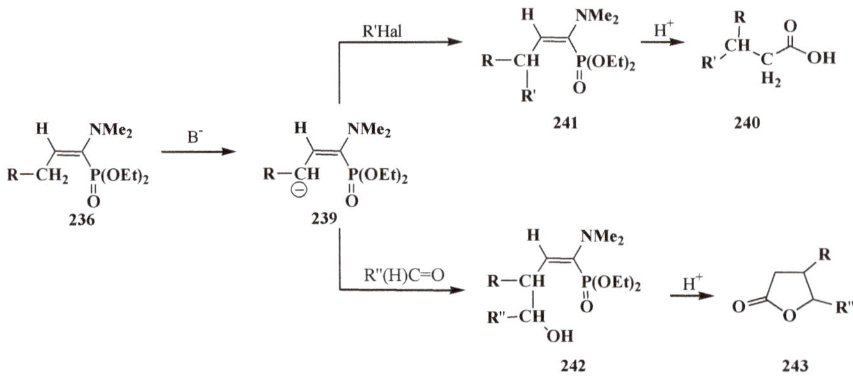

Scheme 73. Syntheses of carboxylic acids **240** and γ-butyrolactones **243** from phosphorylated enamines **236**.

Because only aldehydes react with compound **161** [27,132], this method of synthesis of carboxylic acids is not widely used. However, compound **161** is employed for the preparation of substituted acetic acids as intermediate stages in the synthesis of potential pharmaceuticals [124,145] and pesticides [28], for example, acid **244** (Scheme 74).

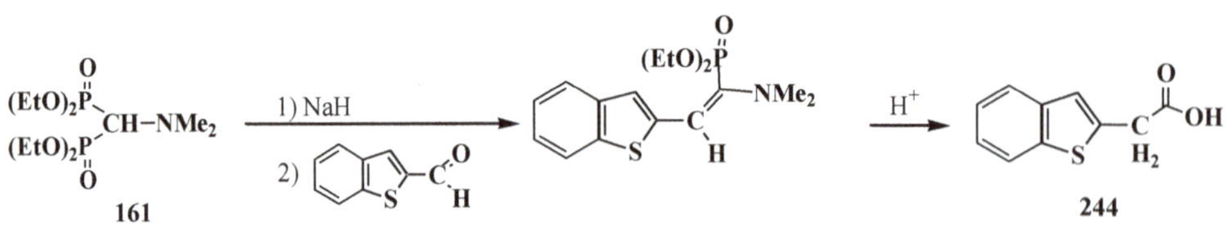

Scheme 74. Synthesis of 2-benzothienylacetic acid **244** by Horner's reaction.

The successful homologation of aldehydes with the use of diphosphonate **161** stimulated further search for the synthetic equivalents of formate carbanion **232** among organophosphorus compounds. In 1976–1977, linear (**107**) and cyclic dialkyl (dialkythiomethyl)phosphonates **113**, **114** were proposed [21, 146]. These compounds can form a stable carbanion **245**, and react with both aldehydes and ketones to form ketene thioacetals **246** and further under subsequent hydrolysis (over thioesters **237**) produce homologous carboxylic acids **235** (Scheme 75), see also section "Alkylation of formacetal carbon atom". Ketene thioacetals **246** were obtained from ketones and aldehydes [102,146], including those unsaturated, in 66–82% and 80–96% yields, respectively, as mixtures of *Z*/*E* isomers [23].

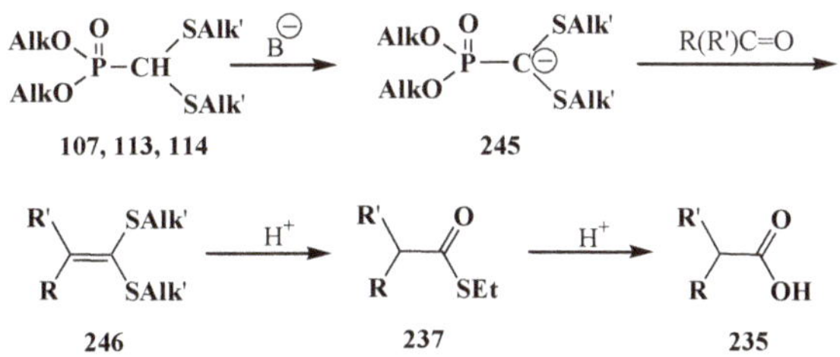

Scheme 75. Syntheses of carboxylic acids **235** by reacting thiomethylphosphonates **107**, **113**, **114** with aldehydes and ketones.

The reaction is observed also for diphenyl([1,3]-dithian-2-yl)phosphine oxide (**122**) [123,124] and diethyl (1,3-benzodithiolylmethyl)phosphonate (**247**) [113,147]. Phosphonate **247** produces benzo-analogs of ketene thioacetals, 1,4-benzodithiafulvenes **248**, in 92–98% yields when reacted with carbonyl compounds (Scheme 76).

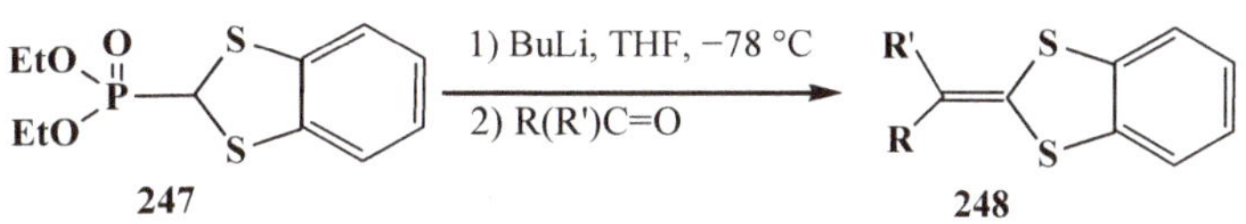

Scheme 76. Syntheses 1,4-benzodithiafulvenes **248** from diphenyl([1,3]-dithian-2-yl)phosphine oxide (**122**) by the reaction with carbonyl compounds.

Phosphorylated thioacetals **107**, **113**, **114** and thioacetals produced from its ketene **246**, **248** have a synthetic importance because, along with carboxylic acid synthesis, they undergo numerous reactions to afford various products [23]. For example, aldehydes **250**, branched out in α-position are formed in the reduction of 1,4-benzodithiafulvenes **248** followed by hydrolysis (over the stage of reduced compounds **249**) [148] (Scheme 77).

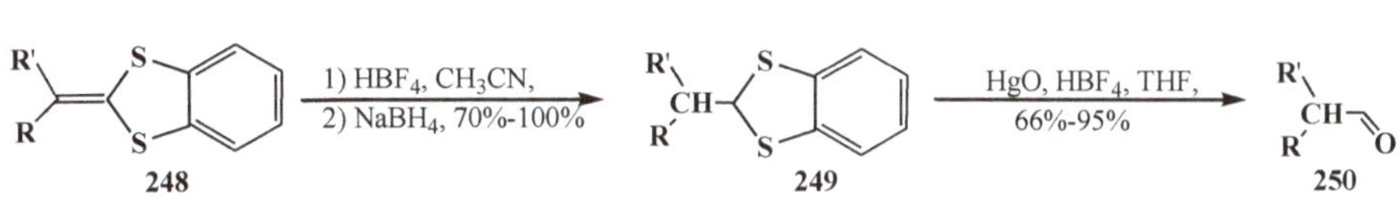

Scheme 77. Obtaining branched aldehydes **250** from 1,4-benzodithiafulvenes **248**.

Methyl esters **251** result from thioacetal **252** methanolysis [149] (Scheme 78).

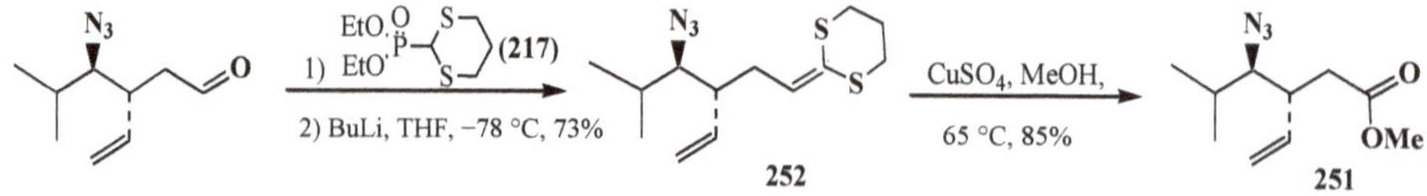

Scheme 78. Synthesis of methyl esters **251** by methanolysis of thioacetal **252**.

Dimethyl (dimethylthiomethyl)phosphonate **253** is used in the practice of contemporary organic chemistry, for example, in the intermediate stages of synthesis of biologically active dipeptide mimetics **254** [149] (Scheme 79), the antibiotic thienamycin (**255**) [150] (Scheme 80), and "organic metals" **256** [116,151–154], for example (Scheme 81).

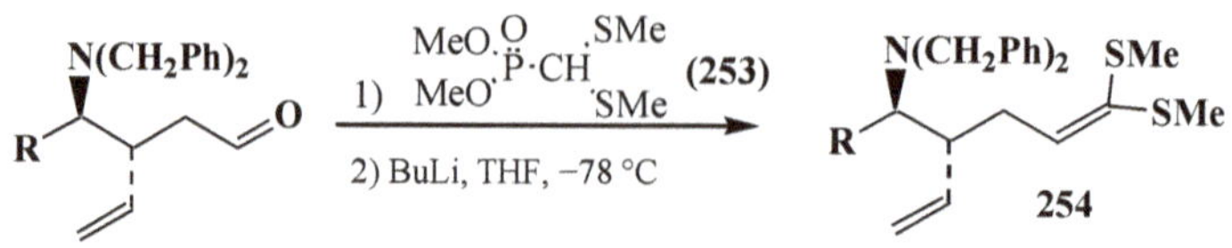

Scheme 79. Synthesis of compound **254**—an intermediate stage of synthesis of dipeptide mimetics.

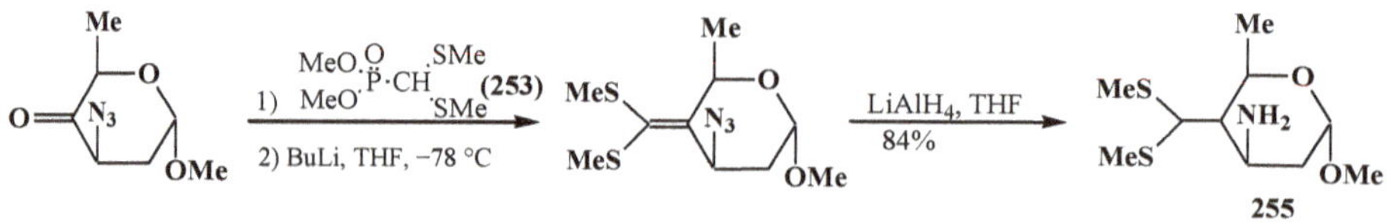

Scheme 80. Synthesis of the antibiotic thienamycin (**255**) by means of thiomethylphosphonate **253**.

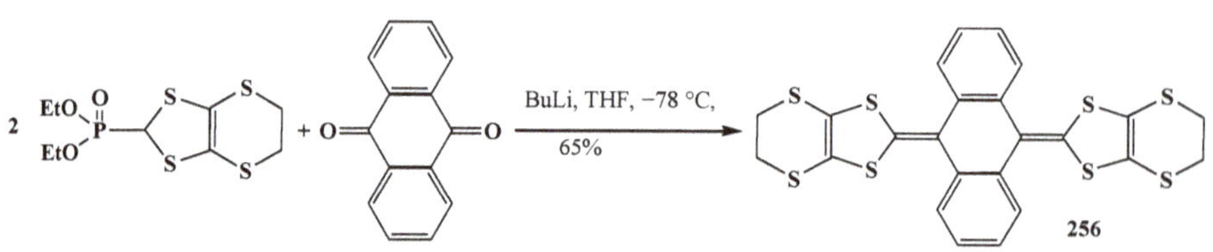

Scheme 81. Example of synthesis of "organic metals" **256**.

However, the long duration of the two-stage conversion of ketene thioacetals **246** into acids **235**, often in the presence of mercury Hg^{2+} [141] or copper Cu^{2+} salts [149] and the necessity of working with mercaptans [23,141,154] limits the application of phosphorylated formaldehyde thioacetals **14** in the Horner reaction.

Therefore, the search for efficient precursors for the synthesis of carboxylic acids **235** from carbonyl compounds by the Horner reaction has continued. From the mid-1970s to the early 1980s, many acetal-like derivatives of diethyl formylphosphonates **13–32** [18,19,21–56], where the negative charge of the carbanion was stabilized by two heteroatoms of the "acetal" group, were obtained [54]. See also Figure 2, where R = OEt. In the case of X = Me_3Si, Peterson olefination prevails over the Horner reaction [54–56,110] and trimethylsyloxy fragment is a leaving group.

Because the majority of compounds **18–32** have no substantial advantages over the phosphorylated formaldehyde thioacetals **14**, the study of Horner reaction with their participation was confined mainly to academic interest. More detailed studies of reactivity of the majority of these compounds were not conducted.

Among the compounds synthesized over this period, diethyl [(*N,N*-dialkylamino)cyano-methyl]phosphonate (**153**) and diphenyl(dialkoxymethyl)phosphine oxides **50** were involved in the practice of organic synthesis.

Compound **153** proposed in 1982 [24] reacts like dialkyl (dialkylthiomethyl)phosphonates **107** with aldehydes, in 50–69% yields, and acetophenone as ketone example, in 24% yield [24,132]. The products of Horner reaction in this case are cyanoenamines **257**, whose acid hydrolysis produces linear carboxylic acids **238** homologous to the initial carbonyl compounds [24,25,132] (Scheme 82). See also Scheme 72.

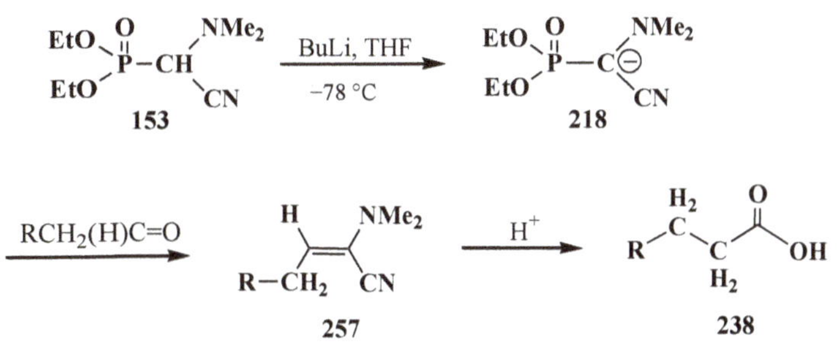

Scheme 82. Carboxylic acids **238** synthesis by means of diethyl [(*N,N*-dialkylamino)cyanomethyl]phosphonates **153**, where R = Alk, Ph.

Like phosphorylated enamines **236** (Scheme 73), compound **257** contains an anion-stabilizing CN-group in the α-position. Resulting cyanoaminoallyl anions **258** combine with alkyl halides to form carboxylic acids **240** through cyanoenamines **259**, while the reaction with aldehydes leads to γ-hydroxy acids that undergo cyclization to give β,γ-disubstituted γ-butyrolactones **243** through cyano aminoallylalcohols **260** [24,25,132,144] (Scheme 83).

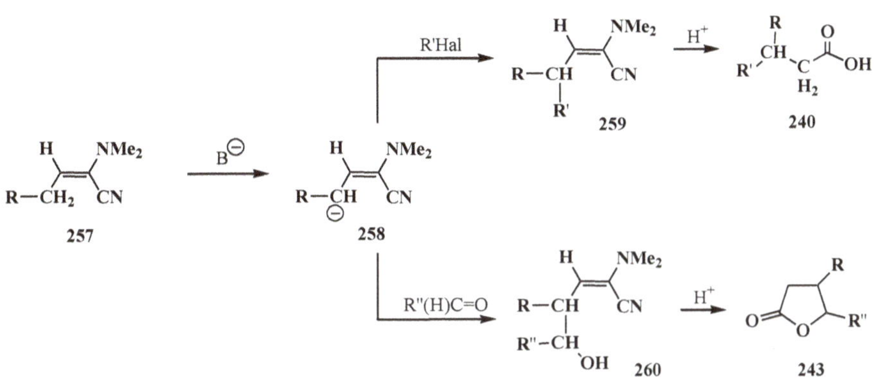

Scheme 83. Enamines **257** alkylation by means of alkyl halides or aldehydes leads to obtaining branched acids **240** or butyrolactones **243**.

Diethyl [(*N*-morpholino)cyanomethyl]phosphonate (**154**) was used in the synthesis of the colerofragarone fragment—terminator of fungi *Collerotrichum fragariac* [155] (Scheme 84, compound **261**).

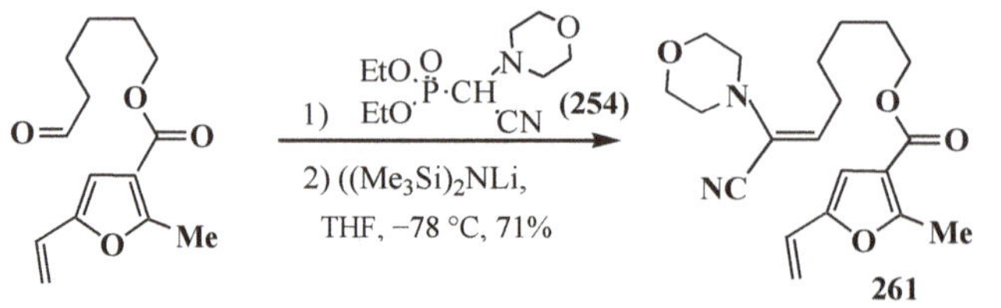

Scheme 84. Synthesis of colerofragarone fragment **262** using diethyl [(*N*-morpholino)cyanomethyl]phosphonate (**154**).

It was shown that compound **153** can react also with cyclic semiacetal **262**, which was used in one of the stages of synthesis of prostaglandin analog cloprosterol PGF_2 [155,156] (Scheme 85, compound **263**).

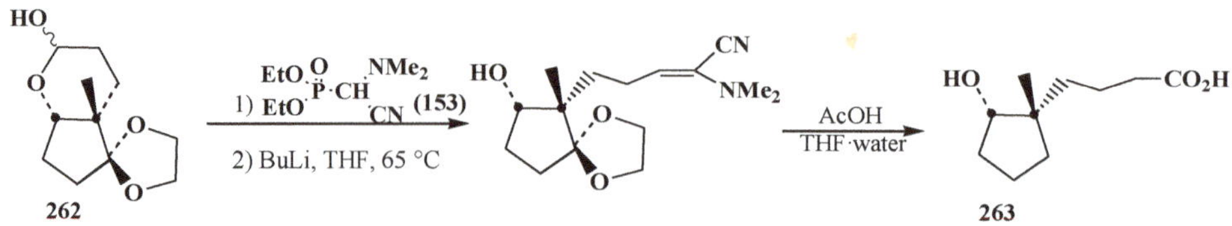

Scheme 85. Synthesis of **263**, a semi-product of synthesis of the analog of cloprosterol PGF_2 from cyclic semiacetal **262** by means of phosphonate **153**.

However, compound **153** was not widely used because of the low yields in its reactions with ketones [25] and formation of hydrogen cyanide on hydrolysis of cyanoenamines. Almost simultaneously with **153**, a successful synthesis of ketene acetal at −110 °C starting from diphenyl(dialkoxymethyl)phosphine oxides **50**, over carbanion **195** and both aldehydes and ketones was reported in 1983 [133]. Intermediate β-hydroxydiphenylphosphinoyl derivatives **264** were isolated in almost quantitative yield that further react with potassium *tert*-butoxide (*t*-BuOK) to give ketene acetals **265** in 45–85% yields [133] (Scheme 86).

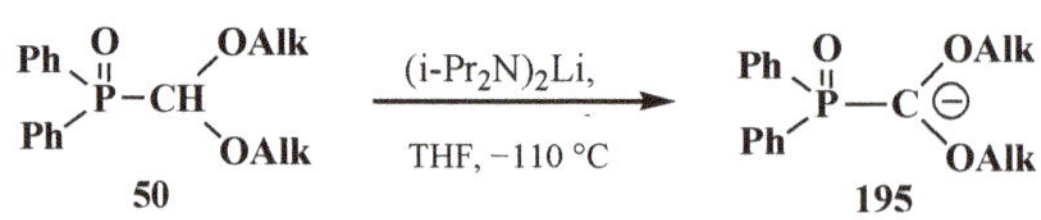

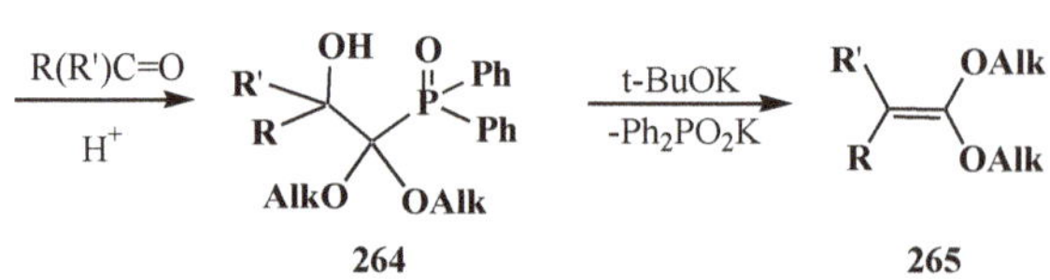

Scheme 86. Synthesis of ketene acetals **265** from diphenyl(dialkoxymethyl)phosphine oxides **50** through conpounds **264** as intermediates.

This publication attracted no attention for a long time, although in 1993 it was shown that the reaction of diphenyl(diethoxymethyl)phosphine oxide (**266**) with substituted cyclohexadienone in the presence of lithium diisopropylamide leads to the formation of carboxylic acid **267** through the stages of ketene acetal **268** and ethyl ester **269** formation [20,157] (Scheme 87).

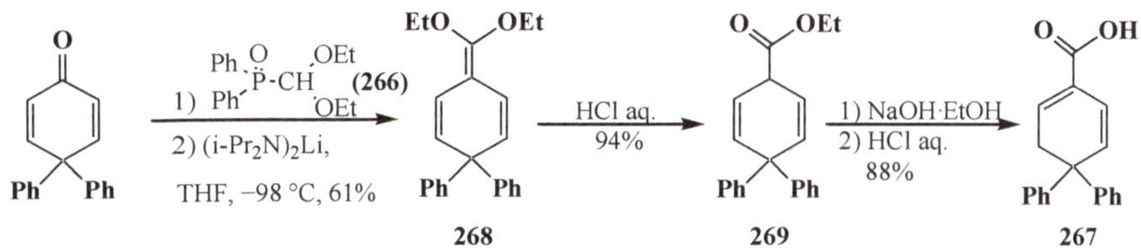

Scheme 87. Synthesis of carboxylic acid **267** from (diethoxymethyl)phosphine oxide (**266**) through ketene acetal **268** and ethyl ester **269**.

It was found in 2002, however, that the reaction of diphenyl(dimethoxymethyl)phosphine oxide (**209**) with aliphatic aldehydes, in contrast to other phosphorylated equivalents of formate anion—compounds **107**, **112–115**, **153** (**154**), **161** (**166**), may also result in derivatives of α-hydroxycarboxylic acids **270** [140,158,159]. Initially formed β-hydroxydiphenylphosphinoyl derivatives **271** in wet acidified dichloromethane readily decompose with the cleavage of the P–C bond to afford methyl esters of α-hydroxycarboxylic acids **272** in 41–89% yield and diphenylphosphine oxide (**53**) (Scheme 88).

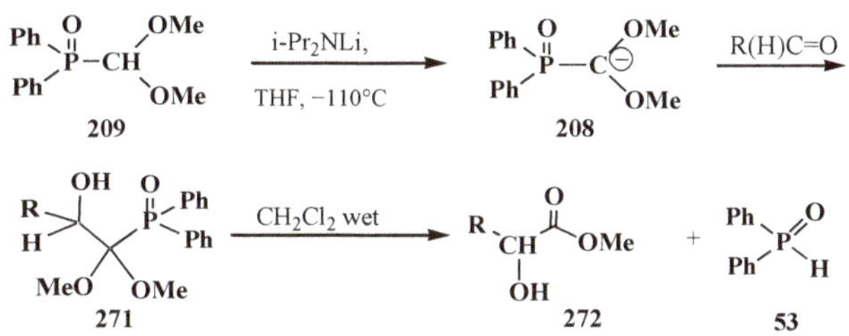

Scheme 88. Synthesis and decomposition of β-hydroxydiphenylphosphinoyl compounds **271** lead to methyl esters of α-hydroxycarboxylic acids **272**.

It was further shown that ketene acetals **273** resulting from aldehydes and phosphine oxide **266** may be successfully oxidized [20,128,160] in the Sharpless asymmetric dihydroxylation—AD reaction [161]. As a result, the esters of chiral α-hydroxycarboxylic acids **274**, **275** were obtained in 49–94% yields and enantiomeric excess up to 98% (Scheme 89).

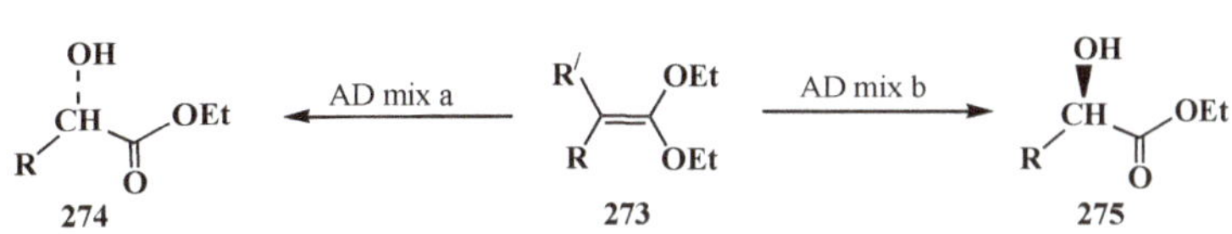

Scheme 89. Using of Sharpless asymmetric dihydroxylation (AD reaction) for synthesis of chiral α-hydroxycarboxylic acids **274**, **275** from ketene acetals **273**.

It should be noted that ketene thioacetals **246** do not undergo the AD reaction. Mixed *O*,*S*-acetal **18** reacts but yields and enantiomeric excess of α-hydroxycarboxylic acids do not exceed 7–37% and 80%, correspondingly [127,128]. The combination of Horner and AD reactions was successfully used on one of stages of synthesis of the diterpenoid tonantzitlolone (**276**) [162] (Scheme 90).

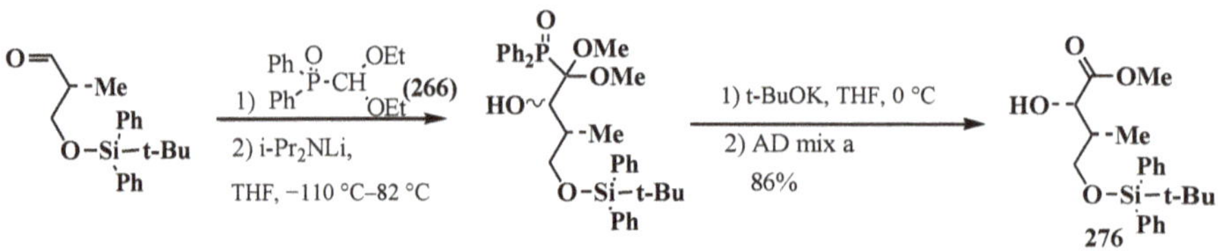

Scheme 90. Using of combination of Horner and AD reactions to obtain compound **276**—intermediate stage of synthesis of the diterpenoid tonantzitlolone.

The involvement of chiral α-alkylaminoaldehydes in the Horner reaction, correspondingly, leads to the synthesis of esters of chiral β-amino acids **277** [20] (Scheme 91).

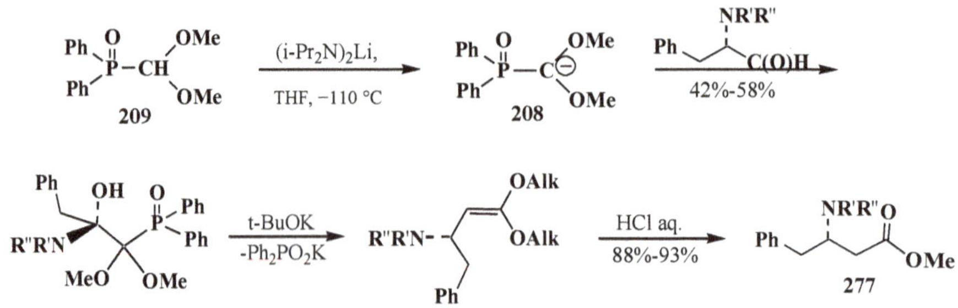

Scheme 91. Synthesis of chiral β-amino acids **277** by the Horner reaction.

Sharpless asymmetric dihydroxylation followed of Horner reaction also provides an opportunity to synthesize α-hydroxy-β-amino acid diastereomers **278, 279** [20,128] (Scheme 92).

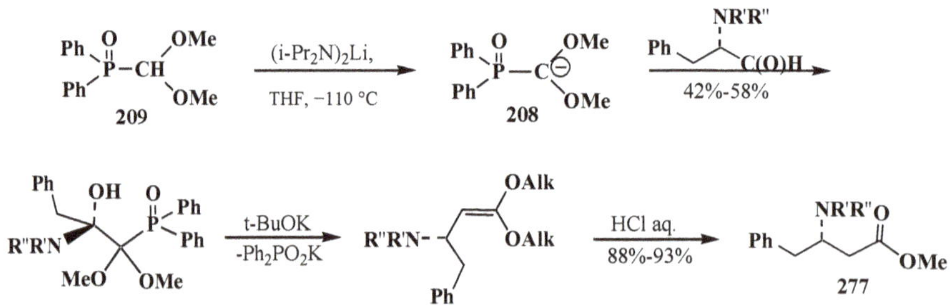

Scheme 92. Using of combination of Horner and AD reactions to form α-hydroxy-β-amino acid diastereomers **278, 279** (R = *t*-BuOC(O)).

4. Phosphorylated Formaldehyde Halogenoaminals (Phosphorylated Vilsmeier–Haak Reagents) 17

At present, four types of phosphorus compounds are known that can be related to phosphorylated Vilsmeier–Haak reagents: *N,N,N′,N′*-tetraisopropyl [(*N″,N″*-diisopropylamino)methylydeniminium] phosphondiamide dichlorophosphate (**8**) [12], dialkyl [(*N,N*-dimethylamino)chloromethyl]phosphonates **171** [31,32], diphenyl[(*N,N*-dialkylamino)chloromethyl]phosphine oxides **280** [163], and structurally related to the phosphorylated Vilsmeier–Haak reagents, diethyl [(*N*-acylamino)bromomethyl]phosphonates **281** [164] (Figure 6).

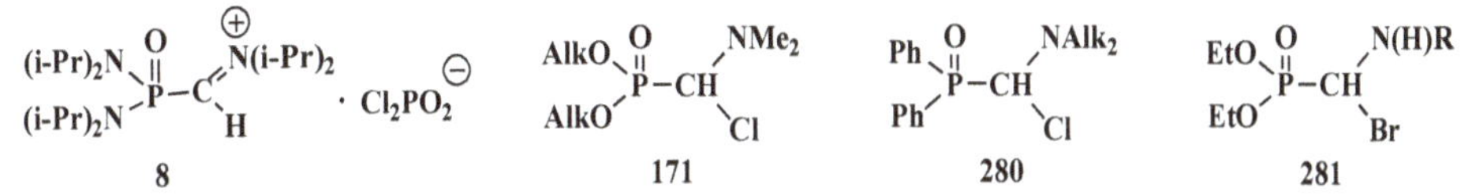

Figure 6. Representatives of phosphorylated Vilsmeier–Haak reagents **8**, **171**, **280**, **281** synthesized to date.

No general methods of synthesis have been found for the compounds of this type **8**, **171**, **280**, **281**. Their chemical properties also differ significantly from each other.

4.1. Synthesis of N,N,N′,N′-Tetraisopropyl [(N″,N″-Diisopropylamino)methylydeniminium] Phosphondiamide Dichlorophosphate (**8**)

A sole paper was published in 1999 [12] on the synthesis of *N,N,N′,N′*-tetraisopropyl [(*N″,N″*-diisopropylamino)methylydeniminium]phosphondiamide dichlorophosphate (**8**) by the oxidation of *N,N,N′,N′*-tetraisopropyl [(*N″,N″*-diisopropylamino)methylydeniminium]phosphindiamide dichlorophosphate (**282**) with dimethyl sulfoxide DMSO (Scheme 93). However, the chemical properties of **8** are insufficiently studied (for some chemical properties of **8** and its derivatives, see Section 1. Introduction).

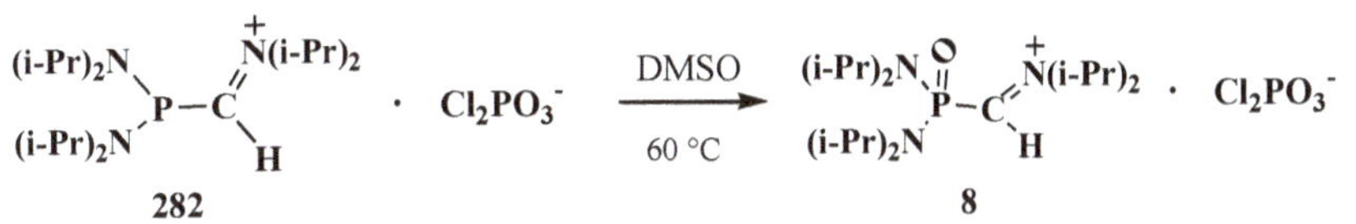

Scheme 93. Phosphindiamide dichlorophosphate **282** oxidation by means of DMSO leads to the formation of phosphondiamide dichlorophosphate **8**.

4.2. Synthesis and Chemical Properties of Dialkyl [(N,N-Dimethylamino)chloromethyl]phosphonates **171**

The first compounds of this type, dimethyl [(*N,N*-dimethylamino)chloromethyl] phosphonate (**172**) and diethyl [(*N,N*-dimethylamino)chloromethyl]phosphonate (**173**) were synthesized in 1969 [31,32]. For the first time, these compounds were obtained as intermediates in the synthesis of tetraalkyl (*N,N*-dimethylaminomethyl)diphosphonates **174** by reaction of N,N-dimethyl(chloromethylideneiminium) chloride (**171**) with trialkyl phosphites **58** [32]—see Scheme 49, section "Diphosphinoyl *N,N*-dialkylaminomethanes".

Preparative amounts of these compounds were prepared by a two-step scheme [30,31]. First, the reaction of dimethylformamide dimethylacetal (**78**) with dialkyl phosphites **51** resulted in dialkyl [(*N,N*-dimethylamino)methoxymethyl]phosphonates **162**, which were further reacted with thionyl chloride ($SOCl_2$) at 0 °C to yield the final products **172**, **173** (Scheme 94). See also Schemes 49 and 50.

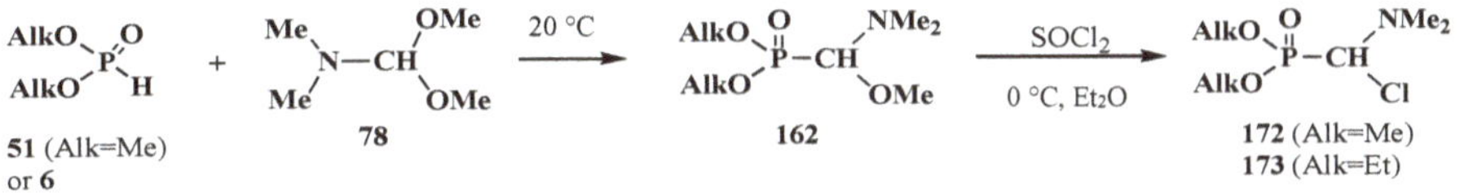

Scheme 94. Reaction dialkyl phosphites **51** and dimethylformamide dimethylacetal (**78**) leads to phosphorilated halogenoaminals **172** (Alk = Me), **173** (Alk = Et).

This scheme is currently used [33,165]; phosphorus trichloride can be used instead of thionyl chloride [83]. Dimethyl [(*N,N*-dimethylamino)chloromethyl]phosphonate (**172**) was found to undergo spontaneous dealkylation on storage [31], therefore only diethyl [(*N*,*N*-dimethylamino)chloromethyl]phosphonate (**173**) is used in organic synthesis. Phosphonate **173**, as a phosphorylated Vilsmeier–Haak reagent where one chlorine atom is substituted by a diethoxyphosphoryl group, show electrophilic properties [166] inherent in compounds of such kind and readily reacts with nucleophiles. Thus, compound **173** reacts vigorously with secondary amines, carbazole, alcohols, thiols, and hydrophosphinoyl compounds **51**, **53** in the presence of equimolar amount of triethylamine. The reaction of α-phosphono-α-aminomethylation leads to the corresponding diethoxyphosphinoylformaldehyde derivatives: asymmetric aminals **283**, **284**, aminoacetals **285**, aminothioacetals **286**, and diphosphorylated *N*,*N*-dimethylaminomethanes **164** [30], including unsymmetrical **165** [165] (Scheme 95). The reaction with aromatic and aliphatic amides RC(O)NH_2, urea and thiourea $(NH_2)_2C$=X [30] proceeds in a similar manner. Products of reactions are phosphorylated amidoaminals **287**, **288** and symmetrically disubstituted derivatives of urea and thiourea **289**, **290** (Scheme 95). See also Scheme 53, section "Diphosphinoyl *N*,*N*-dialkylaminomethanes".

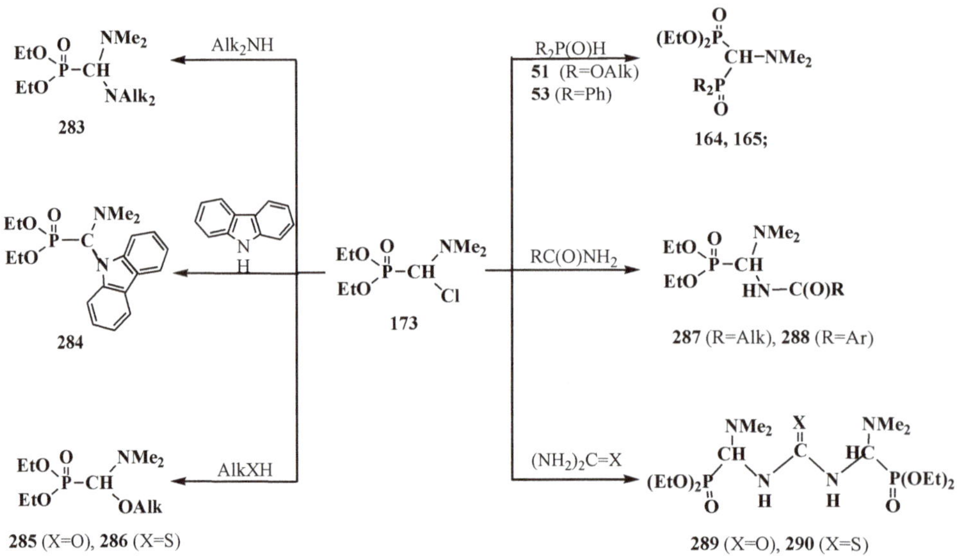

Scheme 95. Reactions of diethyl [(*N*,*N*-dimethylamino)chloromethyl]phosphonate (**173**) with nucleophilic co-reagents.

Diethyl [(*N*,*N*-dimethylamino)chloromethyl]phosphonate (**173**) reacts with arylsulfonamides to lead to the cleavage of the phosphorus–carbon bond and give *N*-sulfonyl-substituted derivatives of formamidines **291** and diethyl phosphite (**6**) [30] (Scheme 96). See also Section 3.1. Cleavage of phosphorus–carbon bond.

Scheme 96. Cleavage of phosphorus–carbon bond by means of the reaction of compound **173** with arylsulfonamides.

Compound **173** reacts with methyl aryl ketones [30] and dialkyl or aryl alkyl ketones [33] to give diethyl [(*N*,*N*-dimethylamino)(aroyl(or alkanoyl)alkylmethyl)methyl]phosphonates **292** and **293** [30,33], phosphorylated analogs of natural α-amino acids potentially possessing biological activity [33] (Scheme 97). However, reaction stereoselectivity is low, and the *anti*/*syn* ratio is 2:1.

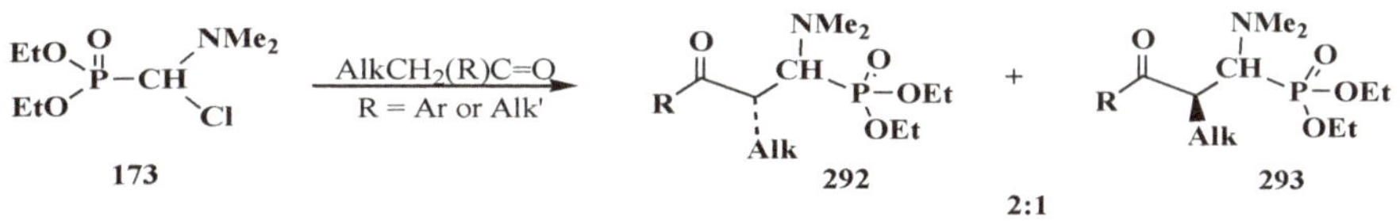

Scheme 97. Reaction of compound **173** with ketones leads to diethyl [(*N*,*N*-dimethylamino)(aroyl(or alkanoyl)alkylmethyl)methyl]phosphonates **292** and **293**.

Like ketones, the reaction of **173** proceeds also with aldehydes branched at the α-position to the carbonyl group to give phosphorylated aminoaldehydes **294** and **295** (Scheme 98), but the reaction is not stereoselective: the stereoisomer ratio of the resulting α-formyl-α-methylaminophosphonates is 1:1 [33].

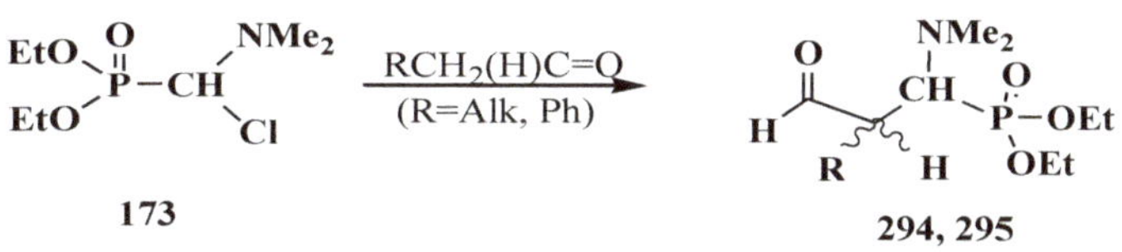

Scheme 98. Reaction of compound **173** with aldehydes leads to phosphorylated aminoaldehydes **294** and **295**.

The involvement of enamines instead of ketones as their synthetic equivalents in the reaction with phosphonate **173** provides stereoselective synthesis of diethyl (α-methylamino-α-aroyl(or alkanoyl))methylphosphonates **292**; the reaction leads to the products with only *anti* configuration [33] (Scheme 99).

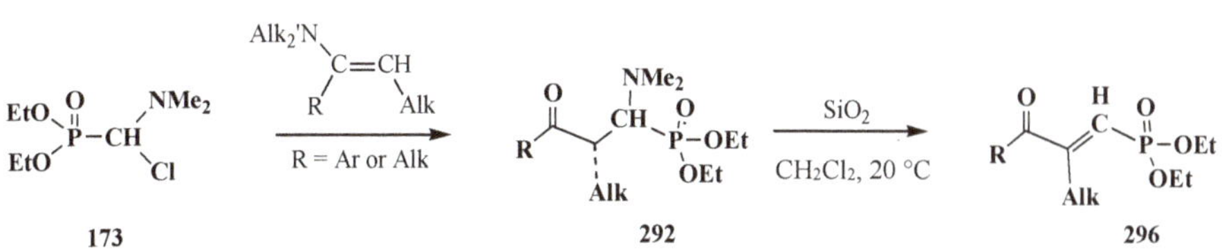

Scheme 99. Synthesis of diethyl (α-methylamino-α-aroyl(or alkanoyl))methylphosphonates **292** and phosphorylated vinyl ketones **296**.

Diethyl [(*N*,*N*-dimethylamino)(aroyl(or alkanoyl)alkylmethyl)methyl]phosphonates **292**, **293** (Schemes 101–103) are obtained as hydrochlorides in the above-stated syntheses. They are unstable and on storage undergo spontaneous β-elimination of dimethylamino group as dimethylammonium chloride—with conversion up to 40% over 24 h [30] to form the corresponding phosphorylated vinyl ketones **296**. The same result was achieved on the heating of hydrochlorides of compounds **292** and **293** to 100 °C over 45 min [30] or stirring their solutions in methylene chloride with silica gel at 20 °C for 15 h [33]. The reaction of amino group elimination with the use of silica gel is also stereoselective: the anti stereoisomers **292** produce only *E* isomers of vinyl compounds **297** [33] (Scheme 99).

The reaction of β-enaminonitriles **298** with compound **173** was used in the synthesis of substituted nitriles of nicotinic acid **299** [167] (Scheme 100).

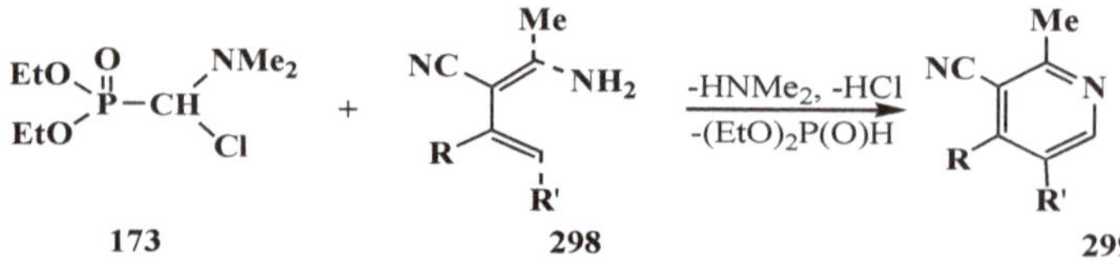

Scheme 100. The reaction of compound **173** with β-enaminonitriles **298** leads to nitriles of nicotinic acid **299**.

As a phosphorylated Vilsmeier–Haak reagent, phosphonate **173** undergoes typical reactions with activated aromatic compounds: *N*,*N*-dimethylaniline, triethylammonium salts of β-naphthol and *p*-cresol, and heteroaromatic compounds: *N*-methylindole, *N*-methylpyrrole, α-methylfuran. The synthesis of aromatic α-aminomethylphosphonates **300** is regiospecific: only one addition product forms in each case [33] (Scheme 101).

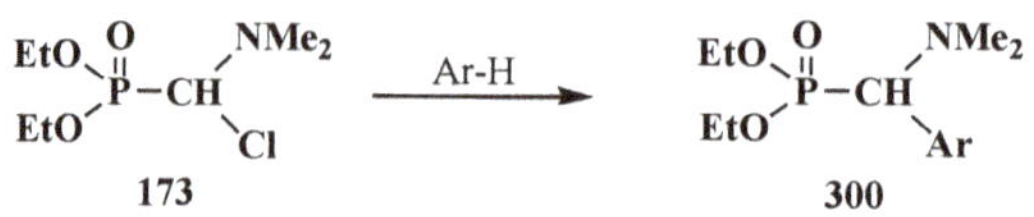

Scheme 101. Interactions of phosphonate **173** with aromatic compounds (Ar = 4-dimethylaminophenyl; 2-hydroxynaphthyl-1; 2-hydroxy(5-methyl)phenyl; 1-methylaminoindol-3-yl; 1,5-dimethylaminopyrrol-2-yl; 5-methylfur-2-yl).

4.3. *Syntheses and Chemical Properties of Diphenyl[(N,N-dialkylamino)chloromethyl]phosphine Oxides* **280**

Diphenyl[(*N*,*N*-dialkylamino)chloromethyl]phosphine oxides **280** are the diphenylphosphinoyl analogs of dialkyl [((*N*,*N*-dialkylamino)chloromethyl]phosphonates. They are obtained by the reaction of chlorodiphenylphosphine (**68**) with *N*,*N*-dialkylformamides in the presence of 10–20% of *N*,*N*-dialkylchloromethyleniminium chloride $[Alk_2N{=}C(H)Cl]^+Cl^-$ (**168**)—Vilsmeier–Haak reagent [168,169] according to Scheme 102.

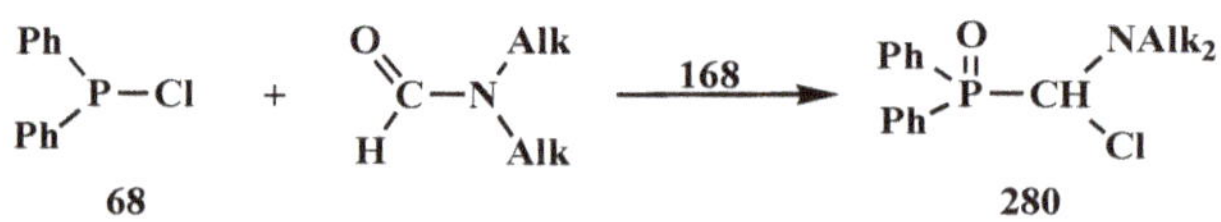

Scheme 102. Synthesis of diphenyl[(*N*,*N*-dialkylamino)chloromethyl]phosphine oxides **280**.

Although phosphine oxides **280** are the formal analogs of dialkyl [(*N*,*N*-dialkylamino)chloromethyl]phosphonates **171**, they differ substantially in reactivity. It was shown by the example of diphenyl[(*N*,*N*-dimethylamino)chloromethyl]phosphine oxide (**301**) that compounds **280** can dissociate in solutions with the cleavage of C–Cl and C–P bonds. See also section "Cleavage of phosphorus–carbon bond". The dissociation in solution results in the formation of both diphenyl[(*N*,*N*-dimethylamino)chloromethyliden]phosphine oxide cation (**302**) and diphenylphosphinite anion (**193**) [170] (Scheme 103).

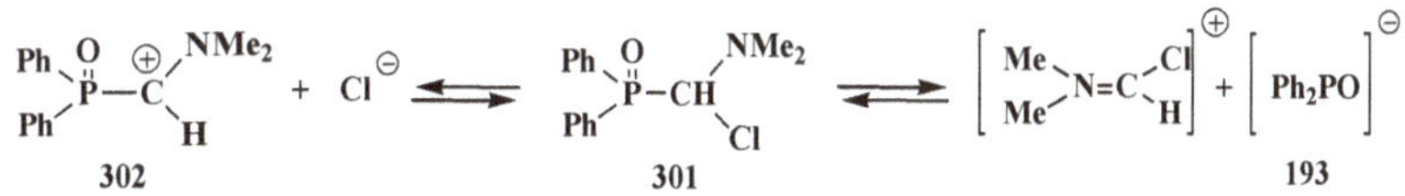

Scheme 103. Diphenyl[(*N,N*-dimethylamino)chloromethyl]phosphine oxide (**301**) dissociation in solutions with a cleavage as bond **C-Cl** and bond **C-P**.

Therefore there are two kinds of reactivity for **301** (and accordingly **280**) [163,170]: electrophilic diphenyl[(*N,N*-dimethylamino)chloromethylidene]phosphine oxide cation (**302**) reacts with nucleophiles such as hydrophosphinoyl compounds **6**, **53**, mixed diethyl trimethylsilyl phosphite $(EtO)_2POSiMe_3$ and *N,N*-dimethylamino aniline $PhNMe_2$, to give symmetrical **303**, R=Ph and unsymmetrical **304**, R=OEt diphosphorylated *N,N*-dimethylaminomethanes and diphenyl[(bis(4-*N,N*-dimethylamino)phenyl)methyl] phosphine oxide (**305**), the product of substitution of both chloro and amino groups at the carbon atom of **301**. However, nucleophilic diphenylphosphinite anion (**193**) reacts with electrophiles: acetic aldehyde MeC(O)H, acetone $Me_2C(O)$, phenyl isocyanate PhNCO, triethyl orthoformate $(EtO)_3CH$, dimethylformamide dimethylacetal (**79**), and bis(diethylamino)methane $(Et_2N)_2CH_2$ to form addition or substitution products, α-phosphorylated compounds: alcohols **306** and **307**, carbamoyl compound **308**, acetal **309**, aminoacetal **310** and amine **311** (Scheme 104).

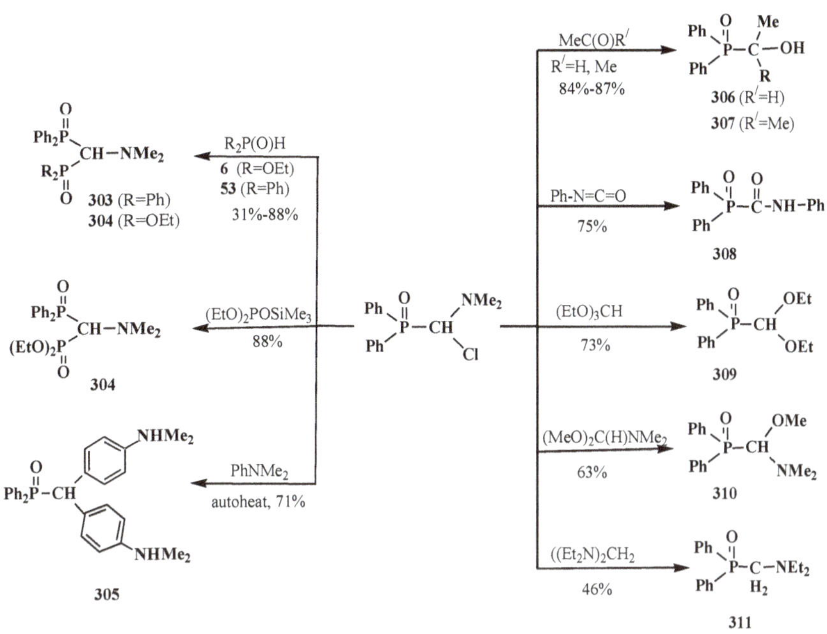

Scheme 104. Reactions of diphenyl[(*N,N*-dimethylamino)chloromethyl]phosphine oxide (**280**) with nucleophilic and electrophilic coreagents.

4.4. *Synthesis and Chemical Properties of Diethyl [(N-Acylamino)bromomethyl]phosphonates* 281

Diethyl [(*N*-acylamino)bromomethyl]phosphonates **281** were prepared in high yields by the bromination of diethyl (*N*-acylamino)methylphosphonates **312** with N-bromosuccinimide (NBS) in carbon tetrachloride [164,171] (Scheme 105).

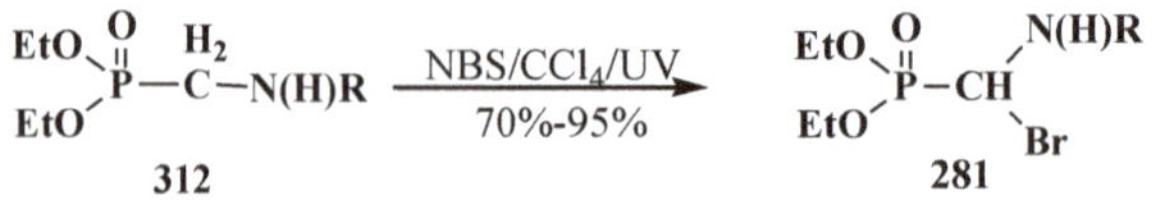

Scheme 105. Synthesis of diethyl [(*N*-acylamino)bromomethyl]phosphonates **281** where R = AlkC(O), ArC(O), AlkS(O)$_2$, ArS(O)$_2$.

To date diethyl [(*N*-acylamino)bromomethyl]phosphonates **281** remain poorly studied, and their chemical properties are insufficiently studied. Nonetheless, compounds **281**, structurally related to the phosphorylated Vilsmeier–Haak reagents, behave as electrophiles similarly to dialkyl [(*N,N*-dimethylamino)chloromethyl]phosphonates **171** [166]. They react readily with such nucleophiles as primary aliphatic and aromatic amines and secondary cyclic amines in the presence of ethyldiisopropylamine—Hünig's base as well as triethylamine (Scheme 106). Reaction products are diethyl [(*N*-acylamino)aminomethyl]phosphonates **313**, unsymmetrical amidoaminals of diethyl (formyl)phosphonate (**3**) [30] showing biological and pharmacological activity [172].

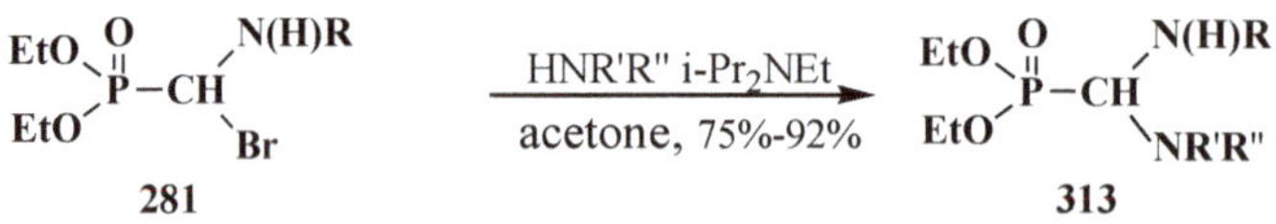

Scheme 106. Transformation of bromomethylphosphonates **281** into unsymmetrical amidoaminals **313**, where HNR'R" are morpholine, piperidine, pyrrole, 2-aminomethyltetrahydrofuran, *N*-methylbenzylamine, and aniline.

The reaction of compounds **282** with sodium azide leads to diethyl [(*N*-acylamino)azidomethyl]-phosphonates **314**, valuable precursors in the synthesis of phosphinoyl-substituted 1,2,3-triazoles, the products of 1,3-dipolar addition to disubstituted alkynes [173,174]. A mixture of the resultant two regioisomers of diethyl [(*N*-acylamino)(1-(1,2,3-triazolyl))methyl]phosphonates (**315**) and diethyl [(*N*-acylamino)(2-(1,2,3-triazolyl))methyl]phosphonates (**316**) can be separated by chromatography on silica gel (Scheme 107).

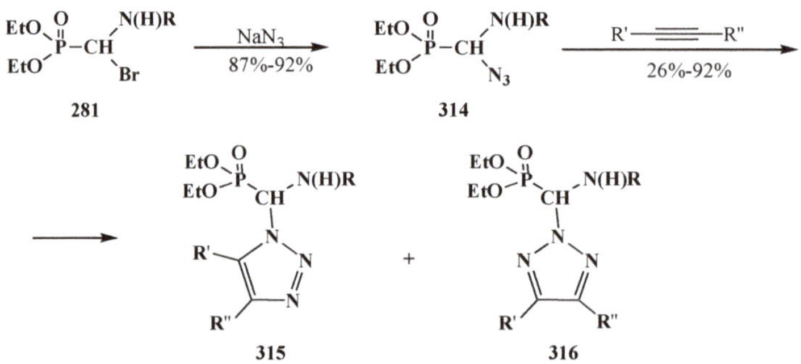

Scheme 107. Synthesis of (triazolylmethyl)phosphonates **315** and **316** from bromomethylphosphonates **281**, where R', R" = AlkC(O), ArC(O), AlkS(O)$_2$, ArS(O)$_2$.

Prolonged heating of acetonitrile solutions of **315** leads to migration of [(*N*-acylamino)(diethylphosphinoyl)]methyl group from the first to the second nitrogen atom (N1–N2 migration) of 1,2,3-triazole group to give **316**. In the presence of nucleophiles, azide ion or imidazole, the triazole fragment is displaced to yield diethyl [(*N*-acylamino)azidomethyl]phosphonates **314** or diethyl [(*N*-acylamino)(1-imidazolyl)methyl]phosphonates **317** (Scheme 108) [174].

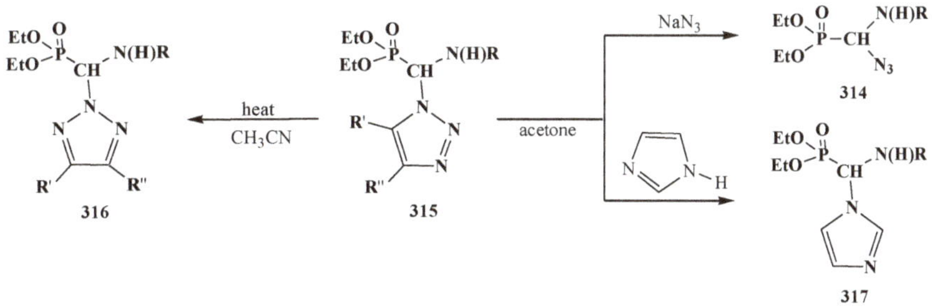

Scheme 108. Reaction replacement of 1,2,3-triazole group of compound **315** with nucleophiles to give compounds **314** and **317** and isomerisation of **315** into compound **316**.

The treatment of phosphonates **281** with Hünig's base i-Pr_2NEt [164] or triethylamine Et_3N [171] in tetrahydrofuran at −78 °C was shown to result in diethyl [(N-acyl)iminomethyl]phosphonates **318** in yields up to 98% (Scheme 109).

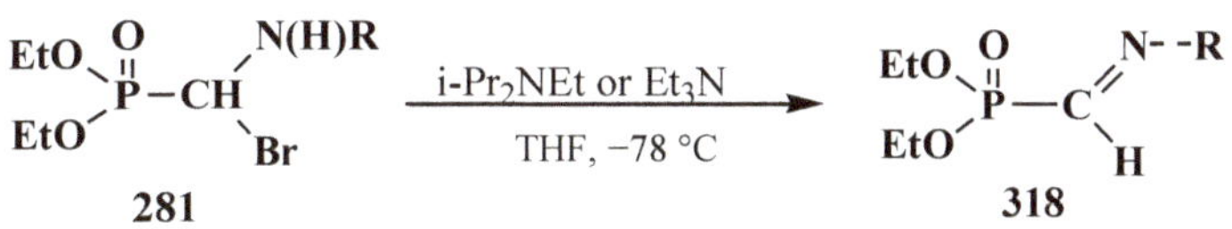

Scheme 109. Synthesis of diethyl [(*N*-acyl)iminomethyl]phosphonates (**318**) from phosphonates **281**.

5. Alkyl (dialkoxymethyl)phosphinates—H-Phosphinates 34. Syntheses and Chemical Properties

The synthesis of alkyl (dialkoxymethyl)phosphinates—H-phosphinates **34** in yields up to 96% by the reaction of hypophosphorous acid (**319**) with orthoformates **54** in the presence of catalytic amounts amounts of *p*-toluenesulfonic acid (p-$TolSO_3H$) [57,175] or trifluoroacetic acid (CF_3SO_3H) [58] was reported in 1977 (Scheme 110). The method is still used at present.

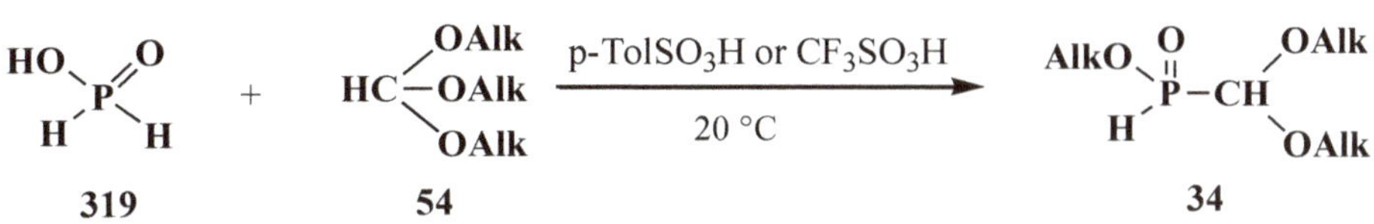

Scheme 110. Synthesis of alkyl (dialkoxymethyl)phosphinates **34**.

The chemical properties of phosphinates **34** were studied almost exclusively by the example of ethyl (diethoxymethyl)phosphinate (**86**). Ethyl (diethoxymethyl)phosphinate (**86**) retains properties typical for both phosphorus esters and hydrophosphoryl compounds and retains the general ability of phosphorylated formaldehyde acetals to undergo the cleavage of the phosphorus–carbon bond.

Phosphinate **86** is readily alkylated under the action of alkyl halides in the presence of bases: Na, NaH, BuLi (B^-) (via anion **320**) to give ethyl alkyl(diethoxymethyl)phosphinates **321** [58,176] (Scheme 111).

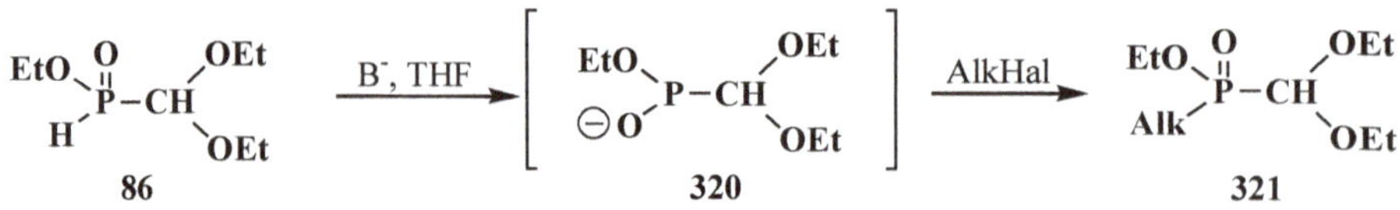

Scheme 111. Phosphinate **86** alkylation with alkylhalydes.

Compound **86** also readily undergoes addition to activated double bonds, or it can transform into three-coordinated phosphorus compounds [59,176,177], for example, in Scheme 112, compounds **322** and ethyl trimethylsilyl (diethoxymethyl)phosphinite (**323**), respectively.

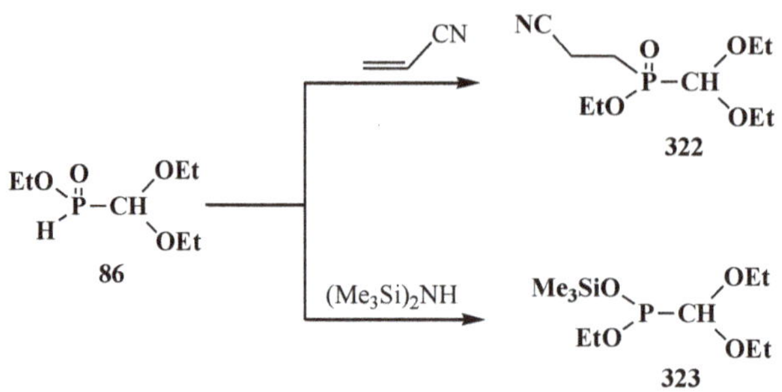

Scheme 112. Transformations of compound **86** into phosphinate **322** and a three-coordinated phosphorus compound (phosphinite **323**).

It also undergoes Todd–Atherton reaction with phenols and cross-coupling reactions with aryl bromides [106] in the presence of tetrakis(triphenylphosphine)palladium(0), $Pd(Ph_3P)_4$, see Scheme 17. In **86**, one of two P–H bonds of initial hypophosphorous acid **319** is protected by a diethoxymethyl group and can be restored in subsequent stages after deprotection [58,176]. When treated with two equivalents of organomagnesium or organolithium compounds, phosphinate **86** produces secondary phosphine oxides **324** [178], a hidden form of unstable primary phosphine oxides **325** apt to disproportionation [179], that can be easily obtained by subsequent acid hydrolysis. Further alkylation of secondary phosphine oxides **324** in the presence of a base leads to unsymmetrical tertiary phosphine oxides **326**, which in turn are the hidden form of unsymmetrical secondary phosphine oxides **328**. Similarly to primary phosphine oxides **325**, compounds **327** can be also obtained by subsequent acid hydrolysis of tertiary phosphine oxides **326**. Further oxidation leads to the corresponding phosphonic **328** and unsymmetrical phosphinic acids **329**. Unsymmetrical tertiary phosphine oxides **330** can be obtained by the following alkylation of secondary phosphine oxides **327** (Scheme 113).

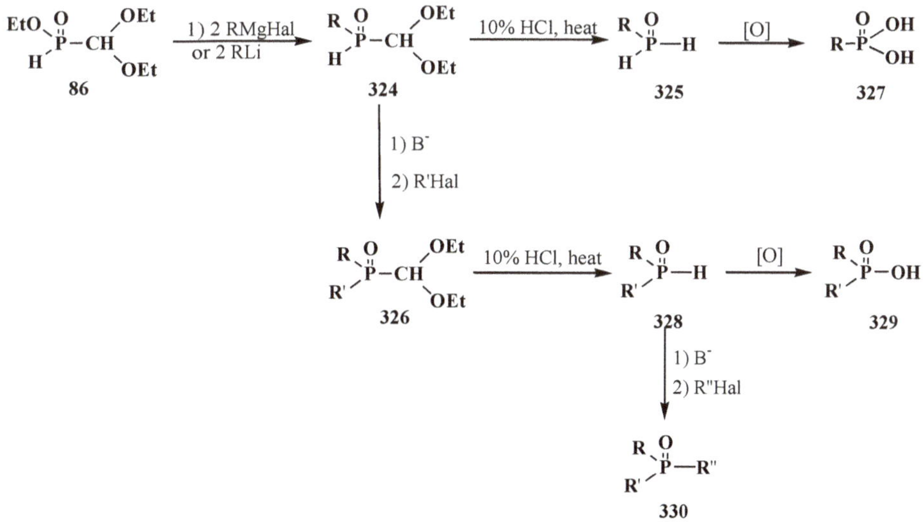

Scheme 113. Using phosphinate **86** for the synthesis of phosphine oxides **325**, **328**, **330**, phosphonic **328** and unsymmetrical phosphinic acids **329**.

Phosphonic and unsymmetrical phosphinic acids **328** and **329** can also be obtained by the scheme that begins with the alkylation reaction by introducing **86** in the reaction of one or sequentially both P–H bonds, thus solving the problem of selectivity [58,59,176,178], over phosphinates **331** and H-phosphinic acids **332**, respectively. The scheme allows one to avoid the use of highly toxic hypophosphorous acid in the syntheses [177] (Scheme 114).

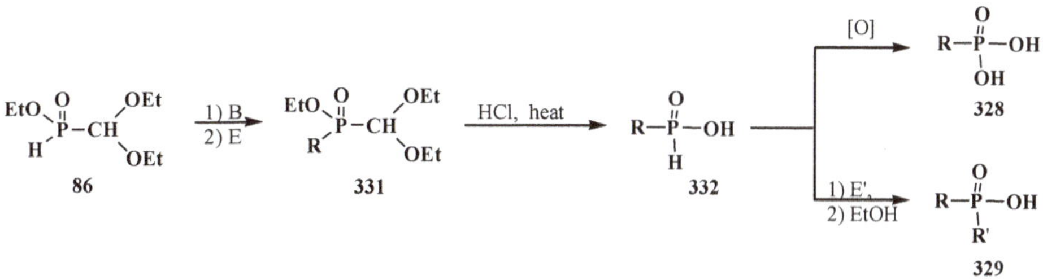

Scheme 114. Phosphonic **328**, H-phosphinic **332** and unsymmetrical phosphinic acids **329** syntheses starting with the reaction of **86** with electrophiles, where E = RHal or olefins with the activated double bond.

Due to the unique combination of chemical properties, phosphinate **86** is used in modern organic synthesis and synthesis of biologically active compounds. H-phosphinic analogs of natural α-amino acids **333** were obtained with >95% enantiomeric excess by the reaction of **86** with chiral (*S*)-*N*-*tert*-butylsulfinylketimines (**334**) [180] (Scheme 115).

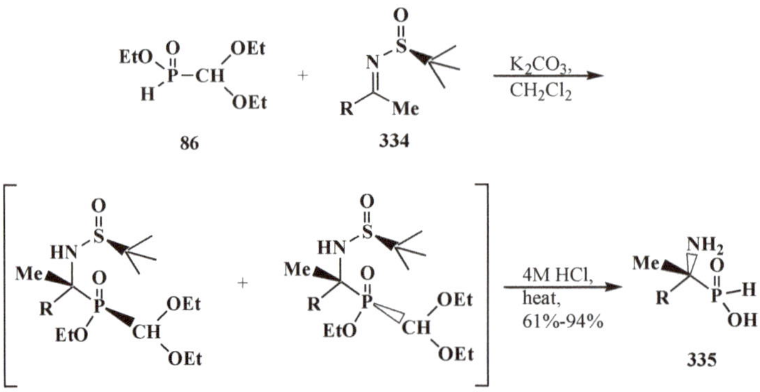

Scheme 115. Synthesis of H-phosphinic analogs of natural α-amino acids **333** by means of the reaction of phosphinate **86** with chiral (*S*)-*N*-*tert*-butylsulfinylketimines **334**.

3-Amino-2-fluoropropyl-*H*-phosphinic acid (**335**), a γ-aminobutyric acid (GABA) analog, a potential pharmaceutical for the treatment of central nervous system diseases, was prepared from compound **323** as the silylated form of ethyl (diethoxymethyl)phosphinate **86** [181] (Scheme 116).

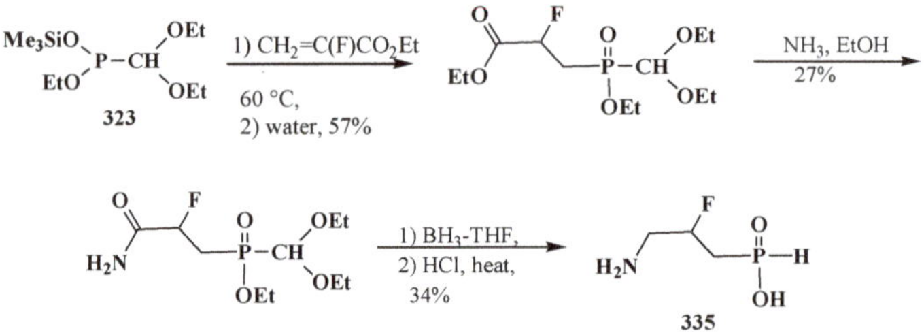

Scheme 116. Silylated compound **323** used for the synthesis of γ-aminobutyric acid analog **335**.

Disubstituted 3-aminopropyl-4-pyridylphosphinic acid (**336**), a potential neurotropic pharmaceutical and GABA antagonist, was obtained in a similar manner [182] (Scheme 117).

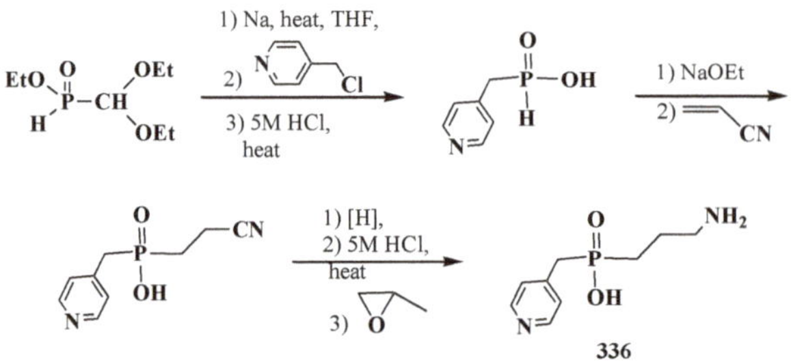

Scheme 117. Synthesis of potential neurotropic pharmaceutical, GABA antagonist **336**.

6. Phosphorylated Formaldehyde Hydrates—Geminal Diols 35. Syntheses and Chemical Properties

The chemistry of phosphorylated formaldehyde hydrates **35** began to develop since the mid 1990s when diethyl (dihydroxymethyl)phosphonate (**337**) was obtained in quantitative yield by the reaction of diethyl (diazomethyl)phosphonate (**338**) with 3,3-dimethyldioxirane—acetone peroxide at 20 °C [60,61] (Scheme 118).

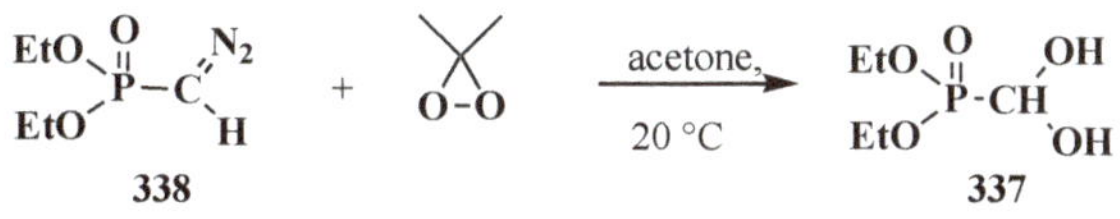

Scheme 118. Synthesis of diethyl (dihydroxymethyl)phosphonate (**337**) by the reaction of diazomethylphosphonate (**338**) with 3,3-dimethyldioxirane.

It was shown that **337** exists in an equilibrium with the hydrated form of diethyl formylphosphonate **3** [60,61] (Scheme 119).

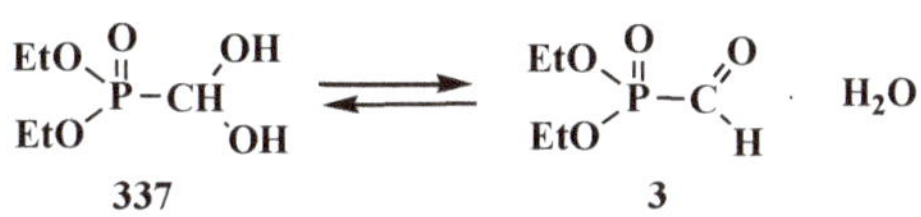

Scheme 119. Equilibrium between diethyl (dihydroxymethyl)phosphonate (**337**) and the hydrated form of diethyl formylphosphonate **3**.

The chemical properties of phosphonate **337** provide the possibility of considering it as a hidden form of phosphorylated formaldehyde. Compound **337** reacts with secondary amines R'_2NH with the cleavage of the phosphorus–carbon bond and formation of the corresponding *N,N*-disubstituted formamides and diethyl phosphite (**6**) [60,61] (it behaves as a source of formyl group $[C(O)H]^+$ cation, synthon type a[1] [183,184].

Compound **337** also shows properties of a typical aldehyde when reacted with primary amines RNH_2, trimethylsilyl cyanide Me_3SiCN, and hydroxylamine NH_2OH to give phosphorylated formaldimines **339**, formalcyanohydrin **340**, and formaldoxime **341**, respectively [60,61] (Scheme 120).

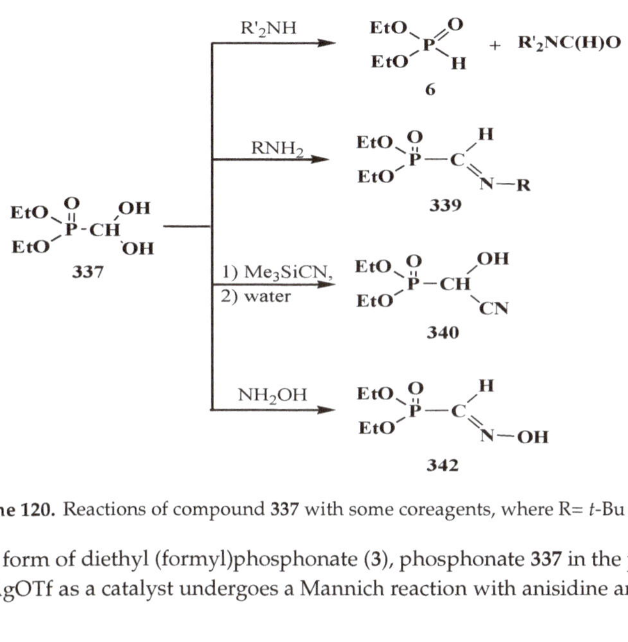

Scheme 120. Reactions of compound **337** with some coreagents, where R= *t*-Bu or Ph.

As a hidden form of diethyl (formyl)phosphonate (**3**), phosphonate **337** in the presence of silver trifluoroacetate AgOTf as a catalyst undergoes a Mannich reaction with anisidine and terminal alkyl-

or aryl-substituted alkynes. Reaction products, α-aminopropargylphosphonates **342**, are valuable precursors in the synthesis of α-aminophosphonic acids showing high biological activity as α-amino acid mimetics [63] (Scheme 121).

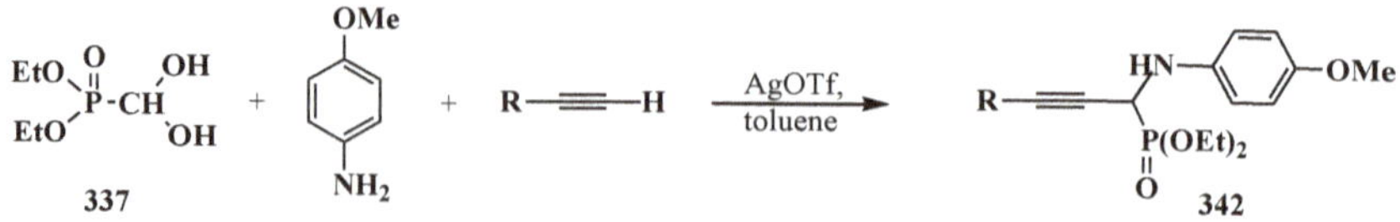

Scheme 121. Catalytical synthesis of α-aminopropargylphosphonates **342** by a Mannich reaction.

In the presence L-proline amide (**343**) as a catalyst, **337** undergoes an asymmetrical cross aldol condensation with aliphatic ketones. Diastereomerically pure α-hydroxyphosphonates **344**, precursors of α-hydroxyphosphonic acids, result from chirality induction with diastereomeric excesses greater that 99% [185] (Scheme 122).

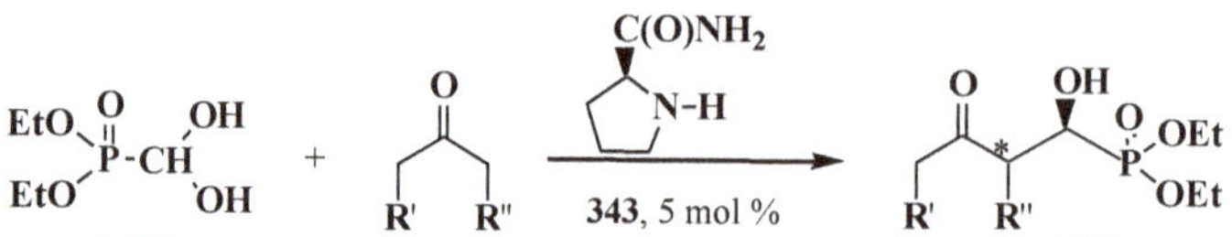

Scheme 122. Synthesis of diastereomeric α-hydroxyphosphonates **344** by asymmetrical cross-aldol condensation.

Like α-aminophosphonic acids, α-hydroxyphosphonic acids are also α-amino acid mimetics and exhibit high biological activity [185]. Similar results were obtained in the aldol condensation of ketones with racemic ethyl phenyl(dihydroxymethyl)phosphinate (**345**). In this case, the final products are α-hydroxyphosphinic acids **346** [62].

7. Conclusions

Phosphorylated formaldehyde derivatives, *i.e.*, acetals and related compounds, are a group of largely underinvestigated species. The experimental data accumulated since the early 1960s confirm that these compounds can be used in a wide variety of syntheses that have not been fully realized to date. For this reason, the growth of interest to these compounds will allow investigating their chemical properties in more detail and potentially enrich organic and organoelement chemistry with new synthetic methods.

Conflicts of Interests

The author declares no conflict of interests.

References

1. Mitsuo, S.; Masaki, S.; Hikaru, Y.; Tsujiaki, H. Acylphosphonates: P-C Bond Cleavage of Dialkyl Acylphosphonates by Means of Amines. Substituent and Solvent Effects for Acylation of Amines. *J. Org. Chem.* **1980**, *45*, 4162–4167.
2. Miller, A.; Stewart, D. Reactions of Carbonyl Compounds with Tervalent Phosporus Reagents. Part 8.1, Acetyldiphenylphosphine Oxide. *J. Chem. Soc. Perkin Trans. I* **1977**, *17*, 1898–1901. [CrossRef]
3. Frey, G.; Lesiecki, H.; Lindner, E.; Vordermaier, G. Synthese und reaktives Verhalten von Acyldiorganylphosphanoxiden. *Chem. Ber.* **1979**, *112*, 763–772. [CrossRef]
4. Sekine, M.; Kume, A.; Nakajima, M.; Hata, T. A new method for acylation of enolates by means of dialkyl acylphosphonates as acylating agents. *Chem. Lett.* **1981**, *10*, 1087–1090. [CrossRef]

5. Shagidullin, R.R.; Plyamovatyi, A.K.; Mukhamadeeva, R.M.; Khairullin, V.K.; Pudovik, A.N. Fourier IR—Spectroscopic study of the mechanism of anomalous Shiff base reaction with phosphorus-containing CH—Acids. *Z. Obshch. Khim.* **1994**, *64*, 926–930.
6. Conti, P.; Pinto, A.; Tamborini, L.; Rizzo, V.; de Micheli, C. A regioselective route to 5-substituted pyrazole- and pyrazoline-3-phosphonic acids and esters. *Tetrahedron* **2007**, *63*, 5554–5560. [CrossRef]
7. Conti, P.; Pinto, A.; Tamborini, L.; Dunkel, P.; Gambaro, V.; Visconti, G.L.; de Micheli, C. A regioselective route to 5-substituted isoxazole- and isoxazoline-3-phosphonates. *Synthesis* **2009**, 591–596. [CrossRef]
8. Vasella, A.; Voeffray, R. Asymmetric Synthesis of α- Aminophosphonic Acids by Cycloaddition of *N*-Glycosyl-*C*- dialkoxyphosphonoylnitrones. *Helv. Chim. Acta* **1982**, *65*, 1953–1964. [CrossRef]
9. Moskva, V.V.; Mavrin, V.Y. *O,O*-Diethylformylphosphonate. *J. Gen. Chem. USSR* **1988**, *57*, 2492.
10. Iorga, B.; Eumery, F.; Mouriès, V.; Savignac, P. Phosphorylated aldehydes. *Tetrahedron* **1998**, *54*, 14637–14677. [CrossRef]
11. Groß, H. Zur Existenz von Estern der Formylphosphonsäure. *Z. Chem.* **1977**, *17*, 131–132.
12. Amsallem, D.; Gornitzka, H.; Baceiredo, A.; Bertrand, G. New types of stable aldehydes: Formylphosphane and formylphosphane oxide. *Angew. Chem. Int. Ed. Engl.* **1999**, *38*, 2201–2203. [CrossRef]
13. Razumov, A.I.; Liorber, B.G.; Moskva, V.V.; Sokolov, M.P. Phosphorylated aldehydes. *Russ. Chem. Rev.* **1973**, *42*, 538–550. [CrossRef]
14. Wagenknecht, J. An electrochemical method for the preparation of iminodimethylenediphosphonic Acid. *Synth. React. Inorg. Met. Org. Chem.* **1974**, *4*, 567–572. [CrossRef]
15. Firestone, R.A. Hydroxy and Imino Containing Phosphonic Acid Diesters. U.S. Patent 3784590, 8 January 1974.
16. Mührle, H.; Vetter, W. Reaktionsbeteiligung von Phosphoester-Nachbargruppen bei Aminodehydrierungen. *Z. Naturforsch. B* **1988**, *43*, 1663–1671.
17. Franczyk, T.S., II. Preparation of Formylphosphonic Acid from Tertiary Aminomethylphosphonic Acid N-oxides. U.S. Patent 6274760, 14 August 2001.
18. Costisella, B.; Gross, H. Synthese und NMR-Spektren offenkettiger und cyclischer Acetale von Formylphosphonsäureestern und Formylphosphinoxid. *J. Prakt. Chem.* **1977**, *319*, 8–16. [CrossRef]
19. Livantsov, M.V.; Proskurina, M.V.; Prishchenko, A.A.; Lutsenko, I.F. Synthesis and some properties of phosphorylated formals. *J. Gen. Chem. USSR* **1984**, *54*, 2504–2517.
20. Brunjes, M.; Kujat, C.; Monenschein, H.; Kirschning, A. Acylation of Alkyl Halides and Amino Aldehydes with Phosphane Oxide-Based d^1-Synthon. *Eur. J. Org. Chem.* **2004**, *5*, 1149–1160. [CrossRef]
21. Gross, H.; Keitel, I.; Costisella, B.; Mikołajczyk, M.; Midura, W. Zur Synthese von bis-alkylmercaptomethanphosphonsäuredialkylestern. *Phosphorus Sulfur Silicon Relat. Elem.* **1983**, *16*, 257–262. [CrossRef]
22. Młotkowska, B.; Gross, H.; Costisella, B.; Mikołajczyk, M.; Grejszczak, S.; Zatorski, A. Synthese offenkettiger und cyclischer *S,S*-Acetale von Formyl phosphonsäurestern. *J. Prakt. Chem.* **1977**, *319*, 17–22. [CrossRef]
23. Mikołajczyk, M.; Grejszczak, S.; Zatorski, A.; Młotkowska, B.; Gross, H.; Costisella, B. A new and general synthesis of ketene *S,S*- and *S,O*-Acetals based on the Horner-Wittig reaction. *Tetrahedron* **1978**, *34*, 3081–3088. [CrossRef]
24. Costisella, B.; Gross, H. 1-Dimethylamino-1-cyano methanphosphonsдurediethylester, ein neues Edukt zur Darstellung von Carbonsäuren, 1-Cyanoenaminen und Homoenolaten. *Tetrahedron* **1982**, *38*, 139–145. [CrossRef]
25. Costisella, B.; Gross, H. 1-Cyano-1-dimethylamino methanphosphonsäurediethylester-Lithium—Ein "Instant-Horner-Reagent". *Z. Chem.* **1987**, *27*, 143–144. [CrossRef]
26. Marrero, Y.; Harwood, L.M. Towards the total synthesis of colletofragaranes: Constructing the macrocyclic lactone by high pressure-mediated intramolecular Diels-Alder reaction. *Tetrahedron Lett.* **2009**, *50*, 3574–3576. [CrossRef]
27. Gross, H.; Costisella, B. Synthesis of carboxylic acids via PO-activated olefination of tetraethyl dimethilamino-methylene-diphosphonate. *Angew. Chem. Int. Ed. Engl.* **1968**, *7*, 391–392. [CrossRef]
28. Qian, D.Q.; Shi, X.D.; Cao, K.Z.; Liu, L.Z. The Synthesis and Reactivity of Alkylaminosubstitutedmethylenediphosphonates. *Heteroat. Chem.* **1999**, *10*, 271–276. [CrossRef]
29. McNulty, J.; Das, P. Development of one-pot method for the homologation of aldehides to carboxylic acids. *Tetrahedron* **2009**, *65*, 7794–7800. [CrossRef]

30. Gross, H.; Costisella, B. Über α-substituerte Phosphonate, XI. *Liebiegs Ann. Chem.* **1971**, *750*, 44–52. [CrossRef]
31. Gross, H.; Costisella, B. Derivate des Formylphosphonsäurediäthylestern. *J. Prakt. Chem.* **1969**, *311*, 925–929. [CrossRef]
32. Gross, H.; Costisella, B.; Haase, L. Zur Reaction von Dimethylformamidederivaten mit Phosphorigsäurediestern und Phosphinoxiden. *J. Prakt. Chem.* **1969**, *311*, 577–585. [CrossRef]
33. Risch, N.; Piper, S.; Winter, A.; Lefarth-Risse, A. An Efficient Synthesis of Novel α-Aminophosphonates Based on a Mannich-Type Reaction. *Eur. J. Org. Chem.* **2005**, *2*, 387–394. [CrossRef]
34. Schrader, T.; Steglich, W. Phosphorus analogs of amino acids. IV. Syntheses of unusual 1-aminophosphonic acids via Diels-Alder reactions of diethyl (*N*-acyliminomethyl)phosphonates. *Synthesis* **1990**, 1153–1156. [CrossRef]
35. Mikołajczyk, M.; Costisella, B.; Grejszczak, S.; Zatorski, A. Synthesis of *O,S*-thioacetals of formylphosphonates. *Tetrahedron Lett.* **1976**, *17*, 477–480. [CrossRef]
36. Kim, T.H.; Uh, D.Y. New synthetic method of *O,S*-thioacetals of formylphosphonates. *Tetrahedron Lett.* **1985**, *26*, 3479–3482.
37. Costisella, B.; Keitel, I. Synthese und Reaktionen von 1-Acetoxy-1-methylthio-methanphosphorylverbindungen. *Phosphorus Sulfur Silicon Relat. Elem.* **1988**, *40*, 161–165. [CrossRef]
38. Mavrin, V.Y.; Moskva, V.V. Reaction of trihetero-substituted carbonium salts with diethyl phosphite. *J. Gen. Chem. USSR* **1988**, *58*, 1930–1931.
39. Mavrin, V.Y.; Moskva, V.V. Condensation of diethyl phosphite with mixed derivatives of orthocarboxylic acid. *J. Gen. Chem. USSR* **1988**, *58*, 1490.
40. Gross, H.; Seibt, H. Synthese von α-chlor-α-methoxy-methanphosphonsäureestern und α-chlor-α-methylthio-methanphosphonsäureestern. *J. Prakt. Chem.* **1970**, *312*, 475–482. [CrossRef]
41. Petrov, K.A.; Chauzov, V.A.; Agafonov, S.V.; Pazhitkova, I.V. Methoxymethanediphosphonic acid esters. *J. Gen. Chem. USSR* **1980**, *50*, 1525–1526.
42. Galli, R.; Scaglioni, L.; Palla, O.; Gozzo, F. Synthesis of 2-alkylthyo(or trifluoromethylthio)-2-halogenothenyl deryvatives by of Wittig (under phase transfer conditions)or Wittig-Horner reactions. Applications in the field of pyretroids. *Tetrahedron* **1984**, *40*, 1523–1532. [CrossRef]
43. Kim, T.H.; Uh, D.Y. Synthesis of *S,S*-thioacetals of formylphosphonate from chloro(arylthio)methanephosphonate. *Synth. Commun.* **1988**, *18*, 1611–1614. [CrossRef]
44. Otten, P.A.; Davies, H.M.; van der Gen, A. A Horner-Wittig Synthesis of 1-Chlorovinyl Sulfoxides. *Tetrahedron Lett.* **1995**, *36*, 781–784. [CrossRef]
45. Otten, P.A.; Davies, H.M.; van der Gen, A. A Horner-Wittig Synthesis of 1-Chlorovinyl Sulfoxides. *Phosphorus Sulfur Silicon Relat. Elem.* **1996**, *109*, 449–452.
46. Dinizo, S.E.; Freerksen, R.W.; Rabst, W.E.; Watt, D.S. Synthesis of α-Alkoxyacrylonitriles Using Substituted Diethyl Cyanomethylphoaphonates. *J. Org. Chem.* **1976**, *41*, 2846–2849. [CrossRef]
47. Dinizo, S.E.; Freerksen, R.W.; Rabst, W.E.; Watt, D.S. A One-Carbon Homologation of Carbonyl Compounds to Carboxylyc Acids, Esters and Amides. *J. Am. Chem. Soc.* **1977**, *99*, 182–186. [CrossRef]
48. Diamond, P.M.; Dinizo, S.E.; Freerksen, R.W.; Curtis, R.; Haltiwanger, R.C.; Watt, D.S. Ferric Chloride-catalysed Conversion of α-t-butoxy- or α-acetoxy-acrylonitryles into Imides. *J. Chem. Soc. Chem. Commun.* **1977**, 298–299. [CrossRef]
49. Koidan, G.N.; Marchenko, A.P.; Oleinik, V.A.; Pinchuk, A.M. 2-Methyl-2-tribromophosphazopropane. *J. Gen. Chem. USSR* **1988**, *58*, 1304–1309.
50. Savignac, P.; Coutrot, P. Preparation of 1,1-dibromoalkanes by Halogen Exchange. *Synthesis* **1976**, 197–199. [CrossRef]
51. Majewski, P.; Koszuk, J.F. Synthesis of Dichloromethylphosphonates. *Phosphorus Sulfur Silicon Relat. Elem.* **2009**, *184*, 956–962. [CrossRef]
52. Bulpin, A.; Masson, S.; Sene, A. Reaction of Phosphonodithioformates with Nucleophilic Reagents; Potential synthetic Uses. *Phosphorus Sulfur Silicon Relat. Elem.* **1990**, *49–50*, 135–138. [CrossRef]
53. Costisella, B.; Oczegowski, S.; Gross, H. α-Substituierte phosphonate. 70 Untersuchungen über Methylthiomethanbisphosphorylderivate. *Phosphorus Sulfur Silicon Relat. Elem.* **1994**, *86*, 169–175. [CrossRef]
54. Binder, J.; Zbiral, E. A new procedure for homologation of carbonyl compounds to α-hydroxy-carboxylic esters by means of diethyl-[trimethylsilylethoxymethyl]phosphonate. *Tetrahedron Lett.* **1986**, *27*, 5829–5832. [CrossRef]

55. Dufrechon, S.; Combert, J.-C.; Malhiac, C.; Collignon, N. Efficient synthesis of α-enamino phosphonates in the series of piperidine and morpholine. *Phosphorus Sulfur Silicon Relat. Elem.* **1997**, *127*, 1–14. [CrossRef]
56. Mikołajczyk, M. α-Heterosubstituted phosphonates and phosphineoxides. *Pure Appl. Chem.* **1987**, *59*, 983–988. [CrossRef]
57. Gallagher, M.J.; Honegger, H. Dialkoxymethylation of phosphorus with trialkyl orthoformates: Reactions of phosphonic and phosphinic acids via their trivalent tautomers. *Tetrahedron Lett.* **1977**, *18*, 2987–2990. [CrossRef]
58. Coudray, L.; Montchamp, J.-L. Temporary Protection of *H*-Phosphinic Acids as a Synthetic Strategy. *Eur. J. Org. Chem.* **2009**, *27*, 4646–4654. [CrossRef]
59. Dingwall, Y.G.; Ehrenfreund, I.; Hall, R.G. Diethoxymethylphosphonites and phosphinates. Intermediates for the synthesis of α, β and γ-aminoalkylphosphinous acids. *Tetrahedron* **1989**, *45*, 3787–37808. [CrossRef]
60. Hamilton, R.; McKervey, M.A.; Rafferty, M.D.; Walker, B.J. The Reaction of Dimethyl Dioxirane with Diazomethylphosphonates; the First Synthesis of a Formylphosphonate Hydrate. *J. Chem. Soc. Chem. Commun.* **1994**, 37–38. [CrossRef]
61. Cairns, J.; Dunne, C.; Franczyk, T.S.; Hamilton, R.; Hardacre, C.; Stern, M.K.; Treacy, A.; Walker, B.J. The Synthesis and Chemistry of Formylphosphonate. *Phosphorus Sulfur Silicon Relat. Elem.* **1999**, *144*, 385–388. [CrossRef]
62. Samanta, S.; Perera, S.; Zhao, C.-G. Organocatalytic Enantioselective Synthesis of Both Diastereomers of α-Hydroxyphosphinates. *J. Org. Chem.* **2010**, *75*, 1101–1106. [CrossRef]
63. Dodda, R.; Zhao, C.-G. Silver(I) triflate-catalyzed direct synthesis of N-PMP protected α-aminopropargylphosphonates from terminal alkynes. *Org. Lett.* **2007**, *9*, 165–167. [CrossRef] [PubMed]
64. Hirai, T.; Han, L.-B. Palladium-Catalyzed Ibsertion of Isocyanides into P(O)-H Bonds: Selective Formation of phosphinoyl Imines and Bisphosphinoylaminomethanes. *J. Am. Chem. Soc.* **2006**, *128*, 4722–4723. [CrossRef] [PubMed]
65. Piotrowska, D.G. N-Substituted C-diethoxyphosphorylated nitrones as useful synthons for the synthesis of α-aminophosphonates. *Tetrahedron Lett.* **2006**, *47*, 5363–5366. [CrossRef]
66. Piotrowska, D.G.; Balzarini, J.; Glowacka, I.E. Design, synthesis antiviral and cytostatic evaluation of novel isoxalidine nucleotide analogues with a 1,2,3-triazole linker. *Eur. J. Med. Chem.* **2012**, *47*, 501–509. [CrossRef] [PubMed]
67. Seufert, D.; Marmor, R.S.; Hilbert, P. Some reactions of dimethilphosphono-substituted diazoalkanes. $(MeO)_2P(O)CR$ transfer to olefins and 1,3-dipolar additions of $(MeO)_2P(O)CR'$. *J. Org. Chem.* **1971**, *36*, 1379–1386.
68. Maehr, H.; Uskokovic, M.R.; Schaffer, C.P. Concise synthesis of dimethyl(2-Oxopropyl)phosphonate and homologation of aldehydes and to alkynes in a tandem process. *Synth. Commun.* **2009**, *39*, 299–310. [CrossRef]
69. Kosobokov, M.D.; Titnyuk, I.D.; Beletskaya, I.P. An expedient synthesis of diethyl diazomethylphosphonate. *Mendeleev Commun.* **2011**, *21*, 142–143. [CrossRef]
70. Neidlein, R.; Keller, H. Syntheses of 1-Benzyloxyaminoalkylphosphonates. *Heterocycles* **1993**, *36*, 1925–1932. [CrossRef]
71. Carter, W.A. Treatment of Human Viral Infection by Doublestranged RNA Combined with Viral Infections. Eur. Patent 286224, 12 October 1988.
72. Franczyk, T.S., II. Method for Preparing Formylphosphonic Acid. U.S. Patent 7294733, 13 November 2007.
73. Franczyk, T.S., II. Method of Making Phosphorus-containing Compounds and Products Thereof. U.S. Patent 6864218, 8 March 2005.
74. Oediger, H.; Lieb, F.; Disselnkotter, H. Phosphonoformaldehyde, a process for its preparation and its use as an intermediate product for the preparation of medicaments. U.S. Patent 4,348,332, 7 September 1982.
75. Livantsov, M.V.; Boiko, V.I.; Proskurina, M.V.; Lutsenko, I.F. Phosphorylation of orthoformates. *J. Gen. Chem. USSR* **1982**, *52*, 811.
76. Razumov, A.I.; Moskva, V.V. Reaction of dialkylphosphorous acids with orthoformic esters. *Z. Obshch. Khim.* **1964**, *34*, 3125–3126.
77. Razumov, A.I.; Moskva, V.V. Reaction of orthoformates with hydrogen phosphited and hydrogen phosphinites. *J. Gen. Chem. USSR* **1965**, *35*, 1599.

78. Beznosko, B.K.; Usanova, V.M.; Zhuravleva, L.V.; Kharitonov, A.V.; Bondarenko, N.A.; Yarkevich, A.N.; Antoshin, A.E.; Tsvetkov, E.N. Antiinflammatory and analgesic activity of quaternary phosphine oxides. *Pharm. Chem. J.* **1990**, *24*, 244–247. [CrossRef]
79. Gross, H.; Costisella, B. Reaktion von phosphoriger Säure bzw. deren Anhydrid mit Orthoameisensäureestern. *J. Prakt. Chem.* **1974**, *311*, 550–556. [CrossRef]
80. Gross, H.; Costisella, B. *O,O*-and S,S-Acetals of formylphosphoniumsalts. *J. Prakt. Chem.* **1978**, *320*, 128–132.
81. Gross, H.; Freiberg, J.; Costisella, B. Zur existanz von halogendialkoxyalkanen, eine einfache synthese von dialkoxymethanphosphonaten. *Chem. Ber.* **1968**, *101*, 1250–1256. [CrossRef]
82. Malenko, D.M.; Gololobov, Y.G. New method for synthesis of phosphorylated formals. *J. Gen. Chem. USSR* **1981**, *51*, 1214–1215.
83. Groß, H.; Costisella, B. Zur Umsetzung von Dimethylformamid-dimethyl-acetal mit Phosphortrichlorid. *Z. Chem.* **1970**, *10*, 404–405. [CrossRef]
84. Costisella, B.; Gross, H. Zur Reaktion von Orthoameisensäureestern mit Phosphor-III-Verbindungen. *J. Prakt. Chem.* **1969**, *311*, 571–576.
85. Krokhina, S.S.; Pyrkin, R.I.; Levin, Y.A.; Ivanov, B.E. Effect of triethyl othoformate on some trivalent phosphorus derivatives. *Russ. Chem. Bull.* **1968**, *17*, 1349. [CrossRef]
86. Prishchenko, A.A.; Livantsov, M.V.; Novikova, O.P.; Livantsova, L.I.; Petrosyan, V.S. Synthesis and reactivity of substituted α-carbonylphosphonites and their derivatives. *Heteroat. Chem.* **2012**, *23*, 352–372. [CrossRef]
87. Dietsche, W. Darstellung von C-phosphorylierten Formaldehydacetalen. *Liebigs Ann. Chem.* **1968**, *712*, 21–27. [CrossRef]
88. Moskva, V.V.; Maikova, A.I.; Razumov, A.I. Reaction of orthocarboxylates and acetals with trivalent phosphorus acid chlorides. *J. Gen. Chem. USSR* **1969**, *39*, 563–566.
89. Costisella, B.; Gross, H. Notiz zur Synthese unsymmetrisch substituirten derivate der diätoxymethanphosphonsäure. *J. Prakt. Chem.* **1977**, *319*, 343–346. [CrossRef]
90. Krutskaya, L.V.; Safiulina, O.Z.; Voronina, S.G.; Bryukhovetskaya, L.V.; Tsivunin, V.S. On the reaction of 2-chloro-1,3,2-dioxaphospholanes with 2-ethoxy-1,3-dioxolanes. *J. Gen. Chem. USSR* **1989**, *59*, 2408–2411.
91. Tsivunin, V.S.; Krutskii, L.N.; Ernazarov, M.; Kamai, G.K. Reaction of the diethylamido chloride of ethylphosphinous acid and ethyldichlorophosphine with orthoformic esters. *J. Gen. Chem. USSR* **1970**, *40*, 2551–2553.
92. Costisella, B.; Gross, H. Zur reaktion von P(III) amiden mit orthoameisensäureesternchloriden. *Phosphorus Sulfur Silicon Relat. Elem.* **1980**, *8*, 99–103. [CrossRef]
93. Młotkowska, B.; Costisella, B.; Gross, H. Eine einfache Methode zur Synthese cyclischer Acetale des Formylphosphonsäreesters. *J. Prakt. Chem.* **1974**, *316*, 913–916. [CrossRef]
94. Bennett, S.N.L.; Hall, R.G. New syntheses of arylphosphinic acids from the reaction of ethyl diethoxymethylphosphinate with aryl bromides and phenols. *J. Chem. Soc. Perkin Trans. I.* **1995**, 1145–1151. [CrossRef]
95. Costisella, B.; Gross, H. Zur reaktivität von Acetalen des Formylphosphonsäureesters. *J. Prakt. Chem.* **1971**, *313*, 265–276.
96. Niyazymbetov, M.S.; Costisella, B.; Kaitel, I.; Shvatrz, K.K. Cathode-catalyzed transesterification of alkylphosphonates. *Russ. Chem. Bull.* **1991**, *40*, 182–186. [CrossRef]
97. Mikołajczyk, M.; Graczyk, P.P.; Wiechorek, M.W.; Bujacz, G. A solution and solid state conformation of 2-diphenylphosphinoyl-1,3-dioxanes. The nature of O-C-P anomeric interactions. *Tetrahedron* **1992**, *48*, 4209–4230. [CrossRef]
98. Gross, H.; Böck, C.; Costisella, B.; Gloede, J. Entalkylierung von Phosphonoesthers mit labilen funktionellen Gruppen mittels Trimethylsilyl-Bromid. *J. Prakt. Chem.* **1978**, *320*, 344–350. [CrossRef]
99. Rol'nik, L.Z.; Pastushenko, E.V.; Livantsov, M.V.; Proskurnina, M.V.; Zlotskii, S.S.; Rakhmankulov, D.L. Free-radical reactions of dialkyl formylphosphonate phosphonacetals. *J. Gen. Chem. USSR* **1983**, *53*, 1104–1109.
100. Razumov, A.I.; Gurevich, P.A.; Liorber, B.G.; Borisova, T.B. A study in the series of phosphinic and phosphinous acid derivatives. *J. Gen. Chem. USSR* **1969**, *39*, 369–372.
101. Nguyen-Ba, P.; Lee, N.; Mitchell, H.; Chan, L.; Quimpere, M. Design and synthesis of a novel class nucleotide analogs with anti-HCMV activity. *Bioorg. Med. Chem. Lett.* **1998**, *8*, 3555–3560. [CrossRef]

102. Kruse, C.G.; Broekhof, N.L.Y.M.; Wijsman, A.; van der Gen, A. Synthetic application of 2-chloro-1,3-dithiane preparation of ketene dithioacetals. *Tetrahedron Lett.* **1977**, *18*, 885–888. [CrossRef]
103. Ishikawa, K.; Akiba, K.; Inamoto, N. Synthesis of 1, 4-benzodithiofulvenes via Wittig reaction. *Tetrahedron Lett.* **1976**, *17*, 3695–3698. [CrossRef]
104. Mikołajczyk, M.; Graczyk, P.P.; Wieczorek, M.W. Conformational preference in 1,3-dithianes containing 2-phosphoryl, -(thiophosphoryl), and (selenophosphoryl)groups. Chemical and crystallographic implications of the nature of the anomeric effect. *J. Org. Chem.* **1994**, *59*, 1672–1679. [CrossRef]
105. Mikołajczyk, M.; Łuczak, T.; Graczyk, P.P.; Wieczorek, M.W.; Błaszczyk, Y.; Bujacz, G.D.; Majzner, W.R. Solid state conformation of the anomeric effect in 2-phosphoryl, 2- thiophosphoryl and 2-selenophosphoryl-substituted 1,3-dithiolans. *J. Organomet. Chem.* **1997**, *536–537*, 355–360. [CrossRef]
106. Aggarwal, V.K.; Barnell, J.K.; Worrall, J.M.; Alexander, R. Highly diastereoselective epoxidation of ketene dithioacetal dioxides: A new approach to asymmetric synthesis of α-amino amides. *J. Org. Chem.* **1998**, *63*, 7128–7129. [CrossRef]
107. Mikołajczyk, M.; Grejszczak, S.; Chefchyńska, A.; Zatorski, A. Addition of elemental sulfur to phosphonate carbanions and its application for synthesis of α-phosphoryl organosulfur compounds, synthesis of oaromatic ketones. *J. Org. Chem.* **1979**, *44*, 2967–2972. [CrossRef]
108. Mikołajczyk, M.; Bałczewski, P.; Grejszczak, S. Sulphenylation of phosphonates. A facile synthesis of α-phosphoryl sulfides and *S,S*-acetals of oxomethanephosphonates. *Synthesis* **1980**, 127–129.
109. Grayson, J.I.; Warren, S. Acyl Anion Equivalents: Synthesis of Ketones and Enones from α-Phenylthioalkylphosphine Oxides. *J. Chem. Soc. Perkin Trans. I* **1977**, 2263–2272. [CrossRef]
110. Mikołajczyk, M.; Bałczewski, P. Synthesis and reactivity of diethyl(methylthio)(trimethylsilyl)methylphosphonate. *Synthesis* **1989**, 101–106. [CrossRef]
111. Makomo, H.; Masson, S.; Saquet, M. Reduction of phosphonodithioformates: Syntheses of α-phosphoryl thiols and hemidithioacetals. *Tetrahedron Lett.* **1994**, *50*, 10277–10288. [CrossRef]
112. Mikołajczyk, M.; Mikina, M.; Graczyk, P.P.; Bałczewski, P. Synthesis of dithio- and diselenoacetals of formylphosphonates. *Synthesis* **1996**, 1232–1238. [CrossRef]
113. Yuaristi, E.; Cordillo, B.; Valle, L. Relative reactivity of 2-diphenylphosphinoil- and 2-diphenylthiophosphinoil-2-[1,3]-dithianyllithium as reagents Wittig-Horner. *Tetrahedron* **1986**, *42*, 1963–1970. [CrossRef]
114. Yuaristi, E.; Valle, L.; Valenzuela, B.A.; Aguilar, M.A. S-C-P anomeric interactions. 4. Conformational analys of 2-(diphenylphosphinoil)-1,3-dithyane. *J. Am. Chem. Soc.* **1996**, *108*, 2000–2005. [CrossRef]
115. Iorga, B.; Mouriès, V.; Savigniac, P. Carbanionic displacement reactions of phosphorus. Synthesis and reactivity of 5,5-dimethyl-2-oxo-2-(1,1-dithian-2-yl)-1, 3, 2-dioxaphosphorinane. *Bull. Chem. Soc. Fr.* **1997**, *134*, 891–985.
116. El-Wareth, A.; Sarhan, A.O.; Murakami, M.; Izumi, T. Synthesis of functionalized compounds related to π-extended tetrathiafulvenes with quinonoidal bearing a ferrocene moiety. *Monatsh. Chem.* **2002**, *133*, 1055–1066.
117. Costisella, B.; Gross, H. 1-Trimethylammonium-1-diethylphosphono-1-cyanomethylid, ein stabiles N-Ylid. *J. Prakt. Chem.* **1987**, *324*, 545–549. [CrossRef]
118. Costisella, B. ^{1}H, ^{13}C, ^{31}P, ^{15}N, ^{17}O-NMR investigations of α-substituted aminophosphonates. *Phosphorus Sulfur Silicon Relat. Elem.* **1990**, *51/52*, 226. [CrossRef]
119. Prishchenko, A.A.; Livantsov, M.V.; Petrosyan, V.S. Reaction of some CH acids with dimethylformamide dimethylacetal. *J. Gen. Chem. USSR* **1993**, *63*, 1326–1327.
120. Gross, H.; Costisella, B.; Gnauk, T.; Brenneke, L. Derivate der Aminomethan-bis-phosphonsäure. *J. Prakt. Chem.* **1976**, *318*, 116–126. [CrossRef]
121. Nesterov, L.V.; Krepysheva, N.E.; Aleksandrova, N.A. Reaction of dimethylformamide dimethylacetal with silyl phosphites. *J. Gen. Chem. USSR* **1989**, *59*, 641.
122. Prishchenko, A.A.; Livantsov, M.V.; Boganova, N.V.; Zhutskii, P.V.; Lutsenko, I.F. Synthesis of tetraalkyl dimethylaminomethylenediphosphorus-containing acids. *J. Gen. Chem. USSR* **1989**, *59*, 2132–2133.
123. Prishchenko, A.A.; Livantsov, M.V.; Petrosyan, V.S. New types of aminomethyl organophosphorus compounds. *J. Gen. Chem. USSR* **1994**, *63*, 1181–1193.
124. Degenhardt, C.R. Use of tetraethyl dimethylaminomethylene diphosphonate in the synthesis of benzothiophene-2-acetic acid and other carboxylic acids. *Synth. Commun.* **1982**, *12*, 415–421. [CrossRef]

125. Gross, H.; Costisella, B. Bis(diethoxyphosphinyl)(trimethylammonio)-methylide, a Stable *N*-Ylide. *Angew. Chem. Int. Ed. Engl.* **1968**, *7*, 463. [CrossRef]
126. Gross, H.; Costisella, B.; Bürger, W. Synthese von Alkylphosphonbetainen und stabilen N-Ylyden aus α-Trimethylammoniummethylphosphonaten. *J. Prakt. Chem.* **1969**, *311*, 563–570. [CrossRef]
127. Monenschein, H.; Dräger, G. Alexander Jung, Kirschning, A. Asymmetric Nucleophilic Acylation of Aldehydes via 1, 1-Heterodisubstituted Alkenes. *Chem. A Eur. J.* **1999**, *5*, 2270–2280. [CrossRef]
128. Kirschning, A.; Kujat, C.; Luiken, S.; Schaumann, E. Small and Versatile—Formyl Anion and Dianion Equivalents. *Eur. J. Org. Chem.* **2007**, 2387–2400. [CrossRef]
129. Mikołajczyk, M.; Midura, W.; Grejszczak, S. Synthesis of mono- and 1, 4-dicarbonyl compounds based on the oxygenation of phosphonate carbanions. Synthesis of dihydrojasmone, allethone and methylenomycin B. *Tetrahedron Lett.* **1984**, *25*, 2489–2492. [CrossRef]
130. Morita, T.; Okamoto, Y.; Sackurai, H. Dealkylation reaction of acetals, phosphonate, and phosphate esters with chlorotrimethylsilane/metal halide reagent in acetonitrile, and its application to the synthesis of phosphonic acids and vinyl phosphonates. *Bull. Chem. Soc. Jpn.* **1981**, *54*, 267–273. [CrossRef]
131. Morita, T.; Okamoto, Y.; Sackurai, H. A convenient dealkylation of dialkyl phosphonates by chlorotrimethylsilane in the presence of sodium iodide. *Tetrahedron Lett.* **1978**, *28*, 2523–2526. [CrossRef]
132. Gross, H.; Keitel, I. Heterosubstituierte olefine durch horner-reaktion mit α-substituierten Phosphonaten. *Z. Chem.* **1982**, *22*, 117–126.
133. Van Schaik, T.A.M.; Henzen, A.V.; van der Gen, A. A Horner-Wittig solution to the synthesis of ketene, O, O-acetals. *Tetrahedron Lett.* **1983**, *24*, 1303–1306. [CrossRef]
134. Streitwieser, A., Jr.; Juaristi, E. Carbon acidity. Equilibrium ion pair acidies of some phosphorus-substituted carbon acids. *J. Org. Chem.* **1982**, *47*, 768–770. [CrossRef]
135. Afarinkia, K.; Faller, A.; Twist, A.J. A new synthesis of α-ketophosphonates. *Synthesis* **2003**, 357–360. [CrossRef]
136. Coutrot, P.; Grison, C.; Lecouvey, M. Preparation of the phosphonic acid analogue of 3-deoxy-D-manno-2-octulosonic acid (KDO). *Tetrahedron Lett.* **1996**, *37*, 1595–1598. [CrossRef]
137. Horner, L.; Hoffmann, H.; Wippel, H.S. Phosphinoxide als Olefinierungsreagenzien. *Chem. Ber.* **1958**, *91*, 61–63. [CrossRef]
138. Horner, L.; Hoffmann, H.; Wippel, H.S.; Klahre, G. Phosphinoxide als olefinierungsreagenzien. *Chem. Ber.* **1959**, *92*, 2499–2505. [CrossRef]
139. Wittig, G.; Schöllkopf, U. Über Triphenyl-phosphin-methylene als olefinbildende Reagenzien (I Mitteilung). *Chem. Ber.* **1954**, *87*, 1318–1330. [CrossRef]
140. Wardsworth, W.S.; Emmons, W.D. The Utility of Phosphonate Carbanions in Olefins Synthesis. *J. Am. Chem. Soc.* **1961**, *83*, 1733–1738. [CrossRef]
141. Martin, S.F. Synthesis of Aldehydes, Ketones, and Carboxylic Acids from Lower Carbonyl Compounds by C-C Coupling Reactions. *Synthesis* **1979**, 633–665. [CrossRef]
142. Korotchenko, V.N.; Nenaidenko, V.G.; Balenkova, E.S.; Shastin, A.V. Olefination of carbonyl compounds. The newest and classical methods. *Russ. Chem. Rev.* **2004**, *73*, 957–989. [CrossRef]
143. Costisella, B.; Keitel, I.; Gross, H. α-Phosphononoenamine und Acyl phosphonate durch Horner-Olefinierung. *Tetrahedron* **1981**, *37*, 1227–1232. [CrossRef]
144. Gross, H.; Costisella, B.; Schick, H. γ-Butyrolactone aus metallierten Enamine. *Tetrahedron* **1984**, *40*, 733–736.
145. O'Boyle, N.M.; Greene, L.M.; Bergin, O.; Fichet, J.-B.; McCabe, T.; Lloyd, D.G.; Zisterer, D.M.; Meegan, M.J. Synthesis, evaluation and structural studies of antiproliferative tubulin-targeting azetidin-2-ones. *Bioorg. Med. Chem.* **2011**, *19*, 2306–2325. [CrossRef]
146. Mikołajczyk, M.; Grzejszczak, S.; Zatorski, A.; Młotkowska, B. A new general synthesis of ketene thioacetals. *Tetrahedron Lett.* **1976**, *31*, 2731–2734. [CrossRef]
147. Mikołajczyk, M.; Bałczewski, P. Diverse reactivity of α-carbanions derived from α-phosphoryl dithioacetals and α-phosphoryl sulphides towards α, β-unsaturated carbonyl compounds. A general synthesis of conjugated ketene dithioacetals. *Tetrahedron* **1992**, *8*, 8697–8710. [CrossRef]
148. Ceruti, M.; Degani, I.; Fochi, R. A new synthetic application of 1, 2-benzodithiolium cations: Synthesis of aldehydes by 1-carbon homologation of carbonyl compounds. *Synthesis* **1987**, 79–83. [CrossRef]

149. Hanessian, S.; Maji, D.K.; Govindan, S.; Matera, R.; Tintelnot-Blomley, M. Substrate-Controlled and Organocatalytic Asymmetric Synthesis of Carbocyclic Amino Acid Dipeptide Mimetics. *J. Org. Chem.* **2010**, *75*, 2861–2876. [CrossRef] [PubMed]
150. Hanessian, S.; Desilets, D.; Rancourt, G.; Fortin, R. The total stereocontrolled synthesis of a chemical precursor to (+)-thienamycin. *A formal synthesis of the antibiotic. Can. J. Chem.* **1982**, *60*, 2292–2294.
151. Moore, A.J.; Bryce, M.R. Generation and trapping of phosphorus stabilized 4, 5-ethylenedithio-1,3-dithiol-2-ide carbanions: Synthesis of ethylenedithio-1,3-dithiafulvalenes. *Synthesis* **1991**, 26–28. [CrossRef]
152. Misaki, Y.; Nishikawa, H.; Kawakami, K.; Uehara, T.; Yamabe, T. Bis(2-methylidene-1,3-dithiolo[4,5-*d*])tetrathiafulvalene (BDT-TTF): A tetrathiafulvalene condensed with 1,3-dithiol-2-ylidene moieties. *Tetrahedron Lett.* **1992**, *33*, 4321–4324. [CrossRef]
153. Heynderick, A.; Kaou, A.M.; Moustrou, C.; Samat, A.; Guglielmetti, R. Synthesis and photochromic behaviour of new Dipyrrolylperfluorocyclopentenes. *New J. Chem.* **2003**, *27*, 1425–1432. [CrossRef]
154. Seebach, D.; Bürstinghaus, R. S-methyl thiocarboxylates from aldehydes and ketones through ketene thioacetals. Reductive nucleophile thiocarboxylation. *Synthesis* **1975**, 461–462. [CrossRef]
155. Theil, F.; Costisella, B.; Mahrwald, R.; Gross, H.; Schick, H.; Schwarz, S. Prostaglandins and Prostaglandin Intermediates. XVII. Synthesis of the Main Metabolite of the $PGF_{2\alpha}$ Analogue Cloprostenol. *J. Prakt. Chem.* **1986**, *325*, 435–440. [CrossRef]
156. Theil, F.; Costisella, B.; Groß, H.; Schick, H.; Schwarz, S. A Three-step Procedure for the Conversion of γ-Lactones into δ-Lactones. *J. Chem. Soc. Perkin Trans. I* **1987**, 2469–2472. [CrossRef]
157. Zimmerman, H.E.; Baker, M.R.; Bottner, R.C.; Morrissey, M.M.; Murphy, S. Photochemistry of some allenic counterparts of cyclohexenones and 2, 5-cyclohexadienones. *J. Am. Chem. Soc.* **1993**, *115*, 459–466. [CrossRef]
158. Monenschein, H.; Brünjes, M.; Kirschning, A. Lithiated dimethoxymethyl diphenyl phosphine oxide. A versatile formiate carbanion equivalent. *Synlett* **2002**, 525–527. [CrossRef]
159. Okada, H.; Mori, T.; Saikawa, Y.; Nakata, M. Formation of α-hydroxyketones via irregular Wittig reaction. *Tetrahedron Lett.* **2009**, *50*, 1276–1278. [CrossRef]
160. Kirschning, A.; Dräger, G.; Yung, A. A new asymmetric formylation of aldehydes. *Angew. Chem. Int. Ed. Engl.* **1997**, *36*, 253–255. [CrossRef]
161. Becker, H.; Sharpless, B. A new ligand class for the asymmetric dihydroxylation of olefins. *Angew. Chem. Int. Ed. Engl.* **1996**, *35*, 448–451. [CrossRef]
162. Wittenberg, R.; Beier, C.; Dräger, G.; Jas, G.; Jasper, C.; Monenschein, H.; Kirschning, A. Towards the total synthesis of tonantzitlolone—-Preparation of key fragments and the complete carbon backbone. *Tetrahedron Lett.* **2004**, *45*, 4457–4460. [CrossRef]
163. Morgalyuk, V.P.; Strelkova, T.V.; Nifant'ev, E.E. Synthesis of polyfunctionalized methylphosphine oxides. *Russ. Chem. Bull.* **2012**, *61*, 380–385. [CrossRef]
164. Schrader, T.; Kober, R.; Steglich, W. Synthese von 1-aminophosphonsäure-derivaten über acyliminophosphonsäure-ester. *Synthesis* **1986**, 372–375. [CrossRef]
165. Costisella, B.; Keitel, I.; Ozegowski, S. α-Substituierte Phosphonate. 69. Diastereoselektivität bei der Knüpfung der Phosphor-kohlenstoffbindung. *Phosphorus Sulfur Silicon Relat. Elem.* **1993**, *84*, 115–120. [CrossRef]
166. Marson, C.M. Reactions of carbonyl compounds with (monohalo) methyleniminium salts (Vilsmeier reagents). *Tetrahedron* **1992**, *48*, 3659–3726. [CrossRef]
167. Winter, A.; Risch, N. The vinylogous mannich reaction: An efficient access to substituted nicotinonitriles. *Synlett* **2003**, *13*, 1959–1964. [CrossRef]
168. Morgalyuk, V.P.; Petrovskii, P.V.; Lysenko, K.A.; Nifant'ev, E.E. Synthesis of polyfunctionalyzed methylphosphine oxides. *Russ. Chem. Bull.* **2009**, *58*, 248–250. [CrossRef]
169. Morgalyuk, V.P.; Strelkova, T.V. Stepwise scheme for the reaction of chlorodiphenylphosphine with N,N-dialkylformamides in the presence of NaI. *Russ. J. Gen. Chem.* **2011**, *81*, 2096–2101. [CrossRef]
170. Morgalyuk, V.P.; Strelkova, T.V.; Nifant'ev, E.E. Reactivity of Chloro(dialkylamino)(diphenylphosphinoyl)methanes. *Bull. Chem. Soc. Jpn.* **2012**, *85*, 93–100. [CrossRef]

171. Schrader, T.; Steglich, W. Phosphoranaloge von Aminosäuren IV. Synthesen ungevönlicher 1-Aminophosphonsäuren über Diels-Alder-Reaktionen von (*N*-Acyliminomethyl)phosphonsäurendiethylestern. *Synthesis* **1990**, 1153–1156. [CrossRef]
172. Boukallaba, K.; Elachqar, A.; El Hallaoui, A.; Alami, A.; El Hajji, S.; Labriti, B.; Atmani, A. Synthesis of α-Heterocyclic α-Aminophosphonates, Part II: Morpholine, Piperidine, Pyrrolidine, Tetrahydrofurylmethylamine, *N*-Benzyl-*N*-Methylamine, and Aniline Derivatives. *Phosphorus Sulfur Silicon Relat. Elem.* **2007**, *182*, 1045–1052. [CrossRef]
173. Elachqar, A.; El Hallaouiq, A.; Roumestant, M.L.; Viallefont, P. Synthesis of Heterocyclic α-Aminophosphonic Acids. *Synth. Commun.* **1994**, *24*, 1279–1286. [CrossRef]
174. Achamlale, S.; Mabrouk, H.; Elachqar, A.; El Hallaoui, A.; El Hajji, S.; Alami, A.; Bellan, J.; Mazières, M.R.; Wolf, J.G.; Pierrot, M. Synthesis and Thermal Isomerization of Carboxylic and Phosphonic α-Aminoesters Substituted With a Triazole Ring. *Phosphorus Sulfur Silicon Relat. Elem.* **2007**, *182*, 357–367. [CrossRef]
175. Gallagher, M.J.; Honegger, H. Organophosphorus intermediates. VI. The acid-catalysed reaction of trialkyl orthoformates with phosphinic acid. *Aust. J. Chem.* **1980**, *33*, 287–294. [CrossRef]
176. Fougère, C.; Guénin, E.; Hardouin, J.; Lecouve, M. Rapid and Efficient Synthesis of Unsymmetrical Phosphinic Acids R'P(O)OHR". *Eur. J. Org. Chem.* **2009**, 6048–6054. [CrossRef]
177. Baylis, E.K. 1,1-Diethoxyethylphosphinates and Phosphinites. Intermediates for the Synthesis Functional Phosphorus Acids. *Tetrahedron Lett.* **1995**, *36*, 9385–9388.
178. Hall, R.G.; Riebli, P. Preparation of New Phosphine Oxides: Synthesis of an Analogue of Muscarinic Antagonists. *Synlett* **1999**, 1633–1635. [CrossRef]
179. Buckler, S.A.; Epstein, M. The preparation and reactions of primary phosphine oxides. *Tetrahedron* **1962**, *3*, 1221–1230. [CrossRef]
180. Zhang, D.; Yuan, C. A concise and first synthesis of α-aminophosphinates with two stereogenic atoms leading to optically pure α-amino-H-phosphinic Acids. *Chem. A Eur. J.* **2008**, *14*, 6049–6052.
181. Alstermark, C.; Amin, K.; Dinn, S.R.; Elebring, T.; Fjellström, O.; Fitzpatrick, K.; Geiss, W.B.; Gottfries, J.; Guzzo, P.R.; Harding, J.P.; *et al.* Synthesis and pharmacological evaluation of novel γ-aminobutyric acid type B (GABAB) receptor agonists as gastroesophageal reflux inhibitors. *J. Med. Chem.* **2008**, *51*, 4315–4320. [CrossRef]
182. Froestl, W.; Mickel, S.J.; von Sprecher, G.; Diel, P.J.; Hall, R.G.; Maier, L.; Strub, D.; Melillo, V.; Baumann, P.A.; Bernasconi, R.; *et al.* Phosphinic acid analogues of GABA. 2. Selective, orally active GABAB antagonists. *J. Med. Chem.* **1995**, *38*, 3313–3331. [CrossRef]
183. Seebach, D. Methoden und möglichkeiten der nucleophilen acylierung. *Angew. Chem. Int. Ed. Engl.* **1969**, *81*, 690–700. [CrossRef]
184. Seebach, D. Methods of reactivity umpolung. *Angew. Chem. Int. Ed. Engl.* **1979**, *18*, 39–336. [CrossRef]
185. Dodda, R.; Zhao, C.-G. Organocatalytic highly enantioselective synthesis of secondary α-hydroxyphosphonates. *Org. Lett.* **2006**, *8*, 4911–4914. [CrossRef]

molecules

MDPI

Article

The Effect of New Thiophene-Derived Aminophosphonic Derivatives on Growth of Terrestrial Plants: A Seedling Emergence and Growth Test

Jarosław Lewkowski [1,*], Zbigniew Malinowski [1], Agnieszka Matusiak [1], Marta Morawska [1], Diana Rogacz [2] and Piotr Rychter [2,*]

[1] Department of Organic Chemistry, Faculty of Chemistry, University of Łódź, Tamka 12, Łódź 91-403, Poland; zbigmal@uni.lodz.pl (Z.M.); agusk@poczta.fm (A.M.); martaem88@gmail.com (M.M.)

[2] Faculty of Mathematics and Natural Science, Jan Długosz University in Częstochowa, 13/15 Armii Krajowej Av., Częstochowa 42-200, Poland; diana.rogacz@gmail.com

* Correspondence: jlewkow@uni.lodz.pl (J.L.); p.rychter@ajd.czest.pl (P.R.); Tel.: +48-426-355-751 (J.L.); +48-343-615-154 (P.R.)

Academic Editor: György Keglevich

Received: 22 April 2016; Accepted: 23 May 2016; Published: 30 May 2016

Abstract: The aim of this work was to synthesize selected thiophene-derived aminophosphonic systems and evaluate the phytotoxicity of newly obtained products according to the OECD 208 Guideline. Seven new thiophene-derived *N*-substituted dimethyl aminomethylphosphonic acid esters **2a–h** were synthesized by the addition of an appropriate phosphite to azomethine bond of starting Schiff bases **1a–h**, and NMR spectroscopic properties of aminophosphonates were investigated. These eight compounds were analyzed in regard to their phytotoxicity towards two plants, radish (*Raphanus sativus*) and oat (*Avena sativa*). On the basis of the obtained results, it was found that tested aminophosphonates **2a–h** showed an ecotoxicological impact against selected plants, albeit to various degrees.

Keywords: thiophene-derived aminophosphonates; phytotoxicity; growth inhibition; environmental protection; ecotoxicology; OECD standard

1. Introduction

Nowadays, since production of food in agricultural areas is in the phase of intensive growth, the use of herbicides is absolutely necessary to manipulate or control undesirable vegetation of plants. Herbicides are used in landscapes throughout all over the world and are generally accepted as a moderately safe compounds. However, widespread use or overdosage of herbicides over long periods of time can result in residues in crops, soil, and land waters and, as a consequence, may lead to health and environmental risks [1–4]. Permanent applications of a given herbicide can result in resistance of a weed to the agent. Consequently, the weeds, while becoming resistant will not respond to the herbicide's active properties [5,6]. With this respect, there is a necessity of designing new compounds with potential herbicidal properties.

Sulfur-containing compounds, which have been developed as agrochemicals for plant protection, are currently being reviewed from the standpoint of their use and biological properties in each field of fungicides, insecticides, or herbicides [7–9].

Sulfur-containing herbicides are successfully and widely used as a foliar (applied to weed foliage) and a soil-applied herbicide absorbed by the root or shoot of emerging seedlings.

Biodegradability, photodegradation, persistence, soil mobility, and the accumulation in living organisms of sulfur-containing herbicides are currently in extensive research phases [10–14].

Designing new sulfur-containing compounds with potential herbicidal activity in soil, their persistence in soil must be taken into consideration. On the one side, herbicides, which are applied to the soil, typically affect seed emergence or the growth of weed seedlings and must remain in the soil to be effective. Unfortunately, on the other side, some herbicides like thifensulfuron-methyl can persist in the environment for a long time, becoming harmful agents for the surrounding environment. Therefore, planning the synthesis, it is difficult to strike a balance between the activity of a compound and its biodegradability in soil [15].

Biological properties of thiophene-deriving aminophosphonates have been largely studied [16–20]. Scientists found several of them to be promising plant protection agents [16] or to have rather strong antimicrobial [17], cytotoxic [18], antifungal [19,20], or even antiviral [20] properties. Therefore, working on new variations of methods for synthesis of aminophosphonic systems, chemists include derivatives bearing thiophene moiety as examples [17,21–26]. It is to stress that all those methodologies are mostly based on the Kabachnik–Fields reaction [27–29].

Working on our large project aimed at the search of a new class of herbicidal agents, we feel obliged to perform tests of phytotoxicity of newly synthesized compounds, especially when these compounds are aminophosphonic derivatives, which are known generally to have moderate-to-strong phytotoxic action on higher plants [30–33].

Our previous investigation provided results demonstrating that aminophosphonic acids bearing furfuryl moiety (derivatives of *C*-furfurylphosphonoglycine) are compounds of moderate phytotoxicity being able to kill both monocotyledonous oat (*Avena sativa*) and dicotyledonous radish (*Raphanus sativus*) with an amount of 100–200 mg in 1 kg of soil [34].

The presented research is concerned with a screening test of a synthesized series of variously *N*-substituted dimethyl amino(2-thienyl)methylphosphonates (**2a**–**g**) and dibenzyl *N*-furfurylamino(2-thienyl)methylphosphonate (**2h**) with respect to their phytotoxic properties. Our assumption was that the potential hazard of thiophene-derived aminophosphonates is important enough in the light of their possible applications and therefore should be investigated.

Apart from that, we performed the preliminary evaluation of aminophosphonates **2a**–**h** as potential soil-applied herbicides for agricultural/horticultural purposes. The potential effects of herbicides strongly results from their mechanism of action and the way they are applied. Since some types of herbicides are non-selective, which means the chemicals kill all classes of plants, not only unwanted weeds, for the proposed experiment, but also two types of plants (mono- and dicotyledonous) have been chosen as experimental objects.

2. Results and Discussion

2.1. Chemistry

Schiff bases **1a**–**h** were synthesized following the published and commonly known procedure of simple mixing thiophene-2-carboxaldehyde with appropriate amine in methanol and stirring them at room temperature for 24 h [35]. This procedure produced imines **1a**–**h**, which were isolated and used for further conversions without any purification. ^{1}H-NMR spectra were made only to verify the identity, based on the ^{1}H-NMR diagnostic singlet of a proton of the azomethine group (–CH=N–) above 8 ppm [31,32].

Aminophosphonates **2a**–**h** were synthesized basing on the aza-Pudovik reaction—a slightly modified procedure of a dimethyl or dibenzyl phosphite addition to the azomethine bond of corresponding Schiff bases **1a**–**h** in boiling toluene for 5 h. After the workup described in Section 3, the resulting aminophosphonates **2a**–**h** were obtained in 60%–70% yields (Scheme 1).

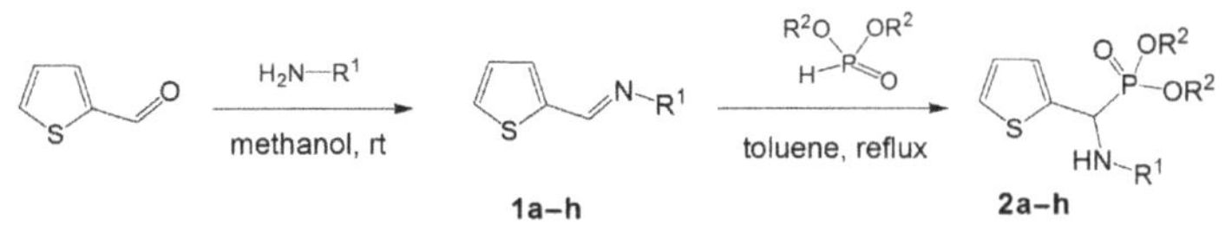

1a: R^1 = 2-$CH_3C_6H_4$; **1b**: R^1 = 3-$CH_3C_6H_4$; **1c**: R^1 = 4-$CH_3C_6H_4$; **1d**: R^1 = 3-$OCH_3C_6H_4$;
1e: R^1 = 4-$OCH_3C_6H_4$; **1f**: R^1 = $CH_2C_6H_5$; **1g**: R^1 = $CH(C_6H_5)_2$; **1h**: R^1 = CH_2Fur.

2a: R^1 = 2-$CH_3C_6H_4$; R^2 = CH_3; **2b**: R^1 = 3-$CH_3C_6H_4$; R^2 = CH_3; **2c**: R^1 = 4-$CH_3C_6H_4$; R^2 = CH_3;
2d: R^1 = 3-$OCH_3C_6H_4$; R^2 = CH_3; **2e**: R^1 = 4-$OCH_3C_6H_4$; R^2 = CH_3; **2f**: R^1 = $CH_2C_6H_5$; R^2 = CH_3;
2g: R^1 = $CH(C_6H_5)_2$; R^2 = CH_3; **2h**: R^1 = CH_2Fur; R^2 = CH_2Ph.

Scheme 1. Synthesis of thiophene-derived aminophosphonates **2a–h**.

2.2. *Evaluation of Phytotoxicity of Aminophosphonates* 2a–h

The rate of growth and development of plants as well as their quality, including increased number and size of leaves and stems, is strongly dependent (as commonly known) on the ground composition. Since plants have a high amount of water, its level depends on the contents of water in its environment, so taking dry mass as a measure of plant growth seems to be more credible.

Water content is given as a percentage of the dry or fresh weight. To assess the phytotoxicity of the analyzed compounds, the emergence and weight (dry and green) of control plant seedlings, with the emergence and mass (dry and green) of plant seedlings growing on the soil with a given admixture of the examined substances **2a–h**, were determined and compared. Dry weight changes of tested plants are shown in Figure 1. The visual assessment of any damage of tested plants concerns mainly the degree of plant growth inhibition, which is evaluated by visible or other signs of necrosis of chlorosis. Necrosis is the symptom of the death of plant tissues or organs caused by infection. Leaf and/or stem deformation may also be observed, while chlorosis is an inhibition of chlorophyll biosynthesis, resulting in yellowing of tissue.

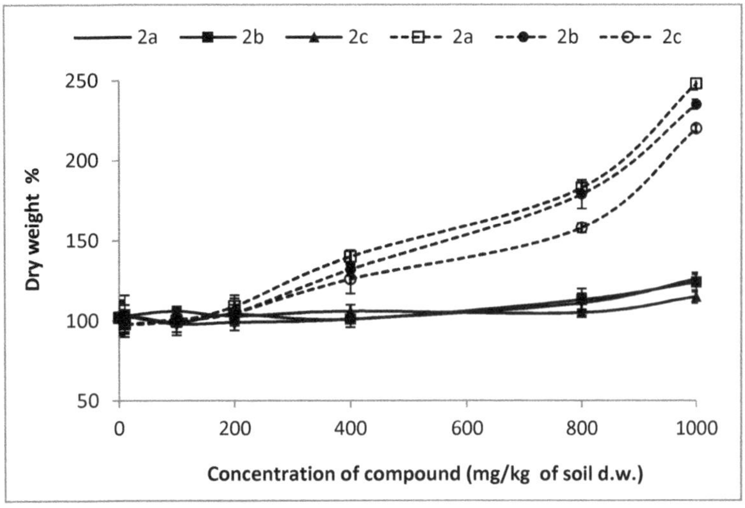

Figure 1. Changes of dry weight of treated plants expressed as percent to the value in untreated plants (control plants = 100% of dry weight). Solid lines represent changes of oat dry weight. Dotted lines represent changes of radish dry weight.

Preliminary tests for aminophosphonates **2a–c** for both oat and radish have shown inhibitive effects for plants. The dicotyledonous radish was noticed as the more sensitive plant when compared to sprouts of oat. Comparing results of toxicity of aminophosphonates **2a–c** for both types of tested plants, it was noticed that Compounds **2a** and **2b** are more toxic as compared with **2c** (Tables 1 and 2 Figures 1–4).

Table 1. Average changes (mean of three replicates) in basic parameters of the plant growth test for oat (Avena sativa) treated with **2a–c**. Least significant difference for samples (LSDS) and concentration (LSDC) is given for each tested parameter. %F.M. refers to plant biomass (fresh weight) expressed as percent of untreated control.

Sample Concentration in Soil (mg/kg of Soil Dry Matter)	Emerged Seedlings Number	% of Germination	Fresh Matter (g/pot)	% F.M.
0	20	100	2.780	100
1	19	97	2.682	96
10	19	95	2.654	95
100	19	97	2.919	105
200	19	95	2.644	95
400	19	95	2.715	98
800	18	88	2.100	76
1000	18	90	1.847	66
0	20	100	2.780	100
1	19	95	2.740	99
10	19	97	2.663	96
100	19	97	2.729	98
200	19	97	2.651	95
400	20	98	2.661	96
800	18	92	2.021	73
1000	18	92	1.659	60
0	20	100	2.780	100
1	19	97	2.759	99
10	20	98	2.704	97
100	20	98	2.801	101
200	19	97	2.862	103
400	20	98	2.762	99
800	19	95	2.643	95
1000	19	95	2.253	81
	LSDS = 1 LSDC = 1		LSDS = 0.401 LSDC = 0.245	

All the other results, to avoid confusion concerned with a large amount of data, have been placed in Supplementary Materials. To describe results, moreover, the demonstrated values of no observed effect concentration (NOEC) and lowest observed effect concentration (LOEC), changes in the fresh mass of plants were expressed as % inhibition compared to control and presented as EC_{50} (Figure 2). Effective concentration EC_{50} was calculated using GraphPad Prism software (Version 7, GraphPad Software, Inc., La Jolla, CA 92037, USA).

Table 2. Average changes (mean of three replicates) in basic parameters of the plant growth test for radish (*Raphanus sativus*) treated with **2a–c**.

Sample Concentration in Soil (mg/kg of Soil Dry Matter)	Emerged Seedlings Number	% of Germination	Fresh Matter (g/pot)	% F.M.
0	20	100	4.951	100
1	20	100	4.908	99
10	19	97	4.831	98
100	19	98	4.728	95
200	18	93	3.928	79
400	16	83	1.752	35
800	18	90	1.136	23
1000	15	75	0.772	16

Table 2. *Cont.*

Sample Concentration in Soil (mg/kg of Soil Dry Matter)	Emerged Seedlings Number	% of Germination	Fresh Matter (g/pot)	% F.M.
0	20	100	4.951	100
1	20	100	4.803	97
10	19	95	4.843	98
100	19	97	4.728	95
200	19	97	3.928	79
400	17	86	2.142	43
800	17	88	1.546	31
1000	15	78	1.106	22
0	20	100	4.951	100
1	19	98	4.914	99
10	19	98	4.798	97
100	19	98	4.873	98
200	19	97	4.787	97
400	18	93	3.934	79
800	18	93	3.328	67
1000	17	86	2.733	55
	$LSD_S = 1$ $LSD_C = 1$		$LSD_S = 0.435$ $LSD_C = 0.267$	

Least significant difference for samples (LSD_S) and concentration (LSD_C) is given for each tested parameter. %F.M. refer to plant biomass (fresh weight) expressed as percent of untreated control.

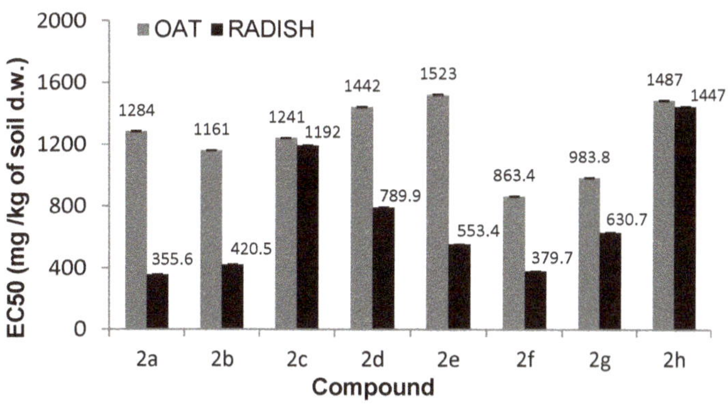

Figure 2. The EC_{50} values of fresh weight inhibition following exposure to tested compounds.

Figure 3. *Cont.*

Figure 3. Digital photographs of oat treated with **2a–c** (concentration in mg/kg of soil dry weight) on the 14th day of growth.

Figure 4. Digital photographs of radish treated with **2a–c** (concentration in mg/kg of soil dry weight) on the 14th day of growth.

N-methylphenyl aminophosphonates with *ortho* and *meta* substituted phenyl ring (**2a**, **2b**) have a significantly stronger influence on the inhibition of plant growth as compared with *N-para*-methylphenyl aminophosphonate (**2c**). In a case of oat, the values of NOEC and LOEC for Compounds **2a** and **2b** were 400 and 800 mg/kg of soil dry weight, respectively. These values for Compound **2c** were higher (NOEC and LOEC 800 and 1000 mg/kg d.w. of soil), indicating lower toxicity and subsequently higher tolerance of oat against this substance (Figure 3).

At the highest concentration (1000 mg), a *ca.* 40% decrease in fresh matter of oat sprouts for Compounds **2a** and **2b** when compared to control plants is observed (Table 1).

To make the description of the obtained results more readable and user-friendly, we only present selected tables and figures to the main text here (Tables 1 and 2, Figures 1–4).

Tested substances were much more toxic for the seedlings of radish, resulting in a decrease in NOEC and LOEC values. The first observed negative symptoms of tested **2a–c** compounds against radish were observed for the concentrations of 200 mg/kg of dry weight of soil (Samples **2a**, **2b**) and of 400 mg/kg of dry weight of soil for Compound **2c** (Figure 4). A very high drop of fresh weight of radish seedling was noticed for all compounds (**2a–c**), but aminophosphonates **2a** and **2b** were more toxic. At the concentration of 1000 mg/kg of dry weight of soil, a decrease in radish fresh weight, resulting from the action of Compounds **2a–c**, was 16%, 22%, and 55%, respectively. High values of dry matter at higher concentrations of the tested substances in pots correspond with a stronger inhibition of water uptake through the root system of plants (Figure 1). The increase in values of oat dry matter when compared to control plants is not as high as in the case of radish, which indicates better tolerance of oat roots against tested substances and moderately observed disturbances in the water uptake process.

The number of emerged seedlings at higher concentrations of tested compounds also decreased and was more visible for radish.

A plant growth test conducted for aminophosphonates bearing methoxy groups substituted in position *meta* (**2d**) and *para* (**2e**) in a phenyl ring have proved their toxic impact against both types of plants. Contrary to Compounds **2b–c** substituted with *meta*-methyl and *para*-methylphenyl groups, respectively, the examined *N-m*-methoxyphenyl derivative **2d** was less toxic for both radish and oat as compared with *N-p*-methoxyphenyl derivative **2e** (Tables S1 and S2, Figures S1–S3). The NOEC and LOEC values in the oat growth test for Compounds **2d** were much higher (800 and 1000 mg respectively) when compared to **2e** (100 and 200 mg respectively). The same values of NOEC and LOEC of Sample **2e** were revealed for radish. Again, as in the case of Compounds **2a–c**, the much more sensitive plant was radish. In comparison, the yield of the fresh matter of oat at the highest concentration (1000 mg) of **2d** and **2e** was 19% and 31% (respectively) lower compared to non-treated plants. The yield of the fresh weight of radish at the highest concentration did not exceed 29% (for **2d**) and 13% (for **2e**) of fresh matter of control plants. The high increase in dry matter values point out the dysfunction of roots (Figure S1).

Comparing the phytotoxicity results of aminophosphonates **2f**, **2g**, and **2h**, Compound **2h** was the most toxic agent for both plants. NOEC and LOEC values of Compound **2h** were the same for radish and oat—40 and 80 mg/kg of dry matter, respectively. Among Compounds **2f** and **2g**, the former had a higher negative impact on treated radish and oat. It was shown again that the more sensitive specimen of plant was radish. A very low yield of fresh matter for radish at the highest concentration (1000 mg) when compared to non-treated plants with simultaneously very high dry matter of plants indicates a strong inhibition of plant growth (both sprouts and roots) (Tables S3–S5, Figures S4–S8)

It is worth comparing the plant growth of Samples **2a–c** with **2f**, the non-aromatic amine derivative bearing the non-substituted phenyl ring. Sample **2f** is much more toxic against oat when compared to **2a–c**. In the case of radish, results are comparable to those obtained for Samples **2a,b** (the same values of NOEC and LOEC). Samples **2a** and **2b**, as well as *N*-benzyl derivative **2f**, in the highest applied concentration (1000 mg/kg), harmfully influenced radish seeds especially, while for oat, at the same concentration, Sample **2f** (without substituted phenyl ring with methyl group) caused almost total inhibition of sprout germination. As a result, contrary to dimethyl *N*-benzylamino(2-thienyl)methylphosphonate (**2f**), which kills both plants in a non-selective way, dimethyl *N*-(2-methylphenyl)amino(2-thienyl)methylphosphonate (**2a**) and dimethyl *N*-(3-methylphenyl)amino(2-thienyl)methylphosphonate (**2b**) were found to be selectively toxic towards radish, slightly harming oat.

3. Materials and Methods

3.1. Chemistry

All solvents (POCh, Gliwice, Poland) were routinely distilled and dried prior to use. Amines, dimethyl and dibenzyl phosphites, as well as 2-thiophenecarboxaldehyde (Aldrich, Poznań, Poland) were used as received. Melting points were measured on a MelTemp II apparatus (Bibby Scientific Limited, Staffordshire, UK) and were not corrected. NMR spectra were recorded on a Bruker Avance III 600 MHz (Billerica, MA, USA) operating at 600 MHz (^{1}H-NMR), 150 MHz (^{13}C-NMR), and 243 MHz (^{31}P-NMR). TMS was used as the internal standard for ^{1}H- and ^{13}C-NMR, and phosphoric acid was used as the external standard for ^{31}P-NMR. The following abbreviations were used for listing NMR signals: s—singlet, d—doublet, dd—doublet of doublets, ddd—doublet of doublet of doublets, and m—multiplet. Elemental analyses were carried out at the Centre for Molecular and Macromolecular Science of the Polish Academy of Science in Łódź, Poland.

General Procedures of Preparaing Amino(2-thienyl)methylphosphonates **2a–h**

2-Thiophenecarboxaldehyde (1.12 g, 10 mmol) was dissolved in methanol (15 mL), and a solution of an appropriate amine (10 mmol) in methanol (15 mL) was added. The obtained solution was stirred at room temperature for 24 h. Then, a small portion of anhydrous potassium carbonate was added, the mixture stirred for additional 15 min. Then, the inorganic salt was filtered off, and the filtrate was evaporated to achieve imines in quantitative yields, which were used for the further reaction without any purification.

Thus, the obtained imine (10 mmol) was dissolved in acetonitrile (15 mL), and an appropriate phosphite (10 mmol) in acetonitrile (15 mL) was added. The mixture was refluxed during the day and stirred at room temperature during the night. Total time of the reaction was 72 h. Then, the solvent was evaporated, and the crude product was isolated and purified using various procedures.

Dimethyl aminophosphonates **2a–g** were isolated as follows: crude products were dissolved in DCM (30 mL), and the solution was washed 5 times with saturated aqueous sodium hydrogen carbonate. After the usual workup, DCM was evaporated, and the product was purified by column chromatography on silica gel using ethyl acetate-hexane in a 4:1 ratio.

Product **2h** was chromatographed after evaporating acetonitrile, eluted using ethyl acetate—hexane in a 3:2 ratio.

Dimethyl N-(2-methylphenyl)amino(2-thienyl)methylphosphonate (**2a**): Yield = 83% (2.58 g) yellow oil. ^{1}H-NMR ($CDCl_3$, 600 MHz): δ 7.27–7.25 (m, H_5^{thioph}, 1H), 7.21–7.19 (m, H_3^{thioph}, 1H), 7.11 (d, $^{3}J_{HH}$ = 7.4 Hz, *o*-C_6H_4, 1H), 7.07 (ddd, $^{3}J^{(1)}_{HH}$ = $^{3}J^{(2)}_{HH}$ = 7.4 Hz and $^{4}J_{HH}$ = 1.7 Hz, *o*-C_6H_4, 1H), 7.01 (ddd, $^{3}J_{HH}$ = 5.0 and 4.2 Hz and $^{4}J_{HH}$ = 0.7 Hz, H_4^{thioph}, 1H), 6.74 (ddd, $^{3}J^{(1)}_{HH}$ = $^{3}J^{(2)}_{HH}$ = 7.4 Hz and $^{4}J_{HH}$ = 1.0 Hz, *o*-C_6H_4, 1H), 6.65 (d, $^{3}J_{HH}$ = 7.4 Hz, *o*-C_6H_4, 1H), 5.16 (d, $^{2}J_{PH}$ = 23.7 Hz, CHP, 1H), 3.82 (d, $^{3}J_{PH}$ = 10.6 Hz, $POCH_3$, 3H), 3.67 (d, $^{3}J_{PH}$ = 10.6 Hz, $POCH_3$, 3H), 2.28 (s, CH_3, 3H). ^{13}C-NMR ($CDCl_3$, 150 MHz): δ 143.98 (d, $^{2}J_{CP}$ = 12.3 Hz, C^2_{thioph}), 139.81 (C_{arom}), 130.41 (C_{arom}), 127.19 (d, *J* = 2.5 Hz, C_{thioph}), 127.03 (C_{arom}), 126.19 (d, *J* = 7.2 Hz, C_{thioph}), 125.43 (d, *J* = 3.4 Hz, C_{thioph}), 123.39 (C_{arom}), 118.86 (C_{arom}), 111.48 (C_{arom}), 54.17 (d, $^{2}J_{CP}$ = 7.0 Hz, POC), 53.88 (d, $^{2}J_{CP}$ = 7.3 Hz, POC), 51.84 (d, $^{1}J_{CP}$ = 157.8 Hz, PC), 17.47 (ArC). ^{31}P-NMR (243 MHz, $CDCl_3$): δ 23.33. Elem. Anal. Calcd. for $C_{14}H_{18}NO_3PS$: C, 54.01, H, 5.83, N, 4.50. Found: C, 54.18, H, 5.97, N, 4.59.

Dimethyl N-(3-methylphenyl)amino(2-thienyl)methylphosphonate (**2b**): Yield = 94% (2.92 g) yellow crystals, m.p.: 81–83 °C. ^{1}H-NMR ($CDCl_3$, 600 MHz): δ 7.26–7.25 (m, H_5^{thioph}, 1H), 7.21–7.20 (m, H_3^{thioph}, 1H), 7.07 (dd, $^3J_{HH}$ = 7.2 and 7.0 Hz, *m*-C_6H_4, 1H), 7.01 (m, H_4^{thioph}, 1H), 6.62 (d, $^3J_{HH}$ = 7.0 Hz, *m*-C_6H_4, 1H), 6.55 (s, *m*-C_6H_4, 1H), 6.52 (dd, $^3J_{HH}$ = 7.2 Hz and $^4J_{HH}$ = 1.9 Hz, *m*-C_6H_4, 1H), 5.11 (d, $^2J_{PH}$ = 24.4 Hz, CHP, 1H), 3.82 (d, $^3J_{PH}$ = 10.4 Hz, $POCH_3$, 3H), 3.65 (d, $^3J_{PH}$ = 10.8 Hz, $POCH_3$, 3H), 2.28 (s, CH_3, 3H). ^{13}C-NMR ($CDCl_3$, 150 MHz): δ 145.94 (d, $^2J_{CP}$ = 13.1 Hz, C^2_{thioph}), 139.68 (C_{arom}), 139.10 (C_{arom}), 129.15 (C_{arom}), 127.17 (d, *J* = 3.1 Hz, C_{thioph}), 126.25 (d, *J* = 7.3 Hz, C_{thioph}), 125.41 (d, *J* = 4.0 Hz, C_{thioph}), 120.09 (C_{arom}), 114.94 (C_{arom}), 111.01 (C_{arom}), 54.13 (d, $^2J_{CP}$ = 6.7 Hz, POC), 53.83 (d, $^2J_{CP}$ = 7.4 Hz, POC), 51.73 (d, $^1J_{CP}$ = 158.4 Hz, PC), 21.54 (ArC). ^{31}P-NMR (243 MHz, $CDCl_3$): δ 23.20. *Elem. Anal.* Calcd. for $C_{14}H_{18}NO_3PS$: C, 54.01, H, 5.83, N, 4.50. Found: C, 54.17, H, 5.79, N, 4.51.

Dimethyl N-(4-methylphenyl)amino(2-thienyl)methylphosphonate (**2c**): Yield = 54% (1.68 g) yellow crystals, m.p.: 98–100 °C. ^{1}H-NMR ($CDCl_3$, 600 MHz): δ 7.25–7.24 (m, H_5^{thioph}, 1H), 7.20–7.19 (m, H_3^{thioph}, 1H), 7.05–6.98 (m, H_5^{thioph}, *p*-C_6H_4, 3H), 6.64 (part of AA'XX' system, $^3J_{HH}$ = 9.0 and $^4J_{HH}$ = 1.2 and 1.1 Hz, *p*-C_6H_4, 2H), 5.09 (d, $^2J_{PH}$ = 24.0 Hz, CHP, 1H), 3.82 (d, $^3J_{PH}$ = 10.8 Hz, $POCH_3$, 3H), 3.65 (d, $^3J_{PH}$ = 10.8 Hz, $POCH_3$, 3H), 2.24 (s, CH_3, 3H). ^{13}C-NMR ($CDCl_3$, 150 MHz): δ 143.63 (d, $^2J_{CP}$ = 13.4 Hz, C^2_{thioph}), 139.77 (C_{arom}), 129.79 (C_{arom}), 128.40 (C_{arom}), 127.15 (d, *J* = 2.4 Hz, C_{thioph}), 126.26 (d, *J* = 7.4 Hz, C_{thioph}), 125.41 (d, *J* = 3.4 Hz, C_{thioph}), 114.25 (C_{arom}), 54.13 (d, $^2J_{CP}$ = 7.1 Hz, POC), 53.81 (d, $^2J_{CP}$ = 7.1 Hz, POC), 52.12 (d, $^1J_{CP}$ = 157.9 Hz, PC), 20.40 (ArC). ^{31}P-NMR (243 MHz, $CDCl_3$): δ 23.27. *Elem. Anal.* Calcd. for $C_{14}H_{18}NO_3PS$: C, 54.01, H, 5.83, N, 4.50. Found: C, 54.13, H, 5.88, N, 4.63.

Dimethyl N-(3-methoxyphenyl)amino(2-thienyl)methylphosphonate (**2d**): Yield = 72% (2.35 g) yellow oil ^{1}H-NMR ($CDCl_3$, 600 MHz): δ 7.26–7.25 (m, H_5^{thioph}, 1H), 7.20–7.19 (m, H_3^{thioph}, 1H), 7.08 (dd, $J_1 = J_2$ = 7.8 Hz, *m*-C_6H_4, 1H), 7.01–6.99 (m, H_4^{thioph}, 1H), 6.35–6.34 (m, *m*-C_6H_4, 1H), 6.33–6.31 (m, *m*-C_6H_4, 1H), 6.28–6.26 (m, *m*-C_6H_4, 1H), 5.10 (d, $^2J_{PH}$ = 24.0 Hz, CHP, 1H), 3.81 (d, $^3J_{PH}$ = 10.8 Hz, $POCH_3$, 3H), 3.74 (s, OCH_3, 3H), 3.65 (d, $^3J_{PH}$ = 10.8 Hz, $POCH_3$, 3H). ^{13}C-NMR ($CDCl_3$, 150 MHz): δ 160.77 (C_{arom}), 147.34 (d, $^2J_{CP}$ = 13.5 Hz, C^2_{thioph}), 139.46 (C_{arom}), 130.12 (C_{arom}), 127.2 (d, *J* = 3.0 Hz, C_{thioph}), 126.3 (d, *J* = 6.9 Hz, C_{thioph}), 125.5 (d, *J* = 3.5 Hz, C_{thioph}), 106.91 (C_{arom}), 104.38 (C_{arom}), 100.23 (C_{arom}), 55.08 (ArOC), 54.12 (d, $^2J_{CP}$ = 6.7 Hz, POC), 53.90 (d, $^2J_{CP}$ = 6.7 Hz, POC), 51.69 (d, $^1J_{CP}$ = 158.4 Hz, PC).^{31}P-NMR (243 MHz, $CDCl_3$: δ 23.06. *Elem. Anal.* Calcd. for $C_{14}H_{18}NO_4PS \times {}^1/_5\, C_6H_{14}$: C, 52.98, H, 6.08, N, 4.06. Found: C, 52.97, H, 5.96, N, 4.38.

Dimethyl N-(4-methoxyphenyl)amino(2-thienyl)methylphosphonate (**2e**): Yield = 80% (2.62 g) yellow oil. ^{1}H-NMR ($CDCl_3$, 600 MHz): δ 7.25–7.23 (m, H_5^{thioph}, 1H), 7.18–7.16 (m, H_3^{thioph}, 1H), 6.99–6.98 (m, H_4^{thioph}, 1H), 6.75 and 6.66 (AA'XX' system, $^3J_{HH}$ = 9.0 and $^4J_{HH}$ = 1.2 and 1.1 Hz, *p*-C_6H_4, 4H), 4.94 (d, $^2J_{PH}$ = 23.4 Hz, CHP, 1H), 4.54 (d, *J* = 8.4 Hz, NH, 1H), 3.81 (d, $^3J_{PH}$ = 10.8 Hz, $POCH_3$, 3H), 3.72 (s, OCH_3, 3H), 3.64 (d, $^3J_{PH}$ = 10.8 Hz, $POCH_3$, 3H). ^{13}C-NMR ($CDCl_3$, 150 MHz): δ 153.26 (C_{arom}), 139.91 (d, $^2J_{CP}$ = 14.1 Hz, C^2_{thioph}), 139.73 (C_{arom}), 127.1 (d, *J* = 2.8 Hz, C_{thioph}), 126.3 (d, *J* = 6.9 Hz, C_{thioph}), 125.4 (d, *J* = 4.0 Hz, C_{thioph}), 115.69 (C_{arom}), 114.83 (C_{arom}), 55.63 (ArOC), 54.13 (d, $^2J_{CP}$ = 6.7 Hz, POC), 53.81 (d, $^2J_{CP}$ = 6.7 Hz, POC), 52.85 (d, $^1J_{CP}$ = 158.4 Hz, PC). ^{31}P-NMR (243 MHz, $CDCl_3$): δ 23.33. *Elem. Anal.* Calcd. for $C_{14}H_{18}NO_4PS \times {}^1/_6\, C_7H_8$: C, 53.16, H, 5.69, N, 4.09. Found: C, 53.21, H, 5.78, N, 4.26.

Dimethyl N-benzylamino(2-thienyl)methylphosphonate (**2f**): Yield = 57% (1.77 g) yellow oil. ^{1}H-NMR ($CDCl_3$, 600 MHz): 7.26–7.20 (m, PhH, H_5^{thioph}, 5H), 7.18–7.15 (m, PhH, 1H), 7.02–7.01 (m, H_3^{thioph}, 1H), 6.93 (dd, $^3J_{HH}$ = 4.8 and 3.6 Hz, H_4^{thioph}, 1H), 4.24 (d, $^2J_{PH}$ = 23.8 Hz, CHP, 1H), 3.82 (d, $^2J_{HH}$ = 13.8 Hz, CH_2Ph, 1H), 3.68 (d, $^3J_{PH}$ = 10.2 Hz, $POCH_3$, 3H), 3.57 (d, $^2J_{HH}$ = 13.8 Hz, CH_2Ph, 1H), 3.55 (d, $^3J_{PH}$ = 10.2 Hz, $POCH_3$, 3H). ^{13}C-NMR ($CDCl_3$, 150 MHz): δ 139.30 (d, $^2J_{CP}$ = 4.3 Hz, C^2_{thioph}), 138.99 (C_{arom}), 128.47 (d, *J* = 7.5 Hz, C_{thioph}), 128.38 (C_{arom}), 128.23 (C_{arom}), 127.28 (C_{arom}), 126.97 (d, *J* = 3.5 Hz, C_{thioph}), 125.68 (d, *J* = 3.3 Hz, C_{thioph}), 54.74 (d, $^1J_{CP}$ = 159.7 Hz, PC), 53.94 (d, $^2J_{CP}$ = 6.8 Hz, POC),

53.63 (d, $^2J_{CP}$ = 6.8 Hz, POC), 51.24 (d, $^3J_{CP}$ = 15.6 Hz, PCC). ^{31}P-NMR (243 MHz, $CDCl_3$): δ 23.98. *Elem. Anal.* Calcd. for $C_{14}H_{18}NO_3PS$: C, 54.01, H, 5.83, N, 4.50. Found: C, 53.77, H, 5.67, N, 4.46.

Dimethyl N-benzhydrylamino(2-thienyl)methylphosphonate (**2g**): Yield = 41% (1.59 g) yellow oil. ^{1}H-NMR ($CDCl_3$, 600 MHz): 7.44–7.34 (m, PhH, H_5^{thioph}, 6H), 7.32–7.22 (m, PhH, 5H), 7.07–7.06 (m, H_3^{thioph}, H_4^{thioph}, 2H), 4.91 (s, CH, 1H), 4.29 (d, $^2J_{PH}$ = 22.8 Hz, CHP, 1H), 3.91 (d, $^3J_{PH}$ = 10.8 Hz, $POCH_3$, 3H), 3.64 (d, $^3J_{PH}$ = 10.8 Hz, $POCH_3$, 3H), 2.57–2.48 (bs, NH, 1H). ^{13}C-NMR ($CDCl_3$, 150 MHz): δ 139.36 (d, $^2J_{CP}$ = 6.4 Hz, C^2_{thioph}), 139.73 (C_{arom}), 128.75 (C_{arom}), 128.56 (C_{arom}), 127.89 (C_{arom}), 127.17 (d, *J* = 7.9 Hz, C_{thioph}), 127.07 (d, *J* = 2.3 Hz, C_{thioph}), 125.72 (d, *J* = 3.4 Hz, C_{thioph}), 63.81 (d, $^3J_{CP}$ = 16.0 Hz, PCC), 54.12 (d, $^2J_{CP}$ = 6.7 Hz, POC), 53.53 (d, $^2J_{CP}$ = 6.7 Hz, POC), 53.19 (d, $^1J_{CP}$ = 161.5 Hz, PC). ^{31}P-NMR (243 MHz, $CDCl_3$): δ 24.32. *Elem. Anal.* Calcd. for $C_{20}H_{22}NO_3PS \times {}^1/_5\, C_7H_8$: C, 63.33, H, 5.86, N, 3.45. Found: C, 63.54, H, 5.69, N, 3.75.

Dibenzyl N-furfurylamino(2-thienyl)methylphosphonate (**2h**): Yield = 62% (2.81 g) yellow oil. ^{1}H-NMR ($CDCl_3$, 600 MHz): δ 7.36–7.32 (m, PhH, 10H), 7.25–7.23 (m, H_5^{thioph}, H_5^{fur}, 1H), 7.11–7.09 (m, H_3^{thioph}, 1H), 7.03–7.02 (m, H_4^{thioph}, 1H), 6.33–6.31 (m, H_3^{fur}, 1H), 6.15–6.14 (m, H_4^{fur}, 1H), 5.10–5.02 (Part AB of ABX system, $^3J_{PH}$ = 7.5 and 8.8 Hz, $^2J_{HH}$ = 11.7 Hz, $POCH_2Ph$, CHP, 2H), 5.00 (Part A of AMX system, $^3J_{PH}$ = 7.2 and $^2J_{HH}$ = 11.8 Hz, $POCH_2Ph$, 1H), 4.90 (Part M of AMX system, $^3J_{PH}$ = 8.2 and $^2J_{HH}$ = 11.8 Hz, $POCH_2Ph$, 1H), 4.43 (d, $^2J_{PH}$ = 19.3 Hz, CHP, 1H), 3.92 and 3.69 (2d, $^2J_{HH}$ = 14.6 Hz, CH_2Fur, 3H). ^{13}C-NMR (150 MHz, $CDCl_3$): δ 152.60 (C_{arom}), 142.12 (C_{arom}), 138.67 (d, *J* = 4.8 Hz, C_{arom}), 136.32 (d, $^3J_{CP}$ = 6.0 Hz, C_{arom}), 136.19 (d, $^3J_{CP}$ = 6.3 Hz, C_{arom}), 128.48 (d, *J* = 6.5 Hz, C_{arom}), 128.34 (C_{arom}), 128.28 (C_{arom}), 128.00 (C_{arom}), 127.84 (C_{arom}), 127.36 (C_{arom}), 127.30 (C_{arom}), 126.92 (d, *J* = 2.4 Hz, C_{arom}), 125.82 (d, *J* = 3.5 Hz, C_{arom}), 110.14 (C_{arom}), 107.92 (C_{arom}), 68.58 and 68.18 (2d, $^2J_{CP}$ = 40.6 Hz, POC), 55.11 (d, $^1J_{CP}$ = 159.8 Hz, PC), 43.72 (d, $^3J_{CP}$ = 16.4 Hz, NC). ^{31}P-NMR (243 MHz, $CDCl_3$): δ 22.19. *Elem. Anal.* Calcd. for $C_{24}H_{24}NO_4PS \times {}^1/_{10}\, CH_2Cl_2$: C, 62.66, H, 5.28, .N, 3.03. Found: C, 62.77, H, 5.03, N, 3.05.

3.2. Plant Growth Test of New Synthesized Compounds

The plant growth test of thiophene-derived aminophosphonates **2a–h** was performed in laboratory conditions following the OECD 208 Guideline Terrestrial Plants Growth Test for oat (*Avena sativa*) as a monocotyledonous plant and radish (*Raphanus sativus* L. subvar. radicula Pers.), a dicotyledonous plant.

According to the mentioned OECD 208 standard, the plant growth test of Compounds **2a–h** was carried out in sandy soil having the following parameters: granulometric composition of soil—77% sand, 14% dust and loam, organic carbon content of approx. 1.2%, pH (KCl) equal to 6.3.

Tests were carried out in polypropylene pots (diameter of 90 mm and capacity of 300 cm^3), which were filled with the control soil or with the soil mixed with the tested Compounds **2a–h** added at a specific concentration. Twenty identical seeds of each of the selected plant species were sown into the soil. Seeds originated from the same source. Plants were grown for 14 days under controlled conditions: a constant humidity content at the level required for the plants (70% field water capacity), light intensity (7000 lux), and temperature (20 ± 2 °C) were maintained. Then, seedlings were counted, and the dry as well as fresh weight of the plants was determined. The parts of plants above the soil surface were analyzed.

The performed plant growth was evaluated using a preliminary test; subsequently, the final test was conducted afterward based on the obtained screening results. Preliminary tests were performed to determine the range of concentrations of compounds affecting the soil quality; therefore, the used concentrations of Samples **2a–h** were as follows: 0 mg (control), 1 mg, 10 mg, 100 mg, and 1000 mg/kg of soil dry weight.

As already mentioned, based on the obtained results from preliminary test, the final, more precise test to find values of the no observed effect concentration (NOEC) and the lowest observed effect concentration (LOEC) of the compounds under study **2a–h** was performed.

The evaluation of phytotoxicity of the studied aminophosphonates **2a–h** at applied concentrations was made by comparing the germination and (dry and fresh) weight of control plant sprouts (seedlings) with germination and of (dry and fresh) plants sprouts grown in the soil with an admixture of given amounts of the tested compounds.

The dry weights of tested plants were measured after drying at 75 °C until the constant weight.

The visual evaluation of phytotoxicity of aminophosphonates **2a–h** at applied concentrations was performed by digital photography. Obtained pictures were analyzed in terms of any type of damage to tested plants, such as their growth inhibition, chlorosis, and necrosis. Tests were carried out three times for each sample.

The significance of the obtained results was evaluated using the analysis of variance (ANOVA). The least significant difference (LSD) values at a confidence level of 95% were computed using the Tukey test.

4. Conclusions

To conclude, the toxicity of *N*-methylphenyl aminophosphonates **2a–c** was diagnosed to be promising as a potential soil-applied herbicide, especially for their selectivity—they are evidently toxic for dicotyledonous radish and not so harmful for monocotyledonous oat. Moreover, their herbicidal efficiency is stronger for the *N*-methylphenyl aminophosphonates substituted with methyl group at position *ortho* (**2a**) and *meta* (**2b**) in the phenyl ring.

Dimethyl *N*-benzylamino(2-thienyl)methylphosphonate (**2f**) exhibited non-selective, harmful effects against both tested plants.

Among *N*-methoxyphenyl aminophosphonates **2d** and **2e**, the latter showed an inhibitive effect especially against dicotyldenous plants. Its impact, however, was significantly weaker than impact of **2a**,**b**. This is to point out that the sole structural difference between *N*-methoxyphenyl aminophosphonates **2d**,**e** and *N*-methylphenyl aminophosphonates **2a–c** is the substituent of the phenyl moiety, *i.e.*, the methoxy and methyl group, respectively, which indicates that the type and position of those groups in the phenyl ring play a key role in the toxicity effect against tested plants.

Among all tested samples, dibenzyl *N*-furfurylamino(2-thienyl)methylphosphonate (**2h**) was found to be the most toxic for radish and oat but without significant selectivity. The investigation of Compounds **2a**,**b**, **2f**, and **2h**, such as their potential applications and mechanistic approaches, will be continued.

Some thiophene-derived aminophosphonic derivatives were found to have various biological properties (*vide supra*); therefore, important microbiological, herbicidal, and cytotoxic properties of the studied aminophosphonic systems bearing thiophene moiety **2a–h** cannot be excluded. If so, the synthesis of such compounds for potential agrochemical or pharmacological application must be taken into consideration from the environmental protection point of view. Our results call attention to the necessity of further phytotoxicological investigation of any new synthesized aminophosphonic derivatives bearing thiophene moiety.

Supplementary Materials: Supplementary materials can be accessed at: http://www.mdpi.com/1420-3049/21/6/694/s1.

Acknowledgments: Studies were founded by the National Centre of Science of Polish State (NCN), grant no. 2014/13/B/NZ9/02418, where funds for publishing in open-access mode were scheduled. Special thanks are addressed to Paulina Tomczyk and Justyna Biernacka, former M.Sc. students at the University of Łódź, for their fruitful assistance in the preparation of the studied compounds.

Author Contributions: J.L. and P.R. conceived and designed the experiments; A.M., Z.M., M.M., and D.R. performed the experiments; J.L. and P.R. analyzed the data; J.L. and P.R. wrote the paper.

Conflicts of Interest: The authors declare no conflict of interest. The founding sponsors had no role in the design of the study; in the collection, analyses, or interpretation of data; in the writing of the manuscript; or in the decision to publish the results.

References

1. Andersen, S.M.; Hertz, P.B.; Holst, T.; Bossi, R.; Jacobsen, C.S. Mineralisation studies of C14-labelled metsulfuron-methyl, tribenuron-methyl, chlorsulfuron and thifensulfuronmethyl in one Danish soil and groundwater sediment profile. *Chemosphere* **2001**, *45*, 775–782. [CrossRef]
2. Polati, S.; Bottaro, M.; Frascarolo, P.; Gosetti, F.; Gianotti, V.; Gennaro, M.C. HPLC-UV and HPLC-MSn multiresidue determination of amidosulfuron, azimsulfuron, nicosulfuron, rimsulfuron, thifensulfuron methyl, tribenuron methyl and azoxystrobin in surface waters. *Anal. Chim. Acta* **2006**, *579*, 146–151. [CrossRef] [PubMed]
3. EFSA (European Food Safety Authority). Conclusion on the peer review of the pesticide risk assessment of the active substance thifensulfuron-methyl. *EFSA J.* **2015**, *13*, 4201. [CrossRef]
4. Rosenkrantz, R.Y.; Baun, A.; Kusk, O. Growth inhibition and recoveryof *Lemna gibba* after pulse exposure to sulfonylurea herbicides. *Ecotoxicol. Environ. Saf.* **2013**, *89*, 89–94. [CrossRef] [PubMed]
5. Beckie, H.J.; Tardif, F.J. Herbicide cross resistance in weeds. *Crop Prot.* **2012**, *35*, 15–28. [CrossRef]
6. Brown, D. *2,4-D and Dicamba-Resistant Crops and Their Implications for Susceptible Non-Target Crops*; Michigan State University Extension: East Lansing, MI, USA, 2013.
7. Filipe, O.M.S.; Santos, S.A.O.; Domingues, M.R.M.; Vidal, M.M.; Silvestre, A.J.D.; Neto, C.P.; Santos, E.B.H. Photodegradation of the fungicide thiram in aqueous solutions. Kinetic studies and identification of the photodegradation products by HPLC-MS/MS. *Chemosphere* **2013**, *91*, 993–1001. [CrossRef] [PubMed]
8. Sanchirico, R.; Pinto, G.; Pollio, A.; Cordella, M.; Cozzani, V. Thermal degradation of Fenitrothion: Identification and eco-toxicity of decomposition products. *J. Hazard. Mater.* **2012**, *199*, 390–400. [CrossRef] [PubMed]
9. Baghestani, M.A.; Zand, E.; Soufizadeh, S.; Jamali, M.; Mighany, F. Evaluation of sulfosulfuron for broadleaved and grass weed control in wheat (*Triticum aestivum* L.) in Iran. *Crop Prot.* **2007**, *26*, 1385–1389. [CrossRef]
10. Zhao, W.; Xu, L.; Li, D.; Li, X.; Wang, C.; Zheng, M.; Pan, C.; Qiu, L. Biodegradation of thifensulfuron-methyl by *Ochrobactrum* sp. in liquid medium and soil. *Biotechnol. Lett.* **2015**, *37*, 1385–1392. [CrossRef] [PubMed]
11. Cessna, A.J.; Donald, D.B.; Bailey, J.; Waiser, M. Persistence of the Sulfonylurea Herbicides Sulfosulfuron, Rimsulfuron, and Nicosulfuron in Farm Dugouts (Ponds). *J. Environ. Qual.* **2015**, *44*, 1948–1955. [CrossRef] [PubMed]
12. Elliott, J.A.; Cessna, A.J. Variability in the distribution and dissipation of the herbicide thifensulfuron-methyl in a prairie wetland. *J. Soil Water Conserv.* **2014**, *69*, 151–159. [CrossRef]
13. Aziz, S.; Dumas, S.; el Azzouzi, M.; Sarakham, M.; Chovelon, J.-M. Photophysical and photochemical studies of thifensulfuron-methyl herbicide in aqueous solution. *J. Photochem. Photobiol. A Chem.* **2010**, *209*, 210–218. [CrossRef]
14. Lazartigues, A.; Thomas, M.; Banas, D.; Brun-Bellut, J.; Cren-Olivé, C.; Feidt, C. Accumulation and half-lives of 13 pesticides in muscle tissue of freshwater fishes through food exposure. *Chemosphere* **2013**, *91*, 530–535. [CrossRef] [PubMed]
15. Brown, H.M.; Joshi, M.M.; van Ailien, T.; Carski, T.H.; Dulka, J.J.; Patrick, M.C.; Reiser, R.W.; Livingston, R.S.; Doughty, J. Degradation of thifensulfuron methyl in soil: Role of microbial carboxyesterase activity. *J. Agric. Food Chem.* **1997**, *45*, 955–961. [CrossRef]
16. Boduszek, B. Synthesis and Biological Activity of Heterocyclic Amino-phosphonates. *Phosphorus Sulfur Silicon Relat. Elem.* **1999**, *144*, 433–436. [CrossRef]
17. Janardhan Rao, A.; Visweswara Rao, P.; Koteswara Rao, V.; Mohan, C.; Naga Raju, C.; Suresh Reddy, C. Microwave Assisted One-pot Synthesis of Novel α-Amino-phosphonates and Their Biological Activity. *Bull. Korean Chem. Soc.* **2010**, *31*, 1863–1868.
18. Maheswara Rao, K.U.; Namkoong, S.; Yu, H.-C.; Park, J.; Chung, C.-M.; Oh, S.Y. Green Synthesis and Biological Evaluation of New Di-aminophosphonate Derivatives as Cytotoxic Agents. *Arch. Pharm. Chem. Life Sci.* **2013**, *346*, 851–859. [CrossRef] [PubMed]

19. Rao Devineni, S.; Doddaga, S.; Donka, R.; Raju Chamarthi, N. $CeCl_3 \cdot 7H_2O$-SiO_2: Catalyst promoted microwave assisted neat synthesis, antifungal and antioxidant activities of α-diaminophosphonates. *Chin. Chem. Lett.* **2013**, *24*, 759–763. [CrossRef]
20. Rasheed, S.; Venkataramana, K.; Chandrasekhar, K.; Fareeda, G.; Naga Raju, C. Ultrasound-Assisted Synthesis of Novel-Aminophosphonates and Their Biological Activity. *Arch. Pharm. Chem. Life Sci.* **2012**, *345*, 294–301. [CrossRef] [PubMed]
21. Banik, A.; Batta, S.; Bandyopadhyay, D.; Banik, B.K. A Highly Efficient Bismuth Salts-Catalyzed Route for the Synthesis of α-Aminophosphonates. *Molecules* **2010**, *15*, 8205–8213. [CrossRef] [PubMed]
22. Boduszek, B. An efficient synthesis of 1-aminophosphonic acids and esters bearing heterocyclic moiety. *Phosphorus Sulfur Silicon Relat. Elem.* **1995**, *104*, 63–70. [CrossRef]
23. Demir, A.S.; Tanyeli, C.; Sesenoglu, O.; Demic, S. A Simple Synthesis of 1-Amino-phosphonic Acids from 1-Hydroxyiminophosphonates with $NaBH_4$ in the Presence of Transition Metal Compounds. *Tetrahedron Lett.* **1996**, *37*, 407–410. [CrossRef]
24. Manjula, A.; Vittal Rao, B.; Neelakantan, P. One-Pot Synthesis of α-aminophosphonates: An Inexpensive Approach. *Synth. Commun.* **2003**, *33*, 2963–2969. [CrossRef]
25. Mu, X.-J.; Lei, M.-Y.; Zoua, J.-P.; Zhang, W. Microwave-assisted solvent-free and catalyst-free Kabachnik–Fields reactions for α-amino phosphonates. *Tetrahedron Lett.* **2006**, *47*, 1125–1127. [CrossRef]
26. Saito, B.; Egami, H.; Katsuki, T. Synthesis of an Optically Active Al(salalen) Complex and Its Application to Catalytic Hydrophosphonylation of Aldehydes and Aldimines. *J. Am. Chem. Soc.* **2007**, *129*, 1978–1986. [CrossRef] [PubMed]
27. Cherkasov, R.A.; Galkin, V.I. The Kabachnik–Fields reaction: Synthetic potential and the problem of the mechanism. *Russ. Chem. Rev.* **1998**, *67*, 857–882. [CrossRef]
28. Keglevich, G.; Bálint, E. The Kabachnik–Fields reaction: Mechanism and synthetic use. *Molecules* **2012**, *17*, 12821–12835. [CrossRef] [PubMed]
29. Kafarski, P.; Górniak, M.G.; Andrasiak, I. Kabachnik–Fields reaction under green conditions-A critical overview. *Curr. Green Chem.* **2015**, *5*, 218–222. [CrossRef]
30. Kafarski, P.; Lejczak, B. Biological Activity of Aminophosphonic Acids. *Phosphorus Sulfur Silicon Relat. Elem.* **1991**, *63*, 193–215. [CrossRef]
31. Forlani, G.; Berlicki, Ł.; Duò, M.; Dziędzioła, G.; Giberti, S.; Bertazzini, M.; Kafarski, P. Synthesis and Evaluation of Effective Inhibitors of Plant δ^1-Pyrroline-5-carboxylate Reductase. *J. Agric. Food Chem.* **2013**, *61*, 6792–6798. [CrossRef] [PubMed]
32. Occhipinti, A.; Berlicki, Ł.; Giberti, S.; Dziędzioła, G.; Kafarski, P.; Forlani, G. Effectiveness and mode of action of phosphonate inhibitors of plant glutamine synthetase. *Pest Manag. Sci.* **2010**, *66*, 51–58. [CrossRef] [PubMed]
33. Giberti, S.; Bertazzini, M.; Liboni, M.; Berlicki, Ł.; Kafarski, P.; Forlani, G. Phytotoxicity of aminobisphosphonates targeting both δ^1-pyrroline-5-carboxylate reductase and glutamine synthetase. *Pest Manag. Sci.* **2016**. [CrossRef] [PubMed]
34. Matusiak, A.; Lewkowski, J.; Rychter, P.; Biczak, R. Phytotoxicity of New Furan-derived Aminophosphonic Acids, *N*-Aryl Furaldimines and 5-Nitrofuraldimine. *J. Agric. Food Chem.* **2013**, *61*, 7673–7678. [CrossRef] [PubMed]
35. Klimczak, A.A.; Kuropatwa, A.; Lewkowski, J.; Szemraj, J. Synthesis of *N*-aryl, furan-derived aminophosponates and studies of their *in vitro* cytotoxicity against esophageal cancer cells. *Med. Chem. Res.* **2013**, *21*, 852–860. [CrossRef]

Sample Availability: Samples of the compounds **2a–h** are available from the authors.

molecules

MDPI

Article

Synthesis, Spectral Characterization of Several Novel Pyrene-Derived Aminophosphonates and Their Ecotoxicological Evaluation Using *Heterocypris incongruens* and *Vibrio fisheri* Tests

Jarosław Lewkowski [1,*], Maria Rodriguez Moya [1], Marta Chmielak [1,†], Diana Rogacz [2], Kamila Lewicka [2] and Piotr Rychter [2,*]

1 Department of Organic Chemistry, Faculty of Chemistry, University of Łódź, Tamka 12, 91-403 Łódź, Poland; MRM_chem@wp.pl (M.R.M.); martachmielak1@gmail.com (M.C.)

2 Institute of Chemistry, Environmental Protection and Biotechnology, Jan Długosz University in Częstochowa, 13/15 Armii Krajowej Av., 42-200 Częstochowa, Poland; diana.rogacz@gmail.com (D.R.); lewickakamilla@gmail.com (K.L.)

* Correspondence: jlewkow@uni.lodz.pl (J.L.); p.rychter@ajd.czest.pl (P.R.); Tel.: +48-426-355-751 (J.L.); +48-343-615-154 (P.R.)

† M.Sc. student at the Faculty of Chemistry, University of Łódź (Academic Year 2015–2016).

Academic Editor: György Keglevich
Received: 26 June 2016; Accepted: 14 July 2016; Published: 19 July 2016

Abstract: Four diphenyl pyrene-derived aminophosphonates were synthesized. Attempts were made to synthesize diphenyl *N*-(*R*)-α-methylbenzylamino(pyren-1-yl)methylphosphonate (**3e**) in order to obtain the chiral aminophosphonate bearing a pyrene moiety. Because these attempts failed, dimethyl and dibenzyl *N*-(*R*)-α-methylbenzyl substituted aminophosphonates **4** and **5** were synthesized and the predominant diastereoisomer of dimethyl aminophosphonate **4** was isolated. The resolution of the diastereomeric mixture of **5** failed. Aminophosphonates **3a–d** and the predominant diastereoisomer of **4** were investigated in terms of their ecotoxicity using tests performed on the ostracode *Heterocypris incongruens* and the fluorescent bacterium *Vibrio fisheri*. The tests confirmed the moderate-to-high ecotoxicity of aminophosphonates **3a–d** and **4**, but no evident correlation between the structure and toxicity has been found.

Keywords: pyrene-1-carboxaldehyde; aminophosphonates; ecotoxicity; ostracode test; *Vibrio fisheri* test

1. Introduction

Aminophosphonic derivatives are well known for their biological activity. They are described as effective inhibitors of various enzymes and, therefore they find applications as crop protective [1], antibacterial [2] or anticancer [3] agents. Analogues of *C*-phenylphosphonoglycine, have been demonstrated to show activity in various fields as herbicides and plant growth regulators [4], agrochemical fungicides [5,6] and glutamate receptor agonists and antagonists [7].

In a sort of logical consequence, a lot of compounds from this group have been synthesized, including those bearing benzene and its polycyclic aromatic analogues [8], five- and six-membered heteroaromatic rings [9,10], and even the ferrocene [11] moiety. Their syntheses are generally based on the Kabachnik-Fields and the aza-Pudovik reactions [12–14].

In contrast, aminophosphonic derivatives of 1-pyrene, i.e., amino(1-pyrene)methylphosphonic compounds have been described only twice in the chemical literature. Hudson et al. [15] reported the synthesis of diethyl *N*-benzhydrylamino(1-pyrene)methyl phosphonate, and Jayaprakash and co-workers [16] described the synthesis of a series of *N*-aryl substituted, diethyl amino(1-pyrene)methyl phosphonates and results of preliminary studies on their fluorescence properties.

We have tried to fill this void and very recently have published results of our studies on the synthesis, photophysical properties and cytotoxic action of a series of α-aminophosphonates bearing pyren-1-yl moieties [17]. They demonstrated moderate-to-satisfactory cytotoxicities against two colon cancer cell lines. Such interesting biological properties may result in potential future applications and their biological hazards should be evaluated using typical biotoxicological tests.

Herein, we wish to report the synthesis of a new series of *N*-substituted-*C*-(pyren-1-yl)-phosphonoglycine derivatives and their ecotoxicological properties evaluated using ostracode (*Heterocypris incongruens*) and fluorescence bacteria *Vibrio fisheri* tests.

Nowadays, ecotoxicology faces the challenge of evaluating and predicting the effects of a variety of chemicals on aquatic species and ecosystems. Since a chemical pollutant may follow various routes from a source to the target, there is an urgent need to follow their potential ecotoxicological impact using different biotests [18–21].

Toxkit microbiotests are presently used in many countries worldwide for biomonitoring of contaminated waters and soils. These small-scale bioassays are becoming more and more popular in laboratories all over the world because of their various useful advantages, such as high degree of standardization and precision, simple testing procedures which are very clearly documented down to the last detail, operation with small sample volumes, the repeatability as well as giving comparable results to equivalent ISO and OECD full-scale tests. However, the major advantage of Toxkit microbiotests in comparison to conventional bioassays is that the tested microorganisms are incorporated in the kits in a dormant or immobilized form, from which they can be hatched or activated "on demand" prior to carrying out the toxicity tests [22].

The Microtox® system test with the luminescent bacterium *Vibrio fischeri* and the crustacean *Heterocypris incongruens*, when used as bioindicator allows one to monitor potential toxicity of soil or water contaminants in a rapid, simple and inexpensive way. This bioanalytical technique allows one to overcome some disadvantages of the traditional physicochemical methods such as concentration of toxic substances below analytical limits or potential chemical interactions leading to additive, synergistic or antagonistic effect which cannot be identified using the abovementioned methods. The *Vibrio fischeri* bioassay can be applied in all types of water, including surface and ground water, wastewater, drilling sump fluids and many other aqueous solutions [23]. Both *Vibrio fischeri* and *Heterocypris incongruens* bioassay tests have been already successfully used for ecotoxicity determination of various pharmaceuticals, pesticides, ionic liquids, detergents, heavy metals, road dust, eluates and organic extracts of wastes, etc. [24–30].

As it was mentioned above, the aim of the study was the preliminary evaluation of the potential toxicity of newly synthesized aminophosphonates **3a–d** and **4**, if introduced to the environment. In order to perform it, the Microtox® test using the bioluminescent bacterium *Vibrio fischeri* was employed to examine the acute toxicity of sediment eluates loaded with tested compounds. Due to the cost effective, simple operation and time-savings, the biotest based on the bioluminescence inhibition of *Vibrio fischeri* is probably the most widely applied bacterial test used for environmental pollution monitoring.

However, due to fact that the obtained substances are almost insoluble in water, the Microtox® solid phase test (MSPT), as a one of the most popular tests for sediment toxicity assessments due to its simplicity, reproducibility, ecological relevance and sensitivity, has been used [26,31–33].

Additionally, the standardized crustacean biotest with *Heterocypris incongruens* was selected as a representative benthic invertebrate species for screening the toxicity of sediments containing the tested substances. Crustaceans are promising representative species of the benthic environment as they spend their lives swimming in water and in direct contact with the sediment. Therefore, the Ostracodtoxkit® test has been used for toxicity evaluation. The toxicity test using mentioned freshwater ostracodes is a sub-chronic static test that exposes individuals to whole sediments over a period of 6 days, the endpoints being mortality and growth changes [34].

2. Results and Discussion

2.1. *Synthesis of Aminophosphonic Derivatives* **3a–e**, **4** *and* **5** *Bearing 1-Pyrene Moieties*

Schiff bases **1a–e** were synthesized following the published and commonly known procedure by simple mixing pyrene-1-carboxaldehyde with an appropriate amine in hexane (in a case of *p*-anisidine, *p*-toluidine and aniline) or methanol (for benzylamine and (*R*)-α-methylbenzylamine) and stirring them at room temperature for 24 h; the reaction with *p*-anisidine required 48 h [14]. This procedure gave imines **1a–e**, which were isolated and used for further conversions without any purification. ^{1}H-NMR spectra were recorded only to verify the identity, based on the diagnostic ^{1}H-NMR singlet corresponding to the proton of the azomethine group (–CH=N–) above 8 ppm [17].

Aminophosphonates **3a–d**, **4** and **5** were synthesized basing on the aza-Pudovik reaction, i.e., the addition of diphenyl phosphite to imines **1a–d**. However, important modifications were necessary to be introduced, whereby reactions were carried out in a small amount of dichloromethane and diphenyl phosphite was used in three-fold excess. The reaction mixtures were heated for 5 min then stirred for 24 h at room temperature. After workup as described in the Experimental Section, the resulting aminophosphonates **3a–d** were obtained in 60%–70% yields (Scheme 1). They were purified by column chromatography on neutral aluminum oxide, because diphenyl amino(1-pyrene)methylphosphonates **3a–d** decomposed on silica gel.

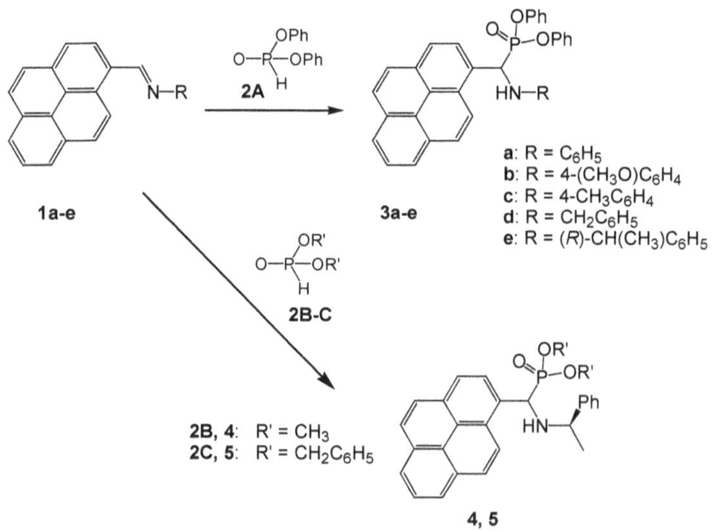

Scheme 1. Synthesis of *N*-substituted amino(1-pyrene)methylphosphonates **3a–e**, **4** and **5**.

The identity of synthesized compounds was verified by means of ^{1}H-, ^{31}P- and ^{13}C-NMR as well as by FT-IR spectroscopy. The proton NMR spectra of diphenyl pyrene-derived aminophosphonates **3a–d** showed the characteristic signals of all compounds, i.e., the doublet corresponding to the proton from the CH-P bond and the system of signals attributable to the pyrene ring protons. The ^{13}C-NMR spectra revealed the presence of two doublets at around 150 ppm with a coupling constant close to 10 Hz, which are attributed to the phenyl ring carbons from the P-O-C groups. Doublets in the 50–55 ppm range with coupling constants between 150 and 160 Hz are easily recognizable as C-P bond signals. IR bands at around 1250 cm^{-1} correspond to the bending vibrations of P=O bonds, while the presence of aromatic carbon-carbon bonds was confirmed by characteristic bands at around 1600, 1570, 1500 and 1450 cm^{-1}. Scans of NMR spectra are collected in the Supplementary Materials as Figures S1–S7.

In order to obtain the chiral, non-racemic diphenyl ester of a phosphonoglycine derivative bearing a pyren-1-yl moiety, the synthesis of diphenyl *N*-(*R*)-α-methylbenzylamino(pyren-1-yl)-methylphosphonate (**3e**) was performed using the method described above. The resulting mixture contained two diastereomers in a 100:15 ratio, but all attempts to purify the product by column chromatography on neutral aluminum oxide caused extensive decomposition of the product. From among available adsorbents, silica gel was totally inconvenient (*vide supra*), and chromatography on microcrystalline cellulose with chloroform allowed us to get rid of impurities only roughly, i.e., crude diphenyl *N*-(*R*)-α-methylbenzylamino(pyren-1-yl)methylphosphonate (**3e**) was obtained.

Thus, in order to investigate the stereochemistry of phosphite addition to the azomethine bond of pyrene-1-carboxaldimine **1e**, addition of dimethyl and dibenzylphosphites has been performed. Dimethyl *N*-(*R*)-α-methylbenzylamino(pyren-1-yl)methylphosphonate (**4**) was obtained as a mixture of two diastereoisomers in a 100:13 ratio. We succeeded in separation of the predominant diastereoisomer using column chromatography on silica gel. Its absolute configuration however could not be determined, because the predominant isomer of aminophosphonate **4** is a dense oily liquid, thus an X-ray study could not be performed. Isolation of the minor diastereoisomer of **4** failed, it was always obtained in a mixture with the major one.

Dibenzyl *N*-(*R*)-α-methylbenzylamino(pyren-1-yl)methylphosphonate (**5**) was obtained as a mixture of two diastereoisomers in a 100:26 ratio. Unfortunately, the obtained mixture of diastereoisomers was not separable by column chromatography, as all applied solvent systems led to mixtures with different ratios of isomers. However, the crude diastereomeric mixture of **5** was purified by column chromatography on silica gel to give satisfactory elemental analysis results.

Thus the obtained diphenyl aminophosphonates **3a–d** and the isolated predominant diastereoisomer of **4** were tested in terms of their ecotoxicity using two popular ecotoxicological tests, the ostracode *Heterocypris incongruens* and fluorescence bacteria *Vibrio fisheri*.

2.2. *Ecotoxicological Properties of Aminophosphonates* **3a–d** *and* **4**

2.2.1. Microtox Toxicity Assay

Bioassays are the complementary tool for characterizing the biological effects and hazards of contaminated sediments or soils. During the solid phase Microtox® assay the bioluminescent bacteria are exposed to a sediment suspension and their response reflects totally the action of toxic agents along with synergists and antagonists present in a given sample. The effects on light emission are evaluated in the liquid phase that remains after removal of the sediments by filtration. Effective concentration (EC_{50}) values (Table 1) were calculated from the concentration–response curves (Figure 1).

Obtained EC_{50} values ranged between 92 and 533 mg/kg of dry weight of soil, depending on the tested compound. Among the compounds **3a–c** with aromatic substituents on the amine groups, diphenyl *N*-(4-methoxyphenyl)amino(pyren-1-yl)methylphosphonate (**3b**) was found to be the most toxic, with the lowest EC_{50} value (92.63 mg/kg). Much less harmful action against *Vibrio fischeri* was found for the *N*-(4-methylphenyl) derivative **3c** (EC_{50} = 214.5 mg/kg), while diphenyl *N*-phenylamino(pyren-1-yl)methylphosphonate (**3a**) demonstrated nearly a lack of toxicity. Thus, the presence of a substituent at the phenyl group linked to nitrogen seems to be crucial in terms of toxicological properties of the studied compounds **3a–c**. The results also showed that a methoxy substituent increases the toxicity of the compound **3b** in comparison with the methyl substituted aminophosphonate **3c**.

Among the aminophosphonates **3d** and **4** having the amino groups substituted with non-aromatic groups, results were indeed surprising. Diphenyl *N*-benzylamino(pyren-1-yl)methylphosphonate (**3d**) with a *N*-benzyl substituent demonstrated five times higher toxicity (EC_{50} = 102.7 mg/kg) as compared to the *N*-(*R*)-α-methylbenzylamino derivative **4** (EC_{50} = 533.1 mg/kg). Therefore we can conclude that the presence of a methyl substituent in the benzyl moiety dramatically decreases the toxicity of compound **4**.

Table 1. Microtox® EC_{50} values (mg/kg of soil dry weight) of exposure of the luminescent marine bacteria *Vibrio fischeri* to tested aminophosphonates **3a–d** and **4**, with respective 95% confidence limits (in brackets) obtained in the fit of the data.

Compounds	EC_{50} (Lower Limit; Upper Limit)	Coefficient of Determination (R^2)
3a	495.9 (203.4; 1209)	0.7632
3b	92.63 (65.27; 131.4)	0.8912
3c	214.5 (165.0; 279.0)	0.9452
3d	102.7 (81.83; 128.9)	0.9571
4	533.1 (439.4; 646.8)	0.9516

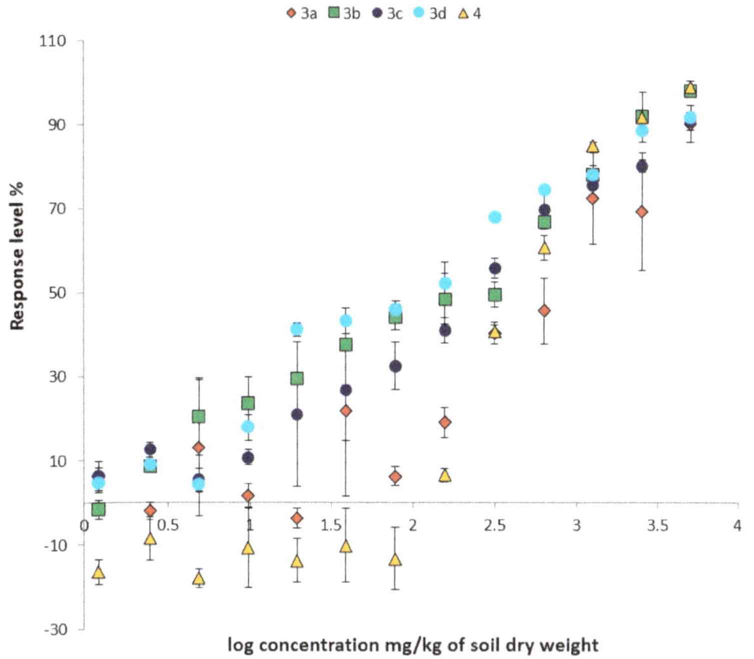

Figure 1. The EC_{50} graded dose-response curves for the five tested compounds **3a–d** and **4**. Overlapped curves are plotted by the Microtox® Analyzer 500 software. Error bars represent standard deviation error (SD, n = 3 determinations).

2.2.2. Ostracod Test Kit

Ostracodtoxkit F™ is the very first sediment-contact microbiotest with a crustacean test species for the assessment of the total toxicity of sediments, hence covering the toxic hazards of both dissolved and not dissolved pollutants. Considering the very low solubility of synthesized aminophosphonates **3a–d** and **4**, it seemed justified to evaluate reliably their potential toxicity.

The first criterion of the effect on tested crustaceans is the percent of mortality. Toxicity was expressed in % of mortality, i.e., the lethal effect of the sample on the number of ostracods compared with the control plate. As it was expected, the mortality of ostracodes increased with growing concentration of tested substance in soil (Figure 2). The concentration 500 mg/kg of soil was found to be lethal for all samples. However, among all tested compounds, *N*-benzylamino (pyren-1-yl)methylphosphonate (**3d**) demonstrated the highest toxicity at every concentration when compared to other samples (with the sole exception of aminophosphonate **4** at a concentration of

100 mg/kg of soil). Even at a concentration of 10 mg/kg of dry weight of soil, compound **3d** caused 10% mortality of crustaceans, while other samples at this concentration did not show any mortal effect.

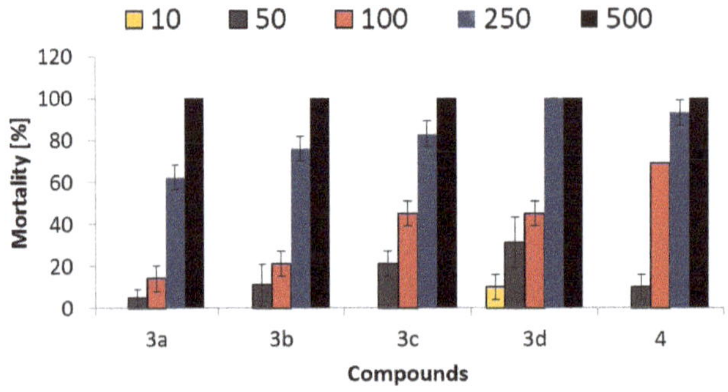

Figure 2. The mortality (%) of *Heterocypris incongruens* living in soil contaminated with pyrene-derived aminophosphonates **3a–d** and **4** at concentrations: 10, 50, 100 and 500 mg/kg of soil dry weight. Error bars represent standard deviation error (SD, n = 6 determinations).

Growth inhibition is the second criterion of the toxic effect indicated by the Ostracodtoxkit F™ microbiotest. This criterion allows one to evaluate the sub-lethal toxicity of sediments. The growth inhibition is determined by comparing the size of surviving ostracods living in the test sediment with the size of ostracods living in the reference sediment at the end of the test. Determination of the sub-lethal impact of sediment toxicants is justified only for sediments which do not cause a high ostracod mortality. It is advised that the growth inhibition should only be determined for sediments which mortality was found to be less than 30%. Therefore, the growth inhibition was chosen as a criterion of sub-lethal effects to determine toxicity not inducing substantial mortality in the test organisms. Consequently, measurements of lengths were only performed when the mortality was less than 30%. In this respect, the growth inhibition rates have not been measured for aminophosphonates **3c**, **3d** and **4** at concentrations of 100, 250 and 500 mg per kg of soil dry weight. Crustaceans were more resistant against compounds **3a** and **3b** and their growth inhibition reached ca. 20% for a concentration of 100 mg/kg of soil, nevertheless it was not measured at 250 and 500 mg per kg of soil dry weight. Values of growth inhibition for compounds **3c**, **3d** and **4** at 50 mg/kg of soil concentration varied between 16% and 23% (Figure 3).

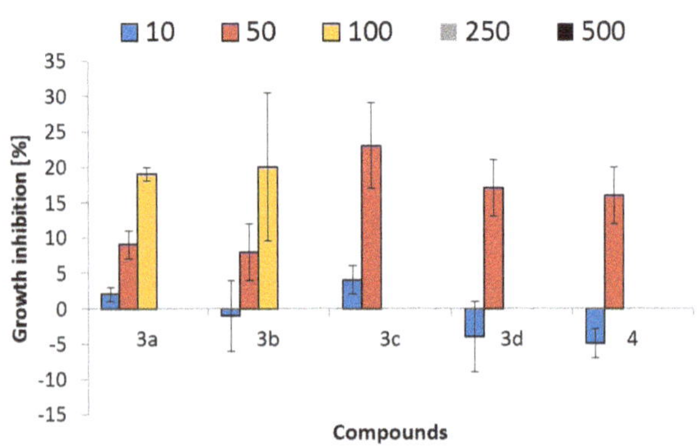

Figure 3. Growth inhibition (%) of *Heterocypris incongruens* living in soil contaminated with pyrene-derived aminophosphonates **3a–d** and **4** at concentrations of 10, 50, 100 and 500 mg/kg of soil dry weight. Error bars represent standard deviation error (SD, n = 6 determinations).

Comparison of results of luminescence inhibition for *Vibrio fischeri* tests with growth inhibition of crustaceans for all concentration of compounds did not show any reasonable relationship. Only diphenyl *N*-benzylamino(pyren-1-yl)methylphosphonate (**3d**), which was strongly toxic for *Vibrio fischeri* (EC_{50} ~103 mg/kg of soil) also revealed the highest mortality of ostracods even at a concentration of 10 mg/kg of soil.

However, diphenyl *N*-(4-methoxyphenyl)amino(pyren-1-yl) methylphosphonate (**3b**) that induced a very strong luminescence inhibition in *V. fischeri* was only moderately toxic for crustaceans considering both the mortality and the growth inhibition index. On the other hand, diphenyl *N*-phenylamino(pyren-1-yl)methylphosphonate (**3a**), being nearly non-toxic for *V. fischeri* (EC_{50} close to 500 mg/kg of soil) is also very weakly toxic for ostracods (with respect to both criteria, the mortality and the growth inhibition).

Thus, based on the obtained results, *Vibrio fischeri* bacteria were very sensitive to aminophosphonates **3b** and **3d**, since relatively low effective concentration (EC_{50}) values were determined, while for ostracods, the highest negative impact considering both mortality and inhibition growth have been found for diphenyl *N*-benzylamino(pyren-1-yl)methylphosphonate (**3d**). Therefore, being the most toxic against both tested organisms, the aminophosphonate **3d** should be considered as a potentially hazardous compound for environment, which may be toxic against many other living species. Its high toxicity may be related with its *N*-benzyl group, which is more flexible than other substituents and may bind more easily to enzymes or receptors of the tested organisms. Nevertheless, this is only a very preliminary hypothesis, which requires further verification.

The usefulness of bioassays for the assessment of areas affected by phosphate industry wastes has already been reported and discussed [35,36]. Authors performing risk assessments in such contaminated areas have proven that the development of sensitive and observable organisms as bio-indicators for monitoring and management of the environmental pollution is a very useful tool to complement chemical analyses. Biological analysis allows one to reflect the harmful effects of contaminants on organisms and provides data to reveal the mechanisms of the toxic effects including formation, development and removal of contaminants.

Since the soil is generally the reservoir for agrochemical residues, the evaluation of any xenobiotics including pesticides using a simple battery of assay tests is strongly needed to control level and fate of their residues. In a successful study focused on the ecotoxicological evaluation of select pesticides (chlorpyrifos, glyphosate, vinclozolin, endosulfan) present in agricultural soils, Antunes et al have reported the use of standard aquatic bioassays for testing of soil eluates [30].

Our obtained effective concentration values are very close to the inhibition concentration (IC_{50}) values for herbicides examined by Joly et al. [28]. Toxicity values of the pesticides *S*-metolachlor, nicosulfuron and benoxacor were 178.4, 167.8 and 93.3 mg/L, respectively. In this respect, the toxicity of the first two herbicides are close to that of the tested aminophosphonate **3c**, while the last IC_{50} value matches the values of compounds **3b** and **3d**. Nevertheless, the action of aminophosphonates **3a–d** and **4** against *V. fischeri* demonstrated much lower toxicity when compared to the herbicides tested by the mentioned authors.

Ostracodtoxit tests using *Heterocypris incongruens* are also a very useful tool for environmental toxicity determination. Results related to mortality and growth inhibition of ostracods correlated with various types of polycyclic aromatic hydrocarbons (PAHs) have been reported by Cvancarova et al. [37]. They found that *H. incongruens* was strongly sensitive to bioavailable PAHs, while luminescence inhibition of *V. fischeri* varied widely between ca. 15%–90% depending on the type of soil examined.

Oleszczuk and Hollert [38], in their studies devoted to determination of the influence of different soils on sewage sludge toxicity, have found that the mortality values of *Heterocypris incongruens* ranged from 0.26% to 11.5% depending on the sludge tested [39]. Growth inhibition values ranged from 10.7% to 36.2% depending on the type of sludge. Płaza et al. [39] applied Microtox®, Ostracodtoxkit F™, to assay bioremediation processes in soils heavily contaminated with petroleum. The test species demonstrated varying sensitivity to soils and the effects on test organisms exposed to tested soils correlated with the soil contaminants' concentrations [39].

Results of toxicity against *H. incongruens* obtained for the studied aminophosphonates **3a–d** and **4** were at a similar level. The mortality varied from ca. 5% until 70% for the concentration range 10–100 mg per kg of soil. At higher concentrations, the tested compounds were practically lethal for crustaceans. The growth inhibition varied from 16% up to 23% for a concentration of 50 mg per kg of soil. All the data presented above lead us to stress the necessity of handling pyrene-derived aminophosphonic derivatives with special care, as they may be potentially hazardous for the environment.

3. Materials and Methods

3.1. Chemistry

All solvents (POCh, Gliwice, Poland) were applied routinely and dried prior to use. Commercial reagents (Aldrich, Poznań, Poland) were generally used as received. NMR spectra were recorded on an Avance III 600 MHz apparatus (Bruker, Billerica, MA, USA) operating at 600 MHz (^{1}H-NMR), 150 MHz (^{13}C-NMR) and 243 MHz (^{31}P-NMR). IR spectra were recorded on a Nexus FT-IR spectrometer (Thermo Nicolet, Waltham, MA, USA). Rotations were measured using a 241 MC polarimeter (Perkin-Elmer, Waltham, MA, USA). Melting points were measured using a MelTemp II apparatus (Bibby Scientific Limited, Staffordshire, UK), in a capillary and were not corrected. Elemental analyses were performed in the Laboratory of Microanalysis, the Center of Molecular and Macromolecular Studies PAS in Łódź and the Laboratory of Molecular Spectroscopy at the Faculty of Chemistry, University of Łódź.

3.1.1. General Procedure for Synthesis of Phosphonates **3a–d**, **4** and **5**

Equimolar quantities of 1-pyrenecarboxaldehyde and amine (2 mmol each) were dissolved in hexane (**1a–c**) or methanol (**1d–e**) (20 mL) and stirred at reflux for 24 h (**1b–c**, **1d–e**) or 48 h (**1a**). The reaction was monitored by ^{1}H-NMR and after the completion, the solvent was removed under reduced pressure, and obtained products were used directly for further conversion. Imine (1 mmol) and the appropriate phosphite **2A–C** (3 mmol or 5–6 mmol in a case of dimethyl phosphite due to the low solubility of the imine in it) were placed in a 25 mL round-bottom flask, which was heated with stirring on a water bath for 5 min until all of the imine was dissolved in the phosphite. In the case of derivatives **3a–d**, dichloromethane (2 mL) was added in order to improve the solubility. Then the mixture was stirred at room temperature for 24 h. The crude aminophosphonates were purified by column chromatography on neutral aluminum oxide using chloroform as eluent.

Diphenyl N-phenylamino(pyren-1-yl)methylphosphonate (**3a**). Y = 0.41 g (76%), yellow solid, m.p. 184–185 °C. ^{1}H-NMR (CDCl$_3$): δ 8.38 (d, $^3J^{HH}$ = 9.3 Hz, H_{pyr}, 1H); 8.28 (dd, $^3J^{HH}$ = 8.0 and $^4J_{HH}$ = 2.6 Hz, H_{pyr}, 1H); 8.20 (t, H_{pyr}, $^3J^{HH}$ = 8.0 Hz, 2H); 8.12 (d, $^3J^{HH}$ = 8.0 Hz, H_{pyr}, 1H); 8.10 (d, $^3J^{HH}$ = 9.2 Hz, H_{pyr}, 1H); 8.07–8.06 (m, H_{pyr}, 1H); 8.04–8.01 (m, H_{pyr}, 2H); 7.30–7.29 (m, H_{Ph}, 5H); 6.94 (t, $^3J^{HH}$ = 7.4 Hz, H_{Ph}, 1H); 6.84 (dd, $^3J^{HH(1)}$ = $^3J^{HH(2)}$ = 7.8 Hz, H_{Ph}, 2H); 6.79 (app. d, $^3J^{HH}$ = 8.5 Hz, *p*-C_6H_4, 2H); 6.67 (d, $^3J^{HH}$ = 7.4 Hz, H_{Ph}, 2H); 6.46 (app. d, $^3J^{HH}$ = 8.5 Hz, *p*-C_6H_4, 2H); 5.92 (d, $^2J_{PH}$ = 24.1 Hz, CHP, 1H); 5.10–5.08 (m, POCH$_2$, 2H); 4.65 and 4.22 (Part of AMX system, $^2J_{HH}$ = 11.6 Hz and $^3J_{PH}$ = 7.8 and 8.5 Hz, POCH$_2$, 2H). ^{13}C-NMR (CDCl$_3$): δ 150.7 (d, $^2J_{CP}$ = 9.9 Hz, C_{POC}); 150.3 (d, $^2J_{CP}$ = 9.5 Hz, C_{POC}); 146.1 (d, $^3J_{CP}$ = 15.27 Hz, C_{Ph}); 131.6 (C_{pyr}); 131.5 (d, $^4J_{CP}$ = 3.1 Hz, C_{pyr}); 130.8 (C_{pyr}); 130.2 (C_{pyr}) 130.0 (C_{Ph}); 129.6 (C_{pyr}); 129.5 (d, $^4J_{CP}$ = 2.0 Hz, C_{Ph}); 128.7 (C_{pyr}); 128.6 (d, $^4J_{CP}$ = 2.1 Hz, C_{pyr}); 127.9 (C_{pyr}); 127.7 (C_{pyr}); 127.4 (C_{pyr}); 127.0 (C_{pyr}); 126.8 (C_{pyr}); 126.3 (C_{pyr}); 125.8 (C_{Ph}); 125.7 (C_{Ph}); 125.6 (d, $^4J_{CP}$ = 3.4 Hz, C_{pyr}); 125.5 (C_{pyr}); 125.2 (C_{pyr}); 125.0 (C_{pyr}); 122.1 (C_{pyr}); 120.9 (d, $^3J_{CP}$ = 4.0 Hz, C_{Ph}); 119.9 (d, $^3J_{CP}$ = 4.4 Hz, C_{Ph}); 119.1 (C_{Ph}); 114.2 (C_{Ph}); 52.2 (d, $^1J_{CP}$ = 154.6 Hz, PC). ^{31}P-NMR (CDCl$_3$): δ 16.16. IR (KBr): 3313 (νCH); 3037; 2923 (CH$_3$); 1600, 1581; 1524, 1483 (C=C); 1249 (P=O); 1214; 1185; 1062; 1021 (P-O); 941; 853;

824; 622.Elemental analysis: Calcd. for $C_{35}H_{26}NO_3P$: C, 77.91; H, 4.86; N, 2.60. Found: C, 77.76, H, 4.86, N, 2.79.

Diphenyl N-(4-methoxyphenyl)amino(pyren-1-yl)methylphosphonate (**3b**). Y = 0.3 g (52%), yellow solid, m.p. 161 °C. ^{1}H-NMR ($CDCl_3$): δ 8.49 (d, $^3J^{HH}$ = 9.2 Hz, H_{pyr}, 1H); 8.34 (dd, $^3J^{HH}$ = 7.7 and $^4J_{HH}$ = 1.9 Hz, H_{pyr}, 1H); 8.24-8.21 (m, H_{pyr}, 3H); 8.14 (d, H_{pyr}, $^3J^{HH}$ = 8.1 Hz, 1H); 8.08 (d, $^3J^{HH}$ = 8.9 Hz, H_{pyr}, 1H); 8.04 (d, $^3J^{HH}$ = 7.7 Hz, H_{pyr}, 1H); 8.02–8.01 (m, H_{pyr}, 2H); 7.31–7.28 (m, H_{Ph}, 2H); 7.19–7.16 (m, H_{Ph}, 3H); 6.95–6.92 (m, H_{Ph}, 2H); 6.89–6.87 (m, H_{Ph}, 1H); 6.62–6.58 (m, H_{Ph}, p-C_6H_4, 6H); 6.23 (d, $^2J_{PH}$ = 24.4 Hz, CHP, 1H); 3.61 (s, CH_3, 3H). ^{13}C-NMR ($CDCl_3$): δ 153.2 (d, $^2J_{CP}$ = 5.4 Hz, C_{para}); 150.7 (d, $^2J_{CP}$ = 9.9 Hz, C_{POC}); 150.3 (d, $^2J_{CP}$ = 9.7 Hz, C_{POC}); 131.5 (C_{pyr}); 131.2 (d, $^4J_{CP}$ = 3.2 Hz, C_{pyr}); 130.8 (C_{pyr}); 130.0 (C_{Ph}); 129.7 (C_{pyr}); 129.6 (d, $^3J_{CP}$ = 7.2 Hz, C_{pyr}); 129.5 (C_{Ph}); 128.8 (C_{pyr}); 128.6 (C_{pyr}); 127.9 (C_{pyr}); 127.7 (C_{pyr}); 126.3 (C_{Ph}); 125.8 (C_{Ph}); 125.6 (C_{pyr}); 125.6 (C_{pyr}); 125.5 (C_{pyr}); 125.2 (d, $^4J_{CP}$ = 1.8 Hz, C_{pyr}); 125.0 (C_{pyr}); 124.9 (C_{pyr}); 122.2 (C_{pyr}); 121.0 (d, $^3J_{CP}$ = 4.2 Hz, C_{Ph}); 120.0 (d, $^4J_{CP}$ = 4.4 Hz, C_{Ph}); 115.7 (d, $^4J_{CP}$ = 9.7 Hz, C_{Ph}); 115.1 (C_{para}); 55.7 (C_{Ar}-OC); 53.0 (d, $^1J_{CP}$ = 154.3 Hz, PC). ^{31}P-NMR ($CDCl_3$): δ 16.32. IR (KBr): 3383 (νNH); 3294 (νCH); 3037; 2958 (CH_3); 1591, 1511, 1486; 1451 (C=C); 1245 (P=O); 1211; 1068; 1024 (P-O); 941; 834; 815; 618. Elemental analysis: Calcd. for $C_{36}H_{28}NO_4P$: C, 75.91; H, 4.95; N, 2.46. Found: C, 75.76, H, 5.06, N, 2.54.

Diphenyl N-(4-methylphenyl)amino(pyren-1-yl)methylphosphonate (**3c**). Y = 0.44 g (79%), yellow solid, m.p. 181–182 °C. ^{1}H-NMR ($CDCl_3$): δ 8.51 (d, $^3J^{HH}$ = 9.3 Hz, H_{pyr}, 1H); 8.34 (dd, $^3J^{HH}$ = 8.2 and $^4J_{HH}$ = 2.5 Hz, H_{pyr}, 1H); 8.25-8.21 (m, H_{pyr}, 3H); 8.13 (d, H_{pyr}, $^3J^{HH}$ = 8.2 Hz, 1H); 8.08 (d, $^3J^{HH}$ = 8.9 Hz, H_{pyr}, 1H); 8.04 (t, $^3J^{HH}$ = 7.6 Hz, H_{pyr}, 1H); 8.01 (d, $^3J^{HH}$ = 8.8 Hz, H_{pyr}, 2H); 7.31–7.29 (m, H_{Ph}, 2H); 7.20–7.15 (m, H_{Ph}, 3H); 7.05–7.02 (m, H_{Ph}, 2H); 6.94–6.91 (m, H_{Ph}, 2H); 6.88-6.86 (m, H_{Ph}, 1H); 6.68-6.64 (m, H_{Ph}, 3H); 6.58–6.57 (m, H_{Ph}, 2H); 6.31 (d, $^2J_{PH}$ = 24.6 Hz, CHP, 1H). ^{13}C-NMR ($CDCl_3$): δ 150.7 (d, $^2J_{CP}$ = 9.8 Hz, C_{POC}); 150.3 (d, $^2J_{CP}$ = 9.7 Hz, C_{POC}); 143.7 (d, $^2J_{CP}$ = 15.9 Hz, C_{para}); 131.6 (C_{pyr}); 131.2 (d, $^4J_{CP}$ = 3.2 Hz, C_{pyr}); 130.8 (C_{pyr}); 130.0 (C_{Ph}); 129.8 (C_{pyr}); 129.6 (d, $^3J_{CP}$ = 7.3 Hz, C_{pyr}); 129.5 (C_{para}); 128.8 (d, $^4J_{CP}$ = 2.1 Hz, C_{pyr}); 128.6 (C_{pyr}); 128.3 (C_{pyr}); 127.8 (C_{para}); 127.7 (d, $^4J_{CP}$ = 1.6 Hz, C_{pyr}); 126.3 (C_{pyr}); 125.8 (C_{pyr}); 125.6 (C_{Ph}); 125.5 (d, $^4J_{CP}$ = 2.3 Hz, C_{pyr}); 125.5 (C_{Ph}); 125.2 (d, $^4J_{CP}$ = 1.9 Hz, C_{pyr}); 125.0 (C_{pyr}); 124.9 (C_{pyr}); 122.2 (C_{pyr}); 121.0 (d, $^3J_{CP}$ = 4.0 Hz, C_{Ph}); 120.0 (d, $^3J_{CP}$ = 4.4 Hz, C_{Ph}); 114.3 (C_{para}); 52.5 (d, $^1J_{CP}$ = 155.1 Hz, PC); 20.5 (C_{Ar}-OC). ^{31}P-NMR ($CDCl_3$): δ 15.23. IR (KBr): 3303 (νCH); 2923 (CH_3); 1591, 1489, 1458 (C=C); 1268 (P=O); 1024 (P-O); 926, 840; 770; 688. Elemental analysis: Calcd. for $C_{36}H_{28}NO_3P$: C, 78.11; H, 5.10; N, 2.53. Found: C, 77.86; H, 5.31; N, 2.47.

Diphenyl N-benzylamino(pyren-1-yl)methylphosphonate (**3d**). Y = 0.43 g (77%), yellow solid, m.p. 118–119 °C. ^{1}H-NMR ($CDCl_3$): δ 8.51–8.49 (m, H_{pyr}, 1H); 8.26 (d, $^3J^{HH}$ = 8.0 Hz, H_{pyr}, 1H); 8.24–8.21 (m, H_{pyr}, 3H); 8.13–8.09 (m, H_{pyr}, 3H); 8.04 (t, $^3J^{HH}$ = 7.6 Hz, H_{pyr}, 1H); 7.29–7.25 (m, H_{Ph}, 5H); 7.20–7.19 (m, H_{Ph}, 2H); 7.15–7.13 (m, H_{Ph}, 3H); 7.02–7.00 (m, H_{Ph}, 2H); 6.93 (app. t, $^3J^{HH}$ = 7.4 Hz, H_{Ph}, 1H); 6.73 (app. d, $^3J^{HH}$ = 8.2 Hz, H_{Ph}, 2H); 5.58 (d, $^2J_{PH}$ = 20.8 Hz, CHP, 1H); 3.93 (d, $^3J^{HH}$ = 13.3 Hz, NCH_2, 1H); 3.66 (d, $^3J^{HH}$ = 13.3 Hz, NCH_2, 1H). ^{13}C-NMR ($CDCl_3$): δ 151.0 (d, $^2J_{CP}$ = 9.9 Hz, C_{POC}); 150.5 (d, $^2J_{CP}$ = 9.7 Hz, C_{POC}); 139.1 (C_{Ph}); 131.6 (C_{pyr}); 131.4 (d, $^4J_{CP}$ = 3.0 Hz, C_{pyr}); 130.8 (C_{pyr}); 130.4 (C_{pyr}); 130.3 (C_{pyr}) 130.0 (C_{pyr}); 129.9 (C_{pyr}); 129.8 (C_{Ph}); 129.5 (C_{Ph}); 128.8 (C_{Ph}); 128.6 (C_{Ph}); 128.2 (C_{pyr}); 128.0 (C_{pyr}); 127.7 (d, $^4J_{CP}$ = 1.5 Hz, C_{pyr}); 127.5 (C_{pyr}); 126.3 (C_{pyr}); 125.7 (C_{Ph}); 125.5 (d, $^4J_{CP}$ = 2.8 Hz, C_{pyr}); 125.4 (C_{Ph}); 125.2 (C_{pyr}); 125.0 (C_{pyr}); 122.7 (C_{pyr}); 120.8 (d, $^3J_{CP}$ = 4.3 Hz, C_{Ph}); 120.3 (d, $^3J_{CP}$ = 4.3 Hz, C_{Ph}); 51.6 (d, $^3J_{CP}$ = 17.8 Hz, PCNC). ^{31}P-NMR ($CDCl_3$): δ 16.22. IR (KBr): 3386 (νNH); 3041, 2917 (CH_3); 1616, 1587, 1518, 1451 (C=C); 1245 (P=O); 1066; 1024 (P-O); 941, 843; 777; 618. Elemental analysis: Calcd. for $C_{36}H_{28}NO_3P$: C, 78.11; H, 5.10; N, 2.53. Found: C, 77.81; H, 5.24; N, 2.40.

3.1.2. *Dimethyl N-(R)-α-Methylbenzylamino(pyren-1-yl)methylphosphonate* (**4**)

The reaction was carried out as it is described above, but the post-reaction mixture was dissolved in dichloromethane and washed three times with a saturated aqueous sodium bicarbonate solution. The organic layer was dried, the solvent was removed in vacuo and the residual oil was purified by column chromatography on silica gel with chloroform as eluent. The procedure gave 440 mg of **4** as a dense yellow oil. *Isolated predominant diastereomer*: Y = 0.4 g (92%, after column chromatography), dense yellow oil. $[\alpha]_D^{20} = -48.3°$ (c = 0.1, $CHCl_3$). ^{1}H-NMR ($CDCl_3$): δ 8.29–8.27 (m, H_{pyr}, 1H); 8.22 (d, $^3J^{HH}$ = 8.0 Hz, H_{pyr}, 1H); 8.21–8.18 (m, H_{pyr}, 3H); 8.10–8.06 (m, H_{pyr}, 3H); 8.02 (t, $^3J^{HH}$ = 7.6 Hz, H_{pyr}, 1H); 7.25–7.23 (m, PhH, 2H); 7.22–7.20 (m, PhH, 2H); 7.19–7,16 (m, PhH, 1H); 5.33 (d, $^2J_{PH}$ = 21.1 Hz, CHP, 1H); 3.91 (q, $^3J^{HH}$ = 6.5 Hz, *C*HPh, 1H); 3.86 (d, $^3J_{PH}$ = 10.5 Hz, $POCH_3$, 3H); 3.26 (d, $^3J_{PH}$ = 10.5 Hz, $POCH_3$, 3H); 1.39 (d, $^3J^{HH}$ = 6.5 Hz, $CHCH_3$, 3H). ^{13}C-NMR ($CDCl_3$): δ 145.2 (C_{Ph}); 131.6 (C_{pyr}); 131.1 (d, $^4J_{CP}$ = 2.3 Hz, C_{pyr}); 130.8 (C_{pyr}); 130.3 (C_{pyr}); 129.4 (d, $^3J_{CP}$ = 7.7 Hz, C_{pyr}); 128.6 (C_{Ph}); 128.2 (C_{pyr}); 127.8 (C_{pyr}); 127.7 (C_{pyr}); 127.3 (C_{Ph}); 127.0 (C_{Ph}); 126.2 (C_{pyr}); 126.1 (C_{pyr}); 125.6 (C_{pyr}); 125.4 (d, $^4J_{CP}$ = 2.8 Hz, C_{pyr}); 125.3 (C_{pyr}); 125.1 (C_{pyr}); 125.0 (C_{pyr}); 55.8 (d, $^3J_{CP}$ = 11.9 Hz, PCNC); 54.0 (d, $^2J_{CP}$ = 7.4 Hz, POC); 53.6 (d, $^2J_{CP}$ = 7.2 Hz, POC); 29.9 (C_{aliph}); 27.8 (C_{aliph}). ^{31}P-NMR ($CDCl_3$): δ 26.70. IR (KBr): 3440 (νNH); 3309 (νCH); 2949 (CH_2); 1600, 1584, 1511, 1454 (C=C); 1242 (P=O); 1055; 1027 (P-O); 846; 815; 761; 609. Elemental analysis: Calcd. for $C_{27}H_{26}NO_3P$: C, 73.12; H, 5.91; N, 3.16. Found: C, 73.15, H, 5.72, N, 2.96. *Mixture of diastereomers—the minor diastereomer is denoted with an asterisk.* ^{1}H-NMR ($CDCl_3$): δ 8.29–8.28 (m, H_{pyr}, 1H); 8.23–8.22 (m, H_{pyr}, 1H); 8.21–8.17 (m, H_{pyr}, 3H); 8.17–8.16* (m, H_{pyr}, 1H); 8.10–8.06 (m, H_{pyr}, 3H); 8.03–8.00 (m, H_{pyr}, 1H); 7.95–7.94* (m, H_{pyr}, 1H); 7.30–7.26* (m, PhH); 7.24–7.22 (m, PhH, 2H); 7.22–7.20 (m, PhH, 2H); 7.19–7.16 (m, PhH, 1H); 7.12–7.10 (m, PhH); 5.33 (d, $^2J_{PH}$ = 21.0 Hz, CHP, 1H); 4.98* (d, $^2J_{PH}$ = 21.7 Hz, CHP, 1H); 3.91 (q, $^3J^{HH}$ = 6.5 Hz, *C*HPh, 1H); 3.86 (d, $^3J_{PH}$ = 10.6 Hz, $POCH_3$, 3H); 3.83* (d, $^3J_{PH}$ = 10.5 Hz, $POCH_3$, 3H); 3.50* (q, $^3J^{HH}$ = 6.6Hz, *C*HPh, 1H); 3.26 (d, $^3J_{PH}$ = 10.4 Hz, $POCH_3$, 3H); 3.23* (d, $^3J_{PH}$ = 10.4 Hz, $POCH_3$, 3H); 1.39 (d, $^3J^{HH}$ = 6.5 Hz, $CHCH_3$, 3H); 1.34* (d, $^3J^{HH}$ = 6.6 Hz, $CHCH_3$, 3H). ^{31}P-NMR ($CDCl_3$): δ 26.70 and 26.18* (10:4). Elemental analysis: Calctd for $C_{27}H_{26}NO_3P$: C, 73.12; H, 5.91; N, 3.16. Found: C, 73.17, H, 5.85, N, 3.10.

3.1.3. *Dibenzyl N-(R)-α-Methylbenzylamino(pyren-1-yl)methylphosphonate* (**5**)

The procedure was identical as in a case of aminophosphonate **4**. *Mixture of diastereoisomers—the predominant diastereoisomer is denoted with an asterisk* (^{13}C-NMR data are quoted with two decimal digits due to the large number of signals situated very close one to another). Y = 0.12 g, (20%, after column chromatography), dense yellow oil (mixture). ^{1}H-NMR ($CDCl_3$): δ 8.21–8.20 (m, H_{pyr}, 2H); 8.16* (d, $^3J^{HH}$ = 7.5 Hz, H_{pyr}, 1H); 8.15–8.12 (m, H_{pyr}); 8.10–8.06 (m, H_{pyr}, 4H); 8.03–8.01 (m, H_{pyr}, 1H); 7.98* (d, $^3J^{HH}$ = 9.3 Hz, H_{pyr}, 1H); 7.38–7.36 (m, PhH, 2H); 7.34–7.32 (m, PhH, 3H); 7.22–7.17 (m, PhH, 5H); 6.97–6.94* (m, PhH, 2H); 6.92–6.90 (m, PhH); 6.89–6.86* (m, PhH, 2H); 6.83–6.81 (m, PhH); 6.76–6.74 (m, PhH, 2H); 6.68–6.66 (m, PhH); 5.38* (d, $^2J_{PH}$ = 21.0 Hz, CHP, 1H); 5.26* and 5.10* (Part of ABX system, $^2J_{HH}$ = 11.8 and $^3J_{PH}$ = 8.6 and 7.0 Hz, OCH_2Ph, 2H); 5.24–5.22 (m, CHP); 5.13–5.03 (m, OCH_2Ph); 4.72* and 4.39* (part of AMX system, $^2J_{HH}$ = 11.8 and $^3J_{PH}$ = 8.4 and 7.3 Hz, OCH_2Ph, 2H); 4.67-4.64 (m, OCH_2Ph); 4.40–4.35 (m, OCH_2Ph); 3.93* (q, $^3J^{HH}$ = 6.4 Hz, *C*HPh, 1H); 1.33* (d, $^3J^{HH}$ = 6.4 Hz, $CHCH_3$, 3H); 1.30–1.29 (m, $CHCH_3$). ^{13}C-NMR ($CDCl_3$): δ 145.21 (C_{Ar}); 144.28 (C_{Ar}); 136.83 (C_{Ar}); 136.79 (C_{Ar}); 136.01 and 135.98 (C_{Ar}); 131.58 (C_{Ar}); 131.06 and 131.05 (C_{Ar}); 130.87 (C_{Ar}); 130.41 (C_{Ar}); 129.57 and 129.52 (C_{Ar}); 128.67 and 128.66 (C_{Ar}); 128.63 (C_{Ar}); 128.45 (C_{Ar}); 128.41 (C_{Ar}); 128.12 (C_{Ar}); 128.03 (C_{Ar}); 127.98 (C_{Ar}); 127.85 (C_{Ar}); 127.75, 127.74, 127.71 and 127.70 (C_{Ar}); 127.49 and 127.48 (C_{Ar}); 127.43 (C_{Ar}); 127.29 (C_{Ar}); 126.98 (C_{Ar}); 126.13 (C_{Ar}); 125.49 (C_{Ar}); 125.40 (C_{Ar}); 125.32 and 125.30 (C_{Ar}); 125.19 (C_{Ar}); 125.16 (C_{Ar}); 125.09 (C_{Ar}); 125.00 (C_{Ar}); 124.97 (C_{Ar}); 68.79 and 68.75 (POC); 68.64 and 68.60 (POC); 68.34 and 68.30 (POC); 68.10 and 68.06 (POC); 55.94 (C_{aliph}); 55.86 (C_{aliph}); 55.58 (C_{aliph}); 55.46 (C_{aliph}); 24.92 (C_{aliph}); 22.85 (C_{aliph}). ^{31}P-NMR ($CDCl_3$): δ 25.31*; 24.64 (4:1).Elemental analysis: Calcd. for $C_{39}H_{34}NO_3P$: C, 78.64; H, 5.75; N, 2.35. Found: C, 78.48, H, 5.96, N, 2.16.

3.1.4. *Diphenyl N-(R)-α-Methylbenzylamino(pyren-1-yl)methylphosphonate* (**3e**)

To a solution of the imine **1e** (1mmol) and diphenyl phosphite (1 mmol) in dichloromethane (ca. 2 mL) was slowly added $BF_3 \cdot OEt_2$ (24 mg, 0.67 mmol) via syringe. The mixture was stirred at room temperature for 24 h while monitoring the course of the reaction using ^{31}P-NMR spectroscopy or TLC. The solvent was removed in vacuo and the residual oil was roughly purified by column chromatography on microcrystalline cellulose with chloroform as eluent to give **3e** as a brown oil in 0.9 g (72%) yield of a crude product. Further attempts to purify and to analyze the product resulted in its decomposition.

3.2. Toxicity Tests

3.2.1. Microtox® Toxicity Assay

The method is based on the analysis of light emission reduction of luminescent bacteria (*Vibrio fischeri*) under toxic stress. The tests were carried out in a Microtox® M500 analyzer according to the 1992 Microtox® Manual. The Microtox® Solid-Phase Test (MSPT) was adopted to report of Doe et al. [31]. The MSPT procedure allows the test organisms to come direct contact with the solid sample in an aqueous suspension of the test sample. Thus it is possible to detect toxicity which is due to the insoluble solids that are not in the solution.

All materials and reagents were purchased from MODERNWATER (New Castle, DE, USA). Toxicity was determined by using the marine luminescent bacterium, *Vibrio fischeri*, naturally adapted to a saline environment. Briefly, bacteria were regenerated with 1 mL of Reconstitution Solution (0.01%) and placed in the reagent well of the Microtox®. A suspension of 7 g of the sediment was prepared in 35 mL of a Solid Phase Diluent (3.5% NaCl) and was magnetically stirred for 10 min. Then a series of dilutions were made and bacteria (approx. 1×10^6 cell·mL^{-1} per assay) were exposed to these dilutions and to a blank (3.5% NaCl solution) for 20 min. Next, after filtration, the light output of supernatants containing exposed bacteria was measured after 5 min with a Microtox® Analyzer 500. Inhibition was calculated as the concentration of compound loaded to sediment (mg/L) that caused a 50% reduction in the light emitted by the bacteria, and EC_{50} along with 95% confidence limit determined by the software provided with the Analyzer.

3.2.2. Ostracod Test Kit

Ecotoxicity evaluation of synthesized compounds was performed in a short term contact test using Ostracodtoxkit F™ provided by MicroBiotests Inc. (Gent, Belgium). This direct sediment contact bioassay was performed in multiwell test plates using neonates of the benthic ostracod crustacean *Heterocypris incongruens* hatched from cysts [35]. After 6 days of contact with the sediment (or soil) the percentage mortality and the growth of the crustaceans were determined and compared to the results obtained in a non-treated reference sediment (soil).

Briefly, according to manual of Ostracodtoxkit test, the cysts (*Heterocypris incongruens*) were transferred into a Petri dish filled with 10 mL standard fresh water (reconstituted water) and were incubated at 25 °C for 52 h under continuous illumination (approx. 3000–4000 lux).

After 48 h of cysts incubation, pre-feeding of the freshly hatched ostracods was performed with algae (spirulina-powder) provided in the test kit. Next, after hatching, before feeding with algal food suspension, the length measurements of ostracod neonates has been done. Algae (*Selenastrum capricornutum*) used as feed in the test plate were reconstituted according to the manufacturer's procedure. Each well of a test plate was filled in the following order: 2 mL standard freshwater, 2 × 500 µL of sediment (soil) treated and non-treated for comparison (blank), 2 mL already prepared algal suspension, 10 ostracods. The test plates were covered with Parafilm® and closed by a lid. Then multiwall plates were incubated at 25 °C in darkness for 6 days. After 6 days of exposure, the ostracods have been recovered from the multiwells to determine the percentage mortality. To calculate the growth inhibition of survived organisms, their length measurements have been also done. Mortality of test organisms was determined in six replicates. The measurement of length was carried out by means of a micrometric strip placed on the bottom of a glass microscope plate. Growth inhibition (GI) of *H. incongruens* in the test sediment was calculated as follows:

$$\% \text{growth inhibition} = 100 - [(\text{growth in test sed./growth in ref. sed.}) \times 100] \quad (1)$$

Statistical differences between variables were analyzed with ANOVA.

4. Conclusions

To conclude, a series of diphenyl pyrene-derived aminophosphonates **3a–d**, **4** and **5** has been prepared. Our previous study [17] revealed that pyrene-derived aminophosphonates have interesting cytotoxic properties against colon cancer cell lines, therefore, the ecotoxicity of the newly synthesized compounds has been evaluated using the crustacean *Heterocypris incongruens* and bacterium *Vibrio fischeri* biotests. The evaluation demonstrated clearly that all compounds **3a–d** and **4** are toxic in a moderate to high degree and that diphenyl *N*-benzylamino(pyren-1-yl) methylphosphonate (**3d**) shows the highest degree of ecotoxicity. It has been also revealed that a weak correlation between the toxicity against *H. incongruens* and against *V. fischeri* has been found and the structure-toxicity correlation is rather problematic.

Although no evident correlations between the structure of the tested compounds and their toxicity have been observed, the combination of the different bioassays—the Microtox® and Ostracodtoxkit F™ tests—has been proved to be a very useful tool for determining and comparing the potential toxicities of examined substances **3a–d** and **4** as well as for the assessment of their toxicity interactions.

Supplementary Materials: The following are available online at: http://www.mdpi.com/1420-3049/21/7/936/s1, scans of ^{1}H, ^{13}C and ^{31}P NMR of compounds **3a–d**, **4** and **5**.

Acknowledgments: We thank the Polish National Centre of Science (NCN) for funding the work in the framework of the grant no. 2014/13/B/NZ9/02418. Within the framework of this grant, funds for covering the costs to publish in open access were scheduled.

Author Contributions: Jarosław Lewkowski and Piotr Rychter conceived and designed the experiments; Maria Rodriguez Moya and Marta Chmielak performed the chemical experiments; Diana Rogacz and Kamila Lewicka performed the ecotoxicological experiments; Jarosław Lewkowski and Piotr Rychter analyzed the data (Jarosław Lewkowski chemistry, Piotr Rychter. ecotoxicity); Jarosław Lewkowski and Piotr Rychter wrote the paper.

Conflicts of Interest: The authors declare no conflict of interest. The founding sponsors had no role in the design of the study; in the collection, analyses, or interpretation of data; in the writing of the manuscript, and in the decision to publish the results.

References

1. Kafarski, P.; Lejczak, B.; Tyka, R.; Koba, L.; Pliszczak, E.; Wieczorek, P. Herbicidal activity of phosphonic, phosphinic, and phosphonous acid analogues of phenylglycine and phenylalanine. *J. Plant Growth Regul.* **1995**, *14*, 199–203. [CrossRef]
2. Kafarski, P.; Lejczak, B. Biological Activity of Aminophosphonic Acids. *Phosphorus Sulfur Silicon Relat. Elem.* **1991**, *63*, 193–215. [CrossRef]
3. Kafarski, P.; Lejczak, B. Aminophosphonic Acids of Potential Medical Importance. *Curr. Med. Chem. Anti-Cancer Agents* **2001**, *1*, 301–312. [CrossRef] [PubMed]
4. Hudson, H.R. Aminophosphonic and Aminophosphinic Acids and their Derivatives as Agrochemicals. In *Aminophosphonic* and Aminophosphinic Acids: Chemistry and Biological Activity; Kukhar, V.P., Hudson, H.R., Eds.; Wiley and Sons: Chichester, UK, 2000.
5. Maier, L.; Diel, P.J. Organic Phosphorus Compounds 94. Preparation, Physical and Biological Properties of Aminoarylmethylphosphonic- and -Phosphonous Acids. *Phosphorus Sulfur Silicon Relat. Elem.* **1991**, *57*, 57–64. [CrossRef]
6. Hudson, H.R. Phosphorus-Containing Fungicides: A Review of Current Research and Prospects. *Phosphorus Sulfur Silicon Relat. Elem.* **1999**, *144–146*, 441–444. [CrossRef]
7. Jane, D.E. Neuroactive aminophosphonic and aminophosphinic acid derivatives. In *Aminophosphonic and Aminophosphinic Acids: Chemistry and Biological Activity*; Kukhar, V.P., Hudson, H.R., Eds.; Wiley and Sons: Chichester, UK, 2000.

8. Mohd-Pahmi, S.H.; Hussein, W.M.; Schenk, G.; McGeary, R.P. Synthesis, modelling and kinetic assays of potent inhibitors of purple acid phosphatase. *Bioorg. Med. Chem. Lett.* **2011**, *21*, 3092–3094. [CrossRef] [PubMed]
9. Klimczak, A.A.; Kuropatwa, A.; Lewkowski, J.; Szemraj, J. Synthesis of new *N*-arylamino(2- furyl) methylphosphonic acid diesters, and in vitro evaluation of their cytotoxicity against esophageal cancer cells. *Med. Chem. Res.* **2013**, *22*, 852–860. [CrossRef]
10. Boduszek, B. The Acidic Cleavage of Pyridylmethyl(amino)phosphonates. Formation of the Corresponding Amines. *Tetrahedron* **1996**, *52*, 12483–12494. [CrossRef]
11. Lewkowski, J.; Rzeźniczak, M.; Skowroński, R.; Zakrzewski, J. The first synthesis of ferrocenyl aminophosphonic esters. *J. Organomet. Chem.* **2001**, *631*, 105–109. [CrossRef]
12. Keglevich, G.; Bálint, E. The Kabachnik–Fields Reaction: Mechanism and Synthetic Use. *Molecules* **2012**, *17*, 12821–12835. [CrossRef] [PubMed]
13. Ali, T.E.; Abdel-Kariem, S.M. Methods for the synthesis of α-heterocyclic/heteroaryl-α-aminophosphonic acids and their esters. *ARKIVOC* **2015**, *6*, 246–287. [CrossRef]
14. Kafarski, P.; Górny vel Górniak, M.; Andrasiak, I. Kabachnik-Fields Reaction under Green Conditions—A Critical Overview. *Curr. Green Chem.* **2015**, *2*, 218–222. [CrossRef]
15. Hudson, H.R.; Lee, R.J.; Matthews, R.W. 1-Amino-1-aryl- and 1-Amino-1-heteroaryl-methanephosphonicacids and Their *N*-Benzhydryl–Protected Diethyl Esters: Preparation and Characterization. *Phosphorus Sulfur Silicon Relat. Elem.* **2004**, *179*, 1691–1709. [CrossRef]
16. Jayaprakash, S.H.; Uma Maheswara Rao, K.; Satheesh Krishna, B.; Siva Prasad, S.; Syama Sundar, C.; Suresh Reddy, C. PAA-SIO2 Catalyzed Synthesis, Uv Absorption, and Fluorescence Emission Studies of Diethyl (aryl/hetero aryl amino)(pyren-1-yl)-methylphosphonates. *Phosphorus Sulfur Silicon Relat. Elem.* **2015**, *190*, 449–460. [CrossRef]
17. Lewkowski, J.; Rodriguez Moya, M.; Wrona-Piotrowicz, A.; Zakrzewski, J.; Kontek, R.; Gajek, G. Synthesis, fluorescent properties and the promising cytotoxicity of pyrene-derived aminophosphonates. *Beilstein J. Org. Chem.* **2016**, *12*, 1229–1235. [CrossRef]
18. Ma, X.Y.; Wang, X.C.; Ngo, H.H.; Guo, W.; Wu, M.N.; Wang, N. Bioassay based luminescent bacteria: Interferences, improvements, and applications. *Sci. Total. Environ.* **2014**, *468–469*, 1–11. [CrossRef] [PubMed]
19. Connon, R.E.; Geist, J.; Werner, I. Effect-Based Tools for Monitoring and Predicting the Ecotoxicological Effects of Chemicals in the Aquatic Environment. *Sensors* **2012**, *12*, 12741–12771. [CrossRef] [PubMed]
20. Lewkowski, J.; Malinowski, Z.; Matusiak, A.; Morawska, M.; Rogacz, D.; Rychter, P. The Effect of New Thiophene-Derived Aminophosphonic Derivatives on Growth of Terrestrial Plants: A Seedling Emergence and Growth Test. *Molecules* **2016**, *21*, 694. [CrossRef] [PubMed]
21. Matusiak, A.; Lewkowski, J.; Rychter, P.; Biczak, R. Phytotoxicity of New Furan-derived Aminophosphonic Acids, N^{-}Aryl Furaldimines and 5^{-}Nitrofuraldimine. *J. Agric. Food Chem.* **2013**, *61*, 7673–7678. [CrossRef] [PubMed]
22. MicroBioTests. Toxkit Advantages/Assets. Available online: http://www.microbiotests.be/information/toxkit-advantagesassets/ (accessed on 19 July 2016).
23. Baran, A.; Tarnawski, M. Phytotoxkit/Phytotestkit and Microtoxs as tools for toxicity assessment of sediments. *Ecotox. Environ. Safe.* **2013**, *98*, 19–27. [CrossRef] [PubMed]
24. Weltens, R.; Deprez, K.; Michiels, L. Validation of Microtox as a first screening tool for waste classification. *Waste Manage.* **2014**, *34*, 2427–2433. [CrossRef] [PubMed]
25. Yu, X.; Zuo, J.; Tang, X.; Li, R.; Li, Z.; Zhang, F. Toxicity evaluation of pharmaceutical wastewaters using the alga *Scenedesmus obliquus* and the bacterium *Vibrio fischeri*. *J. Hazard. Mat.* **2014**, *266*, 68–74. [CrossRef] [PubMed]
26. Rosado, D.; Usero, J.; Morillo, J. Assessment of heavy metals bioavailability and toxicity toward *Vibrio fischeri* in sediment of the Huelva estuary. *Chemosphere* **2016**, *153*, 10–17. [CrossRef] [PubMed]
27. Montalban, M.G.; Hidalgo, J.M.; Collado-Gonzalez, M.; Banos, F.G.D.; Víllora, G. Assessing chemical toxicity of ionic liquids on *Vibrio fischeri*: Correlation with structure and composition. *Chemosphere* **2016**, *155*, 405–414. [CrossRef] [PubMed]
28. Joly, P.; Bonnemoy, F.; Charvy, J.-C.; Bohatier, J.; Mallet, C. Toxicity assessment of the maize herbicides S-metolachlor, benoxacor, mesotrione and nicosulfuron, and their corresponding commercial formulations, alone and in mixtures, using the Microtox test. *Chemosphere* **2013**, *93*, 2444–2450. [CrossRef] [PubMed]

29. Pedrazzani, R.; Ceretti, E.; Zerbini, I.; Casale, R.; Gozio, E.; Bertanza, G.; Gelatti, U.; Donato, F.; Feretti, D. Biodegradability, toxicity and mutagenicity of detergents: Integrated experimental evaluations. *Ecotoxicol. Environ. Saf.* **2012**, *84*, 274–281. [CrossRef] [PubMed]
30. Antunes, S.C.; Pereira, J.L.; Cachada, A.; Duarte, A.C.; Goncalves, P.; Sousa, J.P.; Pereira, R. Structural effects of the bioavailable fraction of pesticides in soil: Suitability of elutriate testing. *J. Hazard. Mater.* **2010**, *184*, 215–225. [CrossRef] [PubMed]
31. Burga Perez, K.F.; Charlatchka, R.; Sahli, L.; Ferard, J.-F. New methodological improvements in the Microtox® solid phase assay. *Chemosphere* **2012**, *86*, 105–110. [CrossRef] [PubMed]
32. Stronkhorst, J.; Ciarelli, S.; Schipper, C.A.; Postma, J.F.; Dubbeldam, M.; Vangheluwe, M.; Brils, J.M.; Hooftman, R. Inter-laboratory comparison of five marine bioassays for evaluating the toxicity of dredged material. *Aquat. Ecosyst. Health Manag.* **2004**, *7*, 147–159. [CrossRef]
33. Doe, K.; Scroggins, R.; Mcleay, D.; Wohlgeschaffen, G. Solid-phase test for sediment toxicity using the luminescent bacterium *Vibrio fischeri*. In *Small-Scale Freshwater Toxicity Investigations*; Blaise, C., Férard, J.F., Eds.; Springer: Dordrecht, Holland, 2005; Volume 1, pp. 107–136.
34. ISO. *Water Quality—Determination of Fresh Water Sediment Toxicity to Heterocyprisincongruens (Crustacea, Ostracoda)*, 1st ed.; ISO 14371:2012; The International Organization for Standardization: Geneva, Switzerland, 2012.
35. Martínez-Sánchez, M.J.; Pérez-Sirvent, C.; García-Lorenzo, M.L.; Martínez-López, S.; Bech, J.; García-Tenorio, R.; Bolívar, J.P. Use of bioassays for the assessment of areas affected by phosphate industry wastes. *J. Geochem. Explor.* **2014**, *147*, 130–138. [CrossRef]
36. Gouider, M.; Feki, M.; Sayadi, S. Bioassay and use in irrigation of untreated and treated wastewaters from phosphate fertilizer industry. *Ecotoxicol. Environ. Saf.* **2010**, *73*, 932–938. [CrossRef] [PubMed]
37. Cvancarova, M.; Kresinova, Z.; Cajthaml, T. Influence of the bioaccessible fraction of polycyclic aromatic hydrocarbons on the ecotoxicity of historically contaminated soils. *J. Hazard. Mater.* **2013**, *254–255*, 116–124. [CrossRef] [PubMed]
38. Oleszczuk, P.; Hollert, H. Comparison of sewage sludge toxicity to plants and invertebrates in three different soils. *Chemosphere* **2011**, *83*, 502–509. [CrossRef] [PubMed]
39. Płaza, G.; Nałęcz-Jawecki, G.; Ulfig, K.; Brigmon, R.L. The application of bioassays as indicators of petroleum-contaminated soil remediation. *Chemosphere* **2005**, *59*, 289–296. [CrossRef] [PubMed]

Sample Availability: Samples of the compounds **2a–d**, **4** and **5** are available from the authors.

molecules

MDPI

Review

Synthesis of Hydroxymethylenebisphosphonic Acid Derivatives in Different Solvents

Dávid Illés Nagy [1], Alajos Grün [1], Sándor Garadnay [2], István Greiner [2] and György Keglevich [1,*]

[1] Department of Organic Chemistry and Technology, Budapest University of Technology and Economics, 1521 Budapest, Hungary; duwa24@gmail.com (D.I.N.); agrun@mail.bme.hu (A.G.)
[2] Gedeon Richter Plc., 1475 Budapest, Hungary; garadnay@richter.hu (S.G.); i.greiner@richter.hu (I.G.)
* Correspondence: gkeglevich@mail.bme.hu; Tel.: +36-1-463-1111 (ext. 5883); Fax: +36-1-463-3648

Academic Editor: Roman Dembinski
Received: 15 July 2016; Accepted: 3 August 2016; Published: 11 August 2016

Abstract: The syntheses of hydroxymethylenebisphosphonic acid derivatives (dronic acid derivatives) starting from the corresponding substituted acetic acids and P-reagents, mainly phosphorus trichloride and phosphorous acid are surveyed according to the solvents applied. The nature of the solvent is a critical point due to the heterogeneity of the reaction mixtures. This review sheds light on the optimum choice and ratio of the P-reactants, and on the optimum conditions.

Keywords: hydroxymethylenebisphosphonic acid (dronic acid) derivatives; solvent; P-reagents; synthesis

1. Introduction

The hydroxymethylenebisphosphonic acid derivatives (dronic acid derivatives) form an important group within organophosphorus pharmaceutics. It is known that the two phosphonic functions are capable of forming complex with calcium ions, hence, the resorption of these ions is prevented. Bisphosphonic derivatives are used in the treatment of osteoporosis, Paget disease and tumor-induced hypercalcaemia, but they also show direct antitumor and antiparasitic activity [1–6].

Although the synthesis of dronic acid derivatives is widely discussed in the literature, it can be considered a sort of "black box". They were prepared in many solvents, but the optimum circumstances, and the molar ratios of the P-reagents were not explored. The role of the reagents and the reaction mechanism were not clarified. In most cases, the purity of the products was not reported, or when crude products were prepared, unrealistically high, and hence misleading yields were claimed. We wish to summarize the preparation of the most important hydroxymethylenebisphosphonic acid derivatives according to solvents, hoping to make the syntheses more transparent. The most important therapeutic agents discussed in this review are listed in Table 1.

Table 1. The dronic acid derivatives discussed in this review.

General Formula	Y	Commercial Name	Abbreviation
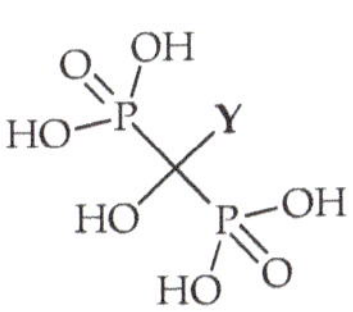	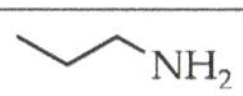	pamidronic acid	**PD**
	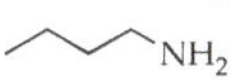	alendronic acid	**AD**
	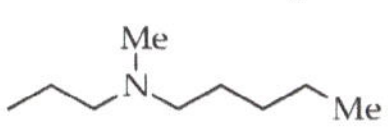	ibandronic acid	**ID**
	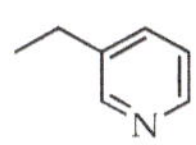	risedronic acid	**RD**
	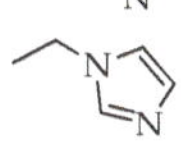	zoledronic acid	**ZD**

According to the general scheme for the preparation of hydroxymethylenebisphosphonic acid derivatives, the corresponding carboxylic acid, or its chloride, ester and anhydride derivative was used as the starting material, and reacted with phosphorus trichloride and/or phosphorous acid, phosphoric acid or phosphoryl chloride in a wide variety of solvents, for example methanesulfonic acid (MSA), sulfolane, chlorobenzene, toluene, xylene, *n*-octane, methyl cyclohexane, 1,4-dioxane, acetonitrile, phenol and its derivatives, alkyl carbonates, phosphates, silicon or sunflower oil, ionic liquids, and also in the absence of any solvent (Scheme 1).

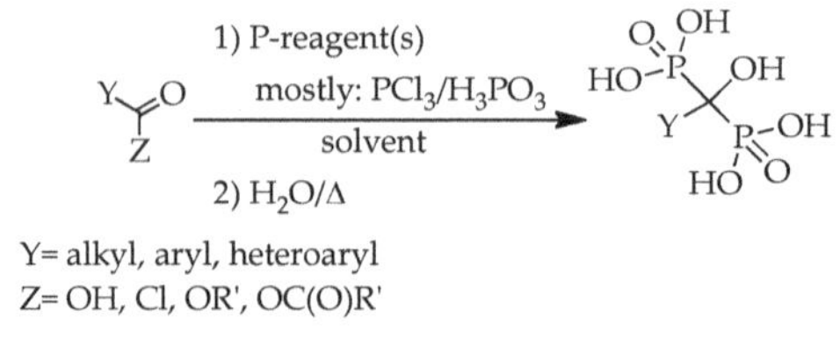

Scheme 1. General scheme for the preparation of hydroxymethylenebisphosphonic acids.

In many publications, water was applied as the solvent, and phosphorus trichloride as the reagent, which is absolutely pointless, as the water reacts fast with the P-reagent to provide phosphorous acid [7–11]. The concept of generating phosphorous acid "in situ" may be questioned, as this approach is expensive, pollutes the environment, and rather dangerous on the industrial scale. We did not deal with this not too practical method.

2. Reactions in Methanesulfonic Acid (MSA)

MSA may be considered the best solvent during the synthesis of dronic acid derivatives, as it helps to overcome the problems of heterogeneity during the reaction. However, the MSA is likely to participate in the reactions themselves. It can activate the carboxylic acid to form a carboxylic acid-MSA mixed anhydride, which may be formed in two ways, as shown in Scheme 2 [12].

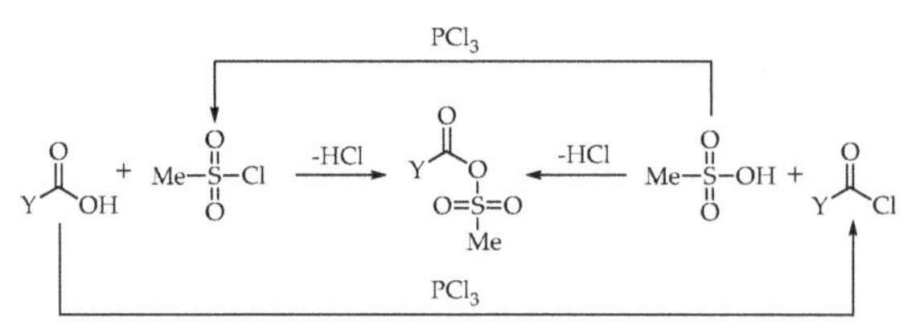

Scheme 2. Possible formation of carboxylic acid-MSA mixed anhydride as an intermediate.

MSA may also react with phosphorus trichloride to provide a more reactive P-species (Scheme 3).

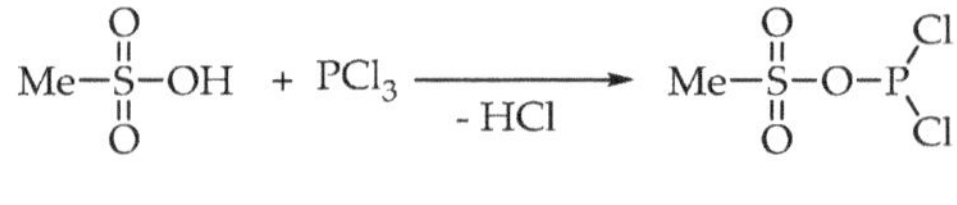

Scheme 3. Activation of phosphorus trichloride by reaction with MSA.

In all syntheses of pamidronate described, β-alanine was applied as the starting material in MSA (Scheme 1, Y = $-CH_2CH_2NH_2$, Z = $-OH$, solvent: MSA). Szűcs, Vepsäläinen and his co-workers used 2 equivalents of phosphorus trichloride and 1 equivalent of phosphoric acid as the P-reagents (Table 2/Entries 1 and 2) [13,14]. Shinkai et al. applied phosphorus trichloride and phosphorous acid in ratio of 2.1:1 (Table 2/Entry 3) [15]. In the lack of purification, no purities were provided. The best result was achieved by Keglevich et al. who proved that the phosphorous acid did not participate in the reaction under discussion due its low nucleophilicity. For this, only 3.2 equivalents of phosphorus trichloride were measured in. The yield of pamidronic acid monosodium salt trihydrate was 57%, and the purity was 99% (Scheme 4) (Table 2/Entry 4) [16].

Table 2. Synthesis of pamidronate from β-alanine in MSA.

Entry	Reagent (Equiv.)	T (°C)	t (h)	Product Form	Purity (%)	Yield (%)	Ref.
1	PCl_3:H_3PO_4 (2:1)	65	16	Na	-	55 [1]	[13]
2	PCl_3:H_3PO_4 (2:1)	65	18	Na	-	22 [1]	[14]
3	PCl_3:H_3PO_3 (2.1:1)	65	16–20	Na	-	57 [1]	[15]
4	PCl_3 (3.2)	75	24	Na·$2H_2O$	99 [2]	57	[16]

[1] For the crude product; [2] On the basis of acid-base titration.

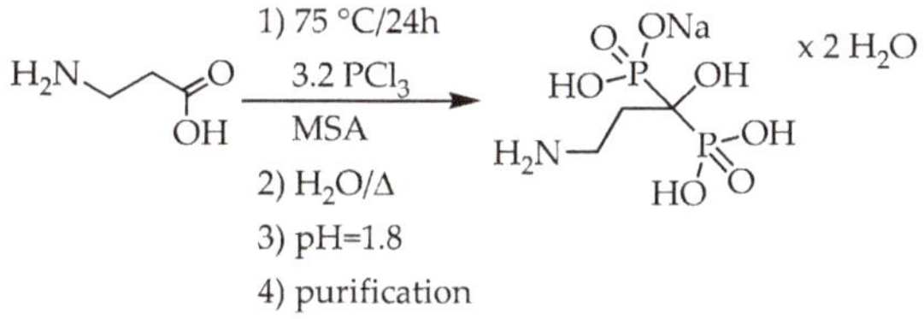

Scheme 4. Preparation of pamidronate in MSA by an optimized method [16].

The preparations of alendronic acid derivatives were also studied in details using MSA. In most cases γ-aminobutyric acid (GABA) was treated with the P-reagents (Table 3/Entries 1–9) [14,15,17–23]. In one instance, 2-pyrrolidone was the starting material (Table 3/Entry 10) [24].

Table 3. Synthesis of alendronic acid derivatives in MSA.

Entry	Starting Material	Reagent (Equiv.)	T (°C)	t (h)	Product form	Purity (%)	Yield (%)	Ref.
1	GABA	PCl_3:H_3PO_3 (2:1)	65	18	Na	-	91 [1]	[19]
2	GABA	PCl_3:H_3PO_3 (2.4:1.5)	65	20	Na·$3H_2O$	99.7 [2]	90 [1]	[23]
3	GABA	PCl_3:H_3PO_3 (2.1:1)	65	16–20	Na·$3H_2O$	99.7 [2]	89 [1]	[15]
4	GABA	PCl_3:H_3PO_4 (2:1)	65	18	acid	-	85 [1]	[14]
5	GABA	PCl_3:H_3PO_3 (2:1)	65	18	-	-	82 [1]	[18]
6	GABA	PCl_3:H_3PO_3 (2.1:1)	65	18	-	-	82 [1]	[20]
7	GABA	PCl_3 (3.2)	75	12	Na·$3H_2O$	93 [2]	58	[21]
8	GABA	$POCl_3$:H_3PO_3 (2:3 or 3:3)	75	12	Na·$3H_2O$	97 [2]	60	[22]
9	GABA	P_2O_5:H_3PO_3 (1:2)	70	24	Na·$3H_2O$	>99.9 [3]	48	[17]
10	2-Pyrrolidone	PCl_3 (3.4)	100–105	6	Na·$3H_2O$	-	81 [1]	[24]

[1] For the crude product; [2] On the basis of acid-base or complexometric titration; [3] On the basis of HPLC.

The majority of the researchers applied phosphorus trichloride and phosphorous acid as reagents in a ratio of ca. 2:1 (Scheme 1, Y = $-CH_2(CH_2)_2NH_2$, Z = $-OH$, solvent: MSA) (Table 3/Entries 1–3, 5 and 6) [15,18–20,23]. Yields around 80%–90% were claimed. As in the case of pamidronate, Vepsäläinen and his co-workers used 2 equivalents of phosphorus trichloride and 1 equivalent of phosphoric acid (Table 3/Entry 4) [14]. The best reproducible results were reported by Keglevich et al., furnishing alendronate in a yield of 58% (Scheme 5) (Table 3/Entry 7) [21]. It was clarified that phosphorous acid is unnecessary, when phosphorus trichloride is the P-reagent and MSA is the solvent.

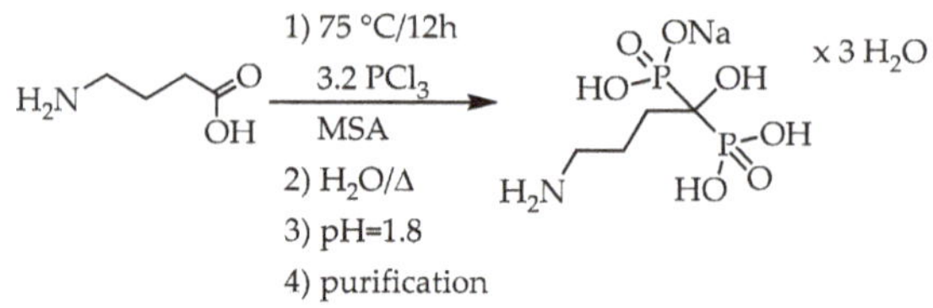

Scheme 5. Preparation of alendronate in MSA by an optimized method [21].

Typically, the yields were higher than 80% for the crude product (Table 3/Entries 1–6 and 10) [14,15,18–20,23,24]. In only two cases were the purities reported, however, the values of 99.7% seem to be excessive (Table 3/Entries 2 and 3) [15,23]. Our experience is that a crude product may never be pure, and such high values are not realistic.

In a special combination, phosphoryl chloride and phosphorous acid were used in a proportion of 2:3 or 3:3 in MSA (Table 3/Entry 8) [22]. In one case, 2 equivalents of phosphorous acid and 1 equivalent of phosphorus pentoxide were the P-reactants, and the target dronate was obtained in a yield of 48% (Table 3/Entry 9) [17]. Using 2-pyrrolidone as the starting material together with 3.4 equivalents of phosphorus trichloride, the crude alendronate was isolated in a yield of 81% (Table 3/Entry 10) [24].

The synthesis of ibandronate in MSA was discussed in only a few publications. In these preparations, *N*-Methyl-*N*-pentyl-β-alanine (MPA) was treated with phosphorus trichloride and phosphorous acid under different conditions (Scheme 1, Y = $-CH_2CH_2N(Me)(Pen)$, Z = −OH, solvent: MSA). Soni et al. applied 5 equivalents of phosphorus trichloride together with 2 equivalents of phosphorous acid. According to their description, pure alendronate was isolated in a yield of 86% (Table 4/Entry 1) [25]. The use of phosphorus trichloride is completely unnecessary. Applying 2.5 equivalents of phosphorus trichloride, and the same amount of phosphorous acid in the mixture of MSA and chlorobenzene, the yield was 60% (Table 4/Entry 2) [26]. The optimum case was, when 3.2 equivalents of phosphorus trichloride were used. The realistic and reproducible yield was 46% (Scheme 6) (Table 4/Entry 3) [21].

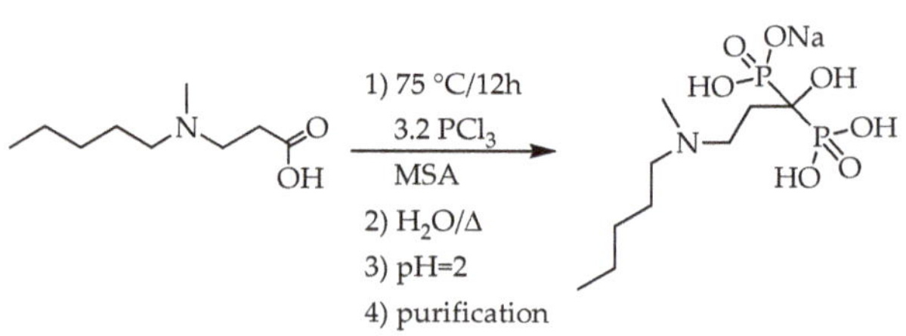

Scheme 6. Preparation of ibandronate in MSA by an optimized method [21].

Table 4. Synthesis of ibandronate from MPA in MSA.

Entry	Reagent (Equiv.)	T (°C)	t (h)	Product Form	Purity (%)	Yield (%)	Ref.
1	PCl_3:H_3PO_3 (5:2)	65–67	24	Na·H_2O	99.6 [1]	86	[25]
2	PCl_3:H_3PO_3 [2] (2.5:2)	96	3	Na·H_2O	99.8 [1]	60	[26]
3	PCl_3 (3.2)	75	12	Na	99 [3]	46	[21]

[1] On the basis of HPLC; [2] in the mixture of MSA and PhCl; [3] On the basis of acid-base titration.

For risedronic acid derivatives, yields of 38%–74% were reported. In the basic syntheses, 3-pyridylacetic acid (PAA) was reacted with phosphorus trichloride and phosphorous acid in MSA (Scheme 1, Y = 3-pyridylmethylene-, Z = −OH, solvent: MSA).

A yield of 73% was obtained, when 3-3 equivalents of phosphorus trichloride and phosphorous acid were used in the mixture of MSA and diethyl carbonate (Table 5/Entry 1) [27]. Another combination, a 2.1:1 ratio of the P-reagents led to the dronic acid in a yield of 38%. No purity was reported (Table 5/Entry 2) [15]. When the mixture of 3 equivalents of phosphorous acid and 2 equivalents of phosphoryl chloride, or 2 equivalents of phosphorous acid and 1 equivalent of phosphorus pentoxide were reacted with the corresponding carboxylic acid, yields of ca. 55% were reported (Table 5/Entries 3 and 4) [17,22]. The application of phosphorus trichloride (3.1 equivalents) alone led to the highest yield (74%), and the unnecessity of phosphorous acid was again proved (Scheme 7) (Table 5/Entry 5) [28,29].

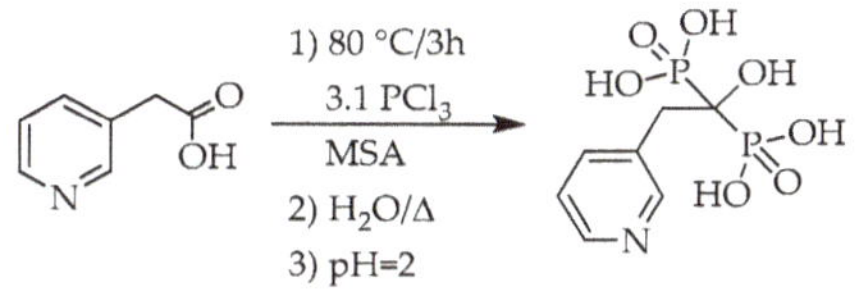

Scheme 7. Preparation of risedronic acid in MSA by an optimized method [28,29].

Table 5. Synthesis of risedronic acid derivatives from PAA in MSA.

Entry	Reagent (Equiv.)	T (°C)	t (h)	Product Form	Purity (%)	Yield (%)	Ref.
1	PCl_3:H_3PO_3 [1] (3:3)	70–72	5	Na·2.5H_2O	99.9 [2]	73	[27]
2	PCl_3:H_3PO_3 (2.1:1)	65	16–20	acid	-	38 [3]	[15]
3	$POCl_3$:H_3PO_3 (2:3)	75	12	acid	98 [4]	55	[22]
4	P_2O_5:H_3PO_3 (1:2)	80–100	72	acid·H_2O	99.9 [2]	54	[17]
5	PCl_3 (3.1)	75	12	acid	92 [4] (100 [2])	74 [3]	[28,29]

[1] in the mixture of MSA and diethyl carbonate; [2] On the basis of HPLC; [3] For the crude product; [4] On the basis of acid-base titration.

During the preparation of zoledronic acid in MSA, imidazol-1-yl-acetic acid (IAA) was treated with only phosphorus trichloride, or with the mixture of this reagent and phosphorous acid (Scheme 1, Y = 1-imidazolylmethylene-, Z = −OH, solvent: MSA). Kieczykowski and co-workers applied phosphorus trichloride and phosphorous acid in a quantity of 2.1:1 equivalents. The yield of the target dronic acid was 31% (Table 6/Entry 1) [15]. Using phosphorus trichloride as the reactant in amounts of 3.1 or 3.2, yields of 53% and 46%, respectively, were reported by the Keglevich group in comparable purity (Scheme 8) (Table 6/Entries 2 and 3) [28,29].

Table 6. Synthesis of zoledronic acid from IAA in MSA.

Entry	Reagent (Equiv.)	T (°C)	t (h)	Product Form	Purity (%)	Yield (%)	Ref.
1	PCl_3:H_3PO_3 (2.1:1)	65	16–20	acid	-	31 [1]	[15]
2	PCl_3 (3.1)	80	3	acid	99 [2]	53	[29]
3	PCl_3 (3.2)	80	3	acid	98 [2]	46	[28]

[1] For the crude product; [2] On the basis of acid-base titration.

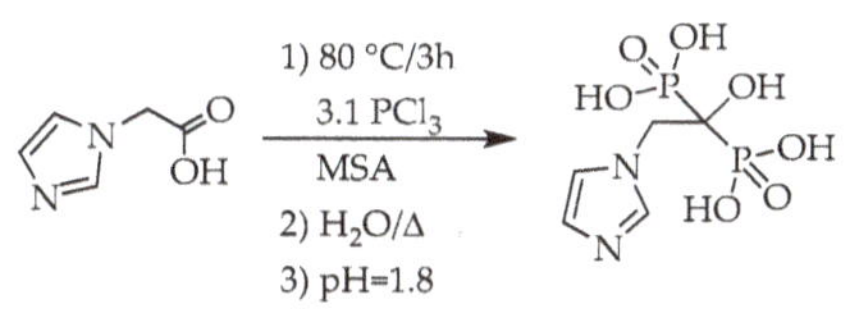

Scheme 8. Preparation of zoledronic acid in MSA by an optimized method [28,29].

In overall, it can be said that the lower yields of the Keglevich group are better, than the higher ones published for the crude products neglecting the purity. The reliable yields were obtained after purifying the products and analyzing the purity by potentiometric titration.

3. Synthesis of Dronic Acid Derivatives in Chlorobenzene

The second frequently used solvent is chlorobenzene. Chlorobenzene has a few advantages in comparison with MSA. It has a lower viscosity, can be removed by distillation, and can be reused. The real problem with MSA is that during the pH adjustment sodium methanesulfonate may be formed from MSA, which is difficult to remove. In the procedures reported, the corresponding carboxylic acids were reacted with phosphorus trichloride and phosphorous acid/phosphoric acid. None of the reactions were performed with solely phosphorus trichloride, as, in this case, it is unreactive without another P-containing partner (Scheme 9). In most reports no data on the purity of dronic acid derivatives was provided, for this, the yields are not reliable.

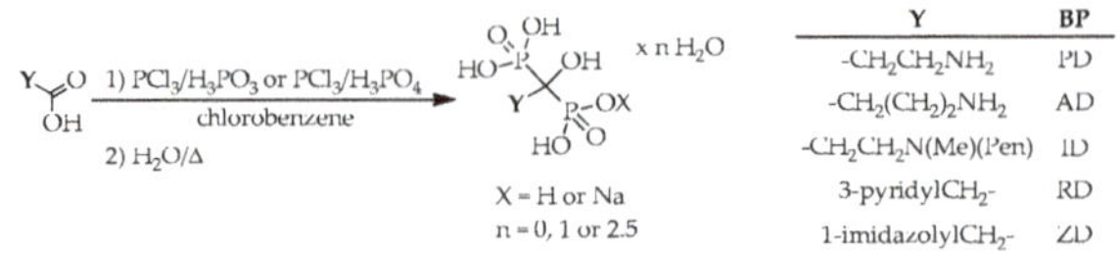

Scheme 9. General scheme for the preparation of dronic acid derivatives in chlorobenzene.

Reacting β-alanine or GABA with 1.5 equivalents of phosphorus trichloride and 1.5 equivalents of phosphorous acid at 100 °C for 3 h, the yields were in the range of 45%–59% (Table 7/Entries 1 and 3–5) [30–32]. Starting from β-alanine or GABA, and measuring in phosphorus trichloride and phosphoric acid in molar equivalents of 2.4:1.5 or 2:3, respectively, the pamidronic acid or alendronic acid was obtained in a yield of 53% and 18%, respectively (Table 7/Entries 2 and 6) [33,34].

Table 7. Synthesis of pamidronic acid and alendronic acid in chlorobenzene.

Entry	BP	Starting Material	Reagent (Equiv.)	T (°C)	t (h)	Product Form	Purity (%)	Yield (%)	Ref.
1	PD	β-alanine	PCl_3:H_3PO_3 (1.5:1.5)	100	3	acid	-	59 [1]	[30]
2	PD	β-alanine	PCl_3:H_3PO_4 (2.4:1.5)	105–110	2	acid	-	53	[33]
3	PD	β-alanine	PCl_3:H_3PO_3 (1.5:1.5)	100	3	acid	-	51	[31]
4	AD	GABA	PCl_3:H_3PO_3 (1.5:1.5)	100	3	acid	-	46	[31]
5	AD	GABA	PCl_3:H_3PO_3 (1.5:1.5)	100	3	acid	-	45	[32]
6	AD	GABA	PCl_3:H_3PO_4 (2:3)	100–110	3	acid	-	18 [1]	[34]

[1] For the crude product.

In two variations for the syntheses of ibandronate, a yield of 60% was reached. Phosphorus trichloride and phosphorous acid were applied in a 1.5:1.5 or 2.5:2 molar equivalent quantity (in a mixture of MSA and chlorobenzene) (Table 8/Entries 1 and 2) [26,35]. Using phosphorus trichloride and phosphoric acid in amounts of 3:3 and 2:3, the yield was 37% and 6%, respectively (Table 8/Entries 3 and 4) [34,36]. In one instance, the proportion of phosphorous acid was not provided, and a peripheral result (a yield of 3%) was reported (Table 8/Entry 5) [26].

Table 8. Synthesis of ibandronic acid derivatives from MPA in chlorobenzene.

Entry	Reagent (Equiv.)	T (°C)	t (h)	Product Form	Purity (%)	Yield (%)	Ref.
1	PCl_3:H_3PO_3 (1.5:1.5)	80	4	Na·H_2O	-	60	[35]
2	PCl_3:H_3PO_3 [1] (2.5:2)	96	3	Na·H_2O	99.8 [2]	60	[26]
3	PCl_3:H_3PO_4 (3:3)	110	11	Na	-	37 [3]	[36]
4	PCl_3:H_3PO_4 (2:3)	100–110	3	acid	-	6 [1]	[34]
5	PCl_3 [4]:H_3PO_3	85–90	4–5	acid	95 [2]	3	[26]

[1] in the mixture of MSA and PhCl; [2] On the basis of HPLC; [3] For the crude product; [4] Two and a half equivalents.

The preparation of risedronic acid derivatives in chlorobenzene as the solvent was in all reactions implemented using phosphorus trichloride and phosphorous acid. An unrealistically high yield of 96% was claimed for risedronic acid using 3.5 equivalents of phosphorus trichloride, and the same amounts of phosphorous acid without specifying the purity (Table 9/Entry 1) [37]. Applying the P-reagents in a quantity of 1.5 equivalents, the yield was 68% (Table 9/Entry 2) [35]. The most reliable result was described by Wadhwa, but the reaction time was not given. The reactants were used in amount of 2.4 equivalents, and the risedronic acid was obtained in a yield of 58% with a purity of 84% (Table 9/Entry 3) [38]. Applying phosphorus trichloride and phosphorous acid in a molar ratio of 3:3.5, the dronate was isolated in a yield of 52%. However, no temperature was reported. A similar yield (52%) was described, when 2 equivalents of phosphorus trichloride was used with phosphorous acid. However, the quantity of the latter species was not provided (Table 9/Entries 4 and 5) [39,40].

Table 9. Synthesis of risedronic acid derivatives from PAA in chlorobenzene.

Entry	Reagent (Equiv.)	T (°C)	t (h)	Product Form	Purity (%)	Yield (%)	Ref.
1	PCl_3:H_3PO_3 (3.5:3.5)	90–95	2.5	acid	-	96 [1]	[37]
2	PCl_3:H_3PO_3 (1.5:1.5)	80	4	acid	-	68 [1]	[35]
3	PCl_3:H_3PO_3 (2.4:2.4)	85–90	-	acid	84 [2]	58 [1]	[38]
4	PCl_3:H_3PO_3 (3:3.5)	-	2.5	Na·2.5H_2O	-	52	[39]
5	PCl_3 [3]:H_3PO_3	100	6	acid	-	52 [1]	[40]

[1] For the crude product; [2] On the basis of HPLC; [3] Two equivalents.

For the synthesis of zoledronic acid in chlorobenzene, the purity remains unclear in all cases, therefore, the yields are not reliable at all. Carrying out the synthesis with 3.7 equivalents of phosphorus trichloride and with the same amount of phosphorous acid, quantitative yield was claimed that is entirely unrealistic (Table 10/Entry 1) [41]. In other instances, phosphorus trichloride and phosphoric acid were applied in different, such as 4.7:2.9, 2:3 and 3:1.9 ratios to lead to yields of 79%, 67% and 41%, respectively (Table 10/Entries 2–4) [34,42,43].

The need for the joint application of phosphorus trichloride and phosphorous acid will be explained in the next sub-chapter.

Table 10. Synthesis of zoledronic acid from IAA in chlorobenzene.

Entry	Reagent (Equiv.)	T (°C)	t (h)	Product Form	Purity (%)	Yield (%)	Ref.
1	PCl_3:H_3PO_3 (3.7:3.7)	100	1	acid	-	100 [1]	[41]
2	PCl_3:H_3PO_4 (4.7:2.9)	80	2–2.5	acid	-	79 [1]	[42]
3	PCl_3:H_3PO_4 (2:3)	100–110	3	acid	-	67 [1]	[34]
4	PCl_3:H_3PO_4 (3:1.9)	100	3	acid	-	41	[43]

[1] For the crude product.

4. Methods for the Preparation of Dronic Acid Derivatives in Sulfolane

Sulfolane is also a preferred solvent in the preparation of hydroxymethylenebisphosphonic acid derivatives, but less data were published. In all such cases, phosphorus trichloride and phosphorous acid were the P-reagents (Scheme 10).

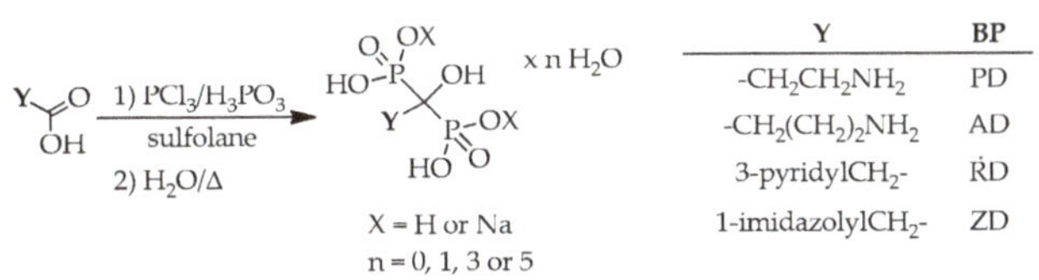

Scheme 10. General scheme for the preparation of dronic acid derivatives in sulfolane.

Approximately in half of the cases, the purities were not described, therefore the data may be misleading. The synthesis of a part of the dronic acid derivatives was attempted also under MW conditions. The synthesis of ibandronic acid derivatives was not attempted in sulfolane.

Applying phosphorus trichloride and phosphorous acid in a molar ratio of 3:3, under conventional and MW-assisted conditions at 65 °C for 3.5 h and 3 min, respectively, yields of 72% and 64%, respectively, were reported for crude pamidronic acid derivatives (Table 11/Entries 1 and 2) [44]. (The forms of product were not provided.) Alendronate was obtained under similar conditions in yields of 38% and 41%, respectively (the forms of product were not provided) (Table 11/Entries 7 and 8) [44]. It can be seen that MW irradiation was useful in respect of reaction time in these reactions. Keglevich et al. used 2-2 equivalents of the P-reagents in reaction with β-alanine to afford pure pamidronic acid in a yield of 63% (Table 11/Entry 3) [16]. When phosphorus trichloride and phosphorous acid were used in a ratio of 3.4:1.5, pure sodium pamidronate was obtained in a lower yield (Table 11/Entry 4) [45]. Applying phosphorus trichloride and phosphorous acid in ratios of 3.4:1.5 and 2.5:3.5, the yield of sodium alendronate was 69% (crude product), and 55% (pure dronate), respectively (Table 11/Entries 5 and 6) [45,46].

Table 11. Synthesis of pamidronic and alendronic acid derivatives in sulfolane.

Entry	BP	Starting Material	Reagent (Equiv.)	T (°C)	t (h)	Product Form	Purity (%)	Yield (%)	Ref.
1	PD	β-alanine	PCl_3:H_3PO_3 (3:3)	65	3.5	-	-	72 [1]	[44]
2	PD	β-alanine	PCl_3:H_3PO_3 (3:3)	65 [2]	0.05	-	-	64 [1]	[44]
3	PD	β-alanine	PCl_3:H_3PO_3 (2:2)	75	12	acid	100 [3]	63	[16]
4	PD	β-alanine	PCl_3:H_3PO_3 (3.4:1.5)	63–67	3	2Na·5H_2O	>99	48	[45]
5	AD	GABA	PCl_3:H_3PO_3 (3.4:1.5)	63–67	3	Na·3H_2O	>99	69 [1]	[45]
6	AD	GABA	PCl_3:H_3PO_3 (2.5:3.5)	60–65	12	Na·3H_2O	>99 [4]	55	[46]
7	AD	GABA	PCl_3:H_3PO_3 (3:3)	65 [2]	0.1	-	-	41 [1]	[44]
8	AD	GABA	PCl_3:H_3PO_3 (3:3)	65	3.5	-	-	38 [1]	[44]

[1] For the crude product; [2] MW heating; [3] On the basis of acid-base titration; [4] On the basis of HPLC.

McKenna and his co-workers investigated the synthesis of risedronic and zoledronic acid derivatives (the forms of product were not provided) also under MW conditions, using 3 equivalents of both the phosphorus trichloride and phosphorous acid at 65 °C for 3 min. The dronic acid derivatives, risedronic acid and zoledronic acid derivatives were obtained in yields of 74% and 70%, respectively (Table 12/Entries 1 and 3) [44]. In case of conventional heating, the yield of zoledronic acid derivative was 67% (Table 12/Entry 4) [44]. The reaction of 1 equivalent of PAA or IAA with 3.4 equivalents of phosphorus trichloride and 1.5 equivalents of phosphorous acid on conventional heating at around 65 °C provided risedronate and zoledronic acid in a yield of 54% and 71%, respectively (Table 12/Entries 2 and 5) [45]. No attempts were made to prepare pure zoledronic acid derivatives.

Table 12. Synthesis of risedronic and zoledronic acid derivatives in sulfolane.

Entry	BP	Starting Material	Reagent (Equiv.)	T (°C)	t (h)	Product Form	Purity (%)	Yield (%)	Ref.
1	RD	PAA	PCl_3:H_3PO_3 (3:3)	65 [1]	0.05	acid·H_2O	-	74 [2]	[44]
2	RD	PAA	PCl_3:H_3PO_3 (3.4:1.5)	63–67	3	Na	>99	54	[45]
3	ZD	IAA	PCl_3:H_3PO_3 (3:3)	65 [1]	0.05	-	-	70 [2]	[44]
4	ZD	IAA	PCl_3:H_3PO_3 (3:3)	65	3.5	-	-	67 [2]	[44]
5	ZD	IAA	PCl_3:H_3PO_3 (3.4:1.5)	63–67	3	acid·H_2O	>99	71 [2]	[45]

[1] Under MW irradiation; [2] For the crude product.

It was shown in Chapter 2 that using MSA as the solvent, there was no need to use phosphorous acid together with phosphorus trichloride (Tables 2–6) [16,21,28,29]. However, using a solvent other than MSA, it is necessary to apply phosphorus trichloride and phosphorous acid jointly. A logical explanation may be that, before phosphorus trichloride would react with the carboxylic acid or its derivatives, it reacts with 1 or 2 equivalents of phosphorous acid resulting in the formation of activated P-species, such as $(HO)_2P\text{-}O\text{-}PCl_2$ and/or $(HO)_2P\text{-}O\text{-}PCl\text{-}O\text{-}P(OH)_2$ (Scheme 11). In the next step, the reactive intermediates formed may react with the substituted acetic acid, providing the target dronic acid derivative after several steps [16].

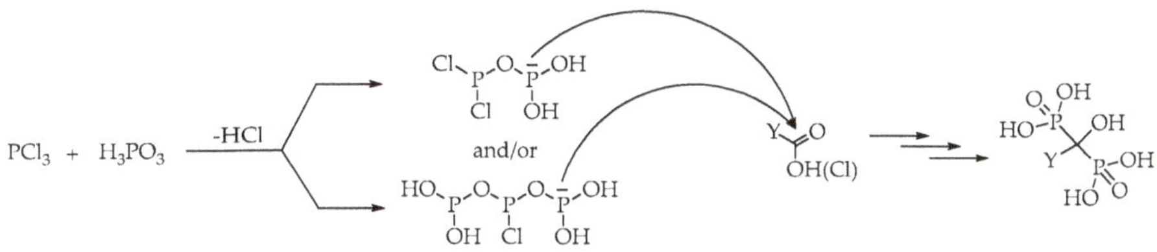

Scheme 11. Formation of the reactive P-species by the condensation of phosphorus trichloride and phosphorous acid and their reaction with the substituted acetic acids.

5. Synthesis of Dronic Acid Derivatives in Toluene

The preparations of alendronic acid derivatives, ibandronate and risedronic acid have also been investigated in toluene (Scheme 12).

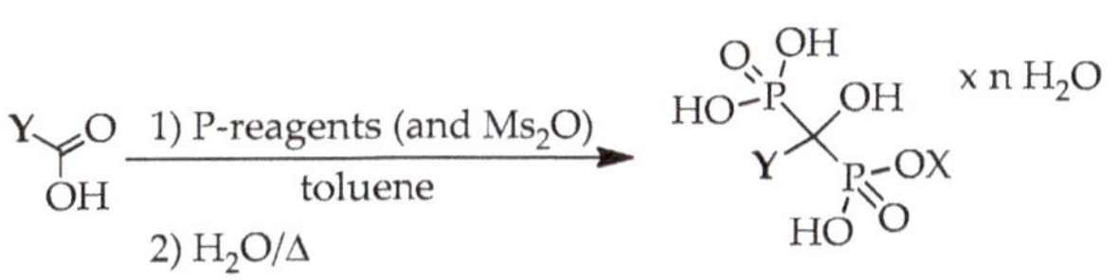

X = H or Na
n = 0, 1 or 3

Y	BP
$-CH_2(CH_2)_2NH_2$	AD
$-CH_2CH_2N(Me)(Pen)$	ID
3-pyridylCH_2-	RD

Scheme 12. General scheme for the preparation of dronic acid derivatives in toluene.

In one approach towards alendronic acid, a mixture of toluene and PEG 400 was used as the solvent, and 1.6 equivalents of phosphorus trichloride along with 1 equivalent of phosphorous acid as the P-reagents. The yield was 56% for crude alendronic acid (Table 13/Entry 1) [47]. Using phosphorus trichloride and methanesulfonic anhydride (Ms_2O) in a molar ratio of 3:3 in toluene, a yield of 66% was reported (Table 13/Entry 2) [48]. The application of 3.7 equivalents of phosphorus trichloride and 4.2 equivalents of phosphorous acid led to a moderate yield (44%) of sodium ibandronate (Table 13/Entry 3) [49].

Table 13. Synthesis of alendronic acid derivatives and ibandronate in toluene.

Entry	BP	Starting Material	Reagent (Equiv.)	T (°C)	t (h)	Product Form	Purity (%)	Yield (%)	Ref.
1	AD	GABA	$PCl_3:H_3PO_3$ [1] (1.6:1)	75	4	acid	-	56 [2]	[47]
2	AD	GABA	$PCl_3:Ms_2O$ (3:3)	70	9.5	Na·$3H_2O$	-	66	[48]
3	ID	MPA	$PCl_3:H_3PO_3$ (3.7:4.2)	80–85	7–8	Na	-	44	[49]

[1] in the mixture of toluene and PEG 400; [2] For the crude product.

Regarding the synthesis of risedronic acid, in most instances unusual reagents for example, propylphosphonic anhydride (T3P), Ms_2O or PCl_5 were used together with phosphorous acid, and purities of >90% were reported according to HPLC (Table 14/Entries 1–6). Unrealistically high yields of 90% and 81% were reported, when 3.1 equivalents phosphorous acid were reacted with 2.5 equivalents of T3P or 3.1 equivalents of phosphoryl chloride (Table 14/Entries 1 and 2) [50]. In similar reactions, with ca. 3 equivalents of reactants (phosphorous acid and Ms_2O) a lower yield of 34% was obtained (Table 14/Entry 3) [50]. Other interesting combinations, such as a 3.1:2:1 ratio of H_3PO_3:Ms_2O:$POCl_3$, or a 3.1:1:1 ratio of H_3PO_3:Ms_2O:PCl_5 afforded the risedronic acid in yields of ca. 60% (Table 14/Entries 4 and 5) [50]. The application of 3.2 equivalents of phosphorus trichloride alone led to the lowest yield

(15%) (Table 14/Entry 6) [12]. It seems to have been confirmed that phosphorus trichloride alone is not enough if not MSA is the solvent. In the cases discussed, it was necessary to apply an activating agent.

Table 14. Synthesis of risedronic acid from PAA in toluene

Entry	Reagent (Equiv.)	T (°C)	t (h)	Product Form	Purity (%)	Yield (%)	Ref.
1	H_3PO_3:T3P (3.1:2.5)	95	20.5	acid·H_2O	99.6 [1]	90 [2]	[50]
2	H_3PO_3:$POCl_3$ (3.1:3.1)	95	20.5	acid·H_2O	98.5 [1]	81 [2]	[50]
3	H_3PO_3:Ms_2O (3.1:3)	95	20.5	acid·H_2O	95.9 [1]	34 [2]	[50]
4	H_3PO_3:Ms_2O:$POCl_3$ (3.1:2:1)	95	20.5	acid·H_2O	99.2 [1]	62 [2]	[50]
5	H_3PO_3:Ms_2O:PCl_5 (3.1:1:1)	95	20.5	acid·H_2O	98.3 [1]	58 [2]	[50]
6	PCl_3 (3.2)	110	6	acid	90.1 [3]	15 [2]	[12]

[1] On the basis of HPLC; [2] For the crude product; [3] On the basis of acid-base titration.

6. Reactions in the Absence of Solvent

In many publications, the hydroxymethylenebisphosphonic acid derivatives were prepared in the absence of any solvent. These reactions are more heterogeneous, and in most cases they cannot be stirred, due to the (almost) solid consistency. Although the yields reported are not wrong, the purities were provided only in a few instances. For this, the conclusions from the yields may be misleading. It can be seen that beside phosphorus trichloride or phosphoryl chloride, there was need for another reactant as well, for example for phosphorous acid or Ms_2O.

Reacting β-alanine with an unnecessarily large amount of phosphorus trichloride (5 equivalents) and phosphorous acid (3 equivalents) under MW conditions, sodium pamidronate was isolated in a yield of 67% (Table 15/Entry 1) [51]. A rather similar yield (61%) was reported applying the above P-reagents in a 2:1.5 equivalents quantity (Table 15/Entry 2) [52]. Keglevich et al. obtained pure pamidronic acid in a yield of 44%, when both phosphorus trichloride and phosphorous acid were used in a quantity of 2 equivalents [53]. According to our experiences, in most cases it is not possible to prepare the hydroxymethylenebisphosphonic acid derivatives in the absence of solvents.

Table 15. Synthesis of pamidronic acid derivatives from β-alanine in neat.

Entry	Reagent (Equiv.)	T (°C)	t (h)	Product Form	Purity (%)	Yield (%)	Ref.
1	PCl_3:H_3PO_3 (5:3)	80 [1]	0.05	Na	-	67 [2]	[51]
2	PCl_3:H_3PO_3 (2:1.5)	100	4	acid	-	61 [2]	[52]
3	PCl_3:H_3PO_3 (2:2)	75	3	acid	>99 [3]	44	[53]

[1] Under MW conditions; [2] For the crude product; [3] On the basis of acid-base titration.

Alendronic acid derivatives were prepared starting from GABA, *N*-phthalimido-GABA, or *N*-phthalimido-GABA-chloride, but the purities were not reported at all. Using GABA as the starting material, and phosphorus trichloride along with Ms_2O or phosphorous acid as the other reagent in a molar ratio of 3:3 (PCl_3:Ms_2O) and 5:3 (PCl_3:H_3PO_3 under MW conditions), or 2:1.5 (PCl_3:H_3PO_3), the yields were 78%, 78% and 59%, respectively (Table 16/Entries 1–3) [48,51,52]. Starting from *N*-phthalimido-GABA and applying phosphorus trichloride along with phosphorous acid, or this combined with phosphoric acid, crude alendronic acid was obtained in yields of 38% and 57% (Table 16/Entries 4–5) [54]. In one instance, *N*-phthalimido-GABA-chloride was reacted with 2 equivalents of phosphorous acid alone, leading to the target dronic acid in a poor yield of 19% (Table 16/Entry 6) [54]. According to our earlier results [16,21,29,53], phosphorous acid alone is completely unreactive in the synthesis of hydroxymethylenebisphosphonic acid derivatives. Therefore, even the low yield (19%) reported is questionable (Table 16/Entry 6) [54].

Table 16. Synthesis of alendronic acid derivatives in the absence of solvents.

Entry	Starting Material	Reagent (Equiv.)	T (°C)	t (h)	Product Form	Purity (%)	Yield (%)	Ref.
1	GABA	PCl_3:Ms_2O (3:3)	70–100	7	Na·3H_2O	-	78	[48]
2	GABA	PCl_3:H_3PO_3 (5:3)	80 [1]	0.05	Na	-	78 [2]	[51]
3	GABA	PCl_3:H_3PO_3 (2:1.5)	100	4	acid	-	59 [2]	[52]
4	*N*-phthalimido-GABA	PCl_3:H_3PO_3 (2:5)	80	3	acid	-	38 [2]	[54]
5	*N*-phthalimido-GABA	PCl_3:H_3PO_3:H_3PO_4 (2:1.5:4)	80	3	acid	-	57 [2]	[54]
6	*N*-phthalimido-GABA-chloride	H_3PO_3 (2)	130	4	acid	-	19 [2]	[54]

[1] Under MW conditions; [2] For the crude product.

The application of 2.9 equivalents of phosphorus trichloride and 1.5 equivalents of phosphorous acid under somewhat different conditions (at 60–65 °C for 1.5 h or at 70–75 °C for 20 h) led to ibandronate in yields of 89% and 82% (Table 17/Entries 1 and 2) [55,56]. A similar result (82%) was reported using a larger excess of the P-reagents (4.1 equivalents of phosphorus trichloride and 2.5 equivalents of phosphorous acid). However, no criterions of purity were provided (Table 17/Entry 3) [57]. Measuring in, even more reagents (5 equivalents of phosphorus trichloride and 3 equivalents of phosphorous acid), and performing the reaction under MW conditions, a somewhat lower yield (72%) was obtained (Table 17/Entry 4) [51]. With a not defined quantity of phosphorus trichloride and 1.3 equivalents of phosphorous acid, sodium ibandronate was prepared in a yield of 68% (Table 17/Entry 5) [58]. The lowest yield of 59% was observed using 3.2 equivalents of phosphoryl chloride and a large excess (9.7 equivalents) of phosphorous acid (Table 17/Entry 6) [59]. The use of this large excess is completely needless, even makes the purification more difficult.

Table 17. Synthesis of ibandronic acid derivatives from MPA in the absence of any solvent.

Entry	Reagent (Equiv.)	T (°C)	t (h)	Product Form	Purity (%)	Yield (%)	Ref.
1	PCl_3:H_3PO_3 (2.9:1.5)	60–65	1.5	Na	-	89	[55]
2	PCl_3:H_3PO_3 (2.9:1.5)	70–75	20	Na·H_2O	-	82	[56]
3	PCl_3:H_3PO_3 (4.1:2.5)	75	5	Na	-	82 [1]	[57]
4	PCl_3:H_3PO_3 (5:3)	80 [2]	0.05	Na	-	72 [1]	[51]
5	PCl_3:H_3PO_3 [3]	75	6	Na·H_2O	-	68 [1]	[58]
6	$POCl_3$:H_3PO_3 (3.2:9.7)	60–70	24	acid	-	59 [1]	[59]

[1] For the crude product; [2] Under MW conditions; [3] 1.3 equivalents of phosphorous acid were used.

Risedronic acid derivatives were synthesized from PAA. Using phosphorus trichloride and phosphorous acid in a molar ratio of 5:3, the yield was 86% (Table 18/Entry 1) [51]. In one case, morpholine was also added to the reaction mixture beside the P-reagents, and the crude risedronic acid was prepared in a yield of 78% (Table 18/Entry 2) [60]. Yields around 66% were claimed using ca. 2 equivalents of both phosphorus trichloride and phosphorous acid at a temperature of 70–100 °C (Table 18/Entries 3 and 4) [52,61]. Carrying out the reaction applying 3.2 equivalents of phosphoryl chloride and a large excess (9.7 equivalents) of phosphorous acid, the yield was 60% (Table 18/Entry 5) [59].

Table 18. Synthesis of risedronic acid derivatives from PAA in the absence of solvent.

Entry	Reagent (Equiv.)	T (°C)	t (h)	Product Form	Purity (%)	Yield (%)	Ref.
1	PCl_3:H_3PO_3 (5:3)	80 [1]	0.05	Na	-	86 [2]	[51]
2	PCl_3:H_3PO_3:morpholine (2:5:1)	68–75	2.5–4	acid	-	78 [2]	[60]
3	PCl_3:H_3PO_3 (2:1.5)	100	4	acid	-	67 [2]	[52]
4	PCl_3:H_3PO_3 (2.5:2.2)	70–75	8	Na·2.5H_2O	99.9 [3]	65	[61]
5	$POCl_3$:H_3PO_3 (3.2:9.7)	60–70	24	acid	-	60 [2]	[59]

[1] Under MW conditions; [2] For the crude product; [3] On the basis of HPLC.

Reacting IAA, phosphorus trichloride and phosphorous acid in a molar ratio of 1:5:3 under MW conditions, or in a ratio of 1:3:5, zoledronate and zoledronic acid were obtained in a yield of 80% and 61%, respectively. The first yield was related on crude product (Table 19/Entries 1–2) [51,62]. Using 3 equivalents of phosphoryl chloride and 5 equivalents of phosphorous acid, the yield was 79% for the crude, but pure zoledronic acid (Table 19/Entry 3) [62]. The application of phosphoryl chloride and phosphorous acid measured in an excessive quantity (reported also for ibandronic and risedronic acid) led to a yield of 62% (Table 19/Entry 4) [59].

Table 19. Synthesis of zoledronic acid derivatives from IAA under solvent-free conditions.

Entry	Reagent (Equiv.)	T (°C)	t (h)	Product Form	Purity (%)	Yield (%)	Ref.
1	PCl_3:H_3PO_3 (5:3)	80 [1]	0.05	Na	-	80 [2]	[51]
2	PCl_3:H_3PO_3 (3:5)	75	6	acid	99.9 [3]	61	[62]
3	$POCl_3$:H_3PO_3 (3:5)	75	5	acid	99.9 [3]	79 [2]	[62]
4	$POCl_3$:H_3PO_3 (3.2:9.7)	60–70	24	acid	-	62 [2]	[59]

[1] Under MW conditions; [2] For the crude product; [3] On the basis of HPLC.

7. Methods in ILs

The ILs are considered green solvents, because of their low vapor pressure, high thermal stability, and, as they can be recycled and reused. Although their use in organic syntheses is spreading, hydroxymethylenebisphosphonic acid derivatives were prepared in ILs only in few cases (Table 20) (Scheme 13).

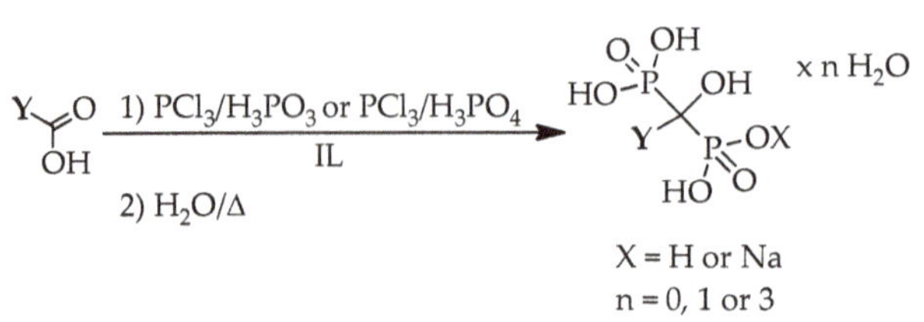

Y	BP
$-CH_2CH_2NH_2$	PD
$-CH_2(CH_2)_2NH_2$	AD
3-pyridylCH_2-	RD
1-imidazolylCH_2-	ZD

Scheme 13. General scheme for the preparation of dronic acid derivatives in ILs.

De Ferra and co-workers investigated the preparation of pamidronic and alendronic acid derivatives (in two cases, the form of the products was not provided), as well as risedronic and zoledronic acid in tributylmethylammonium chloride ($[Bu_3NH][Cl]$). Reacting the mixture of 2 equivalents of phosphorus trichloride and 1 equivalent of phosphorous acid with the corresponding carboxylic acid, low yields of the respective dronic acid derivatives were reported (12%–31%) (Table 20/Entries 2 and 4–6) [63]. The Keglevich group reached better results, when synthesizing pamidronic acid in the presence of 1-butyl-3-methylimidazolium hexafluorophosphate ($[bmim][PF_6]$). The IL was not used a solvent, only as an additive. Reacting β-alanine with 2:2 or 3:2 equivalents of phosphorus trichloride and phosphorous acid, the optimum amount of the IL was found to be around 0.3–0.6 equivalents. Under such conditions, record yields of 70%–72% were reached (Table 20/Entry 1) [53]. GABA was reacted with 2 equivalents of phosphorus trichloride and 1.5 equivalents of phosphorous acid, in the presence of 0.5 equivalents of various ILs at 60 °C for about 6 h. Depending on the ILs used, the yield of pure alendronate was 92%–94% (Table 20/Entry 3) [64]. The same research group also prepared zoledronic acid in different ILs. For crude zoledronic acid, yields of 90%–92% were reported. In the basic syntheses, IAA was reacted with 2.5 equivalents of phosphorus trichloride and 1.7 equivalents of phosphoric acid in 1.6 equivalents of ILs (Table 20/Entry 7). Carrying out the reaction with the same amount of P-reactants but with $[bmim][BF_4]$, the yield was 60% for sodium zoledronate (Table 20/Entry 8) [65].

Table 20. Synthesis of hydroxymethylenebisphosphonic acid derivatives in the presence of ILs.

Entry	BP	Ils (Equiv.)	Starting Material	Reagent (Equiv.)	T (°C)	t (h)	Product Form	Purity (%)	Yield (%)	Ref.
1	PD	[bmim][PF_6] (0.3 or 0.6)	β-alanine	PCl_3:H_3PO_3 (2:2 or 3:2)	75	3	acid	>99 [1]	70–72	[53]
2	PD	[Bu_3NH][Cl]	β-alanine	PCl_3:H_3PO_3 (2:1)	60	2	-	-	26	[63]
3	AD	various Ils [2] (0.5)	GABA	PCl_3:H_3PO_3 (2:1.5)	60	6	Na·$3H_2O$	>99 [3]	92–94	[64]
4	AD	[Bu_3NH][Cl]	GABA	PCl_3:H_3PO_3 (2:1)	60	2	-	-	31	[63]
5	RD	[Bu_3NH][Cl]	PAA	PCl_3:H_3PO_3 (2:1)	80	2	acid	-	12	[63]
6	ZD	[Bu_3NH][Cl]	IAA	PCl_3:H_3PO_3 (2:1)	80	2	acid	-	26	[63]
7	ZD	various Ils [4] (1.6)	IAA	PCl_3:H_3PO_4 (2.5:1.7)	65	4	acid	-	90–92 [5]	[65]
8	ZD	[bmim][BF_4] (1.6)	IAA	PCl_3:H_3PO_4 (2.5:1.7)	65	4	Na·H_2O	99.8 [3]	60	[65]

[1] On the basis of acid-base titration; [2] [bmim][BF_4], [bmim][PF_6], 1-hydroxyethyl-2,3-dimethyl-imidazolium chloride (LOH), 1-propyl-3-carbonitrile-imidazolium chloride (LCN), 1-carboxyethyl-3-methyl-imidazolium chloride (LOOH); [3] On the basis of HPLC; [4] *N*-ethylpyridine tetrafluoroborate [EPy][BF_4], [bmim][PF_6], LOH, LCN, LOOH; [5] For the crude product.

8. Processes for the Preparation of Hydroxymethylenebisphosphonic Acid Derivatives in Other Solvents

In a few instances, the synthesis of hydroxymethylenebisphosphonic acid derivatives was carried out in other solvents, from among *p*-cresol, acetonitrile and *n*-octane should be mentioned. In these cases again the corresponding carboxylic acids were reacted with phosphorus trichloride and phosphorous acid. In the first series let us see the examples carried out in *p*-cresol (Scheme 14).

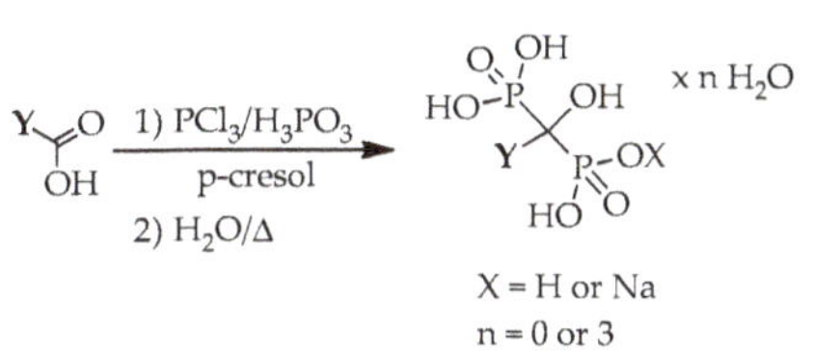

Y	BP
$-CH_2CH_2NH_2$	PD
$-CH_2(CH_2)_2NH_2$	AD
3-pyridylCH_2-	RD
1-imidazolylCH_2-	ZD

Scheme 14. General scheme for the preparation of dronic acid derivatives in *p*-cresol.

Applying β-alanine or IAA together with phosphorus trichloride and phosphorous acid in a molar ratio of 3.5:3 and using *p*-cresol as the solvent, crude pamidronic acid was obtained in a yield of 57%, while crude zoledronic acid in a yield of 80% (Table 21/Entries 1 and 5) [66]. Carrying out the syntheses from GABA, and using phosphorus trichloride and phosphorous acid in a ratio of 3.2:1.5 or 3.5:2, alendronate was obtained in a yield of ca. 40% (Table 21/Entries 2 and 3) [67,68]. When PAA was reacted with the mixture of 3.4 equivalents of phosphorus trichloride and 3 equivalents of phosphorous acid, a yield of 59% was reported for risedronic acid (Table 21/Entry 4) [68].

Table 21. Synthesis of hydroxymethylenebisphosphonic acid derivatives in *p*-cresol.

Entry	BP	Starting Material	Reagent (Equiv.)	T (°C)	t (h)	Product Form	Purity (%)	Yield (%)	Ref.
1	PD	β-alanine	PCl_3:H_3PO_3 (3.5:3)	65–70	5	acid	-	57 [1]	[66]
2	AD	GABA	PCl_3:H_3PO_3 (3.2:1.5)	75	5	Na·3H_2O	99.9 [2]	42	[67]
3	AD	GABA	PCl_3:H_3PO_3 (3.5:2)	50–55	2	Na·3H_2O	99 [2]	39	[68]
4	RD	PAA	PCl_3:H_3PO_3 (3.4:3)	65–70	4	acid	-	59 [1]	[68]
5	ZD	IAA	PCl_3:H_3PO_3 (3.5:3)	65–70	5	acid	-	80 [1]	[66]

[1] For the crude product; [2] On the basis of HPLC.

From among the basic hydroxymethylenebisphosphonates, only pamidronate, alendronate and risedronate was synthesized in acetonitrile (Scheme 15).

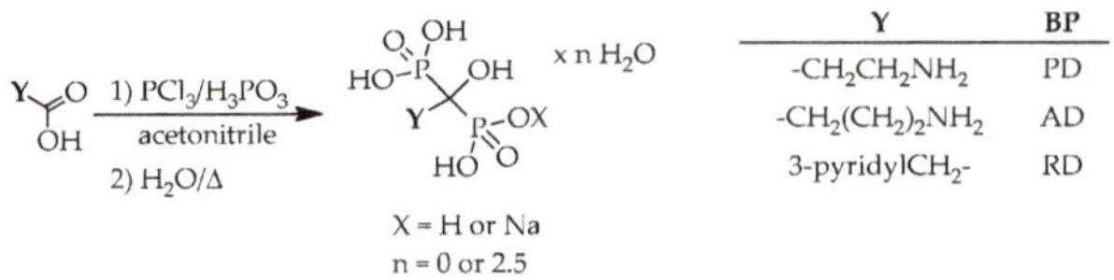

Scheme 15. General scheme for the preparation of dronic acid derivatives in acetonitrile.

In the synthesis of pamidronate, β-alanine and the usual P-reactants were used in a ratio of 1:2:1.5, and the crude product was obtained in a yield of 60% (Table 22/Entry 1) [69]. Applying both phosphorus trichloride and phosphorous acid in amount of 2 equivalents in reaction with GABA, crude alendronate was obtained in a yield of 70% (Table 22/Entry 2) [69]. Risedronic acid was synthesized in a yield of 85% from PAA using phosphorus trichloride and phosphorous acid in a ratio of 2.7:2.4 (Table 22/Entry 3) [38]. When the mixture of 2.7 equivalents of phosphorus trichloride and 2.4 equivalents of phosphorous acid, or 2 equivalents of phosphorus trichloride and 1.2 equivalent of phosphorous acid was reacted with PAA, yields of 77% and 45% were reported for sodium risedronate, respectively (Table 22/Entries 4 and 5) [38,69].

Table 22. Synthesis of hydroxymethylenebisphosphonates in acetonitrile.

Entry	BP	Starting Material	Reagent (Equiv.)	T (°C)	t (h)	Product Form	Purity (%)	Yield (%)	Ref.
1	PD	β-alanine	PCl_3:H_3PO_3 (2:1.5)	70–75	6–9	Na	-	60 [1]	[69]
2	AD	GABA	PCl_3:H_3PO_3 (2:2)	70–75	6–9	Na	-	70 [1]	[69]
3	RD	PAA	PCl_3:H_3PO_3 (2.7:2.4)	82	5	acid	98.6 [2]	85 [1]	[38]
4	RD	PAA	PCl_3:H_3PO_3 (2.7:2.4)	82	5	Na	99.3 [2]	77 [1]	[38]
5	RD	PAA	PCl_3:H_3PO_3 (2:1.2)	70–75	6–9	Na·2.5H_2O	99.7 [2]	45	[69]

[1] For the crude product; [2] On the basis of HPLC.

In a few instances, *n*-octane was also described as the solvent (Scheme 16).

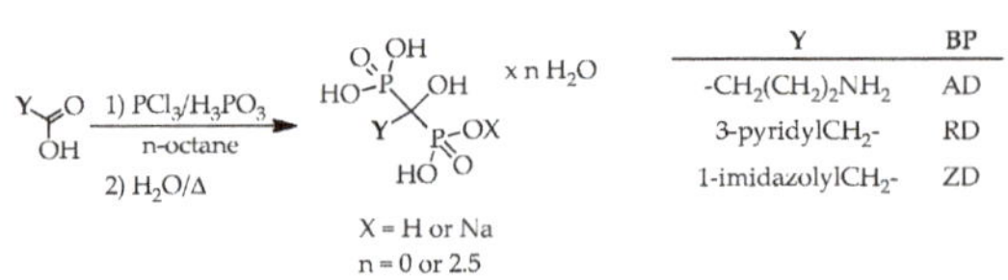

Scheme 16. General scheme for the preparation of dronic acid derivatives in *n*-octane.

Starting from GABA and PAA, and applying phosphorus trichloride and phosphorous acid in a quantity of ca. 3.5 equivalents, yields of 43% and 96% were reported for alendronic acid and risedronic acid, respectively (Table 23/Entries 1 and 2) [70]. In another combination, the use of phosphoryl chloride and phosphorous acid in a ratio of 3:5 led the risedronate in a yield of 64% (Table 23/Entry 3) [71]. Starting from IAA, and measuring in both phosphorus trichloride and phosphorous acid in a 4.6 molar equivalent quantity, zoledronic acid was obtained in a yield of 65% (Table 23/Entry 4) [70].

Table 23. Synthesis of hydroxymethylenebisphosphonic acid derivatives in *n*-octane.

Entry	BP	Starting Material	Reagent (Equiv.)	T (°C)	t (h)	Product Form	Purity (%)	Yield (%)	Ref.
1	AD	GABA	PCl_3:H_3PO_3 (3.6:3.5)	90–95	-	acid	-	43 [1]	[70]
2	RD	PAA	PCl_3:H_3PO_3 (3.5:3.5)	90–95	-	acid	-	96 [1]	[70]
3	RD	PAA	$POCl_3$:H_3PO_3 (3:5)	85–90	3–4	Na·2.5H_2O	99.9 [2]	64	[71]
4	ZD	IAA	PCl_3:H_3PO_3 (4.6:4.6)	90–95	-	acid	-	65	[70]

[1] For the crude product; [2] On the basis of HPLC.

The syntheses of the hydroxymethylenebisphosphonates discussed were studied only sporadically in other solvents. For the sake of completeness, we list the special cases in Table 24.

Table 24. Synthesis of hydroxymethylenebisphosphonic acid derivatives in other solvents.

Solvent	PD	AD	ID	RD	ZD
benzenesulfonic acid	[72]	[72]			
trifluoromethanesulfonic acid		[67]			
4-chlorotoluene			[26]		
ethyl acetate				[50]	
xylene	[73]				
cyclohexane			[26]		[42]
methylcyclohexane				[74]	
decalin				[74]	
tetralin				[74]	
anisole		[75,76]	[26]		
dimethylformamide	[77]				
dioxane		[77]		[70]	[70]
diphenyl ether		[78]		[78]	[78]
poly(ethylene glycol) methyl ether 350		[47]			
tetrahydrofuran		[79]			
phenol		[66,68]		[68]	
4-nitrophenol		[68]		[66,68]	
2,6-di-terc-butil-4-metilfenol (BHT)		[68]			
nonylphenol ethoxylate				[80]	
triethyl phosphate		[67]			
tricresyl phosphate		[67]			
diethyl carbonate			[81–83]	[27]	
propylene carbonate					[84]
in the mixture of propylene carbonate and PEG 600					[84]
PEG 400					[41]
dimethoxymethane					[41,85]
diethoxymethane			[86]		
diglyme					[85]
1,2-dichloroethane					[42]
N,N′-dimethylurea				[87]	[87]
sunflower oil			[88,89]	[80]	
olive oil				[80]	
in the presence of paraffin				[80]	
silicon oil			[90]		[41]

9. Conclusions

The methods for the preparation of the most important dronic acid derivatives starting from the corresponding substituted acetic acids and P-reagents, especially phosphorus trichloride and phosphorous acid were surveyed according to the solvents applied. Methanesulfonic acid is a special solvent, as it may enter into reaction also with phosphorus trichloride. The formation of Cl_2P-OSO_2Me may mean some kind of activation, and hence there is no need to use phosphorous acid. However, in other cases (typically with sulfolane as the solvent), phosphorus trichloride should be applied together with phosphorous acid to provide more reactive species formed by condensation. The optimum molar ratios of the P-reactants were pointed out. In general, it was a problem during the evaluation of the literature cases that not much stress was placed on purity criterions.

Acknowledgments: This project was funded by Gedeon Richter Plc., and the Hungarian Research Development and Innovation Fund (K119202).

Conflicts of Interest: The authors declare there is no conflict of interest.

References

1. Russell, R.G.G. Bisphosphonates: The first 40 years. *Bone* **2011**, *49*, 2–19.
2. Russell, R.G.G. Bisphosphonates: Mode of action and pharmacology. *Pediatrics* **2007**, *119*, 150–162. [CrossRef]
3. Massey, A.S.; Pentlavalli, S.; Cunningham, R.; McCrudden, C.M.; McErlean, E.M.; Redpath, P.; Ali, A.A.; Annett, S.; McBride, J.W.; McCaffrey, J.; et al. Potentiating the anti-cancer properties of bisphosphonates by nanocoplexation with the cationic amphipathic peptide, RALA. *Mol. Pharm.* **2016**, *13*, 1217–1228. [PubMed]
4. Hudson, H.R.; Wardle, N.J.; Blight, S.W.A.; Greiner, I.; Grün, A.; Keglevich, G. N-heterocyclic dronic acids; applications and synthesis. *Mini Rev. Med. Chem.* **2012**, *12*, 313–325. [CrossRef]
5. Bortolini, O.; Fantin, G.; Fogagnolo, M.; Rossetti, S.; Maiuolo, L.; di Pompo, G.; Avnet, S.; Granchi, D. Synthesis, characterization and biological activity of hydroxyl-bisphosphonic analogs of bile acids. *Eur. J. Med. Chem.* **2012**, *52*, 221–229.
6. Rogers, M.J.; Gordon, S.; Benford, H.L.; Coxon, F.P.; Luckman, S.P.; Monkkonen, J.; Frith, J.C. Cellular and molecular mechanisms of action of bisphosphonates. *Cancer* **2000**, *88*, 2961–2978. [PubMed]
7. Kumar, A.; Dike, S.; Nijasure, A.; Bhaware, V. Process for Manufacture of 4-Amino-hydroxybutylidene-1,1-bisphosphonic Acid and Its Salts. US2007/149486, 28 June 2007.
8. Szulc, M.; Slisewski, T.; Dembkowski, L.; Jastrzebska, B.; Rachon, J.; Makowiec, S. A Process for the Synthesis of 1-Hydroxy-3-(n-methylpentylamino) Propylidene Bisphosphonic Acid Monosodium Salt, Monohydrate. WO2011/16738A8, 24 November 2011.
9. Mihaljevic, K.; Bajic, B.; Cavuzic, D.; Oreskovic, R. Preparation of Risedronic Acid and Its Salts for Uses in the Treatment of Bone and Calcium Diseases. WO2006/129056, 7 December 2006.
10. Dembkowski, L.; Rynkiewicz, R.; Rachon, J.; Makowiec, S.; Przychodzen, W.; Witt, D. Process for the Preparation of [1-Hydroxy-2-(3-pyridinyl)ethyliden] Bisphosphonic Acid and Hemi-Pentahydrate Monosodium Salt Thereof. WO2006/71128 A1, 6 July 2006.
11. Dembkowski, L.; Krzyzanowski, M.; Rynkiewicz, R.; Szramka, R.; Roznerski, Z.; Źyla, D.; Rachon, J.; Makowiec, S. Process for the Preparation of [1-Hydroxy-2-(1*H*-imidazol-1-yl)-ethylidene]bisphosphonic Acid. WO2010/50830, 6 May 2010.
12. Keglevich, G.; Grün, A.; Kovács, R.; Koós, K.; Szolnoki, B.; Garadnay, S.; Neu, J.; Drahos, L.; Greiner, I. Heteroacetyl Chlorides and mixed anhydrides as intermediates in the synthesis of heterocyclic dronic acids. *Lett. Drug. Des. Discov.* **2012**, *9*, 345–351.
13. Zeevaart, J.R.; Mbianda, X.Y.; Szűcs, Z. Preparation of Carbon Nanotube Bonded Bisphosphonates Useful as Anticancer Agent. WO2013/005154, 10 January 2013.
14. Alanne, A.L.; Hyvönen, H.; Lahtinen, M.; Ylisirniö, M.; Turhanen, P.; Kolehmainen, E.; Peräniemi, S.; Vepsäläinen, J. Systematic study of the physicochemical properties of a homologous series of aminobisphosphonates. *Molecules* **2012**, *17*, 10928–10945. [CrossRef] [PubMed]
15. Kieczykowski, G.R.; Jobson, R.B.; Melillo, D.G.; Reinhold, D.F.; Grenda, V.J.; Shinkai, I. Preparation of (4-amino-1-hydroxybuty1idene)bisphosphonic acid sodium salt, MK-217 (alendronate sodium). An improved procedure for the preparation of 1-hydroxy-1,l-bisphosphonic acids. *J. Org. Chem.* **1995**, *60*, 8310–8312.
16. Kovács, R.; Grün, A.; Németh, O.; Garadnay, S.; Greiner, I.; Keglevich, G.Y. The synthesis of pamidronic derivatives in different solvents: An optimization and a mechanistic study. *Heteroat. Chem.* **2014**, *25*, 186–193. [CrossRef]

17. Neu, J.; Fischer, J.; Fodor, T.; Törley, J.; Gizur, T.; Lévai, S.; Demeter, Á.; Perényi, É. Industrial Process for the Synthesis of 2-Substituated 1-(Hydroxy-ethylidene)-1,1-bisphosphonic Acids of High Purity and the Salts Thereof. WO2004/067541, 12 August 2004.
18. Benyettou, F.; Rezgui, R.; Ravaux, F.; Jaber, T.; Blumer, K.; Jouiad, M.; Motte, L.; Olsen, J.-C.; Platas-Iglesias, C.; Magzouba, M.; et al. Synthesis of silver nanoparticles for the dual delivery of doxorubicin and alendronate to cancer cells. *J. Mater. Chem. B* **2015**, *3*, 7237–7245. [CrossRef]
19. Migianu-Griffoni, E.; Chebbi, I.; Kachbi, S.; Monteil, M.; Sainte-Catherine, O.; Chaubet, F.; Oudar, O.; Lecouvey, M. Synthesis and Biological Evaluation of New Bisphosphonate–Dextran Conjugates Targeting Breast Primary Tumor. *Bioconjugate Chem.* **2014**, *25*, 224–230. [CrossRef] [PubMed]
20. Gref, R.; Agostoni, V.; Daoud-Mahammed, S.; Rodriguez-Ruiz, V.; Malanga, M.; Jicsinszky, L.; Horcajada-Cortes, P.; Serre, C. Organic-Inorganic Hybrid Solid Having a Modified Outer Surface. WO2013/178954, 5 December 2013.
21. Kovács, R.; Grün, A.; Garadnay, S.; Greiner, I.; Keglevich, G. Rational synthesis of ibandronate and alendronate. *Curr. Org. Synth.* **2013**, *10*, 640–644.
22. Grün, A.; Kovács, R.; Garadnay, S.; Greiner, I.; Keglevich, G. The synthesis of risedronic acid and alendronate applying phosphorus oxychloride and phosphorous acid in methanesulfonic acid. *Lett. Drug. Des. Discov.* **2015**, *12*, 253–258. [CrossRef]
23. Kieczykowski, G.R.; Melillo, D.G.; Jobson, R.B. Process for Preparing 4-Amino-1-hydroxybutylidene-1,1-bisphosphonic Acid or Salts Thereof. US4922007, 1 May 1990.
24. Guangyu, X.; Yuyuan, X.; Xihan, W. A facile and direct synthesis of alendronate from pyrrolidone. *Org. Prep. Proc. Int.* **2004**, *36*, 185–187.
25. Koftis, T.V.; Menisiou, A.; Soni, R.R. Process for the Preparation of 3-(n-Methyl-n-pentyl)amino-1-hydroxypropane-1,1-diphosphonic Acid Salt or Derivatives Thereof. WO2012/007021, 19 January 2012.
26. Bolugoddu, V.; Dahyabhai, J.L.; Kammili, R.V.; More, Y.P.; Thakur, P. Crystalline form a of Ibandronic Acid and proCess for the Preparation. WO2008/14510, 31 January 2008.
27. Bhimavarapu, S.R.; Bolugoddu, V.B.; Donthula, A.; Elati, C.S.R.R.; Sait, S.; Tondepu, N.; Vakamudi, V.P.S.N. Preparation of Risedronate Sodium Hemi-Pentahydrate. WO2009/003001, 31 December 2008.
28. Garadnay, S.; Grün, A.; Keglevich, G.; Neu, J. Novel Process for the Preparation of Dronic Acids. WO2012/107787, 16 August 2012.
29. Keglevich, G.; Grün, A.; Aradi, K.; Garadnay, S.; Greiner, I. Optimized synthesis of N-heterocyclic dronic acids; closing a black-box era. *Tetrahedron Lett.* **2011**, *52*, 2744–2746.
30. Krueger, F.; Bauer, L.; Michel, W. 1-Hydroxy-3-aminopropane-1,1-diphosphonic Acid Preparation from Beta-Alanine and Phosphorus Trihalide, Used as Complexant. DE2130794, 11 January 1973.
31. Kabachnik, M.I.; Medved, T.Y.; Dyaglova, N.M.; Polikarpov, Y.M.; Shcherbakov, B.K.; Bel'skii, F.I. Synthesis and acid-base and complexing properties of amino-substituted α-hydroxyalkylidenediphosphonic acids. *Bull. Acad. Sci. USSR Div. Chem. Sci.* **1978**, *27*, 374–377.
32. Blum, H.; Worms, K.-H. 1-Hydroxy-3-amino-alkane-1,1-diphosphonic Acids and Salts. US4054598, 18 October 1977.
33. Padioukova, N.; Mikhailov, S.; Dixon, H.B.F.; Tzeitline, G. Bisphosphonate Conjugates and Methods of Making and Using the Same. US6750340, 15 June 2004.
34. Widler, L.; Jaeggi, K.A.; Glatt, M.; Müller, K.; Bachmann, R.; Bisping, M.; Born, A.-R.; Cortesi, R.; Guiglia, G.; Jeker, H.; et al. Highly Potent Geminal Bisphosphonates. From Pamidronate Disodium (Aredia) to Zoledronic Acid (Zometa). *J. Med. Chem.* **2002**, *45*, 3721–3738.
35. Lerstrup, K.; Preikschat, H.F.; Fischer, E. Reagent and Use Thereof for the Production of Bisphosphonates. WO2008/58722, 21 May 2008.
36. Pulla, R.D.; Usharani, V.; Venkaiah, C.N. Improved Process for the Preparation of Ibandronate Sodium. WO2007/13097, 1 February 2007.
37. Pandey, S.C.; Ram, K.; Singh, M.K.; Thaper, R.K.; Dubey, S.K. Process for Preparing a Pure Polymorphic form of 3-Pyridyl-1-hydroxyethylidine-1,1-bisphosphonic acid Sodium Salt. WO2006/51553, 18 May 2006.
38. Saxena, R.; Jain, A.K.; Srinivasan, C.V.; Wadhwa, L. Process for the Preparation of Pure Risedronic Acid or Salts. WO2007/132478, 22 November 2007.
39. Lee, A.G. Pharmaceutical Compositions and Methods Comprising Combinations of 2-Alkylidene-19-nor-Vitamin D Derivatives and a Bisphosphonate. US2005/65117, 24 March 2005.

40. Zhoua, L.-S.; Yanga, K.W.; Fenga, L.; Xiaoa, J.-M.; Liua, C.-C.; Zhanga, Y.-L.; Crowder, M.W. Novel fluorescent risedronates: Synthesis, photodynamic inactivation and imaging of *Bacillus subtilis*. *Bioorg. Med. Chem. Lett.* **2013**, *23*, 949–954.
41. Aronhime, J.; Lifshitz-Liron, R. Zoledronic acid Crystal Forms, Zoledronate Sodium salt Crystal Forms, Amorphous Zoledronate Sodium Salt, and Processes for Their Preparation. CA2530193, 20 January 2005.
42. Pulla, R.M.; Usha, R.V.; Venkaiah, C.N. An Improved Process for the Preparation of Zoledronic Acid. WO2005/63717, 14 July 2005.
43. Jaeggi, K.A.; Widler, L. Substituted Alkanediphosphonic Acids and Pharmaceutical Use. US4939130, 3 July 1990.
44. Mustafa, D.A.; Kashemirov, B.A.; McKenna, C.E. Microwave-assisted synthesis of nitrogen-containing 1-hydroxymethylenebisphosphonate drugs. *Tetrahedron Lett.* **2011**, *52*, 2285–2287. [CrossRef]
45. Patel, V.M.; Chitturi, T.R.; Thennati, R. A Process for Preparation of Bisphosphonic Acid Compounds. WO2005/44831, 19 May 2005.
46. Singh, G.P.; Jadhav, H.S.; Maddireddy, N.V.; Srivastava, D. Process for the Production of 4-Amino-1-hydroxybutylidene-1,1-bisphosphonic Acid or Salts Thereof. WO2007/10556, 25 January 2007.
47. Kubela, R.; Tao, Y. Process for the Production of 4-Amino-1-hydroxybutylidene-1,1-bisphosphonic Acid or Salts Thereof. US5908959, 1 June 1999.
48. Lladó, J.B.; Lista, E.P.; Miguel, M.C.O. Process for Producing 4-Amino-1-hydroxybutylidene-1,1-bisphosphonic acid and its Trihydrated Monosodium Salt. US6573401, 3 June 2003.
49. Rao, D.R.; Kankan, R.N.; Ghagare, M.G. Process for the Synthesis of Ibandronate Sodium. CA2652420, 27 March 2008.
50. Serrano, J.P.; Illado, J.B. Process of Making Geminal Bisphosphonic Acids and Pharmaceutically Acceptable Salts and/or Hydrates Thereof. US2008/194525, 14 August 2008.
51. Lenin, R.; Raju, R.M.; Rao, D.V.N.S.; Ray, U.K. Microwave-assisted efficient synthesis of bisphosphonate libraries: A useful procedure for the preparation of bisphosphonates containing nitrogen and sulfur. *Med. Chem. Res.* **2013**, *22*, 1624–1629.
52. Qua, Z.; Chena, X.; Qub, C.; Qua, L.; Yuanc, J.; Weia, D.; Lia, H.; Huanga, X.; Jianga, Y.; Zhao, Y. Fragmentation pathways of eight nitrogen-containing bisphosphonates (BPs) investigated by ESI-MSn in negative ion mode. *Int. J. Mass Spectrom.* **2010**, *295*, 85–93.
53. Grün, A.; Nagy, D.I.; Garadnay, S.; Greiner, I.; Keglevich, G. Efficient synthesis of pamidronic acid using an ionic liquid additive. *Lett. Drug. Des. Discov.* **2016**, *13*, 475–478.
54. Lidor-Hadas, R.; Lifshitz, R. Process for Preparing Alendronic Acid. US6201148, 13 March 2001.
55. Parthasaradhi, R.B.; Rathnakar, R.K.; Muralidhara, R.D.; Subash, C.D.K.; Vamsi, K.B. Ibandronate Sodium Solid Dispersion. WO2013/179305, 5 December 2013.
56. Balasubramaniam, R.; Polsani, P.R.; Tammireddy, G.N. A New and Improved Process for the Preparation of Ibandronate Sodium Monohydrate. WO2009/93258, 20 January 2011.
57. Reddy, V.V.R.M.K.; Chintamani, U.S.; Udaykiran, D.; Madhusudhan, G.; Mukkanti, K. A facile preparation of *N*-methylpentan-1-amine: A key intermediate for ibandronate sodium. *Indian J. Chem. Sect. B* **2010**, *49*, 1257–1260.
58. Pulla, R.M.; Usharani, V.; Venkaiah, C.N. Novel Polymorphic Forms of Ibandronate. WO2007/74475, 5 July 2007.
59. Grassi, S.; Volante, A. A Process for the Preparation of Alkyl- and Aryl-Diphosphonic Acids and Salts Thereof. WO2005/63779, 29 September 2005.
60. Cazer, F.; Cramer, W.; Billings, D.; Parry, G. Process for Making Geminal Bisphosphonates. US2001/041690, 15 November 2001.
61. Balasubramaniam, R.; Polsani, P.R.; Tammireddy, G.N. Improved Process for the Preparation of Risedronate Sodium Hemipentahydrate. WO2009/34580, 19 March 2009.
62. Yadav, R.P.; Shaikh, Z.G.; Mukarram, S.M.J.; Kumar, Y. Processes for the Preparation of Pure Zoledronic Acid. WO2007/69049, 21 June 2007.
63. De Ferra, L.; Turchetta, S.; Massardo, P.; Casellato, P. Preparation of Bisphosphonic Acids and Salts Thereof. WO2003/093282, 7 September 2007.
64. Hao, E.; Jiang, X.; Liu, Y.; Zhang, Q.; Wang, D.; Xie, M.; Wang, H.; Guo, H.; Li, G. A Method for Preparing Alendronate Sodium. CN104558028, 29 April 2015.
65. Hao, E.; Jiang, X.; Liu, Y.; Zhang, Q.; Wang, D.; Xie, M.; Wang, H.; Guo, H.; Li, G. Method for Preparing Sodium Zoledronate. CN104610357, 13 May 2015.

66. Divvela, S.R.V.N.; Dandalaa, R.; Narayanana, G.K.A.S.S.; Lenina, R.; Sivakumarana, M.; Naidub, A. Novel Procedure for the Synthesis of 1-Hydroxy-1,1-bisphosphonic Acids using Phenols as Medium. *Synth. Commun.* **2007**, *37*, 4359–4365.
67. Mandava, V.N.B.; Setty, R.K.S.; Manne, N. Process for Preparing Bisphosphonic Acids. US2007/142636, 21 June 2007.
68. Danda, S.R.; Garimella, N.K.A.S.S.; Divvela, S.R.V.N.; Dandala, R.; Meenakshisunderam, S. Process for the Preparation of Biphosphonic Acids. US2007/173645, 26 July 2007.
69. Vinayak, G.G.; Vinay, K.S.; Manoj, M.G.; Rekha, M.A. Novel Process for the Preparation of Bisphosphonic Acids. WO2008/04000, 10 January 2008.
70. Pandey, S.C.; Haider, H.; Saxena, S.; Singh, M.K.; Thaper, R.K.; Dubey, S.K. Process for Producing Biphosphonic Acids and Forms Thereof. US2009/312551, 17 December 2009.
71. Kumar, N.P.; Singare, D.; Pradhan, N.S.; Valgeirsson, J. Process for the Preparation of Risedronate Sodium. US2010/317859, 16 December 2007.
72. Yanvarev, D.V.; Korovina, A.N.; Usanov, N.N.; Kochetkov, S.N. Non-hydrolysable analogues of inorganic pyrophosphate as inhibitors of hepatitis C virus RNA-dependent RNA-polymerase. *Russ. J. Bioorg. Chem.* **2012**, *38*, 224–229.
73. Ham, W.H.; Jung, Y.H.; Oh, C.Y.; Lee, K.Y.; Kim, Y.H. Preparation of 3-Amino-1-hydroxypropane-1,1-diphosphonic Acid. US5792885, 11 August 1998.
74. Telang, R.B.; Kumar, R.; Tavhare, A.M. Novel Process for Preparing Risedronic Acid. WO2009/50731, 23 April 2009.
75. Senthilkumar, U.; Arulmoli, T.; Lakshmipathi, V.; Rao, S. Process for the Preparation of Bisphosphonic Acid. US2007/066569, 22 March 2007.
76. Senthilkumar, U.; Arulmoli, T.; Lakshmipathi, V.; Rao, S. An Improved Process for the Preparation of Bisphosphonic Acid. WO2008/35131, 27 March 2008.
77. Gibson, A.M.; Mendizabal, M.; Pither, R.; Pullan, S.E.; Griffiths, V.; Duncanson, P. Radiolabelled Bisphosphonates and Method. US6534488, 18 March 2003.
78. Deshpande, P.B.; Luthra, P.K. Process for the Preparation of Biphosphonic Derivatives. US7439385, 21 October 2006.
79. Egorov, M.; Aoun, S.; Padrines, M.; Redini, F.; Heymann, D.; Lebreton, J.; Mathé-Allainmat, M. A One-Pot Synthesis of 1-Hydroxy-1,1-bis(phosphonic acid)s Starting from the Corresponding Carboxylic Acids. *Eur. J. Org. Chem.* **2011**, *35*, 7148–7154.
80. Adiyaman, M.; Keskin, H.; Karliga, B. Process for the Preparation of 3-pyridyl-1-hydroxyethylidene-1,1-biphosphonic Acid and Hydrated Forms Thereof. WO2007/68678, 21 June 2007.
81. Eiermann, U.; Junghans, B.; Knipp, B.; Sattelkau, T. Ibandronate Polymorph. US2006/172976, 3 August 2006.
82. Baetz, F.; Junghans, B. Method for Synthesizing Bisphosphonate. US2006/094898, 4 May 2006.
83. Junghans, B. Process for Preparing Ibandronate. US2008/255386, 16 October 2008.
84. Kas, M.; Benes, M.; Pis, J. Process for Making Zoledronic Acid. US2010/130746, 27 May 2010.
85. Samsel, E.G.; Wu, T.-C. Process for Manufacturing Bisphosphonic Acids. WO2007/109542, 1 November 2007.
86. Bartl, J. Process for Making 1-Hydroxyalkylidene-1,1-biphosphonic Acids. US2010/010258, 14 January 2010.
87. Baptista, J.; Mendes, Z. Process for the Preparation of Biphosphonic Acids and Salts Thereof. WO2008/56129, 15 May 2008.
88. Aslan, T.; Kocak, E. Processes for the Preparation of Sodium Ibandronate Monohydrate Polymorphs a, b and Mixture of Polymorphs a with b. WO2013/109198, 25 July 2013.
89. Havlicek, J.; Jirman, J.; Karliga, B.; Richter, J. Preparation of Ibandronate Trisodium. EP2180003, 28 April 2010.
90. Avhar-Maydan, S.; Gilboa, E.; Koltai, T.; Lifshitz-Liron, R. Crystalline Forms of Ibandronic Acid and Processes for Preparation Thereof. WO2007/127249, 8 November 2007.

molecules

MDPI

Article

Effect of Mono- and Poly-CH/P Exchange(s) on the Aromaticity of the Tropylium Ion

Ankita Puri and Raakhi Gupta *

Department of Chemistry, The IIS University, Jaipur 302020, India; ankita.puri@iisuniv.ac.in
* Correspondence: raakhi.gupta@iisuniv.ac.in; Tel.: +91-141-240-0160

Academic Editor: György Keglevich
Received: 27 June 2016; Accepted: 11 August 2016; Published: 20 August 2016

Abstract: In view of the fact that the phosphorus atom in its low co-ordination state (coordination numbers 1 and 2) has been termed as the carbon copy, there have been attempts to investigate, theoretically as well as experimentally, the effect of the exchange(s) of CH- moiety with phosphorus atom(s) (CH/P) on the structural and other aspects of the classical carbocyclic and heterocyclic systems. Tropylium ion is a well-known non-benzenoid aromatic system and has been studied extensively for its aromatic character. We have now investigated the effect of mono- and poly-CH/P exchange(s) on the aromaticity of the tropylium ion. For this purpose, the parameters based on the geometry and magnetic properties, namely bond equalization, aromatic stabilization energies (ASE), Nucleus-Independent Chemical Shift (NICS) values, (NICS(0), NICS(1), NICS(1)$_{zz}$), proton nucleus magnetic resonance (^{1}H-NMR) chemical shifts, magnetic susceptibility exaltation and magnetic anisotropic values of mono-, di-, tri- and tetra-phosphatropylium ions have been determined at the Density Functional Theory (DFT) (B3LYP/6-31+G(d)) level. Geometry optimization reveals bond length equalization. ASEs range from −46.3 kcal/mol to −6.2 kcal/mol in mono- and diphospha-analogues which are planar. However, the ions having three and four phosphorus atoms lose planarity and their ASE values approach the values typical for non-aromatic structures. Of the three NICS values, the NICS(1)$_{zz}$ is consistently negative showing aromatic character of all the systems studied. It is also supported by the magnetic susceptibility exaltations and magnetic anisotropic values. Furthermore, ^{1}H-NMR chemical shifts also fall in the aromatic region. The conclusion that mono-, di-, tri- and tetra-phosphatropylium ions are aromatic in nature has been further corroborated by determining the energy gap between the Highest Occupied Molecular Orbital (HOMO) and Lowest Unoccupied Molecular Orbital (LUMO) (HOMO − LUMO gap), which falls in the range, ca. 3×10^{-19}–9×10^{-19} J. The systems having more than four phosphorus atoms are not able to sustain their monocyclic structure.

Keywords: phosphatropylium ions; aromaticity; NICS; magnetic susceptibility exaltation; HOMO − LUMO gap

1. Introduction

The concept of aromaticity is of fundamental importance for explaining the structure, stability and reactivity of many molecules, due to which it continues to attract attention of many eminent theoretical chemists and forms the basis of many interesting monographs [1–3], research papers [4–9] and thematic issues [10–12]. Hückel's predictions based on the molecular orbital theory that conjugated monocyclic planar systems containing (4n + 2) π electrons should be aromatic, i.e., conjugatively stabilized [13], was modified by Platt [14], who broadened the scope of the postulation by including neutral as well as charged polycyclic systems in this category. The term "antiaromatic" was used in substitution of "pseudoaromatic" after 1965 by Breslow for those compounds that exhibited significant destabilization characteristics and possessed (4n) π conjugated electrons [15]. The possibility of verifying theoretical predictions with experimental isolations is limited to aromatic compounds only. Post-Hückel, there have been many attempts to propose newer and more reliable indicators to determine aromaticity/antiaromaticity of a particular system or a class of compounds. Schleyer et al. [16] endeavoured to divide various theoretical parameters into four categories, each based on a particular property:

Structure—tendency towards bond length equalization and planarity (if applicable).
Energy—enhanced stability.
Reactivity—lowered reactivity, electrophilic aromatic substitution (if applicable).
Magnetic properties—proton nucleus magnetic resonance (^{1}H-NMR) chemical shift, magnetic susceptibility exaltation and anisotropies, nucleus-independent chemical shift (NICS), ring current plots.

In a recent review [17], geometry-based various aromaticity indices have been discussed elegantly. Bond length equalization has been perceived as one of the important manifestations of aromaticity, but it cannot be used as the sole criterion to characterize it unambiguously, as there are examples like borazine, which has equalized bond length but is not aromatic. The quantitative indices have been developed by Francois and Julg [18] and Bird [19], but extension of these indicators to the heteroaromatic compounds is not straightforward.

The aromatic stabilization energies (ASEs) and the enhanced resonance energies (REs) have been recognized to be the keystone of aromaticity since long ago. However, these energy estimates vary significantly, depending strongly on the equations used and on the choice of the reference molecules. As such, the other effects like hyperconjugative, etc. are also not taken into consideration or are neglected while calculating the ASEs and REs, and, therefore, in some specific cases, large discrepancy is noted. Schleyer et al. [20] have reported improved results by calculating isomerization stabilization energies (ISE) instead of ASEs.

The most reliable and commonly used indicators for validating aromaticity of different systems including the transition states are the magnetic criteria. The exalted diamagnetic susceptibility has been used to quantify aromaticity ever since it was proposed by Pascal in 1910 [21]. ^{1}H-NMR chemical shifts are also used frequently for characterizing aromatic compounds [22]. The effect of the diatropic ring current induced in an aromatic molecule by an external magnetic field is much stronger inside the ring in comparison to outside. Schleyer et al. in 1996 proposed another indicator of aromaticity, namely NICS based on the magnetic shielding at the center of the aromatic ring caused by the induced diatropic current [23]. Subsequently, it was refined and magnetic shielding 1Å above the center of the aromatic ring (NICS(1)), the zz component of NICS(1) ($NICS(1)_{zz}$), and dissected NICS were considered to be better indicators [16].

An impressive analogy between P=C and C=C double bonds in the molecules having several PC double bonds was emphasized for the first time by Appel [24] while describing Cope-rearrangement of several diphospha- and tetraphospa-hexadienes. Thereafter, similar conjugative ability of the P=C and C=C double bonds was reiterated by analyzing π-ionization energies and ring-fragmentation/isodesmic energies of a series of heterocyclic conjugated systems containing σ^2,λ^3-P atoms [25–28]. The analogy turned out to be so prolific that besides a few research papers [29–33] on this theme, a monograph with the title, "Phosphorus: The Carbon Copy" [34] has been published. The effect of the exchange of the CH moiety with phosphorus atom (CH/P exchange) on the aromaticity of the carbocyclic rings has been discussed in two recent reviews [35,36], which include detailed experimental and theoretical discussion of the three- to six-membered rings having one or more phosphorus atoms. It was concluded that due to the small angle at the P atom (ca. 95 degrees), aromaticity is sustained on CH/P exchange in the small rings, but in large rings, it results in reduced aromatic character and planarity is disturbed.

On the basis of the recent studies carried out by our research group [37,38], it was demonstrated that mono- and poly-phospha analogues of carbocyclic cations and anions, namely, cyclopropenium, cyclopropenide, cyclopentadienium and cyclopentadienide, display aromatic or antiaromatic character comparable with that of the respective carbocyclic ion. However, the results are not uniform in the case of antiaromatic systems.

The aromatic character of the tropylium cation and its monophospha- and other hetero-analogues has been probed earlier at the hybrid functional level using various structural and magnetic descriptors [39–41]. The tropylium cation was found to have a planar structure with completely delocalized 6π-electrons. The CH/P exchange in the tropylium cation was accompanied by lowering of the aromaticity only marginally.

As no attempt has been made so far to study the effect of more than one CH/P exchanges on the aromaticity of the tropylium cation, the present study concerns the investigation of the effect of mono- and poly-CH/P exchange(s) and to compare the results obtained on the basis of geometry, energetic and magnetic criteria. For this purpose, we computed the optimized geometries and calculated ASEs, ^{1}H-NMR chemical shifts, NICS(0), NICS(1), NICS(1)$_{zz}$, magnetic susceptibility exaltation and magnetic susceptibility anisotropy values and the energy difference of the highest occupied molecular orbital and lowest unoccupied molecular orbital (HOMO − LUMO energy gap) for the tropylium ion and its mono-, di-, tri- and tetra-phospha analogues at the Density Functional Theory (DFT) (B3LYP/6-31+G(d)) level. As discussed later, phosphatropylium ions having five or more phosphorus atoms do not sustain their monocyclic structures. These results are presented here. The corresponding anions will be discussed in a separate communication.

2. Results and Discussion

The following model cations (Scheme 1) were computed at the B3LYP/6-31+G(d) level to study the effect of the CH/P exchange on the aromaticity of the tropylium ion.

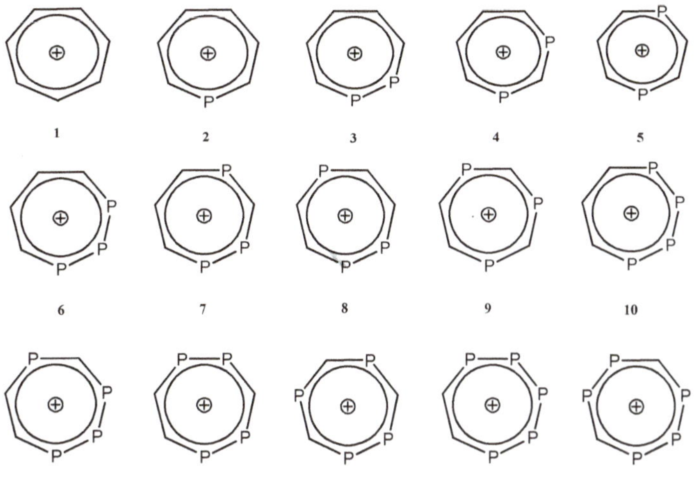

1 2 3 4 5 6 7 8 9 10 11 12 13 14 15

Scheme 1. Tropylium ion and its mono- and poly-phospha analogues.

2.1. Optimized Geometries

The Julg concept [18] is based on bond length equalization in the aromatic systems. However, it could not be extended to the heteroaromatic systems. To overcome this difficulty, the Bird index [19] was developed, but that too involved cumbersome calculations involving several aromaticity indices as constants. In the present study, our discussion is limited to the equalization or localization of the bonds without resorting to the calculation of the Julg or Bird indices.

The geometries of different systems optimized at the B3LYP/6-31+G(d) level along with bond lengths and Wiberg bond indices are reproduced in Figure 1.

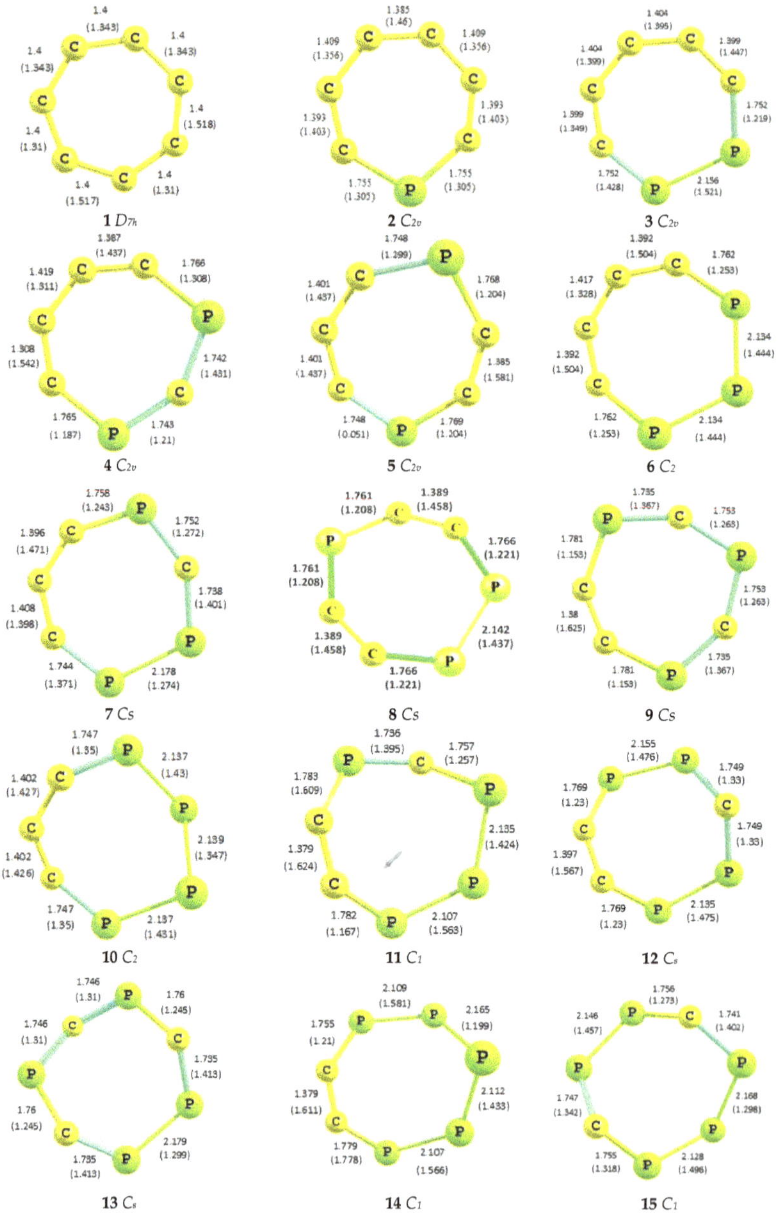

Figure 1. Optimized geometries of tropylium ion and its mono- and poly-phosha analogues at the B3LYP/6-31+G(d) level with bond distances (in Å) and Wiberg bond indices (in parenthesis). For better visualization, H atoms have been omitted.

The compounds **1–15** exhibit significant structural delocalization with CC, CP and PP bond lengths approaching the average of the respective double and single bond lengths of the prototype systems (Table 1).

Table 1. Average bond lengths (in Å) at the B3LYP/6-31+G(d) level.

Species	C-C (1.53 Å)	C=C (1.33 Å)	P-C (1.87 Å)	P=C (1.67 Å)	P-P (2.72 Å)	P=P (2.04 Å)
1	1.40	1.40	-	-	-	-
2	1.41	1.39	1.75	-	-	-
3	1.40	1.40	1.75	1.75	2.15	-
4	1.41	1.35	1.75	1.75	-	-
5	1.40	1.40	1.75	1.75	-	-
6	1.40	1.41	-	1.76	2.13	-
7	1.40	1.40	1.75	1.75	2.17	-
8	-	1.38	1.76	1.74	-	-
9	-	1.40	1.76	-	2.13	2.14
10	1.40	1.39	1.75	1.75	-	2.14
11	-	1.38	1.76	1.74	2.14	2.11
12	-	1.40	1.75	1.75	2.14	2.15
13	-	-	1.75	1.74	2.17	-
14	-	1.38	1.76	-	2.14	2.11
15	-	-	1.75	1.75	2.15	2.13

It is further noted that the planarity of the ring is not disturbed with one and two CH/P exchange(s) in the tropylium ion. Thus, systems **2–5** are planar with C_{2v} symmetry. It is in accordance with the earlier results, according to which incorporation of two σ^2, λ^3-P atoms in the five-membered ring reduces the ring strain and makes it planar even if a σ^3-P atom is also present in the α-position [42]. Likewise, pentaphosphole having a σ^3-P atom was found to be planar [43,44]. Experimentally also, the first aromatic 1*H*-1,2,4-triphosphole having a fully planar structure could be prepared [45]. Three or more CH/P exchanges in the tropylium cation, leading to **6, 10, 11, 12, 13, 14** and **15** are accompanied by loss of planarity, with the exception of 1,2,4- (**7**) and 1,3,5-triphosphacycloheptatrienyl (**9**) cations, though this deviation is not much, as evident from the dihedral angles D_h C_a 30°.

As mentioned earlier, on making five or more CH/P exchanges in the tropylium ion, monocyclic structure is no more stable and it changes into bicyclic (**7P** atoms, **18**) or polycyclic (**5P** atoms, **16**; **6P** atoms, **17**) structures as shown in Figure 2.

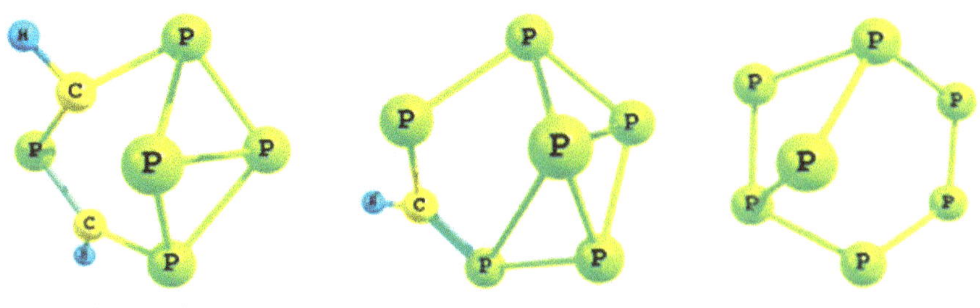

16 **17** **18**

Figure 2. 5P (**16**), 6P (**17**) and 7P (**18**) analogues of tropylium ion.

2.2. Frontier Molecular Orbitals

The tropylium cation (**1**) has doubly degenerate HOMOs, whereas, in the case of mono- and poly-phosphatropylium cations, the *p* orbital on the phosphorus atom(s) containing the lone pair constitutes the HOMO. As discussed later, the HOMO − LUMO gap in a molecule can be used as a measure of its aromatic character. In view of this, the frontier molecular orbitals (FMOs) of the tropylium cation (**1**) and two representative phosphatropylium cations (**2** and **3**) along with their energies are shown in Figure 3. It may be noted that, as compared to the tropylium cation, the HOMO − LUMO gap in the phosphatropylium cations is smaller.

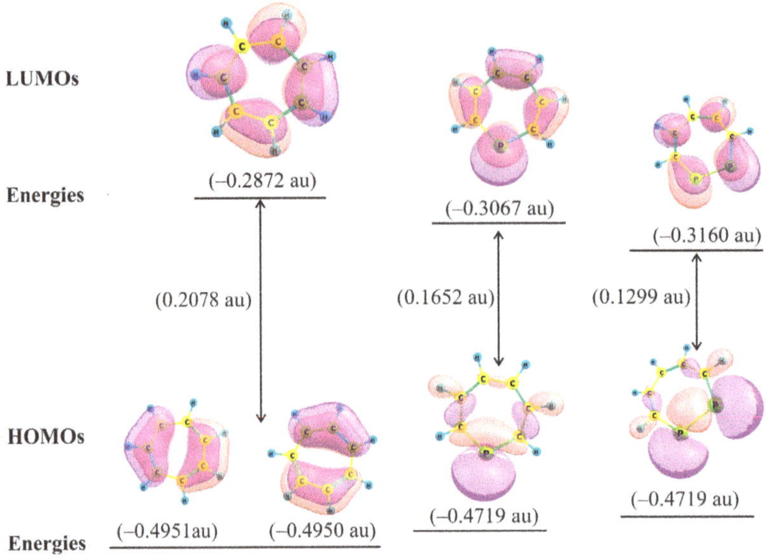

Figure 3. Frontier molecular orbitals (FMOs) of the tropylium cation (**1**); phosphatropylium cation (**2**) and 1,2-diphosphatropylium cation (**3**).

Further FMO analysis reveals that the HOMO-4 orbital of the species **1**, **2**, **3**, HOMO-5 orbital of **4**, **5**, **7**, **8** and HOMO-6 orbital of **9** are the delocalized orbitals of the π symmetry as in their non-phosha analogue i.e., compound **1** (Figure 4). This, in particular, indicates that the aromatic behaviour of these ions is comparable to that of the tropylium ion. However, in the case of the compounds **6**, **10**, **11**, **12**, **13**, **14** and **15**, loss of planarity of the ring reduces the interaction between some of the π orbitals, resulting in diminished delocalization of the π electrons.

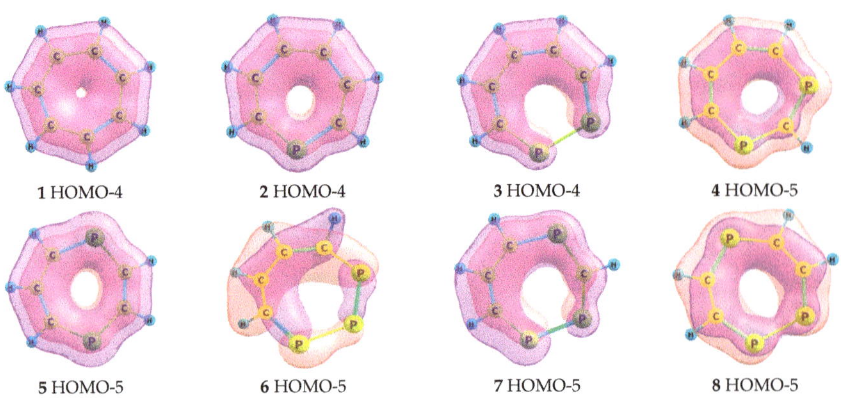

Figure 4. *Cont.*

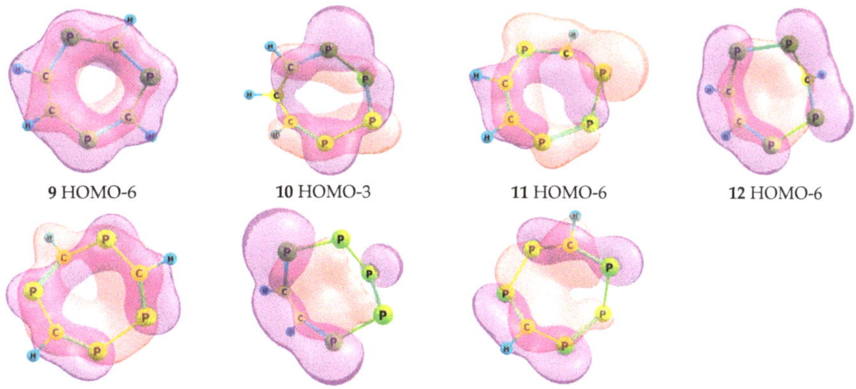

9 HOMO-6 **10** HOMO-3 **11** HOMO-6 **12** HOMO-6

13 HOMO-6 **14** HOMO-6 **15** HOMO-7

Figure 4. HOMOs of π-symmetry of the tropylium ion and its mono- and poly-phospha analogues at the density functional theory (DFT) (B3LYP/6-31+G(d)) level.

2.3. *HOMO − LUMO Energy Gap*

The correlation of the HOMO − LUMO energy gap with kinetic stability and chemical reactivity has been well established [46–54]. It can therefore be an important parameter to determine or explain aromaticity in molecules. A large HOMO − LUMO gap means high kinetic stability and low chemical reactivity associated with the aromatic behaviour. It can be well understood on the basis of the fact that it is energetically unfavourable to add electrons to a high-lying LUMO and to extract electrons from a low-lying HOMO to form the activated complex in a potential reaction [46].

This concept was for the first time used to rationalize aromatic character of the polycyclic aromatic hydrocarbons [55]. We extended application of this concept to the present study and have attempted to explain the aromatic behaviour of the tropylium cation and its phospha analogues on the basis of the HOMO − LUMO energy separation and chemical hardness. The chemical hardness has been calculated using Pearson formula (Equation (1)) [56,57]:

$$\eta = (\varepsilon\text{LUMO} - \varepsilon\text{HOMO}) \ / \ 2 \tag{1}$$

The values of the HOMO − LUMO energy gap ($\Delta\varepsilon$) and chemical hardness (η) are given in Table 2.

Table 2. Frontier molecular orbital energies and hardness of the tropylium ion and its mono- and poly-phospha analogues.

Species	HOMO (a.u.)	LUMO (a.u.)	LUMO − HOMO (a.u.)	$\Delta\varepsilon$ ($\times 10^{-19}$ J)	η ($\times 10^{-19}$ J)
1	−0.4951	−0.2872	0.2078	9.05	4.53
2	−0.4719	−0.3067	0.1652	7.19	3.60
3	−0.4459	−0.3160	0.1299	5.65	2.83
4	−0.4485	−0.3179	0.1306	5.65	2.84
5	−0.4650	−0.3176	0.1474	6.52	3.20
6	−0.4376	−0.3287	0.1089	4.74	2.37
7	−0.4372	−0.3259	0.1113	4.85	2.43
8	−0.4431	−0.3264	0.1167	5.09	2.55
9	−0.4384	−0.3258	0.1126	4.90	2.45
10	−0.4284	−0.3328	0.0956	4.16	2.08
11	−0.4332	−0.3341	0.0991	4.32	2.16
12	−0.4357	−0.3335	0.1022	4.45	2.27
13	−0.4331	−0.3261	0.1070	4.66	2.33
14	−0.4271	−0.3412	0.0859	3.75	1.87
15	−0.4272	−0.3394	0.0878	3.83	1.91

A low HOMO − LUMO energy gap (<2.08 × 10^{-19} J) was associated with the antiaromatic character of the dianions of the polycyclic hydrocarbons [58]. From these results, it was extrapolated that the aromatic systems would have a large HOMO - LUMO energy gap. It may be noted that the value of the HOMO − LUMO energy gap for **1** to **15** lies in the range of 3.75 × 10^{-19}–9.05 × 10^{-19} J, which indicates aromatic character of these systems. The decrease in HOMO − LUMO energy gap in the case of **6** and **10–15** can be attributed to the loss of planarity in these compounds.

It has been shown earlier that an excellent correlation exists between absolute hardness and aromaticity, both increasing almost parallel to each other [56]. In the present case, chemical hardness also varies parallel to the HOMO − LUMO energy gap vis-à-vis aromaticity as shown in Figure 5.

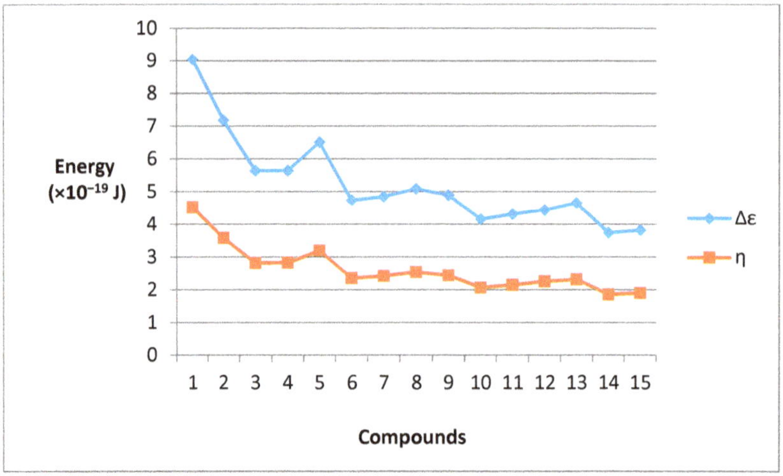

Figure 5. Variation of $\Delta_{HOMO\text{–}LUMO}$ and chemical hardness in tropylium cation and its mono- and poly-phospha analogues.

2.4. ASE

The ASEs of different systems were computed using the following homodesmotic reactions (Scheme 2).

The ASEs of the tropylium ion (**1**) and phosphatropylium ion (**2**) are in conformity with their aromatic character. However, on further CH/P exchange(s), the systems lose planarity as a result of which, ASEs do not change consistently. It is obvious that for the phosphatropylium ions having two or more phosphorus atoms, ASE is no more a reliable indicator of the aromaticity.

2.5. Magnetic Criteria

It is well known that a diatropic current is induced in an aromatic system when it is placed in an external magnetic field. This generates a secondary magnetic field causing a deshielding effect at the periphery and a shielding effect inside the ring. Based on this effect, three indicators, namely ^{1}H-NMR chemicals shifts [22], NICS [23] and magnetic susceptibility exaltation [59,60] have been used extensively to understand aromaticity. A closely related parameter, magnetic susceptibility anisotropy has also been used to validate aromatic character of different compounds. The aromatic character of the species **1–15** is discussed on the basis of all these criteria.

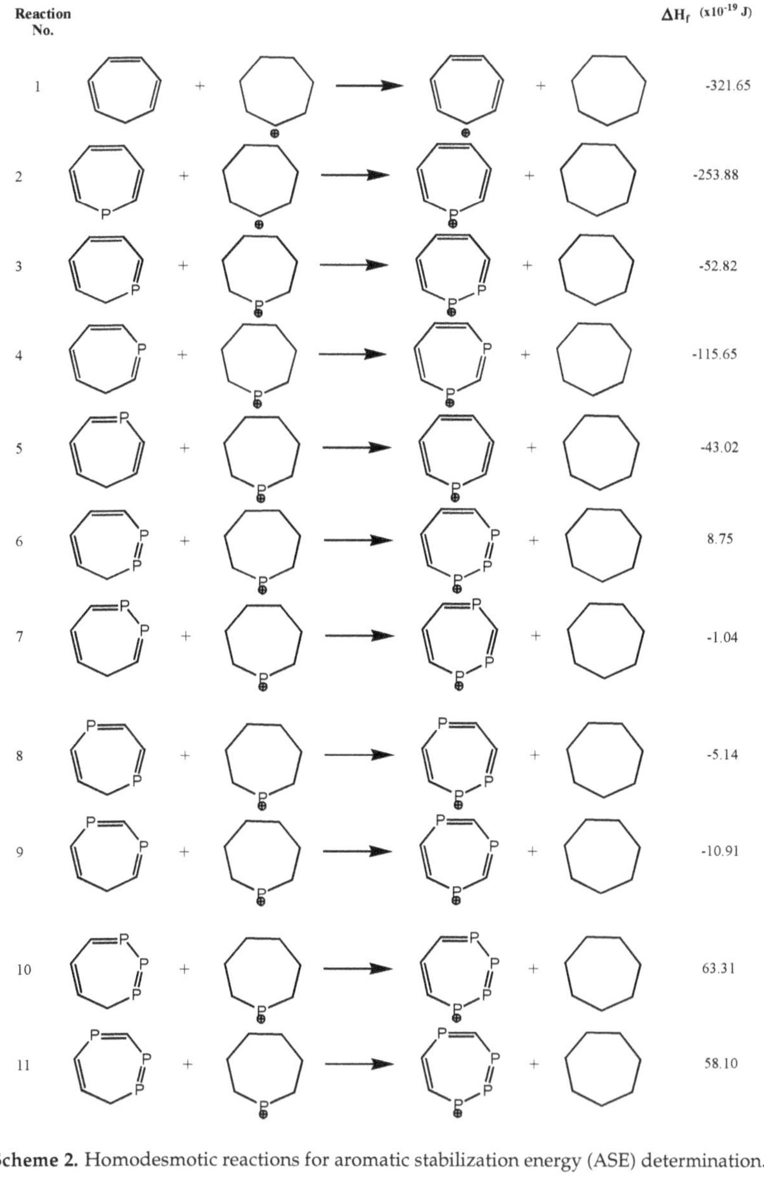

Scheme 2. Homodesmotic reactions for aromatic stabilization energy (ASE) determination.

2.5.1. ^{1}H-NMR Chemical Shifts

The most common measure of the ring current is the ^{1}H-NMR chemical shift, although Schleyer et al. reminded the chemical community of the limitations of proton shifts in the measurement of a ring current [16]. The chemical shift of a proton is the difference of the magnetic shielding of a reference (Tetramethylsilane in the present case) and the proton in question calculated at the same level of theory (Equation (2)):

$$\delta 1\text{H} = \sigma\text{TMS} - \sigma 1\text{H ppm} \tag{2}$$

The ^{1}H-NMR chemical shift values of different species calculated are given in Table 3. It is noted that ^{1}H-NMR chemical shifts of protons in all of the species **1–15**, fall in the aromatic region, confirming their aromatic character. In tropylium ion (**1**), all protons are equivalent and the calculated chemical shift value is δ 9.3. It is noteworthy that on successive CH/P exchanges, the NMR shift values of the protons adjacent to the phosphorus atom(s) appear further downfield, which may be attributed to the stronger electron-acceptor character of the P=C- moiety.

Table 3. ^{1}H-NMR chemical shifts, nucleus-independent chemical shift (NICS), magnetic susceptibility exaltation, magnetic anisotropy and aromatic stabilization energy ASE values of tropylium ion and its phospha analogues.

Species	^{1}H-NMR Chem. Shift (δ)		NICS(0)	NICS(1)	NICS(1)$_{zz}$	Λ	χ_{anis}	ASE (×10^{-19}J)
1	H1	9.39	−6.30	−9.52	−26.42	−20.72	−79.89	−321.65
	H2	9.34						
	H3	9.38						
	H4	9.38						
	H5	9.35						
	H6	9.34						
	H7	9.36						
2	H2	11.14	−5.28	−8.72	−23.78	−34.05	−102.65	−253.88
	H3	9.39						
	H4	9.24						
	H5	9.24						
	H7	11.14						
	H6	9.39						
3	H3	11.53	−4.32	−7.78	−21.73	−32.72	−116.79	−52.82
	H4	9.31						
	H5	9.37						
	H6	9.31						
	H7	11.53						
4	H2	13.16	−4.19	−7.78	−20.70	−35.95	−105.56	−115.65
	H4	10.86						
	H5	9.18						
	H7	10.86						
	H6	9.15						
5	H2	11.11	−4.34	−7.92	−21.31	−36.66	−109.68	−43.02
	H3	11.11						
	H5	11.18						
	H6	9.24						
	H7	11.18						
6	H4	10.98	−4.69	−8.00	−21.26	−39.93	−121.50	8.75
	H5	9.36						
	H6	9.36						
	H7	10.98						
7	H3	13.59	−3.26	−6.97	−18.62	−33.49	−119.21	−1.04
	H5	11.19						
	H6	9.09						
	H7	11.56						
8	H3	11.27	−3.59	−7.31	−19.91	−35.64	−121.98	−5.14
	H4	11.18						
	H6	11.18						
	H7	11.27						
9	H2	13.02	−2.99	−6.75	−17.54	−37.45	−107.94	−10.91
	H4	13.02						
	H6	10.78						
	H7	10.78						
10	H5	10.84	−5.06	−8.06	−20.06	−44.93	−115.3	63.31
	H6	9.12						
	H7	10.84						
11	H4	12.73	−4.68	−8.13	−20.46	−45.94	−109.43	58.10
	H6	10.61						
	H7	10.45						
12	H3	12.31	−5.18	−8.96	-23.86	−47.65	−114.03	89.09
	H6	11.03						
	H7	11.03						
13	H3	13.06	−2.70	−6.99	−17.83	−38.43	−105.24	45.17
	H5	12.74						
	H7	13.06						
14	H7	11.06	−7.83	−9.79	−22.83	−54.31	−107.81	93.69
	H6	10.12						
15	H3	12.63	−3.25	−7.54	−19.96	−42.88	−105.24	173.95
		H7	11.54					

2.5.2. NICS Values

Among the widely used magnetic criteria for aromaticity and anti-aromaticity, NICS proposed by Schleyer and co-workers in 1996 is a simple efficient probe [23]. Subsequently, there have been many refinements and at present NICS$(1)_{zz}$ value is considered to be the most reliable indicator of aromaticity [61]. NICS(0) is defined as the negative value of the absolute shielding computed at the ring centre determined by the average of the heavy atoms coordinates in the ring. NICS(1) is the negative value of the absolute shielding measured 1 Å above the center of the ring, while NICS$(1)_{zz}$ is the out-of-plane component of the absolute shielding estimated in the same position as NICS(1). Rings with highly negative values of NICS are qualified as aromatic by definition, whereas those with positive values are anti-aromatic.

The calculated NICS(0), NICS(1) and NICS$(1)_{zz}$ tensor component values for all species are given in Table 3.

It is noteworthy that these values are negative ranging from −17.54 to −26.42, confirming the aromatic character of the species **1** to **15**.

2.5.3. Magnetic Susceptibility Exaltation

The magnetic susceptibility exaltation (Λ) evaluates the effect of a ring current by comparing the bulk magnetic susceptibility (χ) to the susceptibility of a localized ring system [59,60].

The magnetic susceptibility exaltation with CSGT (continuous set gauge transformations) method values is determined by subtracting the sum of the magnetic susceptibilities of the fragments of cyclic electron delocalization from the values of the respective species (Equation (3)):

$$\Lambda = \chi M - \chi M' \tag{3}$$

The Λ values calculated for different species are given in Table 3. As expected, the exaltation values (Λ) of all these compounds are negative ranging from −20.7 to −54.3 revealing their aromatic character.

2.5.4. Magnetic Anisotropy

Magnetic anisotropy is defined as the difference between out-of-plane and the average in-plane diamagnetic susceptibilities for a ring lying in the (xy) plane [62,63] (Equation (4)):

$$\Delta\chi = \chi zz - (1/2)\,[\chi xx + \chi yy] \tag{4}$$

The values of magnetic anisotropies so calculated are given in Table 3. It may be noted that these values are negative confirming aromatic character of the species.

3. Computational Method and Models

All calculations were carried out using a Gaussian 03 suite of programmes [64]. A diffuse function was added to the heavy atoms and geometry optimization of all the systems was done in the gas phase at the B3LYP/6-31+G(d) level of theory. Frequency calculations were done at the same level to determine zero-point energies and to characterize the energy minima.

^{1}H-NMR chemical shifts were calculated at the GIAO-B3LYP/6-311++G(d,p)//B3LYP/6-31+G(d) level. NICS values were calculated at the (3,+1) ring critical point of the electron density topology as defined by Bader [65] at the GIAO-B3LYP/6-311++G(d,p)//B3LYP/6-31+G(d) level. Magnetic susceptibilities were calculated at the csgt-B3LYP/6-31+G(d)//B3LYP/6-31+G(d) level and exaltation values were determined by subtracting the sum of the magnetic susceptibilities of the fragments (as explained in the supporting information) calculated at the same level from the value of the respective species.

4. Conclusions

On CH/P- exchange in the tropylium ion, aromaticity is sustained up to four carbon atoms. However, with further exchange(s), the system does not remain monocyclic, and it changes to bicyclic and polycyclic systems. Of the various indicators for the aromaticity, the ones based on the magnetic field effect, namely ^{1}H-NMR chemical shifts, NICS(1)$_{zz}$, magnetic susceptibility exaltation and magnetic anisotropies are more consistent and confirm the aromatic character of these species.

Acknowledgments: We are thankful to the authorities of The IIS University, Jaipur (India) for providing us with the computational facilities. Thanks are also due to Raj Kumar Bansal for helpful discussions. We would like to express our gratitude to the anonymous reviewer for making very constructive suggestions, thus contributing indirectly to the improvement of the paper.

Author Contributions: The research problem was conceptualized by the corresponding author (Raakhi Gupta), and the computational work was carried out by Ankita Puri.

Conflicts of Interest: The authors declare no conflict of interest.

References

1. Garratt, P.J. *Aromaticity*; Wiley-Inc.: Hoboken, NJ, USA, 1986.
2. Minkin, V.J.; Glukhovtsev, M.N.; Simkin, B.Y. *Aromaticity and Antiaromaticity; Electronic and Structural Aspects*; John Wiley & Sons, Inc.: New York, NY, USA, 1994.
3. Bergmann, E.D.; Pullman, B. Aromaticity, Pseudo-Aromaticity, Anti-Aromaticity, Proceedings of Jerusalem Symposium on Quantum Chemistry and Biochemistry, Jerusalem, Israel, 31 March–3 April 1970; Academic Press Inc.: Jerusalem, Israel, 1971.
4. Cyranski, M.K.; Krygowski, T.M.; Katritzky, A.R.; Schleyer, P.v.R. To what extent can aromaticity be defined uniquely. *J. Org. Chem.* **2002**, *67*, 1333–1338. [CrossRef] [PubMed]
5. Schleyer, P.v.R.; Jiao, H. What is aromaticity? *Pure Appl. Chem.* **1996**, *68*, 209–218. [CrossRef]
6. Schleyer, P.v.R.; Freeman, P.K.; Jiao, H.; Goldfuss, B. Aromaticity and Antiaromaticity in Five-Membered C_4H_4X Ring Systems: Classical and Magnetic Concepts May Not Be Orthogonal. *Angew. Chem. Int. Ed. Engl.* **1995**, *34*, 337–340. [CrossRef]
7. Mills, N.S.; Benish, M. The Aromaticity/Antiaromaticity Continuum. I. Comparison of the aromaticity of the dianion and the antiaromaticity of the dication of tetrabenzo[5.5]fulvalene via magnetic measures. *J. Org. Chem.* **2006**, *71*, 2207–2213. [CrossRef] [PubMed]
8. Katritzky, A.R.; Karelson, M.; Wells, A.P. Aromaticity as a Quantitative Concept. 6. Aromaticity Variation with Molecular Environment. *J. Org. Chem.* **1996**, *61*, 1619–1623. [CrossRef] [PubMed]
9. Katritzky, A.R.; Karelson, M.; Sild, S.; Krygowski, T.M.; Jug, K. Aromaticity as a Quantitative Concept. 7. Aromaticity Reaffirmed as a Multidimensional Characteristic. *J. Org. Chem.* **1998**, *63*, 5228–5231. [CrossRef]
10. Thematic issue on Aromaticity. *Chem. Rev.* **2001**, *101*, 1115–1566.
11. Thematic issue on Delocalization of pi- and sigma. *Chem. Rev.* **2005**, *105*, 3433–3947.
12. Thematic issue on Challenges in Aromaticity: 150 years after Kekule's benzene. *Chem. Soc. Rev.* **2015**, *44*, 6397–6643.
13. Hückel, E. Grundzüge der Theorie Ungesättiger und Aromatischer Verbindungen. *Z. Elektrochem. Angew. Phys. Chem.* **1937**, *43*, 752–788.
14. Platt, J.R. The box model and electron densities in conjugated systems. *J. Chem. Phys.* **1954**, *22*, 1448–1455.
15. Breslow, R. Aromatic character. *Chem. Eng. News* **1965**, *43*, 90–100. [CrossRef]
16. Schleyer, P.v.R.; Chen, Z.; Wannere, C.S.; Corminboeuf, C.; Puchta, R. Nucleus-independent chemical shifts (NICS) as an aromaticity criterion. *Chem. Rev.* **2005**, *105*, 3842–3888.
17. Krygowski, T.M.; Szatylowicz, H.; Stasyuk, O.A.; Dominikowska, J.; Palusiak, M. Aromaticity from the viewpoint of molecular geometry: Application to planar systems. *Chem. Rev.* **2014**, *114*, 6383–6422. [CrossRef] [PubMed]
18. Julg, A.; Francois, P. Researches sur la géométrie de quelques hydrocarbures non-alternants: Son influence sur les énergies de transition, une nouvelle définition de l'aromaticité. *Theor. Chim. Acta* **1967**, *8*, 249–259. [CrossRef]

19. Bird, C.W. A new aromaticity index and its application to five-membered ring heterocycles. *Tetrahedron* **1985**, *41*, 1409–1414. [CrossRef]
20. Schleyer, P.v.R.; Pülhofer, F. Recommendations for the evaluation of aromatic stabilization energies. *Org. Lett.* **2002**, *4*, 2873–2876. [CrossRef] [PubMed]
21. Pascal, P. Magnetochemical Researches. *Ann. Chim. Phys.* **1910**, *19*, 5–70.
22. Mitchell, R.H. Measuring aromaticity by NMR. *Chem. Rev.* **2001**, *101*, 1301–1315. [CrossRef] [PubMed]
23. Schleyer, P.v.R.; Maerker, C.; Dransfeld, A.; Jiao, H.; Hommes, N.J.R.v.E. Nucleus-independent chemical shifts: A simple and efficient aromaticity probe. *J. Am. Chem. Soc.* **1996**, *118*, 6317–6318. [CrossRef]
24. Appel, R. pπ-Double bonds between phosphorus and carbon—A challenge. *Pure Appl. Chem.* **1987**, *59*, 977–982. [CrossRef]
25. Nyulászi, L.; Veszprémi, T.; Réffy, J. A new look at the similarities of the conjugative ability and reactivity of P=C and C=C double bonding. *J. Phys. Chem.* **1993**, *97*, 4011–4015. [CrossRef]
26. Nyulászi, L.; Várnai, P.; Krill, S.; Regitz, M. Aromaticity of thia- and selena-phospholes: A photoelectron spectroscopic and quantum chemical study. *J. Chem. Soc. Perkin Trans.* **1995**, *2*, 315–318. [CrossRef]
27. Nyulászi, L.; Várnai, P.; Veszprémi, T. About the aromaticity of five-membered heterocycles. *J. Mol. Struct. THEOCHEM* **1995**, *358*, 55–61. [CrossRef]
28. Cloke, F.G.N.; Hitchcock, P.B.; Nixon, J.F.; Wilson, D.J.; Tabellion, F.; Fischbeck, U.; Preuss, F.; Regitz, M. Synthetic, structural and theoretical studies on new aromatic 1,2,4-diazaphosphole ring systems: Crystal and molecular structure of $P_2C_2Bu^t{}_2NPh$. *Chem. Commun.* **1999**. [CrossRef]
29. Mathey, F. Expanding the analogy between phosphorus-carbon and carbon-carbon double bonds. *Acc. Chem. Res.* **1992**, *25*, 90–96. [CrossRef]
30. Schmidpeter, A. Molecules that we made: An essay on phosphorus chemistry. *Heteroat. Chem.* **1999**, *10*, 529–537. [CrossRef]
31. Bansal, R.K.; Heinicke, J. Anellated heterophospholes and phospholides and analogies with related non-phosphorus systems. *Chem. Rev.* **2001**, *101*, 3549–3578. [CrossRef] [PubMed]
32. Simpson, M.C.; Protasiewicz, J.D. Phosphorus as a carbon copy and as a photocopy: New conjugated materials featuring multiply bonded phosphorus. *Pure Appl. Chem.* **2013**, *85*, 801–815. [CrossRef]
33. Bansal, R.K.; Gupta, R.; Maheshwari, P.; Kour, M. Analogy of phosphaalkenes and azaphospholes with their respective non-phosphorus analogues. *Curr. Org. Chem.* **2016**, *20*, 2099–2108. [CrossRef]
34. Dillon, K.B.; Mathey, F.; Nixon, J.F. *Phosphorus: The Carbon Copy*; Wiley: New York, NY, USA, 1988; pp. 203–226.
35. Nyulászi, L. Aromaticity of phosphorus heterocycles. *Chem. Rev.* **2001**, *101*, 1229–1246. [CrossRef] [PubMed]
36. Nyulászi, L.; Benkő, Z. Aromatic phosphorus heterocycles. In *Aromaticity in Heterocyclic Compounds*; Krygowski, T.M., Cyranski, M.K., Eds.; Springer-Verlag: Berlin/Heidelberg, Germany, 2009; Volume 19, pp. 27–81.
37. Gupta, R.; Bansal, R.K. Aromaticity/antiaromaticity of phospha-analogues of carbocyclic ions: A DFT investigation. *Comp. Theor. Chem.* **2016**, *1076*, 1–10. [CrossRef]
38. Gupta, R.; Maheshwari, P.; Kour, M. Reinvestigation of the aromaticity of mono- and polyphosphacyclopentadienide ions at the DFT level. *Comp. Theor. Chem.* **2015**, *1060*, 10–16. [CrossRef]
39. Ghiasi, R. Theoretical study of classical isomers tropylium, azatropylium, phosphatropylium, and arsatropylium cations: Structure, properties and aromaticity. *Main Group Chem.* **2008**, *7*, 147–154. [CrossRef]
40. Lin, L.; Lievens, P.; Nguyen, M.T. In search of aromatic seven-membered rings. *J. Mol. Struct.* **2010**, *943*, 23–31. [CrossRef]
41. Firouzi, R.; Ardani, S.S. Description of heteroaromaticity on the basis of π-electron density anisotropy. *Phys. Chem. Chem. Phys.* **2014**, *16*, 11538–11548. [CrossRef] [PubMed]
42. Nyulászi, L. Toward a planar σ^3–phosphorus. *J. Phys. Chem.* **1996**, *100*, 6194–6198. [CrossRef]
43. Nyulászi, L. Pentaphosphole: An aromatic ring with a planar σ^3–phosphorus. *Inorg. Chem.* **1996**, *35*, 4690–4693. [CrossRef]
44. Glukhovtsev, M.; Dransfeld, A.; Schleyer, P.v.R. Why pentaphosphole, P_5H, is planar in contrast to phosphole, $(CH)_4PH$. *J. Phys. Chem.* **1996**, *100*, 13447–13454. [CrossRef]
45. Cloke, F.G.N.; Hitchcock, P.B.; Hunnable, P.; Nixon, J.F.; Nyulászi, L.; Niecke, E.; Thelen, V. The First Delocalized Phosphole Containing a Planar Tricoordinate Phosphorus Atom: 1-[Bis (trimethylsilyl) methyl]-3, 5-bis (trimethylsilyl)-1,2,4-triphosphole. *Angew. Chem. Int. Ed.* **1998**, *37*, 1083–1086. [CrossRef]

46. Manolopoulos, D.E.; May, J.C.; Down, S.E. Theoretical studies of the fullerenes: C34 to C70. *Chem. Phys. Lett.* **1991**, *181*, 105–111. [CrossRef]
47. Haddon, R.C.; Fukunaga, T. Unified theory of the thermodynamic and kinetic criteria of aromatic character in the [4n + 2] annulenes. *Tetrahedron Lett.* **1980**, *21*, 1191–1192. [CrossRef]
48. Schmalz, T.G.; Seitz, W.A.; Klein, D.J.; Hite, G.E. Elemental carbon cages. *J. Am. Chem. Soc.* **1988**, *110*, 1113–1127. [CrossRef]
49. Zhou, Z.; Parr, R.G.; Garst, J.F. Absolute hardness as a measure of aromaticity. *Tetrahedron Lett.* **1988**, *29*, 4843–4846. [CrossRef]
50. Zhou, Z.; Parr, R.G. New measures of aromaticity: Absolute hardness and relative hardness. *J. Am. Chem. Soc.* **1989**, *111*, 7371–7379. [CrossRef]
51. Zhou, Z.; Parr, R.G. Activation hardness: New index for describing the orientation of electrophilic aromatic substitution. *J. Am. Chem. Soc.* **1990**, *112*, 5720–5724. [CrossRef]
52. Liu, X.; Schmalz, T.G.; Klein, D.J. Favorable structures for higher fullerenes. *Chem. Phys. Lett.* **1992**, *188*, 550–554. [CrossRef]
53. Parr, R.G.; Zhou, Z. Absolute hardness: Unifying concept for identifying shells and subshells in nuclei, atoms, molecules, and metallic clusters. *Acc. Chem. Res.* **1993**, *26*, 256–258. [CrossRef]
54. Aihara, J.; Oe, S.; Yoshida, M.; Osawa, E.J. Further test of the isolated pentagon rule: Thermodynamic and kinetic stabilities of C84 fullerene isomers. *Comput. Chem.* **1996**, *17*, 1387–1394. [CrossRef]
55. Aihara, J. Circuit resonance energy: A key quantity that links energetic and magnetic criteria of aromaticity. *J. Am. Chem. Soc.* **2006**, *128*, 2873–2879. [CrossRef] [PubMed]
56. Pearson, R.G. Absolute electronegativity and hardness correlated with molecular orbital theory. *Proc. Natl. Acad. Sci. USA* **1986**, *83*, 8440–8441. [CrossRef] [PubMed]
57. Pearson, R.G. Recent Advances in the Concept of Hard and Soft Acids and Bases. *J. Chem. Educ.* **1987**, *64*, 561–567. [CrossRef]
58. Minsky, A.; Meyer, A.Y.; Rabinovitz, M. Paratropicity and antiaromaticity: Role of the homo-lumo energy gap. *Tetrahedron* **1985**, *41*, 785–791. [CrossRef]
59. Dauben, H.J.; Wilson, J.D.; Laity, J.L. Diamagnetic susceptibility exaltation as a criterion of aromaticity. *J. Am. Chem. Soc.* **1968**, *90*, 811–813. [CrossRef]
60. Dauben, H.J.; Wilson, J.D.; Laity, J.L. Diamagnetic susceptibility in hydrocarbons. *J. Am. Chem. Soc.* **1969**, *91*, 1991–1998. [CrossRef]
61. Schleyer, P.v.R.; Jiao, H.J.; Hommes, N.J.R.v.E.; Malkin, V.G.; Malkina, O.L. An evaluation of the aromaticity of inorganic rings: Refined evidence from magnetic properties. *J. Am. Chem. Soc.* **1997**, *119*, 12669–12670. [CrossRef]
62. Benson, R.C.; Flygare, W.H. Molecular Zeeman effect of cyclopentadiene and isoprene and comparison of the magnetic susceptibility anisotropies. *J. Am. Chem. Soc.* **1970**, *92*, 7523–7529.
63. Schmalz, T.G.; Norris, C.L.; Flygare, W.H. Localized magnetic susceptibility anisotropies. *J. Am. Chem. Soc.* **1973**, *95*, 7961–7967. [CrossRef]
64. Frisch, M.J.; Trucks, G.W.; Schlegel, H.B.; Scuseria, G.E.; Robb, M.A.; Cheeseman, J.R.; Montgomery, J.A., Jr.; Vreven, T.; Kudin, K.N.; Burant, J.C.; et al. *Gaussian 03*, Revision B.05. Gaussian, Inc.: Wallingford, CT, USA, 2003.
65. Bader, R.F.W. *Atoms in Molecules: A Quantum Theory*; Clarendon Press: Oxford, UK, 1990.

Sample Availability: Not Available.

molecules

MDPI

Review

Stereoselective Synthesis of α-Amino-C-phosphinic Acids and Derivatives

José Luis Viveros-Ceballos [1], Mario Ordóñez [2,*], Francisco J. Sayago [3] and Carlos Cativiela [3,*]

1 Secretaría Académica, Universidad Autónoma del Estado de Morelos, Av. Universidad 1001, 62209 Cuernavaca, Morelos, Mexico; jlvc@uaem.mx

2 Centro de Investigaciones Químicas-IICBA, Universidad Autónoma del Estado de Morelos, Av. Universidad 1001, 62209 Cuernavaca, Morelos, Mexico

3 Departamento de Química Orgánica, Universidad de Zaragoza-CSIC, ISQCH, 50009 Zaragoza, Spain; jsayago@unizar.es

* Correspondence: palacios@uaem.mx (M.O.); cativiela@unizar.es (C.C.); Tel.: +52-777-329-7997 (M.O.); +34-976-761-210 (C.C.)

Academic Editor: György Keglevich

Received: 27 July 2016; Accepted: 25 August 2016; Published: 29 August 2016

Abstract: α-Amino-C-phosphinic acids and derivatives are an important group of compounds of synthetic and medicinal interest and particular attention has been dedicated to their stereoselective synthesis in recent years. Among these, phosphinic pseudopeptides have acquired pharmacological importance in influencing physiologic and pathologic processes, primarily acting as inhibitors for proteolytic enzymes where molecular stereochemistry has proven to be critical. This review summarizes the latest developments in the asymmetric synthesis of acyclic and phosphacyclic α-amino-C-phosphinic acids and derivatives, following in the first case an order according to the strategy used, whereas for cyclic compounds the nitrogen embedding in the heterocyclic core is considered. In addition selected examples of pharmacological implications of title compounds are also disclosed.

Keywords: α-amino-C-phosphinic acids; α-amino-C-phosphinates; α-amino acids; stereoselective synthesis; biological activity

1. Introduction

Optically active α-aminoalkylphosphonic acids **1** are probably the most important analogues of the α-amino acids **2**, obtained by isosteric substitution of the planar and less bulky carboxylic acid (CO_2H) by a tetrahedral phosphonic acid functionality (PO_3H_2). These classes of compounds are currently attracting great interest in organic and medicinal chemistry, as well as in agriculture, due to their important biological and pharmacological properties [1–5]. The great importance of this type of compounds has prompted organic chemists to report numerous procedures for their racemic or stereoselective synthesis [6–11]. On the other hand, the optically active α-amino-C-phosphinic acids **3** are also considered as analogues of α-amino acids **2** where now the carboxylic group (CO_2H) is replaced by a tetrahedral and sterically more demanding phosphinic acid functionality (PO_2HR'). As part of peptides, α-amino-C-phosphinic acids **3** are much closer analogues to natural α-amino acids **2** than their α-aminophosphonic counterparts **1**, due to their mono acidic character, and the higher stability of the P-C bond of **3** compared to the P-O bond of **1** (Figure 1) [12].

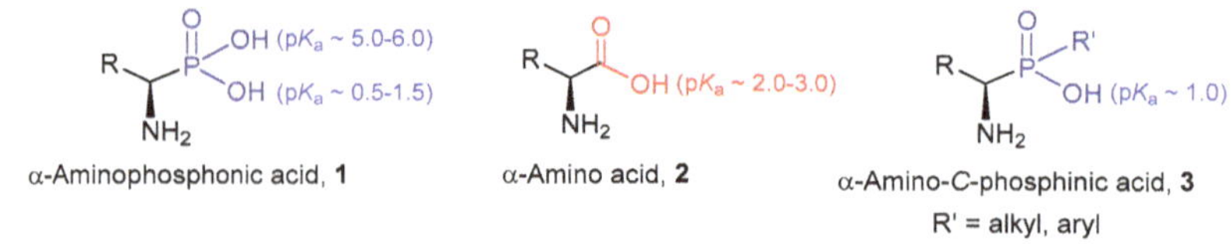

Figure 1. General structure for α-amino acids (**2**) and their phosphonic (**1**) and *C*-phosphinic (**3**) analogues.

Phosphinic, phosphonic and carboxylic acids differ in several characteristics. The carboxylic group is planar at carbon while their phosphorus analogues are tetrahedral and considerably larger in size (the phosphorus atom has a much larger atomic radius than carbon). Phosphinic and phosphonic acids exist mainly in their tetracoordinate form, and they are considered as mono- and diacids, respectively. The phosphinic acids (pK_a ~1.0) are slightly stronger than the corresponding phosphonic acids (pK_a ~0.5–1.5) and both more acidic than the carboxylic acid equivalent (pK_a ~2.0–3.0); the former are mono acids and no evidence for the second acidity due to the P-H bond is reported, while for phosphonic acids, it is well-known that the second acidity is about 5 pK units higher than the first P-OH acidity. Similar to α-amino acids, the aminophosphonic and aminophosphinic acids have a zwitterionic form due to the internal hydrogen transfer between the P-OH and amino groups [13,14]. In addition, a characteristic feature of tetracoordinated phosphorus (III) compounds such as *H*-phosphonate and *H*-phosphinate derivatives, is the configurational stability of the phosphorus center at room temperature, and their tautomeric equilibria with trivalent species that occurs without epimerization at the phosphorus center. This is the basis for synthetic applications of these compounds as chiral precursors in several stereospecific transformations [15]. Although, different structures of phosphonic and phosphinic groups compared with the carboxylic function could a priori disrupt enzyme-ligand interactions, they frequently are recognized by enzymes or receptors as inhibitors or false substrates [16,17].

Phosphinopeptides containing *C*-terminal α-aminophosphinic acids have been prepared by similar methods to those used with their phosphonic analogues, for example the coupling of *N*-protected amino acids or their active esters with α-aminophosphinates [18], or with free α-aminophosphinic acids in organic or aqueous-organic media has been reported [19]. Additionally, the synthesis of phosphinopeptides via the Mannich-type condensation of *N*-Cbz protected alkanamides/peptide amides, aldehydes and aryldichlorophosphines, followed by hydrolysis, has also been reported [20]. Therefore, the α-amino-*C*-phosphinic acids are currently attracting interest in medicinal chemistry due to their relevance mainly as pseudopeptides, which have acquired pharmacological importance in influencing physiologic and pathologic processes, primarily acting as inhibitors for proteolytic enzymes. The success of these compounds is based on the resemblance of phosphinic acids to the sp^3 transition state of the hydrolysis of peptide bonds. Considering that lengths of O-P and C-P bonds are significantly longer than the corresponding O-C and C-C bonds, the phosphinic fragment might be considered as an extended tetrahedral intermediate and thus can be treated similar to the transition state of the hydrolysis (Figure 2) [21]. Moreover, they can act as simple analogues of amino acids replacing the carboxylic group by the phosphinic moiety, as non-hydrolysable phosphate analogues or as chelating agents of metal ions present in the active site of the enzymes [22]. A number of excellent reviews on various aspects of their activity in natural systems have been published [16,17].

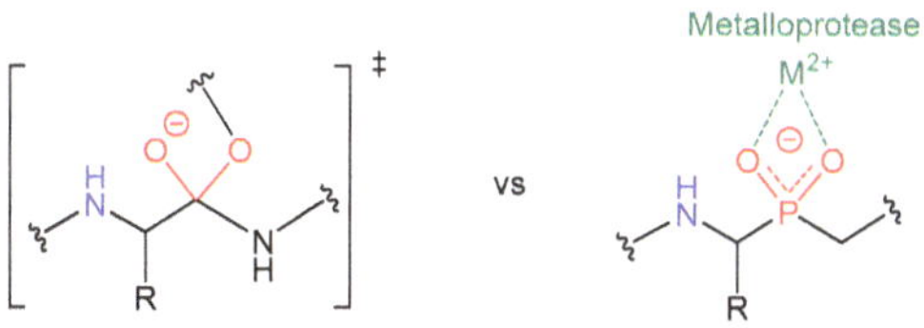

Figure 2. Similarity between the transition state of peptide hydrolysis and the α-amino-C-phosphinic motif. The additional interaction with the metal ion in the metalloproteases active site is included.

Simple pseudopeptides containing α-amino-C-phosphinate moieties replacing the scissile peptide bonds, rank among most potent inhibitors of metalloproteinases, where their tumorigenesis and invasion-related functions makes these enzymes molecular targets for the development of potential anticancer drugs [23]. The role of neutral aminopeptidase in the pathogenesis of hypertension provides an opportunity for regulating arterial blood pressure through its inhibition with this class of compounds [24]. Furthermore, these pseudopeptides appear to be excellent inhibitors to *Plasmodium falciparum* aminopeptidases, promising targets for a novel treatment of malaria [25,26]. α-Amino-C-phosphinic pseudopeptides have also revealed their potential activity in the regulation of matrix metalloproteinases (MMPs), whose overexpression or inadequate levels leads to pathological states such as osteoarthritis, rheumatoid arthritis and inflammation, but also it is associated with tumor growth, invasion and metastasis [27,28]. Additionally, is noteworthy that in these derivatives the stereochemistry affects their biological activity, hence the importance of the synthesis of these compounds in enantiomerically pure form [29,30].

In view of the broad biological applications of the optically active α-amino-C-phosphinic acids and their peptidic derivatives, in the last years the development of suitable synthetic methodologies for their preparation in optically pure form has been a topic of great interest in several research groups [31,32]. In this context, now we would like to report herein a summary of the stereoselective synthesis of α-amino-C-phosphinic acids and their derivatives covering the last 20 years. In this review article, the compounds have been classified in acyclic and phosphacyclic α-amino-C-phosphinic derivatives. In the first case an order is established according to the strategy used, in this way all procedures can be classified into C-P bond formation using a Pudovik or Kabachnik-Fields like reaction, C-H bond formation derived from catalytic asymmetric hydrogenation of α,β-dehydroaminophosphinates, resolution methodologies, preparation from chiral α-amino-phosphonates and also P-C bond formation from chiral α-amino-*H*-phosphinic derivatives is included (Scheme 1). Otherwise, in the second case involving the phosphacyclic derivatives, the classification will be based in the presence or lack of the nitrogen atom in the heterocycle.

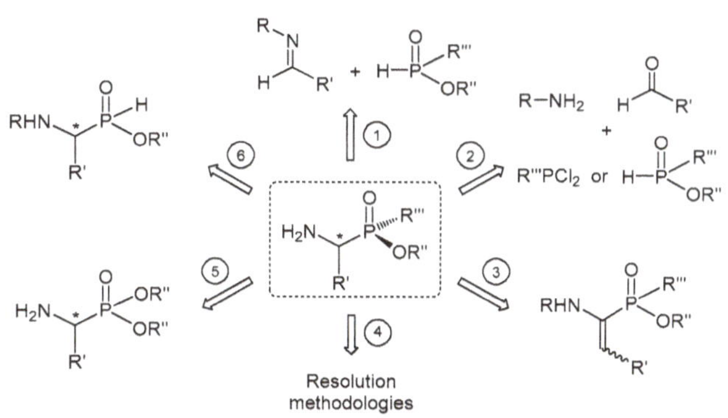

Scheme 1. General overview of the synthetic strategies leading to α-amino-C-phosphinic acids.

2. Stereoselective Synthesis of Acyclic α-Amino-*C*-phosphinic Acids and Derivatives

2.1. *Stereoselective C-P Bond Formation (Addition of Phosphorus Compounds to Imines)*

A strategy widely used for the stereoselective synthesis of α-amino-C-phosphinic acids is the addition of a nucleophile HX (X = CN, PO_2R, PO_2RR') to a preformed *N*-substituted chiral imine in the presence or absence of a chiral catalyst [33], leading to optically active α-amino carboxylic, α-aminophosphonic and α-aminophosphinic acids precursors, respectively. The extent of asymmetric induction at the C_α depends on the nature of the substituents (steric or hydrophobic interactions) and nucleophile–substrate interactions, which are favored in organocatalytic processes [34].

Indeed, the nucleophilic addition of phosphorus compounds (alkyl or aryl phosphinates) to imines, a Pudovik-like reaction [35], is an appropriate procedure for the preparation of α-amino-C-phosphinates. Generally, the high diastereoselective ratio can be explained on the basis of the most stable conformation of the Schiff bases, in which the C-H bond of the imine is eclipsed with the N–C–H fragment, as would be expected from the 1,3-allylic strain model [36,37], and the nucleophilic attack of the alkyl or aryl phosphinate takes place on the less hindered side (in the example below, R′ is bulkier than the methyl group) (Scheme 2).

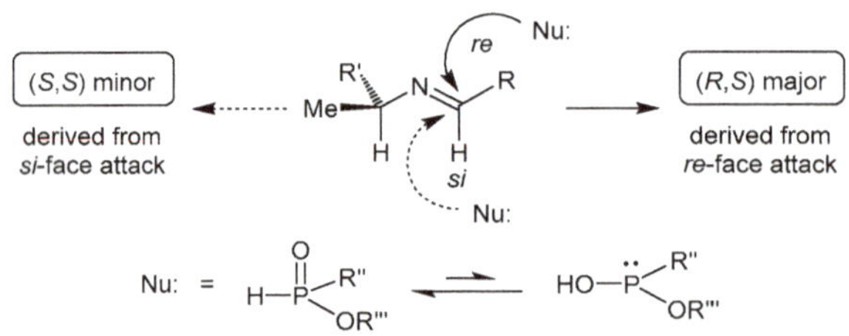

Scheme 2. Model to explain the high diastereoselectivity in the nucleophilic addition to chiral imine compounds.

Contrary to the phosphonate synthesis, the addition of *H*-phosphinates to imines results in the generation of two new stereogenic centers due to the attack of the chiral phosphinates at the prochiral C=N bond [38]; however, the phosphorus chirality is lost during the hydrolysis through the delocalization of P=O bond due to d-p π bond (Scheme 3) [28].

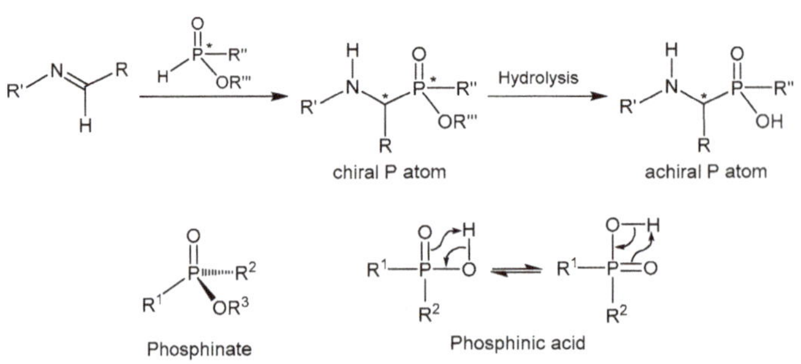

Scheme 3. Structure of α-amino-C-phosphinates and α-amino-C-phosphinic acids.

2.1.1. Chiral Imine Compounds

Petneházy et al. [39] reported the first enantioselective synthesis of α-amino-C-phosphinic acids by the addition of phosphinates to chiral imines in the absence of a catalyst. In this regard, the (*S*)-α-methylbenzylamine and (*S*)-α-methoxymethylbenzylamine derived imines **4a–g** were reacted with ethyl phenylphosphinate in toluene at 70 °C, to obtain the α-amino-*C*-phosphinates **5a–g** in 40%–84% yield and moderate diastereoisomeric ratio. The hydrolysis of the phosphinate moiety in **5a–g** using concd. HCl or HBr/AcOH, followed by treatment with propylene oxide and hydrogenolysis using Pd/C in MeOH, provided the optically active α-amino-*C*-phosphinic acids **6a–d** in 20%–60% yield and with an enantiomeric ratio of good to excellent (Scheme 4).

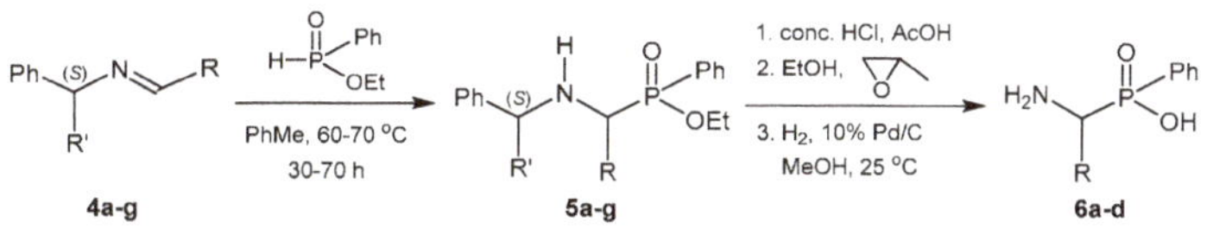

5a; R = 3-MeOC$_6$H$_4$, R' = Me, 68%, 45:13:35:9 d.r.
5b; R = 2-MeC$_6$H$_4$, R' = Me, 48%, 30:14:40:16 d.r.
5c; R = *i*-Bu, R' = Me, 60%, 45:39:11:5 d.r.
5d; R = 3-MeOC$_6$H$_4$, R' = CH$_2$OMe, 62%, 30:33:19:18 d.r.
5e; R = 2-MeC$_6$H$_4$, R' = CH$_2$OMe, 40%, 42:4:41:14 d.r.
5f; R = *i*-Bu, R' = CH$_2$OMe, 60%, 42:10:35:13 d.r.
5g; R = Ph, R' = Me, 84%, 47:6:40:7 d.r.

6a; R = 3-MeOC$_6$H$_4$, 23%, 90:10 e.r.
6b; R = 2-MeC$_6$H$_4$, 30%, 88:12 e.r.
6c; R = *i*-Bu, 20%, 80:20 e.r.
6a; R = 3-MeOC$_6$H$_4$, 40%, 79:21 e.r.
6b; R = 2-MeC$_6$H$_4$, 42%, 86:14 e.r.
6c; R = *i*-Bu, 50%, 70:30 e.r.
6d; R = Ph, 60%, 100:0 e.r.

Scheme 4. First enantioselective synthesis of α-amino-*C*-phosphinic acids **6a–d**.

Recently, Zhao et al. [40] described the synthesis of a *P*-chiral compound, whose R_P chirality, was fully confirmed by X-ray analysis. In this context, the hydrophosphinylation of the *N*-(*R*)-α-methylbenzyl Schiff base (*R*)-**4g** with the (R_P)-*O*-(−)-menthyl phenylphosphinate at 80 °C, gave after crystallization the optically pure *O*-menthyl phosphinate (*R*,*S*,*R*)-**7** (Scheme 5).

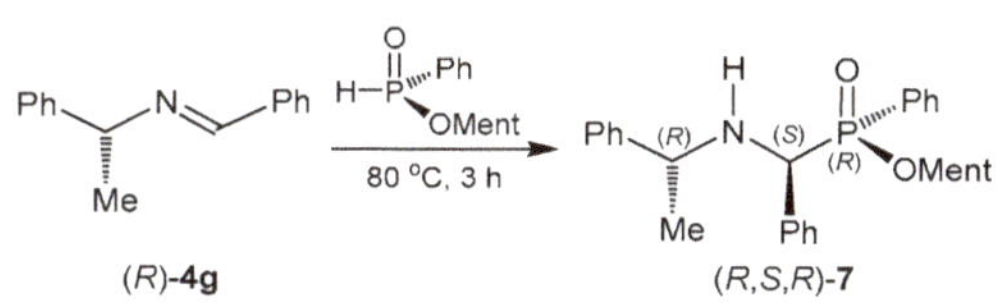

Scheme 5. Synthesis of the *P*-chiral *O*-menthyl phosphinate (*R*,*S*,*R*)-**7**.

Rossi et al. [41] carried out the nucleophilic addition of the racemic ethyl phenylphosphinate onto the chiral imines **8a**,**b** in chloroform at 60–70 °C, to obtain the α-amino-*C*-phosphinates **9a**,**b** in 96% and 80% yield, respectively, and good diastereoselectivities (Scheme 6).

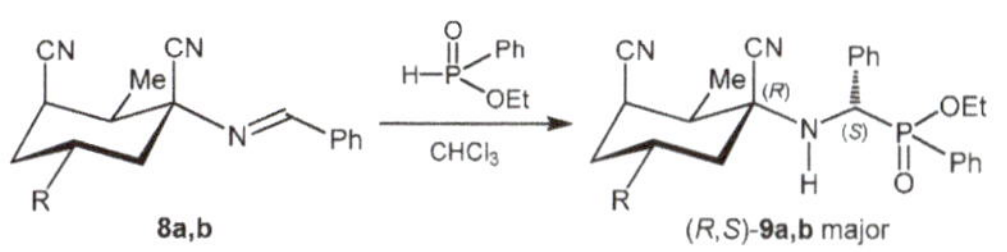

9a; R = HO(CH$_3$)$_2$C-, 96%, 8:46:3:43
9b; R = CH$_2$=C(Me)-, 80%, 7:51:5:37

Scheme 6. Nucleophilic addition of ethyl phenylphosphinate onto the chiral imines **8a**,**b**.

The (*S*)-configuration at C_α of the major diastereoisomers **9a** resulted from the preferred addition of ethyl phenylphosphinate onto the *si* face of the imine **8a**, which adopts the most stable conformation according to the 1,3-allylic strain model [37], as shown in the Scheme 7 [41].

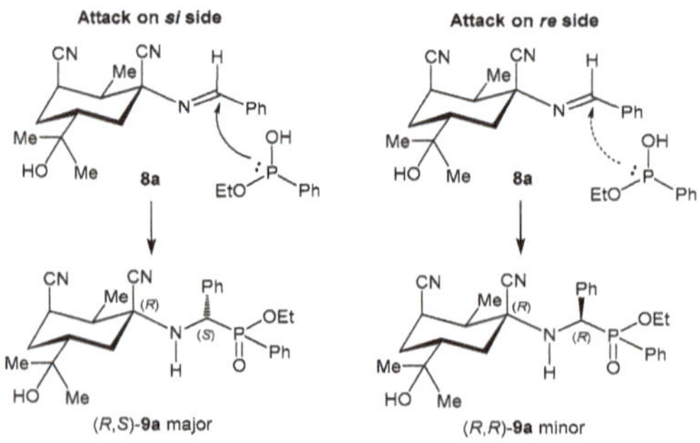

Scheme 7. Preferential addition of ethylphenylphosphinate onto the *si* face of the imine **8a**.

The reaction of (*R*,*S*)-**9a** with aqueous $AgNO_3/HNO_3$ in methanol at 50 °C, produced the α-amino-*C*-phosphinate **10a** in 63% yield and with 78:22 diastereoisomeric ratio, which by treatment with HBr/AcOH, afforded the enantiomerically pure (*S*)-α-aminobenzyl phenylphosphinic acid **6d** (Scheme 8) [41].

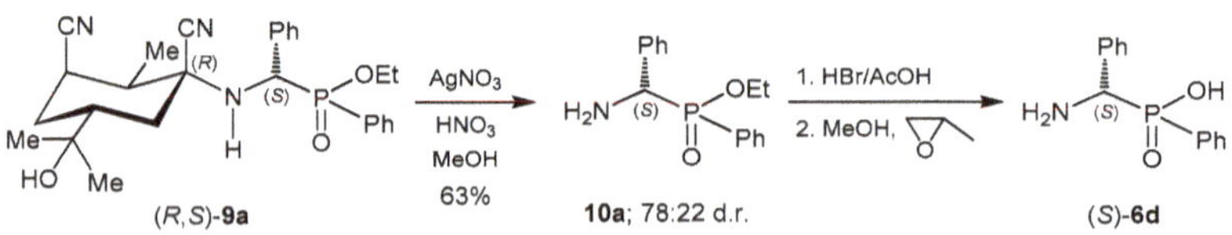

Scheme 8. Synthesis of the (*S*)-α-aminobenzyl phenylphosphinic acid **6d**.

Carbohydrate derivatives are efficient auxiliaries in many stereoselective chiral syntheses [42–44]. For example, Chen et al. [45] achieved the nucleophilic addition of ethyl phenylphosphinate to *N*-galactosylaldimines **11a–g** in the presence of catalytic amounts of $SnCl_4$ in THF at room temperature to obtain the *N*-galactosyl α-aminoalkyl-*C*-phosphinates **12a–g** in good yield and moderate diastereoselectivity (Scheme 9). Diastereoisomerically pure compounds **12a–c,e,g** were obtained by recrystallization.

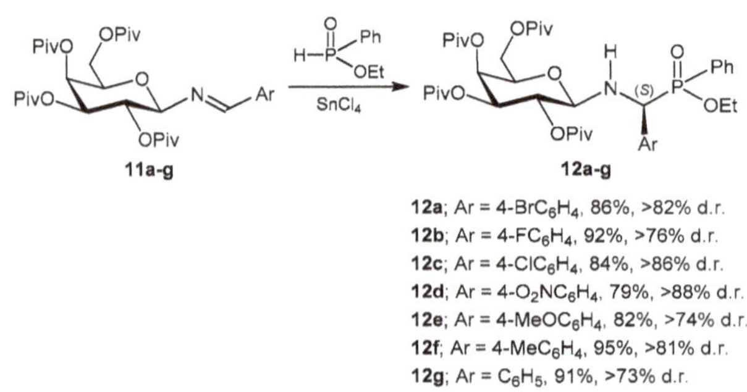

Scheme 9. Nucleophilic addition of ethyl phenylphosphinate to *N*-galactosylaldimines **11a–g**.

The (*S*)-configuration at the α-carbon of the major diastereoisomers **12a–g** can be explained by the attack of ethyl phenylphosphinate from *si* face of (*E*)-imines **11a–g**. The imine-$SnCl_4$ complex and simultaneous chelation with the auxiliary's pivaloyl group inhibits free rotation along the *N*-anomeric carbon bond. According to this rationalization, the attack of the phosphinate in its P^{III} phosphonous tautomeric form proceeds from the back face of the imine **11a–g**, as shown in Scheme 10 [45].

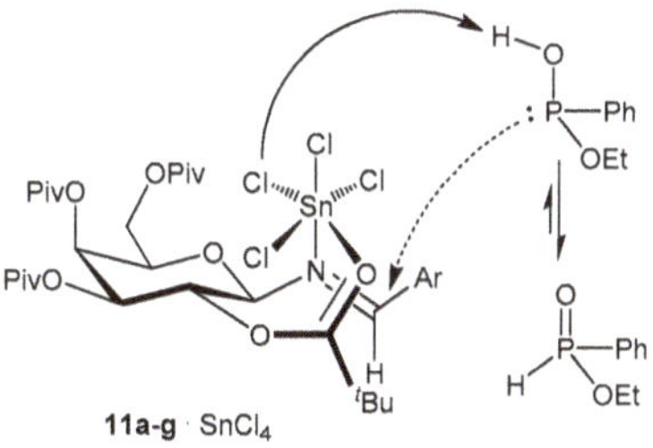

11a-g · $SnCl_4$

Scheme 10. Plausible reaction mechanism for the addition of ethyl phenylphosphinate to **11a–g**.

With the aim to assign the configuration, the reaction of *N*-galactosyl α-aminoalkyl-*C*-phosphinates **12a–g** with 1 M HCl/MeOH at room temperature produced the α-amino-*C*-phosphinate hydrochlorides **13a–g** in 78%–95% yield, which by hydrolysis with 1.5 M HCl under reflux followed by treatment with propylene oxide in ethanol at reflux, afforded the α-amino-*C*-phosphinic acids **6d** and **14a–f** in 82%–93% yield and with 73%–97% enantiomeric excess (Scheme 11) [45].

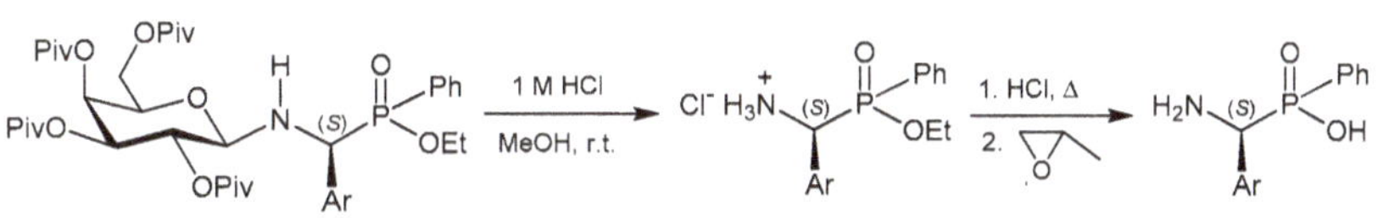

12a-g (*S*)-**13a-g** (*S*)-**6d**,**14a-f**

13a; Ar = 4-BrC_6H_4, 90%
13b; Ar = 4-FC_6H_4, 93%
13c; Ar = 4-ClC_6H_4, 78%
13d; Ar = 4-$O_2NC_6H_4$, 87%
13e; Ar = 4-MeOC_6H_4, 95%
13f; Ar = 4-MeC_6H_4, 85%
13g; Ar = C_6H_5, 87%

14a; Ar = 4-BrC_6H_4, 87%, 78% e.e.
14b; Ar = 4-FC_6H_4, 82%, 90% e.e.
14c; Ar = 4-ClC_6H_4, 90%, 94% e.e.
14d; Ar = 4-$O_2NC_6H_4$, 93%, 73% e.e.
14e; Ar = 4-MeOC_6H_4, 90%, 88% e.e.
14f; Ar = 4-MeC_6H_4, 91%, 97% e.e.
6d; Ar = C_6H_5, 88%, 84% e.e.

Scheme 11. Synthesis of the α-amino-*C*-phosphinic acids **6d** and **14a–f**.

Readily available enantiopure sulfinylimines also constitute valuable molecules in asymmetric synthesis [46]. Indeed, Petneházy et al. [47] reported the synthesis of optically enriched α-amino-*C*-phosphinic acids using ethyl phenylphosphinate and chiral *N*-sulfinylaldimines. In this context, the nucleophilic addition of ethyl phenylphosphinate to the chiral imines (*S*)-**15a–c** in toluene at 70 °C, furnished the α-amino-*C*-phosphinates (R_C,S_S)-**16a–c** in 19%–35% yield, which by simultaneous removal of the *N*-sulfinyl auxiliary and hydrolysis of the ethyl phosphinate with concentrated HCl at reflux, gave the α-amino-*C*-phosphinic acids (*R*)-**6a,d** and (*R*)-**17a** in 48%–58% yield and 20%–40% enantiomeric excesses (Scheme 12).

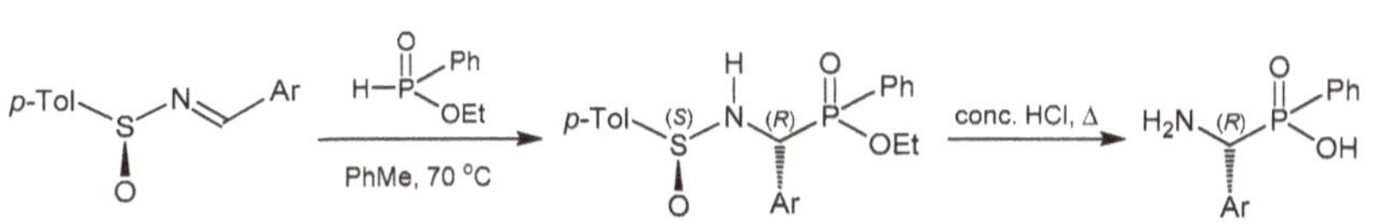

(*S*)-**15a-c** (R_C,S_S)-**16a-c** (*R*)-**6a,d,17a**

16a; Ar = 2-ClC_6H_4, 22%
16b; Ar = 3-MeOC_6H_4, 35%
16c; Ar = C_6H_5, 19%

17a; Ar = 2-ClC_6H_4, 58%, 30% e.e.
6a; Ar = 3-MeOC_6H_4, 48%, 40% e.e.
6d; Ar = C_6H_5, 52%, 20% e.e.

Scheme 12. Synthesis of α-amino-*C*-phosphinic acids from chiral *N*-sulfinylaldimines (*S*)-**15a–c**.

2.1.2. Chiral Catalyst

The stereoselective hydrophosphinylation of aldimines under organocatalytic conditions is emerging as a suitable method for the synthesis of optically active α-amino-C-phosphinates, in this regard the guanidines and guanidinium salts have shown to be highly efficient catalysts for enantioselective reactions [48]. For example, following the principle of activation of imines through hydrogen bonding, Tan et al. [49] carried out the hydrophosphinylation reaction of the N-tosyl imines **18a–g** with arylphosphinates and alkylphosphinates in the presence of catalytic amounts of the guanidinium salt **19**·HBArF$_4$ in a CH_2Cl_2:toluene mixture and K_2CO_3 as a base at −40 °C, obtaining the α-amino-C-phosphinates **20a–k** in 71%–93% yield and 3:1–16:1 diastereoisomeric ratio, with preference for the diastereoisomers (R_C,S_P)-**20a–k** with 82%–94% enantiomeric excess (Scheme 13).

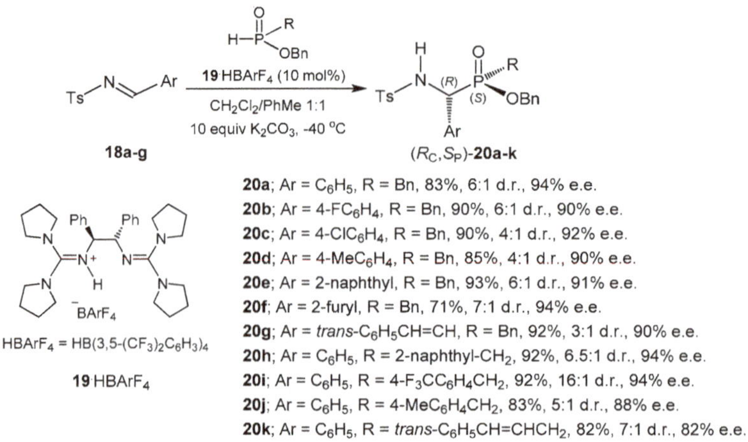

Scheme 13. Hydrophosphinylation of aldimines **18a–g** under organocatalytic conditions.

2.2. *Stereoselective C-P Bond Formation (One-Pot Three-Component Reaction)*

2.2.1. Chiral Amino Compounds

Possibly the simplest experimental proposal to prepare optically active α-amino-C-phosphinates is the "one-pot" three-component reaction, where the reactants are placed all together with or without solvent and catalyst, which is identical to the Kabachnik-Fields reaction widely studied in the synthesis of α-aminophosphonates [50–52].

Meng and Xu [19] reported a direct synthetic route for the preparation of phosphinopeptides containing C-terminal α-aminoalkyl-C-phosphinic acids, valuable peptide mimics. Thus, the "one-pot" three-component reaction of N-Cbz-aminoalkanamides (S)-**21a–c** obtained from N-protected amino acids, with benzaldehyde and phenyldichlorophosphine in dry acetonitrile at 80 °C followed by hydrolysis, produced the phosphinopeptides **22a–c** in 65%–75% yield and low degree of asymmetric induction, obtaining the (S,S)-**22a–c** diastereoisomers as major products (Scheme 14).

According to the authors, the reaction of the N-Cbz-aminoalkanamide, benzaldehyde and phenyldichlorophosphine produce the N-acylimine **23b** and phenylchlorophosphonous acid, which attacks to the imine favorably from the *si* face because the amino acid substituent blocks the opposite *re* face. Finally, a proton transfer affords the intermediate aminoalkylphosphinic chloride, which undergoes hydrolysis gave the (S,S)-phosphinopeptide **22b** as major product (Scheme 15) [19].

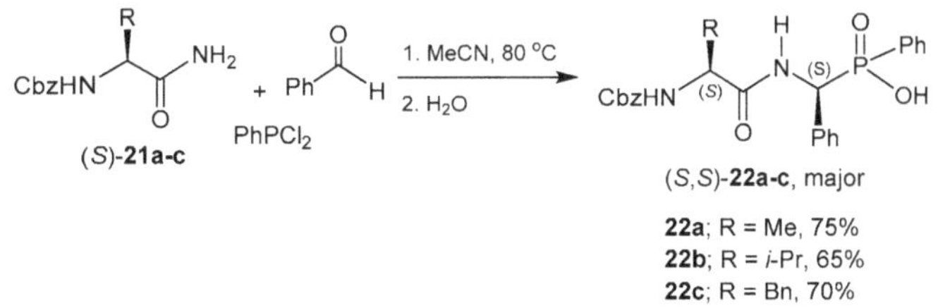

Scheme 14. Preparation of phosphinopeptides (*S*,*S*)-**22a–c**.

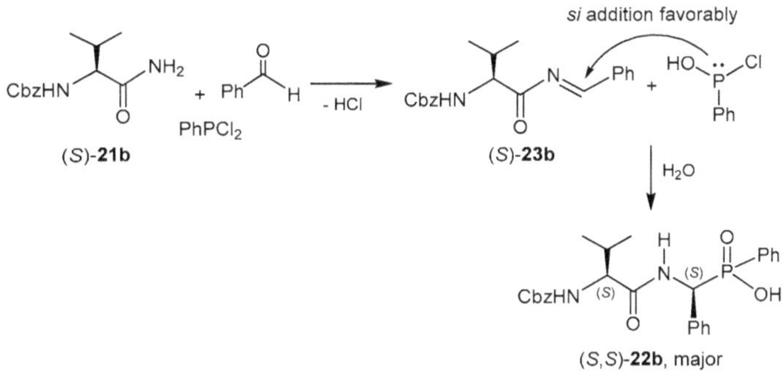

Scheme 15. Mechanism for the formation of diastereoisomeric phosphinopeptide (*S*,*S*)-**22b**.

Similarly, a series of phosphinodepsipeptides were synthesized via a pseudo-four-component condensation reaction, in a convergent and atom economic synthetic strategy [53]. In this context, the "one-pot" reaction of *N*-Cbz-amino amides derived from amino acids with different aldehydes and aryldichlorophosphines in dry acetonitrile at 80 °C followed by alcoholysis with α-hydroxy esters or ethanol, produced the phosphinodepsipeptides **24a–k** in 50%–66% yield and low diastereo selectivities, where the (R_C*,R_P*)-products were assumed as major diastereoisomers (Scheme 16).

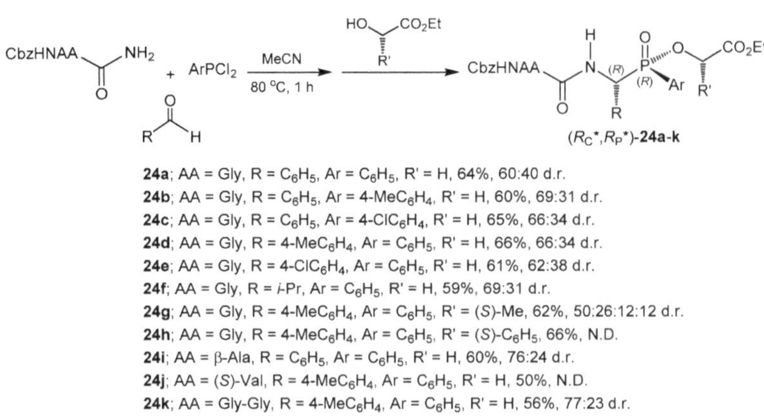

Scheme 16. Synthesis of phosphinodepsipeptides (R_C*,R_P*)-**24a–k**.

In a similar way, Meng et al. [54] reported the preparation of hybrid sulfonophosphinopeptides incorporating α-aminoalkyl-*C*-phosphinic acids. The compounds of interest were synthesized by reaction of the *N*-Cbz protected aminoalkanesulfonamides, benzaldehyde and phenyldichloro phosphine in dry acetonitrile at 45 °C and subsequent hydrolysis at room temperature, obtaining the sulfonophosphinodipeptides **26a–d** in good yield, but low diastereoselectivity. The diastereo isomers (*S*,*S*)-**26a–d** were assumed as the major products (Scheme 17).

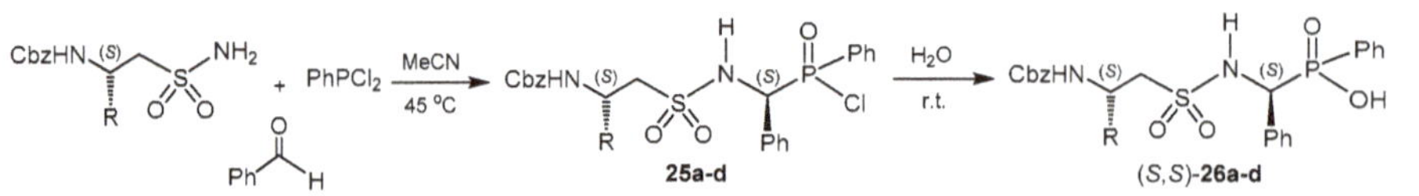

26a; R = Me, 75%, (*S*,*S*):(*S*,*R*) 54:46 d.r.
26b; R = Bn, 72%, (*S*,*S*):(*S*,*R*) 57:43 d.r.
26c; R = *i*-Pr, 71%, (*S*,*S*):(*S*,*R*) 59:41 d.r.
26d; R = *i*-Bu, 80%, (*S*,*S*):(*S*,*R*) 66:34 d.r.

Scheme 17. Synthesis of the sulfonophosphinodipeptides **26a–d**.

2.3. Stereoselective C-H Bond Formation

2.3.1. Reduction of C=C Bond in α,β-Dehydroaminophosphinates

Catalytic asymmetric hydrogenation of α,β-dehydroaminophosphinates is another nice method for the synthesis of optically pure α-amino-C-phosphinic acids and its derivatives. For example, Drauz et al. [55] described for the first time the catalytic asymmetric hydrogenations of the α,β-dehydroaminophosphinates **27a–e** [56] using the rhodium complex **28** in methanol at room temperature and 0.1 MPa H_2-pressure, obtaining the α-amino-C-phosphinates **29a–e** in quantitative yield and 81%–87% enantiomeric excesses (Scheme 18).

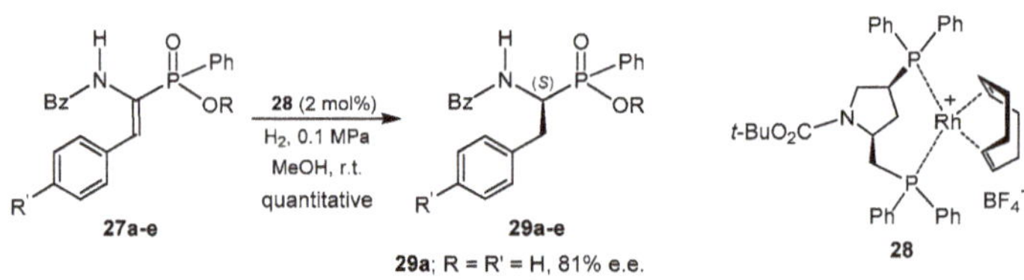

29a; R = R' = H, 81% e.e.
29b; R = Me, R' = H, 86% e.e.
29c; R = Et, R' = H, 87% e.e.
29d; R = Et, R' = F, 87% e.e.
29e; R = Et, R' = *i*-Pr, 87% e.e.

Scheme 18. Catalytic asymmetric hydrogenation of the α,β-dehydroaminophosphinates **27a–e**.

Additionally, Oehme et al. [57] conducted a comprehensive study on the hydrogenation of α,β-dehydroaminophosphinates evaluating different chiral rhodium complexes, substrate-catalyst ratio, temperatures, hydrogen pressures, modified substrates and both organic and aqueous solvents. Thus, under optimized conditions, they carried out the catalytic hydrogenation of the α,β-dehydroaminophosphinates **27a–g** [56] in the presence of catalytic amounts of [Rh(cod)(*S*,*S*)-bppm]BF_4 **28** in aqueous micellar media at room temperature and 1 bar H_2-pressure, obtaining the α-amino-C-phosphinates **29a–g** in 66%–100% yield and with 53%–97% enantiomeric excess (Scheme 19).

Furthermore, Darses et al. [58] reported an alternative approach for the synthesis of α-amino-C-phosphinates through the asymmetric rhodium-catalyzed addition of organoboron derivatives to α,β-dehydroaminophosphinates. Indeed, the addition of potassium phenyltrifluoroborate to the α,β-dehydroaminophosphinate **30** [59] in the presence of $NaHCO_3$, $[RhCl(CH_2CH_2)_2]_2$ as catalyst precursor and (*S*)-DifluorPhos **31** as chiral ligand in isopropyl alcohol at 90 °C, afforded the diastereoisomeric α-amino-C-phosphinates **32** and **33** in moderate yield and with 94% and 92% enantiomeric excesses, respectively (Scheme 20).

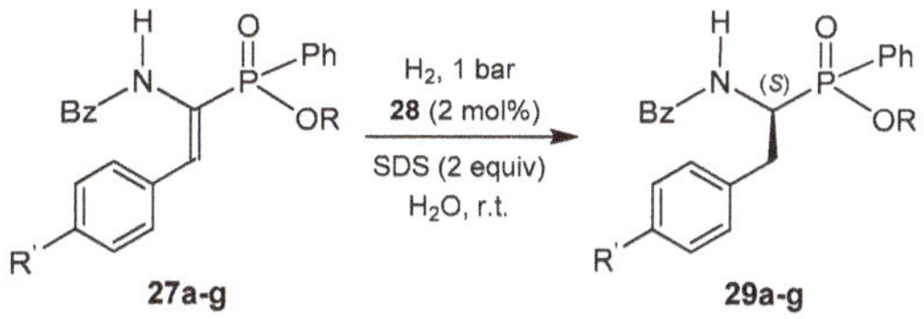

27a-g → **29a-g**

29a; R = R' = H, n.d., 53% e.e.
29b; R = Me, R' = H, 96%, 97% e.e.
29c; R = Et, R' = H, 100%, 96% e.e.
29d; R = Et, R' = F, 66%, 96% e.e.
29e; R = Et, R' = *i*-Pr, 95%, 93% e.e.
29f; R = Me, R' = Me, 80%, 97% e.e.
29g; R = Et, R' = Me, 100%, 96% e.e.

Scheme 19. Optimized conditions for the catalytic hydrogenation of **27a–g**.

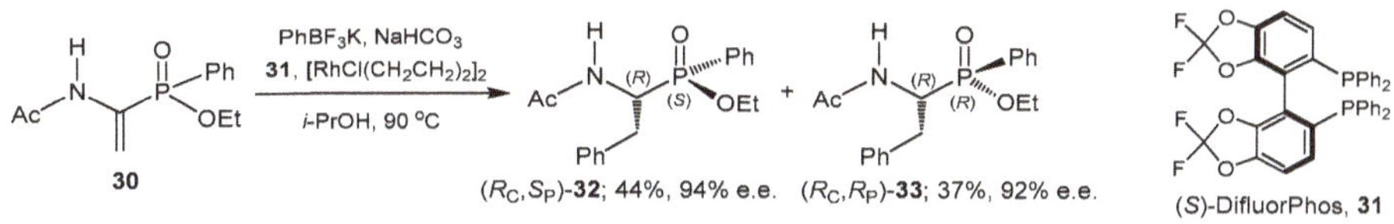

Scheme 20. Asymmetric rhodium-catalyzed addition of phenyltrifluoroborate to **30**.

2.4. Resolution Methodologies

The resolution strategies of racemic compounds have also demonstrated its usefulness for the preparation of optically enriched α-amino-*C*-phosphinic acids. For example, Lämmerhofer et al. [60] carried out the chiral HPLC separation of *N*-Cbz phosphinic pseudopeptide esters **34a–d** as well as their free C-terminal carboxylic derivatives **34e–h** using a set of cinchona alkaloid-derived chiral anion-exchangers **35**,**36** (Scheme 21). Semi-preparative scale chromatography supplied single enantiomers in 100 mg quantities allowing the biological activity assays.

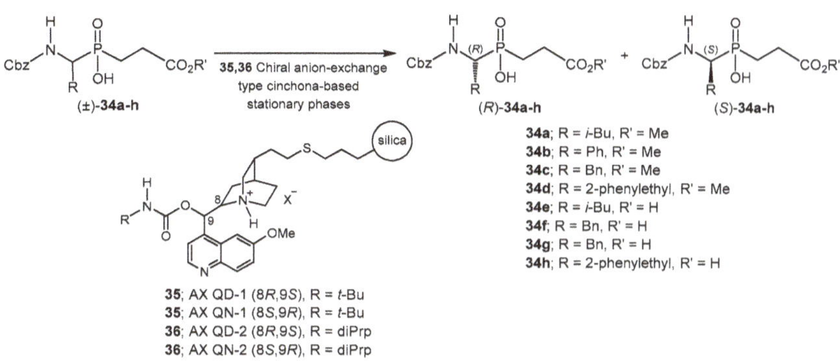

Scheme 21. Chiral HPLC separation of *N*-Cbz phosphinic pseudopeptides **34a–h**.

In a related work, Lämmerhofer et al. [61] developed a capillary electrochromatography (CEC) method for the separation of the stereoisomers of *N*-Cbz phosphinic pseudodipeptide methyl ester **37a** and **38a** and its *N*-DNP derivative with free *C*-terminal carboxylic group **37b** and **38b**, using a monolithic silica capillary column modified with a cinchona alkaloid-derived anion-exchange-type chiral selector **39** (Scheme 22). This method proved superiority compared to the HPLC separation, which was attributed to significantly enhanced plate numbers.

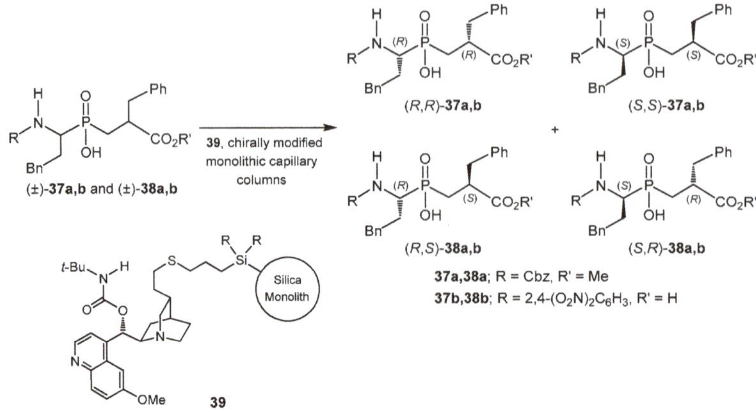

Scheme 22. Capillary electrochromatography (CEC) method for the separation of (±)-**37a,b** and (±)-**38a,b**.

Taking into account that phosphinic dipeptides, although tested as stereoisomeric mixtures, are considered amongst the most potent inhibitors of bizinc cytosolic leucine aminopeptidase (LAP) [62], Mucha, et al. [63] carried out the chiral HPLC separation of all stereoisomers of phosphinic dipeptide homophenylalanyl-phenylalanine derivative **40** and **41** on a quinidine carbamate modified silica stationary phase **42** (Scheme 23). Significant differences in inhibition of the unprotected derivatives **43** and **44** show the importance of the molecular stereochemistry in spatial interaction with the enzyme active site.

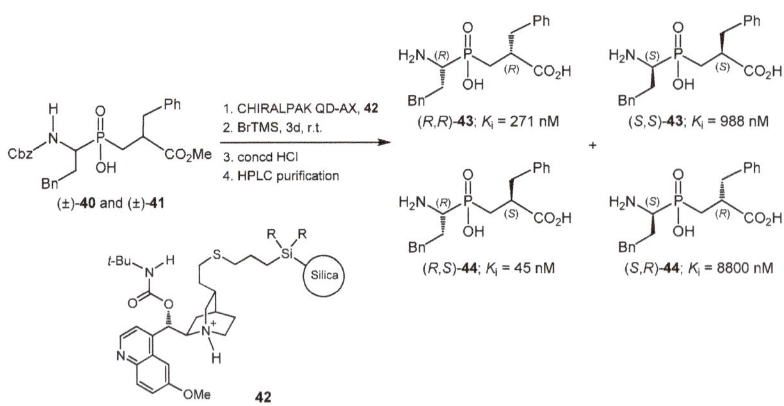

Scheme 23. Chiral HPLC separation of phosphinic dipeptide derivatives **40** and **41** and activities of unprotected derivatives **43** and **44** as inhibitors of leucine aminopeptidase (LAP).

In another example, Ragulun and Rozhko [64] reported the preparation of the optically pure α-aminobenzyl hydroxymethyl phosphinic acid (*R*)- and (*S*)-**45**, where the key step was a biocatalytic resolution. Thus, the racemic α-aminobenzyl hydroxymethyl phosphinic acid (±)-**45**, obtained from the hydroxymethyl phosphinic acid, was acylated with phenylacetyl chloride and KOH in water to obtain the *N*-phenylacetyl derivative **46** in 63% yield, which by hydrolysis using penicillin amidase (PcAm) as the biocatalyst, afforded the enantiomerically pure (*R*)-**46** in 62% yield and the deacylated product (*S*)-**45** in 57% yield. Finally, hydrolysis of compounds (*R*)-**46** with 6 N HCl at reflux followed by treatment with propylene oxide, provided the α-amino-*C*-phosphinic acid (*R*)-**45** in 84% yield (Scheme 24).

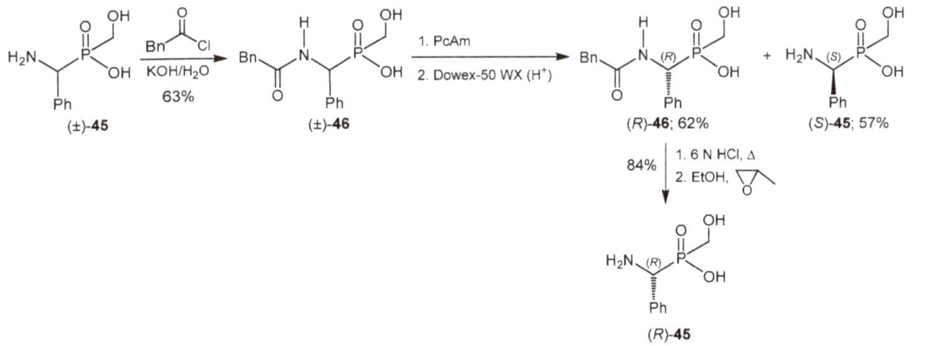

Scheme 24. Synthesis of optically pure α-aminobenzyl hydroxymethyl phosphinic acid (*R*)- and (*S*)-**45**.

2.5. Conversion from Chiral α-Aminophosphonates

Undoubtedly, a useful but not as explored strategy for the synthesis of optically enriched α-amino-*C*-phosphinic acids is their conversion from the corresponding enantiomerically pure α-aminophosphonic counterparts [8]. Considering this possibility, Pyun et al. [65,66] carried out the transformation of chiral phosphonate (1*S*,2*S*)-**47** into *C*-phosphinates (1*S*,2*S*)-**50a–l**. Thus, the reaction of the optically pure diethyl (1*S*,2*S*)-1-amino-2-vinylcyclopropanephosphonate **47** [67] with benzyl chloroformate and $NaHCO_3$ followed by treatment with NaI in pyridine at reflux, gave the phosphonate monoester (1*S*,2*S*)-**48** in 88% yield. The reaction of (1*S*,2*S*)-**48** with oxalyl chloride and DMF in MeCN, afforded the phosphonomonochloridate (1*S*,2*S*)-**49**, that without additional purification was reacted with different organolithium or Grignard reagents, to obtain the α-amino-*C*-phosphinates (1*S*,2*S*)-**50a–d** in 37%–56% yield, which have been used in the design and synthesis of new enzyme inhibitors (Scheme 25).

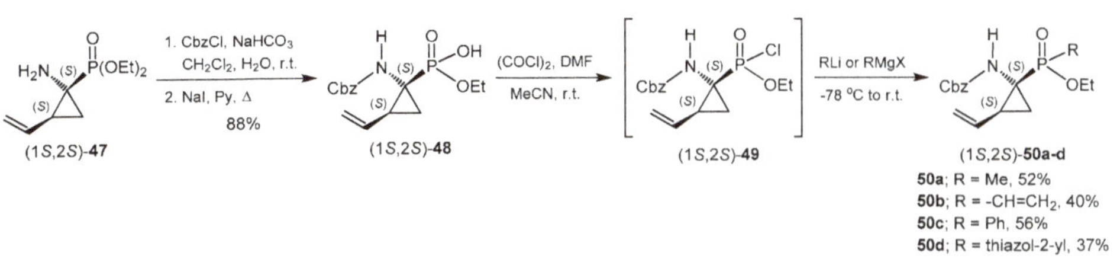

Scheme 25. Transformation of chiral phosphonate (1*S*,2*S*)-**47** into *C*-phosphinates (1*S*,2*S*)-**50a–d**.

On the other hand, the reduction of the phosphonomonochloridate (1*S*,2*S*)-**49** with $LiAlH(Ot\text{-}Bu)_3$ in THF at −78 °C, produced the *H*-phosphinate (1*S*,2*S*)-**51** in 78% yield, which by alkylation either by activation with chlorotrimethylsilane (TMSCl) and diisopropyl ethylamine (DIPEA) followed by addition of an alkyl halide (method A), or using NaHMDS and subsequent addition of an alkyl halide (method B), furnished the *P*-alkylated α-amino-*C*-phosphinates (1*S*,2*S*)-**50e–l** in 27%–63% yield (Scheme 26) [65,66].

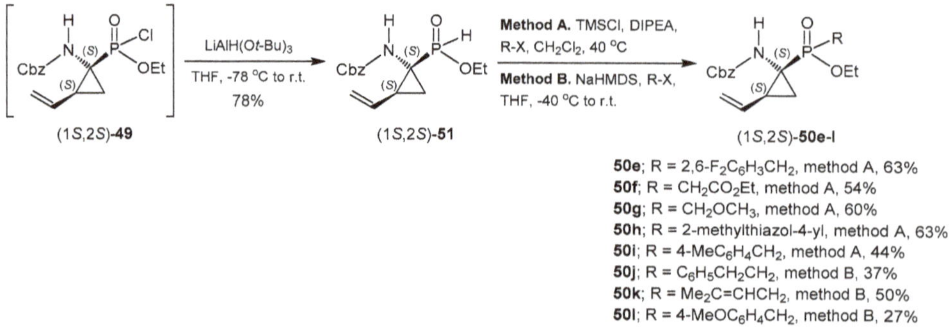

Scheme 26. Transformation of chiral phosphonate (1*S*,2*S*)-**47** into *C*-phosphinates (1*S*,2*S*)-**50e–l**.

Additionally, the reaction of (1*S*,2*S*)-**50a–l** with iodotrimethylsilane in MeCN produced the simultaneous deprotection of both the Cbz group and the ethyl ester, to obtain the α-amino-*C*-phosphinic acids (1*S*,2*S*)-**52a–l**. On the other hand, cleavage of *N*-Cbz bond in (1*S*,2*S*)-**50a–l** with TFA-SMe_2 [68] gave the α-amino-*C*-phosphinates (1*S*,2*S*)-**53a–l**. Both, the α-amino-*C*-phosphinic acids (1*S*,2*S*)-**52a–l** and *C*-phosphinates (1*S*,2*S*)-**53a–l** were used in the synthesis of the acyclic **54a–l** and cyclic **55a–l** *C*-phosphinate analogs of BI-2061 [66], respectively, and their inhibition of the HCV NS3 protease was tested (Scheme 27).

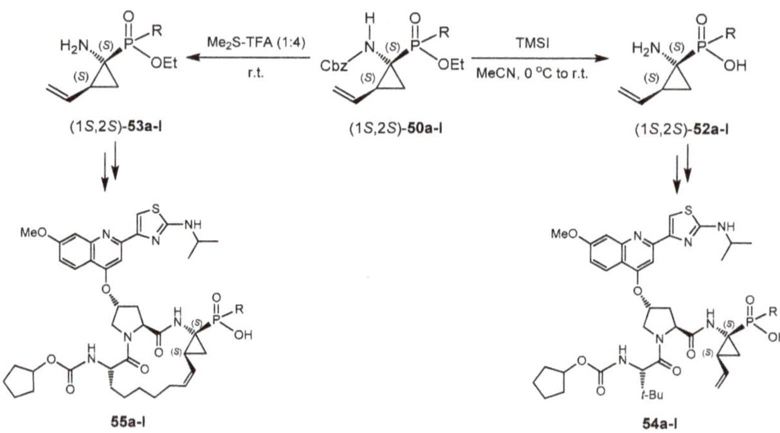

Scheme 27. Synthesis of the acyclic **54a–l** and cyclic **55a–l** *C*-phosphinate analogs of BI-2061.

The acyclic *C*-phosphinate analog **54a** showed an IC_{50} of 6 nM, whereas the carboxylic **56** and phosphonic **57** derivatives were found to have an IC_{50} of 3 and 0.9 nM, respectively. The lower inhibition of the *C*-phosphinic compared to its phosphonic analog was attributed to the loss of *H*-bonding interaction (Figure 3) [66].

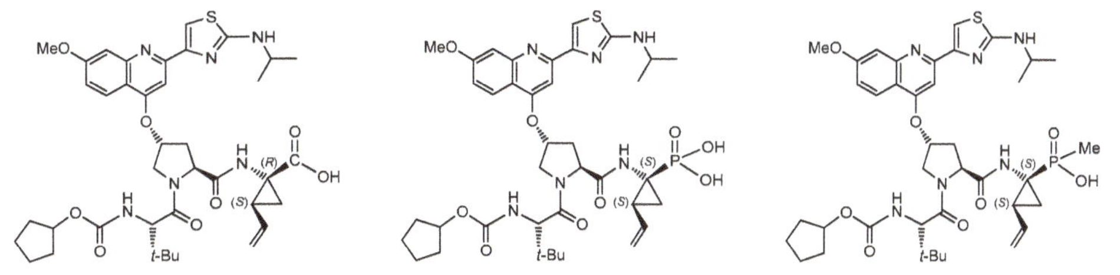

Figure 3. Carboxylic **56**, phosphonic **57** and *C*-phosphinic **54a** analogs of BI-2061 and activities as inhibitors of the HCV NS3 protease.

Recently, Hammerschmidt et al. [69] reported the *s*-BuLi or LiTMP induced phosphonate-phosphinate rearrangement of *N*-Boc α-aminophosphonates through deprotonation- metallation at the methoxy group. In this context, the treatment of enantiomerically pure *N*-Boc α-aminophosphonate **58** with 2.5 equiv of LiTMP in dry THF at −78 °C followed by acidic workup, provided the *C*-phosphinates (R,S_P)-**64** and (R,R_P)-**65** in 12% yield and with 22:78 diastereoisomeric ratio, via the metallated phosphonates (R_C,S_P)-**60** and (R_C,R_P)-**61** and the ensuing rearrangement to *C*-phosphinates (R_C,S_P)-**62** and (R_C,R_P)-**63**, respectively (Scheme 28).

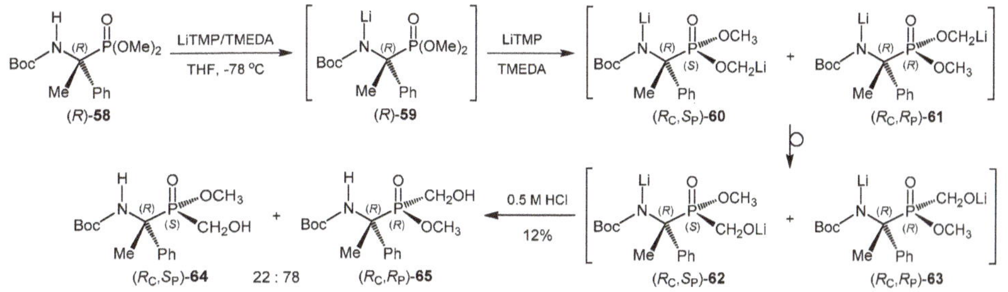

Scheme 28. LiTMP induced phosphonate-phosphinate rearrangement of the α-aminophosphonate (*R*)-**58**.

2.6. P-C Bond Formation from Chiral α-Amino-H-phosphinic Derivatives

There is a continuously growing interest in phosphorus-carbon bond formation due to the great applications of organophosphorus compounds in synthetic organic and bioorganic chemistry [5,70]. In this context, the chiral α-amino-*H*-phosphinates have also proved efficiency as *P*-nucleophiles in Michael additions. In this regard, Yiotakis et al. [27,71] reported the synthesis of dehydroalaninyl phosphinic dipeptide analogues, which were tested as zinc-metalloproteases MMP-8 inhibitors. For this purpose, the optically pure α-amino-*H*-phosphinic acid (*R*)-**66a** [72] was reacted with TMSCl/DIPEA in CH_2Cl_2 followed by addition of *tert*-butyl 2-(bromomethyl)acrylate, obtaining the dehydroalanine derivative (*R*)-**67** in 94% yield, presumably via an allylic rearrangement. Cleavage of the *O-tert*-butyl bond in (*R*)-**67** with trifluoroacetic acid (TFA) and coupling with L-tryptophanyl-amide, gave the pseudotripeptide (*R*,*S*)-**68** in 69% yield. Finally, the Michael-type addition of benzyl thiols to (*R*,*S*)-**68**, produced the cysteine analogues **69a–c** in good yield (Scheme 29).

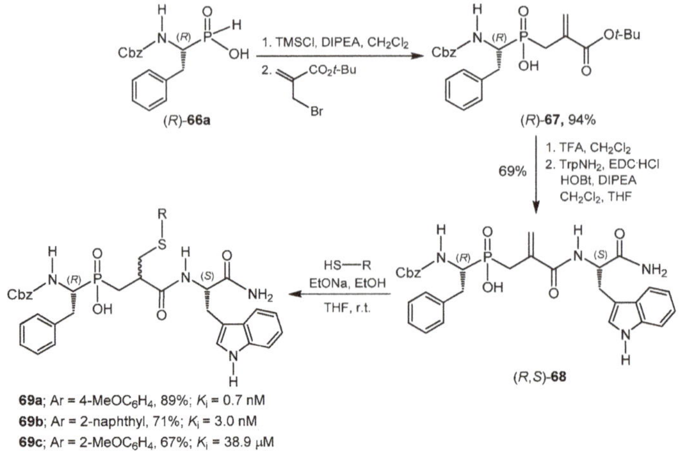

Scheme 29. Synthesis of phosphinic pseudotripeptides **69a–c** and their activities toward MMP-8 (human neutrophile collagenase, HNC).

On the other hand, McKittrick et al. [73] reported the synthesis of the phosphinic pseudotripeptide (*R*,*S*)-**71**, which not only inhibits the metalloprotease endothelin converting enzyme (ECE), but also the angiotensin converting enzyme (ACE) and neutral endopeptidase (NEP) with IC_{50} values <100 nM. Thus, the hydroxyphosphinyl group in enantiopure Cbz-protected phenylalanine phosphinic analogue (*R*)-**66a** [72] was reacted with diazomethane, followed by treatment with LDA in THF at −78 °C and subsequent addition of the triflate, to afford the *P*-alkylated α-amino-*C*-phosphinate (*R*)-**70** in 43% yield, which was further transformed into pseudotripeptide (*R*,*S*)-**71** (Scheme 30).

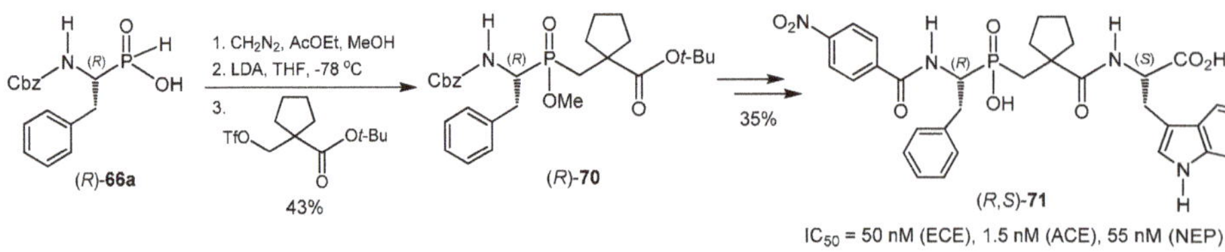

Scheme 30. Synthesis of the phosphinic pseudotripeptide (*R*,*S*)-**71** and its activities as inhibitor of endothelin converting enzyme (ECE), angiotensin converting enzyme (ACE) and neutral endopeptidase (NEP).

Hamilton et al. [74] reported the synthesis of phenyl phosphinate derivatives **73a,b** from Cbz-protected phenylalanine and valine phosphinic analogues [72] and their inhibitory properties against human neutrophile elastase (HNE) (valine analogue) and chymotrypsin (phenylalanine analogue) were evaluated, determining specific inhibitory activity for analogues having (*R*) configuration at α-carbon. Thus, the reaction of the α-amino-*H*-phosphinic acids (*R*)-**66a,b** with TMSCl and Et_3N in CH_2Cl_2 at 0 °C followed by the addition of ethyl acrylate, produced the Michael adducts (*R*)-**72a,b**, which by reaction with thionyl chloride ($SOCl_2$) in CH_2Cl_2 followed by addition of phenol, provided the phenyl *C*-phosphinates (*R*)-**73a,b** with 91:9 and 89:11 diastereoisomeric ratio, respectively (Scheme 31).

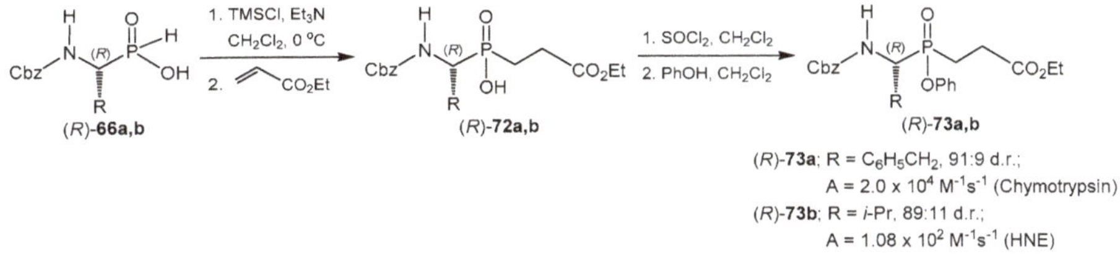

Scheme 31. Synthesis of phenyl phosphinate derivatives **73a,b** and second-order rate constants [A = k/K$_i$] of inhibition of chymotrypsin and human neutrophile elastase (HNE).

Additionally, Lloyd et al. [75] synthesized the *C*-phosphinic acid **75** and tested its effect on ECE inhibition. In this context, the reaction of α-amino-*H*-phosphinic acid (*R*)-**66a** obtained by resolution [72], with *N*,*O*-bis(trimethylsilyl)trifluoroacetamide (BSTFA) followed by the Michael addition to ethyl 2-benzylacrylate and subsequent basic hydrolysis of the ethoxycarbonyl group, produced the *C*-phosphinic acid **74**. Finally, the reaction of **74** with oxalyl chloride, followed by coupling with (*S*)-Trp-OMe and subsequent saponification with LiOH, afforded the phosphinic pseudopeptide **75** as a 4:1 diastereoisomeric mixture (Scheme 32).

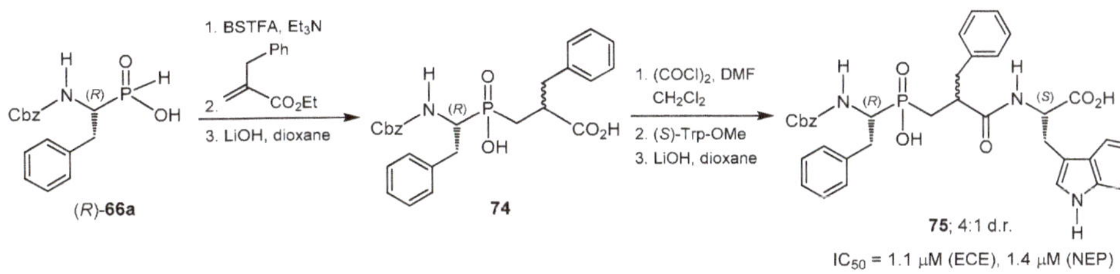

Scheme 32. Synthesis of the *C*-phosphinic acid **75** and its activities as inhibitor of endothelin converting enzyme (ECE) and neutral endopeptidase (NEP).

Roques et al. [76] reported the synthesis of the iodinated tripeptide (*R*,*S*,*S*)-**78** analogue of a highly efficient aminopeptidase N inhibitor, as a tool for complete characterization of the biochemical and pharmacological properties of this enzyme. The Michael addition of *N*-Cbz alanine *H*-phosphinic analogue (*R*)-**66c** obtained by resolution [72], to the methyl benzylacrylate in the presence of *N*,*O*-bistrimethylsilylacetamide (BSA), followed by alkaline hydrolysis and subsequent coupling with the L-tyrosine benzyl ester, produced the tripeptide **76** in 57% yield as a diastereoisomeric mixture, which by saponification of benzyl ester and Cbz removal under acidic conditions, gave the (*R*,*S*,*S*)-**77** diastereoisomer in 47% yield after HPLC separation. Finally, the radioiodination of (*R*,*S*,*S*)-**77** was performed by using $Na^{125}I$ in sodium hydroxide solution and chloramine-T, providing the radiolabeled compound (*R*,*S*,*S*)-**78** (Scheme 33).

In a similar way, Roques et al. [77] reported the synthesis of the phosphinic pseudopeptide (*R*,*S*,*S*)-**81**, which proved to be a potent dual inhibitor of the zinc metallopeptidases neprilysin and aminopeptidase N. For this purpose, the Michael addition of *N*-Cbz alanine *H*-phosphinic acid (*R*)-**66c** obtained by resolution [72], to ethyl *p*-bromobenzylacrylate followed by a Suzuki coupling using phenylboronic acid and subsequent alkaline hydrolysis of the ethyl ester, gave the *C*-phosphinic acid derivative **79** in 75% yield, which by the coupling with (*S*)-alanine methyl ester afforded the tripeptide **80** in 59% yield. Finally, the saponification of the methyl ester followed by the cleavage of the *N*-Cbz bond with BBr_3 and the semipreparative HPLC purification, provided the enantiomerically pure phosphinic pseudopeptide (*R*,*S*,*S*)-**81** in 92% yield (Scheme 34).

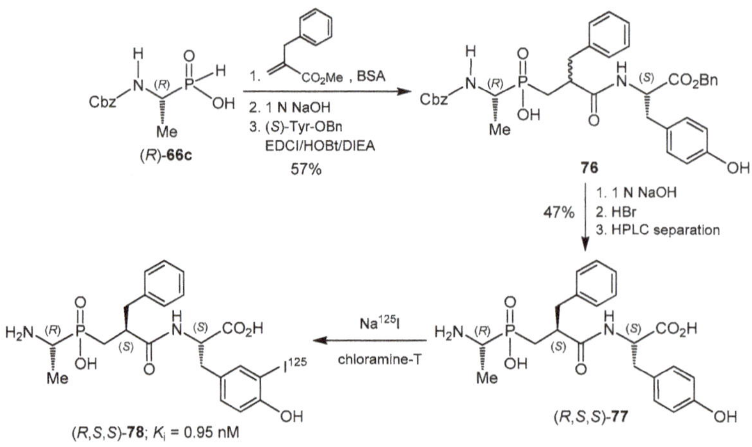

Scheme 33. Synthesis of the iodinated tripeptide (*R*,*S*,*S*)-**78** and its activity as inhibitor of aminopeptidase N.

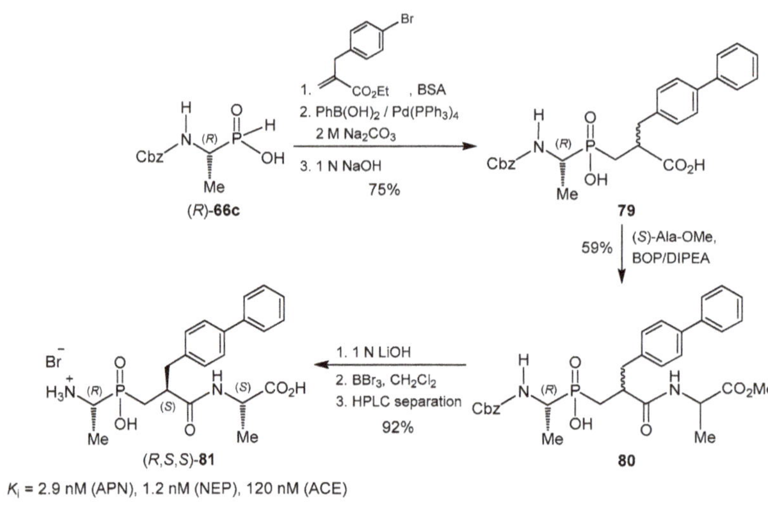

Scheme 34. Synthesis of the phosphinic pseudopeptide (*R*,*S*,*S*)-**81** and its activities as inhibitor of aminopeptidase N (APN), neutral endopeptidase (NEP) and angiotensin converting enzyme (ACE).

In a related work, Samios et al. [78] developed an efficient synthetic methodology for the preparation of the phosphinic pseudotripeptide known as RXP03, which has been widely studied as MMPs inhibitor [79,80]. Initially, reaction of the *N*-Cbz *H*-phosphinic acid analogue of phenylalanine (*R*)-**66a** [72] with bis(trimethylsilyl)amine (HMDS) followed by the addition of ethyl phenylpropyl-acrylate and subsequent hydrolysis of the ethyl ester group, gave *C*-phosphinate **82** in 90% yield, which by coupling with (*S*)-TrpNH$_2$, afforded the phosphinic pseudotripeptides (*R*,*S*,*S*)-**83** and (*R*,*R*,*S*)-**84** in 77% yield, which were separated in 99% purity by preferential precipitation in EtOH (Scheme 35). This methodology allows the separation of the diastereoisomers based in their solubility differences, which was explained by conformational and solvation theoretical studies in terms of their intra- and intermolecular structure.

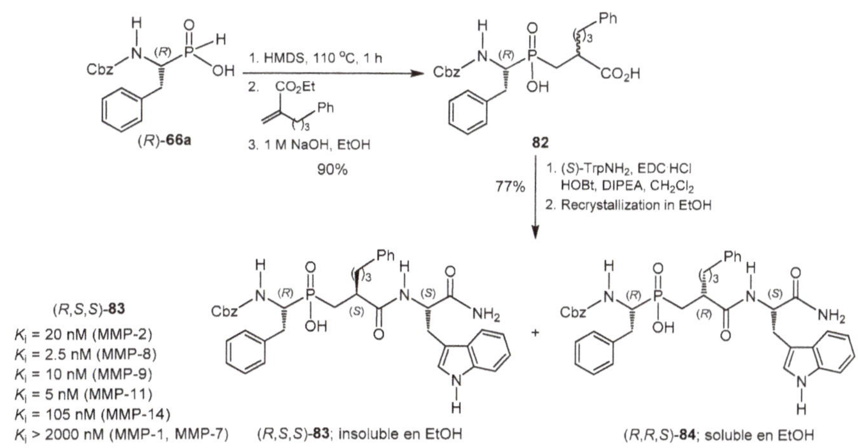

Scheme 35. Synthesis of the phosphinic pseudotripeptides (*R,S,S*)-**83** and (*R,R,S*)-**84** and its activities of the most potent diastereoisomer of RXP03 [(*R,S,S*)-**83**] as inhibitor of matrix metalloproteinases.

Similarly, the phosphinic pseudotripeptides (*S,S,S*)-**85** and (*S,R,S*)-**86** were obtained from the enantiomeric pure *N*-Cbz *H*-phosphinic acid analogue of phenylalanine (*S*)-**66a** in 64% yield. As in the previous example, each diastereoisomer showed different solubility properties in ethyl ether allowing their complete separation (Scheme 36) [78].

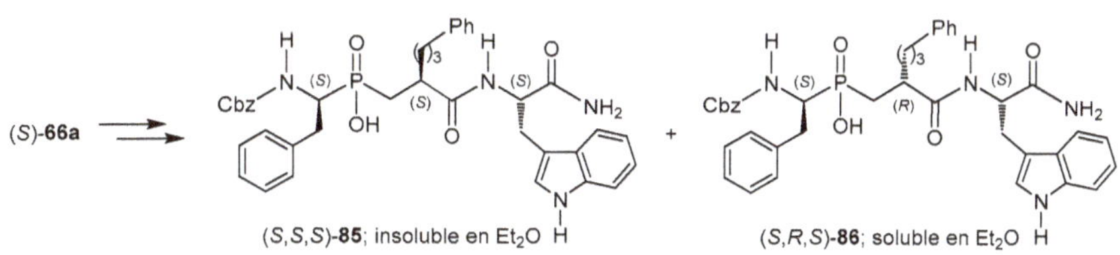

Scheme 36. Synthesis of the phosphinic pseudotripeptides (*S,S,S*)-**85** and (*S,R,S*)-**86**.

Considering that phosphinic peptides are well recognized as peptide isosters and powerful inhibitors of many classes of enzymes, mainly zinc proteases, Yiotakis et al. [81,82] reported the synthesis of phosphinopeptidic building blocks incorporating a triple bond, which through 1,3-dipolar cycloaddition process, gives access to novel class of isoxazole-containing phosphinic peptides. Thus, the reaction of *N*-Cbz *H*-phosphinic acid analogue of phenylalanine (*R*)-**66a** [72] with TMSCl/DIPEA followed by the addition of ethyl α-propargylic acrylate and saponification of the ethyl ester, gave the *C*-phosphinates **87** in 96% yield and 3.5:1 diastereoisomeric ratio. Coupling of **87** with (*S*)-TrpNH$_2$ afforded the tripeptide **88** as a diastereoisomeric mixture, from which the diastereoisomer (*R,S,S*)-**88** was isolated by means of simple crystallization. Finally, the 1,3-dipolar cycloaddition reaction of (*R,S,S*)-**88** with different nitrile oxides generated in situ from the chlorination of the corresponding oxime, furnished the isoxazoles (*R,S,S*)-**89a–d** in good yields (Scheme 37).

On the other hand, the 1,3-dipolar cycloaddition reaction of propargylic *C*-phosphinic acid **87** (1:1 diastereoisomeric mixture) [81] with benzonitrile oxide, gave the isoxazole-phosphinate **90** in 83% yield, which by coupling with di-*tert*-butyl protected (*S*)-tyrosine followed by the cleavage of the *tert*-butyl group and semipreparative RP-HPLC isolation, provided the pure diastereoisomeric phosphinic tripeptide (*R,R,S*)-**91** in 90% yield (Scheme 38) [83]. These compounds were administered intravenously and lowered mean arterial blood pressure in spontaneously hypertensive rats.

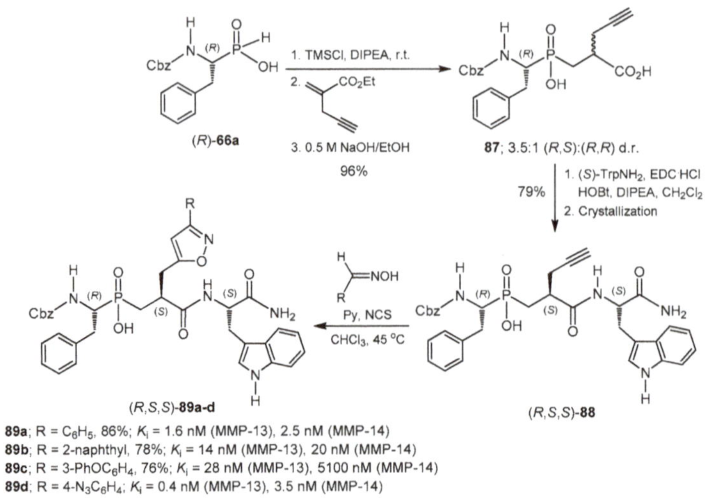

Scheme 37. Preparation of a novel class of isoxazole-containing phosphinic peptides **89a–d** and its activities as inhibitors of matrix metalloproteinases MMP-13 and MMP-14.

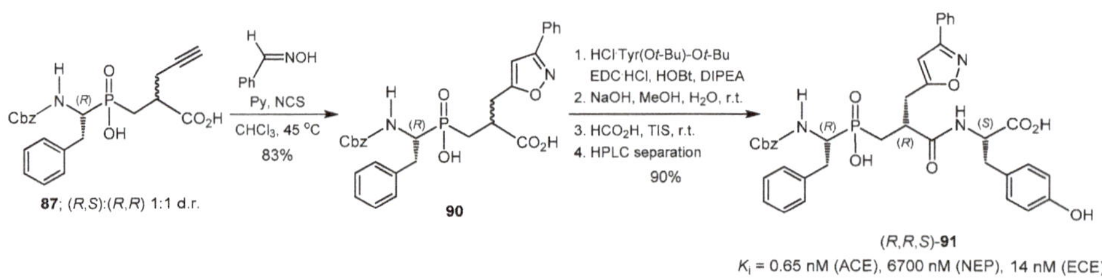

Scheme 38. Synthesis of the phosphinic tripeptide (*R*,*R*,*S*)-**91** and its activities as inhibitor of angiotensin converting enzyme (ACE), neutral endopeptidase (NEP) and endothelin converting enzyme (ECE).

In a similar way, Stratikos et al. [84] reported the synthesis of the phosphinic tripeptide (*R*,*S*,*S*)-**94**, which was tested in the inhibition of the endoplasmic reticulum aminopeptidases 1 and 2 (ERAP1 and ERAP2), involved in regulation of cytotoxic cellular responses. Thus, the reaction of *N*-Boc *H*-phosphinic acid analogue of homophenylalanine (*R*)-**92** [72] with HMDS followed by the addition of ethyl isobutylacrylate and saponification of the ethyl ester, gave the C-phosphinate **93** in 72% yield, which by coupling with (*S*)-TrpNH_2, deprotection and HPLC separation, afforded the phosphinic tripeptide (*R*,*S*,*S*)-**94** in 87% (Scheme 39) [78].

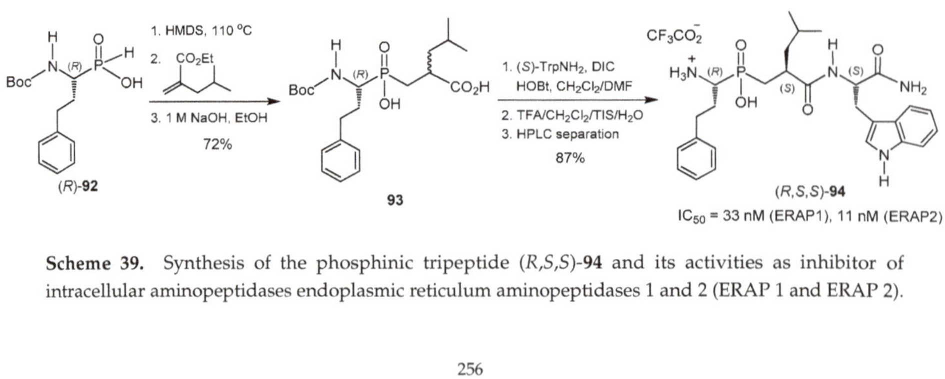

Scheme 39. Synthesis of the phosphinic tripeptide (*R*,*S*,*S*)-**94** and its activities as inhibitor of intracellular aminopeptidases endoplasmic reticulum aminopeptidases 1 and 2 (ERAP 1 and ERAP 2).

In order to functionalize the phosphinic pseudopeptide framework, Yiotakis et al. [85] developed a simple and efficient method for $P_1{}'$ diversification obtaining a variety of dehydrophosphinic peptides. According to this strategy, the treatment of *N*-Cbz phenylalanine *H*-phosphinic analogue (*S*)-**66a** [72] with TMSCl/DIPEA followed by addition of diethyl 2-methylenemalonate and subsequent alkaline hydrolysis, afforded the phosphinic pseudodipeptide (*S*)-**95** in 67% yield, which was subjected to a Knoevenagel-type condensation with several aldehydes, to give the dehydrophosphinic peptides **96a–i** in good yields (Scheme 40).

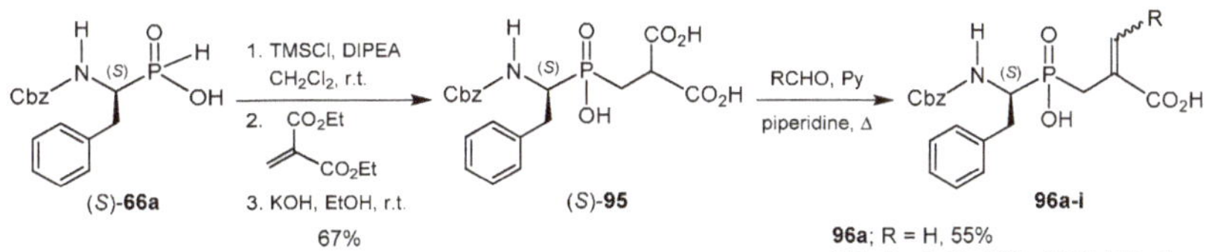

96a; R = H, 55%
96b; R = C_6H_5, 66%, 100:0 *E*/*Z* ratio
96c; R = $C_6H_5(CH_2)_2$, 57%, 67:33 *E*/*Z* ratio
96d; R = $C_6H_5CH_2OCH_2$, 55%, 56:44 *E*/*Z* ratio
96e; R = *i*-Bu, 62%, 54:46 *E*/*Z* ratio
96f; R = 4-imidazolyl, 75%, 77:23 *E*/*Z* ratio
96g; R = 4-ClC_6H_4, 53%, 100:0 *E*/*Z* ratio
96h; R = 4-*t*-$BuO_2CC_6H_4$, 56%, 100:0 *E*/*Z* ratio
96i; R = 4-$HO_2CC_6H_4$, 71%, 100:0 *E*/*Z* ratio

Scheme 40. Synthesis of the dehydrophosphinic peptides **96a–i**.

Gegnas et al. [86] described the synthesis of the phosphinic pseudopeptide **99**, which proved to be an inhibitor of the D-glutamic acid-adding enzyme (MurD) of bacterial peptidoglycan biosynthesis. The nucleophilic addition of compound (*R*)-**97** [72] to 2-methylene pentanedioate in the presence of sodium methoxide provided the dipeptide isostere as a mixture of four diastereoisomers, which under hydrogenolysis over Pd/C catalyst in methanol provided the derivative **98** with the free amino group in 56% yield. Compound **98** was further transformed into the target pseudopeptide **99** in 2% overall yield through a six steps sequence, whose biological activity lied below the sensitivity of the enzyme assay used (<1 nM) (Scheme 41).

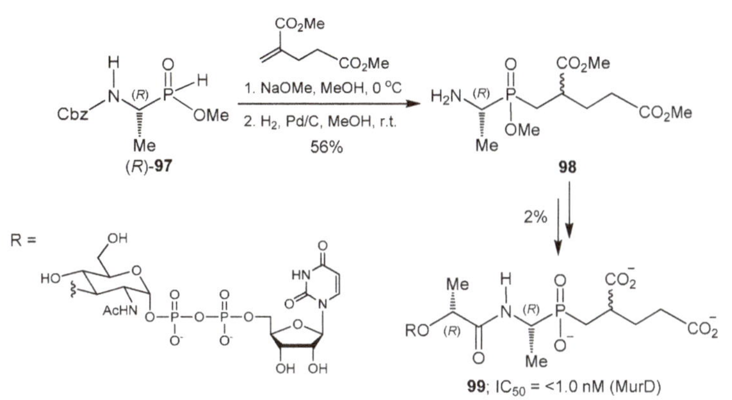

Scheme 41. Synthesis of the phosphinic pseudopeptide **99** and its activity as inhibitor of the D-glutamic acid-adding enzyme (MurD) isolated from *E. coli*.

Additionally, Yiotakis et al. [87] developed an Ireland-Claisen rearrangement triggered by the phospha-Michael addition of silyl phosphonites to allyl acrylates, as a new strategy to access pharmacologically relevant *C*-phosphinic derivatives. In this context, the reaction of Cbz-protected phenylalanine *H*-phosphinic analogue (*R*)-**66a** [72] with the acrylate in the presence of DIPEA in CH_2Cl_2 followed by addition of TMSCl at −78 °C, provided the *P*-alkylated α-amino-*C*-phosphinic acid (*R*)-**100** in 52% yield with retention of configuration at phosphorus atom (Scheme 42).

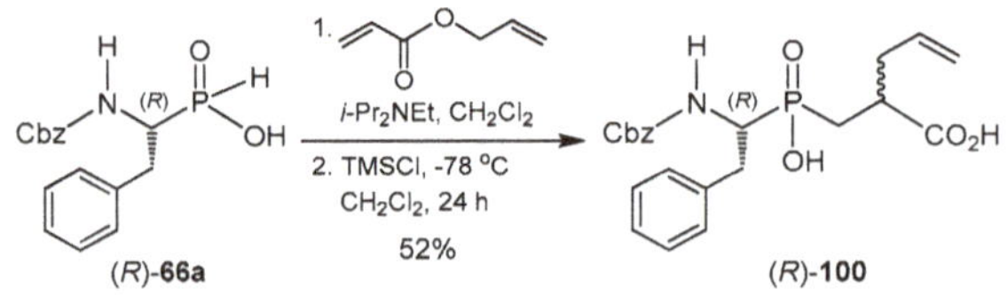

Scheme 42. Synthesis of the α-amino-C-phosphinic acid (*R*)-**100**.

On the other hand, the diastereoselective synthesis of Pro-Phe phosphinyl dipeptide isosteres was accomplished from optically active prolinephosphinate derivative (*R*,R_P)-**101** [88], which was obtained in 13 steps from 1,1-diethoxyethyl(hydroxymethyl)phosphinate via a lipase-catalyzed acylation [89]. Thus, the Michael addition of the *H*-phosphinate (*R*,R_P)-**101** to *t*-butyl acrylate in the presence of *t*-BuOMgBr in THF at 0 °C, provided the C-phosphinate (*R*,R_P)-**102** in 95% yield, which by treatment with LiHMDS followed by the addition of benzyl bromide in THF at −78 °C, produced the Pro-Phe phosphinyl dipeptide (*R*,R_P,*R*)-**103** in 81% yield and 4.8:1 diastereoisomeric ratio (Scheme 43).

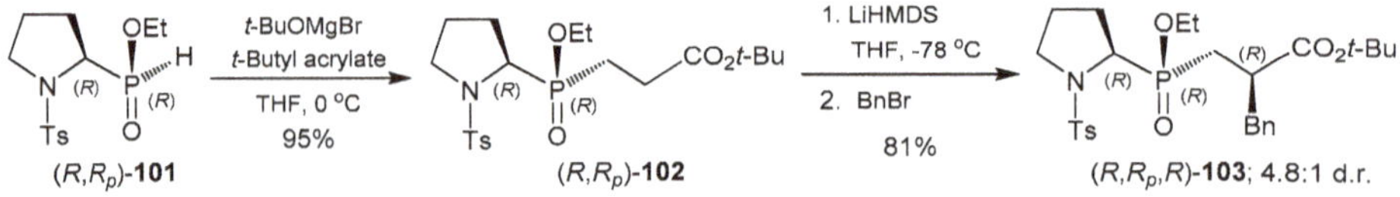

Scheme 43. Synthesis of the Pro-Phe phosphinyl dipeptide (*R*,R_P,*R*)-**103**.

3. Synthesis of Phosphacyclic α-Amino-C-phosphinates

Several strategies for the preparation of *P*-heterocycles have been described over the last years, and excellent reviews have been published [90–93]. In the past 20 years, significant effort has been devoted to synthetic and reactivity studies of this particular class of compounds, here we only report the stereoselective methods recently reported in the literature where the α-amino-*C*-phosphinate motif is incorporated.

3.1. 1,4,2-Oxazaphosphacycles

The strategies described for the stereoselective synthesis of 1,4,2-oxazaphosphacycles typically involve the diastereoselective nucleophilic addition-cyclization reaction from hypophosphorous acid (H_3PO_2) or methyl hypophosphite (H_2PO_2Me) and chiral imino alcohols (Scheme 44).

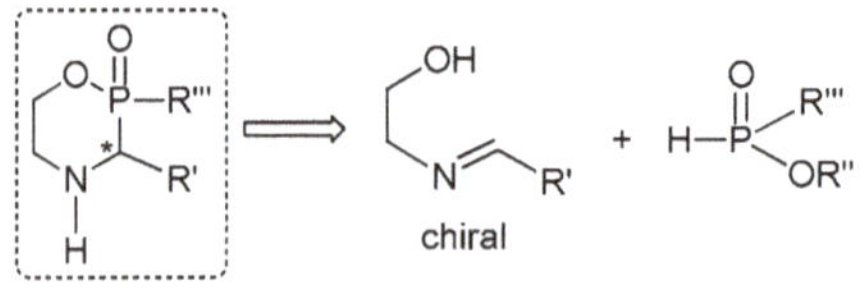

Scheme 44. Retrosynthetic analysis of the 1,4,2-oxazaphosphacycle scaffold.

With previously acquired knowledge in the preparation and reactivity of oxazaphosphinanes, Pirat et al. [94] addressed the enantioselective synthesis of phosphinic analogues of 2-aryl-morpholinols, which have shown strong activity on the noradrenergic systems [95] and therefore constitute new therapeutic means for depression treatment. Thus, the palladium catalyzed coupling of (2*R*,3*R*,5*R*)-**104** with aryl halides and catalytic amounts of tetrakis(triphenylphosphine) palladium and Et_3N in PhMe at reflux, afforded the 2-aryl-1,4,2-oxazaphosphinanes (2*R*,3*R*,5*R*)-**105a–e** in 68%–74% yield as single diastereoisomers (Scheme 45). The retention of configuration at phosphorus atom was confirmed through X-ray analysis of final products.

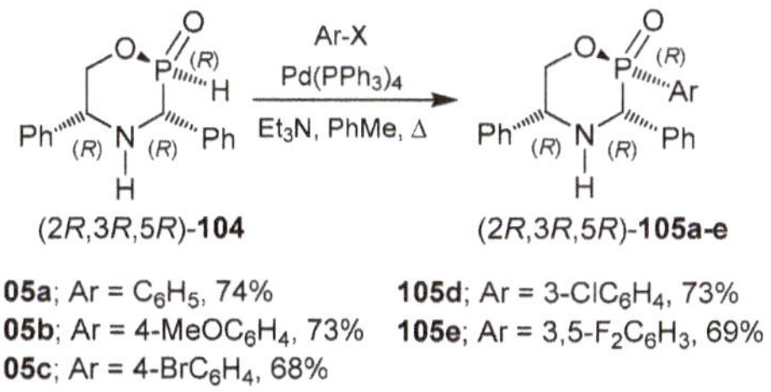

Scheme 45. Palladium catalyzed coupling of (2*R*,3*R*,5*R*)-**104** with aryl halides.

Treatment of (2*R*,3*R*,5*R*)-**105a** with 12 M HCl at 70 °C and subsequent addition of propylene oxide, produced the (2*S*,3*R*,5*R*)-2,3,5-triphenyl-1,4,2-oxazaphosphinane (−)-**106a** in 80% yield through a selective inversion of configuration at phosphorus center (Scheme 46) [94].

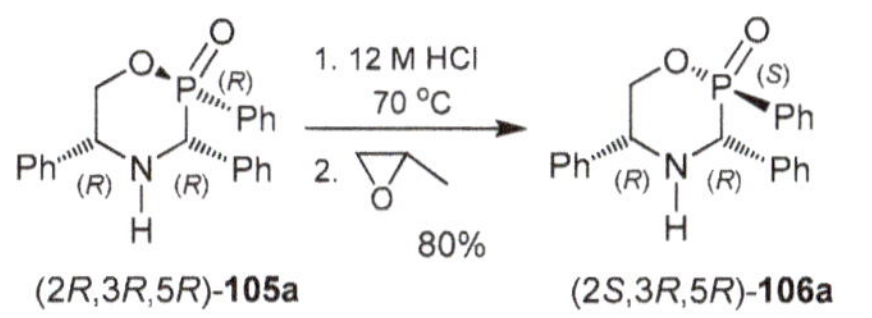

Scheme 46. Inversion of configuration at the phosphorus of (2*R*,3*R*,5*R*)-**105a**.

In a complementary work, Pirat et al. [96] evaluated the diastereoselectivity in the Michael addition of the enantiomerically pure oxazaphosphinane (2*S*,3*S*,5*S*,6*R*)-**107** to methyl cinnamate. Thus, reaction of (2*S*,3*S*,5*S*,6*R*)-**107** with catalytic amounts of potassium *tert*-butoxide in CH_2Cl_2 at room temperature followed by the addition of methyl cinnamate, gave the cyclic α-amino-C-phosphinate (2*S*,3*S*,5*S*,6*R*)-**108** in 92% yield and with 70% diastereoisomeric excess (Scheme 47).

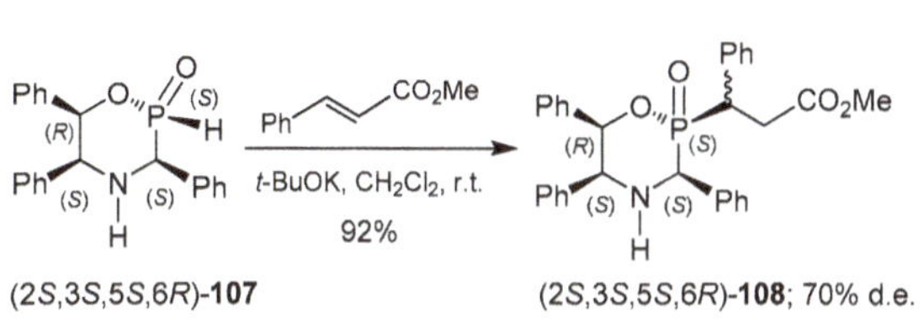

Scheme 47. Michael addition of the oxazaphosphinane (2*S*,3*S*,5*S*,6*R*)-**107** to methyl cinnamate.

On the other hand, Ordóñez et al. [97] carried out the stereoselective synthesis of novel 1,4,2-oxazaphosphepines from chiral 1,3-benzoxazines. For this purpose, the reaction of the chiral 1,3-benzoxazines **109a–d** with dichlorophenylphosphine in the presence of Et_3N in CH_2Cl_2 at room temperature, gave the 1,4,2-oxazaphosphepines **110a–d** in 50:50 to 100:0 diastereoisomeric ratio. In a similar way, the reaction of **111a,b**, afforded the 1,4,2-oxazaphosphepines **112a,b** in 74:26 and 84:16 diastereoisomeric ratio, respectively (Scheme 48). The configuration assignment was stablished by X-ray analysis of final products.

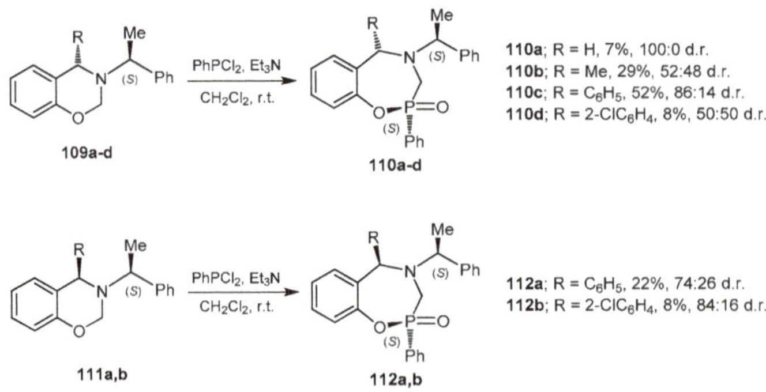

Scheme 48. Synthesis of novel 1,4,2-oxazaphosphepines from chiral 1,3-benzoxazines.

3.2. 1,2-Oxaphosphacycles

The stereoselective synthesis of 1,2-oxaphosphacycles with an exocyclic amino group has been performed, mainly through two pathways: (a) diastereoselective nucleophilic addition-cyclization reaction from arylphosphinates and chiral cyclic hemiaminals; or (b) substitution with *N*-nucleophiles on derivatized chiral 3-hydroxy-1,2-oxaphosphinanes (Scheme 49).

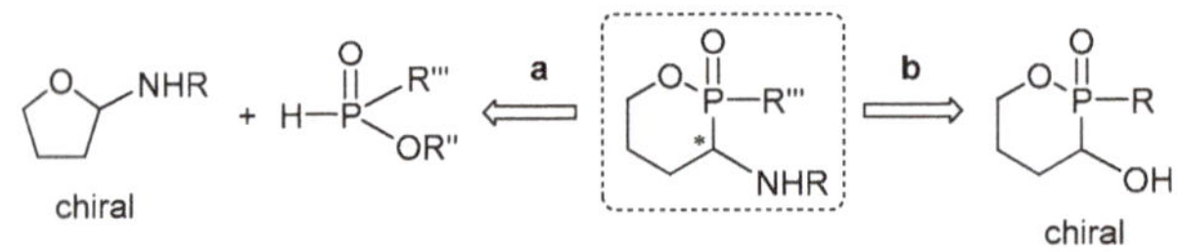

Scheme 49. Retrosynthetic analysis of the 1,2-oxaphosphacycle scaffold.

For example, the reaction of tris(benzyloxy)-D-arabinofuranose **113** with ethyl phenylphosphinate in the presence of catalytic amounts of potassium *tert*-butoxide followed by preparative reverse phase HPLC separation, led to the cyclic α-amino-*C*-phosphinate (2*S*,3*R*,4*S*,5*S*,6*R*)-**114** in 17% yield. The reaction of (2*S*,3*R*,4*S*,5*S*,6*R*)-**114** with triflic anhydride in the presence of pyridine in CH_2Cl_2 at 0 °C, produced the triflate, which without additional purification it was reacted with trimethylsilylazide in the presence of TBAF in THF at 65 °C, obtaining the azido derivative (2*S*,3*S*,4*S*,5*S*,6*R*)-**115** in 28% yield. Finally, reduction of azide group in the compound **115** with Ph_3P/THF/H_2O, furnished the glucose like oxaphosphinane (2*S*,3*S*,4*S*,5*S*,6*R*)-**116** in 42% yield, which was tested for their antiproliferative activity in the search for new therapeutic agents against glioblastoma (Scheme 50) [98].

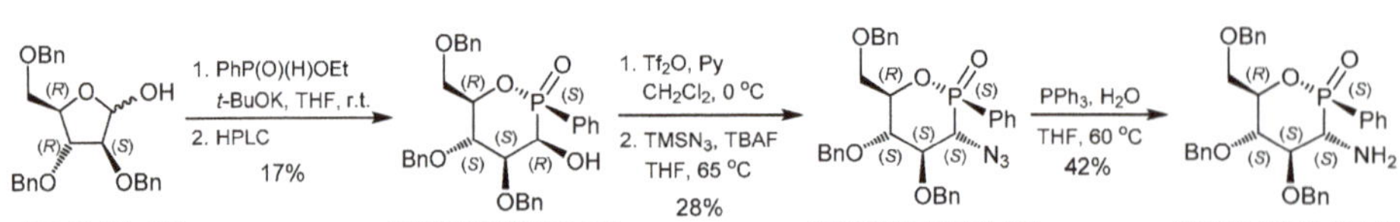

Scheme 50. Synthesis of (2*S*,3*S*,4*S*,5*S*,6*R*)-**116** and its antiproliferative activity on glioblastoma cell line C6.

Additionally, the reaction of tris(benzyloxy)-D-arabinofuranose **113** with allylamine or benzylamine in the presence of $MgSO_4$ in EtOH at reflux, provided the cyclic hemiaminal ethers **117a**,**b** in 70% yield, which by nucleophilic addition of ethyl phenylphosphinate and cyclization in the presence of catalytic amounts of potassium *tert*-butoxide, afforded the 3-alkylamino-1,2-oxaphosphinanes **118a,b** in 12% and 7% yield, respectively (Scheme 51) [98].

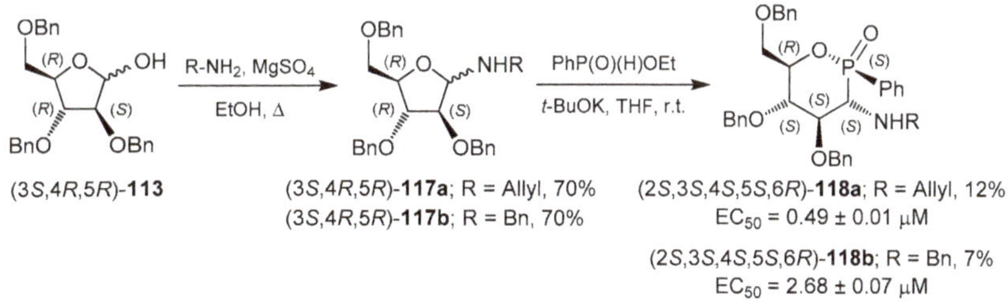

Scheme 51. Synthesis of **118a,b** and its antiproliferative activities against glioblastoma cell line C6.

Finally, Bakalara et al. [99] carried out the synthesis of 3-amino-1,2-oxaphosphinanes as *C*-glycoside mimetics and new pharmacological agents against glioma stem cell proliferation, migration and invasion. In this respect, the condensation of the 2,3,5,6-di-*O*-isopropylidene-α-D-manno-furanose (3*S*,4*S*,5*R*)-**119** with different amines in the presence of $MgSO_4$ in EtOH at reflux, afforded the cyclic hemiaminal ethers (3*S*,4*S*,5*R*)-**120a–d** in 58%–90% yield, which by reaction with ethyl phenylphosphinate and cyclization in the presence of catalytic amounts of potassium *tert*-butoxide, produced the 3-aryl or 3-alkylamino-1,2-oxaphosphinanes (2*S*,3*S*,4*S*,5*R*,6*R*)-**121a–d** in 53%–73% yield (Scheme 52).

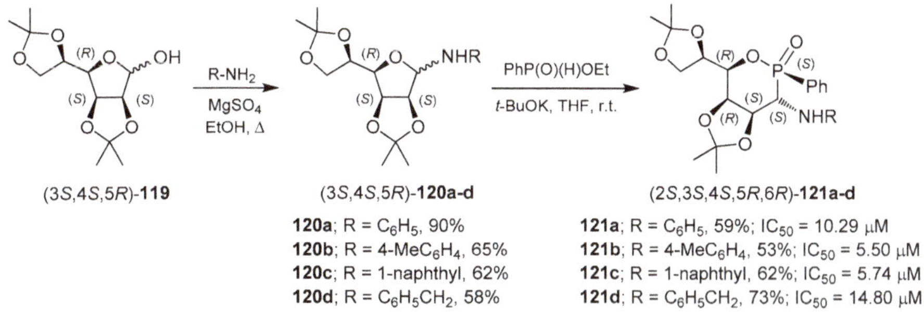

Scheme 52. Synthesis of the 3-aryl or 3-alkylamino-1,2-oxaphosphinanes (2*S*,3*S*,4*S*,5*R*,6*R*)-**121a–d** and its antiproliferative activities toward glioblastoma cell line C6.

4. Conclusions

In this review, we have covered recent progress in the development of stereoselective methodologies for the synthesis of both acyclic and cyclic α-amino-*C*-phosphinic acids and derivatives. As we have shown, a number of optically active α-amino-*C*-phosphinic derivatives are now accessible through different strategies; however, there is still a long way to be covered, considering the poor availability of stereoselective methods for the preparation of pure stereoisomers of phosphinic peptides, taking into account that diastereoisomers with a particular three dimensional structure are able to interact efficiently with the receptor binding site, while other configurations are discriminated. Authors believe that much effort must be made in this direction and further advances in the search and improvement of synthetic procedures, new chemical applications and biological activities of these interesting compounds will be a very rewarding task in coming years.

Acknowledgments: The authors gratefully acknowledge the CONACYT (grant 181816, 248868) and PRODEP (project UAEMOR-PTC-379) of Mexico, Ministerio de Economía y Competitividad (grant CTQ2013-40855-R) and Gobierno de Aragón-FSE (research group E40) for their financial support.

Conflicts of Interest: The authors declare no conflict of interest.

Abbreviations

The following abbreviations are used in this manuscript: Ac: acetyl; ACE: angiotensin converting enzyme; APN: alanyl aminopeptidase; Ar: aryl; Boc: *t*-butyloxycarbonyl; BOP: (benzotriazol-1-yloxy)tris(dimethylamino)-phosphonium hexafluorophosphate; BPPM: (2*S*,4*S*)-*N*-butoxycarbonyl-4-diphenylphosphino-2-diphenyl-phosphino methylpyrrolidine; BSA: *N*,*O*-bis(trimethylsilyl)acetamide; BSTFA: *N*,*O*-bis(trimethylsilyl)trifluoro acetamide; Cbz: benzyloxycarbonyl; CEC: capillary electrochromatography; cod: 1,5-cyclo octadiene; d.e.: diastereomeric excess; DCC: *N*,*N*′-dicyclohexylcarbodiimide; DIPEA: diisopropyl ethylamine; DMAP: 4-dimethylaminopyridine; DMF: dimethylformamide; DNP: 2,4-dinitro phenol; e.e.: enantiomeric excess; ECE: endothelin converting enzyme; EDC: 1-ethyl-3-(3′-dimethylaminopropyl)carbodiimide; ERAP1: endoplasmic reticulum aminopeptidases 1; ERAP2: endoplasmic reticulum aminopeptidases 2; Et: ethyl; HCV NS3: hepatitis C virus nonstructural protein 3; HMDS: bis(trimethylsilyl)amine; HNC: human neutrophile collagenase; HNE: human neutrophile elastase; HOBt: 1-hydroxybenzotriazole; HPLC: high performance liquid chromatography; *i*-Bu: *iso*-butyl; IC_{50}: half maximal inhibitory concentration; *i*-Pr: *iso*-propyl; LAP: leucine aminopeptidase; LDA: lithium diisopropylamide; LiTMP: lithium tetramethylpiperidide; Me: methyl; MMPs: matrix metalloproteinases; MPa: megapascal; MurD: D-glutamic acid-adding enzyme; NaHMDS: sodium hexamethyl-disilazide; *n*-Bu: *n*-butyl; NCS: *N*-chlorosuccinimide; NEP: neutral endopeptidase; nM: nanomolar; PcAm: penicillin amidase; Ph: phenyl; Phth: phthaloyl; Py: pyridine; r.t.: room temperature; RP-HPLC: reversed phase high performance liquid chromato-graphy; SDS: sodium dodecyl sulfate; *sec*-Bu: *sec*-butyl; TBAF: tetra-*n*-butyl-ammonium fluoride; *t*-Bu: *tert*-butyl; Tf: trifluoromethanesulfonate; TFA: trifluoroacetic acid; THF: tetrahydrofuran; TMEDA: tetramethylethylenediamine; TMSCl: chlorotrimethylsilane; $TMSN_3$: trimethyl-silylazide; Ts: tosyl.

References

1. Orsini, F.; Sello, G.; Sisti, M. Aminophosphonic acids and derivatives. Synthesis and biological applications. *Curr. Med. Chem.* **2010**, *17*, 264–289. [CrossRef] [PubMed]
2. Naydenova, E.D.; Todorov, P.T.; Troev, K.D. Recent synthesis of aminophosphonic acids as potential biological importance. *Amino Acids* **2010**, *38*, 23–30. [CrossRef] [PubMed]
3. Sienczyk, M.; Oleksyszyn, J. Irreversible inhibition of serine proteases—Design and in vivo activity of diaryl α-aminophosphonate derivatives. *Curr. Med. Chem.* **2009**, *16*, 1673–1687. [CrossRef] [PubMed]
4. Lejczak, B.; Kafarski, P. Biological activity of aminophosphonic acids and their short peptides. *Top. Heterocycl. Chem.* **2009**, *20*, 31–63.
5. Kukhar, V.P.; Hudson, H.R. *Aminophosphonic and Aminophospinic Acids: Chemistry and Biological Activity*; Wiley: Chichester, UK, 2000.
6. Ordóñez, M.; Viveros-Ceballos, J.L.; Cativiela, C.; Sayago, F.J. An update on the stereoselective synthesis of α-aminophosphonic acids and derivatives. *Tetrahedron* **2015**, *71*, 1745–1784. [CrossRef]
7. Ordóñez, M.; Sayago, F.J.; Cativiela, C. Synthesis of quaternary α-aminophosphonic acids. *Tetrahedron* **2012**, *68*, 6369–6412. [CrossRef]
8. Ordóñez, M.; Viveros-Ceballos, J.L.; Cativiela, C.; Arizpe, A. Stereoselective synthesis of α-aminophosphonic acids analogs of the 20 Proteinogenic α-amino acids. *Curr. Org. Synth.* **2012**, *9*, 310–341. [CrossRef]
9. Kudzin, Z.H.; Kudzin, M.H.; Drabowicz, J.; Stevens, C.V. Aminophosphonic acids—Phosphorus analogues of natural amino acids. Part 1. Syntheses of α-aminophosphonic acids. *Curr. Org. Chem.* **2011**, *15*, 2015–2071. [CrossRef]
10. Gulyukina, N.S.; Makukhin, N.N.; Beletskaya, I.P. Synthesis methods of (1-aminocyclopropyl)phosphonic acids. *Russ. J. Org. Chem.* **2011**, *47*, 633–649. [CrossRef]
11. Ordóñez, M.; Rojas-Cabrera, H.; Cativiela, C. An overview of stereoselective synthesis of α-aminophosphonic acids and derivatives. *Tetrahedron* **2009**, *65*, 17–49. [CrossRef] [PubMed]

12. Moore, W.J.; van der Marel, G.A.; van Boom, J.H.; Liskamp, R.M.J. Peptides containing the novel methylphosphinamide transition-state isostere. *Tetrahedron* **1993**, *49*, 11055–11064. [CrossRef]
13. Albouy, D.; Brun, A.; Munoz, A.; Etemad-Moghadam, G. New (α-hidroxyalkyl)phosphorus amphiphiles: Synthesis and dissociation constants. *J. Org. Chem.* **1998**, *63*, 7223–7230. [CrossRef] [PubMed]
14. Kukhar, V.P.; Romanenko, V.D.; Hughes, A.B. *Chemistry of Aminophosphonic Acids and Phosphonopeptides*; Kukhar, V.P., Romanenko, V.D., Eds.; Wiley-VCH Verlag GmbH & Co. KGaA: Weinheim, Germany, 2009.
15. Kraszewski, A.; Stawinski, J. H-Phosphonates: Versatile synthetic precursors to biologically active phosphorus compounds. *Pure Appl. Chem.* **2007**, *79*, 2217–2227. [CrossRef]
16. Mucha, A.; Kafarski, P.; Berlicki, Ł. Remarkable potential of the α-aminophosphonate/phosphinate structural motif in medicinal chemistry. *J. Med. Chem.* **2011**, *54*, 5955–5980. [CrossRef] [PubMed]
17. Dive, V.; Georgiadis, D.; Matziari, M.; Makaritis, A.; Beau, F.; Cuniasse, P.; Yiotakis, A. Phosphinic peptides as zinc metalloproteinase inhibitors. *Cell. Mol. Life Sci.* **2004**, *61*, 2010–2019. [CrossRef] [PubMed]
18. Vassiliou, S.; Grabowiecka, A.; Kosikowska, P.; Yiotakis, A.; Kafarski, P.; Berlicki, L. Design, synthesis, and evaluation of novel organophosphorus inhibitors of bacterial ureases. *J. Med. Chem.* **2008**, *51*, 5736–5744. [CrossRef] [PubMed]
19. Lukas, M.; Vojtisek, P.; Hermann, P.; Rohovec, J.; Lukes, I. Synthesis of phosphinic acid analogues of glycyl-glycine and crystal structure of *N*-glycyl-aminomethyl-(phenylphosphinic) acid. *Synth. Commun.* **2002**, *32*, 79–88. [CrossRef]
20. Meng, F.-H.; Xu, J.-X. Direct synthesis of phosphinopeptides containing *C*-terminal α-aminoalkylphosphinic acids. *Amino Acids* **2010**, *39*, 533–538. [CrossRef] [PubMed]
21. Gluza, K.; Kafarski, P. Transition state analogues of enzymatic reaction as potential drugs. In *Drug Discovery*; El-Shemy, H., Ed.; InTech: Rijeka, Croatia, 2013; pp. 325–372.
22. Berlicki, L.; Kafarski, P. Computer-aided analysis and design of phosphonic and phosphinic enzyme inhibitors as potential drugs and agrochemicals. *Curr. Org. Chem.* **2005**, *9*, 1829–1850. [CrossRef]
23. Fournié-Zaluski, M.-C.; Poras, H.; Roques, B.P.; Nakajima, Y.; Ito, K.; Yoshimoto, T. Structure of aminopeptidase N from escherichia coli complexed with the transition-state analogue aminophosphinic inhibitor PL250. *Acta Crystallogr. Sect. D Biol. Crystallogr.* **2009**, *65*, 814–822. [CrossRef] [PubMed]
24. Banegas, I.; Prieto, I.; Vives, F.; Alba, F.; de Gasparo, M.; Segarra, A.B.; Hermoso, F.; Duran, R.; Ramirez, M. Brain aminopeptidases and hypertension. *J. Renin Angiotensin Aldosterone Syst.* **2006**, *7*, 129–134. [CrossRef] [PubMed]
25. Skinner-Adams, T.S.; Stack, C.M.; Trenholme, K.R.; Brown, C.L.; Grembecka, J.; Lowther, J.; Mucha, A.; Drag, M.; Kafarski, P.; McGowan, S.; et al. Plasmodium falciparum neutral aminopeptidases: New targets for anti-malarials. *Trends Biochem. Sci.* **2010**, *35*, 53–61. [CrossRef] [PubMed]
26. Stack, C.M.; Lowther, J.; Cunningham, E.; Donnelly, S.; Gardiner, D.L.; Trenholme, K.R.; Skinner-Adams, T.S.; Teuscher, F.; Grembecka, J.; Mucha, A.; et al. Characterization of the plasmodium falciparum M17 leucyl aminopeptidase: A protease involved in amino acid regulation with potential for antimalarial drug development. *J. Biol. Chem.* **2007**, *282*, 2069–2080. [CrossRef] [PubMed]
27. Yiotakis, A.; Georgiadis, D.; Matziari, M.; Makaritis, A.; Dive, V. Phosphinic peptides: Synthetic approaches and biochemical evaluation as Zn-metalloprotease inhibitors. *Curr. Org. Chem.* **2004**, *8*, 1135–1158. [CrossRef]
28. Matziari, M.; Beau, F.; Cuniasse, P.; Dive, V.; Yiotakis, A. Evaluation of P_1'-diversified phosphinic peptides leads to the development of highly selective inhibitors of MMP-11. *J. Med. Chem.* **2004**, *47*, 325–336. [CrossRef] [PubMed]
29. Yamagishi, T. Stereocontrolled synthesis of phosphinyl dipeptide isosteres using an asymmetric center at the phosphorus atom. *J. Pharm. Soc. Jpn.* **2013**, *134*, 915–924. [CrossRef]
30. Drag, M.; Pawelczak, M.; Kafarski, P. Stereoselective synthesis of 1-aminoalkanephosphonic acids with two chiral centers and their activity towards leucine aminopeptidase. *Chirality* **2003**, *15*, S104–S107. [CrossRef] [PubMed]
31. Mucha, A. Synthesis and modifications of phosphinic dipeptide analogues. *Molecules* **2012**, *17*, 13530–13568. [CrossRef] [PubMed]
32. Virieux, D.; Volle, J.N.; Pirat, J.L. Acyclic to cyclic aminophosphonic and phosphinic acids. *ARKIVOC-Online J. Org. Chem.* **2012**, *11*, 264–277.
33. Kobayashi, S.; Ishitani, H. Catalytic enantioselective addition to imines. *Chem. Rev.* **1999**, *99*, 1069–1094. [CrossRef] [PubMed]

34. Gröger, H. Catalytic enantioselective Strecker reactions and analogous syntheses. *Chem. Rev.* **2003**, *103*, 2795–2828. [CrossRef] [PubMed]
35. Pudovik, A.N. Addition of dialkyl phosphites to imines. New method of synthesis of esters of amino phosphonic acids. *Dokl. Akad. Nauk SSSR* **1952**, *83*, 865–869.
36. Tibhe, G.D.; Lagunas-Rivera, S.; Vargas-Díaz, E.; García-Barradas, O.; Ordóñez, M. Uncatalyzed one-pot diastereoselective synthesis of α-amino phosphonates under solvent-free conditions. *Eur. J. Org. Chem.* **2010**, *2010*, 6573–6581. [CrossRef]
37. Hoffmann, R.W. Allylic 1,3-strain as a controlling factor in stereoselective transformations. *Chem. Rev.* **1989**, *89*, 1841–1860. [CrossRef]
38. Afarinkia, K.; Cadogan, J.I.G.; Rees, C.W. Diastereoselectivity in the formation of phosphorus-carbon bonds. *Synlett* **1992**, *1992*, 124–125. [CrossRef]
39. Szabó, A.; Jászay, Z.M.; Hegedűs, L.; Tőke, L.; Petneházy, I. The first enantioselective synthesis of α-aminophosphinates. *Tetrahedron Lett.* **2003**, *44*, 4603–4606. [CrossRef]
40. Yang, M.; Sun, Y.-M.; Hou, Q.-G.; Zhao, C.-Q. (R_P)-5-Methyl-2-(propan-2-yl)cyclohexyl phenyl{phenyl [(1-phenylethyl)amino]methyl}phosphinate. *Acta Crystallogr. Sect. E Struct. Rep. Online* **2012**, *68*. [CrossRef] [PubMed]
41. Rossi, J.-C.; Marull, M.; Larcher, N.; Taillades, J.; Pascal, R.; van der Lee, A.; Gerbier, P. A recyclable chiral auxiliary for the asymmetric syntheses of α-aminonitriles and α-aminophosphinic derivatives. *Tetrahedron Asymmetry* **2008**, *19*, 876–883. [CrossRef]
42. Katritzky, A.R.; Narindoshvili, T.; Draghici, B.; Angrish, P. Regioselective syntheses of β-*N*-linked glycoaminoacids and glycopeptides. *J. Org. Chem.* **2008**, *73*, 511–516. [CrossRef] [PubMed]
43. Zhou, G.; Zheng, W.; Wang, D.; Zhang, P.; Pan, Y. Practical stereo- and regioselective, copper(I)-promoted Strecker synthesis of sugar-modified α,β-unsaturated imines. *Helv. Chim. Acta* **2006**, *89*, 520–526. [CrossRef]
44. Knauer, S.; Kranke, B.; Krause, L.; Kunz, H. Amino sugars and glycosylamines as tools in stereoselective synthesis. *Curr. Org. Chem.* **2004**, *8*, 1739–1761. [CrossRef]
45. Wang, Y.; Wang, Y.; Yu, J.; Miao, Z.; Chen, R. Stereoselective synthesis of α-amino(phenyl)methyl(phenyl) phosphinic acids with *O*-pivaloylated D-galactosylamine as chiral auxiliary. *Chem. Eur. J.* **2009**, *15*, 9290–9293. [CrossRef] [PubMed]
46. Edupuganti, R.; Davis, F.A. Synthesis and applications of masked oxo-sulfinamides in asymmetric synthesis. *Org. Biomol. Chem.* **2012**, *10*, 5021–5031. [CrossRef] [PubMed]
47. Szabó, A.; Jászay, Z.M.; Töke, L.; Petneházy, I. Interesting by-products in the synthesis of chiral α-aminophosphinates from enantiopure sulfinimines. *Tetrahedron Lett.* **2004**, *45*, 1991–1994. [CrossRef]
48. Leow, D.; Tan, C.-H. Chiral guanidine catalyzed enantioselective reactions. *Chem. Asian J.* **2009**, *4*, 488–507. [CrossRef] [PubMed]
49. Fu, X.; Loh, W.; Zhang, Y.; Chen, T.; Ma, T.; Liu, H.; Wang, J.; Tan, C.-H. Chiral guanidinium salt catalyzed enantioselective phospha-Mannich reactions. *Angew. Chem. Int. Ed.* **2009**, *48*, 7387–7390. [CrossRef] [PubMed]
50. Cherkasov, R.A.; Galkin, V.I. The Kabachnik-Fields reaction: Synthetic potential and the problem of the mechanism. *Russ. Chem. Rev.* **1998**, *67*, 857–882. [CrossRef]
51. Fields, E.K. The synthesis of esters of substituted amino phosphonic acids. *J. Am. Chem. Soc.* **1952**, *74*, 1528–1531. [CrossRef]
52. Kabachnik, M.I.; Medved, T.Y. New synthesis of aminophosphonic acids. *Dokl. Akad. Nauk SSSR* **1952**, *83*, 689–692.
53. Meng, F.; Xu, J. Synthesis of phosphinodepsipeptides via the pseudo-four-component condensation reaction. *Tetrahedron* **2013**, *69*, 4944–4952. [CrossRef]
54. Meng, F.; He, F.; Song, X.; Zhang, L.; Hu, W.; Liu, G.; Xu, J. Facile synthesis of hybrid sulfonophosphinodipeptides composing of taurines and 1-aminoalkylphosphinic acids. *Amino Acids* **2012**, *43*, 423–429. [CrossRef] [PubMed]
55. Dwars, T.; Schmidt, U.; Fischer, C.; Grassert, I.; Kempe, R.; Fröhlich, R.; Drauz, K.; Oehme, G. Synthesis of optically active α-aminophosphinic acids by catalytic asymmetric hydrogenation in organic solvents and aqueous micellar media. *Angew. Chem. Int. Ed.* **1998**, *37*, 2851–2853. [CrossRef]
56. Brovarets, V.S.; Zyuz, K.V.; Budnik, L.V.; Solodenko, V.A.; Drach, B.S. New approach to the synthesis of 1-acylaminoalkenylphosphonic acids, their analogs and derivatives. *Zh. Obs. Khim.* **1993**, *63*, 1259–1265.

57. Dwars, T.; Schmidt, U.; Fischer, C.; Grassert, I.; Krause, H.-W.; Michalik, M.; Oehme, G. Synthesis of enantiomerically enriched α-aminophosphonic acid derivatives via asymmetric hydrogenation. *Phosphorus Sulfur Silicon Relat. Elem.* **2000**, *158*, 209–240. [CrossRef]
58. Lefevre, N.; Brayer, J.-L.; Folléas, B.; Darses, S. Chiral α-amino phosphonates via rhodium-catalyzed asymmetric 1,4-addition reactions. *Org. Lett.* **2013**, *15*, 4274–4276. [CrossRef] [PubMed]
59. Zoń, J. A simple preparation of diethyl 1-acylamino-1-ethenephosphonates. *Synthesis* **1981**, *1981*, 324–324. [CrossRef]
60. Lämmerhofer, M.; Hebenstreit, D.; Gavioli, E.; Lindner, W.; Mucha, A.; Kafarski, P.; Wieczorek, P. High-performance liquid chromatographic enantiomer separation and determination of absolute configurations of phosphinic acid analogues of dipeptides and their α-aminophosphinic acid precursors. *Tetrahedron Asymmetry* **2003**, *14*, 2557–2565. [CrossRef]
61. Preinerstorfer, B.; Lubda, D.; Mucha, A.; Kafarski, P.; Lindner, W.; Lämmerhofer, M. Stereoselective separations of chiral phosphinic acid pseudodipeptides by CEC using silica monoliths modified with an anion-exchange-type chiral selector. *Electrophoresis* **2006**, *27*, 4312–4320. [CrossRef] [PubMed]
62. Grembecka, J.; Mucha, A.; Cierpicki, T.; Kafarski, P. The most potent organophosphorus inhibitors of leucine aminopeptidase. Structure-based design, chemistry, and activity. *J. Med. Chem.* **2003**, *46*, 2641–2655. [CrossRef] [PubMed]
63. Mucha, A.; Lämmerhofer, M.; Lindner, W.; Pawełczak, M.; Kafarski, P. Individual stereoisomers of phosphinic dipeptide inhibitor of leucine aminopeptidase. *Bioorg. Med. Chem. Lett.* **2008**, *18*, 1550–1554. [CrossRef] [PubMed]
64. Rozhko, L.F.; Ragulun, V.V. α-Hydroxy-α-aminophosphinic acids: I. Synthesis of a new analog of phenylglycine and its enantiomers. *Russ. J. Gen. Chem.* **2005**, *75*, 533–536. [CrossRef]
65. Pyun, H.; Clarke, M.O.; Cho, A.; Casarez, A.; Ji, M.; Fardis, M.; Pastor, R.; Sheng, X.C.; Kim, C.U. Synthesis of 1-amino-2-vinylcyclopropane-1-phosphinates. Conversion of a phosphonate to phosphinates. *Tetrahedron Lett.* **2012**, *53*, 2360–2363. [CrossRef]
66. Clarke, M.O.; Chen, X.; Cho, A.; Delaney, W.E.; Doerffler, E.; Fardis, M.; Ji, M.; Mertzman, M.; Pakdaman, R.; Pyun, H.; et al. Novel, potent, and orally bioavailable phosphinic acid inhibitors of the hepatitis C virus NS3 protease. *Bioorg. Med. Chem. Lett.* **2011**, *21*, 3568–3572. [CrossRef] [PubMed]
67. Pyun, H.-J.; Chaudhary, K.; Somoza, J.R.; Sheng, X.C.; Kim, C.U. Synthesis and resolution of diethyl (1*S*,2*S*)-1-amino-2-vinylcyclopropane-1-phosphonate for HCV NS3 protease inhibitors. *Tetrahedron Lett.* **2009**, *50*, 3833–3835. [CrossRef]
68. Njoroge, F.G.; Venkatraman, S.; Girijavallabhan, V.M. Depeptidized Inhibitors of Hepatitis C Virus NS3 protease. U.S. Patent 2005164921 (A1), 28 July 2005.
69. Qian, R.; Roller, A.; Hammerschmidt, F. Phosphonate–phosphinate rearrangement. *J. Org. Chem.* **2015**, *80*, 1082–1091. [CrossRef] [PubMed]
70. Collinsova, M.; Jiracek, J. Phosphinic acid compounds in biochemistry, biology and medicine. *Curr. Med. Chem.* **2000**, *7*, 629–647. [CrossRef] [PubMed]
71. Matziari, M.; Georgiadis, D.; Dive, V.; Yiotakis, A. Convenient synthesis and diversification of dehydroalaninyl phosphinic peptide analogues. *Org. Lett.* **2001**, *3*, 659–662. [CrossRef] [PubMed]
72. Baylis, E.K.; Campbell, C.D.; Dingwall, J.G. 1-Aminoalkylphosphonous acids. Part 1. Isosteres of the protein amino acids. *J. Chem. Soc. Perkin Trans. 1* **1984**, 2845–2853. [CrossRef]
73. McKittrick, B.A.; Stamford, A.W.; Weng, X.; Ma, K.; Chackalamannil, S.; Czarniecki, M.; Cleven, R.; Fawzi, A.B. Design and synthesis of phosphinic acids that triply inhibit endothelin converting enzyme, angiotensin converting enzyme and neutral endopeptidase 24.11. *Bioorg. Med. Chem. Lett.* **1996**, *6*, 1629–1634. [CrossRef]
74. Hamilton, R.; Wharry, S.; Walker, B.; Walker, B.J. The synthesis of phosphinic acid based proteinase inhibitors. *Phosphorus Sulfur Silicon Relat. Elem.* **1999**, *144*, 761–764. [CrossRef]
75. Lloyd, J.; Schmidt, J.B.; Hunt, J.T.; Barrish, J.C.; Little, D.K.; Tymiak, A.A. Solid phase synthesis of phosphinic acid endothelin converting enzyme inhibitors. *Bioorg. Med. Chem. Lett.* **1996**, *6*, 1323–1326. [CrossRef]
76. Chen, H.; Bischoff, L.; Fournie-Zaluski, M.-C.; Roques, B.P. Synthesis of 2(*S*)-benzyl-3-[hydroxy(1′(*R*)-amino ethyl)phosphinyl]propanoyl-L-3-[^{125}I]-iodotyrosine: A radiolabelled inhibitor of aminopeptidase N. *J. Label. Compd. Radiopharm.* **2000**, *43*, 103–111. [CrossRef]

77. Chen, H.; Noble, F.; Mothé, A.; Meudal, H.; Coric, P.; Danascimento, S.; Roques, B.P.; George, P.; Fournié-Zaluski, M.-C. Phosphinic derivatives as new dual enkephalin-degrading enzyme inhibitors: Synthesis, biological properties, and antinociceptive activities. *J. Med. Chem.* **2000**, *43*, 1398–1408. [CrossRef] [PubMed]
78. Matziari, M.; Dellis, D.; Dive, V.; Yiotakis, A.; Samios, J. Conformational and solvation studies via computer simulation of the novel large scale diastereoselectively synthesized phosphinic MMP inhibitor RXP03 diluted in selected solvents. *J. Phys. Chem. B* **2010**, *114*, 421–428. [CrossRef] [PubMed]
79. Vassiliou, S.; Mucha, A.; Cuniasse, P.; Georgiadis, D.; Lucet-Levannier, K.; Beau, F.; Kannan, R.; Murphy, G.; Knäuper, V.; Rio, M.-C.; et al. Phosphinic pseudo-tripeptides as potent inhibitors of matrix metalloproteinases: A structure-activity study. *J. Med. Chem.* **1999**, *42*, 2610–2620. [CrossRef] [PubMed]
80. Dive, V.; Andarawewa, K.L.; Boulay, A.; Matziari, M.; Beau, F.; Guerin, E.; Rousseau, B.; Yiotakis, A.; Rio, M.-C. Dosing and scheduling influence the antitumor afficacy of a phosphinic peptide inhibitor of matrix metalloproteinases. *Int. J. Cancer* **2005**, *113*, 775–781. [CrossRef] [PubMed]
81. David, A.; Steer, D.; Bregant, S.; Devel, L.; Makaritis, A.; Beau, F.; Yiotakis, A.; Dive, V. Cross-linking yield variation of a potent matrix metalloproteinase photoaffinity probe and consequences for functional proteomics. *Angew. Chem. Int. Ed.* **2007**, *46*, 3275–3277. [CrossRef] [PubMed]
82. Makaritis, A.; Georgiadis, D.; Dive, V.; Yiotakis, A. Diastereoselective solution and multipin-based combinatorial array synthesis of a novel class of potent phosphinic metalloprotease inhibitors. *Chem. Eur. J.* **2003**, *9*, 2079–2094. [CrossRef] [PubMed]
83. Jullien, N.; Makritis, A.; Georgiadis, D.; Beau, F.; Yiotakis, A.; Dive, V. Phosphinic tripeptides as dual angiotensin-converting enzyme C-domain and endothelin-converting enzyme-1 inhibitors. *J. Med. Chem.* **2010**, *53*, 208–220. [CrossRef] [PubMed]
84. Zervoudi, E.; Saridakis, E.; Birtley, J.R.; Seregin, S.S.; Reeves, E.; Kokkala, P.; Aldhamen, Y.A.; Amalfitano, A.; Mavridis, I.M.; James, E.; et al. Rationally designed inhibitor targeting antigen-trimming aminopeptidases enhances antigen presentation and cytotoxic T-cell responses. *Proc. Natl. Acad. Sci. USA* **2013**, *110*, 19890–19895. [CrossRef] [PubMed]
85. Matziari, M.; Nasopoulou, M.; Yiotakis, A. Active methylene phosphinic peptides: A new diversification approach. *Org. Lett.* **2006**, *8*, 2317–2319. [CrossRef] [PubMed]
86. Gegnas, L.D.; Waddell, S.T.; Chabin, R.M.; Reddy, S.; Wong, K.K. Inhibitors of the bacterial cell wall biosynthesis enzyme Mur D. *Bioorg. Med. Chem. Lett.* **1998**, *8*, 1643–1648. [CrossRef]
87. Rogakos, V.; Georgiadis, D.; Dive, V.; Yiotakis, A. A modular rearrangement approach toward medicinally relevant phosphinic structures. *Org. Lett.* **2009**, *11*, 4696–4699. [CrossRef] [PubMed]
88. Yamagishi, T.; Kinbara, A.; Okubo, N.; Sato, S.; Fukaya, H. Diastereoselective synthesis of Pro-Phe phosphinyl dipeptide isosteres. *Tetrahedron Asymmetry* **2012**, *23*, 1633–1639. [CrossRef]
89. Yamagishi, T.; Mori, J.; Haruki, T.; Yokomatsu, T. A chemo-enzymatic synthesis of optically active 1,1-diethoxyethyl(aminomethyl)phosphinates: Useful chiral building blocks for phosphinyl dipeptide isosteres. *Tetrahedron Asymmetry* **2011**, *22*, 1358–1363. [CrossRef]
90. Berger, O.; Montchamp, J.-L. Phosphinate-containing heterocycles: A mini-review. *Beilstein J. Org. Chem.* **2014**, *10*, 732–740. [CrossRef] [PubMed]
91. Mathey, F. *Phosphorus-Carbon Heterocyclic Chemistry: The Rise of a New Domain*; Elsevier Health Sciences: Oxford, UK, 2001.
92. Quin, L.D. *A Guide to Organophosphorus Chemistry*; John Wiley & Sons: New York, NY, USA, 2000.
93. Mathey, F. Chemistry of 3-membered carbon-phosphorus heterocycles. *Chem. Rev.* **1990**, *90*, 997–1025. [CrossRef]
94. Volle, J.-N.; Virieux, D.; Starck, M.; Monbrun, J.; Clarion, L.; Pirat, J.-L. Chiral phosphinyl analogues of 2-C-arylmorpholinols: 2-Aryl-3,5-diphenyl-[1,4,2]-oxazaphosphinanes. *Tetrahedron Asymmetry* **2006**, *17*, 1402–1408. [CrossRef]
95. Kelley, J.L.; Musso, D.L.; Boswell, G.E.; Soroko, F.E.; Cooper, B.R. (2*S*,3*S*,5*R*)-2-(3,5-Difluorophenyl)-3,5-dimethyl-2-morpholinol: A novel antidepressant agent and selective inhibitor of norepinephrine uptake. *J. Med. Chem.* **1996**, *39*, 347–349. [CrossRef] [PubMed]
96. Monbrun, J.; Dayde, B.; Cristau, H.-J.; Volle, J.-N.; Virieux, D.; Pirat, J.-L. Diastereoselective Michael addition of 2*H*-2-oxo-1,4,2-oxaza phosphinanes to olefins. *Tetrahedron* **2011**, *67*, 540–545. [CrossRef]

97. Salgado-Escobar, O.; Chavelas-Hernández, L.; Domínguez-Mendoza, B.; Linzaga-Elizalde, I.; Ordoñez, M. Synthesis of chiral 1,4,2-oxazaphosphepines. *Molecules* **2015**, *20*, 13794–13813. [CrossRef] [PubMed]
98. Clarion, L.; Jacquard, C.; Sainte-Catherine, O.; Loiseau, S.; Filippini, D.; Hirlemann, M.-H.; Volle, J.-N.; Virieux, D.; Lecouvey, M.; Pirat, J.-L.; et al. Oxaphosphinanes: New therapeutic perspectives for glioblastoma. *J. Med. Chem.* **2012**, *55*, 2196–2211. [CrossRef] [PubMed]
99. Clarion, L.; Jacquard, C.; Sainte-Catherine, O.; Decoux, M.; Loiseau, S.; Rolland, M.; Lecouvey, M.; Hugnot, J.-P.; Volle, J.-N.; Virieux, D.; et al. *C*-Glycoside mimetics inhibit glioma stem cell proliferation, migration, and invasion. *J. Med. Chem.* **2014**, *57*, 8293–8306. [CrossRef] [PubMed]

molecules

MDPI

Article

Practical and Efficient Synthesis of α-Aminophosphonic Acids Containing 1,2,3,4-Tetrahydroquinoline or 1,2,3,4-Tetrahydroisoquinoline Heterocycles

Mario Ordóñez [1,*], Alicia Arizpe [2], Fracisco J. Sayago [2], Ana I. Jiménez [2] and Carlos Cativiela [2,*]

1. Centro de Investigaciones Químicas-IICBA, Universidad Autónoma del Estado de Morelos, Av. Universidad 1001, Cuernavaca 62209, Morelos, Mexico
2. Departamento de Química Orgánica, Universidad de Zaragoza-CSIC, ISQCH, Zaragoza 50009, Spain; aliceiw@gmail.com (A.A.); jsayago@unizar.es (F.J.S.); anisjim@unizar.es (A.I.J.)

* Correspondence: palacios@uaem.mx (M.O.); cativiela@unizar.es (C.C.); Tel.: +52-777-329-7997 (M.O.); +34-976-761-210 (C.C.)

Academic Editor: György Keglevich
Received: 25 July 2016; Accepted: 25 August 2016; Published: 31 August 2016

Abstract: We report here a practical and efficient synthesis of α-aminophosphonic acid incorporated into 1,2,3,4-tetrahydroquinoline and 1,2,3,4-tetrahydroisoquinoline heterocycles, which could be considered to be conformationally constrained analogues of pipecolic acid. The principal contribution of this synthesis is the introduction of the phosphonate group in the *N*-acyliminium ion intermediates, obtained from activation of the quinoline and isoquinoline heterocycles or from the appropriate δ-lactam with benzyl chloroformate. Finally, the hydrolysis of phosphonate moiety with simultaneous cleavage of the carbamate afforded the target compounds.

Keywords: α-aminophosphonic acids; *N*-acyliminium ions; conformationally constrained

1. Introduction

There is a continuously growing interest in the development of new peptidomimetics, compounds that mimic the bioactive conformation and action of therapeutic peptides while possessing greater bioavailability and stability and less undesirable effects. In this regard, the incorporation of rigid unusual secondary α-amino acids, where the nitrogen is involved in a ring, may result in significant consequences for the conformation of peptidomimetics as synthetic tools for drug discovery [1,2]. Some of the most important molecules are the 1,2,3,4-tetrahydroquinoline-2-carboxylic acid **1** [3,4] and 1,2,3,4-tetrahydroisoquinoline-1-carboxylic acid **2** [5,6], which are conformationally constrained analogues of unnatural pipecolic acid, and 1,2,3,4-tetrahydroisoquinoline-3-carboxylic acid (Tic) **3** [7,8], which is considered a conformationally constrained analogue of phenylalanine (Phe). These compounds are important unnatural α-amino acids, and they are used as key intermediates in organic synthesis for the preparation of biologically active compounds. Therefore, much effort has been dedicated to the preparation of these compounds.

On the other hand, the α-aminoalkylphosphonic acids are probably the most important analogues of the α-amino acids, obtained by isosteric substitution of the planar and less bulky carboxylic acid (CO_2H) by a tetrahedral phosphonic acid functionality (PO_3H_2). This class of compounds is currently attracting interest in organic and medicinal chemistry, due to their important biological and pharmacological properties [9–12]. The great importance of this type of compound has prompted organic chemists to report numerous procedures for their racemic or stereoselective synthesis [13–18], principally using the diastereoselective and enantioselective

Pudovik [19,20] and Kabachnik–Fields [21–25] reactions for acyclic α-aminoalkylphosphonates, and through *N*-acyl iminium ions for the synthesis of cyclic derivatives [26–34]. However, to the best of our knowledge, the synthesis of 1,2,3,4-tetrahydroquinoline-2-phosphonic acid **4** and 1,2,3,4-tetrahydroisoquinoline-1-phosphonic acid **5** analogues [35–38] has not yet been described in the literature, whereas the synthesis of 1,2,3,4-tetrahydroisoquinoline-3-phosphonic acid **6** has been recently described by our research group [39,40] (Figure 1).

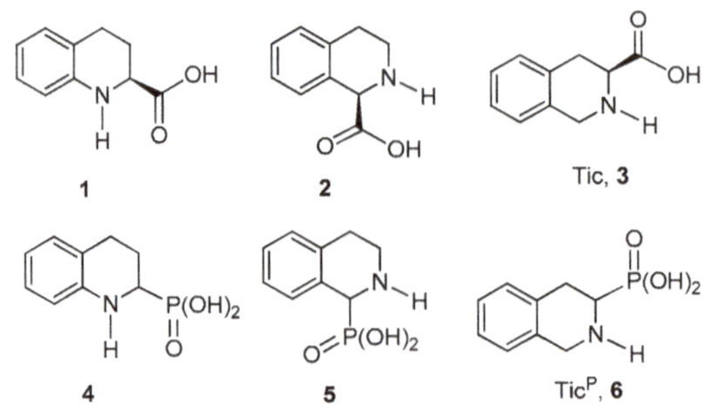

Figure 1. α-Amino acids characterized by a tetrahydroquinoline or tetrahydroisoquinoline heterocyles and their α-amino phosphonic analogues.

Considering the high value of these non-coded compounds in connection with our current research interest in the synthesis of novel conformationally restricted α-aminophosphonic acids [26–28,41–43], we now report herein the practical and efficient synthesis of α-aminophosphonic acids incorporating 1,2,3,4-tetrahydroquinoline **4**, and 1,2,3,4-tetrahydroisoquinoline rings **5** and **6**, which could be considered to be conformationally constrained analogues of pipecolic acid. The principal contribution of this work is the regioselective introduction of the phosphonate group in the *N*-acyliminium ions derivatives obtained from activation of the quinoline and isoquinoline nucleous or from the appropriate δ-lactams with benzyl chloroformate.

2. Results and Discussion

For the synthesis of 1,2,3,4-tetrahydroquinoline-2-phosphonic acid **4**, we proposed the *N*-acyliminium ions as suitable precursors, considering that the well-known reduction of the carbonyl group of lactams such as the 3,4-dihydro-2(1*H*)-quinolinone **7**, and their subsequent transformation into a *N*-acyliminium ion, is one of the methods that allows the incorporation of the phosphonate functionality into the α position of the nitrogen atom such as we have previously described [26–28]. These proposals prompted us to explore further application of the *N*-acyliminium strategy for the synthesis of the target compound. In this context, the commercially available 3,4-dihydro-2(1*H*)-quinolinone **7** was reacted with lithium bis(trimethylsilyl)amide (LiHMDS) as a base and benzyl chloroformate in THF at −78 °C, to obtain the *N*-Cbz-3,4-dihydro-2(1*H*)-quinolinone **8** in 91% yield. Reduction of the carbonyl group in **8** with diisobutylaluminium hydride (DIBAL-H) and subsequent reaction with methanol and catalytic amounts of pyridinium *p*-toluenesulfonate (PPTS), gave the methoxyaminal **9**, which was treated immediately with trimethyl phosphite in the presence of $BF_3{\cdot}OEt_2$, obtaining the dimethyl *N*-Cbz-1,2,3,4-tetrahydroquinoline-2-phosphonate **11** in 62% via the *N*-acyliminium ion **10** in 62% yield from **8**. Treatment of this *N*-Cbz-protected phosphonate with a 33% solution of hydrogen bromide in acetic acid, afforded the 1,2,3,4-tetrahydroquinoline-2-phosphonic acid **4** as hydrobromide in quantitative yield (Scheme 1).

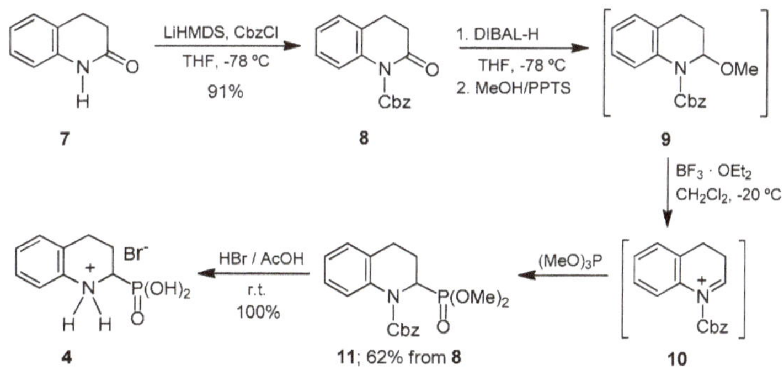

Scheme 1. Synthesis of 1,2,3,4-tetrahydroquinoline-2-phosphonic acid **4** from lactam **7**.

The literature has described the hydrophosphonylation of quinoline derivatives to obtain the 1,2-dihydroquinolin-2-ylphosphonate and 2,4-diphosphono-1,2,3,4-tetrahydroquinoline derivatives using activating agents [44–48]. These reactions proceed via quinolinium ions that can be viewed as counterparts of the above-mentioned iminium species. Thus, the addition of an acyl chloride to quinoline can be considered as a way to generate the iminium ion necessary for subsequent regioselective incorporation of the phosphonate functionality α to the nitrogen atom.

Based on this precedent, we decided to compare the efficiency of the *N*-acylquinolinium salts as intermediates in the synthesis of α-aminophosphonic acid **4**. For this purpose, benzyl chloroformate was added to quinoline and the *N*-Cbz-quinolinium chloride formed **12** was reacted with trimethyl phosphite in acetonitrile at 50 °C, obtaining the dimethyl *N*-Cbz-1,2-dihydroquinoline-2-phosphonate derivative **13** in 74% yield from quinolone. The diphosphonylation products at the 2- and 4-positions of the quinoline ring were not detected. The regioselectivity observed in the addition of trimethyl phosphite to the *N*-acylquinolinium ion **12** may be attributed to the electron-withdrawing character of the benzyloxycarbonyl group, which increases the electrophilic nature of the carbon adjacent to the nitrogen atom. Additionally, the benzyloxycarbonyl group was selected for its easy incorporation and removal under mild conditions. Thus, the catalytic hydrogenation resulted in simultaneous removal of the Cbz group and reduction of the double bond at positions 3,4 to afford the α-aminophosphonate **14**, which, by treatment with hydrogen bromide in acetic acid, gave the 1,2,3,4-tetrahydroquinoline-2-phosphonic acid **4** as hydrobromide. The latter two steps proceeded quantitatively and obviated the need for purification (Scheme 2).

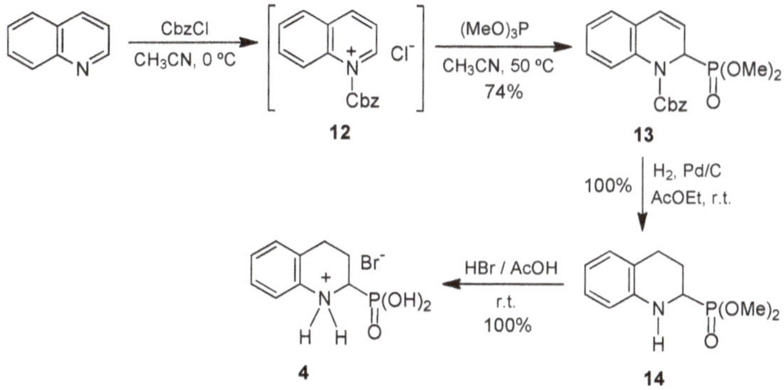

Scheme 2. Synthesis of 1,2,3,4-tetrahydroquinoline-2-phosphonic acid **4** from quinoline.

The 1,2,3,4-tetrahydroquinoline-2-phosphonic acid **4** was obtained in 74% overall yield from quinoline following the synthetic strategy in Scheme 2, which compares favorably with the 56% global yield achieved in Scheme 1, when starting from quinolinone **1**. Accordingly, the superior global yield, together with the much lower price of quinoline in comparison with 3,4-dihydro-2(1*H*)-quinolinone **7**, makes the methodology in Scheme 2 more advantageous than the lactam-based one for the preparation of α-aminophosphonic acid **4**. From an operational viewpoint, both routes required few purification steps—by column chromatography—of intermediate compounds.

We next addressed the preparation of 1,2,3,4-tetrahydroisoquinoline-1-phosphonic acid **5**. Similarly to that described above for the quinoline counterpart, the preparation of *N*-substituted-1,2-dihydroisoquinoline-1-phosphonates through the isoquinolinium salts formed by reaction of isoquinoline with acyl chlorides and subsequent addition of trialkyl phosphites has been described [49,50]. Thus, the synthesis of dimethyl *N*-Cbz-1,2-dihydroisoquinoline-1-phosphonate should be straightforward, as shown above for the preparation of dimethyl *N*-Cbz-1,2-dihydroquinoline-2-phosphonate **13**.

On the other hand, the lactam to be used as a starting product for the preparation of the target amino acid **5** is not easily available from commercial sources in this case. This fact together with the advantageous preparation of amino acid **4** via the formation of quinolinium salt prompted us to undertake the synthesis of 1,2,3,4-tetrahydroisoquinoline-1-phosphonic acid following an analogous synthetic route.

Thus, using the reaction conditions described above, the isoquinoline was reacted with benzyl chloroformate followed by the addition of trimethyl phosphite in acetonitrile at 50 °C. However, under these reaction conditions, the desired compound **16** was not obtained. It is noteworthy that during the addition of benzyl chloroformate to isoquinoline, a yellow solid was formed, which was later identified as the isoquinolinium salt **15**. This compound remained unaltered upon addition of trimethyl phosphite so that the formation of the expected 1,2-dihydroisoquinoline-1-phosphonate **16** did not take place. We reasoned that if the trimethyl phosphite was present in the reaction medium before benzyl chloroformate was added, the isoquinolinium ion formed **16** would immediately react with it, thus preventing precipitation. To our delight, when the order of addition of these reagents was exchanged (that is, trimethyl phosphite prior to benzyl chloroformate), the dimethyl *N*-Cbz-1,2-dihydroisoquinoline-1-phosphonate **16** was obtained at 92% yield. In the next step, we carried out the catalytic hydrogenation of the double bond in **16** under an atmospheric pressure of hydrogen gas and using Pd/C as a catalyst, to generate the tetrahydroisoquinoline moiety. However, the Cbz group in **16** was readily eliminated at room temperature, whereas the double bond at positions 3,4 was only partially hydrogenated even after long reaction times. As a consequence, mixtures of the desired tetrahydroisoquinoline derivative **17** and the analogous dihydroisoquinoline **18** were obtained. Attempts to improve this result by changing the solvent as well as by increasing the reaction temperature or hydrogen gas pressure proved unsuccessful and a mixture of **17** and **18** was always obtained (Scheme 3).

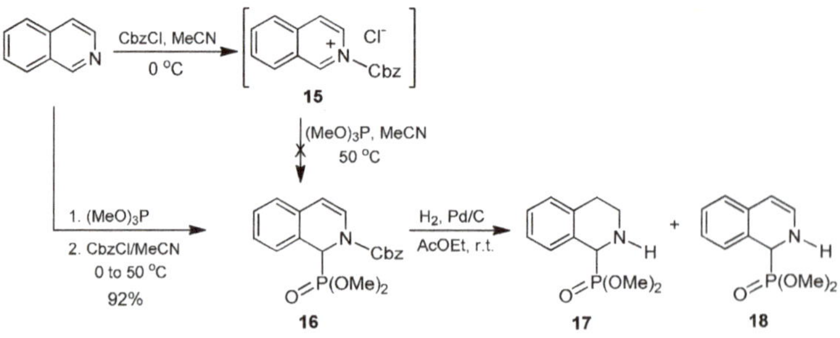

Scheme 3. First attempt to synthetize 1,2,3,4-tetrahydroisoquinoline-1-phosphonic acids derivatives.

To circumvent this problem, we synthesized the compound analogous to **16** that bears the less acid-sensitive diethyl phosphonate group **19**. A procedure identical to that established for the dimethyl derivative but using triethyl phosphite as the phosphorus source provided the desired diethyl *N*-Cbz-1,2-dihydroisoquinoline-1-phosphonate **19** in 98% yield from isoquinoline. Taking into account the fact that compound **19** can be considered not only as an *N*-acyl-α-aminophosphonate, but also as an enamide, we decided to perform the reduction of the double bond using trifluoroacetic acid and triethylsilane as the reducing agent, following the conditions described by Jacobsen et al. [51] for cyclic enamides. In this case, the diethyl *N*-Cbz-1,2,3,4-tetrahydroisoquinoline-1-phosphonate **20** was readily obtained in quantitative yield by reaction with triethylsilane and trifluoroacetic acid. Subsequent cleavage of the protecting groups in **20** by treatment with hydrogen bromide in acetic acid, afforded the 1,2,3,4-tetrahydroisoquinoline-1-phosphonic acid **5** as hydrobromide in quantitative yield. This compound was isolated in 97% overall yield from isoquinoline (Scheme 4).

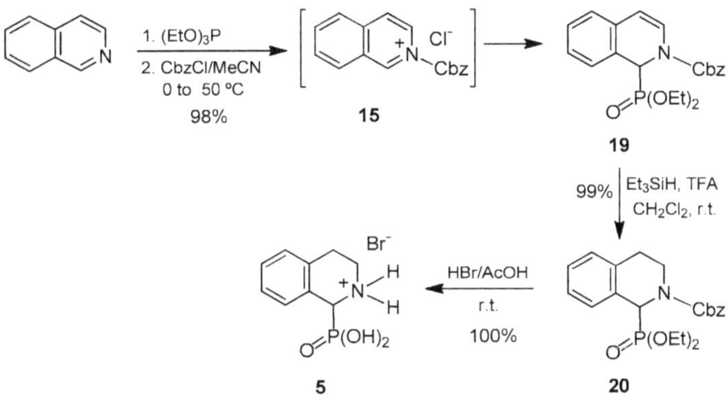

Scheme 4. Synthesis of 1,2,3,4-tetrahydroisoquinoline-1-phosphonic acid **5** from isoquinoline.

Finally, we focused on the preparation of 1,2,3,4-tetrahydroisoquinoline-3-phosphonic acid **6**, that is, the phosphonic analogue of Tic (Tic^P). This compound exhibits a tetrahydroisoquinoline core, as does the aminophosphonic acid **5**. However, in the case of Tic^P, the phosphonate group is at the 3-position. However, the isoquinoline selectively provides the phosphonate group at the 1-position of the ring. Therefore, we believe that the generation of a suitable iminium ion for the introduction of the phosphonate moiety at the desired 3-position, the lactam precursor was the only route to consider in this case. It should be noted that, following this strategy, the position of the carbonyl group in the starting lactam determines with complete regiocontrol the introduction of the phosphonate substituent, which is a distinctive advantage of this methodology. The adequate lactam **21** used in the synthesis of the target amino acid was easily prepared following reported procedures that involved reaction of 2-phenylacetyl chloride with aqueous ammonia to give 2-phenylacetamide [52] and subsequent condensation with formaldehyde [53]. Lactam **21** was readily transformed into 1,2,3,4-tetrahydroisoquinoline-3-phosphonic acid (Tic^P, **6**) following the synthetic route strictly similar to that reported above for the preparation of the α-aminophosphonic acid **4** from lactam **7** (Scheme 1). The 1,2,3,4-tetrahydroisoquinoline-3-phosphonic acid **6** was thus obtained as hydrobromide in 52% global yield from the starting substrate **21**, with isolation and purification of only two synthetic intermediates **22**, **25** (Scheme 5).

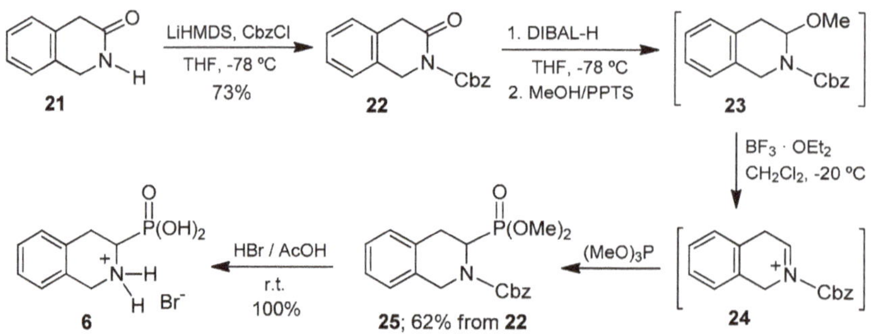

Scheme 5. Synthesis of 1,2,3,4-tetrahydroisoquinoline-3-phosphonic acid, Tic^P **6**.

The results obtained show the successful use of the lactam **21** to generate the key intermediate *N*-acyliminium ion, which by phosphonylation, afforded the desired diethyl *N*-Cbz-1,2,3,4-tetrahydroisoquinoline-3-phosphonate **25**, demonstrating thus the versatility of this methodology in the preparation of α-aminophosphonic acids structurally related to pipecolic acid.

3. Materials and Methods

3.1. General

All reagents were used as received from commercial suppliers (Sigma-Aldrich Chemie GmbH, Buchs, Switzerland) without further purification. Thin-layer chromatography (TLC) was performed on Macherey-Nagel Polygram® SIL G/UV_{254} (Macherey-Nagel, Duren, Germany) precoated silica gel polyester plates. The products were visualized by exposure to UV light (254 nm), iodine vapour or ethanolic solution of phosphomolybdic acid. Column chromatography was performed using 60 M (0.04–0.063 mm) silica gel from Macherey-Nagel. Melting points were determined on a Gallenkamp apparatus (Weiss Gallenkamp, Loughborough, UK). IR spectra were registered on a Nicolet Avatar 360 FTIR spectrophotometer (Thermo Electron Corporation, Madison, WI, USA); ν_{max} is given for the main absorption bands. ^{1}H-, ^{13}C- and ^{31}P-NMR spectra were recorded on Bruker AV-400 or AV-300 instruments (Bruker BioSpin GmbH, Rheinstetten, Germany) at room temperature, unless otherwise indicated, using the residual solvent signal as the internal standard; chemical shifts (δ) are expressed in ppm and coupling constants (*J*) in Hertz. High-resolution mass spectra were obtained on a Bruker Microtof-Q spectrometer. Compound **22** was prepared by reaction of 2-phenylacetamide [52] with formaldehyde [53]. ^{1}H-, ^{13}C- and ^{31}P- NMR spectra of all final compounds are showed in the supplementary material (Figures S1–S34).

3.2. Synthesis of N-Benzyloxycarbonyl-3,4-dihydro-2-quinolinone **8**

A 1 M solution of lithium bis(trimethylsilyl)amide in tetrahydrofuran (2.80 mL, 2.80 mmol) was slowly added to a solution of 3,4-dihydro-2(1*H*)-quinolinone **7** (400 mg, 2.72 mmol) in anhydrous tetrahydrofuran (10 mL) kept at −78 °C under argon. After 30 min, benzyl chloroformate (0.39 mL, 464 mg, 2.72 mmol) was added dropwise and stirring was continued for additional 3 h. The reaction mixture was then treated with saturated aqueous ammonium chloride (10 mL) and allowed to warm to room temperature. The two layers were separated and the aqueous phase was extracted with dichloromethane (2 × 20 mL). The combined organic extracts were dried, filtered, and concentrated. Purification by column chromatography (eluent:hexanes/ethyl acetate 4:1) afforded **8** as a colourless oil (694 mg, 2.47 mmol, 91% yield). IR (neat) ν_{max} 1773, 1699 cm^{-1}. ^{1}H-NMR (400 MHz, $CDCl_3$): δ = 7.48–7.33 (m, 5H, Ar), 7.20–7.13 (m, 2H, Ar), 7.10–7.04 (m, 1H, Ar), 6.92–6.88 (m, 1H, Ar), 5.41 (s, 2H, CH_2Ph), 2.98–2.92 (m, 2H, H-4), 2.72–2.67 (m, 2H, H-3) ppm. ^{13}C-NMR (100 MHz, $CDCl_3$): δ = 169.90 (CO), 153.46 (COO), 136.94 (Ar), 134.52 (Ar), 128.85 (Ar), 128.76 (Ar), 128.75 (Ar), 127.95 (Ar),

127.39 (Ar), 126.95 (Ar), 124.78 (Ar), 118.65 (Ar), 70.05 (CH_2Ph), 33.05 (C-3), 25.55 (C-4) ppm. HRMS (ESI): calcd. for $C_{17}H_{15}NNaO_3$ $[M + Na]^+$ 304.0944; found 304.0947.

3.3. Synthesis of Dimethyl N-benzyloxycarbonyl-1,2,3,4-tetrahydroquinoline-2-phosphonate **11**

A 1 M solution of diisobutylaluminium hydride in hexanes (2.40 mL, 2.40 mmol) was slowly added to a solution of **8** (448 mg, 1.59 mmol) in anhydrous tetrahydrofuran (8 mL) kept at −78 °C under argon. After stirring at this temperature for 2 h, the reaction was treated with saturated aqueous sodium acetate (5 mL) and warmed to room temperature. A 3:1 mixture of diethyl ether and saturated aqueous ammonium chloride (16 mL) was then added and the resulting mixture was stirred at room temperature until a suspension was formed. The solid was filtered off under reduced pressure and washed with diethyl ether (2 × 10 mL). The organic layer was separated and the aqueous phase was extracted with diethyl ether (2 × 20 mL). The combined organic extracts were washed with brine (20 mL), dried, filtered, and evaporated to provide the hemiaminal as an oil. It was dissolved in methanol (6 mL) and treated with pyridinium *p*-toluenesulfonate (40 mg, 0.16 mmol). After stirring at room temperature for 2 h, triethylamine (0.10 mL, 74 mg, 0.73 mmol) was added. The solvent was evaporated and the crude methoxyaminal **9** thus obtained was dissolved in anhydrous dichloromethane (7 mL) and kept under argon. Trimethyl phosphite (0.19 mL, 198 mg, 1.59 mmol) was added and the resulting solution was cooled to −20 °C. Boron trifluoride-diethyl ether (0.20 mL, 226 mg, 1.59 mmol) was added dropwise and the reaction mixture was slowly warmed to room temperature and stirred for 12 h. After quenching with water (2 mL), the two layers were separated and the aqueous phase was extracted with dichloromethane (2 × 10 mL). The combined organic extracts were dried, filtered, and concentrated. Purification by column chromatography (eluent:ethyl acetate/hexanes 4:1) afforded **11** as a colourless oil (372 mg, 0.99 mmol, 62% yield). IR (neat) ν_{max} 1702, 1279, 1029 cm^{-1}. ^1H-NMR (400 MHz, $CDCl_3$): δ = 7.53–7.46 (m, 1H, Ar), 7.37–7.28 (m, 5H, Ar), 7.22–7.16 (m, 1H, Ar), 7.14–7.06 (m, 2H, Ar), 5.31 (d, *J* = 12.5 Hz, 1H, CH_2Ph), 5.16 (d, *J* = 12.5 Hz, 1H, CH_2Ph), 5.04 (ddd, *J* = 13.4, 8.7, 7.3 Hz, 1H, H-2), 3.63 (d, *J* = 10.6 Hz, 3H, OMe), 3.51 (d, *J* = 10.6 Hz, 3H, OMe), 2.85–2.75 (m, 1H, H-4), 2.67–2.57 (m, 1H, H-4′), 2.54–2.40 (m, 1H, H-3), 2.17–2.03 (m, 1H, H-3′) ppm. ^{13}C-NMR (100 MHz, $CDCl_3$): δ = 154.68 (CO), 136.84 (Ar), 136.08 (Ar), 132.78 (Ar), 128.54 (Ar), 128.16 (Ar), 127.88 (Ar), 127.56 (Ar), 126.36 (Ar), 125.97 (Ar), 125.22 (Ar), 68.11 (CH_2Ph), 53.07 (d, *J* = 6.2 Hz, OMe), 52.82 (d, *J* = 7.2 Hz, OMe), 49.33 (d, *J* = 158.2 Hz, C-2), 25.75 (C-3), 25.66 (C-4) ppm. ^{31}P-NMR (162 MHz, $CDCl_3$): δ = 26.39 ppm. HRMS (ESI): calcd. for $C_{19}H_{22}NNaO_5P$ $[M + Na]^+$ 398.1128; found 398.1145.

3.4. Synthesis of Dimethyl N-Benzyloxycarbonyl-1,2-dihydroquinoline-2-phosphonate **13**

Benzyl chloroformate (0.61 mL, 726 mg, 4.26 mmol) was added dropwise to a solution of quinoline (0.46 mL, 500 mg, 3.87 mmol) in anhydrous acetonitrile (6 mL) kept at 0 °C under argon. After stirring at this temperature for 10 min, trimethyl phosphite (0.50 mL, 528 mg, 4.26 mmol) was slowly added followed by sodium iodide (853 mg, 5.69 mmol). The mixture was heated at 50 °C for 10 min. The solvent was evaporated and the crude product was partitioned between dichloromethane (15 mL) and saturated aqueous sodium bicarbonate (15 mL). The organic phase was separated and the aqueous layer was further extracted with dichloromethane (3 × 15 mL). The combined organic extracts were washed with brine (10 mL), dried, filtered, and concentrated. Purification by column chromatography (eluent:ethyl acetate/hexanes 4:1) afforded **13** as a colourless oil (1.07 g, 2.87 mmol, 74% yield). IR (neat) ν_{max} 1706, 1653, 1603, 1263, 1125, 1026 cm^{-1}. ^1H-NMR (400 MHz, DMSO-d_6, 80 °C): δ = 7.58–7.52 (m, 1H, Ar), 7.44–7.30 (m, 5H, Ar), 7.25–7.07 (m, 3H, Ar), 6.71–6.65 (m, 1H, H-4), 6.07 (ddd, *J* = 9.5, 6.3, 4.5 Hz, 1H, H-3), 5.61 (ddd, *J* = 21.4, 6.3, 1.2 Hz, 1H, H-2), 5.27 (s, 2H, CH_2Ph), 3.47 (d, *J* = 10.6 Hz, 6H, OMe) ppm. ^{13}C-NMR (100 MHz, DMSO-d_6, 80 °C): δ = 152.90 (d, *J* = 4.7 Hz, CO), 135.68 (Ar), 134.66 (Ar), 128.05 (Ar), 127.70 (Ar), 127.40 (Ar), 127.35 (Ar), 127.01 (d, *J* = 9.7 Hz, C-4), 126.77 (d, *J* = 4.0 Hz, Ar), 126.06 (d, *J* = 1.5 Hz, Ar), 124.35 (Ar), 123.66 (Ar), 122.55 (d, *J* = 3.0 Hz, C-3), 67.44 (CH_2Ph), 52.70 (d, *J* = 7.0 Hz, OMe), 52.39 (d, *J* = 6.6 Hz, OMe), 50.84 (d, *J* = 151.5 Hz, C-2) ppm.

^{31}P-NMR (162 MHz, DMSO-d_6, 80 °C): δ = 20.41 ppm. HRMS (ESI): calcd. for $C_{19}H_{20}NNaO_5P$ [M + Na]$^+$ 396.0971; found 396.0991.

3.5. Synthesis of Dimethyl 1,2,3,4-Tetrahydroquinoline-2-phosphonate **14**

A mixture of **13** (200 mg, 0.54 mmol) and 10% Pd/C (20 mg) in ethyl acetate (10 mL) was stirred at room temperature under an atmospheric pressure of hydrogen gas for 12 h. Filtration of the catalyst and evaporation of the solvent provided pure **14** as a colourless oil (130 mg, 0.54 mmol, 100% yield). IR (neat) ν_{max} 3316, 1231, 1057, 1029 cm^{-1}. ^{1}H-NMR (400 MHz, $CDCl_3$): δ = 7.02–6.93 (m, 2H, Ar), 6.70–6.63 (m, 1H, Ar), 6.58–6.53 (m, 1H, Ar), 4.16 (br s, 1H, NH), 3.82 (d, *J* = 10.3 Hz, 3H, OMe), 3.81 (d, *J* = 10.5 Hz, 3H, OMe), 3.71 (ddd, *J* = 10.1, 6.7, 3.4 Hz, 1H, H-2), 2.88–2.75 (m, 2H, H-4), 2.29–2.19 (m, 1H, H-3), 2.12–1.98 (m, 1H, H-3′) ppm. ^{13}C-NMR (100 MHz, $CDCl_3$): δ = 143.08 (d, *J* = 10.9 Hz, Ar), 129.40 (Ar), 127.11 (Ar), 121.05 (Ar), 118.26 (Ar), 114.97 (Ar), 53.83 (d, *J* = 6.7 Hz, OMe), 53.08 (d, *J* = 7.3 Hz, OMe), 49.05 (d, *J* = 161.5 Hz, C-2), 26.09 (d, *J* = 13.7 Hz, C-4), 22.39 (d, *J* = 5.2 Hz, C-3) ppm. ^{31}P-NMR (162 MHz, $CDCl_3$): δ = 27.39 ppm. HRMS (ESI): calcd. for $C_{11}H_{16}NNaO_3P$ [M + Na]$^+$ 264.0760; found 264.0752.

3.6. Synthesis of 1,2,3,4-Tetrahydroquinoline-2-phosphonic Acid Hydrobromide **4**

From **11**: A 33% solution of hydrogen bromide in acetic acid (2 mL) was added to compound **11** (110 mg, 0.29 mmol) and the reaction mixture was stirred at room temperature for 3 h. The solvent was evaporated and the residue was taken up in water and lyophilised to afford **4** as a white solid (86 mg, 0.29 mmol, 100% yield). M.p. 94–96 °C (dec.). IR (nujol) ν_{max} 3381, 1171, 1077, 1057 cm^{-1}. ^{1}H-NMR (400 MHz, CD_3OD): δ = 7.40–7.28 (m, 4H, Ar), 3.78 (ddd, *J* = 13.1, 12.1, 2.5 Hz, 1H, H-2), 3.08–3.01 (m, 2H, H-4), 2.52–2.43 (m, 1H, H-3), 2.19–2.06 (m, 1H, H-3′) ppm. ^{13}C-NMR (100 MHz, CD_3OD): δ = 132.21 (d, *J* = 0.7 Hz, Ar), 131.87 (Ar), 131.78 (Ar), 130.25 (Ar), 128.64 (Ar), 124.49 (Ar), 52.52 (d, *J* = 153.2 Hz, C-2), 25.94 (d, *J* = 12.4 Hz, C-4), 22.37 (d, *J* = 2.2 Hz, C-3) ppm. ^{31}P-NMR (162 MHz, CD_3OD): δ = 14.61 ppm. HRMS (ESI): calcd. for $C_9H_{13}NO_3P$ [M − Br]$^+$ 214.0628; found 214.0632.

From **14**: A 33% solution of hydrogen bromide in acetic acid (2 mL) was added to compound **14** (130 mg, 0.54 mmol) and the reaction mixture was stirred at room temperature for 3 h. The solvent was evaporated and the residue was taken up in water and lyophilised to afford **4** as a white solid (158 mg, 0.54 mmol, 100% yield). Spectroscopic data were identical to those described above.

3.7. Synthesis of Dimethyl N-Benzyloxycarbonyl-1,2-dihydroisoquinoline-1-phosphonate **16**

Benzyl chloroformate (0.61 mL, 726 mg, 4.26 mmol) was added dropwise to a solution of isoquinoline (500 mg, 3.87 mmol) and trimethyl phosphite (0.50 mL, 528 mg, 4.26 mmol) in anhydrous acetonitrile (6 mL) kept at 0 °C under argon. Sodium iodide (853 mg, 5.69 mmol) was slowly added and the mixture was heated at 50 °C for 10 min. The solvent was evaporated and the crude product was partitioned between dichloromethane (15 mL) and saturated aqueous sodium bicarbonate (15 mL). The organic phase was separated and the aqueous layer was further extracted with dichloromethane (3 × 15 mL). The combined organic extracts were washed with brine (10 mL), dried, filtered, and concentrated. Purification by column chromatography (eluent: ethyl acetate/hexanes 4:1) afforded **16** as a colourless oil (1.33 g, 3.56 mmol, 92% yield). IR (neat) ν_{max} 1715, 1342, 1295, 1238, 1121, 1028 cm^{-1}. ^{1}H-NMR (300 MHz, DMSO-d_6, 50 °C): δ = 7.51–7.11 (m, 9H, Ar), 6.91 (d, *J* = 7.7 Hz, 1H, H-3), 6.08–5.94 (m, 1H, H-4), 5.83 (d, *J* = 15.8 Hz, 1H, H-1), 5.27 (s, 2H, CH_2Ph), 3.57–3.38 (m, 6H, OMe) ppm. ^{13}C-NMR (100 MHz, DMSO-d_6): δ = (duplicate signals are observed for some carbons; asterisks indicate those corresponding to the minor rotamer) 152.37* (CO), 151.88 (CO), 136.01 (Ar), 135.91* (Ar), 131.03* (d, *J* = 3.6 Hz, Ar), 130.84 (d, *J* = 3.6 Hz, Ar), 128.75* (d *J* = 3.0 Hz, Ar), 128.68 (d, *J* = 3.1 Hz, Ar), 128.54 (Ar), 128.47* (Ar), 128.26 (Ar), 128.11* (Ar), 127.92 (Ar), 127.55 (d, *J* = 5.2 Hz, Ar), 127.30 (d, *J* = 2.3 Hz, Ar), 127.18 (d, *J* = 2.0 Hz, Ar), 125.43 (Ar), 125.35 (Ar), 124.86 (d, *J* = 2.9 Hz, C-3), 124.79* (C-3), 109.86 (C-4), 67.86* (CH_2Ph), 67.77 (CH_2Ph), 53.40* (d, *J* = 148.4 Hz, C-1), 53.16* (d, *J* = 6.1 Hz,

OMe), 53.12 (d, J = 7.0 Hz, OMe), 52.64 (d, J = 149.6 Hz, C-1) ppm. ^{31}P-NMR (122 MHz, DMSO-d_6, 50 °C): δ = 21.34 ppm. HRMS (ESI): calcd. for $C_{19}H_{20}NNaO_5P$ $[M + Na]^+$ 396.0971; found 396.0982.

3.8. Synthesis of Diethyl N-Benzyloxycarbonyl-1,2-dihydroisoquinoline-1-phosphonate 19

Benzyl chloroformate (0.61 mL, 726 mg, 4.26 mmol) was added dropwise to a solution of isoquinoline (500 mg, 3.87 mmol) and triethyl phosphite (0.73 mL, 708 mg, 4.26 mmol) in anhydrous acetonitrile (6 mL) kept at 0 °C under argon. Sodium iodide (853 mg, 5.69 mmol) was slowly added and the mixture was heated at 50 °C for 10 min. The solvent was evaporated and the crude product was partitioned between dichloromethane (15 mL) and saturated aqueous sodium bicarbonate (15 mL). The organic phase was separated and the aqueous layer was further extracted with dichloromethane (3 × 15 mL). The combined organic extracts were washed with brine (10 mL), dried, filtered, and concentrated. Purification by column chromatography (eluent:ethyl acetate/hexanes 3:2) afforded **19** as a colourless oil (1.52 g, 3.79 mmol, 98% yield). IR (neat) ν_{max} 1715, 1294, 1253, 1120, 1023 cm^{-1}. ^{1}H-NMR (300 MHz, DMSO-d_6, 70 °C): δ = 7.50–7.10 (m, 9H, Ar), 6.91 (d, J = 7.8 Hz, 1H, H-3), 5.98 (d, J = 7.8 Hz, 1H, H-4), 5.77 (d, J = 15.9 Hz, 1H, H-1), 5.26 (s, 2H, CH_2Ph), 3.95–3.69 (m, 4H, OCH_2), 1.08 (t, J = 7.0 Hz, 3H, Me), 1.07 (t, J = 7.0 Hz, 3H, Me) ppm. ^{13}C-NMR (100 MHz, DMSO-d_6): δ = (duplicate signals are observed for some carbons; asterisks indicate those corresponding to the minor rotamer) 152.48* (CO), 151.91 (CO), 136.05 (Ar), 135.88* (Ar), 131.15* (d, J = 3.6 Hz, Ar), 130.98 (d, J = 3.7 Hz, Ar), 128.69* (d, J = 3.3 Hz, Ar), 128.61 (d, J = 3.3 Hz, Ar), 128.56 (Ar), 128.47* (Ar), 128.28 (Ar), 128.25* (Ar), 128.13* (Ar), 127.97 (Ar), 127.53 (d, J = 5.2 Hz, Ar), 127.22 (d, J = 2.6 Hz, Ar), 127.11* (d, J = 2.7 Hz, Ar), 125.59 (d, J = 1.8 Hz, Ar), 125.56* (d, J = 2.4 Hz, Ar), 125.42 (Ar), 124.84* (C-3), 124.81 (C-3), 109.95 (C-4), 67.84* (CH_2Ph), 67.71 (CH_2Ph), 62.54 (d, J = 7.1 Hz, OCH_2), 62.42* (d, J = 6.6 Hz, OCH_2), 62.38* (d, J = 7.1 Hz, OCH_2), 53.97* (d, J = 148.9 Hz, C-1), 53.15 (d, J = 150.2 Hz, C-1), 16.17 (d, J = 5.3 Hz, Me), 16.15* (d, J = 5.7 Hz, Me) ppm. ^{31}P-NMR (122 MHz, DMSO-d_6, 70 °C): δ = 18.77 ppm. HRMS (ESI): calcd. for $C_{21}H_{25}NO_5P$ $[M + H]^+$ 402.1465; found 402.1466.

3.9. Synthesis of Diethyl N-Benzyloxycarbonyl-1,2,3,4-tetrahydroisoquinoline-1-phosphonate 20

Triethylsilane (1.0 mL, 730 mg, 6.28 mmol) and trifluoroacetic acid (0.48 mL, 716 mg, 6.28 mmol) were added to a solution of **19** (300 mg, 0.75 mmol) in anhydrous dichloromethane (15 mL) kept at 0 °C under argon. The solution was allowed to warm to room temperature and stirred for 18 h. Evaporation of the solvent followed by column chromatography (eluent:hexanes/ethyl acetate 1:1) afforded **20** as a colourless oil (298 mg, 0.74 mmol, 99% yield). IR (neat) ν_{max} 1701, 1294, 1249, 1230, 1051, 1022 cm^{-1}. ^{1}H-NMR (300 MHz, DMSO-d_6, 70 °C): δ = 7.44–7.16 (m, 9H, Ar), 5.53 (d, J = 20.4 Hz, 1H, H-1), 5.17 (s, 2H, CH_2Ph), 4.14–3.77 (m, 5H, H-3, OCH_2), 3.72–3.52 (m, 1H, H-3′), 2.98–2.78 (m, 2H, H-4), 1.17 (t, J = 7.0 Hz, 3H, Me), 1.08 (t, J = 6.9 Hz, 3H, Me) ppm. ^{13}C-NMR (100 MHz, DMSO-d_6): δ = (duplicate signals are observed for some carbons; asterisks indicate those corresponding to the minor rotamer) 154.60 (d, J = 3.9 Hz, CO), 154.18* (d, J = 2.4 Hz, CO), 136.66 (Ar), 136.46* (Ar), 134.73 (d, J = 5.6 Hz, Ar), 134.63* (d, J = 5.6 Hz, Ar), 129.22 (Ar), 129.13* (d, J = 2.2 Hz, Ar), 128.95 (d, J = 2.3 Hz, Ar), 128.85 (Ar), 128.38 (Ar), 128.33* (Ar), 127.94* (Ar), 127.91 (Ar), 127.86* (Ar), 127.67 (d, J = 4.0 Hz, Ar), 127.58 (Ar), 127.38 (Ar), 125.88 (d, J = 2.8 Hz, Ar), 125.83* (d, J = 2.8 Hz, Ar), 66.92* (CH_2Ph), 66.82 (CH_2Ph), 62.58 (d, J = 7.2 Hz, OCH_2), 62.19 (d, J = 7.1 Hz, OCH_2), 52.76* (d, J = 149.9 Hz, C-1), 52.34 (d, J = 152.0 Hz, C-1), 39.05* (C-3), 38.84 (C-3), 27.43 (C-4), 27.15* (C-4), 16.08* (d, J = 5.6 Hz, Me), 16.05 (d, J = 5.3 Hz, Me). ^{31}P-NMR (122 MHz, DMSO-d_6, 70 °C): δ = 20.75. HRMS (ESI): calcd. for $C_{21}H_{27}NO_5P$ $[M + H]^+$ 404.1621; found 404.1611.

3.10. Synthesis of 1,2,3,4-Tetrahydroisoquinoline-1-phosphonic Acid Hydrobromide 5

A 33% solution of hydrogen bromide in acetic acid (2 mL) was added to **20** (160 mg, 0.40 mmol) and the reaction mixture was stirred at room temperature for 3 h. The solvent was evaporated and the residue was taken up in water and lyophilised to afford **5** as a white solid (117 mg, 0.40 mmol, 100% yield). M.p. 85–87 °C (dec.). IR (nujol) ν_{max} 3421, 1212, 1118, 1019 cm^{-1}. ^{1}H-NMR (400 MHz,

D_2O): δ = 7.45–7.41 (m, 1H, Ar), 7.37–7.27 (m, 3H, Ar), 4.70 (d, *J* = 17.6 Hz, 1H, H-1), 3.78 (ddd, *J* = 12.7, 9.2, 6.5 Hz, 1H, H-3), 3.58–3.51 (m, 1H, H-3′), 3.18–3.12 (m, 2H, H-4) ppm. ^{13}C-NMR (100 MHz, CD_3OD): δ = 132.88 (d, *J* = 5.4 Hz, Ar), 130.40 (d, *J* = 2.2 Hz, Ar), 129.47 (d, *J* = 2.7 Hz, Ar), 128.94 (d, *J* = 3.6 Hz, Ar), 127.92 (d, *J* = 2.6 Hz, Ar), 127.40 (d, *J* = 5.1 Hz, Ar), 54.00 (d, *J* = 147.3 Hz, C-1), 41.15 (d, *J* = 2.1 Hz, C-3), 25.88 (C-4) ppm. ^{31}P-NMR (162 MHz, D_2O): δ = 10.02 ppm. HRMS (ESI): calcd. for $C_9H_{13}NO_3P$ [M − Br]$^+$ 214.0628; found 214.0633.

3.11. *Synthesis of N-Benzyloxycarbonyl-1,4-dihydro-3-isoquinolinone* **22**

A 1 M solution of lithium bis(trimethylsilyl)amide in tetrahydrofuran (0.83 mL, 0.83 mmol) was slowly added to a solution of 1,4-dihydro-3(2*H*)-isoquinolinone **21** (122 mg, 0.83 mmol) in anhydrous tetrahydrofuran (5 mL) kept at −78 °C under argon. After 30 min, benzyl chloroformate (0.12 mL, 142 mg, 0.83 mmol) was added dropwise and stirring was continued for additional 3 h. The reaction was then treated with saturated aqueous ammonium chloride (10 mL) and allowed to warm to room temperature. The two layers were separated and the aqueous phase was extracted with dichloromethane (2 × 10 mL). The combined organic extracts were dried, filtered, and concentrated. Purification by column chromatography (eluent:hexanes/ethyl acetate 7:3) afforded **22** as a white solid (172 mg, 0.61 mmol, 73% yield). M.p. 63–65 °C. IR (nujol) ν_{max} 1692 cm^{-1}. ^{1}H-NMR (400 MHz, $CDCl_3$): δ = 7.49–7.44 (m, 2H, Ar), 7.40–7.25 (m, 6H, Ar), 7.23–7.19 (m, 1H, Ar), 5.34 (s, 2H, CH_2Ph), 4.92 (s, 2H, H-1), 3.73 (s, 2H, H-4) ppm. ^{13}C-NMR (100 MHz, $CDCl_3$): δ = 169.43 (CO), 153.62 (COO), 135.38 (Ar), 132.25 (Ar), 132.13 (Ar), 128.74 (Ar), 128.50 (Ar), 128.46 (Ar), 128.22 (Ar), 127.35 (Ar), 126.99 (Ar), 125.81 (Ar), 68.97 (CH_2Ph), 48.75 (C-1), 41.53 (C-4) ppm. HRMS (ESI): calcd. for $C_{17}H_{15}NNaO_3$ [M + Na]$^+$ 304.0944; found 304.0943.

3.12. *Synthesis of Dimethyl N-Benzyloxycarbonyl-1,2,3,4-tetrahydroisoquinoline-3-phosphonate* **25**

A 1 M solution of diisobutylaluminium hydride in hexanes (0.78 mL, 0.78 mmol) was slowly added to a solution of **22** (146 mg, 0.52 mmol) in anhydrous tetrahydrofuran (5 mL) kept at −78 °C under argon. After stirring at this temperature for 2 h, the reaction was treated with saturated aqueous sodium acetate (10 mL) and allowed to warm to room temperature. A 3:1 mixture of diethyl ether and saturated aqueous ammonium chloride (16 mL) was then added and the resulting mixture was stirred at room temperature until a suspension was formed. The solid was filtered off under reduced pressure and washed with diethyl ether (2 × 10 mL). The organic layer was separated and the aqueous phase was extracted with diethyl ether (2 × 10 mL). The combined organic extracts were washed with brine (10 mL), dried, filtered, and evaporated to provide the hemiaminal as an oil. It was dissolved in methanol (5 mL) and treated with pyridinium *p*-toluenesulfonate (13 mg, 0.05 mmol). After stirring at room temperature for 2 h, triethylamine (31 µL, 22 mg, 0.22 mmol) was added. The solvent was evaporated and the crude methoxyaminal obtained **23** was dissolved in anhydrous dichloromethane (5 mL) and kept under argon. Trimethyl phosphite (61 µL, 65 mg, 0.52 mmol) was added and the resulting solution was cooled to −20 °C. Boron trifluoride-diethyl ether (65 µL, 74 mg, 0.52 mmol) was added and the reaction mixture was slowly warmed to room temperature and stirred for 12 h. After quenching with water (10 mL), the two layers were separated and the aqueous phase was extracted with dichloromethane (2 × 10 mL). The combined organic extracts were dried, filtered, and concentrated. Purification by column chromatography (eluent:ethyl acetate/hexanes 9:1) afforded **25** as a colourless oil (138 mg, 0.37 mmol, 71% yield). IR (neat) ν_{max} 1703, 1410, 1246, 1055, 1031 cm^{-1}. ^{1}H-NMR (400 MHz, $CDCl_3$): δ = (duplicate signals are observed for some protons; asterisks indicate those corresponding to the minor rotamer) 7.42–7.30 (m, 5H, Ar), 7.23–7.04 (m, 4H, Ar), 5.31–5.12 (m, 2H, CH_2Ph), 5.11–5.03 (m, 1H, H-3), 5.00–4.86 (m, 1H, H-1) overlapped with 4.97*–4.89* (m, 1H, H-3), 4.52 (d, *J* = 16.6 Hz, 1H, H-1′), 4.45* (d, *J* = 16.6 Hz, 1H, H-1′), 3.65 (d, *J* = 10.7 Hz, 3H, OMe), 3.51* (d, *J* = 10.7 Hz, 3H, OMe), 3.35 (d, *J* = 10.7 Hz, 3H, OMe), 3.33–3.16 (m, 2H, H-4) overlapped with 3.24* (d, *J* = 10.7 Hz, 3H, OMe) ppm. ^{13}C-NMR (100 MHz, $CDCl_3$): δ = (duplicate signals are observed for some carbons; asterisks indicate those corresponding to the minor rotamer)

155.72 (CO), 155.05* (CO), 136.37 (Ar), 136.18* (Ar), 132.61* (Ar), 132.46 (Ar), 131.41 (Ar), 131.00* (Ar), 128.64 (Ar), 128.53 (Ar), 128.45* (Ar), 128.24 (Ar), 128.17* (Ar), 127.95 (Ar), 126.70 (Ar), 126.64* (Ar), 126.60 (Ar), 126.09* (Ar), 125.86 (Ar), 67.79 (CH_2Ph), 52.74 (d, *J* = 6.4 Hz, OMe), 52.58 (d, *J* = 6.6 Hz, OMe), 52.34* (d, *J* = 6.9 Hz, OMe), 47.07* (d, *J* = 154.2 Hz, C-3), 46.29 (d, *J* = 154.1 Hz, C-3), 44.37 (C-1), 28.56* (C-4), 28.32 (C-4) ppm. ^{31}P-NMR (162 MHz, $CDCl_3$): δ = (a duplicate signal is observed; an asterisk indicates that corresponding to the minor rotamer) 27.67, 27.07* ppm. HRMS (ESI): calcd. for $C_{19}H_{22}NNaO_5P$ $[M + Na]^+$ 398.1128; found 398.1144.

3.13. *Synthesis of 1,2,3,4-Tetrahydroisoquinoline-3-phosphonic Acid Hydrobromide* 6

A 33% solution of hydrogen bromide in acetic acid (2 mL) was added to **25** (100 mg, 0.27 mmol) and the reaction mixture was stirred at room temperature for 3 h. The solvent was evaporated and the residue was taken up in water and lyophilised to afford **6** as a white solid (79 mg, 0.27 mmol, 100% yield). M.p. 102–103 °C (dec.). IR (nujol) ν_{max} 3404, 1232, 1010 cm^{-1}. ^{1}H-NMR (400 MHz, CD_3OD): δ = 7.35–7.22 (m, 4H, Ar), 4.52 (d, *J* = 15.7 Hz, 1H, H-1), 4.42 (dd, *J* = 15.7, 3.1 Hz, 1H, H-1′), 3.94–3.83 (m, 1H, H-3), 3.34–3.25 (m, 2H, H-4) ppm. ^{13}C-NMR (100 MHz, CD_3OD): δ = 131.71 (d, *J* = 12.2 Hz, Ar), 129.97 (Ar), 129.27 (Ar), 128.69 (Ar), 128.32 (Ar), 127.77 (Ar), 51.34 (d, *J* = 153.1 Hz, C-3), 47.01 (d, *J* = 7.7 Hz, C-1), 27.44 (C-4) ppm. ^{31}P-NMR (162 MHz, CD_3OD): δ = 14.06 ppm. HRMS (ESI): calcd. for $C_9H_{13}NO_3P$ $[M - Br]^+$ 214.0628; found 214.0631.

4. Conclusions

The synthesis of α-aminophosphonic acids characterized by a tetrahydroquinoline or tetrahydroisoquinoline heterocycles is described. Specifically, 1,2,3,4-tetrahydroquinoline-2-phosphonic acid **4**, 1,2,3,4-tetrahydroisoquinoline-1-phosphonic acid **5**, and 1,2,3,4-tetrahydro-isoquinoline-3-phosphonic acid **6** have been prepared in high overall yields using efficient methods that make use of easily available substrates. The compounds obtained can be viewed as higher homologues of phosphopipecolic acid that bear a benzene ring fused at different positions of the six-membered piperidine cycle. In particular, the latter is the phosphonic surrogate of Tic, an amino acid of extraordinary value in the design of pharmacologically useful peptides. The synthetic routes developed rely on the addition of trialkyl phosphites to iminium ions generated from quinoline/isoquinoline or from the appropriate δ-lactam bearing a fused benzene ring. Thus, lactams have proven to be versatile starting materials for the efficient and selective synthesis of α-aminophosphonic acids containing an azacyclic skeleton. The presence of the carbonyl group in the starting lactam determines, with full regiocontrol, the position at which the phosphonate substituent is incorporated. Formation of the intermediate *N*-acyliminium ion with the adequate regiochemistry has allowed the synthesis of the phosphonic counterpart of Tic (Tic^P). Importantly, this α-aminophosphonic acid is not accessible through other methodologies traditionally used to generate iminium ions.

Supplementary Materials: ^{1}H-, ^{13}C-, and ^{31}P- (when applicable) NMR spectra of the final compounds and all key intermediates are available online at http://www.mdpi.com/1420-3049/21/9/1140/s1.

Acknowledgments: Financial support to this work was provided by Consejo Nacional de Ciencia y Tecnología (CONACYT-Mexico, grants 181816 and 248868). Ministerio Economia y Competitividad (grant CTQ2013-40855-R), and Gobierno de Aragón-Fondo Social Europeo (research group E40). Consejo Superior de Investigaciones Científicas (grant 2008MX0044; JAE predoctoral fellowship to A.A).

Author Contributions: M.O., C.C. and A.I.J. provided the concepts of the work, interpreted the results and prepared the manuscript, A.A., F.J.S., carried out the experimental work, interpreted the results and prepared the manuscript. All authors read and approved the final manuscript.

Conflicts of Interest: The authors declare no conflict of interest.

References

1. Ko, E.; Liu, J.; Perez, L.M.; Lu, G.; Schaefer, A.; Burgess, K. Universal peptidomimetics. *J. Am. Chem. Soc.* **2011**, *133*, 462–477. [CrossRef] [PubMed]

2. Hruby, V.J. Designing peptide receptor agonists and antagonists. *Nat. Rev. Drug. Discov.* **2002**, *1*, 847–858. [CrossRef] [PubMed]
3. Brindisi, M.; Butini, S.; Franceschini, S.; Brogi, S.; Trotta, F.; Ros, S.; Cagnotto, A.; Salmona, M.; Casagni, A.; Andreassi, M.; et al. Targeting Dopamine D_3 and Serotonin 5-HT_{1A} and 5-HT_{2A} Receptors for developing effective antipsychotics: Synthesis, biological characterization, and behavioral studies. *J. Med. Chem.* **2014**, *57*, 9578–9597. [CrossRef] [PubMed]
4. Mitsunaga, S.; Ohbayashi, T.; Sugiyama, S.; Saitou, T.; Tadokoro, M.; Satoh, T. Asymmetric synthesis of cyclic α-amino acid derivatives by the intramolecular reaction of magnesium carbenoid with an *N*-magnesio arylamine. *Tetrahedron: Asymmetry* **2009**, *20*, 1697–1708. [CrossRef]
5. Kurata, K.; Inoue, K.; Nishimura, K.; Hoshiya, N.; Kawai, N.; Uenishi, J. Synthesis of optically pure (*R*)- and (*S*)-tetrahydroisoquinoline-1- and -3-carboxylic acids. *Synthesis* **2015**, *47*, 1238–1244.
6. Hu, L.; Magesh, S.; Chen, L.; Wang, L.; Lewis, T.A.; Chen, Y.; Khodier, C.; Inoyama, D.; Beamer, L.J.; Emge, T.J.; et al. Discovery of a small-molecule inhibitor and cellular probe of Keap1-Nrf2 protein-protein interaction. *Bioorg. Med. Chem. Lett.* **2013**, *23*, 3039–3043. [CrossRef] [PubMed]
7. Kotha, S.; Deodhar, D.; Khedkar, P. Diversity-oriented synthesis of medicinally important 1,2,3,4-tetrahydroisoquinoline-3-carboxylic acid (Tic) derivatives and higher analogs. *Org. Biomol. Chem.* **2014**, *12*, 9054–9091. [CrossRef] [PubMed]
8. Yolacan, C.; Mavis, M.E.; Aydogan, F. Evaluation of mono- and dipeptides as organocatalysts for enantioselective aldol reaction. *Tetrahedron* **2014**, *70*, 3707–3713. [CrossRef]
9. Mucha, A.; Kafarski, P.; Berlicki, Ł. Remarkable Potential of the α-aminophosphonate/phosphinate structural motif in medicinal chemistry. *J. Med. Chem.* **2011**, *54*, 5955–5980. [CrossRef] [PubMed]
10. Orsini, F.; Sello, G.; Sisti, M. Aminophosphonic acids and derivatives. Synthesis and biological applications. *Curr. Med. Chem.* **2010**, *17*, 264–289. [CrossRef] [PubMed]
11. Naydenova, E.D.; Todorov, P.T.; Troev, K.D. Recent synthesis of aminophosphonic acids as potential biological importance. *Amino Acids* **2010**, *38*, 23–30. [CrossRef] [PubMed]
12. Lejczak, B.; Kafarski, P. Biological activity of aminophosphonic acids and their short peptides. *Top. Heterocycl. Chem.* **2009**, *20*, 31–63.
13. Ordóñez, M.; Viveros-Ceballos, J.L.; Cativiela, C.; Sayago, F.J. An update on the stereoselective synthesis of α-aminophosphonic acids and derivatives. *Tetrahedron* **2015**, *71*, 1745–1784. [CrossRef]
14. Ordóñez, M.; Sayago, F.J.; Cativiela, C. Synthesis of quaternary α-aminophosphonic acids. *Tetrahedron* **2012**, *68*, 6369–6412. [CrossRef]
15. Ordóñez, M.; Viveros-Ceballos, J.L.; Cativiela, C.; Arizpe, A. Stereoselective synthesis of α-aminophosphonic acids analogs of the 20 proteinogenic α-amino acids. *Curr. Org. Synth.* **2012**, *9*, 310–341. [CrossRef]
16. Kudzin, Z.H.; Kudzin, M.H.; Drabowicz, J.; Stevens, C.V. Aminophosphonic acids-phosphorus analogues of natural Amino Acids. Part 1: Syntheses of α-aminophosphonic acids. *Curr. Org. Chem.* **2011**, *15*, 2015–2071. [CrossRef]
17. Gulyukina, N.S.; Makukhin, N.N.; Beletskaya, I.P. Synthesis methods of (1-aminocyclopropyl)phosphonic acids. *Russ. J. Org. Chem.* **2011**, *47*, 633–649. [CrossRef]
18. Ordóñez, M.; Rojas-Cabrera, H.; Cativiela, C. An overview of stereoselective synthesis of α-aminophosphonic acids and derivatives. *Tetrahedron* **2009**, *65*, 17–49. [CrossRef] [PubMed]
19. Pudovik, A.N. Addition of dialkyl phosphites to imines. New method of synthesis of esters of amino phosphonic acids. *Dokl. Akad. Nauk. SSSR* **1952**, *83*, 865–869.
20. Alfonsov, V.A. Diastereoselective Synthesis of Enantiopure α-Aminophosphonic Acid Derivatives: Pudovik Reaction in Stereoselective Synthesis (Dedicated to A.N. Pudovik, 1916–2006). *Phosphorus. Sulfur. Silicon Relat. Elem.* **2008**, *183*, 2637–2644. [CrossRef]
21. Kabachnik, M.I.; Medved, T.Y. New synthesis of aminophosphonic acids. *Dokl. Akad. Nauk SSSR* **1952**, *83*, 689–692.
22. Fields, E.K. The synthesis of esters of substituted amino phosphonic acids. *J. Am. Chem. Soc.* **1952**, *74*, 1528–1531. [CrossRef]
23. Zefirov, N.S.; Matveeva, E.D. Catalytic Kabachnik-Fields reaction: New horizons for old reaction. *Arkivoc* **2008**, *2008*. [CrossRef]
24. Kafarski, P.; Gorny vel Gorniak, M.; Andrasiak, I. Kabachnik-Fields Reaction under Green Conditions—A Critical Overview. *Curr. Green Chem.* **2015**, *2*, 218–222. [CrossRef]

25. Keglevich, G.; Bálint, E. The Kabachnik–Fields Reaction: Mechanism and Synthetic Use. *Molecules* **2012**, *17*, 12821–12835. [CrossRef] [PubMed]
26. Bonilla-Landa, I.; Viveros-Ceballos, J.L.; Ordóñez, M. Diastereoselective synthesis of novel 5-substituted morpholine-3-phosphonic acids: Further exploitation of *N*-acyliminium intermediates. *Tetrahedron: Asymmetry* **2014**, *25*, 485–487. [CrossRef]
27. Arizpe, A.; Sayago, F.J.; Jiménez, A.I.; Ordóñez, M.; Cativiela, C. Synthesis of phosphoproline derivatives with an octahydroisoindole structure. *Eur. J. Org. Chem.* **2011**, *2001*, 6732–6738. [CrossRef]
28. Arizpe, A.; Sayago, F.J.; Jiménez, A.I.; Ordóñez, M.; Cativiela, C. Stereodivergent synthesis of two novel α-aminophosphonic acids characterised by a *cis*-fused octahydroindole system. *Eur. J. Org. Chem.* **2011**, *2001*, 3074–3081. [CrossRef]
29. Boto, A.; Gallardo, J.A.; Hernández, R.; Saavedra, C.J. One-pot synthesis of α-amino phosphonates from α-amino acids and β-amino alcohols. *Tetrahedron Lett.* **2005**, *46*, 7807–7811. [CrossRef]
30. Shono, T.; Matsumura, Y.; Tsubata, K. A new carbon—Phosphorous bond forming reaction and synthesis of aminoalkylphosphonic acid derivatives. *Tetrahedron Lett.* **1981**, *22*, 3249–3252. [CrossRef]
31. Maury, C.; Wang, Q.; Gharbaoui, T.; Chiadmi, M.; Tomas, A.; Royer, J.; Husson, H.-P. Asymmetric synthesis of (*R*)- and (*S*)-piperidin-2-yl-phosphonic acid by diastereoselective addition of trialkyl phosphite onto potential iminium salt. *Tetrahedron* **1997**, *53*, 3627–3636. [CrossRef]
32. Katritzky, A.R.; Qiu, G.; He, H.-Y.; Yang, B. Novel Syntheses of Hexahydro-1*H*-pyrrolo[1,2-*a*]imidazoles and Octahydroimidazo[1,2-*a*]pyridines. *J. Org. Chem.* **2000**, *65*, 3683–3689. [CrossRef] [PubMed]
33. Kaname, M.; Mashige, H.; Yoshifuji, S. Chemical Conversion of Cyclic α-Amino Acids to Cyclic α-Aminophosphonic Acids. *Chem. Pharm. Bull.* **2001**, *49*, 531–536. [CrossRef] [PubMed]
34. Yang, D.; Zhao, D.; Mao, L.; Wang, L.; Wang, R. Copper/DIPEA-Catalyzed, Aldehyde-Induced Tandem Decarboxylation—Coupling of Natural α-Amino Acids and Phosphites or Secondary Phosphine Oxides. *J. Org. Chem.* **2011**, *76*, 6426–6431. [CrossRef] [PubMed]
35. Han, W.; Mayer, P.; Ofial, A.R. Iron-Catalyzed Oxidative Mono- and Bis-Phosphonation of *N*,*N*-Dialkylanilines. *Adv. Synth. Catal.* **2010**, *352*, 1667–1676. [CrossRef]
36. Rueping, M.; Zhu, S.; Koenigs, R.M. Photoredox catalyzed C-P bond forming reactions-visible light mediated oxidative phosphonylations of amines. *Chem. Commun.* **2011**, *47*, 8679–8681. [CrossRef] [PubMed]
37. Wang, H.; Li, X.; Wu, F.; Wan, B. Direct oxidative phosphonylation of amines under metal-free conditions. *Tetrahedron Lett.* **2012**, *53*, 681–683. [CrossRef]
38. Yoo, W.-J.; Kobayashi, S. Efficient visible light-mediated crossdehydrogenative coupling reactions of tertiary amines catalyzed by a polymer-immobilized iridium-based photocatalyst. *Green Chem.* **2014**, *16*, 2438–2442. [CrossRef]
39. Viveros-Ceballos, J.L.; Ordóñez, M.; Sayago, F.J.; Jimenez, A.I.; Cativiela, C. First synthesis of (*R*)- and (*S*)-1,2,3,4-tetrahydroisoquinoline-3-phosphonic acid (Tic^P) using a Pictet-Spengler reaction. *Eur. J. Org. Chem.* **2016**, 2711–2719. [CrossRef]
40. Ramirez-Marroquin, O.A.; Romero-Estudillo, I.; Viveros-Ceballos, J.L.; Cativiela, C.; Ordóñez, M. Convenient synthesis of cyclic α-aminophosphonates by alkylation-cyclization reaction of iminophosphoglycinates using phase-transfer catalysis. *Eur. J. Org. Chem.* **2016**, *2016*, 308–313. [CrossRef]
41. López-Iglesias, M.; Arizpe, A.; Sayago, F.J.; Gotor, V.; Cativiela, C.; Gotor-Fernández, V. Lipase-catalyzed dynamic kinetic resolution of dimethyl (1,3-dihydro-2*H*-isoindol-1-yl)phosphonate. *Tetrahedron* **2016**. [CrossRef]
42. Arizpe, A.; Rodríguez-Mata, M.; Sayago, F.J.; Pueyo, M.J.; Gotor, V.; Jiménez, A.I.; Gotor-Fernández, V.; Cativiela, C. Enzymatic and chromatographic resolution procedures applied to the synthesis of the phosphoproline enantiomers. *Tetrahedron: Asymmetry* **2015**, *26*, 1469–1477. [CrossRef]
43. Viveros-Ceballos, J.L.; Sayago, F.J.; Cativiela, C.; Ordóñez, M. First practical and efficient synthesis of 3-phosphorylated β-carboline derivatives using the Pictet-Spengler reaction. *Eur. J. Org. Chem.* **2015**, *2015*, 1084–1091. [CrossRef]
44. Liu, K.; Liu, L.-L.; Gu, C.-Z.; Dai, B.; He, L. Aryne-induced dearomatized phosphonylation of electron-deficient azaarenes. *RSC Adv.* **2016**, *6*, 33606–33610. [CrossRef]
45. Zhang, Q.; Wei, D.; Cui, X.; Zhang, D.; Wang, H.; Wu, Y. Direct diphosphonylation of quinolines with *H*-phosphonates under metal-free conditions. *Tetrahedron* **2015**, *71*, 6087–6093. [CrossRef]

46. Shaabani, A.; Sarvary, A.; Mousavi-Faraz, S.; Ng, S.W. Synthesis of 1,2-dihydroquinolin-2-ylphosphonates and 1,2-dihydroisoquinolin-1-ylphosphonates via three-component reactions. *Monatsh. Chem.* **2012**, *143*, 1061–1065. [CrossRef]
47. De Blieck, A.; Masschelein, K.G.R.; Dhaene, F.; Rozycka-Sokolowska, E.; Marciniak, B.; Drabowicz, J.; Stevens, C.V. One-pot tandem 1,4–1,2-addition of phosphites to quinolines. *Chem. Commun.* **2010**, *46*, 258–260. [CrossRef] [PubMed]
48. De Blieck, A.; Catak, S.; Debrouwer, W.; Drabowicz, J.; Hemelsoet, K.; Verstraelen, T.; Waroquier, M.; van Speybroeck, V.; Stevens, C.V. Diphosphonylation of Aromatic Diazaheterocycles and Theoretical Rationalization of Product Yields. *Eur. J. Org. Chem.* **2013**, *2013*, 1058–1067. [CrossRef]
49. Hamada, Y.; Sugiura, M.; Asai, K.; Hamada, Y. Simple synthetic method of dialkyl 1,2-dihydro(iso)quinoline (1 or 2)-phosphonates. *Heterocycles* **1996**, *43*, 953–958. [CrossRef]
50. Akiba, K.; Negishi, Y.; Inamoto, N. A facile general synthesis of dimethyl 1-acyl-1,2-dihydroquinoline-2-phosphonates and 2-acyl-1,2-dihydroisoquinoline-1-phosphonates. *Synthesis* **1979**, 55–56. [CrossRef]
51. Taylor, M.S.; Tokunaga, N.; Jacobsen, E.N. Enantioselective thiourea-catalyzed acyl-Mannich reactions of isoquinolines. *Angew. Chem. Int. Ed.* **2005**, *44*, 6700–6704. [CrossRef] [PubMed]
52. Ilayaraja, N.; Noel, M. A Comparative study of anodic fluorination of *N*-alkyl and *N,N*-dialkyl phenylacetamides in $Et_3N{\cdot}4HF$ medium. *J. Electroanal. Chem.* **2009**, *632*, 45–54. [CrossRef]
53. Kamochi, Y.; Watanabe, Y. Studies on the synthesis of benzolactam rings. II. Synthesis of 1,4-dihydro-3(2*H*)-isoquinolinone derivatives. *Heterocycles* **1987**, *26*. [CrossRef]

Sample Availability: Samples of the compounds are not available from the authors.

molecules

MDPI

Article

Palladium-Catalyzed Allylation/Benzylation of *H*-Phosphinate Esters with Alcohols

Anthony Fers-Lidou, Olivier Berger and Jean-Luc Montchamp *

Department of Chemistry and Biochemistry, TCU Box 298860, Texas Christian University, Fort Worth, TX 76129, USA; a.fers-lidou@outlook.fr (A.F.-L.); o.berger@tcu.edu (O.B.)

* Correspondence: j.montchamp@tcu.edu; Tel.: +1-817-257-5851

Academic Editor: György Keglevich

Received: 10 September 2016; Accepted: 19 September 2016; Published: 28 September 2016

Abstract: The Pd-catalyzed direct alkylation of *H*-phosphinic acids and hypophosphorous acid with allylic/benzylic alcohols has been described previously. Here, the extension of this methodology to *H*-phosphinate esters is presented. The new reaction appears general, although its scope is narrower than with the acids, and its mechanism is likely different. Various alcohols are examined in their reaction with phosphinylidene compounds $R^1R^2P(O)H$.

Keywords: palladium; cross-coupling; alcohols; *H*-phosphinate esters; allylation; benzylation

1. Introduction

In the classic Pd-catalyzed Tsuji-Trost reaction, the allylic electrophile is an alcohol derivative (most often the acetate) and numerous nucleophiles can be employed easily [1]. More recently, the use of allylic alcohols has emerged [1,2]. Using alcohols for the direct allylation of nucleophiles is desirable because water is the only byproduct. In the context of carbon-phosphorus bond-formation, we disclosed 10 years ago the reaction between hypophosphorous acid (HPA) and allylic alcohols catalyzed by Pd to directly afford the corresponding allylic *H*-phosphinic acids (Scheme 1a) [3]. The reaction was subsequently extended to include *H*-phosphinic acids instead of HPA to form disubstituted phosphinic acids under slightly more forcing conditions (Scheme 1b) [4]. Around the same time, we also described the benzylation of HPA and *H*-phosphinic acids with benzylic alcohols under similar conditions (Scheme 1c) [5]. These reactions were discovered based on mechanistic reasoning that was supported by model studies [6]. It was thought that the reaction required a PO_2H motif for Fisher esterification and tautomerization. More recently, we decided to reexamine this type of reaction, but using *H*-phosphinic esters as starting materials, and were surprised to observe a successful allylation, suggesting that a different mechanism could be operative. This manuscript describes these findings (Scheme 1d).

2. Results and Discussion

Based on our prior work in this area, we selected cinnamyl alcohol and butyl phenyl-*H*-phosphinate in equimolar amounts as the reacting partners for the initial investigation (Table 1). Cinnamyl alcohol was identified previously as a very reactive partner in our Pd-catalyzed allylation [3–6]. Additionally, based on our prior findings, *t*-amyl alcohol was selected as the solvent for the azeotropic removal of the water byproduct (reflux, Dean-Stark trap) [4,5]. Using palladium acetate as the catalyst, and without added ligand, the reaction failed to produce any detectable amount of product (entry 1). As we had found in our other couplings, Xantphos (4,5-bis(diphenylphosphino)-9,9-dimethylxanthene) performed superbly affording the desired product in nearly quantitative isolated yield (entry 2). Switching the solvent to toluene (still with a Dean-Stark trap) gave a satisfactory yield, albeit lower (entry 3). This was not entirely unexpected as *t*-amyl alcohol was initially identified for its ability to promote P(V) to P(III) tautomerization through hydrogen-bonding. Changing the salt from palladium acetate to palladium chloride resulted in a slightly lower yield (entry 4 vs. 2). On the other hand, tris(dibenzylideneacetone)dipalladium(0) ($Pd_2(dba)_3$) (as we had done in most of our prior work on allylation/benzylation) also gave a nearly quantitative yield of product (entry 5). Using 1,1′-bis(diphenylphosphino)ferrocene (Dppf) as the ligand did afford the desired product, but in significantly lower yield (entry 6). Thus, the results shown in Table 1 confirmed the catalyst system we had identified earlier for the cross-coupling of phosphinic acids.

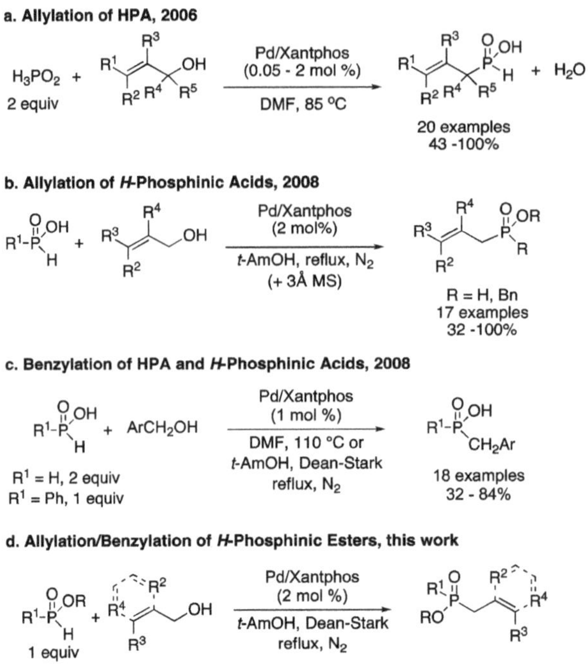

Scheme 1. Carbon-phosphorus bond formation by allylation/benzylation with alcohols. Xantphos, 4,5-bis(diphenylphosphino)-9,9-dimethylxanthene; MS, molecular sieves; *t*-AmOH, *t*-amyl alcohol.

Table 1. Test reaction between butyl phenyl-*H*-phosphinate and cinnamyl alcohol.

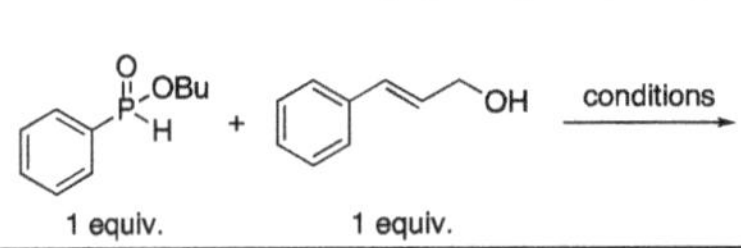

Entry	Solvent [a]	Temp. (°C) [a]	Pd Salt (mol %)	Ligand (mol %)	Yield (%) [b]
1	*t*-AmOH	102	$Pd(OAc)_2$ (2)	None (-)	0 [c]
2	*t*-AmOH	102	$Pd(OAc)_2$ (2)	Xantphos (2) [d]	99
3	toluene	111	$Pd(OAc)_2$ (2)	Xantphos (2) [d]	88
4	*t*-AmOH	102	$PdCl_2$ (2)	Xantphos (2) [d]	90
5	*t*-AmOH	102	$Pd_2(dba)_3$ (1) [e]	Xantphos (2) [d]	99
6	*t*-AmOH	102	$Pd_2(dba)_3$ (1) [e]	Dppf (2) [f]	67

[a] Reflux, Dean-Stark trap; [b] Isolated yield after column chromatography over silica gel (Note: the ^{31}P-NMR yield of the crude before purification is comparable); [c] No product in the ^{31}P-NMR spectrum; [d] Xantphos, 4,5-bis(diphenylphosphino)-9,9-dimethylxanthene; [e] $Pd_2(dba)_3$, tris(dibenzylideneacetone)dipalladium(0); dba, dibenzylideneacetone; [f] Dppf, 1,1′-bis(diphenylphosphino)ferrocene.

Next, the scope of the allylation with cinnamyl alcohol was investigated with a variety of phosphinylidene compounds (Table 2). Not surprisingly, changing the ester group from *n*-butyl to cyclohexyl gave a good result (entry 2), while a benzyl ester gave a lower yield (entry 3), presumably because of competing transesterification. The less reactive cyclohexyl-octyl-*H*-phosphinate [7] reacted uneventfully (entry 4), while butyl cinnamyl-*H*-phosphinate afforded a quantitative yield of butyl bis(cinnamyl)phosphinate (entry 5). Other functionalized *H*-phosphinate esters were tested (entries 6–9) giving generally good results. However, the Ciba-Geigy reagent (entry 10) was unsatisfactory. The acetal moiety is acid sensitive and this result may point out to the presence of acidic species along the reaction coordinates. The rather special *H*-phosphinate DOPO (6*H*-dibenzo[*c*,*e*][1,2λ^5]oxaphosphinine 6-oxide) [7] gave an excellent yield of product (entry 11). Other types of phosphinylidene were tested: diethyl *H*-phosphonate and diphenyl phosphine oxide both afforded the desired products in excellent yields (entries 12–13).

Table 2. Allylation of various phosphinylidene compounds with cinnamyl alcohol.

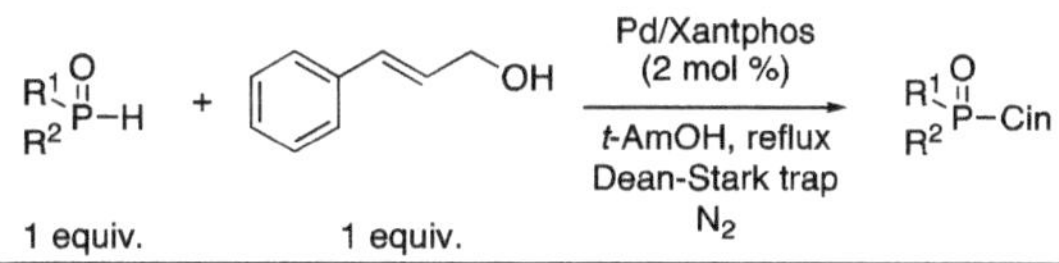

Entry	R^1	R^2	Isolated Yield (%) [a]
1	Ph	OBu	99
2	Ph	OCy	91
3	Ph	OBn	77
4	Oct	OCy	87
5	$PhCH{=}CHCH_2$ (Cin)	OBu	100
6	$4\text{-}MeOC_6H_4CH_2CH_2$	OBu	70
7	$PhtNCH_2CH_2$	OBu	100
8	$AcOCH_2$	OBu	77
9	$HOCH_2$	OMen [b,c]	91
10	$MeC(OEt)_2$	OEt	47 [d,e]
11	DOPO [f]		96
12	OEt	OEt	94
13	Ph	Ph	92

[a] Isolated yield after column chromatography over silica gel. Unless otherwise noted, the palladium salt is $Pd_2(dba)_3$ (1 mol %); [b] Men, (−)-menthyl; [c] 50:50 diastereoisomeric mixture; [d] $Pd(OAc)_2$ (2 mol %) was used; [e] ^{31}P-NMR yield; [f] DOPO, 6*H*-dibenzo[*c*,*e*][1,2λ^5]oxaphosphinine 6-oxide.

Whereas the cinnamyl moiety is a versatile functional group and its introduction appears quite general (Table 2), we next examined other allylic alcohols as well as benzylic alcohols. The results are gathered in Table 3.

Table 3. Allylation and benzylation of various phosphinylidene compounds.

Entry	Phosphorus P(O)H Compound	Alcohol	Product	Isolated Yield (%) [a]
1	PhP(O)(OBu)H	$H_2C{=}CHCH_2OH$ (2 equiv.)	PhP(O)(OBu)$CH_2CH{=}CH_2$	98
2	PhP(O)(OCy)H	$H_2C{=}CHCH_2OH$ (2 equiv.)	PhP(O)(OCy)$CH_2CH{=}CH_2$	94
3	CinP(O)(OBu)H	$H_2C{=}CHCH_2OH$ (2 equiv.)	CinP(O)(OBu)$CH_2CH{=}CH_2$	91
4	PhP(O)(OCy)H	2-Methyl-3-phenyl-2-propen-1-ol (1 equiv.)	PhP(O)(OCy)CH_2C(Me)=CHPh	93
5	CinP(O)(OBu)H	2-Methyl-3-phenyl-2-propen-1-ol (1 equiv.)	CinP(O)(OBu)CH_2C(Me)=CHPh	77
6	CinP(O)(OBu)H	Methallyl alcohol (1 equiv.)	CinP(O)(OBu)$CH_2C(Me){=}CH_2$	51
7	PhP(O)(OBu)H	Myrtenol (1 equiv.)	PhP(O)(OBu)$C_{10}H_{15}$	74
8	OctP(O)(OCy)H	$PhCH_2OH$ (1 equiv.)	OctP(O)(OCy)CH_2Ph	65
9	CinP(O)(OBu)H	$PhCH_2OH$ (1 equiv.)	CinP(O)(OBu)CH_2Ph	57
10	PhP(O)(OCy)H	1-NpCH_2OH (5 equiv.)	PhP(O)(OCy)CH_2-1-Np	73
11	CinP(O)(OBu)H	1-NpCH_2OH (1 equiv.)	CinP(O)(OBu)CH_2-1-Np	49
12	OctP(O)(OCy)H	Furfuryl alcohol (2 equiv.)	OctP(O)(OCy)$CH_2(OC_4H_3)$	24 [b]
13	OctP(O)(OBn)H	$PhCH_2OH$ (1 equiv.)	OctP(O)(OBn)CH_2Ph	18 [b]
14	$(EtO)_2$P(O)H	$PhCH_2OH$ (2 equiv.)	$(EtO)_2P(O)CH_2Ph$	23 [b]

[a] Isolated yield after column chromatography over silica gel. The palladium salt is $Pd_2(dba)_3$ (1 mol %); [b] ^{31}P-NMR yield.

Reactions with the simple allyl alcohol (2 equiv.) gave excellent results (>90% isolated yield) as shown in entries 1–3. The 2-methyl substituted version of cinnamyl alcohol, 2-methyl-3-phenyl-2-propen-1-ol (1 equiv.) also gave satisfactory results (entries 4–5). Methallyl alcohol, on the other hand, gave only a moderate yield of product (entry 6), but this result was not unexpected, as this alcohol also had given poor results with HPA [3]. Myrtenol also reacted successfully (entry 7). Next, the benzylation was investigated (entries 8–11) and products were obtained in moderate to good yields. Unfortunately, some other combinations of reactants did not afford the desired product in acceptable yields (for example: entries 12–14), thereby showing some limitations in scope. Electron-rich furfuryl alcohol (entry 12) may lead to slow oxidative-addition and an overall inefficient transformation. Acid sensitivity may become an issue: while this alcohol was successful with hypophosphorous acid [5], the more difficult the desired reaction, the more side reactions will be competing. Perhaps for a similar reason, none of the secondary allylic alcohols we tried reacted successfully. In entry 13, transesterification of the benzyl ester is a greater problem than it was in entry 8 because the reaction is slower than with cinnamyl alcohol (Table 2, entry 4). Similarly, diethyl *H*-phosphonate, which reacted satisfactorily with cinnamyl alcohol (Table 2, entry 12) also gave little product with benzyl alcohol (entry 14). Thus, marginal results are obtained when the reaction is slowed due to any of the following parameters (or combinations): less reactive allylic/benzylic electrophile (aromatics, of course, being much less reactive than alkenes), unfavorable tautomerization profile [7], transesterification of the phosphorus ester, or acid sensitivity of a reactant.

Although the scope of this reaction seems more limited than the corresponding reaction of phosphinic acids, it can offer significant synthetic advantages. Since disubstituted phosphinic acids cannot be esterified easily through Fischer-like reactions with azeotropic water-removal [8,9], formation of their esters requires prior activation of the acid (P(O)OH to P(O)LVG + ROH, P(O)OH to P(O)(OAg) + RX, where LVG is a leaving group and RX an alkyl halide) or a diazoalkane. Therefore, the intermediacy of disubstituted phosphinic acids implies atom-wasteful procedures (Scheme 2). On the other hand, *H*-phosphinic acids can be esterified easily with an alcohol and, therefore, the resulting synthetic sequence is more convenient and environmentally friendly (Scheme 2).

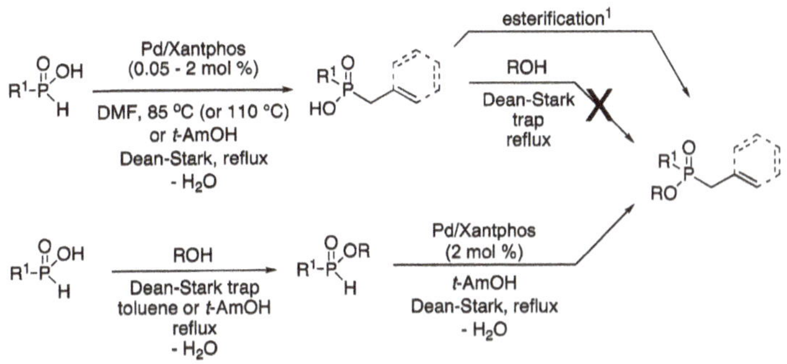

Scheme 2. *H*-phosphinic acids versus *H*-phosphinic esters.[1] $SOCl_2$ then ROH; carbodiimide + ROH; $RCHN_2$; R^1COCl then ROH; Ag_2O then alkyl halide (RX), etc.

An illustrative example is shown in Scheme 3 for the preparation of 3-phospholenic esters. The "classic" approach is illustrated by Miokowski's work (Scheme 3a) [10]. The double allylation of bis-(trimethylsiloxy)phosphine afforded the symmetrical phosphinic acid. Subsequent esterification via P(O)Cl then provided the benzyl ester, which was converted to the phospholene ester by ring-closing metathesis (RCM). It should be noted that the attempted RCM reaction of diallylphosphinic acid failed completely. The overall sequence produced the phospholene in 31% isolated yield.

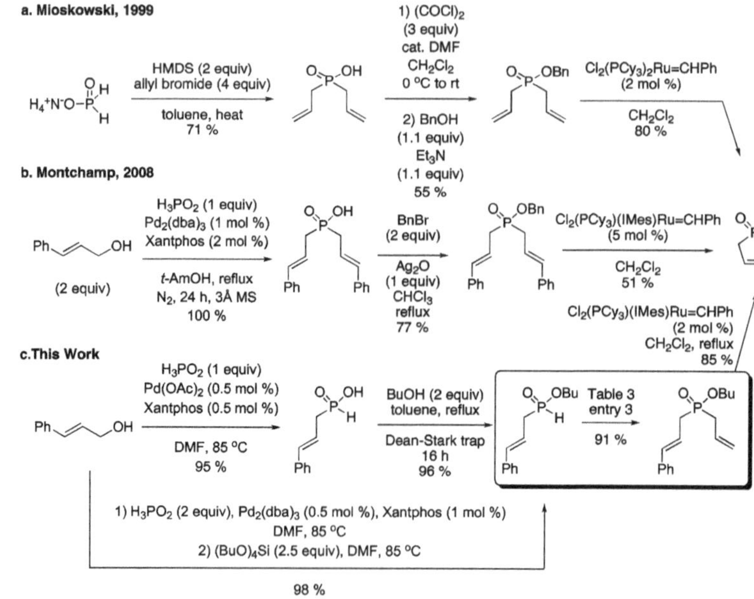

Scheme 3. Comparison of methods for the synthesis of 3-phospholenic esters. HMDS: hexamethyldisilazane; IMes: 1,3-bis(2,4,6-trimethylphenyl)imidazol-2-ylidene.

Next, our allylation methodology was used to produce bis(cinnamyl)phosphinic acid in quantitative yield [4]. Silver-promoted esterification gave the corresponding benzyl ester in 77% yield. Ring-closing metathesis afforded the desired heterocycle in 51% isolated yield. The moderate yield of this reaction may be attributed to the fact that both alkenes are substituted. In spite of this, the overall sequence gave the phospholene derivative in 39% yield.

Cinnamyl-*H*-phosphinic acid is easily synthesized, as we reported [11]. Low loading of Pd (0.5 mol %) can be employed to still deliver a very high 95% yield (Scheme 2c) [3]. Esterification with *n*-butanol under Dean-Stark conditions proceeded in 96% yield. The present allylation reaction (Table 3, entry 3) gave the disubstituted ester in 91% yield (Scheme 2c, in box). Subsequent RCM afforded the phospholene in 85% yield. Overall, the sequence in Scheme 3c produced the phospholene derivative in four steps and 70% overall yield, through catalytic reactions and with only water as the byproduct. An alternative using our silicate esterification [9] directly gives butyl cinnamyl-*H*-phosphinate from cinnamyl alcohol in 98% isolated yield [6]. Using this sequence, the phospholenic acid is still produced inexpensively in an outstanding 76% overall yield.

In addition to the advantages of Dean-Stark processes over the use of wasteful stoichiometric reagents, the present allylation of *H*-phosphinate esters can offer unique opportunities. For example (Scheme 4), the cinnamylation of optically-active menthyl hydroxymethyl-*H*-phosphinate proceeds with complete stereoselectivity in a nearly quantitative isolated yield [12].

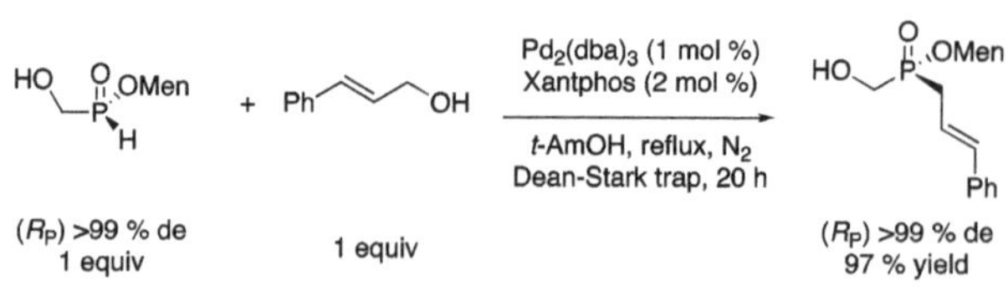

Scheme 4. Stereospecific cinnamylation. Men: (−)-menthyl.

Mechanistically, the present reaction must proceed through a pathway different from the one we have proposed for the allylation of phosphinic acids [6]. Scheme 5 shows a proposed mechanism for the Pd-catalyzed allylation/benzylation of *H*-phosphinate esters.

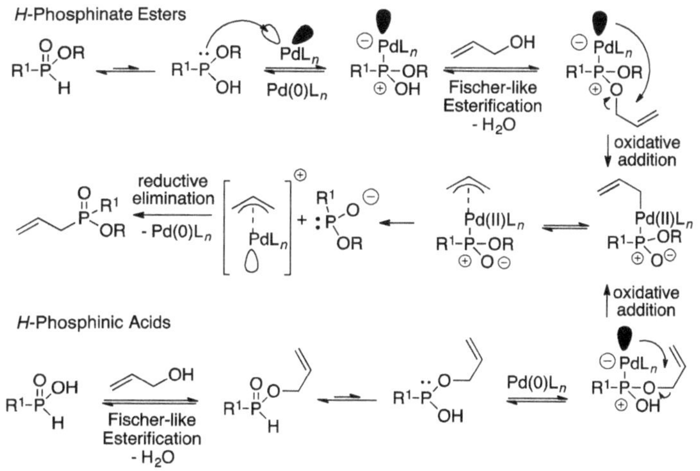

Scheme 5. Proposed Mechanisms for the allylation of *H*-phosphinate esters (**top**) and phosphinic acids (**bottom**).

The key difference between the two mechanisms (Scheme 5) is in the order of tautomerization and esterification. *H*-phosphinic esters are not esterified to the P(III) phosphonite under Dean-Stark conditions. Furthermore, while transesterification is possible in principle, this reaction is inefficient (in the absence of catalysts, particularly bases) and very difficult on esters like cyclohexyl. This would also result in the formation of disubstituted phosphinic acids. The fact that the reaction takes place with Pd(0) complexes but not with Pd(II) (Table 1) also points to a mechanism in which Lewis acidity is not key. Given those facts and some of the limitations in the scope discussed earlier, we propose the following (Scheme 5, top): the *H*-phosphinate ester must tautomerize to the P(III) form, which can then act as a ligand to the Pd(0) catalyst. The resulting putative intermediate would be sufficiently acidic to undergo Fischer-like esterification (with water being removed azeotropically). The esterification of the phosphinite-Pd complex would be a key step. The resulting Pd complex of the mixed phosphonite ester would then undergo oxidative addition (allylO to allylPd migration) to produce the classic Pd(II) intermediate in the Tsuji-Trost reaction. Subsequent formation of a π-allyl complex and attack of the phosphorus nucleophile would afford the allylated product with concomitant reductive elimination to regenerate the Pd(0) catalyst.

In the case of phosphinic acids (HPA or *H*-phosphinic, Scheme 5, bottom), first esterification to produce the allyl ester takes place in a well-established process. Tautomerization of the allyl-*H*-phosphinate ester leads to a phosphonite, which then complexes the Pd(0). Oxidative addition in this complex produces the same type of Pd(II) intermediate (R = H). The mechanism of phosphinic acid allylation (Scheme 5, bottom) is fully consistent with all our prior results [6]. Overall, the success or failure of the allylation depends on various parameters, like the tautomeric equilibrium of the phosphinylidene species, the nucleophilicity of the P(III) tautomer to complex the Pd, and the reactivity of the allylic/benzylic moiety towards oxidative addition.

3. Materials and Methods

3.1. General Procedure for the Allylation/Benzylation of H-Phosphinates and Related Compounds

To a solution of the appropriate *H*-phosphinate ester (1 equiv.) in *t*-amyl alcohol (10 mL), tris(dibenzylideneacetone)dipalladium(0) $Pd_2(dba)_3$ (1 mol %), Xantphos (2 mol %), and the corresponding alcohol (1 equiv.) were added. The reaction mixture was stirred at reflux for 24 h under N_2 in a flask equipped with a Dean-Stark trap. After cooling down the reaction to room temperature (rt), the solvent was removed under vacuum and the residue obtained was purified by column chromatography on silica gel using a mixture of hexane/ethyl acetate to afford the different products. The NMR spectra of the products can be found in the Supplementary Materials.

Butyl cinnamyl phenylphosphinate (Table 2, Entry 1) [13]. General procedure was used with cinnamyl alcohol (0.13 mL, 1 mmol, 1 equiv.) and butyl phenyl-*H*-phosphinate (198 mg, 1 mmol, 1 equiv.). The crude obtained was purified by column chromatography (hexane/ethyl acetate 100:0 to 0:100) to afford the product as a yellow oil (310 mg, 99%). ^{31}P-NMR ($CDCl_3$, 162 MHz) δ = 39.1 (s); ^{1}H-NMR ($CDCl_3$, 400 MHz) δ = 7.73–7.80 (m, 2H), 7.47–7.54 (m, 1H), 7.40–7.46 (m, 2H), 7.21–7.27 (m, 4H), 7.14–7.20 (m, 1H), 6.33 (dd, *J* = 5.0 and 15.8 Hz, 1H), 6.04–6.15 (m, 1H), 3.93 (dm, *J* = 95 Hz, 2H), 2.89 (dd, *J* = 7.6 and 18.5 Hz, 2H), 1.63 (quint., *J* = 6.8 Hz, 2H), 1.37 (dsextuplet, *J* = 1.7 and 7.5 Hz, 2H), 0.87 (t, *J* = 7.4 Hz, 3H).

Cyclohexyl cinnamyl phenylphosphinate (Table 2, Entry 2). General procedure was used with cinnamyl alcohol (0.13 mL, 1 mmol, 1 equiv.) and cyclohexyl phenyl-*H*-phosphinate (224 mg, 1 mmol, 1 equiv.). The crude obtained was purified by column chromatography (hexane/ethyl acetate 100:0 to 0:100) to afford the product as a yellow oil (311 mg, 91%). ^{31}P-NMR ($CDCl_3$, 162 MHz) δ = 37.7 (s); ^{1}H-NMR ($CDCl_3$, 400 MHz) δ = 7.76–7.84 (m, 2H), 7.49–7.55 (m, 1H), 7.41–7.47 (m, 2H), 7.23–7.30 (m, 4H), 7.16–7.22 (m, 1H), 6.35 (dd, *J* = 5.0 and 15.8 Hz, 1H), 6.06–6.17 (m, 1H), 4.30–4.41 (m, 1H), 2.89 (ddd, *J* = 1.0, 7.6 and 18.4 Hz, 2H), 1.97–2.07 (m, 1H), 1.57–1.79 (m, 4H), 1.39–1.52 (m, 2H), 1.15–1.36 (m, 3H); ^{13}C-NMR (101 MHz, $CDCl_3$): δ = 137.0 (d, J_{PCCCC} = 3.3 Hz), 134.9 (d, J_{PCC} = 13.2 Hz), 132.2 (d, J_{PCCCC} = 2.7 Hz), 131.7 (d, J_{PCCC} = 9.6 Hz, 2C), 131.5 (d, J_{PC} = 128 Hz), 128.5 (2C), 128.4 (d, J_{PCC} = 12.5 Hz, 2C), 127.4, 126.1, 126.1, 118.9 (d, J_{PCCC} = 10.3 Hz), 74.7 (d, J_{POC} = 6.9 Hz), 35.9 (d, J_{PC} = 97.3 Hz), 34.2 (d, J_{POCC} = 2.9 Hz), 33.7 (d, J_{POC} = 4.1 Hz), 25.1, 23.6, 23.6; HRMS (EI+) *m*/*z* calcd for $C_{21}H_{26}O_2P$ ($[M + H]^+$) 341.1665, found 341.1675.

Benzyl cinnamyl phenylphosphinate (Table 2, Entry 3). General procedure was used with cinnamyl alcohol (0.13 mL, 1 mmol, 1 equiv.) and benzyl phenyl-*H*-phosphinate (232 mg, 1 mmol, 1 equiv.). The crude obtained was purified by column chromatography (hexane/ethyl acetate 100:0 to 0:100) to afford the product as an orange oil (248 mg, 71%). ^{31}P-NMR ($CDCl_3$, 162 MHz) δ = 41.4 (s); ^{1}H-NMR ($CDCl_3$, 400 MHz) δ = 7.80–7.87 (m, 2H), 7.54–7.60 (m, 1H), 7.45–7.52 (m, 2H), 7.31–7.40 (m, 5H), 7.26–7.31 (m, 4H), 7.20–7.25 (m, 1H), 6.38 (dd, *J* = 5.2 and 15.9 Hz, 1H), 6.08–6.19 (m, 1H), 5.18 (dd, *J* = 7.3 and 11.8 Hz, 1H), 4.88 (dd, *J* = 7.2 and 11.8 Hz, 1H), 2.99 (dd, *J* = 7.6 and 18.7 Hz, 2H); ^{13}C-NMR (101 MHz, $CDCl_3$): δ = 136.9 (d, J_{PCCCC} = 3.5 Hz), 136.4 (d, J_{POCC} = 6.8 Hz), 135.3 (d, J_{PCC} = 13.3 Hz), 132.6 (d, J_{PCCCC} = 2.5 Hz), 131.9 (d, J_{PCCC} = 9.7 Hz, 2C), 130.1 (d, J_{PC} = 125 Hz), 128.7 (d, J_{PCC} = 12.3 Hz, 2C), 128.6 (2C), 128.5 (2C), 128.4, 127.9 (2C), 127.6, 126.3, 126.2, 118.4 (d, J_{PCCC} = 10.4 Hz), 66.3 (d, J_{POC} = 6.4 Hz), 35.3 (d, J_{PC} = 96.3 Hz); HRMS (EI+) *m*/*z* calcd for $C_{22}H_{22}O_2P$ ($[M + H]^+$) 349.1357, found 349.1379.

Cyclohexyl cinnamyl octylphosphinate (Table 2, Entry 4). General procedure was used with cinnamyl alcohol (0.13 mL, 1 mmol, 1 equiv.) and cyclohexyl octyl-*H*-phosphinate (260 mg, 1 mmol, 1 equiv.). The crude obtained was purified by column chromatography (hexane/ethyl acetate 50:50 to 0:100) to afford the product as an orange oil (328 mg, 87%). ^{31}P-NMR ($CDCl_3$, 162 MHz) δ = 51.6 (s); ^{1}H-NMR ($CDCl_3$, 400 MHz) δ = 7.24–7.35 (m, 4H), 7.16–7.22 (m, 1H), 6.47 (dd, *J* = 4.5 and 15.8 Hz, 1H), 6.10–6.22

(m, 1H), 4.34–4.46 (m, 1H), 2.71 (dd, J = 7.7 and 17.3 Hz, 2H), 1.83–1.94 (m, 2H), 1.42–1.75 (m, 8H), 1.15–1.37 (m, 14 H), 0.84 (t, J = 6.8 Hz, 3 H); ^{13}C-NMR (101 MHz, $CDCl_3$): δ = 136.8 (d, J_{PCCCC} = 3.0 Hz), 134.4 (d, J_{PCC} = 12.6 Hz), 128.5 (2C), 127.5, 126.1, 126.1, 119.7 (d, J_{PCCC} = 9.4 Hz), 73.7 (d, J_{POC} = 7.0 Hz), 34.6 (d, J_{PC} = 82.2 Hz), 34.2 (d, J_{POCC} = 5.6 Hz), 34.1, 31.8, 30.8 (d, J_{PCC} = 15.1 Hz), 29.0, 29.0, 28.3 (d, J_{PC} = 93.1 Hz), 25.2, 23.7 (2C), 22.6, 21.6 (d, J_{PCCC} = 4.2 Hz), 14.1; HRMS (EI+) m/z calcd for $C_{23}H_{38}O_2P$ ([M + H]$^+$) 377.2609, found 377.2531.

Butyl bis cinnamylphosphinate (Table 2, Entry 5). General procedure was used with cinnamyl alcohol (0.13 mL, 1 mmol, 1 equiv.) and butyl cinnamyl-*H*-phosphinate (238 mg, 1 mmol, 1 equiv.). The crude obtained was purified by column chromatography (hexane/ethyl acetate 50:50 to 0:100) to afford the product as an orange oil (354 mg, 100%). ^{31}P-NMR ($CDCl_3$, 162 MHz) δ = 49.9 (s); ^{1}H-NMR ($CDCl_3$, 400 MHz) δ = 7.21–7.33 (m, 8H), 7.14–7.20 (m, 2H), 6.45 (dd, J = 4.6 and 15.8 Hz, 2H), 6.11–6.22 (m, 2H), 4.02 (dt, J = 6.8 and 7.0 Hz, 2H), 2.73 (dd, J = 7.6 and 17.5 Hz, 4H), 1.61 (quint., J = 7.2 Hz, 2H), 1.36 (sextuplet, J = 7.5 Hz, 2H), 0.88 (t, J = 7.4 Hz, 3H); ^{13}C-NMR (101 MHz, $CDCl_3$): δ = 136.7 (d, J_{PCCCC} = 3.0 Hz, 2C), 135.0 (d, J_{PCC} = 13.1 Hz, 2C), 128.6 (4C), 127.7 (2C), 126.2 (2C), 126.2 (2C), 118.9 (d, J_{PCCC} = 9.6 Hz, 2C), 64.6 (d, J_{POC} = 7.2 Hz), 33.3 (d, J_{PC} = 88.7 Hz, 2C), 32.8 (d, J_{POCC} = 5.6 Hz), 18.8, 13.7; HRMS (EI+) m/z calcd for $C_{22}H_{28}O_2P$ ([M + H]$^+$) 355.1827, found 355.1831.

Butyl 2-[(4-methoxyphenyl)ethyl] cinnamylphosphinate (Table 2, Entry 6). General procedure was used with cinnamyl alcohol (0.13 mL, 1 mmol, 1 equiv.) and butyl 2-[4-methoxyphenyl)ethyl]-*H*-phosphinate (256 mg, 1 mmol, 1 equiv.). The crude obtained was purified by column chromatography (hexane/ethyl acetate 100:0 to 0:100) to afford the product as an orange oil (261 mg, 70%). ^{31}P-NMR ($CDCl_3$, 162 MHz) δ = 52.3 (s); ^{1}H-NMR ($CDCl_3$, 400 MHz) δ = 7.29–7.38 (m, 4H), 7.22–7.28 (m, 1H), 7.10–7.15 (m, 2H), 6.81–6.87 (m, 2H), 6.45 (dd, J = 4.6 and 15.8 Hz, 1H), 6.11–6.22 (m, 1H), 4.06 (dt, J = 6.6 and 6.7 Hz, 2H), 3.78 (s, 3H), 2.83–2.99 (m, 2H), 2.73 (ddd, J = 0.7, 7.2 and 16.8 Hz, 2H), 2.01–2.12 (m, 2H), 1.68 (quint., J = 7.5 Hz, 2H), 1.43 (sextuplet, J = 7.5 Hz, 2H), 0.95 (t, J = 7.4 Hz, 3H); ^{13}C-NMR (101 MHz, $CDCl_3$): δ = 158.2, 136.7 (d, J_{PCCCC} = 3.2 Hz), 134.8 (d, J_{PCCC} = 12.9 Hz), 133.0 (d, J_{PCC} = 14.6 Hz), 129.1 (2C), 128.6 (2C), 127.7, 126.2, 126.2, 119.1 (d, J_{PCCC} = 9.4 Hz, 2C), 114.0 (2C), 64.4 (d, J_{POC} = 7.0 Hz), 55.3, 34.0 (d, J_{PC} = 86.0 Hz), 32.8 (d, J_{POCC} = 5.7 Hz), 29.7 (d, J_{PC} = 90.4 Hz), 27.0 (d, J_{PCC} = 3.5 Hz), 18.9, 13.7; HRMS (EI+) m/z calcd for $C_{22}H_{29}O_3P$ ([M + H]$^+$) 373.1933, found 373.1855.

Butyl (2-ethylphthalimide) cinnamylphosphinate (Table 2, Entry 7). General procedure was used with cinnamyl alcohol (1.7 mL, 12.94 mmol, 1 equiv.) and butyl (2-ethylphthalimide)-*H*-phosphinate (3.3 g, 12.94 mmol, 1 equiv.). The crude obtained was purified by column chromatography (dichloromethane/acetone 100:0 to 90:10) to afford the product as an orange oil (4.87 g, 100%). ^{31}P-NMR ($CDCl_3$, 162 MHz) δ = 49.1 (s); ^{1}H-NMR ($CDCl_3$, 400 MHz) δ = 7.82–7.90 (m, 2H), 7.68–7.75 (m, 2H), 7.20–7.44 (m, 5H), 6.62 (dd, J = 4.7 and 15.8 Hz, 1H), 6.16–6.27 (m, 1H), 3.95–4.12 (m, 4H), 2.88 (dd, J = 7.5 and 17.5 Hz, 2H), 2.17–2.34 (m, 2H), 1.58 (quint., J = 7.1 Hz, 2H), 1.35 (sextuplet, J = 7.4 Hz, 2H), 0.89 (t, J = 7.4 Hz, 3H); ^{13}C-NMR (101 MHz, $CDCl_3$): δ = 167.8 (2C), 136.6 (d, J_{PCCCC} = 3.2 Hz), 135.3 (d, J_{PCC} = 13.1 Hz), 134.1 (2C), 132.0 (2C), 128.6 (2C), 127.7, 126.3, 126.3, 123.4 (2C), 118.5 (d, J_{PCCC} = 9.6 Hz), 64.6 (d, J_{POC} = 7.0 Hz), 33.7 (d, J_{PC} = 88.4 Hz), 32.6 (d, J_{PCC} = 5.9 Hz), 31.7, 26.5 (d, J_{PC} = 89.7 Hz), 18.8, 13.6; HRMS (EI+) m/z calcd for $C_{23}H_{27}NO_4P$ ([M + H]$^+$) 412.1678, found 412.1676.

Butyl (acetoxymethyl) cinnamylphosphinate (Table 2, Entry 8). General procedure was used with cinnamyl alcohol (0.13 mL, 1 mmol, 1 equiv.) and butyl (acetoxymethyl)-*H*-phosphinate (194 mg, 1 mmol, 1 equiv.). The crude obtained was purified by column chromatography (hexane/ethyl acetate 100:0 to 0:100) to afford the product as an orange oil (239 mg, 77%). ^{31}P-NMR ($CDCl_3$, 162 MHz) δ = 43.6 (s); ^{1}H-NMR ($CDCl_3$, 400 MHz) δ = 7.27–7.32 (m, 2H), 7.21–7.27 (m, 2H), 7.14–7.20 (m, 1H), 6.50 (dd, J = 5.0 and 15.9 Hz, 1H), 6.06–6.17 (m, 1H), 4.43 (dd, J = 7.8 and 14.5 Hz, 1H), 4.27 (dd, J = 4.2 and 14.5 Hz, 1H), 4.03 (dm, J = 39.6 Hz, 2H), 2.79 (ddd, J = 1.0, 7.7 and 18.5 Hz, 2H), 2.03 (s, 3H), 1.60 (quint., J = 7.5 Hz,

2H), 1.34 (sextuplet, J = 7.5 Hz, 2H), 0.86 (t, J = 7.4 Hz, 3H); ^{13}C-NMR (101 MHz, $CDCl_3$): δ = 170.0 (d, J_{PCOC} = 6.8 Hz), 136.5 (d, J_{PCCCC} = 3.5 Hz), 135.5 (d, J_{PCC} = 13.3 Hz), 128.6 (2C), 127.8, 126.3, 126.2, 117.5 (d, J_{PCCC} = 10.2 Hz), 65.2 (d, J_{POC} = 6.9 Hz), 57.9 (d, J_{PC} = 108 Hz), 32.6 (d, J_{POCC} = 5.8 Hz), 32.5 (d, J_{PC} = 93.1 Hz), 20.5 (d, J_{POCCC} = 8.8 Hz), 18.7, 13.6; HRMS (EI+) m/z calcd for $C_{16}H_{24}O_4P$ ($[M + H]^+$) 311.1407, found 311.1401.

(R_p)/(S_p)-Menthyl cinnamyl(hydroxymethyl)phosphinate (Table 2, Entry 9) [14]. To a solution of (R_p)/(S_p)-menthyl (hydroxymethyl)-*H*-phosphinate (2.34 g, 10 mmol, 1 equiv, 56:44 diastereoisomeric ratio) in *t*-amyl alcohol (30 mL), $Pd_2(dba)_3$ (92 mg, 0.1 mmol, 1 mol %), Xantphos (116 mg, 0.2 mmol, 2 mol %), and cinnamyl alcohol (1.3 mL, 10 mmol, 1 equiv.) were added. The reaction mixture was stirred at reflux for 20 h under N_2 in a flask equipped with a Dean-Stark trap. After cooling down the reaction to rt, the solvent was removed under vacuum and the residue obtained was purified by column chromatography (dichloromethane/acetone 100:0 to 90:10) to afford the product as a white solid (3.19 g, 91%, 56:44 diastereoisomeric ratio). ^{31}P-NMR (162 MHz, $CDCl_3$): δ = 48.4 (s, 56%) and 48.1 (s, 44%).

6-(Cinnamyl)-6H-dibenzo[c,e][1,2λ^5]oxaphosphinine-6-oxide (Table 2, Entry 11). General procedure was used with cinnamyl alcohol (0.13 mL, 1 mmol, 1 equiv.) and DOPO (216 mg, 1 mmol, 1 equiv.). The crude obtained was purified by column chromatography (hexane/ethyl acetate 50:50 to 0:100) to afford the product as a yellow solid (318 mg, 96%). ^{31}P-NMR ($CDCl_3$, 162 MHz) δ = 27.4 (s); ^{1}H-NMR ($CDCl_3$, 400 MHz) δ = 7.89–7.99 (m, 2H), 7.68–7.74 (m, 1H), 7.49–7.55 (m, 1H), 7.36–7.42 (m, 1H), 7.22–7.30 (m, 6H), 7.16–7.21 (m, 2H), 6.35 (dd, J = 5.3 and 15.8 Hz, 1H), 6.02–6.13 (m, 1H), 3.08 (ddt, J = 1.3, 7.6 and 18.2 Hz, 2H); ^{13}C-NMR (101 MHz, $CDCl_3$): δ = 149.6 (d, J_{POC} = 8.2 Hz), 136.5 (d, J_{PCCCC} = 3.6 Hz), 135.9 (d, J_{PCCC} = 6.1 Hz), 135.8 (d, J_{PCC} = 13.8 Hz), 133.4 (d, J_{PCCCC} = 2.3 Hz), 130.7, 130.5 (d, J_{PCCC} = 10.2 Hz), 128.5 (2C), 128.4 (d, J_{PCC} = 13.0 Hz), 127.7 (d, J_{PCCCC} = 2.2 Hz), 126.2, 126.2, 125.2, 124.6, 124.2 (d, J_{PC} = 120 Hz), 123.8 (d, J_{PCCC} = 9.6 Hz), 122.3 (d, J_{PCCC} = 10.5 Hz), 120.4 (d, J_{PCCC} = 6.3 Hz), 117.2 (d, J_{PCC} = 11.0 Hz), 34.5 (d, J_{PC} = 93.8 Hz); HRMS (EI+) m/z calcd for $C_{21}H_{18}O_2P$ ($[M + H]^+$) 333.1039, found 333.1048.

Diethyl cinnamylphosphonate (Table 2, Entry 12) [15]. General procedure was used with cinnamyl alcohol (0.13 mL, 1 mmol, 1 equiv.) and diethylphosphite (138 mg, 1 mmol, 1 equiv.). The crude obtained was purified by column chromatography (hexane/ethyl acetate 50:50 to 0:100) to afford the product as an orange oil (240 mg, 94%). ^{31}P-NMR ($CDCl_3$, 162 MHz) δ = 34.8 (s); ^{1}H-NMR ($CDCl_3$, 400 MHz) δ = 7.34–7.40 (m, 2H), 7.28–7.34 (m, 2H), 7.21–7.26 (m, 1H), 6.53 (dd, J = 5.1 and 15.8 Hz, 1H), 6.12–6.23 (m, 1H), 4.08–4.18 (m, 4H), 2.76 (dd, J = 7.6 and 22.2 Hz, 2H), 1.32 (t, J = 6.1 Hz, 6H).

Cinnamyl diphenylphosphine oxide (Table 2, Entry 13) [16]. General procedure was used with cinnamyl alcohol (0.13 mL, 1 mmol, 1 equiv.) and diphenyl phosphine oxide (202 mg, 1 mmol, 1 equiv.). The crude obtained was purified by column chromatography (hexane/ethyl acetate 100:0 to 0:100) to afford the product as a yellow oil (292 mg, 92%). ^{31}P-NMR ($CDCl_3$, 162 MHz) δ = 28.9 (s); ^{1}H-NMR ($CDCl_3$, 400 MHz) δ = 7.75–7.82 (m, 4H), 7.46–7.58 (m, 6H), 7.18–7.29 (m, 5H), 6.44 (dd, J = 4.5 and 15.8 Hz, 1H), 6.14–6.25 (m, 1H), 3.31 (ddd, J = 1.2, 7.5 and 15.0 Hz, 2H).

Butyl allyl phenylphosphinate (Table 3, Entry 1) [17]. General procedure was used with allyl alcohol (0.13 mL, 2 mmol, 2 equiv.) and butyl phenyl-*H*-phosphinate (198 mg, 1 mmol, 1 equiv.). The crude obtained was purified by column chromatography (hexane/ethyl acetate 100:0 to 0:100) to afford the product as a yellow oil (234 mg, 98%). ^{31}P-NMR ($CDCl_3$, 162 MHz) δ = 35.1 (s); ^{1}H-NMR ($CDCl_3$, 400 MHz) δ = 7.66–7.75 (m, 2H), 7.45–7.51 (m, 1H), 7.36–7.43 (m, 2H), 5.61–5.75 (m, 1H), 4.95–5.09 (m, 1H), 3.93–4.03 (m, 1H), 3.69–3.80 (m, 1H), 3.31 (dd, J = 7.0 and 18.5 Hz, 2H), 1.58 (quint., J = 7.1 Hz, 2H), 1.32 (sextuplet, J = 1.9 and 7.4 Hz, 2H), 0.83 (t, J = 7.4 Hz, 3H).

Cyclohexyl allyl phenylphosphinate (Table 3, Entry 2). General procedure was used with allyl alcohol (0.13 mL, 2 mmol, 2 equiv.) and cyclohexyl phenyl-*H*-phosphinate (224 mg, 1 mmol, 1 equiv.). The crude obtained was purified by column chromatography (hexane/ethyl acetate 100:0 to 0:100) to afford the product as an orange oil (247 mg, 94%). ^{31}P-NMR ($CDCl_3$, 162 MHz) δ = 37.7 (s); ^{1}H-NMR ($CDCl_3$, 400 MHz) δ = 7.70–7.79 (m, 2H), 7.46–7.52 (m, 1H), 7.38–7.45 (m, 2H), 5.64–5.78 (m, 1H), 4.97–5.11 (m, 2H), 4.22–4.33 (m, 1H), 2.65–2.76 (m, 2H), 1.92–2.02 (m, 1H), 1.51–1.75 (m, 4H), 1.34–1.47 (m, 2H), 1.12–1.32 (m, 3H); ^{13}C-NMR (101 MHz, $CDCl_3$): δ = 132.1 (d, J_{PCCCC} = 2.7 Hz), 131.7 (d, J_{PCCC} = 9.6 Hz, 2C), 131.4 (d, J_{PC} = 126 Hz), 128.3 (d, J_{PCC} = 12.5 Hz, 2C), 127.4 (d, J_{PCCC} = 9.2 Hz), 120.1 (d, J_{PCC} = 13.0 Hz), 74.6 (d, J_{POC} = 6.9 Hz), 36.6 (d, J_{PC} = 97.5 Hz), 34.2 (d, J_{POCC} = 3.1 Hz), 33.7 (d, J_{POCC} = 4.3 Hz), 25.1, 23.6, 23.6; HRMS (EI+) m/z calcd for $C_{15}H_{22}O_2P$ ($[M + H]^+$) 265.1352, found 265.1359.

Butyl allyl cinnamylphosphinate (Table 3, Entry 3) [6]. General procedure was used with allyl alcohol (0.15 mL, 2 mmol, 2 equiv.) and butyl cinnamyl-*H*-phosphinate (238 mg, 1 mmol, 1 equiv.). The crude obtained was purified by column chromatography (hexane/ethyl acetate 100:0 to 0:100) to afford the product as an orange oil (254 mg, 91%). ^{31}P-NMR ($CDCl_3$, 162 MHz) δ = 48.0 (s); ^{1}H-NMR ($CDCl_3$, 400 MHz) δ = 7.23–7.34 (m, 4H), 7.15–7.21 (m, 1H), 6.46 (dd, J = 4.7 and 15.8 Hz, 1H), 6.08–6.19 (m, 1H), 5.72–5.86 (m, 1H), 5.14–5.23 (m, 2H), 4.00 (dt, J = 6.8 and 6.9 Hz, 2H), 2.72 (ddd, J = 0.7, 7.6 and 17.6 Hz, 2H), 2.60 (dd, J = 7.5 and 17.2 Hz, 2H), 1.60 (quint., J = 7.1 Hz, 2H), 1.35 (sextuplet, J = 7.5 Hz, 2H), 0.88 (t, J = 7.4 Hz, 3H); ^{13}C-NMR (101 MHz, $CDCl_3$): δ = 136.7 (d, J_{PCCCC} = 3.3 Hz), 135.0 (d, J_{PCC} = 12.9 Hz), 128.6 (2C), 127.7 (d, J_{PCCC} = 8.0 Hz), 127.6, 126.2, 126.1, 120.3 (d, J_{PCC} = 12.6 Hz), 118.8 (d, J_{PCCC} = 9.6 Hz), 64.5 (d, J_{POC} = 7.2 Hz), 33.8 (d, J_{PC} = 88.7 Hz), 32.9 (d, J_{PC} = 89.0 Hz), 32.7 (d, J_{POCC} = 5.7 Hz), 18.8, 13.6.

Cyclohexyl 3-[(2-methyl-2-propene)phenyl] phenylphosphinate (Table 3, Entry 4). General procedure was used with *trans*-2-methyl-3-phenyl-2-propen-1-ol (0.15 mL, 1 mmol, 1 equiv.) and cyclohexyl phenyl-*H*-phosphinate (224 mg, 1 mmol, 1 equiv.). The crude obtained was purified by column chromatography (hexane/ethyl acetate 100:0 to 0:100) to afford the product as an orange oil (330 mg, 93%). ^{31}P-NMR ($CDCl_3$, 162 MHz) δ = 39.5 (s); ^{1}H-NMR ($CDCl_3$, 400 MHz) δ = 7.77–7.85 (m, 2H), 7.48–7.54 (m, 1H), 7.40–7.47 (m, 2H), 7.22–7.29 (m, 2H), 7.11–7.17 (m, 1H), 7.03–7.09 (m, 2H), 6.09 (d, J = 5.5 Hz, 1H), 4.29–4.40 (m, 1H), 2.76–2.91 (m, 2H), 1.98–2.07 (m, 1H), 1.92 (dd, J = 1.3 and 3.5 Hz, 3H), 1.59–1.79 (m, 4H), 1.39–1.49 (m, 2H), 1.17–1.36 (m, 3H); ^{13}C-NMR (101 MHz, $CDCl_3$): δ = 137.8 (d, J_{PCCCC} = 3.7 Hz), 132.1 (d, J_{PCCCC} = 2.7 Hz), 131.9 (d, J_{PCCC} = 9.5 Hz, 2C), 131.6 (d, J_{PC} = 125 Hz), 130.0 (d, J_{PCC} = 11.5 Hz), 129.3 (d, J_{PCCC} = 10.3 Hz), 128.7 (d, J_{PCCCCC} = 2.9 Hz, 2C), 128.3 (d, J_{PCC} = 12.4 Hz, 2C), 128.0 (2C), 126.3, 74.7 (d, J_{POC} = 7.0 Hz), 42.8 (d, J_{PC} = 95.2 Hz), 34.3 (d, J_{POCC} = 2.9 Hz), 33.7 (d, J_{POCC} = 4.3 Hz), 25.2, 23.6 (2C), 19.5 (d, J_{CCC} = 2.1 Hz); HRMS (EI+) m/z calcd for $C_{22}H_{28}O_2P$ ($[M + H]^+$) 355.1827, found 355.1849.

Butyl 3-[(2-methyl-2-propene)phenyl] cinnamylphosphinate (Table 3, Entry 5). General procedure was used with *trans*-2-methyl-3-phenyl-2-propen-1-ol (0.15 mL, 1 mmol, 1 equiv.) and butyl cinnamyl-*H*-phosphinate (238 mg, 1 mmol, 1 equiv.). The crude obtained was purified by column chromatography (hexane/ethyl acetate 50:50 to 0:100) to afford the product as an orange oil (283 mg, 77%). ^{31}P-NMR ($CDCl_3$, 162 MHz) δ = 48.2 (s); ^{1}H-NMR ($CDCl_3$, 400 MHz) δ = 7.19–7.38 (m, 10H), 6.54 (dd, J = 4.4 and 15.8 Hz, 1H), 6.43 (d, J = 4.8 Hz, 1H), 6.19–6.30 (m, 1H), 4.10 (dt, J = 6.7 and 6.8 Hz, 2H), 2.84 (dd, J = 7.6 and 17.1 Hz, 2H), 2.77 (d, J = 17.3 Hz, 2H), 2.07 (d, J = 3.1 Hz, 3H), 1.68 (quint., J = 7.4 Hz, 2H), 1.42 (sextuplet, J = 7.5 Hz, 2H), 0.93 (t, J = 7.4 Hz, 3H); ^{13}C-NMR (101 MHz, $CDCl_3$): δ = 137.5 (d, J_{PCCCC} = 3.5 Hz), 136.8 (d, J_{PCCCC} = 3.1 Hz), 135.0 (d, J_{PCC} = 12.7 Hz), 129.9 (d, J_{PCCC} = 11.7 Hz), 129.8 (d, J_{PCC} = 10.0 Hz), 128.9, 128.8, 128.6 (2C), 128.2 (2C), 127.7, 126.6, 126.2, 126.2, 119.2 (d, J_{PCCC} = 9.9 Hz), 64.6 (d, J_{POC} = 7.2 Hz), 40.6 (d, J_{PC} = 86.7 Hz), 33.5 (d, J_{PC} = 88.0 Hz), 32.9 (d, J_{POCC} = 5.7 Hz), 19.6 (d, J_{PCCC} = 1.9 Hz), 18.9, 13.7; HRMS (EI+) m/z calcd for $C_{23}H_{30}O_2P$ ($[M + H]^+$) 369.1983, found 369.1906.

Butyl (2-methylprop-2-ene) cinnamylphosphinate (Table 3, Entry 6). General procedure was used with methylallyl alcohol (0.154 mL, 1 mmol, 1 equiv.) and butyl cinnamyl-*H*-phosphinate (238 mg, 1 mmol, 1 equiv.). The crude obtained was purified by column chromatography (hexane/ethyl acetate 100:0 to 0:100) to afford the product as an orange oil (148 mg, 51%). ^{31}P-NMR ($CDCl_3$, 162 MHz) δ = 47.4 (s); ^{1}H-NMR ($CDCl_3$, 400 MHz) δ = 7.30–7.39 (m, 4H), 7.22–7.28 (m, 1H), 6.52 (dd, *J* = 4.6 and 15.8 Hz, 1H), 6.14–6.25 (m, 1H), 4.97–5.01 (m, 1H), 4.95 (d, *J* = 4.5 Hz, 1H), 4.06 (dt, *J* = 6.8 Hz, 2H), 2.81 (dd, *J* = 7.6 and 17.3 Hz, 2H), 2.62 (d, *J* = 17.4 Hz, 2H), 1.92 (t, *J* = 1.2 Hz, 3H), 1.66 (quint., *J* = 7.5 Hz, 2H), 1.41 (sextuplet, *J* = 7.6 Hz, 2H), 0.93 (t, *J* = 7.4 Hz, 3H); ^{13}C-NMR (101 MHz, $CDCl_3$): δ = 136.8 (d, J_{PCCCC} = 3.3 Hz), 136.7 (d, J_{PCCC} = 9.0 Hz), 134.9 (d, J_{PCC} = 13.1 Hz), 128.6 (2C), 127.6, 126.2, 126.2, 119.2 (d, J_{PCCC} = 9.7 Hz), 115.7 (d, J_{PCC} = 10.9 Hz), 64.6 (d, J_{POC} = 7.2 Hz), 37.6 (d, J_{PC} = 87.4 Hz), 33.1 (d, J_{PC} = 88.3 Hz), 32.8 (d, J_{POCC} = 5.8 Hz), 24.1 (d, J_{PCCC} = 2.2 Hz), 18.8, 13.7; HRMS (EI+) *m/z* calcd for $C_{17}H_{26}O_2P$ ($[M + H]^+$) 293.1670, found 293.1693.

Butyl myrtenyl phenylphosphinate (Table 3, Entry 7). General procedure was used with myrtenol (0.17 mL, 1 mmol, 1 equiv.) and butyl phenyl-*H*-phosphinate (198 mg, 1 mmol, 1 equiv.). The crude obtained was purified by column chromatography (hexane/ethyl acetate 100:0 to 0:100) to afford the product as an orange oil (246 mg, 74%). ^{31}P-NMR ($CDCl_3$, 162 MHz) δ = 40.3 (s); ^{1}H-NMR ($CDCl_3$, 400 MHz) δ = 7.64–7.72 (m, 2H), 7.41–7.47 (m, 1H), 7.33–7.40 (m, 2H), 5.15–5.21 (m, 1H), 3.88–3.97 (m, 1H), 3.61–3.71 (m, 1H), 2.52–2.76 (m, 2H), 2.17–2.26 (m, 1H), 2.00–2.12 (m, 3H), 1.87–1.96 (m, 1H), 1.53 (quint., *J* = 7.2 Hz, 2H), 1.30 (sextuplet, *J* = 7.5 Hz, 2H), 1.14 (d, *J* = 10.4 Hz, 3H), 0.99 (dd, *J* = 8.7 and 20.1 Hz, 1H), 0.81 (t, *J* = 7.6 Hz, 3H), 0.72 (d, *J* = 4.0 Hz, 3H); ^{13}C-NMR (101 MHz, $CDCl_3$): δ = 137.8 (d, J_{PCC} = 10.4 Hz), 132.0 (d, J_{PCCCC} = 2.8 Hz, 0.5C), 132.0 (d, J_{PCCCC} = 2.8 Hz, 0.5C), 131.8 (d, J_{PCCC} = 9.7 Hz), 131.8 (d, J_{PCCC} = 9.8 Hz), 131.0 (d, J_{PC} = 123 Hz), 131.0 (d, J_{PC} = 123 Hz), 128.2 (d, J_{PCC} = 12.3 Hz, 2C), 122.2 (d, J_{PCCC} = 12.6 Hz, 0.5C), 122.2 (d, J_{PCCC} = 12.6 Hz, 0.5C), 64.1 (d, J_{POC} = 7.0 Hz, 0.5C), 64.0 (d, J_{POC} = 7.0 Hz, 0.5C), 46.9 (d, J_{POCC} = 2.8 Hz, 0.5C), 46.7 (d, J_{POCC} = 2.4 Hz, 0.5C), 40.1 (0.5C), 40.1 (0.5C), 38.7 (d, J_{PC} = 97.6 Hz), 37.9 (d, J_{PCCCC} = 2.2 Hz, 0.5C), 37.9 (d, J_{POCC} = 2.2 Hz, 0.5C), 32.5 (0.5C), 32.5 (0.5C), 31.6 (d, J_{PCCCC} = 2.4 Hz, 0.5C), 31.5 (d, J_{PCCCC} = 2.0 Hz, 0.5C), 26.1 (0.5C), 26.1 (0.5C), 21.0 (0.5C), 20.9 (0.5C), 18.7, 13.6; HRMS (EI+) *m/z* calcd for $C_{20}H_{30}O_2P$ ($[M + H]^+$) 333.1983, found 333.1906.

Cyclohexyl benzyl octylphosphinate (Table 3, Entry 8) [1]. General procedure was used with benzyl alcohol (0.11 mL, 1 mmol, 1 equiv.) and cyclohexyl octyl-*H*-phosphinate (260 mg, 1 mmol, 1 equiv.). The crude obtained was purified by column chromatography (hexane/ethyl acetate 50:50 to 0:100) to afford the product as an orange oil (226 mg, 65%). ^{31}P-NMR ($CDCl_3$, 162 MHz) δ = 52.1 (s); ^{1}H-NMR ($CDCl_3$, 400 MHz) δ = 7.22–7.37 (m, 5H), 4.30–4.42 (m, 1H), 3.13 (d, *J* = 16.7 Hz, 2H), 1.86–1.95 (m, 2H), 1.40–1.84 (m, 10H), 1.17–1.37 (m, 12H), 0.89 (t, *J* = 6.9 Hz, 3H).

Butyl benzyl cinnamylphosphinate (Table 3, Entry 9). General procedure was used with benzyl alcohol (0.11 mL, 1 mmol, 1 equiv.) and butyl cinnamyl-*H*-phosphinate (238 mg, 1 mmol, 1 equiv.). The crude obtained was purified by column chromatography (hexane/ethyl acetate 100:0 to 0:100) to afford the product as an orange oil (187 mg, 57%). ^{31}P-NMR ($CDCl_3$, 162 MHz) δ = 48.5 (s); ^{1}H-NMR ($CDCl_3$, 400 MHz) δ = 7.21–7.39 (m, 10H), 6.45 (dd, *J* = 4.2 and 15.8 Hz, 1H), 6.08–6.20 (m, 1H), 3.93–4.07 (m, 2H), 3.20 (d, *J* = 16.8 Hz, 2H), 2.70 (dd, *J* = 7.6 and 17.2 Hz, 2H), 2.01–2.12 (m, 2H), 1.62 (quint., *J* = 7.3 Hz, 2H), 1.37 (sextuplet, *J* = 7.5 Hz, 2H), 0.91 (t, *J* = 7.4 Hz, 3H); ^{13}C-NMR (101 MHz, $CDCl_3$): δ = 136.8 (d, J_{PCCCC} = 3.1 Hz), 135.1 (d, J_{PCCC} = 13.0 Hz), 131.6 (d, J_{PCC} = 7.3 Hz), 129.9 (d, J_{PCCC} = 5.8 Hz, 2C), 128.7 (d, J_{PCCCC} = 2.6 Hz, 2C), 128.6 (2C), 127.7, 127.0 (d, J_{PCCCCC} = 3.0 Hz), 126.2, 126.2, 118.9 (d, J_{PCCC} = 9.8 Hz), 64.8 (d, J_{POC} = 7.0 Hz), 36.0 (d, J_{PC} = 87.1 Hz), 33.2 (d, J_{PC} = 88.0 Hz), 32.8 (d, J_{POCC} = 5.9 Hz), 27.0 (d, J_{PCC} = 3.5 Hz), 18.8, 13.7; HRMS (EI+) *m/z* calcd for $C_{20}H_{26}O_2P$ ($[M + H]^+$) 329.1665, found 329.1671.

Cyclohexyl 1-(naphthylmethyl) phenylphosphinate (Table 3, Entry 10). General procedure was used with 1-naphthalenemethanol (158 mg, 1 mmol, 1 equiv.) and cyclohexyl phenyl-*H*-phosphinate (224 mg,

1 mmol, 1 equiv.). The crude obtained was purified by column chromatography (hexane/ethyl acetate 100:0 to 0:100) to afford the product as an orange oil (266 mg, 73%). ^{31}P-NMR ($CDCl_3$, 162 MHz) δ = 35.5 (s); ^{1}H-NMR ($CDCl_3$, 400 MHz) δ = 7.95–8.02 (m, 1H), 7.77–7.84 (m, 1H), 7.69–7.75 (m, 1H), 7.56–7.65 (m, 2H), 7.39–7.48 (m, 3H), 7.29–7.37 (m, 3H), 7.22–7.28 (m, 1H), 4.22–4.33 (m, 1H), 3.76 (d, *J* = 17.8 Hz, 2H), 1.77–1.86 (m, 1H), 1.45–1.67 (m, 4H), 1.09–1.44 (m, 5H); ^{13}C-NMR (101 MHz, $CDCl_3$): δ = 133.8 (d, J_{PCCCC} = 2.1 Hz), 132.2 (d, J_{PCCCC} = 4.3 Hz), 132.1 (d, J_{PCCCC} = 2.2 Hz), 132.0 (d, J_{PCCC} = 9.5 Hz, 2C), 131.4 (d, J_{PC} = 124 Hz), 128.6 (d, J_{PCCC} = 6.7 Hz), 128.4, 128.4 (d, J_{PCC} = 8.0 Hz), 128.2 (d, J_{PCC} = 12.4 Hz, 2C), 127.5 (d, J_{PCCCC} = 3.8 Hz), 125.7, 125.5, 125.2 (d, J_{PCCCC} = 3.6 Hz), 124.6, 74.9 (d, J_{POC} = 6.8 Hz), 35.9 (d, J_{PC} = 96.1 Hz), 34.0 (d, J_{POCC} = 2.6 Hz), 33.5 (d, J_{POCC} = 4.0 Hz), 25.1, 23.4 (2C); HRMS (EI+) *m/z* calcd for $C_{22}H_{28}O_2P$ ($[M + H]^+$) 355.1827, found 355.1849.

Butyl (1-methylnaphthalene) cinnamylphosphinate (Table 3, Entry 11). General procedure was used with 1-naphthalenemethanol (316 mg, 2 mmol, 2 equiv.) and butyl cinnamyl-*H*-phosphinate (238 mg, 1 mmol, 1 equiv.). The crude obtained was purified by column chromatography (hexane/ethyl acetate 100:0 to 0:100) to afford the product as an orange oil (186 mg, 49%). ^{31}P-NMR ($CDCl_3$, 162 MHz) δ = 47.5 (s); ^{1}H-NMR ($CDCl_3$, 400 MHz) δ = 8.09–8.14 (m, 1H), 7.85–7.90 (m, 1H), 7.76–7.82 (m, 1H), 7.49–7.58 (m, 2H), 7.42–7.48 (m, 1H), 7.29–7.36 (m, 5H), 7.23–7.29 (m, 1H), 6.40 (dd, *J* = 4.5 and 15.8 Hz, 1H), 6.09–6.20 (m, 1H), 3.83–4.01 (m, 2H), 3.62–3.78 (m, 2H), 2.75 (dd, *J* = 7.6 and 16.8 Hz, 2H), 1.51 (quint., *J* = 7.4 Hz, 2H), 1.27 (sextuplet, *J* = 7.6 Hz, 2H), 0.85 (t, *J* = 7.4 Hz, 3H); ^{13}C-NMR (101 MHz, $CDCl_3$): δ = 136.7 (d, J_{PCCCC} = 3.1 Hz), 135.0 (d, J_{PCC} = 12.8 Hz), 134.0 (d, J_{PCCCC} = 2.4 Hz), 132.1 (d, J_{PCCC} = 4.6 Hz), 128.8, 128.6 (2C), 128.5 (d, J_{PCCCC} = 6.7 Hz), 128.1 (d, J_{PCC} = 8.2 Hz), 127.9 (d, J_{PCCCC} = 3.8 Hz), 127.7, 126.2 (3C), 125.9, 125.4 (d, J_{PCCCC} = 3.6 Hz), 124.4 (d, J_{PCCCCC} = 1.2 Hz), 118.9 (d, J_{PCCC} = 9.9 Hz), 64.9 (d, J_{POC} = 7.3 Hz), 33.7 (d, J_{PC} = 88.2 Hz), 33.2 (d, J_{PC} = 87.7 Hz), 32.7 (d, J_{POCC} = 5.9 Hz), 18.7, 13.6; HRMS (EI+) *m/z* calcd for $C_{24}H_{28}O_2P$ ($[M + H]^+$) 379.1827, found 379.1750.

3.2. 1-Butoxy-3-Phospholene 1-Oxide (Scheme 3c) [6]

To a solution of cinnamyl phosphinic acid [13] (3.64 g, 20 mmol, 1 equiv.) in toluene (40 mL), 1-butanol (3.7 mL, 40 mmol, 2 equiv.) was added and the reaction was stirred for 16 h at reflux in a flask equipped with a Dean-Stark trap. After cooling down the reaction to room temperature, the solvent was removed under vacuum. The residue obtained was dissolved in ethyl acetate and the organic layer was washed with an aqueous saturated solution of $NaHCO_3$ and brine, dried over $MgSO_4$, filtered, and concentrated under vacuum. The residue obtained was purified by column chromatography (hexane/ethyl acetate 80:0 to 20:80) to afford butyl cinnamyl-*H*-phosphinate as a yellow oil (4.59 g, 96%) [1]. ^{31}P-NMR ($CDCl_3$, 162 MHz) δ = 36.2 (dm, *J* = 543 Hz); ^{1}H-NMR ($CDCl_3$, 400 MHz) δ = 7.31–7.41 (m, 4H), 7.23–7.29 (m, 1H), 7.07 (dt, *J* = 1.9 and 543 Hz, 1H), 6.57 (dd, *J* = 6.0 and 15.9 Hz, 1H), 6.06–6.18 (m, 1H), 4.12–4.21 (m, 1H), 4.00–4.09 (m, 1H), 2.83 (dd, *J* = 7.6 and 16.8 Hz, 2H), 1.71 (quint., *J* = 7.1 Hz, 2H), 1.43 (sextuplet, *J* = 7.5 Hz, 2H), 0.95 (t, *J* = 7.4 Hz, 3H).

General procedure was used with allyl alcohol (0.15 mL, 2 mmol, 2 equiv.) and butyl cinnamyl-*H*-phosphinate (238 mg, 1 mmol, 1 equiv.). The crude obtained was purified by column chromatography (hexane/ethyl acetate 100:0 to 0:100) to afford butyl allyl cinnamylphosphinate as an orange oil (254 mg, 91%) [6]. ^{31}P-NMR ($CDCl_3$, 162 MHz) δ = 48.0 (s); ^{1}H-NMR ($CDCl_3$, 400 MHz) δ = 7.23–7.34 (m, 4H), 7.15–7.21 (m, 1H), 6.46 (dd, *J* = 4.7 and 15.8 Hz, 1H), 6.08–6.19 (m, 1H), 5.72–5.86 (m, 1H), 5.14–5.23 (m, 2H), 4.00 (dt, *J* = 6.8 and 6.9 Hz, 2H), 2.72 (ddd, *J* = 0.7, 7.6 and 17.6 Hz, 2H), 2.60 (dd, *J* = 7.5 and 17.2 Hz, 2H), 1.60 (quint., *J* = 7.1 Hz, 2H), 1.35 (sextuplet, *J* = 7.5 Hz, 2H), 0.88 (t, *J* = 7.4 Hz, 3H); ^{13}C-NMR (101 MHz, $CDCl_3$): δ = 136.7 (d, J_{PCCCC} = 3.3 Hz), 135.0 (d, J_{PCC} = 12.9 Hz), 128.6 (2C), 127.7 (d, J_{PCCC} = 8.0 Hz), 127.6, 126.2, 126.1, 120.3 (d, J_{PCC} = 12.6 Hz), 118.8 (d, J_{PCCC} = 9.6 Hz), 64.5 (d, J_{POC} = 7.2 Hz), 33.8 (d, J_{PC} = 88.7 Hz), 32.9 (d, J_{PC} = 89.0 Hz), 32.7 (d, J_{POCC} = 5.7 Hz), 18.8, 13.6.

To a solution of butyl allyl cinnamylphosphinate (92 mg, 0.33 mmol, 1 equiv.) in dichloromethane (50 mL), [1,3-bis(2,4,6-trimethylphenyl)-2-imidazolidinylidene] dichloro(phenylmethylene)-

(tricyclohexylphosphine)ruthenium (5.6 mg, 0.0066 mmol, 0.02 equiv.) was added and the mixture was placed under N_2. The reaction was heated at reflux for 12 h and then allowed to cool down to rt and treated with activated charcoal (0.1 g). The resulting suspension was stirred for 12 h at rt, suction-filtered through a Celite pad in a Buchner funnel, and concentrated under vacuum. The crude obtained was purified by column chromatography (hexane/ethyl acetate 100:0 to 0:100) to afford an orange oil (48 mg, 85%). ^{31}P-NMR ($CDCl_3$, 162 MHz) δ = 74.2 (s); ^{1}H-NMR ($CDCl_3$, 400 MHz) δ = 5.91 (d, J = 33.0 Hz, 2H), 4.04 (dt, J = 6.9 and 7.0 Hz, 2H), 2.35–2.50 (m, 4H), 1.67 (quint., J = 7.1 Hz, 2H), 1.40 (sextuplet, J = 7.5 Hz, 2H), 0.93 (t, J = 7.4 Hz, 3H); ^{13}C-NMR (101 MHz, $CDCl_3$): δ = 127.0 (d, J_{PCC} = 15.4 Hz, 2C), 64.7 (d, J_{POC} = 6.7 Hz), 32.6 (d, J_{POCC} = 6.0 Hz), 29.2 (d, J_{PC} = 91.2 Hz, 2C), 18.8, 13.6.

3.3. *(R_p)-Menthyl Cinnamyl(hydroxymethyl)phosphinate (Scheme 4) [14]*

To a solution of (R_p)-menthyl (hydroxymethyl)-*H*-phosphinate (468 mg, 2 mmol, 1 equiv., >99% diastereoisomeric excess) in *t*-amyl alcohol (10 mL), $Pd_2(dba)_3$ (18.3 mg, 0.02 mmol, 1 mol %), Xantphos (23.2 mg, 0.04 mmol, 2 mol %) and cinnamyl alcohol (0.26 mL, 2 mmol, 1 equiv.) were added. The reaction mixture was stirred at reflux for 20 h under N_2 in a flask equipped with a Dean-Stark trap. After cooling down the reaction to rt, the solvent was removed under vacuum and the residue obtained was purified by column chromatography (dichloromethane/acetone 100:0 to 90:10) to afford the product as a white solid (681 mg, 97%, >99% diastereoisomeric excess). Mp = 145–146 °C; ^{31}P-NMR (162 MHz, $CDCl_3$): δ = 48.8 (s); ^{1}H-NMR (400 MHz, $CDCl_3$): δ = 7.19–7.39 (m, 5H), 6.55 (dd, J = 4.7 and 15.8 Hz, 1H), 6.12–6.27 (m, 1H), 4.20–4.34 (m, 1H), 3.87 (s, 2H), 3.64 (s, 1H), 2.85 (dd, J = 7.6 and 17.6 Hz, 2H), 2.06–2.22 (m, 2H), 1.60–1.71 (m, 2H), 1.28–1.54 (m, 2H), 1.15 (q, J = 11.7 Hz, 1H), 0.74–1.07 (m, 2H), 0.91 (d, J = 6.4 Hz, 3H), 0.86 (d, J = 6.8 Hz, 3H), 0.77 (d, J = 7.0 Hz, 3H); ^{13}C-NMR (101 MHz, $CDCl_3$): δ = 136.8 (d, J_{PCCCC} = 3.3 Hz), 135.0 (d, J_{PCC} = 12.2 Hz), 128.5 (2C), 127.5, 126.2 (d, J_{PCCCCC} = 1.7 Hz, 2C), 118.4 (d, J_{PCCC} = 10.5 Hz), 76.7 (d, J_{POC} = 8.3 Hz), 59.5 (d, J_{PC} = 106 Hz), 48.6 (d, J_{POCC} = 5.6 Hz, 2C), 43.5, 34.0, 31.6 (d, J_{PC} = 87.3 Hz), 31.5, 25.5, 22.7, 22.1, 21.0, 15.5; HRMS (EI+) m/z calcd for $C_{20}H_{31}O_3P$ ($[M]^+$) 350.2011, found 350.2012; $[\alpha]_D^{24}$ −51.6° (c 1 g/100 mL, chloroform).

4. Conclusions

We have described the Pd-catalyzed allylation/benzylation of *H*-phosphinate esters using alcohols as the electrophilic partner. While the scope of this reaction is somewhat narrower than when phosphinic acids are used, it still constitutes a useful synthetic methodology for the synthesis of disubstituted phosphinic acid derivatives. In addition, because *H*-phosphinate esters can be produced easily by Dean-Stark esterification of the corresponding acids, it avoids the need for cumbersome esterification procedures on disubstituted phosphinic acids. Therefore, when the present reaction is successful, the sequence esterification-allylation/benzylation is superior to the reverse sequence allylation/benzylation-esterification. A possible mechanism is also proposed.

Supplementary Materials: Supplementary materials can be accessed at: http://www.mdpi.com/1420-3049/21/10/1295/s1.

Acknowledgments: This material is based in part upon work supported by the National Science Foundation under grant CHE-1262254.

Author Contributions: Jean-Luc Montchamp conceived and designed the experiments. Anthony Fers-Lidou conducted the majority of the experiments and Olivier Berger performed some of the experiments and wrote the experimental procedures. Jean-Luc Montchamp prepared the manuscript.

Conflicts of Interest: The authors declare no conflict of interest. The funding sponsors had no role in the design of the study; in the collection, analyses, or interpretation of data; in the writing of the manuscript, and in the decision to publish the results.

Abbreviations

The following abbreviations are used in this manuscript:

HPA	hypophosphorous acid (H_3PO_2)
MS	molecular sieves
t-AmOH	*t*-amyl alcohol

$Pd_2(dba)_3$	tris(dibenzylideneacetone)dipalladium(0)
Dba	dibenzylideneacetone
Dppf	1,1′-bis(diphenylphosphino)ferrocene
Xantphos	4,5-bis(diphenylphosphino)-9,9-dimethylxanthene
Cin	cinnamyl
Cy	cyclohexyl
Np	naphthyl
Men	(-)-menthyl
DOPO	6*H*-dibenzo[*c*,*e*][1,2λ^5]oxaphosphinine 6-oxide
HMDS	hexamethyldisilazane
IMes	1,3-bis(2,4,6-trimethylphenyl)imidazol-2-ylidene
RCM	ring closing metathesis

References

1. Ferraccioli, R.; Pignataro, L. Tsuji-Trost Type Functionalization of Allylic Substrates with Challenging Leaving Groups: Recent Developments. *Curr. Org. Chem.* **2015**, *19*, 106–120. [CrossRef]
2. Gumrukcu, Y.; de Bruin, B.; Reek, J.N.H. Hydrogen-Bond-Assisted Activation of Allylic Alcohols for Palladium-Catalyzed Coupling Reactions. *ChemSusChem* **2014**, *7*, 890–896. [CrossRef] [PubMed]
3. Bravo-Altamirano, K.; Montchamp, J.-L. Palladium-Catalyzed Dehydrative Allylation of Hypophosphorous Acid with Allylic Alcohols. *Org. Lett.* **2006**, *8*, 4169–4171. [CrossRef] [PubMed]
4. Coudray, L.; Bravo-Altamirano, K.; Montchamp, J.-L. Allylic Phosphinates via Palladium-Catalyzed Allylation of *H*-Phosphinic Acids with Allylic Alcohols. *Org. Lett.* **2008**, *10*, 1123–1126. [CrossRef] [PubMed]
5. Coudray, L.; Montchamp, J.-L. Green, Palladium-Catalyzed Synthesis of Benzylic *H*-Phosphinates from Hypophosphorous Acid and Benzylic Alcohols. *Eur. J. Org. Chem.* **2008**, 4101–4103. [CrossRef] [PubMed]
6. Bravo-Altamirano, K.; Abrunhosa-Thomas, I.; Montchamp, J.-L. Palladium-Catalyzed Reactions of Hypophosphorous Compounds with Alkenes, Dienes, and Allylic Electrophiles: Methodology for the Synthesis of Allylic *H*-Phosphinates. *J. Org. Chem.* **2008**, *73*, 2292–2301. [CrossRef] [PubMed]
7. Janesko, B.J.; Fisher, H.C.; Bridle, M.J.; Montchamp, J.-L. P(=O)H to P-OH Tautomerism: A Theoretical and Experimental Study. *J. Org. Chem.* **2015**, *80*, 10025–10032. [CrossRef] [PubMed]
8. Keglevich, G.; Kiss, N.Z.; Mucso, Z.; Körtvélyesi, T. Insights into a surprising reaction: The microwave-assisted direct esterification of phosphinic acids. *Org. Biomol. Chem.* **2012**, *10*, 2011–2018. [CrossRef] [PubMed]
9. Dumond, Y.R.; Baker, R.L.; Montchamp, J.-L. Orthosilicate-Mediated Esterification of Monosubstituted Phosphinic Acids. *Org. Lett.* **2000**, *2*, 3341–3344. [CrossRef] [PubMed]
10. Bujard, M.; Gouverneur, V.; Mioskowski, C. A Highly Efficient and Practical Synthesis of Cyclic Phosphinates Using Ring-Closing Metathesis. *J. Org. Chem.* **1999**, *64*, 2119–2123. [CrossRef] [PubMed]
11. Bravo-Altamirano, K.; Montchamp, J.-L. Palladium-Catalyzed Dehydrative Allylation of Hypophosphorous Acid with Allylic Alcohols. Preparation of Cinnamyl-*H*-Phosphinic Acid. *Org. Synth.* **2008**, *85*, 96–105.
12. Berger, O.; Montchamp, J.-L. General Synthesis of *P*-Stereogenic Compounds: The Menthyl Phosphinate Approach. *Org. Biomol. Chem.* **2016**, *14*, 7552–7562. [CrossRef] [PubMed]
13. Berger, O.; Petit, C.; Deal, E.L.; Montchamp, J.-L. Phosphorus-Carbon Bond Formation: Palladium-Catalyzed Cross-Coupling of *H*-Phosphinates and other P(O)-Containing Compounds. *Adv. Synth. Catal.* **2013**, *355*, 1361–1373. [CrossRef]
14. Berger, O.; Montchamp, J.-L. A General Strategy for the Synthesis of P-Stereogenic Compounds. *Angew. Chem. Int. Ed.* **2013**, *52*, 11377–11380. [CrossRef] [PubMed]
15. Loy, N.S.Y.; Singh, A.; Xu, X.; Park, C.-M. Synthesis of Pyridines by Carbenoid-Mediated Ring Opening of 2*H*-Azirines. *Angew. Chem. Int. Ed.* **2013**, *52*, 2212–2216. [CrossRef] [PubMed]
16. Nelson, A.; O'Brien, P.; Warren, S. Asymmetric dihydroxylations of allylic phosphine oxides. *Tetrahedron Lett.* **1995**, *36*, 2685–2688. [CrossRef]
17. Stiles, A.R.; Harman, D. Esters of Aromatic Olefinic Phosphinic Acids. U.S. Patent US2711403, 21 June 1955.

Sample Availability: Not available.

molecules

Review

Synthesis, Properties and Stereochemistry of 2-Halo-1,2λ⁵-oxaphosphetanes

Anastasy O. Kolodiazhna and Oleg I. Kolodiazhnyi *

Institute of Bioorganic Chemistry and Petrochemistry, National Academy of Sciences of Ukraine, 1, Murmanska Street, Kiev 02094, Ukraine; anastasik7@rambler.ru
* Correspondence: olegkol321@gmail.com; Tel.: +380-445-732-555

Academic Editors: György Keglevich and Derek J. McPhee
Received: 31 August 2016; Accepted: 8 October 2016; Published: 17 October 2016

Abstract: Results of research into four-membered 2-halo-1,2λ^5-oxaphosphetane phosphorus(V)-heterocycles are presented. The preparation of 2-halo-1,2λ^5-oxaphosphetanes by reaction of *P*-haloylides with carbonyl compounds is described. The mechanism of asynchronous [2+2]-cycloaddition of ylides to aldehydes was proposed on the base of low-temperature NMR investigations. 2-Halo-1,2λ^5-oxaphosphetanes were isolated as individual compounds and their structures were confirmed by ^{1}H-, ^{13}C-, ^{19}F- and ^{31}P-NMR spectra. These compounds are convenient reagents for preparing of various organic and organophosphorus compounds hardly available by other methods. Chemical and physical properties of the 2-halo-1,2λ^5-oxaphosphetanes are reviewed. The 2-chloro-1,2λ^5-oxaphosphetanes, rearrange with formation of 2-chloroalkyl-phosphonates or convert into *trans*-phosphorylated alkenes depending on the substituents at the α-carbon atom. Prospective synthetic applications of 2-halo-1,2λ^5-oxaphosphetanes are analyzed. The 2-halo-1,2λ^5-oxaphosphetanes may be easily converted to various alkenylphosphonates: allyl- or vinylphosphonates, phosphorus ketenes, thioketenes, ketenimines.

Keywords: 2-halo-1,2λ^5-oxaphosphetanes; allylphosphonates; vinylphosphonates; phosphorus ketenes

1. Introduction

One of the most interesting and intriguing classes of organophosphorus compounds are the 1,2-oxaphosphetanes—four-membered heterocycles containing pentacoordinated phosphorus [1–7]. Since 1,2-oxaphosphetanes are well-known intermediates in the Wittig reaction, a number of efforts have been made for their structural characterization both in solution and the solid state [8–13].

In 1967, Birum and Matthews had already reported the structural characterization (NMR and X-ray study) of the first isolated 1,2-oxaphosphetane **1**. Compound **1** was prepared in 76% yield by allowing hexaphenylcarbodiphosphorane to react with hexafluoroacetone in dry diglyme [14] (Scheme 1).

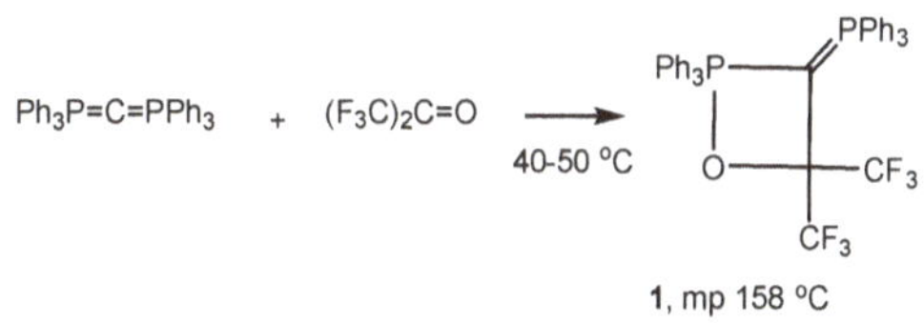

Scheme 1. First representative of stable 1,2-oxaphosphetanes isolated by Birum and Matthews.

Vedejs [3,4] succeeded in detecting of 1,2-oxaphosphetanes **2–4** by low temperature NMR spectroscopy during typical Wittig reactions and observed that these intermediates readily decompose upon warming to room temperature into alkenes and phosphine oxides (Scheme 2).

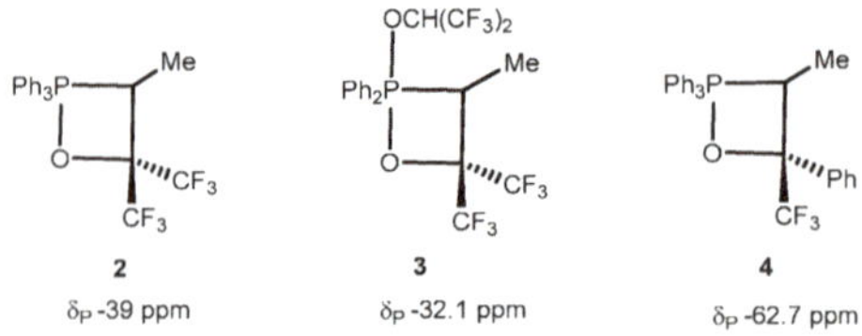

Scheme 2. 1,2-oxaphosphetanes **2–4** registered by low temperature NMR spectroscopy.

Schmutzler and co-workers reported several stabilized bis(trifluoromethylated) oxaphosphetanes **5**, **6** (Scheme 3) [15] which were characterized by NMR, MS spectra and X-ray analysis. At room temperature Berry-pseudorotation was fast on the NMR time scale, impeding one from distinguishing apical and equatorial P-CF_3 groups. Decreasing the temperature to −60 °C in toluene-d_8 allowed resolving the signals for all CF_3 groups of molecule **7** [16].

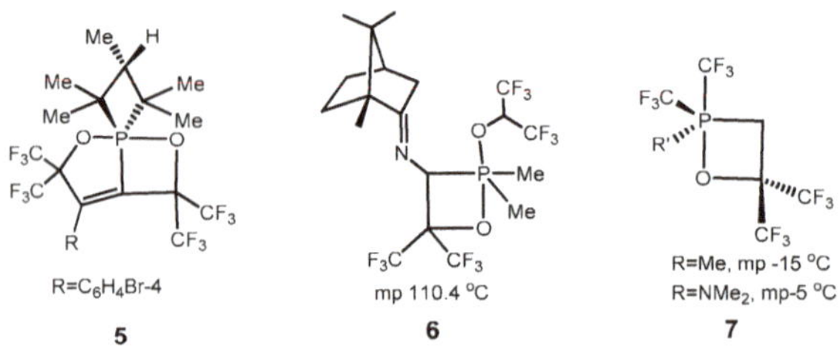

Scheme 3. Stabilized bis(trifluoromethylated) oxaphosphetanes **5**,**6**.

Kojima [17] reported the interesting anti-apicophilic spirophosphorane **8** bearing an oxaphosphetane ring. The structure of compounds **8** was confirmed by X-ray diffraction. Crystallization from hexane gave the pure anti-apicophilic derivative. Stereomutation of compound **8** was observed in the presence of acids and slowed down when DBU was present, suggesting that the isomerization into **9** is rather the result of a P–O bond breaking-recombination process. Evidently, this conversion represents an example of a thermodynamically stable oxaphosphetanes, in which pseudorotation is faster than alkene formation (Scheme 4).

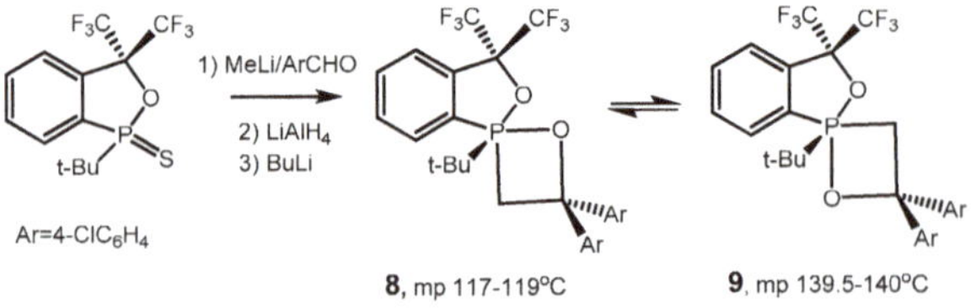

Scheme 4. Stereomutation of anti-apicophilic oxaphosphetane **8**.

Most of the previously reported stable oxaphosphetane structures contain fluorine-bearing or bicyclic phosphole-type ligands either at the phosphorus position or at the 4 position in the oxaphosphetane ring **10**, **11** [15–20]. Recently, Streubel and coworkers have prepared the first 1,2-oxaphosphetane complexes **12** formally similar to traditional oxaphosphetanes, using low-temperature ring expansion of epoxides with a Li/Cl phosphinidenoid complex [18] (Scheme 5).

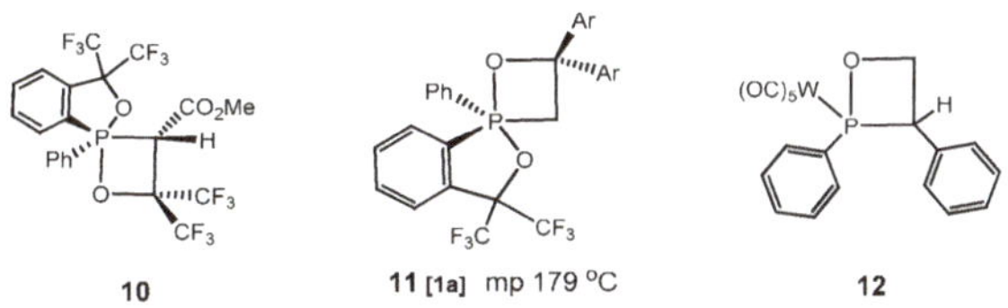

Scheme 5. Oxaphosphetanes stabilized by CF_3 groups **10,11** and 1,2-oxaphosphetane complex **12**.

Gilheany studied oxaphosphetane intermediates in the Wittig reaction by variable-temperature NMR spectroscopy [9–11]. Compound **13** was obtained by low-temperature acid quenching of the Wittig reaction of ylide with benzaldehyde, a suitable representative aromatic aldehyde (see Scheme 6). The major diastereomer was the *syn*-**13** on the basis that the unquenched Wittig reaction gives the (*Z*)-alkene as the major product. In this manner, the *syn/anti* ratio of **13** was 89:11.

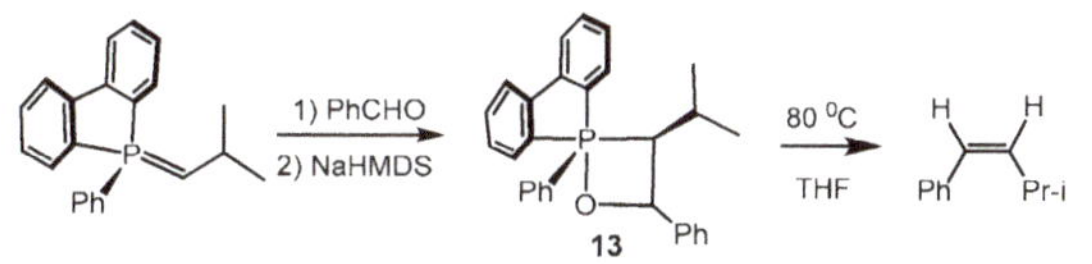

Scheme 6. Stereospecific decomposition of oxaphosphetane **13**.

Keglevich reported detection of enantiomers of *P*-stereogenic pentacoordinated phosphorus compounds [20]. Detailed ^{31}P-NMR investigations of oxaphosphetes in optically active solvents have clearly shown that the most electronegative substituents (e.g., oxygen) prefers the apical position in a trigonal bipyramidal structure and that the pentacoordinated phosphorus atom is in a dynamic condition due to pseudorotation. Berger and coworkers found that 2-furyl groups on the phosphorus atom increase the thermal stabilities of oxaphosphetanes and succeeded in isolation and determination of the X-ray structure of tris(2-furyl) substituted oxaphosphetane, the stability of which is attributed to the electron-withdrawing properties of the 2-furyl group [21].

Among the stable oxaphosphetanes [1–22], 2-halo-1,2λ^5-oxaphosphetanes **14**, which possess relatively high stability and diverse reactivity, attract particular interest [23–38] (Scheme 7). These oxaphosphetanes containing fluorine, chlorine or bromine atoms bonded to phosphorus are an interesting class of pentacoordinated phosphorus heterocycles possessing peculiar properties. The chemical properties of 2-halo-1,2λ^5-oxaphosphetanes, first of all of *P*-chloro- and *P*-fluoroylides, due to the presence of a labile halogen atom on phosphorus, are very specific and differ from the properties of triphenylphosphonium ylides. For example, reactions and conversions of 2-halo-1,2λ^5-oxaphosphetanes proceed with preservation of the P-C bond and leads to the formation of different organophosphorus compounds. In addition, this type of compounds exhibit also physico-chemical properties uncharacteristic for traditional *P,P,P*-triorganosubstituted oxaphosphetanes. For the first time the 2-halo-1,2λ^5-oxaphosphetanes were prepared in our laboratory almost twenty years ago and up to today we and some other authors are still studying their chemistry. In this article, we summarize the synthesis and properties of this type of organophosphorus compounds. The chemistry of 2-halo-1,2λ^5-oxaphosphetanes was not previously analyzed, generalized or reviewed.

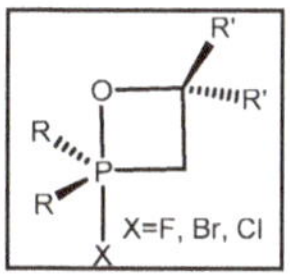

14

Scheme 7. Examples of stable oxaphosphetanes.

2. Results and Discussion

2.1. General Description

Since pentacoordinated phosphoranes formally have 10 electrons in the valence shell, they display a specific bonding model. Therefore pentacoordinated phosphoranes take a trigonal bipyramidal structure and there are two ligating sites, apical and equatorial sites. The apical bond consists of three-center four-electron bond using the *p* orbital of the central phosphorus atom, while the equatorial bond is a typical *s* bond using sp^2 hybrid orbital of the phosphorus atom. This three-center four-electron bond forms three molecular orbitals.

It is generally known that pentacoordinated phosphoranes rapidly undergo intramolecular positional isomerization without bond cleavage. A very rapid non-dissociative intramolecular site exchange is usually explained by the Berry pseudorotation mechanism [19,22,23].

The 2-halo-1,2λ^5-oxaphosphetanes (halogen = chlorine, bromine, fluorine) are the most stable representatives of this type of compounds. They can be purified by distillation under vacuum and stored in a refrigerator. At the same time they possess interesting chemical properties and participate in various chemical transformations [24–28].

The stability of 2-halo-oxaphosphetanes changes in the same sequence of substituents R^4 at the endocyclic carbon atoms at position 4. The most stable are compounds containing strong electron-accepting groups at C(4), drawing off electron density from the oxygen atom as a result of which the three-centre apical bond O—P—Hal is strengthened. Chloro-oxaphosphetanes, having less electron-accepting substituents at C(4), are dissociated to a large extent and correspondingly are converted into vinylphosphine oxides at room temperature.

2.2. Synthesis of 2-Halo-1,2λ^5-oxaphosphetane

Available methods for the synthesis of 2-halo-1,2λ^5-oxaphosphetanes can be used for investigation of the reaction mechanism of phosphorus ylides with carbonyl compounds as well as for preparing stable oxaphosphetanes that can be used as reactants for organic synthesis. The 2-halo-1,2λ^5-oxaphosphetanes were prepared by reaction of *P*-fluoro-, chloro- or bromoylides with carbonyl compounds. *P*-Chloro- and *P*-bromoylides react with active ketones, containing a trifluoromethyl group, with the formation of stable [2+2]-cycloaddition products, 2-chloro- or 2-bromo-1,2λ^5-oxaphosphetenes **14** were isolated in yields close to quantitative as crystalline substances or as liquids distillable in vacuum (Scheme 8, Table 1) [29–42].

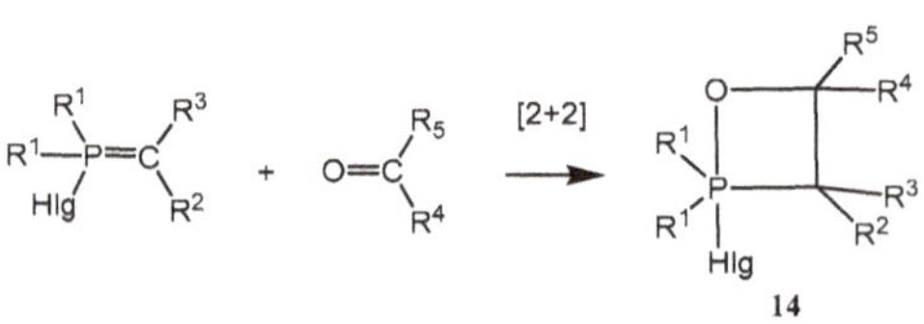

Scheme 8. Synthesis of 2-halo-1,2λ^5-oxaphosphetanes. Hlg = F, Cl, Br; R^1 = Alk, Ph; CR^2R^3 = CH_2, CHAlk, $CAlk_2$, CHPh, $CHSiMe_3$, CCl_2, CBr_2; CR^4R^5 = C=O, CNPh, CHAlk, $CAlk_2$, CHPh, CPh_2, $CH_2CH = CH_2$.

The addition of *P*-halogen-ylides to ketones proceeded stereoselectively and led predominantly to the formation of one of the possible 2-halo-1,2λ^5-oxaphosphetane diastereomers. 2-Halo-1,2λ^5-oxaphosphetanes dissociate at the *P*-halogen bond in solution with the formation of cyclic phosphonium salts, as a result of which an equilibrium is established between the forms with five- and four-coordinate phosphorus atoms. Dissociation of 2-halo-1,2λ^5-oxaphosphetanes is enhanced by reducing the electron-accepting properties of substituents and also by increasing priority solvent. The 2-halo-1,2λ^5-oxaphosphetanes containing electron-accepting groups at C(4) are distinctly stabler than oxaphosphetanes with alkyl groups in this position [25–27].

The reaction of *P*-bromomethylides with fluorinated acetophenone afforded in high yield oxaphosphetanes **15**, which were isolated as crystalline compounds (Scheme 9 and Table 1, entries 10–13). The compounds **15** exist in solution as cyclic phosphonium salt.

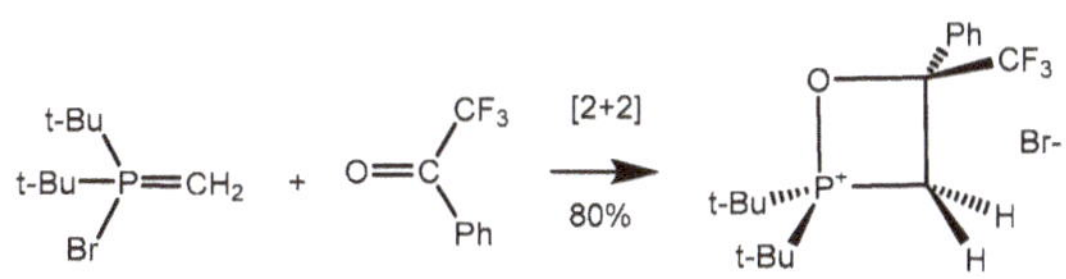

15

δ_P +46 ppm
mp 132 °C (hexane)

Scheme 9. Ionized form of 2-bromo-1,2λ^5-oxaphosphetanes **15**.

At the same time the 2-chloro-1,2λ^5-oxaphosphetanes **16** exist as mixture of P(IV) and P(V)-forms. These compounds can be distilled under vacuum and dissolved in non-polar solvents (benzene) (Scheme 10 and Table 1, entries 1–9). Reaction of *P*-Fluoroylids with aldehydes and ketones proceeds in ether or pentane at −40–−20 °C and leads to the formation of stable 2-fluoro-1,2λ^5-oxaphosphetanes **17** (Scheme 11 and Table 2).

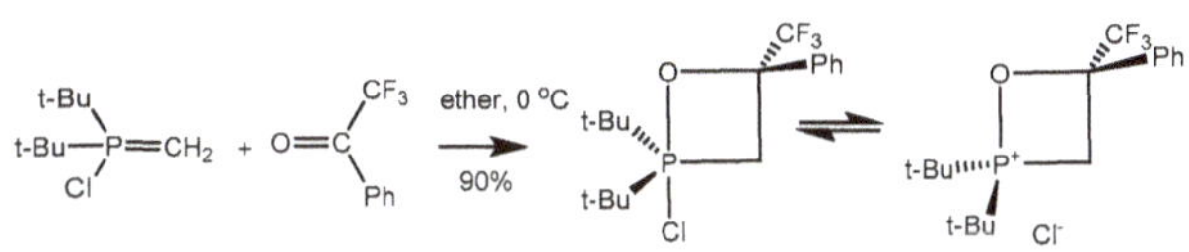

16

mp 82-84 °C, δ_P 9.94 ppm

Scheme 10. P(IV) and P(V)-forms of 2-chloro-1,2λ^5-oxaphosphetane **16**.

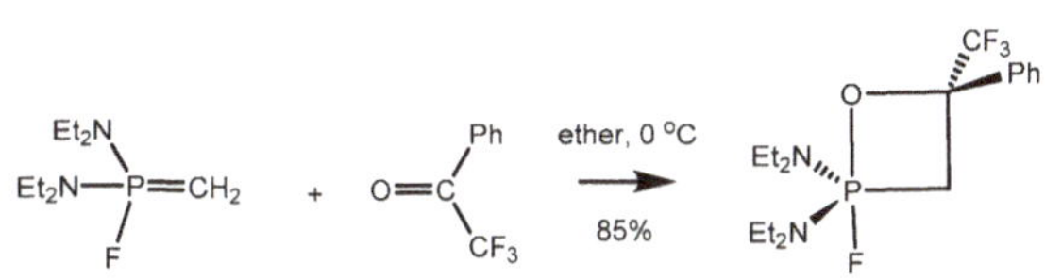

17

bp 110 °C (0.06 mmHg), mp 20 °C,
δ_P -38.96 ppm

Scheme 11. 2-Fluoro-1,2λ^5-oxaphosphetane **17**.

The compounds **17** are liquids distilling in vacuo, the structure of which was proved by means of mass and NMR spectra. The ^{31}P-NMR spectra of 2-fluoro-oxaphosphetanes **17** present doublets with 800 Hz $^1J_{PF}$ constants in the high magnetic field of a NMR spectrum at −37–−8 ppm. This corresponds to a pentacoordinate state of compounds **17** [24,26–31]. Tetracoordinated forms of 2-fluoro-oxaphosphetanes **17** were not registered by ^{19}F- and ^{31}P-NMR spectroscopy (Scheme 11).

C-Silyl-*P*-chloroylides **18** react with carbonyl compounds to afford 2-chloro-1,2λ^5-oxaphosphetanes **19**. The oxaphosphetanes **19** bearing an electronegative CF_3 substituent at C-4 are relatively stable and can be isolated and analyzed by NMR (Scheme 12). The NMR spectra of these compounds reveal signals at 0.2 ppm, singlet (Me_3Si), at 2 ppm, doublet, $^3J_{PH}$ 18.0 Hz (C^3H), and at 4.5 ppm (C^4H). ^{13}C-NMR signals of C-3 and C-4 carbons were found at 30 and 90 ppm, correspondingly. The ^{31}P-NMR signals of 2-chloro-1,2λ^5-oxaphosphetanes **19** at δ_P +48 ppm (R^1 = *i*-PrO) and at δ_P = 60 ppm (R^1 = Et_2N) correspond to tetracoordinate phosphorus included in four-membered phosphetane cycle (Table 1) [25,32].

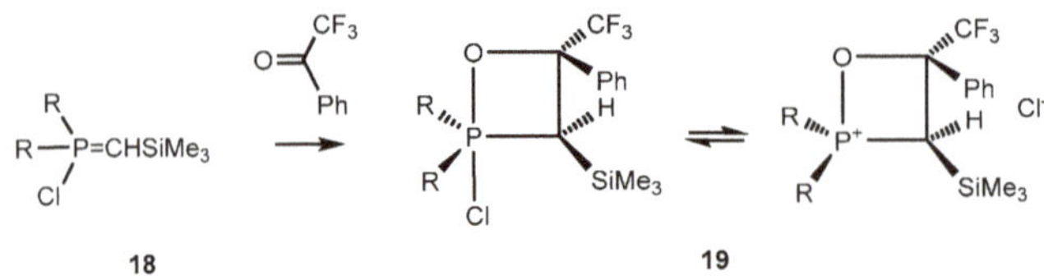

Scheme 12. Silylated 2-halogene-1,2λ^5-oxaphosphetanes **19**.

Table 1. 2-Chloro- and 2-bromo-1,2λ^5-oxaphosphetanes.

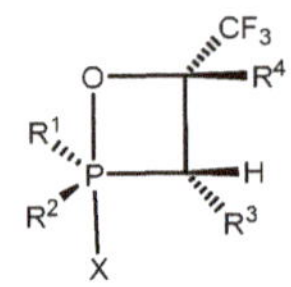

Entry	R^1	R^2	R^3	R^4	X	δ_P	δ_F	References
1	*t*-Bu	*t*-Bu	H	C_6H_4F-4	Cl	+ 9.0	+7.2 c (CF_3); −32.9 (C_6H_4F)	[36,42]
2	*i*-Pr	*i*-Pr	Ph	Ph	Cl	+1.23; +0.53	+1.08; +1.64	[42]
3	*t*-Bu	*t*-Bu	H	Ph	Cl	+ 9.94	+1.0	[42]
4	*t*-Bu	*t*-Bu	H	An-4	Cl	+13.6	+7.35 c	[36,42]
5	*t*-Bu	*t*-Bu	H	CF_3	Cl	+0.25	6.0	[36,42]
6	*t*-Bu	*t*-Bu	Me	Ph	Cl	+13.8; +11.65	+6.1	[42]
7	*t*-Bu	Et_2N	H	Ph	Cl	+5.01; +9.13	+6.06; +6.09	[42]
8	*t*-Bu	*t*-Bu	Me	CF_3	Cl	+2.53	+5.38;	[42]
9	*t*-Bu	Et_2N	H	CF_3	Cl	−2.6	+2.65 q; +2.94 q, $^1J_{FF}$ 8	[42]
10	*t*-Bu	*t*-Bu	H	CF_3	Br	+24.4	+6.25	[42]
11	*t*-Bu	*t*-Bu	H	Ph	Br	+46	+7.35	[42]
12	*t*-Bu	*t*-Bu	H	C_6H_4F-4	Br	+46	+5.6 (CF_3); −34.2 (C_6H_4F)	[42]
13	*t*-Bu	*t*-Bu	H	An-4	Br	+60	+6.1	[42]
14	*t*-Bu	*t*-Bu	H	Ph	OMe	−14.6	+1.15	[42]
15	*t*-Bu	*t*-Bu	H	Ph	OPh	−12.8	+2.0	[42]

Sotiropulos and Bertrand [33] reported the addition of phosphacumulene ylides **20** bearing the diazo group to isocyanates, leading to the formation of products **22**. It was proposed that initial nucleophilic attack of the ylide carbon atom on the carbonyl carbon gives a oxaphosphetane **21**, which depending on the relationship of the oxygen atom (or NR group) to nitrogen or phosphorus rearranges into products of 1,4- or 1,5-cyclisation **22** (Scheme 13). Benzaldehyde gives the 2-chlorooxaphosphetane **23** with the ylide **20** which readily eliminates hydrogen chloride and a nitrogen molecule being converted into an acetylene phosphonate **24** (Scheme 14).

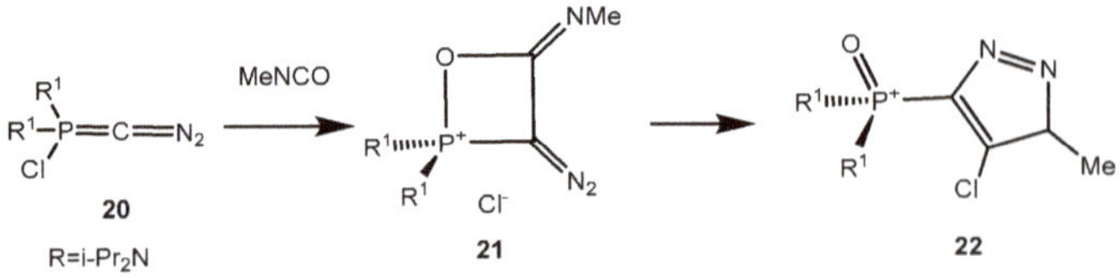

Scheme 13. Adduct of phosphacumulene-ylide **20** with isocyanate.

A number of 2-fluoro-1,2λ^5-oxaphosphetanes **25** were prepared by reaction of *P*-fluoroylides with aldehydes and ketones (Scheme 15, Table 2). The 2-fluoro-1,2λ^5-oxaphosphetanes **25** are stable compounds, which can be isolated and purified by distillation under vacuum or by crystallization from non-polar solvents. These compounds are much distinguished from unstable adducts of carbonyl compounds with triphenylphosphonium ylides. Cycloadducts of *P*-fluoroylides with carbonyl compounds, 2-fluoro-1,2λ^5-oxaphosphetanes, are also much more stable than 2-chloro- or 2-bromo-1,2λ^5-oxaphosphetanes [25–31].

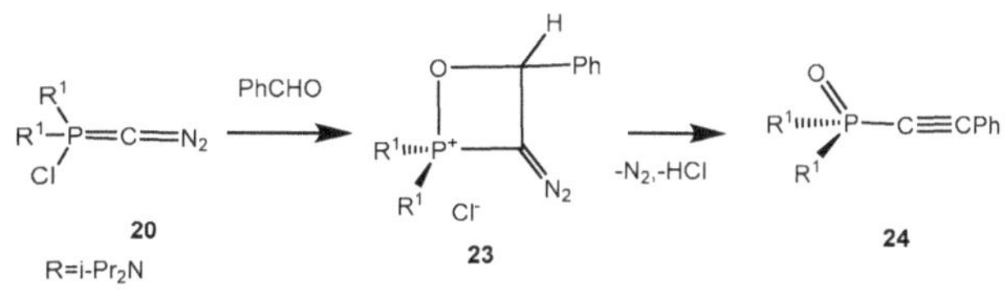

Scheme 14. Reaction of phosphacumulene-ylide **20** with benzaldehyde.

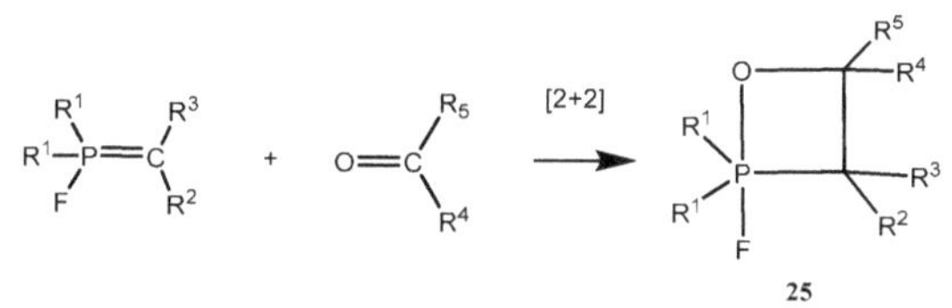

Scheme 15. Synthesis of 2-fluoro-1,2λ^5-oxaphosphetanes **25**. R^1 = Alk, Ph; CR^2R^3 = CH_2, CHAlk, $CAlk_2$, CHPh, $CHSiMe_3$, CCl_2, CBr_2; CR^4R^5 = C=O, CNPh, CHAlk, $CAlk_2$, CHPh, CPh_2, $CH_2CH = CH_2$.

The stability of 2-fluoro-oxaphosphetanes is explained by the high electronegativity of the fluorine atom, compared to the electronegativities of chlorine and bromine. The P-F bond in 2-fluoro-1,2λ^5-oxaphosphetanes is very strong, and, therefore, these compounds do not dissociate with formation of cyclic phosphonium salts, what is observed, for example, with 2-chloro-1,2λ^5-oxaphosphetanes. Various stable 2-fluoro-1,2λ^5-oxaphosphetanes were synthesized, isolated as pure specimens, and characterized (Table 2). Typical representatives of such compounds **26–29** are shown in Schemes 16–18 [26,29].

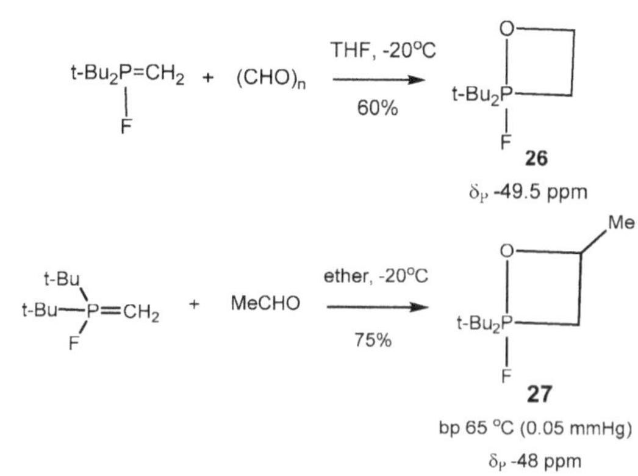

Scheme 16. The simplest representatives of 2-Fluoro-1,2λ^5-oxaphosphetanes **26**,**27**.

The ^{31}P-NMR spectra of compounds **25** show a doublet at −6 to −45 ppm with 760–850 Hz PF coupling constants appropriate for axial fluorine atoms. The ^{31}P-NMR spectra of 2-fluoro-1,2λ^5-oxaphosphetanes exhibit a doublet in the range −6 to −45 ppm, belonging to the five-coordinate phosphorus with corresponding coupling constants on axial fluorine atoms (760–850 Hz). The ^{19}F-NMR spectra of 2-fluoro-1,2λ^5-oxaphosphetanes exhibit doublets with the same PF coupling constants.

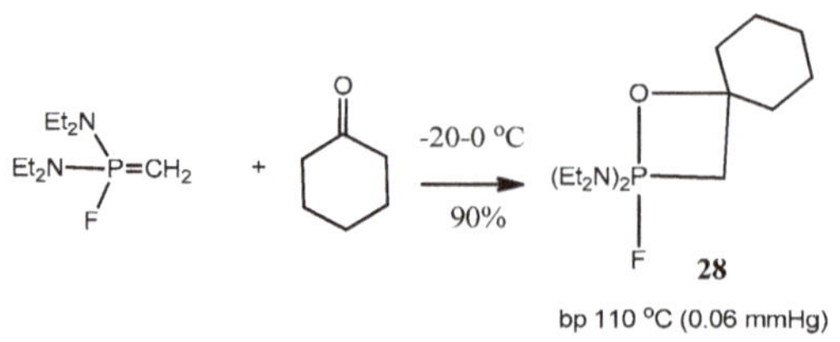

Scheme 17. 2-Fluoro-1,2λ^5-oxaphosphetane **28**.

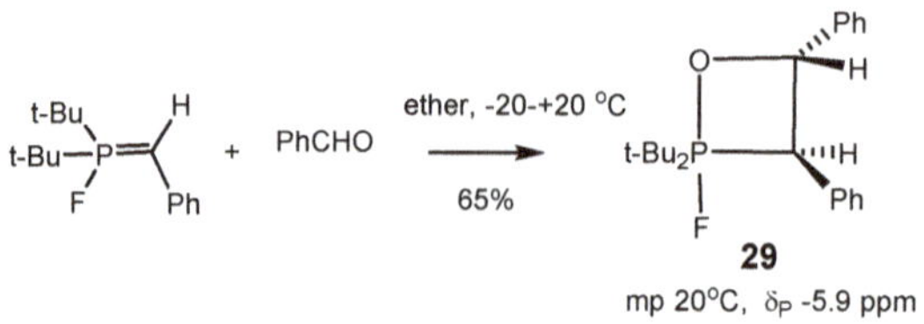

Scheme 18. 2-Fluoro-1,2λ^5-oxaphosphetane **29**.

The *P,P*-difluoroylides **30** react with aldehydes and active ketones to afford the stable oxaphosphetanes **31** bearing two fluorine atoms at the phosphorus (Scheme 19) [34–36]. The compounds **31** (R=CF_3) were distilled under reduced pressure without decomposition and were characterized by means of NMR spectroscopy.

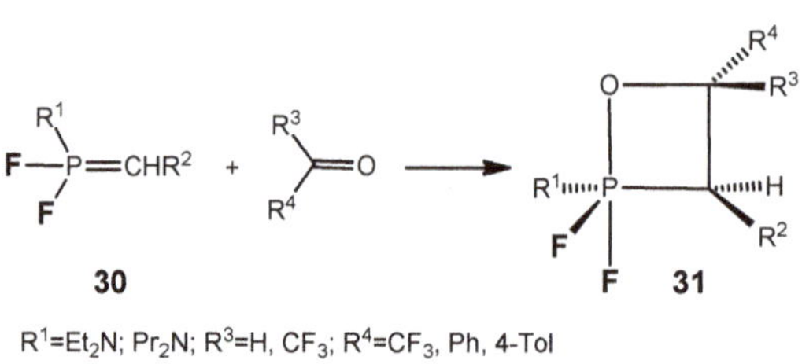

Scheme 19. 2,2-Difluoro-1,2λ^5-oxaphosphetanes **31**.

The ^{19}F-NMR spectra contain two double doublets at −47 and −65 ppm with expected coupling constants for the axial and equatorial fluorine atoms: $^1J_{PFa}$ = 915 Hz, $^1J_{PFe}$ = 1025 Hz, and $^2J_{FaFe}$ = 62 Hz. Apparently [2+2]-cycloaddition of the C=O group to the ylide **15** proceeds with high stereoselectivity, because the ^{19}F- and ^{31}P-NMR spectra show the signals belonging to the single diastereomer of the compounds **31**. The signals of the second diastereomer, the existence of which one can suppose as a consequence of the presence of two asymmetric endocyclic C-3 and C-4 carbon atoms are absent. The ^{13}C-NMR spectra reveal the presence of the signals at 62.5 ppm ($^1J_{CP}$ = 150 Hz, $^2J_{CF}$ = 53 Hz) and 75.6 ppm due to the C-3 and C-4 carbon atoms in complete accordance with the assigned structure of oxaphosphetane (Figures 1 and 2).

P-Fluoroylides add easily isocyanates, carbon dioxide and carbon disulfide with formation of oxaphosphetanes **32–35**. For example, the reaction of *P*-fluoroylides with phenyl isocyanate resulted in the formation of 2-fluorooxaphosphetane 34 which was stable during several hours at ambient temperature. Compound **34** was purified by crystallization in pentane and isolated as colorless crystalline matter. In IR spectra of compounds was found the strong band at 1730 cm^{-1} belonging to the C=O bond in a four-membered cycle. The ^{31}P-NMR spectra disclosed a doublet with constant $^1J_{PF}$ = 780 Hz and a doublet with the same constant in ^{19}F-NMR spectra. The oxaphosphetanes **32–35** convert at room temperature slowly and at heating quickly into phosphorylated heterocumulenes (ketenes, thioketenes, ketenimines, see Scheme 20) [29,37–39].

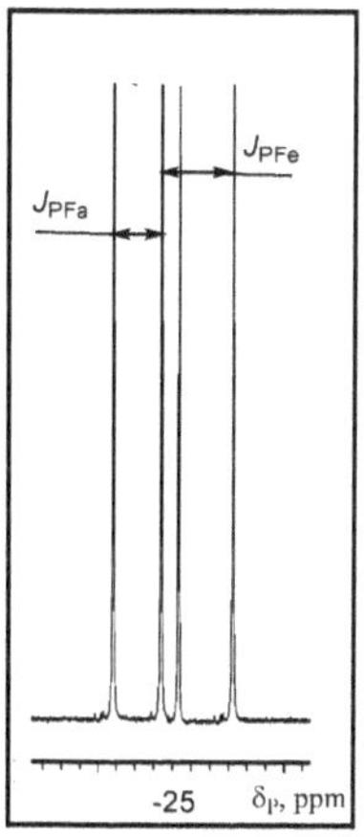

Figure 1. ^{31}P-NMR spectrum of 2-(diethylamino)-2,2-difluoro-4-tolyl-3-propyl-4-(trifluorophenyl-1,2λ^5-oxaphosphetane **31** (See Table 2, entry 15).

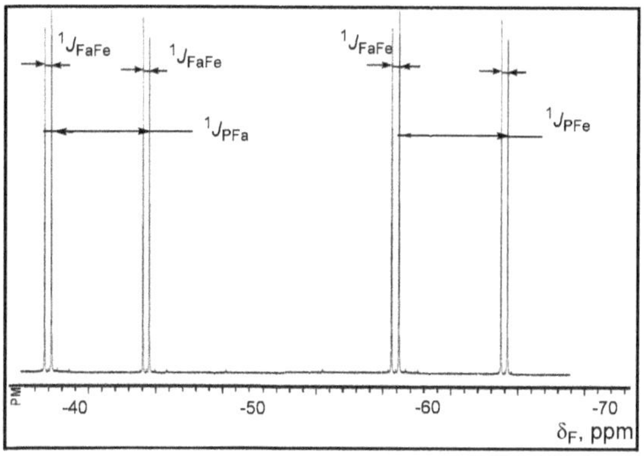

Figure 2. ^{19}F-NMR spectra of 2-(diethylamino)-2,2-difluoro-4-tolyl-3-propyl-4-(trifluorophenyl-1,2λ^5-oxaphosphetane **31** (see Table 2, entry 15).

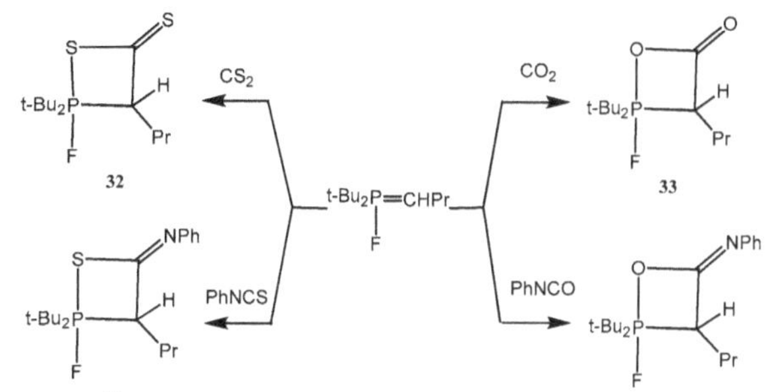

Scheme 20. Addition of heterocumulenes (X=C=Y) to *P*-fluoroylides.

Table 2. 2-Fluoro-1,2λ⁵-oxaphosphetanes [26,32,37].

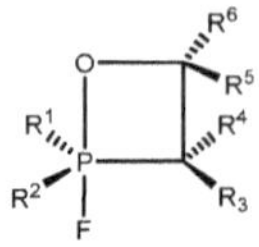

Entry	R^1	R^2	R^3	R^4	R^5	R^6	Yield, %	b.p., °C (P, mmHg)	δ_P, ppm.	δ_F, ppm.	$^1J_{PF}$, Hz
1	Et_2N	Et_2N	H	H	H	Bu	85	90 (0.06)	−42.15	31.30	767
2	Et_2N	Et_2N	H	H	H	C_6H_{13}	85	100 (0.06)	−42.25	31.12	766
3	Et_2N	Et_2N	H	H	Me	Me	70	75 (0.02)	−47.53	31.04	765
4	Et_2N	Et_2N	H	H	Me	Et	85	85 (0.06)	−49.20	31.04	765
5	Et_2N	Et_2N	H	H	Me	Ph	99	[a]	−44.00	32.00	766
6	Et_2N	Et_2N	H	H	$(CH_2)_5$		90	110 (0.06)	−44.60	34.90	766
7	Et_2N	Et_2N	H	Me	$(CH_2)_5$		99	[a]	−44.00	21.00	853
8	Et_2N	Et_2N	H	H	Ph	Ph	99	[a]	−41.60	34.90	766
9	Et_2N	Et_2N	H	H	H	Ph	99	[a]	−42.13	31.18	768
10	Et_2N	Et_2N	Cl	Cl	H	Pr	90	[a]	−47.00	5.40	853
11	Et_2N	Et_2N	Br	Br	H	Pr		[a]	−56.03	5.80	842
12	Et_2N	Et_2N	H	Me	CF_3	Ph	80	120 (0.06)	−36.39; −38.27 (7:1) [b]	−0.4	795; 795
13	Ph	Ph	Ph	H	CH_3	H	95	[a]	−43.50	47.8	670
14	Et_2N	F	H	Pr	CF_3	Ph	85	105 (0.06)	−26.7 (99% dr)	−46.9, −65 (CF_3)	915; 1025
15	Et_2N	F	H	Pr	CF_3	4-Tol	80	108 (0.06)	−26.5 (99% dr)	−46.6, −65 (CF_3)	915; 1025
16	Et_2N	Et_2N	H	Ph	CF_3	Ph	85	<20 [c]	−38.96	27	792.6
17	*t*-Bu	*t*-Bu	H	Pr	CF_3	Ph	70	125 (0.06)	−7.73, −8.79 (9:1) [b]	−1.03	768; 770
18	*t*-Bu	*t*-Bu	H	Pr	H	Ph	75	120 (0.08)	−10.1; −9.62 (15:1) [b]	9.9	762; 762
19	*t*-Bu	*t*-Bu	H	H	H	Ph	70	<20 [c]	−5.9	13	750
20	Et_2N	Et_2N	H	*i*-Pr	H	Ph	70	110 (0.06)	−33.7; −33.54 (15:1) [b]	7.2	827; 827
21	*t*-Bu	*t*-Bu	H	Pr	C=NPh		50	84–86 [c]	−18.85	5.55	820
22	*t*-Bu	*t*-Bu	H	Pr	C=O			[a]	20.01	−26.23	785
23	Ph	Ph	Ph	$SiMe_3$	Ph	H	85	[a]	−43.5	28.76	670
24	Et_2N	Et_2N	H	$SiMe_3$	Ph	H	90	[a]	−39.18; −38.83 (3:2) [b]	−40.19; −8.84	670
25	*t*-Bu	*t*-Bu	H	$SiMe_3$	Ph	H	90	[a]	−11.0	-	768
26	Et_2N	Et_2N	H	$SiMe_3$	Pr	H	90	[a]	−39.27; −42.11 (4:1) [b]	−43.71; −42.61	768
27	Et_2N	Et_2N	H	$SiMe_3$	Bu	H	95	[a]	−39.22; −42.32 (4:1) [b]	−43.71; −42.61	768
28	Et_2N	Et_2N	H	$SiMe_3$	C_6H_{13}	H	90	[a]	−39.20; −42.10 (3:1) [b]	−43.71; −42.61	763
29	Et_2N		H	$SiMe_3$	C_8H_{17}	H	90	[a]	−39.88; 42.20 (4:1) [b]	−42.85; −42.07	763
30	Et-N		H	$SiMe_3$	CF_3	Ph	95	[a]	−38.60; −37.45 (3:1) [b]	−41.11 −49.01	777

[a] Oil; [b] Diastereomers; [c] melting point (m.p.).

2.3. Properties of Oxaphosphetanes

The cycloaddition of *P*-halogenylides to aldehydes and ketones proceeds stereoselectively and leads predominantly to the formation of one of the possible diastereomers of 2-halo-1,2λ⁵-oxaphosphetanes (Tables 1 and 2). *P*-Haloylides (chloro- and bromo) react stereoselectively with aldehydes to give predominantly single diastereoisomers of 2-halo-oxaphosphetanes, bearing asymmetric endocyclic C-3 and C-4 carbon atoms (dr ~7:1–15:1). It was established by NMR that the ratio of diastereoisomers of 2-halooxaphosphetanes containing asymmetric atoms at C(3) and C(4) was within the limits ~99:1–90:10. The ratio of diastereomers after completion of the reaction was approximately 6:4–9:1. However on heating, as a result of permutational changes in the molecule, the ratio of diastereomers grew in favour of the thermodynamically more stable diastereomer. 2-Fluoro- and 2-chloroxaphosphetanes containing chiral phosphorus or endocyclic carbon atoms exist as mixtures of diastereomers whose ratio depends on the nature of the starting reagents (Scheme 21). The diastereomeric purity of the compounds, assessed by NMR spectroscopy, was 80%–96%. The reaction of α,α,α-trifluoroacetophenone with *P,P*-difluoroylides proceeded with very high stereo-selectivity to furnish only one diastereomer of compound. At the same time 2-halo-1,2λ^5-oxa-phosphetanes bearing asymmetric phosphorus atom are formed with low stereoselectivity (dr~3:1).

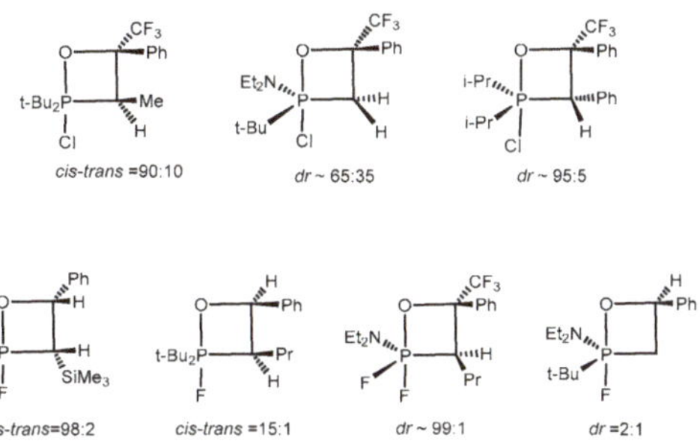

Scheme 21. Diastereomeric composition of 2-chloro- and 2-fluoro-1,2λ^5-oxaphosphetanes.

The reaction of *P*-halogenylides with aldehydes leads to the formation of an *erythro*-oxaphosphetanes with the oxygen in the apical position, because of the tendency of electron-acceptor atoms to occupy this position [40]. The preferred orientation of the transition state leading to the *erythro*-oxaphosphetane is stereoselective approach of the ylide nucleophilic center to the carbonyl group at an angle of 107°, with the double bonds of both reagents arranged in one plane. This was associated with the Burgi-Dunitz trajectory concept (Scheme 22) [41]. The first step of reaction stereoselectively leads to the formation of the betaine **B**. The conversion of reagents into oxaphosphetane requires minimum energy when it proceeds with the non-synchronous formation of bonds between the carbonyl and ylidic carbon atoms. On the second step the betaine converts into oxaphosphetane **C** with low stereoselectivity, because of free rotation around of the P-C bond [28,42]. This mechanism of asynchronous addition of *P*-haloylides to carbonyl compounds was confirmed by low temperature NMR investigations. The reaction of di-*tert*-butylchlorphosphonium methylide with $Ph(CF_3)C{=}O$ was studied by ^{31}P-NMR at low temperature (−70–0 °C) in diethyl ether solution. At −70 °C was found only a decreasing signal of *P*-chloroylide (+114 ppm) and a growing signal of 2-chloro-1,2λ^5-oxaphosphetane (+9.9 ppm). The signal which could be referred to betaine was not found, probably, because of the high speed of betaine cyclisation into 2-chloroxaphosphetane. However the reaction of *tert*-butyl(diethylamino)chlorophosphonium methylide with benzaldehyde began at −70 °C and led to the formation of *erythro*-betaine which was registered by increasing signal δ_P +106 ppm, located in more weaker field than the signal of initial *P*-chloroylide (+102 ppm) (Figure 3). The betaine formation was proceeded stereoselectively as in the ^{31}P-(^{1}H-)NMR spectrum only one diastereomer signal was registered. The rise of temperature to −50 °C led to the formation of two signals of *threo* and *erythro*-oxaphosphetane diastereomers δ_P + 92.6 and +84.6 ppm in the ratio of 2:1. Evidently, the formation of diastereomers proceeded via S_N2@P substitution with inversion of configuration at the phosphorus atom [43]. The formation of isomers indicates on the nucleophilic substitution at asymmetric tetracoordinate phosphorus atom, proceeding with partial inversion of configuration. At −50–−40 °C the oxaphosphetane was converted into 2-chloralkylphosphine oxide, which was also formed as two *threo*- and *erythro*-diastereomers in the ratio of 2:1 (δ_P +48.7 and +49.2 ppm). The conversion was completed at −20 °C. The *threo*- and *erythro* diastereomers were separated and isolated in pure state by chromatography and crystallization (m.p. 100 and 123 °C, see Table 3, entries 13, 14). Although at present many of the puzzling features about the reaction mechanism of ylides phosphorus with carbonyl compounds have been clarified, it seems there is no general mechanism, which could explain the progress of the reaction, transition states and stereochemistry. Nevertheless in case of *P*-chloroylide addition to aldehydes, on the basis of presented above stereochemical researches the asynchronous [2+2]-cycloaddition seems to be the most likely mechanism.

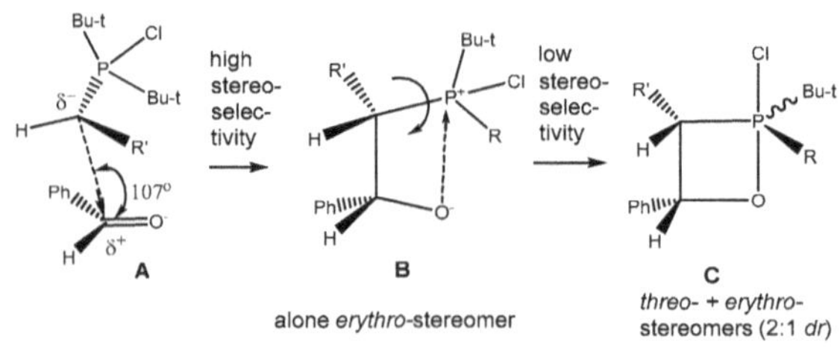

Scheme 22. Mechanism of asynchronous [2+2]-cycloaddition of *P*-chloroylide to benzaldehyde.

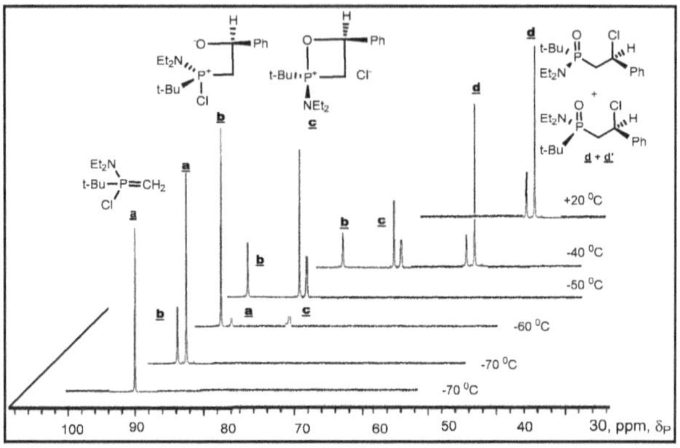

Figure 3. ^{31}P-(^{1}H-)NMR monitoring of the reaction of ylide <u>a</u> with PhCHO leading via the formation of betaine <u>b</u>, and oxaphosphetane <u>c</u> to the formation of β-chloroalkylphosphonates <u>d</u> and <u>d′</u>.

2-Halo-oxaphosphetanes exist in pentacoordinate form, but ionize under certain conditions with formation of tetracoordinate forms. In the mass spectra of 2-chlorooxaphosphetanes the peak of molecular ion was observed that indicated the covalent character of the P–Cl bond. All oxaphosphetanes show methyl carbons with $^{1}J_{PC}$ values characteristic of equatorial placement in the trigonal bipyramid. The resonances of the methylene carbons of the phosphetane ring system lay downfield of the methyl carbons and the large measured $^{1}J_{PC}$ values indicate equatorial placement. The four-membered ring is occupying an apical-equatorial plane where the O-P-C angle is 90°, and never a diequatorial plane or a diapical plane where the O-P-C angle would have to be 120 or 180°, respectively.

In the ^{1}H-NMR spectra of the 2-halooxaphosphetanes possessing asymmetric carbon atom C^4 the magnetic nonequivalence of protons CH^aH^b in the four-membered ring became apparent because the geminal spin-spin interaction arose between them. Each of the signals CH^a and CH^b was the double doublet with the coupling constant with phosphorus nucleus $^{2}J_{PH}$ = 20–22 Hz and a constant of geminal interaction $^{2}J_{HH}$ = 16–17 Hz. The chemical shifts ^{31}P of 2-halogenoxaphosphetanes depended on the polarity of the solvent. Thus, in nonpolar solvents (diethyl ether, pentane) the signals of δ_P are located in the strong field of the ^{31}P-NMR spectrum (from −3 to 10 ppm) that corresponds to the pentacoordinated state of the phosphorus atom. In polar solvents and especially in the presence of Lewis acids, like $AlCl_3$, the values of δ_P were shifted downfield. For example, for 2-chloro-oxaphosphetanes the chemical shift of phosphorus, δ_P was as follows (ppm): 9 (pentane), 22 (CH_2C1_2), 30 ($CHC1_3$), 35 (CH_3CN), 45 ($CHC1_3$ + $A1C1_3$ (traces)). With the increase in amount of aluminum chloride up to equimolar level the value of δ_P 100 ppm was registered for 2-chlorooxaphosphetanes that was in accordance with the values of chemical shifts for the known alkoxyphosphonium salts and apparently indicated complete ionization of chlorophosphorane with the formation of phosphonium salt. The dependence of the chemical shift value of phosphorus in 2-chlorooxaphosphetanes on the solvent polarity and on the presence of Lewis acid is in accordance with the published data showing that chlorophosphoranes can be ionized with the formation of phosphonium structures. Herewith in the ^{31}P-NMR spectrum the resultant signal is registered due to fast exchange in the phosphorus coordination **15A**$\rightleftharpoons$**15B**. The value of δ_P is shifted downfield proportionally to the increase in the phosphonium structure content, which in its turn depends on the solvent polarity. Substitution of electron withdrawing CF_3 group at C4 atom by hydrogen atom destabilizes the oxaphosphetane cycle and considerably reinforces the ionization of P-C1 bond. The stability of 2-halogenoxaphosphetanes decreases, and positive values δ_P and ionization of P-C1 bond increases in the sequence of substituents at C4 atom: CF_3 > C_6H_4F-4 >-C_6N_5 > C_6H_4OMe > H (Table 4) [25,30,31]. Chemical shifts of phosphorus in 2-bromo-1,2λ^5-oxaphosphetanes (δ_P = +20–+75 ppm) were in the weaker fields relatively to chemical shifts of chloro-oxaphosphetanes. Evidently high positive values δ_P of 2-bromoxaphosphetanes is undoubtedly explained by bigger, than in case of 2-chlorooxaphosphetanes, contribution of phosphonium forms **15B** in the equilibrium 15**A**$\rightleftharpoons$15**B**.

Table 3. 2-Chloroalkylphosphine oxides **35** (Scheme 23) [36,42].

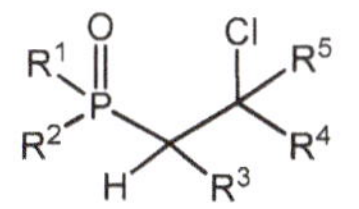

N	R¹	R²	R³′	R⁴	R⁵	Yield, %	m.p./b.p. °C	δ_P
1	*t*-Bu	*t*-Bu	H	H	CF_3	60	62 (pentane)	57.8
2	*t*-Bu	*t*-Bu	H	Me	CF_3	60	86 (hexane)	57.0
3	*t*-Bu	*t*-Bu	H	Pr	CF_3	60	88 (hexane)	57.0
4	*t*-Bu	*t*-Bu	H	Ph	CF_3	75	112.5 (benzene)	57.2
5	*t*-Bu	*t*-Bu	H	$C_6H_4NO_2$	CF_3	70	139 (hexane)	59
6	*t*-Bu	*t*-Bu	Me	$C_6H_4NO_2$	CF_3	70	131 (hexane)	60.1
7	*t*-Bu	*t*-Bu	Pr	$C_6H_4NO_2$	CF_3	60	143.5 (hexane)	60.0
8	Me_2N	Me_2N	H	Pr	CF_3	60	b.p. 95 (0.06 mmHg)	45.8
9	Et_2N	Et_2N	H	Ph	CF_3	40	123 (ether-pentane)	46.0
10	Et_2N	Et_2N	H	Ph	CF_3	20	160 (hexane)	46.3
11	*i*-Pr_2N	*i*-Pr_2N	Me	Ph	H	50	107–111	34.4
12	*t*-Bu	Ph	H	Ph	H	75	oil	57.9; 58.3 (10:1)
13	*t*-Bu	Et_2N	H	H	Ph	20	100 (hexane)	46.3
14	*t*-Bu	Et_2N	H	Ph	H	40	123 (ether-pentane)	46.0

Table 4. Effect of solvent on δ_P of 2-chloro and 2-bromo-1,2λ^5-oxaphosphetanes.

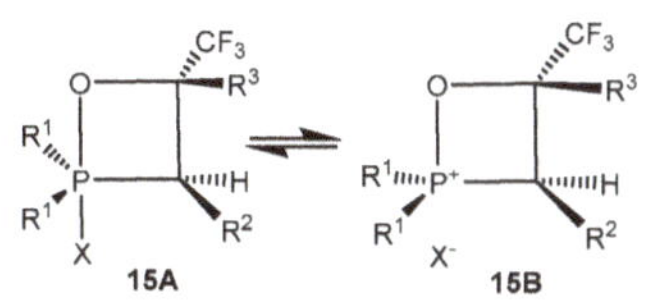

R³	CF_3	C_6H_4F	C_6H_5	C_6H_4OMe
δ_P ($CHCl_3$), Hlg = C1	+0.25	+9.0	+10.0	+13.6
δ_P (ether), Hlg = Br	+25.0	+60.0	+65.0	+75.0

2-Chloro-1,2λ^5-oxaphosphetanes enter into a number of interesting chemical conversions. Thus, 2-chloro-1,2λ^5-oxaphosphetanes as a result of [1,3]-migration of chlorine atom to carbon atom underwent 2-chlorooxaphosphetane-2-chlorooxyphosphine oxide rearrangement to convert into the 2-chlorooxyphosphine oxides **35**. Thermal stability of 2-chloro-1,2λ^5-oxaphosphetanes was decreased in case of oxaphosphetanes which not contain at C4 atom strong acceptor substituents, which rearranged into chloroalkylphosphonates at room temperature (Table 3). At heating 2-chloralkylphosphine oxides yield vinylphosphine oxides **38**. Hydrolysis of 2-chloro- and bromo-oxaphosphetanes led to the formation of 2-hydroxyalkylphosphine oxides **36** (Table 5). The chlorine atom of 2-chloro-1,2λ^5-oxaphosphetanes is easily substituted on methoxy- or phenoxy groups by reaction with methanol or phenol in the presence of triethylamine with formation of 2-alkoxyoxaphosphetanes **37**, which at heating were converted into alkenes (Scheme 23, Table 1, entries 14, 15) [27,42].

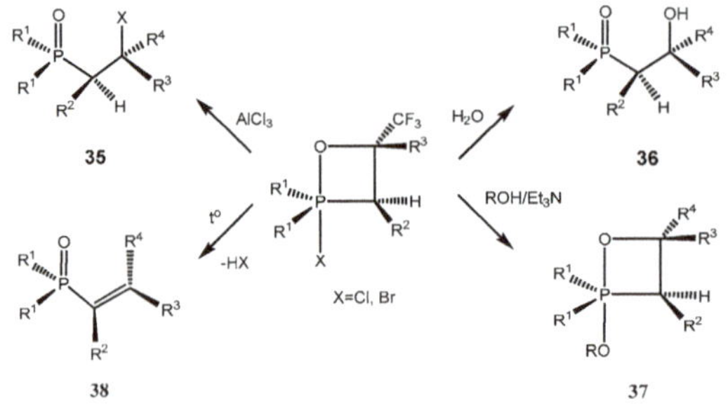

Scheme 23. Chemical reactions of 2-chloro and 2-bromo-1,2λ^5-oxaphosphetanes.

The 2-chloro-1,2λ^5-oxaphosphetanes not containing electron accepting groups at C-3 are unstable and rearrange easily into 2-cloroalkylphosphonates at temperature below +20 °C. The oxaphophetanes bearing at C4 electronegative CF_3 groups convert into vinylphosphonates with elimination of hydrogen chloride at heading up to 150–160 °C. 2-Bromooxaphosphetanes are less stable than 2-chlorooxaphosphetanes and are converted into vinylphosphine oxides at room temperature. At heating (160–190 °C) 2-chloralkylphosphine oxides eliminate HCl to convert into vinylphosphine oxides (Scheme 24) [27–29,42].

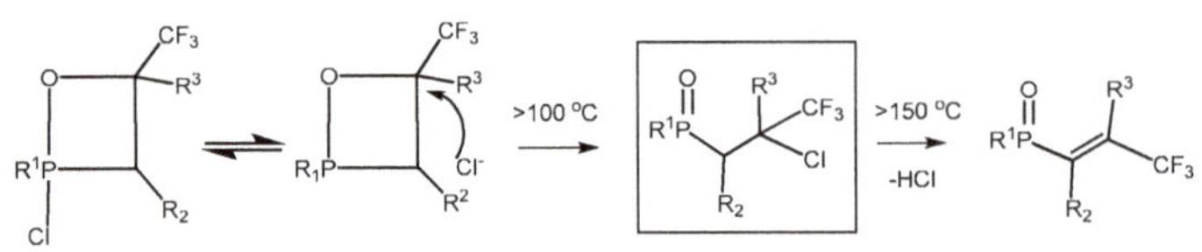

Scheme 24. 2-Chloro-1,2λ^5-oxaphosphetane-2-chloroalkylphosphonate rearrangement.

Table 5. 2-Hydroxyalkylphosphine oxides **23** (Scheme 23) [39,42].

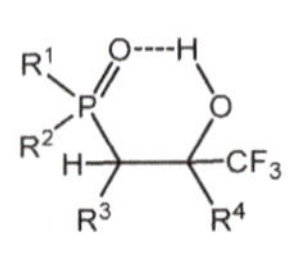

Entry	R^1	R^2	R^3	R^4	Yield, %	m.p., °C	δ_P	δ_F
1	*i*-Pr	*i*-Pr	Ph	Ph	90	131–132	+66.0	+5.63
2	*t*-Bu	*t*-Bu	H	Ph	90	155–156	+68.25	−1.78
3	*t*-Bu	*t*-Bu	H	C_6H_4F-*p*	85	132	+68.7	−2.02; −34.5
4	*t*-Bu	*t*-Bu	H	An-*p*	70	140–141	+68.5	+4.12
5	*t*-Bu	*t*-Bu	Me	Ph	50	152	+73.4	+3.13
6	*t*-Bu	Et_2N	Me	CF_3	50	108	+53.4	+8.1
7	*t*-Bu	Et_2N	H	Ph	65	150	+57.4	−3.06
8	*t*-Bu	Et_2N	H	CF_3	90	46–48	+56.6	+6.1
9	*t*-Bu	*t*-Bu	H	CF_3	90	88-89	+67.5	+6.25

The oxaphosphetanes bearing a trimethylsilyl group at C-3 atom eliminate the trimethylchlorosilane moiety and convert into the phosphorylated alkenes **38**. The conversion of oxaphosphetanes into alkenephosphonates proceeds at room temperature slowly, and at heating faster to give the alkenes **38** in good yields (Scheme 25). The phosphorylated alkenes **38** were purified by distillation under vacuum and isolated as pure compounds. The reaction of ylides **14** with aldehydes is regioselective. For example, the oxaphosphetanes obtained by reaction of silylated *P*-chloroylide **5** with terephthalic aldehyde depending on a ratio of initial reactants led to the formation of 1,4-bis-vinylphosphonobenzene or phosphonovinylbenzaldehyde, that represent interest as reactants for organic synthesis (See Table 6, entries 10, 11) [13,18–20].

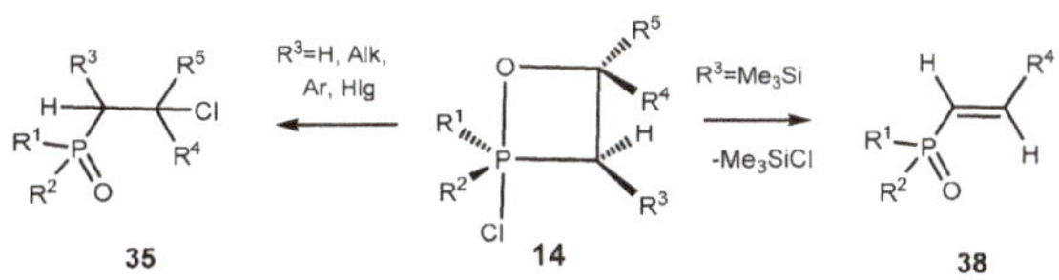

Scheme 25. Conversions of 2-chloro-1,2λ^5-oxaphosphetanes **14**.

The 2-fluoro-1,2λ^5-oxaphosphetanes enter readily into a number of interesting organophosphorus compounds proceeding without P—C bond cleavage [26]. Thus, the treatment of 2-fluorooxaphosphetanes **25** with ether solution of HCl led to the formation of 2-chloro-oxaphosphetanes **40**, which was isolated in good yield. At heating the 2-fluorooxaphosphetanes **25**, in contrast to triphenyloxaphosphetanes, convert into phosphorylated alkenes: vinylphosphonates or allylphosphonates. The direction of reaction depended on substituents R^2 and R^3 at C-3 and C-4 of oxaphosphetane cycle. The 2-fluorooxaphosphetanes bearing at C-3 R^2 = H, Alk, Ar, and at C-4 R^3 = Alkyl eliminated HF to convert into the allylphosphonates **42** [39,44,45]. The reaction was catalyzed by boron trifluoride etherate. At the same time the 2-fluorooxaphosphetanes **25** bearing at C-3 substituent R = Me_3Si eliminated Me_3SiF and afforded the vinylphosphonates **41** (Scheme 26, Table 7). This reaction represent a convenient method for the preparation of phosphorylated alkenes that are versatile building blocks for organic synthesis [45–48]. 2-Fluoro-1,2λ^5-oxaphosphetanes containing two alkyl groups at C3 as well eliminated hydrogen fluoride and afforded the allylphosphonates **42** (Scheme 26, Table 8).

Table 6. Vinylphophonates **38** (Scheme 25).

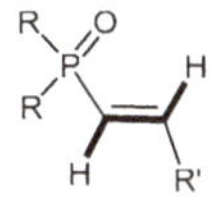

Entry	R	R′	Reaction Conditions			Yields, %	References
			Time, h	t °C	Solvent		
1	*t*-Bu	Ph	0.5	150	a	80	[25,32]
2	EtO	Ph	2	25–100	ether	80	[25,32]
3	EtO	2-BrC_6H_4	6–8	25–100	ether	60	[25,32]
4	*i*-PrO	Ph	2	25–100	ether	60	[25,32]
5	Et_2N	Ph	14	25	ether	80	[25,32]
6	Et_2N	2-FC_6H_4	24	25	ether	60	[25,32]
8	Et_2N	CH=CHPh	18	25	ether	80	[25,32]
9	Et_2N	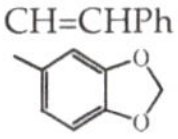	24	25	THF	70	[25,32]
9	EtO	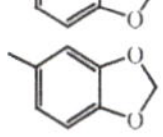	48	35	THF	65	[25,32]
10	Et_2N	C_6H_4CHO-*p*	25	18	THF	50	[25,32]
11	Et_2N	$R_2P(O)CH{=}CHC_6H_4$	25	24	THF	50	[25,32]
12	*i*-PrO	$R_2P(O)CH{=}CHC_6H_4$	25	24	THF	50	[25,32]

a without solvent.

The reaction occurs by the 1,4-elimination as shown in Scheme 26 [25]. Opposite to 2,2,2-triphenyloxaphosphetanes, upon heating, the decyclization of 2-fluorooxaphosphetanes led to the formation of phosphorylated alkenes **41**, **42**.

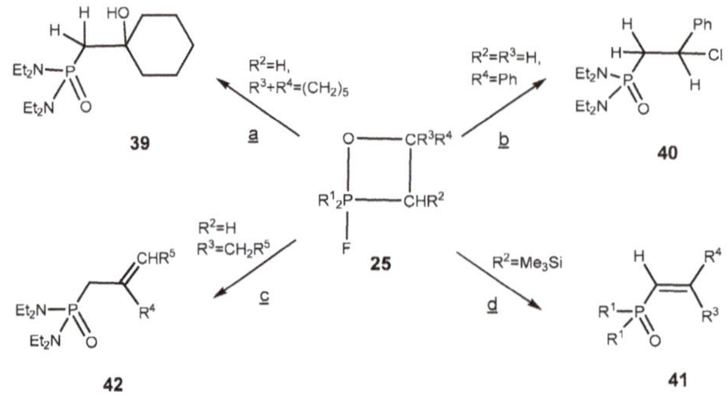

Scheme 26. Chemical properties of 2-fluoro-1,2λ^5-oxaphosphetanes. a = H_2O (-HF); b = HCl, ether (-HF); c = 100–120 °C, ~0.5 h (-HF) or $BF_3{\cdot}Et_2O$, +20 °C, 168 h; d = 60–80 °C (-Me_3SiF).

At heating to +100 °C the 2-fluoro-3-silyloxaphosphetanes **25**, bearing CF_3 group at C-4, afforded a mixture of *E*- and *Z*-vinylphosphonates **29**, **30** in the ratio 2:1 with elimination of Me_3SiF. However the slow conversion of oxaphosphetane at +20 °C during several days provided almost pure vinylphosphonates *E*-**29**, containing only 2%–3% of *Z*-isomer (Scheme 27) [26]. This effect was explained by formation of carbocation intermediate and rotation of substituents around the C-C bond (Scheme 28).

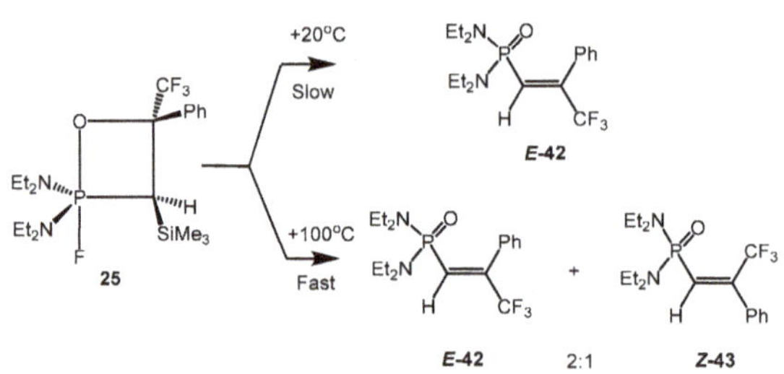

Scheme 27. Effect of temperature on the stereoselectivity of 2-fluoro-1,2λ^5-oxaphosphetane conversion into vinylphosphonates.

The conversion of 2-fluoro-1,2λ^5-oxaphosphetanes bearing alkyl groups at C-3 into allylphosphonates **42** represents an interesting example of 1,4-elimination as Scheme 28 and Table 8 show [45]. The study of the reaction mechanism showed that Lewis and Broensted acids actively catalyze the conversion of 2-fluorooxaphosphetanes into allylphosphine oxides. The reaction is autocatalytic because the evolving hydrogen fluoride catalyzes the transition of 2-fluorooxaphosphetanes into allylphosphine oxides. The decomposition of protonated 2-fluorooxaphosphetanes leads to the formation of oxonium salts and carbocation intermediates under conditions of E_N1 elimination. We suppose that the 2-fluorooxaphosphetanes under condition of acid catalysis (with HF or BF_3) via the formation of an oxonium intermediate F, convert to carbocation intermediate **G** which has a planar configuration [46]. The removal of a proton from the carbocation intermediate **G** depends on electronic effects of the substituents. Alkyl groups possessing the +*I*-effect and the effect of hyperconjugation stabilize the positive charge and reduce the energy of intermediate **G** formation. Therefore, the intermediate **G′** leading to allylphosphonates is energetically more favorable than this one that is converted into vinylphosphonates.

Table 7. Vinylphosphonates **41** (Scheme 26) [26].

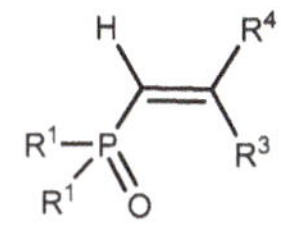

Entry	R^1	R^3	R^4	Yield, % [a]	b.p. °C (*p* mmHg)/ m.p. °C (Solvent)	δ_P, ppm
1	*t*-Bu	H	Ph	50	138 (heptane)	35.0
2	Ph	H	Ph	40	165 (heptane)	22.0
3	Et_2N	H	Ph	85	103.5 (hexane)	24.7
4	Et_2N	H	Me	80	120 (0.05)	24.7
5	Et_2N	H	Pr	72	120 (0.05)	24.3
6	Et_2N	H	Bu	68	120–123 (0.04)	23.95
7	Et_2N	H	C_6H_{13}	80	135 (0.06)	24.1
8	Et_2N	H	C_8H_{17}	79	145–150 (0.05)	24.1
9	Et_2N	H	PhCH=CH-	60	170 (0.06)	23.30
10	Et_2N	H	$Me_2C{=}CHCH_2CH_2C(Me)CH{=}C$-	35	160 (0.08)	25.0
11	Et_2N	CF_3	Ph	74	140 (0.03)	17.0; 18.7 [b]
12	Et_2N	H	$-C_6H_4CHO$-4	50	98 (hexane)	23.7
13	Et_2N	H	$-C_6H_4CH{=}CHP(O)(NEt_2)_2$	50	190 (heptane)	24.8

[a] yield of the isolated product; [b] mixture of *E/Z*-diastereomers in 97:3 ratio.

In addition, even in case when the initial 2-fluorooxaphosphetanes **25** exist as a mixture of *threo*- and *erythro*-diastereomers, they converted into pure *E*-vinylphosphonates. Probably, this effect can be explained by rotation of substituents around the C-C bond in carbocation intermediate G. However, the presence of Me_3Si at C-3 in **G″** leads to the elimination of Me_3SiF, which has a high energy of formation, that creates a preference for the formation of vinylphosphonates **45** (Scheme 28) [26,40].

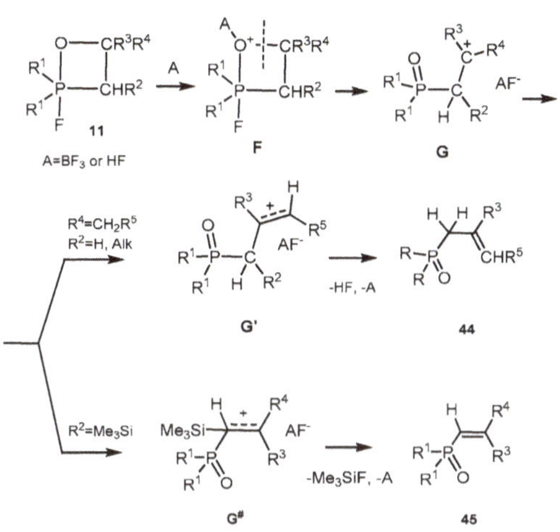

Scheme 28. The mechanism for the 2-fluoro-1,2λ^5-oxaphosphetane conversion into allyl- or vinylphosphonates **44**, **45**.

The formation of carbocation intermediate **G** was experimentally confirmed (Scheme 29). The treatment of 2-hydroxyphosphonate **46** with trifluoracetic acid and refluxing for several hours, generated the carbocation intermediate, as a result of acid-catalyzed dehydration of alcohol.

After that the carbocation intermediate is converted into allylphosphonate **47**, which is identical to the one obtained from 2-fluorooxaphosphetane **25** [26,40,45].

Table 8. Allylphosphonates **28**.

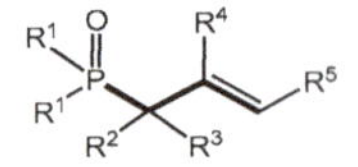

Entry	R^1	R^2	R^3	R^4	R^5	Yield, %	b.p. °C (*P*, mmHg)	References
1	Et_2N	H	H	H	Pr	70	120 (0.06)	[40,45]
2	Et_2N	H	H	H	C_5H_{11}	70	140 (0.08)	[40,45]
3	Et_2N	H	H	Me	H	55	105 (0.08)	[40,45]
4	Et_2N	H	H	Me	Me	60	95 (0.06)	[40,43]
5	Et_2N	H	H	Ph	H	65	145 (0.08)	[40]
6	Et_2N	H	H	$(CH_2)_4$		85	145 (0.08)	[26,45]
7	Et_2N	H	Me	$(CH_2)_4$		75	110 (0.06)	[40,45]
8	Et_2N	Me	Me	$(CH_2)_4$		70	110 (0.06)	[40,45]

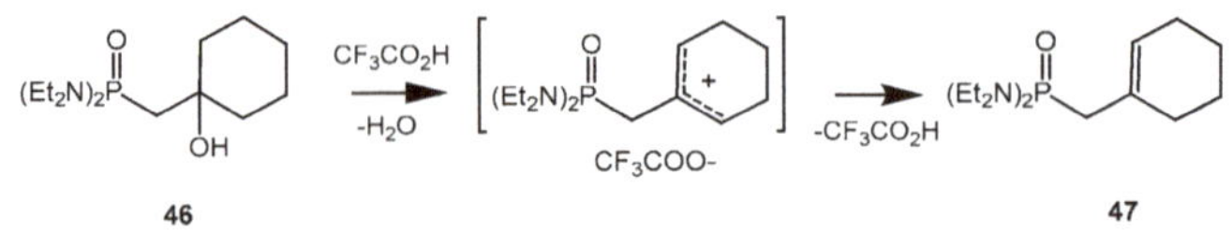

Scheme 29. The conversion of hydroxyphosphonate **46** into the allylphosphonate **47**.

The reaction of *P*-fluoroylides with carbonyl compounds is a convenient method for the synthesis of allylphosphonates having various applications in the synthesis of naturally occurring compounds (Scheme 30) [45–47]. In this case, the carbonyl compounds can be used twice for the constructing of diene structures: first in reaction with *P*-fluoroylide and then in the Wittig reaction with allylphosphonate. This reaction was used for preparing analogs of the juvenile hormone.

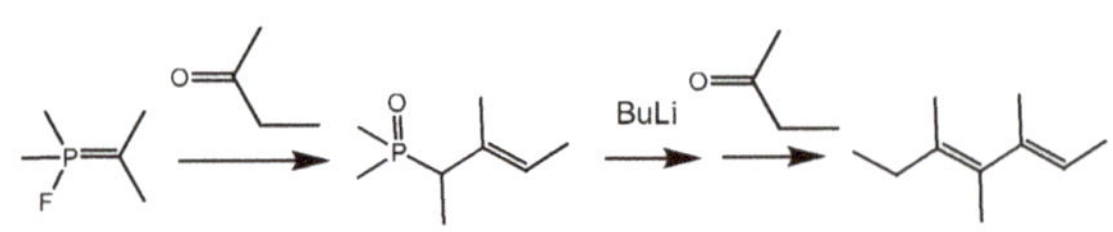

Scheme 30. Synthetic application of allylphosphonates.

The cycloadducts of *P*-fluoroylides with carbon dioxide or with carbon disulfide were isolated as colorless liquids or crystalline substances. Their structure was confirmed by the NMR spectra. Upon gentle heating or at room temperature these cycloadducts **32–35** (Table 2, entries 21, 22) are converted to phosphorylated ketenes or thioketenes **48** (Scheme 31). Stable [2+2]-cycloadducts of 2-fluoroylides with alkyl and aryl isothiocyanates were also synthesized and converted into phosphorylated ketenimines. The reaction of *P*-fluoroylides with carbon disulfide gives unstable cycloadducts that at temperatures above 0 °C converted completely into phosphorylated thioketenes in high yields. Phosphorylated thioketenes are red liquids distillable in vacuum and susceptible for various transformations [29,36,37]. The same phosphorylated ketenes or thioketenes **29** were prepared by reaction of *P*-cloroylides corresponding with CO_2 and CS_2 (See Scheme 31 and Table 9) [37–39].

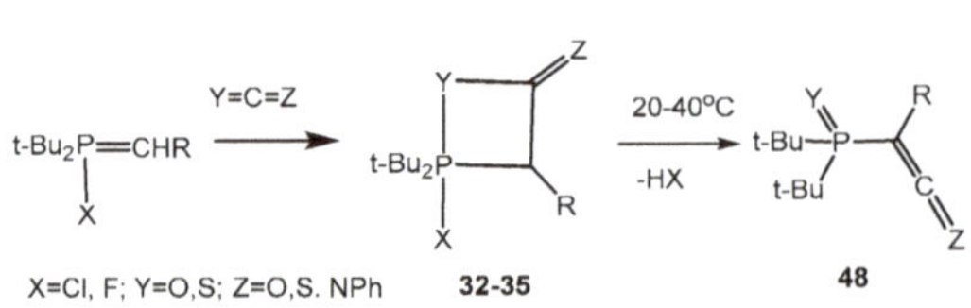

Scheme 31. Preparation of phosphorylated ketenes, thioketenes, ketenimines **48** (Table 9).

Table 9. Preparation of phosphorylated ketenes, thioketenes, ketenimines **48** (Scheme 31).

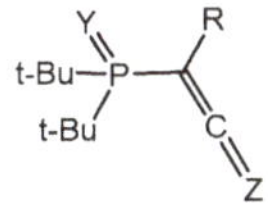

Entry	R	X	Y	Z	Yield, %	b.p. °C (mmHg)	References
1	H	Cl	O	O	80	a	[36,39]
2	Me	F	O	O	70	98 (0.06)	[37]
3	Me	Cl	O	O	85	98 (0.06)	[36,39]
4	Pr	F	O	O	60	102 (0.06)	[37]
5	Pr	Cl	O	O	69	102 (0.06)	[36,39]
6	Me	F	O	NPh	45	150 (0.06)	[37]
7	Pr	F	O	NPh	45	150 (0.06)	[37]
8	Me	Cl	S	S	75	95 (0.06)	[38]
9	Pr	Cl	S	S	80	115 (0.06)	[38]
10	Pr	F	S	S	70	115–120 (0.06)	[37,38]
11	*i*-Pr	Cl	S	S	80	105 (0.06)	[38]

a unstable compound.

3. Conclusions

In conclusion, this review has summarized the achievements in the synthesis and properties of stable four-membered phosphorus heterocycles-2-chloro-, 2-bromo- and 2-fluoro-1,2λ^5-oxaphosphetanes. These interesting compounds were obtained by reaction of *P*-halogenylides with various carbonyl compounds (aldehydes, ketones, isocyanates, carbon dioxide, and others). The 2-chloro and 2-bromo-1,2λ^5-oxaphosphetanes, depending on the halogen nature and substituents at the α-carbon atom, underwent the rearrangement into 2-haloalkylphosphonates or with elimination of hydrogen halides were converted into *trans*-phosphorylated alkenes. Hydrolysis of 2-halo-1,2λ^5-oxaphosphetanes led to the formation of 2-hydroxyalkylphosphonates. The 2-fluoro-oxaphosphetanes, bearing alkyl groups at the C-4 atom at heating as a result of 1,4-elimination of hydrogen fluoride are turned to allylphosphonates. Upon heating 3-silyl-2-fluoro-1,2λ^5-oxaphosphetanes easily eliminate the trimethylsilylfluoride to convert into *E*-vinylphosphonates in high yields. These reactions provide efficient protocols for the preparation of various phosphorylated alkenes (vinylphosphonates, allylphosphonates, phosphorus ketenes, ketenimines, thioketenes) and can be useful for fine organic synthesis.

Acknowledgments: This work was supported by grants of the State Foundation for Basic Research of Ukraine (Projects No 03.07/00047, No. F40.3/034 and No. F53.3/016), grant of the Science and Technology Center in Ukraine, STCU (Project No. 3558).), the research projects of the National Academy of Sciences of Ukraine (No. 0109U001795 and No. 0106U004307).

Author Contributions: O.I.K and A.O.K. conceived and designed the review. O.I.K. wrote the text of review. A.O.K. have prepared reaction schemes and did literature research on relevant articles for this review. Both authors discussed and approved the final version.

Conflicts of Interest: The authors declare no conflict of interest.

References

1. López-Ortiz, F.; López, J.G.; Álvarez Manzaneda, R.; Pérez Álvarez, I.J. Isolable 1,2-Oxaphosphetanes: From curiosities to starting materials for the synthesis of olefins. *Mini-Rev. Org. Chem.* **2004**, *1*, 65–76. [CrossRef]
2. Kawashima, T.; Kato, K.; Okazaki, R. Novel Synthetic Route to Isolable Pentacoordinate 1,2-Oxaphosphetanes and Mechanism of Their Thermolysis, the Second Step of the Wittig Reaction. *J. Am. Chem. Soc.* **1992**, *114*, 4008–2010. [CrossRef]
3. Vedejs, E. Studies in Heteroelement-Based Synthesis. *J. Org. Chem.* **2004**, *69*, 5159–5167. [CrossRef] [PubMed]

4. Vedejs, E.; Marth, C.F. *Phosphorus-31 NMR Spectral Properties in Compound Characterization and Structural Analysis*; Quin, L.D., Verkade, J.G., Eds.; VCH: New York, NY, USA, 1994; Chapter 23; pp. 297–313.
5. Hamaguchi, M.; Iyama, Y.; Mochizuki, E.; Oshima, T. First isolation and characterization of 1,2-oxaphosphetanes with three phenyl groups at the phosphorus atom in typical Wittig reaction using cyclopropylidenetriphenylphosphorane. *Tetrahedron Lett.* **2005**, *46*, 8949–8952. [CrossRef]
6. Robiette, R.; Richardson, J.; Aggarwal, V.K.; Harvey, J.N. Reactivity and selectivity in the Wittig reaction: A computational study. *J. Am. Chem. Soc.* **2006**, *128*, 2394–2409. [CrossRef] [PubMed]
7. Kobayashi, J.; Kawashima, T. Chemistry of pentacoordinated anti-apicophilic phosphorus compounds. *C. R. Chim.* **2010**, *13*, 1249–1259. [CrossRef]
8. Garcia-López, J.; Morán-Ramallal, A.; Gonzalez, J.; Roces, L.; García-Granda, S.; Iglesias, M.J.; Ona-Burgos, P.; López-Ortiz, F. Mechanisms of stereomutation and thermolysis of spiro-1,2-oxaphosphetanes: New insights into the second step of the Wittig reaction. *J. Am. Chem. Soc.* **2012**, *134*, 19504–19507. [CrossRef] [PubMed]
9. Byrne, P.A.; Muldoon, J.; Ortin, Y.; Müller-Bunz, H.; Gilheany, D.G. Investigations on the Operation of Stereochemical Drift in the Wittig Reaction by NMR and Variable-Temperature NMR Spectroscopy of Oxaphosphetane Intermediates and Their Quench Products. *Eur. J. Org. Chem.* **2014**, *2014*, 86–98. [CrossRef]
10. Byrne, P.A.; Gilheany, D.G. The Mechanism of phosphonium ylide alcoholysis and hydrolysis: Concerted addition of the OH bond across the P=C bond. *Chem. Eur. J.* **2016**, *22*, 9140–9154. [CrossRef] [PubMed]
11. Byrne, P.A.; Gilheany, D.G. Unequivocal experimental evidence for a unified lithium salt-free Wittig reaction mechanism for all phosphonium ylide types: Reactions with β-heteroatom-substituted aldehydes are consistently selective for *cis*-oxaphosphetane-derived products. *J. Am. Chem. Soc.* **2012**, *134*, 9225–9239. [CrossRef] [PubMed]
12. Byrne, P.A.; Gilheany, D.G. The modern interpretation of the Wittig reaction mechanism. *Chem. Soc. Rev.* **2013**, *42*, 6670–6696. [CrossRef] [PubMed]
13. Restrepo-Cossio, A.A.; Cano, H.; Mari, F.; Gonzalez, C.A. Theoretical study of the mechanism of the Wittig reaction: Ab initio and MNDO-PM3 treatment of the reaction of unstabilized, semistabilized and stabilized ylides with acetaldehyde. *Heteroatom Chem.* **1997**, *8*, 557–569. [CrossRef]
14. Birum, G.H.; Matthews, C.N. Reactions of Triphenyl-2,2-bis(trifluoromethyl)vinylide-nephosphorane, synthesized from a cyclic ylide-ketone adduct. *J. Org. Chem.* **1967**, *32*, 3554–3559. [CrossRef]
15. Borkenhagen, F.; Neda, I.; Thönnessen, H.; Jones, P.G.; Schmutzler, R. An unusual rearrangement during the oxidative addition of hexafluoroacetone and trifluoroacetophenone to 2-bornanylen (dimethylphosphino)methyl imine-formation of a P=C double-bond. *Z. Anorg. Allg. Chem.* **1998**, *624*, 650–654. [CrossRef]
16. Dieckbreder, U.; Röschenthaler, G.-V.; Kolomeitsev, A.A. P-Bis(trifluoromethyl) ylides: Synthesis and reactions. *Heteroatom Chem.* **2002**, *13*, 650–653. [CrossRef]
17. Kojima, S.; Supino, M.; Matsukawa, S.; Nakamoto, M.; Akiba, K.-Y. First isolation and scheme 4 characterization of an anti-apicophilic spirophosphorane bearing an oxaphosphetane ring: A model for the possible reactive intermediate in the Wittig reaction. *J. Am. Chem. Soc.* **2002**, *124*, 7674–7675. [CrossRef] [PubMed]
18. Kyri, A.W.; Nesterov, V.; Schnakenburg, G.; Streubel, R. Synthesis and Reaction of the First 1,2-Oxaphosphetane Complexes. *Angew. Chem. Int. Ed.* **2014**, *53*, 10809–10812. [CrossRef] [PubMed]
19. Couzijn, E.P.A.; Slootweg, J.C.; Ehlers, A.W.; Lammertsma, K. Stereomutation of pentavalent compounds: Validating the berry pseudorotation. *J. Am. Chem. Soc.* **2010**, *132*, 18127–18140. [CrossRef] [PubMed]
20. Kollar, L.; Berente, Z.; Forintos, H.; Keglevich, G. Detection of the enantiomers of P-stereogenic pentacoordinated phosphorus compounds: ^{31}P-NMR of oxaphosphetes in optically active solvents. *Tetrahedron Asymmetry* **2000**, *11*, 4433–4444. [CrossRef]
21. Appel, M.; Blaurock, S.; Berger, S. A Wittig reaction with 2-furyl substituents at the phosphorus atom: Improved (Z) selectivity and isolation of a stable oxaphosphetane intermediate. *Eur. J. Org. Chem.* **2002**, *2002*, 1143–1148. [CrossRef]
22. Kumara Swamy, K.C.; Satish Kumar, N. New Features in Pentacoordinate Phosphorus Chemistry. *Acc. Chem. Res.* **2006**, *39*, 324–333. [CrossRef] [PubMed]
23. Ugi, I.; Marquarding, D.; Klusacek, H.; Gillespie, P.; Ramirez, F. Berry pseudorotation and turnstile rotation. *Acc. Chem. Res.*, **1971**, *4*, 288–296. [CrossRef]

24. Kolodiazhnyi, O.I.; Schmutzler, R. Synthesis and properties of phosphorus ylides containing fluorine atoms bonded to phosphorus. *Synlett* **2001**, *2001*, 1065–1078. [CrossRef]
25. Kolodiazhna, O.O.; Kolodiazhnyi, O.I. Reaction of *C*-silyl-*P*-chloroalkylidene-phosphoranes with carbonyl compounds. *Phosphorus Sulfur Silicon Relat. Elem.* **2015**, *190*, 322–328.
26. Kolodiazhna, A.O.; Kolodiazhnyi, O.I. Synthesis and properties of four-membered phosphorus heterocycles-2-fluoro-1,2λ^5-oxaphosphetanes. *Phosphorus Sulfur Silicon Relat. Elem.* **2015**, *190*, 2232–2245. [CrossRef]
27. Kolodyazhna, O.O.; Grishkun, E.V.; Sheiko, S.Yu.; Kolodyazhna, A.O.; Kolodyazhnyi, O.I. Synthesis and properties of tert-butylphenylmethylene(chloro)phosphorane. *Russ. J. Gen. Chem.* **2015**, *85*, 1639–1643. [CrossRef]
28. Kolodiazhnyi, O.I. *Phosphorus Ylides. Chemistry and Application in Organic Synthesis*; Wiley-VCH: Weinheim, Germany, 1999; p. 565.
29. Kolodyazhnyi, O.I. Chemistry of P-F ylides. *Russ. J. Gen. Chem.* **2005**, *75*, 1017–1039. [CrossRef]
30. Kolodyazhnyi, O.I.; Kolodyazhnaya, A.O. A new approach towards synthesis of phosphorylated alkenes. *Russ. J. Gen. Chem.* **2015**, *85*, 359–365. [CrossRef]
31. Kolodiazhnyi, O.I. *P*-Halogen-substituted phosphorus ylids. *Russ. Chem. Rev.* **1991**, *60*, 391–409. [CrossRef]
32. Kolodiazhna, A.O.; Kolodiazhnyi, O.I.; Kolodiazhna, O.O. Stereoselective Method for the Synthesis of Phosphorylated Alkenes. U.S. Patent 111025, 10 March 2016.
33. Sotiropoulos, J.-M.; Baceiredo, A.; Bertrand, G. Synthesis and reactivity of diazomethylenephosphoranes ($>P=C=N_2$): New phosphacumulene ylides and first stable pseudounsaturated diazo derivatives. *J. Am. Chem. Soc.* **1987**, *109*, 4711–4712. [CrossRef]
34. Kolodiazhnyi, O.I.; Ustenko, S.N. Synthesis and properties of *P*,*P*-Difluoroylids. *Phosphorus Sulfur Silicon Relat. Elem.* **1991**, *62*, 111–118. [CrossRef]
35. Ustenko, S.N. Synthesis and Properties of New Types of Phosphorus *P*-halogenylides. Ph.D. Thesis, Institute of Organic Chemistry, National Academy of Sciences of Ukraine, Kiev, Ukraine, 12 October 1995.
36. Golokhov, D.B. *P*-Halogenoylides-Synthesis and Properties. Ph.D. Thesis, Institute of Organic Chemistry, Academy of Sciences of Ukraine, Kiev, Ukraine, 20 September 1990.
37. Kolodyazhnyi, O.I. 2-Halogen-1,2λ^5-oxaphosphetanes. Reaction of *P*-Fluoroylids with carbonyl compounds and carbon disulfide. *Russ. J. Gen. Chem.* **1986**, *56*, 821–827.
38. Kolodiazhnyi, O.I. Thiocetenes phosphores. *Tetrahedron Lett.* **1987**, *28*, 881–884. [CrossRef]
39. Kolodiazhnyi, O.I. Reaction de Wittig "anormale". *Tetrahedron Lett.* **1985**, *26*, 439–442. [CrossRef]
40. Kolodiazhnyi, O.I.; Golokhov, D.B. 2-Halogen-1,2λ^5-oxaphosphetanes. III. About conversion of 2-fluorine-1,2λ^5-oxaphosphetanes to alkenes. *J. Gen. Chem.* **1989**, *59*, 293–306.
41. Burgi, H.B.; Dunitz, I.D.; Lehn, J.M.; Wipff, G. Stereochemistry of reaction paths at carbonyl centres. *Tetrahedron* **1974**, *30*, 1563–1572. [CrossRef]
42. Kolodiazhnyi, O.I. 2-Halogen-1,2λ^5-oxaphosphetanes. *Russ. J. Gen. Chem.* **1986**, *56*, 283–298.
43. Van Bochove, M.A.; Swart, M.; Bickelhaupt, F.M. Nucleophilic substitution at phosphorus centers ($S_N2@P$). *Chem. Phys. Chem.* **2007**, *8*, 2452–2463. [CrossRef] [PubMed]
44. Kolodiazhnyi, O.I.; Kovalenko, A.B. Reaction of P-Halogenylides with carbonyl compounds. *Synthesis of allylphosphonates. Russ. J. Gen. Chem.* **1987**, *57*, 2147–2149.
45. Kolodiazhnyi, O.I. Transformation des 2-fluoro-1,2λ^5-oxaphosphetanes en allylphosphonates. *Tetrahedron Lett.* **1988**, *29*, 3663–3666. [CrossRef]
46. Mayr, H.; Ammer, J.; Baidya, M.; Maji, B.; Nigst, T.A.; Ofial, A.R.; Singer, T. Scales of Lewis basicities toward C-centered Lewis acids (Carbocations). *J. Am. Chem. Soc.* **2015**, *137*, 2580–2599. [CrossRef] [PubMed]
47. Corbridge, D.E.C. *Phosphorus: Chemistry, Biochemistry and Technology*, 6th ed.; CRC Press: Boca Raton, FL, USA, 2013; p. 1473.
48. Robert, A.S., Jr. *Practical Functional Group Synthesis*; John Wiley & Sons, Inc.: Hoboken, NJ, USA, 2016; p. 685.

Sample Availability: Samples of the compounds are not available from the authors.

molecules

MDPI

Review

Synthetic Procedures Leading towards Aminobisphosphonates

Ewa Chmielewska * and Paweł Kafarski

Department of Bioorganic Chemistry, Faculty of Chemistry, Wrocław University of Science and Technology, Wrocław 50-370, Poland; pawel.kafarski@pwr.edu.pl

* Correspondence: ewa.chmielewska@pwr.edu.pl; Tel.: +48-71-320-29-77; Fax: +48-71-320-24-27

Academic Editor: György Keglevich

Received: 30 September 2016; Accepted: 2 November 2016; Published: 4 November 2016

Abstract: Growing interest in the biological activity of aminobisphosphonates has stimulated the development of methods for their synthesis. Although several general procedures were previously elaborated to reach this goal, aminobisphosphonate chemistry is still developing quite substantially. Thus, innovative modifications of the existing commonly used reactions, as well as development of new procedures, are presented in this review, concentrating on recent achievements. Additionally, selected examples of aminobisphosphonate derivatization illustrate their usefulness for obtaining new diagnostic and therapeutic agents.

Keywords: bisphosphonates; bisphosphonylation procedures; Vilsmayer-Haack-like reactions; functionalization of aminoalkylbisphosphonates

1. Introduction

Bisphosphonates are a class of compounds that are currently receiving significant attention. Over 50 new papers are seen each week when searching the literature with the keyword "bisphosphonate" via Web of Science. More than 17,000 various bisphosphonate structures have been synthesized and described in the literature [1]. Most papers concern their important subclass, namely aminobisphosphonates. This strong interest results from these compounds acting as strong inhibitors of bone resorption, with several representatives of this class already commercialized as drugs of choice for the treatment of osteoporosis, skeletal complications of malignancy, Paget's disease, multiple myeloma, hypercalcemia and fibrous dysplasia [2–7] Consequently, most of the papers are devoted to various clinical aspects of the anti-resorptive effects that bisphosphonates exert towards bone tissues; however, there is also a growing interest in their applications as anticancer and antibacterial agents [5–7]. Additionally, aminobisphosphonic acids have found important industrial applications, largely as inhibitors of scale formation and as corrosion inhibitors, actions which result from their ability to complex metal ions [8–10].

Thus, simple and effective procedures for their synthesis are becoming increasingly important. However, only a few general reactions leading to these compounds have been described to date and are only partially reviewed in the literature [11,12]. Novel reports are mostly concentrated on modifications and improvement of these procedures, and there are only a few papers aiming at new reactions, which results from the commonly applied procedures being simple, economical and effective.

In this paper, we comprehensively review the recent studies (supplemented by older papers if necessary) on reactions applied to synthesize the most important class of bisphosphonates—aminobisphosphonates—and discuss their scope and limitations.

2. Overview of Synthetic Procedures

There are contradictory reports of the origin of bisphosphonates [13]. Most likely, the first one was obtained in the 19th century by Nikolay Menschutkin and/or Theodor Saltzer as an impurity in reactions designed to obtain different compounds. It was further identified by Hans von Baeyer and Wilhelm Heideprim as 1-hydroxyethanebisphosphonic acid.

There are only a few general methods for the synthesis of aminobisphosphonates. However, there are many individual procedures described for their preparation [11,12]. In this review, the following general reactions are presented: (i) starting from carboxylic acids and (ii) their amides; (iii) reactions using nitriles and (iv) isonitriles as substrates; (v) syntheses based on addition of phosphites to oxophosphonates; (vi) three-component condensation of amines, trialkyl orthoformates and dialkyl phosphites; (vii) addition of amines to vinylidenebisphosphonates; and (viii) functionalization of simple bisphosphonates treated as building blocks for the preparation of more complex structures. Additionally, some specific and non-conventional procedures that have been elucidated will be presented in this review.

2.1. Synthesis from Carboxylic Acids

1-Hydroxyethylidene-1,1-bisphosphonic acids are perhaps the oldest group of bisphosphonates. They are standardly prepared by a large-scale, one-step reaction of carboxylic acids with phosphorus trichloride and phosphorous or phosphoric acids, followed by hydrolysis with water; the procedure was optimized by Kieczykowski et al. [14]. The reaction is carried out in selected solvents (phenylsulphonic acid, various phenols, chlorobenzene, diphenyl ether or ionic liquids) with sulfone and methanesulphonic acid being preferred choices [11,15–20]. Despite many theories [19,21–23], the exact mechanism of this reaction is not fully understood; however, the formation of acid chloride as a first intermediate has been undoubtedly demonstrated (Scheme 1). This intermediate may react with methanesulphonic acid (when used as a solvent), and the formed mixed anhydride is also considered a potential intermediate for the next step [24], which is an Arbuzov-like reaction of phosphorus acid or one of its several derivatives (including anhydrides of variable structure [17]) formed during the reaction. The formed derivative of ketophosphonate is a substrate for the addition reaction of trivalent P-OH species, and bisphosphonate is obtained (Scheme 1). It has also been documented that the use of phosphorous acid could be omitted if water was added to the reaction medium, and thus this compound was formed in situ.

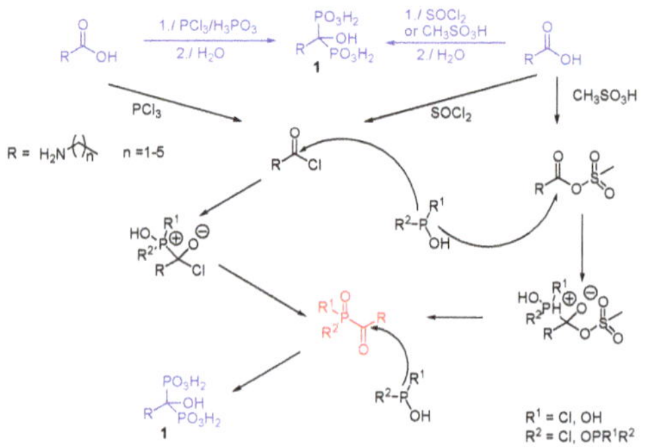

Scheme 1. Presumable mechanism of the synthesis of 1-hydroxy-1,1-bisphosphonic acids.

This reaction was commonly used for the preparation of a wide variety of anti-osteoporotic bisphosphonic acids containing free amino groups (so called dronic acids, compound **1**). In this case, free unblocked amino acids are used as substrates, and the final neutralization of reaction mixtures to a pH of approximately 4 causes precipitation of the desired products, which are formed in satisfactory yields and are of good purity [16,23–26]. Because of its technological importance, this reaction is still quite intensively studied and optimized; however, most of the studies are done in industrial laboratories, and their results are mostly disseminated as patents. Consequently, it is difficult, if not impossible, to determine which conditions are optimal for the synthesis of individual drugs. It is important because, as in the case of any multicomponent reaction, a synthetic course is strongly dependent on applied conditions and molar ratios of reagents. For example, one of the patents reports that the omission of phosphorus chloride in the reaction of 4-aminobutyric acid as the substrate in methanesulphonic acid provides mixtures of anhydrides as final products (compounds **2** and **3**, Scheme 2) [11]. This corresponds well to the known tendency of phosphonic acids to cyclize.

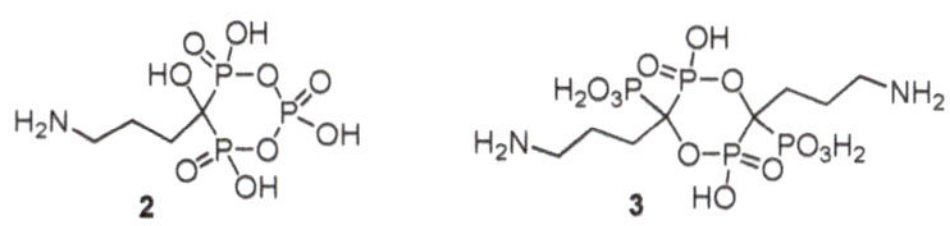

Scheme 2. Cyclic products of the reaction between carboxylic acids and phosphorous acid in methanesulphonic acid.

Among modifications of this procedure are (i) replacement of phosphorus trichloride with thionyl chloride to generate acid chloride in the first reaction step [11]; (ii) replacement of acid by its chloride or anhydride [11] or (iii) t-butyl ester [27]; (iv) blocking the amino moiety in the case of α-amino acids [28]; and (v) application of microwave-assisted procedures [29].

2.2. Synthesis from Amides

Although studies on the biological activity of 1-amino-1,1-bisphosphonic acids are scarce, a significant number of procedures for their preparation have been described. They have recently been reviewed by Romanenko and Kukhar [12]. Amides, being highly stable and easily available compounds, are substrates of choice, and the most commonly applied procedures are simple modifications of those elaborated for carboxylic acids. The most straightforward are reactions of *N*-acylamines and *N*-formylamines with phosphorus trichloride and phosphorous acid [30–32], phosphorus tribromide [33], or triphosgene [34], which provide a wide structural variety of 1-amino-1,1-bisphosphonic acids (compound **4**, Scheme 3). They are based on modification of the first procedure elaborated by Plöger et al. with formamide as a substrate [35].

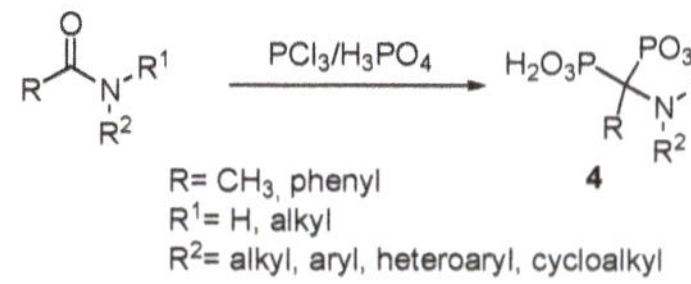

Scheme 3. Bisphosphonylation of amides.

The exact reaction mechanism in this case is also not fully known; however, a Vilsmayer-Haack-like route via iminium ion is postulated here (Scheme 4).

Another possibility is using trialkyl or trimethylsilyl phosphites in reactions prompted by phosphoryl chloride [36–39], trifluoromethanesulphonic anhydride (Tf_2O) [40], zinc chloride [41] or trimethylsilyl trifluoromethanesulfonate [42,43]. Unfortunately, primary amides are not suitable substrates for this reaction [38]. The use of triethyl phosphite and phosphoryl chloride appeared to be especially useful when lactams were used as substrates, resulting in high yields of aminomethylene-*gem*-bisphosphonates (Scheme 5) [36–39]. They were readily hydrolyzed, yielding corresponding bisphosphonic acids (representative example compound **5**). On the other hand, this reaction is not suitable for the conversion of benzoannulated lactams, and mixtures of the desired bisphosphonates and monophosphonates of variable structures were obtained (compounds **6**, **7**, **8** and **9**, Scheme 5). In fact, monophosphonates are usually major products, and the reaction course is dependent on the size of the substrate aliphatic ring [44]. Additionally, these bisphosphonates and monophosphonates appeared to be unstable upon acid hydrolysis and upon storage, and undergo degradation with cleavage of the carbon-to-phosphorus bond (for example compound **10**, Scheme 5). Thus, corresponding acids have not yet been obtained.

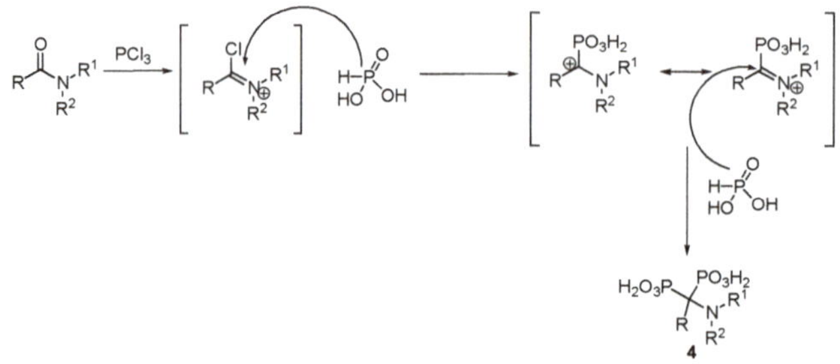

Scheme 4. Bisphosphonylation of amides and the presumable mechanism of this reaction.

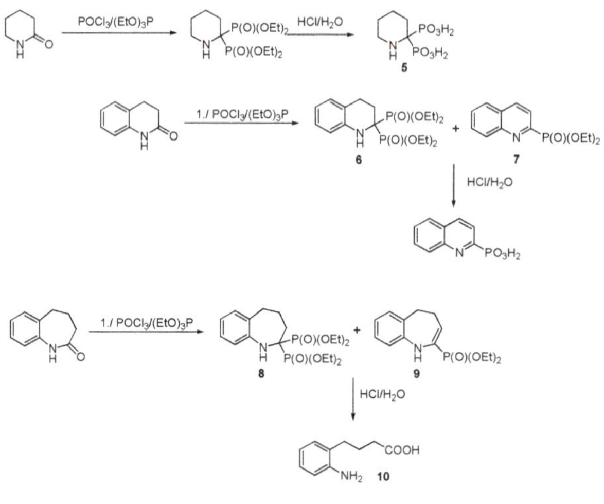

Scheme 5. Representative reactions of lactams with triethyl phosphite prompted by phosphoryl chloride.

The formation of a Vilsmayer-Haack-like intermediate is also proposed in this case with the further addition of a nucleophilic phosphorus reagent. This assumption is additionally supported by studies on the use of structurally variable imine salts as substrates for synthesis of compounds **11** (Scheme 6) [41–45].

The reaction of dibutoxyphosphine or bis(trimethylsiloxy)phosphine with dimethylformamides in the presence of trimethylsilyl trifluoromethanesulfonate affords the corresponding aminomethylenebisphosphonites **12** in high yields (Scheme 7) [41–43]. Similar products were obtained by reacting diethyl pivaloylphosphonite with dialkylformamides in the presence of excess ethanol and catalytic amounts of zinc chloride (Scheme 7). Pivaloylphosphonite decomposes in these reaction conditions, yielding diethoxyphosphine, which is the real substrate of the reaction [41].

N-Octylpyrrolidinone treated with LDA, followed by the addition of diethyl phosphorochloridite and oxidation of the reaction mixture with 30% hydrogen peroxide resulted in bisphosphonylated lactam **13** with excellent yield (Scheme 8). This reaction was then applied for the synthesis of *N*-geranylated lactams **14** and **15** and linear amides, potential inhibitors of farnesyl:protein transferase [46]. In the case of cyclic imides, in which two equivalent positions for enolate formation are present on the imide rings, and thus two carbon atoms may be phosphonylated, the structure of the product was dependent on the mode of reaction. If phosphonylation was carried out in one step, two carbon atoms were phosphonylated and vicinal bisphosphonate was obtained. Step-by-step reaction resulted in the predomination of the phosphonylation of one carbon atom, and the desired gem-bisphophonate was the major product.

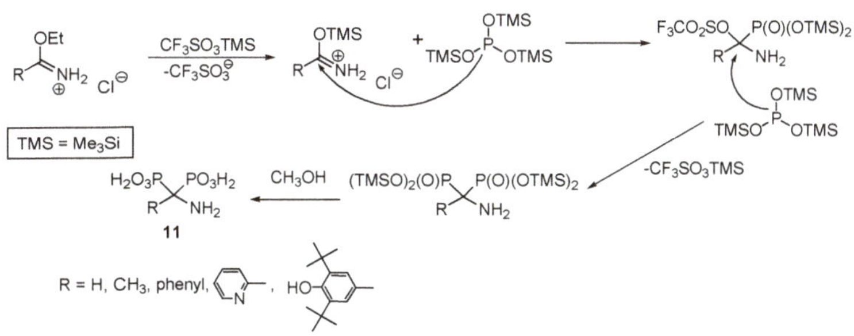

Scheme 6. Imine salts as substrates for preparation of bisphosphonic acids.

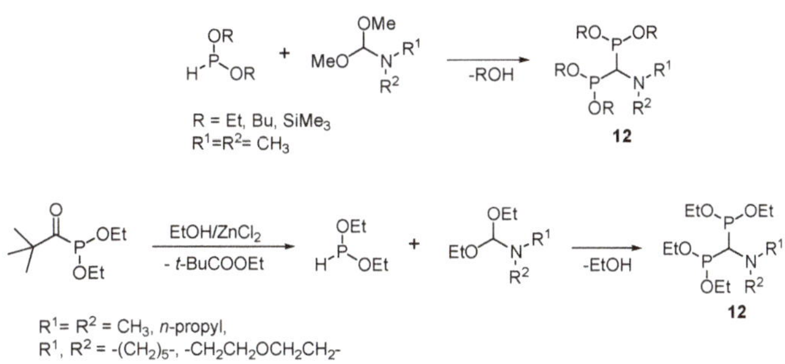

Scheme 7. Reaction of diethoxyphosphine with the acetal of dimethylformamide.

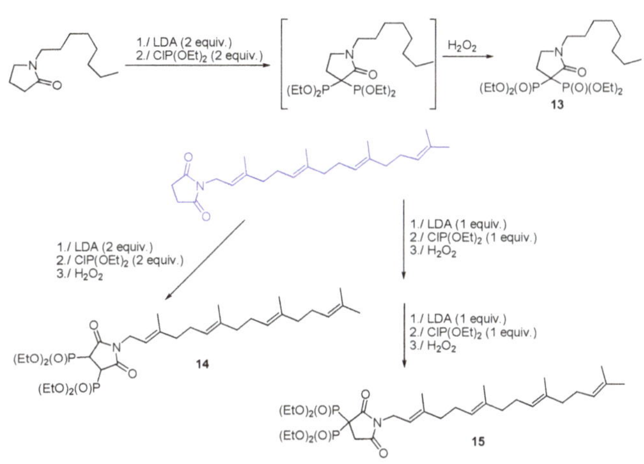

Scheme 8. Synthesis of bisphosphonate lactams.

2.3. *Synthesis from Nitriles*

1-Aminoalkylidene-1,1-bisphosphonic acids could also be prepared from nitriles by applying the same procedures as in syntheses starting from amides. Despite being readily available substrates, there is limited literature on their use for that purpose. The most popular route is the reaction of nitriles with phosphorous acid, in some cases prompted by phosphorus trichloride and phenylsulphonic or methylsulphonic acids [12,31,47,48]. The conditions of this reaction are sufficiently delicate to use peptidyl nitriles as substrates, which was demonstrated using cyanoethyl derivatives of dipeptides (for example, peptide **16**, Scheme 9) [49].

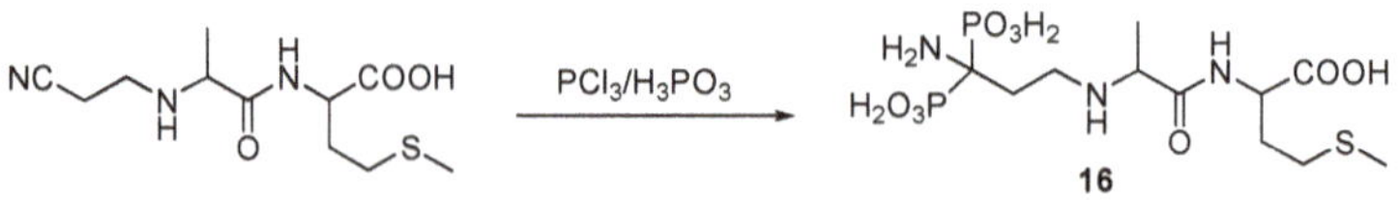

Scheme 9. Representative example of the synthesis of peptidylbisphosphonates from nitriles.

Recently, elegant, mild, and atom economical double phosphonylation of nitriles in the presence of titanocene has been described. This procedure used induced phosphorus-centered radicals mediated by titanocene dichloride (Cp_2TiCl_2) (Scheme 10) and provided a wide structural variety of aminobisphosphonates **17** [50]. This is a double radical transfer reaction initiated by the reaction of titanocene with zinc dust and the transfer of the obtained radical to epoxypropane by cleavage of its oxirane ring, followed by transfer of this radical to diethyl phosphite, which in turn reacts with nitrile (Scheme 10). The reaction carried out without epoxide, under microwave stimulation, also results in the desired bisphosphonates, although it is accompanied by formation of many side products.

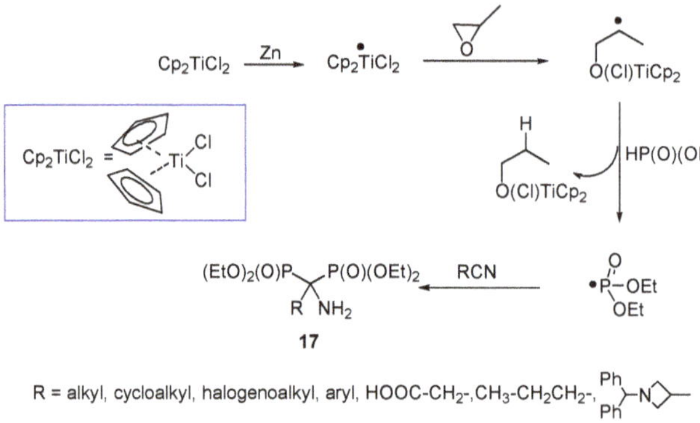

Scheme 10. Radical procedure for the synthesis of bisphosphonates from nitriles.

2.4. *Synthesis from Isonitriles*

Isonitriles, which are quite common substrates, especially in multicomponent reactions [51], have been seldom used for the preparation of aminobisphosphonates **18**. Their reactions with *H*-phosphine oxides carried out in the presence of typical palladium catalysts (Pd_2dba_3) afforded the product **19** of monophosphonylation, whereas the use of various rhodium catalysts afforded products of diphosphonylation (Scheme 11) [52].

Far simpler is the addition of diethyl phosphite to isonitriles. This reaction is carried out with an excess of hydrogen chloride in aprotic solvents. Under these conditions, nitrile is converted to an onium salt, and after addition of the first phosphite molecule, iminium salt is formed, a substrate for the addition of a second phosphite molecule (Scheme 12) [53]. Using this procedure, two distinct libraries of bisphosphonates **20** and **21** have been prepared [54–56].

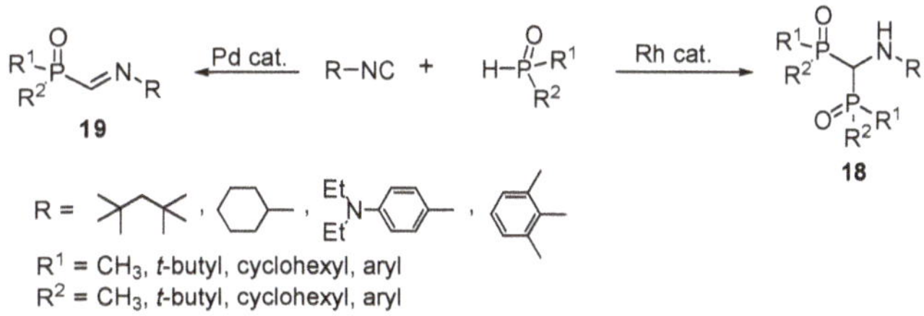

Scheme 11. Addition of *H*-phosphine oxides to isonitriles catalyzed by metallocatalysts.

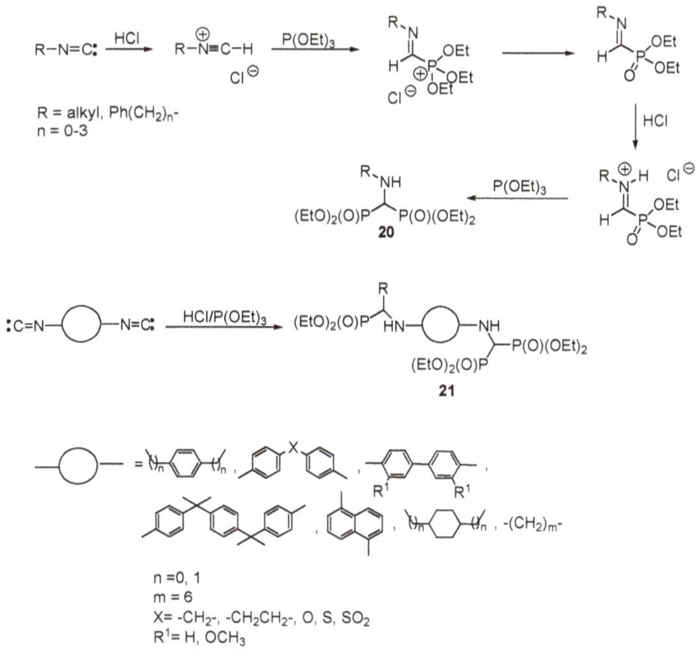

Scheme 12. Addition of phosphites to isonitriles.

2.5. Synthesis via Ketophosphonates

Presumably, the addition of phosphites to ketophosphonates was the first procedure for preparation of dialkyl 1-hydroxy-1,1-bisphosphonates [16,19,57]. Starting ketophosphonates are obtained by the Arbuzov reaction of acyl chlorides with trialkyl phosphites and used as substrates in the next step, namely, the addition of dialkyl phosphite to carbonyl double bonds (a representative example for preparation of tetramethyl pamidronate **22** is shown in Scheme 13). Since ketophosphonates are unstable species [58], the desired bisphosphonates are usually obtained via one-pot procedures applying mixtures of trialkyl- and dialkylphosphonates at elevated temperature [59]. Additionally, in situ generation of dialkyl- from trialkyl phosphites is possible by the addition of a protic solvent to the reaction mixture. A useful modification of this procedure is the application of tris(trimethylsilyl) phosphite, followed by the easy removal of ester groups by methanolysis [23,60,61]. This was used to prepare a wide variety of 1-hydroxy-1,1-bisphosphonates containing amino groups [19,23,61–63]. Additionally, the use of bis((trimethylsilyl)oxy)phosphine generated from ammonium hypophosphite was applied to prepare hydroxybisphosphinic acids (also presented in Scheme 13) [64]. Another modification was the use of acyl phosphonamidates readily prepared from phosphoramidite type reagents and a range of acid chlorides, followed by reaction with trimethyl phosphite in the presence of pyridinium perchlorate [65].

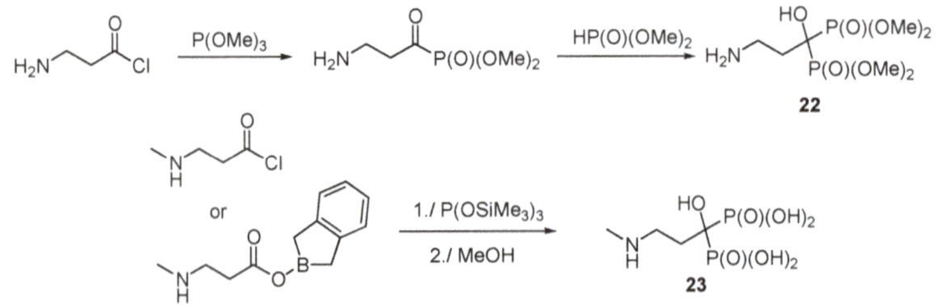

Scheme 13. Representative syntheses via ketophosphonates.

Recently, a one-pot synthesis was described reacting carboxylic acids with catecholborane, followed by the treatment of the formed acyloxy-benzodioxaborolane with tris(trimethylsilyl)phosphite (Scheme 13) [66]. The efficiency of that simple methodology was proved by the syntheses of alendronate and *N*-methyl pamidronate (compound **23**) without additional steps for the protection/deprotection of their amine functions.

2.6. Three-Component Condensation of Amines with Triethyl Orthoformate and Diethylphosphite

Simple three-component condensation of stoichiometric ratios of amines, diethyl phosphite and triethyl orthoformate, first reported in patent literature by Suzuki [67] and further extended by Maier [68], is perhaps the most common procedure for the preparation of a wide variety of aminomethylenebisphosphonic acids (Scheme 14). Since this reaction usually gives a complex mixture of products [56] that are difficult to separate, the resulting esters are not isolated but rather, the crude reaction mixtures are hydrolyzed, yielding bisphosphonic acids that are isolated after the hydrolytic step. Some modifications of this classic procedure have also been reported. They include the use of a solvent-free, microwave-assisted reaction [69–71] and reactions catalyzed by titanium dioxide [72] and by crown ethers (increasing the selectivity of the process by 10%–20%) [73]. Additionally, a reaction carried out in a micellar environment was described [74].

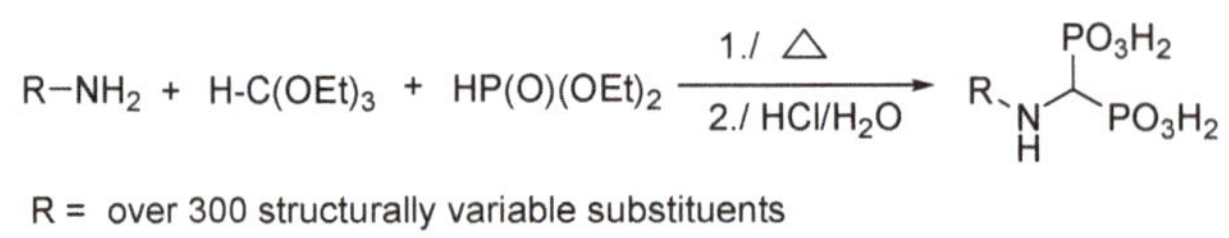

Scheme 14. Condensation of amines with triethyl orthoformate and diethyl phosphite.

The three-component reaction was applied to synthesize a large series of physiologically active compounds, in some cases of very complex chemical structures, including bone antiresorptive drug candidates [75–82]; bone imaging [83,84], antiprotozoal [85–88], antibacterial [72,89–92], anti-HIV [93] and anti-inflammatory [94] agents; herbicides [95–97]; and complex ones for various metals [10,98].

In some cases, this reaction appears to be quite capricious and affords unexpected side-products along with the expected aminomethylenebisphosphonates (see representative examples in Scheme 15), the compositions of which are dependent on the applied conditions (molar ratio of substrates, temperature and reaction time). Most often, alkylation (for example, compounds **24** and **25**) or formylation (compound **26**) of amine moieties is observed [71,99], while in selected cases the formation of aminomethylenebisphosphonic acid (compound **27**) is observed. Because this compound was previously prepared by acid hydrolysis of *N*-benzhydrylaminomethylenebisphosphonic acid [83,100], it may be that it is formed upon hydrolysis of *N*-aryl derivatives.

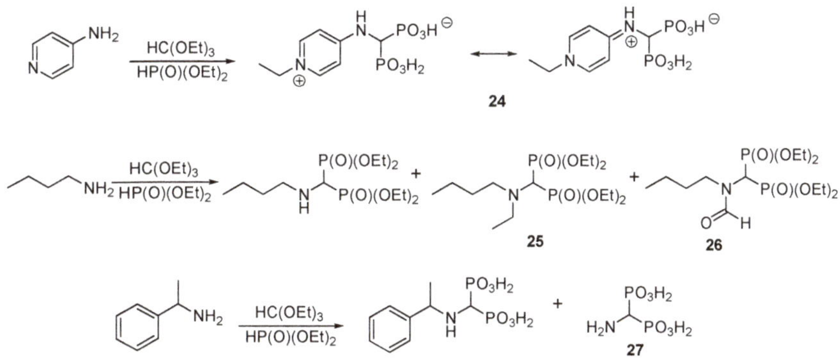

Scheme 15. Representative side products of the three-component procedure for the synthesis of aminomethylenebisphosphonic acids.

The mechanism of this useful reaction had been thoroughly studied by using ^{31}P-NMR and by isolation of all intermediates [71,72,99] and identifying interrelations between them (Scheme 16) [99]. The mechanism appeared to be quite complex because the intermediates exist in thermodynamic equilibrium. Thus, its course is strongly dependent on the properties of the used amine and applied reaction conditions.

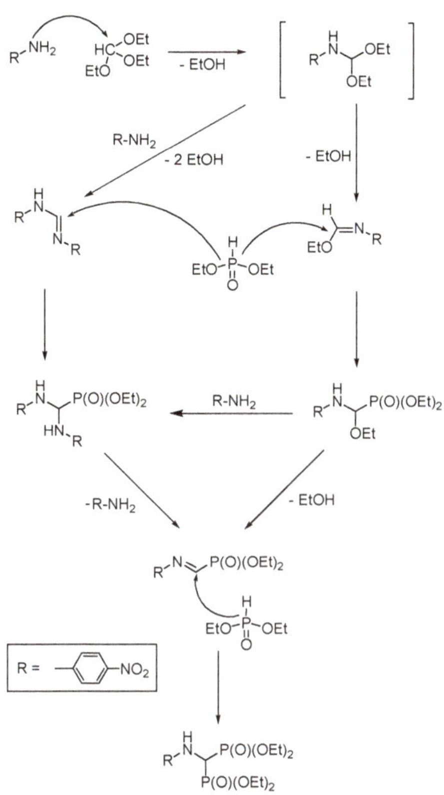

Scheme 16. Mechanism of the three-component reaction of amines with orthoformates and phosphites.

Modification of this procedure and the use of ethyl diethoxymethyl-*H*-phosphinate instead of diethyl phosphite allowed obtaining structurally variable bisphosphinates **28** (Scheme 17) [101]. Better results were obtained if the reaction was carried out under nitrogen (air oxidizes substrates and products). Notably, aromatic amines provided the desired bisphosphinates in high yields (78%–92%), whereas the reaction yields when using cyclic and aliphatic derivatives were lower. Interestingly, of many methods for the synthesis of corresponding acids **29**, acidolysis of the obtained esters **28** appeared to be optimal. The use of other conditions, including mild transesterification with trimethylsilyl bromide followed by methanolysis, resulted in the formation of side products **30** and **31**.

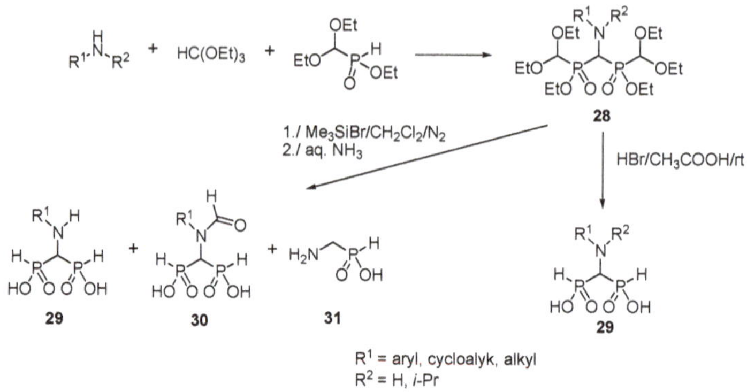

Scheme 17. Three-component synthesis of aminobisphosphonates.

2.7. Addition of Amines to Vinylidenebisphosphonates

Tetraethyl vinylidenebisphosphonate **32** is a versatile synthon useful for preparation of a wide variety of bisphosphonates. Its usefulness is a subject of recent comprehensive review [102]. As an electron-deficient alkene, it can undergo conjugate addition of strong and mild nucleophiles, with amines belonging to the latter class. Primary amines undergo smooth Michael addition (Scheme 18), however, the obtained compounds **33** must be quickly purified and hydrolyzed since they tend to undergo a retro-Michael reaction [103,104]. Fortunately, free acids **34** are substantially more stable.

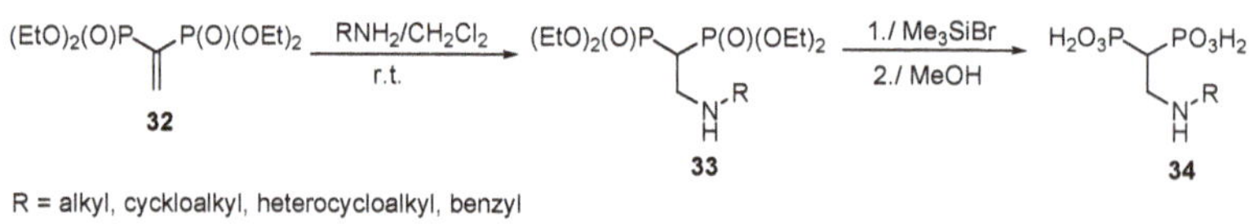

Scheme 18. Michael addition of amines to vinylidenebisphosphonate.

The reactivity of vinylidenebisphosphonates has been used for the synthesis of derivatives of various analogues of fluoroquinolone antibacterial agents [105,106], heteroaromatics with potential pharmaceutical applications [107,108], simple antiplasmodial agents [103,104], HIV reverse transcriptase inhibitors [59,93], potential anti-osteoporotic agents [109,110], and *N*-alkylated antitumor pyridines [111]. Some representatives of these compounds (**35–39**) are shown in Scheme 19.

Another example, albeit far less developed, is the addition of organometallic amines to vinylidenebisphosphonate **22** (Scheme 20) [112]. Huisgen copper-catalyzed 1,3-dipolar cycloaddition of this bisphosphonate to azides was used to obtain substrate **40** for a "click reaction". This reaction yielded compounds that may be considered aza-analogues **41** of zoledronate (Scheme 20) [113]. Similar substrates could also be obtained by addition of propargylamine to the double bond of vinylidenebisphosphonate [114].

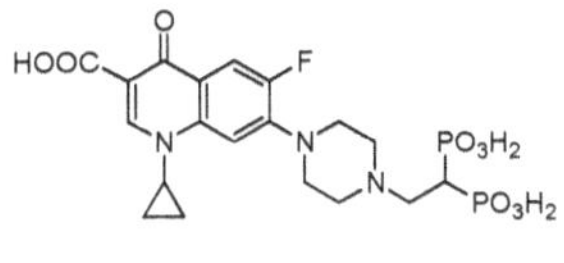

35, analog of Ciprofloxacin

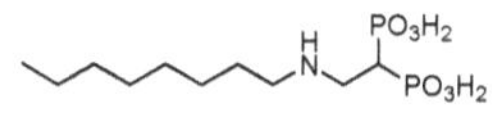

36, antiprotozoal agent

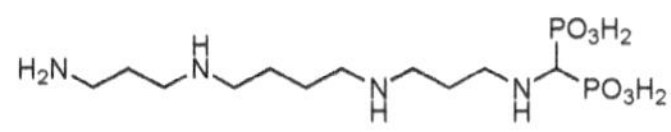

37, potential antiosteoporotic agent

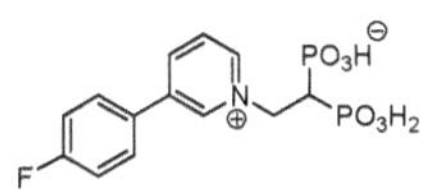

38, anticancer agent

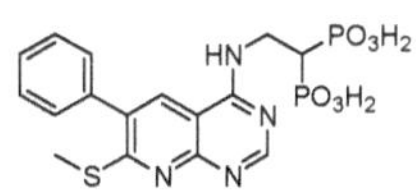

39, inhibitor of HIV reverse transcriptase

Scheme 19. Representatives of useful bisphosphonates obtained via addition of amines to vinylidenebisphosphonate.

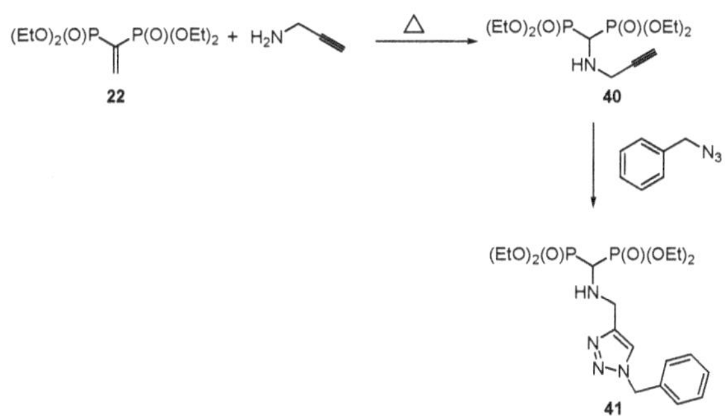

Scheme 20. An example of the synthesis of novel bisphosphonates via "click chemistry".

2.8. Miscellaneous Procedures

The desire to obtain new aminobisphosphonate scaffolds for biological studies has stimulated numerous studies on general methods of their synthesis. The methods discussed here are mostly designed to prepare specific scaffolds or are applicable only to specific, if not unusual, substrates. Only some of them may be considered as novel, general procedures.

Radical addition of sodium hypophosphite to terminal alkynes in the presence of triethylborane, which produced 1-alkyl-1,1-bis-*H*-phosphinates in moderate yields, gives access to a wide structural variety of bisphosphonates [115]. When using propargylamino acids, the corresponding bisphosphinates were obtained in satisfactory yields (a representative example is given in Scheme 21 for analogue **42** of pamidronate). Bisphosphinates are easily converted to bisphosphonates by ozonolysis.

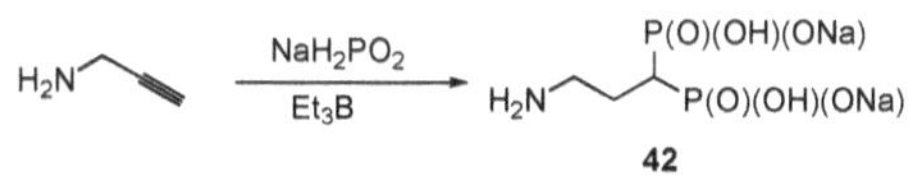

Scheme 21. Procedure for the synthesis of bisphosphinates and their conversion into bisphosphonates.

1-(*N*-acylamino)alkylphosphonates, easily accessible from *N*-acyl-α-amino acids using a two-step transformation, underwent electrophilic activation at the α-carbon by electrochemical α-methoxylation in methanol in a process mediated by NaCl. Attempts to carry out a Michaelis-Arbuzov-like reaction of the obtained diethyl 1-(*N*-acetylamino)-1-methoxy-alkylphosphonates with triethyl phosphite failed; however, they readily reacted with triphenylphosphine (Scheme 22) [116]. The resulting diethyl 1-(*N*-acetylamino)-1-triphenylphosphoniumalkylphosphonate tetrafluoroborates **43** reacted smoothly with trialkyl phosphites, dialkyl phosphonites or alkyl phosphinites in the presence of Hünig's base and methyltriphenylphosphonium iodide as catalysts. This gave bisphosphonates **44,** 1-phosphinylalkylphosphonates or 1-phosphinoylalkylphosphonates in good yields. This reaction has potential to become a general procedure.

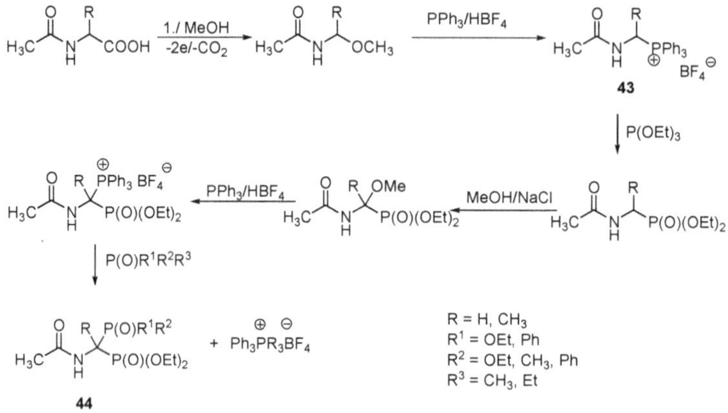

Scheme 22. Transformation of diethyl 1-(*N*-acetylamino)-1-alkylphosphonates into bisphosphoric acid esters via the corresponding phosphonium salts.

Alkylation of tetraalkyl methylenebisphosphonate **45**, a classical method for synthesizing variable bisphosphonic acids, has been rarely used for the synthesis of amino derivatives. Thus, its direct amination with the hydroxylamine ester of diphenylphosphinic acid, followed by its bromoacetylation (Scheme 23), provided substrate **46** for functionalization into glycopeptide antibiotics, in the hope that they will find an application as medication for osteomyelitis [117]. Another classical example is the monoalkylation of tetraisopropyl methylenebisphosphonate with 1,6-dibromohexane, followed by fluorination with *Selectofluor* and conversion of the remaining bromide into amine, which resulted in compound **47** (Scheme 23) [118].

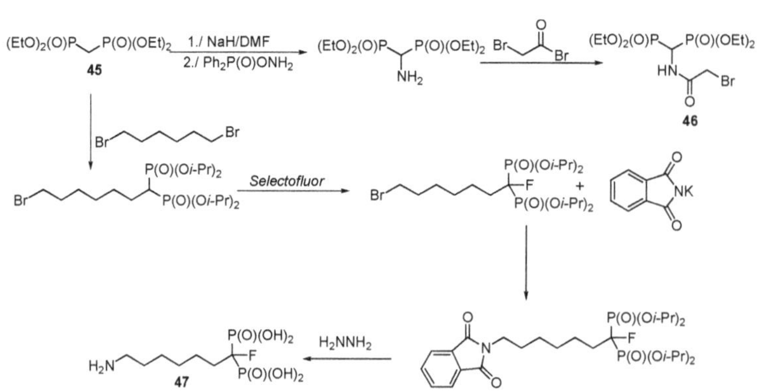

Scheme 23. Synthesis of aminobisphosphonates via alkylation of tetraalkyl methylenebisphosphonates.

An unusual procedure is metallacarbenoid insertion of aromatic amines into tetraethyl diphosphonodiazomethane **48**, which yields corresponding aminomethylenebisphosphonates **49** (Scheme 24) [119]. Among the tested catalysts, $Rh_2(NHCOCF_3)_4$ was found to be the best.

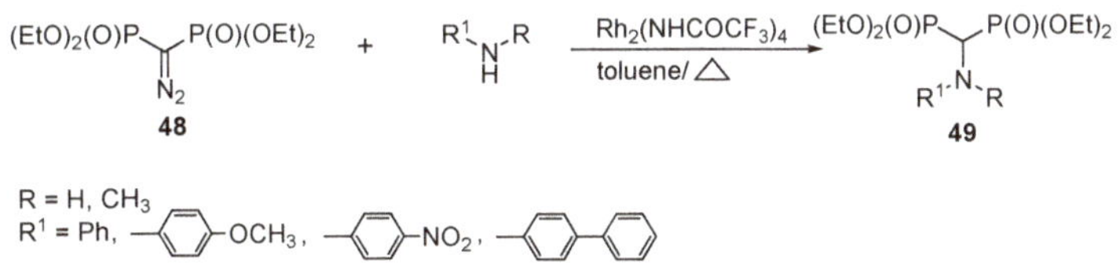

Scheme 24. N-H insertion of the reaction of aromatic amines.

A specific and unexpected reaction was the addition of silylated dialkyl phosphites to 4-phosphono-1-aza-1,3-dienes **50**, which resulted in γ-phosphono-α–aminobisphosphonates **51** (Scheme 25) [120]. This reaction is interesting in that, depending on the steric demand of the substituent present on nitrogen, double 1,2-addition or tandem 1,4-1,2-addition with formation of bisphosphonate **52** occurred (Scheme 25).

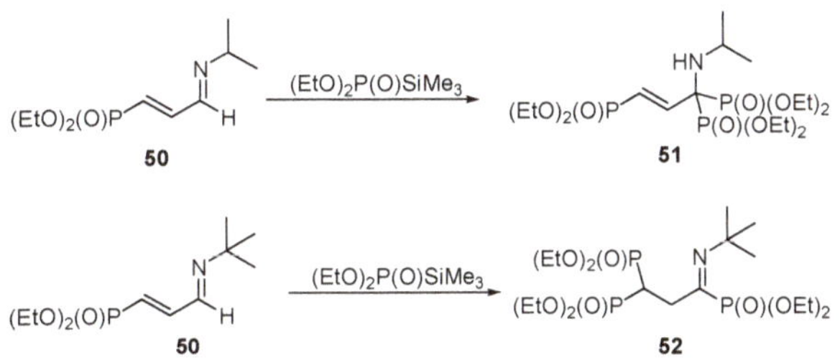

Scheme 25. Addition of phosphites to 4-phosphono-1-aza-1,3-dienes.

Additionally, the addition of phosphites to iminophosphonate esters has been studied. This reaction is limited to specific substrates, and usually the obtained bisphosphonates **53** are unstable, and phosphoryl C-N transfer to compounds **54** is observed (Scheme 26), which may be considered an example of an aza-Perkov reaction [121–123].

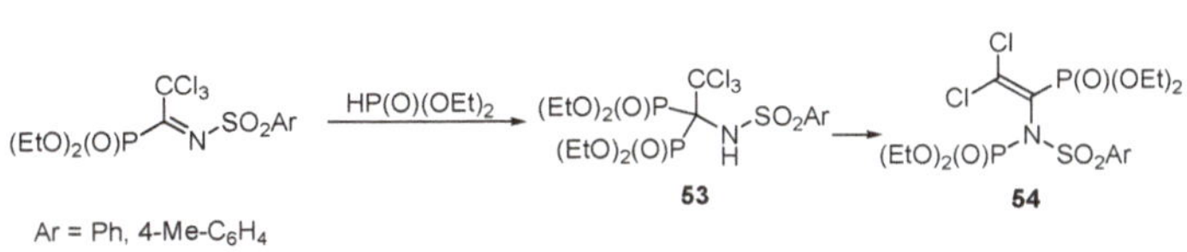

Scheme 26. Addition of phosphites to iminophosphonates.

Finally, syntheses of rare aminomethyltrisphosphonates are presented in a recent review by Romanenko and Kukhar devoted to applications of methylidynetrisphosphonates [124]. Two procedures for the synthesis of bis- and trisphosphonates **55** and **56** taken from this review and described in patent literature are depicted in Scheme 27.

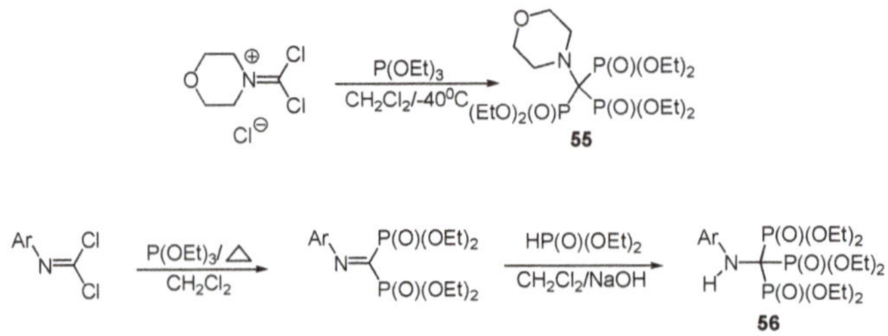

Scheme 27. Syntheses of aminomethyltrisphosphonates.

3. Functionalization of Aminobisphosphonates

Bisphosphonates are known for their affinity to bone tissue, and thus their conjugation to various drugs has been quite intensively studied; they are expected to serve as system-targeting drugs [6]. Because dronic acids are produced at industrial scale and are thus readily available and inexpensive, they are most frequently used for that purpose. However, their extreme hydrophilic character means that they are practically insoluble in most organic solvents, which limits the use of aqueous media or requires their conversion into phosphonate esters prior to functionalization. Unfortunately, the methods for direct esterification of phosphonate moieties are scarce and usually give unsatisfactory results; these esters must generally be synthesized independently.

Methods for functionalizing aminobisphosphonic acids with structurally variable molecules were recently reviewed [12]; therefore, in this review, the methods arbitrarily chosen as the most representative are reported.

3.1. *Direct Acylation of Aminobisphosphonic Acids*

Direct acylation of aminobisphosphonic acids is difficult because this reaction is accompanied by the possible competitive acylation of phosphonic groups [125,126] and hydroxylic groups when amino-1-hydroxy-1,1-bisphosphonic acids are substrates [126,127]. A representative example of this reaction is given in Scheme 28 for alendronic acid **57**.

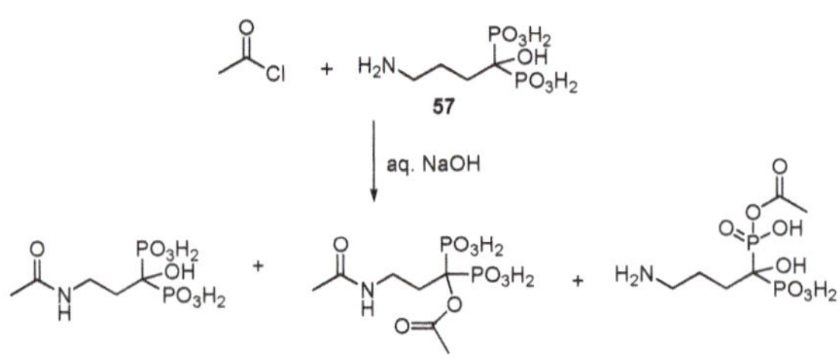

Scheme 28. Direct acylation of alendronic acid with indication of possible side-products.

Acylation of dronic acid salts with acyl chlorides in sodium hydroxide in water or in water/propanol solutions is the simple Schotten-Baumann variant [127,128]. Acidification of the solution results in the precipitation of the desired acids or their monosodium salts. To conjugate aminobisphosphonic acids with molecules bearing carboxylic groups, classical coupling protocols used in peptide synthesis have also been used. These include the activation of carboxylic groups with *N*,*N*′-dicyclohexylcarbodiimide (DCC) [129,130]; the use of previously prepared succinimidate esters to obtain 21 extremely complex fluorescent imaging probes (representative examples of a green fluorescent dye **58** and a red fluorescent dye **59** are shown in Scheme 29) [131]; and *N*,*N*′-dicarbonylimidazole, which was used to conjugate pamidronate to pullan (polysaccharide composed of maltotriose units) [132] to obtain a system for bone regeneration.

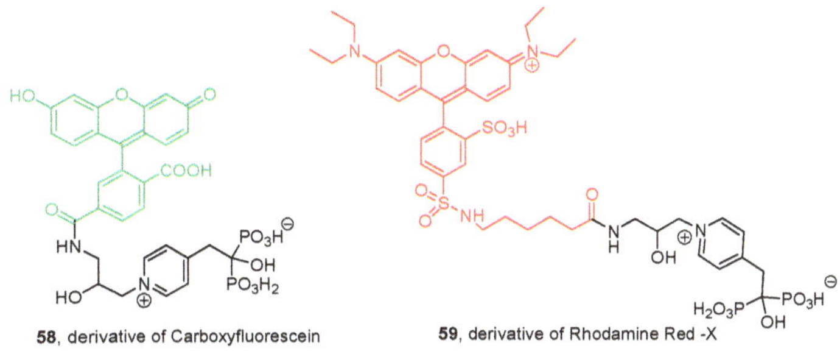

Scheme 29. Representative fluorescent imaging probes obtained via acylation of bisphosphonic acids (green—carboxyfluoresceine fragment, red—rhodamine Red-X fragment).

Acylation of aminobisphosphonic acids with small acids, such as acrylic acid [133], chloroacetic acid [134] or succinic acid [135], provided useful substrates for further functionalization of larger molecules, such as macrocycles devised for lanthanide ion complexation, chitosan, and hyaluronan.

3.2. Direct Acylation of Tetraethyl Aminobisphosphonates

Esters are far more suitable substrates for acylation than free phosphonic acids, with tetraethyl aminomethylenebisphosphonate being the most popular substrate. It has been used to obtain structurally variable conjugates with estradiol (compounds **60**, **61** and **62** in Scheme 30) using classical peptide synthesis coupling agents such as DCC and DPPA (diphenylphosphoryl azide) [136]. These conjugates were synthesized as bone-specific estrogens in the hope that they will protect elderly women from bone loss resulting from osteoporosis.

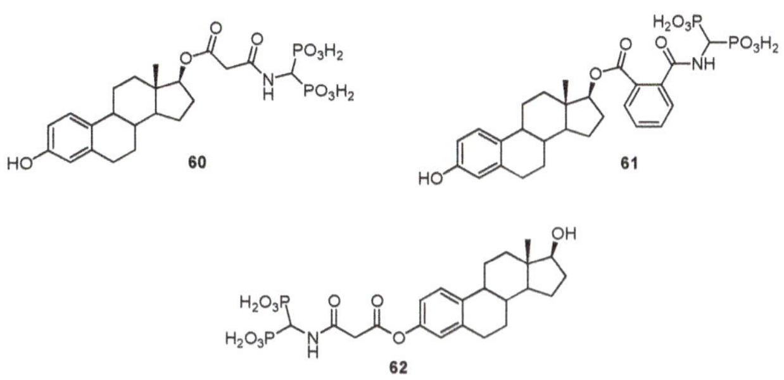

Scheme 30. Representative structures of bone-specific estrogens.

A derivative of raloxifene, a selective estrogen receptor or modulator, was obtained using a suitable acid chloride as substrate [137], whereas radioligands, which are selectively bound to bone tissue, have been synthesized by using DCC/HOBt (hydroxybenzotriazol) activation [138] and antibacterial bisphosphonated benzoxazinorifamycin prodrugs using EDCl (1-ethyl-3-(3-dimethylaminopropyl)carbodiimide) [139].

In this case, acylation of aminobisphosphonic acids with small acids was also applied as a starting step to produce antibacterial agents against osteomyelitis [88] and bone imaging macrocycles [81], using acid chlorides as acylating agents.

Using DCC as a coupling agent, diethyl 1-(2-aminoethylamino)-1,1-ethylbisphosphonate was acylated by structurally variable natural acids, such as folic acid [140,141], ursulonic and betulinic acids [142], and trolox [141,143].

Acylation of pamidronic acid ester **22** by using HBTU (2-(1*H*-benzotriazol-1-yl)-1,1,3,3-tetramethyluronium hexafluorophosphate) was used to prepare a catecholic derivative, which was then bound to the surface of magnetic iron oxide nanoparticles (Scheme 31) [144]. The particle **63** has been designed to remove uranyl ions from blood.

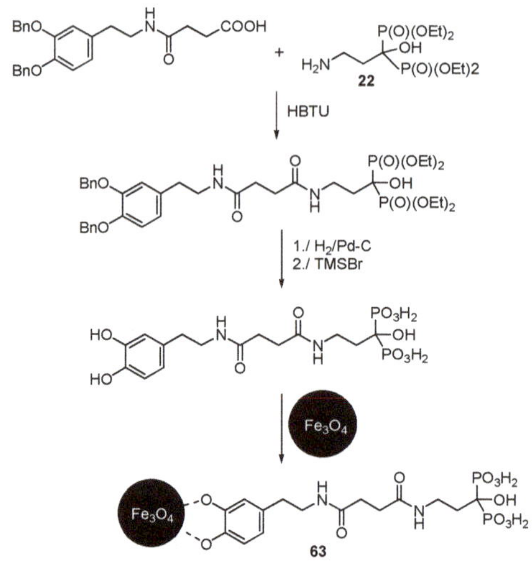

Scheme 31. Synthesis of bisphosphonate functionalized magnetic nanoparticles.

3.3. *Synthesis of Building Blocks for Polymer Chemistry*

Self-etching adhesives are polymeric materials containing phosphonate groups that have become popular in restorative dentistry because they allow strong bonds between dental hard tissues (enamel and dentin). One possibility for their preparation is to obtain monomers containing phosphonic groups [145]. Such methacrylamide monomers **64** and **65** were synthesized by acylation with suitable acryloylchlorides (Scheme 32) [146]. Studies of their photopolymerization indicated that they may be suitable for potential use in dentistry.

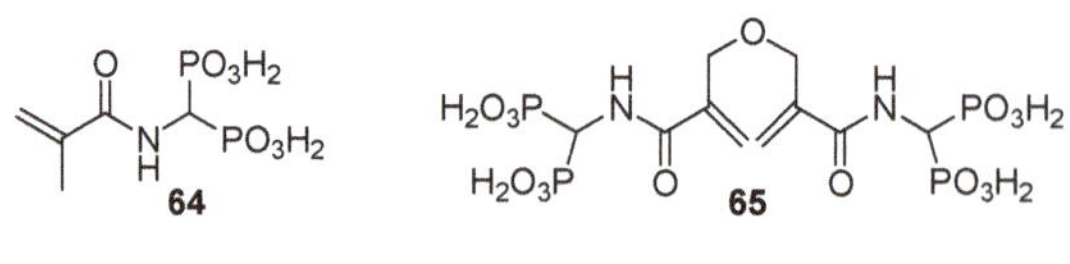

Scheme 32. Bisphosphonylated methacrylamide monomers.

A similar monomer, acryloylated pamidronate, has been used to obtain a hyaluronic acid derivative that is dually polymerized with cross-linkable hydrazide groups and bisphosphonate ligands. By mixing bisphosphonate polymer with calcium ions and aldehyde-derivatized hyaluronic acid, a hybrid hydrogel was obtained, which quickly mineralizes [147]. Such a system is of interest as a mediator for fast bone regeneration.

By dispersion copolymerization of three monomers (methacrylate bisphosphonate **66**, *N*-(3-aminopropyl) methacrylamide **67**, and tetra(ethylene glycol) diacrylate **68**) (Scheme 33), polymeric nanoparticles **69** were obtained [148]. Their size distribution was controlled by changing various polymerization parameters. By covalent attachment of a drug and/or a dye to amino groups (such as a near IR fluorescent dye), a theranostic system may be obtained.

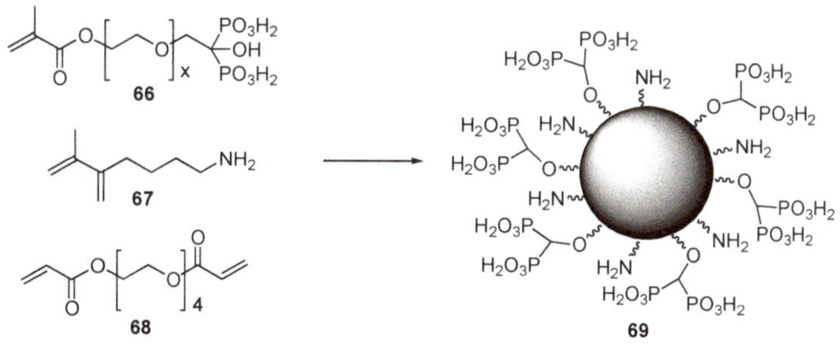

Scheme 33. Synthesis of polymeric nanoparticles as potential dye and drug carriers.

Macromolecular co-conjugates of bisphosphonate and ferrocene were synthesized by means of Michael addition copolymerization of methylenebisacrylamide (MBA) with primary amines-6-amino-1-hydroxyhexylidene-1,1-bisphosphonate and 4-ferrocenyl-butamidopropylamine [149]. The mass percentage incorporation of ferrocene analogues was found to be between 4%–5%, and 10%–12% for bisphosphonate. Such polymers could be selectively bound to bone tissue and slowly release anticancer ferrocene derivatives at this target site.

3.4. Miscellaneous

Other means to functionalize amino moieties are quite scarce and dispersed. Classical ones include the formation of Schiff bases, which are then alkylated [150] or reduced [126], and synthesis of thioureido derivatives as intermediates in the preparation of heterocyclic compounds [151–153].

The functionalization of polysaccharides is the gateway of aminobisphosphonates into nanoscience. For example, phosphonated cellulose was utilized to obtain nanocellulose with good thermal stability and potential intumescent properties. It was synthesized from birch pulp via sequential periodate oxidation and reductive amination using alendronate **57** as a phosphonating reagent (Scheme 34) [154]. After high-pressure homogenization, bisphosphonate cellulose nanofibers or nanocrystals of the general formula **70** were obtained, depending on the initial oxidation degree.

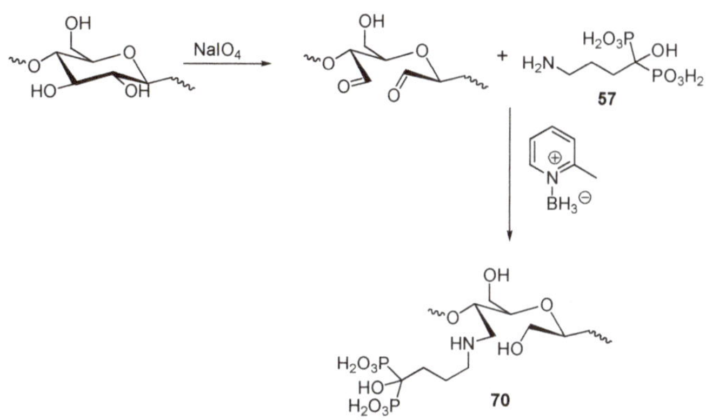

Scheme 34. Periodate oxidation of cellulose followed by reductive amination with sodium alendronate.

Alendronate was also bound to conjugates of pullulan and paclitaxel, which were bound to the polysaccharide by a cathepsin K-sensitive tetrapeptide spacer. This should ensure release of paclitaxel in bones. Then, bisphosphonate was covalently conjugated to the sugar chain via a polyethyleneglycol chain, using a technique similar to that described above (identical to those shown in Scheme 34) [155]. This system exhibited strong antiproliferative action against several cancer cell lines.

4. Conclusions

Aminobisphosphonic acids are gaining significant interest as a class of compounds with promising physiologic activity, with some representatives already commercialized as bone resorption inhibitors, and therefore useful drugs against osteoporosis and related bone disorders. This leads to both the modification of existing and the elaboration of novel procedures for their preparation. There are several commonly used reactions for this purpose; however, they have also been modified in the last decade to be tailored to specific biological needs. A few novel procedures have also been developed. Functionalization of simple aminobisphosphonic acids is a difficult task because of their strongly polar character. However, successful examples of functionalization of the amino moieties of these compounds have provided promising diagnostics and novel therapeutic agents.

Acknowledgments: This work was supported by a statuary activity subsidy from the Polish Ministry of Science and Higher Education.

Conflicts of Interest: The authors declare no conflict of interest.

References

1. Turhanen, P.A. Synthesis of triple bond containing 1-hydroxy-1,1-bisphosphonic acid derivatives to be used as precursors in "click" chemistry: Two examples. *J. Org. Chem.* **2014**, *79*, 6330–6335. [CrossRef] [PubMed]
2. Russell, R.G. Bisphosphonates: The first 40 years. *Bone* **2011**, *49*, 2–19. [CrossRef] [PubMed]
3. Ebetino, F.H.; Hogan, A.M.; Sun, S.; Tsuompra, M.K.; Duan, X.; Triffitt, J.T.; Kwaasi, A.A.; Dunford, J.E.; Barnett, B.L.; Oppermann, U.; et al. The relationship between the chemistry and biological activity of the bisphosphonates. *Bone* **2011**, *49*, 20–33. [CrossRef] [PubMed]
4. Maraka, S.; Kennel, K.A. Bisphosphonates for the prevention and treatment of osteoporosis. *Br. Med. J.* **2015**, *351*, h3783. [CrossRef] [PubMed]
5. Shi, C.G.; Zhang, Y.; Yuan, W. Efficacy of bisphosphonates on bone mineral density and fracture Rate in patients with osteogenesis imperfecta: A systematic review and meta-analysis. *Am. J. Ther.* **2016**, *3*, e894–e904. [CrossRef] [PubMed]
6. Chmielewska, E.; Kafarski, P. Physiologic Activity of Bisphosphonates—Recent Advances. *Open Pharm. Sci. J.* **2016**, *3*, 56–78. [CrossRef]
7. Demkowicz, S.; Rachoń, J.; Daśko, M.; Kozak, W. Selected organophosphorus compounds with biological activity. Applications in medicine. *RSC Adv.* **2016**, *6*, 7101–7112. [CrossRef]
8. Studnik, H.; Liebsch, S.; Forlani, G.; Wieczorek, D.; Kafarski, P.; Lipok, J. Amino polyphosphonates—Chemical features and practical uses, environmental durability and biodegradation. *New Biotechnol.* **2015**, *32*, 1–6. [CrossRef] [PubMed]
9. Turhanen, P.A.; Vepsäläinen, J.J.; Peräniemi, S. Advanced material and approach for metal ions removal from aqueous solutions. *Sci. Rep.* **2015**, *5*. [CrossRef] [PubMed]
10. Gałęzowska, J.; Gumienna-Kontecka, E. Phosphonates, their complexes and bio-applications: A spectrum of surprising diversity. *Coord. Chem. Rev.* **2012**, *256*, 105–124. [CrossRef]
11. Abdou, W.M.; Shaddy, A.A. The development of bisphosphonates for therapeutic uses, and bisphosphonate structure-activity consideration. *Arch. Org. Chem.* **2008**, *2009*, 143–182.
12. Romaneneko, V.D.; Kukhar, V.A. 1-Amino-1,1-bisphosphonates. Fundamental syntheses and new developments. *Arch. Org. Chem.* **2012**, *2012*, 127–166.
13. Petroianu, G.A. Pharmacist Theodor Salzer (1833–1900) and the discovery of bisphosphonates. *Pharmazie* **2011**, *66*, 804–807. [PubMed]

14. Kieczykowski, G.R.; Jobson, R.B.; Melillo, D.G.; Reinhold, D.F.; Grenda, V.J.; Shinkai, I. Preparation of (4-amino-1-hydroxybutylidene)bisphosphonic acid sodium salt, MK-217 (alendronate sodium). An improved procedure for the preparation of 1-hydroxy-1,1-bisphosphonic acids. *J. Org. Chem.* **1995**, *60*, 8310–8312. [CrossRef]
15. Teixeira, F.C.; Antunes, I.F.; Curto, M.J.M.; Neves, M.; Teixeira, A.P.S. Novel 1-hydroxy-1,1-bisphosphonates derived from indazole: Synthesis and characterization. *Arch. Org. Chem.* **2009**, *2009*, 69–84.
16. Yanvarev, D.V.; Koovina, A.N.; Usanov, N.N.; Kochetkov, S.N. Non-hydrolysable analogues of inorganic pyrophosphate as inhibitors of Hepatitis C virus RNA-dependent RNA polymerase. *Russ. J. Bioorg. Chem.* **2012**, *38*, 224–229. [CrossRef]
17. Srinivasa Rao, D.V.N.; Dandala, R.; Narayan, G.K.A.S.S.; Lenin, R.; Sivakumaran, N.; Naidu, A. Novel procedure for the synthesis of 1-hydroxy-1,1-bisphosphonic acids using phenols as medium. *Synth. Commun.* **2007**, *37*, 4359–4365. [CrossRef]
18. Grün, A.; Nagy, D.I.; Németh, O.; Mucsi, Z.; Garadnay, S.; Greiner, I.; Keglevich, G. The Synthesis of 3-Phenylpropidronate Applying Phosphorus Trichloride and Phosphorous Acid in Methanesulfonic Acid. *Curr. Org. Chem.* **2016**, *20*, 1745–1752. [CrossRef]
19. Lecouvey, M.; Leroux, I. Synthesis of 1-hydroxy-1,1-bisphosphonates. *Heteroat. Chem.* **2000**, *11*, 556–561. [CrossRef]
20. Keglevich, G.; Grün, A.; Kovács, R.; Garadnay, S.; Greiner, I. Green chemical synthesis of bisphosphonic/dronic derivatives. *Phosphorus Sulfur Silicon Relat. Elem.* **2015**, *190*, 664–667. [CrossRef]
21. Troev, K.; Todorov, P.; Naydenova, E.; Mitova, V.; Vassiliev, N. A study of the reaction of phosphorus trichloride with paraformaldehyde in the presence of carboxylic acids. *Phosphorus Sulfur Silicon Relat. Elem.* **2013**, *188*, 1147–1155. [CrossRef]
22. Nagy, D.I.; Grün, A.; Garadnay, S.; Greiner, I.; Keglevich, G. Synthesis of hydroxymethylenebisphosphonic acid derivatives in different solvents. *Molecules* **2016**, *21*, 1046. [CrossRef] [PubMed]
23. Kachbi Khelfallah, S.; Monteil, M.; Deschamp, J.; Gager, O.; Migianu-Griffoni, E.; Lecouvey, M. Synthesis of novel polymerisable molecules bearing bisphosphonate. *Org. Biomol. Chem.* **2015**, *13*, 11382–11392. [CrossRef] [PubMed]
24. Grün, A.; Kovács, R.; Nagy, D.L.; Garadnay, S.; Greiner, I.; Keglevich, G. Efficient synthesis of benzidronate applying of phosphorus trichloride and phosphorous acid. *Lett. Drug Des. Discov.* **2015**, *12*, 78–84. [CrossRef]
25. Keglevich, G.; Grün, A.; Aradi, K.; Garadnay, S.; Greiner, I. Optimized synthesis of *N*-heterocyclic dronic acids; closing a black-box era. *Tetrahedron Lett.* **2011**, *21*, 2744–2746. [CrossRef]
26. Keglevich, G.; Grün, A.; Garadnay, S.; Greiner, I. Rational synthesis of dronic acid derivatives. *Phosphorus Sulfur Silicon Relat. Elem.* **2015**, *190*, 2116–2124. [CrossRef]
27. Ratrout, S.S.; Al Sarabi, A.L.; Sweidan, K.A. A one-pot and efficient synthesis of zolendronic acid starting from tert-butyl imodazol-1-yl acetate. *Pharm. Chem. J.* **2015**, *48*, 837–841. [CrossRef]
28. Mizrahi, D.A.; Waner, T.; Segall, Y. α-Amino acid derived bisphosphonates: Synthesis and anti-resorptive activity. *Phosphorus Sulfur Silicon Relat. Elem.* **2001**, *173*, 1–25. [CrossRef]
29. Lenin, R.; Raju, M.R.; Srinisava Rao, D.V.N.; Ray, U.K. Microvawe-assistant efficient synthesis of bisphosphonate libraries: A useful procedure for the preparation of bisphosphonates containing nitrogen and sulfur. *Med. Chem. Res.* **2013**, *22*, 1624–1629. [CrossRef]
30. Roth, A.G.; Drescher, A.; Yang, Y.; Redmer, S.; Uhling, S.; Arenz, C. Potent and selective inhibition of acid sphingomyelinase by bisphosphonates. *Angew. Chem. Int. Ed.* **2009**, *48*, 7560–7563. [CrossRef] [PubMed]
31. Szajnman, S.H.; Ravaschino, E.L.; Docampo, R.; Rodrigues, J.B. Synthesis and biological evaluation of 1-amino-1,1-bisphosphonates derived from fatty acids against *Trypanosoma cruzi* targeting farnesyl pyrophosphate synthase. *Bioorg. Med. Chem. Lett.* **2005**, *15*, 4685–4690. [CrossRef] [PubMed]
32. Wu, M.; Chen, P.; Huang, Y. Convenient synthesis of analogs of aminomethylene gem-diphosphonic acid from amines without catalyst. *Synth. Commun.* **2004**, *34*, 1393–1398. [CrossRef]
33. Fukuda, M.; Okamoto, Y.; Sakurai, H. Synthesis of dialkylaminomethylenediphosphonic acids. *Bull. Chem. Soc. Jpn.* **1975**, *48*, 1030–1031. [CrossRef]
34. Yu, C.M.; Wang, B.; Chen, Z.W. A Novel synthesis of alkylamino substituted methylenediphosphonates using bis(trichloromethyl) carbonate and $RCONR^1R^2$. *Chin. J. Chem.* **2008**, *26*, 1899–1901. [CrossRef]
35. Plöger, W.; Schindler, N.; Wollmann, K.; Worms, K.H. Herstellung von 1-Aminoalkan-1,1-diphosphonsäuren. *Z. Anorg. Allg. Chem.* **1972**, *389*, 119–128. [CrossRef]

36. Olive, G.; le Moigne, F.; Mercier, A.; Rockenbauer, A.; Tordo, P. Synthesis of tetraalkyl (pyrrolidine-2,2-diyl) bisphosphonates and 2,2-bis(diethoxyphosphoryl)-3,4-dihydro-2*H*-pyrrole 1-oxide; ESR study of derived nitroxides. *J. Org. Chem.* **1998**, *63*, 9095–9099. [CrossRef]
37. Olive, G.; Jacques, A. Tetraethyl(pyrrolidine-2,2-diyl)bisphosphonate. *Molecules* **2001**, *6*, M275. [CrossRef]
38. Olive, G.; Jacques, A. Optimization, continuation and lack of the one-step diphosphorylation reaction assay of modification of the tetraethyl(pyrrolidine-2,2-diyl)bisphosphonate. *Phosphorus Sulfur Silicon Relat. Elem.* **2003**, *178*, 33–46. [CrossRef]
39. Olive, G.; Rockenbauer, A.; Rozanska, X.; Jacques, A.; Peeters, D.; German, A. Synthesis of new tetraethyl(*N*-alkyl-1-aminoethan-1,1-diyl)bisphosphonates and ESR analysis of chemical exchange of derived nitroxides of acyclic aminobisphosphonates. *Phosphorus Sulfur Silicon Relat. Elem.* **2008**, *182*, 2359–2369. [CrossRef]
40. Wang, A.-E.; Chang, Z.; Sun, W.-T.; Huang, P.-Q. General and chemoselective bisphosphonylation of secondary and tertiary amides. *Org. Lett.* **2015**, *17*, 732–735. [CrossRef] [PubMed]
41. Prishchenko, A.A.; Livantsov, N.V.; Novikova, O.P.; Livantsova, L.I.; Petrosyan, V.S. Synthesis of the new types of *N*-substituted aminomethylenebisorganophosphorus acids and their derivatives. *Heteroatom Chem.* **2009**, *20*, 319–324. [CrossRef]
42. Prishchenko, A.A.; Livantsov, N.V.; Novikova, O.P.; Livantsova, L.I.; Ershov, I.R.; Petrosyan, V.S. Synthesis and reactivity of the new trimethylsilyl esters of aminomethylenebisorganophosphorus acids. *Heteroatom Chem.* **2013**, *24*, 355–360. [CrossRef]
43. Prishchenko, A.A.; Livantsov, N.V.; Novikova, O.P.; Livantsova, L.I.; Petrosyan, V.S. Synthesis of aminomethylenediphosphonates and their derivatives containing $PCNH_2$ fragments. *J. Gen. Chem.* **2014**, *83*, 608–610. [CrossRef]
44. Chmielewska, E.; Miszczyk, P.; Kozłowska, J.; Prokopowicz, M.; Młynarz, P.; Kafarski, P. Reaction of benzolactams with triethyl phosphite prompted by phosphoryl chloride affords benzoannulated monophosphonates instead of expected bisphosphonates. *J. Organomet. Chem.* **2015**, *785*, 84–91. [CrossRef]
45. Prishchenko, A.A.; Livantsov, N.V.; Novikova, O.P.; Livantsova, L.I.; Ershov, I.R.; Petrosyan, V.S. Synthesis of new types of aminomethylenediphosphorus-containing acids and their derivatives. *Russ. J. Gen. Chem.* **2015**, *85*, 370–379. [CrossRef]
46. Du, Y.; Jung, K.Y.; Wimer, D.F. A one-flask synthesis of α,α-bisphosphonates via enolate chemistry. *Tetrahedron Lett.* **2002**, *43*, 8665–8668. [CrossRef]
47. Srinivasa Rao, D.V.N.; Dandala, R.; Lenin, R.; Sivakumaran, N.; Shivashankar, S.; Naidu, A. A facile one pot synthesis of bisphosphonic acids and their sodium salts from nitriles. *Arch. Org. Chem.* **2007**, *2007*, 34–38.
48. Kaabak, L.V.; Kuz'mina, N.E.; Khudneko, A.V.; Tomilov, A.P. Improved synthesis of 1-aminoethylidenediphosphonic acid. *Russ. J. Gen. Chem.* **2006**, *76*, 1673–1674. [CrossRef]
49. Bandurina, T.A.; Konyukhov, V.N.; Panomareva, O.A.; Barybin, O.S.; Pushkareva, Z.N. Synthesis and antitumor activity of aminophosphonic acids. *Pharm. Chem. J.* **1978**, *12*, 1428–1431. [CrossRef]
50. Midier, C.; Lantsoght, M.; Volle, J.-N.; Pirat, J.L.; Virieux, D.; Stevens, C.V. Hydrophosphonylation of alkenes or nitriles by double radical transfer mediated by titanocene/propylene oxide. *Tetrahedron Lett.* **2011**, *52*, 6693–6696. [CrossRef]
51. Váradi, A.; Palmer, T.C.; Notis Dardashti, R.; Majumdar, S. Isocyanide-based multicomponent reactions for the synthesis of heterocycles. *Molecules* **2016**, *21*, 19. [CrossRef] [PubMed]
52. Hirai, T.; Han, L.-B. Palladium-catalyzed insertion of isocyanides into P(O)−H bonds: Selective formation of phosphinoyl imines and bisphosphinoylaminomethanes. *J. Am. Chem. Soc.* **2006**, *128*, 7422–7423. [CrossRef] [PubMed]
53. Goldemen, W.; Kluczyński, A.; Soroka, M. The preparation of *N*-substituted aminomethylidenebisphosphonates and their tetraalkyl esters via reaction of isonitriles with trialkyl phosphites and hydrogen chloride. Part 1. *Tetrahedron Lett.* **2012**, *53*, 5290–5292.
54. Goldeman, W.; Nasulewicz-Goldeman, A. Synthesis and antiproliferative activity of aromatic and aliphatic bis(aminomethylidene(bisphosphonic)) acids. *Bioorg. Med. Chem. Lett.* **2014**, *24*, 3475–3479. [CrossRef] [PubMed]
55. Kurzak, B.; Goldeman, W.; Szpak, M.; Matczak-Jon, E.; Kamecka, E. Synthesis of *N*-methyl alkylaminomethane-1,1-diphosphonic acids and evaluation of their complex-formation abilities towards copper(II). *Polyhedron* **2015**, *85*, 675–684. [CrossRef]

56. Goldeman, W.; Nasulewicz-Goldeman, A. Synthesis and biological evaluation of aminomethylidenebisphosphonic derivatives of β-arylethylamines. *Tetrahedron* **2015**, *71*, 3282–3289. [CrossRef]
57. Bhushan, K.R.; Tanaka, E.; Frangioni, J.V. Synthesis of conjugatable bisphosphonates for molecular imaging of large animals. *Angew. Chem. Int. Ed.* **2007**, *46*, 7969–7971. [CrossRef] [PubMed]
58. Ziora, Z.; Maly, A.; Lejczak, B.; Kafarski, P.; Holband, J.; Wójcik, G. Reactions of *N*-phthalylamino acid chlorides with trialkyl phosphites. *Heteroatom Chem.* **2000**, *11*, 232–239. [CrossRef]
59. Yanvarev, D.V.; Korovina, A.N.; Usanov, N.N.; Khomich, O.A.; Vepsäläinen, J.; Puljula, E.; Kukhanova, M.K.; Kochetkov, S.N. Data on synthesis of methylene bisphosphonates and screening of their inhibitory activity towards HIV reverse transcriptase. *Data Brief* **2016**, *8*, 1157–1167. [CrossRef] [PubMed]
60. Guenin, E.; Degache, E.; Liquier, J.; Lecouvey, M. Synthesis of 1-hydroxymethylene-1,1-bis(phosphonic acids) from acid anhydrides: Preparation of a new cyclic 1-acyloxymethylene-1,1-bis(phosphonic acid). *Eur. J. Org. Chem.* **2004**, 2983–2987. [CrossRef]
61. Kachbi-Khelfallah, S.; Monteil, M.; Cortes-Clerget, M.; Migianu-Griffoni, E.; Pirat, J.-L.; Gager, O.; Deschamp, J.; Lecouvey, M. Towards potential nanoparticle contrast agents: Synthesis of new functionalized PEG bisphosphonates. *Beilstein J. Org. Chem.* **2016**, *12*, 1366–1371. [CrossRef] [PubMed]
62. Teixeira, F.C.; Rangel, C.M.; Teixeira, A.P.S. Synthesis of new azole phosphonate precursors for fuel cells proton exchange membranes. *Heteroatom Chem.* **2015**, *26*, 238–248. [CrossRef]
63. Teixeira, F.C.; Lucas, C.; Curto, M.J.M.; Neves, M.; Duarte, M.T.; André, V.; Teixeira, A.P.S. New 1-Hydroxy-1,1-bisphosphonates derived from 1*H*-pyrazolo[3,4-*b*]pyridine: Synthesis and characterization. *J. Braz. Chem. Soc.* **2013**, *24*, 1295–1306. [CrossRef]
64. Kaboudin, B.; Ezzati, A.; Faghihi, M.R.; Barati, A.; Kazemi, F.; Abdollahi, H.; Yokomatsu, T. Hydroxy-bisphosphinic acids: Synthesis and complexation properties with transition metals and lanthanide ions in aqueous solution. *J. Iran. Chem. Soc.* **2016**, *13*, 747–752. [CrossRef]
65. Crossey, K.; Migaud, M.E. Solventless synthesis of acyl phosphonamidates, precursors to masked bisphosphonates. *Chem. Commun.* **2015**, *51*, 11088–11091. [CrossRef] [PubMed]
66. Egorov, M.; Aoun, S.; Padrines, M.; Redini, F.; Heymann, D.; Lebreton, J.; Mathé-Allainmat, M.A. One-pot synthesis of 1-hydroxy-1,1-bis(phosphonic acid)s starting from the corresponding carboxylic acids. *Eur. J. Org. Chem.* **2011**, 7148–7154. [CrossRef]
67. Suzuki, F.; Fujikawa, Y.; Yamamoto, S.; Mizutani, H.; Funabashi, C.; Ohya, T.; Ikai, T.; Oguchi, T. Pharmaceutical Compositions Containing Geminal Diphosphonates. US 5583122 A, 1 February 1979.
68. Maier, L. Organishe phosphorverbindungen 75: Herstellung und Eigenschaften von Aminomethylendiphosphinaten und -diphosphonaten, $RR^1NCH[P(O)R^2(OR^3)]_2$ und Derivaten. *Phosphorus Sulfur Silicon Relat. Elem.* **1981**, *11*, 311–332. [CrossRef]
69. Kaboudin, B.; Kalipour, S. A microwave-assisted solvent- and catalyst-free synthesis of aminomethylene bisphosphonates. *Tetrahedron Lett.* **2009**, *50*, 4243–4245. [CrossRef]
70. Minaeva, L.; Patrikeeva, L.S.; Kabachnik, M.M.; Beletskaya, I.P.; Orlinson, B.S.; Novakov, I.A. Synthesis of novel aminomethylenebisphosphonates and bisphosphonic acids, containing adamantyl fragment. *Heteroatom Chem.* **2011**, *22*, 55–58. [CrossRef]
71. Bálint, E.; Tajti, A.; Dziełak, A.; Hägele, G.; Keglevich, G. Microwave-assisted synthesis of (aminomethylene) bisphosphine oxides and (aminomethylene)bisphosphonates by a three-component condensation. *Beilstein J. Org. Chem.* **2016**, *12*, 1493–1502. [CrossRef] [PubMed]
72. Reddy, N.B.; Sundar, C.S.; Krishna, B.S.; Reddy, K.M.K.; Reddy, C.S. Synthesis and in-vitro antimicrobial activity of aminomethylene bisphosphonates. *Iran. J. Org. Chem.* **2014**, *6*, 1227–1234.
73. Krutikov, V.I.; Erkin, A.V.; Pautov, P.A.; Zolotukhina, M.M. Heteryl- and arylaminomethylenebisphosphonates: Synthesis and biologic activity. *Russ. J. Gen. Chem.* **2003**, *73*, 187–191. [CrossRef]
74. Siva Prasad, S.; Jayaprakash, S.H.; Syamasundar, Ch.; Sreelakshmi, P.; Bhuvaneswar, C.; Vijaya Bhaskar, B.; Rajendra, W.; Nayak, S.K.; Suresh Reddy, C. Tween 20-/H_2O promoted green synthesis, computational and antibacterial activity of amino acid substituted methylene bisphosphonates. *Phosphorus Sulfur Silicon Relat. Elem.* **2015**, *190*, 2040–2050. [CrossRef]
75. Mimura, M.; Hayashida, M.; Nomiyama, K.; Ikegami, S.; Iida, Y.; Tamura, M.; Hiyama, Y.; Ohishi, Y. Synthesis and evaluation of (piperidinomethylene)bis(phosphonic acid) derivatives as anti-osteoporosis agents. *Chem. Pharm. Bull.* **1993**, *41*, 1973–1986. [CrossRef]

76. De Schutter, J.W.; Shaw, J.; Lin, Y.-S.; Tsantrizos, Y.S. Design of potent bisphosphonate inhibitors of the human farnesyl pyrophosphate synthase via targeted interactions with the active site "capping" phenyls. *Bioorg. Med. Chem.* **2012**, *20*, 5583–5591. [CrossRef] [PubMed]
77. Leung, C.Y.; Langille, A.M.; Mancuso, J.; Tsantrizos, Y.S. Discovery of thienopyrimidine-based inhibitors of the human farnesyl pyrophosphate synthase—Parallel synthesis of analogs via a trimethylsilyl ylidene intermediate. *Bioorg. Med. Chem.* **2013**, *21*, 2229–2240. [CrossRef] [PubMed]
78. Leung, C.Y.; Park, J.; de Schutter, J.W.; Sebag, M.; Berghuis, A.M.; Tsantrizos, Y.S. Thienopyrimidine bisphosphonate (ThPBP) inhibitors of the human farnesyl pyrophosphate synthase: Optimization and characterization of the mode of inhibition. *J. Med. Chem.* **2013**, *56*, 7939–7950. [CrossRef] [PubMed]
79. Rubino, M.T.; Agamennone, M.; Campestre, C.; Campiglia, P.; Cremasco, V.; Faccio, R.; Laghezza, A.; Loiodice, F.; Maggi, D.; Panza, E.; et al. Biphenyl sulfonylamino methyl bisphosphonic acids as inhibitors of matrix metalloproteinases and bone resorption. *ChemMedChem* **2011**, *6*, 1258–1268. [CrossRef] [PubMed]
80. Tauro, M.; Lagezza, A.; Loiodice, F.; Agamennone, M.; Campestre, C.; Tortorella, P. Arylamino methylene bisphosphonate derivatives as bone seeking matrix metalloproteinase inhibitors. *Bioorg. Med. Chem.* **2013**, *21*, 6456–6465. [CrossRef] [PubMed]
81. Chmielewska, E.; Mazur, Z.; Kempińska, K.; Wietrzyk, J.; Piątek, A.; Kuryszko, J.J.; Kiełbowicz, Z.; Kafarski, P. *N*-Arylaminomethylenebisphosphonates bearing fluorine atoms: Synthesis and antiosteoporotic activity. *Phosphorus Sulfur Silicon Relat. Elem.* **2016**, *190*. [CrossRef]
82. Chmielewska, E.; Kempińska, K.; Wietrzyk, J.; Piątek, A.; Kuryszko, J.J.; Kiełbowicz, Z.; Kafarski, P. Novel Bisphosphonates and Their Use. WO2015159153 A1, 22 October 2015.
83. Kubiček, V.; Rudovský, J.; Kotek, J.; Hermann, P.; Elst, L.V.; Muller, R.N.; Kolar, Z.I.; Wolterbeek, H.Th.; Peters, J.A.; Lukeš, I. A Bisphosphonate monoamide analogue of DOTA: A potential agent for bone targeting. *J. Am. Chem. Soc.* **2005**, *127*, 16477–16485. [CrossRef] [PubMed]
84. Årstad, E.; Hoff, P.; Skattebøl, L.; Skretting, A.; Breistøl, K. Studies on the synthesis and biological properties of non-carrier-added [^{125}I and ^{131}I]-labeled arylalkylidenebisphosphonates: Potent bone-seekers for diagnosis and therapy of malignant osseous lesions. *J. Med. Chem.* **2003**, *46*, 3021–3032. [CrossRef] [PubMed]
85. Martin, M.B.; Sanders, J.M.; Kendrick, H.; de Luca-Fradley, K.; Lewis, J.C.; Grimley, J.S.; van Brussel, E.M.; Olsen, J.R.; Meints, G.A.; Burzyńska, A.; et al. Activity of Bisphosphonates against *Trypanosoma brucei rhodesiense*. *J. Med. Chem.* **2002**, *45*, 2904–2914. [CrossRef] [PubMed]
86. Sanders, J.M.; Gomez, A.O.; Mao, J.; Meints, G.A.; van Brussel, E.M.; Burzyńska, A.; Kafarski, P.; Gonzalez-Pacanowska, D.; Oldfield, E. 3-D QSAR investigations of the inhibition of *Leishmania major* farnesyl pyrophosphate synthase by bisphosphonates. *J. Med. Chem.* **2003**, *46*, 5171–5183. [CrossRef] [PubMed]
87. Ghosh, S.; Chan, J.M.; Lea, C.R.; Meints, G.A.; Lewis, J.C.; Tovian, Z.S.; Flessner, R.M.; Loftus, T.C.; Bruchhaus, I.; Kendrick, H.; et al. Effects of bisphosphonates on the growth of *Entamoeba histolytica* and Plasmodium species in vitro and in vivo. *J. Med. Chem.* **2004**, *47*, 175–187. [CrossRef] [PubMed]
88. Ling, Y.; Sahota, D.; Odeh, S.; Chan, J.M.; Araujo, F.G.; Moreno, S.N.; Oldfield, E. Bisphosphonate inhibitors of Toxoplasma gondi growth: In vitro, QSAR, and in vivo investigations. *J. Med. Chem.* **2005**, *48*, 3130–3140. [CrossRef] [PubMed]
89. Tanaka, K.S.E.; Houghton, T.J.; Kang, T.; Dietrich, E.; Delorme, D.; Ferreira, S.S.; Caron, L.; Viens, F.; Arhin, S.S.; Sarmiento, I.; et al. Bisphosphonated fluoroquinolone esters as osteotropic prodrugs for the prevention of osteomyelitis. *Bioorg. Med. Chem.* **2008**, *16*, 9217–9229. [CrossRef] [PubMed]
90. Forlani, G.; Petrolino, D.; Fusetti, M.; Romanini, L.; Nocek, B.; Joachimiak, A.; Berlicki, L.; Kafarski, P. δ^1-Pyrroline-5-carboxylate reductase as a new target for therapeutics: Inhibition of the enzyme from *Streptococcus pyogenes* and effects in vivo. *Amino Acids* **2012**, *42*, 2283–2291. [CrossRef] [PubMed]
91. Brel, V.K. Synthesis of *gem*-bisphosphonates with (3-aryl-4,5-dihydroxisoxazol-5-yl)methylamino moiety. *Mendeleev Commun.* **2015**, *25*, 234–235. [CrossRef]
92. Kosikowska, P.; Bochno, M.; Macegoniuk, K.; Forlani, G.; Kafarski, P.; Berlicki, Ł. Bisphosphonic acids as effective inhibitors of *Mycobacterium tuberculosis* glutamine synthetase. *J. Enzyme Inhib. Med. Chem.* **2016**, *31*, 931–938. [CrossRef] [PubMed]
93. Lacbay, C.M.; Mancuso, J.; Lin, Y.-S.; Bennett, N.; Götte, M.; Tsantrizos, Y.S. Modular assembly of purine-like bisphosphonates as inhibitors of HIV-1 reverse transcriptase. *J. Med. Chem.* **2014**, *57*, 7435–7449. [CrossRef] [PubMed]

94. Shaddy, A.A.; Kamel, A.A.; Abdou, W.M. Synthesis, quantitative structure-activity relationship, and anti-inflammatory profiles of substituted 5- and 6-*N*-heterocycle bisphosphonate esters. *Chem. Commun.* **2013**, *43*, 236–252. [CrossRef]
95. Kafarski, P.; Lejczak, B.; Forlani, G. Herbicidally active aminomethylenebisphosphonic acids. *Heteroat. Chem.* **2000**, *11*, 449–453. [CrossRef]
96. Forlani, G.; Berlicki, Ł.; Duò, M.; Dzięddzioła, G.; Giberti, S.; Bertazzini, M.; Kafarski, P. Synthesis and Evaluation of Effective Inhibitors of Plant δ^1-Pyrroline-5-carboxylate Reductase. *J. Agric. Food Chem.* **2013**, *61*, 6792–6798. [CrossRef] [PubMed]
97. Giberti, S.; Bertazzini, M.; Liboni, M.; Berlicki, Ł.; Kafarski, P.; Forlani, G. Phytotoxicity of aminobisphosphonates targeting both δ1-pyrroline-5-carboxylate reductase and glutamine synthetase. *Pest Manag. Sci.* **2016**, in press. [CrossRef] [PubMed]
98. Dobosz, A.; Spychała, J.; Ptak, T.; Chmielewska, E.; Maciejewska, G.; Kafarski, P.; Młynarz, P. Electrochemical and spectroscopic investigations of selected *N*-heteroalkylaminomethylene-bisphosphonic acids with Pb(II) ions. *Coord. Chem. Rev.* **2016**, in press. [CrossRef]
99. Dąbrowska, E.; Burzyńska, A.; Mucha, A.; Matczak-Jon, E.; Sawka-Dobrowolska, W.; Berlicki, Ł.; Kafarski, P. Insight into the mechanism of three component condensation leading to aminomethylenebisphosphonates. *J. Organomet. Chem.* **2009**, *6943*, 3806–3813. [CrossRef]
100. Lejczak, B.; Boduszek, B.; Kafarski, P.; Forlani, G.; Wojtasek, H.; Wieczorek, P. Mode of action of herbicidal derivatives of aminomethylenebisphosphonic acid. I. Physiologic activity and inhibition of anthocyanin biosynthesis. *J. Plant Growth Regul.* **1996**, *15*, 109–113. [CrossRef]
101. Bochno, M.; Berlicki, Ł. A three-component synthesis of aminomethylene bis-*H*-phosphinates. *Tetrahedron Lett.* **2014**, *55*, 219–223. [CrossRef]
102. Rodriguez, J.B. Tetraethyl vinylidenebisphosphonate: A versatile synthon for the preparation of bisphosphonates. *Synthesis* **2014**, *46*, 1129–1142. [CrossRef]
103. Rosso, V.S.; Szajman, S.H.; Malayil, L.; Galizzi, M.; Moreno, S.N.J.; Docampo, R.; Rodriguez, J.B. Synthesis and biological evaluation of new 2-alkylaminoethyl-1,1-bisphosphonic acids against *Trypanosoma cruzi* and *Toxoplasma gondii* targeting farnesyl diphosphate synthase. *Bioorg. Med. Chem.* **2011**, *19*, 2211–2217. [CrossRef] [PubMed]
104. Szajnman, S.H.; Liñares, G.E.G.; Li, Z.-H.; Jiang, C.; Galizzi, M.; Bontempi, E.; Ferella, M.; Moreno, S.N.J.; Docampo, R.; Rodriguez, J.B. Synthesis and biological evaluation of 2-alkylaminoethyl-1,1-bisphosphonic acids against *Trypanosoma cruzi* and *Toxoplasma gondii* targeting farnesyl diphosphate synthase. *Bioorg. Med. Chem.* **2008**, *15*, 3283–3290. [CrossRef] [PubMed]
105. Herczegh, P.; Buxton, T.B.; McPherson, J.C., III; Kovács-Kulyassa, A.; Brewer, P.D.; Sztaricskai, F.; Stroebel, G.C.; Plowman, K.M.; Farcasiu, D.; Hartmann, J.F. Osteoadsorptive bisphosphonate derivatives of fluoroquinolone antibacterials. *J. Med. Chem.* **2002**, *45*, 2338–2341. [CrossRef] [PubMed]
106. Houghton, T.J.; Tanaka, K.S.E.; Kang, T.; Dietrich, E.; Lafontaine, Y.; Delorme, D.; Ferreira, S.S.; Viens, F.; Arhin, F.F.; Sarmiento, I.; et al. Linking bisphosphonates to the free amino groups in fluoroquinolones: Preparation of osteotropic prodrugs for the prevention of osteomyelitis. *J. Med. Chem.* **2008**, *51*, 6955–6969. [CrossRef] [PubMed]
107. Dzięgielewski, M.; Hejmanowska, J.; Albrecht, Ł. A convenient approach to a novel group of quaternary amino acids containing a geminal bisphosphonate moiety. *Synthesis* **2014**, *46*, 3233–3239. [CrossRef]
108. Simoni, D.; Gebbia, N.; Invidiata, M.P.; Eleopra, M.; Marchetti, P.; Rondanin, R.; Baruchello, R.; Provera, S.; Marchioro, C.; Tolomeo, M.; et al. Design, synthesis, and biological evaluation of novel aminobisphosphonates possessing an in vivo antitumor activity through a γδ-T lymphocytes-mediated activation mechanism. *J. Med. Chem.* **2008**, *51*, 6800–6807. [CrossRef] [PubMed]
109. Sankala, E.; Weisell, J.M.; Huhtala, T.; Närvänen, A.T.O.; Vepsäläinen, J.J. Synthesis of novel bisphosphonate polyamine conjugates and their affinity to hydroxyapatite. *Arch. Org. Chem.* **2012**, *2012*, 233–241.
110. Chen, K.M.C.; Hudock, M.P.; Zhang, Y.; Guo, R.T.; Cao, R.; No, J.H.; Liang, P.H.; Ko, T.-P.; Chang, T.H.; Chang, S.; et al. Inhibition of geranylgeranyl diphosphate synthase by bisphosphonates: A crystallographic and computational investigation. *J. Med. Chem.* **2008**, *51*, 5594–5607.
111. Zhang, Y.; Leon, A.; Song, Y.; Studer, D.; Haase, C.; Kościelski, Ł.A.; Oldfield, E. Activity of nitrogen-containing and non-nitrogen-containing bisphosphonates on tumor cell lines. *J. Med. Chem.* **2006**, *49*, 5804–5814. [CrossRef] [PubMed]

112. Makarov, M.V.; Rybalkina, E.Y.; Röschenthaler, G.-V. New 3,5-bis(arylidene)-4-piperidones with bisphosphonate moiety: Synthesis and antitumor activity. *Russ. Chem. Bull. Int. Ed.* **2014**, *63*, 1181–1186. [CrossRef]
113. Skarpos, H.; Osipov, S.N.; Vororb'eva, D.V.; Odinets, I.L.; Lork, E.; Röschenthaler, G.-V. Synthesis of functionalized bisphosphonates via click chemistry. *Org. Biomol. Chem.* **2007**, *5*, 2361–2367. [CrossRef] [PubMed]
114. Liu, S.; Bi, W.; Li, X.; Chen, X.; Qu, N.; Zhao, Y. A Practical method to synthesize 1,2,3-triazole-amino-bisphosphonate derivatives. *Phosphorus Sulfur Relat. Elem.* **2015**, *190*, 1735–1742. [CrossRef]
115. Gouault-Bironneau, S.; Deprèle, S.; Sutor, A.; Montchamp, J.-L. Radical reaction of sodium hypophosphite with terminal alkynes: Synthesis of 1,1,-bis-*H*-phosphinates. *Org. Lett.* **2005**, *7*, 5909–5912. [CrossRef] [PubMed]
116. Kuźnik, A.; Mazurkiewicz, R.; Grymel, M.; Zielińska, K.; Adamek, J.; Chmielewska, E.; Bochno, M.; Kubica, S. A new method for the synthesis of α-aminoalkylidenebisphosphonates and their asymmetric phosphonyl-phosphinyl and phosphonyl-phosphinoyl analogues. *Beilstein J. Org. Chem.* **2015**, *11*, 1418–1424. [CrossRef] [PubMed]
117. Tanaka, K.S.; Dietrich, E.; Ciblat, S.; Métayer, C.; Arhin, F.F.; Sarmiento, I.; Moeck, G.; Parr, T.R., Jr.; Far, A.R. Synthesis and in vitro evaluation of bisphosphonate glycopeptide prodrugs for the treatment of osteomyelitis. *Bioorg. Med. Chem. Lett.* **2010**, *20*, 1355–1359. [CrossRef] [PubMed]
118. Bala, J.L.F.; KAshemirov, B.A.; McKenna, C.E. Synthesis of novel bisphosphonic acid alkene monomer. *Synth. Commun.* **2010**, *40*, 3577–3584. [CrossRef]
119. Lecerclé, D.; Gabillet, S.; Gomis, J.-M.; Taran, F. A facile synthesis of aminomethylene bisphosphonates through rhodium carbenoid mediated N-H insertion reaction. Application to the preparation of powerful uranyl ligands. *Tetrahedron Lett.* **2008**, *49*, 2083–2087. [CrossRef]
120. Masschelein, K.G.R.; Stevens, C.V. Double nucleophilic 1,2-addition of silylated dialkyl phosphites to 4-phosphono-1-aza-1,3-dienes: Synthesis of γ-phosphono-α–aminobisphosphonates. *J. Org. Chem.* **2007**, *72*, 9248–9252. [CrossRef] [PubMed]
121. Rassukana, Y.V.; Davydova, K.O.; Onys'ko, P.P.; Sinitsa, A.D. Synthesis and rearrangements of *N*-trichloroacetylfluoroacetimidolyl chloride and its phosphorylation products. *J. Fluor. Chem.* **2002**, *117*, 107–113. [CrossRef]
122. Rassukana, Y.V.; Onys'ko, P.P.; Davydova, K.O.; Sinitsa, A.D. A new reaction of phosphorylated *N*-sulfonylimines with hydrophosphoryl agents involving C-N transfer of phosphoryl groups. *Tetrahedron Lett.* **2004**, *45*, 3899–3902. [CrossRef]
123. Kolotilo, N.V.; Sinitsa, A.D.; Rassukanaya, Y.V.; Onys'ko, P.P. *N*-Sulfonyl- and *N*-phosphorylbenzimidoylphosphonates. *Russ. J. Gen. Chem.* **2006**, *76*, 1210–1218. [CrossRef]
124. Romanenko, V.D.; Kukhar, V.P. Methylidynetrisphosphonates: Promising C1 building block for the design of phosphate mimetics. *Beilstein J. Org. Chem.* **2013**, *9*, 991–1001. [CrossRef] [PubMed]
125. Kafarski, P.; Lejczak, B. A facile conversion of aminoalkanephosphonic acids into their diethyl esters. The use of unblocked aminophosphonic acids in phosphono peptide synthesis. *Synthesis* **1988**, *4*, 307–310. [CrossRef]
126. Turhanen, P.A.; Weisell, J.; Vepsäläinen, J.J. Preparation of mixed trialkyl alkylcarbonate derivatives of etidronic acid via an unusual route. *Beilstein J. Org. Chem.* **2012**, *8*, 2019–2024. [CrossRef] [PubMed]
127. Turhanen, P.A.; Vepsäläinen, J.J. Synthesis of novel (1-alkanoyloxy-4-alkanoylaminobutylidene)-1,1-bisphosphonic acid derivatives. *Beilstein J. Org. Chem.* **2006**, *2*, 2. [CrossRef] [PubMed]
128. Xu, G.; Yang, C.; Bo, L.; Wu, X.; Xie, Y. Synthesis of new potential chelating agents: Catechol-bisphosphonate conjugates for metal intoxication therapy. *Heteroatom Chem.* **2004**, *15*, 251–257. [CrossRef]
129. Agyin, J.K.; Santhamma, B.; Roy, S.S. Design, synthesis, and biological evaluation of bone-targeted proteasome inhibitors for multiple myeloma. *Bioorg. Med. Chem.* **2013**, *23*, 5455–6458. [CrossRef] [PubMed]
130. Ezra, A.; Hoffman, A.; Breuer, E.; Alferiev, A.S.; Mönkkönen, J.; El Hanany-Rozen, N.; Weiss, G.; Stepensky, D.; Gati, I.; Cohen, H.; et al. A Peptide prodrug approach for improving bisphosphonate oral absorption. *J. Med. Chem.* **2000**, *43*, 3641–3652. [CrossRef] [PubMed]
131. Sun, S.; Błażewska, K.; Kadina, A.P.; Kashemirov, B.A.; Duan, X.; Triffitt, J.T.; Dunford, J.E.; Russell, R.G.R.; Ebetino, F.H.; Roelofs, A.J.; et al. Fluorescent bisphosphonate and carboxyphosphonate probes: A versatile imaging toolkit for applications in bone biology and biomedicine. *Bioconj. Chem.* **2016**, *27*, 329–340. [CrossRef] [PubMed]

132. Liu, J.; Jo, J.; Kawai, Y.; Aoki, I.; Tanaka, C.; Yamamoto, M.; Tabata, Y. Preparation of polymer-based multimodal imaging agent to visualize the process of bone regeneration. *J. Control. Release* **2012**, *157*, 398–405. [CrossRef] [PubMed]
133. Varghese, O.P.; Sun, W.; Hilborn, J.; Ossipov, D.A. In situ cross-linkable high molecular weight hyaluronan-bisphosphonate conjugate for localized delivery and cell-specific targeting: A hydrogel linked prodrug approach. *J. Am. Chem. Soc.* **2009**, *131*, 8781–8783. [CrossRef] [PubMed]
134. Vitha, T.; Kubíček, V.; Hermann, P.; Elst, L.V.; Muller, R.N.; Kollar, Z.I.; Wolterbeek, H.T.; Breeman, W.A.P.; Lukeš, I.; Peters, J.A. Lanthanide(III) Complexes of Bis(phosphonate) Monoamide Analogues of DOTA: Bone-Seeking Agents for Imaging and Therapy. *J. Med. Chem.* **2008**, *51*, 677–683. [CrossRef] [PubMed]
135. Shao, X.; Xu, Y.; Jiao, Y.; Zhou, C. Synthesis and characterization of an alendronate-chitosan conjugate. *Appl. Mech. Mater.* **2012**, *140*, 53–57. [CrossRef]
136. Morioka, M.; Kamizono, A.; Takikawa, H.; Mori, A.; Ueno, H.; Kadowaki, S.; Nakano, Y.; Kato, K.; Umezawa, K. Design, synthesis, and biological evaluation of novel estradiol-bisphosphonate conjugates as bone-specific estrogens. *Bioorg. Med. Chem.* **2010**, *18*, 1143–1148. [CrossRef] [PubMed]
137. Dadiboyena, S. Recent advances in the synthesis of raloxifene: A selective estrogen receptor modulator. *Eur. J. Med. Chem.* **2012**, *51*, 17–34. [CrossRef] [PubMed]
138. Palma, E.; Oliveira, B.M.; Correira, J.D.G.; Gano, L.; Maria, L.; Santos, I.C.; Santos, I. A new bisphosphonate-containing ^{99m}Tc(I) tricarbonyl complex potentially useful as bone-seeking agent: Synthesis and biological evaluation. *J. Biol. Inorg. Chem.* **2007**, *12*, 667–679. [CrossRef] [PubMed]
139. Reddy, R.; Dietrich, E.; Lafontaine, Y.; Houghton, T.J.; Belanger, O.; Dubois, A.; Arhin, F.F.; Sarmiento, I.; Fadhil, I.; Laquerre, K.; et al. Bisphosphonated benzoxazinorifamycin prodrugs for the prevention and treatment of osteomyelitis. *ChemMedChem* **2008**, *3*, 1863–1868. [CrossRef] [PubMed]
140. Bekker, K.S.; Chukanov, N.V.; Grigor'ev, I.A. Synthesis of a bisphosphonate derivative of folic acid. *Chem. Nat. Prod.* **2013**, *49*, 495. [CrossRef]
141. Bekker, K.S.; Chukanov, N.V.; Popov, S.A.; Grigor'ev, I.A. New Amino-Bisphosphonate Building Blocks in the Synthesis of Bisphosphonic Derivatives Based on Lead Compounds. *Phosphorus Sulfur Silicon Relat. Elem.* **2015**, *190*, 1209–1212. [CrossRef]
142. Bekker, K.S.; Chukanov, N.V.; Popov, S.A.; Grigor'ev, I.A. Conjugates of bisphosphonates with derivatives of ursolic, betulonic, and betulinic acids. *Chem. Nat. Prod.* **2013**, *49*, 581–582. [CrossRef]
143. Wang, L.; Yang, Z.; Gao, J.; Xu, K.; Gu, H.; Zhang, B.; Zhang, X.; Xu, B. A biocompatible method of decorporation: Bisphosphonate-modified magnetite nanoparticles to remove uranyl ions from blood. *J. Am. Chem. Soc.* **2006**, *131*, 13358–13359. [CrossRef] [PubMed]
144. Bekker, K.S.; Chukanov, N.V.; Grigor'ev, I.A. Synthesis of a bis-phosphonic acid derivative of trolox, a new potential antioxidant. *Chem. Nat. Prod.* **2013**, *49*, 785–786. [CrossRef]
145. Altin, A.; Akgun, B.; Bilgici, Z.S.; Turker, S.B.; Avci, D. Synthesis, photopolymerization, and adhesive properties of hydrolytically stable phosphonic acid-containing (meth)acrylamides. *J. Polym. Sci. A Polym. Chem.* **2014**, *52*, 512–522. [CrossRef]
146. Akgun, B.; Avci, D. Synthesis and evaluations of bisphosphonate-containing monomers for dental materials. *J. Polym. Sci. A Polym. Chem.* **2012**, *50*, 4854–4863. [CrossRef]
147. Yang, X.; Akhtar, X.; Rubino, S.; Leifer, K.; Hilborn, J.; Ossipov, D. Direct "click" synthesis of hybrid bisphosphonate—Hyaluronic acid hydrogel in aqueous solution for biomineralization. *Chem. Mater.* **2012**, *24*, 1690–1697. [CrossRef]
148. Gluz, E.; Grinberg, I.; Corem-Salkmon, E.; Mizrahi, D.; Margel, S. Engineering of new crosslinked near-infrared fluorescent polyethylene glycol bisphosphonate nanoparticles for bone targeting. *J. Polym. Sci. A Polym. Chem.* **2013**, *51*, 4282–4291. [CrossRef]
149. Mukaya, H.E.; Mbianda, X.Y. Macromolecular co-conjugate of ferrocene and bisphosphonate: Synthesis, characterization and kinetic drug release study. *J. Inorg. Organomet. Polym.* **2015**, *25*, 411–418. [CrossRef]
150. Beck, J.; Gharbi, S.; Herteg-Fernea, A.; Vercheval, L.; Berbone, C.; Lassaux, P.; Zervosen, A.; Marchant-Bryanert, J. Aminophosphonic acids and aminobis(phosphonic acids) as potential inhibitors of penicillin-binding proteins. *Eur. J. Org. Chem.* **2009**, *2009*, 85–97. [CrossRef]
151. Chuiko, A.L.; Filonenko, L.P.; Borisevich, A.M.; Lozinskii, M.O. pH-Dependent isomerism of (iso)thioureidomethylenebisphosphonates. *Russ. J. Gen. Chem.* **2009**, *79*, 72–77. [CrossRef]

152. Chuiko, A.L.; Filonenko, L.P.; Lozinskii, M.O. Reaction of thioureidomethylenebisphosphonic acids with α-haloketones: II. Effect of pH on the reaction pathway. *Russ. J. Gen. Chem.* **2009**, *79*, 57–66. [CrossRef]
153. Chuiko, A.L.; Filonenko, L.P.; Lozinskii, M.O. Synthesis of ((thiazol-2-ylamino)-methylene)bisphosphonic acids. *Chem. Heterocycl. Compd.* **2011**, *47*, 1137–1140. [CrossRef]
154. Sirviö, J.A.; Hasa, T.; Ahola, J.; Liimatainen, H.; Niinimäki, J.; Hormi, O. Phosphonated nanocelluloses from sequential oxidative-reductive treatment—Physicochemical characteristics and thermal properties. *Carbohydr. Polym.* **2015**, *133*, 524–532. [CrossRef] [PubMed]
155. Bonzi, G.; Salmaso, S.; Scomparin, A.; Edar-Boock, A.; Satchi-Fainaro, R.; Caliceti, P. Novel pullulan bioconjugate for selective breast cancer bone metastases treatment. *Bioconj. Chem.* **2015**, *26*, 489–501. [CrossRef] [PubMed]

molecules

MDPI

Review

Chiral Hypervalent, Pentacoordinated Phosphoranes

Dorota Krasowska [1,*,†], Jacek Chrzanowski [1,*,†], Piotr Kiełbasiński [1,†] and Józef Drabowicz [1,2,*,†]

1 Centre of Molecular and Macromolecular Studies, Polish Academy of Sciences, Sienkiewicza 112, 90-363 Lodz, Poland; piokiel@cbmm.lodz.pl
2 Department of Chemistry, Environment Protection and Biotechnology, Jan Dlugosz University in Czestochowa, Armii Krajowej Ave. 13/15, 42-200 Czestochowa, Poland
* Correspondence: dkrasowska@gmail.com (D.K.); jacekchrzanowski1@gmail.com (J.C.); drabow@gmail.com (J.D.); Tel.: +48-42-680-32-34 (J.D.)
† These authors contributed equally to this work.

Academic Editor: György Keglevich
Received: 29 September 2016; Accepted: 16 November 2016; Published: 21 November 2016

Abstract: This review presents synthetic procedures applied to the preparation of chiral (mainly optically active) pentacoordinated, hypervalent mono and bicyclic phosphoranes. The mechanisms of their stereoisomerization and their selected interconversions are also presented.

Keywords: hypervalency; chirality; phosphoranes; Berry pseudorotation; turnstile rotation

1. Introduction

The phenomenon of chirality plays a very important role in organic chemistry and biological processes [1–5]. Therefore an easy access to chiral compounds, especially optically active species, constitutes a vital challenge in modern synthetic organic chemistry. At least two reasons are responsible for this fact. Firstly, single stereoisomers are used very often as active components in formulations which serve as drugs, food additives or substances having the requested flavor or fragrance. The second reason stems from their application as chiral catalysts in various asymmetric reactions [6,7]. Moreover, detailed mechanistic description of a number of reactions would not be possible without taking into account their stereochemical aspects. These can be studied only with the use of optically active species which allow simultaneous application of kinetic and polarimetric measurements carried out with the use of the suitably designated model substrates (enantiomerically or diastereomerically pure or enriched samples) containing an appropriate chirality element [8]. In the organophosphorus chemistry the phenomenon of chirality is very common among organic derivatives having different valency and/or coordination number [9,10]. Their optical activity can be related to the presence of the following elements of chirality:

(a) a stereogenic center (for trivalent, tricoordinated, tetravalent tetracoordinated, pentavalent tetracoordinated derivatives)
(b) "trigonal bipyramidal chirality"
(c) "tetragonal bipyramidal chirality"

Organophosphorus derivatives in which optical activity results from the presence of a phosphorus atom at a center of trigonal bipyramid, are commonly named phosphoranes [11,12]. They have the general formula $P(L_a)_2(L_e)_3$, where "a" means an axial position and "e" means an equatorial position (Figure 1).

Figure 1. General formula of a phosphorus-containing trigonal bipyramid.

They constitute the family of hypervalent derivatives, showing "trigonal bipyramidal chirality" in which the central phosphorus atom has expanded its valence shell from 8 to 10 electrons (the concept was for the first time proposed by Musher in 1969) [13]. According to general systematic scheme proposed by Martin and coworkers [14,15] [N-P-L ($_nA_mB$) coding system, in which N stands for the number of valence electrons associated formally with a central phosphorus atom and L shows the number of ligands (A and B stand for the bonding element)]—this group should be considered as the family of 10-P-5 derivatives. The definition of hypervalency is also formally fulfilled by the corresponding phosphonium ylides (in which a phosphorus atom has also expanded its formal valence shell from 8 to 10 electrons). However, when one considers that the important resonance structures of these compounds are polarized from phosphorus to carbon it becomes evident that they represent 8-P-4 species (Figure 2). Therefore they do not meet the condition of hypervalency and they are excluded from this review.

Figure 2. Resonance structure of phosphonium ylides.

The nature of bonding of hypervalent derivatives is presented in details in a few chapters of "*The Chemistry of Hypervalent Compounds*" [16,17] and more recently has again been studied using the topological analysis of the electron localization function (ELF)" [18]. Their structure and reactivity are dominated by the weak three-center-four-electron (3c-4e) bond. Its weakness results from the fact that two of the four electrons that participate in the bonding are in a nonbonding molecular orbital.

Phosphoranes, like other compounds having a trigonal bipyramide (TBP) geometry can exist in an enantiomeric form when the number of different ligands is sufficiently large to induce chirality of such a structure. For 10-P-5 derivatives having the general structure with five substituents (Figure 1) bonded to the central phosphorus atom the numbers of achiral and chiral structures are given in Table 1. According to the Mutterties rule [19] in hypervalent structures the more electronegative ligands tend to occupy the apical positions whereas the lone electron pair "should be located at the equatorial position" [19]. Moreover, "five- and six-membered rings, which stabilize the hypervalent molecules, span both the axial and equatorial positions" [14].

Table 1. Number of chiral and achiral species of 10-P-5 (TBP geometry).

General Structure	Number of Different Structures and Their Chirality
PL_5	1 achiral
$PL^1{}_4L^2$	2 achiral
$PL^1{}_3L^2{}_2$	3 achiral
$PL^1{}_3L^2L^3$	2 achiral and 2 pairs of enantiomers
$PL^1{}_2L^2{}_2L^3$	2 achiral and 3 pairs of enantiomers
$PL^1{}_2L^2L^3L^4$	1 achiral and 6 pairs of enantiomers
$PL^1L^2L^3L^4L^5$	10 pairs of enantiomers

It is evident from this table that phosphoranes containing at least three different ligands can be chiral and for structures $PL^1{}_2L^2{}_2L^3$ chirality may appear in the symmetric bicyclic spiro derivatives

(Figure 3). Therefore, from the synthetic point of view, hypervalent structures exhibiting chirality due to the presence of one or two bidentate ligands should be more easily available than its acyclic analogues.

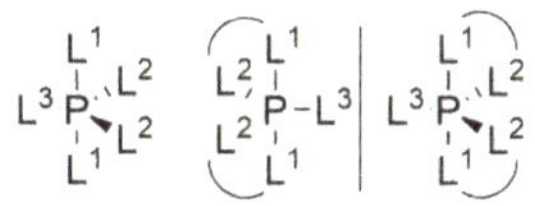

Figure 3. Chiral and achiral structure of hypervalent phosphoranes.

In hypervalent phosphoranes 10-P-5 the apical and equatorial ligands can be interchanged with each other according to the Berry pseudorotation mechanism (BPR) [20–22] (Scheme 1) or by the "turnstile" rotation mechanism (TR) proposed by Ugi and Ramirez [23–26] (Scheme 2). It should be noted here that the very recent DFT calculations of the fluxional behavior of experimentally known pentavalent molecules have suggested the equivalence of the turnstile rotation with the Berry pseudorotation. This suggestion is based on identification of three principal mechanisms by which the substituents interchange can be achieved (Berry pseudorotation, threefold cyclic permutation, and half-twist axial-equatorial interchange) [27].

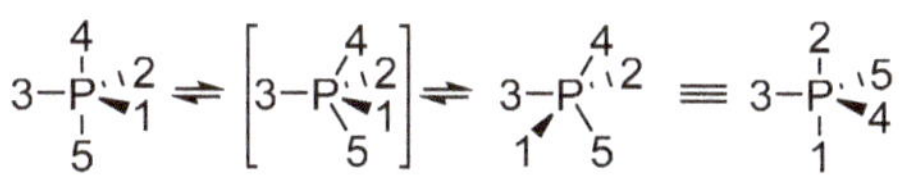

Scheme 1. Berry pseudorotation mechanism.

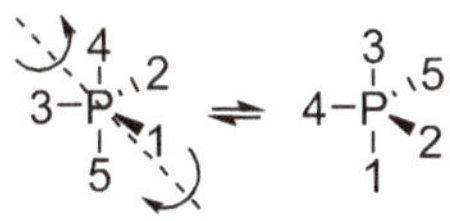

Scheme 2. "Turnstile" rotation mechanism (TR).

As early as 1949, Wittig and Rieber [28] reported isolation of pentaphenylphosphorane as the first example of a pentacoordinated organophosphorus derivative. However, a real opening in the hypervalent organophosphorus chemistry can be ascribed to a much later finding that the Ramirez reaction affords oxyphosphoranes by an oxidative addition of α-diketones to trivalent phosphines [29,30] and Westheimer's proposal that a pentacoordinate phosphorane (able to stereomutate by pseudorotation) is involved in the hydrolysis of phosphoric esters [31]. Since that time, studies devoted to the synthesis, structural determinations, stereochemistry and reactivity of this group of the organophosphorus derivatives, have been carried out in a number of academic and industrial laboratories. Their results are regularly published in renowned chemical journals and, fortunately for those who work on these topics, are regularly presented in the chapters devoted to penta- and hexacoordinated organophosphorus derivatives included in the subsequent volumes of RSC monographic series "*Organophosphorus Chemistry*" [32–43]. In this respect, a very recent review highlighting methods for the asymmetric synthesis of P-chiral pentacoordinated spirophosphoranes should also be mentioned [44].

2. Chiral Hypervalent Phosphoranes and Their Anions

2.1. Chiral Phosphoranes as Reactive Intermediates

In most cases the papers describing the involvement of unstable chiral phosphoranes as reactive intermediates tackle the problem of the mechanism of S_N-P reactions. This is connected with the fact that a nucleophilic substitution at phosphorus can occur either synchronously according to an S_N2-P mechanism or in a stepwise manner by an addition-elimination mechanism (A-E). The later involves formation of a phosphorane as an intermediate, by the addition of a nucleophile$_s$ (Nu) to a chiral substrate. It is now generally accepted that diaxial or diequatorial disposal of entering "Nu" and leaving "L" groups in a trigonal bipyramidal structure of a transient or intermediary phosphorane should lead to inversion of the configuration at phosphorus while the steric course of axial-equatorial substitution (which is possible if a phosphorane intermediate undergoes pseudorotation) is predicted to be a retention (Scheme 3).

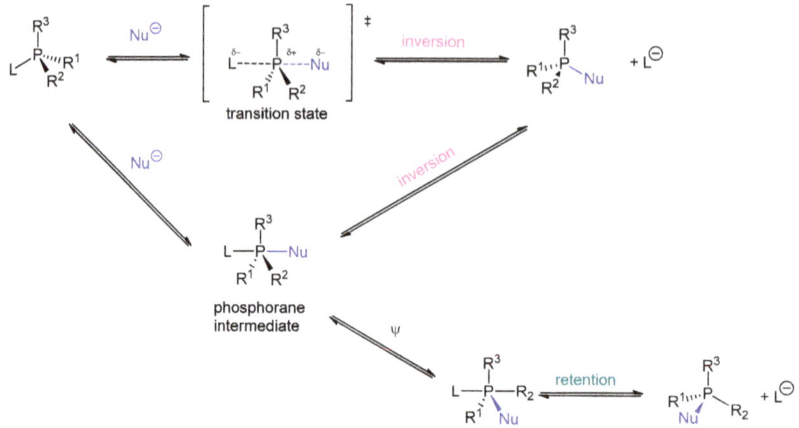

Scheme 3. Possible mechanism for nucleophilic substitution reaction at phosphorus.

The stereochemical consequences (retention or inversion) of such substitution reactions may be conveniently discussed with the aid of the Desargues–Levi graph [45], which originally was proposed by Mislow [46] for stereoisomerization of pentacoordinate phosphoranes.

The participation of such chiral phosphoranes **I–V** as intermediates and their pseudorotation was suggested by Juge and coworkers [47] to explain retention of configuration during the rearrangement of the lithium salts **2a–c** derived from the 2-bromophenylphosphinite boranes **1a–c** to the *o*-hydroxyphenylphosphine boranes **3a–c** induced by halogen–metal exchange (Scheme 4) [47].

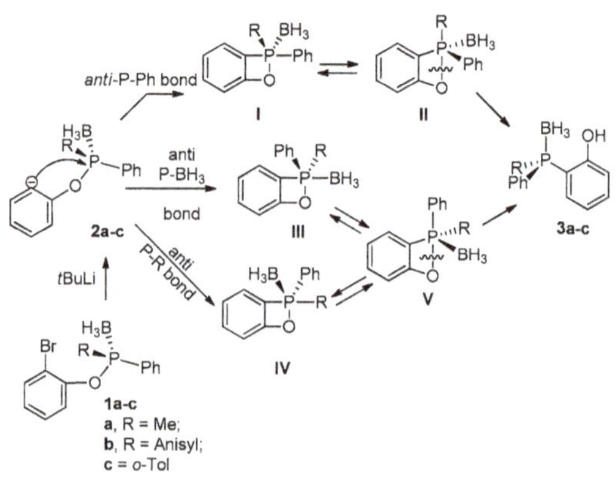

Scheme 4. The rearrangement of the lithium salts **2a–c** derived from the 2-bromophenylphosphinite boranes **1a–c** to the *o*-hydroxyphenylphosphine boranes **3a–c**.

Earlier the same group proposed that pseudorotation of the intermediate phosphoranes **VI**, **VII** was responsible for predominant retention of configuration during the regiospecific ring opening reaction of the diastereomerically pure dioxaphospholane borane **4** with organolithium reagents which afforded phosphinite boranes **7a,b** (Scheme 5) [48].

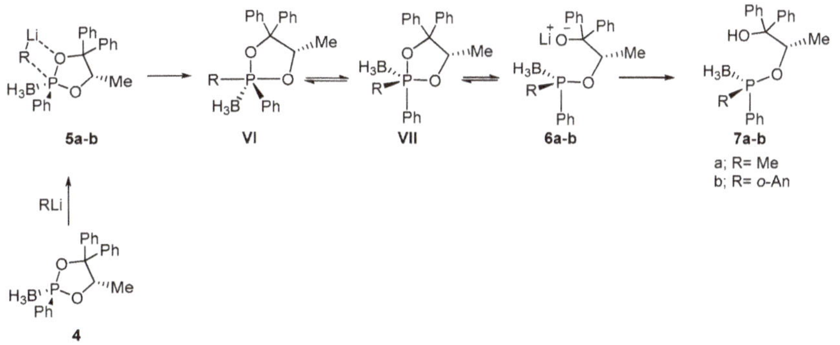

Scheme 5. Stereochemical outcome of the regiospecific ring opening reaction of the diastereomerically pure dioxaphospholane borane **4** with organolithium reagents.

As another example of such a transformation, hydrolysis of quasiphosphonium salts, *cis*- and *trans*-3-methoxy-2,2,6-trimethyl-3-phenyl-1,3-oxaphosphorinanium tetrafluoroborate salts **8a** and **8b** can be presented, by which the corresponding phosphine oxides **9a** and **9b** were formed with complete retention of configuration at phosphorus. The detailed NMR and stereochemical studies allow rationalization of the stereochemical outcome of this reaction in terms of the addition-elimination mechanism, A-E, involving two hydroxyphosphorane intermediates **10a** and **10b** (that are able to undergo pseudorotation) in which at least one oxygen is apical as shown in Scheme 6 [49].

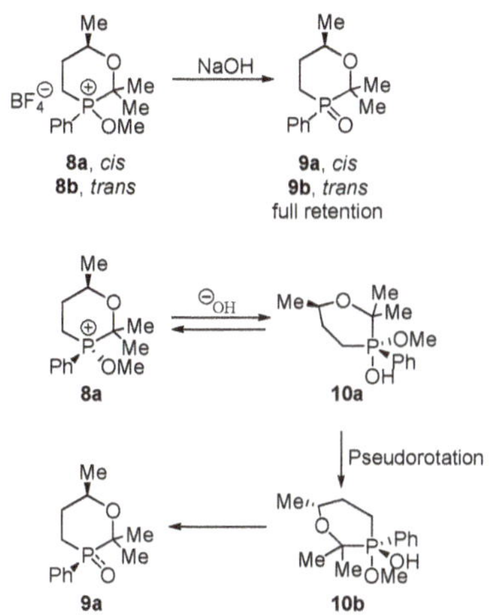

Scheme 6. The hydrolysis of quasiphosphonium salts **8a** and **8b** involving two hydroxyphosphorane intermediates **10a** and **10b**.

Similarly, the participation of the chiral hydridophosphorane **12** as an intermediate in the intramolecular transesterification of the 2-hydroxyphenyl phosphite **11** to the 3-hydroxypropyl phosphite **13** (Scheme 7) was supported by low temperature ^{31}P-NMR measurements [50].

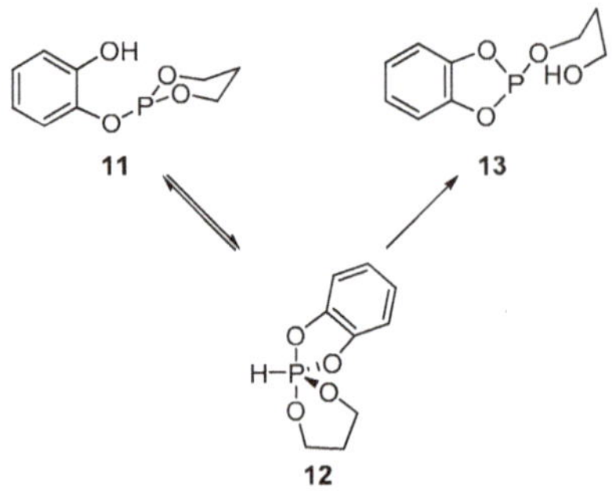

Scheme 7. The intramolecular transesterification of the 2-hydroxyphenyl phosphite **11**.

Earlier, the participation of chiral hydridophosphoranes was proposed in the intermolecular transesterification of chiral asymmetric phenyl phosphites [51] and in the addition of achiral alcohols to compounds having a stereogenic tricoordinated phosphorus atom [52] and very recently in the hydrolysis of a trinucleoside monophosphate by the intramolecular 2α-hydroxy group neighbouring the scissile phosphodiester linkage [53]. The regioselective formation of chiral 3-hydroxypropylphosphinates **17** from cyclic oxaphospholane **14** and Grignard reagents was explained by assuming that during the reactions a pentavalent TBP phosphorane intermediate **15** is formed, which could not undergo pseudorotation to generate another phosphorane **16** due to a high energetic barrier. Therefore, the endocyclic P–O bond in **15** is cleaved almost exclusively to form the phosphinates **17** in high yields (Scheme 8) [54].

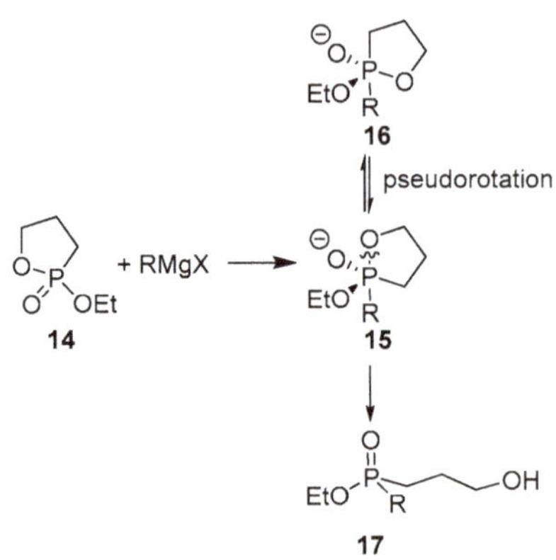

Scheme 8. Reaction of cyclic oxaphospholane **14** and Grignard reagents.

2.2. *Chiral Phosphoranes and Their Anions as Isolable Species*

As it was mentioned above, from the synthetic point of view, hypervalent structures exhibiting chirality due to the presence of one or two bidentate ligands should be more easily available than their acyclic analogues. Therefore, the majority of isolated hypervalent phosphoranes constitute mono- and especially bicylic derivatives in which a cyclic frame include pentacoordinated phosphorus atom bonded to a carbon and/or heteroatoms.

2.2.1. Chiral Monocyclic Phosphoranes and Their Anions as Isolable Species

The non-concerted [2 + 2] cycloaddition reaction of P-haloylides **18** with trifluorometyl ketones led to the stereoselective formation of monocyclic halophosphoranes **19** (Scheme 9) [55].

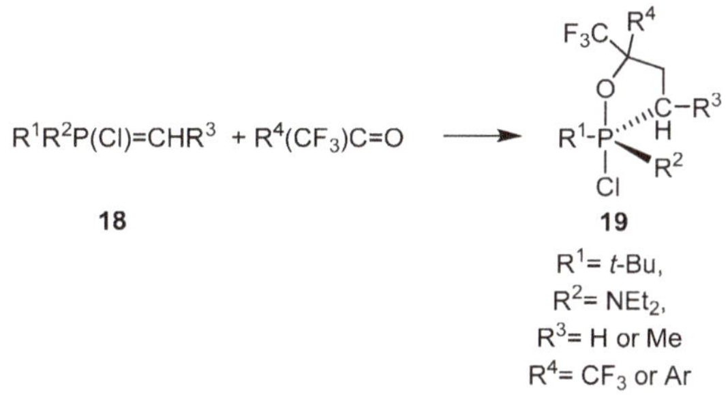

Scheme 9. Monocyclic halophosphoranes **19**.

The monocyclic vinyl phosphoranes **21** were prepared by the reaction of cyclic phosphite **20** with dimethyl acetylenedicarboxylate in the presence of the appropriate phenol (Scheme 10) [56].

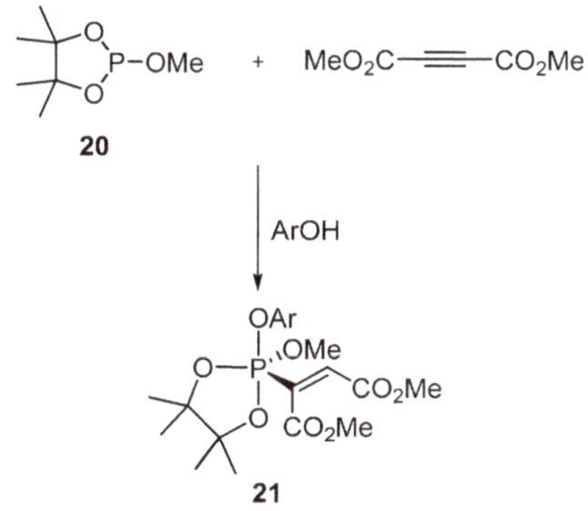

Scheme 10. Reaction of cyclic phosphite **20** with dimethyl acetylenedicarboxylate.

The formation of monocyclic phosphorane **23** was observed in the reaction of ylide **22** with hexafluoroacetone (Scheme 11) [57].

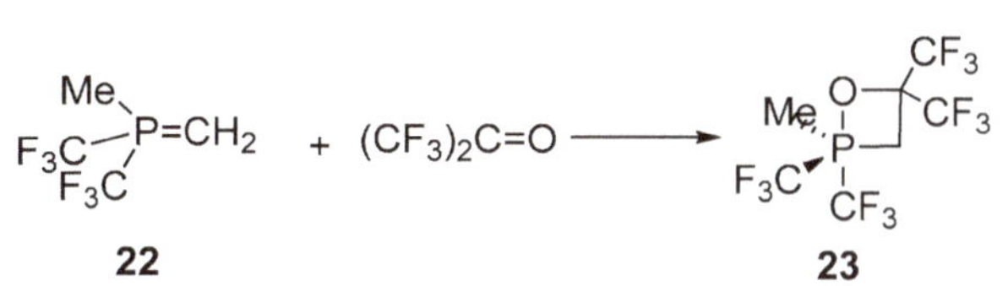

Scheme 11. Reaction of ylide **22** with hexafluoroacetone.

A similar reaction of ylide **24** with hexafluoroacetone gave monocyclic phosphorene **25** (Scheme 12) [57].

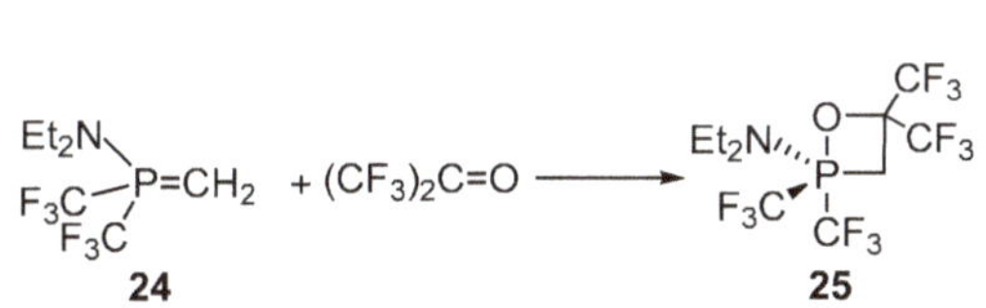

Scheme 12. Reaction of ylide **24** with hexafluoroacetone.

The tricoordinate organophosphorus derivatives **26a–c** were found to react with hexafluoroacetone to give chiral monocyclic phosphoranes **27a–c** (Scheme 13) in which no ligand exchange processes were observed at room temperature. However, the chiral phosphonic acid ester **29** was isolated upon hydrolysis of the chlorophosphorane **27a**, which proceeds most probably via hydroxyphosphorane **28** (Scheme 13) [58].

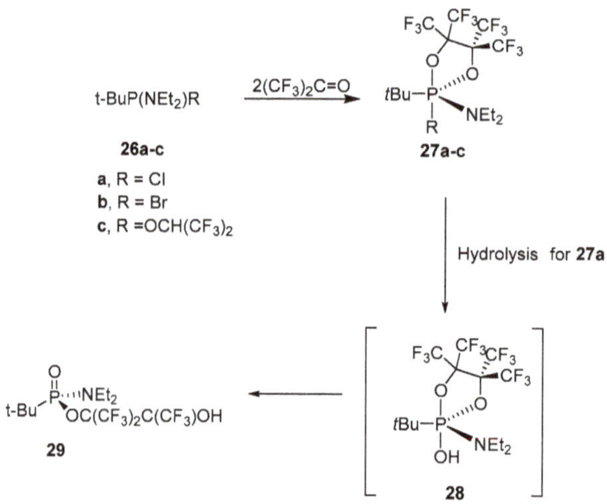

Scheme 13. Reaction of organophosphorus derivatives **26a–c** with hexafluoroacetone.

Preparation of monocyclic halophosphoranes **31a,b** and **31c** was based on the reaction of the tricoordinate phosphinites **30a,b** with benzyl chloride or benzyl bromide. The corresponding fluorophosphoranes **31d** or **31e** were isolated upon the chloride-fluoride or bromide-fluoride exchange reactions when the chloride **31a** or bromide **31c** were used as a substrate respectively (Scheme 14) [59].

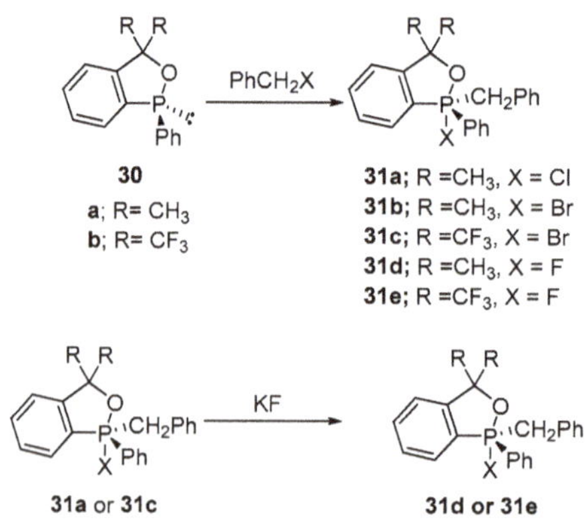

Scheme 14. Preparation of monocyclic halophosphoranes **31a–e**.

Monocyclic (ethylthio)fluorophosphorane **34** was isolated in quantitative yield upon fluoride anion abstraction from BF_4 counter anion of phosphonium salt **33** easily generated by the reaction of cyclic phosphinothionate **32** with triethyloxonium tetrafluoroborate. Defluorination of the fluorophosphorane **34** with trimethylsilyl trifluoromethanesulfonate gave phosphonium salt **35** [60] (Scheme 15).

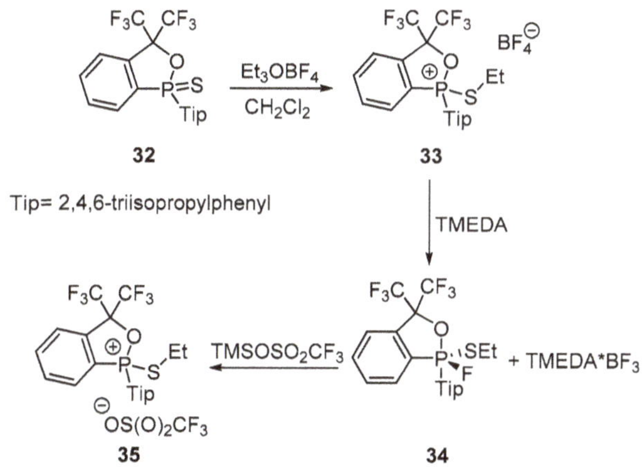

Scheme 15. Preparation of monocyclic (ethylthio)fluorophosphorane **34**.

2.2.2. Chiral Bicyclic Phosphoranes and Their Anions as Isolable Species

A relatively rich number of isolable bicyclic hypervalent phosphoranes have been well documented in the older and more recent chapters by Hellwinkel [61], Burger [62], Holmes [63], Husband and McNab [64], Burgada and Setton [65], Akiba [16], and Kawashima [66]. These chapters mention, more or less extensively, chiral bicyclic phosphoranes, but without deeper comments devoted to the phenomenon of chirality in such derivatives. Therefore, in this part of our compilation we are going to present, as comprehensively as possible, isolable hypervalent derivatives in which their chirality results from the presence of two heterocyclic units forming a P-spiro system. They are divided in accord with the commonly accepted Martin's N-P-L ($_{n}A_{m}B$) coding system, in which N stands for the number of valence electrons associated formally with a central phosphorus atom and L shows the number of ligands (A and B stand for the bonding element) [14].

10P-5C Phosphoranes

Hellwinkel as early as 1966 reported on the isolation of the first optically active pentaarylphosphorane. Initially, hexacoordinate phosphorane potassium salt **37** was synthesized in good yield in racemic form by the reaction of bis(biphenyl)phosphonium iodide **36** with 2,2′-dilithio-4-methyl-biphenyl. The optical resolution of the racemic salt **37** was based on its conversion into two diastereomeric ammonium salts by the treatment with *N*-methylammonium iodide derived from brucine. The isolation of the levorotatory diastereoisomer ($[\alpha]_{578}$ = −1200) was achieved in moderate yield by recrystallization of a crude crystalline fraction from acetone. The dextrorotatory diastereoisomer of the ammonium salt was isolated from a mother liquor in poor yield and with a lower optical purity ($[\alpha]_{578}$ = +986). The isolated diastereoisomers were reconverted into the enantiomeric potassium salts upon the treatment with potassium iodide in acetone followed by repeating crystallizations. Their maximum specific rotation reached values $[\alpha]_{578}$ = + and −1870. Acidification of optically active potassium salts of hexacoordinate phosphorus derivatives (+) and (−)-**37** with HCl in methanol/acetone solution provided the mixture of spirophosphoranes from which enantiomeric spirophosphoranes **38** having $[\alpha]_{578}$ = + and −94, respectively were isolated by multiple recrystallization (Scheme 16). Additionally, the structure of **38** as a racemate was confirmed by the alternative method for its preparation using biphenyl-biphenylyl-2-phosphine **39** as the starting material (Scheme 17) [67].

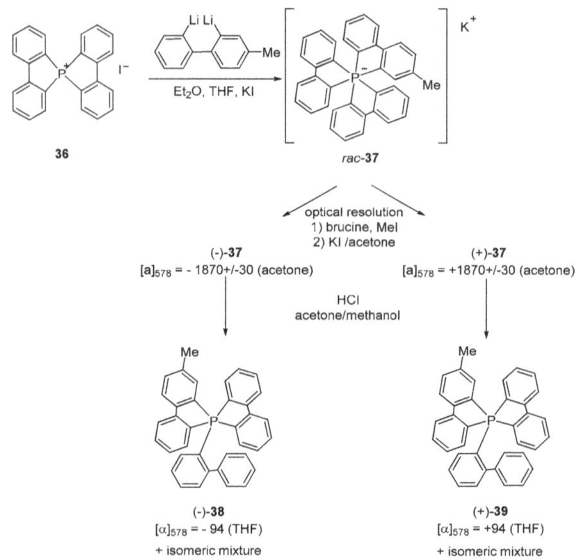

Scheme 16. Preparation of enantiomeric spirophosphoranes **38**.

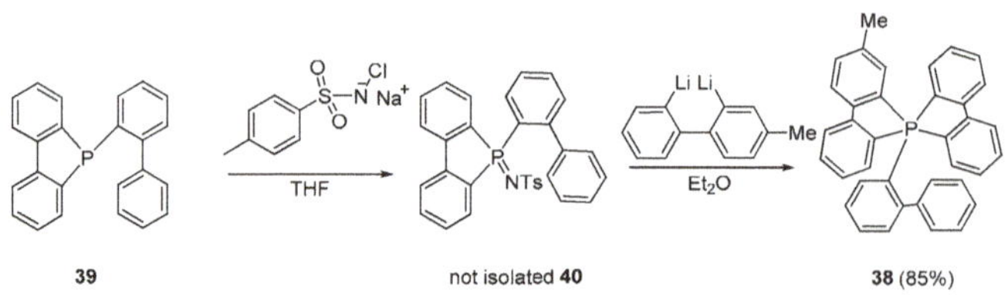

Scheme 17. Preparation of **38** using biphenyl-biphenylyl-2-phosphine **39**.

More recently the reaction of spirocyclotetraalkylphosphonium salts **41** with a series of organolithium reagents RLi, was found to give spirobicyclophosphoranes **42a–e** in good to low yields (Scheme 18). Single crystal X-ray analysis of **42a** showed a TBP geometry with the axial-equatorial rings and the methyl group in an equatorial position. Low temperature NMR measurements indicated that all the pentacoordinate structures **42** show fluxional behaviour in a solution with a very low pseudorotation barrier [68].

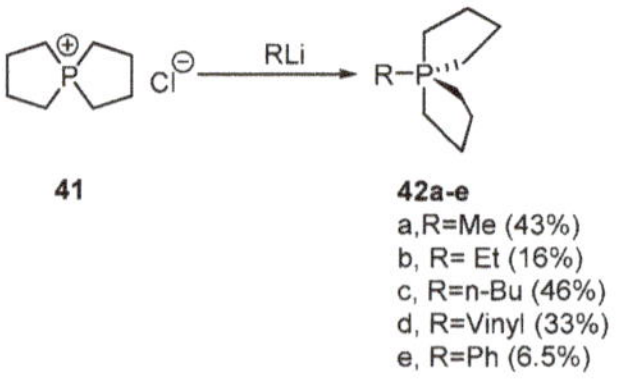

Scheme 18. Reaction of spirocyclotetraalkylphosphonium salts **41** with a series of organolithium reagents.

10P-4C-1O Phosphoranes

Oxaphosphetane (OPA) is a widely recognized intermediate formed via an irreversible [2 + 2] cycloaddition of an aldehyde or ketone with the phosphonium ylide in the Wittig reactions.

The stereoselectivity of the reaction is decided by the course of the first irreversible step. Then, the oxaphosphetane undergoes irreversible [2 + 2] *syn* cycloreversion to give an alkene and phosphine oxide. In most cases the diastereomeric ratio of the oxaphosphetane (OPA) intermediate exactly corresponds to the final alkene, except the cases when a stereochemical drift occurs in the formation of the alkene product. The process "stereochemical drift" refers to the nonstereospecific decomposition of the oxaphosphetane (OPA) intermediate in reactions of certain alkylides with certain aldehydes. The particular investigations on the stereochemical drift in the Wittig reactions (alkylidenephosphonium ylide with aldehydes) under Li-salt-free conditions were undertaken by D. G. Gilheany et al. [69], and based on Variable-Temperature NMR measurements. A particularly suitable candidate for the study of stereochemical drift was intermediate **44a** formed in the reaction of biphenylphenylethylidene-phosphonium ylide **43** with 2-bromobenzaldehyde. It is due to its stability at room temperature (Scheme 19). It undergoes significant stereochemical drift on decomposition in THF at reflux: A sample containing *cis*- and *trans*-**44a** in a ratio of 94:6 gave alkenes of a (*Z*)/(*E*) ratio 82:18 under reflux conditions. On the other hand, the decomposition conducted at 50 °C was not complete after 6d, but the maximum possible (*Z*)/(*E*) ratio that could have been achieved in this reaction was 88:12, revealing that a smaller amount of the stereochemical drift had occurred. A very similar experiment supported with VTNMR spectra analysis over the temperature range −20 to +40 °C on **44b** (produced in the reaction of biphenylphenylethylidenephosphonium ylide **43** and benzaldehyde) was performed. Oxaphosphetane intermediate **44b** was formed in a diastereomeric ratio 71:29 (*cis*/*trans*). Again, the ratio of OPA, ylide and phosphine oxide was invariant within the temperature range from −20 to 30 °C, which indicates that no decomposition to alkene and phosphine oxide occurred. On heating the OPA **44b** was decomposed to produce the alkene in 53:47 (*Z*)/(*E*) isomeric ratio. In the ^{31}P NMR spectra recorded for **44a** and **44b** in the presence of an excess aldehyde additional signals appeared in the pentavalent region (δ = −50 to −80 ppm). The observation was consistent with that reported earlier by Vedejs et al. [70]. Thus, the apparent interaction between OPA and aldehyde was expected to have an impact on the occurrence of the stereochemical drift.

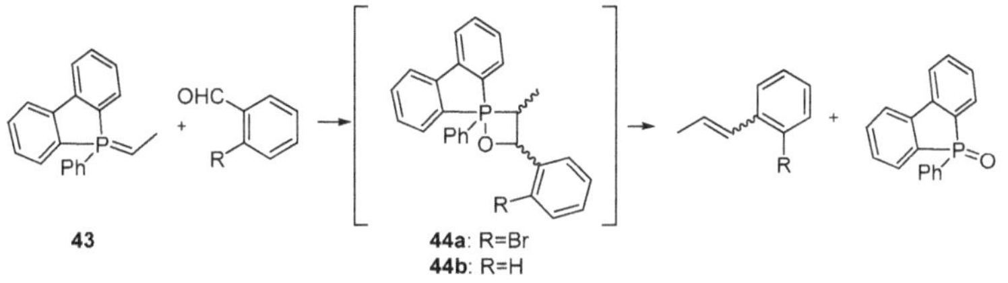

Scheme 19. Reaction of biphenylphenylethylidene-phosphonium ylide **43** with benzaldehydes.

10P-3C-1N-1O Phosphoranes

In 1996 Kawashima and Okazaki succeeded in preparing pentacoordinate *N*-apical 1,2-azaphosphetidines, whose structures were fully characterized. They also investigated their thermolysis. The first observation on the occurrence of the *N*-equatorial pseudorotamers was also reported [71]. The synthetic pathway was as follows: the treatment of 2-(methylphenylphosphinyl)-α,α-bis(trifluoromethyl)phenylmethanol **45** with *n*-BuLi led to the generation of methylene carbanion species, which subsequently underwent nucleophilic addition to a Schiff base providing the β-aminoethyl derivative **46**. 1,2-Azaphosphetidine **47** was formed in 93% yield by intramolecular cyclization followed by dehydration under the Mitsunobu reaction condition (Scheme 20).

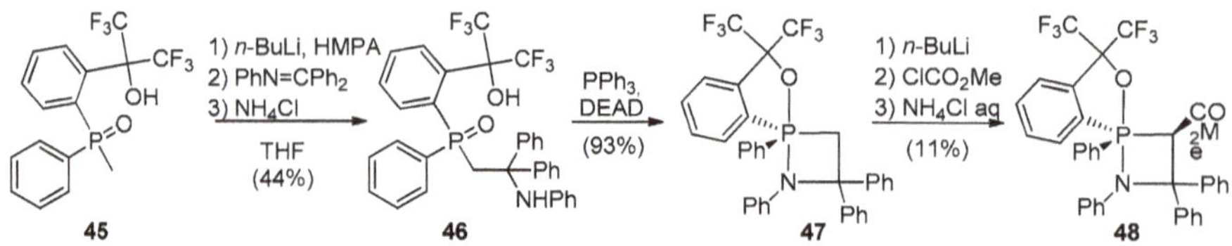

Scheme 20. Synthesis of 1,2-Azaphosphetidine **47** and **48**.

A typical chemical shift in the NMR spectra was observed for compounds with a TBP geometry in which an apical position was occupied with O or N atom. X-ray single crystal analysis revealed the distorted TBP geometry with phosphorus as a central core. Hydrolysis of **47** on silica gel resulted in its cycloreversion to the starting material **46**. Finally, 1,2-azaphosphetidine **47** was converted into the 3-methoxycarbonyl derivative **48**, likewise it had been reported, for the corresponding 1,2-λ^5-oxaphosphetanes [72,73]. In the ^{31}P-NMR spectrum recorded for the solution of **47** in deuterated toluene, apart from the appropriate resonance signal ($\delta_P = -29.9$), another upfielded peak ($\delta_P = -50.9$), was observed which was assigned to the *N*-equatorial pseudorotamer **49** (Scheme 21). A similar pseudorotamer ($\delta_P = -51$) was observed also in the case of **48**.

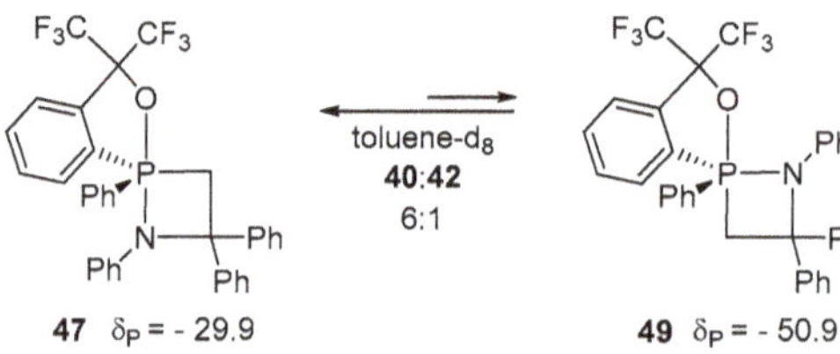

Scheme 21. The equilibrium between phosphoranes **47** and **49**.

Thermolysis of **47** or **48** proceeded with the quantitative formation of the corresponding alkenes and iminophosphorane (Scheme 22), which implies that azaphosphetidines should be considered as aza-Wittig intermediates.

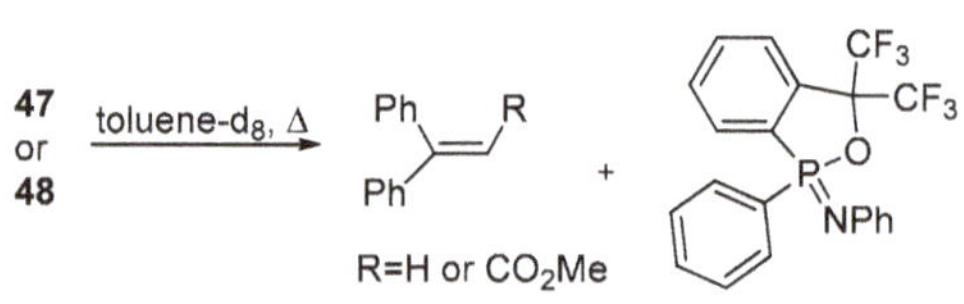

Scheme 22. Thermolysis of **47** or **48**.

The reaction of an iminophosphorane, the nitrogen analogue of a phosphonium ylide, with a reactive alkyne has been intensely studied because it constitutes a useful synthetic pathway for the corresponding α-iminoalkylidenephosphoranes via a [2 + 2]-cycloadduct. The intermediate had neither been observed nor isolated until the first stable pentacoordinate 1,2-λ^5-azaphosphetine **51** was synthesized by Kawashima et al. [74]. Its preparation was achieved by the cycloaddition reaction of the Martin ligand-based iminophosphorane **50** with an alkyne (Scheme 23). X-ray crystallography of **51** showed a distorted TBP with N and O atoms at the apical positions. The variable temperature ^{31}P NMR spectra of **51** in C_7D_8 or CD_3CN showed a shift to a lower field with decreasing temperature indicating that (**51**) was in an equilibrium with the corresponding ylide structure **52**.

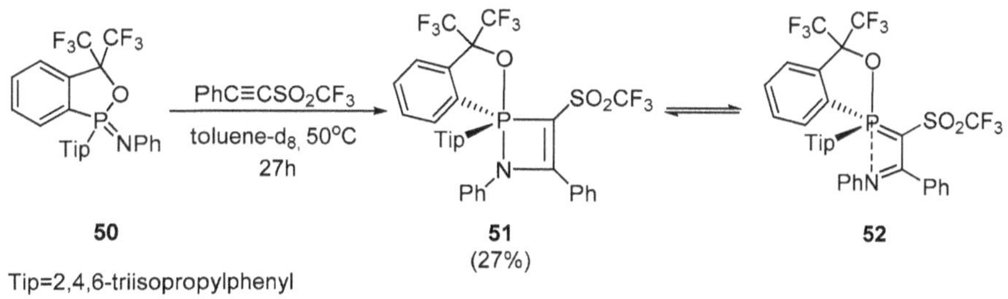

Scheme 23. Cycloaddition reaction of the iminophosphorane **50** with an alkyne.

The isomerization and thermal decomposition of isolable spiro-1,2-oxaphosphetanes has been studied in details with the aim to recognize a stereomutation mechanism of spiro-1,2-oxaphosphetanes, the Wittig intermediate products. Spiro-oxaphosphetanes **54–58**, stabilized by the presence of the *o*-benzamide moiety, were found to be good candidates for the studies [75]. The OPA's **54–56** were obtained according to the previously reported procedure [76] (Scheme 24). The reactions with the hindered ketone, L-(−)-camphor, gave products **57** and **58** (in a 45:55 ratio) arising from the endo attack at the CO group of L-camphor.

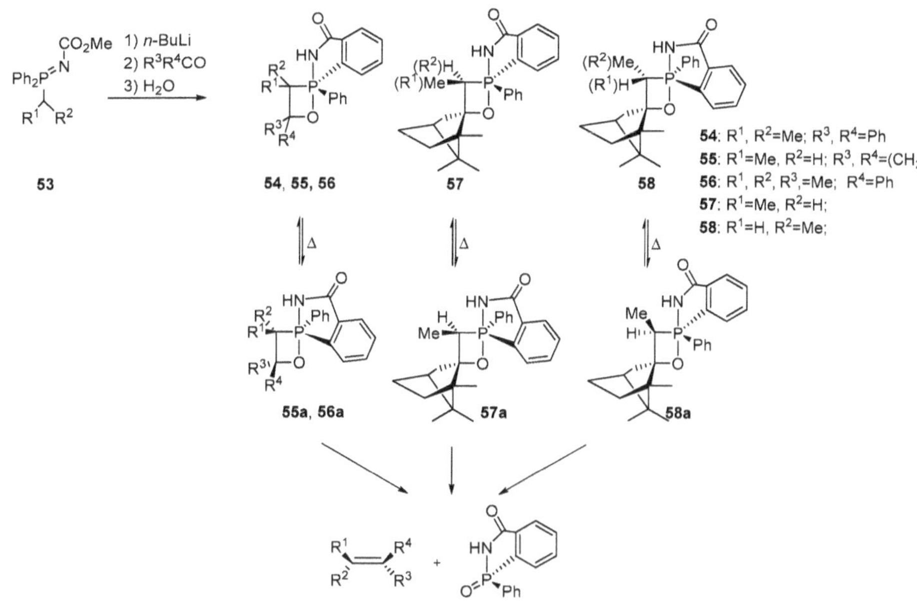

Scheme 24. Synthesis of spiro-oxaphosphetanes **54–58** and their thermal decompositions.

All thermal decompositions of OPAs **54–58** furnished the appropriate alkenes and a phosphinamide, quantitatively. On the thermolysis of **55–58**, partial isomerization to **55a–58a** was observed, which resulted in the inverted configuration at phosphorus atom. It has been found that oxaphosphetane decomposition took place in a single step via a polar transition state. For the first time the stereomutation through three possible mechanisms MB_2, MB_3, and MB_4 involving two, three, and four Berry pseudorotations (at phosphorus atom), respectively, supported by DFT calculations has been evidenced [75].

10P-3C-2O Phosphoranes

The first configurationally stable enantiomeric pair of 10-P-3C-2O phosphoranes with well-defined sole stereogenic centre at phosphorus atom was successfully obtained and characterized by Akiba et al. [77]. The synthesis of diastereomeric **60**-(R_P) and **60**-(S_P) was achieved via a facile alkylation of the in situ generated phosphoranide anion by (−)-menthyl chloroacetate in 87% yield (Scheme 25). The diastereomeric mixture could be resolved effectively by fractional crystallization to furnish **60**-(R_P) in 25% and **60**-(S_P) in 22% yield. Thus, the separation and the selection of suitable crystals allowed for the determination of absolute configurations for both diastereoisomers by X-ray diffraction. The alkylation of P–H phosphorane derivatives was recognized to proceed with complete retention of configuration at the phosphorus strereogenic centre, most probably due to no detectable racemization at the step of phosphoranide intermediate formation.

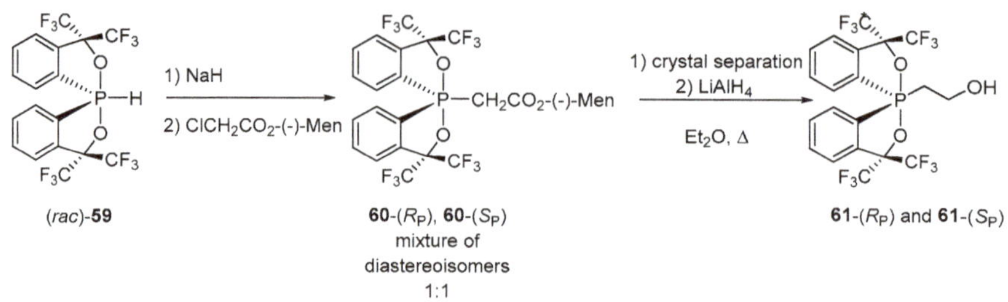

Scheme 25. Synthesis of diastereomeric phosphoranes **60** and enantiomeric phosphoranes **61**.

The removal of the menthyl unit by treating the diastereoisomers **60**-(R_P) and **60**-(S_P) with an excess of $LiAlH_4$ provided the enantiomerically pure phosphoranes **61**-(R_P) and **61**-(S_P), respectively, with a single chirality center at the pentacoordinated phosphorus atom. In order to verify enantiomeric purity of the obtained *P*-chiral 2-hydroxyethylphosphoranes **61**, they were converted into the (*R*)-(+) Mosher esters **62** using (*R*)-(+)-2-methoxy-2-(trifluoromethyl)phenylacetic acid chloride ((+)-MTPA-Cl)) (Scheme 26) [78].

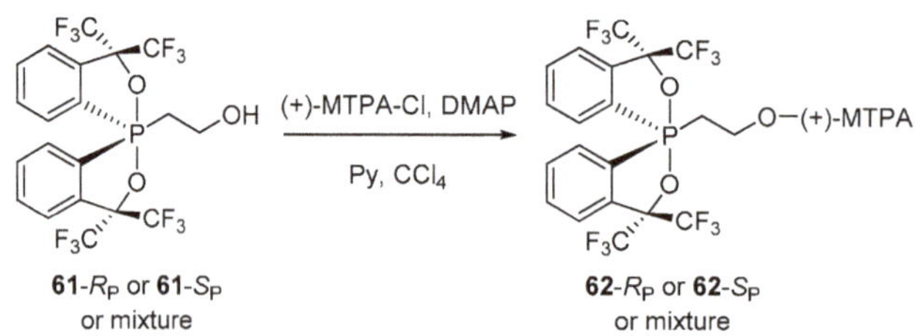

Scheme 26. Conversion of phosphoranes **61** to esters **62**.

Independently, the H-phosphorane **59** was converted (by the treatment with *n*-BuLi) into an anion, which was reacted with CH_2I_2 affording the alkylating product **63** in 82% yield. Subsequently α-iodomethyl phosphorane derivative was treated with α-methylbenzylamine to give a diastereomeric mixture of **64** in 56% yield (Scheme 27). The pure diastereoisomers were isolated by recrystallization and their absolute configuration was determined by X-ray analysis.

It was also found that deprotonation of the nitrogen-bonded proton in each single diastereomer **64** afforded enantiomeric phosphoroanidates **65**-(R_P) and **65**-(S_P) (via imine elimination). They, after acidification, gave the desired optically active P-H phosphoranes **59**-(R_P) and **59**-(S_P) (Scheme 28).

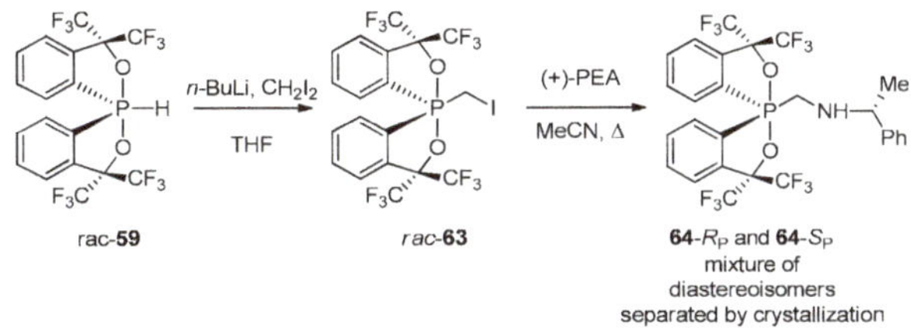

Scheme 27. Synthesis of diastereomeric phosphoranes **64**.

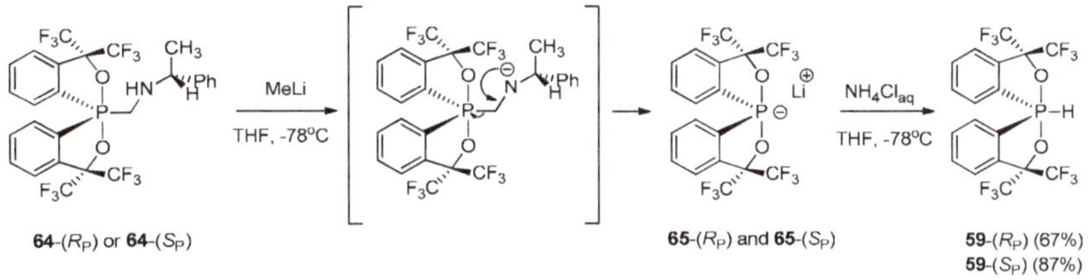

Scheme 28. Conversion of diastereomeric phosphoranes **64** to enantiomeric phosphoranes **59**.

Later on, Akiba and co-workers reported the first example of an "anti-apicophilic" (*O-cis*) spirophosphoranes **67** [79–81], in which the oxygen atom occupies an equatorial position and the carbon atom is located in an apical position of a five membered ring, which is in contrast to the general concept of apicophilicity. According to the procedure shown in Scheme 29 (*O-cis*) phosphoranes **66** were obtained. The treatment of P–H phosphorane **59** with an excess of alkyl(aryl)lithium resulted in the formation of a dianionic intermediate which upon treatment with I_2, as an oxidizing agent, afforded P–I equatorial phosphorane. Its cyclization, initiated by the nucleophilic attack of the alkoxide anion at the phosphorous centre and the simultaneous extrusion of I^- produced *O-cis*-alkyl-or aryl spirophosphoranes **66**.

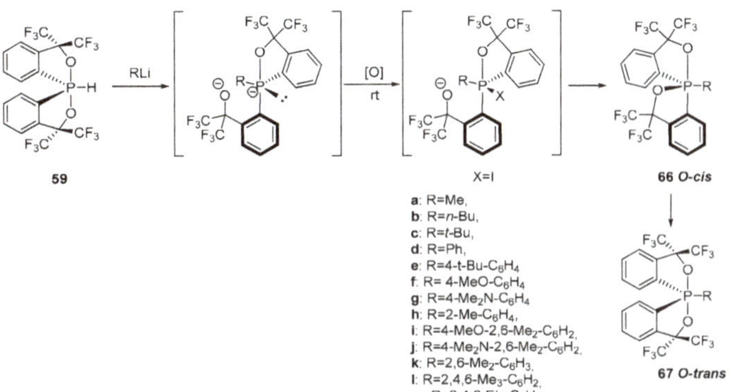

Scheme 29. Synthesis of (*O-cis*) spirophosphoranes **66** and their stereomutation to (*O-trans*) phosphoranes **67**.

The effects of substituents in the monodentate aryl ring of *O-cis* arylphosphoranes on the stereomutation process, along with a kinetic study on the stereomutation of *O-cis* **66k–n** to *O-trans* **67k–n** were studied in details. The exclusive generation of *O-cis* **66** could be observed for the alkyl derivatives **66a–c** right after oxidation, however, in the case of the phenyl derivative **66d** the

O-cis isomer was the minor product. The pseudorotation to O-trans **67d** isomer was complete in 30 min at room temperature, which could be explained by the higher apicophilicity of aryl groups in comparison to alkyl groups. For compounds **66e–g**, bearing electron donating groups (*t*Bu, OMe, NMe_2, respectively) in the 4-position of the monodentate aryl groups *O-cis* isomers were the major products. Finally, the use of bulky aryl organometallic reagents containing two substituents in *ortho* positions led to the exclusive formation of the *O-cis* isomers (**66i–n**), having the sufficient lifetimes for characterization. From the relative stability of **66i–n**, it is evident that steric effects contributed mainly to the stabilization against pseudorotation in these isomers that exhibited reversed apicophilicity. In the case of **66n**, pseudorotation was slow enough to allow for its isolation (purification on silica gel in 71% yield) and the crystal structure determination by X-ray diffraction. The differences in the reactivity of *O-cis*-**66b** and *O-trans*-**67b** toward nucleophiles were also indicated (Scheme 30). The reaction of *O-cis*-**66** with TBAF as a nucleophile afforded the hexacoordinated phosphate **68** with the newly formed P–F bond while the *trans*-isomer did not react. Rapid decomposition of **68** caused by the presence of traces of water was observed. *O-Cis* **66b** likewise was reacted with MeLi to give product **69**, while *O-trans*-**67b** treated with organolithium reagent remained unreacted under the same conditions; however by increasing MeLi molar equivalency and temperature the product **69** was successfully isolated.

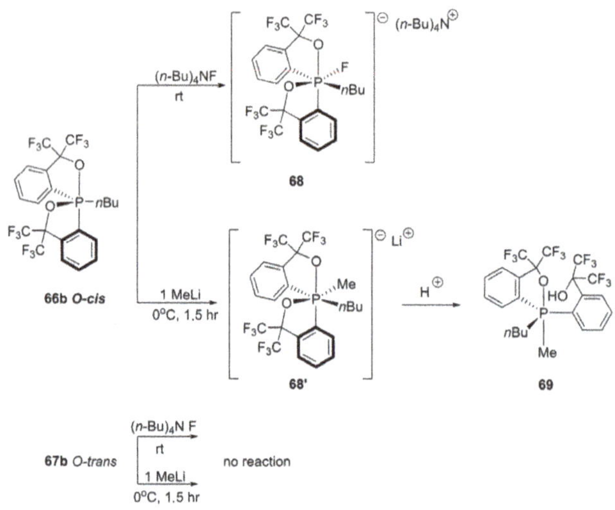

Scheme 30. Reactivity of *O-cis*-**66b** and *O-trans*-**67b** toward nucleophiles.

A three-component reaction of phosphoranide **70**, silane **71**, and THF was applied for the preparation of hypervalent compound **72** containing simultaneously pentacoordinated phosphorus and silicon atoms (Scheme 31) [82,83]. The ^{31}P-NMR spectrum indicated the presence of two diastereomeric products in a ratio of 1:1.

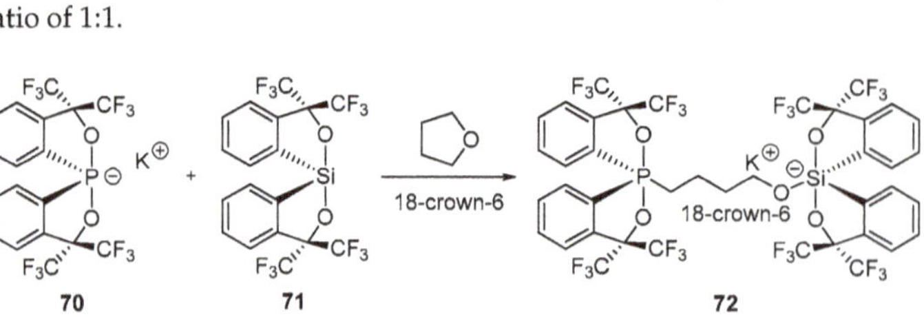

Scheme 31. Reaction of phosphoranide **70** with silane **71**.

The synthesis of other phosphoranylalkoxysilicates was achieved by the nucleophilic addition of phosphoranylalkoxides generated from hydroxy(alkyl)phosphoranes **61** and **73a–c** to silane **74** (Scheme 32). Hydroxy(alkyl)phosphoranes **61** and **73a–c** were prepared by deprotonation of hydrophosphorane **59** with DBU and subsequent treatment with an electrophile, such as hydrogen peroxide, formaldehyde, 2-bromoethanol, and 4-bromobutanol, respectively. Their treatment with KH in the presence of 18-crown-6 followed by the addition of silane **71** in THF afforded the appropriate phosphoranyloxysilicate **74a** (56%), and phosphoranylalkoxysilicates **74b,c** and **72** (53%–96%). Potassium silicates **74a–c** were obtained as a mixture of two diastereomers (in ratios of 66:34 for **74a**, 53:47 for **74b**, 50:50 for **74c**, respectively), as estimated by the integrals in the ^{31}P-NMR spectra [82,83].

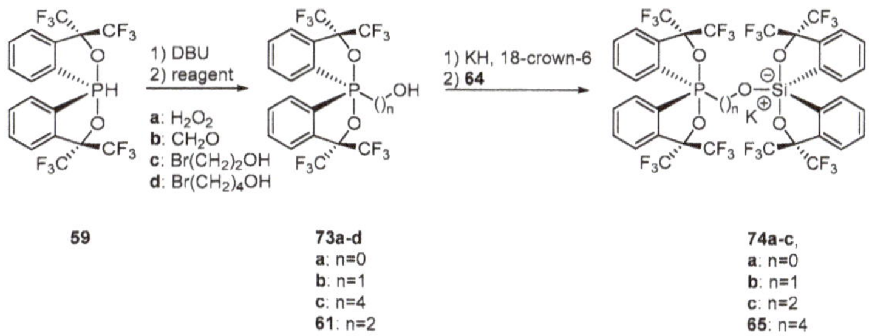

Scheme 32. Reactivity of phosphoranes **59**, **61** and **73**.

More recently, studies on spirophosphoranes, prepared from a new bulkier analogue of the Martin ligand containing two decafluoroetyl units, have been carried out by the Yamamoto group. This new ligand was expected to suppress BPR more efficiently than the Martin ligand [84]. The synthesis of P–H-spirophosphoranes **75a,b** was based on a two-step reaction involving dimetallation of perfluorocumyl alcohol derivatives leading to the corresponding dianion and its subsequent condensation with PCl_3, followed by acidification with 6 molar HCl. The *O*-equatorial spirophosphoranes **76a,b** were obtained by the reaction of **75a,b** with 3 molar equivalents of RLi followed by the oxidation with I_2. Their conversion to the corresponding *O*-apical stereoisomers **70a,b** proceeded in the solution at high temperatures. The stereomutation of **76a-1** to **77a-1** was found to proceed with first–order kinetics. The free energy of activation for stereomutation was higher by 3.6 kcal/mol in comparison with the CF_3 analogue indicating that steric hindrance of pentafluoroethyl group is the major factor for freezing pseudorotation. Recently, Jiang and coworkers [85] obtained a new stable anti-apicophilic phosphorene **76b** with four bulky *n*-C_3F_7 groups, which was converted to more stable *O*-apical phosphorane **77b** only by heating, (Scheme 33).

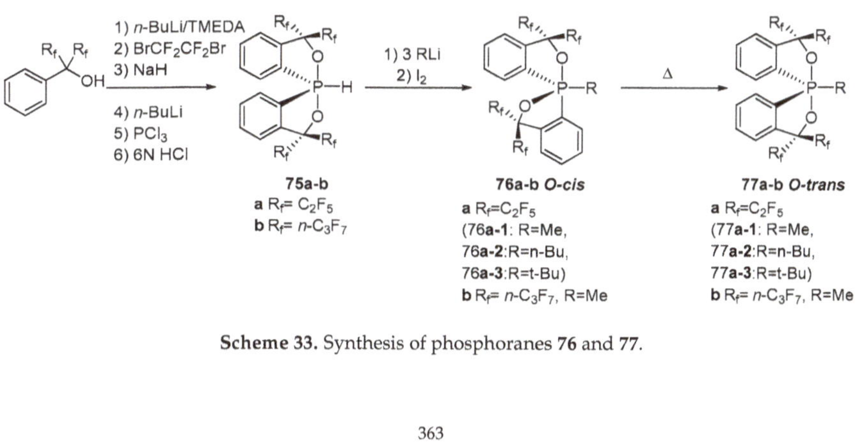

Scheme 33. Synthesis of phosphoranes **76** and **77**.

The same research group elaborated the synthesis of a series of anti-apicophilic phosphoranes bearing a *para*-substituted aryl group (–C_6H_4(*p*-X); X = H, CF_3, F, OMe) or a mesityl group as a more electronegative monodentate ligand than an alkyl group. The *O*-equatorial arylphosphoranes **69c–g** were successfully synthesized by the reaction of the P-H spirophosphorane **68a** with an excess of ArLi, followed by treatment with I_2 (Scheme 34) [86]. These phosphoranes were found to be stable at r.t. Their isolation indicates that the steric effect of the C_2F_5 groups for freezing the isomerization is remarkable. By heating in an organic solvent the *O*-equatorial arylphosphoranes **69c–g** were quantitatively converted into the more stable *O*-apical isomers **70c–g**. The effect of a solvent and a para-substituent on the rate of the isomerization were also investigated in order to provide an insight into the $\pi \rightarrow \sigma^*$ P–O interaction in the *O*-equatorial arylphosphoranes. The kinetic study showed a small para-substituent effect on the stereomutations. However, the multi-regression analysis for the para substituents revealed the 1.3 times greater contribution of the resonance effect on the on the isomerization rate in C_6D_6 than the inductive effect suggesting that the $\pi \rightarrow \sigma^*$ P–O stabilizing interaction in the *O*-equatorial isomer plays some role in the isomerization.

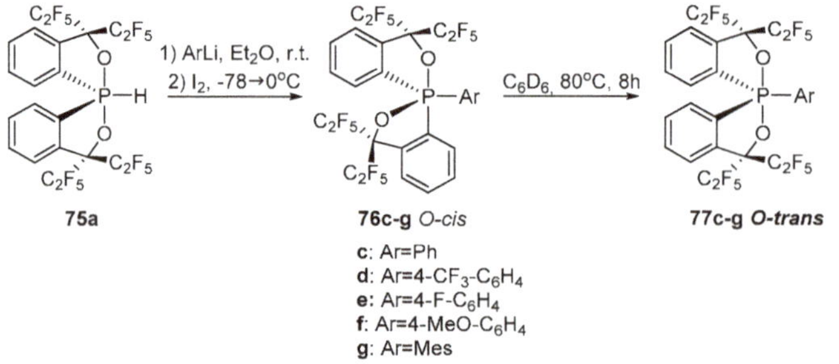

Scheme 34. Synthesis of phosphoranes **76** and their interconversion into **77**.

In this context it is interesting to note that the *O*-equatorial methylphosphorane **76a** treated with MeLi was converted to the more stable *O*-apical isomer **77a**, but on the contrary, the *O*-apical phosphorane did not react with MeLi at all (Scheme 35) [87].

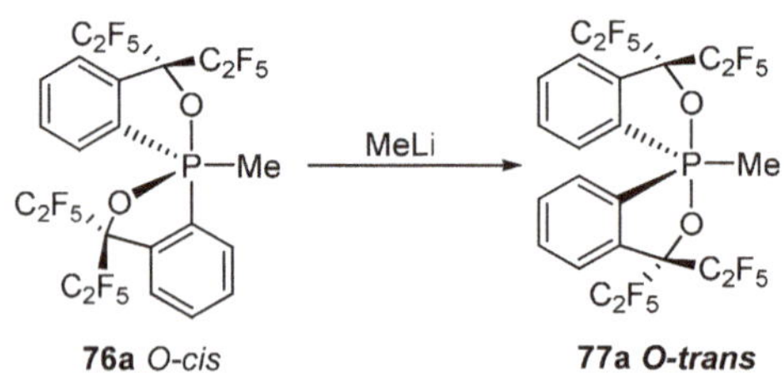

Scheme 35. Methyllithium induced epimerization of *O*-equatorial methylphosphorane **76a**.

Deprotonation of methylphosphorane **76a** at the methyl group using superbase (*t*-BuOK/*n*-BuLi) led to the corresponding α-anion, which upon treatment with electrophilic agents gave new phosphorane derivatives **78** and/or **79** (Scheme 36). The *O*-equatorial and *O*-apical phosphoranes having a β-hydroxyethyl group were synthesized according to this procedure using paraformaldehyde and applied as monodentate ligands [87].

Akiba et al. reported the preparation and characterization of an anti-apicophilic spirophosphorane bearing an oxaphosphetane ring and the Martin ligand [88]. Although the substitution pattern on the phosphorus atom slightly differs from that of oxaphosphetane in the typical Wittig reaction they can be considered as a model for the possible reactive intermediate for the Wittig reaction.

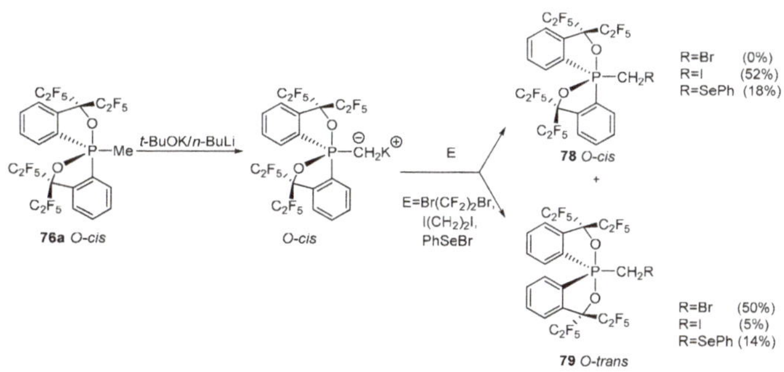

Scheme 36. Synthesis of phosphorane derivatives **78** and/or **79**.

The synthesis of isomeric mixture of oxaphosphetane **80a** and **80b** in 1:1 ratio was achieved by the oxidation procedure using *n*-BuLi followed by iodine in ether (Scheme 37). The sole *anti*-apicophilic **80a** isomer was isolated from the reaction mixture by crystallization upon addition of hexane. Under acidic conditions **80a** underwent complete stereomutation to **80b**. A similar conversion took place within minutes upon dissolution of **80a** in anhydrous $CDCl_3$ at room temperature. A ^{31}P-NMR measurement of the **80b** sample heated to 120 °C in *p*-xylene did not show the presence of **80a** in equilibrium, indicating that **80a** is thermodynamically much less favorable than its *O*-apical isomer. Moreover, heating a solid sample of **80a** at 120 °C for 5 min. gave only **80b** as a result of pseudorotation.

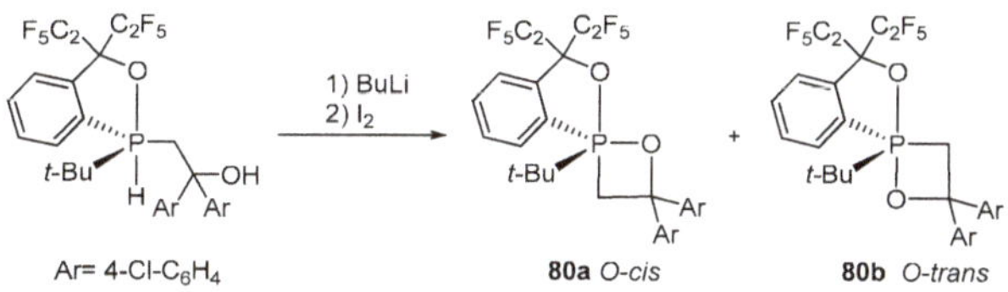

Scheme 37. Synthesis of isomeric mixture of oxaphosphetane **80a** and **80b**.

The synthesis of pentacoordinate phosphirenes **82a** and **82b** bearing a phenyl group on the phosphorus atom was achieved by the reaction of tricoordinate phosphirenes **81a** and **81b** with *o*-chloranil (Scheme 38) [89].

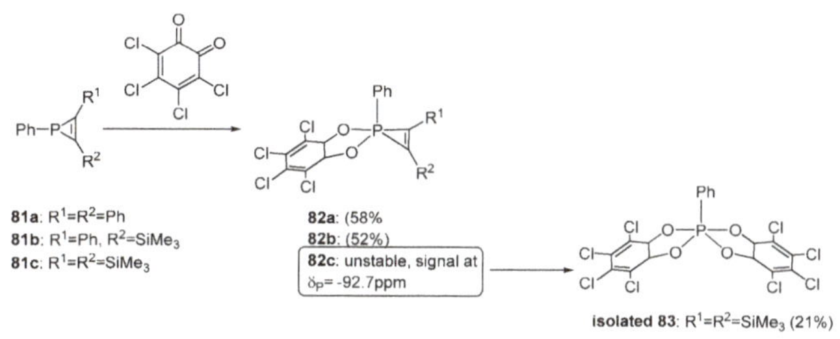

Scheme 38. Reaction of phosphirenes **81a** and **81b** with *o*-chloranil.

The formation of **82c** was evidenced by ^{31}P-NMR spectrum that showed a suitable signal at δ_P −92.7. It was converted rapidly to phosphorane **83** bearing two tetrachlorocatecholate ligands. For relatively more thermally stable phosphirene **82a** the crystal structure was determined by X-ray analysis. It showed that phosphirene **82a** has a highly distorted square pyramidal SP arrangement at the phosphorus atom as a core. The basal positions are occupied by two oxygen atoms from the tetrachlorocatecholate ligand and two carbon atoms belonging to the three-membered ring of phosphirene, when the phenyl group is located at the apical position. No decomposition was observed for **82a** after heating at 60 °C for 40 h. On the other hand, phosphirene **82b** was thermally less stable, and decomposed at room temperature to give the *o*-chloranil derived phenylphosphinite and phenyl(trimethylsilyl)acetylene. These results indicate that an attachment of a trimethylsilyl group to the endocyclic carbon lowers the thermal stability of the pentacoordinate phosphirenes.

10P-3C-1O-1Se; 10P-3C-1O-1S; 10P-3C-1O-1F; 10P-3C-1O-1H Phosphoranes

The synthesis of the first 1,2-σ^{5-}selenaphosphirane and 1,2-σ^5-thiaphosphirane involving a pentacoordinate phosphorus atom was based on the reaction of the Martin ligand based phosphorus ylide with elemental selenium [90], $CpTiSe_5$ [91] or elemental sulfur [92]. Treatment of ylide **84** with 1.6 equiv. of elemental selenium in THF at room temperature or with Cp_2TiSe_5 under the same conditions resulted in the formation of selenaphosphirane **85** as a solid in 96% and 85% yield, respectively (Scheme 37). Although the selenaphosphirane **85** is highly moisture-sensitive, its isolation was successful under an argon atmosphere and its structure was crystallographically characterized. Both, the solid-state $^{31}P\{^1H\}$ NMR spectrum and the $^{31}P\{^1H\}$ NMR spectrum of the C_6D_6 solution showed a single peak with chemical shifts at −26.1 and −26.6, respectively. This was considered as evidence for the pentacoordinate state of the phosphorus atom of the selenaphosphirane. It is interesting to note that in the $^{77}Se\{^1H\}$-NMR spectrum (the resonance signal at δ_{Se} 147.5 in C_6D_6), the coupling constant between the phosphorus and the selenium nuclei was not observed. This selenaphosphirane upon treatment with methyl triflate in $CDCl_3$ afforded the highly air-sensitive α-(methylseleno)-α-methylethyl phosphonium triflate **86** in 76% yield (Scheme 39). The formation of **86** was initiated by the electrophilic attack of methyl triflate on the negatively charged selenium atom in **85** and proceeded with the subsequent cleavage of the polarized P–Se bond. In a similar manner the thiaphosphirane **87** was prepared in 68% yield, by the reaction of the ylide **84** with 0.96 equivalent of elemental sulfur in THF at −30 °C for 7 h (Scheme 39). Thiaphosphirane **87** was found to be air-sensitive and its single crystal was obtained by recrystallization from hexane under an argon atmosphere. Its X-ray crystallographic analysis showed that the oxygen and sulfur atom are located at apical positions.

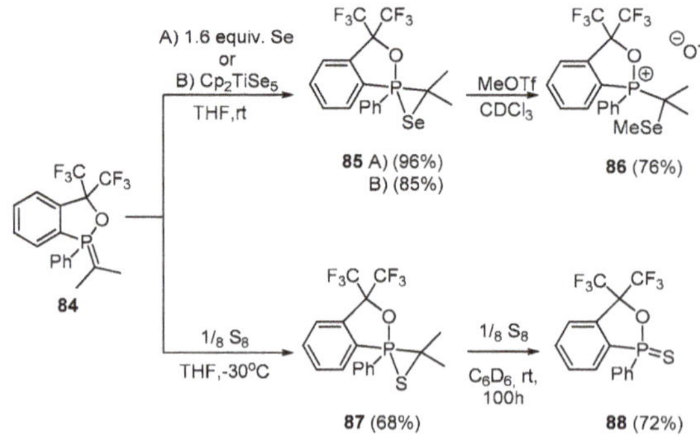

Scheme 39. Reactions of monocyclic phosphorus ylide **84**.

Thiaphosphirane **87** showed inertness toward triaryl- or trialkylphosphines. This is in a contrast to tri- and tetracoordinate thiaphosphiranes which underwent desulfurization to form phosphaalkene derivarives and phosphine sulfides [93]. When the thiaphosphirane **87** was treated with additional portion of elemental sulfur the corresponding phosphinothionate **88** was isolated in 72% yield. The independent experiment showed that the phosphorus ylide **84** treated with 2 equiv. of elemental sulfur provided phosphinothionate **88** in 31% yield.

10P-2C-2N-1O; 10P-2C-1N-2O Phosphoranes

The reactions of iminophosphorane bearing the Martin ligand and a bulky 2,4,6-triisopropyl-phenyl group with a ketone, an isothiocyanate and an alkyne has been found to proceed with the formation of the corresponding cycloadducts as novel heterocycles bearing pentacoordinated phosphorus atom [94,95]. Cyclic iminophosphinate **89** reacted with carbonyl compounds, phenyl isothiocyanate and phenylethynyl trifluoromethyl sulfone affording the appropriate cycloadducts: 1,3,2λ^5-oxazaphosphetidines **90a–c**, 1,3,2λ^5-diazaphosphetidine-4-thione **91**, 1,2λ^5-azaphosphetine **92**, respectively. The reaction with hexafluoroacetone proceeded at room temperature yielding cyclic phosphorane which could be even purified by silica gel chromatography. With dimethyl acetylenedicarboxylate (DMAD) and water the reaction provided 1,2λ^5-oxaphosphol-5(2*H*)-one **93** instead of the expected 1,2-azaphosphetine (Scheme 40).

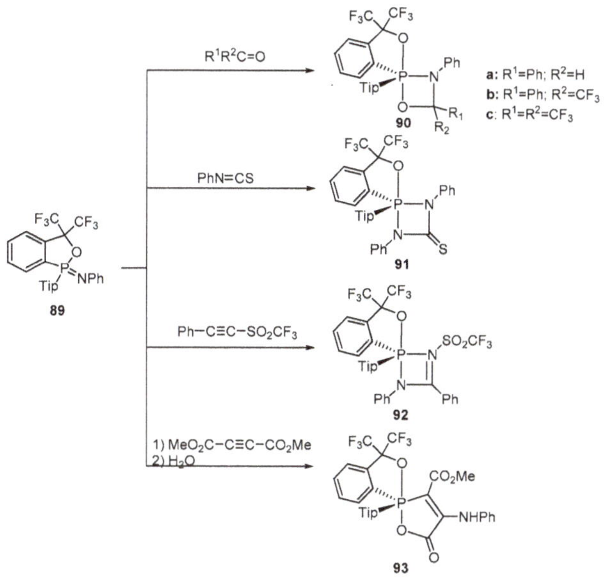

Scheme 40. Syntheses of bicyclic phosphoranes **90–93**.

Thermolysis of **90c** at 140 °C in a sealed tube gave the corresponding imine **94** and cyclic phosphinate **95** (Scheme 41), indicating that 1,3,2λ^5-oxazaphosphetidine **90c** is regarded as an intermediate of the aza-Wittig reaction [95].

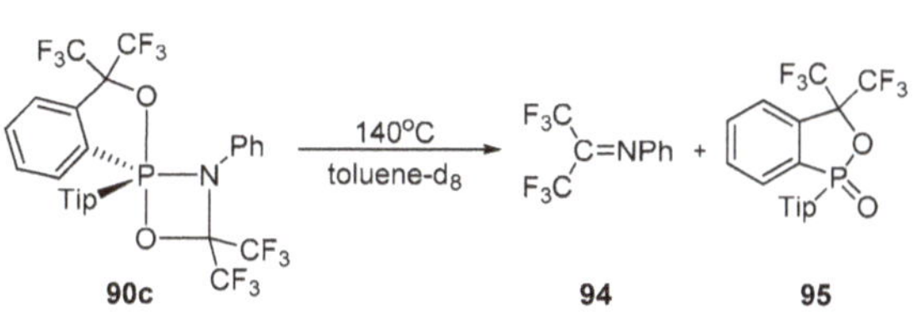

Scheme 41. Thermolysis of spirophosphorane **90c**.

The detailed studies on the formation of 1,2λ^5-oxaphosphol-5(2*H*)-one **93** were conducted on the basis of ^{31}P-NMR analysis [96]. In the course of the reaction a decrease of the resonance signal at +25.6 ppm along with an increase of the new one at +47.3 ppm corresponding to **96** was observed.

The latter was found to be unstable in the presence of moisture which resulted in the slow conversion to the final product. The proven synthetic pathway was concluded in Scheme 42.

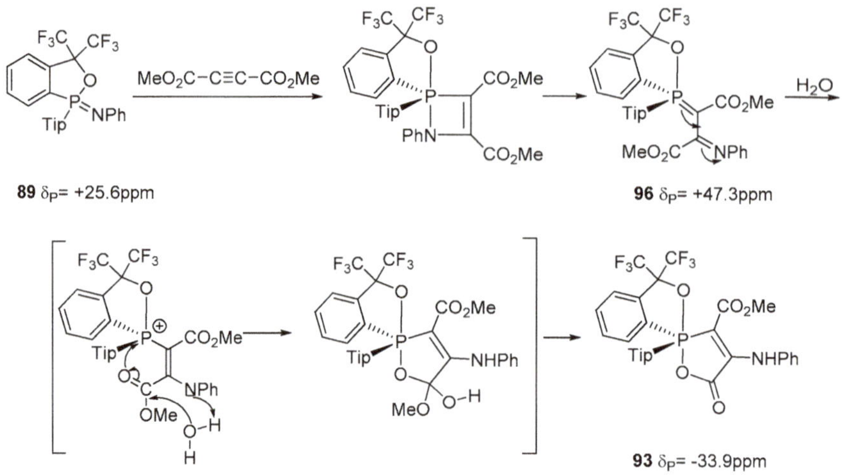

Scheme 42. The mechanism of formation of the bicyclic phosphorane **93** from cyclic iminophosphinate **89**.

10P-2C-2O-1Z (Z = N, O, etc.) Phosphoranes

It is already well known that the preparation of pentacoordinate phosphoranes can be accomplished by the reaction of phosphorus (III) compounds with ylidene derivatives of β-carbonyl compounds. This approach was exemplified by the reaction of 2-phenyl-1,3,2-benzodioxaphosphorin-4-one **97** with diethyl benzylidenemalonate **98** which occurred along two pathway, yielding pentacoordinate phosphorane **99** and phosphinate derivative **100** containing tetracoordinate phosphorus atom [97]. When the reaction was allowed to heat for a short time the formation of the seven-membered phosphonate derivative with relatively high stereoselectivity equal to 50% was observed (Scheme 43).

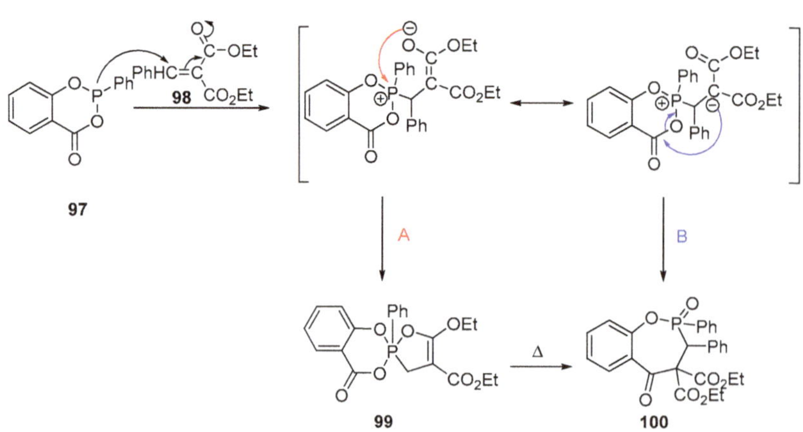

Scheme 43. Synthesis of pentacoordinate phosphorane **99** and phosphinate derivative **100**.

The plausible mechanism involves the nucleophilic attack of the phosphorus atom at the β-carbon atom in α,β-unsaturated carbonyl compounds to produce an intermediate with delocalized π-electrons along three atoms, as is presented by its two resonance structures. In case of synthetic pathway A the attack of negatively charged oxygen on phosphorus atom occurs with reversible formation of pentacoordinate phosphorane. The second synthetic pathway (B) assumes nucleophilic substitution at the carbonyl carbon atom to yield the seven-membered ring, 1,2-benzoxaphosphepine. In further investigations Akiba et al. [98] succeeded in preparing spirophosphoranes bearing Martin ligand units and a monodentate primary amino substituent. These *O*-equatorial isomers were found to be surprisingly stable, taking into account the fact that the equatorial substituents, which are more electronegative than carbon, are expected to facilitate pseudorotation. The *O*-equatorial aminophosphoranes **102** along with *O*-apical isomers **103** were obtained from the in-situ generated chlorophosphorane **101** which was allowed to react with the corresponding amines (Scheme 44). However, when methylamine or aniline were used, the corresponding *O*-equatorial isomer were not produced.

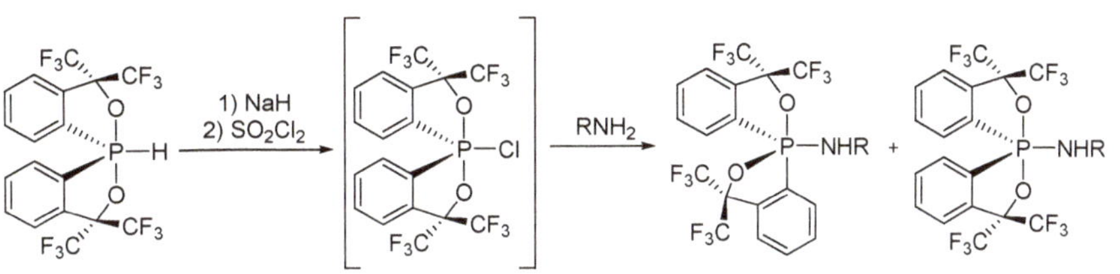

59 **101** **102** **103**

	102/103
a: R=Me,	2/98
b: R=*n*-Pr,	36/64
c: R=$PhCH_2$,	62/38
d: R=*i*-Pr,	80/20
e: R=*t*-Bu	98/2
f: R= Ph,	2/98

Scheme 44. Synthesis of aminophosphoranes **102** and **103**.

The *O*-equatorial phosphoranes **102** were stable at room temperature and could still be converted to their more stable *O*-apical pseudorotamers **103** when they were heated in a solution, indicating that the phosphoranes **102** are kinetic products, as in the case of *O*-equatorial alkylphosphoranes. The structures of **102c** and **102e** were determined by single-crystal X-ray analyses to be in the *O*-equatorial configuration. The unusual stability of the *O*-equatorial phosphorane **102** could be attributed to the stabilizing $n_N \rightarrow \sigma^*_{P\text{-}O}$ orbital interaction. More detailed studies on the experimental determination of the $n_N \rightarrow \sigma^*_{P\text{-}O}$ interaction energy of *O*-equatorial *C*-apical phosphorane supported with DFT calculations have been also reported providing an additional proof for this assumption.

Juge et al. described the formation of chiral bicyclic phosphoranes in the Arbuzov-type reaction of diastereomerically pure cyclic tricoordinated oxaphosphacycloalkanes and very reactive Koshland reagent I **104** [99]. They found that both enantiomers of 2-phenyl-1.3.2-oxazaphospholidine [(−)-**105** derived from (+)-ephedrine or (+)-**105** from (−)-ephedrine], treated with 2-hydroxy-5-nitrobenzyl bromide **104a** afforded the stable pentacoordinated phosphoranes [(−)-**106** from (−)-**105** or (+)-**106** from (+)-**105**] which were isolated by simple chromatography. The minor isomeric spirophosphoranes with the opposite configuration at the phosphorus atom were also formed (Scheme 45). 2-Hydroxy-5-nitrobenzyl halides **104a**,**b** reacted also with five- or six-membered cyclic tricoordinated phosphorus compounds (−)-**107** (derived from (+)-2-methyl-2,3-butanediol) or (−)-**110** (derived from (−)-chloramphenicol), yielding spirophosphoranes, **108** and **109** or **110**, respectively. However, these spirophosphoranes were too unstable to be isolated. Their formation was confirmed only by the ^{31}P-NMR spectra of a crude reaction product.

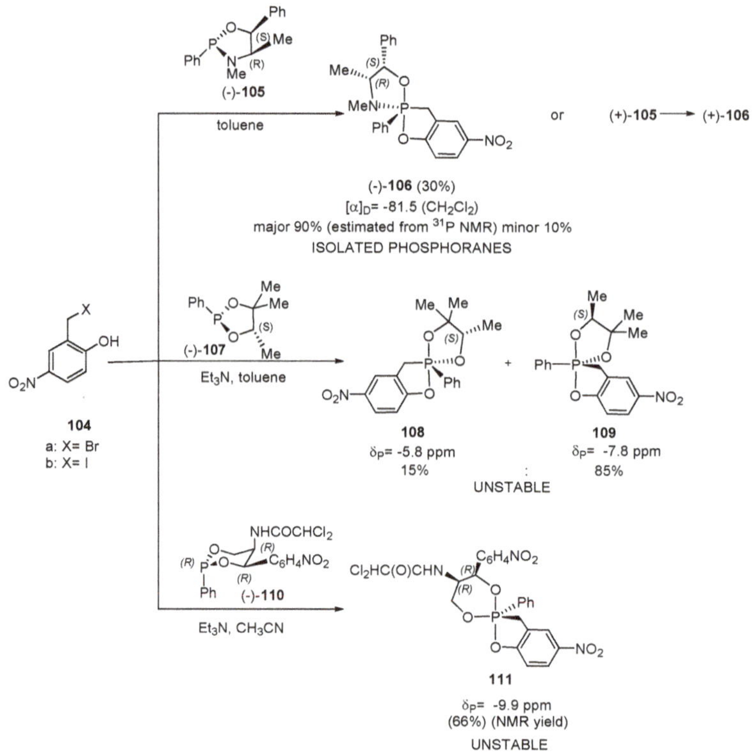

Scheme 45. Reactions of cyclic five or six-membered tricoordinated phosphorus compounds with 2-hydroxy-5-nitrobenzyl halides.

10P-1C-1N-3O Phosphoranes

The synthesis of a pentacoordinated phosphorane containing tricyclic system was reported by Sevenard et al. [100]. 2-Fluoroacetylcycloalkanones **111a,b** reacted with diethyl isocyanatophosphite diastereospecifically leading to phosphoranes **112a,b** via intermediate species formed by the addition of phosphorus to a trifluoromethyl substituted carbonyl atom which subsequently underwent two additional heterocyclizations (Scheme 46).

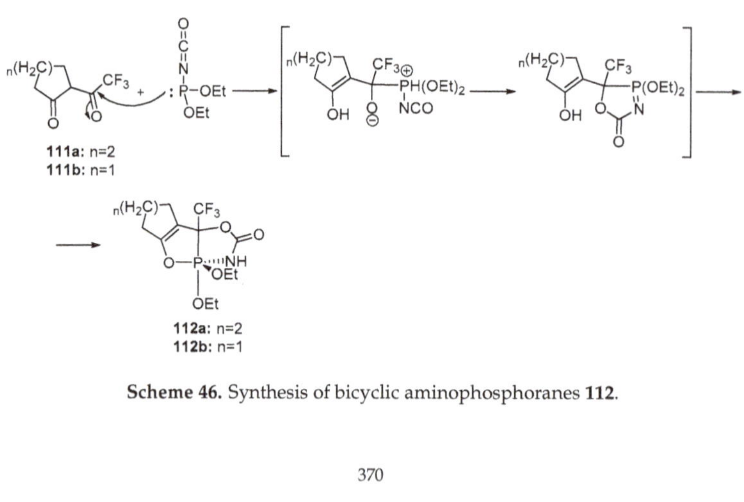

Scheme 46. Synthesis of bicyclic aminophosphoranes **112**.

The phosphorane derived from 2-trifluoroacetylcyclopentanone **112b** was moisture-sensitive, unlike the cyclohexane derivative **112a**, probably due to the greater ring strains. The proposed structure was strictly confirmed by means of NMR data on the basis of which it was found that carbon atom occupies an axial position and two annulated five membered rings are arranged in the equatorial-axial-equatorial order.

10P-1C-2N-2O Phosphoranes

Pudovik et al. reported the synthesis and crystal structure analysis of spirophosphoranes derived from benzoxazaphospholidines and/or 2-aminophenol [101]. It has been found that the condensation reaction of 2-aminophenol with chloromethylphosphonic dichloride proceeded with the formation of the expected benzoxazaphospholidine-2-oxide **99** along with the unexpected one: spirophosphorane **100** (Scheme 47).

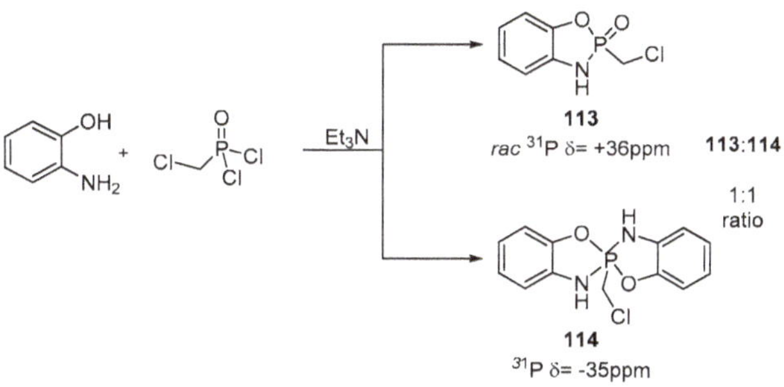

Scheme 47. Synthesis of bicyclic aminophosphorane **114**.

As indicated by the authors, the crucial step in a plausible pathway for their synthesis is an initial phosphorylation of a hydroxy group in 2-aminophenol with phosphonic dichloride leading to chloromethylphosphonochloridate intermediate (I). Subsequent nucleophilic attack of the amino group at the phosphorus atom leads to the cyclization product **99**. Simultaneously, the second pathway is realized when chloromethylphosphonochloridate intermediate reacts with the next molecule of the starting aminophenol to give the OH condensation product, which undergoes cyclization to provide unstable intermediate (II, III). It is then converted to the spirophosphorane as a result of intramolecular nucleophilic attack of the amino group at electrophilic phosphorus atom with the elimination of water (Scheme 48).

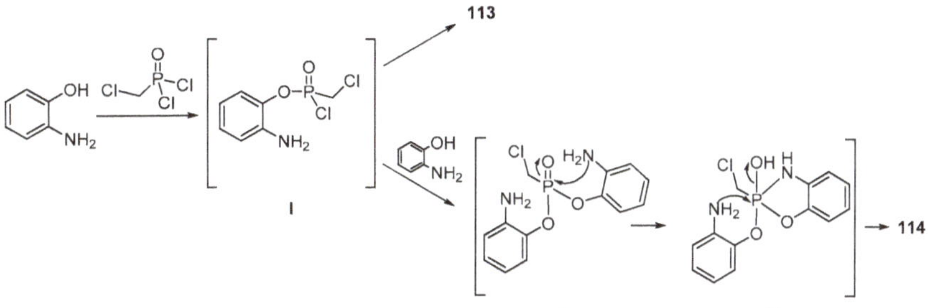

Scheme 48. The mechanism for the formation of spirophosphorane **114**.

The alkylation of 2-alkoxy-1,3,2-benzoxazaphospholidine with bis(dialkylamino)methanes afforded N-alkylated product **115** which underwent tautomerization to phosphonimidate **116** and further transformation to polycyclic phosphoranes **117** in 30 days [102]. It is worth to mention that both benzoxazaphospholidine **115** and polycyclic phosphorane **117** were used as phosphorylating agent for 2-aminophenol yielding spirophosphorane **118** (Scheme 49). The crystal and molecular structure of the synthesized spirophosphoranes **114** and **118** were determined by single-crystal X-ray diffraction [101,103,104].

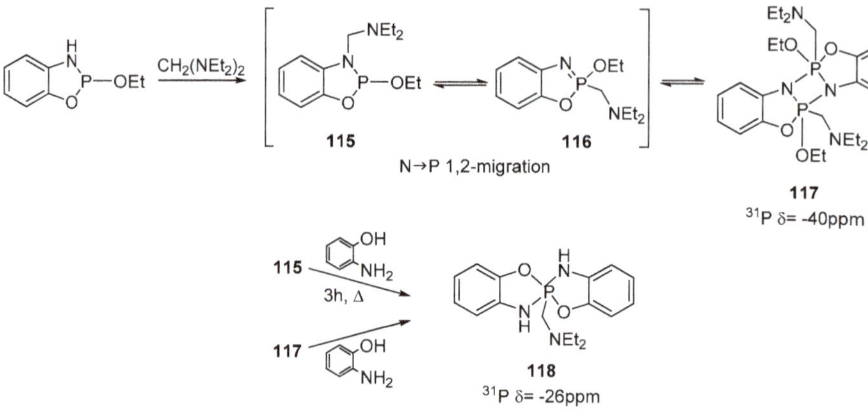

Scheme 49. Syntheses of bicyclic diaminophosphoranes **117** and **118**.

Amino alcohol based monocyclic hydrophosphorane **119** was prepared in the reaction of 2-aminophenol with tetraalkylphenylphosphonous diamide and used for the preparation of bicyclic spirophosphorane **120** [105] based on the oxidation procedure or on heating above its melting point (Scheme 50).

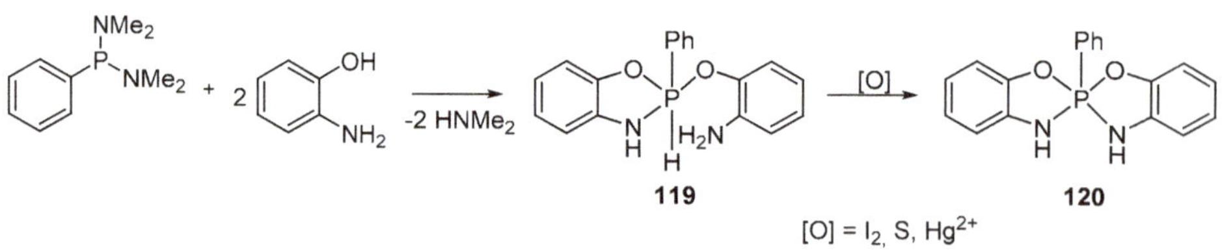

Scheme 50. Synthesis of bicyclic diaminophosphorane **120**.

The alternative approach to synthesize bicyclic pentacoordinate spirophosphoranes proposed by Gholivand et al. [106] is based on the condensation of simple hydrazides (i.e., benzhydrazide **123a** and 4-pyridinecarboxylic acid hydrazide **123b**) with phosphoryl reagents followed by a dehydration–cyclization rearrangement of the resulting phosphorylated hydrazides. Such spiro-bicyclophosphorane with a trigonal bipyramidal structure is formed if phosphoryl reagent contains at least two appropriate leaving groups such as chlorine atoms (in $PhPOCl_2$ or $POCl_3$). Benzhydrazide reacted with $POCl_3$ in refluxing acetonitrile to give the P–Cl phosphorane intermediate **121** which treated with the appropriate amine was converted to the P-N spirophosphorane **122a,b** (Scheme 51). The mechanism proposed for the formation of the intermediate involved β-amidic proton elimination in phosphorylated hydrazides **A** upon which the cyclization product with new C=N imine bond **B** was formed. The intermediate was finally converted into **121** by dehydration.

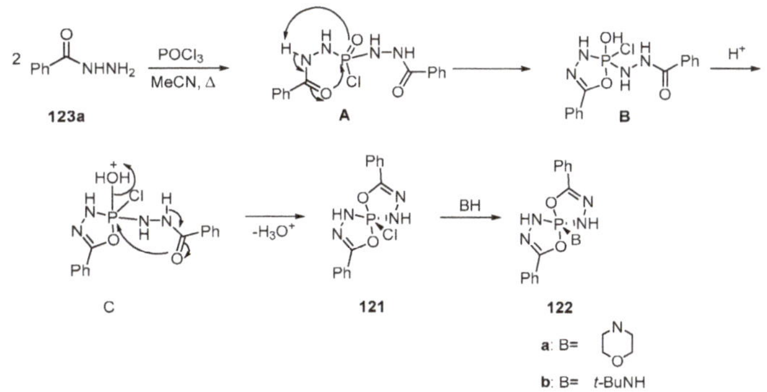

Scheme 51. Syntheses of bicyclic diaminophosphoranes **121** and **122**.

Similarly, several pentacoordinate phosphoranes **122c–g** were easily prepared in the reaction of benzhydrazide **123a** or 4-pyridinecarboxylic acid hydrazide **123b** with $PhPOCl_2$, $PhNHPOCl_2$ or $POCl_3$ in the presence of triethylamine [106] (Scheme 52).

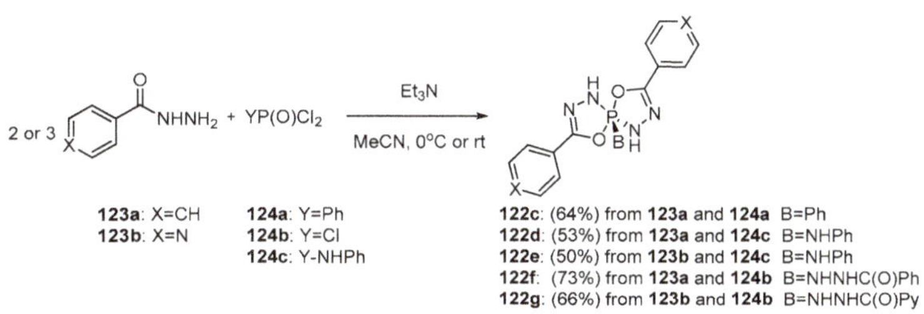

Scheme 52. Synthesis of bicyclic diaminophosphoranes **122**.

10P-1C-4O Phosphoranes

The reaction of (*E*)-bis-(2,4,6-tri-*t*-butylphenyl)diphosphane **125** with a fourfold excess of tetrachloro-*o*-benzoquinone **126** was found to give the chiral perchlorinated spirophosphorane **127**. It was suggested that the reaction, proceeding according to an electron transfer mechanism, occurred via **128** with a cleavage of the P-P bond and indeed, **127** was isolated in low yield even with two equivalents of **126** (Scheme 53) [107].

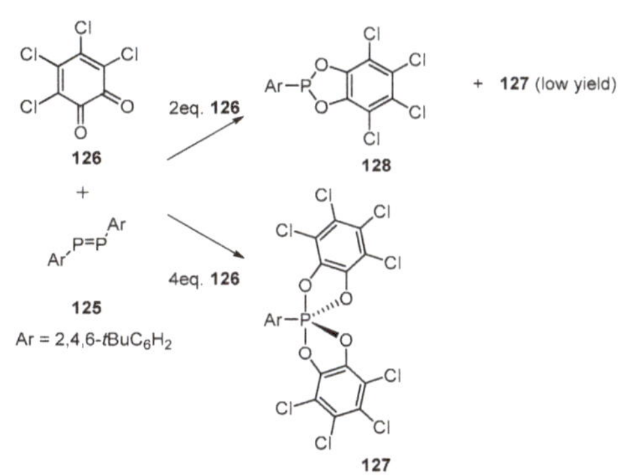

Scheme 53. Synthesis of bicyclic phosphoranes **127**.

Similarly, the reactions of tricoordinated organophosphorus precursors **129–131** with an appropriate equivalent of ethylene glycol was found to give the spirophosphorane **132**. All the three reactions occurred via the bis-ester **133** formed in a nucleophilic exchange reaction at the tricoordinated phosphorus center of precursors **129–131** [108] (Scheme 54).

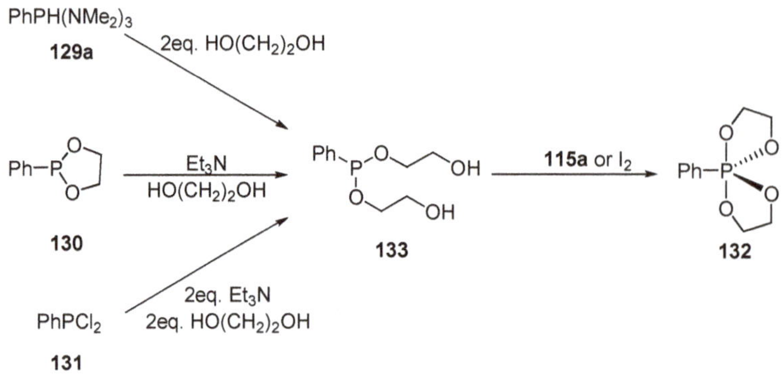

Scheme 54. Synthesis of bicyclic phosphorane **132**.

10P-1C-1N-2O-1H Phosphoranes

The reactions of tricoordinated phosphonous diamides **115a–d** with an appropriate equivalent of aminodiols **120** led to a series of chiral enantiomeric or diastereomeric bicyclic spirophosphoranes **121a–f** (Scheme 55) [109].

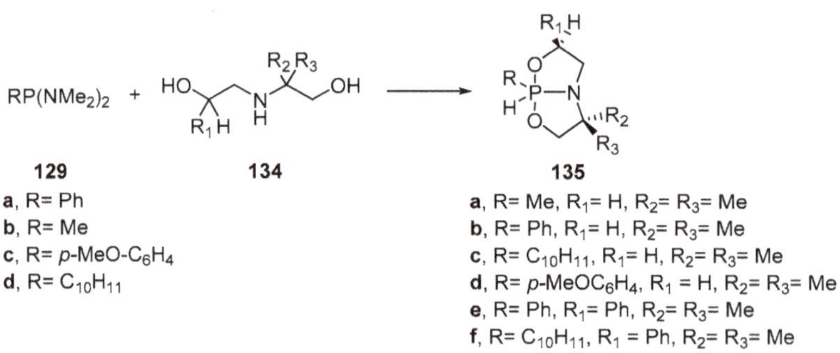

Scheme 55. Synthesis of bicyclic azaphosphoranes **135**.

10P-2N-2O-1H and 10P-2N-3O Phosphoranes

Chiral hydrospirophosphoranes derived from *L*-amino acids were for the first time obtained by Koenig et al. as early as 1979 [110]. More recently this approach based on the reaction of phosphorus trichloride with amino acids was used by Zhao and coworkers for the preparation of spirophosphoranes from enantiomerically pure valine **136a**, isoleucine **136b** and phenylalanine **136c**. [111–114]. The spirophosphoranes formed as mixtures of diastereoisomers were separated to the pure stereoisomers **137a–c** and **138a–c** by reverse-phase HPLC or recrystallization (Scheme 56).

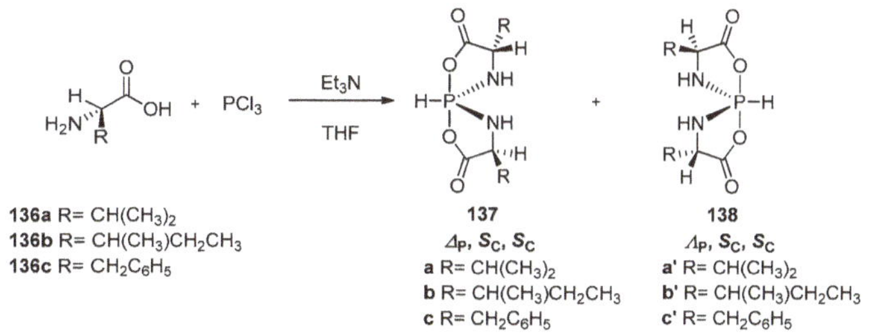

Scheme 56. Synthesis of hydrospirophosphoranes **137** and **138** derived from L-amino acids.

The pure stereoisomers **137** were used for the preparation of pentacoordinate spirophosphorane carbamates **139** [115] via the two-steps Atherton–Todd-type reaction (Scheme 57). Initially, the spirophosphoranes **137** reacted with CCl_4 yielding chlorospirophosphoranes **140** (with retention of configuration at the phosphorus atom). In a second step chlorophosphoranes, upon the reaction with carbamate anions, (formed in situ from CO_2 and secondary amines), afforded the carbamates **139** with inversion of configuration (Scheme 58).

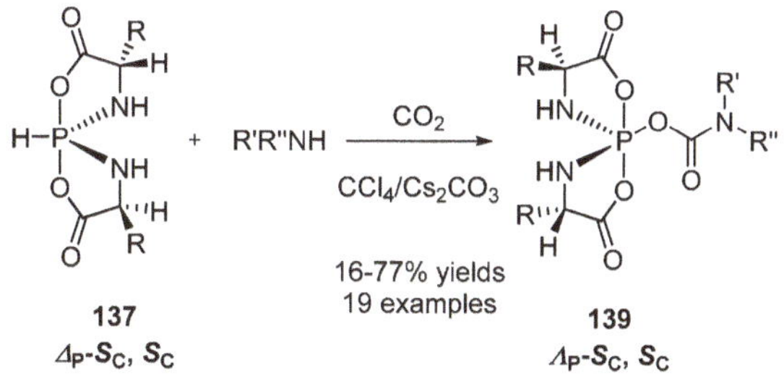

Scheme 57. Preparation of pentacoordinate spirophosphorane carbamates **139**.

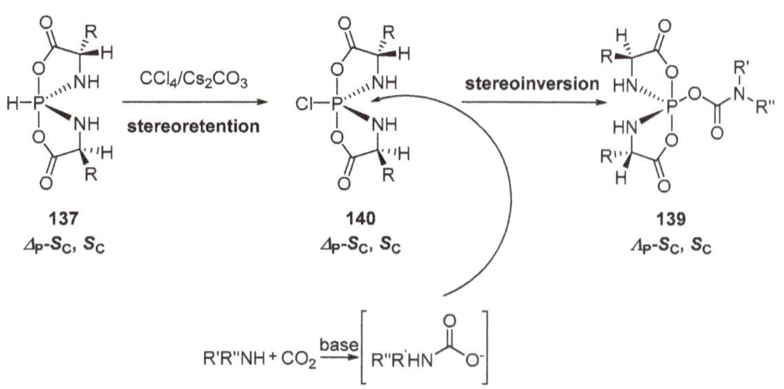

Scheme 58. The mechanism of formation spirophosphorane carbamates **139**.

Similarly, the spirophosphoranes **137a–e** and **138a–e** were used as substrates for the synthesis of pentacoordinate pyrospirophosphoranes containing a P–O–P bond **141–143** under modified Atherton–Todd conditions [116] (Scheme 59). It was found, upon optimization of the reaction conditions, that the spirophosphoranes **137a,b** gave the pure diasteroisomer of pyrospirophosphoranes **142a,b** whereas a mixture of diastereoisomers **141c–e** and **143c–e** were isolated if hydrospirophosphoranes **137c–e** were used as substrates.

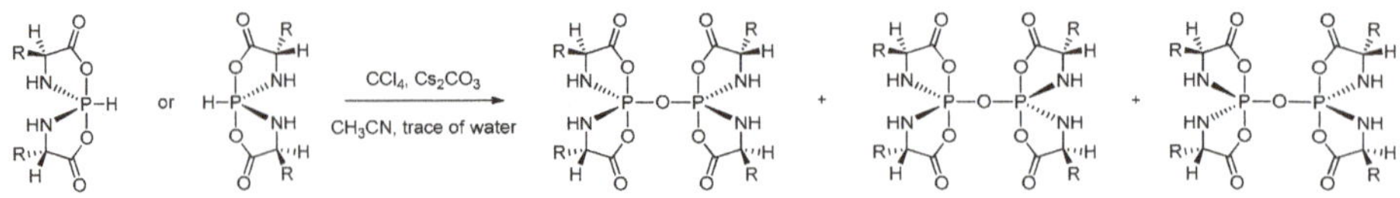

138 Λ_P, S_C, S_C / 137 Δ_P, S_C, S_C	141 Λ_P:Δ_P	142 Λ_P:Λ_P	143 Δ_P:Δ_P
a L-Val-Δ_P R= $CH(CH_3)_2$	a 0%	a' 100%	a 0%
b L-Ile-Δ_P R= $CH(CH_3)CH_2CH_3$	b 0%	b' 100%	b 0%
c L-Phe-Λ_P R= $CH_2C_6H_5$	c 94%	c' 0%	c 6%
d L-Met-Δ_P R= $CH_2CH_2SCH_3$	d 67%	d' 33%	d 0%
e L-Leu-Δ_P R= $CH_2CH(CH_3)_2$	e 40%	e' 60%	e 0%

Scheme 59. Synthesis of pentacoordinate pyrospirophosphoranes **141–143**.

A series of the alkoxy spirophosphoranes **144,145** was also prepared by this approach (Scheme 60) but unfortunately as a mixture of diastereoisomers (in most cases in 1:1 ratio) [117].

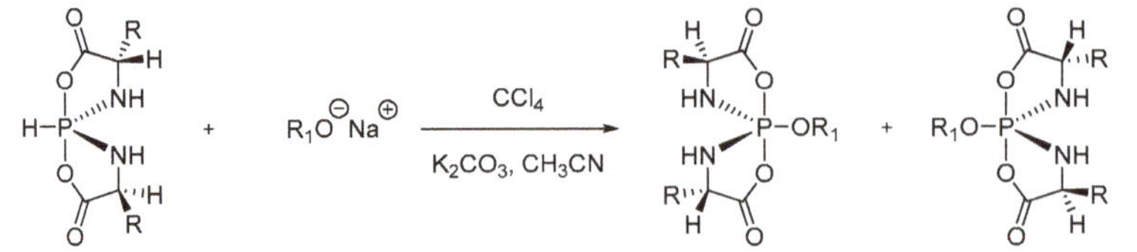

137 Δ_P, S_C, S_C — **144** Λ_P, S_C, S_C — **145** Δ_P, S_C, S_C

a R= $CH(CH_3)_2$ R_1= Me
b R= $CH(CH_3)CH_2CH_3$ R_1'= Et

R_1= Me, R= $CH(CH_3)_2$; $dr(\Lambda_P:\Delta_P)$= 50:50 yield 18/25%
R_1= Et, R= $CH(CH_3)_2$; $dr(\Lambda_P:\Delta_P)$= 44:56 yield 21/46%
R_1= Et, R= $CH(CH_3)CH_2CH_3$; $dr(\Lambda_P:\Delta_P)$= 44:56 yield 14/23%

Scheme 60. Synthesis of alkoxy spirophosphoranes **144,145**.

Bicyclic hydrophosphoranes **147** and **148** were synthesized as a 3:2 mixture of two epimers by the reaction of tris-(*N*,*N*-diethyl)phosphorus amide with isoleucinol **146**. On the other hand, the tricyclic phosphorane **149** was isolated as a single stereoisomer in a similar reaction with the diaminodiole **151** (Scheme 61) [118]. The isolated spirophosphoranes **147–149** were used as chiral ligands in the Pd-catalyzed alkylation of 1,3-diphenyl allyl acetate with acceptable stereoselectivity (up to 74% *ee*) (Scheme 61) [119]. Moreover, these hydrophosphorane derivatives were used to obtain complexes with $[Pt(COD)Cl_2]$, $[Rh(CO)_2Cl]_2$ $[Rh(THF)_2(COD)]^+BF_4^-$.

Chiral triquinphosphoranes **152–154** were easily synthesized from chiral enantiopure diaminodiols with a C_2 symmetry axis. It was shown that their structure is best represented by two trigonal bipyramid structures (TBP) with opposite absolute configurations at the phosphorus atom, R_P and S_P, being in a fast equilibrium by a Berry pseudorotation process (Scheme 62) [120]. They reacted with various activated carbonyl compounds: trifluoroacetophenone, ketopantolactone and aromatic aldehydes to afford two diastereomeric hydroxyphosphoranes with a *de* up to 90% depending on the nature of the electrophile (Scheme 62). Diastereomeric mixture of hydroxyphosphoranes formed by the addition of ketopantolactone **155–156d–f** were quantitatively converted into alkoxyphosphoranes **157** with diastereomeric excesses decreasing from 86% to 8% for R = Me, from 90% to 50% for R = iPr and from 84% to 2% for R = Bn.

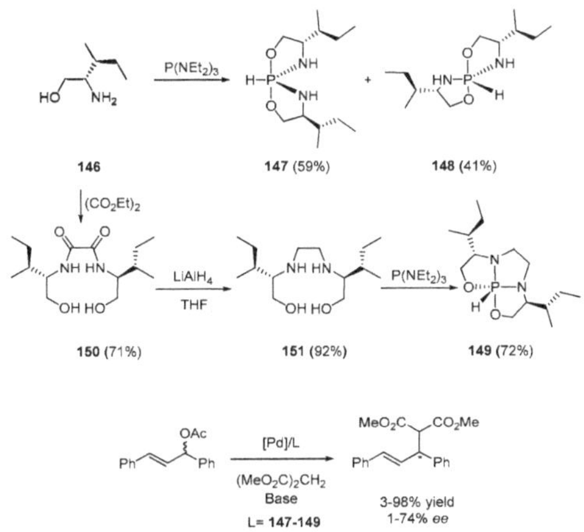

Scheme 61. Syntheses of bicyclic hydrophosphoranes **147–149** and their use as chiral ligands.

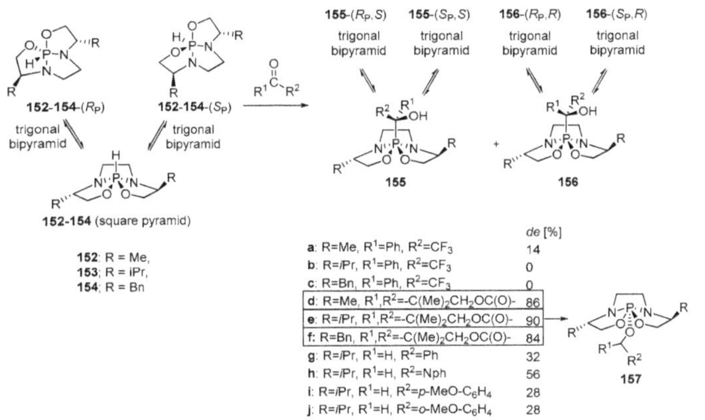

Scheme 62. Chiral triquinphosphoranes **152–154** and their reactions with various activated carbonyl compounds.

Similarly, the parent triquinphosphorane **158** was found to react with methyl, *n*-butyl, or *t*-butyl disulfide to produce the corresponding alkylthiophosphoranes **159a–c**, although *t*-butylthio-phosphorane **159c** was only a minor product and the major one was identified as the thiophosphoramide **160** (Scheme 63). Under irradiation at −50 °C *t*-butylthiophosphorane **159c** was produced in high yield, however, by increasing the reaction temperature its slow conversion to **160** was observed. The reactivity of the triquinphosphorane **158** with alkyl disulfides was compared both under irradiation and in the dark at room temperature. The reactivity increased significantly in the case of *n*-butyl disulfide and *t*-butyl disulfide when the reactions were conducted under irradiation. This observation led to the conclusion that this type of the reaction does involve free radical species [121].

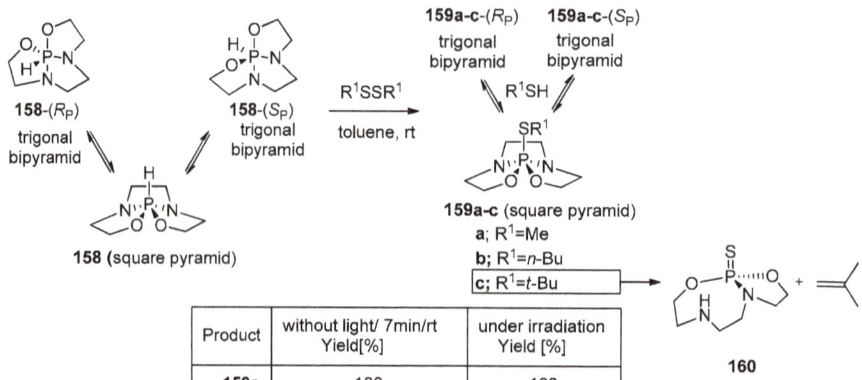

Product	without light/ 7min/rt Yield[%]	under irradiation Yield [%]
159a	100	100
159b	20	100
159c	0	80 (**159c+160**)

Scheme 63. Reaction of chiral hydrido-triquinphosphorane **158** with alkyl disulfides.

10P-1N-3O-1H and 10P-2N-2O-1H Phosphoranes

Three synthetic protocols were applied to the preparation of this type of hydrophosphoranes with the use of amidooxime derivatives as substrates [122]. The first procedure started with the reaction of amidooxime **161a–e** with PCl_3 leading to 5-chloro-1,2,4,5-oxadiazaphospholines **162a–e**. Their reactions with ethylene glycol or *ortho*-hydroxyphenol afforded the unsymmetrical spirophosphoranes **163** or **164** whereas the treatment with another dose of the amidooxime **161** led to symmetrically substituted spirophosphoranes **165a–e** (Scheme 64).

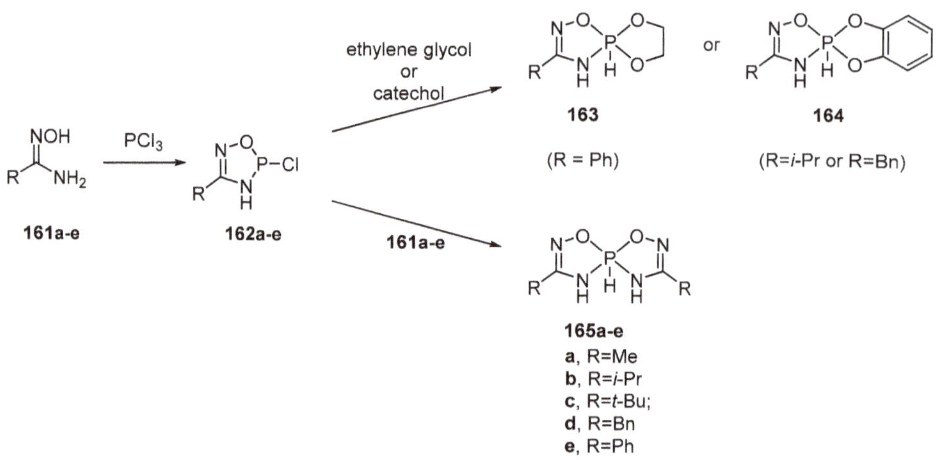

Scheme 64. Syntheses of bicyclic hydrophosphoranes **163–165**.

In the second approach, amidooxime **161a,c,d** upon reaction with 2-dimethylamino-1,3,2-dioxaphospholane **166** provided unsymmetrical spirophosphoranes **163a,c,d** and in the reaction with 2-dimethylamino-(4,5)benzo-1,3,2-dioxaphospholane **167** in acetonitrile yielded unexpected symmetrically substituted spirophosphoranes **165a–e** instead of unsymmetrically substituted phosphorane **164a–e** (Scheme 65).

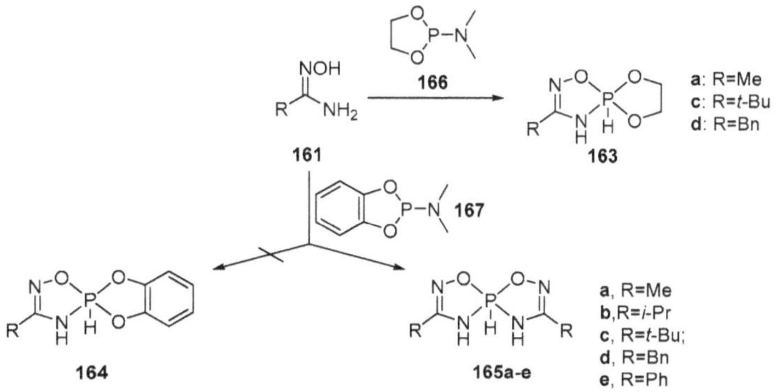

Scheme 65. Syntheses of bicyclic hydrophosphoranes **163** and **165**.

The third synthetic pathway involved the treatment of amidooximes **161b,d,e** with 2-chloro-(4,5)benzo-1,3,2-dioxaphospholane **168** leading, as it was expected, to the corresponding unsymmetrical spirophosphoranes **164**. The latter were converted into the corresponding unsymmetrical spirophosphorane **170** by transesterification with pinacol **169** taking place in the presence of triethylamine (Scheme 66).

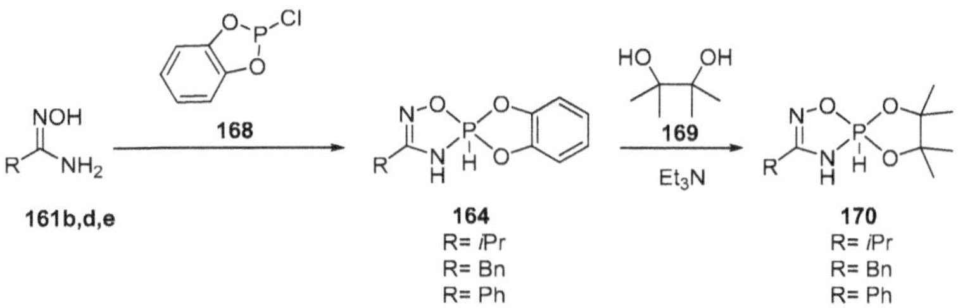

Scheme 66. Synthesis of bicyclic hydrophosphoranes **164** and **170**.

10P-2N-3O and 10P-1N-4O Phosphoranes

Amidooximes **161d,e** were used also as substrates for the preparation of the spirophosphoranes containing the P–OMe grouping **173**, **175**. They were isolated upon application of the oxidation procedure carried out under basic conditions with iodine as an oxidant [122] (Scheme 67).

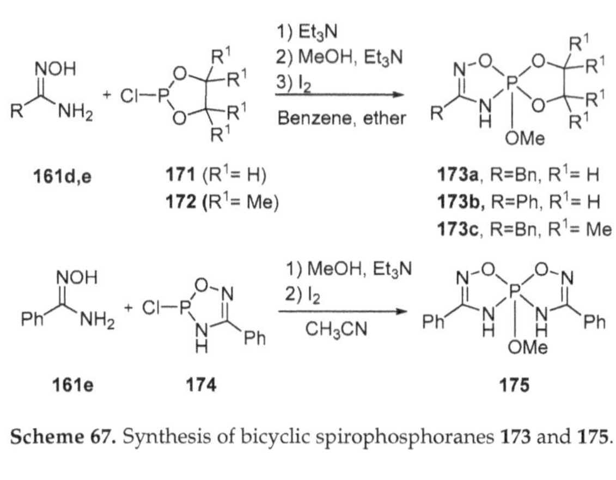

Scheme 67. Synthesis of bicyclic spirophosphoranes **173** and **175**.

O-Trimethylsilyl derivatives of spirophosphoranes **177** containing an amino acid residue were formed by the cyclisation of the corresponding cyclic amides **176** [123] (Scheme 68). The resulting spirophosphoranes **177** were used for the formation of peptides from amino acids such as histidine, serine, threonine and α-alanine but not β-alanine.

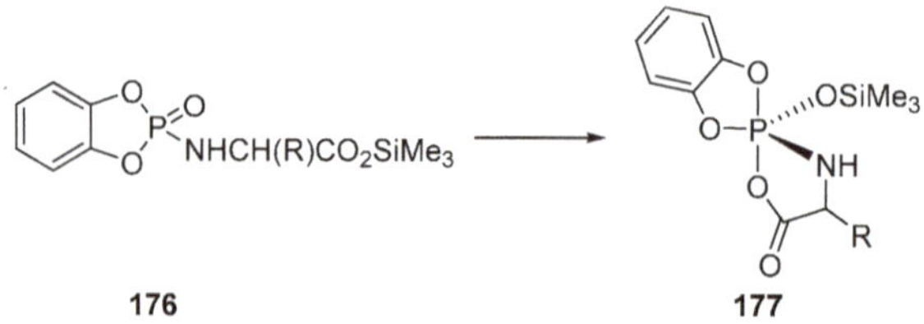

Scheme 68. Synthesis of bicyclic phosphoranes **177**.

Later on, spirophosphoranes **177** were also used as substrates in a new method for the solid phase synthesis of oligopeptides. This approach was based on their rapid reaction with hydroxymethyl polystyrene resin leading to unstable hydroxyphosphoranes **178** which after mild hydrolysis gave the phosphate **180** and a solid phase bound amino-acids **179**. Coupling of **179** with further phosphorane **177**, followed by hydrolysis afforded a solid phase bound dipeptide. The process was repeated to give oligopeptides. The final oligopeptide was liberated from the resin by treatment with TFA [124] (Scheme 69).

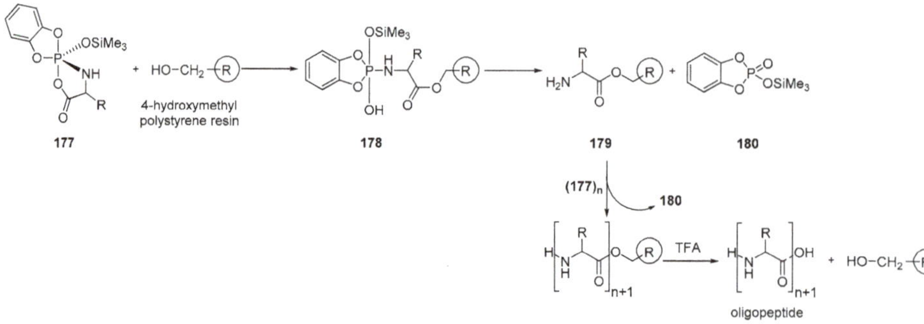

Scheme 69. Synthesis of oligopeptides with the use of spirophosphoranes **177** as substrates.

10P-4O-1H Phosphoranes

The synthesis and reactivity of the hydrophosphorane **181** derived from perfluoropinacol was described as early as 1983 [125]. It was found that it could be rapidly oxidized by dimethyl sulfoxide to form the hydroxyphosphorane **182**, and then silylated to the derivative **183**. Chlorine and bromine reacted with **181** to give the corresponding halospirophosphoranes **184a,b**. In this context it is interesting to note that the corresponding fluorospirophosphorane **184c** was obtained from the dichlorofluoro-perfluoropinacolophosphorane $FCl_2P(CH_2Ph)[OC(CF_3)C(CF_3)O]$ **185** and dilithium perfluoropinacolate $LiOC(CF_3)C(CF_3)OLi$ **186**. In the presence of triethylamine the hydrophosphorane **181** reacted with benzyl bromide or acetyl chloride to form the phosphoranes **187** or **188**. Moreover, the treatment of the hydrophosphorane **181** with trimethylphosphine gave a thermally unstable phosphonium salts **189**. Hydrophosphorane **181** undergoes fast hydrolysis affording phosphorous acid and perfluoropinacol [125] (Scheme 70).

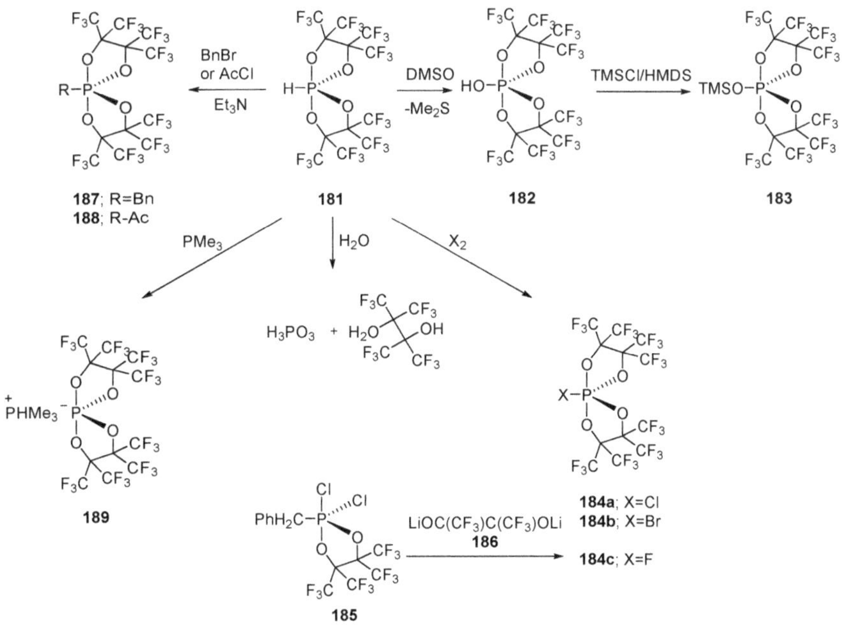

Scheme 70. Interconversions of spirophosphoranes **181** and **182**.

10P-4O-1Metal and 10P-3N-1O-1Metal Phosphoranes

Another group of chiral pentacoordinate organophosphorus compounds are metallophosphoranes **192a–e**. Nakazawa et al. [126] developed a synthetic method for the preparation of metallophosphoranes **192**, which was based on the nucleophilic attack of the, in situ generated, anion **191a–e** on the complexed trivalent phosphorus atom **190a,b** (Scheme 71). For example, ferrocenyl complex **190a** reacted with in situ generated anions **191a–e** to form chiral metallophosphoranes **192a–e**. The valency of the phosphorus atom increased from III to V without breaking the Fe-P bond and the nature of Fe-P bond changed from coordinative to covalent.

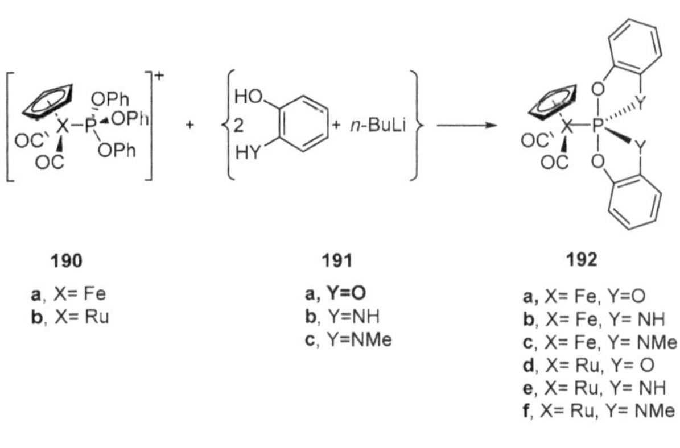

Scheme 71. Preparation of chiral metallophosphoranes **192**.

It is worthy to note that treatment of $HP(C_6H_4NH)_2$-**193** with *n*BuLi led to unexpected deprotonation on nitrogen atom (giving **194b**) instead on phosphorus (to afford **194a**). Moreover, an amide anion **194b** is in an equilibrium with phosphoranide **194a**. The reaction of the in situ generated anion **194a** with [Cp(CO)LFeCl] (**195a** or **195b**) gave a P-metallated phosphorane **196a** or **196b**. On the contrary, the reaction the equilibrated mixture of **194** with [Cp(CO)CoI_2]-**197** led to *N*-metallated compound **198**. Interestingly, in the reaction of the phosphoranide **194a** with MeI the P-methylated product **199** was formed. On the other hand, electrophiles such as Me_3SiCl, Me_3GeCl or Me_3SnCl, provided *N*-substituted products (**200a–c**) [127,128] (Scheme 72).

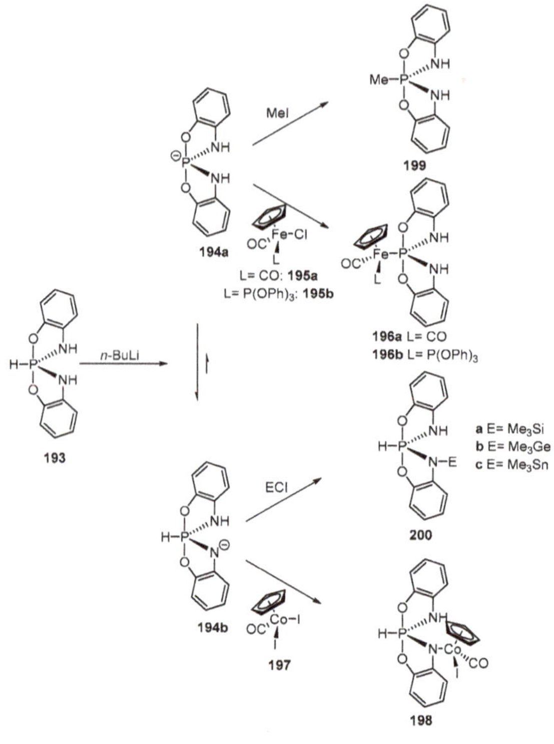

Scheme 72. Functionalization of hydrophosphorane **193** leading to phosphoranes **196** and **198–200**.

Moreover, it was found that in the presence of a base a migration of pentacoordinate phosphorane fragment occured from the transition metal in **196** to the carbon atom in the Cp ring in **201** [129] (Scheme 73).

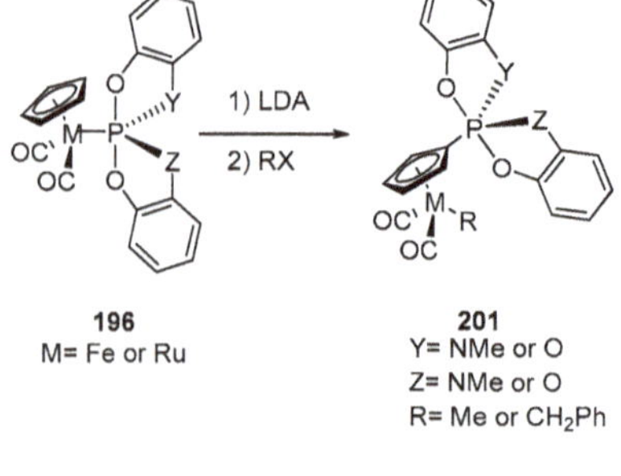

Scheme 73. Alkylation of metallophosphoranes **196**.

10P-5O Phosphoranes

The hydroxyphosphorane **203** was prepared by N_2O_4 oxidation of the parent hydrophosphorane **202** (Scheme 74). Its single crystal analysis showed an almost perfect trigonal bipyramidal structure with the unit cell containing two molecules of the same helicity connected by H-bonds between the P–OH and carbonyl groups [130].

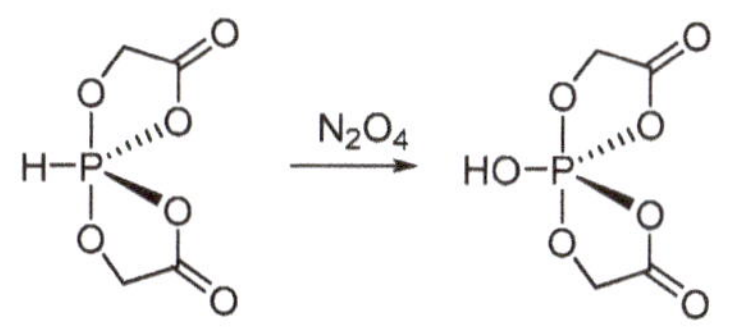

202 **203**

Scheme 74. Oxidation of hydrophosphorane **202**.

The triethylammonium salt of the hydroxyspirophosphorane **205** derived from *n*-butyl tartrate was prepared on treatment of the very acidic [pK_a = 4.4(DMSO)] monocyclic phosphorus ester **204** with triethylamine (Scheme 75) [131].

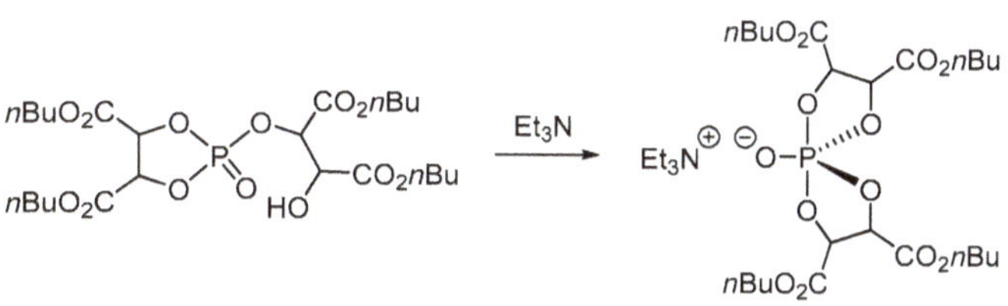

204 **205**

Scheme 75. Formation of salt **205**.

The radical reaction of the bicyclic hydrophosphorane **206** derived from glycol with ethyl vinyl ether provided the corresponding P–C spirophosphorane **207** along with the monocyclic phosphite **208**. The reaction of the hydrospirophosphorane **206** with ethylene glycol and enamine **209** afforded the pentaoxyspirophosphorane **210** with the simultaneous reduction of the enamine **209** to amine **211** [132] (Scheme 76).

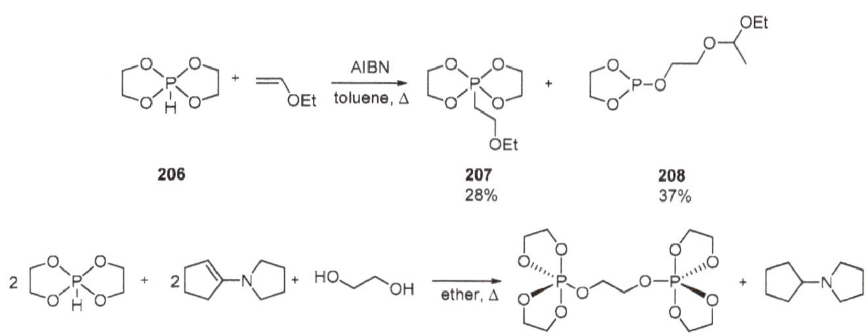

206 **207** 28% **208** 37%

206 **209** **210** 100% **211**

Scheme 76. Synthesis of spirophosphoranes **207** and **210**.

3. Conclusions

Enantiomerically enriched pentavalent phosphoranes constitute an interesting group of heterocyclic chiral auxiliaries. Their still limited applications in asymmetric synthesis are however hampered mainly by a limited access to diastereomerically pure and especially enantiomerically pure (or at least enantiomerically enriched) samples. It is our hope that this review, which describes briefly the basic procedures used for the preparation of these derivatives, their stereoisomerization mechanisms and their selected interconversions, will encourage wider interest in the synthesis, structural determinations and applications of these groups of chiral, hypervalent organophosphorus derivatives.

Acknowledgments: Preparation of this review was financially supported by the National Science Center, Poland fund awarded based on the decisions (grant Maestro UMO-2012/06/A/ST5/00227 to (Józef Drabowicz).

Conflicts of Interest: The authors declare no conflict of interest.

References

1. Eliel, E.L.; Wilen, S.H. *Stereochemistry of Organic Compounds*; John Wiley & Sons: New York, NY, USA, 1994.
2. Wolf, C. *Dynamic Stereochemistry of Chiral Compounds: Principles and Applications*; The Royal Society of Chemistry: Cambridge, UK, 2008.
3. Singh, J.; Hagen, T.J. Chirality and Biological Activity. In *Burger's Medicinal Chemistry and Drug Discovery*, 7th ed.; Abraham, D., Rotella, D., Eds.; John Wiley & Sons: New York, NY, USA, 2010; pp. 127–166.
4. Reddy, I.K.; Mehvar, R. *Chirality in Drug Design and Development*; CRC Press: New York, NY, USA; Basel, Switzerland, 2004.
5. Mori, K. Bioactive Natural Products and Chirality. *Chirality* **2011**, *23*, 449–462. [CrossRef] [PubMed]
6. Ojima, I. (Ed.) *Catalytic Asymmetric Synthesis*, 3rd ed.; Wiley-VCH: New York, NY, USA, 2010.
7. Christmann, M.; Brase, S. (Eds.) *Asymmetric Synthesis: More Methods and Applications*; Wiley-VCH: New York, NY, USA, 2012.
8. Bruckner, R. *Organic Mechanisms: Reactions, Stereochemistry and Synthesis*; Harmata, M., Ed.; Springer: Berlin/Heidelberg, Germany, 2010.
9. Montchamp, J.-L. Phosphorus Chemistry I: Asymmetric Synthesis and Bioactive Compounds. In *Topics in Current Chemistry*; Springer International Publishing: Cham, Switzerland, 2015; Volume 360.
10. Dutartre, M.; Bayardon, J.; Juge, S. Applications and stereoselective syntheses of P-chirogenic phosphorus compounds. *Chem. Soc. Rev.* **2016**, *45*, 5771–5794. [CrossRef] [PubMed]
11. International Union of Pure and Applied Chemistry (IUPAC). *Compendium of Chemical Terminology*, 2nd ed.; (the "Gold Book"); McNaught, A.D., Wilkinson, A., Eds.; Blackwell Scientific Publications: Oxford, UK, 1997.
12. Quin, L.D. *A Guide to Organophosphorus Chemistry*; John Wiley & Sons: New York, NY, USA, 2000.
13. Musher, J.I. The Chemistry of Hypervalent Molecules. *Angew. Chem. Int. Ed. Engl.* **1969**, *8*, 54–68. [CrossRef]
14. Perkins, C.W.; Martin, J.C.; Arduengo, A.J.; Lau, W.; Alergia, A.; Kochi, J.K. An Electrically Neutral sigma-Sulfuranyl Radical from the Homolysis of a Perester with Neighboring Sulfenyl Sulfur: 9-S-3 Species. *J. Am. Chem. Soc.* **1980**, *102*, 7753–7759. [CrossRef]
15. Hayes, R.A.; Martin, J.C. *Organic Sulfur Chemistry: Theoretical and Experimental Advances*; Bernardi, F., Csizmadia, J.G., Mangini, A., Eds.; Elsevier: Amsterdam, The Netherlands, 1985; Chapter 8; pp. 408–458.
16. Akiba, K. Hypervalent Compounds. In *Chemistry of Hypervalent Compounds*; Akiba, K., Ed.; Wiley-VCH: New York, NY, USA, 1999; Chapter 1; pp. 1–8.
17. Akiba, K. Structure and Reactivity of Hypervalent Organic Compounds: General Aspects. In *Chemistry of Hypervalent Compounds*; Akiba, K., Ed.; Wiley-VCH: New York, NY, USA, 1999; Chapter 2; pp. 9–47.
18. Noury, S.; Silyi, B.; Gillespie, J.R. Chemical Bonding in Hypervalent Molecules: Is the Octet Rule Relevant? *Inorg. Chem.* **2002**, *41*, 2164–2172. [CrossRef] [PubMed]
19. Mutterties, E.L.; Schunn, R.A. Pentacoordination. *Q. Rev.* **1966**, *20*, 245–299.
20. Berry, R.S. Correlation of Rates of Intramolecular Tunneling Processes, with Application to Some Group V Compounds. *J. Chem. Phys.* **1960**, *32*, 933–938. [CrossRef]
21. Berry, J.S. Time-Dependent Measurements and Molecular Structure: Ozone. *Rev. Mod. Phys.* **1960**, *32*, 447–454. [CrossRef]
22. Sheppard, W.A. Alkyl- and Arylsulfur Trifluorides. *J. Am. Chem. Soc.* **1962**, *84*, 3058–3063. [CrossRef]

23. Ugi, I.; Marquarding, D.; Klusacek, H.; Gillespie, P.; Ramirez, F. Berry Pseudorotation and Turnstile Rotation. *Acc. Chem. Res.* **1971**, *4*, 288–296. [CrossRef]
24. Gillespie, P.; Hoffmann, P.; Klusacek, H.; Marquarding, D.; Pfohl, S.; Ramirez, F.; Tsolis, E.A.; Ugi, I. Bewegliche Molekülgerüste—Pseudorotation und Turnstile-Rotation Pentakoordinierter Phosphorverbindungen und Verwandte Vorgänge. *Angew. Chem.* **1971**, *83*, 691–721. [CrossRef]
25. Ugi, I.; Marquading, D.; Klusacek, H.; Gokel, G.; Gillespie, P. Chemie und Logische Strukturen. *Angew. Chem.* **1970**, *82*, 741–771. [CrossRef]
26. Ramirez, F.; Ugi, I. Turnstile Rearrangement and Pseudorotation in the Permutational Isomerization of Pentavalent Phosphorus Compounds. *Adv. Phys. Org. Chem.* **1971**, *9*, 25–126.
27. Couzijn, E.P.A.; Slootweg, J.C.; Ehlers, A.W.; Lammertsma, K. Stereomutation of Pentavalent Compounds: Validating the Berry Pseudorotation, Redressing Ugi's Turnstile Rotation, and Revealing the Two- and Three-Arm Turnstiles. *J. Am. Chem. Soc.* **2010**, *132*, 18127–18140. [CrossRef] [PubMed]
28. Wittig, G.; Rieber, M. Darstellung und Eigenschaften des Pentaphenyl-phosphors. *Justus Liebigs Ann. Chem.* **1949**, *562*, 187–192. [CrossRef]
29. Ramirez, F. Condensations of carbonyl compounds with phosphite esters. *Pure Appl. Chem.* **1964**, *9*, 337–370. [CrossRef]
30. Ramirez, F. Recent developments in the chemistry of hydroxyphosphoranes. *Bull. Soc. Chem. Fr.* **1966**, *8*, 2443–2450.
31. Westheimer, F.H. Pseudo-rotation in the hydrolysis of phosphate esters. *Acc. Chem. Res.* **1968**, *1*, 70–78. [CrossRef]
32. Hall, C.D. Pentacoordinated and hexacoordinated compounds. In *Organophosphorus Chemistry*; Hutchinson, D.W., Walker, B.J., Eds.; The Royal Society of Chemistry: London, UK, 1986; Volume 16, pp. 51–78.
33. Hall, C.D. Pentacoordinated and hexacoordinated compounds. In *Organophosphorus Chemistry*; Walker, B.J., Hobbs, J.B., Eds.; The Royal Society of Chemistry: London, UK, 1988; Volume 19, pp. 47–69.
34. Hall, C.D. Pentacoordinated and hexacoordinated compounds. In *Organophosphorus Chemistry*; Walker, B.J., Hobbs, J.B., Eds.; The Royal Society of Chemistry: London, UK, 1991; Volume 21, pp. 51–72.
35. Hall, C.D. Pentacoordinated and hexacoordinated compounds. In *Organophosphorus Chemistry*; Allen, D.W., Tebby, J.C., Eds.; The Royal Society of Chemistry: London, UK, 2003; Volume 33, pp. 68–83.
36. Hall, C.D. Pentacoordinated and hexacoordinated compounds. In *Organophosphorus Chemistry*; Allen, D.W., Tebby, J.C., Eds.; The Royal Society of Chemistry: London, UK, 2006; Volume 35, pp. 265–303.
37. Tebby, J.C. Pentacoordinated and hexacoordinated compounds. In *Organophosphorus Chemistry*; Allen, D.W., Tebby, J.C., Eds.; The Royal Society of Chemistry: London, UK, 2007; Volume 36, pp. 184–196.
38. Roschenthaler, G.-V. Pentacoordinated and hexacoordinated compounds. In *Organophosphorus Chemistry*; Allen, D.W., Tebby, J.C., Eds.; The Royal Society of Chemistry: London, UK, 2009; Volume 38, pp. 318–331.
39. Pajkert, R.; Roschenthaler, G.-V. Pentacoordinated and hexacoordinated compounds. In *Organophosphorus Chemistry*; Allen, D.W., Tebby, J.C., Loakes, D., Eds.; The Royal Society of Chemistry: London, UK, 2010; Volume 39, pp. 290–307.
40. Pajkert, R.; Roschenthaler, G.-V. Pentacoordinated and hexacoordinated compounds. In *Organophosphorus Chemistry*; Allen, D.W., Tebby, J.C., Loakes, D., Eds.; The Royal Society of Chemistry: London, UK, 2011; Volume 40, pp. 297–315.
41. Pajkert, R.; Roschenthaler, G.-V. Pentacoordinated and hexacoordinated compounds. In *Organophosphorus Chemistry*; Allen, D.W., Tebby, J.C., Loakes, D., Eds.; The Royal Society of Chemistry: London, UK, 2012; Volume 41, pp. 322–348.
42. Pajkert, R.; Roschenthaler, G.-V. Pentacoordinated and hexacoordinated compounds. In *Organophosphorus Chemistry*; Allen, D.W., Tebby, J.C., Loakes, D., Eds.; The Royal Society of Chemistry: London, UK, 2013; Volume 42, pp. 197–215.
43. Pajkert, R.; Roschenthaler, G.-V. Pentacoordinated and hexacoordinated compounds. In *Organophosphorus Chemistry*; Allen, D.W., Tebby, J.C., Loakes, D., Eds.; The Royal Society of Chemistry: London, UK, 2014; Volume 43, pp. 348–365.
44. Kolodiazhnyi, O.I. Recent developments in the asymmetric synthesis of *P*-chiral phosphorus compounds. *Tetrahedron Asymmetry* **2012**, *23*, 1–46. [CrossRef]
45. Randić, M. Symmetry properties of graphs of interest in chemistry. II. Desargues Levi graph Quantum Chemistry. *Int. J. Quantum Chem.* **1979**, *15*, 663–682. [CrossRef]

46. DeBruin, K.E.; Naumann, K.; Zon, G.; Mislow, K. Topological representation of the stereochemistry of displacement reactions at phosphorus in phosphonium salts and cognate systems. *J. Am. Chem. Soc.* **1969**, *91*, 7031–7040. [CrossRef]
47. Moulin, D.; Bago, S.; Bauduin, C.; Darcel, C.; Juge, S. Asymmetric synthesis of P-stereogenic o-hydroxyarylphosphine (borane) and phosphine-phosphinite ligands. *Tetrahedron Asymmetry* **2000**, *11*, 3939–3956. [CrossRef]
48. Uziel, J.; Stéphan, M.; Kaloun, El B.; Genet, J.P.; Juge, S. Regio- and stereochemistry of nucleophilic attack at the P-chiral center of a dioxaphospholane-borane complex: A model of study for the P-O bond cleavage. *Bull. Soc.Chim. Fr.* **1997**, *134*, 379–389.
49. Lopez-Cortina, S.; Basiulis, D.I.; Marsi, K.; Munoz-Hernandez, M.A.; Fernandez-Zertuche, M. Synthesis of New 1,3-Oxaphosphorinanium Salts. Stereochemistry of Hydroxide-Induced Displacement of Methoxide Ion. *J. Org. Chem.* **2005**, *70*, 7473–7478. [CrossRef] [PubMed]
50. Van Lier, J.J.C.; Hermans, R.J.M.; Buck, H.M. Evidence in support of tetraoxaspirophosphorane intermediates with a six-membered ring and P[sbnd]H bond in intramolecular transesterification reactions. *Phosphorus Sulfur Silicon Relat. Elem.* **1984**, *19*, 173–188. [CrossRef]
51. Boisdon, M.T.; Malavaud, C.; Mathis, F.; Barrans, J. Investigations of phosphoranes with P–H bond during the reaction of aminophosphine with alcohols. *Tetrahedron Lett.* **1977**, *18*, 3501–3505. [CrossRef]
52. Chung, F.-L.; Earl, R.A.; Townsend, L.B. The novel synthesis of a [6:5:6] linear isoguanosine type tricyclic nucleoside using carbonyl sulfide. *Tetrahedron Lett.* **1980**, *21*, 1599–1602. [CrossRef]
53. Lönnberg, T.; Laine, M. Phosphorane intermediate vs. leaving group stabilization by intramolecular hydrogen bonding in the cleavage of trinucleoside monophosphates: Implications for understanding catalysis by the large ribozymes. *Org. Biomol. Chem.* **2010**, *8*, 349–356. [CrossRef] [PubMed]
54. Binyamin, I.; Meidan-Shani, S.; Ashkenaz, N. Synthesis of γ-hydroxypropyl P-chirogenic ($\pm$)-phosphorus oxide derivatives by regioselective ring-opening of oxaphospholane 2-oxide precursors. *Beilstein J. Org. Chem.* **2015**, *11*, 1332–1339. [CrossRef] [PubMed]
55. Kolodyazhnyi, O.I. 2-Halo-1,2λ^5-oxaphosphetanes. *Zh. Obsh Kh.* **1986**, *56*, 283–298.
56. Labaudiniere, L.; Burgada, R. Evolution thermique de vinylphosphoranes. *Tetrahedron* **1986**, *42*, 3521–3536. [CrossRef]
57. Dieckbreder, U.; Roschenthaler, G.-V.; Kolomeitsev, A.A. *P*-bis(trifluoromethyl) ylides: Synthesis and reactions. *Heteroat. Chem.* **2002**, *13*, 650–653. [CrossRef]
58. Francke, R.; Roschenthaler, G.-V.; Di Giacomo, R.; Dakternieks, D. Oxidative addition von hexafluoraceton an die phosphor(III)-verbindungen tBuP(X)NEt_2 (X = F, Cl, OCH(CF_3)$_2$. *Phosphorus Sulfur Silicon Relat. Elem.* **1984**, *20*, 107–115. [CrossRef]
59. Yamamoto, Y.; Nakao, K.; Hashimoto, T.; Matsukawa, S.; Suzukawa, N.; Kojima, S.; Akiba, K.-Y. Crystallographic and NMR studies on species intermediate between haloalkoxyphosphoranes and alkoxyphosphonium halides. *Heteroat. Chem.* **2011**, *22*, 523–530. [CrossRef]
60. Sase, S.; Kano, N.; Kawashima, T. Novel Synthetic Method of Fluorophosphoranes by Fluoride Ion Abstraction from Tetrafluoroborate. *Phosphorus Sulfur Silicon Relat. Elem.* **2002**, *177*, 2041. [CrossRef]
61. Hellwinkel, D. *Organic Phosphorus Chemistry*; Kosolapoff, G.M., Maier, L., Eds.; Wiley Interscience: New York, NY, USA, 1973; Volume 3, pp. 186–339.
62. Burger, K. *Organophosphorus Reagents in Organic Synthesis*; Cadogan, J.I.G., Ed.; Academic Press: London, UK, 1979.
63. Holmes, R.R. *Structure and Spectroscopy and Reaction Mechanisms in Pentacoordinated Phosphorus*; ACS Monograph Series 175 and 176; American Chemical Society: Washington, DC, USA, 1980; Volume 1–2.
64. Husband, J.B.; McNab, H. The thermolysis of pentacoordinate phosphorus heterocycles. *Phosphorus Sulfur Silicon Relat. Elem.* **1984**, *20*, 207–230. [CrossRef]
65. Burgada, R.R.; Setton, R. *The Chemistry of Organophosphorus Compounds*; Hartley, F.R., Ed.; Wiley-Interscience: Chichester, UK, 1994; Volume 3, pp. 185–277.
66. Kawashima, T. *Chemistry of Hypervalent Compounds*; Akiba, K.-Y., Ed.; Wiley: New York, NY, USA, 1999.
67. Hellwinkel, D. Die Stereochemie organischer Derivate des fünf-und sechsbindigen Phosphors, II. Über ein erstes optisch aktives Pentaarylphosphoran. *Chem. Ber.* **1966**, *99*, 3642–3659. [CrossRef]
68. Monkowius, U.N.; Mitzel, W.; Schier, A.; Schmidbaur, H. 5-Organyl-5-phosphaspiro[4.4]nonanes: A Contribution to the Structural Chemistry of Spirocyclic Tetraalkylphosphonium Salts and Pentaalkylphosphoranes. *J. Am. Chem. Soc.* **2002**, *124*, 6126–6132. [CrossRef] [PubMed]

69. Byrne, P.A.; Muldoon, J.; Ortin, Y.; Müller-Bunz, H.; Gilheany, D.G. Investigations on the Operation of Stereochemical Drift in the Wittig Reaction by NMR and Variable-Temperature NMR Spectroscopy of Oxaphosphetane Intermediates and Their Quench Products. *Eur. J. Org. Chem.* **2014**, 86–98. [CrossRef]
70. Vedejs, E.; Meier, G.P.; Snoble, K.A.J. Low-temperature characterization of the intermediates in the Wittig reaction. *J. Am. Chem. Soc.* **1981**, *103*, 2823–2831. [CrossRef]
71. Kawashima, T.; Soda, T.; Okazaki, R. Synthesis, Structure, and Thermolysis of N-Apical 1,2λ^5-Azaphosphetidines with a Pentacoordinate P Center and the First Observation of Their *N*-Equatorial Pseudorotamers. *Angew. Chem. Int. Ed. Engl.* **1996**, *35*, 1096–1098. [CrossRef]
72. Kawashima, T.; Kato, K.; Okazaki, R. Synthese, Struktur und Thermolyse eines 3-Methoxycarbonyl-1,2λ^5-oxaphosphetans. *Angew. Chem.* **1993**, *105*, 941–942. [CrossRef]
73. Kawashima, T.; Kato, K.; Okazaki, R. Synthesis, Structure, and Thermolysis of a 3-Methoxycarbonyl-1,2λ^5-oxaphosphetane. *Angew. Chem. Int. Ed. Eng.* **1993**, *32*, 869–870. [CrossRef]
74. Kano, N.; Kikuchi, A.; Kawashima, T. The first isolable pentacoordinate 1,2λ^5-azaphosphetine: Synthesis, X-ray crystallographic analysis, and dynamic behaviour. *Chem. Commun.* **2001**, 2096–2097. [CrossRef]
75. López, J.G.; Ramallal, A.M.; González, J.; Roces, L.; Garcia-Ganda, S.; Iglesias, M.J.; Oña-Burgos, P.; Ortis, F.L. Mechanisms of Stereomutation and Thermolysis of Spiro-1,2-oxaphosphetanes: New Insights into the Second Step of the Wittig Reaction. *J. Am. Chem. Soc.* **2012**, *134*, 19504–19507. [CrossRef] [PubMed]
76. García-López, J.; Peralta-Pérez, E.; Forcén-Acebal, A.; García-Granda, S.; López-Ortiz, F. Dilithiated phosphazenes: Scaffolds for the synthesis of olefins through a new class of bicyclic 1,2-oxaphosphetanes. *Chem. Commun.* **2003**, 856–857. [CrossRef]
77. Kojima, S.; Kajiyama, K.; Akiba, K.-Y. Characterization of Enantiomeric Pairs of Optically Active 10-P-5 Phosphoranes with Asymmetry Only at Phosphorus. *Bull. Chem. Soc. Jpn.* **1995**, *68*, 1785–1797. [CrossRef]
78. Kojima, S.; Kajiyama, K.; Akiba, K.-Y. Characterization of an optically active pentacoordinate phosphorane with asymmetry only at phosphorus. *Tetrahedron Lett.* **1994**, *35*, 7037–7040. [CrossRef]
79. Kajiyama, K.; Yoshimune, M.; Nakamoto, M.; Matsukawa, S.; Kojima, S.; Akiba, K.-Y. Highly Selective One-Pot Synthesis of Spirophosphoranes Exhibiting Reversed Apicophilicity by Oxidation of Dianions Generated from P–H Spirophosphorane. *Org. Lett.* **2001**, *3*, 1873–1875. [CrossRef] [PubMed]
80. Akiba, K.-Y.; Matsukawa, S.; Kajiyama, K.; Nakamoto, M.; Kojima, S.; Yamamoto, Y. Novel results obtained by freezing berry pseudorotation of phosphoranes (10-P-5). *Heteroat. Chem.* **2002**, *13*, 390–396. [CrossRef]
81. Kajiyama, K.; Yoshimune, M.; Kojima, S.; Akiba, K.-Y. A New Method for the Formation of Anti-apicophilic (*O-cis*) Spirophosphoranes—Kinetic Studies on the Stereomutation of *O-cis* Arylspirophosphoranes to Their *O-trans* Isomers. *Eur. J. Org. Chem.* **2006**, 2739–2746. [CrossRef]
82. Kano, N.; Miyaka, H.; Kawashima, T. Hypervalent Silicon and Phosphorus Atoms in Single Molecules: Synthesis and Properties of Phosphoranylalkoxysilicates and a Phosphoranyloxysilicate. *Chem. Lett.* **2007**, *36*, 1260–1261. [CrossRef]
83. Miyaka, H.; Kano, N.; Kawashima, T. Synthesis and Properties of Pentacoordinate Phosphorus Compounds Containing a Pentacoordinate Silicon Atom. *Phosphorus Sulfur Silicon Relat. Elem.* **2008**, *183*, 673–674. [CrossRef]
84. Jiang, X.-D.; Kakuda, K.; Matsukawa, S.; Yamamichi, H.; Kojima, S.; Yamamoto, Y. Synthesis and application of a bidentate ligand based on decafluoro-3-phenyl-3-pentanol: Steric effect of pentafluoroethyl groups on the stereomutation of *O*-equatorial C-apical spirophosphoranes. *Chem. Asian J.* **2007**, *2*, 314–323. [CrossRef] [PubMed]
85. Jiang, X.-D.; Matsukawa, S.; Kakuda, K.; Fukuzaki, Y.; Zhao, W-L.; Li, L.-S.; Shen, H.-B.; Kojima, S.; Yamamoto, Y. Efficient synthesis of tetradecafluoro-4-phenylheptan-4-ol by a Cannizzaro-type reaction and application of the alcohol as a bulky Martin ligand variant for a new anti-apicophilic phosphorane. *Dalton Trans.* **2010**, *39*, 9823–9829. [CrossRef] [PubMed]
86. Jiang, X.-D.; Matsukawa, S.; Yamamoto, Y. Synthesis, structure and isomerization of arylphosphoranes with anti-apicophilic bonding modes using a novel bidentate ligand with two C2F5 groups. *Dalton Trans.* **2008**, 3678–3687. [CrossRef] [PubMed]
87. Jiang, X.D.; Matsukawa, S.; Yamamichi, H.; Yamamoto, Y. Some Reactions of *O*-Equatorial Spirophosphoranes Bearing the Bidentate Ligand Based on Decafluoro-3-phenyl-3-pentanol. *Heterocycles* **2007**, *73*, 805–824.
88. Kojima, S.; Sugino, M.; Matsukawa, S.; Nakamoto, M.; Akiba, K.-Y. First Isolation and Characterization of an Anti-Apicophilic Spirophosphorane Bearing an Oxaphosphetane Ring: A Model for the Possible Reactive Intermediate in the Wittig Reaction. *J. Am. Chem. Soc.* **2002**, *124*, 7674–7675. [CrossRef] [PubMed]

89. Sase, S.; Kano, N.; Kawashima, T. Pentacoordinate 1H-Phosphirenes: Reactivity, Bonding Properties, and Substituent Effects on Their Structures and Thermal Stability. *J. Org. Chem.* **2006**, *71*, 5448–5456. [CrossRef] [PubMed]
90. Sase, S.; Kano, N.; Kawashima, T. Synthesis and Structure of the First 1,2σ^5-Selenaphosphirane. *J. Am. Chem. Soc.* **2002**, *124*, 9706–9707. [CrossRef] [PubMed]
91. Sase, S.; Kano, N.; Kawashima, T. Synthesis of the First Stable Pentacoordinate Selenaphosphirane. *Phosphorus Sulfur Silicon Relat. Elem.* **2002**, *177*, 2039–2040. [CrossRef]
92. Sase, S.; Kano, N.; Kawashima, T. Isolation of a Cyclic Intermediate in the Reaction of a Phosphorus Ylide with Elemental Sulfur: Synthesis, Structure, and Reactivity of a 1,2σ^5-Thiaphosphirane. *Chem. Lett.* **2004**, *33*, 1434–1435. [CrossRef]
93. Ken Hirotsu, K.; Akihiro Okamoto, A.; Kozo Toyota, K.; Masaaki Yoshifuji, M. X-ray structure of 3-diphenylmethylene-2-(2,4,6-tri-*t*-butylphenyl)thiaphosphirane 2-sulfide: The first thiaphosphirane with *exo*-methylene. *Heteroat. Chem.* **1990**, *1*, 251–254. [CrossRef]
94. Kano, N.; Xing, J.-H.; Kikuchi, A.; Kawa, S.; Kawashima, T. Synthesis of Four- and Five-Membered Heterocycles Derived from an Iminophosphorane. *Phosphorus Sulfur Silicon Relat. Elem.* **2002**, *177*, 1685–1687. [CrossRef]
95. Kano, N.; Xing, J.-H.; Kawa, S.; Kawashima, T. Synthesis, structure, and thermolysis of pentacoordinate 1,3,2λ^5-oxazaphosphetidines: The intermediates of aza-Wittig reactions. *Tetrahedron Lett.* **2000**, *41*, 5237–5241. [CrossRef]
96. Kano, N.; Xing, J.-H.; Kikuchi, A.; Kawashima, T. Formation and X-ray crystallographic analysis of a 1,2λ^5-oxaphosphol-5(2*H*)-one. *Heteroat. Chem.* **2001**, *12*, 282–286. [CrossRef]
97. Burnaeva, L.M.; Mironov, V.F.; Romanov, S.V.; Ivkova, G.A.; Shulaeva, I.L.; Konovalova, I.V. Reaction of 2-Phenyl-1,3,2-benzodioxaphosphinan-4-one with Diethyl Benzylidenemalonate. *Russ. J. Gen. Chem.* **2001**, *71*, 488–489. [CrossRef]
98. Adachi, T.; Matsukawa, S.; Nakamoto, M.; Kajiyama, K.; Kojima, S.; Yamamoto, Y.; Akiba, K.-Y.; Re, S.; Nagase, S. Experimental Determination of the $n_N \rightarrow \sigma^*_{P\text{-}O}$ Interaction Energy of *O*-Equatorial C-Apical Phosphoranes Bearing a Primary Amino Group. *Inorg. Chem.* **2006**, *45*, 7269–7277. [CrossRef] [PubMed]
99. Acher, F.; Juge, S.; Wakselman, M. Chiral bicyclic spirophosphoranes in an Arbuzov-type reaction. *Tetrahedron* **1987**, *43*, 3721–3728. [CrossRef]
100. Sevenard, D.V.; Kazimir, E.L.; Pashkevich, I.; Röschenthaler, G.-V. Polyfluoroacylcycloalkanones in reactions with selected phosphorus(III) compounds. *Heteroat. Chem.* **2002**, *13*, 97–107. [CrossRef]
101. Terent'eva, S.A.; Pudovik, M.A.; Gubaidullin, A.T.; Litvinov, I.A.; Pudovik, A.N. Synthesis and crystal and molecular structures of 2-diethylaminomethyl- and 2-chloromethyl-2,2′-spirobi[benzo-1,3,2-oxazaphospholines]. *Russ. J. Gen. Chem.* **2001**, *71*, 330–336. [CrossRef]
102. Terent'eva, S.A.; Pudovik, M.A.; Pudovik, A.N. Aminoalkylation of 2-substituted 4,5-benzo-1,3,2-oxazaphospholanes. *Zh. Obshch. Khim.* **1987**, *57*, 496–499.
103. Chekhlov, A.N.; Bovin, A.N.; Tsvetkov, E.N. Structure of 2-chloromethyl-2λ5–2,2′(3*H*,3′*H*)-spirobis[1,3,2-benzoxazaphosphole]. *Izv. Akad. Nauk. SSSR Ser. Khim.* **1991**, *7*, 1523–1526.
104. Terent'eva, S.A.; Nikolaeva, I.L.; Burilov, A.R.; Kharitonov, D.I.; Popova, E.V.; Pudovik, M.A.; Litvinov, I.A.; Gubaidullin, A.T.; Konovalov, A.I. Reaction of 2-Ethyl-1,3,2-bezoxaphospholine with Calix[4]resorcinarenes. *Russ. J. Gen. Chem.* **2001**, *71*, 389–395. [CrossRef]
105. Malavaud, C.; Barrans, J. Les phosphoranes monocycliques a ligand hydrogene intermediaires dans la synthese de spirophosphoranes. *Tetrahedron Lett.* **1975**, *16*, 3077–3080. [CrossRef]
106. Gholivand, K.; Mahzouni, H.R.; Molaei, F.; Kalateh, A.A. A phosphoryl to spiro-bicyclophosphorane transformation via β-amidic proton elimination in phosphorylated hydrazides. *Tetrahedron Lett.* **2012**, *53*, 5944–5947. [CrossRef]
107. Freytag, M.; Jones, P.G.; Schmutzler, R.; Yoshifuji, M. Reaction of (*E*)-bis-(2,4,6-tri-*t*-butylphenyl)diphosphane with tetrachloro-*o*-benzoquinone. *Heteroat. Chem.* **2001**, *12*, 300–308. [CrossRef]
108. Malavaud, C.; Charbonnel, Y.; Barrans, J. Synthese de P-methyl spirophosphoranes a partir de diols ou D-aminoalcohols. *Tetrahedron Lett.* **1975**, *16*, 497–498. [CrossRef]
109. Bonningue, C.; Brazier, J.F.; Houalla, D.; Osman, F.H. Phosphoranes par fermeture *trans* annulaire IV. Quelques Exemples de Phosphoranes Bicycliques Chiraux. *Phosphorus Sulfur Silicon Relat. Elem.* **1979**, *5*, 291–298. [CrossRef]

110. Koenig, M.; Munoz, A.; Garrigues, B.; Wolf, R. Spirophosphoranes a Liaison P–H D'α-Hydroxyacidies et de Tartrates D'Alcoyle: Preparation, Properties-Stereochimie. *Phosphorous Sulfur Relat. Elem.* **1979**, *6*, 435–451. [CrossRef]
111. Hou, J.-B.; Tang, G.; Guo, J.-N.; Liu, Y.; Zhang, H.; Zhao, Y.-F. Stereochemistry of chiral pentacoordinate spirophosphoranes correlated with solid-state circular dichroism and ^{1}H NMR spectroscopy. *Tetrahedron Asymmetry* **2009**, *20*, 1301–1307. [CrossRef]
112. Hou, J.-B.; Zhang, H.; Guo, J.-N.; Liu, Y.; Xu, P.-X.; Zhao, Y.-F.; Blackburn, G.M. Chirality at phosphorus in pentacoordinate spirophosphoranes: Stereochemistry by X-ray structure and spectroscopic analysis. *Org. Biomol. Chem.* **2009**, *7*, 3020–3023. [CrossRef]
113. Yang, G.; Xu, Y.; Hou, J.; Zhang, H.; Zhao, Y. Determination of the Absolute Configuration of Pentacoordinate Chiral Phosphorus Compounds in Solution by Using Vibrational Circular Dichroism Spectroscopy and Density Functional Theory. *Chem. Eur. J.* **2010**, *16*, 2518–2527. [CrossRef] [PubMed]
114. Yang, G.; Xu, Y.; Hou, J.; Zhang, H.; Zhao, Y. Diastereomers of the pentacoordinate chiral phosphorus compounds in solution: Absolute configurations and predominant conformations. *Dalton Trans.* **2010**, *39*, 6953–6959. [CrossRef] [PubMed]
115. Cao, S.; Gao, P.; Gua, Y.; Zhao, H.; Wang, J.; Liu, Y.; Zhao, Y. Unexpected Insertion of CO_2 into the Pentacoordinate P–N Bond: Atherton–Todd-Type Reaction of Hydrospirophosphorane with Amines. *J. Org. Chem.* **2013**, *78*, 11283–11293. [CrossRef] [PubMed]
116. You, X.H.; Qi, L.; Zheng, J.; Dai, W.; Guo, Y.C.; Zhao, Y.F.; Cao, S.X. Synthesis and Characterization of New Pyrospirophosphoranes Containing a P–O–P Bond by the Atherton–Todd Reaction. *Heteroat. Chem.* **2015**, *26*, 168–174. [CrossRef]
117. Dai, W.; Liu, Q.; You, X.; Zhou, Z.; Guo, Y.; Zhao, Y.; Cao, S. Synthesis and Characterization of Alkoxy Spirophosphoranes Prepared from Hydrospirophosphoranes and Sodium Alcoholates. *Heteroat. Chem.* **2016**, *27*, 63–71. [CrossRef]
118. Bondarev, O.G.; Mikhel, I.S.; Tsarev, P.V.; Petrovskii, P.V.; Davankov, V.A.; Gavrilov, K.N. Platinum and rhodium complexes with isoleucinol-based bi- and tricyclic hydrophosphoranes. *Russ. Chem. Bull. Int. Ed.* **2003**, *52*, 116–121. [CrossRef]
119. Gavrilov, K.N.; Polosukhin, A.I.; Bondarev, O.G.; Lyubimov, S.E.; Lyssenko, K.A.; Petrovski, P.; Davankov, V.A. Synthesis and properties of pentacoordinated phospha derivatives of *iso*-leucinol. A rare example of using of hydrophosphoranes as ligands in asymmetric catalysis. *J. Mol. Catal. A Chem.* **2003**, *196*, 39–53. [CrossRef]
120. Marchi, C.; Buono, G. Asymmetric Addition of Chiral Triquinphosphoranes on Activated Carbonyl Compounds. *Tetrahedron Lett.* **1999**, *40*, 9251–9254. [CrossRef]
121. Marchi, C.; Buono, G. Alkylthiylation of triquinphosphoranes by disulfides: An entry to chiral thiatriquinphosphoranes. *Tetrahedron Lett.* **2000**, *41*, 3073–3076. [CrossRef]
122. Lopez, L.; Fabas, C.; Barrans, J. Heterocycles phosphores derivant des amidoximes. II. Composes du phosphore penta et hexacoordine. *Phosphorus Sulfur Silicon Relat. Elem.* **1979**, *7*, 81–87. [CrossRef]
123. Zhao, Y.-F.; Han, B.; Chen, J.; Jiang, Y. Penta-Coordinate Phosphorus Compounds and Biochemistry. *Phosphorus Sulfur Silicon Relat. Elem.* **2002**, *177*, 1391–1396. [CrossRef]
124. Li, Z.; Fu, H.; Gong, H.; Zhao, Y. Convenient Solid Phase Synthesis of Oligopeptides Using Pentacoordinated Phosphoranes with Amino Acid Residue as Building Blocks. *Bioorg. Chem.* **2004**, *32*, 170–177. [CrossRef] [PubMed]
125. Schenthaler, G.-V.R.; Bohlen, R.; Storzer, W.; Sopchik, A.E.; Bentrude, W.G. Reaktionen eines Tetraalkoxyhydro-spirophosphorans. *Z. Anorg. Allg. Chem.* **1983**, *507*, 93–99. [CrossRef]
126. Nakazawa, H.; Kubo, K.; Miyoshi, K. Metallaphosphorane Chemistry: Preparations, Structures, and Reactivities. *Bull. Chem. Soc. Jpn.* **2001**, *74*, 2255–2267. [CrossRef]
127. Nakazawa, H.; Ogawa, T.; Kawamura, K.; Miyoshi, K. Syntheses of Metallaphosphorane Complexes and Berry Pseudorotation. *Phosphorus Sulfur Silicon Relat. Elem.* **2002**, *177*, 2163–2164. [CrossRef]
128. Nakazawa, H.; Kawamura, K.; Ogawa, T.; Myoshi, K. Syntheses, structure, and Berry pseudorotation of ruthenium-, iron-, and cobalt-phosphorane complexes. *J. Organomet. Chem.* **2002**, *646*, 204–211. [CrossRef]
129. Kubo, K.; Nakazawa, H.; Kawamura, K.; Mizuta, T.; Miyoshi, K. Migration Reaction of a Hypervalent Fragment: Base-Induced Migration of a Phosphorane Fragment from Iron to the Cyclopentadienyl Ring in $Cp(CO)_2Fe\{P(OC_6H_4Y)(OC_6H_4Z)\}$ (Y, Z = NMe, NH, O). *J. Am. Chem. Soc.* **1998**, *120*, 6715–6721. [CrossRef]

130. Munoz, A.; Gornitzka, H. Phosphorus, Structure Cristalline D'un Hydroxyspirophosphorane Modele D'intermediaire de Reactions Impliquant Des Esters Cycliques de L'acide Phosphorique. *Phosphorus Sulfur Silicon Relat. Elem.* **2003**, *178*, 55–60. [CrossRef]
131. Boyer, D.G.; Boisdon, M.-T.; Rochal, A.; Munoz, A. Equilibre Ester Phosphorique, Hydroxyphosphorane Role des Liaisons Hydrogene, Acidite de Bronsted. *Phosphorus Sulfur Silicon Relat. Elem.* **2003**, *178*, 2117–2125. [CrossRef]
132. Laurenco, C.; Burgada, R. Mecanisme de formation et de transformation des spirophosphoranes-IX: Reactions des spirophosphoranes a liaison P–H avec les ethers vinyliques et les enamines. *Tetrahedron* **1976**, *22*, 2253–2255. [CrossRef]

molecules

MDPI

Article

1-(Acylamino)alkylphosphonic Acids—Alkaline Deacylation

Marek Cypryk [1], Jozef Drabowicz [1,2,*], Bartlomiej Gostynski [1], Marcin H. Kudzin [3], Zbigniew H. Kudzin [4,*] and Pawel Urbaniak [4]

1 Centre of Molecular and Macromolecular Studies, Polish Academy of Sciences, Sienkiewicza 120a, Lodz 90-363, Poland; mcypryk@cbmm.lodz.pl (M.C.); bgostyns@cbmm.lodz.pl (B.G.)
2 Department of Chemistry and Environment Protection, Jan Dlugosz University, Czestochowa 42-200, Poland
3 Textile Research Institute, Brzezinska 5/15, Lodz 92-103, Poland; kudzin@iw.lodz.pl
4 Faculty of Chemistry, University of Lodz, Tamka 12, Lodz 91-403, Poland; pawelurb@chemia.uni.lodz.pl
* Correspondence: drabowicz@gmail.com (J.D.); zhkudzin@uni.lodz.pl (Z.H.K.)

Academic Editor: György Keglevich
Received: 5 March 2018; Accepted: 28 March 2018; Published: 9 April 2018

check for updates

Abstract: The alkaline deacylation of a representative series of 1-(acylamino)alkylphosphonic acids [(AC)-AA^P: (AC) = Ac, TFA, Bz; AA^P = Gly^P, Ala^P, Val^P, Pgl^P and Phe^P] in an aqueous solution of KOH (2M) was investigated. The results suggested a two-stage reaction mechanism with a quick interaction of the hydroxyl ion on the carbonyl function of the amide R-C(O)-N(H)- group in the first stage, which leads to instant formation of the intermediary acyl-hydroxyl adducts of R-C$(O^-)_2$-N(H)-, visible in the ^{31}P NMR spectra. In the second stage, these intermediates decompose slowly by splitting of the RC$(O^-)_2$-N(H)- function with the subsequent formation of 1-aminoalkylphosphonate and carboxylate ions.

Keywords: amino acids; aminophosphonic acids; 1-aminoalkylphosphonic acids; 1-(acylamino) alkylphosphonic acids; ^{31}P NMR spectra of intermediates; hydrolytic deacylation

1. Introduction

Aminoalkylphosphonic acids (AA^P) are structural analogues of amino acids (AA^C) [1], some of which are of natural origin [2]. Due to the structural analogy, they present similar biological properties to the class and are important inhibitors of enzymes of the amino acids metabolism [1,3] (Table 1).

Table 1. Structures of aminophosphonic acids (AA^P)—phosphonic analogs of amino acids (AA^C) and 1-(acylamino)alkylphosphonic acids (AC)-AA^P and phosphonopeptides AA^C-AA^P and AA^P-AA^C [4–14].

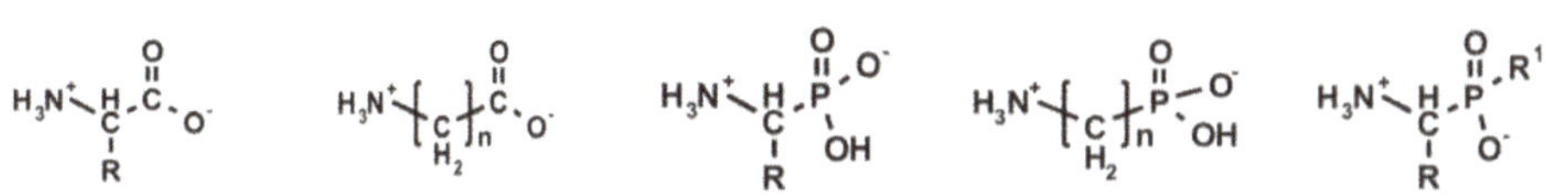

AA^C	ω-AA^C	AA^P	ω-AA^P	AA^{P-R}
Naturally occurring aminophosphonic acids.				

Table 1. *Cont.*

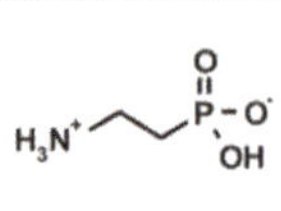	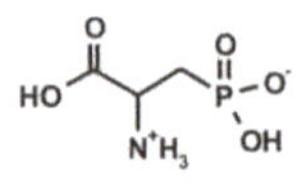	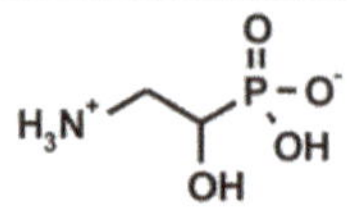	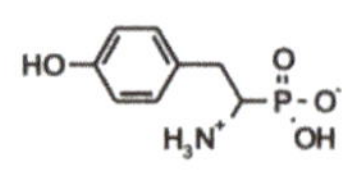	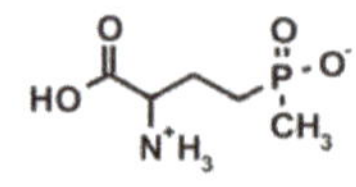
β-Ala^{P} [4]	Asp$^{\beta-P}$ [5]	Iser^{P} [6]	Tyr^{P} [7]	Glu$^{-\gamma-P(Me)}$ [8]
Representative, biologically active aminophosphonates.				
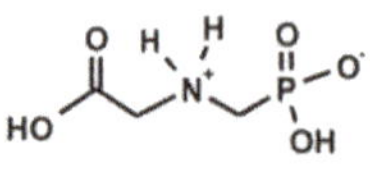	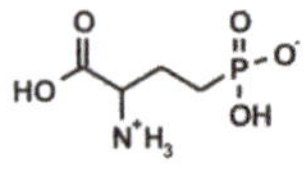	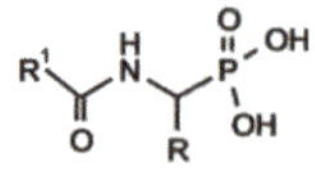	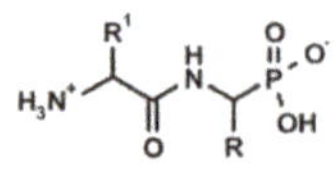	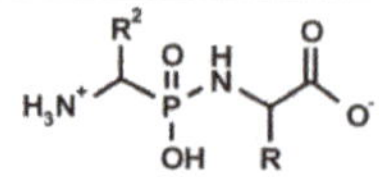
PMG [9]	Glu$^{\gamma-P}$ [10]	(AC)-AA^{P}[R=H, (AC)-Gly^{P}] [11]	AA^{C}-AA^{P}(R=R^{1}=Me, Ala-Ala^{P}) [12]	AA^{P}-AA^{C} [13,14]

Abbreviation of AA^{P} and (AC)-AA^{P} follow the general rules elaborated by Kudzin et al. [15].

Several papers reflected the complexing abilities of this class of compounds [16–18] and their pharmacological [19–22], agro-chemical [23] and industrial [11,24,25] applications. Therefore, the studies on the synthesis [26–30] and physico-chemical properties [31–40] of AA^{P} and their derivatives constitute an important topic in chemistry and biochemistry also in addition to material science (e.g., Self-assembled monolayers agents, SAMs [41]).

The 1-(acylamino)alkylphosphonic acids (AC)-AA^{P} belong to the interesting group of compounds that is of potential pharmacological importance since the corresponding 1-*N*-aminoacylamino derivatives of AA^{C}-AA^{P} (mixed *P*-terminal phosphono-dipeptides, e.g., Alaphosphaline) exhibit antibacterial activity [1,23] (Table 1).

1-(Acylamino)alkylphosphonic acids [36,37] and esters [42] are easily formed from starting aminoalkylphosphonic acids, the latter being substrates for selective deacylation to *O,O*-dialkyl aminoalkylphosphonates [43]. Both of these groups are substrates for stereoselective enzymatic hydrolysis [44–48] (Figure 1). 1-(Acylamino)alkylphosphonic acids are also formed in so-called Engelman-Pikl-Oleksyszyn method synthesis of 1-aminoalkylphosphonic and 1-aminoalkylphosphonic acids, which starts from appropriate amides and phosphorus chlorides [10,27,49,50] (Figure 2). As some secondary 1-aminoalkanephosphonates are unstable in the process of acidic degradation [40], alkaline hydrolysis can present an alternative method in such procedures.

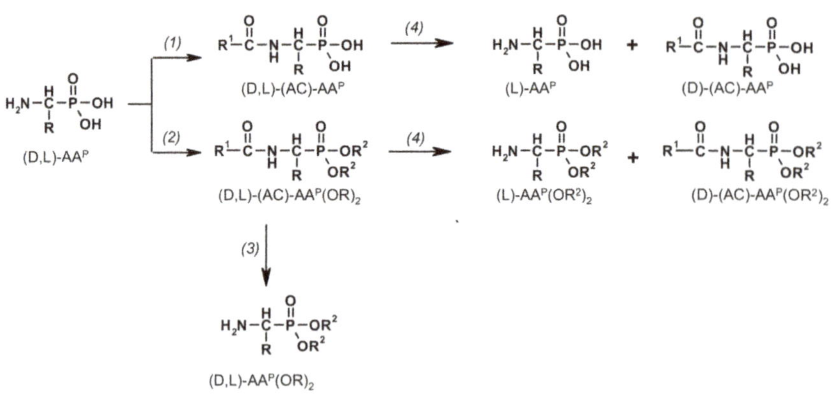

Figure 1. Transformation of 1-aminoalkylphosphonic acids via *N*-acyl-derivatives [(*1*) R^{1}-C(O)OH/[R^{1}-C(O)]$_2$O [36–38]; (*2*) R^{1}-C(O)OH/[R^{1}-C(O)]$_2$O/HC(OR^{2})$_3$ [42]; (*3*) NaBH$_4$/MeOH [43]; (*4*) enzymatic hydrolysis of racemic (D,L)-(AC)-AA^{P} and (D,L)-(AC)-AA^{P}(OR)$_2$ [44–48]].

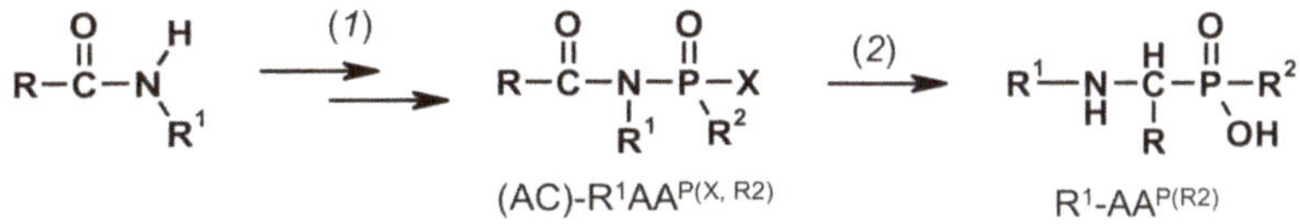

Figure 2. Scheme of Engelmann-Pikl-Oleksyszyn methods for synthesis of secondary 1-aminoalkylphosphonic (R^2=OH) and 1-aminoalkylphosphinic (R^2=H, alkyl) acids (*1*) aldehyde, R-PCl_2/AcOH; and (2) 5M HCl-H_2O, Δ8–10 h.

The susceptibility of (AC)-AA^P to hydrolysis/solvolysis constitutes an important factor influencing their biological activity, especially during penetration through cell membranes. Therefore, the detailed studies on deacylation of these compounds should be helpful in obtaining a deeper understanding of this phenomenon. To tackle this problem, we have decided to carry out studies devoted to deacylation of (AC)-AA^P under different reaction conditions. Our preliminary results of experiments were carried out in aqueous media in the pH range of 0–6.5, which occurs namely in an aqueous 2M HCl and buffer solutions. These are briefly described in a recently published article [51]. As an obvious and necessary continuation of this topic, this paper presents our results on the deacylation of representative types of 1-(acylamino)alkylphosphonic acids (AC)-AA^P in aqueous 2M KOH solution. These include 1-(acetylamino)alkylphosphonic Ac-AA^P, 1-(trifluoroacetylamino)alkylphosphonic acids TFA-AA^P and 1-(benzoylamino)alkylphosphonic Bz-AA^P, derived from representative 1-aminoalkylphosphonic acids AA^P (Gly^P, Ala^P, Val^P, Pgl^P and Phe^P).

2. Results and Discussion

It is generally known that amides can be hydrolyzed with either acidic or basic catalysis, with products being the free acid and the ammonium/substituted ammonium ions or the salts of the acid and ammonia/amine, respectively. Both the acid- and base-catalyzed hydrolyses are essentially irreversible, since salts are formed in both cases [52–55]. Water alone is not sufficient to hydrolyze most of amides [56]. The very low rate of amide hydrolysis by water has been measured by Kahne & Still [57].

A kinetic study has been conducted on the alkaline hydrolyses of several types of amides, including formanilides [58,59], *N*-methylformanilides and *N*-acetanilides [60], 1,8-bis(trifluoroacetylamino)-naphthalene [61] also in addition to *N*-methylacetamide, *N*-methylbenzamide and acetanilide [62]. A mild protocol for the alkaline hydrolysis of secondary and tertiary amides in non-aqueous conditions at room temperature or under reflux has been recently described [63].

The structural analogy of 1-(acylamino)alkylphosphonic acids (AC)-AA^P and 1-(acylamino)-alkanoic acids (AC)-AA^C or amides causes the mechanism of hydrolytic scission of the amide linkage R-C(O)-N in common amides and in 1-(acylamino)alkylphosphonic acids. These should be ruled by similar if not the same mechanisms.

2.1. *Investigations on the Deacylation Course of 1-(acylamino)alkylphosphonic Acids*

The (AC)-AA^P contain amidic functions, which has a hydrolytic sensitivity that should be dependent on the structure of their acyl moieties and on the type of applied hydrolytic medium [49]. For conducting an inquiry into the hydrolytic stability of (AC)-AA^P, we undertook deacylation investigations of these compounds in aqueous 2M KOH solutions.

The application of ^{31}P NMR monitoring of the reaction course enabled identification and quantification of phosphoric components of the reaction mixtures formed. Essentially, monitoring of the deacylation course of Ac-Gly^P in 2M KOH solution reveals the formation of Gly^P (Figure 3). The results of ^{31}P NMR monitoring of the deacylation of the various types of (AC)-AA^P (Ac-AA^P, Bz-AA^P and TFA-AA^P) at a temperature of 25 °C and exposition period up to 240 h are presented in Figure 4 (Ac-AA^P), Figure 5 (TFA-AA^P) and Figure 6 (TFA-AA^P), respectively.

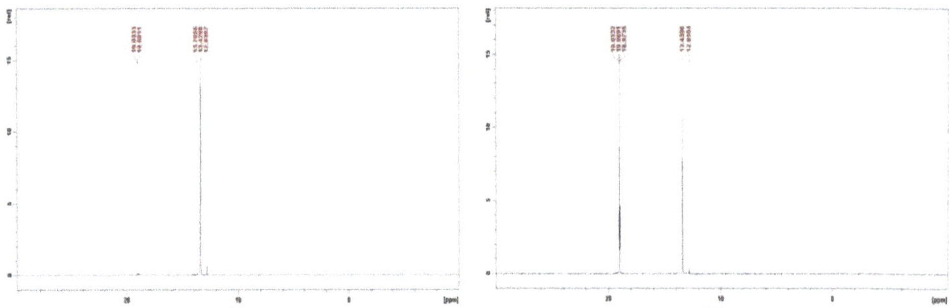

(A) (B)

Figure 3. ^{31}P NMR spectra of Ac-Gly^P recorded in 2 M KOH. (**A**) Immediately after mixing the reaction mixture, and (**B**) After 192 h of exposition of the reaction mixture. [Ac-Gly^P: 13.4 ppm; Gly^P: 19.0 ppm].

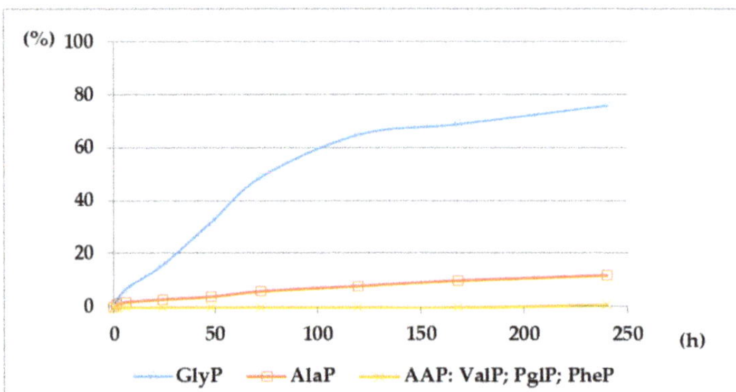

Figure 4. A solution of 0.1 M 1-(acetylamino)alkylphosphonic acids Ac-AA^P in 2 M KOH reacted at 25 °C, which was monitored by ^{31}P NMR [DD (%) vs. time (h)].

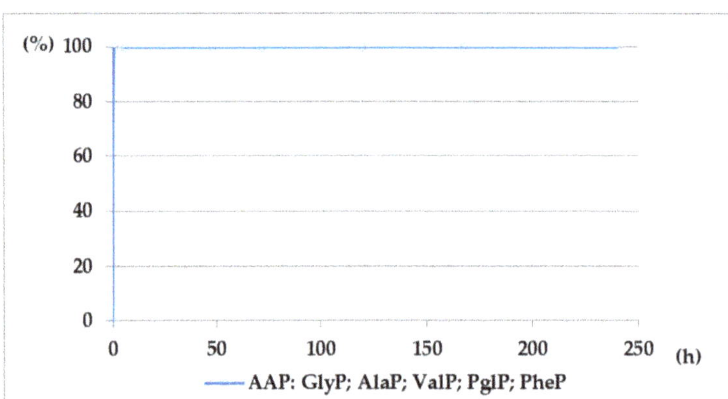

Figure 5. A solution of 0.1 M 1-(trifluoroacetylamino)alkylphosphonic acids TFA-AA^P in 2 M KOH reacted at 25 °C, which was monitored by ^{31}P NMR [DD (%) vs. time (h)].

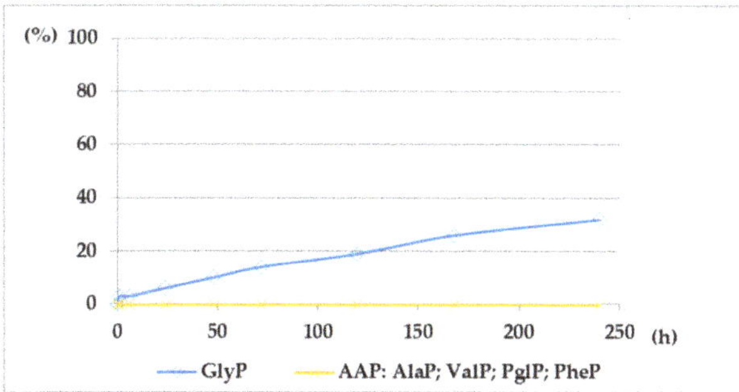

Figure 6. A solution of 0.1 M 1-(benzoylamino)alkylphosphonic acids Bz-AA^P in 2 M KOH reacted at 25 °C, which was monitored by ^{31}P NMR [DD (%) vs. time (h)]. Thus, the presented graphs illustrate much higher stability of benzoyl-derivatives Bz-AA^P (Bz-Gly^P: DD_{200h} ~30%; for Bz-Ala^P and Bz-Pgl^P: DD_{200h} < 3%) in comparison with acetyl-derivatives Ac-AA^P (Ac-Gly^P: DD_{200h} ~70%; for Ac-Ala^P: DD_{200h} ~20%; and Ac-Pgl^P: DD_{200h} < 7%).

The comparison of hydrolytic stability of TFA-Gly^P, Ac-Gly^P and Bz-Gly^P in 2M KOH is presented in Figure 7.

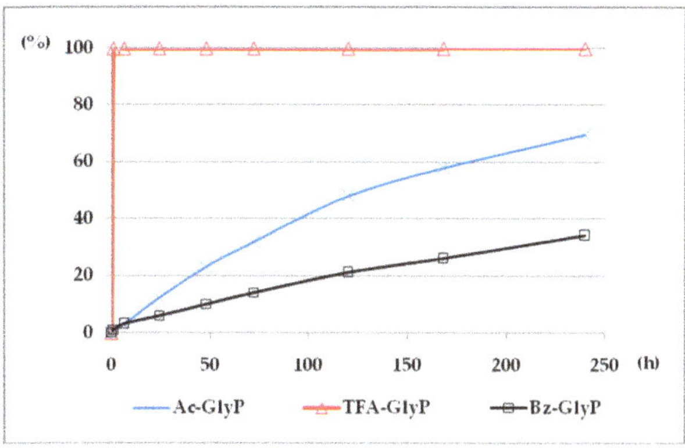

Figure 7. A solution of 0.1 M 1-(acylamino)methylphosphonic acids (AC)-Gly^P (AC: Ac, TFA, Bz) in 2M KOH reacted at 25 °C, which was monitored by ^{31}P NMR [DD (%) vs. time (h)]. Points on the plots corresponds to Gly^P released from the parent (AC)-Gly^P, which means Gly^P from Ac-Gly^P, Gly^P from Bz-Gly^P and Gly^P from TFA-Gly^P.

Since no products of the scission of the C-P bond in all investigated cases were observed, the deacylation mechanism of (AC)-AA^P in alkaline solutions can be illustrated by an analogy to the corresponding basic deacylation of 1-(acylamino)alkanoic acids (Scheme 1) [54,62,63].

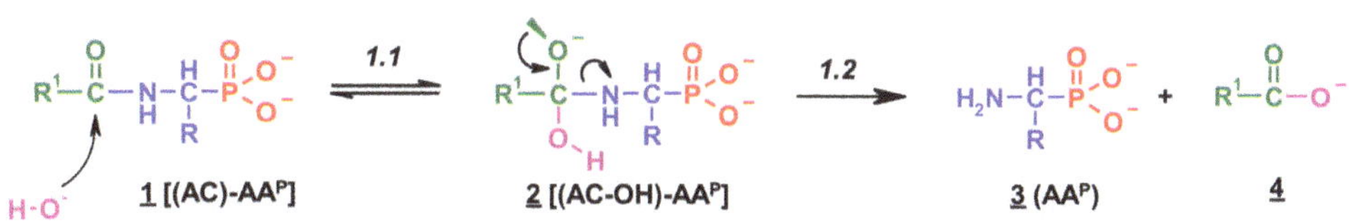

Scheme 1. Hypothetical mechanism of deacylation of (AC)-AA^P in 2 M KOH solution.

This includes the prior interaction of hydroxyl ion on the carbonyl carbon of 1, which occurs with the formation of the tetrahedral intermediate state 2 (Scheme 1). These intermediates were observed on ^{31}P NMR spectra as signals down-shifted in comparison with the parent signals of AC-AA^P (e.g., Figure 2). The ^{31}P NMR chemical shifts of (AC)-AA^P (1) and the corresponding adduct compounds AC(OH)-AA^P (2) formed are listed in Table 2.

Table 2. ^{31}P NMR chemical shifts of (AC)-AA^P and the corresponding adduct compounds (AC)(OH)-AA^P formed immediately after mixing the reaction mixture. (**A**) ^{31}P NMR chemical shifts of Ac-AA^P and the corresponding adduct compounds Ac(OH)-AA^P formed. (**B**) ^{31}P NMR chemical shifts of (AC)-Gly^P and the corresponding adduct compounds [AC(OH)]-Gly^P formed.

(A)

	Ac-Gly^P		**Ac-Ala^P**		**Ac-Nva^P**		**Ac-Val^P**		**Ac-Pgl^P**	
	Ac-Gly^P	Ac(OH)-Gly^P	Ac-Ala^P	Ac(OH)-Ala^P	Ac-Nva^P	Ac(OH)-Nva^P	Ac-Val^P	Ac(OH)-Val^P	Ac-Pgl^P	Ac(OH)-Pgl^P
δ	13.4	*12.8*	17.0	*16.5*	15.9	*15.2*	16.8	*16.1*	13.3	*12.7*
%	97.1	*2.9*	97.1	*2.9*	96.3	*3.7*	97.0	*3.0*	97.0	*3.0*

(B)

	Ac-Gly^P			**Prp-Gly^P**			**Bz-Gly^P**		
	Gly^P	Ac-Gly^P	Ac(OH)-Gly^P	Gly^P	Prp-Gly^P	Prp(OH)-Gly^P	Gly^P	Bz-Gly^P	Bz(OH)-Gly^P
δ	19.1	13.4	12.8	19.3	13.4	12.8	19.3	13.5	13.2
%	-	97.1	2.9	-	98.3	1.7	-	99.5	<0.5

These compounds [AC(OH)]-AA^P] slowly disappeared during the deacylation progress (Figure 8).

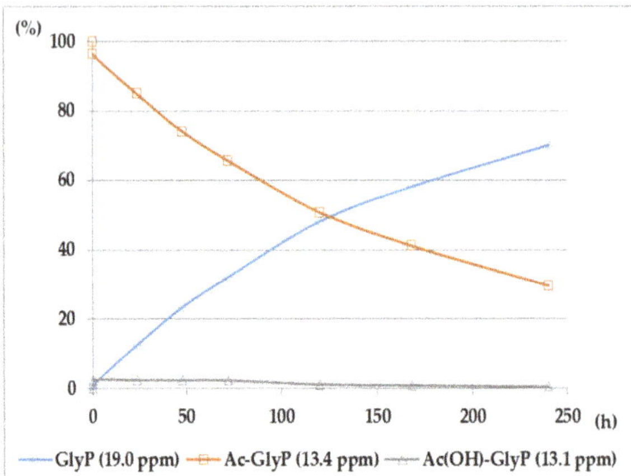

Figure 8. Ac-Gly^P hydrolysis in 2M KOH [DD (%) vs. time (h)].

The rehybridization ($sp^3 \rightarrow sp^2$) of the carbonyl carbon in 2 enforces the splitting of the amide bond R-C(O)-N and the subsequent formation of the aminophosphonate 3.

The comparison of (AC)-AA^P stability in aqueous 2M KOH and 2M HCl solutions is given in Figure 9.

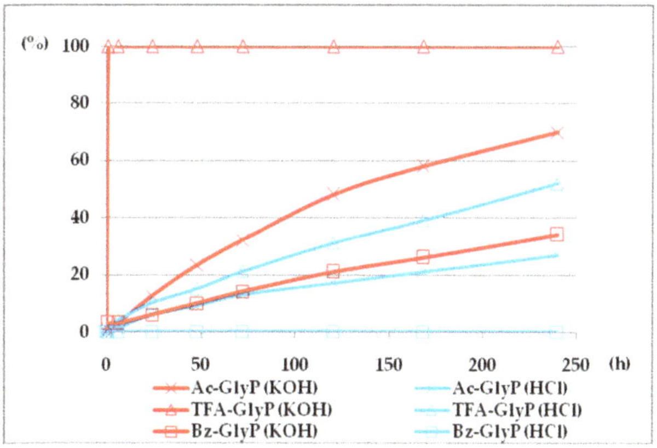

Figure 9. (AC)-Gly^P(Ac-Gly^P, TFA-Gly^P, TFA-Gly^P) hydrolysis in aqueous solutions of 2M KOH and 2M HCl [DD (%) vs. time (h)].

2.2. Quantum Chemical Calculations for Deacylation of (AC)-AA^P in Basic Conditions

The reaction illustrated in Scheme 1 was divided into two stages (reactions *1.1* and *1.2*, respectively), which were studied separately. Stage *1.1* concerns the addition of hydroxide to amidophosphonate anion $\underline{1}$ with the formation of trianion $\underline{2}$ (the example structure of the addition product $\underline{2}$ for Ac-Ala^P (R=R^1=Me) is given in Figure 10). Stage *1.2* involves dissociation of the trianion $\underline{2}$ into aminophosphonate dianion $\underline{3}$ and carboxylate anion $\underline{4}$. We assumed that the transfer of proton from OH to the amide nitrogen in $\underline{2}$ proceeds simultaneously with dissociation. The thermodynamics of this two-stage process was studied by density functional theory (DFT) methods in the gas phase and in aqueous solution. Gas phase calculations were performed only for comparison with the aqueous system, which was the real reaction system to illustrate the essential role of the solvation effect.

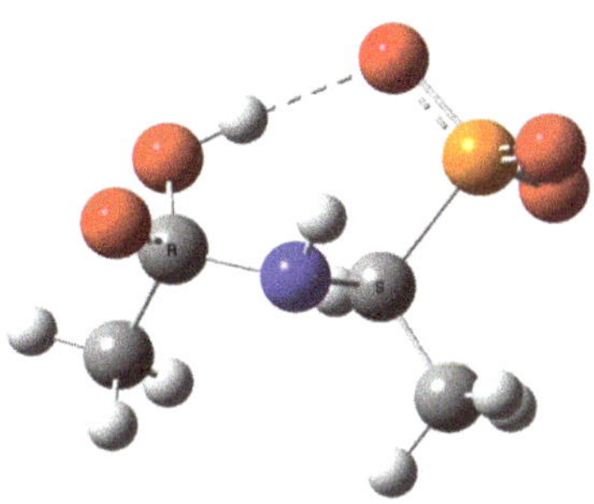

Figure 10. The structure of the addition product **2** (R=R^1=Me, Ac(OH)-Ala^P) with intramolecular hydrogen bond between OH and phosphoryl group (orange—phosphorus, blue—nitrogen, red—oxygen).

Stage 1: Hydroxide anion attack on the carbonyl moiety of (AC)-AA^P dianion $\underline{1}$ (Scheme 1, r. *1.1*).

This reaction stage presumably proceeds by the attack of a lone pair of negatively charged oxygen atoms on sp^2 carbon in the carbonyl group. This process is highly unfavorable since it involves the reaction of two anions with the formation of the molecule with a total charge of −3. The energy of this reaction in the gas phase is very high, so this process seems to be thermodynamically very unlikely (Table 3).

Table 3. Gibbs free energies ΔG (298 K) for reaction *1.1* in gas phase (kcal/mol).

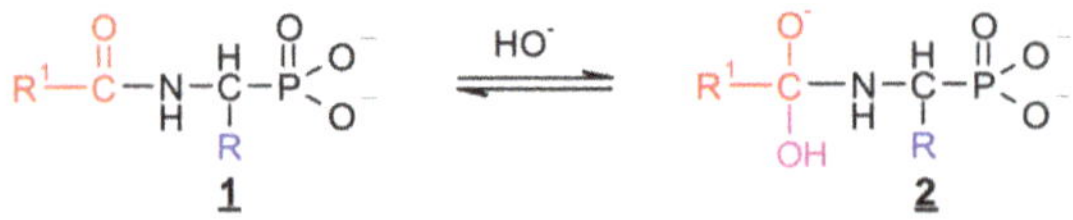

(AC)-AA^P (1Xx)	R	R^1-C(O)- (1Xx)		
		CH_3- (1Ax)	CF_3- (1Bx)	Ph- (1Cx)
(AC)-Gly^P (1Xa)	H-	139.43	123.16	126.87
(AC)-Ala^P (1Xb)	CH_3-	135.19	119.26	124.82
(AC)-Val^P (1Xc)	$(CH_3)_2CH$-	133.61	117.48	132.28
(AC)-Pgl^P (1Xd)	Ph-	125.63	121.43	117.08
(AC)-Phe^P (1Xe)	$PhCH_2$-	139.31	122.06	129.57

However, when carried out in aqueous solutions, this reaction is much less unfavorable due to the stabilization of ionic structures by the polar solvent (Table 2). Similarly, this reaction should also be facilitated by counterions, although these were omitted in order to keep the model system simple. The values of the Gibbs free energy ΔG^{298} (kcal/mol) for reaction *1.1* in gas phase and in aqueous phase are listed in Tables 3 and 4, respectively.

Table 4. Gibbs free energies ΔG (298 K) for reaction *1.1* in aqueous phase (kcal/mol).

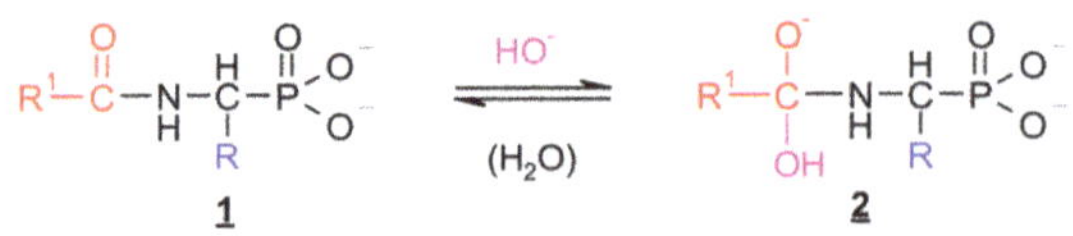

(AC)-AA^P (1Xx)	R	R^1-C(O)- (1Xx)		
		CH_3- (1Ax)	CF_3- (1Bx)	Ph- (1Cx)
(AC)-Gly^P (1Xa)	H-	26.70	11.77	25.70
(AC)-Ala^P (1Xb)	CH_3-	26.26	10.54	25.45
(AC)-Val^P (1Xc)	$(CH_3)_2CH$-	26.15	10.90	26.57
(AC)-Pgl^P (1Xd)	Ph-	25.47	12.27	24.96
(AC)-Phe^P (1Xe)	$PhCH_2$-	24.22	7.73	24.53

The free energy barriers, $\Delta G^{\ddagger}$ (298 K), calculated for reaction *1.1* in a water solution are shown in Table 5. The lowest energy barriers for OH^- addition (Scheme 1, r. *1.1*) occur when $R^1 = CF_3$, which results from stabilization of the negative charge by the electron-withdrawing CF_3 group.

Table 5. Free energy barriers ($\Delta G^{\ddagger}$ (298 K), kcal/mol) for OH^- addition (reaction *1.1*) in water.

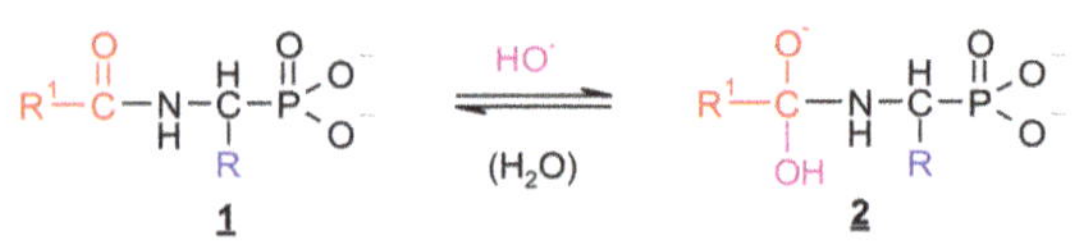

(AC)-AA^P (1Xx)	R	R^1-C(O)- (1Xx)		
		CH_3- (1Ax)	CF_3- (1Bx)	Ph- (1Cx)
(AC)-Gly^P (1Xa)	H-	29.99	19.51	25.70
(AC)-Ala^P (1Xb)	CH_3-	30.85	19.12	26.75
(AC)-Val^P (1Xc)	$(CH_3)_2CH$-	31.40	18.47	29.68
(AC)-Pgl^P (1Xd)	Ph-	28.57	17.60	28.44
(AC)-Phe^P (1Xe)	$PhCH_2$-	28.85	18.66	30.23

Stage 2: Dissociation of trianion **2** into aminophosphonate dianion **3** and carboxylate anion **4** (Scheme 1, reaction *1.2*).

This reaction stage presumably involves the proton transfer from the geminal hydroxyl group to nitrogen with dissociation of the C-N bond. This hypothesis is supported by the transition structures found for this reaction. The example of this structure for the proton transfer/dissociation of the C-N bond in Ac-Ala$^{P(3-)}$ is shown in Figure 11. Vibrational analysis indicates that the imaginary vibration is associated with proton transfer from OH to NH. In the structure of the transition state shown in Figure 11, the geminal C-O bond distance is 1.284 Å, which corresponds well to C=O double bond. Furthermore, the C-N distance is 1.802 Å compared to 1.503 Å in the starting amide, which indicates that the bond breaking in the transition state is significantly advanced.

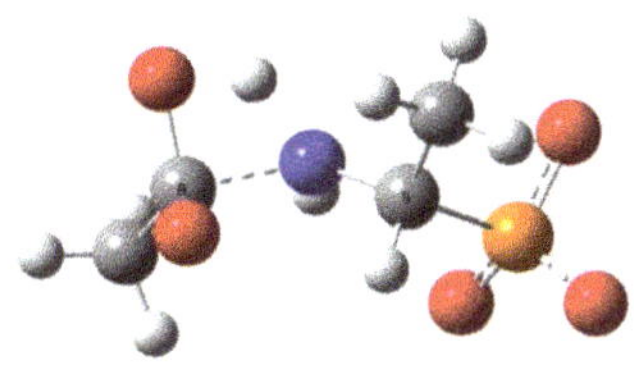

Figure 11. The transition state for proton transfer/dissociation of the C-N bond in Ac(OH)-Ala$^{P(3-)}$.

The Gibbs free energies for the dissociation reaction *1.2* in the gas phase and in water are presented in Tables 6 and 7, respectively.

Table 6. Gibbs free energy ΔG (298 K) for the reaction *1.2* in gas phase (kcal/mol).

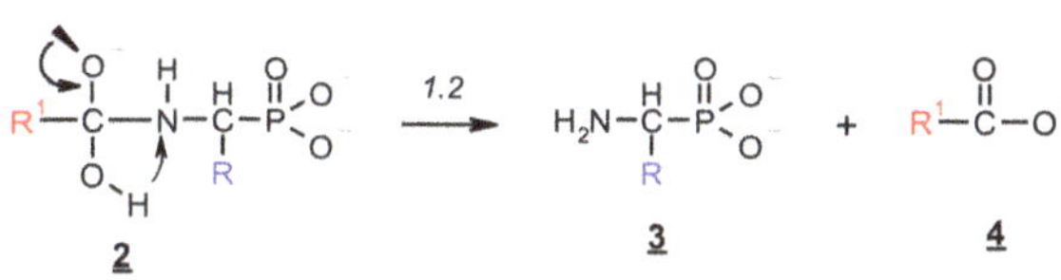

(AC-OH)-AA^P (2-Xx)	R	R^1-C(O$^-$)(OH)- (2Xx)		
		CH_3- (2Ax)	CF_3- (2Bx)	Ph- (2Cx)
(AC-OH)-Gly^P (2Xa)	H-	−165.22	−153.75	−152.91
(AC-OH)-Ala^P (2Xb)	CH_3-	−159.50	−149.37	−149.31
(AC-OH)-Val^P (2Xc)	$(CH_3)_2CH$-	−162.42	−152.74	−161.46
(AC-OH)-Pgl^P (2Xd)	Ph-	−157.24	−160.78	−150.14
(AC-OH)-Phe^P (2Xe)	$PhCH_2$-	−168.00	−158.21	−159.56

Table 7. Gibbs free energy ΔG (298 K) for the reaction *1.2* in an aqueous phase (kcal/mol).

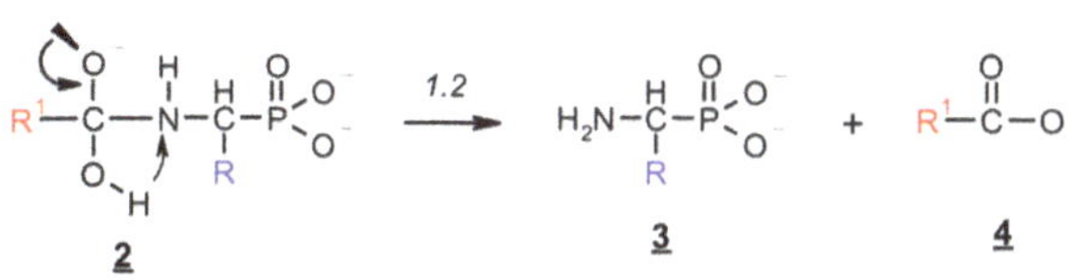

(AC-OH)-AA^P (2-Xx)	R	R^1-C(O$^-$)(OH)- (2Xx)		
		CH_3- (2Ax)	CF_3- (2Bx)	Ph- (2Cx)
(AC-OH)-Gly^P (2Xa)	H-	−46.41	−40.32	−48.51
(AC-OH)-Ala^P (2Xb)	CH_3-	−46.62	−38.58	−48.56
(AC-OH)-Val^P (2Xc)	$(CH_3)_2CH$-	−44.45	−37.28	−47.29
(AC-OH)-Pgl^P (2Xd)	Ph-	−45.60	−40.90	−47.44
(AC-OH)-Phe^P (2Xe)	$PhCH_2$-	−47.74	−39.80	−50.89

Comparing the thermodynamic quantities in Tables 3 and 6, it is evident that the total free energy of the reaction *1* is negative in all cases by approximately −30 kcal/mol and thus, the entire process *1* is thermodynamically feasible. However, the very high energy barrier for the first step of reaction *1.1* (Table 3) prevents this reaction from proceeding in the gas phase.

Analogously, the reaction 1 in water is equally thermodynamically favorable (ΔG = −20 to −30 kcal/mol, Tables 4 and 7), but the energy barrier associated with the first step *1.1* is much lower in this case (Table 4). Table 8 presents the results of free energy barrier calculations for reaction *1.2* in water.

Table 8. Free energy barriers (ΔG‡ (298 K), kcal/mol) for proton transfer/dissociation of the C-N bond in water (reaction *1.2*).

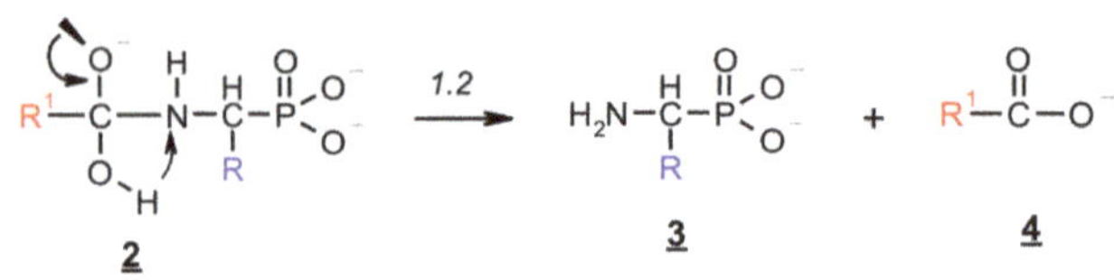

(AC-OH)-AA^P (2-Xx)	R	*R*^{1}*-C(O$^-$)(OH)-* (2Xx)		
		CH$_3$- (2Ax)	*CF$_3$-* (2Bx)	*Ph-* (2Cx)
(AC-OH)-Gly^P (2Xa)	H-	20.76	21.54	16.91
(AC-OH)-Ala^P (2Xb)	CH$_3$-	21.47	23.43	19.00
(AC-OH)-Val^P (2Xc)	(CH$_3$)$_2$CH-	23.73	24.91	20.56
(AC-OH)-Pgl^P (2Xd)	Ph-	20.05	21.74	18.51
(AC-OH)-Phe^P (2Xe)	PhCH$_2$-	22.50	25.00	18.99

The total Gibbs free energy profiles of entire process *1* for two selected pairs of 1-(acylamino)alkylphosphonic acids [Ac-Ala$^{P(2-)}$, TFA-Ala$^{P(2-)}$, Bz-Pgl$^{P(2-)}$ and TFA-Pgl$^{P(2-)}$] in water solution are shown in Figure 12.

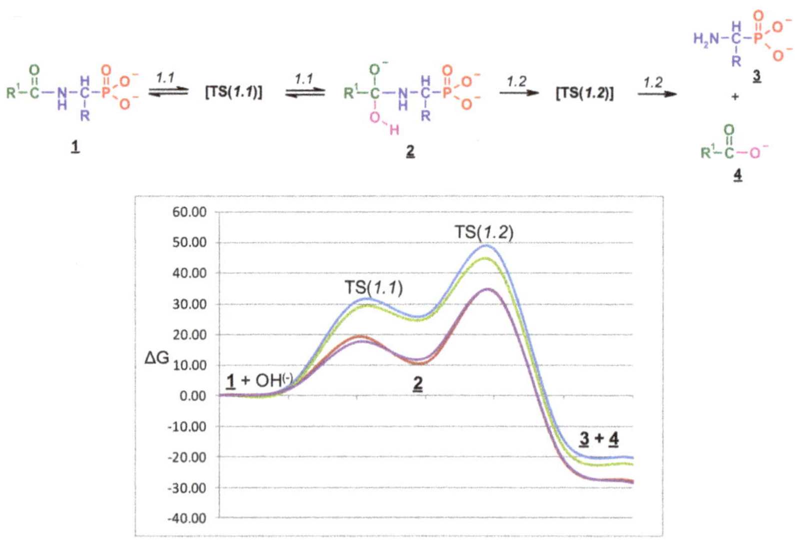

Figure 12. Free energy profiles ΔG^{298} (kcal/mol) for the reaction of Ac-Ala$^{P(2-)}$ (blue), TFA-Ala$^{P(2-)}$ (red), Bz-Pgl$^{P(2-)}$ (green), and TFA-Pgl$^{P(2-)}$ (purple) with hydroxyl ions in water solution (the general reaction scheme is given above).

The total Gibbs free energies and reaction barriers for the entire process *1* are shown in Table 9. Moreover, Figure 12 reveals that the energy profiles for both TFA-substituted phosphonic acids are essentially identical and all stationary points are significantly lower in energy than Ac-substituted phosphonic acids.

Table 9. Total Gibbs free energies$\Delta G^{298}(1.1) + \Delta G^{298}(1.2)$ for the overall deacylation process in the aqueous phase and total Gibbs free energy barriers ($\Delta G^{298}(1.1) + \Delta G^{\ddagger}(1.2)$, in parentheses) (kcal/mol).

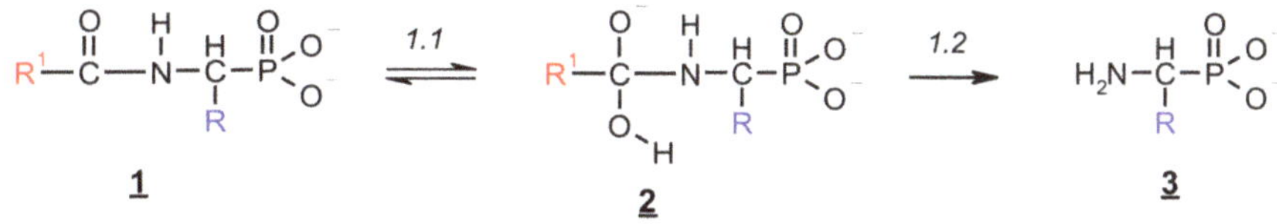

(AC)-AA^P (1Xx)	R	R^1-C(O)- (1Xx)		
		CH_3- (1Ax)	CF_3- (1Bx)	Ph- (1Cx)
(AC)-Gly^P (1Xa)	H-	−19.72 (47.5)	−28.56 (31.3)	−22.82 (42.6)
(AC)-Ala^P (1Xb)	CH_3-	−20.36 (47.7)	−28.04 (34.0)	−23.11 (44.5)
(AC)-Val^P (1Xc)	$(CH_3)_2CH$-	−18.30 (49.9)	−26.38 (35.8)	−20.72 (47.1)
(AC)-Pgl^P (1Xd)	Ph-	−20.13 (45.5)	−28.63 (34.0)	−22.48 (43.5)
(AC)-Phe^P (1Xe)	$PhCH_2$-	−23.53 (46.7)	−32.08 (32.7)	−26.36 (43.5)

Therefore, the presence of a TFA group is the only factor determining the free energy barrier of the reaction. From the energy profile, it is obvious that the overall energy barrier in water $\Delta G^{\ddagger}$ is G^{298}[TS (*1.2*)] − (G^{298}(**1**) + $G^{298}(OH^-)$), while the net Gibbs energy effect for the reaction in water solution can be calculated as $\Delta G_r{}^{298} = G^{298}(\mathbf{3}) + G^{298}(\mathbf{4}) - [(G^{298}(\mathbf{1}) + G^{298}(OH^-)] = \Delta G^{298}(1.1) + \Delta G^{\ddagger}(1.2)$.

The total energetic effect of the process *1* in water is negative, which means that the overall reaction is thermodynamically feasible (Table 8). The free energy of the process when R^1=CF_3 is also approximately 8 kcal/mol more favorable than for R^1=Me. The comparison of values in Table 9 also reveals that the transformation *1* of substrates with R^1=Ph are thermodynamically more favored by approximately 3 kcal/mol than R^1=Me. Thus, the thermodynamic effect of reaction *1* depends on the R^1 substituents in the order of: CH_3 < Ph < CF_3. Table 9 also leads to the conclusion that the substrates with a trifluoromethyl group in the R^1 position systematically show free energy barriers that are approximately 10–15 kcal/mol lower than those for other substituents R^1, due to the additional stabilization of the trianion **2** by the inductive effect of the strongly electron-withdrawing CF_3 group. This suggests that trifluoromethyl-substituted anions should be more reactive than the other derivatives.

3. Materials and Methods

3.1. General Information

The ^{31}P NMR spectra were recorded on a Bruker AV 200 spectrometer (Rheinstetten, Germany) operating at 81.01 MHz and on a Bruker Avance III 600 spectrometer (Bruker BioSpin, Rheinstetten, Germany) operating at 242.9 MHz. The1H NMR spectra were recorded on a Bruker Avance III 600 spectrometer operating at 600 MHz. The positive chemical shift values of ^{31}P were reported for compounds that were absorbing at lower fields than H_3PO_4. The purity of the 1-aminoalkylphosphonic acids and corresponding 1-(acylamino)alkylphosphonic acids were determined by pH-metric titration using a computer aided automatic titrator connected to the EMU-meter (Wroclaw University of Science and Technology, Wrocław, Poland), which was fitted with a combined glass microelectrode Crison 5028.

3.2. Investigated Compounds

Phosphonoglycine (Gly^P) was obtained according to a previous study [64]. The 1-Aminoalkylphosphonic acids of phosphonoalanine (Ala^P); phosphonovaline (Val^P); phosphonophenylglycine (Pgl^P) and phosphonophenylalanine (Phe^P) have been prepared according to a previous study [65]. 1-(Acetylamino)alkylphosphonic acids (Ac-AA^P: Ac-Gly^P; Ac-Ala^P; Ac-Val^P; Ac-Pgl^P and Ac-Phe^P) and 1-(benzoylamino)alkylphosphonic acids (Bz-AA^P: Bz-Gly^P; Bz-Ala^P; Bz-Val^P; Bz-Pgl^P and Bz-Phe^P) were synthesized according to a previous study [36]. 1-(Trifluoroacetylamino)alkylphosphonic acids (TFA-AA^P: TFA-Gly^P; TFA-Ala^P; TFA-Val^P; TFA-Pgl^P and TFA-Phe^P) were synthesized according to a previous study [37]. (Structures, names and abbreviations of aminophosphonic acids and corresponding 1-(acylamino)phosphonic acids discussed in this work are given in Table 10). Purity of all (AC)-AA^P synthesized was determined by ^{31}P NMR and ^{1}H NMR in addition to potentiometric titration. Analytical standards (fixanals) and other reagents were purchased from Aldrich–Sigma (Poznań, Poland).

Table 10. Structures, names and abbreviations of aminophosphonic acids and corresponding 1-(acylamino)phosphonic acids discussed in this work [a].

		Name	Trivial Name	Abbr./(No)	Ref.
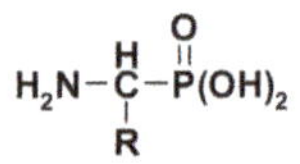		Aminoalkyl-phosphonic acid	Phosphono-Amino Acid	AA^P (**3**)	
	R				
	H	Aminomethyl-phosphonic acid	Phosphono-glycine	Gly^P (**3a**)	[64]
	Me	1-Aminoethyl-phosphonic acid	Phosphono-alanine	Ala^P (**3b**)	[65]
	iPr	1-Amino-1-methylethyl-phosphonic acid	Phosphono-valine	Val^P (**3c**)	
	Ph	1-Amino-1-phenylmethyl-phosphonic acid	Phosphono-phenyl-glycine	Pgl^P (**3d**)	
	$PhCH_2$	1-Amino-1-phenylethyl-phosphonic acid	Phosphono-phenyl-alanine	Phe^P (**3e**)	
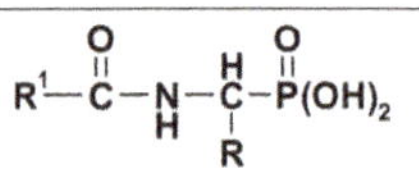		1-(Acylamino)alkyl-phosphonic acid	Acylamino-Phosphono Acid	(AC)-AA^P (**2**)	
R^1	R				
CH_3	H, Me, iPr, Ph, $PhCH_2$	1-(Acetylamino)alkyl -phosphonic acid	Acetylamino-Phosphono Acid	Ac-AA^P (**2A**)	[36]
CH_3	H	1-(Acetylamino)methyl-phosphonic acid	Phosphono-Acetylamino Glycine	Ac-Gly^P (**2Aa**)	
CH_3	Ph	1-(Acetylamino)-1-phenyl-methyl-phosphonic acid	Phosphono-Acetyl-amino-PhenylGlycine	Ac-Pgl^P (**2Ad**)	
CF_3	H, Me, iPr, Ph, $PhCH_2$	1-(Trifluoroacetylamino)-alkyl-phosphonic acid	Phosphono-Trifluoroacetyl-amino Acid	TFA-AA^P (**2B**)	[37]
CF_3	iPr	1-(Trifluoroacetylamino)-2-methylethyl-phosphonic acid	Phosphono-Trifluoroacetyl-Valine	TFA-Val^P (**2Bc**)	
Ph	H, Me, iPr, Ph, $PhCH_2$	1-(Benzoylamino)alkyl-phosphonic acid	Phosphono-Benzoylamino Acid	Bz-AA^P (**2C**)	[36]
Ph	$PhCH_2$	1-(Benzoylamino)-2-phenyl-ethyl-phosphonic acid	Phosphono-Benzoyl-Phenyl-alanine	Bz-Phe^P (**2Ce**)	

[a] Applied names were in accordance with the IUPAC rules, and the abbreviations were in agreement with the general rules elaborated by Kudzin et al. [15].

3.3. Reaction of Deacylation of 1-(acylamino)alkylphosphonic Acids

A sample of (AC)-AA^P (0.5 mmol) was placed into 5-mL V-vials, before being dissolved in 5 mL of appropriate solution fortified with D_2O (2 M KOH). The vials were maintained at 25 °C ± 0.2 °C. At various time intervals, the vials were removed from the baths and the samples of 0.5 mL were taken for ^{31}P NMR analysis. The concentration of the substrate and formed products were determined from the integration of its NMR signal.

3.4. Computational Methods

All quantum mechanical calculations were performed using the Gaussian 09 suite of programs [66]. Geometries of model compounds in the gas phase and in water solution were optimized using the B3LYP hybrid functional containing three-parameter Becke (B3) exchange and Lee-Yang-Parr (LYP) correlation functional and the 6–31+G(d) basis set. All stationary points were identified as stable minima by frequency calculations. The vibrational analysis also provided the thermal enthalpy and entropy corrections at 298 K within the rigid rotor/harmonic oscillator/ideal gas approximation. Computed frequencies were scaled by a factor of 0.98. Transition states were verified by the intrinsic reaction path (IRC) procedure to confirm if they did link the substrates and products along the reaction path.

Free energies of the reaction in water were calculated within a continuum (implicit) solvent approximation using the Conductor-like Polarizable Continuum Model (CPCM) with UFF (Universal Force Field) cavities (SCRF = CPCM option as defined in Gaussian 09 program) [67]. This approximation assumes that molecule of the solute is placed in a cavity within the solvent, which is treated as a dielectric continuum (self-consistent reaction field). Thermodynamic functions in solution were calculated based on vibrational analysis, which was described above.

4. Conclusions

The analysis of ^{31}P NMR results presented above leads to the following conclusions:

- The alkaline deacylation of (AC)-AA^P occurs through the hydroxyl adduct intermediates, which was observed on ^{31}P NMR spectra of the reaction mixtures;
- The deacylation ability of investigated (AC)-AA^P derivatives exhibited strong dependence of electron-acceptor character of the acyl group, with CF_3-C(O)- > CH_3-C(O)-;
- The 1-(acylamino)alkylphosphonic acids are present in aqueous solutions with substantial stability at ambient temperatures, which decreases with temperature elevation;
- The deacylation of (AC)-AA^P increases substantially in basic (2 M KOH) solutions;
- The lowest deacylation ability (highest stability) was found for the 1-(benzoylamino acids Bz-AA^P;
- For the same type of 1-(acylamino)-derivatives (AC)-AA^P, the highest stability was found in(AC)-Val^P and (AC)-Phe^P; lower stability in (AC)-Ala^P; and the lowest in (AC)-Gly^P and (AC)-Pgl^P;
- All examined ^{31}P NMR spectra of deacylation mixtures did not reveal any trace of H_3PO_3 and/orH_3PO_4, which represent the products of dephosphonylation or oxidative dephosphonylation of (AC)-AA^P.

The theoretical calculations lead to the following conclusions:

The substrates with the trifluoromethyl group in R^1 position are predicted to be significantly more reactive than the other phosphonic amino acids studied, which is consistent with the experimental results. The intermediates and products when R^1=CF_3 are also more thermodynamically favored than the other derivatives. This difference in reactivity is due to additional stabilization of the trianion **2** by the inductive effect of the strongly electron-withdrawing CF_3 group. The calculations did not provide unequivocal information about relative reactivities of other AA derivatives as the calculated reaction energy barriers are similar (within a computational error), which is shown in

Table 9 and Figure 12. The calculation results are consistent with those reported for acidic deacylation of 1-(acylamino)alkylphosphonic acids [51]. As CF_3 is strongly electron-withdrawing, this group plays an activating role in basic media, increasing the acidity of carbonyl moiety. In contrast, it has a deactivating effect in acidic media, decreasing the basicity of carbonyl oxygen.

Acknowledgments: The studies at the Textile Research Institute, Łódź, in 2017 were partly financed by the Polish Ministry of Science and Higher Education within statutory research founds (for M.H.K.). DFT calculations were supported by the PL-Grid infrastructure. We would like to thank the National Science Center (Krakow, Poland) for the support in the frame of the grant UMO 2014/15/B/ST5/05329 (for J.D.). Dedicated to the memory of Professor Andrzej Okruszek.

Author Contributions: J.D. and Z.H.K. designed the research study and contributed to the data interpretation and to the manuscript drafting and revisions. M.C. and B.G. carried out the Quantum Chemical Calculations and contributed to writing the manuscript. M.H.K. performed the synthesis and purification of investigated 1-(acylamino)alkylphosphonic acids and was involved in the conception of the research study, analyzed the data, and contributed to writing the manuscript. P.U. recorded NMR spectra and analyzed the experimental data.

Conflicts of Interest: The authors declare no conflict of interest.

References

1. Kukhar, V.P.; Hudson, H.R. (Eds.) *Aminophosphonic and Aminophosphinic Acids: Chemistry and Biological Activity*; J. Wiley & Sons, Ltd.: Chichester, UK; New York, NY, USA; Weinheim, Germany; Brisbane, Australia; Singapore; Toronto, ON, Canada, 2000; ISBN 0-471-89149-5.
2. Kittredge, J.S.; Roberts, E. A carbon-phosphorus compounds in nature. *Science* **1969**, *164*, 37–42. [CrossRef] [PubMed]
3. Kafarski, P.; Mastalerz, P. *Aminophosphonates, Natural Occurrence, Biochemistry and Biological Properties*; Beiträge zur Wirkstofforschung; Institut für Wirkstofforschung: Berlin, Germany, 1984; Volume 24, pp. 1–110.
4. Horiguchi, M.; Kandatsu, M. Isolation of 2-aminoethyl phosphonic amid from rumen protozoa. *Nature* **1959**, *184*, 901–902. [CrossRef] [PubMed]
5. Kittredge, J.S.; Hughes, R.R. Occurence of α-amino-β-phosphono-propionic acid in the zoanthid, zoanthussociatus, and the ciliate, Tetrahymena pyrifornis. *Biochemistry* **1964**, *3*, 991–996. [CrossRef] [PubMed]
6. Korn, E.D.; Deaborn, D.G.; Falles, H.M.; Sokoloski, E.A. A major polysaccharide constituents of the amoeba plasma membrane contains 2-aminoethylphosphonic acid and 1-hydroxy-2-aminoethyl-phosphonic acid. *J. Biol. Chem.* **1973**, *248*, 2257–2259. [PubMed]
7. Kido, Y.; Hamakado, T.; Anno, M.; Miyagawa, E.; Motoki, Y.; Wakamiya, T.; Shiba, T. Isolation and characterization of 15112, a new phosphorus containing inhibitor of angiotensin I converting enzyme produced by *Actinomadura* sp. *J. Antibiot.* **1984**, *37*, 965–969. [CrossRef] [PubMed]
8. Bayer, E.; Gugel, K.H.; Haegele, K.; Hagenmaier, H.; Jessipov, S.; Koenig, W.A.; Zaehner, H. Metabolic products of microorganisms. 98. Phosphinothricine and phosphinothricylo-alanylo-alanine. *Helv. Chim. Acta* **1972**, *55*, 224–239. [CrossRef] [PubMed]
9. Franz, J.E. Herbicidal compositions and methods employing esters of *N*-phosphonoglycine. U.S. Patent US 3997860, 14 December 1976.
10. Mastalerz, P. Inhibition of glutamine synthetase by phosphonic analogs of glutamic acid. *Arch. Immun. Ter. Dośw.* **1959**, *7*, 201–210.
11. Engelmann, M.; Pikl, J. Phosphonic Acids Derived from Organic Acylamidomethyl Compounds. U.S. Patent 2304156, 8 December 1942.
12. Allen, J.G.; Atherton, F.R.; Hall, M.J.; Hassal, C.H.; Holmes, S.W.; Lambert, R.W.; Nisbet, L.J.; Ringrose, P.S. Phosphonopeptides, a new class of synthetic antibacterial agents. *Nature* **1978**, *272*, 56–58. [CrossRef] [PubMed]
13. Jacobsen, N.E.; Bartlett, P.A. A phosphonamidate dipeptide analog as an inhibitor of carboxypeptidase A. *J. Am. Chem. Soc.* **1981**, *103*, 654–657. [CrossRef]
14. McLeod, D.A.; Brinkworth, R.; Ashley, J.A.; Janda, K.D.; Wirsching, P. Phosphonamidates and phosphonamidate esters as HIV-1 protease inhibitors. *Bioorg. Med. Chem. Lett.* **1991**, *1*, 653–658. [CrossRef]

15. Drabowicz, J.; Jakubowski, H.; Kudzin, M.H.; Kudzin, Z.H. The nomenclature of 1-aminoalkylphosphonic acids and derivatives: Evolution of the code system. *Acta Biochim. Pol.* **2015**, *62*, 139–150. [CrossRef] [PubMed]
16. Kabachnik, M.I.; Medved, T.Y.; Dyatlova, N.M.; Archipova, O.G.; Rudomino, M.W. Phosphoroorganic complexones. *Usp. Khim.* **1974**, *43*, 1554–1574. [CrossRef]
17. Rizkalla, E.N. Metal chelates of phosphonate-containing ligands. *Rev. Inorg. Chem.* **1983**, *5*, 223–304.
18. Nowack, B. Environmental chemistry of phosphonates. *Water Res.* **2003**, *37*, 2533–2546. [CrossRef]
19. Kafarski, P.; Lejczak, B. Aminophosphonic acids of potential medical importance. *Curr. Med. Chem. Anti-Cancer Agents* **2001**, *1*, 301–312. [CrossRef] [PubMed]
20. Lejczak, B.; Kafarski, P. Biological activity of aminophosphonic acids and their short peptides. *Top. Heterocycl. Chem.* **2009**, *20*, 31–63. [CrossRef]
21. Orsini, F.; Sello, G.; Sisti, M. Aminophosphonic acids and derivatives. Synthesis and biological applications. *Curr. Med. Chem.* **2010**, *17*, 264–289. [CrossRef] [PubMed]
22. Mucha, A.; Kafarski, P.; Berlicki, L. Remarkable potential of the α-amino-phosphonate/phosphinate structural motif in medicinal chemistry. *J. Med. Chem.* **2011**, *54*, 5955–5980. [CrossRef] [PubMed]
23. Sikorski, J.A.; Logush, E.W. Aliphatic carbon-phosphorus compound as herbicides. In *Handbook in Organophosphorus Chemistry*; Engel, R., Ed.; Marcel Dekker Inc.: New York, NY, USA, 1988; Volume Chapter 15, pp. 737–806.
24. Maier, L. Phosphoroorganic detergents. *Chimia* **1969**, *23*, 323–330.
25. Petrov, K.A.; Chauzov, V.A.; Erokhina, T.E. Aminoalkyl organo-phosphorus compounds. *Usp. Khim.* **1974**, *43*, 2045–2087. [CrossRef]
26. Palacios, F.; Alonso, C.; de los Santos, J.M. Synthesis of β-amino-phosphonates and -phosphinates. *Chem. Rev.* **2005**, *105*, 899–931. [CrossRef] [PubMed]
27. Kudzin, Z.H.; Kudzin, M.H.; Drabowicz, J.; Stevens, C. Aminophosphonic acids—Phosphorus analogues of natural amino acids. P. 1: Syntheses of α-amino-phosphonic acids. *Curr. Org. Chem.* **2011**, *15*, 2015–2071. [CrossRef]
28. Kudzin, Z.H.; Kudzin, M.H.; Drabowicz, J. Thioureidoalkylphosphonates in the synthesis of 1-aminoalkylphosphonic acids. The Ptc-aminophosphonate method. *Arkivoc* **2011**, *6*, 227–269. [CrossRef]
29. Ma, J.-N. Catalytic asymmetric synthesis of α- and β-amino phosphonic acid derivatives. *Chem. Soc. Rev.* **2006**, *35*, 630–636. [CrossRef] [PubMed]
30. Ordonez, M.; Rojas-Cabrera, H.; Cativiela, C. An overview of stereoselective synthesis of α-aminophosphonic acids and derivatives. *Tetrahedron* **2009**, *65*, 17–49. [CrossRef] [PubMed]
31. Kukhar, V.P.; Solodenko, V.A. The phosphorus analogs of aminocarboxylic acids. *Usp. Khim.* **1987**, *56*, 1504–1532. [CrossRef]
32. Kudzin, Z.H.; Mokrzan, J.; Skowroński, R. Long chain 1-aminothia-alkanephosphonates, their sulphinyl and sulphonyl derivatives. A new class of complexane type surfactants. *Phosphorus Sulfur Silicon Relat. Elem.* **1989**, *42*, 41–46. [CrossRef]
33. Kudzin, Z.H.; Skrzypek, S.W.; Skowroński, R.; Ciesielski, W.; Cristau, H.J.; Plenat, F. Polarographic investigations of functionalized alkanephosphonic acids. Part II. *Phosphorus Sulfur Silicon Relat. Elem.* **1996**, *119*, 201–207. [CrossRef]
34. Chęcińska, L.; Kudzin, Z.H.; Małecka, M.; Nazarski, R.B.; Okruszek, A. [(Diphenoxyphosphinyl) methylidene]triphenylphosphorane—The double P^+-stabilisedcarboanion: A crystallographic, computational and solution NMR comperative study on the ylidic bonding. *Tetrahedron* **2003**, *59*, 7681–7693. [CrossRef]
35. Kudzin, Z.H.; Saganiak, M.; Andrijewski, G.; Drabowicz, J. Oxidation of phosphonocysteine and phosphonohomocysteine. Synthesis of phosphonocysteic and phosphonohomocysteic acids. *Pol. J. Chem.* **2005**, *79*, 529–539.
36. Kudzin, Z.H.; Depczyński, R.; Andrijewski, G.; Drabowicz, J.; Łuczak, J. 1-(*N*-acylamino)alkanephosphonates. IV. *N*-acylation of 1-aminoalkanephosphonic acids. *Pol. J. Chem.* **2005**, *79*, 499–513.
37. Kudzin, Z.H.; Depczyński, R.; Kudzin, M.H.; Drabowicz, J.; Łuczak, J. 1-(*N*-Trifluoroacetylamino) alkylphosphonic acids. Synthesis and properties. *Amino Acids* **2007**, *33*, 663–667. [CrossRef] [PubMed]

38. Kudzin, Z.H.; Depczyński, R.; Kudzin, M.H.; Drabowicz, J. 1-(*N*-Chloroacetylamino)alkylphosphonic acids—Synthetic precursors of glycylophosphono-peptides and related compounds. *Amino Acids* **2008**, *34*, 163–168. [CrossRef] [PubMed]
39. Drabowicz, J.; Jordan, F.; Kudzin, M.H; Kudzin, Z.H.; Urbaniak, P.; Stevens, C. Reactivity of aminophosphonic acids. oxidative dephosphonylation of 1-aminoalkylphosphonic acids by aqueous halogens. *Dalton Trans.* **2016**, *45*, 2308–2317. [CrossRef] [PubMed]
40. Gancarz, R.; Kiersnowska, A.D. Associative vs. dissociative mechanism of P–C bond breaking in α-aminophosphonates leading to phosphoric acid [P(V)] derivatives. *Arkivoc* **2017**, *2*, 285–302. [CrossRef]
41. Aldrich-Sigma Catalog 2016. Aminophosphonic Self-Assembled Monolayers Agents (e.g., 795798, SIK 7701-10; SIK 7701-101). Available online: https://www.sigmaaldrich.com/china-mainland.html/ (accessed on 9 April 2018).
42. Kudzin, Z.H.; Łuczak, J. A facile conversion of aminoalkanephosphonic acids into *O,O*-dialkyl 1-(*N*-acylamino)alkanephosphonate derivatives. *Synthesis* **1995**, 509–511. [CrossRef]
43. Kudzin, Z.H.; Łyżwa, P.; Łuczak, J.; Andrijewski, G. Aminoalkanephosphonates. P.II. Facile conversion of 1-aminoalkanephosphonic acids their *O,O*-diester derivatives. *Synthesis* **1997**, 44–47. [CrossRef]
44. Svedas, V.K.; Kozlova, E.V.; Mironenko, D.A.; Kukhar, V.P.; Kasheva, T.N.; Solodenko, V.A.; Belozersky, A.N. Enzymatic hydrolysis of *N*-acylated 1-aminophosphonic acids. *Phosphorus Sulfur Silicon Relat. Elem.* **1990**, *51*, 411. [CrossRef]
45. Solodenko, V.; Kasheva, T.; Kukhar, V.; Kozlova, E.; Mironenko, D.; Svedas, V. Preparation of optically active 1-aminoalkylphosphonic acids by stereoselective enzymatic hydrolysis of racemic *N*-acylated 1-aminoalkyl-phosphonic acids. *Tetrahedron* **1991**, *47*, 3989–3998. [CrossRef]
46. Solodenko, V.; Belik, M.; Galushko, S.; Kukhar, V.; Kozlova, E.; Mironenko, D.; Svedas, V. Enzymatic preparation of both L- and D-enantiomers of phosphonic and phosphonous analogues of alanine using penicillin acylase. *Tetrahedron Asymmetry* **1993**, *4*, 1965–1968. [CrossRef]
47. Zielinska, K.; Mazurkiewicz, R.; Szymanska, K.; Jarzebski, A.; Magierac, S.; Erfurt, K. Penicillin G acylase-mediated kinetic resolution of racemic 1-(*N*-acylamino)alkylphosphonic and 1-(*N*-acylamino)alkylphosphinic acids and their esters. *J. Mol. Catal. B Enzym.* **2016**, *132*, 31–40. [CrossRef]
48. Zielinska, K.; Szymanska, K.; Mazurkiewicz, R.; Jarzebski, A. Batch and in-flow kinetic resolution of racemic 1-(*N*-acylamino)alkylphosphonic and 1-(*N*-acylamino)alkylphosphinic acids and their esters using immobilized penicillin G acylase. *Tetrahedron Asymmetry* **2017**, *28*, 146–152. [CrossRef]
49. Oleksyszyn, J. An amidoalkylation of trivalent phosphorous compounds with P(O)H functions including acetic acid solutions of PCl_3, $RPCl_2$ or R_2PCl, diesters of phosphorous acid and phosphorous-III-acids. *J. Prakt. Chem.* **1987**, *329*, 19–29. [CrossRef]
50. Soroka, M. The synthesis of 1-aminoalkylphosphonic acids. A revised mechanism of the reaction of phosphorus trichloride, amides and aldehydes or ketones in acetic acid solution (Oleksyszyn reaction). *Liebigs Ann. Chem.* **1990**, *4*, 331–334. [CrossRef]
51. Cypryk, M.; Drabowicz, J.; Gostynski, B.; Kudzin, M.H.; Kudzin, Z.H.; Urbaniak, P. 1-(*N*-Acylamino)-alkylphosphonic acids. Deacylation in aqueous solutions. *Phosphorus Sulfur Silicon Relat. Elem.* **2017**, *192*, 651–658. [CrossRef]
52. Smith, M.B.; March, J. *March's Advanced Organic Chemistry: Reactions, Mechanisms, and Structure*, 6th ed.; John Wiley & Sons, Inc.: Hoboken, NJ, USA, 2007; Volume Chapters 16–60, pp. 1408–1411. ISBN 13-978-0-471-72091-1.
53. O'Connor, C. Acidic and basic amide hydrolysis. *Q. Rev. Chem. Soc.* **1970**, *24*, 553–564. [CrossRef]
54. Brown, R.S.; Bennet, A.J.; Slebocka-Tilk, H. Recent perspectives concerning the mechanism of H_3O^+ and OH^- promoted amide hydrolysis. *Acc. Chem. Res.* **1992**, *25*, 481–488. [CrossRef]
55. Bagno, A.; Lovato, G.; Scorrano, G. Thermodynamics of protonation and hydration of aliphatic amides. *J. Chem. Soc. Perkin Trans.* **1993**, *2*, 1091–1098. [CrossRef]
56. Zahn, D. On the role of water in amide hydrolysis. *Eur. J. Org. Chem.* **2004**, 4020–4023. [CrossRef]
57. Kahne, D.; Still, W.C. Hydrolysis of a peptide bond in neutral water. *J. Am. Chem. Soc.* **1988**, *110*, 7529–7534. [CrossRef]
58. DeWolfe, R.H.; Newcombe, R.C. Hydrolysis of formanilides in alkaline solutions. *J. Org. Chem.* **1971**, *36*, 3870–3878. [CrossRef]

59. Desai, S.D.; Kirsch, L.E. The ortho effect on the acidic and alkaline hydrolysis of substituted formanilides. *Int. J. Chem. Kinet.* **2015**, *47*, 471–488. [CrossRef]
60. Bowden, K.; Bromley, K. Reactions of carbonyl compounds in basic solutions. Part 14. The alkaline hydrolysis of substituted *N*-methylformanilides, *N*-methylacetanilides, 1-phenylazetidin-2-ones, 1-phenyl-2-pyrrolidones, and 1-phenyl-2-piperidones. *J. Chem. Soc. Perkin Trans.* **1990**, *2*, 2103–2109. [CrossRef]
61. Hibbert, F.; Malana, M.A. Kinetics of the alkaline hydrolysis of 1,8-bis(trifluoroacetylamino)-naphthalene to 1-amino-8-trifluoroacetylaminonaphthalene in 70%, 80% and 90%(v/v) $Me_2SO–H_2O$. *J. Chem. Soc. Perkin Trans.* **1992**, *2*, 755–759. [CrossRef]
62. Cheshmedzhieva, D.; Ilieva, S.; Hadjieva, B.; Galabov, B. The mechanism of alkaline hydrolysis of amides: A comparative computational and experimental study of the hydrolysis of *N*-methylacetamide, *N*-methylbenzamide, and acetanilide. *J. Phys. Org. Chem.* **2009**, *22*, 619–631. [CrossRef]
63. Theodorou, V.; Paraskevopoulos, G.; Skobridis, K. A mild alkaline hydrolysis of *N*- and *N,N*-substituted amides and nitriles. *Arkivoc* **2015**, *7*, 101–112. [CrossRef]
64. Soroka, M. Comments on the synthesis of aminomethylphosphonic acid. *Synthesis* **1989**, 547–548. [CrossRef]
65. Kudzin, Z.H.; Stec, W.J. Synthesis of 1-aminoalkanephosphonic acids via thioureidoalkanephosphonates. *Synthesis* **1978**, 469–472. [CrossRef]
66. Frisch, M.J.; Trucks, G.W.; Schlegel, H.B.; Scuseria, G.E.; Robb, M.A.; Cheeseman, J.R.; Scalmani, G.; Barone, V.; Mennucci, B.; Petersson, G.A.; et al. *Gaussian 09, Revision D.01*; Gaussian, Inc.: Wallingford, CT, USA, 2009.
67. Cossi, M.; Rega, N.; Scalmani, G.; Barone, V. Energies, structures, and electronic properties of molecules in solution with the C-PCM solvation model. *J. Comput. Chem.* **2003**, *24*, 669–681. [CrossRef] [PubMed]

Sample Availability: Samples of all 1-(acylamino)phosphonic acids discussed in this paper are available from the authors. Calculated geometries of all modeled species are available upon request.

molecules

MDPI

Review

Synergy between Experimental and Theoretical Results of Some Reactions of Annelated 1,3-Azaphospholes

Raj K. Bansal *, Raakhi Gupta and Manjinder Kour

Department of Chemistry, The IIS University, Jaipur 302020, India; raakhi.gupta@iisuniv.ac.in (R.G.); manjinderkour1992@gmail.com (M.K.)

* Correspondence: bansal56@gmail.com; Tel.: +91-9057-242266

Received: 5 May 2018; Accepted: 24 May 2018; Published: 27 May 2018

Abstract: Computational calculations have been used successfully to explain the reactivity of the >C=P- functionality in pyrido-annelated 1,3-azaphospholes. Theoretical investigation at the Density Functional Theory (DFT) level shows that the lone pair of the bridgehead nitrogen atoms is involved in extended conjugation, due to which electron density increases considerably in the five-membered azaphosphole ring. The electron density in the azaphosphole is further enhanced by the presence of an ester group at the 3-position making the >C=P- functionality electron-rich. Thus, 1,3-azaphospholo[5,1-*a*]pyridine, i.e., 2-phosphaindolizine having ester group at the 3-position only does not undergo Diels-Alder (DA) reaction with an electron rich diene, such as 2,3-dimethyl-1,3-butadiene (DMB). However, an ester group at 1-position acts as electron-sink, due to which transfer of the electron density to the >C=P- moiety is checked and DA reaction occurs across the >C=P- functionality. The coordination of the Lewis acid to the carbonyl group at the 3-position raises the activation barrier, while it is lowered remarkably when it is coordinated to the P atom. Furthermore, the attack of 1,3-butadiene on the *Si* face of the P-coordinated (*o*-menthoxy)aluminum dichloride complex is a lower activation energy path. Fukui functions could be used to explain relative reactivities of indolizine and 2-phosphaindolizine towards electrophilic substitution reactions.

Keywords: synergy; 1,3-azaphospholes; Diels-Alder reaction; electrophilic substitution; DFT calculations

1. Introduction

Almost a century ago, Paul Dirac [1] could foresee the possibility of applying quantum mechanical concepts to solving the experimental problems in chemistry when he remarked, "The underlying physical laws necessary for the mathematical theory of a large part of physics and the whole of chemistry are thus completely known, and the difficulty is only that the exact application of these laws leads to equations much too complicated to be soluble. It therefore becomes desirable that approximate practical methods of applying quantum mechanics should be developed, which can lead to an explanation of the main features of complex atomic systems without too much computation". Dirac's dream inched towards fruition when Kohn, Hohenberg, and Sham [2,3] developed the density functional theory (DFT). It was followed by dramatic progress when many DFT codes were made available commercially, which enabled even less experienced persons with modest computational facilities to do calculations with much accuracy and look into the electronic structures and many other properties of the substances of interest. During the last two decades, the availability of a variety of new and more accurate DFT functionals incorporating specific features has brought DFT computations to the stage where they are used advantageously as a tool not only to explain the experimental results but also to plan and modify experimental conditions for achieving the desired objectives [4–9].

In recent years, there has been a perceptible increase in the number of research papers in chemistry reporting both experimental and theoretical results together. Fristrup and co-workers in a recent article emphasized the synergy between experimental and theoretical methods in the exploration of homogeneous transitional metal catalysis [10]. In another interesting review article titled, "Towards the computational design of solid catalysts", Norskov et al. [11] commented, "*Theoretical methods can be used to describe surface chemical reactions in detail and to understand variations in catalytic activity from one catalyst to another*." In spite of many challenges in the field, they added "Computational approaches for the discovery and development of catalysts hold great promise for the future." Similar views have been expressed in other reports [12–14].

In this perspective, we shall present mainly our own results highlighting the application of DFT calculations to plan and accomplish successfully some reactions of annelated 1,3-azaphospholes. It will also be illustrated how computational calculations helped to resolve an experimental log jam and paved the way for the use of catalysts, leading to a successful reaction and asymmetric synthesis.

2. Synthesis of Annelated 1,3-Azaphospholes

Before coming to the reactions, it will be appropriate to describe briefly two general methods of synthesis of annelated 1,3-azaphospholes that we developed.

2.1. [4+1] Cyclocondensation Method

The reaction of 2-ethyl-1-phenacylpyridinium bromide (**1**) with PCl_3 in the presence of Et_3N at room temperature (r.t.) afforded 3-benzoyl-1-methyl-1,3-azaphospholo[5,1-*a*]pyridine (**2**) in very good yield (Scheme 1). The general skeleton was named as 2-phosphaindolizine perceiving it to result from a CH/P exchange at the 2-position of indolizine nucleus [15].

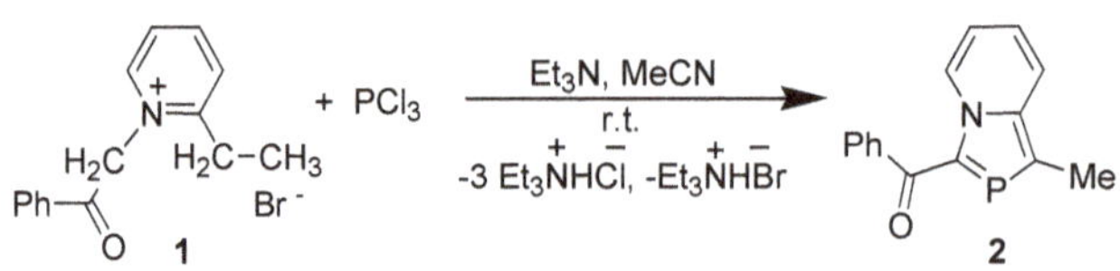

Scheme 1. Synthesis of 2-phosphaindolizine derivative.

Later, this synthetic strategy made accessible not only variously substituted 2-phosphaindolizines [15–17], but also 1-aza-2-phosphaindolizines [18], 3-aza-2-phosphaindolizines [19], and 1,3-diaza-2-phosphaindolizines [20]. Besides, the method could be extended to the synthesis of 1,3-azaphospholes annelated to pyrazine [15], thiazole [21], benzothiazole [21], and oxazole [22] skeletons also (Scheme 2). These reactions have been compiled in a mini-review [23].

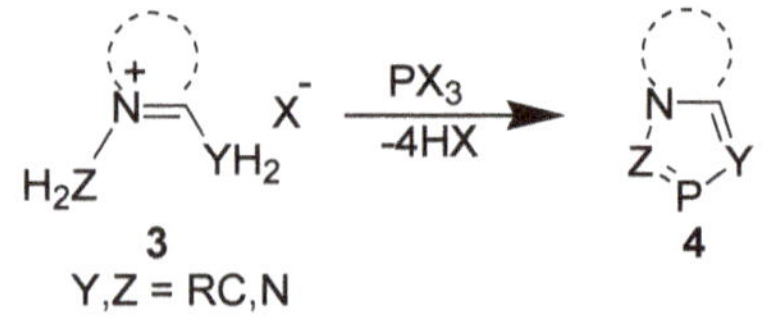

Scheme 2. Synthesis of annelated 1,3-azaphospholes.

2.2. Synthesis Involving Tandem Pyridinium Ylide generation, Disproportionation, and 1,5-Electrocyclization

We obtained 1,3-bis(alkoxycarbonyl)-1,3-azaphospholo[5,1-*a*]pyridines (**9**) serendipitously on reacting (alkoxycarbonyl)methylpyridinium bromide with Et_3N and PCl_3. The reaction involved tandem pyridinium alkoxycarbonyl-dichlorophosphinomethylide (**6**) generation, disproportionation, 1,5-electrocyclization, and 1,2-elimination to furnish the final product (Scheme 3) [24,25].

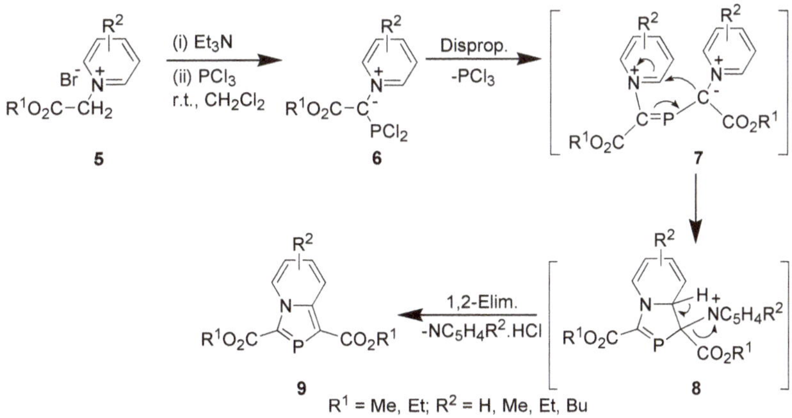

Scheme 3. Synthesis of 1,3-bis(alkoxycarbonyl)-1,3-azaphospholo[5,1-*a*]pyridines.

Subsequently, isoquinoline [26], and phenanthridine [27] analogues of **9** could also be prepared.

3. Reactions Response

3.1. Diels-Alder Reaction

There have been earlier reports about the Diels-Alder (DA) reaction across the >C=P- functionality as dienophile [28,29]. We investigated theoretically the model DA reactions of ethene and phosphaethene with 1,3-butadiene at the DFT(B3LYP/6-31+G**) level [30]. The Frontier Molecular Orbitals (FMO) of the reactants are shown in Figure 1.

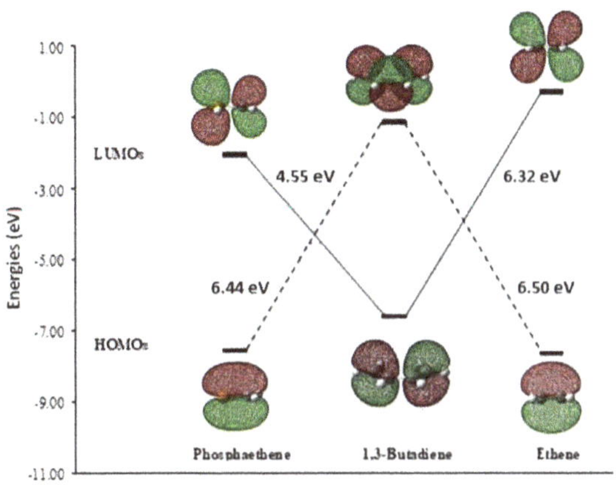

Figure 1. The HOMO-LUMO energy differences between FMOs of phosphaethene, 1,3-butadiene, and ethene calculated at the B3LYP/6-31+G** level.

It may be noted that the $HOMO_{diene}$–$LUMO_{phosphaethene}$ gap (4.55 eV) is much lower than the gap $HOMO_{diene}$–$LUMO_{ethene}$ (6.32 eV). Furthermore, the activation energy barriers (ΔEact.) for the two DA reactions are shown in Scheme 4 [30].

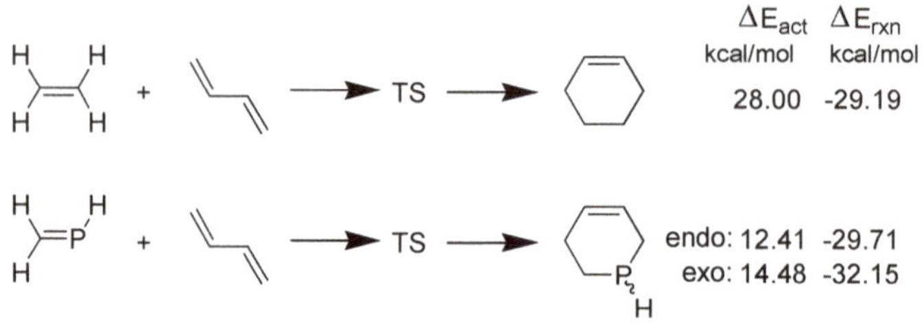

Scheme 4. Model DA reactions of ethene and phosphaethene with 1,3-butadiene computed at B3LYP/6-31+G** in the gas phase.

Encouraged by these theoretical results, we attempted DA reactions of 1,3-azaphospholo[5,1-*a*]pyridines, i.e., 2-phosphaindolizines prepared by two methods. However, to our great surprise, the compounds 3-ethoxycarbonyl-1-methyl-1,3-azaphospholo[5,1-*a*]pyridine **10a** (**10**, Z=Me, $R^1=R^2=R^3=H$) obtained through [4+1] cyclocondensation and 1,3-bis(ethoxylcarbonyl)-1,3-azaphospholo[5,1-*a*]pyridine **10b** (**10**, Z=CO_2Et, $R^1=R^2=R^3=H$) showed remarkable difference in their reactivity with 2,3-dimethylbuta-1,3-diene (DMB). It was found that 3-(ethoxycarbonyl)-1-methyl-1,3-azaphospholo[5,1-*a*]pyridine (**10a**) did not undergo DA reaction with DMB, even upon boiling in toluene in the presence of sulfur (used to oxidize phosphorus atom of the initially formed cycloadduct to push the reaction in the forward direction, see later) [31]. But, 1,3-bis(ethoxylcarbonyl)-1,3-azaphospholo[5,1-*a*]pyridine (**10b**) underwent DA reaction with DMB at room temperature (r.t.). However, the reaction at room temperature (ca 25 °C) was very slow and could be completed (δ ^{31}P NMR = 14.1 ppm) after heating at reflux temperature in chloroform for four days. On carrying out the reaction in the presence of sulfur or methyl iodide, it was complete at r.t. in 4 days [31]. Thus, the DA reactions of 1,3-bis(ethoxycarbonyl)-1,3-azaphospholo[5,1-*a*]pyridines and also of an isoquinoline analogue with DMB and isoprene could be accomplished successfully (Scheme 5) [31,32].

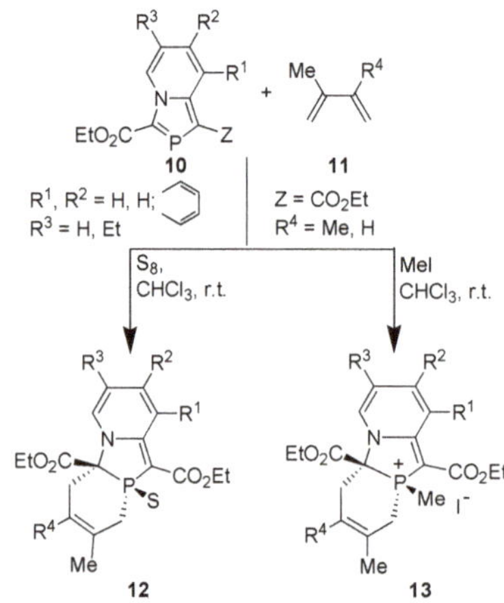

Scheme 5. DA reactions of 1,3-bis-(ethoxycarbonyl)-1,3-azaphospholo[5,1-*a*]pyridines with 2,3-dimethyl-1,3-butadiene (DMB) and isoprene.

All the reactions occurred with complete stereo- and regioselectivity, except the reaction of 1,3-bis(ethoxycarbonyl)-1,3-azaphospholo[5,1-*a*]isoquinoline (**14**) with isoprene in the presence of methyl iodide which afforded two regioisomers, the major (62%) being **15** (Scheme 6) [32].

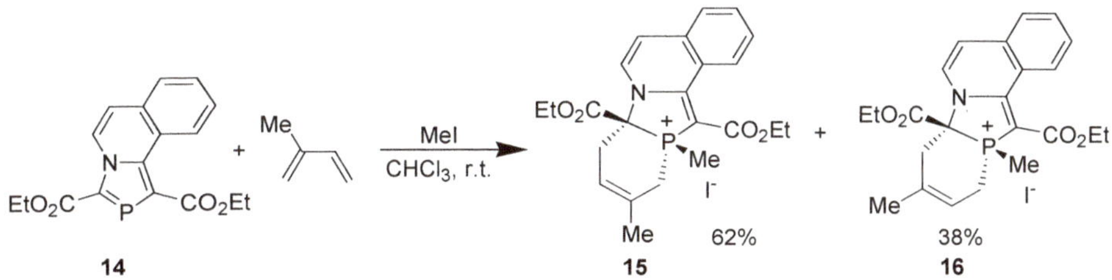

Scheme 6. Regioselectivity in the DA reaction of 1,3-bis(ethoxycarbonyl)-1,3-azaphospholo[5,1-*a*]isoquinoline.

The structure of the cycloadduct **12** (R^1R^2= CH=CH-CH=CH-; R^3=R^4=H) was confirmed by single crystal X-ray diffraction studies [32]. That the [2+4]cycloaddition of annelated 1,3-azaphosphole with DMB precedes the oxidation of the phosphorus atom of the initially formed cycloadduct by sulfur or methyl iodide could be proved unambiguously by carrying out the DA reaction of 1,4,2-diazaphospholo[4-*a*]pyridine derivative **17** in the presence of methyl iodide, when N-methylated product **18** (R=Me) was formed (Scheme 7) [33]. It has been reported that [1,4,2] diazaphospholo[4-*a*]pyridines did not react with methyl iodide [34].

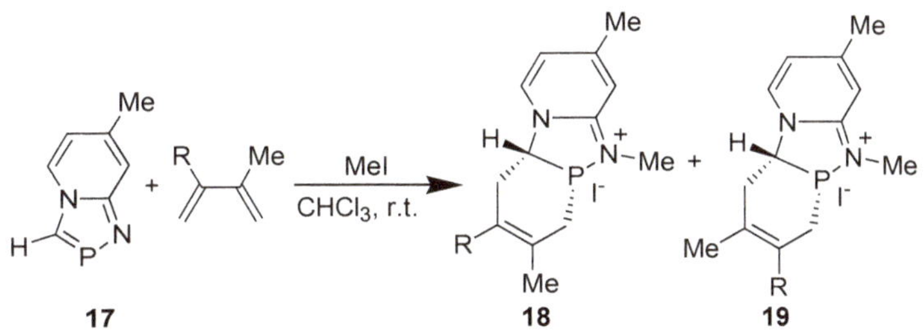

Scheme 7. DA reaction of 7-methyl-1,4,2-diazaphospholo[4,5-*a*]pyridine in the presence of methyl iodide.

The regioselectivity observed in the reaction of 1,3-bis(ethoxycarbonyl)- 1,3-azaphospholo[5,1-*a*] pyridine with isoprene was investigated theoretically at the DFT (B3LYP/6-311++G**// B3LYP/6-31G**) level. Following model reactions were computed (Figure 2). For studying the solvent effect, single point energy was computed using gas phase-optimized geometry at the B3LYP/6-311++G** level with polarizable continuum model (PCM).

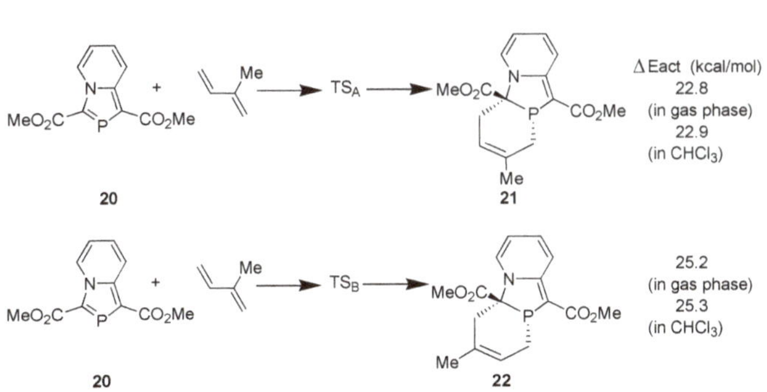

Figure 2. Model DA reactions computed at B3LYP/6-311++G**//B3LYP/6-31G**.

The regioselectivity was calculated by using the formula, $k_1/k_2 = e^{-\Delta E/RT}$ [35]. A difference of 2.4 kcal/mol in the activation energies corresponds to only 2% regioselectivity. It is obvious that the observed regioselectivity of 100% cannot be accounted for at this theoretical level.

We investigated the reason for the difference in the dienophilic reactivities of 3-ethoxycarbonyl -1-methyl-1,3-azaphospholo[5,1-*a*]pyridine (**10a**) and 1,3-bis(ethoxycarbonyl)-1,3-azaphospholo[5,1-*a*] pyridine (**10b**), theoretically. For this purpose, following four model reactions were computed at the DFT (B3LYP/6-31G**) level (Scheme 8) [36].

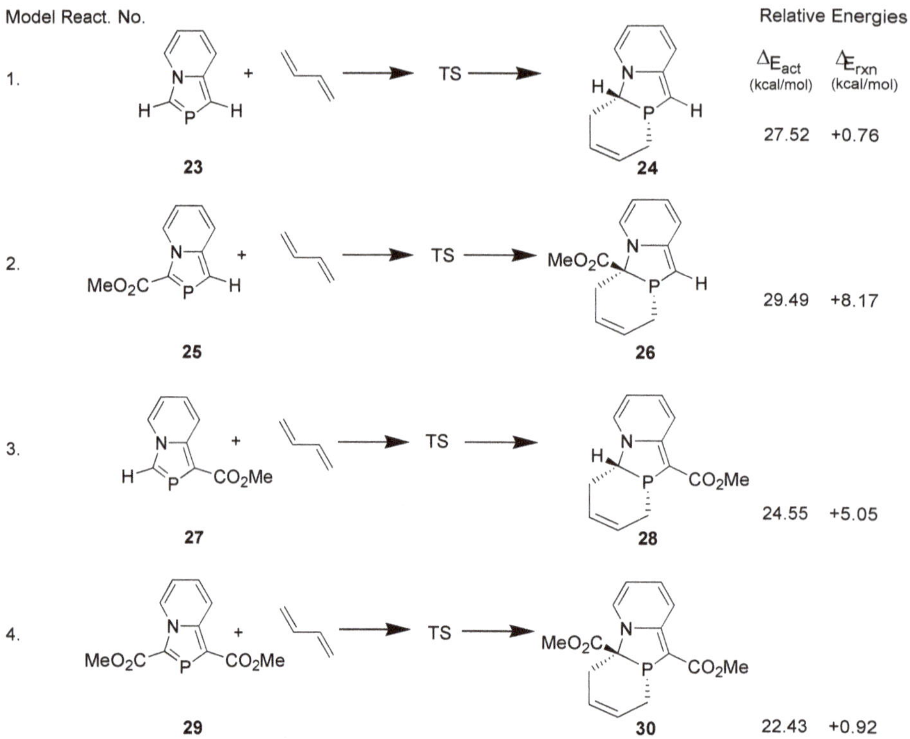

Scheme 8. Model DA reactions computed at the B3LYP/6-31G** level.

As expected, the activation energy barrier for the DA reaction with 1,3-bis(methoxycarbonyl)-1, 3-azaphospholo[5,1-*a*]pyridine (**29**) is lowest. However, it is interesting to find that the activation barrier for the DA reaction with 3-methoxycarbonyl-1,3-azaphospholo[5,1-*a*]pyridine (**25**) is highest (29.49 kcal/mol), higher than even for the DA reaction with the unsubstituted 1,3-azaphospholo[5,1-*a*]pyridine (**23**). It is rather an unexpected effect of an electron-withdrawing group (EWG). To look into the real cause of this unusual effect, we carried out the natural bond orbital (NBO) analysis calculations, which revealed the existence of an interesting phenomenon: due to conjugation of the lone pair of the bridgehead nitrogen in **23**, negative charge density is effectively transferred to the five-membered azaphosphole ring, which is further enhanced by the presence of an EWG at the 3-position. It is shown in Figure 3A. It makes the >C=P- functionality electron-rich, which fails to react with electron-rich diene, DMB. However, a CO_2Me group at the 1-position acts as a trap for the negative charge, the latter being delocalized over to the carbonyl group. It is shown in Figure 3B. Thus in this case, the >C=P- functionality remains unaffected and is able to undergo the DA reaction with DMB.

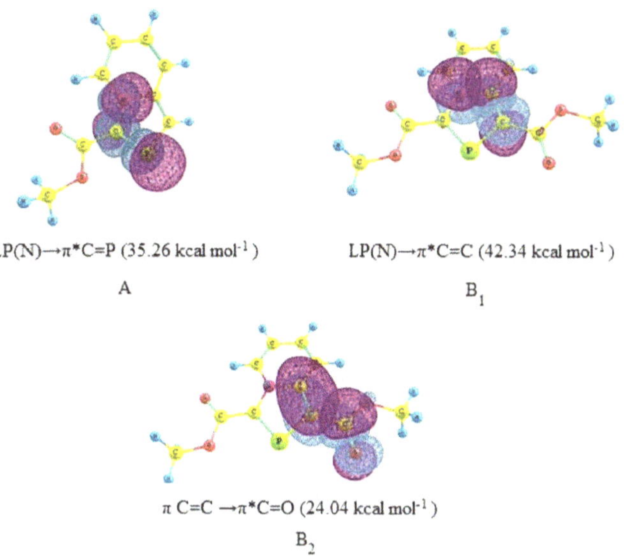

Figure 3. Natural Bond Orbital (NBO) interactions in 3-methoxycarbonyl-1,3-azaphospholo[5,1-*a*]pyridine (**A**) and 1,3-bis(methoxycarbonyl)- 1,3-azaphospholo[5,1-*a*]pyridine (**B_1** and **B_2**) with second order perturbation energies (in parenthesis).

3.2. Lewis Acid Catalyzed Diels-Alder Reaction

Then, we explored the possibility of using a Lewis acid as catalyst to accomplish DA reaction with 3-(alkoxycarbonyl)-1-methyl-1,3-azaphospholo[5,1-*a*]pyridines and computed the following model reactions at the DFT (B3LYP/6-31+G**) level (Scheme 9) [37]. For studying the solvent effect, single point energy was computed using gas phase-optimized geometry with PCM solvation model.

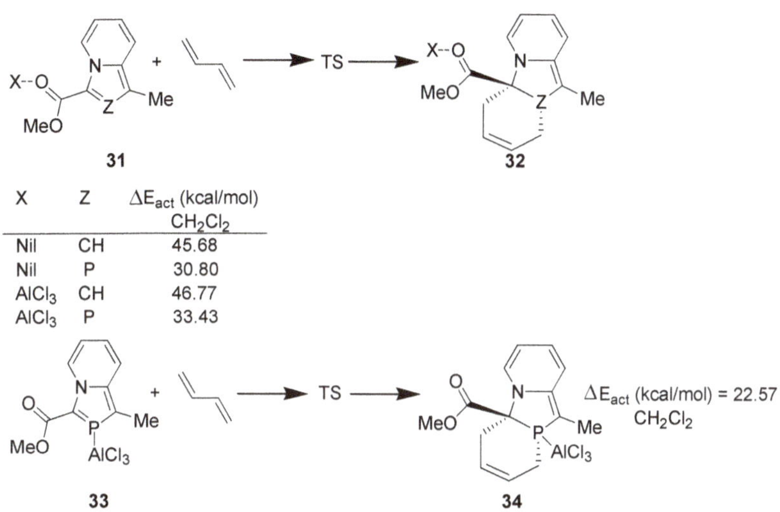

Scheme 9. Model DA reactions computed at the B3LYP/6-31+G** level.

There was another surprise in store for us. When the catalyst, $AlCl_3$, is coordinated to the carbonyl group of indolizine or 1,3-azaphospholo[5,1-*a*]pyridine, the activation energy barrier is increased further as compared with that in the absence of the catalyst. The NBO studies reveal that coordination of $AlCl_3$ with the carbonyl group enhances transfer of the negative charge in these cases, making the >C=C< and >C=P- functionality in indolizine and 1,3-azaphospholo[5,1-*a*]pyridine, respectively, electron-rich. This appears to be the reason why no DA reaction across the >C=C< moiety of indolizine has been reported even in the presence of a catalyst.

In 2-phosphaindolizine, however, there is another site for the coordination of $AlCl_3$: the P atom. It may be noted that on coordination of the catalyst to the P atom, the activation energy barrier is lowered remarkably, to 22.57 kcal/mol (in CH_2Cl_2) only. These results indicated that in contrast to indolizine, DA reactions across the >C=P- moiety of 2-phosphaindolizine having EWG at 3-position only should be possible in the presence of a Lewis acid catalyst. However, in this context a question arises: if coordination of $AlCl_3$ to the carbonyl group is thermodynamically more favorable than its coordination to the P atom by 3.43 kcal/mol, why should it coordinate to the P atom at all? In our opinion, in solution, an equilibrium may exist between the C=O- → $AlCl_3$ and P → $AlCl_3$ coordinated species, making a small concentration of the latter available that undergoes DA reaction with the diene to produce irreversibly the cycloadduct, thereby shifting the equilibrium to the right (Scheme 10).

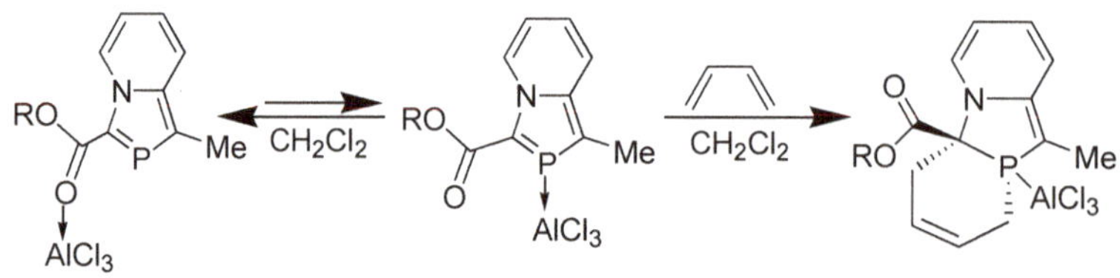

35 **36** **37**

Scheme 10. Equilibrium between C=O → and P → coordinated species, and DA reaction with the latter.

Guided by the above results, we accomplished successfully the DA reaction of 3-(alkoxycarbonyl/acyl)-1-methyl-2-phosphaindolizines (**38**) with DMB in the presence of ethylaluminium dichloride as a promoter (i.e., using its 1 equiv.) in dichloromethane at r.t. under nitrogen atmosphere (Scheme 11) [38]. The occurrence of a clean reaction was revealed by ^{31}P NMR signal at $\delta = -16.6$ to -5.3 ppm, which is in accordance with the ^{31}P NMR chemical shifts reported for the [2+4] cycloadducts obtained from the DA reaction of P-W(Co)$_5$ complexes of λ^3-phosphinines [39].

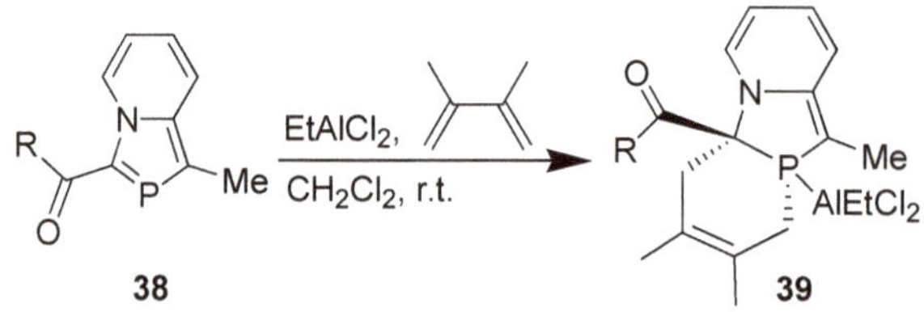

38, 39	**a**	**b**	**c**	**d**
R	OMe	OEt	CMe_3	Ph
δ ^{31}P (**39**)	-5.8	-5.3	-16.6	-10.1

Scheme 11. DA reaction of 3-(alkoxycarbonyl/acyl)-1-methyl-2-phosphaindolizines in the presence of $EtAlCl_2$.

3.3. *Asymmetric Diels-Alder Reaction with 2-Phosphaindolizines*

The successful results described above motivated us to investigate asymmetric DA reaction across the >C=P- moiety of 2-phosphaindolizines by using a chiral organoaluminium catalyst, namely, o-menthoxyaluminium dichloride [40].

We computed the following model reactions at the B3LYP/6-31+G* level (Scheme 12).

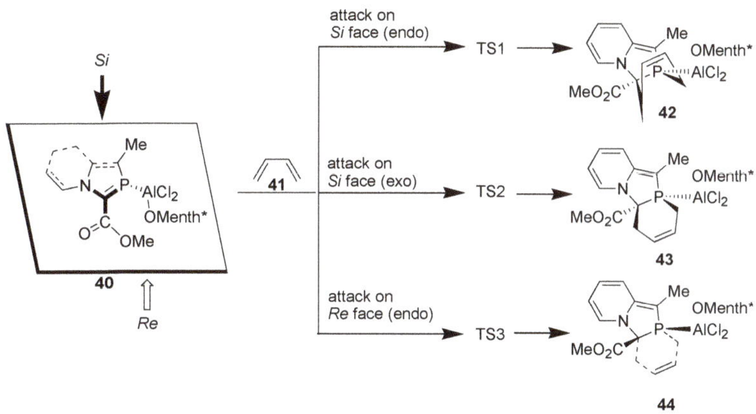

Scheme 12. Attack of 1,3-butadiene on *Si* and *Re* faces of the >C=P- functionality of 2-phosphaindolizine complex computed at the B3LYP/6-31+G* level.

Computational calculations reveal that attack of the diene occurs preferentially from the *Si* face. Furthermore, during attack on the *Si* face, activation energy barrier for the *exo* approach is smaller than for the *endo* approach by ca. 0.3 kcal/mol. Presence of the bulky o-menthoxy group possibly makes the *endo* approach comparatively less accessible. All the reactions are moderately exothermic, but due to decrease in entropy, they are endergonic, Gibbs free energies values being +11.65 to +13.08 kcal/mol.

In accordance with the theoretical results, DA reaction of 2-phosphaindolizines (**47**) could be accomplished with complete diastereoselectivity, as indicated by ^{31}P NMR, with a single product being formed in each case (Scheme 13) [40].

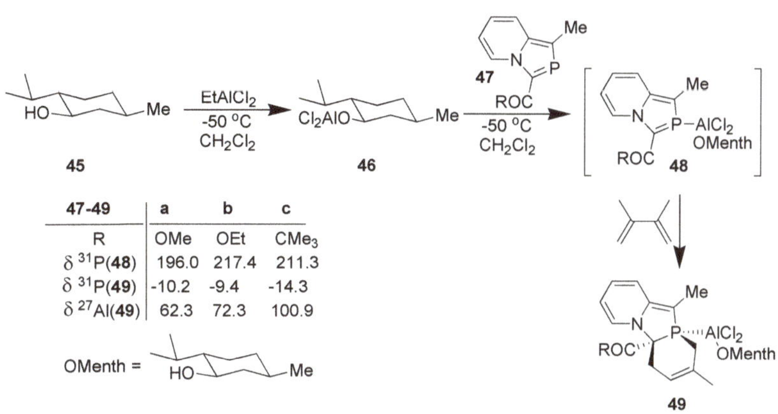

47-49	**a**	**b**	**c**
R	OMe	OEt	CMe$_3$
δ ^{31}P(**48**)	196.0	217.4	211.3
δ ^{31}P(**49**)	-10.2	-9.4	-14.3
δ ^{27}Al(**49**)	62.3	72.3	100.9

Scheme 13. Diels-Alder reaction of 2-phosphaindolizine-η^1-P-aluminium(*o*-menthoxy) dichloride with DMB.

3.4. Electrophilic Substitution

In contrast to indolizines [41], 1-unsubstituted 2-phosphaindolizines failed to react with MeCOCl, PhCOCl, or Me_3SiCl, even on prolonged heating in the presence of Et_3N [42]. However, like indolizines [43], 1-unsubstituted 2-phosphaindolizines reacted with PCl_3 or $PhPCl_2$ (but not with Ph_2PCl) in the presence of Et_3N to afford the phosphynated products (Scheme 14) [44]. It is noteworthy that on reacting with PCl_3, the initially formed dichlorophosphino product (**51**, R=Cl) undergoes dismutation accompanied by the loss of PCl_3 to afford a PCl bridged bis(2-phosphaindolizine) (**52**) as the final product.

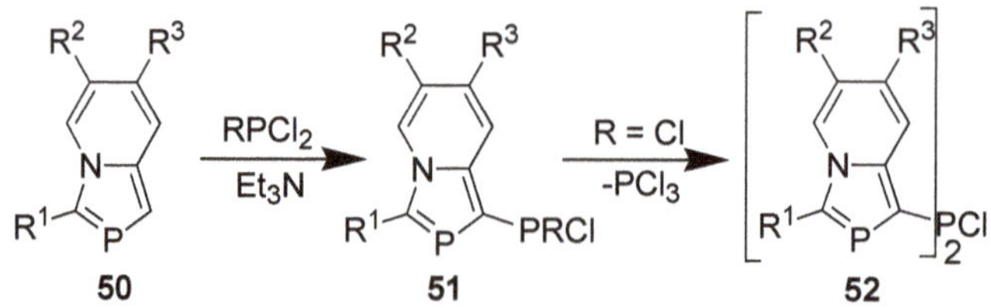

	a	b	c	d	e	f
R^1	COPh	COPh	COPh	CO_2Et	CO_2Et	p-$NO_2C_6H_4$
R^2	H	Me	H	H	Me	H
R^3	H	H	^{n}Bu	H	H	H

Scheme 14. Reaction of 1-unsubstituted 2-phosphaindolizines with chlorophosphines.

The concept of hard-soft acid-base (HSAB) theory was extended to explain the global and local reactivity of organic molecules towards nucleophilic and electrophilic reagents [45]. The chemical reactivity at a particular molecular site could be rationalized using a quantitative descriptor, the Fukui function (f(r)). We recently used Fukui function to rationalize 1,2- versus 1,4-addition of amines to maleic anhydride [46]. The difference in the behavior of indolizine and 2-phosphaindolizine towards electrophilic reagents can be rationalized on the basis of the Fukui functions calculated at the B3LYP/6-31+G* level (Figure 4) [47].

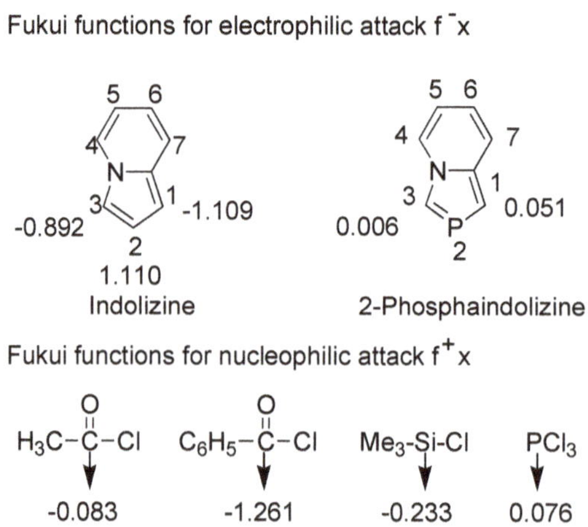

Figure 4. Fukui functions calculated at the B3LYP/6-31+G* level.

Fukui functions reveal that C1 and C3 in indolizine are much harder than C1 and C3 in 2-phosphaindolizine. Thus, hard electrophilic reagents such as acyl chlorides and chlorotrimethylsilane attack at C1 and C3 positions of indolizine, but they fail to react with 2-phosphaindolizine. On the other hand, phosphorus trichloride is comparatively softer and hence reacts with 2-phosphaindolizine at C1 (C3 being substituted in our compounds). Being at the borderline, PCl_3 reacts with indolizine also.

4. Conclusions

The computational calculations could be used as a useful tool to understand and solve problems encountered during reactions of annelated 1,3-azaphospholers. Furthermore, this strategy worked well while planning the reactions such as Lewis acid catalyzed and asymmetric DA reactions across the >C=P- functionality of 2-phosphaindolizines.

Author Contributions: All authors contributed equally in the compilation of this perspective.

Funding: This research received no external funding. APC was sponsored by MDPI.

Acknowledgments: We are thankful to the authorities of The IIS University, Jaipur (India) for the facilities.

Conflicts of Interest: The authors declare no conflict of interest.

References

1. Dirac, P.A.M. Quantum mechanics of many-electrons systems. *Proc. R. Soc. Lond. Ser. A.* **1929**, *123*, 714–733. [CrossRef]
2. Hohenberg, P.; Kohn, W. Inhomogenous electron gas. *Phys. Rev.* **1964**, *136*, B867–B871. [CrossRef]
3. Kohn, W.; Sham, L.J. Self-consistent equations including exchange and correlation effects. *Phys. Rev.* **1965**, *140*, A1133–A1138. [CrossRef]
4. Zangwill, A. A half century of density functional theory. *Phys. Today* **2015**, *68*, 34–39. [CrossRef]
5. Jain, A.; Shin, Y.; Persson, K.A. Computational predictions of energy materials using density functional theory. *Nat. Rev. Mater.* **2016**, *1*, 1–13. [CrossRef]
6. Cohen, A.J.; Mori-Sánchez, P.; Yang, W. Challenges for density functional theory. *Chem. Rev.* **2012**, *112*, 289–320. [CrossRef] [PubMed]
7. Jones, R.O. Density functional theory: Its origins, rise to prominence, and future. *Rev. Mod. Phys.* **2015**, *87*, 897–923. [CrossRef]
8. Tsipis, A.C. DFT flavor of coordination chemistry. *Coord. Chem. Rev.* **2014**, *272*, 1–29. [CrossRef]
9. Cole, D.J.; Hine, N.D.M. Applications of large-scale density functional theory in biology. *J. Phys. Condens. Matter* **2016**, *28*, 393001. [CrossRef] [PubMed]
10. Lupp, D.; Christensen, N.J.; Fristrup, P. Synergy between experimental and theoretical methods in the exploration of homogeneous transition metal catalysis. *Dalton Trans.* **2014**, *43*, 11093–11105. [CrossRef] [PubMed]
11. Nørskov, J.K.; Bligaard, T.; Rossmeisl, J.; Christensen, C.H. Towards the computational design of solid catalysts. *Nat. Chem.* **2009**, *1*, 37–46. [CrossRef] [PubMed]
12. Lineberger, W.C.; Borden, W.T. The synergy between qualitative theory, quantitative calculations, and direct experiments in understanding, calculating, and measuring the energy differences between the lowest singlet and triplet states of organic diradicals. *Phys. Chem. Chem. Phys.* **2011**, *13*, 11792–11813. [CrossRef] [PubMed]
13. Sierka, M. Synergy between theory and experiment in structure resolution of low-dimensional oxides. *Prog. Surf. Sci.* **2010**, *85*, 398–434. [CrossRef]
14. Mcgrady, J. Guest Editor The themed collection, "The synergy between theory and experiment. *Dalton Trans.* **2009**, 5805–6064. [CrossRef]
15. Bansal, R.K.; Gupta, N.; Karaghiosoff, K.; Schmidpeter, A.; Spindler, C. 2-Phosphaindolizines. *Chem. Ber.* **1991**, *124*, 475–480. [CrossRef]
16. Bansal, R.K.; Kabra, V.; Gupta, N.; Karaghiosoff, K. Synthesis of some new 2-phosphaindolizines. *Indian J. Chem.* **1992**, *31*, 254–256.
17. Gupta, N.; Jain, C.B.; Heinicke, J.; Bharatiya, N.; Bansal, R.K.; Jones, P.G. 2-Phosphaindolizines. *Heteroat. Chem.* **1998**, *9*, 333–339. [CrossRef]

18. Karaghioshoff, K.; Bansal, R.K.; Gupta, N. 1,4,2-Diazaphospholo[4,5-*a*]pyridines. *Z. Naturforsch. B* **1992**, *47b*, 373–378.
19. Bansal, R.K.; Pandey, G.; Karaghiosoff, K.; Schmidpeter, A. 1,2,3-Diazaphospholo[1,5-*a*]pyridines. *Synthesis* **1995**, 173–175. [CrossRef]
20. Bansal, R.K.; Gandhi, N.; Karaghiosoff, K.; Schmidpeter, A. Synthesis of [1,2,4,3]triazaphospholo[1,5-*a*]pyridines. *Z. Naturforsch.* **1995**, *50b*, 558–562. [CrossRef]
21. Bansal, R.K.; Mahnot, R.; Sharma, D.C.; Karaghiosoff, K.; Schmidpeter, A. 1,3-Azaphospholo[5,1-b]thiazolines and benzothiazoles. *Heteroat. Chem.* **1992**, *3*, 351–357. [CrossRef]
22. Bansal, R.K.; Jain, C.B.; Gupta, N.; Kabra, V.; Karaghiosoff, K.; Schmidpeter, A. Synthesis and properties of a 5,6-dihydro-1,3-azaphospholo[5,1-b]oxazole. *Phosphorus Sulfur Silicon* **1994**, *86*, 139–143. [CrossRef]
23. Bansal, R.K.; Karaghiosoff, K.; Gandhi, N.; Schmidpeter, A. 2-Substituted cycloiminium salts in azaphosphole synthesis. *Synthesis* **1995**, 361–369. [CrossRef]
24. Bansal, R.K.; Surana, A.; Gupta, N. 2-Phosphaindolizines via 1,5-electrocyclization. *Tetrahedron Lett.* **1999**, *40*, 1565–1568. [CrossRef]
25. Bansal, R.K.; Gupta, N.; Baweja, M.; Hemrajani, L.; Jain, V.K. Pyridinium dichlorophosphinomethylides. *Heteroat. Chem.* **2001**, *12*, 602–609. [CrossRef]
26. Bansal, R.K.; Hemrajani, L.; Gupta, N. 1,3-Azaphospholo[5,1-*a*]isoquinolines. *Heteroat. Chem.* **1999**, *10*, 598–604. [CrossRef]
27. Singh, D.; Sinha, P.; Gupta, N.; Bansal, R.K. Synthesis of 1,3-azaphospholo[1,5-*f*]phenanthridines through 1,5-electrocyclization. *Phosphorus Sulfur Silicon Relat. Elem.* **2016**, *191*, 488–492. [CrossRef]
28. Appel, R.; Knoll, F.; Ruppert, I. Phospha-alkenes and phospha-alkynes, genesis and properties of the (*p*-*p*)π-multiple bond. *Angew. Chem. Int. Ed.* **1981**, *20*, 731–744. [CrossRef]
29. Mathey, F.; Mercier, F.; Charrier, C.; Fischer, J.; Mitschler, A. Dicoordinated 2H-phospholes as transient intermediates in the reactions of tervalent phospholes at high temperature. One-step synthesis of 1-phosphabornadienes and phosphorins from phospholes. *J. Am. Chem. Soc.* **1981**, *103*, 4595–4597. [CrossRef]
30. Wannere, C.S.; Bansal, R.K.; Schleyer, P.v.R. Diels-Alder reaction of phosphaethene with 1,3-dienes: An ab-initio study. *J. Org. Chem.* **2002**, *67*, 9162–9174. [CrossRef] [PubMed]
31. Bansal, R.K.; Gupta, N.; Kumawat, S.K.; Dixit, G. Diastereo- and regioselective Diels-Alder reactions of 2-phosphaindolizines. *Tetrahedron* **2008**, *64*, 6395–6401. [CrossRef]
32. Bansal, R.K.; Jain, V.K.; Gupta, N.; Gupta, N.; Hemrajani, L.; Baweja, M.; Jones, P.G. Stereo- and regioselectivity in Diels-Alder reactions of 1,3-azaphospholo[5,1-a]isoquinoline and –[5,1-*a*]pyridine. *Tetrahedron* **2002**, *58*, 1573–1579. [CrossRef]
33. Bansal, R.K.; Karaghiosoff, K.; Gupta, N.; Gandhi, N.; Kumawat, S.K. Diastereo- and regioselectivity in Diels-Alder reaction of [1,4,2]diazaphospholo[4,5-*a*]pyridines. *Tetrahedron* **2005**, *61*, 10521–10528. [CrossRef]
34. Karaghiosoff, K.; Cleve, C.; Schmidpeter, A. Chloromethyl-dichlorophosphane: A useful reagent for the synthesis of new heterocycles with dicoordinated phosphorus. *Phosphorus Sulfur Silicon Relat. Elem.* **1986**, *28*, 289–296. [CrossRef]
35. Pellegriner, S.C.; Silva, M.A.; Goodman, J.M. Theoretical evaluation of the origin of the regio- and stereoselectivity in the Diels-Alder reactions of dialkylvinylboranes: Studies on the reactions of vinylborane, dimethylvinylborane, and vinyl-9-BBN with *trans*-piperylene and isoprene. *J. Am. Chem. Soc.* **2001**, *123*, 8832–8837. [CrossRef]
36. Bansal, R.K.; Gupta, N.; Dixit, G.; Kumawat, S.K. Theoretical investigation of an unusual substituent effect on the dienophilicity of >C=P– Functionality in 2-phosphaindolizines. *J. Phys. Org. Chem.* **2009**, *22*, 125–129. [CrossRef]
37. Gupta, N.; Jangid, R.K.; Bansal, R.K.; Hopffgarten, M.V. Catalytic effect of organoaluminium chloride reagents on the dienophilic reactivity of indolizine and 2-phosphaindolizine towards [2+4] cycloaddition: A DFT investigation. *Curr. Cat.* **2012**, *1*, 93–99. [CrossRef]
38. Jangid, R.K.; Gupta, N.; Bansal, R.K.; Hopffgarten, M.V.; Frenking, G. Diels-Alder reaction of 2-phosphaindolizines catalyzed by organoaluminium reagent: Theoretical and experimental results. *Tetrahedron Lett.* **2011**, *52*, 1721–1724. [CrossRef]
39. Alcaraz, J.P.; Mathey, F. Accroissement de la reactivite des phosphorines en tant que dienes et philodienes par complexation du phosphore. *Tetrahedron Lett.* **1984**, *25*, 207–210. [CrossRef]

40. Jangid, R.K.; Sogani, N.; Gupta, N.; Bansal, R.K.; Hopffgarten, M.V.; Frenking, G. Asymmetric Diels-Alder reaction with >C=P– functionality of the 2-phosphaindolizine-η^1-P-aluminium(*o*-menthoxy) dichloride complex: Experimental and theoretical results. *Beilstein J. Org. Chem.* **2013**, *9*, 392–400. [CrossRef] [PubMed]
41. Swinbourne, F.J.; Hunt, J.H.; Klinkert, G. *Advances in Indolizine Chemistry in the Series: Advances in Heterocyclic Chemistry*; Katritzky, A.R., Boulton, A.J., Eds.; Academic Press: New York, NY, USA, 1979; Volume 23, pp. 103–170.
42. Bansal, R.K.; Heinicke, J. Anellated heterophospholes and phospholides and analogies with related non-phosphorus systems. *Chem. Rev.* **2001**, *101*, 3549–3578. [CrossRef] [PubMed]
43. Tolmachev, A.A.; Yurchenko, A.A.; Kozlov, E.S.; Shulezhko, V.A.; Pinchuk, A.M. Phosphorylated indolizines. *Heteroat. Chem.* **1993**, *4*, 343–360. [CrossRef]
44. Bansal, R.K.; Gupta, N.; Kabra, V.; Spindler, C.; Karaghiosoff, K.; Schmidpeter, A. Substitution of 2-phosphaindolizines by bromine and by chlorophosphines. *Heteroat. Chem.* **1992**, *3*, 359–366. [CrossRef]
45. Mendez, F.; Gázquez, J.L. Chemical reactivity of enolate ions: The local hard and soft-acids and bases principle viewpoint. *J. Am. Chem. Soc.* **1994**, *116*, 9298–9301. [CrossRef]
46. Kour, M.; Gupta, R.; Bansal, R.K. Experimental and theoretical investigation of the reaction of secondary amines to maleic anhydride. *Aust. J. Chem.* **2017**, *70*, 1247–1253. [CrossRef]
47. Gupta, R.; Paul, B.; Bansal, R.K. Application of Fukui functions for comparing reactivities of indolizine and 2-phosphaindolizine towards electrophilic substitution. 2018, unpublished work.

molecules

MDPI

Review

Synthesis and Reactions of α-Hydroxyphosphonates

Zita Rádai and György Keglevich *

Department of Organic Chemistry and Technology, Budapest University of Technology and Economics, 1521 Budapest, Hungary; radai.zita@mail.bme.hu

* Correspondence: gkeglevich@mail.bme.hu; Tel.: +36-1-463-1111 (ext. 5883)

Received: 8 May 2018; Accepted: 16 June 2018; Published: 20 June 2018

Abstract: This review summarizes the main synthetic routes towards α-hydroxyphosphonates that are known as enzyme inhibitors, herbicides and antioxidants, moreover, a number of representatives express antibacterial or antifungal effect. Special attention is devoted to green chemical aspects. α-Hydroxyphosphonates are also versatile intermediates for other valuable derivatives. *O*-Alkylation and *O*-acylation are typical reactions to afford α-alkoxy-, or α-acyloxyphosphonates, respectively. The oxidation of hydroxyphosphonates leads to ketophosphonates. The hydroxy function at the α carbon atom of hydroxyphosphonates may be replaced by a halogen atom. α-Aminophosphonates formed in the nucleophilic substitution reaction of α-hydroxyphosphonates with primary or secondary amines are also potentially bioactive compounds. Another typical reaction is the base-catalyzed rearrangement of α-hydroxy-phosphonates to phosphates. Hydrolysis of the ester function of hydroxyphosphonates leads to the corresponding phosphonic acids.

Keywords: α-hydroxyphosphonate; Pudovik reaction; *O*-derivatization; oxidation; substitution; rearrangement; hydrolysis

1. Synthetic Routes towards α-Hydroxyphosphonates

Due to the bioactivity of α-hydroxyphosphonates **1**, the synthesis of these derivatives is an evergreen topic in organophosphorus chemistry [1–8]. The main approaches to obtain α-hydroxyphosphonates **1** are shown in Scheme 1. The most commonly studied method is the addition of a dialkyl phosphite to an oxo compound (method "A") [9]. The good atom economy makes this method the most appealing way to synthesize α-hydroxyphosphonates **1**. In the majority of cases, the addition is carried out in the presence of a base catalyst, but a few acid-catalyzed variations are also known. An alternative route is the condensation of an oxo compound and a trialkyl phosphite (method "B") [10]. In contrast to method "A", this reaction is usually catalyzed by different acids. In the literature, the denomination of the reactions is not consistent, as both methods "A" and "B" are referred to as the Pudovik and Abramov reaction, and also as the phospha-aldol reaction. The third main approach towards α-hydroxyphosphonates **1** involves the reaction of α-ketophosphonates **2** with Grignard reagents [11] or other organometallic compounds [11,12], followed by hydrolysis (method "C"). α-Alkyl phosphonates **3** may also be converted to the corresponding α-hydroxy derivatives by oxidation (method "D") [13,14]. In this review, methods "A" and "B" are discussed in detail. Although the asymmetric synthesis of α-hydroxyphosphonates **1** is of special importance [4–8], in this review the discussion is limited to racemic derivatives.

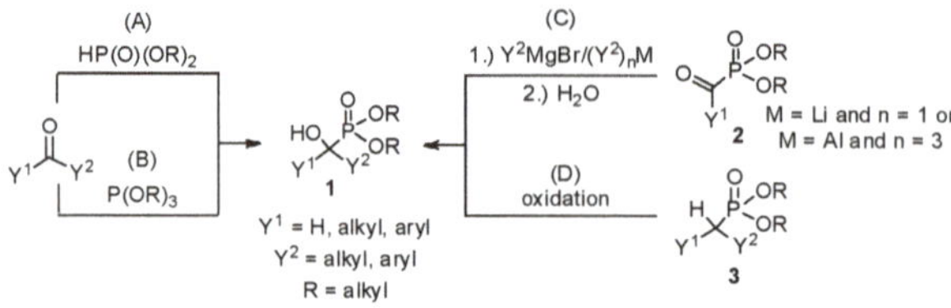

Scheme 1. Synthetic pathways towards α-hydroxyphosphonates **1**.

1.1. Synthesis of α-Hydroxyphosphonates by the Reaction of Aldehydes/Ketones and Dialkyl Phosphites

The reaction of oxo compounds and dialkyl phosphites catalyzed by alkali alcoholates (Scheme 1, method "A") was first reported by Pudovik [15]. Since then, a number of variants involving different catalysts and conditions were elaborated to obtain α-hydroxyphosphonates by the reaction under discussion (Table 1). Recent methods have targeted the use of inexpensive and simple catalysts, and mild reaction conditions in the spirit of green chemistry. In the great majority of the cases, the addition was carried out without using any solvent. It is worth mentioning that starting from ketones, the accomplishment of the reaction is more challenging than in the cases applying aldehydes. The synthesis of α-alkyl-α-hydroxyphosphonates often requires the use of "exotic" catalysts [16–20], or an excess of a base. Procedures also applicable for the conversion of ketones are marked by an asterisk in Table 1.

A base-catalyzed variation of the Pudovik reaction was carried out in the presence of 5 mol % potassium phosphate as the catalyst (Table 1/Entry 1) [21]. This method has the advantage of using an inexpensive catalyst, requiring short reaction times and an easy work-up procedure that comprises simple extraction, and, in most cases, provides the products in high yields. Barium hydroxide was also an efficient catalyst in the addition [22,23]. One method employed 10 mol % $Ba(OH)_2$ (Table 1/Entry 2) [22], while according to another protocol, $Ba(OH)_2 \cdot 8H_2O$ was used (Table 1/Entry 3) [23]. The first $Ba(OH)_2$-catalyzed variation suffers from the drawback of a tedious work-up procedure involving extraction, and washing, followed by crystallization [22], while in the latter case, the use of THF as the solvent is a disadvantage [23]. One equivalent of magnesium oxide efficiently catalyzed the addition of diethyl phosphite to substituted benzaldehydes (Table 1/Entry 4) [24]. The method could be extended to the conversion of γ-cyanoketones by applying the base in a twofold quantity (Table 1/Entry 5) [25]. The reaction of dimethyl phosphite with oxo compounds including ketones afforded the corresponding α-hydroxyphosphonates **1** in the presence of three equivalents of aluminum oxide after stirring for 72 h (Table 1/Entry 6) [26]. The reaction time could be shortened dramatically by using potassium fluoride with only one equivalent of Al_2O_3. However, this latter method was efficient only in case of aldehydes as the starting compounds (Table 1/Entry 7) [27]. The addition was also performed in the presence of one equivalent of triethylamine as the catalyst (Table 1/Entry 8) [28]. A simple crystallization afforded the desired products in good to excellent yields. Another method employed three equivalents of triethylamine together with one equivalent of magnesium chloride (Table 1/Entry 9) [29]. In the latter case, an extraction was inserted before the crystallization step in the work-up procedure [29]. In agreement with the requirements of environmentally friendly approaches, green activation methods, such as microwave (MW) irradiation [30,31] and grinding [32,33] were also employed in the synthesis of α-hydroxyphosphonates **1**. According to a MW-assisted procedure, the reaction of benzaldehydes and diethyl phosphite was carried out without the use of any catalyst or solvent (Table 1/Entry 10) [30], while another method combined MW irradiation with sodium carbonate as the base (Table 1/Entry 11) [31]. Na_2CO_3 was also used efficiently in another protocol, when the contact among the reaction components was promoted by grinding (Table 1/Entry 12) [32]. The drawback of the method is the complicated work-up procedure comprising washing, extraction, and finally recrystallization. A similar protocol utilizes the piperazine-catalyzed reaction of aldehydes and diethyl phosphite in a mill (Table 1/Entry 13) [33]. In the latter case, the work-up involved washing, extraction and column chromatography. The synthesis of α-hydroxyphosphonates **1** was also carried out in the presence of 10 mol % choline hydroxide as an IL catalyst (Table 1/Entry 14) [34]. According to a plausible mechanism suggested by the authors, the IL promotes the reaction by activation through *H*-bridges. An interesting protocol is a special fluoroapatite-catalyzed Pudovik reaction (Table 1/Entry 15) [35]. A mixture of the starting dialkyl phosphite and oxo compound was stirred with a spatula, and then left standing on the catalyst over 1–1.5 min. The work-up involved extraction with dichloromethane, and recrystallization of the crude product. Acid-catalyzed variations of the Pudovik reaction occur rarely in the literature [36,37]. Potassium hydrogensulfate proved to be an efficient catalyst in an amount of 20 mol % in addition of diethyl phosphite to substituted aldehydes (Table 1/Entry 16) [36]. According to another method, silica-supported tungstic acid was applied as the catalyst (Table 1/Entry 17) [37]. Organometallic compounds may also play the role of the catalyst during the synthesis of α-hydroxyphosphonates **1** [38–40]. Butyllithium was applied in the amount of 0.1 mol % in hexane as the solvent (Table 1/Entry 18) [38]. The method not only bears the disadvantages of using an exotic catalyst in an inert atmosphere, but also the need for a complicated work-up involving quenching, washing, and finally column chromatography. As for the advantages, mild reaction conditions and wide substrate scope including ketones can be mentioned. Titanium tetraisopropylate was also applied to catalyze the reaction of ketones and dimethyl phosphite (Table 1/Entry 19) [39]. Column chromatography of the crude product afforded α-alkyl α-hydroxyphosphonates **1** in good to excellent yields. Molybdenum dioxide dichloride (MoO_2Cl_2) was also tested as the catalyst in the Pudovik reaction (Table 1/Entry 20) [40]. Using it in an amount of 5 mol % under solvent-free conditions at 80 °C, the hydroxyphosphonates **1** were obtained good yields.

1.2. Synthesis of α-Hydroxyphosphonates by the Reaction of Aldehydes/Ketones and Trialkyl Phosphites

Besides the reaction of oxo compounds with dialkyl phosphites discussed in Section 1.1 (Scheme 1, method "A"), the other widespread method for the synthesis of α-hydroxyphosphonates is the condensation of aldehydes or ketones with trialkyl phosphites (Scheme 1, method "B") [10]. This way is of somewhat less importance due to the reduced atom economy, as compared to the major route outlined. While the reaction involving dialkyl phosphites is usually catalyzed by different bases, the typical catalysts for the condensation of oxo compounds with trialkyl phosphites exhibit acidic character (Table 2).

A method of choice is a solvent- and catalyst-free accomplishment under ultrasonic irradiation (Table 2/Entry 1) [41]. The reaction of aldehydes and trialkyl phosphites at 25 °C was complete after 10–35 min and provided the α-hydroxyphosphonates **1** in excellent yields after crystallization (Table 2/Entry 1). The method was successfully applied to ketones as starting materials as well. Another ultrasound-assisted variation is the potassium dihydrogen phosphate-catalyzed reaction (Table 2/Entry 2) [42]. According to a plausible mechanism, the role of the acid catalyst is the protonation of the carbonyl function of the aldehyde making the C=O bond more electrophilic, and hence facilitating the nucleophilic attack of the $(RO)_3P$ reagent. A similar procedure employed sulphamic acid as the catalyst (Table 2/Entry 3) [43]. Again, the catalytic effect of the sulphamic acid was attributed to its capability to protonate the oxo compound. The camphorsulfonic acid-catalyzed syntheses of α-hydroxyphosphonates **1** from aldehydes and triethyl phosphite was carried out under ultrasonic irradiation on stirring (Table 2/Entry 4) [44]. Applying the ultrasound technique, the reaction times could be shortened from 30–75 min to 8–15 min after comparing the results with those obtained without ultrasonication. At the same time, the yields were similar (85–93%) for the two procedures [44]. Oxalic acid also proved to be an efficient catalyst in the condensation under discussion at 80 °C (Table 2/Entry 5) [45]. A series of organic solvents were tested as the reaction medium, but the desired products were obtained in low yields (<30%). Under neat conditions, the α-hydroxyphosphonates **1** were formed efficiently as suggested by the yields of 83–98% obtained after quenching, extraction and column chromatography. In the hope of synthesizing a new class of antibiotics, α-hydroxy-phosphonates **1** containing a β-lactam scaffold were prepared in the presence of tartaric acid as the catalyst (Table 2/Entry 6) [46].

Table 1. Selected examples for the synthesis of α-hydroxyphosphonates (**1**) from aldehydes or ketones and dialkyl phosphites (method "A").

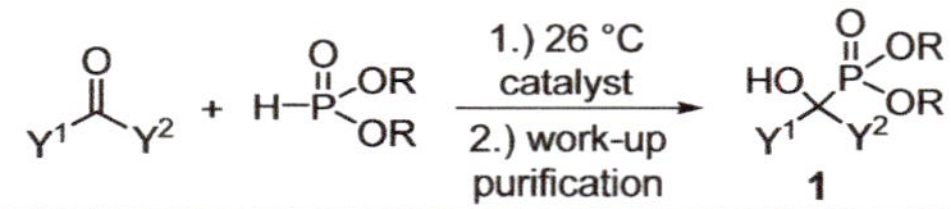

Entry	Y^1	Y^2	R	Catalyst	Amount of Catalyst	Conditions	Yield (%)	Ref.
1	^{i}Pr, ^{c}Hex, C_6H_5, 4-MeC_6H_4, 4-$^iPrC_6H_4$, 4-$MeOC_6H_4$, 3,4-diMeOC_6H_3, 4-ClC_6H_4, 2-$NO_2C_6H_4$, 4-CNC_6H_4, C_6H_5CH=CH, 3,4-OCH$_2$OC_6H_3, 4-CH_2=CHCH$_2$OC_6H_4, 4-PhCH$_2$OC_6H_4, 2-furyl, 2-thienyl	H	Me, Et, ^{i}Pr	K_3PO_4	5 mol %	25 °C, 4–8 min for aromatic aldehydes, 24 h for aliphatic aldehydes	20–98	[21]
2	C_6H_5, 4-MeOC_6H_4, 3,4-diMeOC_6H_3, 2-ClC_6H_4, 4-ClC_6H_4, 2,3-diClC_6H_3, 2,4-diClC_6H_3, 2,6-diClC_6H_3, 2-BrC_6H_4, 4-BrC_6H_4, 3-FC_6H_4, 4-FC_6H_4, 2-$NO_2C_6H_4$, 3-$NO_2C_6H_4$, 4-$NO_2C_6H_4$, 4-CNC_6H_4, C_6H_5CH=CH, 2-thienyl, 4-pyridyl	H	Et	$Ba(OH)_2$	10 mol %	25 °C, 4–10 min	70–98	[22]
3	^{i}Bu, C_6H_5, 2-MeC_6H_4, 3-MeC_6H_4, 4-MeC_6H_4, 4-EtC_6H_4, 4-MeOC_6H_4, 2,3,4-triMeOC_6H_2, 2-ClC_6H_4, 4-ClC_6H_4, 2,6-diClC_6H_3, 3-BrC_6H_4, 4-BrC_6H_4, 3-FC_6H_4, 4-FC_6H_4, 2-$NO_2C_6H_4$, 3-$NO_2C_6H_4$, 4-$NO_2C_6H_4$, 3-CNC_6H_4, 4-CNC_6H_4, 2-furyl, 2-thienyl, 2-pyridyl, 3-pyridyl, 4-pyridyl, 1-naphthyl, 2-naphthyl, 9-anthryl, C_6H_5CH=CH	H	Et, Bu, Bn	$Ba(OH)_2 \cdot 8H_2O$	2–7 mol %	25 °C, 15 min, THF	72–99	[23]
4	C_6H_5, 4-MeC_6H_4, 3-MeOC_6H_4, 4-MeOC_6H_4, 3-HOC_6H_4, 2-ClC_6H_4, 4-ClC_6H_4, 2,6-diClC_6H_3, 2-$NO_2C_6H_4$, 3-$NO_2C_6H_4$, 4-$NO_2C_6H_4$, 4-$Me_2NC_6H_4$, C_6H_5CH=CH, 2-furyl, 1-naphthyl, 2-naphthyl	H	Et	MgO	1 equiv.	25 °C, 2 min–4 h	80–100	[24]
5 *	C_6H_5, 4-MeC_6H_4, 4-MeOC_6H_4	$(CH_2)_2CN$, MeCHCH$_2$CN	Me, Et	MgO	2 equiv.	25 °C, 1–6 h	70–82	[25]
6 *	Pr, Bu, C_6H_5, 4-MeC_6H_4, 4-MeOC_6H_4, 4-$^iPrC_6H_4$, 2,4,6-triMeC_6H_2, 2-OHC_6H_4, 4-ClC_6H_4, 4-BrC_6H_4, 3-FC_6H_4, 4-FC_6H_4, 4-Br-2-OHC_6H_3, 3,4-diMeOC_6H_3, 3,4-OCH$_2$OC_6H_3, 2-$NO_2C_6H_4$, 4-$NO_2C_6H_4$, 2-furyl, 2-thienyl, 3-thienyl, 1-naphthyl, 2-naphthyl	H, Me	Me	Al_2O_3	3 equiv.	25 °C, 72 h	52–98	[26]
7	C_6H_5, 4-MeC_6H_4, 4-MeOC_6H_4, 4-ClC_6H_4, 4-FC_6H_4, 4-$NO_2C_6H_4$	H	Me, Et	Al_2O_3 + KF	1 + 2 equiv.	25 °C, 30 min	53–88	[27]
8 *	Me, ClCH$_2$, C_6H_5, 4-MeC_6H_4, 4-MeOC_6H_4, 2-ClC_6H_4, 3-ClC_6H_4, 4-ClC_6H_4, 3-BrC_6H_4, 4-BrC_6H_4, 2-FC_6H_4, 4-FC_6H_4, 2,4-diClC_6H_3, 2-furyl, 2-thienyl	Me, Pr, Ph	Me	Et_3N	1 equiv.	40 °C, 2 h	63–89	[28]

Table 1. *Cont.*

Entry	Y^1	Y^2	R	Catalyst	Amount of Catalyst	Conditions	Yield (%)	Ref.
9 *	Pr, C_6H_5, 4-MeC_6H_4, 4-$MeOC_6H_4$, 2-ClC_6H_4, 4-ClC_6H_4, 3-$NO_2C_6H_4$, 4-$NO_2C_6H_4$, 4-CNC_6H_4, $(CH_2)_2C_6H_5$, C_6H_5CH=CH, 2-furyl, 2-naphthyl, 9-anthryl	H, Me	Me	Et_3N + $MgCl_2$	3 + 1 equiv.	50 °C, 1–2 h	85–98	[29]
10	C_6H_5, 4-MeC_6H_4, 4-$MeOC_6H_4$, 4-OHC_6H_4, 2-ClC_6H_4, 4-ClC_6H_4, 4-BrC_6H_4, 2-$NO_2C_6H_4$, 3-$NO_2C_6H_4$, 4-$NO_2C_6H_4$	H	Et	–	–	MW, 90–100 °C, 10–20 min	79–95	[30]
11	C_6H_5, 4-MeC_6H_4, 4-$MeOC_6H_4$, 4-ClC_6H_4, 4-$NO_2C_6H_4$	H	Me, Et	Na_2CO_3	0.75 equiv.	MW, 110 °C, 20 min	62–88	[31]
12	C_6H_5, 4-MeC_6H_4, 4-$MeOC_6H_4$, 4-ClC_6H_4, 3-$NO_2C_6H_4$, 4-$NO_2C_6H_4$	H	Et	Na_2CO_3	1 equiv.	Grinding, 25 °C, 10 min	75–83	[32]
13	C_6H_5, 4-MeC_6H_4, 4-$^i PrC_6H_4$, 4-$MeOC_6H_4$, 3,4-di$MeOC_6H_3$, 2-ClC_6H_4, 4-ClC_6H_4, 2-$NO_2C_6H_4$, 3-$NO_2C_6H_4$, 4-$NO_2C_6H_4$, 4-$Me_2NC_6H_4$, 4-$PhCH_2OC_6H_4$, 4-$C_5H_4NC_6H_4$, 9-anthryl	H	Et	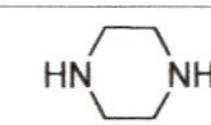	1 equiv.	Grinding, 25 °C, 2–10 min	78–96	[33]
14	C_6H_5, 2-MeC_6H_4, 4-MeC_6H_4, 2-$^i PrC_6H_4$, 3,5-diMeC_6H_3, 2-$MeOC_6H_4$, 4-$MeOC_6H_4$, 3,5-di$MeOC_6H_3$, 3,4,5-tri$MeOC_6H_2$, 4-$EtOC_6H_4$, 2-ClC_6H_4, 2-BrC_6H_4, 3-BrC_6H_4, 4-BrC_6H_4, 2-Cl-6-furylC_6H_3, 2-F-4-BrC_6H_3, 2-$NO_2C_6H_4$, 4-$NO_2C_6H_4$, 9-anthryl, 1-pyrenyl, 2-furyl, 1-C_6H_5-4-pyrazolyl	H	Et	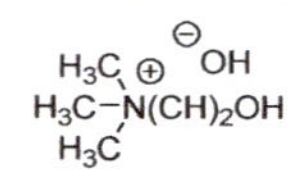	10 mol %	25 °C, 5–10 min	90–98	[34]
15 *	C_6H_5, 3-$MeOC_6H_4$, 4-$MeOC_6H_4$, 4-ClC_6H_4, 3-$NO_2C_6H_4$, 4-$NO_2C_6H_4$, C_6H_5CH=CH	H, Me	Me, Et	Na-modified fluoroapatite	1 g/1.25 mol acetophenone	20–25 °C, 1–1.5 min	75–98	[35]
16	4-$^i PrC_6H_4$, 4-$MeOC_6H_4$, 2,6-di$MeOC_6H_3$, 3,5-di$MeOC_6H_3$, 4-OHC_6H_4, 3-$NO_2C_6H_4$, 4-$Me_2NC_6H_4$, 2-$PhCH_2OC_6H_4$, 2-imidazyl, 3-indolyl	H	Et	$KHSO_4$	20 mol %	25 °C, 2–4 h	82–91	[36]
17	3-FC_6H_4, 4-$NO_2C_6H_4$, 3,4-$OCH_2OC_6H_3$, 4-$MeSC_6H_4$, $C_5H_{10}N$, 3-MeO-4-OHC_6H_3, 4-$C_4H_8NC_6H_4$, 2-furyl, 2-thienyl, 4-imidazyl, 2-pyrrolyl, 4-pyridyl	H	Me	silica-supported tungstic acid	20 mol %	25 °C, 30 min	85–96	[37]
18 *	Me, Pr, ^{i}Bu, Pent, C_6H_5, 2-MeC_6H_4, 4-MeC_6H_4, 2-$MeOC_6H_4$, 3-$MeOC_6H_4$, 4-$MeOC_6H_4$, 2-ClC_6H_4, 3-ClC_6H_4, 4-ClC_6H_4, 2-BrC_6H_4, 4-BrC_6H_4, 4-FC_6H_4, 3-$NO_2C_6H_4$, 4-$NO_2C_6H_4$, 1-naphthyl, 2-naphthyl, 2-furyl, 2-thienyl, 3-pyridyl	H, Me, Ph, CF_3, $(CH_2)_{10}CH_3$, C(O)Ph, $CH_2C(O)Ph$	Et, ^{i}Pr, Ph	BuLi	0.1 mol %	10–25 °C, 5 min, hexane	35–99	[38]
19 *	Et, ^{c}Hex, C_6H_5, 3-MeC_6H_4, 4-MeC_6H_4, 3-$MeOC_6H_4$, 4-ClC_6H_4, 3,4-diClC_6H_3, 2,3,4-triClC_6H_2, 2-FC_6H_4, 4-FC_6H_4, 4-$NO_2C_6H_4$, 3-$CF_3C_6H_4$CH2=CH, 3-$CF_3C_6H_4CH(OH)CH_2$, 2-thienyl, 2-naphthyl, EtOC(O)CHBn	Me, Et, Ph, $CH(OEt)_2$, COOMe, $(CH_2)_2Cl$	Me	$Ti(O^iPr)_4$	5 mol %	30 °C, 15 min	74–98	[39]
20	C_6H_5, 4-$MeOC_6H_4$, 4-FC_6H_4, 4-$NO_2C_6H_4$, 4-$CF_3C_6H_4$, 4-CNC_6H_4, 4-MeOC(O)C_6H_4, $(CH_2)_2C_6H_5$, C_6H_5CH=CH	H	Et	MoO_2Cl_2	5 mol %	80 °C, 1–24 h	70–96	[40]

* The procedure is also suitable starting from ketones.

Table 2. Selected procedures to synthesize α-hydroxyphosphonates (**1**) from aldehydes or ketones and trialkyl phosphites (method "B").

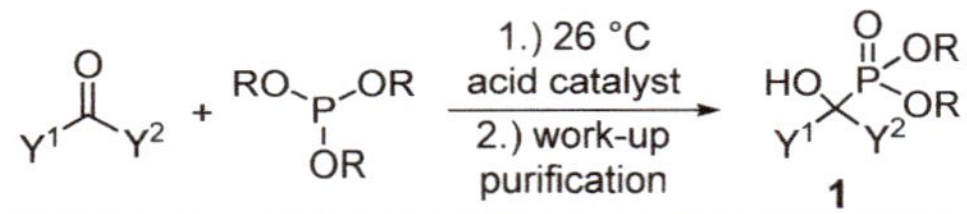

Entry	Y^1	Y^2	R	Catalyst	Amount of Catalyst	Conditions	Yield (%)	Ref.
1 *	C_6H_5, 4-MeC_6H_4, 4-$^iPrC_6H_4$, 4-$MeOC_6H_4$, 2-ClC_6H_4, 4-ClC_6H_4, 4-BrC_6H_4, 4-FC_6H_4, 4-$NO_2C_6H_4$, 4-$CF_3C_6H_4$, 4-$Me_2NC_6H_4$, $CH_2C_6H_5$, $(CH_2)_2C_6H_5$	H, Me, Et, Ph	Me, Et	–	–	Sonication, 25 °C, 10–35 min	84–94	[41]
2	Pr, ^iBu, C_6H_5, 4-MeC_6H_4, 4-$MeOC_6H_4$, 3-OHC_6H_4, 2-ClC_6H_4, 4-ClC_6H_4, C_6H_5CH=CH, 3-chromonyl, 6-Cl-7-Me-3-chromonyl, 2-Cl-3-quinolinyl	H	Et	KH_2PO_4	5 mol %	Sonication, 25 °C, 5–45 min	48–92	[42]
3	C_6H_5, 4-MeC_6H_4, 4-$MeOC_6H_4$, 4-ClC_6H_4, 3-pyridyl, 3-chromonyl, 6,8-diMe-3-chromonyl, 2-Cl-3-quinolinyl, 6-Cl-3-chromonyl, 6,7-diCl-3-chromonyl, 6,8-diCl-3-chromonyl, 6-Br-3-chromonyl, 6-Cl-7-Me-3-chromonyl, 2-Cl-6-Me-3-quinolinyl, 2-Cl-7-Me-3-quinolinyl, 2-Cl-8-Me-3-quinolinyl, 2-Cl-6-MeO-3-quinolinyl, 2-Cl-6-EtO-3-quinolinyl, 2-Cl-8-Et-3-quinolinyl	H	Me, Et	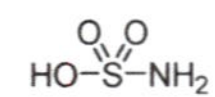	25 mol %	Sonication, 25 °C, 1–60 min	78–98	[43]
4	C_6H_5, 4-MeC_6H_4, 4-$MeOC_6H_4$, 4-OHC_6H_4, 4-ClC_6H_4, 4-$NO_2C_6H_4$, 3,4-$OCH_2OC_6H_3$, C_6H_5CH=CH, 2-furyl, 2-thienyl, 2-Cl-3-quinolinyl, 4-tetrazolo[1,5-*a*]quinolinyl	H	Et	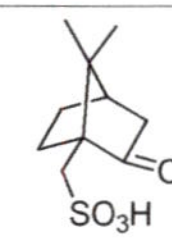	10 mol %	Sonication, 25 °C, 8–20 min	85–93	[44]
5	Pr, ^iPr, Pent, ^cHex, C_6H_5, 4-MeC_6H_4, 4-$MeOC_6H_4$, 4-OHC_6H_4, 2-ClC_6H_4, 4-ClC_6H_4, 4-CNC_6H_4, C_6H_5CH=CH, 2-furyl	H	Me	$(COOH)_2$	10 mol %	80 °C, 3 h	83–98	[45]
6	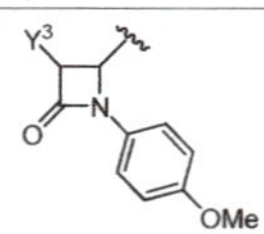 Y^3 = Me, Bu, Ph, OPh	H	Me, Et	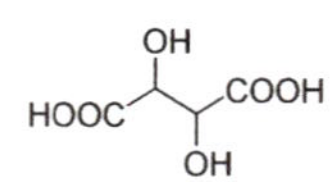	10 mol %	Δ, 30 min, acetonitrile	41–69	[46]
7 *	Pr, C_6H_5, 4-MeC_6H_4, 4-$MeOC_6H_4$, 2-ClC_6H_4, 4-ClC_6H_4, 3-$NO_2C_6H_4$, 4-$NO_2C_6H_4$, 4-CNC_6H_4, $(CH_2)_2C_6H_5$, C_6H_5CH=CH, 2-furyl	H, Me	Me	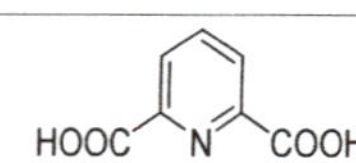	10 mol %	50 °C, 1.3–3 h, H_2O	60–95	[47]

Table 2. *Cont.*

Entry	Y^1	Y^2	R	Catalyst	Amount of Catalyst	Conditions	Yield (%)	Ref.
8	Pr, ^{i}Pr, Bu, ^{t}Bu, Hex, ^{c}Hex, C_6H_5, 4-ClC_6H_4, 2-furyl, 2-pyridyl	H	Me	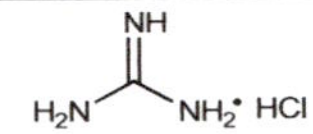	0.5 mol %	50 °C, 2 h, H_2O	60–95	[48]
9	Me, Et, ^{i}Pr, C_6H_5, 4-MeC_6H_4, 4-$MeOC_6H_4$, 2-ClC_6H_4, 4-ClC_6H_4, 2,4-diClC_6H_3, 2-$NO_2C_6H_4$, 3-$NO_2C_6H_4$, 4-$NO_2C_6H_4$, $C_6H_5CH{=}CH$, 2-furyl, 2-thienyl	H	Et	I_2	10 mol %	80 °C, 15–120 min, H_2O	83–97	[49]
10	C_6H_5, 4-MeC_6H_4, 4-$MeOC_6H_4$, 2-OHC_6H_4, 4-ClC_6H_4, 4-BrC_6H_4, 4-FC_6H_4, 3-$NO_2C_6H_4$, 4-$NO_2C_6H_4$, 2-pyridyl, 2-naphthyl, 2-furyl, 2-thienyl	H	Et	β-cyclodextrin	1 equiv.	60–70 °C, 8–12 h, H_2O	80–93	[50]
11	C_6H_5, 4-MeC_6H_4, 4-$MeOC_6H_4$, 3-OHC_6H_4, 2-ClC_6H_4, 4-ClC_6H_4, 4-$NO_2C_6H_4$, $C_6H_5CH{=}CH$, 3-chromonyl, 6-Cl-3-chromonyl, 6,7-diCl-3-chromonyl, 6,8-diCl-3-chromonyl, 6-Cl-7-Me-3-chromonyl, 2-Cl-3-quinolinyl, 2-Cl-6-Me-3-quinolinyl, 2-Cl-7-MeO-3-quinolinyl, 2-Cl-8-Et-3-quinolinyl, 2-Cl-6-EtO-3-quinolinyl	H	Et	NH_4VO_3	10 mol %	25 °C, 5–40 min	80–94	[51]
12	4-ClC_6H_4, 2,4-diClC_6H_3, 4-BrC_6H_4, 4-MeOC(O)C_6H_4, 4-$CF_3C_6H_4$, 2-NO_2-3,6-di$MeOC_6H_2$, 1-naphthyl	H	Et	$ZnBr_2$	10 mol %	25 °C, 10–30 min	68–91	[52]
13	C_6H_5, 4-MeC_6H_4, 4-$MeOC_6H_4$, 4-HOC_6H_4, 2-ClC_6H_4, 4-ClC_6H_4, 4-$NO_2C_6H_4$, 4-$Me_2NC_6H_4$, $C_6H_5CH{=}CH$, 2-Cl-3-quinolinyl, 2-Cl-6-Me-3-quinolinyl	H	Et	$Bi(NO_3)_3{\cdot}5H_2O$	10 mol %	MW, 70 °C, 10–15 min	88–95	[53]
14 *	Et, Pr, ^{i}Pr, ^{t}Bu, ^{c}Hex, $CH_3CH{=}CH$, C_6H_5, 4-$MeOC_6H_4$, 4-ClC_6H_4, $(CH_2)_2C_6H_5$, $C_6H_5CH{=}CH$	H, Me	Me, Et	$NbCl_5$, TMSCl	0.05 mol %	25 °C, 20–90 min	44–96	[54]

* The procedure is also suitable starting from ketones.

Among a series of acids (e.g., acetic acid, trifluoroacetic acid, *p*-toluenesulfonic acid, lactic acid, fumaric acid and tartaric acid) screened as potential catalysts of the reaction, fumaric acid and tartaric acid were found to be the most efficient. It was concluded that weak acids (with a pKa value around 4.5) are the best catalysts of the reaction [46]. According to another procedure, pyridine-2,6-dicarboxylic acid was applied as the catalyst (Table 2/Entry 7) [47]. The reaction of oxo compounds including ketones with trimethyl phosphite was carried out in water at 50 °C to furnish hydroxyphosphonates. In another approach, the pyridine-based catalyst was replaced by guanidine hydrochloride (Table 2/Entry 8) [48]. The drawback of this protocol is the rather complicated work-up comprising quenching, extraction, and then chromatography. An interesting finding is that iodine may also be a catalyst in the reaction of aldehydes and triethyl phosphite (Table 2/Entry 9) [49]. The catalytic effect of I_2 was explained similarly as that of the acid catalysts: iodine is able to interact with the oxygen atom of the carbonyl group, making the C=O bond more electrophilic. β-Cyclodextrin was found to be a promising and recoverable catalyst of the condensation (Table 2/Entry 10) [50]. However, it was necessary to use this catalyst in a one equivalent quantity, and completion of the reaction required longer times (8–12 h) as compared with other protocols. Ammonium metavanadate was tested as the catalyst in the reaction of benzaldehydes with triethyl phosphite, as well as with diethyl phosphite [51]. In the first case, the reaction was complete after 12–25 min (Table 2/Entry 11). However, in the second case, the desired products were only formed in traces. Metal salts were also applied in the reaction of oxo compounds and trialkyl phosphites with success (Table 2/Entries 12–14) [52–54]. The reaction of slightly electron-rich benzaldehydes and triethyl phosphite in the presence of $ZnBr_2$ afforded the corresponding α-hydroxyphosphonates (Table 2/Entry 12) [52]. Bismuth(III) nitrate pentahydrate was applied efficiently as the catalyst in the reaction of aldehydes and triethyl phosphite under MW irradiation at 70 °C (Table 2/Entry 13) [53], while the niobium(V)chloride-trimethylsilyl chloride system was used in an amount of only 0.05 mol % at 25 °C to prepare hydroxyphosphonates (Table 2/Entry 14) [54].

1.3. The "Greenest" Protocol for the Synthesis of α-Hydroxyphosphonates

As it was presented in Sections 1.1 and 1.2, a great number of procedures have been elaborated for the synthesis of α-hydroxyphosphonates **1** from oxo compounds by reaction with dialkyl or trialkyl phosphites. Although, several methods involved the use of inexpensive and easily available catalysts together with mild and solvent-free reaction conditions, the work-up procedures still suffered from drawbacks. In most cases, the crude product was extracted with organic solvents, and then purified by column chromatography or recrystallization. The combination of the purification steps mentioned is also a frequent option [22,32,33,35,38,48].

We aimed at the elaboration of a new, environmentally-friendly, procedure for the synthesis of α-hydroxyphosphonates **1**. As opposed to the procedures already published, we wished to reduce the amount of volatile organic solvents used not only during the reaction, but also in the course of the work-up procedure. According to our method, an equimolar mixture of substituted benzaldehyde and dialkyl phosphite was stirred at reflux in the presence of 10 mol % of triethylamine as the catalyst, in a minimal amount of acetone (1.0 mL/11.0 mmol of the reagents). Then, *n*-pentane was added to the mixture, which was then cooled to 5 °C. The desired product crystallized from the reaction mixture, and the α-hydroxyphosphonates **1** could be obtained easily by a simple filtration in good to excellent yields (78–99%) in a pure form. The main novelty of our protocol is the absence of further purification steps, a consequence of the "one-pot" synthesis and precipitation from the reaction mixture (Scheme 2) [55].

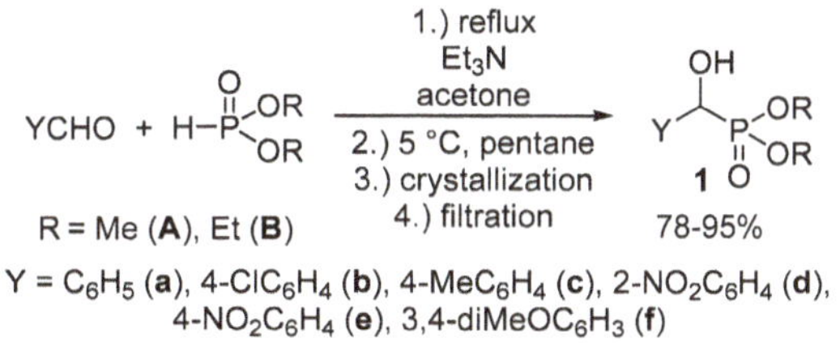

Scheme 2. A green, solvent-economical synthetic method towards α-hydroxyphosphonates (**1**).

To reveal the role of triethylamine in the addition, the reaction of benzaldehyde with diethyl phosphite was investigated by quantum chemical calculations [56]. It was found that in the absence of a catalyst, the enthalpy of activation is around 85.9 kJ/mol, and the addition is thermoneutral. The calculations revealed that triethylamine is able to promote the proton transfer from the >P(O)H reagent to the oxo compound. The amine-catalyzed reaction goes with a decreased enthalpy of activation of 68.8 kJ/mol, and follows an exothermic route ($\Delta H = -17.2$ kJ/mol). The suggested mechanism involving transition state **4** and intermediate **5** is shown in Scheme 3.

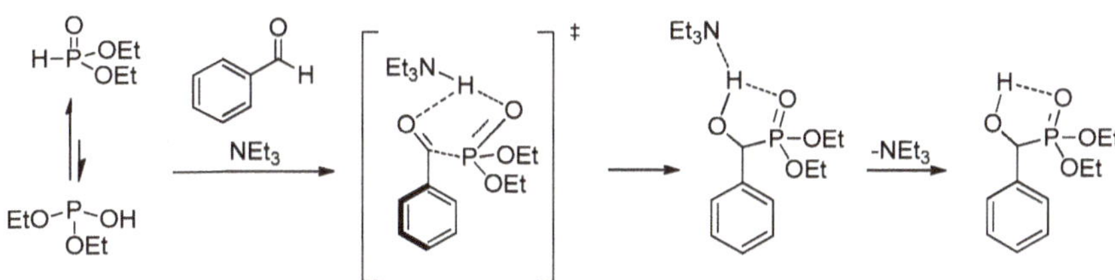

Scheme 3. Calculated mechanism of the triethylamine-catalyzed Pudovik reaction.

In conclusion, we were successful in the elaboration of another green protocol for the synthesis of α-hydroxyphosphonates **1** from substituted benzaldehydes and a series of dialkyl phosphites in the presence of triethylamine as the catalyst [55]. The new method involves the preparation in boiling acetone followed by crystallization of the product in high purity. There was no need for additional purification. The role of triethylamine in the reaction was justified by quantum chemical calculations [56].

2. Reactions of α-Hydroxyphosphonates

α-Hydroxyphosphonates **1** deserve interest not only as potential enzyme inhibitors [57], herbicides [58], antibiotics [59] and antibacterial or antifungal [60] agents, but also as starting materials for other valuable derivatives [61]. Section 2.1 summarizes the alkylation reactions of α-hydroxyphosphonates (Scheme 4, Route "A"). Acylation of the hydroxy function is one of the most extensively studied transformations within this family of compounds (Scheme 4, Route "B", Section 2.2). A series of α-acyloxyphosphonates (**8**) was synthesized applying different acylating agents, like carboxylic and sulfonyl chlorides, as well as carboxylic anhydrides and acids. The oxidation of α-hydroxyphosphonates leads to α-ketophosphonates **9** (Scheme 4, Route "C", Section 2.3). Another thoroughly studied field is nucleophilic substitution at the α carbon atom of α-hydroxyphosphonates to afford α-halo-, and α-amino or α-alkylphosphonates **10** (Scheme 4, Route "D", Section 2.4). The rearrangement of hydroxyphosphonates to benzyl phosphates **11** is an interesting field (Scheme 4, Route "E", Section 2.5). The curiosity of this transformation is the high substrate-dependence. Last but not least, α-hydroxyphosphonic acids **12** may be obtained by the hydrolysis of the ester function of α-hydroxyphosphonates (Scheme 4, Route "F", Section 2.6).

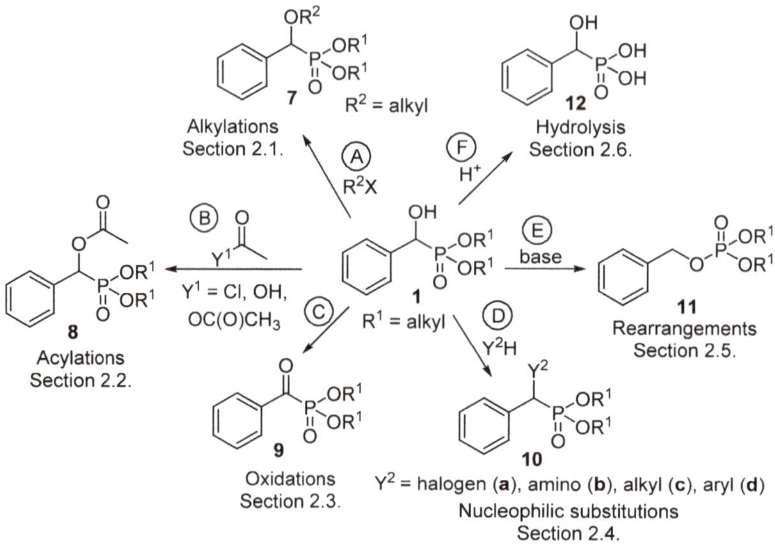

Scheme 4. Family tree of compounds **7–12** derived from α-hydroxyphosphonates (**1**).

The derivatives **7–12** synthesized from α-hydroxyphosphonates were summarized in the family tree shown in Scheme 4. Classes **7–12** represent potentially bioactive compounds.

2.1. Alkylations

Alkylation of the hydroxy function of α-hydroxyphosphonates **1** does not belong to their extensively studied reactions, as only a few articles reported the synthesis of α-alkoxyphosphonates **7**. A related article targeted antimalarial drugs, and the synthesis involved the *O*-alkylation of hydroxyphosphonates **1B** with benzyl bromoacetate in the presence of silver oxide (Scheme 5) [62].

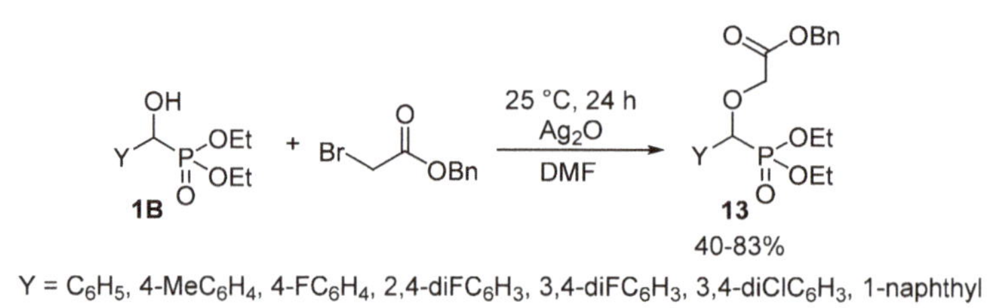

Scheme 5. *O*-alkylation of hydroxyphosphonates **1B** with benzyl bromoacetate.

The protection of the α-hydroxy function may also occur in the literature as another utilization of *O*-alkylation. A tetrahydropyranyl (THP) protecting group was introduced to an α-hydroxyphosphonate **1B** by treatment with 3,4-dihydro-2*H*-pyran applying *p*-toluenesulfonic acid as the catalyst [63]. The silylation of α-hydroxyphosphonates **1** is also a useful tool to protect the hydroxy function [64–67]. The reaction of hydroxyphosphonates **1B** with hexamethyldisilazane (HMDS) was studied using a great variety of catalysts including iodine (Scheme 6, Route "a") [65], copper triflate (Scheme 6, Route "b") [66] or iron(III) trifluoroacetate (Scheme 6, Route "c") [67].

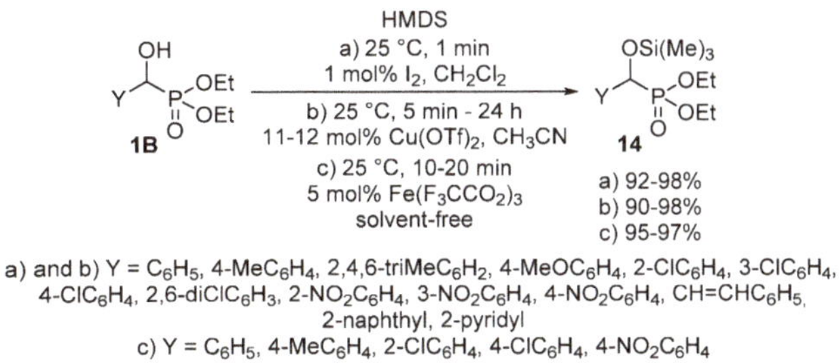

Scheme 6. Possible ways of introducing a silyl protecting group to α-hydroxyphosphonates **1B**.

A unique protocol has been reported for the cyclization of α-hydroxyphosphonates bearing an alkynyl function in the *ortho*-position of the aromatic ring (**15**) [68]. Applying Pd(II) acetate as the catalyst, the nucleophilic attack of the α-*O* atom occurred on the more distant *C*-atom of the triple bond of the alkynyl substituent to form the six-membered product **16** selectively. Surprisingly, changing the $Pd(OAc)_2$ catalyst to DBU used in 2 equivalents led to a five-membered cyclic product **17**. In the latter case, the attack of the hydroxy group occurred on the nearer *C*-atom of the triple bond, followed by isomerization (Scheme 7).

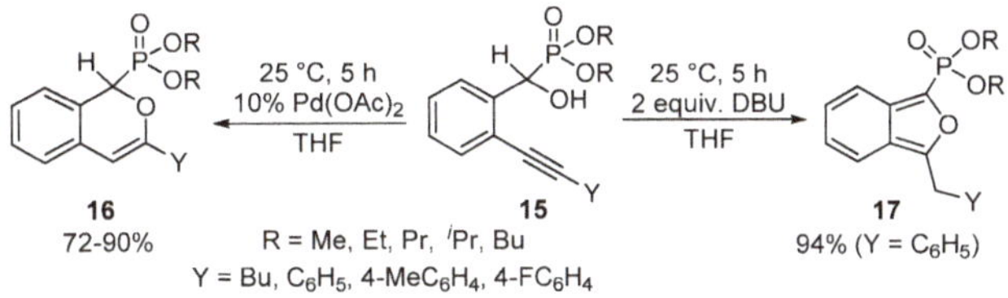

Scheme 7. Selective cyclization of *ortho*-alkynyl α-hydroxyphosphonates **15**.

2.2. *Acylations*

The synthesis of α-acyloxyphosphonates **8** is one of the most thoroughly studied transformations of α-hydroxyphosphonates **1**. α-Acyloxyphosphonates **8** attracted attention as potential herbicides [69–71], fungicides [69], and insecticides [72]. A number of articles reported the acylation of α-hydroxyphosphonates **1** with carboxylic acid chlorides in the presence of triethylamine [73–75] or pyridine [76,77] as the hydrochloric acid scavenger. Among the –C(O)Cl reagents, phenoxyacetyl chloride derivatives **18** are frequently applied acylating agents [70,71,76–84].

In general, the reaction of α-hydroxyphosphonates **1** with phenoxyacetyl chlorides **18** required a reaction time of 3–7 h, and the corresponding α-acyloxyphosphonates **19** were isolated in yields of 53–89% (Scheme 8) [78]. The compounds **19** thus-obtained are popular target molecules, as a series of related derivatives exhibited herbicidal activity [70,76–80,83,84].

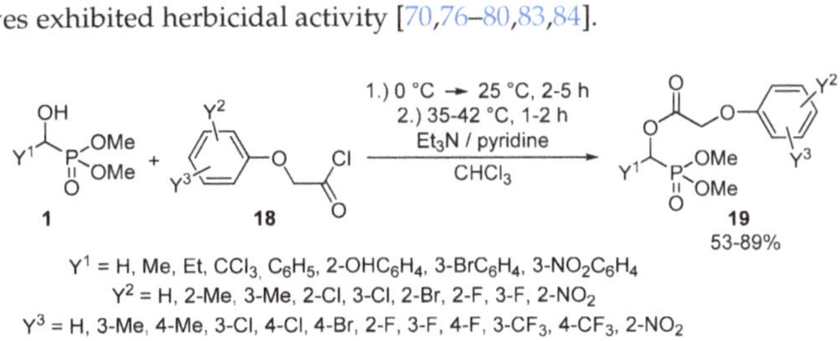

Scheme 8. Synthesis of α-acyloxyphosphonates (**19**) from α-hydroxyphosphonates (**1**).

The reaction of α-hydroxy-phosphonates **1A** with 2,6-pyridinedicarboxylic acid chloride (**20**) led to bis[dimethyl phosphonomethyl pyridine-2,6-dicarboxylates] **21** containing two α-acyloxyphosphonate moieties (Scheme 9). The reaction took place under mild conditions, and the products **21** were isolated with variable yields (38–80%). The derivatives so-obtained **21** also showed herbicidal activity [85].

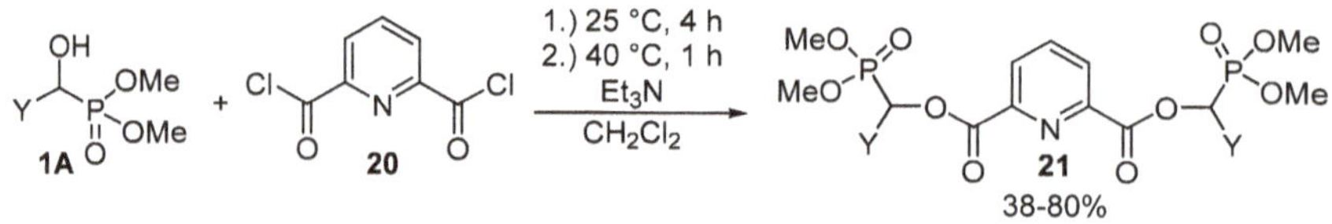

Y = H, Me, Et, Pr, ^{i}Pr, CCl_3, C_6H_5, 4-MeC_6H_4, 4-$MeOC_6H_4$, 2-ClC_6H_4, 4-ClC_6H_4, 2,4-diClC_6H_3, 3-$NO_2C_6H_4$, 3,4-$OCH_2OC_6H_3$, 2-furyl, 2-thienyl

Scheme 9. Acylation of α-hydroxyphosphonates (**1A**) with 2,6-pyridinedicarboxylic acid chloride (**20**).

The acylation of α-hydroxyphosphonates **1B** has been also performed with acetic anhydride [86–88]. This reaction was carried out in the presence of a catalytic amount (1.3–8 mol %) of $Cu(OTf)_2$ at 25 °C to afford the desired α-acetyloxyphosphonates **22** almost quantitatively (Scheme 10, Route "a") [86]. In another instance, 1 mol % of $TiCl_3(OTf)$ was employed as the catalyst (Scheme 10, Route "b"). In this case, completion of the reaction required a somewhat higher temperature (70 °C) [87]. A more benign synthetic route to α-acetyloxyphosphonates **22** is the reaction of α-hydroxyphosphonates (**1B**) and acetic anhydride under MW irradiation (Scheme 10, Route "c") [88]. The acylation was complete after a short reaction time of 5 min, and the products were obtained in excellent yields (90–98%). However, a lack of control of the reaction temperatures is a shortcoming that prevents reproduction.

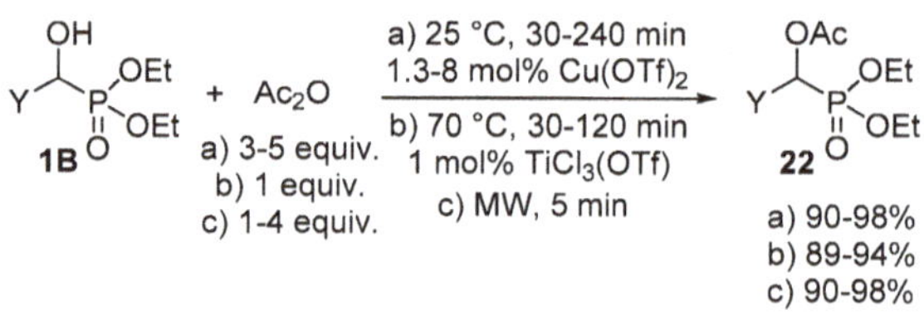

a) and c) Y = C_6H_5, 4-MeC_6H_4, 2,4,6-triMeC_6H_2, 4-$MeOC_6H_4$, 2-ClC_6H_4, 3-ClC_6H_4, 4-ClC_6H_4, 2,6-diClC_6H_3, 2-$NO_2C_6H_4$, 3-$NO_2C_6H_4$, 4-$NO_2C_6H_4$, 2-naphthyl, 2-pyridyl, CH_3CH=CH, C_6H_5CH=CH
b) Y = C_6H_5, 4-MeC_6H_4, 4-ClC_6H_4

Scheme 10. Possible ways for the synthesis of α-acetyloxyphosphonates **22**.

The synthesis of α-acyloxyphosphonates **8** was also carried out starting from substituted benzaldehydes, dimethyl phosphite and acetic anhydride in the presence of magnetic nanoparticle-supported guanidine [89]. In this one-pot procedure, the formation of the α-hydroxyphosphonate **1** in the Pudovik reaction was followed by acylation of the hydroxy function.

Carboxylic acids may also be efficient acylating agents of α-hydroxyphosphonates **1** in the presence of *N*,*N*′-dicyclohexylcarbodiimide (DCC) as the coupling reagent, and 4-dimethyl- aminopyridine (DMAP) as the base [27,90,91]. This is exemplified by the reaction of 2-(4,6-dimethoxypyrimidin- 2-yloxy)benzoic acid (**23**) at 25 °C to give α-acyloxyphosphonates **24** possessing herbicidal activity (Scheme 11) [27].

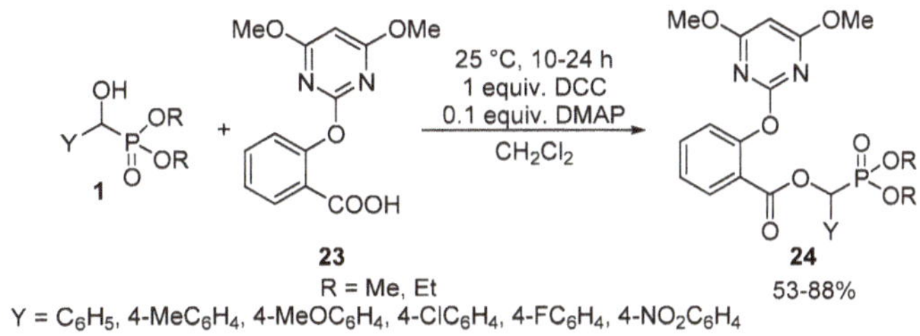

Scheme 11. Coupling reaction of α-hydroxyphosphonates (**1**) with carboxylic acid **23**.

Another method of choice for the acylation of α-hydroxyphosphonates **1B** with carboxylic acids, is the Mitsunobu reaction applying 4,4′-azopyridine (**25**) and triphenylphosphine in acetonitrile as the solvent (Scheme 12) [92]. Completion of the reaction required a reaction time of 9–24 h at 80 °C, and the products **26** were obtained in yields of 65–88%.

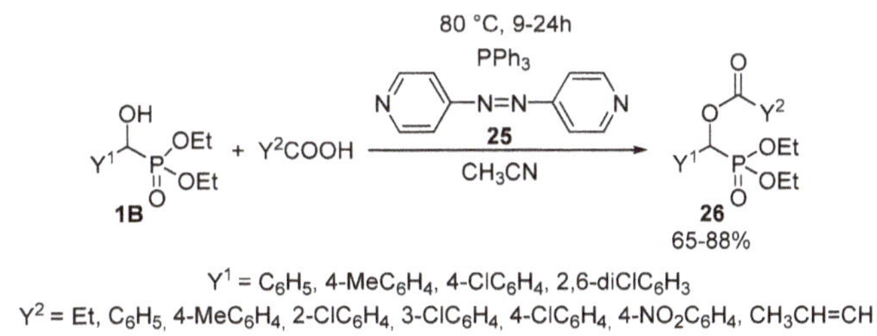

Scheme 12. Synthesis of α-acyloxyphosphonates (**26**) via Mitsunobu reaction.

The reaction of α-hydroxyphosphonates **1** with isocyanates **27** or isothiocyanates **29** resulted in the formation of α-carbamoyloxy-**28** or α-thiocarbamoyloxyphosphonates **30**, respectively [93,94]. One study targeted herbicides by the acylation of α-hydroxyphosphonates (**1**) with pyrimidine-based isocyanates **27** (Scheme 13, Route "A") [93].

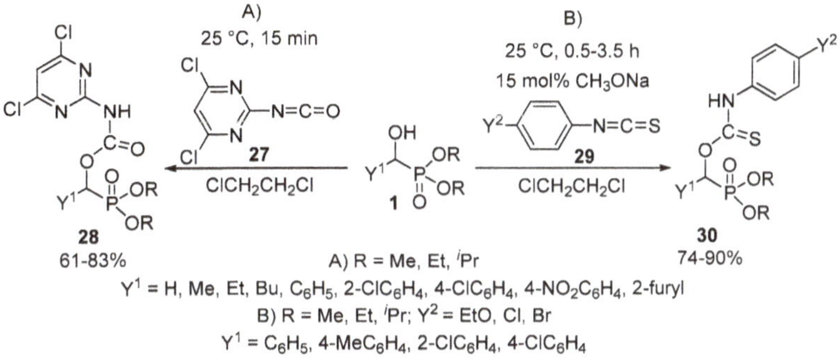

Scheme 13. The reaction of α-hydroxyphosphonates (**1**) with isocyanates **27** and isothiocyanates **29**.

The reaction took place under benign conditions, and the products **28** so-obtained revealed a weak herbicidal activity. On reacting hydroxyphosphonates (**1**) and isothiocyanates **29** in the absence of a catalyst, the corresponding α-thiocarbamoyl-oxyphosphonates **30** were obtained in modest yields of 20–30%. However, the application of 15 mol % of sodium methoxide promoted the acylation, and the products **30** could be isolated in yields of 74–90% (Scheme 13, Route "B") [94]. The authors of the article emphasize the need for mild reaction conditions (25 °C, 0.5–3.5 h), as heating of the reaction mixture resulted in the formation of a by-product through the trimerization of the starting isothiocyanate. The α-thiocarbamoyl-oxyphosphonates **30** obtained in this way expressed plant growth regulating activity.

As shown above, the acylation of α-hydroxyphosphonates (**1B**) was carried out with a wide range of carboxylic acid derivatives. Besides carboxylic chlorides, anhydrides and acids, sulfonic acid derivatives may also function as the acylating agents. By analogy with the reaction of α-hydroxyphosphonates (**1B**) with carboxylic chlorides, the acylation was also attempted with sulfonyl chlorides [95–97]. Methanesulfonyl chloride (Scheme 14, Route "a") [96] and *p*-toluenesulfonyl chloride (Scheme 14, Route "b") [97] worked well as sulfonylating agents in the presence of triethylamine.

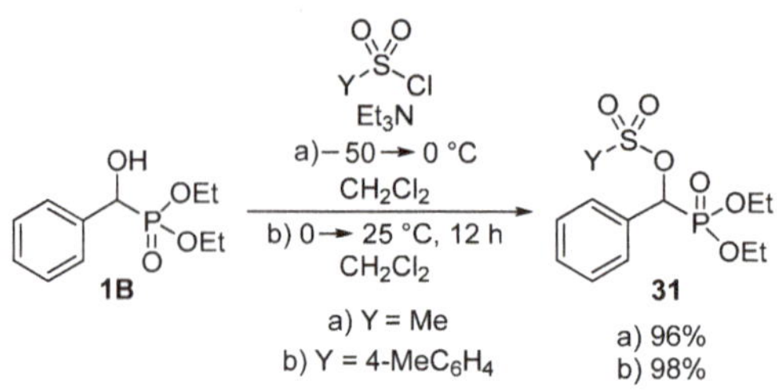

Scheme 14. Sulfonylation of α-hydroxyphosphonates **1B**.

In contrast to the wide range of α-acyloxy and α-sulfonyloxy phosphonates, α-phosphoryloxy analogues appear rarely in the literature. A study from the early 1960's reported the synthesis of a >P(O)OCHP(O)< type compound **33** by a thermo-induced rearrangement (Scheme 15) [98]. However, the phosphorylation of α-hydroxyphosphonates (**1**) with *P*-chlorides presented a challenge.

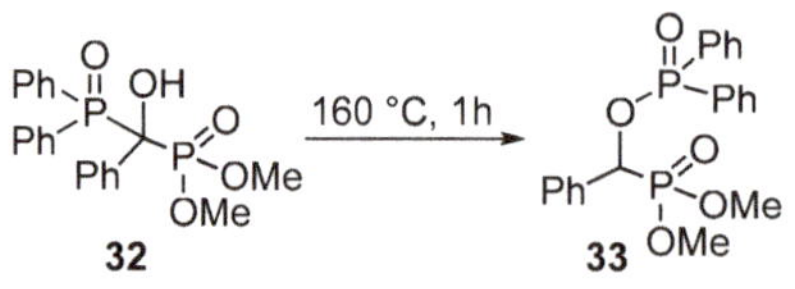

Scheme 15. Synthesis of *O*-phosphoryloxy phosphonate **33** by rearrangement.

Recognizing the lack of *O*-phosphorylated α-hydroxyphosphonates **34** in the literature, we aimed at the synthesis of new classes of derivatives by the acylation of α-hydroxyphosphonates **1** with phosphoryl and phosphinic chlorides. It was observed that by stirring the reaction components at ambient temperature, the *O*-phosphorylated phosphonates **34** were formed with conversions of 55–90% (Scheme 16) [99].

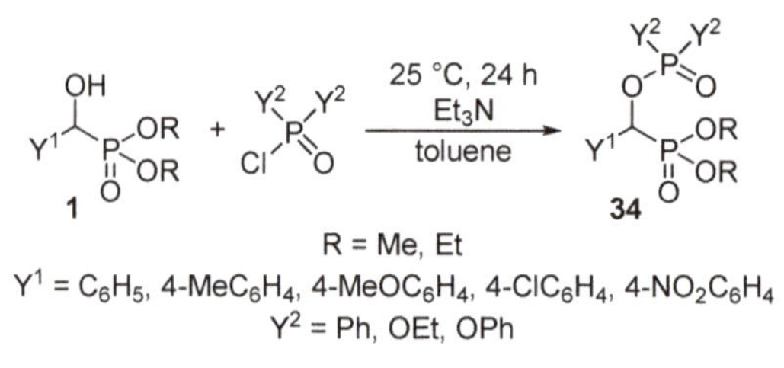

Scheme 16. Phosphorylation of α-hydroxyphosphonates **1** with phosphoryl and phosphinic chlorides.

2.3. Oxidations

The oxidation of α-hydroxyphosphonates **1B** affords the corresponding α-ketophosphonates (**35**) that are versatile precursors of a series of organophosphorus compounds [100–102]. The application of metal compounds with variable valency as the oxidizing agent is predominating in the literature. Alumina-supported chromium(III) oxide (CrO_3/Al_2O_3) (Scheme 17, Route "a") [103], and chromium salts, like zinc dichromate trihydrate ($ZnCr_2O_7{\cdot}3H_2O$) (Scheme 17, Route "b") [104] were found to be efficient reagents in the oxidation of α-hydroxyphosphonates **1B**. In both cases, the reaction took place at ambient temperature under solvent-free conditions. The procedures provided the desired products **35** in good to excellent yields (up to 96%).

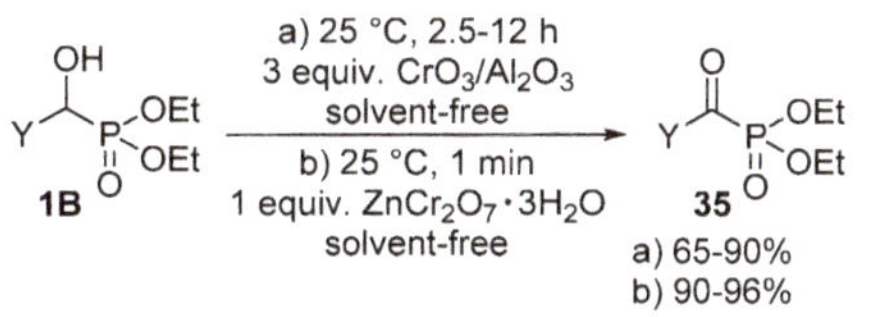

a) Y = Bu, Pent, C_6H_5, 4-MeC_6H_4, 4-MeOC_6H_4, 2-ClC_6H_4, 4-ClC_6H_4, 2,4-diClC_6H_3, 2-$NO_2C_6H_4$, 3-$NO_2C_6H_4$, 4-$NO_2C_6H_4$, 2-furyl, 1-naphthyl, 2-naphthyl, $CH_3CH{=}CH$, $C_6H_5CH{=}CH$, $C_6H_5CHCH_3$, $C_6H_5(CH_2)_2$
b) Y = ^{i}Pr, C_6H_5, 4-MeC_6H_4, 4-MeOC_6H_4, 2,4,6-triMeC_6H_2, 2-ClC_6H_4, 3-ClC_6H_4, 4-ClC_6H_4, 2,6-diClC_6H_3, 2-$NO_2C_6H_4$, 3-$NO_2C_6H_4$, 4-$NO_2C_6H_4$, 2-naphthyl

Scheme 17. Oxidation of α-hydroxyphosphonates **1B** using metal compounds with variable valency.

In other variations, pyridinium dichromate [105] or pyridinium chlorochromate [106] were applied successfully to give α-ketophosphonates **35**. The reaction under discussion was also reported using $KMnO_4$ as the oxidant (Scheme 18). A shortcoming of this latter protocol is the need for dry benzene as the reaction medium [107]. Another problem of the chemical oxidating agents is the low atom efficiency and the lack of "greenness".

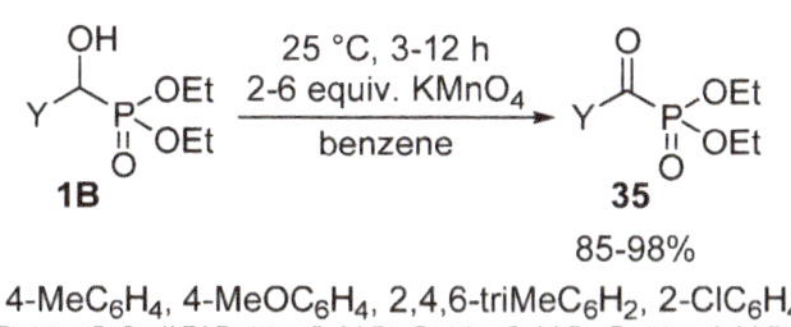

Y = C_6H_5, 4-MeC_6H_4, 4-MeOC_6H_4, 2,4,6-triMeC_6H_2, 2-ClC_6H_4, 3-ClC_6H_4, 4-ClC_6H_4, 2,6-diClC_6H_3, 2-$NO_2C_6H_4$, 3-$NO_2C_6H_4$, 4-$NO_2C_6H_4$, $C_6H_5CHCH_3$, 2-naphthyl, 3-pyridyl

Scheme 18. Synthesis of α-ketophosphonates **35** from α-hydroxyphosphonates **1B** with $KMnO_4$ oxidant.

An elegant achievement is the oxidation of dibenzyl α-hydroxyphosphonates **1C** with oxygen, employing a chiral vanadyl(V) complex (**37**) as the catalyst (Scheme 19) [108]. After a reaction time of 6–150 h, the corresponding α-ketophosphonates **36** were obtained in conversions of 49–51%, while due to the presence of the optically active catalyst, the (*S*) enantiomer of the α-hydroxyphosphonate (*S*)-**1C** remained untouched. This protocol offered an efficient way for the resolution of α-hydroxyphosphonates **1C** with high enantiomeric excesses up to 99% (with two exceptions), in parallel with the formation of α-ketophosphonates **36**.

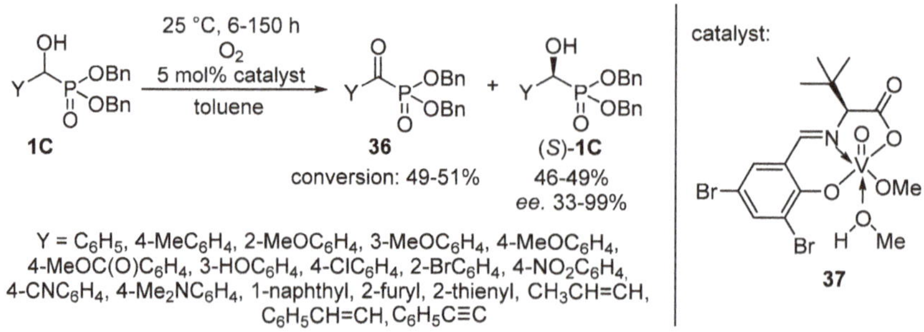

Scheme 19. Tandem resolution and oxidation of α-hydroxyphosphonates **1C**.

2.4. Nucleophilic Substitutions

The nucleophilic substitution of the hydroxy group of α-hydroxyphosphonates **1** represents one of the most widely studied reaction types of this class of compounds. A typical transformation is the change of the hydroxy function to a halogen atom. α-Halobenzylphosphonates **38** deserve interest as potential inhibitors of protein tyrosine phosphatases [109,110]. A chloro or bromo substituent may be introduced into an α-hydroxyphosphonate **1** molecule by treatment with thionyl chloride [111] or thionyl bromide (Scheme 20, Route "a") [110], respectively. Another method of choice to obtain α-halobenzylphosphonates **38** is the joint application of triphenylphosphine and the corresponding carbon tetrahalide [112–114]. One procedure applied this synthetic route to obtain α-chlorobenzylphosphonates (Scheme 20, Route "b") [114], while another paper reported the introduction of a chloro atom to both aliphatic and aromatic hydroxyphosphonates under similar conditions (Scheme 20, Route "c") [113].

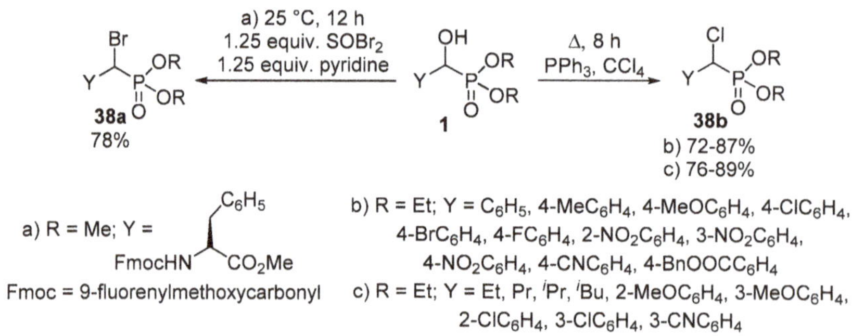

Scheme 20. Synthetic routes to obtain α-halophosphonates **38**.

A further variation involves the use of tetrabutylammonium bromide as the nucleophile in the presence of triphenylphosphine—DDQ (2,3-dichloro-5,6-dicyanobenzoquinone) system, affording α-bromophosphonates [115,116].

In another protocol, hydroxyphosphonates **1** were converted to their corresponding α-halo analogues **38** by reaction with molecular halogens (Br_2 or I_2) applying 4-aminophenyl-diphenylphosphinite **39** as the acid scavenger (Scheme 21) [117]. The amino function of the phosphinite **39** made possible the removal of the forming hydrogen halogenide as an ammonium salt after a simple filtration.

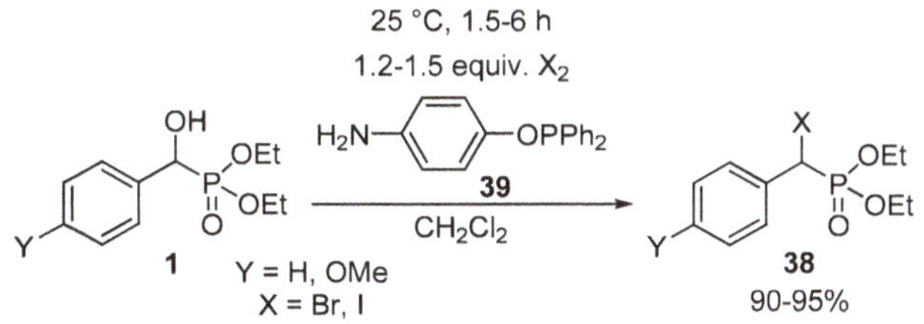

Scheme 21. Synthesis of α-halophosphonates **38** from α-hydroxyphosphonates **1** with molecular halogens.

To synthesize α-fluorophosphonates, the most commonly used fluorinating agent is diethylaminosulfur trifluoride (Et_2NSF_3) [111,114,118–121]. As a variation of this method, a protocol applying morpholinosulfur trifluoride was elaborated for the OH → F change in α-hydroxy-bisphosphonates [122]. *N*-fluorobisbenzenesulfonimide was used efficiently to introduce a fluoro function to the α-carbon atom of α-monohalophosphonates **38** in the presence of sodium hexamethyldisilazane [113,114,116].

Halogens are not the sole nucleophiles that the α-hydroxy function of benzylphosphonates **1B** may be replaced with. One method operated with the above mentioned triphenylphosphine−DDQ system to catalyze the reaction of α-hydroxyphosphonates **1B** with ammonium thiocyanate to obtain α-thiocyanatophosphonates **40** (Scheme 22, Route "A") [115]. Another study reported the Mitsunobu reaction of α-hydroxyphosphonates **1B** with hydrazoic acid as the nucleophile. The treatment of the so-formed α-azidophosphonates **41** with triphenylphosphine and carbon disulfide resulted in the formation of α-isothiocyanatophosphonates **42** (Scheme 22, Route "B") [123].

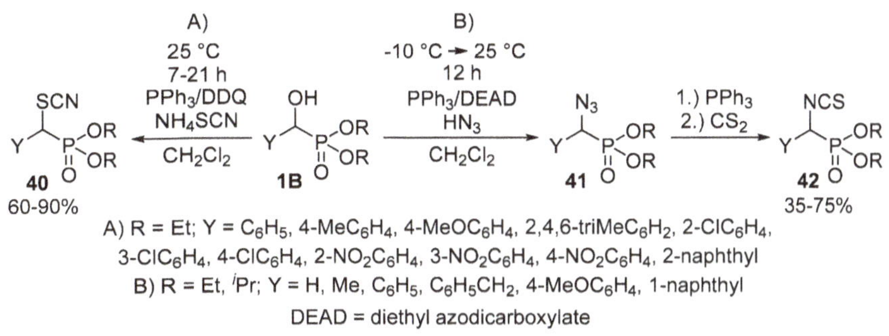

Scheme 22. Nucleophilic substitution of α-hydroxyphosphonates **1B** with NH_4SCN and hydrazoic acid.

α-Aminophosphonates **43** are usually obtained by the three-component condensation (Kabachnik-Fields reaction) of oxo compounds, amines and dialkyl phosphites or secondary phosphine oxides [124–127]. However, these compounds may also be synthesized via the nucleophilic substitution of α-hydroxyphosphonates **1B** with primary or secondary amines. According to a simple protocol, this latter reaction was performed under MW irradiation, in the presence of Al_2O_3 [128]. After 3–7 min, a complete conversion was attained (Scheme 23). However, reaction temperatures are missing from the articles, as the experiments were carried out in a kitchen MW oven.

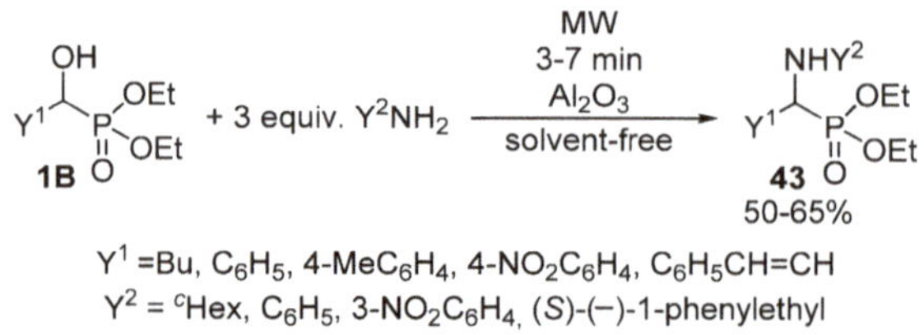

Scheme 23. Synthesis of α-aminophosphonates **43** by nucleophilic substitution of **1B** with amines.

We found that α-hydroxyphosphonates **1B** may be converted to the corresponding α-aminophosphonates **44–46** under MW conditions without the use of any catalyst or solvent (Scheme 24) [56,129]. The starting hydroxyphosphonates **1B** were reacted with three equivalents of a primary amine (propyl-, butyl- or cyclohexyl-amine) applying MW irradiation at 100 °C. Following this method, the substitution took place with surprising ease. Starting from substituted hydroxyphosphonates **1Bb**, **1Bc** and **1Bg**, a short reaction time of 15–30 min was sufficient for complete conversion (Table 3/Entries 4–12). The unsubstituted analogue **1Ba** was even more reactive, and in this case the reaction took place after 10–15 min (Table 3/Entries 1–3).

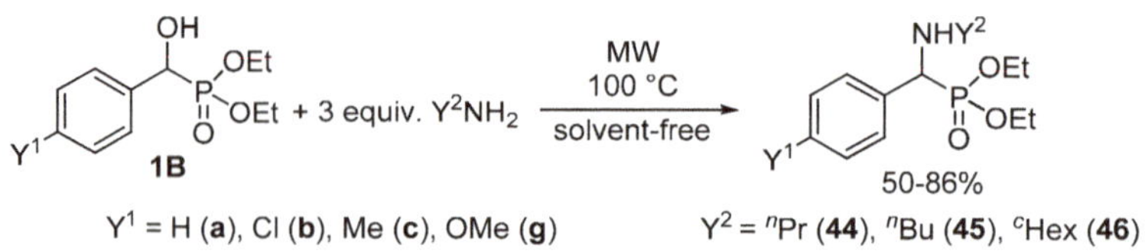

Scheme 24. MW-Assisted synthesis of α-aminophosphonates **44–46** through nucleophilic substitution.

Table 3. Details of the synthesis of α-aminophosphonates **44–46** from α-hydroxyphosphonates **1B**.

Entry	Y^1	Y^2	Reaction Time (min)	Yield (%)	Product	Ref.
1	H	^{n}Pr	10	78	**44a**	[129]
2	H	^{n}Bu	15	86	**45a**	[129]
3	H	^{c}Hex	10	84	**46a**	[129]
4	Cl	^{n}Pr	15	60	**44b**	[50]
5	Cl	^{n}Bu	20	50	**45b**	[56]
6	Cl	^{c}Hex	30	54	**46b**	[56]
7	Me	^{n}Pr	15	58	**44c**	[56]
8	Me	^{n}Bu	15	79	**45c**	[56]
9	Me	^{c}Hex	30	73	**46c**	[56]
10	OMe	^{n}Pr	15	72	**44g**	[56]
11	OMe	^{n}Bu	30	66	**45g**	[56]
12	OMe	^{c}Hex	30	70	**46g**	[56]

Encouraged by the recognition that the reaction of α-hydroxyphosphonates **1B** with primary amines takes place rather easily, we wished to study the mechanism of the substitution by quantum chemical (DFT) calculations [129]. The nucleophilic attack of methylamine on diethyl α-hydroxybenzylphosphonate **1Ba** was chosen as the model reaction. It was found that the reaction follows an S_N2 mechanism. It was also revealed that a favorable neighboring group effect facilitates the transfer of the oxygen atom of the hydroxy group from the α-carbon atom to the phosphorus (transition state **47**) [129]. The proposed mechanism is shown on Scheme 25.

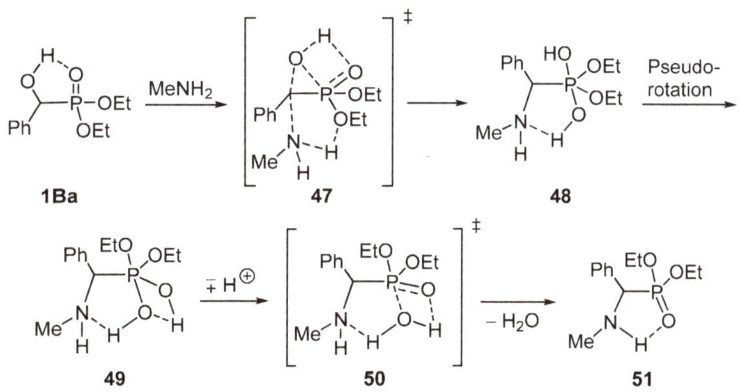

Scheme 25. Calculated mechanism of the substitution of hydroxyphosphonate **1Ba** with methylamine.

α-Sulfonamidophosphonates were obtained through the reaction of α-hydroxyphosphonates and sulfonamides in the presence of HOTf in dioxane. The reaction took place at ambient temperature within 5 h, and the desired products were obtained in yields of 70–94% [130]. The products so-obtained were found to be as promising corrosion inhibitors for mild steel [131].

The Friedel-Crafts alkylation using α-hydroxyphosphonates **1** in the presence of an acid catalyst (e.g., $FeCl_3$ or HOTf) led to α,α-diarylphosphonates **53** [132–134]. Benzene and naphthalene served as the aromatic substrate. According to the proposed mechanism, the first step is the formation of benzylphosphonate carbocation **52** that is followed by its electrophilic attack on the aromatic substrate (Scheme 26) [132].

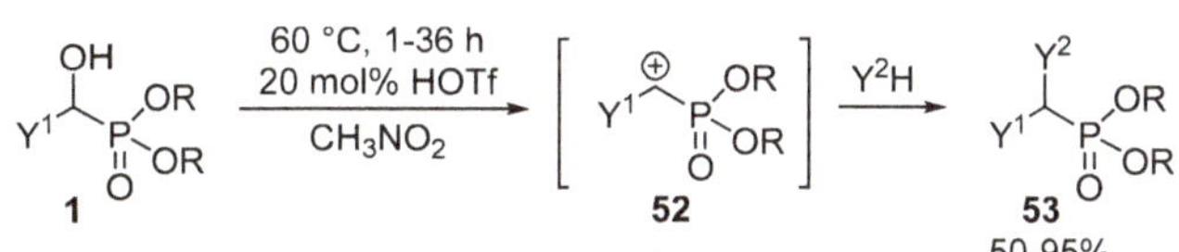

R = Me, Et, ^{i}Pr, Bu

Y^1 = 4-MeC_6H_4, 2,4-diMeC_6H_3, 3,5-diMe-4-OHC_6H_2, 2-$MeOC_6H_4$, 4-$MeOC_6H_4$, 2,4-di$MeOC_6H_3$, 3,4,5-tri$MeOC_6H_2$, 4-$EtOC_6H_4$, 4-$MeSC_6H_4$, 4-$Me_2NC_6H_4$

Y^2 = 2-MeC_6H_4, 4-MeC_6H_4, 4-$^tBuC_6H_4$, 2,4,6-triMeC_6H_2, 3,5-diMe-4-OHC_6H_2, 4-$MeOC_6H_5$, 2,4-di$MeOC_6H_3$, 2,5-di$MeOC_6H_3$, 2,4,6-tri$MeOC_6H_2$, 2-OH-4-$^tBuC_6H_3$, 4-$MeSC_6H_4$, 1-naphthyl, 4-Me-1-naphthyl, 1-Me-3-indolyl, 1-Me-2-pyrrolyl, 2-Me-5-thienyl, 3-benzofuryl

Scheme 26. Friedel-Crafts alkylation of α-hydroxyphosphonates **1** with arene nucleophiles.

An interesting reaction is when α-hydroxyphosphonates **1B** function as an alkylating agent of 1,3-diketones to afford γ-ketophosphonates **54** (Scheme 27) [135]. The reaction of the starting components in organic solvents in the presence of $FeCl_3$ or $Cu(OTf)_2$ Lewis acids resulted in the formation of compound **54** (Scheme 27, Route "A"). The optimization of the alkylations was a real challenge for the authors, as each hydroxyphosphonate—diketone combination required different reaction conditions regarding the Lewis acid, temperature, time and solvent. However, applying $FeCl_3 \cdot 6H_2O$ as the catalyst, a *C–C* bond cleavage of the desired product **54** also occurred due to the presence of the crystal hydrate of the iron salt. In this case, a mixture of compounds **54** and **55** was attained (Scheme 27, Route "B").

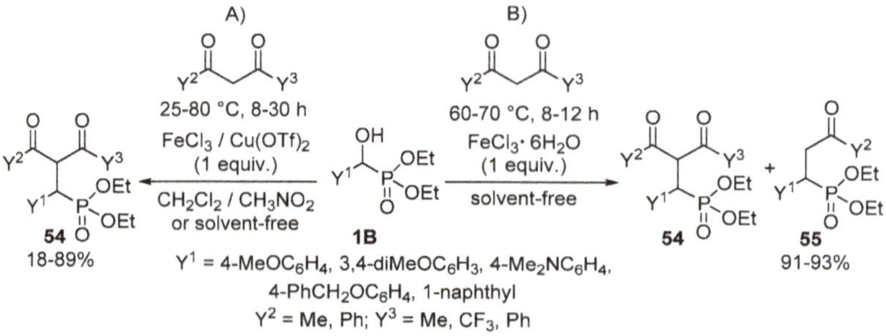

Scheme 27. Alkylation of 1,3-diketones with α-hydroxyphosphonates **1B**.

2.5. Rearrangements

The rearrangement of α-hydroxyphosphonates **1** to benzyl phosphates was first discovered through the example of the widely known insecticide prodrug, trichlorofon (*O,O*-dimethyl (2,2,2-trichloro-1-hydroxyethyl)phosphonate) that is rearranged to 2,2-dichlorovinyl dimethyl phosphate (DDVP) acting as an acetylcholinesterase inhibitor [136]. Phosphonate-phosphate rearrangements were investigated in the presence of strong bases, such as sodium hydroxide [136], sodium ethoxide [137] and sodium hydride [138]. A number of protocols operated with triethylamine as the catalyst [139–141]. However, the Et_3N-catalyzed rearrangement is not a versatile method, as it was reported to apply to only six-membered cyclic phosphonates [139–141].

A new approach of this transformation is when the Pudovik reaction of benzaldehydes and dialkyl phosphites and the subsequent rearrangement of the so-formed α-hydroxyphosphonate **1** take place in "one-pot", under the same conditions. The synthesis of benzyl phosphates **56** directly from an oxo compound and a *P*-reagent was efficiently catalyzed by butyllithium (Scheme 28, Route "a") [142] as well as by DBU (Scheme 28, Route "b") [143]. The tandem Pudovik reaction and rearrangement was also reported in the presence of superbases **57** (Scheme 28, Route "c") [144]. The authors of this latter article emphasized that both the reaction media and the substituent in the aromatic ring had a great impact on the reaction. The rearrangement took place faster in alcohols (EtOH or ^{i}PrOH) than in acetonitrile. It was also shown that the presence of electron-withdrawing substituents facilitated the formation of the benzyl phosphate **56** [144].

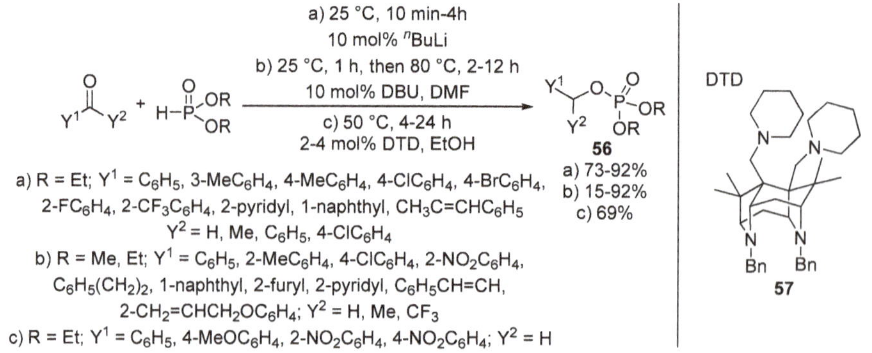

Scheme 28. Methods of the tandem Pudovik reaction and rearrangement affording benzyl phosphates **56**.

In this field, our research targets the accomplishment of the rearrangement of α-hydroxyphosphonates **1** using bases under phase transfer catalytic conditions (Scheme 29) [145].

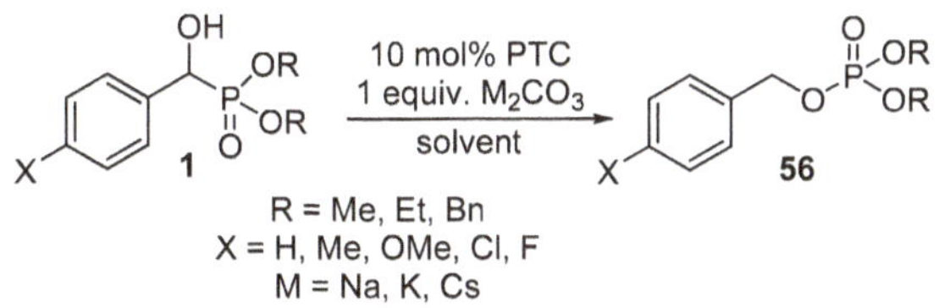

Scheme 29. Rearrangement of α-hydroxyphosphonates **1** under phase-transfer catalytic conditions.

Our preliminary results highlighted that the substituent in the aromatic ring has a significant impact on the reaction. In the presence of electron-withdrawing substituents (Cl and F), the reaction took place at ambient temperature, whereas the rearrangement of α-hydroxyphosphonates bearing electron-releasing substituents required 2–3 h of heating [145].

2.6. Hydrolysis

The hydrolysis of α-hydroxyphosphonates **1** is the most common way for the synthesis of α-hydroxyphosphonic acids **12** that may be inhibitors of a wide variety of enzymes including protein tyrosine phosphatases [110,146,147], mandelate racemase [148], phosphoglycerate kinase [149], plant P5C reductase [150] and plant glutamine synthetase [151]. A series of α-hydroxyphosphonic acid (**12**) and α-hydroxybisphosphonic acid was recognized as antibiotics against Gram-positive bacteria [152]. α-Hydroxyphosphonic acids **12** bearing an amide function in the *para*-position of the aromatic ring were found to be in vivo inhibitors of autotaxin [153].

As for the synthesis of α-hydroxyphosphonic acids **12**, most of the related articles reported the acidic hydrolysis of the corresponding methyl or ethyl ester [149,150,152,154]. A number of protocols applied hydrochloric acid in an excess at reflux, and the hydrolysis usually required a prolonged reaction time of 6–24 h [149,150,152]. According to another method, beside the excess of 6 M hydrochloric acid, dioxane was also added as a co-solvent [154].

In the hope of selective hydrolysis of one of the ester functions, the hydrolytic reaction of dimethyl 1-hydroxy-1-phenylmethylphosphonate **1Aa** was carried out in aqueous sodium hydroxide at reflux for 5 h (Scheme 30, Route "A") [155]. After pH adjustment, the desired product (**57**) was obtained in a yield of 85%. This method was suitable only for the synthesis of the racemic target compound **57**. In order to maintain the C chirality center of the starting α-hydroxyphosphonate (*R*)-**1Aa**, a milder protocol had to be elaborated. On stirring the starting optically active α-hydroxyphosphonate (*R*)-**1Aa** in the presence of sodium iodide in acetone at reflux for 22 h, racemization of the hydroxyphosphonate (*R*)-**1Aa** could be avoided (Scheme 30, Route "B") [155].

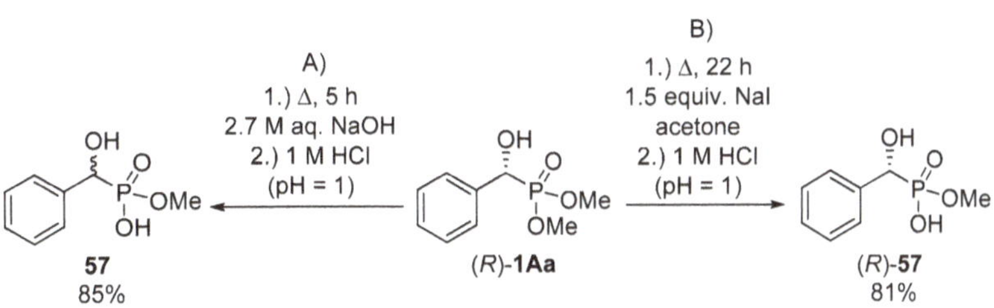

Scheme 30. Selective hydrolysis of one ester function of hydroxyphosphonate (*R*)-**1Aa**.

Another way of synthesizing α-hydroxyphosphonic acids (**12**) from α-hydroxyphosphonates **1** involves the cleavage of the C–O bond of the ester function by trimethylsilyl chloride [77,138] or trimethylsilyl bromide [110]. This reaction is usually carried out at ambient temperature within 2–18 h, generally in acetonitrile as the solvent.

3. Conclusions

This review summarizes the synthesis and reactions of α-hydroxyphosphonates that are of importance due to their bioactivity. The first half of the review presents the most common synthetic routes towards α-hydroxyphosphonates, with special stress on the addition of dialkyl phosphite to an oxo compound, and the condensation of trialkyl phosphites with aldehydes or ketones. The overview is followed by the discussion of the green synthetic protocols for α-hydroxyphosphonates. Then, the reactions of α-hydroxyphosphonates involving *O*-alkylation, *O*-acylation and oxidation of the hydroxy function, substitutions of the hydroxy group by a chloro or amino function, rearrangements to phosphate derivatives and hydrolysis of the ester function are discussed. Beside the literature examples, our own synthetic results are also included.

Acknowledgments: The authors are indebted to David W. Allen for his advice. The above project was supported by the National Research Development and Innovation Fund (K119202). Z. Rádai is grateful for the fellowship provided by the Chinoin-Sanofi Pharmaceuticals and the József Varga Foundation.

Conflicts of Interest: The authors declare no conflict of interest.

References

1. Olszewski, T.K. Environmentally benign syntheses of α-substituted phosphonates: Preparation of α-amino- and α-hydroxyphosphonates in water, in ionic liquids, and under solvent-free conditions. *Synthesis* **2014**, *46*, 403–429. [CrossRef]
2. Failla, S.; Finocchiaro, P.; Consiglio, G.A. Syntheses, characterization, stereochemistry and complexing properties of acyclic and macrocyclic compounds possessing α-amino- or α-hydroxyphosphonate units: A review article. *Heteroatom. Chem.* **2000**, *11*, 493–504. [CrossRef]
3. Rádai, Z.; Kiss, N.Z.; Keglevich, G. *Synthesis of α-Hydroxyphosphonates, an Important Class of Bioactive Compounds*; György, K., Ed.; Organophosphorus Chemistry: Novel Developments; Walter de Gruyter: Berlin, Germany; Boston, MA, USA, 2018; pp. 91–107. 315p.
4. Gröger, H.; Hammer, B. Catalytic concepts for the enantioselective synthesis of α-amino and α-hydroxy phosphonates. *Chem. Eur. J.* **2000**, *6*, 943–948. [CrossRef]
5. Kolodiazhnyi, O.I. Chiral hydroxyl phosphonates: Synthesis, configuration and biological properties. *Russ. Chem. Rev.* **2006**, *75*, 227–253. [CrossRef]
6. Merino, P.; Marqués-López, E.; Herrera, R.P. Catalytic enantioselective hydrophosphonylation of aldehydes and imines. *Adv. Synth. Catal.* **2008**, *350*, 1195–1208. [CrossRef]
7. Spilling, C.D.; Malla, R.K. Synthesis of non-racemic α-hydroxyphosphonates via asymmetric phosphor-aldol reaction. *Top. Curr. Chem.* **2015**, *361*, 83–136. [PubMed]
8. Phillips, A.M.F. Organocatalytic asymmetric synthesis of chiral phosphonates. *Mini Rev. Org. Chem.* **2014**, *11*, 164–185. [CrossRef]
9. Pudovik, A.N.; Konovalova, I.V. Addition reactions of esters of phosphorus(III) acids with unsaturated systems. *Synthesis* **1979**, *2*, 81–96. [CrossRef]
10. Abramov, V.S.; Kiryukhina, L.I.; Kudryavtseva, N. Reaction of dialkyl esters of phosphorus acids with aldehydes and ketones. XXV. Esters of α-hydroxy-β-(2,2,3-trimethyl3-cyclopentenyl)ethylphosphonic and α-hydroxynitrobenzylphosphonic acids. *Zh. Obshch. Khim.* **1964**, *34*, 2235–2238.
11. Maeda, H.; Takahashi, K.; Ohmori, H. Reactions of acyl tributylphosphonium chlorides and dialkyl acylphosphonates with Grignard and organolithium reagents. *Tetrahedron* **1998**, *54*, 12233–12242. [CrossRef]
12. Seven, O.; Polat-Cakir, S.; Hossain, M.S.; Emrullahoglu, M.; Demir, A.S. Reactions of acyl phosphonates with organoaluminum reagents: A new method for the synthesis of secondary and tertiary α-hydroxy phosphonates. *Tetrahedron* **2011**, *67*, 3464–3469. [CrossRef]
13. Gu, L.; Jin, C.; Zhang, H. The catalytic aerobic synthesis of quaternary α-hydroxy phosphonates via direct hydroxylation of phosphonate compounds. *New J. Chem.* **2015**, *39*, 1579–1582. [CrossRef]
14. Li, X.; Jin, C.; Gu, L. C–H hydroxylation of phosphonates with oxygen in [bmIm]OH to produce quaternary α-hydroxy phosphonates. *J. Org. Chem.* **2015**, *80*, 2443–2447. [CrossRef] [PubMed]

15. Pudovik, A.N.; Zametaeva, G.A. New synthesis of esters of phosphonic and thiophosphonic acids. XIII. Addition of diethyl thiophosphite to ketones and aldehydes. *Izv. Akad. Nauk SSSR Ser. Khim.* **1952**, *1952*, 932–939.
16. Zhou, S.; Wu, Z.; Rong, J.; Wang, S.; Yang, G.; Zhu, X.; Zhang, L. Highly efficient hydrophosphonylation of aldehydes and unactivated ketones catalyzed by methylene-linked pyrrolyl rare earth metal amido complexes. *Chem. Eur. J.* **2012**, *18*, 2653–2659. [CrossRef] [PubMed]
17. Zhao, L.; Ding, H.; Zhao, B.; Lu, C.; Yao, Y. Synthesis and characterization of amidate rare-earth metal amides and their catalytic activities toward hydrophosphonylation of aldehydes and unactivated ketones. *Polyhedron* **2014**, *83*, 50–59. [CrossRef]
18. Yang, S.; Zhu, X.; Zhou, S.; Wang, S.; Feng, Z.; Wei, Y.; Miao, H.; Guo, L.; Wang, F.; Zhang, G.; et al. Synthesis, structure, and catalytic activity of novel trinuclear rare-earth metal amido complexes incorporating μ–η^5:η^1 bonding indolyl and μ_3-oxo groups. *Dalton Trans.* **2014**, *43*, 2521–2533. [CrossRef] [PubMed]
19. Miao, H.; Zhou, S.L.; Wang, S.W.; Zhang, L.J.; Wei, Y.; Yang, S.; Wang, F.H.; Chen, Z.; Chen, Y.; Yuan, Q.B. Rare-earth metal amido complexes supported by bridged bis(β-diketiminato) ligand as efficient catalysts for hydrophosphonylation of aldehydes and ketones. *Sci. China Chem.* **2013**, *56*, 329–336. [CrossRef]
20. Liu, B.; Carpentier, J.F.; Sarazin, Y. Highly effective alkaline earth catalysts for the sterically governed hydrophosphonylation of aldehydes and nonactivated ketones. *Chem. Eur. J.* **2012**, *18*, 13259–13264. [CrossRef] [PubMed]
21. Kulkarni, M.A.; Lad, U.P.; Desai, U.V.; Mitragotri, S.D.; Wadgaonkar, P.P. Mechanistic approach for expeditious and solvent-free synthesis of α-hydroxy phosphonates using potassium phosphate as catalyst. *C. R. Chim.* **2013**, *16*, 148–152. [CrossRef]
22. Nandre, K.P.; Nandre, J.P.; Patil, V.S.; Bhosale, S.V. Barium hydroxide catalyzed greener protocol for the highly efficient and rapid synthesis of α-hydroxyphosphonates under solvent free conditions. *Chem. Biol. Interface* **2012**, *2*, 314–321.
23. Pandi, M.; Chanani, P.K.; Govindasamy, S. An efficient synthesis of α-hydroxy phosphonates and 2-nitroalkanols using $Ba(OH)_2$ as catalyst. *Appl. Catal. A* **2012**, *441–442*, 119–123. [CrossRef]
24. Sardarian, A.R.; Kaboudin, B. Surface-mediated solid phase reactions: Preparation of diethyl 1-hydroxyarylmethylphosphonates on the surface of magnesia. *Synth. Commun.* **1997**, *27*, 543–551. [CrossRef]
25. Aouani, I.; Lahbib, K.; Touil, S. Green synthesis and antioxidant activity of novel γ-cyano-α-hydroxyphosphonate derivatives. *Med. Chem.* **2015**, *11*, 206–213. [CrossRef] [PubMed]
26. Hudson, H.R.; Yusuf, R.O.; Matthews, R.W. The preparation of dimethyl α-hydroxyphosphonates and the chemical shift non-equivalence of their diastereotopic methyl ester groups. *Phosphorus Sulfur Silicon Relat. Elem.* **2008**, *183*, 1527–1540. [CrossRef]
27. Jin, C.; He, H. Synthesis and herbicidal activity of novel dialkoxyphosphoryl aryl methyl 2-(4,6-dimethoxypyrimidin-2-yloxy) benzoate derivatives. *Phosphorus Sulfur Silicon Relat. Elem.* **2011**, *186*, 1397–1403. [CrossRef]
28. Wang, C.; Zhou, J.; Lv, X.; Wen, J.; He, H. Solvent-free synthesis of tertiary α-hydroxyphosphates by the triethylamine-catalyzed hydrophosphonylation of ketones. *Phosphorus Sulfur Silicon Relat. Elem.* **2013**, *188*, 1334–1339. [CrossRef]
29. Tajbakhsh, M.; Samad, K.; Zahra, T.; Ahmadreza, B. $MgCl_2/Et_3N$ base system as a new catalyst for the synthesis of α-hydroxyphosphonate. *Chin. J. Chem.* **2012**, *30*, 827–829. [CrossRef]
30. Kumari, S.; Shekhar, A.; Pathak, D.D. A new catalyst and solvent-free green synthesis of α-hydroxy phosphonates and α-aminophosphonates. *Chem. Sci. Trans.* **2014**, *3*, 45–54.
31. Keglevich, G.; Tóth, V.R.; Drahos, L. Microwave-assisted synthesis of α-hydroxy-benzylphosphonates and –benzylphosphine oxides. *Heteroatom Chem.* **2011**, *22*, 15–17. [CrossRef]
32. Kong, D.L.; Liu, R.D.; Li, G.Z.; Zhang, P.W.; Wu, M.S. A rapid, convenient, solventless green approach for the synthesis of α-hydroxyphosphonates by grinding. *Asian J. Chem.* **2014**, *26*, 1246–1248.
33. Kumar, K.S.; Reddy, C.B.; Reddy, M.V.N.; Rani, C.R.; Reddy, C.S. Green chemical synthesis of α-hydroxyphosphonates. *Org. Commun.* **2012**, *5*, 50–57.
34. Kalla, R.M.N.; Zhang, Y.; Kim, I. Highly efficient green synthesis of α-hydroxyphosphonates using a recyclable choline hydroxide catalyst. *New J. Chem.* **2017**, *41*, 5373–5379. [CrossRef]

35. Ramananarivo, H.R.; Solhy, A.; Sebti, J.; Smahi, A.; Zahouily, M.; Clark, J.; Sebti, S. An eco-friendly paradigm for the synthesis of α-hydroxyphosphonates using sodium-modified fluorapatite under solventless conditions. *ACS Sustain. Chem. Eng.* **2013**, *1*, 403–409. [CrossRef]
36. Rao, K.U.M.; Sundar, C.S.; Prasad, S.S.; Rani, C.R.; Reddy, C.S. Neat synthesis and anti-oxidant activity of α-hydroxyphosphonates. *Bull. Korean Chem. Soc.* **2011**, *32*, 3343–3347. [CrossRef]
37. Santhisudha, S.; Sreelakshimi, P.; Jayaprakash, S.H.; Kumar, B.V.; Reddy, C.S. Silica-supported tungstic acid catalyzed synthesis and antioxidant activity of α-hydroxyphosphonates. *Phosphorus Sulfur Silicon Relat. Elem.* **2015**, *190*, 1479–1488. [CrossRef]
38. Liu, C.; Zhang, Y.; Qian, Q.; Yuan, D.; Yao, Y. n-BuLi as a highly efficient precatalyst for hydrophosphonylation of aldehydes and unactivated ketones. *Org. Lett.* **2014**, *16*, 6172–6175. [CrossRef] [PubMed]
39. Zhou, X.; Liu, Y.; Chang, L.; Zhao, J.; Shang, D.; Liu, X.; Lin, L.; Feng, X. Highly efficient synthesis of quaternary α-hydroxy phosphonates via Lewis acid-catalyzed hydrophosphonylation of ketones. *Adv. Synth. Catal.* **2009**, *351*, 2567–2572. [CrossRef]
40. de Noronha, R.G.; Costa, P.J.; Romão, C.C.; Calhorda, M.J.; Fernandes, A.C. MoO_2Cl_2 as a novel catalyst for C–P bond formation and for hydrophosphonylation of aldehydes. *Organometallics* **2009**, *28*, 6206–6212. [CrossRef]
41. Bouzina, A.; Aouf, N.E.; Berredjem, M. Ultrasound assisted green synthesis of α-hydroxyphosphonates under solvent-free conditions. *Res. Chem. Intermed.* **2016**, *42*, 5993–6002. [CrossRef]
42. Mandhane, P.G.; Joshi, R.S.; Nagargoje, D.R.; Gill, C.H. Ultrasound-promoted greener approach to synthesize α-hydroxy phosphonates catalyzed by potassium dihydrogen phosphate under solvent-free condition. *Tetrahedron Lett.* **2010**, *51*, 1490–1492. [CrossRef]
43. Sadaphal, A.S.; Sonar, S.S.; Pokalwar, R.U.; Shitole, N.V.; Shingare, M.S. Sulphamic acid: An efficient catalyst for the synthesis of α-hydroxy phosphonates using ultrasound irradiation. *J. Korean Chem. Soc.* **2009**, *53*, 536–541.
44. Shinde, P.V.; Kategaonkar, A.H.; Shingate, B.B.; Shingare, M.S. An organocatalyzed facile and rapid access to α-hydroxy and α-amino phosphonates under conventional/ultrasound technique. *Tetrahedron Lett.* **2011**, *52*, 2889–2892. [CrossRef]
45. Vahdat, S.M.; Baharfar, R.; Tajbakhsh, M.; Heydari, A.; Baghbanian, S.M.; Khaksar, S. Organocatalytic synthesis of α-hydroxy and α-aminophosphonates. *Tetrahedron Lett.* **2008**, *49*, 6501–6504. [CrossRef]
46. Vangala, V.B.; Pati, H.N. Efficient synthesis of β-lactam containing α-hydroxy phosphonates using tartaric acid and fumaric acid as mild catalysts. *Synth. Commun.* **2016**, *46*, 374–378. [CrossRef]
47. Jahani, F.; Zamenian, B.; Khaksar, S.; Tajbakhsh, M. Pyridine 2,6-dicarboxylic acid as a bifunctional organocatalyst for hydrophosphonylation of aldehydes and ketones in water. *Synthesis* **2010**, *19*, 3315–3318. [CrossRef]
48. Heydari, A.; Arefi, A.; Khaksar, S.; Tajbakhsh, M. Hydrophosphonylation of aldehydes catalyzed by guanidine hydrochloride in water. *Catal. Commun.* **2006**, *7*, 982–984. [CrossRef]
49. Wang, H.S.; Zeng, J.E. Iodine-catalyzed, efficient synthesis of α-hydroxy phosphonates in water. *Phosphorus Sulfur Silicon Relat. Elem.* **2010**, *185*, 1425–1428. [CrossRef]
50. Ramesh, K.; Madhav, B.; Murthy, S.N.; Nageswar, Y.V.D. Aqueous-phase synthesis of α-hydroxyphosphonates catalyzed by β-cyclodextrin. *Synth. Commun.* **2012**, *42*, 258–265. [CrossRef]
51. Sonar, S.S.; Kategaonkar, A.H.; Ware, M.N.; Gill, C.H.; Shingate, B.B.; Shingare, M.S. Ammonium metavanadate: An effective catalyst for synthesis of α-hydroxyphosphonates. *Arkivoc* **2009**, *2*, 138–148.
52. Raju, P.; Rajeshwaran, G.G.; Nandakumar, M.; Mohanakrishnan, A.K. Unusual reactivity of aryl aldehydes with triethyl phosphite and zinc bromide: A facile preparation of epoxides, benzisoxazoles, and α-hydroxy phosphonate esters. *Eur. J. Org. Chem.* **2015**, *2015*, 3513–3523. [CrossRef]
53. Kumar, A.; Jamwal, S.; Khan, S.; Singh, N.; Rai, V.K. $Bi(NO_3)_3{\cdot}5H_2O$ catalyzed phosphonylation of aldehydes: An efficient route to α-hydroxyphosphonates. *Phosphorus Sulfur Silicon Relat. Elem.* **2016**, *192*, 381–385. [CrossRef]
54. Thottempudi, V.; Chung, K.H. Niobium(V) chloride catalyzed Abramov reaction: An efficient protocol for the preparation of α-hydroxy phosphonates. *Bull. Korean Chem. Soc.* **2008**, *29*, 1781–1783. [CrossRef]
55. Keglevich, G.; Rádai, Z.; Kiss, N.Z. To date the greenest method for the preparation of α-hydroxyphosphonates from substituted benzaldehydes and dialkyl phosphites. *Green Process Synth.* **2017**, *6*, 197–201. [CrossRef]

56. Kiss, N.Z.; Rádai, Z.; Mucsi, Z.; Keglevich, G. Synthesis of α-aminophosphonates from α-hydroxyphosphonates; a theoretical study. *Heteroat. Chem.* **2016**, *27*, 260–268. [CrossRef]
57. Lorenz, W.; Henglein, A.; Schrader, G. The new insecticide *O,O*-dimethyl 2,2,2-trichloro-1-hydroxyethylphosphonate. *J. Am. Chem. Soc.* **1955**, *77*, 2554–2556. [CrossRef]
58. Song, H.; Mao, H.; Shi, D. Synthesis and herbicidal activity of α-hydroxy phosphonate derivatives containing pyrimidine moiety. *Chin. J. Chem.* **2010**, *28*, 2020–2024. [CrossRef]
59. Pokalwar, R.U.; Hangarge, R.V.; Maske, P.V.; Shingare, M.S. Synthesis and antibacterial activities of α-hydroxyphosphonates and α-acetyloxyphosphonates derived from 2-chloroquinoline-3-carbaldehyde. *Arkivoc* **2006**, *11*, 196–204.
60. Kategaonkar, A.H.; Pokalwar, R.U.; Sonar, S.S.; Gawali, V.U.; Shingate, B.B.; Shingare, M.S. Synthesis, in vitro antibacterial and antifungal evaluations of new α-hydroxyphosphonate and new α-acetoxyphosphonate derivatives of tetrazolo [1, 5-a] quinolone. *Eur. J. Med. Chem.* **2010**, *45*, 1128–1132. [CrossRef] [PubMed]
61. Sobhani, S.; Tashrifi, Z. Synthesis of α-functionalized phosphonates from α-hydroxyphosphonates. *Tetrahedron* **2010**, *66*, 1429–1439. [CrossRef]
62. Brücher, K.; Illarionov, B.; Held, J.; Tschan, S.; Kunfermann, A.; Pein, M.K.; Bacher, A.; Gräwert, T.; Maes, L.; Mordmüller, B.; et al. α-Substituted β-oxa isosteres of fosmidomycin: Synthesis and biological evaluation. *J. Med. Chem.* **2012**, *55*, 6566–6575. [CrossRef] [PubMed]
63. Guan, Q.; Yang, F.; Guo, D.; Xu, J.; Jiang, M.; Liu, C.; Bao, K.; Wu, Y.; Zhang, W. Synthesis and biological evaluation of novel 3,4-diaryl-1,2,5-selenadiazol analogues of combretastatin A-4. *Eur. J. Med. Chem.* **2014**, *87*, 1–9. [CrossRef] [PubMed]
64. Sekine, M.; Nakajima, M.; Kume, A.; Hashizume, A.; Hata, T. Silyl phosphites. XVIII. Versatile utility of α-(trimethylsilyloxy)-alkylphosphonates as key intermediates for transformation of aldehydes into several carbonyl derivatives. *Bull. Chem. Soc. Jpn.* **1982**, *55*, 224–238. [CrossRef]
65. Firouzabadi, H.; Iranpoor, N.; Sobhani, S. A high yielding preparation of α-trimethylsilyloxyphosphonates by silylation of α-hydroxyphosphonates with HMDS catalyzed by iodine. *Tetrahedron Lett.* **2002**, *43*, 3653–3655. [CrossRef]
66. Firouzabadi, H.; Iranpoor, N.; Sobhani, S.; Ghassamipour, S.; Amoozgar, Z. Copper triflate [$Cu(OTf)_2$] is an efficient and mild catalyst for the silylation of α-hydroxyphosphonates to α-trimethylsilyloxyphosphonates with HMDS at room temperature. *Tetrahedron Lett.* **2003**, *44*, 891–893. [CrossRef]
67. Firouzabadi, H.; Iranpoor, N.; Jafari, A.A.; Jafari, M.R. Iron(III) trifluoroacetate [$Fe(F_3CCO_2)_3$] as an easily available, non-hygroscopic, non-corrosive, highly stable and a reusable Lewis acis catalyst: Efficient *O*-silylation of α-hydroxyphosphonates, alcohols and phenols by hexamethyldisilazane (HMDS) under solvent-free conditions. *J. Organomet. Chem.* **2008**, *693*, 2711–2714.
68. Wang, F.; Miao, Z.; Chen, R. Efficient syntheses of phosphonylated isochromenes by regioselective 6-*endo-dig* addition to carbon-carbon triple bond catalyzed by $Pd(OAc)_2$. *Org. Biomol. Chem.* **2009**, *7*, 2848–2850. [CrossRef] [PubMed]
69. Chen, X.B.; Shi, D.Q. Synthesis and biological activity of novel phosphonate derivatives containing of pyridyl and 1,2,3-triazole rings. *Phosphorus Sulfur Silicon Relat. Elem.* **2008**, *183*, 1134–1144. [CrossRef]
70. Chen, T.; Shen, P.; Li, Y.; He, H. Synthesis and herbicidal activity of *O,O*-dialkyl phenoxyacetoxyalkylphosphonates containing fluorine. *J. Fluorine Chem.* **2006**, *127*, 291–295. [CrossRef]
71. Peng, H.; Wang, T.; Xie, P.; Chen, T.; He, H.W.; Wan, J. Molecular docking and three-dimensional quantitative structure-activity relationship studied on the binding modes of herbicidal 1-(substituted phenoxyacetoxy)alkylphosphonates to the E1 component of pyruvate dehydrogenase. *J. Agric. Food Chem.* **2007**, *55*, 1871–1880. [CrossRef] [PubMed]
72. Wang, W.; Wang, L.P.; Ning, B.K.; Mao, M.Z.; Xue, C.; Wang, H.Y. Synthesis and insecticidal activities of *O,O*-dialkyl-2-[3-bromo-1-(3-chloropyridin-2-yl)-1H-pyrazole-5-carbonyloxy] (aryl) methylphosphonates. *Phosphorus Sulfur Silicon Relat. Elem.* **2016**, *191*, 1362–1367. [CrossRef]
73. Fang, H.; Wang, D.; Chen, W.; Zhao, Y.; Fang, M. Synthesis, crystal structure and fragmentation pathway of arylcarboxy ester of α-hydroxyphosphonate. *Chin. J. Org. Chem.* **2010**, *30*, 1377–1382.
74. Hudson, H.R.; Jászay, Z.M.; Pianka, M. The preparation and properties of some α-acyloxy- and α-carbamoyloxy-phosphonothionates. *Phosphorus Sulfur Silicon Relat. Elem.* **2003**, *178*, 1571–1582. [CrossRef]
75. Shi, D.Q.; Li, X.J.; Wei, J. 5-Fluorouracil derivatives containing α-hydroxy phosphonates. *Phosphorus Sulfur Silicon Relat. Elem.* **2007**, *182*, 405–412. [CrossRef]

76. Peng, H.; Long, Q.; Deng, X.; He, H. Synthesis and herbicidal activities of lithium or potassium hydrogen 1-(substituted phenoxyacetoxy)alkylphosphonates. *Phosphorus Sulfur Silicon Relat. Elem.* **2013**, *188*, 1868–1874. [CrossRef]
77. Long, Q.; Deng, X.; Gao, Y.; Xie, H.; Peng, H.; He, H. Synthesis and herbicidal activities of sodium hydrogen 1-(substituted phenoxyacetoxy)alkylphosphonates. *Phosphorus Sulfur Silicon Relat. Elem.* **2013**, *188*, 819–825. [CrossRef]
78. He, H.W.; Yuan, J.L.; Peng, H.; Chen, T.; Shen, P.; Wan, S.Q.; Li, Y.; Tan, H.L.; He, Y.H.; He, J.B.; et al. Studies of *O,O*-dimethyl α-(2,4-dichlorophenoxyacetoxy)-ethylphosphonate (HW02) as a new herbicide. 1. Synthesis and herbicidal activity of HW02 and analogues as novel inhibitors of pyruvate dehydrogenase complex. *J. Agric. Food Chem.* **2011**, *59*, 4801–4813. [CrossRef] [PubMed]
79. Meng, L.; Joshi, R.; Li, M.; He, H. Synthesis and biological activity of novel 1-(substituted phenoxyacetoxy) alkyl phosphonates and phosphinates. *J. Nepal Chem. Soc.* **2009**, *23*, 11–20.
80. Soung, M.G.; Hwang, T.Y.; Sung, N.D. Synthesis and 3D-QSARs analyses of herbicidal *O,O*-dialkyl-1-phenoxy-1-acetoxy-1-methylphosphonate analogues as a new class of potent inhibitors of pyruvate dehydrogenase. *Bull. Korean Chem. Soc.* **2010**, *31*, 1361–1367. [CrossRef]
81. Wang, T.; He, H.W. An efficient synthesis of α-(2,4-dichloro-phenoxyacetoxy)aryl methyl phosphonate monosodium salts. *Synth. Commun.* **2004**, *34*, 1415–1423. [CrossRef]
82. Wang, T.; He, H.W. Simple and improved preparation of α-oxophosphonate monolithium salts. *Phosphorus Sulfur Silicon Relat. Elem.* **2004**, *179*, 2081–2089. [CrossRef]
83. Wang, W.; He, H.W.; Zuo, N.; Zhang, X.; Lin, J.S.; Chen, W.; Peng, H. Synthesis and herbicidal activity of 2-(substituted phenoxyacetoxy)alkyl-5,5-dimethyl-1,3,2-dioxaphosphinan-2-one containing fluorine. *J. Fluorine Chem.* **2012**, *142*, 24–28. [CrossRef]
84. Wang, W.; He, H.W.; Zuo, N.; He, H.F.; Peng, H.; Tan, X.S. Synthesis and herbicidal activity of 2-(substituted phenoxyacetoxy)alkyl-5,5-dimethyl-1,3,2-dioxaphosphinan-2-one. *J. Agric. Food Chem.* **2012**, *60*, 7581–7587. [CrossRef] [PubMed]
85. Wang, T.; Lei, D.Y.; Huang, Y.; Ao, L.H. Synthesis and biological activity of *O,O*-dimethyl-2,6-pyridinyl diformyloxy alkyl phosphonates. *Phosphorus Sulfur Silicon Relat. Elem.* **2009**, *184*, 2777–2785. [CrossRef]
86. Firouzabadi, H.; Iranpoor, N.; Sobhani, S.; Amoozgar, Z. Copper triflate as a useful catalyst for the high-yielding preparation of α-acetyloxyphosphonates under solvent-free conditions. *Synthesis* **2004**, *2*, 295–297. [CrossRef]
87. Firouzabadi, H.; Iranpoor, N.; Farahi, S. Solid trichlorotitanium(IV) trifluoromethanesulfonate $TiCl_3$(OTf) catalyzed efficient acylation of –OH and –SH: Direct esterification of alcohols with carboxylic acids and transesterification of alcohols with esters under neat conditions. *J. Mol. Catal. A: Chem.* **2008**, *289*, 61–68. [CrossRef]
88. Firouzabadi, H.; Iranpoor, N.; Sobhani, S.; Amoozgar, Z. Facile and high-yielding preparation of α-hydroxyphosphonates assisted by microwave irradiation. *Synthesis* **2004**, *11*, 1771–1774. [CrossRef]
89. Rostami, A.; Atashkar, B.; Moradi, D. Synthesis, characterization and catalytic properties of magnetic nanoparticle supported guanidine in base catalyzed synthesis of α-hydroxyphosphonates and α-acetoxyphosphonates. *Appl. Catal. A* **2013**, *467*, 7–16. [CrossRef]
90. Green, D.; Elgendy, S.; Patel, G.; Skordalakes, E.; Goodwin, C.A.; Scully, M.F.; Kakkar, V.V.; Deadman, J.J. Substrate related O,O-dialkyldipeptidyly ψ carboxybenzylphosphonates, a new type of thrombin inhibitor. *Phosphorus Sulfur Silicon Relat. Elem.* **2000**, *156*, 151–155. [CrossRef]
91. Yang, J.; Ma, J.; Che, W.; Li, M.; Li, G.; Song, B. Microwave-assisted synthesis and antitumor activity of salicyl acyloxy phosphonate derivatives. *Chin. J. Org. Chem.* **2014**, *34*, 2566–2571. [CrossRef]
92. Iranpoor, N.; Firouzabadi, H.; Khalili, D. The first Mitsunobu protocol for efficient synthesis of α-acyloxyphosphonates using 4,4′-azopyridine. *Phosphorus Sulfur Silicon Relat. Elem.* **2011**, *186*, 2166–2171. [CrossRef]
93. Xu, L.; You, G.; Peng, H.; He, H. Synthesis and biological activities of *O,O*-dialkyl 1-((4,6-dichloropyrimidin-2-yl)carbamyloxy) alkylphosphonates. *Phosphorus Sulfur Silicon Relat. Elem.* **2014**, *189*, 812–818. [CrossRef]
94. Li, J.P.; Zhu, J.G.; Liu, R.J.; Cui, F.L.; Liu, P.; Liu, G.S. Straightforward synthesis of a new series of α-(arylamino thiocarbonyloxy) hydrocarbylphosphonates. *S. Afr. J. Chem.* **2008**, *61*, 5–8.
95. Li, Z.G.; Sun, H.K.; Wang, Q.M.; Huang, R.Q. An α-hydrazinoalkylphosphonate as building block for novel *N*-phosphonoalkylheterocycles. *Heteroatom Chem.* **2003**, *14*, 384–386. [CrossRef]

96. Creary, X.; Geiger, C.C.; Hilton, K. Mesylate derivatives of α-hydroxy phosphonates. Formation of carbocations adjacent to the diethyl phosphonate group. *J. Am. Chem. Soc.* **1983**, *105*, 2851–2858. [CrossRef]
97. Kong, D.L.; Li, G.Z.; Liu, R.D. Synthesis and crystal structure of diethyl tosyloxybenzylphosphonate. *Asian J. Chem.* **2014**, *26*, 2138–2140.
98. Davidson, R.S.; Sheldon, R.A.; Trippett, S. The reaction of tetraphenyldiphosphine with aromatic carboxylic acids. *J. Chem. Soc.* **1967**, 1547–1552. [CrossRef]
99. Rádai, Z.; Hodula, V.; Kiss, N.Z.; Kegelevich, G. Unpublished results.
100. Samanta, S.; Zhao, C.G. Organocatalyzed nitroaldol reaction of α-ketophosphonates and nitromethane revisited. *Arkivoc* **2007**, *8*, 218–226.
101. Guang, J.; Zhao, C.G. Organocatalyzed asymmetric Michael reaction of β-aryl-α-ketophosphonates and nitroalkenes. *Tetrahedron Lett.* **2013**, *54*, 5703–5706. [CrossRef] [PubMed]
102. Telan, L.A.; Poon, C.D.; Evans, S.A. Diastereoselectivity in the Mukaiyama–Michael reaction employing α-acyl β,γ-unsaturated phosphonates. *J. Org. Chem.* **1996**, *61*, 7455–7462. [CrossRef] [PubMed]
103. Kaboudin, B. Surface-mediated solid-phase reactions: The preparation of acyl phosphonates by oxidation of 1-hydroxyphosphonates on the solid surface. *Tetrahedron Lett.* **2000**, *41*, 3169–3171. [CrossRef]
104. Firouzabadi, H.; Iranpoor, N.; Sobhani, S.; Sardarian, A.R. High yields preparation of α-ketophosphonates by oxidation of α-hydroxyphosphonates with zinc dichromate trihydrate ($ZnCr_2O_7{\cdot}3H_2O$) under solvent-free conditions. *Tetrahedron Lett.* **2001**, *42*, 4369–4371. [CrossRef]
105. Guliaiko, I.; Nesterov, V.; Sheiko, S.; Kolodiazhnyi, O.I.; Freytag, M.; Jones, P.G.; Schmutzler, R. Synthesis of optically active hydroxyphosphonates. *Heteroatom Chem.* **2008**, *19*, 133–139. [CrossRef]
106. Firouzabadi, H.; Iranpoor, N.; Sobhani, S. Preparation of α-ketophosphonates by oxidation of α-hydroxyphosphonates with pyridinium chlorochromate (PCC). *Phosphorus Sulfur Silicon Relat. Elem.* **2004**, *179*, 1483–1491. [CrossRef]
107. Firouzabadi, H.; Iranpoor, N.; Sobhani, S. Preparation of α-ketophosphonates by oxidation of α-hydroxyphosphonates with neutral alumina supported potassium permanganate (NASPP) under solvent-free conditions and potassium permanganate in dry benzene. *Tetrahedron Lett.* **2002**, *43*, 477–480. [CrossRef]
108. Pawar, V.D.; Bettigeri, S.; Weng, S.S.; Kao, J.Q.; Chen, C.T. Highly enantioselective aerobic oxidation of α-hydroxyphosphonates catalyzed by chiral vanadyl(V) methoxides bearing *N*-salicylidene-α-aminocarboxylates. *J. Am. Chem. Soc.* **2006**, *128*, 6308–6309. [CrossRef] [PubMed]
109. Taylor, W.P.; Zhang, Z.Y.; Widlanski, T.S. Quiescent affinity inactivators of protein tyrosine phosphatases. *Bioorg. Med. Chem.* **1996**, *4*, 1515–1520. [CrossRef]
110. Tulsi, N.S.; Downey, A.M.; Cairo, C.W. A protected L-bromophosphonomethylphenylalanine amino acid derivative (BrPmp) for synthesis of irreversible protein tyrosine phosphatase inhibitors. *Bioorg. Med. Chem.* **2010**, *18*, 8679–8686. [CrossRef] [PubMed]
111. Caplan, N.A.; Pogson, C.I.; Hayes, D.J.; Blackburn, G.M. The synthesis of novel bisphosphonates as inhibitors of phosphoglycerate kinase (3-PGK). *J. Chem. Soc. Perkin Trans. 1* **2000**, 421–437. [CrossRef]
112. Gajda, T. Preparation of diethyl 1-bromoalkylphosphonates. *Phosphorus Sulfur Silicon Relat. Elem.* **1990**, *53*, 327–331. [CrossRef]
113. Guan, Z.; Wu, D.; Fu, J.P.; He, Y.H. A facile and efficient synthesis of diethyl α,α-chlorofluoroalkanephosphonates. *Heteroatom Chem.* **2010**, *21*, 250–255. [CrossRef]
114. Wu, D.; He, Y.; Tang, R.; Guan, Z. The first synthesis of diethyl α,α-chlorofluorobenzylphosphonates. *Synlett* **2009**, *13*, 2180–2182.
115. Firouzabadi, H.; Iranpoor, N.; Sobhani, S. PPh_3/DDQ as a neutral system for the facile preparation of diethyl α-bromo, α-iodo and α-azidophosphonates from diethyl α-hydroxyphosphonates. *Tetrahedron* **2004**, *60*, 203–210. [CrossRef]
116. Fu, J.P.; He, Y.H.; Zhong, J.; Yang, Y.; Deng, X.; Guan, Z. An efficient and general route to the synthesis of diethyl α,α-bromofluorophosphonates. *J. Fluorine Chem.* **2011**, *132*, 636–640. [CrossRef]
117. Iranpoor, N.; Firouzabadi, H.; Gholinejad, M. 4-Aminophenyldiphenylphosphinite (APDPP), a new heterogeneous and acid scavenger phosphinite—Conversion of alcohols, trimethylsilyl, and tetrahydropyranyl ethers to alkyl halides with halogens or *N*-halosuccinimides. *Can. J. Chem.* **2006**, *84*, 1006–1012. [CrossRef]

118. Allmendinger, T.; Fujimoto, R.; Gasparini, F.; Schilling, W.; Satoh, Y. α-Fluoro-benzylphosphonates as reagents for the preparation of 1-fluoro-1-aryl alkenes and α-fluorostilbenes. *Chimia* **2004**, *58*, 133–137. [CrossRef]
119. Blackburn, G.M.; Kent, D.E. A novel synthesis of α- and γ-fluoroalkylphosphonates. *J. Chem. Soc. Chem. Commun.* **1981**, 511–513. [CrossRef]
120. Blackburn, G.M.; Kent, D.E. Synthesis of α- and γ-fluoroalkylphosphonates. *J. Chem. Soc. Perkin Trans.* **1986**, 913–917. [CrossRef]
121. Yokomatsu, T.; Yamagishi, T.; Matsumoto, K.; Shibuya, S. Stereocontrolled synthesis of hydroxymethylene phosphonate analogues of phosphorylated tyrosine and their conversion to monofluoromethylene phosphonate analogues. *Tetrahedron* **1996**, *52*, 11725–11738. [CrossRef]
122. Guzyr, O.I.; Zasukha, S.V.; Vlasenko, Y.G.; Chernega, A.N.; Rozhenko, A.B.; Shermolovich, Y.G. Simple route to adducts of (amino)(aryl)carbene with phosphorus pentafluoride. *Eur. J. Inorg. Chem.* **2013**, *2013*, 4154–4158. [CrossRef]
123. Psurski, M.; Błażewska, K.; Gajda, A.; Gajda, T.; Wietrzyk, J.; Oleksyszyn, J. Synthesis and antiproliferative activity of novel α- and β-dialkoxyphosphoryl isothiocyanates. *Bioorg. Med. Chem. Lett.* **2011**, *21*, 4572–4576. [CrossRef] [PubMed]
124. Keglevich, G.; Bálint, E. The Kabachnik–Fields reaction: Mechanism and synthetic use. *Molecules* **2012**, *17*, 12821–12835. [CrossRef] [PubMed]
125. Bálint, E.; Fazekas, E.; Tripolszky, A.; Tajti, Á.; Kangyal, R.; Milen, M.; Keglevich, G. Synthesis of aminophosphonate derivatives by microwave-assisted Kabachnik–Fields reaction. *Organomet. Chem.* **2012**, *717*, 655–659. [CrossRef]
126. Zefirov, N.S.; Matveeva, E.D. Catalytic Kabachnik–Fields reaction: New horizons for old reaction. *Arkivoc* **2008**, *1*, 1–17.
127. Kafarski, P.; Gorny vel Gorniak, M.; Andrasiak, I. Kabachnik–Fields reaction under green conditions—A critical overview. *Curr. Green Chem.* **2015**, *2*, 218–222. [CrossRef]
128. Kaboudin, B. A convenient synthesis of 1-aminophosphonates from 1-hydroxyphosphonates. *Tetrahedron Lett.* **2003**, *44*, 1051–1053. [CrossRef]
129. Kiss, N.Z.; Kaszás, A.; Drahos, L.; Mucsi, Z.; Keglevich, G. A neighbouring group effect leading to enhanced nucleophilic substitution of amines at the hindered α-carbon atom of an α-hydroxyphosphonate. *Tetrahedron Lett.* **2012**, *53*, 207–209. [CrossRef]
130. Pallikonda, G.; Chakravarty, M. Triflic acid mediated functionalization of α-hydroxyphosphonates: Route for sulfonamide phosphates. *RSC Adv.* **2013**, *3*, 20503–20511. [CrossRef]
131. Verma, C.; Singh, A.; Pallikonda, G.; Chakravarty, M.; Quraishi, M.A.; Bahadur, I.; Ebenso, E.E. Aryl sulfonamidomethylphosphonates as new class of green corrosion inhibitors for mild steel in 1 M HCl: Electrochemical, surface and quantum chemical investigation. *J. Mol. Liq.* **2015**, *209*, 306–319. [CrossRef]
132. Chen, L.; Zou, Y.X.; Fang, X.Y.; Yin, X. Convenient synthesis of α-diarylmethylphosphonates by HOTf catalyzed Friedel–Crafts arylation of α-aryl α-hydroxyphosphonates. *Phosphorus Sulfur Silicon Relat. Elem.* **2018**, *193*, 168–177. [CrossRef]
133. Khalid, M.B.Z.; Pallikonda, G.; Tulichala, R.N.P.; Chakravarty, M. Oxy-Wittig reactions of 1-naphthyl(aryl) methylphosphonates: A new approach to naphthylarylketones. *Tetrahedron* **2016**, *72*, 2094–2101. [CrossRef]
134. Pallikonda, G.; Chakravarty, M. $FeCl_3$-mediated arylation of α-hydroxyphosphonates with unactivated arenes: Pseudo-umpolung in allylic phosphonates. *Eur. J. Org. Chem.* **2013**, *2013*, 944–951. [CrossRef]
135. Pallikonda, G.; Chakravarty, M.; Sahoo, M. An easy access to α-aryl substituted γ-ketophosphonates: Lewis acid mediated reactions of 1,3-diketones with α-hydroxyphosphonates and tandem regioselective C–C bond cleavage. *Org. Biomol. Chem.* **2014**, *12*, 7140–7149. [CrossRef] [PubMed]
136. Hall, L.A.R.; Stephens, C.W.; Drysdale, J.J. A rearrangement to from diethyl 1-cyanoethyl phosphate. *J. Am. Chem. Soc.* **1957**, *79*, 1768–1769. [CrossRef]
137. Yoshino, K.; Kohno, T.; Morita, T.; Tsukamoto, G. Organic phosphorus compounds. 2. Synthesis and coronary vasodilator activity of (benzothiazolylbenzyl)phosphonate derivatives. *J. Med. Chem.* **1989**, *32*, 1528–1532. [CrossRef] [PubMed]
138. McGeary, R.P.; Vella, P.; Mak, J.Y.W.; Guddat, L.W.; Schenk, G. Inhibition of purple acid phosphatase with α-alkoxynaphthylmethylphosphonic acids. *Bioorg. Med. Chem. Lett.* **2009**, *19*, 163–166. [CrossRef] [PubMed]

139. Kumaraswamy, S.; Selvi, R.S.; Swamy, K.C.K. Synthesis of new α-hydroxy-, α-halogeno- and vinylphosphonates derived from 5,5-dimethyl-1,3,2-dioxaphosphinan-2-one. *Synthesis* **1997**, *1997*, 207–212. [CrossRef]
140. Jankowski, S.; Marczak, J.; Olczak, A.; Główka, M.L. Stereochemistry of 1-hydroxyphosphonate–phosphate rearrangement. Retention of configuration at the phosphorus atom. *Tetrahedron Lett.* **2006**, *47*, 3341–3344. [CrossRef]
141. Pallitsch, K.; Roller, A.; Hammerschmidt, F. The stereochemical course of the α-hydroxyphosphonate–phosphate rearrangement. *Chem. Eur. J.* **2015**, *21*, 10200–10206. [CrossRef] [PubMed]
142. Pallikonda, G.; Santosh, R.; Ghosal, S.; Chakravarty, M. BuLi-triggered phospha-Brook rearrangement: Efficient synthesis of organophosphates from ketones and aldehydes. *Tetrahedron Lett.* **2015**, *56*, 3796–3798. [CrossRef]
143. El Kaïm, L.; Gaultier, L.; Grimaud, L.; Santos, A.D. Formation of new phosphates from aldehydes by a DBU-catalysed phospha-brook rearrangement in a polar solvent. *Synlett* **2005**, *15*, 2335–2336. [CrossRef]
144. Galeta, J.; Potáček, M. Applications of caged-designed proton sponges in base-catalyzed transformations. *J. Mol. Catal. A: Chem.* **2014**, *395*, 87–92. [CrossRef]
145. Rádai, Z.; Szabó, R.; Kiss, N.Z.; Keglevich, G. Unpublished results.
146. Baguley, T.D.; Xu, H.C.; Chatterjee, M.; Nairn, A.C.; Lombroso, P.J.; Ellman, J.A. Substrate-based fragment identification for the development of selective, nonpeptidic inhibitors of striatal-enriched protein tyrosine phosphatase. *J. Med. Chem.* **2013**, *56*, 7636–7650. [CrossRef] [PubMed]
147. Beers, S.A.; Malloy, E.A.; Wu, W.; Wachter, M.P.; Gunnia, U.; Cavender, D.; Harris, C.; Davis, J.; Brosius, R.; Pellegrino-Gensey, J.L.; et al. Nitroarylhydroxymethylphosphonic acids as inhibitors of CD45. *Bioorg. Med. Chem.* **1997**, *5*, 2203–2211. [CrossRef]
148. Burley, R.K.M.; Bearne, S.L. Inhibition of mandelate racemase by the substrate–intermediate–product analogue 1,1-diphenyl-1-hydroxymethylphosphonate. *Bioorg. Med. Chem.* **2005**, *15*, 4342–4344. [CrossRef] [PubMed]
149. Caplan, N.A.; Pogson, C.I.; Hayes, D.J.; Blackburn, G.M. Novel bisphosphonate inhibitors of phosphoglycerate kinase. *Bioorg. Med. Chem. Lett.* **1998**, *8*, 515–520. [CrossRef]
150. Forlani, G.; Occhipinti, A.; Berlicki, Ł.; Dziedzioła, G.; Wieczorek, A.; Kafarski, P. Tailoring the structure of aminobisphosphonates to target plant P5C reductase. *J. Agric. Food Chem.* **2008**, *56*, 3193–3199. [CrossRef] [PubMed]
151. Occhipinti, A.; Berlicki, Ł.; Giberti, S.; Dziedzioła, G.; Kafarski, P.; Forlani, G. Effectiveness and mode of action of phosphonate inhibitors of plant glutamine synthetase. *Pest. Manag. Sci.* **2010**, *66*, 51–58. [CrossRef] [PubMed]
152. Desai, J.; Wang, Y.; Wang, K.; Malwal, S.R.; Oldfield, E. Isoprenoid biosynthesis inhibitors targeting bacterial cell growth. *Chem. Med. Chem.* **2016**, *11*, 2205–2215. [CrossRef] [PubMed]
153. Jiang, G.; Madan, D.; Prestwich, G.D. Aromatic phosphonates inhibit the lysophospholipase D activity of autotaxin. *Bioorg. Med. Chem.* **2011**, *21*, 5098–5101. [CrossRef] [PubMed]
154. Nesterov, V.V.; Kolodiazhnyi, O.I. Efficient method for the asymmetric reduction of α- and β-ketophosphonates. *Tetrahedron* **2007**, *63*, 6720–6731. [CrossRef]
155. Colton, I.J.; Yin, D.T.; Grochulski, P.; Kazlauskas, R.J. Molecular basis of chiral acid recognition by *candida rugosa* lipase: X-ray structure of transition state analog and modeling of the hydrolysis of methyl 2-methoxy-2-phenylacetate. *Adv. Synth. Catal.* **2011**, *353*, 2529–2544. [CrossRef]

molecules

MDPI

Article

Continuous Flow Alcoholysis of Dialkyl *H*-Phosphonates with Aliphatic Alcohols

Erika Bálint *, Ádám Tajti, Nóra Tóth and György Keglevich *

Department of Organic Chemistry and Technology, Budapest University of Technology and Economics, 1521 Budapest, Hungary; tajti.adam@mail.bme.hu (Á.T.); toth.nora@mail.bme.hu (N.T.)

* Correspondence: ebalint@mail.bme.hu (E.B.); gkeglevich@mail.bme.hu (G.K.); Tel.: +36-1-463-3653 (E.B.)

Received: 14 June 2018; Accepted: 1 July 2018; Published: 3 July 2018

Abstract: The continuous flow alcoholysis of dialkyl *H*-phosphonates by aliphatic alcohols in the absence of a catalyst was elaborated using a microwave (MW) reactor equipped with a flow cell. By the precise control of the reaction conditions, the synthesis could be fine-tuned towards dialkyl *H*-phosphonates with two different and with two identical alkyl groups. In contrast to the "traditional" batch alcoholysis, flow approaches required shorter reaction times, and the products became available at a larger scale.

Keywords: dialkyl *H*-phosphonates; alcoholysis; transesterification; microwave; continuous flow reactor

1. Introduction

Flow chemistry induced an inevitable revolution in the field of chemical transformations [1]. In modern chemistry, reactions have to be strictly controlled, which requires more and more precise synthetic techniques. In contrast to the "traditional" batch reactions, flow approaches create a significantly different processing environment, which enables a more efficient control of the reaction conditions (such as the reaction time or temperature). Due to this, flow protocols may be significantly better with respect to purity, selectivity and yields, as compared to batch approaches [2].

In microwave (MW) chemistry, the continuous flow technique represents a special importance. Although applications of the MW technique proved to be useful in many chemical transformations, the scale-up of MW-assisted reactions means a challenge due to the limited geometry of the MW devices [3–6]. One possibility to solve this problem is the use of continuous flow MW reactors, where the reaction mixture flows through an irradiated flow cell. The MW unit can be of a similar size used in batch mode. During the last decade, the advantages of the applications of continuous flow MW reactors were reported in certain transformations; however, in most cases, the usefulness of the occasional devices may be questionable, as non-professional MW reactors do not allow reproductions, and what is more important, the temperatures were not reported [7,8].

Esters represent a fundamental family among organic compounds. Esters of carboxylic acids may be important intermediates, solvents or products in organic chemistry [9].

The most common preparations of simple carboxylic esters comprise the acid-catalyzed reaction of a carboxylic acid with alcohol (Fischer esterification) and the reaction of an ester with alcohol (alcoholysis) [10]. Continuous flow Fischer esterifications may be carried out in systems containing a packed catalyst bed [11–20] or a heated coil [21,22], as well as in microreactors [23]. In a few cases, the esterifications were performed in continuous flow MW systems based on professional MW reactors [24–26] or in household MW ovens [27,28]. The continuous flow alcoholysis of carboxylic esters may also be performed in reactors equipped with a catalyst bed [29] or a heated coil [21,30]. The alcoholysis is also of great importance in biodiesel production, the area of which was recently summarized by Lee and co-workers [31].

Organophosphorus esters are also of great importance [32]. Dialkyl esters of phosphorous acid (dialkyl *H*-phosphonates) are widely applied building blocks in syntheses [33]. They are important starting materials of the Kabachnik–Fields condensations and the aza-Pudovik reactions [34] (towards α-aminophosphonates), the Pudovik reactions [35] (resulting in the formation of α-hydroxyphosphonates) and further organophosphorus transformations, such as the Hirao reaction [36] and the phospha-Michael addition [37]. Dialkyl *H*-phosphonates bearing different alkoxy groups on the phosphorus atom are valuable intermediates for P-chiral organophosphorus derivatives [38,39].

The industrial synthesis of dialkyl *H*-phosphonates is comprised of the reaction of phosphorus trichloride with alcohols (Scheme 1) [40–45]. Although this transformation is convenient, efficient and may be scaled up, it requires a solvent, and the liberating HCl decreases the atom efficiency. It is also a disadvantage that the synthesis of derivatives with different alkyl groups is not possible.

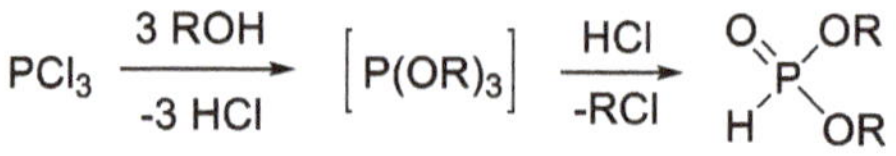

Scheme 1. Synthesis of dialkyl *H*-phosphonates from phosphorus trichloride.

Another possibility for the preparation of dialkyl *H*-phosphonates is alcoholysis (Scheme 2), which provides the products in good yields [46–54]. The only by-product is the leaving alcohol. The relatively high temperature and the need for the high excess of the alcohol mean disadvantages.

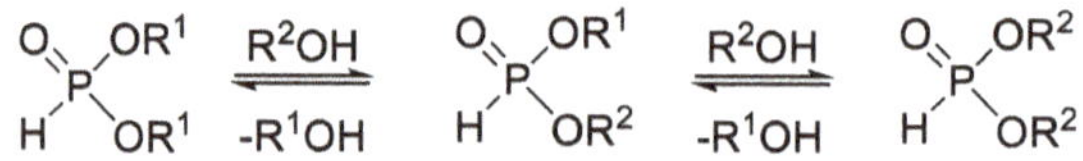

Scheme 2. Synthesis of dialkyl *H*-phosphonates by alcoholysis.

Besides the two main routes mentioned above, there are a few less important methods for the synthesis of dialkyl *H*-phosphonates. Such protocols are the oxidative reaction of elemental phosphorus with alcohols (Scheme 3a) [55–60] and the Ni-catalyzed oxidation of hypophosphorous esters (Scheme 3b) [61].

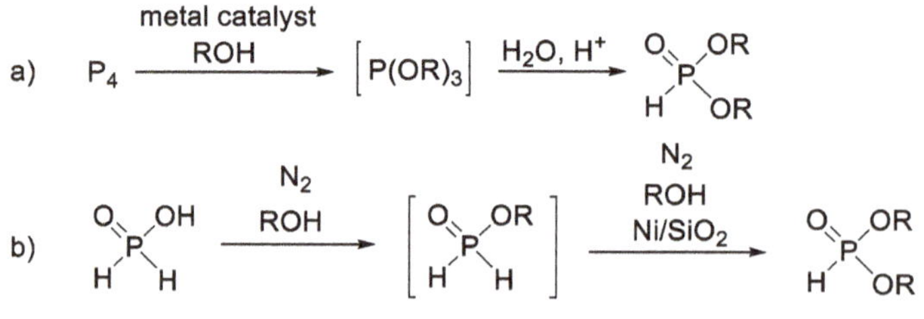

Scheme 3. Miscellaneous synthesis of dialkyl *H*-phosphonates.

The synthesis of dialkyl-*H*-phosphonates bearing different alkyl groups is not easy and requires special methods (Table 1).

Table 1. Synthetic methods for the preparation of dialkyl *H*-phosphonates bearing two different alkyl groups.

Entry	Reaction	Average Yield (%)	Flow Compatible	Ref.
1	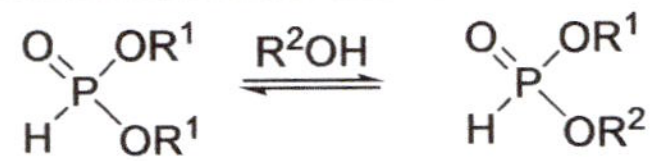	30–50	+	[46,50,52]
2	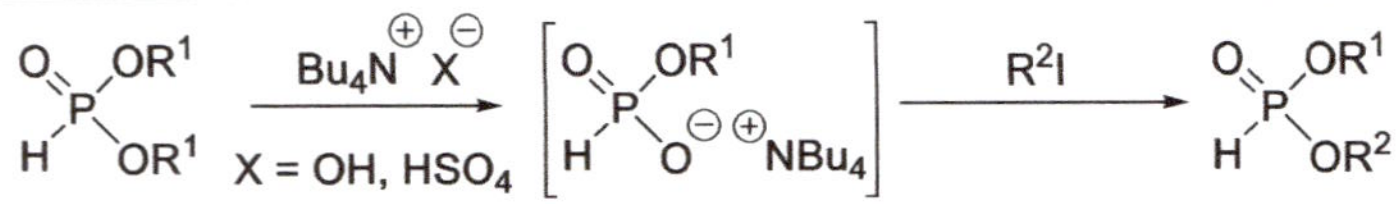	50–70	–	[62,63]
3	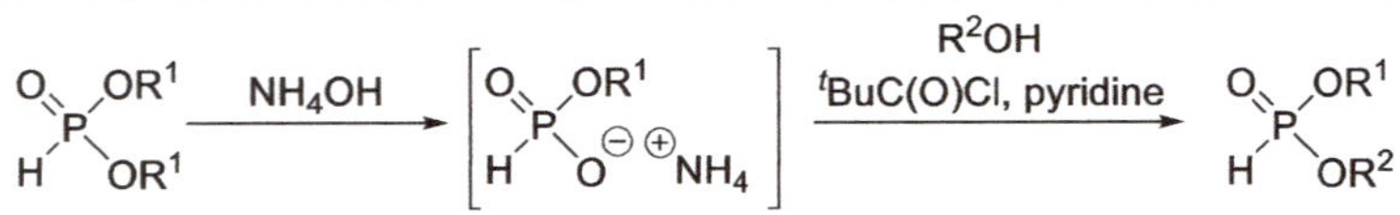	50–85	–	[64]
4	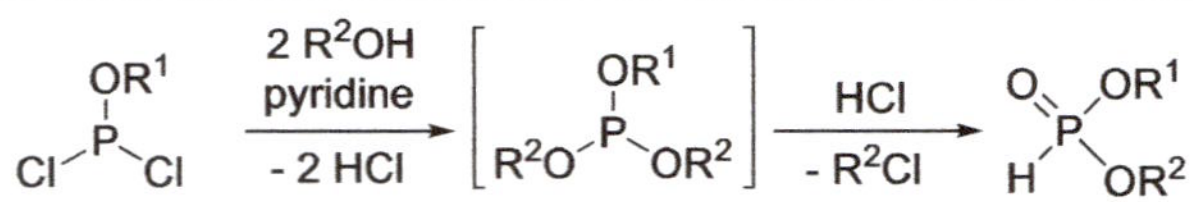	30–60	–	[65]

Phosphites with different alkyl groups may be synthesized by the partial alcoholysis of dialkyl *H*-phosphonates (Table 1, Entry 1). The reaction conditions must be controlled strictly to avoid complete transesterification.

Dialkyl *H*-phosphonates with two different alkyl groups may be obtained by the controlled O–C cleavage of dialkyl *H*-phosphonates (Table 2, Entries 2 and 3). In the first step of the first method, a phosphite-ammonium salt is prepared, which is then alkylated with alkyl iodides (Table 2, Entry 2), or with alcohols in the presence of pivaloyl chloride and pyridine (Table 2 entry 3) to furnish the target compounds selectively. A serious drawback of these transformations is the low atom efficiency.

Table 2. Alcoholysis of dialkyl *H*-phosphonates in a batch microwave (MW) reactor [33].

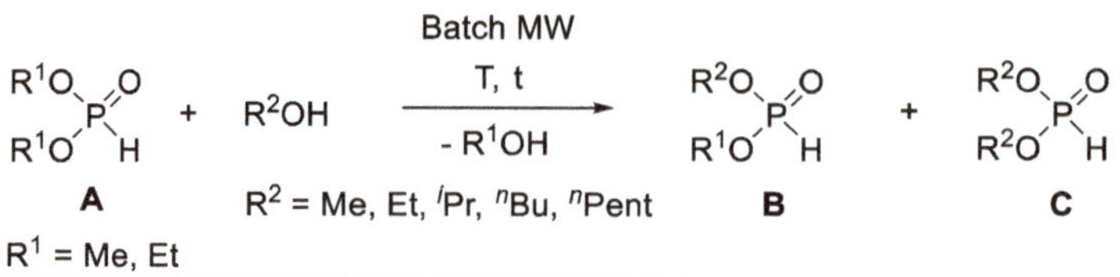

Entry	R^1	R^2	R^2OH (equiv)	T (°C)	t (min)	Composition (%) [a]		
						A	B	C
1	Me	Et	25	100	120	22	**56**	22
2	Me	Et	50	175	40	0	4	**96**
3	Me	^{n}Bu	25	125	60	0	**40**	60
4	Me	^{n}Bu	50	150	60	0	0	**96**
5	Et	Me	25	125	120	49	**38**	13
6	Et	Me	50	175	40	0	21	**79**
7	Et	^{i}Pr	25	125	60	26	**57**	17
8	Et	^{i}Pr	50	175	40	0	6	**94**
9	Et	^{n}Bu	25	125	60	25	**54**	21
10	Et	^{n}Bu	50	175	40	0	2	**98**
11	Et	^{n}Pent	25	125	60	3	**52**	45
12	Et	^{n}Pent	50	175	40	0	8	**92**

[a] Analyzed by GC. Bold numbers indicate the target compounds in the reactions.

Dialkyl *H*-phosphonates bearing different alkoxy groups can also be obtained in the reaction of alkyl dichlorophosphites and alcohols. The primarily formed trialkyl phosphite is cleaved by a part of the hydrochloric acid formed (Table 2, Entry 4).

From these synthetic methods, the alcoholysis may be flow compatible. This solvolytic reaction may be carried out without any catalyst by fine-tuning the reaction conditions, to afford mixed or fully-transesterified dialkyl phosphites as the predominant component.

The alcoholysis of dialkyl *H*-phosphonates in a batch MW reactor was investigated by us previously (Table 2) [52]. The reaction of dialkyl *H*-phosphonates (**A**) with aliphatic alcohols under mild conditions (lower alcohol excess and lower temperature) afforded the derivatives with different alkyl groups (**B**) as the main products, while in the case of applying higher excess of alcohol and higher temperature, the fully-transesterified dialkyl *H*-phosphonates (**C**) predominated in the mixture.

In this paper, we aimed at elaborating an MW-assisted continuous flow method for the alcoholysis of dialkyl *H*-phosphonates. Our purpose was to find the optimum conditions for the formation of dialkyl *H*-phosphonates bearing different alkyl groups and for the fully-transesterified phosphites in a flow MW reactor. Furthermore, we wished to prepare new dialkyl *H*-phosphonates with two different alkyl groups, which may be valuable building blocks of chiral organophosphorus compounds.

2. Results and Discussion

In the first case, the alcoholysis of dimethyl *H*-phosphonate (DMP) (**1**) with *n*-butanol was investigated in the absence of any catalyst, in a CEM® (Matthews, NC, USA) MW reactor equipped with a commercially available CEM® continuous flow cell (Figures 1 and 2). The mixture of DMP (**1**) and a 25-fold excess of *n*-butanol was fed into the reactor by an HPLC pump at a flow rate of 0.1–1.4 mL/min (corresponding to residence times of 60–5 min, respectively). The temperature was monitored and controlled by an IR sensor. The mixture leaving the reactor was cooled down using a spiral-like cooler and was passed through a back pressure regulator operating at 250 psi. Consecutive fractions of the leaving mixture were analyzed by GC measurements in order to determine the composition and to identify the stationary operation.

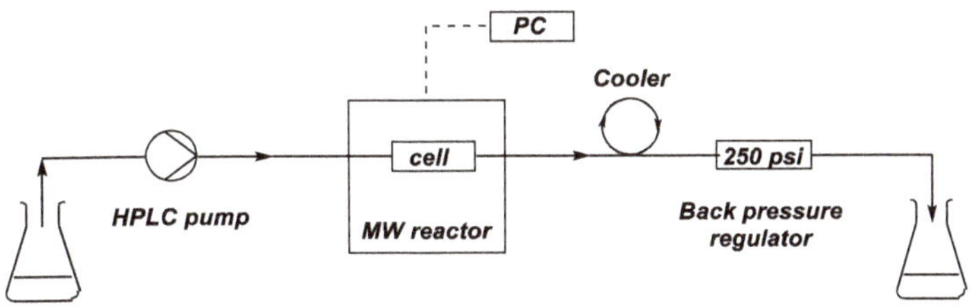

Figure 1. Schematic drawing of the continuous flow system developed.

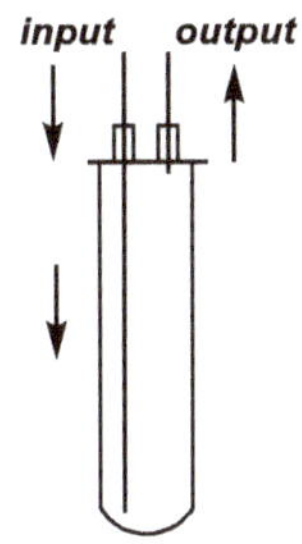

Figure 2. Sketch of the continuous flow cell.

The reaction conditions and the results are summarized in Table 3. To find the optimum conditions for the formation of *n*-butyl methyl *H*-phosphonate (nBMP) (**2a**) and *n*-dibutyl *H*-phosphonate (DnBP) (**3a**), the alcoholysis was carried out applying different temperatures and residence times. Performing the reaction at 100 °C, at a flow rate of 1.4 mL/min (with a residence time of 5 min), the conversion was only 26%, and the proportion of nBMP (**2a**) was 25%, along with 1% of DnBP (**3a**) (Table 3, Entry 1). Using a flow rate of 0.45 mL/min (with a residence time of 15 min), the starting DMP (**1**) was still the main component of the leaving mixture; however, the proportion of nBMP (**2a**) was 44% (Table 3, Entry 2). Increasing the residence time to 30 min (by decreasing the flow rate to 0.25 mL/min), the ratio of mixed *H*-phosphonate (**2a**) was 9% higher (53%) (Table 3, Entry 3). Applying a longer residence time of 45 or 60 min (flow rate of 0.15 or 0.10 mL/min), the proportion of phosphite **2a** was somewhat lower (50% or 47%), and the ratio of DnBP (**3a**) increased to 26% or 35%, respectively (Table 3, Entries 4 and 5). The maximum proportion of nBMP (**2a**) (53%) was obtained applying a reaction time of 30 min at 100 °C (Table 3, Entry 3). Increasing or decreasing the residence time at the same temperature resulted in lower ratio of product **2a** (Table 3, Entries 1, 2 and 4, 5, respectively). In a comparative thermal experiment at 100 °C at a residence time of 30 min, the conversion was only 15% (Table 3, Entry 6). Next, the effect of the reaction temperature was investigated at a residence time of 30 min (Table 3, Entries 6–8) (Figure 3). At 125 °C, the mixture was comprised of 7% of DMP (**1**), 43% of nBMP (**2a**) and

50% of DnBP (**3a**) (Table 3, Entry 7). Increasing the temperature to 150 °C, the conversion was complete, and the fully-transesterified product (**3a**) predominated in a proportion of 78% (Table 3, Entry 8). Performing the alcoholysis at 175 °C, the ratio of DnBP (**3a**) was 91% (Table 3, Entry 9). Carrying out the reaction at a longer residence time (45 min) and/or higher excess of the ^{n}BuOH (50 equiv), the composition did not change. Applying common heating at 175 °C, the reaction was not complete, and the ratio of product **3a** was 8% lower, than under MW conditions (Table 3, Entry 10). It can be observed that the difference between MW-assisted and conventionally-heated experiments was significant at 100 °C, while significantly smaller at 175 °C. After column chromatography, nBMP (**2a**) was obtained in a yield of 48%, while DnBP (**3a**) was isolated in a yield of 88% (Table 3, Entries 3 and 9).

Table 3. Continuous flow alcoholysis of DMP with ^{n}BuOH.

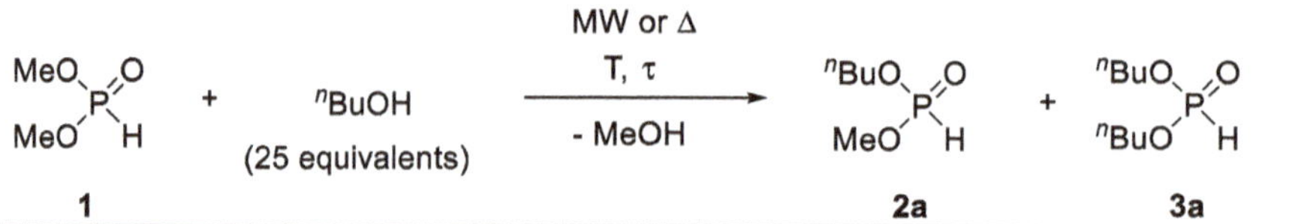

Entry	Mode of Heating	Power (W)	T (°C) [a]	Flow Rate (mL/min) [b]	τ (min)	Conversion (%)	Composition (%) [c]			Yield (%) [d]
							1	2a	3a	
1	MW	22	100	1.4	5	26	74	25	1	-
2	MW	14	100	0.45	15	52	48	44	8	-
3	MW	10	100	0.25	30	65	35	**53**	12	48 (**2a**)
4	MW	8	100	0.15	45	76	24	50	26	-
5	MW	5	100	0.10	60	83	18	47	35	-
6	Δ	-	100	0.25	30	15	85	15	0	-
7	MW	18	125	0.25	30	93	7	43	50	-
8	MW	38	150	0.25	30	100	0	22	78	-
9	MW	59	175	0.25	30	100	0	9	**91** [e,f]	88 (**3a**)
10	Δ	-	175	0.25	30	98	2	15	83	-

[a] The pressure was 17 bar; [b] Based on the HPLC pump; [c] the mixtures from the stationary operation were analyzed by GC; [d] after column chromatography; [e] No change on longer residence time (45 min); [f] No change on higher excess (50 equiv) of ^{n}BuOH. Bold numbers indicate the target compounds in the reactions.

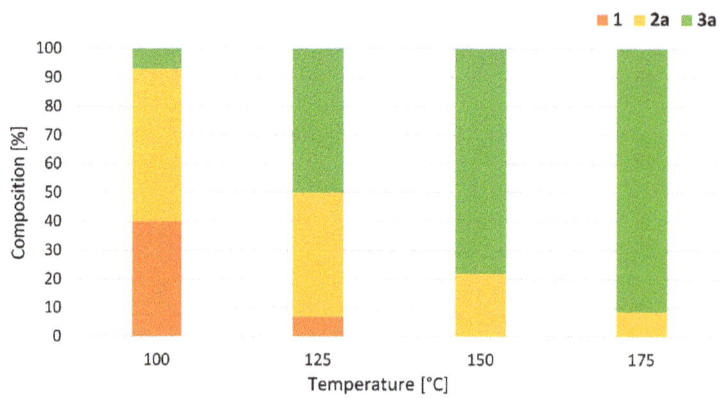

Figure 3. Composition of the reaction mixtures of the MW-assisted alcoholysis of DMP with ^{n}BuOH at different temperatures at a residence time of 30 min (Table 3 entries 3 and 7–9).

Comparative batch experiments were also carried out in the DMP (**1**)—^{n}BuOH model reaction at 100 °C. The alcohol excess (25 equiv) and the reaction time range studied (5–60 min) were the same as in the flow approaches. The batch results are listed in Table 4. Similarly to the flow experiments, the composition was highly reaction time-dependent.

Table 4. Comparative experiments for the alcoholysis of dimethyl *H*-phosphonate (DMP) with ^{n}BuOH in a batch MW reactor.

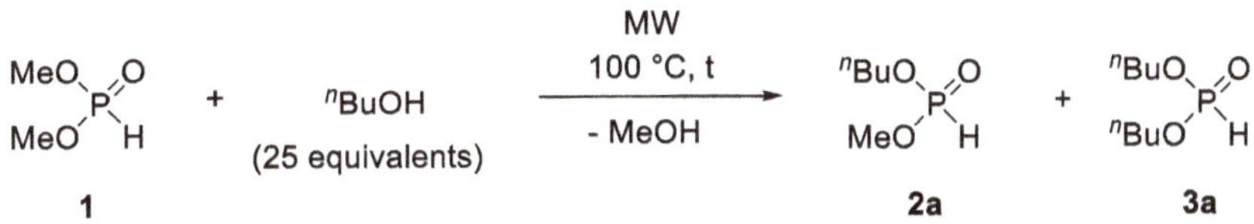

Entry	t (min)	Conversion (%)	Composition (%) [a]		
			1	2a	3a
1	5	24	76	20	4
2	15	46	54	37	9
3	30	59	41	47	12
4	45	67	33	49	18
5	60	73	26	45	29

[a] Based on GC.

Comparing the conversions of the flow and batch processes, the two plots show a similar shape; however, after a reaction time of 15 min, the batch conversions were 5–10% lower than the flow results (Figure 4).

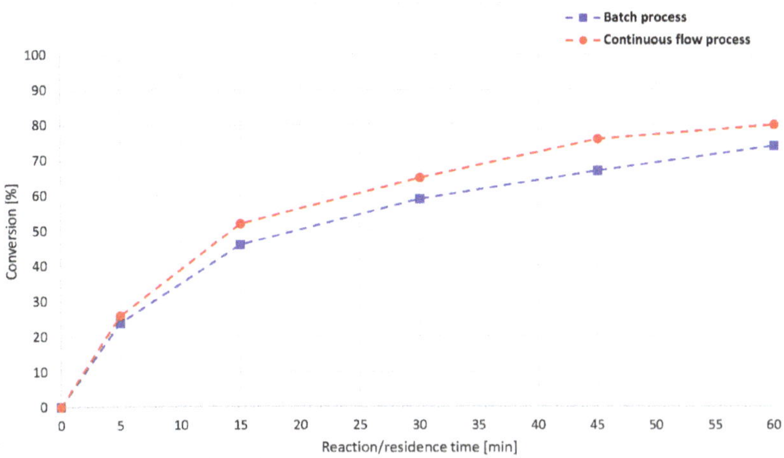

Figure 4. Reaction/residence time-dependence of the conversion of the alcoholysis of DMP with ^{n}BuOH at 100 °C.

A more significant difference may be observed if the ratio of product **2a** is examined separately (Figure 5). Maximum points are present on both curves: at 30 min and 53% in the flow experiments and at 45 min 49% in the case of the batch approaches. Based on the comparison, the flow reaction reaches the optimum point of phosphite **2a** under a shorter time, with a higher ratio.

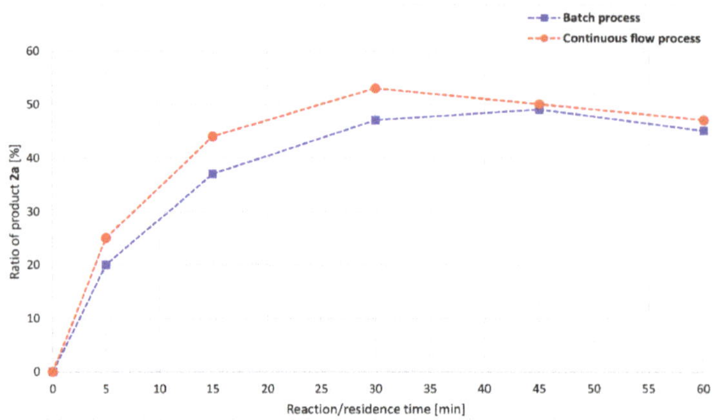

Figure 5. Reaction/residence time-dependence of the product **2a** ratio in the alcoholysis of DMP with ^{n}BuOH at 100 °C.

Based on the experiences above, the alcoholysis of DMP (**1**) was also carried out with *n*-propanol, *i*-butanol, as well as *n*-pentanol at 100 °C and 175 °C with a residence time of 30 min. In all cases, a mixture containing DMP (**1**) and 25 equivalents of the corresponding alcohol was passed through the continuous flow system. The data shown in Table 4 are concerned to the stationary operation. In the reaction of DMP (**1**) with ^{n}PrOH at 100 °C, the leaving mixture contained 45% of unreacted DMP (**1**), 46% of *n*-propyl methyl *H*-phosphonate (PrMP) (**2b**) and 9% of *n*-dipropyl *H*-phosphonate (DPrP) (**3b**) (Table 5, Entry 1). At a higher temperature of 175 °C, the fully-transesterified product (**3b**) was the major component (90%) (Table 5, Entry 2). The outcome of the alcoholysis of DMP (**1**) with ^{i}BuOH and ^{n}PentOH was similar (Table 5, Entries 3–6) However, in the case of the experiments carried out at 100 °C, the ratio of *i*-butyl methyl *H*-phosphonate (iBMP) (**2c**) and *n*-pentyl methyl *H*-phosphonate (PeMP) (**2d**) was above 50% (Table 5, Entries 3 and 5). After column chromatography, the corresponding mixed phosphonates (**2b**, **2c** and **2d**) were obtained in yields of 42–51% (Table 5, Entries 1, 3 and 5), and the fully-transesterified products (**3b**, **3c** and **3d**) were isolated in yields of 85–88% (Table 5, Entries 2, 4 and 6).

Table 5. Continuous flow alcoholysis of DMP with various alcohols.

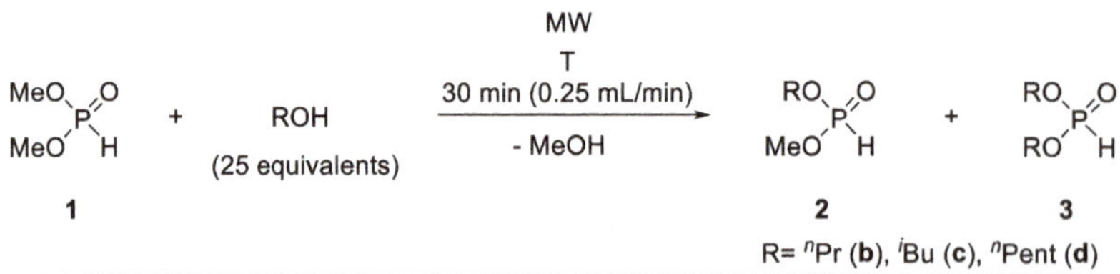

Entry	R	Power (W)	T (°C) [a]	Composition (%) [b]			Yield (%) [c]
				1	2	3	
1	^{n}Pr	10	100	45	**46**	9	42 (**2b**)
2		57	175	0	10	**90**	85 (**3b**)
3	^{i}Bu	10	100	38	**51**	11	46 (**2c**)
4		71	175	0	10	**90**	86 (**3c**)
5	^{n}Pent	10	100	32	**55**	13	51 (**2d**)
6		55	175	0	8	**92**	88 (**3d**)

[a] The pressure was 17 bar; [b] the mixtures from the stationary operation were analyzed by GC; [c] after column chromatography. Bold numbers indicate the target compounds in the reactions.

The next model was the continuous flow alcoholysis of diethyl *H*-phosphonate (DEP) (**4**) (Table 6). Based on the experiences from the transesterification of DMP (**1**), the reactions were carried out using 25 equivalents of the alcohols at a flow rate of 0.25 mL/min (or at a residence time of 30 min). First, the alcoholysis of DEP (**4**) by ^{n}BuOH was investigated. Performing the reaction at 100 °C, the unreacted DEP (**4**) was the main component along with 34% of *n*-butyl ethyl *H*-phosphonate (nBEP) (**5a**) and 8% of DBP (**3a**) (Table 6, Entry 1). At a higher temperature of 125 °C, the ratio of nBEP (**5a**) was 42%, and after column chromatography, compound **5a** could be isolated in a yield of 38% (Table 6, Entry 2). Increasing the temperature to 150 °C and 175 °C, the proportion of **5a** was decreased, while that of DBP (**3a**) increased (Table 6, Entries 3 and 4). In the latter case, the alcoholysis was complete, and the DBP (**3a**) was obtained in a yield of 85% (Table 6, Entry 4). The transesterification of DEP (**4**) by ^{n}PrOH at 125 °C took place similarly as the alcoholysis by ^{n}BuOH, and the *n*-propyl ethyl *H*-phosphonate (PrEP) (**5b**) was obtained in a yield of 36% (Table 6, Entries 2 and 5). At 175 °C, the conversion was not complete, and the resulting mixture contained only 74% of DPrP (**3b**) besides 4% of unreacted DEP (**4**) and 22% of PrEP (**5b**) (Table 6, Entry 6). To obtain a higher proportion of product **3b**, the temperature was increased to 200 °C. Hence, phosphite **3b** was formed in 84% (Table 6, Entry 7). In the case of ^{i}BuOH, the tendency was similar, and the highest amount of mixed phosphonate (**5c**) was formed at 125 °C, while at 175 and 200 °C, the fully-transesterified product (**3c**) predominated (Table 6, Entries 8–10). Using PentOH at 125 °C, the pentyl ethyl *H*-phosphonate (PeEP) (**5d**) was the main component in the departing mixture, from which it was isolated in a yield of 40% (Table 6, Entry 11). For the preparation of dipentyl *H*-phosphonate (DPeP) (**3d**), the application of 175 °C was enough, and compound **3d** was obtained in a yield of 80% (Table 6, Entry 12).

Table 6. Continuous flow alcoholysis of DEP with alcohols.

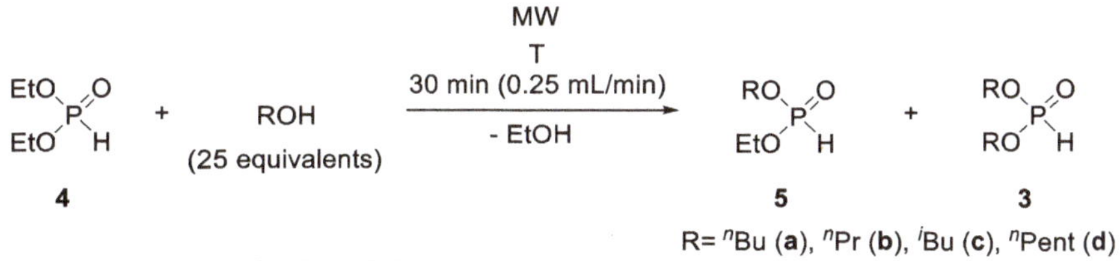

Entry	R	Power (W)	T (°C) [a]	Composition (%) [b]			Yield (%) [c]
				4	5	3	
1	^{n}Bu	10	100	58	34	8	-
2		17	125	44	**42**	14	38 (**5a**)
3		38	150	7	32	61	-
4		57	175	0	11	**89** [d,e]	85 (**3a**)
5	^{n}Pr	18	125	51	**41**	8	36 (**5b**)
6		54	175	4	22	74	–
7		83	200	2	14	**84** [d,e]	78 (**3b**)
8	^{i}Bu	21	125	21	**40**	39	36 (**5c**)
9		63	175	3	26	71	–
10		105	200	0	16	**84** [d,e]	77 (**3c**)
11	^{n}Pent	10	125	31	**44**	25	40 (**5d**)
12		39	175	2	13	**85** [d,e]	80 (**3d**)

[a] The pressure was 17 bar; [b] the mixtures from the stationary operation were analyzed by GC; [c] after column chromatography; [d] no change upon longer residence time (45 min); [e] no change upon higher excess (50 equiv) of ^{n}BuOH. Bold numbers indicate the target compounds in the reactions.

The catalyst-free alcoholysis of dialkyl *H*-phosphonates with various aliphatic alcohols was carried out efficiently in a continuous flow MW reactor. Applying a residence time of 30 min at different temperatures, the alcoholysis could be fine-tuned towards the dialkyl *H*-phosphonates with two different or with two identical alkyl groups. In the case of the reaction studied, the dimethyl phosphite proved to be the more reactive starting material. After almost complete conversions, the $(RO)_2P(O)H$ species formed by transesterification were obtained in yields above 85%. The ratio of (RO)(R′O)P(O)H derivatives with two different alkyl groups was around 50% in all mixtures, affording the target compounds in a yield range of ca. 40–50%. By the method developed, a series of mixed dialkyl *H*-phosphonates, which are valuable building blocks for the preparation of chiral organophosphorus derivatives, was synthesized.

3. Materials and Methods

3.1. General

GC measurements were performed on an HP5890 Series 2 GC-FID chromatograph, using a 15 m × 0.18 mm Restek, Rtx-5 column with a film layer of 0.20 μm. The temperature of the column was initially held at 40 °C for 1 min, followed by programming at 25 °C/min up to 300 °C and a final period at 300 °C (isothermal) for 10 min. The temperature of the injector was 290 °C and of the FID detector was 300 °C. The carrier gas was N_2.

GC-MS measurements were performed on an Agilent 6890 N-GC-5973 N-MSD chromatograph, using a 30 m × 0.25 mm Restek, Rtx-5SILMS column with a film layer of 0.25 μm. The initial temperature of the column was 45 °C for 1 min, followed by programming at 10 °C/min. up to 310 °C and a final period at 310 °C (isothermal) for 17 min. The temperature of the injector was 250 °C. The carrier gas was He, and the operation mode was splitless.

High resolution mass spectrometric measurements were performed using a TripleTOF 5600+ mass spectrometer in positive electrospray mode.

The ^{13}C- and ^{1}H-NMR spectra were obtained in $CDCl_3$ solution on a Bruker DRX-500 spectrometer operating at 125.7 and 500.1 MHz, respectively. The ^{13}C and ^{1}H chemical shifts are referred to TMS. ^{31}P-NMR spectra were obtained on a Bruker AV-300 spectrometer at 121.5 MHz. Chemical shifts are downfield relative to 85% H_3PO_4.

3.2. Equipment

The continuous flow reactions were performed in a self-developed continuous flow system comprising a 300-W CEM® Discover-focused microwave reactor equipped with a CEM® 10-mL Flow Cell Accessory continuous flow unit (irradiated volume 7 mL), a Gilson 305 HPLC pump, an HPLC backpressure regulator with a 250-psi (17.2 bar) cartridge and a cooler. Teflon® (Wilmington, DE, USA) PFA tubes with an outside diameter of 0.125″ (3.175 mm) and an inside diameter of 0.064″ (1.575 mm) were used. The exact lengths and volumes of each tube part are shown in Figure 6. All of the tubes, screws and ferrules applied were fully compatible with a regular HPLC system.

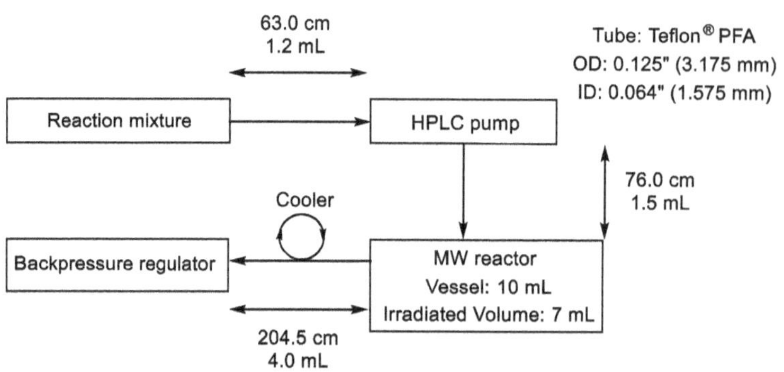

Figure 6. Design parameters of the continuous flow system.

3.3. General Procedure for the Continuous Flow Alcoholysis of Dialkyl H-Phosphonates

A mixture of 25.0 mmol of the dialkyl *H*-phosphonate (2.7 mL of dimethyl *H*-phosphonate, 3.9 mL of diethyl *H*-phosphonate) and 25 equivalents, 0.625 mol of alcohol (47 mL of *n*-propanol, 57 mL of *n*-butanol, 58 mL of *i*-butanol, 68 mL of *n*-pentanol) was homogenized by stirring for 5 min at 25 °C. The reactor was flushed with 20 mL of the mixture with a flow rate of 10 mL/min at 25 °C and 17 bar. Next, under the same pressure, the flow rate was set to the desired value, and the vessel was irradiated with a power of 45–300 W for 3–5 min, until the temperature reached the desired value, then the power was controlled automatically by the software of the MW reactor. The flow rate and the necessary power at stationary operation can be seen in Tables 3, 5 and 6. The operation was regarded as the steady state on the basis of the results of the GC measurements. The products were separated by column chromatography using silica gel as the absorbent and ethyl acetate as the eluent. Yields were calculated on the basis of the weights obtained after separation and evaporation taking into consideration the quantity of the dialkyl *H*-phosphonate fed in during a given time. The ^{31}P-NMR and mass data for the dialkyl *H*-phosphonates are listed in Table 7.

3.4. General Procedure for the Comparative Batch Experiments

A mixture of 0.5 mmol (0.05 mL) of dimethyl *H*-phosphonate and 25 equivalents, 12.5 mmol (1.1 mL) of *n*-butanol was heated at 100 °C in a vial in the MW reactor for 5–60 min as shown in Table 4. The volatile components were removed in vacuum, and the residual oil was analyzed by GC.

Methyl propyl H-phosphonate (**2b**): Yield: 42%; ^{31}P-NMR and HRMS (see Table 7); ^{13}C-NMR ($CDCl_3$) δ 10.0 (CH_3CH_2), 23.8 (J = 6.2, CH_2), 52.0 (J = 5.8, CH_3O), 67.1 (J = 6.0, CH_2O); ^{1}H-NMR ($CDCl_3$) δ 0.98 (t, J_{HH} = 7.4, 3H, CH_3CH_2), 1.65–1.82 (m, 2H, CH_2), 3.78, (d, J_{PH} = 11.9, 3H, CH_3O) 3.97–4.12 (m, 2H, CH_2O), 6.79 (d, J_{PH} = 695.4, 1H, PH).

i-Butyl methyl H-phosphonate (**2c**): Yield: 46%; ^{31}P-NMR and HRMS (see Table 7); ^{13}C-NMR ($CDCl_3$) δ 18.6 (J = 3.1, CH_3CH), 29.2 (J = 6.4, CH), 51.9 (J = 5.7, CH_3O), 71.8 (J = 6.3, CH_2O); ^{1}H-NMR ($CDCl_3$) δ 0.89 (d, J_{HH} = 6.8, 6H, CH_3CH), 1.84–1.96 (m, 1H, CH), 3.71 (d, J_{PH} = 11.9, 3H, CH_3O) 3.74–3.83 (m, 2H, CH_2O), 6.72 (d, J_{PH} = 695.8, 1H, PH).

Methyl pentyl H-phosphonate (**2d**): Yield: 50%; ^{31}P-NMR and HRMS (see Table 7); ^{13}C-NMR ($CDCl_3$) δ 13.9 (CH_3CH_2), 22.2 (CH_3CH_2), 27.6 ($CH_3CH_2CH_2$), 30.1 (J = 6.2, CH_2CH_2O), 51.9 (J = 5.7, CH_3O), 67.0 (J = 6.1, CH_2O); ^{1}H-NMR ($CDCl_3$) δ 0.91 (t, J = 6.9, 3H, CH_3CH_2), 1.29–1.44 (m, 4H, $CH_2CH_2CH_3$), 1.63–1.78 (m, 2H, CH_2CH_2O), 3.78, (d, J_{PH} = 11.9, 3H, CH_3O) 3.99–4.17 (m, 2H, CH_2O), 6.79 (d, J_{PH} = 695.2, 1H, PH).

Ethyl propyl H-phosphonate (**5b**): Yield: 30%; ^{31}P-NMR and HRMS (see Table 7); ^{13}C-NMR ($CDCl_3$) δ 10.0 ($CH_3CH_2CH_2$), 16.3 (J = 6.1, CH_3CH_2O), 23.8 (J = 6.4, $CH_3CH_2CH_2$), 61.8 (J = 5.8, CH_2CH_2O), 67.3 (J = 6.0, CH_3CH_2O); ^{1}H NMR ($CDCl_3$) δ 0.96 (t, J_{HH} = 7.4, 3H, $CH_3CH_2CH_2$), 1.35 (t, J_{HH} = 7.1, 3H, CH_3CH_2O), 1.63–1.78 (m, 2H, $CH_3CH_2CH_2$), 3.94–4.07, (m, 2H, CH_2CH_2O) 4.07–4.21 (m, 2H, CH_3CH_2O), 6.80 (d, J_{PH} = 692.4, 1H, PH).

i-Butyl ethyl H-phosphonate (**5c**): Yield: 36%; ^{31}P-NMR and HRMS (see Table 7); ^{13}C-NMR ($CDCl_3$) δ 16.3 (J = 6.2, CH_3CH_2), 18.7 (J = 1.7, CH_3CH), 29.1 (J = 6.5, CH), 61.8 (J = 5.7, CH_3CH_2O), 71.6 (J = 6.2, $CHCH_2O$); ^{1}H NMR ($CDCl_3$) δ 0.96 (d, J_{HH} = 6.7, 6H, CH_3CH), 1.37 (t, J_{HH} = 7.1, 3H, CH_3CH_2), 1.89–2.06 (m, 1H, CH), 3.77–3.91, (m, 2H, $CHCH_2O$) 4.09–4.22 (m, 2H, CH_3CH_2O), 6.82 (d, J_{PH} = 692.7, 1H, PH).

Table 7. ^{31}P-NMR and mass data for the dialkyl *H*-phosphonates prepared.

Compound	δ_P ($CDCl_3$)	δ_P [Ref]	$[M + H]^+_{found}$	$[M + H]^+_{requires}$
2a	9.3 (695.0)	9.3 [52]	153.0684	153.0681
2b	9.3 (695.4)	-	139.0522	139.0523
2c	6.7 (695.7)	-	153.0676	153.0681
2d	6.8 (695.2)	-	167.0833	167.0837
3a	7.8	7.9 [66]	195.1153	195.1150
3b	7.9	7.8 [67]	167.0833	167.0837
3c	8.1	8.0 [67]	195.1145	195.1150
3d	7.8	8.1 [68]	223.1462	223.1463
5a	7.6 (693.0)	7.6 [52]	167.0840	167.0837
5b	8.6 (692.4)	-	153.0673	153.0681
5c	7.8 (692.8)	-	167.0832	167.0837
5d	7.7 (692.1)	7.7 [52]	181.0992	181.0994

4. Conclusions

In summary, a MW-assisted continuous flow method was developed for the catalyst-free alcoholysis of dialkyl *H*-phosphonates. By the optimization of the reaction parameters, the alcoholysis was fine-tuned towards dialkyl *H*-phosphonates with two different or with two identical alkyl groups. The selectivity of the reaction was similar in the flow and batch approaches; however, the continuous protocol offered several advantages. The dialkyl *H*-phosphonates with two different alkyl groups can be obtained at a shorter reaction time and in a higher ratio. Furthermore, a somewhat higher productivity (0.8–1.0 g/h) can be attained by the continuous flow method. Five new dialkyl *H*-phosphonates with different alkyl groups were isolated and characterized, which may be valuable building blocks in the synthesis of chiral organophosphorus compounds.

Author Contributions: E.B., Á.T. and G.K. conceived of and designed the experiments. Á.T. and N.T. performed the experiments. E.B. and G.K. contributed reagents/materials/analysis tools. E.B., Á.T. and G.K. wrote the paper.

Funding: The project was supported by the Hungarian Research Development and Innovation Fund (FK123961 and K119202), by the National Research, Development and Innovation Fund of Hungary in the frame of the FIEK_16-1-2016-0007 (Higher Education and Industrial Cooperation Center) project and by the János Bolyai Research Scholarship of the Hungarian Academy of Sciences (BO/00278/17/7) (E.B.).

Conflicts of Interest: The authors declare no conflict of interest.

References

1. Plutschack, M.B.; Pieber, B.; Gilmore, K.; Seeberger, P.H. The Hitchhiker's guide to flow chemistry. *Chem. Rev.* **2017**, *117*, 11796–11893. [CrossRef] [PubMed]
2. Glasnov, T. *Continuous-Flow Chemistry in the Research Laboratory*; Springer International Publishing: Basel, Switzerland, 2016; ISBN 978-3-319-32194-3.
3. Bálint, E.; Keglevich, G. The Spread of the Application of the Microwave Technique in Organic Synthesis. In *Milestones in Microwave Chemistry*; Keglevich, G., Ed.; Springer: Basel, Switzerland, 2016; pp. 1–10. ISBN 978-3-319-30632-2.
4. de la Hoz, A.; Loupy, A. (Eds.) *Microwaves in Organic Synthesis*, 3rd ed.; Wiley: Weinheim, Germany, 2012; ISBN 978-3-527-65131-3.
5. Moseley, J.D. Microwave heating as a tool for process chemistry. In *Microwave Heating as a Tool for Sustainable Chemistry*; Leadbeater, N., Ed.; CRC Press: New York, NY, USA, 2010; pp. 105–147. ISBN 9781138111981.
6. Kappe, C.O.; Stadler, A.; Dallinger, D. *Microwaves in Organic and Medicinal Chemistry*, 2nd ed.; Wiley: Weinheim, Germany, 2012; Volume 52, ISBN 978-3-527-33185-7.
7. Estela, L.; Pouxb, M.; Benamaraa, N.; Polaerta, I. Continuous flow-microwave reactor: Where are we? *Chem. Eng. Process.* **2016**, *113*, 56–64. [CrossRef]
8. Baxendale, I.; Hayward, J.; Ley, S. Microwave reactions under continuous flow conditions. *Comb. Chem. High Throughput Screen.* **2007**, *10*, 802–836. [CrossRef] [PubMed]

9. Riemenschneider, W.; Bolt, H.M. Esters, Organic. In *Ullmann's Encyclopedia of Industrial Chemistry*, 7th ed.; Wiley: Weinheim, Germany, 2005; pp. 245–265. ISBN 978-3-527-30673-2.
10. Otera, J.; Nishikido, J. (Eds.) *Esterification: Methods, Reactions, and Applications*, 2nd ed.; Wiley: Weinheim, Germany, 2010; ISBN 978-3-527-32289-3.
11. Woodcock, L.L.; Wiles, C.; Greenway, G.M.; Watts, P.; Wells, A.; Eyley, S. Enzymatic synthesis of a series of alkyl esters using novozyme 435 in a packed-bed, miniaturized, continuous flow reactor. *Biocatal. Biotransform.* **2008**, *26*, 501–507. [CrossRef]
12. Junior, I.I.; Flores, M.C.; Sutili, F.K.; Leite, S.G.F.; Miranda, L.S.D.M.; Leal, I.C.R.; de Souza, R.O.M.A. Fatty acids residue from palm oil refining process as feedstock for lipase catalyzed monoacylglicerol production under batch and continuous flow conditions. *J. Mol. Catal. B Enzym.* **2012**, *77*, 53–58. [CrossRef]
13. Sutili, F.K.; Ruela, H.S.; Leite, S.G.F.; Miranda, L.S.D.M.; Leal, I.C.R.; de Souza, R.O.M.A. Lipase-catalyzed esterification of steric hindered fructose derivative by continuous flow and batch conditions. *J. Mol. Catal. B Enzym.* **2013**, *85–86*, 37–42. [CrossRef]
14. Patil, N.G.; Benaskar, F.; Rebrov, E.V.; Meuldijk, J.; Hulshof, L.A.; Hessel, V.; Schouten, J.C. Scale-up of Microwave Assisted Flow Synthesis by Transient Processing through Monomode Cavities in Series. *Org. Process Res. Dev.* **2014**, *18*, 1400–1407. [CrossRef]
15. Sutili, F.K.; Ruela, H.S.; Nogueira, D.D.O.; Leal, I.C.R.; Miranda, L.S.D.M.; de Souza, R.O.M.A. Enhanced production of fructose ester by biocatalyzed continuous flow process. *Sustain. Chem. Process.* **2015**, *3*, 6. [CrossRef]
16. Okuno, Y.; Isomura, S.; Sugamata, A.; Tamahori, K.; Fukuhara, A.; Kashiwagi, M.; Kitagawa, Y.; Kasai, E.; Takeda, K. Convenient and Simple Esterification in Continuous-Flow Systems using g-DMAP. *ChemSusChem* **2015**, *8*, 3587–3589. [CrossRef] [PubMed]
17. Koreniuk, A.; Maresz, K.; Odrozek, K.; Jarzebski, A.B.; Mrowiec-Bialon, J. Highly effective continuous-flow monolithic silica microreactors for acid catalyzed processes. *Appl. Catal. A* **2015**, *489*, 203–208. [CrossRef]
18. Baek, H.; Minakawa, M.; Yamada, Y.M.A.; Han, J.W.; Uozumi, Y. In-Water and Neat Batch and Continuous-Flow Direct Esterifcation and Transesterifcation by a Porous Polymeric Acid Catalyst. *Sci. Rep.* **2016**, *6*, 25925. [CrossRef] [PubMed]
19. Furuta, A.; Fukuyama, T.; Ryu, I. Efficient Flow Fischer Esterification of Carboxylic Acids with Alcohols Using Sulfonic Acid-Functionalized Silica as Supported Catalyst. *Bull. Chem. Soc. Jpn.* **2017**, *90*, 607–612. [CrossRef]
20. Iemhoff, A.; Sherwood, J.; McElroy, C.R.; Hunt, A.J. Towards sustainable kinetic resolution, a combination of bio-catalysis, flow chemistry and bio-based solvents. *Green Chem.* **2018**, *20*, 136–140. [CrossRef]
21. Razzaq, T.; Glasnov, T.N.; Kappe, C.O. Continuous-Flow Microreactor Chemistry under High-Temperature/Pressure Conditions. *Eur. J. Org. Chem.* **2009**, 1321–1325. [CrossRef]
22. Archambault, C.M.; Leadbeater, N.E. A benchtop NMR spectrometer as a tool for monitoring mesoscale continuous-flow organic synthesis: Equipment interface and assessment in four organic transformations. *RSC Adv.* **2016**, *6*, 101171–101177. [CrossRef]
23. Gumel, A.M.; Annuar, M.S.M. Thermomyces lanuginosus lipase-catalyzed synthesis of natural flavor esters in a continuous flow microreactor. *3 Biotech* **2016**, *6*, 24. [CrossRef] [PubMed]
24. Tajti, Á.; Tóth, N.; Bálint, E.; Keglevich, G. Esterification of benzoic acid in a continuous flow microwave reactor. *J. Flow Chem.* **2017**, *8*, 11–19. [CrossRef]
25. Krull, M.; Moschhaeuser, R. Continuous Method for Producing Esters of Aromatic Carboxylic Acids. U.S. Patent 0088918, 12 April 2012.
26. Cablewski, T.; Faux, A.F.; Strauss, C.R. Development and Application of a Continuous Microwave Reactor for Organic Synthesis. *J. Org. Chem.* **1994**, *59*, 3408–3412. [CrossRef]
27. Chen, S.-T.; Chiou, S.-H.; Wang, K.-T. Preparative scale organic synthesis using a kitchen microwave oven. *J. Chem. Soc. Chem. Commun.* **1990**, 807–809. [CrossRef]
28. Pipus, G.; Plazl, I.; Koloini, T. Esterification of benzoic acid in microwave tubular flow reactor. *Chem. Eng. J.* **2000**, *76*, 239–245. [CrossRef]
29. Asadi, M.; Hooper, J.F.; Lupton, D.W. Biodiesel synthesis using integrated acid and base catalysis in continuous flow. *Tetrahedron* **2016**, *72*, 3729–3733. [CrossRef]

30. Adeyemi, A.; Bergman, J.; Branalt, J.; Savmarker, J.; Larhed, M. Continuous Flow Synthesis under High-Temperature/High-Pressure Conditions Using a Resistively Heated Flow Reactor. *Org. Process Res. Dev.* **2017**, *21*, 947–955. [CrossRef]
31. Tran, D.-T.; Chang, J.-S.; Lee, D.-J. Recent insights into continuous-flow biodiesel production via catalytic and non-catalytic transesterification processes. *Appl. Energy* **2017**, *185*, 376–409. [CrossRef]
32. Corbridge, D.E.C. *Phosphorus: Chemistry, Biochemistry and Technology*, 6th ed.; CRC Press: New York, NY, USA, 2013; ISBN 978-1-439-84088-7.
33. Troev, K.D. *Chemistry and Application of H-Phosphonates*; Elsevier: Amsterdam, The Netherlands, 2006; ISBN 978-0-080-47649-0.
34. Bálint, E.; Tajti, Á.; Tripolszky, A. Synthesis of α-aminophosphonates by the Kabachnik–Fields reaction and by the Pudovik reaction. In *Organophosphorus Chemistry*; Keglevich, G., Ed.; Walter de Gruyter GmbH: Berlin, Germany, 2018; pp. 108–147. ISBN 978-3-11-053453-5.
35. Rádai, Z.; Kiss, N.Z.; Keglevich, G. Synthesis of α-hydroxyphosphonates, an important class of bioactive compounds. In *Organophosphorus Chemistry*; Keglevich, G., Ed.; Walter de Gruyter GmbH: Berlin, Germany, 2018; pp. 91–107. ISBN 978-3-11-053453-5.
36. Henyecz, R.; Keglevich, G. P-C couplings by the Hiaro reaction. In *Organophosphorus Chemistry*; Keglevich, G., Ed.; Walter de Gruyter GmbH: Berlin, Germany, 2018; pp. 158–178. ISBN 978-3-11-053453-5.
37. Enders, D.; Saint-Dizier, A.; Lannou, M.-I.; Lenzen, A. The phospha-Michael addition in organic synthesis. *Eur. J. Org. Chem.* **2006**, 29–49. [CrossRef]
38. Tajti, Á.; Bálint, E.; Keglevich, G. Synthesis of ethyl octyl α-aminophosphonate derivatives. *Curr. Org. Synth.* **2016**, *13*, 638–675. [CrossRef]
39. Bálint, E.; Tajti, Á.; Kalocsai, D.; Mátravölgyi, B.; Konstantin, K.; Czugler, M.; Keglevich, G. Synthesis and utilization of optically active α-aminophosphonate derivatives by Kabachnik-Fields reaction. *Tetrahedron* **2017**, *73*, 5659–5667. [CrossRef]
40. Gerrard, W. The interaction of n-butyl alcohol and the chlorides and oxychloride of phosphorus in the absence and in the presence of pyridine. *J. Chem. Soc.* **1940**, 1466–1469. [CrossRef]
41. Hardy, E.E.; Anniston, A.; Kosolapoff, G.M. Halogenated Compounds and Process for Making Same. U.S. Patent 2409039, 8 October 1946.
42. Foss, O. Di-O-alkylmonothiophosphates and the corresponding pseudohalogens. *Acta Chem. Scand.* **1947**, *1*, 8–31. [CrossRef] [PubMed]
43. Campbell, C.H.; Chadwick, D.H.; Kaufman, S. Continuous process for preparing dialkyl phosphites. *Ind. Eng. Chem. Res.* **1957**, *49*, 1871–1873. [CrossRef]
44. Mitschke, K.-H. Method for the Combined Production of Diethyl Phosphite and Ethyl Chloride. Application Number WO200424742 A1, 25 March 2004.
45. Kendall, A.J.; Salazar, C.A.; Martino, P.F.; Tyler, D.R. Direct conversion of phosphonates to phosphine oxides: An improved synthetic route to phosphines including the first synthesis of methyl JohnPhos. *Organometallics* **2014**, *33*, 6171–6178. [CrossRef]
46. Kosolapoff, G.M. Preparation of some mixed dialkyl phosphites. *J. Am. Chem. Soc.* **1951**, *73*, 4989. [CrossRef]
47. Fields, E. The synthesis of esters of substituted amino phosphonic acids. *J. Am. Chem. Soc.* **1952**, *74*, 1528–1531. [CrossRef]
48. Oswald, A.A. Synthesis of cyclic phosphorus acid esters by transesterification. *Can. J. Chem.* **1959**, *37*, 1499–1504. [CrossRef]
49. Kuskov, V.K.; Gradis, G.K. Reaction of diethyl phosphite with sodium alcoholates. *Dokl. Akad. Nauk SSSR* **1953**, *92*, 323–324.
50. Froneman, M.; Modro, T.A. The titanium-mediated transesterification of phosphorus esters. *Tetrahedron Lett.* **1988**, *27*, 3327–3330. [CrossRef]
51. Aitken, R.A.; Collett, C.J.; Mesher, S.T.E. Convenient preparation of long-chain dialkyl phosphates: Synthesis of dialkyl phosphates. *Synthesis* **2012**, *44*, 2515–2518. [CrossRef]
52. Bálint, E.; Tajti, Á.; Drahos, L.; Ilia, G.; Keglevich, G. Alcoholysis of dialkyl phosphites under microwave conditions. *Curr. Org. Chem.* **2013**, *17*, 555–562. [CrossRef]
53. Zarrougui, R.; Raouafi, N.; Lemordant, D. New series of green cyclic ammonium-based room temperature ionic liquids with alkylphosphite-containing anion: Synthesis and physicochemical characterization. *J. Chem. Eng. Data* **2014**, *59*, 1193–1201. [CrossRef]

54. Salin, A.V.; Il'in, A.V.; Shamsutdinova, F.G.; Fatkhutdinov, A.R.; Galkin, V.I.; Islamov, D.R.; Kataeva, O.N. Phosphine-catalyzed addition of P(O)-H compounds to ethyl phenylpropiolate. *Tetrahedron Lett.* **2015**, *56*, 6282–6286. [CrossRef]
55. Ernsberger, M.L.; Hill, J.W. Preparation of Organic Phosphorus Compounds, and in Particular, of Dialkyl Phosphites. U.S. Patent 2661364, 1 December 1953.
56. Budnikova, Y.G.; Kargin, Y.M. Electrosynthesis of aliphatic esters of phosphorus acids from white phosphorus in alcohol solutions, involving radical cations of phenothiazine and triarylamine. *Russ. J. Gen. Chem.* **1995**, *65*, 504–507.
57. Abdreimova, R.R.; Akbayeva, D.N.; Polimbetova, G.S.; Caminade, A.-M.; Majoral, J.-P. Chlorine free synthesis of organophosphorus compounds based on the functionalization of white phosphorus (P_4). *Phosphorus Sulfur Silicon Relat. Elem.* **2000**, *156*, 239–254. [CrossRef]
58. Trofimov, B.; Timokhin, B.; Gusarova, N.; Kazantseva, M. Golubin, Cu-catalyzed oxidative phosphorylation of alkanols with white phosphorus and H_2O_2. *Phosphorus Sulfur Silicon Relat. Elem.* **2002**, *177*, 2385–2390. [CrossRef]
59. Budnikova, Y.H.; Kafiyatullina, A.G.; Sinyashin, O.G.; Abdreimova, R.R. Electrochemical synthesis of phosphorus esters from white phosphorus in the presence of copper complexes and ethanol. *Russ. Chem. Bull.* **2003**, *52*, 929–938. [CrossRef]
60. Budnikova, Y.H.; Yakhvarov, D.G.; Sinyashin, O.G. Electrocatalytic eco-efficient functionalization of white phosphorus. *J. Organomet. Chem.* **2005**, *690*, 2416–2425. [CrossRef]
61. Fisher, H.C.; Prost, L.; Montchamp, J.L. Organophosphorus chemistry without PCl3: A bridge from hypophosphorous acid to *H*-phosphonate diesters. *Eur. J. Org. Chem.* **2013**, 7973–7978. [CrossRef]
62. Kluba, M.; Zwierzak, A. Alkylation of tetra-n-butylaminium alkyl hydrogen phosphites. A new route to mixed dialkyl phosphites. *Synthesis* **1978**, *2*, 134–137. [CrossRef]
63. Ilia, G.; Kurunczi, L. Synthesis of mixed alkylphosphites and alkylphosphates. *Phosphorus Sulfur Silicon Relat. Elem.* **2003**, *178*, 1513–1519. [CrossRef]
64. Berchel, M.; Haddad, J.; Le Corre, S.S.; Haelters, J.-P.; Jaffrès, P.-A. Synthesis of lipid-based unsymmetrical O,O-dialkylphosphites. *Tetrahedron Lett.* **2015**, *56*, 2345–2348. [CrossRef]
65. Guin, J.; Wang, Q.; van Gemmeren, M.; List, B. The catalytic asymmetric Abramov reaction. *Angew. Chem. Int. Ed.* **2014**, *53*, 1–5. [CrossRef] [PubMed]
66. Peng, W.; Schreeve, J.M. Rapid and high yield oxidation of phosphine, phosphite and phosphinite compounds to phosphine oxides, phosphates and phosphinates using hypofluorous acid–acetonitrile complex. *J. Fluorine Chem.* **2005**, *126*, 1054–1056. [CrossRef]
67. Santschi, N.; Togni, A. Electrophilic Trifluoromethylation of S-Hydrogen Phosphorothioates. *J. Org. Chem.* **2011**, *76*, 4189–4193. [CrossRef] [PubMed]
68. Kers, A.; Kers, I.; Stawinski, J.; Sobkowski, M.; Kraszewski, A. Studies on Aryl *H*-Phosphonates; Part 2: A General Method for the Preparation of Alkyl *H*-Phosphonate Monoesters. *Synthesis* **1995**, 427–430. [CrossRef]

Sample Availability: Samples of the compounds **2a–d**, **3a–d**, **5a–d** are available from the authors.

molecules

MDPI

Article

1-(*N*-Acylamino)alkyltriarylphosphonium Salts with Weakened C_α-P^+ Bond Strength—Synthetic Applications

Jakub Adamek [1,2,*], **Anna Węgrzyk** [1,2], **Justyna Kończewicz** [1,2], **Krzysztof Walczak** [1] and **Karol Erfurt** [3]

[1] Department of Organic Chemistry, Bioorganic Chemistry and Biotechnology, Silesian University of Technology, B. Krzywoustego 4, 44-100 Gliwice, Poland; anna.wegrzyk@polsl.pl (A.W.); j.konczewicz@gmail.com (J.K.); krzysztof.walczak@polsl.pl (K.W.)

[2] Biotechnology Center of Silesian University of Technology, B. Krzywoustego 8, 44-100 Gliwice, Poland

[3] Department of Chemical Organic Technology and Petrochemistry, Silesian University of Technology, B. Krzywoustego 4, 44-100 Gliwice, Poland; karol.erfurt@polsl.pl

* Correspondence: jakub.adamek@polsl.pl; Tel.: +48-032-237-1080; Fax: +48-032-237-2094

Received: 31 August 2018; Accepted: 23 September 2018; Published: 25 September 2018

check for updates

Abstract: The α-amidoalkylating properties of 1-(*N*-acylamino)alkyltriarylphosphonium salts with weakened C_α-P^+ bond strength are discussed and examined. It is demonstrated that such type of phosphonium salts reacts smoothly with a diverse array of carbon- and heteroatom-based nucleophiles, including 1-morpholinocyclohexene, 1,3-dicarbonyl compounds, benzotriazole sodium salt, *p*-toluenesulfinate sodium salt, benzylamine, triarylphosphines, and other P-nucleophiles. Reactions are conducted at room temperature, in a short time (5–15 min) and mostly without catalysts. Simple work-up procedures result in good or very good yields of products. The structures of known compounds were established by spectroscopic methods and all new compounds have been fully characterized using ^{1}H-, ^{13}C-, ^{31}P-NMR, IR spectroscopy, and high-resolution mass spectrometry. Mechanistic aspects of described transformations are also performed and discussed. It was demonstrated that unique properties make 1-(*N*-acylamino)alkyl-triarylphosphonium salts with weakened C_α-P^+ bond strength interesting building blocks with great potential, especially in α-amidoalkylation reactions.

Keywords: organophosphorus chemistry; phosphonium salts; α-amidoalkylating agents; *N*-acyliminium cation; *N*-acylimine

1. Introduction

Specific structural features of 1-(*N*-acylamino)alkylphosphonium salts make them very interesting reagents. The presence of a positively charged nucleofugal phosphonium moiety in the close surroundings of the *N*-acyl group determines its unique chemical properties such as high reactivity in α-amidoalkylations [1–10]. This type of reactions has enjoyed unflagging interest for years as a synthetic method with great potential, especially valuable for C-C and C-heteroatom bond formation [1,11–37].

In general, the reactivity of α-amidoalkylating agents **1** depends on the efficiency of the generation of *N*-acyliminium cation **2** or *N*-acylimine **3** from its precursor and the equilibrium constant of this reaction. Of course, the reactivity of *N*-acyliminium cation **2** or *N*-acylimine **3** toward a nucleophile is also significant [1–3]. To produce the proper α-amidoalkylating agents, for instance *N*-acyliminium cation **2** or *N*-acylimine **3** from the most popular precursors such as α-amido sulfones (Z = SO_2Ar), *N*-(1-benzotriazolil)alkylamides (Z = Bt), and *N*-(1-alkoxyalkyl)amides (Z = OR), it is necessary to use catalysts, mainly Lewis acids (route **a**, Scheme 1 (I)) [18–25,29–37]. In contrast, 1-(*N*-acylamino)alkyltriphenylphosphonium salts **1** (Z = Ph_3P^+ X^-) do not require the use of acidic catalysts because of the permanent positive charge on the phosphonium group. However, the relatively high stability of C_α-P^+ bond forces the use of a base catalyst (e.g., Hünig's base, DBU, TBD; route **b**, Scheme 1 (I)) and sometimes microwave radiation [1,4–10].

Recently, we have proven that the use of 1-(*N*-acylamino)alkyltriarylphosphonium salts **4** derived from the EWG-substituted triarylphosphines facilitates the cleavage of C_α-P^+ bond and thereby the generation of *N*-acyliminium cation (Scheme 1 (II)) [2,3]. This phenomenon significantly increases the reactivity of 1-(*N*-acylamino)alkyltriarylphosphonium salts with weakened C_α-P^+ bond strength and allows us to conduct α-amidoalkylation without the need for any catalyst. In this work, we demonstrate that the abovementioned phosphonium salts **4** react smoothly with various nucleophiles in a short time under mild conditions and create new carbon-carbon or carbon-heteroatom bonds with good or very good yields (Scheme 1 (III)).

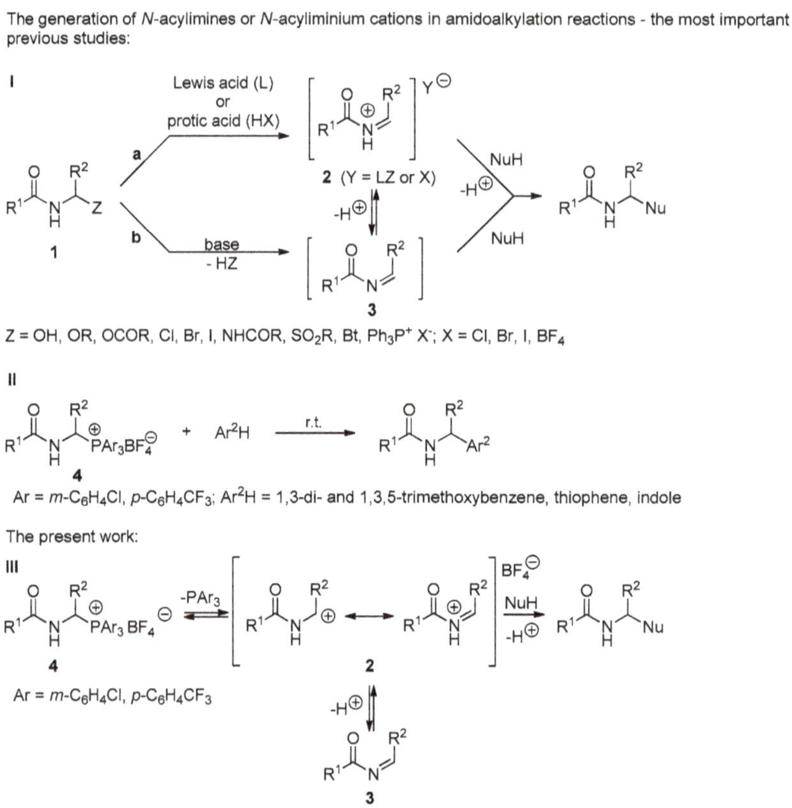

Scheme 1. α-Amidoalkylation reactions—various synthetic routes.

2. Results and Discussion

In this section, we focus our studies on the reactivity of phosphonium salts **4** with weakened C_α-P^+ bond strength in α-amidoalkylation of various types of carbon- and heteronucleophiles, as shown in Scheme 2. 1-Morpholinocyclohexene and 1,3-dicarbonyl compounds, such as dimethyl malonate, diethyl malonate, and ethyl acetoacetate, are used as carbon nucleophiles. In the case of heteronucleophiles, the reaction toward benzotriazole sodium salt, *p*-toluenesulfinate sodium salt, benzylamine, triarylphosphines, trimethyl phosphite, dimethyl phenylphosphonite, and methyl diphenylphosphinite is examined.

The synthesis of 1-(*N*-acylamino)alkyltriarylphosphonium salts **4** from α-amino acids was performed according to our previously described procedure, for which the electrochemical decarboxylative α-methoxylation of *N*-acyl-α-amino acids and substitution of the methoxy group by triarylphosphonium moiety are crucial steps (see Scheme 2) [2,9].

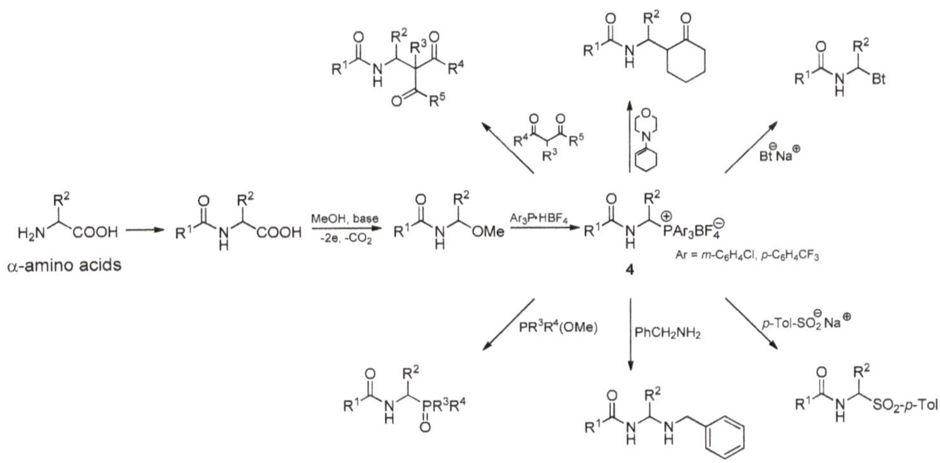

Scheme 2. 1-(*N*-Acylamino)alkyltriarylphosphonium salts with weakened C_α-P^+ bond strength **4**—preparation and synthetic application.

2.1. 1-(N-Acylamino)alkyltriarylphosphonium Salts with Weakened C_α-P^+ Bond Strength in the Selected Reaction of C-C Bond Formation

One of the crucial issues of organic synthesis is the formation of new C-C bonds. The possibility of an effective extension of the carbon skeleton is important in many fields, especially in medicinal chemistry, agrochemical synthesis, or in the synthesis of natural products.

Several years ago, we had proved that 1-(*N*-acylamino)alkyltriphenylphosphonium salts **4** (Ar = Ph) reacted quite easily with 1,3-dicarbonyl compounds in the presence of a base such as DBU (1,8-diazabicyclo[5.4.0]undec-7-ene), DBN (1,5-diazabicyclo[4.3.0]non-5-ene), or TBD (1,5,7-triaza-bicyclo[4.4.0]dec-5-ene) under microwave irradiation at 60 °C. However, the use of organic bases (DBU, DBN, TBD) complicates the course of α-amidoalkylation reaction due to the formation of amidinium and guanidinium salts [10]. To avoid this, in the current work, we used lithium diisopropylamide (LDA) as a base to produce enolate anions from the corresponding 1,3-dicarbonyl compounds. In our protocol, α-amidoalkylation was conducted under argon, in THF, at room temperature. A THF solution of enolate anions generated using LDA was introduced into the THF solution of the phosphonium salt **4**. The reaction was performed for 15 min, and the use of microwave irradiation was not necessary. In addition, it was found that the most favorable molar ratio of phosphonium salt **4**: 1,3-dicarbonyl compound: LDA was 1:8:1 (compare experiments 1 and 2, Table 1).

Under the abovementioned conditions, phosphonium salts with weakened C_α-P^+ bond strength **4** react with 1,3-dicarbonyl compounds, including diethyl and dimethyl malonate and ethyl acetoacetate, to give the corresponding products **6** with good yields, regardless of whether tris(4-trifluoromethylphenyl)- or tris(3-chlorophenyl)phosphonium salts were used as substrates. Only in the case of phosphonium salts with the benzyloxy carbamate protective group, the yields were significantly lower and did not exceed 30% (entry 8, Table 1).

Table 1. Conditions and yields for α-amidoalkylation of 1,3-dicarbonyl compounds by 1-(*N*-acylamino) alkyltriarylphosphonium salts **4**.

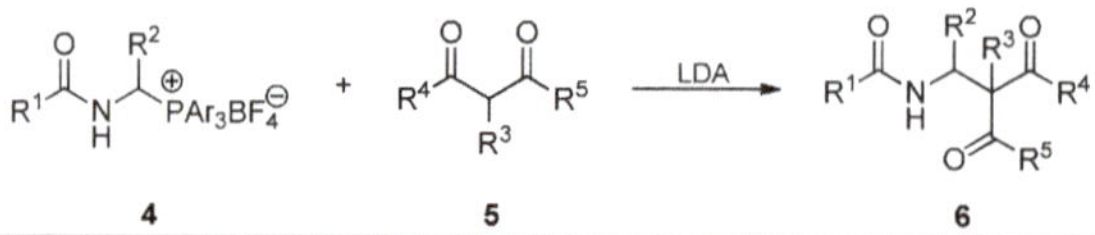

Entry	Phosphonium Salt 4			1,3-Dicarbonyl Compound, 5	Molar Ratio of 4:5:LDA	6	Yield, %
	R[1]	R[2]	Ar				
1	*t*-Bu	Me	*m*-C_6H_4Cl	diethyl malonate	1:2:1	**6a**	40
2	*t*-Bu	Me	*m*-C_6H_4Cl	diethyl malonate	1:8:1	**6a**	65
3	*t*-Bu	Me	*p*-$C_6H_4CF_3$	diethyl malonate	1:8:1	**6a**	67
4	*t*-Bu	Me	*m*-C_6H_4Cl	ethyl acetoacetate	1:8:1	**6b**	62 [a]
5	Bn	*i*-Bu	*m*-C_6H_4Cl	ethyl acetoacetate	1:8:1	**6c**	83 [b]
6	Bn	*i*-Bu	*m*-C_6H_4Cl	dimethyl malonate	1:8:1	**6d**	63
7	Bn	*i*-Bu	*m*-C_6H_4Cl	diethyl malonate	1:8:1	**6e**	52
8	BnO	Bn	*m*-C_6H_4Cl	ethyl acetoacetate	1:8:1	**6f**	30 [c]

[a] A mixture of diastereoisomers in a molar ratio of 1.4:1. [b] A mixture of diastereoisomers in a molar ratio of 1.25:1. [c] Attempts to isolate an analytically pure sample failed.

α-Amidoalkylation of carbon nucleophiles was successfully extended to enamines. Based on the recently described protocol [8], we have demonstrated that 1-(*N*-acylamino)alkyltriarylphosphonium salts **4** with weakened C_α-P^+ bond strength react with 1-morpholinocyclohexene in Stork-type enamination to give the expected products **8** with good yields. We have proved that this reaction can be conducted in acetonitrile, at room temperature and without using any base catalysts. The optimized reaction time was 60 min, and the molar ratio of phosphonium salt to 1-morpholinocyclohexene was 1:2. We were able to separate the major diastereoisomer from the mixture using column chromatography and crystallization technique. Furthermore, the use of phosphonium salts with the benzyloxy carbamate protective group resulted in the decrease in reaction efficiency, as already observed for reactions with 1,3-dicarbonyl compounds (see Table 2).

Table 2. Conditions and yields for α-amidoalkylation of 1-morpholinocyclohexene by 1-(*N*-acylamino) alkyltriarylphosphonium salts **4**.

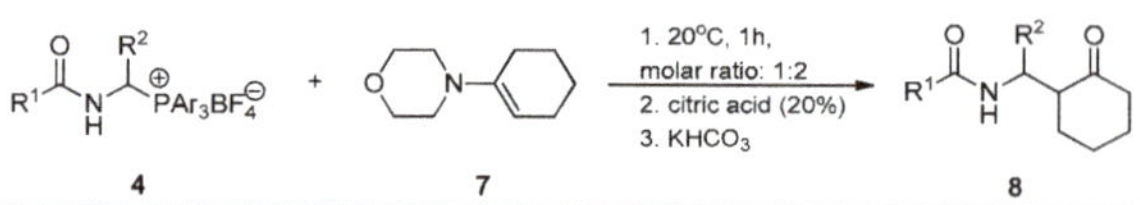

Entry	Phosphonium Salt 4			8	Yield, %
	R[1]	R[2]	Ar		
1	*t*-Bu	Me	*m*-C_6H_4Cl	**8a**	63 [a]
2	Bn	*i*-Bu	*m*-C_6H_4Cl	**8b**	62 [b]
3	BnO	*i*-Bu	*m*-C_6H_4Cl	**8c**	40 [c]
4	BnO	Bn	*m*-C_6H_4Cl	**8d**	33 [d]
5	BnO	CH_2O*t*-Bu	*p*-$C_6H_4CF_3$	**8e**	31 [c]

[a] A mixture of diastereoisomers in a molar ratio of 3.5:1. [b] A mixture of diastereoisomers in a molar ratio of 3.2:1. [c] Only one diastereoisomer was detected and isolated. [d] A mixture of diastereoisomers in a molar ratio of 13.4:1.

2.2. *1-(N-Acylamino)alkyltriarylphosphonium Salts with Weakened C_α-P^+ Bond Strength in α-Amidoalkylation of Selected Heteronucleophiles*

Applications of 1-(*N*-acylamino)alkyltriphenylphosphonium salts **4** (Ar = Ph) in the α-amidoalkylation of heteronucleophiles have been reported many times in the literature [5–8,10]. Usually in these types of reactions, it was necessary to use base catalysts, elevated temperature, and microwave irradiation. The effect of the use of 1-(*N*-acylamino)alkyltriarylphosphonium salts with weakened C_α-P^+ bond strength **4** (R = *m*-C_6H_4Cl, *p*-$C_6H_4CF_3$) on α-amidoalkylation has not been investigated so far. Therefore, we selected several heteronucleophiles such as benzotriazole sodium salt, *p*-toluenesulfinate sodium salt, benzylamine, and triphenylphosphine, and examined their α-amidoalkylation by phosphonium salts with weakened C_α-P^+ bond strength. As it was expected, reactions of salts with a modified structure of the phosphonium group occurred much faster and under mild conditions. In none of the described examples was it necessary to use a base catalyst. The reaction time was only 5–15 min. α-Amidoalkylation of benzotriazole sodium salt and sodium *p*-toluenesulfinate (entries 1–3, Table 3) was conducted at room temperature in $CHCl_3$. Similar conditions were used for reaction with benzylamine (entries 4 and 5, Table 3). The obtained aminal derivatives **10d** and **10e** exhibited limited stability. Therefore, the use of excess amount of the nucleophile was required to make the alkaline environment safe for aminals.

Table 3. Conditions and yields for α-amidoalkylation of selected heteronucleophiles by 1-(*N*-acylamino) alkyltriarylphosphonium salts **4**.

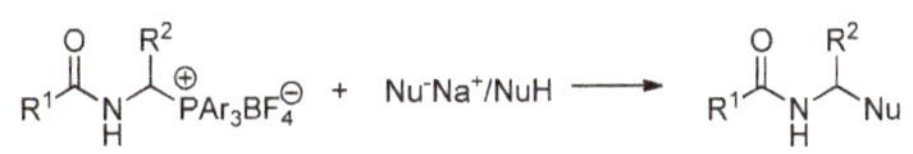

4 **9** **10**

Entry	Phosphonium Salt 4			Nu^-Na^+/ NuH	Temp., °C	Time, Min.	Molar Ratio of 4:9	10	Yield, %
	R^1	R^2	Ar						
1	*t*-Bu	Me	*p*-$C_6H_4CF_3$	Bt^-Na^+	20	15	1:1	**10a**	99
2	BnO	Bn	*m*-C_6H_4Cl	Bt^-Na^+	20	15	1:1	**10b**	70
3	*t*-Bu	Me	*p*-$C_6H_4CF_3$	$TolSO_2^-Na^+$	20	15	1:1	**10c**	88
4	*t*-Bu	Me	*m*-C_6H_4Cl	Bn-NH_2	20	5	1:4	**10d**	91
5	Bn	*i*-Bu	*p*-$C_6H_4CF_3$	Bn-NH_2	20	5	1:4	**10e**	55

1-(*N*-Acylamino)alkyltriarylphosphonium salts **4** react also with triarylphosphines. For example, 1-(*N*-pivaloylamino)ethyltris(4-trifluoromethylphenyl)phosphonium tetrafluoroborate (**4a**) is completely transformed into 1-(*N*-pivaloylamino)ethyltriphenylphosphonium tetrafluoroborate (**4b**) during the reaction with triphenylphosphine. On the other hand, using tris(3-chlorophenyl)- phosphine as the nucleophile, we obtain a reaction mixture in which both 1-(*N*-pivaloylamino)ethyltris(4-trifluoromethylphenyl)phosphonium tetrafluoroborate (**4a**) and 1-(*N*-pivaloylamino)ethyltris(3-chlorophenyl)phosphonium tetrafluoroborate (**4c**) are present in a molar ratio of 1:3. A mixture of the same composition can also be obtained by the reaction of salt **4c** with tris(4-trifluoromethylphenyl)phosphine. These observations may suggest the existence of some equilibrium, as described by the equations in Scheme 3. The equilibrium of the reaction is shifted toward more stable and less reactive phosphonium salts, which is evident for the reaction with triphenylphosphine (equation a, Scheme 3).

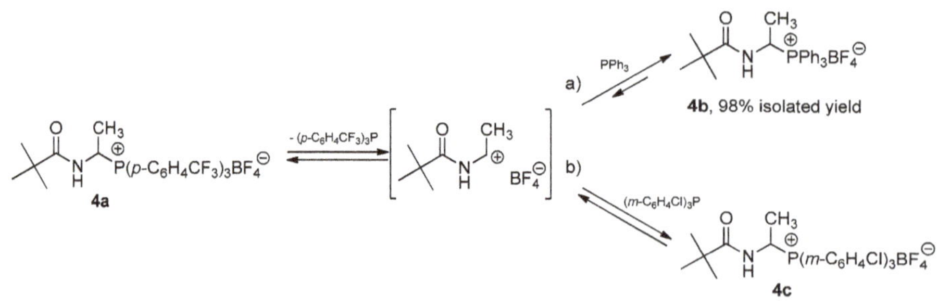

Scheme 3. 1-(*N*-Acylamino)alkyltriarylphosphonium salts with weakened C_{α}-P^{+} bond strength **4**—reactions with triarylphosphines.

The results described above have encouraged us to extend the range of P-nucleophiles by trimethyl phosphite, dimethyl phenylphosphonite, and methyl diphenylphosphinite. It was expected that reactions with these types of phosphorus nucleophiles may occur quickly at room temperature in a non-catalytic environment, as for earlier tested nucleophiles. Surprisingly, the first experiments have shown that reactions are much slower. The analysis of reaction kinetics allowed us to explain these observations. In 2013, we proposed a plausible reaction mechanism wherein the *N*-acyliminium cation or *N*-acylimine, both generated from the 1-(*N*-acylamino)alkyltriphenylphosphonium salt **4** (Ar = Ph), reacts with phosphorus nucleophile to form alkoxyphosphonium salt **13**—the characteristic intermediate of the Michaelis–Arbuzov reaction. The final step of the reaction is the dealkylation of the alkoxyphosphonium salt **13** and may occur directly with triphenylphosphine (see Scheme 4) [7]. Although we have not yet been able to isolate or even observe the formation of postulated intermediate product, we assume that 1-(*N*-acylamino)alkyltriarylphosphonium salts with weakened C_{α}-P^{+} bond strength react with phosphorus nucleophiles in an analogous manner. To prove this, we performed the reaction of 1-(*N*-pivaloylamino)ethyltris(3-chlorophenyl)phosphonium tetrafluoroborate (**4c**) with trimethyl phosphite. The reaction progress was monitored using NMR spectroscopy. In the reaction mixture, besides substrates, two products were also detected. Changes in their concentration as a function of time were measured, as shown in Figure 1. The concentration of the first product quickly reaches its maximum and then rapidly decreases, while the concentration of the second one at the beginning of the reaction is low and then increases. The induction period, which is characteristic for the formation of the final product in the consecutive-type reaction, is very clearly visible. Detailed NMR (^{1}H- and ^{31}P-NMR) and HR-MS analysis confirmed that the first, fast-growing product was 1-(*N*-acylamino)alkyltrimethoxyphosphonium tetrafluoroborate (**13a**) —the postulated intermediate of the Michaelis–Arbuzov reaction. As a result of its demethylation, dimethyl 1-(*N*-pivaloylamino)ethanephosphonate (**12a**) is formed with a much slower reaction. It seems that the second step determines the overall rate of the process.

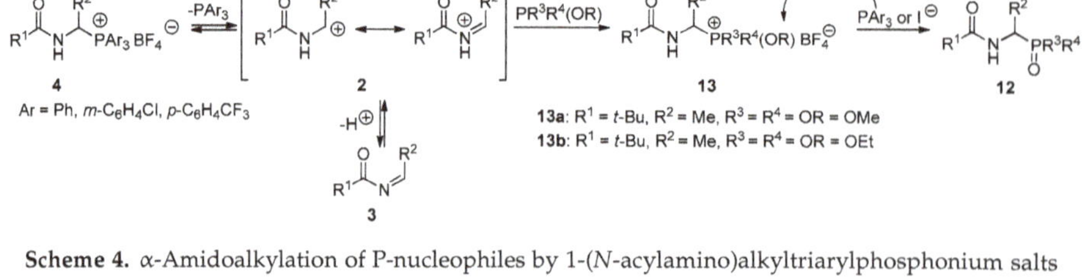

Scheme 4. α-Amidoalkylation of P-nucleophiles by 1-(*N*-acylamino)alkyltriarylphosphonium salts **4**—a plausible mechanism.

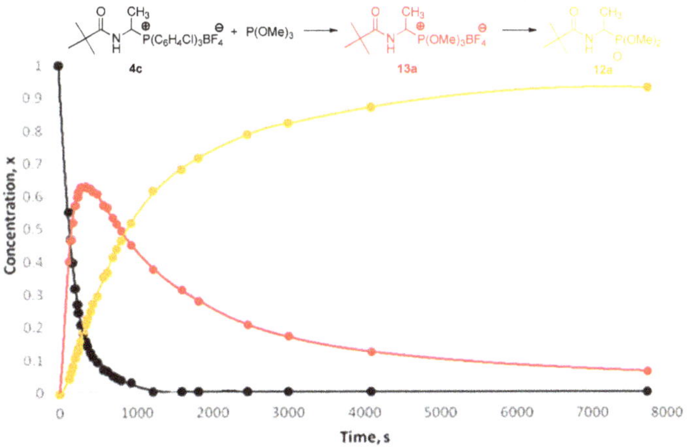

Figure 1. Concentration of the substrate **4c**, intermediate **13a**, and product **12a** as a function of time for the reaction of 1-(*N*-pivaloylamino)ethyltris(3-chlorophenyl)phosphonium tetrafluoroborate (**4c**) with trimethyl phosphite at 26 °C.

A plausible mechanism that explains the described kinetic facts is shown in Scheme 4 and is consistent with the mechanism proposed earlier in 2013. The only difference is that the generation of *N*-acyliminium cation from 1-(*N*-acylamino)alkyltriphenylphosphonium salt **4** (Ar = Ph) is more difficult. On the other hand, the highnucleophilicity of triphenylphosphine facilitates dealkylation, which makes the first step crucial for the course of the entire reaction. In the case of 1-(*N*-acylamino)alkyltriarylphosphonium salts with weakened C_α-P^+ bond strength **4** (R = m-C_6H_4Cl, p-$C_6H_4CF_3$), the first step is easier, which facilitates the formation of alkoxyphosphonium salts **13**. However, due to the lower nucleophilicity of phosphines with electron-withdrawing substituents [$P(m\text{-}C_6H_4Cl)_3$, $P(p\text{-}C_6H_4CF_3)_3$], the rate of dealkylation drops. To overcome this drawback, we decided to add to the reaction mixture the substoichiometric amounts of methyltriphenylphosphonium iodide as the dealkylating agent (molar ratio of 1:0.25). This protocol allowed us to obtain a series of phosphorus analogs of α-amino acids **12** (*N*-acyl-α-aminoalkanephosphonates **12a–c**, *N*-acyl-α-aminoalkanephosphinate (**12d**) and *N*-acyl-α-aminoalkylphosphine oxide (**12e**)) with good or very good yields. Usually, reactions occur efficiently at room temperature. Only in the reaction with dimethyl phenylphosphonite, it was necessary to raise the temperature to 60 °C. At room temperature, the reaction is very slow even after the addition of methyltriphenylphosphonium iodide (see Table 4).

Finally, we tried to isolate and fully characterize the intermediate of the Michaelis–Arbuzov reaction **13**. To this end, we conducted the reaction of 1-(*N*-pivaloylamino)ethyltris(3-chlorophenyl)phosphonium tetrafluoroborate (**4c**) with trimethyl phosphite. Unfortunately, the expected 1-(*N*-pivaloylamino) ethyltrimethoxyphosphonium tetrafluoroborate (**13a**, Scheme 4) was too reactive and attempts to isolate the analytically pure sample failed. To increase the stability of the intermediate **13**, we used triethyl phosphite as the phosphorus nucleophile. This allowed us to isolate 1-(*N*-pivaloylamino)ethyltriethoxyphosphonium tetrafluoroborate (**13b**) and determine its basic physicochemical properties.

Table 4. Conditions and yields for α-amidoalkylation of P-nucleophiles by 1-(*N*-acylamino) alkyltriarylphosphonium salts **4**.

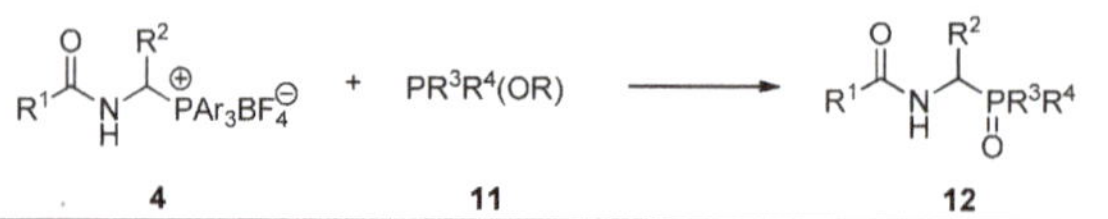

Entry	Phosphonium Salt 4			P-Nu 11			Temp., °C	Time, h	Molar Ratio of 4:11	12	Yield, %
	R[1]	R[2]	Ar	R	R[3]	R[4]					
1	*t*-Bu	Me	*m*-C_6H_4Cl	Me	OMe	OMe	20	3	1:1.5	**12a**	47
2	*t*-Bu	Me	*m*-C_6H_4Cl	Me	OMe	OMe	20	3	1:1.5 [a]	**12a**	85
3	Bn	*i*-Bu	*m*-C_6H_4Cl	Me	OMe	OMe	20	3	1:1.5 [a]	**12b**	77
4	BnO	CH_2O*t*-Bu	*p*-$C_6H_4CF_3$	Me	OMe	OMe	20	3	1:1.5 [a]	**12c**	61
5	*t*-Bu	Me	*m*-C_6H_4Cl	Me	Ph	OMe	60	2	1:1.5 [a]	**12d**	69
6	BnO	*i*-Bu	*p*-$C_6H_4CF_3$	Me	Ph	Ph	20	3	1:1.5 [a]	**12e**	83

[a] Substoichiometric amounts of methyltriphenylphosphonium iodide as a dealkylating agent (molar ratio of 1:0.25) was used.

3. Experimental Section

3.1. General Information

Melting points were determined in capillaries and were uncorrected. IR spectra were measured on an FT-IR spectrophotometer (ATR method). ^{1}H- and ^{13}C-NMR were recorded at operating frequencies of 400 and 100 MHz, respectively, using TMS as the resonance shift standard. ^{31}P-NMR spectra were recorded at an operating frequency of 161.9 MHz without the resonance shift standard, with respect to H_3PO_4 as zero ppm. All chemical shifts (δ) are reported in ppm and coupling constants (*J*) in Hz. High-resolution mass spectrometry (HR-MS) analyses were performed on a Xevo G2 Q-TOF mass spectrometer (Waters, Milford, MA, USA) equipped with an ESI source operating in the positive ion mode. The accurate mass and composition of molecular ion adducts were calculated using the MassLynx software incorporated within the instrument.

^{1}H, ^{13}C, and ^{31}P NMR spectra of all new compounds **6**, **8**, **10**, **12**, **13** as well as the summary table in which we compare conditions and yields for reactions of 1-(*N*-acylamino)alkyltriphenylphosphonium salts (former studies) and 1-(*N*-acylamino)alkyltriarylphosphonium salts (the current work) with selected nucleophiles, are placed in the Supplementary data.

3.2. Syntheses

3.2.1. Substrate Synthesis

Electrochemical decarboxylative α-methoxylation of N-acyl-α-amino acids [2,9]

The electrolysis was conducted in an undivided glass electrolyzer (85 cm^3) equipped with a thermostatic jacket, a magnetic stirrer, a concentrically arranged, cylindrical Pt mesh anode (47 cm^2) and cathode (44 cm^2). To the thus prepared electrolyzer, *N*-acyl-α-amino acid (3.0 mmol), SiO_2-Pip (200 mg), and methanol (30 cm^3) were added. The electrochemical decarboxylative α-methoxylation was executed while stirring, at a constant current of 0.15 A, at 10 °C until a 3.75 F/mol charge had passed. Then, SiO_2-Pip was filtered off, and methanol was evaporated under reduced pressure to obtain *N*-(1-methoxyalkyl)amide, which was used in the next reaction without further purification.

Transformation of N-(1-methoxyalkyl)amides to 1-(N-acylamino)alkyltriarylphosphonium salts **4** [2,10]

To a solution of triarylphosphine (1 mmol) in DCM (2 mL), a tetrafluoroboric acid diethyl ether complex ($HBF_4{\cdot}Et_2O$, 136 μl, 161.9 mg, 1 mmol) was added at 0 °C. The reaction mixture was stirred for 2 h at room temperature. Thereafter, *N*-(1-methoxyalkyl)amide (1 mmol) was added and stirring was continued for 15 min. Evaporation of the solvent yielded the crude 1-(*N*-acylamino) alkyltriarylphosphonium salt **4**, which was used in the next reaction without further purification.

3.2.2. Reactivity of 1-(*N*-acylamino)alkyltriarylphosphonium Salts **4** toward Carbon Nucleophiles

Reaction of 1-(N-acylamino)alkyltriarylphosphonium salts **4** *with C-H acids—general procedure*

A solution of a nucleophile (1.6 mmol), THF (0.5 mmol), and base LDA (2.0 M solution in THF, 0.1 cm^3) was stirred under argon atmosphere. After a few minutes, this mixture was relegated to a solution of 1-(*N*-acylamino)alkyltriarylphosphonium salt (0.2 mmol) in THF (0.5 cm^3) under argon atmosphere. The resulting reaction mixture was stirred for 15 min at room temperature, and then it was evaporated under reduced pressure. The product was separated using column chromatography [toluene/AcOEt 5:1 *v*/*v*].

Diethyl 1-(pivaloylamino)etylpropanedioate (**6a**) [10]. Colorless oil (38.5 mg, 67% yield). ^{1}H-NMR ($CDCl_3$) δ 6.80 (d, *J* = 7.5 Hz, 1H, NH), 4.74–4.62 (m, 1H, $C_\alpha H$), 4.29–4.21 (m, 2H, OCH_2), 4.21–4.12 (m, 2H, OCH_2), 3.57 (d, *J* = 3.4 Hz, 1H, CH), 1.32–1.25 (m, 9H, CH_3 and 2 × $OCH_2\underline{CH_3}$), 1.17 (s, 9H, *t*-Bu) ppm; ^{13}C-NMR ($CDCl_3$) δ 177.6 (NHC = O), 168.7 (C = O), 167.7 (C = O), 61.7 (OCH_2), 61.5 (OCH_2), 55.6 ($C_\alpha H$), 44.2 (CH), 38.6 ($\underline{C}(CH_3)_3$), 27.4 ($C(\underline{C}H_3)_3$), 19.0 (CH_3), 14.0 ($OCH_2\underline{C}H_3$) ppm; IR (ATR) 2977, 1731, 1647, 1510, 1175 cm^{-1}.

Ethyl 2-acetyl-3-(pivaloylamino)butanoate (**6b**). A mixture of two disatereoisomers in a molar ratio of 1.4:1 (31.9 mg, 62% yield). ^{1}H-NMR ($CDCl_3$) δ 6.92 (d, *J* = 8.0 Hz, 1H, NH)[a], 6.41 (d, *J* = 7.8 Hz, 1H, NH)[a], 4.85–4.74 (m, 1H, $C_\alpha H$)[a], 4.68–4.57 (m, 1H, $C_\alpha H$)[a], 4.33–4.10 (m, 2H, OCH_2)[b], 3.72 (d, *J* = 4.9 Hz, 1H, CH)[a], 3.65 (d, *J* = 3.6 Hz, 1H, CH)[a], 2.26 (s, 3H, CH_3CO)[a], 2.25 (s, 3H, CH_3CO)[a], 1.35–1.20 (m, 6H, CH_3 and $OCH_2\underline{CH_3}$)[b], 1.16 (s, 9H, *t*-Bu)[a], 1.15 (s, 9H, *t*-Bu)[a] ppm; ^{13}C-NMR ($CDCl_3$) δ 203.4[a] ($CH_3\underline{C}$ = O), 202.1[a] ($CH_3\underline{C}$ = O), 178.0[a] (NHC = O), 177.7[a] (NHC = O), 169.5[a] (C = O), 168.2[a] (C = O), 63.4[a] ($C_\alpha H$), 62.2[a] ($C_\alpha H$), 61.6[a] (OCH_2), 61.4[a] (OCH_2), 44.3[a] (CH), 43.4[a] (CH), 38.6[a] ($\underline{C}(CH_3)_3$), 38.6[a] ($\underline{C}(CH_3)_3$), 30.4[a] ($\underline{C}H_3CO$), 29.0[a] ($\underline{C}H_3CO$), 27.4[a] ($C(\underline{C}H_3)_3$), 27.3[a] ($C(\underline{C}H_3)_3$), 19.7[a] (CH_3), 18.7[a] (CH_3), 14.1[a] ($OCH_2\underline{C}H_3$), 14.1[a] ($OCH_2\underline{C}H_3$) ppm; IR (ATR) 3318, 2971, 1730, 1635, 1534, 1301, 1190 cm^{-1}. HRMS (TOF-ESI) calcd for $C_{13}H_{24}NO_4$ $[M + H]^+$ 258.1705, found 258.1701. [a] Separate signals from both diastereoisomers. [b] Overlapping signals of both diastereoisomers.

Ethyl 2-acetyl-5-methyl-3-(phenylacetylamino)hexanoate (**6c**). A mixture of two disatereoisomers in a molar ratio of 1.25:1 (55.4 mg, 83% yield). ^{1}H-NMR ($CDCl_3$) δ 7.39–7.19 (m, 5H, Ph)[a], 6.43 (d, *J* = 9.7 Hz, 1H, NH)[b], 6.02 (d, *J* = 9.5 Hz, 1H, NH)[b], 4.77–4.68 (m, 1H, $C_\alpha H$)[b], 4.63–4.54 (m, 1H, $C_\alpha H$)[b], 4.21–4.07 (m, 2H, OCH_2)[a], 3.64 (d, *J* = 4.9 Hz, 1H, CH)[b], 3.60 (d, *J* = 3.7 Hz, 1H, CH)[b], 3.51 (s, 2H, $Ph\underline{CH_2}$)[a], 2.23 (s, 3H, CH_3CO)[b], 2.19 (s, 3H, CH_3CO)[b], 1.56–1.37 (m, 2H, CH_2)[a], 1.28–1.14 (m, 4H, CH and $OCH_2\underline{CH_3}$)[a], 0.95–0.80 (m, 6H, 2 × CH_3)[a] ppm; ^{13}C-NMR ($CDCl_3$) δ 203.2[b] ($CH_3\underline{C}$ = O), 202.2[b] ($CH_3\underline{C}$ = O), 170.6[b] (NHC = O), 170.4[b] (NHC = O), 169.1[b] (C = O), 168.1[b] (C = O), aromatic carbons: 134.7[b], 134.6[b], 129.3[b], 129.3[b], 128.9[b], 128.8[b], 127.2[b], 127.2[b], 62.4[b] ($C_\alpha H$), 61.8[b] ($C_\alpha H$), 61.6[b] (OCH_2), 61.3[b] (OCH_2), 47.1[b] ($Ph\underline{C}H_2$), 46.2[b] ($Ph\underline{C}H_2$), 43.9[b] (CH), 43.8[b] (CH), 42.7[b] (CH_2), 41.9[b] (CH_2), 30.3[b] ($\underline{C}H_3CO$), 29.1[b] ($\underline{C}H_3CO$), 25.1[b] (CH), 25.1[b] (CH), 23.1[b](CH_3), 22.7[b] (CH_3), 22.1[b] (CH_3), 21.7[b] (CH_3), 14.0[b] ($OCH_2\underline{C}H_3$), 14.0[b] ($OCH_2\underline{C}H_3$) ppm; IR (ATR) 3275, 2956, 1740, 1648, 1496, 1260, 1144 cm^{-1}. HRMS (TOF-ESI) calcd for $C_{19}H_{28}NO_4$ $[M + H]^+$ 334.2018, found 334.2009. [a] Overlapping signals of both diastereoisomers. [b] Separate signals from both diastereoisomers.

Dimethyl 3-methyl-1-(phenylacetylamino)butylpropanedioate (**6d**). Colorless crystals (42.3 mg, 63% yield), mp 78–80 °C. ^{1}H-NMR ($CDCl_3$) δ 7.41–7.18 (m, 5H, Ph), 6.25 (d, *J* = 9.6 Hz, 1H, NH), 4.68–4.59 (m, 1H, $C_\alpha H$), 3.69 (s, 3H, OCH_3), 3.62 (s, 3H, OCH_3), 3.55 (d, *J* = 3.9 Hz, 1H, CH), 3.52 (s, 2H, $Ph\underline{CH_2}$), 1.54–1.39 (m, 2H, CH_2), 1.28–1.19 (m, 1H, CH), 0.90 (d, *J* = 6.5 Hz, 3H, CH_3), 0.86 (d, *J* = 6.6 Hz, 3H, CH_3) ppm; ^{13}C-NMR ($CDCl_3$) δ 170.3 (NHC = O), 168.6 (C = O), 168.0 (C = O), aromatic carbons: 134.8, 129.3, 128.8, 127.1, 54.6 ($C_\alpha H$), 52.6 (OCH_3), 52.3 (OCH_3), 46.8 ($Ph\underline{C}H_2$), 43.9 (CH), 42.1 (CH_2), 25.0 (CH), 22.9 (CH_3), 21.9 (CH_3) ppm; IR (ATR) 3283, 2954, 1736, 1648, 1454, 1263 cm^{-1}. HRMS (TOF-ESI) calcd for $C_{18}H_{26}NO_5$ $[M + H]^+$ 336.1811, found 336.1805.

Diethyl 3-methyl-1-(phenylacetylamino)butylpropanedioate (**6e**). Colorless crystals (37.8 mg, 52% yield), mp 93.5–95.5 °C. ^{1}H-NMR ($CDCl_3$) δ 7.42–7.19 (m, 5H, Ph), 6.30 (d, *J* = 9.6 Hz, 1H, NH), 4.68–4.60 (m, 1H, $C_\alpha H$), 4.30–4.00 (m, 4H, 2 × OCH_2), 3.52 (s, 2H, $PhC\underline{H_2}$), 3.52 (d, *J* = 2.8 Hz, 1H, CH), 1.57–1.37 (m, 2H, CH_2), 1.36–1.06 (m, 7H, CH and 2 × $OCH_2C\underline{H_3}$), 0.90 (d, *J* = 6.5 Hz, 3H, CH_3), 0.86 (d, *J* = 6.6 Hz, 3H, CH_3) ppm; ^{13}C-NMR ($CDCl_3$) δ 170.2 (NHC = O), 168.3 (C = O), 167.7 (C = O), aromatic carbons: 134.9, 129.3, 128.8, 127.1, 61.7 (OCH_2), 61.4 (OCH_2), 54.9 ($C_\alpha H$), 46.8 ($Ph\underline{C}H_2$), 43.9 (CH), 42.2 (CH_2), 25.0 (CH), 22.9 (CH_3), 22.0 (CH_3), 14.0 ($OCH_2\underline{C}H_3$), 13.9 ($OCH_2\underline{C}H_3$) ppm; IR (ATR) 3362, 2964, 1745, 1656, 1533, 1345, 1144 cm^{-1}. HRMS (TOF-ESI) calcd for $C_{20}H_{30}NO_5$ $[M + H]^+$ 364.2124, found 364.2122.

Reaction of 1-(N-acylamino)alkyltriarylphosphonium salts **4** *with 1-morpholinocyclohexene—general procedure*

To a solution of 1-(*N*-acylamino)alkyltriarylphosphonium salt **4** (0.25 mmol) in MeCN (1 cm^3), 1-morpholinocyclohexene (84.1 μL, 83.7 mg, 0.5 mmol) was added. After 1 h of stirring at room temperature, an aqueous solution of citric acid (20%) (1.125 cm^3) was added. Stirring was continued for 45 min, and then a saturated solution of $KHCO_3$ was added. Thereafter, the mixture was extracted with DCM (5 × 3 cm^3), and the organic layer was combined and dried over $MgSO_4$. Then, the solvent was evaporated under reduced pressure and the crude product was purified using column chromatography [toluene/EtOAc 2:1 *v*/*v* (**8a**), 5:1 (**8b**, **8c**, **8e**) or 10:1 *v*/*v* (**8d**)].

N-[1-(2-oxocyclohexyl)ethyl]pivalamide (**8a**). A mixture of two disatereoisomers in a molar ratio of 3.5:1 (35.5 mg, 63% yield). The major diastereoisomers was isolated using column chromatography [toluene/EtOAc 2:1] and crystallization from toluene. Colorless crystals, mp 124–126 °C. ^{1}H-NMR ($CDCl_3$) δ 6.51 (d, *J* = 8.1 Hz, 1H, NH), 4.19–4.10 (m, 1H, $C_\alpha H$), 2.56–2.48 (m, 1H, CH), 2.39–2.30 (m, 2H, CH_2), 2.11–2.01 (m, 2H, CH_2), 1.91–1.83 (m, 1H, $C\underline{H}H$), 1.74–1.57 (m, 3H, $CH\underline{H}$ and CH_2), 1.22 (d, *J* = 7.0 Hz, 3H, CH_3), 1.17 (s, 9H, *t*-Bu) ppm; ^{13}C-NMR ($CDCl_3$) δ 214.2 (C = O), 178.0 (NHC = O), 55.4 ($C_\alpha H$), 45.6 (CH), 43.1 (CH_2), 38.7 ($\underline{C}(CH_3)_3$), 32.7 (CH_2), 28.4 (CH_2), 27.6 ($C(\underline{C}H_3)_3$), 24.8 (CH_2), 20.0 (CH_3) ppm; IR (ATR) 3339, 2961, 2869, 1706, 1627, 1526, 1305, 1209, 1117 cm^{-1}. HRMS (TOF-ESI) calcd for $C_{13}H_{24}NO_2$ $[M + H]^+$ 226.1807, found 226.1800.

N-[1-(2-oxocyclohexyl)-3-methylbutyl]phenylacetamide (**8b**). A mixture of two disatereoisomers in a molar ratio of 3.2:1 (46.7 mg, 62% yield). The major diastereoisomers was isolated using column chromatography [toluene/EtOAc 5:1] and crystallization from toluene. Colorless crystals, mp 115.5–117.5 °C. ^{1}H-NMR ($CDCl_3$) δ 7.42–7.09 (m, 5H, Ph), 6.13 (d, *J* = 9.7 Hz, 1H, NH), 4.10–3.99 (m, 1H, $C_\alpha H$), 3.51 (s, 2H, $PhC\underline{H_2}$), 2.49–2.40 (m, 1H, CH), 2.31–2.18 (m, 2H, CH_2), 2.07–1.96 (m, 2H, CH_2), 1.86–1.75 (m, 1H, CH), 1.65–1.37 (m, 5H, CH and 2 × CH_2), 1.21–1.09 (m, 1H, CH), 0.86 (d, *J* = 6.5 Hz, 3H, CH_3), 0.83 (d, *J* = 6.6 Hz, 3H, CH_3) ppm; ^{13}C-NMR ($CDCl_3$) δ 213.5 (C = O), 170.7 (NHC = O), aromatic carbons: 135.3, 129.1, 128.8, 127.1, 54.8 ($C_\alpha H$), 48.7 ($Ph\underline{C}H_2$), 44.2 (CH), 43.1 (CH_2), 43.1 (CH_2), 32.6 (CH_2), 28.3 (CH_2), 25.2 (CH), 25.0 (CH_2), 23.1 (CH_3), 22.0 (CH_3) ppm; IR (ATR) 3273, 2954, 2865, 1702, 1633, 1552, 1455, 1315, 1138 cm^{-1}. HRMS (TOF-ESI) calcd for $C_{19}H_{28}NO_2$ $[M + H]^+$ 302.2120, found 302.2115.

Benzyl N-[1-(2-oxocyclohexyl)-3-methylbutyl]carbamate (**8c**). Only one disatereoisomer was detected and isolated. Colorless oil (31.7 mg, 40% yield). ^{1}H-NMR ($CDCl_3$) δ 7.42–7.27 (m, 5H, Ph), 5.42 (d, *J* = 10.1 Hz, 1H, NH), 5.09 (d, *J* = 12.5 Hz, 1H, $PhC\underline{H}HO$), 5.06 (d, *J* = 12.4 Hz, 1H, $PhCH\underline{H}O$), 3.83–3.69 (m, 1H, $C_\alpha H$), 2.54–2.44 (m, 1H, CH), 2.38–2.23 (m, 2H, CH_2), 2.15–1.99 (m, 2H, CH_2), 1.93–1.82 (m, 1H, $C\underline{H}H$), 1.74–1.55 (m, 5H, $CH\underline{H}$ and 2 × CH_2), 1.28–1.16 (m, 1H, CH), 0.91 (d, *J* = 6.4 Hz, 3H, CH_3), 0.89 (d, *J* = 6.5 Hz, 3H, CH_3) ppm; ^{13}C-NMR ($CDCl_3$) δ 213.1 (C = O), 156.5 (NHC = O), aromatic carbons: 136.8, 128.4, 127.9, 127.9, 66.5 ($Ph\underline{C}H_2O$), 54.7 ($C_\alpha H$), 50.8 (CH), 43.4 (CH_2), 43.2 (CH_2), 32.3 (CH_2), 28.1 (CH_2), 25.1 (CH_2), 25.1 (CH), 23.2 (CH_3), 21.9 (CH_3) ppm; IR (ATR) 3439, 3333, 2952, 2866, 1699, 1499, 1213, 1052 cm^{-1}. HRMS (TOF-ESI) calcd for $C_{19}H_{28}NO_3$ $[M + H]^+$ 318.2069, found 318.2061.

Benzyl N-[1-(2-oxocyclohexyl)-2-phenylethyl]carbamate (**8d**). A mixture of two disatereoisomers in a molar ratio of 13.4:1 (29.0 mg, 33% yield). The major diastereoisomers was isolated using column chromatography [toluene/EtOAc 10:1]. Colorless crystals, mp 99–101 °C. ^{1}H-NMR ($CDCl_3$) δ 7.42–7.06 (m, 10H, 2 × Ph), 5.65 (d, J = 10.0 Hz, 1H, NH), 5.07 (d, J = 12.3 Hz, 1H, PhC$\underline{H}$HO), 5.00 (d, J = 12.4 Hz, 1H, PhCH$\underline{H}$O), 3.95–3.87 (m, 1H, C_αH), 2.99 (dd, J_1 = 13.5, J_2 = 7.3 Hz, 1H, PhC$\underline{H}$H), 2.92 (dd, J_1 = 13.5, J_2 = 8.5 Hz, 1H, PhCH$\underline{H}$), 2.51–2.41 (m, 1H, CH), 2.40–2.31 (m, 1H, CH), 2.29–2.17 (m, 1H, CH), 2.09–1.92 (m, 2H, CH_2), 1.87–1.78 (m, 1H, CH), 1.77–1.51 (m, 3H, CH and CH_2) ppm; ^{13}C-NMR ($CDCl_3$) δ 213.5 (C = O), 156.4 (NHC = O), aromatic carbons: 138.6, 136.7, 129.1, 128.5, 128.4, 127.9, 127.8, 126.4, 66.4 (Ph$\underline{C}H_2$O), 54.2 (C_αH), 51.9 (CH), 43.2 (CH_2), 40.3 (Ph$\underline{C}H_2$), 32.6 (CH_2), 28.1 (CH_2), 25.0 (CH_2) ppm; IR (ATR) 3308, 2929, 1721, 1697, 1541, 1341, 1242, 1094, 1051 cm^{-1}. HRMS (TOF-ESI) calcd for $C_{22}H_{26}NO_3$ [M + H]$^+$ 352.1913, found 352.1913.

Benzyl N-[1-(2-oxocyclohexyl)-2-tert-butoxyethyl]carbamate (**8e**). Only one disatereoisomer was detected and isolated. Colorless oil (26.9 mg, 31% yield). ^{1}H-NMR ($CDCl_3$) δ 7.38–7.28 (m, 5H, Ph), 5.57 (d, J = 9.7 Hz, 1H, NH), 5.11 (d, J = 12.3 Hz, 1H, PhC$\underline{H}$HO), 5.06 (d, J = 12.3 Hz, 1H, PhCH$\underline{H}$O), 3.89–3.79 (m, 1H, C_αH), 3.44 (d, J = 7.1 Hz, 2H, $\underline{CH_2}$O*t*-Bu), 2.90–2.82 (m, 1H, CH), 2.38–2.27 (m, 2H, CH_2), 2.09–2.00 (m, 2H, CH_2), 1.92–1.85 (m, 1H, $\underline{C}$HH), 1.78–1.64 (m, 3H, CH$\underline{H}$ and CH_2), 1.12 (s, 9H, *t*-Bu) ppm; ^{13}C-NMR ($CDCl_3$) δ 213.8 (C = O), 156.4 (NHC = O), aromatic carbons: 136.6, 128.5, 128.0, 73.1 (O$\underline{C}(CH_3)_3$), 66.6 (Ph$\underline{C}H_2$O), 62.1 ($\underline{C}H_2$O*t*-Bu), 52.5 (C_αH), 50.0 (CH), 42.8 (CH_2), 31.9 (CH_2), 28.1 (CH_2), 27.5 (OC($\underline{C}H_3)_3$), 24.9 (CH_2) ppm; IR (ATR) 3432, 2972, 2935, 2867, 1701, 1499, 1363, 1195, 1057 cm^{-1}. HRMS (TOF-ESI) calcd for $C_{20}H_{30}NO_4$ [M + H]$^+$ 348.2175, found 348.2173.

3.2.3. Reactivity of 1-(*N*-Acylamino)alkyltriarylphosphonium Salts 4 toward Heteronucleophiles

Reaction of 1-(N-acylamino)alkyltriarylphosphonium salts **4** *with benzotriazole sodium salt—general procedure*

To a solution of 1-(*N*-acylamino)alkyltriarylphosphonium salt **4** (0.5 mmol) in $CHCl_3$ (1 cm^3), benzotriazole sodium salt (0.5 mmol) was added. The reaction mixture was stirred at room temperature for 15 min. Thereafter, it was filtered through a fluted filter and the filtrate was evaporated under reduced pressure. The residue was crystallized from toluene.

N-[1-(Benzotriazol-1-yl)ethyl]pivalamide (**10a**) [6]. Colorless crystals (121.9 mg, 99% yield), mp 142–144 °C. ^{1}H-NMR ($CDCl_3$) δ 8.06–8.02 (m, 1H, aromatic), 7.85–7.79 (m, 1H, aromatic), 7.54–7.48 (m, 1H, aromatic), 7.41–7.35 (m, 1H, aromatic), 6.88 (dq, J_1 = 9.1, J_2 = 6.7 Hz, 1H, C_αH), 6.63 (d, J = 8.8 Hz, 1H, NH), 2.04 (d, J = 6.7 Hz, 3H, CH_3), 1.15 (s, 9H, *t*-Bu) ppm; ^{13}C-NMR (100 MHz, $CDCl_3$) δ 178.2 (C = O), aromatic carbons: 145.5, 132.4, 127.7, 124.4, 119.5, 110.4, 58.7 (C_αH), 38.7 ($\underline{C}(CH_3)_3$), 27.2 (C($\underline{C}H_3)_3$), 20.7 (CH_3) ppm; IR (ATR) 3346, 2969, 1668, 1512, 1193, 1152, 1065 cm^{-1}.

Benzyl N-[1-(benzotriazol-1-yl)-2-phenylethyl]carbamate (**10b**) [6]. Colorless crystals (130.4 mg, 70% yield), mp 117.5–119.5 °C. ^{1}H-NMR ($CDCl_3$) δ 8.00 (d, J = 8.4 Hz, 1H, aromatic), 7.54 (d, J = 7.5 Hz, 1H, aromatic), 7.42–7.35 (m, 1H, aromatic), 7.33–7.27 (m, 4H, aromatic), 7.27–7.20 (m, 2H, aromatic), 7.19–7.12 (m, 3H, aromatic), 7.11–7.04 (m, 2H, aromatic), 6.71–6.58 (m, 1H, C_αH), 6.02 (d, J = 8.9 Hz, 1H, NH), 5.10 (d, J = 12.4 Hz, 1H, PhC$\underline{H}$HO), 4.97 (d, J = 12.2 Hz, 1H, PhCH$\underline{H}$O), 3.80–3.68 (m, 1H, PhC$\underline{H}$H), 3.63 (dd, J_1 = 13.8, J_2 = 6.6 Hz, 1H, PhCH$\underline{H}$) ppm; ^{13}C-NMR ($CDCl_3$) δ 160.3 (C = O), aromatic carbons: 145.5, 136.0, 134.7, 131.9, 129.1, 128.7, 128.5, 128.4, 128.1, 127.7, 127.3, 124.1, 119.7, 109.7, 67.5 (Ph$\underline{C}H_2$O), 65.8 (C_αH), 41.1 (PH$\underline{C}H_2$) ppm; IR (ATR) 3177, 3008, 1712, 1548, 1280, 1261, 1244, 1195, 1046, 1022 cm^{-1}.

Reaction of 1-(N-pivaloylamino)ethyltris(4-trifluoromethylphenyl)phosphonium tetrafluoroborate **4a** *with sodium p-toluenesulfinate*

To a solution of 1-(*N*-pivaloylamino)ethyltris(4-trifluoromethylphenyl)phosphonium tetrafluoroborate **4a** (340.7 mg, 0.5 mmol) in $CHCl_3$ (1 cm^3), sodium *p*-toluenesulfinate (89.1 mg, 0.5 mmol) was added.

The reaction mixture was stirred at room temperature for 15 min. Thereafter, it was filtered through a fluted filter and the filtrate was evaporated under reduced pressure. The residue was crystallized from toluene.

N-[1-(p-Toluenesulfonyl)ethyl]pivalamide (**10c**) [5]. Colorless crystals (124.7 mg, 88% yield), mp 143–145 °C. ^{1}H-NMR ($CDCl_3$) δ 7.77 (d, J = 8.3 Hz, 2H, aromatic), 7.33 (d, J = 7.9 Hz, 2H, aromatic), 5.98 (d, J = 10.2 Hz, 1H, NH), 5.41 (dq, J_1 = 10.2, J_2 = 7.0 Hz, 1H, $C_{\alpha}H$), 2.42 (s, 3H, CH_3), 1.62 (d, J = 7.0 Hz, 3H, CH_3), 1.01 (s, 9H, *t*-Bu) ppm; ^{13}C-NMR ($CDCl_3$) δ 176.9 (C = O), aromatic carbons: 145.1, 133.5, 129.6, 129.1, 64.3 ($C_{\alpha}H$), 38.7 ($\underline{C}(CH_3)_3$), 27.2 ($C(\underline{C}H_3)_3$), 21.6 (CH_3), 13.2 (CH_3) ppm; IR (ATR) 3372, 2973, 1677, 1516, 1308, 1287, 1135, 1083, 1016, 724 cm^{-1}.

Reaction of 1-(N-acylamino)alkyltriarylphosphonium salts ***4*** *with benzylamine—general procedure*

To a stirred solution of 1-(*N*-acylamino)alkyltriarylphosphonium salt **4** (0.25 mmol) in DCM (1 cm^3), benzylamine (109.4 μl, 110 mg, 1 mmol) was added dropwise. Stirring was continued for 5 min at room temperature. Thereafter, the mixture was evaporated under reduced pressure and dried. The residue was purified using column chromatography [DCM/MeOH/Et_3N, 5:1:0.2 *v*/*v*/*v*].

N-[1-(Benzylamino)ethyl]pivalamide (**10d**). Yellow crystals (53.3 mg, 91% yield), mp 56.5–58.5 °C. ^{1}H-NMR ($CDCl_3$) δ 7.35–7.29 (m, 4H, aromatic), 7.26–7.21 (m, 1H, aromatic), 5.67 (d, J = 6.7 Hz, 1H, NH), 4.90 (qd, J_1 = 6.2, J_2 = 1.6 Hz, 1H, $C_{\alpha}H$), 3.77 (s, 2H, $PhC\underline{H_2}$), 1.92 (br s, 1H, NH), 1.32 (d, J = 6.2 Hz, 3H, CH_3), 1.17 (s, 9H, *t*-Bu) ppm; ^{13}C-NMR ($CDCl_3$) δ 178.1 (C = O), aromatic carbons: 140.1, 128.4, 127.9, 126.9, 60.9 ($C_{\alpha}H$), 49.8 ($Ph\underline{C}H_2$), 38.6 ($\underline{C}(CH_3)_3$), 27.4 ($C(\underline{C}H_3)_3$), 21.6 ppm (CH_3); IR (ATR) 3328, 2971, 1622, 1527, 1475, 1144, 1097 cm^{-1}. HRMS (TOF-ESI) calcd for $C_{14}H_{23}N_2O$ [M + H]$^+$ 235.1810, found 235.1802.

N-[1-(Benzylamino)-3-methylbutyl]phenylacetamide (**10e**). Colorless crystals (42.7 mg, 55% yield), mp 93–95 °C. ^{1}H-NMR ($CDCl_3$) δ 7.59–6.96 (m, 10H, 2 × Ph), 5.33 (d, J = 8.3 Hz, 1H, NH), 4.82–4.75 (m, 1H, $C_{\alpha}H$), 3.68 (s, 2H, $PhC\underline{H_2}NH$), 3.54 (s, 2H, $PhC\underline{H_2}$), 1.66 (br s, 1H, NH), 1.64–1.53 (m, 1H, CH), 1.45–1.28 (m, 2H, CH_2), 0.87 (d, J = 6.6 Hz, 3H, CH_3), 0.84 (d, J = 6.6 Hz, 3H, CH_3) ppm; ^{13}C-NMR ($CDCl_3$) δ 170.9 (C = O), aromatic carbons: 140.3, 135.0, 129.3, 129.0, 128.4, 128.0, 127.3, 126.9, 63.3 ($C_{\alpha}H$), 49.6 ($Ph\underline{C}H_2$), 44.5 ($Ph\underline{C}H_2$), 44.1 (CH_2), 24.9 (CH), 22.6 (CH_3), 22.5 (CH_3) ppm; IR (ATR) 3306, 2962, 2946, 1632, 1518, 1144, 1009 cm^{-1}. HRMS (TOF-ESI) calcd for $C_{20}H_{27}N_2O$ [M + H]$^+$ 311.2123, found 311.2121.

Reaction of 1-(N-pivaloylamino)ethyltris(4-trifluoromethylphenyl)phosphonium tetrafluoroborate ***4a*** *with triphenylphosphine*

To a solution of 1-(*N*-pivaloylamino)ethyltris(4-trifluromethylphenyl)phosphonium tetrafluoroborate **4a** (102.2 mg, 0.15 mmol) in DCM (1 cm^3), triphenylphosphine (39.3 mg, 0.15 mmol) was added. The homogeneous mixture was allowed to react at room temperature for 5 min, and 1-(*N*-pivaloylamino)ethyltriphenylphosphonium tetrafluoroborate (**4b**) [9] was precipitated with Et_2O, separated by decantation, and dried under reduced pressure.

1-(*N*-Pivaloylamino)ethyltriphenylphosphonium tetrafluoroborate (**4b**). Colorless crystals (70.2 mg, 98% yield), mp 160.5–162.5 °C. ^{1}H-NMR ($CDCl_3$) δ 7.89–7.64 (m, 16H, 3 × Ph and NH), 5.82–5.73 (m, 1H, $C_{\alpha}H$), 1.72 (dd, J_1 = 17.8, J_2 = 7.4 Hz, 3H, CH_3), 0.91 (s, 9H, *t*-Bu) ppm; ^{13}C-NMR ($CDCl_3$) δ 179.7 (d, J = 2.3 Hz, C = O), aromatic carbons: 134.8 (d, J = 3.0 Hz), 134.5 (d, J = 9.2 Hz), 130.0 (d, J = 12.3 Hz), 118.4 (d, J = 82.5 Hz), 45.0 (d, J = 53.4 Hz, $C_{\alpha}H$), 38.5 ($\underline{C}(CH_3)_3$), 26.8 ($C(\underline{C}H_3)_3$), 17.4 (d, J = 4.6 Hz, CH_3) ppm; ^{31}P NMR (161.9 MHz, $CDCl_3$) δ 29.2 ppm; IR (ATR) 3373, 1684, 1516, 1447, 1136, 1040 cm^{-1}.

Reaction of 1-(N-acylamino)alkyltriarylphosphonium salts **4** *with triarylphosphines—NMR scale*

To a solution of 1-(*N*-acylamino)alkyltriarylphosphonium salt (0.025 mmol) in $CDCl_3$ (0.65 cm^3), the corresponding triarylphosphine (0.025 mmol) was added. Reactions were conducted in NMR tubes at 26 °C, and their course was monitored using NMR spectroscopy.

Reaction of 1-(N-acylamino)alkyltriarylphosphonium salts **4** *with phosphorus nucleophiles in Michaelis–Arbuzov type reaction—general procedure*

To a stirred solution of 1-(*N*-acylamino)alkyltriarylphosphonium salt **4** (0.25 mmol) and methyltriphenylphosphonium iodide (25.3 mg, 0.0625 mmol) in DCM (2 cm^3), phosphorus nucleophile (0.375 mmol) was added, and the mixture was stirred for 3 h at room temperature (**12a–c** and **12e**) or for 2 h at 60 °C (**12d**). The mixture was evaporated under reduced pressure and the residue was extracted with toluene (3 × 1 cm^3) at 50 °C. The solvent was evaporated under reduced pressure, and the product was isolated using column chromatography [DCM/MeOH 20:1 *v*/*v*].

Dimethyl 1-(N-pivaloylamino)ethanephosphonate (**12a**) [7]. Colorless crystals (50.4 mg, 85% yield), mp 125.5–127.5 °C. ^{1}H-NMR ($CDCl_3$) δ 5.96 (d, *J* = 8.1 Hz, 1H, NH), 4.65–4.49 (m, 1H, C_αH), 3.78 (d, *J* = 4.4 Hz, 3H, OCH_3), 3.75 (d, *J* = 4.3 Hz, 3H, OCH_3), 1.38 (dd, J_1 = 16.8, J_2 = 7.4 Hz, 3H, CH_3), 1.21 (s, 9H, *t*-Bu) ppm; ^{13}C-NMR ($CDCl_3$) δ 177.6 (d, *J* = 5.1 Hz, C = O), 53.1 (d, *J* = 7.0 Hz, OCH_3), 52.9 (d, *J* = 6.5 Hz, OCH_3), 40.1 (d, *J* = 156.4 Hz, C_αH), 38.6 ($\underline{C}(CH_3)_3$), 27.3 ($C(\underline{C}H_3)_3$), 15.5 (CH_3) ppm; ^{31}P-NMR ($CDCl_3$) δ 28.5 ppm; IR (ATR) 3281, 2956, 1659, 1529, 1224, 1205, 1057, 1026, 1007 cm^{-1}.

Dimethyl 1-(N-phenylcetylamino)-3-methylbutanephosphonate (**12b**) [4]. Colorless crystals (60.3 mg, 77% yield), mp 117–119 °C. ^{1}H-NMR ($CDCl_3$) δ 7.39–7.23 (m, 5H, Ph), 5.72 (d, *J* = 9.9 Hz, 1H, NH), 4.62–4.48 (m, 1H, C_αH), 3.73 (d, *J* = 10.6 Hz, 3H, OCH_3), 3.63 (d, *J* = 10.6 Hz, 3H, OCH_3), 3.60 (s, 2H, $PhC\underline{H_2}$), 1.61–1.44 (m, 3H, CH_2 and CH), 0.88 (d, *J* = 5.9 Hz, 6H, 2 × CH_3) ppm; ^{13}C-NMR ($CDCl_3$) δ 170.5 (d, *J* = 4.3 Hz, C = O), aromatic carbons: 134.6, 129.2, 128.9, 127.4, 53.1 (d, *J* = 7.2 Hz, OCH_3), 53.0 (d, *J* = 6.6 Hz, OCH_3), 43.7 ($Ph\underline{C}H_2$), 43.2 (d, *J* = 155.5 Hz, C_αH), 38.1 (d, *J* = 1.8 Hz, CH_2), 24.5 (d, *J* = 13.4 Hz, CH), 23.2 (CH_3), 21.1 (CH_3) ppm; ^{31}P-NMR ($CDCl_3$) δ 27.6 ppm; IR (ATR) 3243, 2959, 1673, 1540, 1217, 1027, 836, 742 cm^{-1}.

Dimethyl 1-(N-benzyloxycarbonylamino)-2-tert-butoxyethanephosphonate (**12c**) [4]. Colorless crystals (54.8 mg, 61% yield), mp 59.5–61.5 °C. ^{1}H-NMR ($CDCl_3$) δ 7.41–7.28 (m, 5H, Ph), 5.29 (d, *J* = 8.8 Hz, 1H, NH), 5.13 (s, 2H, $PhC\underline{H_2}O$), 4.33–4.19 (m, 1H, C_αH), 3.76 (d, *J* = 10.6 Hz, 3H, OCH_3), 3.75 (d, *J* = 10.3 Hz, 3H, OCH_3), 3.61 (dd, J_1 = 9.4, J_2 = 3.8 Hz, 1H, C$\underline{H}$HO*t*-Bu), 3.54 (dd, J_1 = 9.4, J_2 = 3.8 Hz, 1H, CH$\underline{H}$O*t*-Bu), 1.18 (s, 9H, *t*-Bu) ppm; ^{13}C-NMR ($CDCl_3$) δ 155.7 (C = O), aromatic carbons: 136.2, 128.5, 128.2, 128.1, 73.7 ($O\underline{C}(CH_3)_3$), 67.2 ($Ph\underline{C}H_2O$), 60.6 ($\underline{C}H_2O$*t*-Bu), 53.3 (d, *J* = 6.0 Hz, OCH_3), 52.6 (d, *J* = 5.8 Hz, OCH_3), 48.3 (d, *J* = 156.0 Hz, C_αH), 27.3 ($OC(\underline{C}H_3)_3$) ppm; ^{31}P-NMR ($CDCl_3$) δ 26.1 ppm; IR (ATR) 3293, 2974, 1699, 1532, 1278, 1235, 1039, 1023, 756, 697 cm^{-1}.

Methyl phenyl(1-pivaloylaminoethyl)phosphinate (**12d**). Colorless crystals (48.9 mg, 69% yield), mp 131–133 °C. ^{1}H-NMR ($CDCl_3$) δ 7.84–7.75 (m, 2H, aromatic), 7.59–7.43 (m, 3H, aromatic), 5.76 (d, *J* = 9.6 Hz, 1H, NH), 4.79–4.67 (m, 1H, C_αH), 3.67 (d, *J* = 10.8 Hz, 3H, OCH_3), 1.45 (dd, J_1 = 14.6, J_2 = 7.3 Hz, 3H, CH_3), 0.92 (s, 9H, *t*-Bu) ppm; ^{13}C-NMR ($CDCl_3$) δ 177.3 (d, *J* = 4.7 Hz, C = O), aromatic carbons: 132.8 (d, *J* = 2.8 Hz), 132.4 (d, *J* = 9.4 Hz), 128.5 (d, *J* = 12.5 Hz), 127.5 (d, *J* = 124.2 Hz), 51.8 (d, *J* = 7.1 Hz, OCH_3), 42.3 (d, *J* = 115.5 Hz, C_αH), 38.5 ($\underline{C}(CH_3)_3$), 27.1 ($C(\underline{C}H_3)_3$), 14.4 (CH_3) ppm; ^{31}P-NMR ($CDCl_3$) δ 42.9 ppm. IR (ATR) 3264, 2954, 1660, 1531, 1198, 1139, 1027, 799 cm^{-1}; HRMS (TOF-ESI) calcd for $C_{14}H_{23}NO_3P$ $[M + H]^+$ 284.1416, found 284.1405.

Diphenyl 1-(N-benzyloxycarbonylamino)-3-methylbutylphosphine oxide (**12e**) [8]. Colorless crystals (87.5 mg, 83% yield), mp 173.5–175.5 °C. ^{1}H-NMR ($CDCl_3$) δ 7.87–7.09 (m, 15H, 3 × Ph), 5.40 (d, *J* = 10.6 Hz, 1H, NH), 5.02 (d, *J* = 12.5 Hz, 1H, PhC$\underline{H}$HO), 4.89 (d, *J* = 12.5 Hz, 1H, PhCH$\underline{H}$O), 4.83–4.72 (m, 1H, C_αH), 1.83–1.66 (m, 2H, CH_2), 1.36–1.21 (m, 1H, CH), 0.90 (d, *J* = 6.5 Hz, 3H, CH_3), 0.86 (d, *J* = 6.6 Hz, 3H, CH_3) ppm; ^{13}C-NMR ($CDCl_3$) δ 156.1 (d, *J* = 4.5 Hz, C = O), aromatic carbons: 136.4, 132.0 (d,

J = 2.8 Hz), 131.9 (d, J = 2.8 Hz), 131.5, 131.3, 131.1 (d, J = 9.2 Hz), 130.9 (d, J = 9.0 Hz), 128.8 (d, J = 11.4 Hz), 128.4 (d, J = 11.6 Hz), 128.4, 127.9, 127.6, 66.8 (Ph$\underline{C}H_2$O), 48.1 (d, J = 79.5 Hz, C_αH), 37.5 (d, J = 3.5 Hz, CH_2), 24.5 (d, J = 11.0 Hz, CH), 23.4 (CH_3), 21.0 (CH_3) ppm; ^{31}P-NMR ($CDCl_3$) δ 33.5 ppm; IR (ATR) 3181, 2954, 1702, 1545, 1436, 1263, 1187, 1120 cm^{-1}.

Reaction of 1-(N-pivaloylamino)ethyltris(3-chlorophenyl)phosphonium tetrafluoroborate **4c** *with trimethyl phosphite—measurement of changes in concentrations of the substrate* **4c**, *intermediate* **13a** *and product* **12a** *by NMR*

To a solution of 1-(*N*-pivaloylamino)ethyltris(3-chlorophenyl)phosphonium tetrafluoroborate **4c** (29.0 mg, 0.05 mmol) in $CDCl_3$ (0.65 cm^3), trimethyl phosphite (8.9 μL, 9.3 mg, 0.075 mmol) was added. Dimethyldiphenylsilane (5 mg) was used as the internal standard. The reaction mixture was placed directly into the NMR tube. Changes in the concentrations of substrate **4c**, intermediate **13a**, and product **12a** were monitored using ^{1}H-NMR spectroscopy.

Reaction of 1-(N-pivaloylamino)ethyltris(3-chlorophenyl)phosphonium tetrafluoroborate **4c** *with triethyl phosphite–synthesis of 1-(N-pivaloylamino)ethyltriethoxyphosphonium tetrafluoroborate* **13b**

To a stirred solution of 1-(*N*-pivaloylamino)ethyltris(3-chlorophenyl)phosphonium tetrafluoroborate (**4c**, 145.2 mg, 0.25 mmol) in DCM (1 cm^3), triethyl phosphite (64.6 μL, 62.3 mg, 0.375 mmol) was added. Stirring was continued for 30 min at room temperature. Thereafter the intermediate product **13b** was precipitated with Et_2O, separated by decantation, and dried under reduced pressure. In order to obtain the triethoxyphosphonium salt **13** with a higher purity, the reaction should be carried out under argon atmosphere.

1-(N-Pivaloylamino)ethyltriethoxyphosphonium tetrafluoroborate (**13b**). Colorless oil (47.6 mg, 50% yield). ^{1}H-NMR ($CDCl_3$) δ 7.53 (t, J = 8.2 Hz, 1H, NH), 4.69 (dq, J_1 = 19.7, J_2 = 7.0 Hz, 1H, C_αH), 4.62–4.46 (m, 6H, 3 × OCH_2), 1.56 (dd, J_1 = 20.2, J_2 = 7.3 Hz, 3H, CH_3), 1.48 (td, J_1 = 7.0, J_2 = 0.9 Hz, 9H, 3 × $OCH_2\underline{CH_3}$), 1.23 (s, 9H, *t*-Bu) ppm; ^{13}C-NMR ($CDCl_3$) δ 179.9 (C = O), 70.2 (d, J = 9.1 Hz, OCH_2), 42.3 (d, J = 148.6 Hz, C_αH), 38.5 ($\underline{C}(CH_3)_3$), 27.1 ($C(\underline{C}H_3)_3$), 15.9 (d, J = 6.0 Hz, $OCH_2\underline{C}H_3$), 14.0 (CH_3) ppm; ^{31}P-NMR ($CDCl_3$) δ 37.0 ppm; IR (ATR) 3375, 2973, 1664, 1515, 1030 cm^{-1}. HRMS (TOF-ESI) calcd for $C_{13}H_{29}NO_4P$ $[M]^+$ 294.1834, found 294.1829.

4. Conclusions

Modification of the phosphonium group by introducing electron-withdrawing substituents results in the weakening of the C_α-P^+ bond and makes it susceptible to cleavage. This phenomenon is the cause of the high reactivity of 1-(*N*-acylamino)alkyltriarylphosphonium salts with weakened C_α-P^+ bond strength. As we have demonstrated, these types of phosphonium salts react smoothly, usually at room temperature with various types of nucleophiles such as 1-morpholinocyclohexene, 1,3-dicarbonyl compounds, benzotriazole sodium salt, *p*-toluenesulfinate sodium salt, benzylamine, triarylphosphines, and other P-nucleophiles. Only in the case of 1,3-dicarbonyl compounds, it was necessary to use a strong base to generate enolate anions. Reactions with 1-morpholinocyclohexene, benzotriazole sodium salt, *p*-toluenesulfinate sodium salt, benzylamine, and triarylphosphines do not require the use of any catalysts and occur quite fast (5–60 min). Other examined P-nucleophiles also react efficiently with 1-(*N*-acylamino)alkyltriarylphosphonium salts. However, the quickly formed intermediate, in the absence of any dealkylating agent, slowly transforms into a final product. Therefore, to facilitate the reaction, we used a substoichiometric amount of the dealkylating agent in the form of methyltriphenylphosphonium iodide.

The use of 1-(*N*-acylamino)alkyltriarylphosphonium salts with weakened C_α-P^+ bond strength allowed us to discover interesting mechanistic aspects of the examined reactions. Detection, isolation, and characterization of 1-(*N*-acylamino)alkyltrialkoxyphosphonium salt **13b**—the reactive intermediate in the Michaelis–Arbuzov type reaction, were particularly important.

Futher studies on expanding the range of nucleophiles, which can be used in α-amidoalkylation by 1-(*N*-acylamino)alkyltriarylphosphonium salts with weakened C_{α}-P^{+} bond strength, are in progress.

Supplementary Materials: The following are available online at http://www.mdpi.com/1420-3049/23/10/2453/s1, Supporting information includes a summary table (comparison of conditions and yields for reactions of 1-(*N*-acylamino)alkyltriphenylphosphonium salts (former studies) and 1-(*N*-acylamino)alkyltriarylphosphonium salts (the current work) with selected nucleophiles), ^{1}H, ^{13}C, and ^{31}P NMR spectra of all new compounds **6**, **8**, **10**, **12**, **13**. Supplementary data associated with this article can be found in the online version.

Author Contributions: Conceptualization, J.A.; Formal analysis, J.A., A.W., J.K. and K.E.; Investigation, J.A., A.W., J.K. and K.E.; Methodology, J.A. and A.W.; Supervision, J.A. and K.W.; Writing—original draft, J.A.; Writing—review & editing, J.A. and A.W.

Funding: This research received no external funding.

Acknowledgments: This work received financial support from the National Science Center, Poland (NCN) under Grant No. 2015/19/D/ST5/00733.

Conflicts of Interest: The authors declare no conflict of interest.

References

1. Mazurkiewicz, R.; Październiok-Holewa, A.; Adamek, J.; Zielińska, K. α-Amidoalkylating agents: Structure, synthesis, reactivity and application. *Adv. Heterocycl. Chem.* **2014**, *111*, 43–94. [CrossRef]
2. Adamek, J.; Węgrzyk, A.; Krawczyk, M.; Erfurt, K. Catalyst-free Friedel-Crafts reaction of 1-(*N*-acylamino)alkyltriarylphosphonium salts with electron-rich arenes. *Tetrahedron* **2018**, *74*, 2575–2583. [CrossRef]
3. Adamek, J.; Mazurkiewicz, R.; Węgrzyk, A.; Erfurt, K. 1-Imidoalkylphosphonium salts with modulated C_{α}-P^{+} bond strength: Synthesis and application as new active α-imidoalkylating agents. *Beilstein J. Org. Chem.* **2017**, *13*, 1446–1455. [CrossRef] [PubMed]
4. Zielińska, K.; Mazurkiewicz, R.; Szymańska, K.; Jarzębski, A.; Magiera, S.; Erfurt, K. Penicillin G acylase-mediated kinetic resolution of racemic 1-(*N*-acylamino)alkylphosphonic and 1-(*N*-acylamino)alkylphosphinic acids and their esters. *J. Mol. Catal. B Enzym.* **2016**, *132*, 31–40. [CrossRef]
5. Adamek, J.; Mazurkiewicz, R.; Październiok-Holewa, A.; Grymel, M.; Kuźnik, A.; Zielińska, K. 1-(*N*-Acylamino)alkyl Sulfones from *N*-Acyl-α-amino Acids or *N*-Alkylamides. *J. Org. Chem.* **2014**, *79*, 2765–2770. [CrossRef] [PubMed]
6. Adamek, J.; Mazurkiewicz, R.; Październiok-Holewa, A.; Kuźnik, A.; Grymel, M.; Zielińska, K.; Simka, W. *N*-[1-(Benzotriazol-1-yl)alkyl] amides from *N*-acyl-α-amino acids or *N*-alkylamides. *Tetrahedron* **2014**, *70*, 5725–5729. [CrossRef]
7. Adamek, J.; Październiok-Holewa, A.; Zielińska, K.; Mazurkiewicz, R. Comparative Studies on the Amidoalkylating Properties of *N*-(1-Methoxyalkyl)Amides and 1-(*N*-Acylamino) Alkyltriphenylphosphonium Salts in the Michaelis–Arbuzov-Like Reaction: A New One-Pot Transformation of *N*-(1-Methoxyalkyl)Amides into Phosphonic or Phosphinic Analogs of *N*-Acyl-α-Amino Acids. *Phosphorus Sulfur Silicon Relat. Elem.* **2013**, *188*, 967–980. [CrossRef]
8. Październiok-Holewa, A.; Adamek, J.; Mazurkiewicz, R.; Zielińska, K. Amidoalkylating Properties of 1-(*N*-Acylamino)Alkyltriphenylphosphonium Salts. *Phosphorus Sulfur Silicon Relat. Elem.* **2013**, *188*, 205–212. [CrossRef]
9. Mazurkiewicz, R.; Adamek, J.; Październiok-Holewa, A.; Zielińska, K.; Simka, W.; Gajos, A.; Szymura, K. α-Amidoalkylating Agents from *N*-Acyl-α-amino Acids: 1-(*N*-Acylamino)alkyltriphenylphosphonium Salts. *J. Org. Chem.* **2012**, *77*, 1952–1960. [CrossRef] [PubMed]
10. Październiok-Holewa, A.; Adamek, J.; Zielińska, K.; Piernikarczyk, K.; Mazurkiewicz, R. *N*-(1-acyloaminoalkyl) amidinium salts derived from DBU or related bases as reactive intermediates in α-amidoalkylation reactions. *Arkivoc* **2012**, 314–329. [CrossRef]
11. Zaugg, H.E. Recent Synthetic Methods Involving Intermolecular alpha-Amidoalkylation at Carbon. *Synthesis* **1970**, 49–73. [CrossRef]
12. Zaugh, H.E. α-Amidoalkylation at Carbon: Recent Advances–Part I. *Synthesis* **1984**, 85–110. [CrossRef]

13. Zaugg, H.E. α-Amidoalkylation at Carbon: Recent Advances–Part II. *Synthesis* **1984**, 181–212. [CrossRef]
14. Speckamp, W.N.; Hiemstra, H. Intramolecular reactions of *N*-acyliminium intermediates. *Tetrahedron* **1985**, *41*, 4367–4416. [CrossRef]
15. Hiemstra, H.; Speckamp, W.N. *N*-Acyliminium Ions as Intermediates in Alkaloid Synthesis. *Alkaloids* **1988**, *32*, 271–339. [CrossRef]
16. Hiemstra, H.; Speckamp, W.N. *Comprehensive Organic Synthesis*; Throst, B.M., Flemming, I., Eds.; Oxford: Pergamon, Turkey, 1991; Volume 2, pp. 1047–1082. [CrossRef]
17. Kleinman, E.F. The Bimolecular Aliphatic Mannich and Related Reactions. In *Comprehensive Organic Synthesis*; Trost, B.M., Ed.; Pergamon Press: Oxford, UK, 1991; Volume 2, pp. 893–951.
18. Katritzky, A.R.; Lan, X.; Yang, J.Z.; Denisko, O.V. Properties and Synthetic Utility of *N*-Substituted Benzotriazoles. *Chem. Rev.* **1998**, *98*, 409–548. [CrossRef] [PubMed]
19. Speckamp, W.N.; Moolenaar, M.J. New developments in the chemistry of *N*-acyliminium ions and related intermediates. *Tetrahedron* **2000**, *56*, 3817–3856. [CrossRef]
20. Mecozzi, T.; Petrini, M.; Profeta, R. Reactivity of chiral exocyclic *N*-acyliminium ions with aromatic derivatives. *Tetrahedron Asymmetry* **2003**, *14*, 1171–1178. [CrossRef]
21. Maryanoff, B.E.; Zhang, H.C.; Cohen, J.H.; Turchi, I.J.; Maryanoff, C.A. Cyclizations of *N*-acyliminium ions. *Chem. Rev.* **2004**, *104*, 1431–1628. [CrossRef] [PubMed]
22. Katritzky, A.R.; Manju, K.; Singh, S.K.; Meher, N.K. Benzotriazole mediated amino-, amido-, alkoxy- and alkylthioalkylation. *Tetrahedron* **2005**, *61*, 2555–2581. [CrossRef]
23. Petrini, M. α-Amido Sulfones as Stable Precursors of Reactive *N*-Acylimino Derivatives. *Chem. Rev.* **2005**, *105*, 3949–3977. [CrossRef] [PubMed]
24. Ollevier, T.; Li, Z. The first catalytic Sakurai reaction of *N*-alkoxycarbonylamino sulfones with allyltrimethylsilane. *Org. Biomol. Chem.* **2006**, *4*, 4440–4443. [CrossRef] [PubMed]
25. Ballini, R.; Palmieri, A.; Petrini, M.; Torregiani, E. Solventless Clay-Promoted Friedel−Crafts Reaction of Indoles with α-Amido Sulfones: Unexpected Synthesis of 3-(1-Arylsulfonylalkyl) Indoles. *Org. Lett.* **2006**, *8*, 4093–4096. [CrossRef] [PubMed]
26. Marianacci, O.; Micheletti, G.; Bernardi, L.; Fini, F.; Fochi, M.; Pettersen, D.; Sgarzani, V.; Ricci, A. Organocatalytic Asymmetric Mannich Reactions with *N*-Boc and *N*-Cbz Protected α-Amido Sulfones. *Chem. Eur. J.* **2007**, *13*, 8338–8351. [CrossRef] [PubMed]
27. Yazici, A.; Pyne, S.G. Intermolecular addition reactions of *N*-acyliminium ions (Part I). *Synthesis* **2009**, 339–368. [CrossRef]
28. Yazici, A.; Pyne, S.G. Intermolecular addition reactions of *N*-acyliminium ions (Part II). *Synthesis* **2009**, 513–541. [CrossRef]
29. Thirupathi, P.; Kim, S.S. $InBr_3$: A Versatile Catalyst for the Different Types of Friedel−Crafts Reactions. *J. Org. Chem.* **2009**, *74*, 7755–7761. [CrossRef] [PubMed]
30. Das, B.; Damodar, K.; Bhunia, N. A Simple and Efficient Access to α-Amino Phosphonates from *N*-Benzyloxycarbonylamino Sulfones Using Indium(III) Chloride. *J. Org. Chem.* **2009**, *74*, 5607–5609. [CrossRef] [PubMed]
31. Martínez-Estibalez, U.; Gómez-SanJuan, A.; García-Calvo, O.; Aranzamendi, E.; Lete, E.; Sotomayor, N. Strategies based on aryllithium and *N*-acyliminium ion cyclizations for the stereocontrolled synthesis of alkaloids and related systems. *Eur. J. Org. Chem.* **2011**, 3610–3633. [CrossRef]
32. Schneider, A.E.; Manolikakes, G. $Bi(OTf)_3$-Catalyzed Multicomponent α-Amidoalkylation Reactions. *J. Org. Chem.* **2015**, *80*, 6193–6212. [CrossRef] [PubMed]
33. Aranzamendi, E.; Arrasate, S.; Sotomayor, N.; González-Díaz, H.; Lete, E. Chiral Brønsted Acid Catalyzed Enantioselective α-Amidoalkylation Reactions: A Joint Experimental and Predictive Study. *ChemistryOpen* **2016**, *5*, 540–549. [CrossRef] [PubMed]
34. Touati, B.; El Bouakher, A.; Azizi, M.S.; Taillier, C.; Othman, R.B.; Trabelsi-Ayadi, M.; Antoniotti, S.; Dunach, E.; Dalla, V. Enolizable Carbonyls and N,O-Acetals: A Rational Approach for Room-Temperature Lewis Superacid-Catalyzed Direct α-Amidoalkylation of Ketones and Aldehydes. *Chem. Eur. J.* **2016**, *22*, 6012–6022. [CrossRef] [PubMed]

35. Vinogradov, M.G.; Turova, O.V.; Zlotin, S.G. The progress in the chemistry of *N*-acyliminium ions and their use in stereoselective organic synthesis. *Russ. Chem. Rev.* **2017**, *86*, 1–17. [CrossRef]
36. Aranzamendi, E.; Sotomayor, N.; González-Díaz, H.; Lete, E. Phenolic Activation in Chiral Brønsted Acid-Catalyzed Intramolecular α-Amidoalkylation Reactions for the Synthesis of Fused Isoquinolines. *ACS Omega* **2017**, *2*, 2706–2718. [CrossRef]
37. Touati, B.; El Bouakher, A.; Azizi, M.S.; Taillier, C.; Othman, R.B.; Trabelsi-Ayadi, M.; Antoniotti, S.; Dunach, E.; Dalla, V. Atom-Economic Catalytic Direct Substitution of N,O-Acetals with Simple Ketones. *Eur. J. Org. Chem.* **2017**, 4445–4460. [CrossRef]

Sample Availability: Samples of the compounds **6**, **8**, **10** and **12** are available from the authors.

molecules

MDPI

Article

Computational Characterization of Bidentate P-Donor Ligands: Direct Comparison to Tolman's Electronic Parameters

Tímea R. Kégl [1,†,‡], Noémi Pálinkás [2,†,‡], László Kollár [1,†,‡] and Tamás Kégl [1,*,†,‡]

1 Department of Inorganic Chemistry and MTA-PTE Research Group for Selective Chemical Syntheses, University of Pécs, H-7622 Pécs, Hungary; timea.kegl@gmail.com (T.R.K.); kollar@gamma.ttk.pte.hu (L.K.)

2 Department of Inorganic Chemistry, University of Pécs, H-7622 Pécs, Hungary; noemi.palinkas@gmail.com

* Correspondence: tkegl@gamma.ttk.pte.hu; Tel.: +36-72-501-500-24585

† Current address: Ifjúság útja 6., H-7624 Hungary.

‡ These authors contributed equally to this work.

Academic Editor: György Keglevich

Received: 30 September 2018; Accepted: 28 November 2018; Published: 1 December 2018

check for updates

Abstract: The applicability of two types of transition-metal carbonyl complexes as appropriate candidates for computationally derived Tolman's ligand electronic parameters were examined with density functional theory (DFT) calculations employing the B97D3 functional. Both $Pd^{(0)}L_2(CO)$ and $HRh^{(I)}L_2(CO)$ complexes correlated well with the experimental Tolman Electronic Parameter scale. For direct comparison of the electronic effects of diphosphines with those of monophosphines, the palladium-containing system is recommended. The *trans* influence of various phosphines did not show a major difference, but the decrease of the H-Rh-P angle from linear can cause a significant change.

Keywords: diphosphines; electronic parameters; DFT; QTAIM

1. Introduction

Phosphines, phosphites, and other P-donor ligands are of vital importance in the wide range of reactions catalyzed by transition-metal (TM) compounds [1–3]. Changing the coordinated ligands is an effective strategy for modifying the properties of transition metal containing catalysts. The chemical properties of transition-metal complexes, such as their activity in various catalytic reactions, are principally determined by the steric and electronic properties of the ligands bound to the metal center [4–8]. Their structural variation opens the possibility to tune catalytic activity, as well as chemo-, regio-, and enantioselectivity [9,10]. It has long been known that variation of the substituents can cause changes upon coordination in the behavior of the ligands as well as their TM complexes. Information about the nature of the transition metal–phosphorus bond and its influence on other bonds in the molecule is crucial for the characterization of catalytically active compounds and for the fine-tuning their properties in order to develop more efficient catalysts.

In the 1960s, Strohmeier et al. investigated the σ-donor ability and π-acceptor strength of various ligand classes (nitriles, isonitriles, sulfoxides, phosphines). The phosphines were further divided into four categories designated from I to IV, classified by the frequency regions of 1927–1996, 1946–1982, 2045–2101, and 2060–2110 cm^{-1} for complexes $C_5H_5Mn(CO)_2L$, $C_5H_5V(CO)_3L$, $Fe(CO)_4L$, and $Ni(CO)_3L$, respectively, and sorted by wavenumber into increasing order [11]. These classes constitute the basis for Tolman's famous formula.

In the 1970s, Tolman reported the A_1 carbonyl-stretching frequencies of $NiL(CO)_3$ complexes, with monodentate P-donor ligands ($L = PR^1R^2R^3$) [2,12]. The electronic effects of 70 phosphorus-containing ligands were compared. It was found that a contribution could be assigned to each substituent on phosphorus, and the sum of those contributions were equal to the entire influence of the ligand on ν(CO). This contribution was designated as χ_i (cm^{-1}), and the most basic ligand (tri-*t*-butylphosphine) was chosen as reference on the Tolman electronic parameter (TEP) scale recognizing that PBu^t_3 is the best σ-donor and the worst π-acceptor ligand (Equation (1)). The $NiL(CO)_3$ scale, that is, the TEP, can be correlated with Strohmeier's $CpMnL(CO)_2$ system [7,13] (Equation (2)).

$$\nu(\text{CO})_{Ni} = 2056.1 + \Sigma\chi_j \tag{1}$$

$$\nu(\text{CO})_{Ni} = 0.711 \cdot \nu(\text{CO})_{Mn} + 692 \quad (\text{cm}^{-1}) \qquad \text{R} = 0.970 \tag{2}$$

Chelating ligands, such as diphosphines, however, cannot be described within the framework of the TEP scale. Crabtree et al. suggested more suitable model complexes for chelating ligands, for example, $MoL_2(CO)_4$ (L_2: bidentate phosphine ligand or two monodentate phosphines) [14]. They compared 11 phosphorus ligands of various types and correlated to the existing TEP scale using Equation (3). Thus, Crabtree proved Tolman's statement that the choice of transition-metal carbonyl system is arbitrary and interchangable.

$$\nu(\text{CO})_{Ni} = 0.593 \cdot \nu(\text{CO})_{Mo} + 871 \qquad \text{R} = 0.996 \tag{3}$$

In a recent study comparing the Rh–Vaska complexes and $NiL(CO)_3$, Otto and Roodt established a simple quadratic equation [15]. Later, the relationship between the carbonyl-stretching frequencies of the two reference complexes were revised and, by the basicities of the ligands, the ν(CO) range was divided into two sections [16]. The slope for the more basic phosphines showed a difference to those of less basicity.

Arsines and stibines were investigated in order to determine their electronic effects as well [16]. The contributions to the electronic effect of As and Sb, relative to P, were determined, and an additional term compared to Equation (1) was suggested addressing the effect of the donor atom (C_L; L = P, Sb or As). (Equation (4), C_P = 0)

$$\nu(\text{CO})_{Ni} = 2056.1 + \Sigma\chi_j + C_L \tag{4}$$

In the past three decades several attempts have been made for invoking theoretical methods to characterize the donor and acceptor properties of phosphines and other, mainly P-donor, ligands. These methods can be divided into two categories. The first approximation deals only with the isolated ligand, focusing on its electronic and steric properties, while neglecting the influence of the metal-containing fragment. As a prominent example, the molecular electrostatic potential at the lone pair of the phosphorus atom should be mentioned, which correlates reasonably well with the TEP scale, according to Suresh and Koga [17]. Moreover, the method known as quantitative analysis of ligand effects (QALE) relies on experimental data of known ligands and provides the resolution of net donating ability into QALE parameters [18,19]. It should be noted that the pnictogen interactions, as important secondary interactions, of a wide variety of P-donor ligands were studied as well [20–22]. The second category uses approaches that focus on the entire transition-metal complex, thereby including the possibility to scrutinize ligand–ligand effects as well [23]. Various electronic structure methods, such as Extended Transition State theory combined with Natural Orbitals of Chemical Valence (ETS-NOCV) [24–27] and Quantum Theory of Atoms in Molecules (QTAIM) [26] have been employed as well.

Crabtree and coworkers investigated the CEP, that is, the computationally derived ligand electronic parameter using the $NiL(CO)_3$ model complex with various neutral, cationic, and anionic ligands [28]. They found that the CEP correlated well with the Tolman electronic parameter and, for certain anionic ligands, even with the Hammett *meta* substituent constant.

Although the TEP scale achieved a widely accepted status, the limitations of this approximation should be mentioned as well. The coordinating properties of a ligand are governed together by its σ-donor and π-acceptor abilities, whereas TEP (or CEP) only gives the *net* donor strength. Moreover, through-space ligand interactions may affect the carbonyl-stretching frequencies as reported by Sierra and coworkers for manganese half-sandwich complexes [29]. The Tolman electronic parameters failed for linear Au–carbonyl complexes [30].

The goal of this computational study is to establish a relationship between the ligand electronic effects of late transition-metal–diphosphine complexes with the Tolman's scale, that is, the totalsymmetric carbonyl-stretching frequencies of $Ni(CO)_3L$ complexes, thereby providing a direct comparison of the electronic effects of diphosphines and monophosphines. By the selection of appropriate ligands, the set of ligands employed by Crabtree [14] and Tolman [12] has been adopted, that is, PEt_3 (**1**), PEt_2Ph (**2**), PMe_3 (**3**), PMe_2Ph (**4**), PPh_3 (**5**), $P(OMe)_3$ (**6**), $P(OPh)_3$ (**7**), $PCl_2(OEt)$ (**8**), PCl_3 (**9**), PF_3 (**10**), and $PF_2(CF_3)$ (**11**).

The secondary objective of the study is to find a dependence of the *trans* influence upon the bite angle of the diphosphines by scrutinizing the Rh–H bond distance, the rhodium–hydride stretching frequencies as well as quantum chemical descriptors within the framework of the QTAIM methodology. Those diphosphines that are involved in this study are depicted in Figure 1. As test complexes, the $Pd^{(0)}L_2(CO)$ and the $HRh^{(I)}L_2(CO)$ complexes were selected. The former is supposed to be sterically less strained in comparison to the $Mo(CO)_4L_2$ complexes with a natural bite angle of about 90°, thereby reducing the possibility of coupling steric and electronic effects and through-space interactions, whereas the rhodium-containing species are of high importance as the potential catalysts of Rh-catalyzed hydroformylation. It should be noted that only the theoretical treatment of the selected test complexes is possible because of their high reactivity, rendering them unsuitable for the experimental spectroscopical measurements.

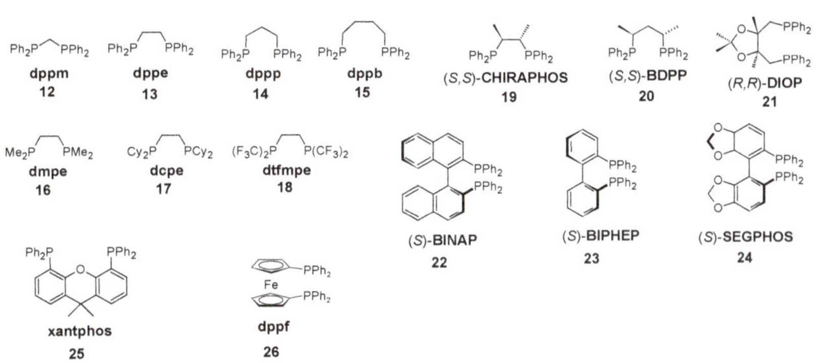

Figure 1. Chelating phosphines considered in this study.

2. Computational Details

All the structures were optimized without symmetry constraints with tight convergence criteria using the program ORCA 4.0.1 [31] with the exchange and correlation functionals developed by Grimme [32] containing the D3 empirical dispersion correction with Becke and Johnson damping [33]. For palladium and rhodium atoms the def2-TZVP basis set [34]. was employed with the respective pseudopotentials [34]. For the other atoms, the def2-SVP basis set [34] was used. The B97D3 functional was selected for the calculations involving diphosphines as it gave slightly better linear correlations for monophosphines as compared to other GGA (BP86 [35,36], PBEPBE [37]) and meta-GGA (TPSS [38], and M06L [39]) functionals (*vide infra*).

For the complexes containing conformationally flexible ligands, conformational analyses have been performed in a similar manner reported earlier [26], and the lowest energy species were considered. QTAIM analyses of the wave function [40] were carried out with AIMAll software [41]. Topological analyses of the Electron Localization Function (ELF) were completed using the Topmod09 package [42]. For electronic structure calculations, the B97D3/def2-TZVP level of theory was employed on the equilibrium geometries obtained by the B97D3 calculations with the mixed (def2-TZVP/ def2-SVP) basis set.

3. Results and Discussion

In Table 1, the experimental nickel ν(CO) parameters, as well as the computed palladium, rhodium ν(CO) parameters were collected for various phosphorus ligands, the same ligands that had also been used by Crabtree et al. In Figure 2, the 11 calculated (B97D3) carbonyl-stretching frequencies of $PdL_2(CO)$ and $HRhL_2(CO)$ training sets are plotted against the experimental ν(CO) frequencies for $NiL(CO)_3$ [2,12]. The high values of the correlation coefficients $r^2 = 0.971$ and $r^2 = 0.975$, obtained for $PdL_2(CO)$ and $HRhL_2(CO)$, respectively, shows that the computed ν(CO) for these systems (CEP parameters) have reasonable linear correlation with the experimental $NiL(CO)_3$ ν(CO) values (TEP) according to Equations (5) and (6):

$$\nu(\mathrm{CO})_{\mathrm{Ni,exp}} = \mathrm{CEP_{Pd}} = 0.5144 \cdot \nu(\mathrm{CO})_{\mathrm{PdL_2(CO)}} + 1002.2 \quad (5)$$

$$\nu(\mathrm{CO})_{\mathrm{Ni,exp}} = \mathrm{CEP_{Rh}} = 0.5744 \cdot \nu(\mathrm{CO})_{\mathrm{HRhL_2(CO)}} + 874.0 \quad (6)$$

Table 1. Experimental CO-stretching frequencies (Reference [12]) for $NiL(CO)_3$, and computed CO-stretching frequencies for $PdL_2(CO)$ and $HRhL_2(CO)$ complexes at the B97D3 level of theory. Carbonyl-stretching frequencies are given in cm^{-1}.

Ligand	$\nu(\mathrm{CO})_{\mathrm{NiL(CO)_3}}$	$\nu(\mathrm{CO})_{\mathrm{PdL_2(CO)}}$	$\nu(\mathrm{CO})_{\mathrm{HRhL_2(CO)}}$
PEt_3 (**1**)	2061.7	2059.6	2065.3
PEt_2Ph (**2**)	2063.7	2062.2	2066.5
PMe_3 (**3**)	2064.1	2068.2	2074.7
PMe_2Ph (**4**)	2065.3	2068.4	2074.6
PPh_3 (**5**)	2068.9	2068.9	2078.8
$P(OMe)_3$ (**6**)	2079.5	2096.5	2107.2
$P(OPh)_3$ (**7**)	2085.3	2100.2	2109.7
$PCl_2(OEt)$ (**8**)	2092.5	2127.6	2124.5
PCl_3 (**9**)	2097.0	2140.4	2134.5
PF_3 (**10**)	2110.8	2148.9	2152.8
$P(CF_3)F_2$ (**11**)	2112.1	2149.0	2144.4

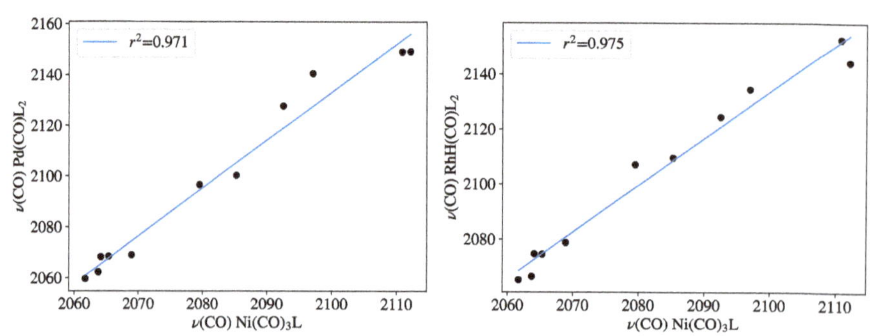

Figure 2. Carbonyl-stretching frequencies of $PdL_2(CO)$ and $HRhL_2(CO)$ against Tolman's parameters at the B97D3 level. Frequency values are given in Table 1.

To clarify the role of the theoretical method employed for obtaining CEP parameters, the training set was recomputed with various GGA and meta-GGA functionals (Figure 3). The B97D3 results were compared with those obtained with BP86 [35,36], PBEPBE [37], TPSS [38], and M06L [39] functionals for both the palladium and the rhodium complexes. The r^2 correlation coefficients remained within quite a narrow range, as the worst correlation was found to be $r^2 = 0.949$ for the PBEPBE functional in the case of the Rh–hydride model systems. For both complexes, the B97D3 functional proved to be slightly superior; therefore, it was chosen for describing the chelating ligands as well.

The chelating ligands shown in Table 2 were selected to address various geometrical as well as electronic parameters. Ligands **12–15** were used to illustrate the influence of the P-M-P bite angle while ligands **16–18** show the effect of the substituents on the phosphorus with ethanediphos backbone. The chiral ligands **19–21** reveal the substituent effects on the backbone. To address the structural changes of ligands with axial chirality **22–24** were selected. Finally, **25** and **26** represent ligands possessing rigid structure with wide bite angles.

Table 2. Computed CO-stretching frequencies and P-M-P bite angles (θ_{PPdP} and θ_{PRhP}, in degrees) for diphosphine-containing complexes $PdL_2(CO)$, $HRhL_2(CO)$, and their respective Computed Tolman Parameters (CEPs), determined according to Equations (5) and (6). Vibrational frequencies are in cm^{-1}.

Ligand	$\nu(CO)_{PdL_2(CO)}$	CEP_{Pd}	θ_{PPdP}	$\nu(CO)_{HRhL_2(CO)}$	CEP_{Rh}	θ_{PRhP}
dppm (**12**)	2080.8	2072.6	69.8	2069.8	2062.9	71.0
dppe (**13**)	2076.1	2070.1	85.1	2071.9	2064.1	84.7
dppp (**14**)	2068.4	2066.2	90.5	2074.6	2065.7	88.7
dppb (**15**)	2064.5	2064.2	96.2	2079.0	2068.2	94.6
dmpe (**16**)	2073.5	2068.8	86.0	2075.1	2065.9	85.5
dcpe (**17**)	2058.6	2061.1	86.2	2060.6	2057.6	85.3
dtfmpe (**18**)	2123.1	2094.3	86.3	2125.8	2095.1	84.9
CHIRAPHOS (**19**)	2074.3	2069.2	85.1	2076.3	2066.6	84.6
BDPP (**20**)	2067.7	2065.8	89.0	2071.2	2063.7	87.7
DIOP (**21**)	2069.6	2066.8	101.5	2077.4	2067.3	95.5
BINAP (**22**)	2069.0	2066.5	93.7	2075.9	2066.4	92.8
BIPHEP (**23**)	2069.6	2066.8	96.4	2076.8	2066.9	93.8
SEGPHOS (**24**)	2067.5	2065.7	94.3	2073.9	2065.2	92.8
xantphos (**25**)	2065.0	2064.4	105.9	2070.1	2063.1	102.8
dppf (**26**)	2069.3	2066.6	103.6	2078.6	2067.9	97.8

The CEP determined on two different types of carbonyl complexes reveals a remarkable difference (near 10 cm^{-1}) for the narrow bite-angle case **12**. The increase of the bite angle by going in the direction of dppm → dppe → dppp → dppb also caused a decrease in ν(CO) for the palladium and an increase in ν(CO) for the rhodium complexes. The replacement of the phenyl substituent in dppe to the electron-withdrawing trifluoromethyl group resulted in a similar blue-shift for the two systems. On the other hand, the presence of electron-donating substituents, such as methyl or cyclohexyl, on the phosphorus results in the expected red-shift in the palladium carbonyls and a mixed effect for the Rh carbonyls as, in HRh(CO)(dmpe), virtually no change occurs in the carbonyl-stretching frequency. When the electron-donating methyl groups are introduced in the chiral backbone, such as for the ligand CHIRAPHOS, a slight blue-shift occurs for HRh(**19**)(CO), whereas an almost negligible red-shift takes place for Pd(**19**)(CO).

The change or the neglect of the condensed ring in the ligands with axial chirality has practically no effect on ν(CO). Moreover, the predicted CEP is almost identical for both model systems. This difference is also minor (below 2 cm^{-1}) for the xantphos and dppf ligands. Not surprisingly, in the three-coordinate palladium systems, all the P-M-P bite angles were somewhat larger than in the HRh(LL)(CO) complexes, except for the dppm ligand.

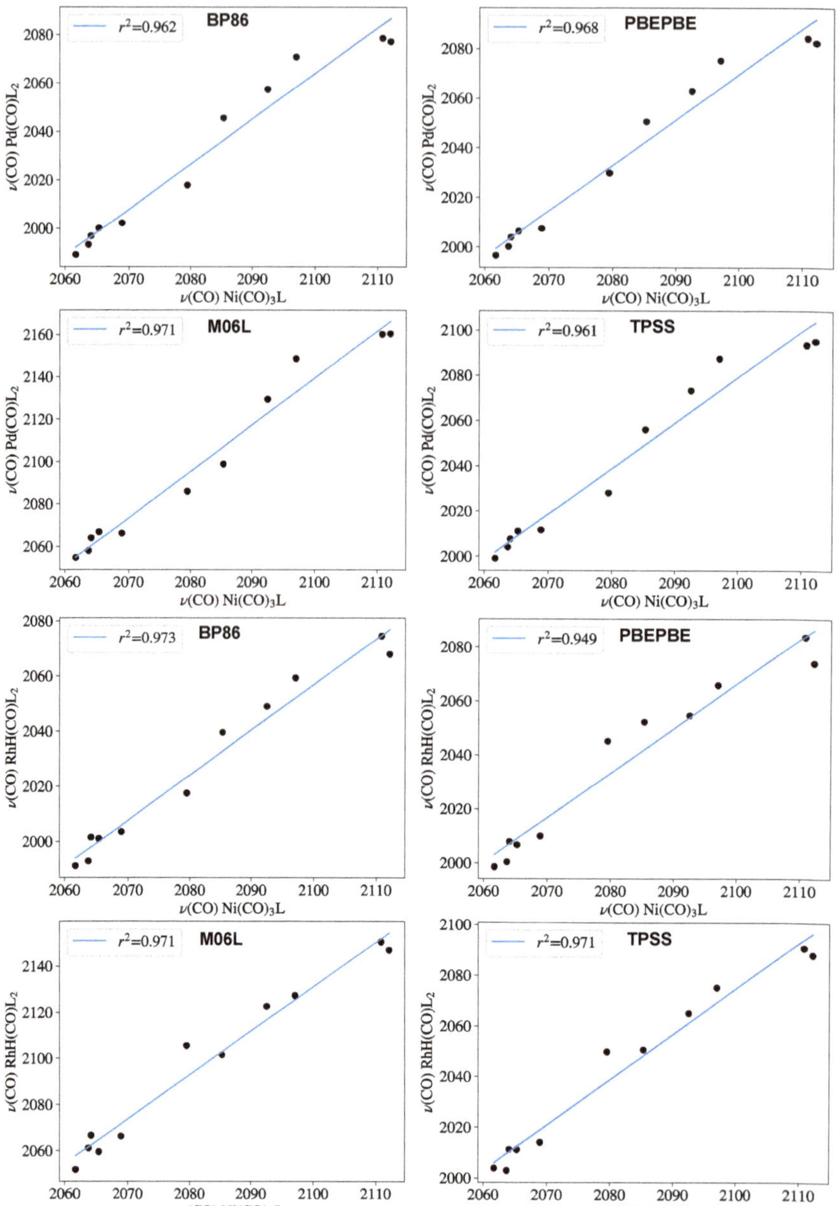

Figure 3. Carbonyl-stretching frequencies of $PdL_2(CO)$ (top row) and $HRhL_2(CO)$ against Tolman's parameter computed with various DFT functionals. Frequency values are given in Table 1.

The *trans* influence in the four-coordinate, distorted square-planar $HRhL_2(CO)$ complexes is formally based on a $n_P \rightarrow \sigma^*_{RhH}$ donor–acceptor interaction. Here, it is followed by the metal–hydrogen stretching frequency (ν(RhH)), the Rh–H distance, and two QTAIM parameters. Both the electron density at the bond critical point (ρ_{BCP}) and the δ(Rh,H) delocalization index, which provides the number of electron pairs delocalized between the atomic basins of Rh and H [43], were somewhat proportional with the bond strength. The delocalization index is, to some extent, related to formal bond orders for an equally shared pair between two atoms in a polyatomic molecule.

According to the ρ_{BCP}(RhH) values in Table 3 the difference in electronic effects caused by the P-ligand *trans* to the hydride is rather subtle. As the trend in the delocalization indices is quite the opposite compared to that obtained from the geometrical parameters, it is likely that the electron sharing between atomic basins did not reflect the overall bond strengths in these cases.

Table 3. Computed Rh-stretching frequencies, Rh–H bond lengths (in Å), P–Rh–P bond angles (in degrees), delocalization indices, and electron densities at bond critical points for $HRhL_2(CO)$ complexes.

Ligand	ν(RhH)	r_{RhH}	θ_{PRhP}	δ(Rh,H)	ρ_{BCP}(RhH)
PEt_3 (**1**)	1920.8	1.618	100.4	0.858	0.131
PEt_2Ph (**2**)	1931.6	1.615	95.6	0.873	0.132
PMe_3 (**3**)	1927.3	1.618	100.9	0.860	0.131
PMe_2Ph (**4**)	1948.3	1.611	98.4	0.876	0.133
PPh_3 (**5**)	1964.7	1.606	102.6	0.849	0.135
$P(OMe)_3$ (**6**)	1949.0	1.618	100.7	0.844	0.132
$P(OPh)_3$ (**7**)	1984.5	1.607	106.9	0.831	0.135
$PCl_2(OEt)$ (**8**)	2014.5	1.598	96.4	0.866	0.139
PCl_3 (**9**)	1993.0	1.602	101.2	0.837	0.135
PF_3 (**10**)	1970.6	1.614	99.9	0.841	0.134
$P(CF_3)F_2$ (**11**)	1967.5	1.617	100.2	0.829	0.133

The ν(RhH) frequencies, however, tend to follow the computed electronic parameters, as the largest red-shift occurs for the more electron donating ligands and the frequencies are increased for the electron withdrawing ones. Thus, the *trans* influence increases in the order of the substituent donor strength.

To address the effect of the P–Rh–P bite angle upon the *trans* influence, the structures and the Laplacians of the electron density were compared and are illustrated in Figure 4. The structural distortion caused by the dppm ligand resulted in distortion in the electron-density distribution as well. The stronger bond in the dppm complex, as compared to the HRh(dppe)(CO) complex, shows that even when the basicities of the phosphorus atoms of the standalone dppm and dppe ligands should be very similar, the slight decrease of the H–Rh–P angle causes a notable decrease in the *trans* influence. It should be noted that, in the dppm complex, a red-shift can be observed in ν(CO) in comparison to that of the dppe complex, which may be explained with the less preferred orientation to a $n_P \rightarrow \pi^*_{CO}$ orbital interaction.

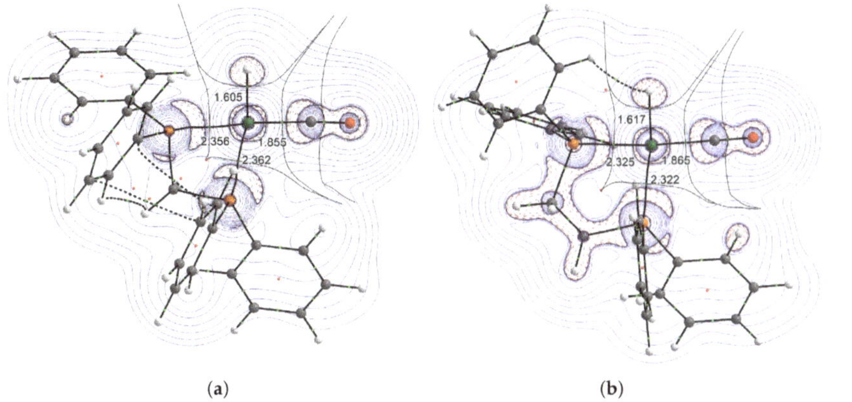

(**a**) (**b**)

Figure 4. Laplacian ($\nabla^2\rho(\mathbf{r})$) of the electron density of complexes HRh(dppm)(CO) (**a**) and HRh(dppe)(CO) (**b**). Charge concentration regions ($\nabla^2\rho(\mathbf{r}) < 0$) are designated with dashed lines.

It was found previously that, among QTAIM parameters, CO delocalization index gave good performance, whereas the electron density of the CO bond critical point gives moderate linear correlation with the computed carbonyl symmetric stretching frequencies in Ni-tricarbonyl phosphine complexes [26]. To the best of our knowledge, the relationship between the electron localization function (ELF) and the stretching frequencies in transition-metal carbonyl complexes is unexplored.

The ELF, denoted as $\eta(\mathbf{r})$ introduced by Becke and Edgecombe [35] and reinterpreted by Savin et al. [44] using the mathematical scheme from Thom's catastrophe theory [45] is based on the conditional pair probability function. It takes values in a range $0 \leqq \eta \leqq 1$ between 0 and 1, where values close to 1 are located in regions of space with strong electron localization, whereas values close to 0 correspond to electron delocalization, and η = 0.5 is the value that would be expected for a homogeneous electron gas. It is worth mentioning that Putz [46,47] proposed general formulations for the ELF on the basis of the path integral formulation of Markovian open systems.

The electronic effect of the monodentate ligands was followed by the population of the V(C,O) and V(Rh,H) disynaptic basins. The population of an ELF basin is defined according to Equation (7).

$$\bar{N}(\Omega_i) = \int\limits_{\Omega_i} \rho(\mathbf{r})d\mathbf{r} \tag{7}$$

The ELF basins, which were the subject of our study, are depicted for complexes $HRh(PMe_3)_2(CO)$ and $HRh(PF_3)_2(CO)$ in Figure 5. The spacial distribution of the V(Rh,H) basin resembles those representing the H–C bonds, that is, it is localized around the hydrogen and not along the H–Rh bond path. On the other hand, the V(C,O) basins are situated in the middle between the carbon and oxygen of the carbonyl ligands.

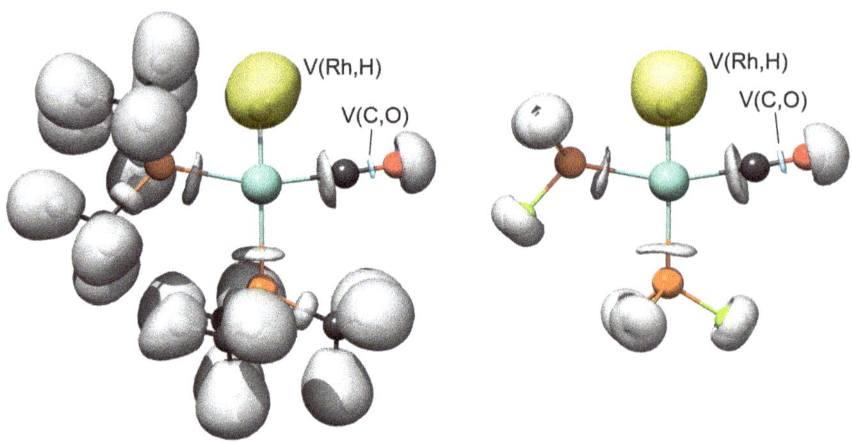

(**a**) (**b**)

Figure 5. Selected electron localization function (ELF) valence basins for complexes $HRh(PMe_3)_2(CO)$ (**a**) and $HRh(PF_3)_2(CO)$ (**b**) at isovalue of η = 0.84. Yellow and blue envelopes designate the disynaptic V(Rh,H) and V(C,O) basins, respectively.

The ELF was computed for all the $HRh(monophosphine)_2(CO)$ type of complexes. The basin population associated with the C≡O bond revealed fairly good linear correlation (r^2 = 0.915) with the experimental TEP parameters (Figure 6a). The extrema of the $\bar{N}(\Omega)$ values are between 2.62 (for ligand PEt_2Ph) and 2.83 (for PF_3), indicating the somewhat depleted triple-bond character of the bound CO. The population change of the Rh–hydride bond, however, shows moderate linear correlation

(r^2 = 0.823) with TEP; only showing that the trans influence generally increases with the greater donor strength of the P-donor ligand.

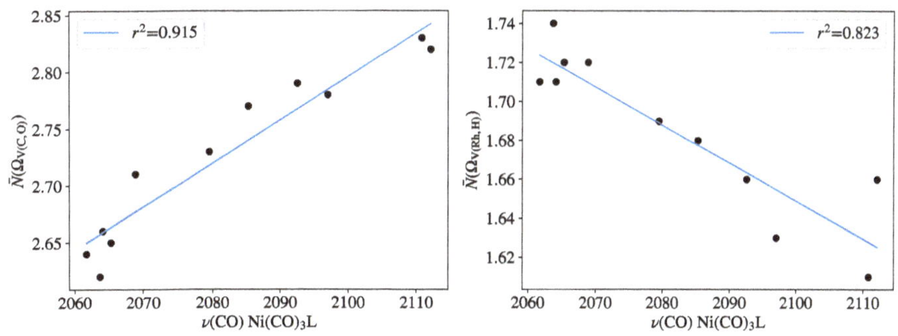

Figure 6. Population of ELF basins (**a**) V(C,O) and (**b**) V(Rh,H) for complexes $HRhL_2(CO)$ against Tolman's parameters at the B97D3 level. Frequency values are given in Table 1.

4. Conclusions

In the present study, two classes of model compounds were scrutinized to check their applicability as appropriate candidates for computationally derived ligand electronic parameters (CEPs) for bidentate P-donor ligands employing the B97D3/def2-TZVP(def2-SVP) level of theory. The results of this work can be summarized as follows:

- For both model systems, no major difference in accuracy can be expected for various DFT methods in terms of the linear regression coefficient.
- For monophosphines, both $Pd^{(0)}L_2(CO)$ and $HRh^{(I)}L_2(CO)$ complexes showed high linearity with the experimental TEP scale.
- For monophosphines, the population of the disynaptic ELF basin V(C,O), associated with the carbonyl ligand, showed good linearity with the experimental TEP scale.
- For diphosphine-containing systems, the bite angle effect showed the opposite trend, with $PdL_2(CO)$ complexes revealing a decrease in ν(CO) by the increase of bite angle.
- In the case of the Pd complexes, the change of substituents on phosphorus caused a change in ν(CO) being consistent with the donor character of the substituent.
- For diphosphines possessing axial chirality, the nature of the condensed ring does not alter the electronic parameter of the ligand.
- The *trans* effect estimated in the $HRhL_2(CO)$ complexes is very sensitive to the H–Rh–P angle.

Supplementary Materials: The following are available online. Cartesian coordinates of palladium-diphosphine complexes occuring in this study.

Author Contributions: The research problem was conceptualized by the corresponding author (T.K.), the computational work was carried out by T.R.K. and N.P., the writing, review and editing is the joint work of all authors. All authors read and approved the final manuscript.

Funding: This research received no external funding.

Acknowledgments: The authors thank the Hungarian Research Fund (K113177) for the financial support. This work was supported by the GINOP-2.3.2-15-2016-00049 grant. N.P. thanks Gedeon Richter's Talentum Foundation for the scholarship. The present scientific contribution is dedicated to the 650th anniversary of the foundation of the University of Pécs, Hungary. The project has been supported by the European Union, cofinanced by the European Social Fund Grant no. EFOP-3.6.1.-16-2016-00004 entitled by Comprehensive Development for Implementing Smart Specialization Strategies at the University of Pécs. The study was also financed by the Higher Education Institutional Excellence Program of the Ministry of Human Capacities in Hungary, within the framework of the 20765-3/2018/FEKUTSTRAT thematic program of the University of Pécs.

Conflicts of Interest: The authors declare no conflict of interest.

References

1. McAuliffe, C.A. *Transition Metal Complexes of Phosphorus, Arsenic and Antimony Ligands*; Halsted Press: New York, NY, USA, 1973.
2. Tolman, C.A. Steric effects of phosphorus ligands in organometallic chemistry and homogeneous catalysis. *Chem. Rev.* **1977**, *77*, 313–348. [CrossRef]
3. Hartwig, J.F.; Collman, J.P. *Organotransition Metal Chemistry: From Bonding to Catalysis*; University Science Books: Sausalito, CA, USA, 2010.
4. Brown, T.L.; Lee, K.J. Ligand steric properties. *Coord. Chem. Rev.* **1993**, *128*, 89–116. [CrossRef]
5. Dias, P.B.; de Piedade, M.E.M.; Simões, J.A.M. Bonding and energetics of phosphorus (III) ligands in transition metal complexes. *Coord. Chem. Rev.* **1994**, *135*, 737–807. [CrossRef]
6. Kamer, P.C.; van Leeuwen, P.W.; Reek, J.N. Wide bite angle diphosphines: Xantphos ligands in transition metal complexes and catalysis. *Acc. Chem. Res.* **2001**, *34*, 895–904. [CrossRef] [PubMed]
7. Kühl, O. Predicting the net donating ability of phosphines do we need sophisticated theoretical methods? *Coord. Chem. Rev.* **2005**, *249*, 693–704. [CrossRef]
8. de Vries, J.G.; Lefort, L. The combinatorial approach to asymmetric hydrogenation: Phosphoramidite libraries, ruthenacycles, and artificial enzymes. *Chem. A Eur. J.* **2006**, *12*, 4722–4734. [CrossRef] [PubMed]
9. Botteghi, C.; Paganelli, S.; Schionato, A.; Marchetti, M. The asymmetric hydroformylation in the synthesis of pharmaceuticals. *Chirality* **1991**, *3*, 355–369. [CrossRef] [PubMed]
10. Kégl, T.; Kollár, L. Chiral Phosphorous Ligands in Asymmetric Catalysis. In *Comprehensive Inorganic Chemistry II (Second Edition): From Elements to Applications*; Elsevier Ltd.: Amsterdam, The Netherlands, 2013.
11. Strohmeier, W.; Müller, F.J. Klassifizierung phosphorhaltiger Liganden in Metallcarbonyl-Derivaten nach der π-Acceptorstärke. *Chem. Ber.* **1967**, *100*, 2812–2821. [CrossRef]
12. Tolman, C.A. Electron donor-acceptor properties of phosphorus ligands. Substituent additivity. *J. Am. Chem. Soc.* **1970**, *92*, 2953–2956. [CrossRef]
13. Strohmeier, W.; Müller, F.J. π-Acceptorstärke von Phosphinen als Liganden in Cyclopentadienylmangantricarbonyl und Nickelcarbonyl. *Z. Naturforschg.* **1967**, *22b*, 451–452. [CrossRef]
14. Anton, D.R.; Crabtree, R.H. Metalation-resistant ligands: Some properties of dibenzocyclooctatetraene complexes of molybdenum, rhodium and iridium. *Organometallics* **1983**, *2*, 621–627. [CrossRef]
15. Roodt, A.; Otto, S.; Steyl, G. Structure and solution behaviour of rhodium(I) Vaska-type complexes for correlation of steric and electronic properties of tertiary phosphine ligands. *Coord. Chem. Rev.* **2003**, *245*, 121–137. [CrossRef]
16. Otto, S.; Roodt, A. Quantifying the electronic cis effect of phosphine, arsine and stibine ligands by use of rhodium(I) Vaska-type complexes. *Inorg. Chim. Acta* **2004**, *357*, 1–10. [CrossRef]
17. Suresh, C.H.; Koga, N. Quantifying the Electronic Effect of Substituted Phosphine Ligands via Molecular Electrostatic Potential. *Inorg. Chem.* **2002**, *41*, 1573–1578. [CrossRef]
18. Liu, H.Y.; Eriks, K.; Prock, A.; Giering, W.P. Quantitative analysis of ligand effects (QALE). Systematic study of iron-phosphorus bond lengths and their relationship to steric thresholds. *Organometallics* **1990**, *9*, 1758–1766. [CrossRef]
19. Bartholomew, J.; Fernandez, A.L.; Lorsbach, B.A.; Wilson, M.R.; Prock, A.; Giering, W.P. Comments on Coupling Graphical and Regression Analyses of Ligand Effect Data. *Organometallics* **1996**, *15*, 295–301. [CrossRef]
20. Scheiner, S. The Pnicogen Bond: Its Relation to Hydrogen, Halogen, and Other Noncovalent Bonds. *Acc. Chem. Res.* **2012**, *46*, 280–288. [CrossRef]
21. Sánchez-Sanz, G.; Trujillo, C.; Alkorta, I.; Elguero, J. Intramolecular pnicogen interactions in phosphorus and arsenic analogues of proton sponges. *Phys. Chem. Chem. Phys.* **2014**, *16*, 15900–15909. [CrossRef]
22. Molloy, A.D.; Sánchez-Sanz, G.; Gilheany, D.G. PP-Rotation, P-Inversion and Metathesis in Diphosphines Studied by DFT Calculations: Comments on Some Literature Conflicts. *Inorganics* **2016**, *4*, 36. [CrossRef]
23. Gusev, D.G. Donor properties of a series of two-electron ligands. *Organometallics* **2009**, *28*, 763–770. [CrossRef]
24. Mitoraj, M.; Michalak, A. Donor-Acceptor Properties of Ligands from the Natural Orbitals for Chemical Valence. *Organometallics* **2007**, *26*, 6576–6580. [CrossRef]
25. Mitoraj, M.P.; Michalak, A. σ-donor and π-acceptor properties of phosphorus ligands: An insight from the natural orbitals for chemical valence. *Inorg. Chem.* **2010**, *49*, 578–582. [CrossRef] [PubMed]

26. Kégl, T.R.; Kollár, L.; Kégl, T. Relationship of QTAIM and NOCV Descriptors with Tolman's Electronic Parameter. *Adv. Chem.* **2016**, *2016*, 4109758. [CrossRef]
27. Couzijn, E.P.; Lai, Y.Y.; Limacher, A.; Chen, P. Intuitive Quantifiers of Charge Flows in Coordinate Bonding. *Organometallics* **2017**, *36*, 3205–3214. [CrossRef]
28. Perrin, L.; Clot, E.; Eisenstein, O.; Loch, J.; Crabtree, R.H. Computed ligand electronic parameters from quantum chemistry and their relation to Tolman parameters, lever parameters, and Hammett constants. *Inorg. Chem.* **2001**, *40*, 5806–5811. [CrossRef] [PubMed]
29. Valyaev, D.A.; Brousses, R.; Lugan, N.; Fernández, I.; Sierra, M.A. Do ν(CO) Stretching Frequencies in Metal Carbonyl Complexes Unequivocally Correlate with the Intrinsic Electron-Donicity of Ancillary Ligands? *Chem. A Eur. J.* **2011**, *17*, 6602–6605. [CrossRef]
30. Ciancaleoni, G.; Scafuri, N.; Bistoni, G.; Macchioni, A.; Tarantelli, F.; Zuccaccia, D.; Belpassi, L. When the Tolman Electronic Parameter Fails: A Comparative DFT and Charge Displacement Study of $[(L)Ni(CO)_3]^{0/-}$ and $[(L)Au(CO)]^{0/+}$. *Inorg. Chem.* **2014**, *53*, 9907–9916. [CrossRef]
31. Frank, N. The ORCA program system. *WIREs Comput. Mol. Sci.* **2012**, *2*, 73–78.
32. Grimme, S. Accurate description of van der Waals complexes by density functional theory including empirical corrections. *J. Comput. Chem.* **2004**, *25*, 1463–1473. [CrossRef]
33. Grimme, S.; Ehrlich, S.; Goerigk, L. Effect of the damping function in dispersion corrected density functional theory. *J. Comput. Chem.* **2011**, *32*, 1456–1465. [CrossRef]
34. Weigend, F.; Ahlrichs, R. Balanced basis sets of split valence, triple zeta valence and quadruple zeta valence quality for H to Rn: Design and assessment of accuracy. *Phys. Chem. Chem. Phys.* **2005**, *7*, 3297–3305. [CrossRef] [PubMed]
35. Becke, A.D. Density-functional exchange-energy approximation with correct asymptotic behavior. *Phys. Rev. A* **1988**, *38*, 3098–3100. [CrossRef]
36. Perdew, J.P. Density-functional approximation for the correlation energy of the inhomogeneous electron gas. *Phys. Rev. B* **1986**, *33*, 8822–8824. [CrossRef]
37. Perdew, J.P.; Burke, K.; Ernzerhof, M. Generalized gradient approximation made simple. *Phys. Rev. Lett.* **1996**, *77*, 3865–3868. [CrossRef] [PubMed]
38. Tao, J.M.; Perdew, J.P.; Staroverov, V.N.; Scuseria, G.E. Climbing the density functional ladder: Nonempirical meta-generalized gradient approximation designed for molecules and solids. *Phys. Rev. Lett.* **2003**, *91*, 146401. [CrossRef] [PubMed]
39. Zhao, Y.; Truhlar, D. A new local density functional for main-group thermochemistry, transition metal bonding, thermochemical kinetics, and noncovalent interactions. *J. Chem. Phys.* **2006**, *125*, 194101–194118. [CrossRef]
40. Bader, R.F.W. *Atoms Moleculs—A Quantum Theory*; Oxford University Press: Oxford, UK, 1990.
41. Keith, T.A. *AIMAll (Version 15.05.18)*; TK Gristmill Software: Overland Park, KS, USA, 2015.
42. Noury, S.; Krokidis, X.; Fuster, F.; Silvi, B. Computational tools for the electron localization function topological analysis. *Comput. Chem.* **1999**, *23*, 597–604. [CrossRef]
43. Bader, R.F.W.; Stephens, M.E. Spatial localization of the electronic pair and number distributions in molecules. *J. Am. Chem. Soc.* **1975**, *97*, 7391–7399. [CrossRef]
44. Savin, A.; Nesper, R.; Wengert, S.; Fässler, T.F. ELF: The electron localization function. *Angew. Chem. Int. Ed. Engl.* **1997**, *36*, 1808–1832. [CrossRef]
45. René, T. *Stabilité Structurelle et Morphogénese: Essai D'une Théorie Générale des Modeles*; WA Benjamin: New York, NY, USA, 1972.
46. Putz, M.V. Markovian approach of the electron localization functions. *Int. J. Quantum Chem.* **2005**, *105*, 1–11. [CrossRef]
47. Putz, M. Density functionals of chemical bonding. *Int. J. Mol. Sci.* **2008**, *9*, 1050–1095. [CrossRef] [PubMed]

molecules

MDPI

Communication

The Triple-Decker Complex $[Cp^*Fe(\mu,\eta^5{:}\eta^5\text{-}P_5)Mo(CO)_3]$ as a Building Block in Coordination Chemistry

Mehdi Elsayed Moussa, Stefan Welsch, Luis Dütsch, Martin Piesch, Stephan Reichl, Michael Seidl and Manfred Scheer *

Department of Inorganic Chemistry, University of Regensburg, 93040 Regensburg, Germany; Mehdi.Elsayed-Mousa@chemie.uni-regensburg.de (M.E.M.); stefan-welsch@gmx.de (S.W.); Luis.duetsch@chemie.uni-regensburg.de (L.D.); Martin.piesch@chemie.uni-regensburg.de (M.P.); Stephan.reichl@chemie.uni-regensburg.de (S.R.); michael1.seidl@chemie.uni-regensburg.de (M.S.)

* Correspondence: manfred.scheer@ur.de or mascheer@chemie.uni-regensburg.de; Tel.: +49-(0)941-943-4440; Fax: +49-(0)941-943-4439

Academic editor: György Keglevich

Received: 19 December 2018; Accepted: 12 January 2019; Published: 17 January 2019

check for updates

Abstract: Although the triple-decker complex $[Cp^*Fe(\mu,\eta^5{:}\eta^5\text{-}P_5)Mo(CO)_3]$ (**2**) was first reported 26 years ago, its reactivity has not yet been explored. Herein, we report a new high-yielding synthesis of **2** and the isolation of its new polymorph (**2'**). In addition, we study its reactivity towards Ag^I and Cu^I ions. The reaction of **2** with $Ag[BF_4]$ selectively produces the coordination compound $[Ag\{Cp^*Fe(\mu,\eta^5{:}\eta^5\text{-}P_5)Mo(CO)_3\}_2][BF_4]$ (**3**). Its reaction with Ag[TEF] and Cu[TEF] ($[TEF]^- = [Al\{OC(CF_3)_3\}_4]^-$) leads to the selective formation of the complexes $[Ag\{Cp^*Fe(\mu,\eta^5{:}\eta^5\text{-}P_5)Mo(CO)_3\}_2][TEF]$ (**4**) and $[Cu\{Cp^*Fe(\mu,\eta^5{:}\eta^5\text{-}P_5)Mo(CO)_3\}_2][TEF]$ (**5**), respectively. The X-ray structures of compounds **3–5** each show an M^I ion ($M^I = Ag^I$, Cu^I) bridged by two P atoms from two triple-decker complexes (**2**). Additionally, four short $M^I\cdots CO$ distances (two to each triple-decker complex **2**) participate in stabilizing the coordination sphere of the M^I ion. Evidently, the X-ray structure for compound **3** shows a weak interaction of the Ag^I ion with one fluorine atom of the counterion $[BF_4]^-$. Such an $Ag\cdots F$ interaction does not exist for compound **4**. These findings demonstrate the possibility of using triple-decker complex **2** as a ligand in coordination chemistry and opening a new perspective in the field of supramolecular chemistry of transition metal compounds with phosphorus-rich complexes.

Keywords: triple-decker; *cyclo*-P_5; weakly coordinating; molybdenum; silver; copper

1. Introduction

For more than two decades, the chemistry of transition metal complexes bearing polyphosphorus ligands has been an active area of research for both inorganic and organometallic chemists [1]. The interest in this class of compounds does not only originate from the large variety of bonding patterns in the formed products, but is also due to their potential as building blocks in supramolecular chemistry [1,2]. Until now, the most successful candidates in this field have been polyphosphorus (P_n) complexes, such as the Cp* derivative of pentaphosphaferrocene $[Cp^*Fe(\eta^5\text{-}P_5)]$ (**1**, $Cp^* = \eta^5\text{-}C_5Me_5$) [3]. Its reaction with Ag^I, Cu^I and group 13 metal salts (Tl^I, In^I, Ga^I) of the weakly coordinating anion $[Al\{OC(CF_3)_3\}_4]^-$ ($[TEF]^-$) allows the formation of one-dimensional polymers with various coordination modes of the P_5 unit [4–8]. With Cu^I halides and **1**, depending on the reaction conditions, a set of one- and two-dimensional coordination polymers [9,10], fascinating fullerene-like spherical aggregates [11–16] as well as an organometallic nanosized capsule [17] are accessible. Moreover, when ditopic organic linkers are added to **1** and Cu^I halides, unprecedented organometallic–organic hybrid polymers with various dimensionalities can be realized [18,19].

Besides its exciting supramolecular chemistry, the *cyclo*-P_5 complex **1** was used as a suitable starting compound for the synthesis of a number of triple-decker complexes with *cyclo*-P_5 middle-decks. Based on **1**, the Scherer group prepared the 30-electron triple-decker complexes [CpFe(μ,η^5:η^5-P_5)FeCp*]$^{+3}$ and [Cp*Fe(μ,η^5:η^5-P_5)M(CO)$_3$] (M = Cr, Mo) [20,21]. Kudinov et al. reported a few iron-, ruthenium- and molybdenum-containing triple-decker complexes such as [Cp*Fe(μ,η^5:η^5-P_5)MCp*]$^+$ (M = Fe, Ru) and [(C_7H_7)Mo(μ,η^5:η^5-P_5)FeCp*]$^+$ [22–24]. More recently, our group used **1** to synthesize several unique neutral triple-decker sandwich complexes with functionalized *cyclo*-P_5 middle-decks [25]. In addition to these examples, some others continuously appear in the literature with *cyclo*-P_5 middle-decks synthesized from starting materials other than **1** [26–28].

Regardless of the ongoing progress in the synthesis of this class of triple-decker compounds, their reactivity and especially their coordination chemistry are totally unexplored. In fact, to the best of our knowledge, apart from the few coordination compounds of the [(Cp*Mo)$_2$(μ,η^6:η^6-P_6)] triple-decker complex recently reported by our group [29], no coordination chemistry of any other triple-deck with any *cyclo*-P_n middle-deck has been reported as yet. Accordingly, we studied the potential of triple-decker complexes with *cyclo*-P_5 middle-decks as multidentate ligands in supramolecular chemistry. The goal of such studies is to understand the coordination chemistry of these complexes with transition metal salts and compare it to that of the complex [Cp*Fe(η^5-P_5)] (**1**). Herein, we present a new high-yielding synthesis of the triple-decker complex [Cp*Fe(μ,η^5:η^5-P_5)Mo(CO)$_3$] (**2**), and we show that its reaction with the coinage metal salts Ag[BF_4], Ag[TEF] and Cu[TEF] allows the formation of the three unprecedented coordination compounds: [Ag{Cp*Fe(μ,η^5:η^5-P_5)Mo(CO)$_3$}$_2$][BF_4] (**3**), [Ag{Cp*Fe(μ,η^5:η^5-P_5)Mo(CO)$_3$}$_2$][TEF] (**4**) and [Cu{Cp*Fe(μ,η^5:η^5-P_5)Mo(CO)$_3$}$_2$][TEF] (**5**), respectively.

2. Results and Discussion

2.1. Synthesis and X-Ray Structure of [Cp*Fe(μ,η^5:η^5-P_5)Mo(CO)$_3$] (2)

The previously reported synthesis of the triple-decker complex **2** from the *cyclo*-P_5 complex **1** and [Mo(CO)$_3$(CH_3CN)$_4$] requires an extended chromatographic purification. Moreover, it is only partially reproducible, as several attempts isolated **2** with very different yields (10–40%) that were all lower than the initially reported one (51%) [21], which is probably one reason why the chemistry of this compound has not been studied despite it having been discovered almost three decades ago. Therefore, we developed another method for the synthesis of **2**, involving the 1:1 reaction of the *cyclo*-P_5 complex **1** with [(η^6-C_7H_8)Mo(CO)$_3$] in THF instead of [Mo(CO)$_3$(CH_3CN)$_4$] in CH_2Cl_2 (Scheme 1). Although this new method requires a higher temperature than the one previously reported, it only takes 30 min instead of one night and has a higher yield (57%) and better reproducibility (for further details, see Section 3.2.1). In addition, chromatographic purification can be avoided by removing the THF solvent and recrystallizing the product from the crude mixture using the appropriate amount of CH_2Cl_2. Interestingly, this procedure additionally allowed the crystallization of a new polymorph, **2′**, of the triple-decker complex **2**.

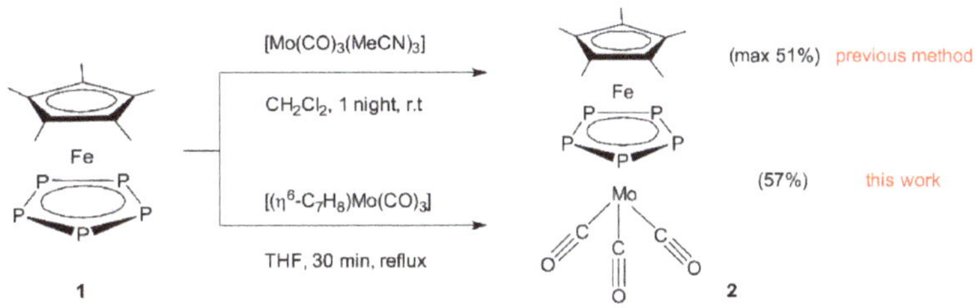

Scheme 1. Previous and current methods used for the synthesis of triple-decker complex **2**.

Compound **2′** crystallizes in the space group $C2/c$ of the orthorhombic crystal system (**2**; monoclinic $P2_1/n$). The geometry of **2′** is mostly similar to that of **2** (Figure 1). The P–P bond lengths in **2′** are slightly longer (2.1453(1)–2.1611(1) Å) than those in **2** (2.116(6)–2.138(8) Å). The average Fe–P distance is also slightly longer in **2′** (2.388(1) Å) than in **2** (2.370(4) Å). The mean Mo–P distances are identical within the margin of error (2.634(1) Å, 2.630(4) Å). The *cyclo*-P_5 ring in **2′** is approximately planar as in **2** (maximum deviation: 0.064(1) Å). Due to the slightly longer P–P distances, a slightly shorter intermetallic distance between Fe and Mo is found for **2′** (3.428 Å) than for **2** (3.443 Å). The new structure determination of **2′** was carried out at a lower temperature (101 K) than that of **2** (r.t.) and showed better quality factors and a lower residual electron density. For this reason, the metric data of **2′** are used in this paper for geometric comparisons.

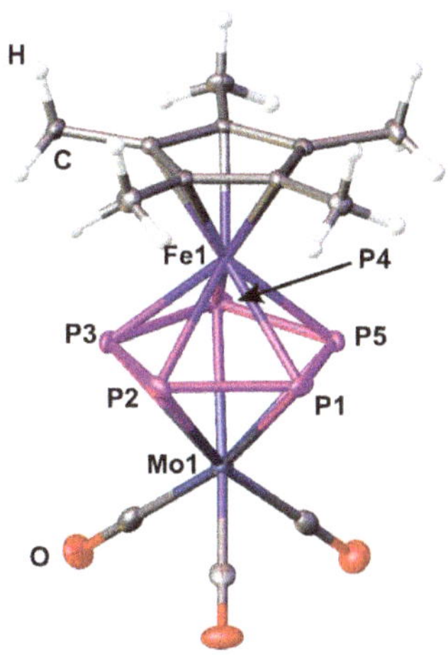

Figure 1. Molecular structure of **2′** in the solid state. Selected bond lengths [Å] and angles [°]: Fe1–P1 2.3822(8), Fe1–P2 2.3812(8), Fe1–P3 2.4010(8), Fe1–P4 2.3665(8), Fe1–P5 2.4097(7), Mo1–P1 2.6548(8), Mo1–P2 2.6579(7), Mo1–P3 2.5843(8), Mo1–P4 2.7036(7), Mo1–P5 2.5694(7), P1–P2 2.1611(1), P2–P3 2.1453(1), P3–P4 2.1582(1), P4–P5 2.1561(1), P5–P1 2.1475(1), Mo1–P1-Fe1 85.59(2), Mo1–P2–Fe1 85.53(2), Mo1–P3–Fe1 86.79(2), Mo1–P4–Fe1 84.80(2), Mo1–P5–Fe1 86.95(2).

2.2. *Synthesis and X-Ray Structure of the Coordination Compound [Ag{Cp*Fe(μ,η⁵:η⁵-P₅)Mo(CO)₃}₂][BF₄]* (3)

As a first approach, the triple-decker complex **2** was reacted with Ag[BF$_4$]. This reaction was carried out in CH_2Cl_2 at room temperature with a 2:1 stoichiometric ratio and led to the selective isolation of complex [Ag{Cp*Fe(μ,η^5:η^5-P_5)Mo(CO)$_3$}$_2$][BF$_4$] (**3**) in a good (62%) yield (Scheme 2).

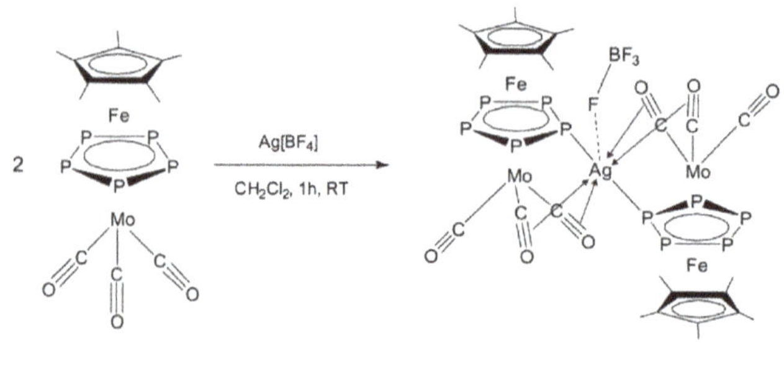

2 **3**, 62%

Scheme 2. The reaction of triple-decker complex **2** with Ag[BF_4]. Synthesis of the coordination compound **3**.

Dark red crystals of **3**, which are suitable for single crystal X-ray diffraction, were obtained after layering the crude reaction mixture with *n*-pentane (Figure 2). Compound **3** crystallizes in the monoclinic space group $P2_1/c$. The asymmetric unit contains one Ag^I cation, two complexes of **2**, and one BF_4^- anion. Two P atoms coordinate to the Ag^I ion, one from each of the two complexes **2** (Figure 2), with Ag–P distances of 2.7149(1) Å and 2.7683(1) Å, respectively. Additionally, four short distances to the C atoms of the two CO ligands of both complexes **2** were identified ($\bar{d}$(Ag···CO) = 2.726 Å). A possible bonding mode for these CO ligands makes them bind to one Mo atom in a common terminal fashion and also donate electron density from the CO π bonds to the Ag atom. This bonding pattern of CO ligands has only been rarely reported in the literature [30]. Moreover, the Ag1–F1 distance is short (2.622(3) Å) and falls within the limits of van der Waals interactions (2.30–2.90 Å) [31]. Overall, the Ag^I cation shows a distorted pentagonal bipyramidal coordination geometry. The average P–P bond length in **3** is slightly longer than that in the free complex **2'** (2.160 Å compared to 2.154(1) Å; Table 1). The Mo–C distances of the semi-bridging CO ligands [32] are slightly longer than the ones of the terminal CO ligands (mean values: 2.008 Å and 1.981 Å). The C–O bond lengths range between 1.136(5) and 1.147(5) Å. The Mo–C–O angles of the semi-bridging CO ligands deviate more from a linear geometry (173.6(3)–176.3(3)°) than those of the terminal ones (179.0(4)° and 179.2(3)°). The CO bands in the IR spectra of **3** are comparable to those of the free ligand. Therefore, it is supposed that the interaction between the cation Ag^+ and the bridging CO ligands is weak. The Ag···Mo distances are relatively short at 2.8301(4) Å and 2.8442(4) Å. The room temperature ^1H and $^{13}C\{^1H\}$ NMR spectra of **3** in CD_2Cl_2 show characteristic signals for the Cp* and CO ligands. In the $^{31}P\{^1H\}$ spectrum, in CD_2Cl_2, a broad singlet at 25.2 ppm can be detected, which is significantly shifted to lower field compared to the free ligand (9.7 ppm). The signal for the $[BF_4]^-$ anion in the $^{19}F\{^1H\}$ NMR spectrum in CD_2Cl_2 at −151.7 ppm is broadened due to the interaction with the Ag^+ cation ($\omega_{1/2}$ = 84 Hz).

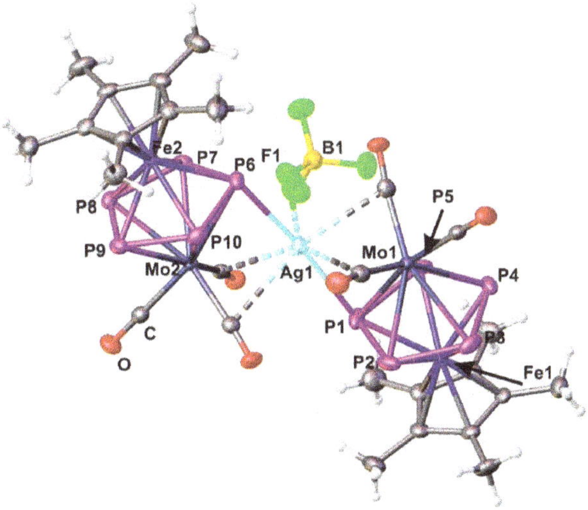

Figure 2. Molecular structure of **3** in the solid state. Selected bond lengths [Å] and angles [°]: Ag1–P1 2.7683(1), Ag1–P6 2.7149(1), Ag1–F1 2.622(3), P1–P2 2.1670(2), P2–P3 2.1465(2), P3–P4 2.1668(2), P4–P5 2.1368(2), P5–P1 2.1660(2), P6–P7 2.1711(1), P7–P8 2.1429(1), P8–P9 2.1721(1), P9–P10 2.1531(1), P10–P6 2.1745(1), P1–Ag1–P6 175.09(4), P1–Ag1–F1 99.07(7), P6–Ag1–F1 83.45(6), Mo1–Ag1–Mo2 155.93(3).

Table 1. Comparison of structural and spectroscopic data of the compounds $[M(\mathbf{2})_2][X]$ and the triple-decker complex **2′**.

	2′	M = Ag [X] = [BF$_4$] 3	M = Ag [X] = [TEF] 4	M = Cu [X] = [TEF] 5′	M = Cu [X] = [TEF] 5″
$\bar{d}$(P,P)/Å	2.154(1)	2.160	2.158	2.156	2.154
$\bar{d}$(M,P)/Å	—	2.742	2.695	2.687	2.662
$\bar{d}$(Mo,M)/Å	—	2.837	2.824	2.680	2.673
∡(P,M,P)/°	—	175.09(3)	177.12(7)	179.61(9)	176.89(8)
∡(Mo,M,Mo)/°	—	155.93(3)	173.56(3)	174.92(4)	173.93(4)
$\delta\ ^{31}P\{^1H\}$/ppm CD$_2$Cl$_2$, RT	9.7	25.2	30.1	31.7	
$\tilde{\nu}_{CO}$/cm^{-1} KBr	1963, 1955, 1894, 1881	1963, 1955, 1892, 1880	1980 (br), 1928, 1914	1993, 1961, 1923, 1910	

2.3. *Synthesis and X-Ray Structure of the Coordination Compound [Ag{Cp*Fe(μ,η^5:η^5-P$_5$)Mo(CO)$_3$}$_2$][TEF]* (**4**)

The result obtained from the reaction of complex **2** with Ag[BF$_4$] raised the question of the possibility of expanding this chemistry to other Ag^I salts. Accordingly, we selected the Ag^I salt of the weakly coordinating anion [TEF]$^-$ because it offers a high solubility, thus allowing a more detailed study of the formed compounds in solution [33]. The reaction of **2** with Ag[TEF], performed by applying reaction conditions similar to those used for the reaction with Ag[BF$_4$], afforded the selective isolation of the complex [Ag{Cp*Fe(μ,η^5:η^5-P$_5$)Mo(CO)$_3$}$_2$][TEF] (**4**) with an excellent yield (80%, Scheme 3).

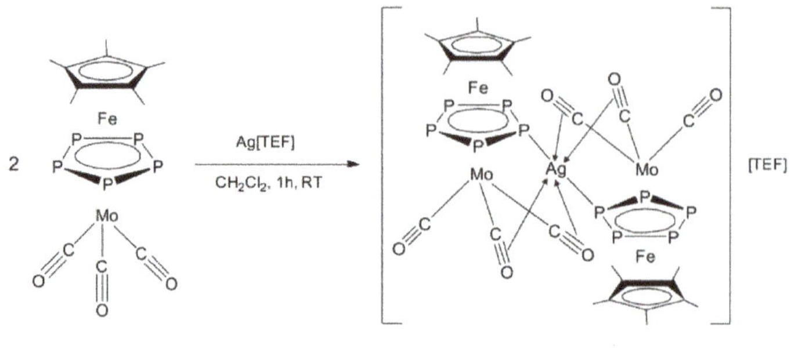

2   4, 80%

Scheme 3. The reaction of the triple-decker complex **2** with Ag[TEF]. Synthesis of complex **4**.

Single crystals of **4** were obtained by layering the CH_2Cl_2 crude reaction mixture with *n*-pentane (Figure 3). Compound **4** crystallizes in the monoclinic space group $P2_1/c$. Similar to what was observed for **3**, the crystal structure of **4** shows an Ag^I cation coordinated by two P atoms with Ag–P distances of 2.6809(2) Å and 2.7097(2) Å. Additionally, it exhibits four weak interactions with the C atoms of four CO ligands ($\bar{d}$(Ag···CO) = 2.748 Å). In contrast to **3**, no additional interactions are observed between the Ag^I cation and any fluorine atom from the $[TEF]^-$ ion, probably due to the non-coordinating property of the $[TEF]^-$ anion. As a consequence, the Ag^I ion adopts a distorted square bipyramidal coordination sphere in contrast to the pentagonal bipyramidal one observed for the Ag^I ion in **3**. Interestingly, the Mo–Ag–Mo angle in **4** (173.56(3)°) deviates much less from linearity than that in **3** (155.93(3)°), and the Ag···Mo distances in **4** (2.8131(7) Å and 2.8353(7) Å) are slightly shorter than those found in **3**.

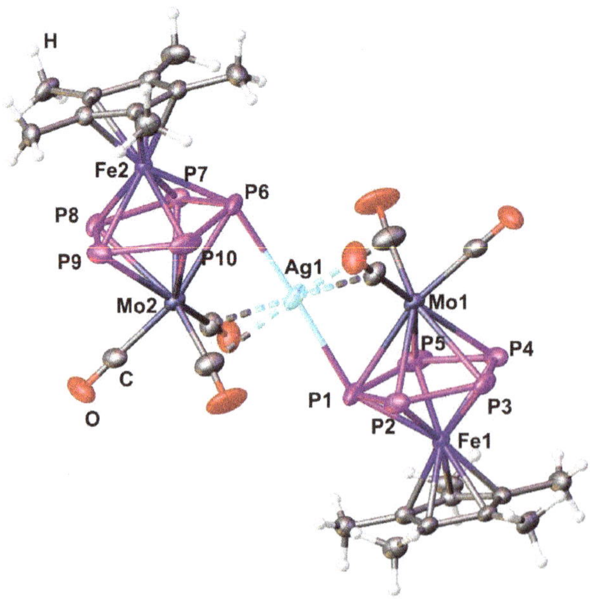

Figure 3. Molecular structure of the cation of **4** in the solid state. Selected bond lengths [Å] and angles [°]: Ag1–P1 2.7097(2), Ag1–P6 2.6809(2), P1–P2 2.173(3), P2–P3 2.139(3), P3–P4 2.159(3), P4–P5 2.147(3), P5–P1 2.168(3), P6–P7 2.166(3), P7–P8 2.137(3), P8–P9 2.164(3), P9–P10 2.159(3), P10–P6 2.169(3), P1–Ag1–P6 177.12(7), Mo1–Ag1–Mo2 173.56(3).

As expected, compound **4** is excellently soluble in CH_2Cl_2 due to its $[TEF]^-$ anion, whereas it is moderately soluble in toluene and insoluble in *n*-alkanes. The ^{1}H, $^{13}C\{^{1}H\}$, and ^{19}F NMR spectra of **4** in CD_2Cl_2 at r.t. show characteristic signals for the Cp* and CO ligands as well as for the $[TEF]^-$ anion. In the $^{31}P\{^{1}H\}$ NMR spectrum, a singlet at 30.1 ppm is observed, which is significantly shifted to lower field compared to the free ligand (9.7 ppm). Upon cooling to 183 K, no broadening or significant shift of the signal occur. The $^{31}P\{^{1}H\}$ MAS-NMR spectrum of solid **4** at room temperature shows a singlet at a chemical shift (28.3 ppm) similar to that in solution. Thus, molecular dynamic processes such as the rotation of the *cyclo*-P_5 ring take place at low temperatures in solution and in the solid state at room temperature as well. However, the single-crystal X-ray analysis performed at 100 K did not indicate a rotation of the *cyclo*-P_5 ring because of the low temperature needed for an appropriate structure solution. The molecular ion peak can be detected in the cation mode of the ESI mass spectrum as well as peaks for the intact and partially decarbonylated ligands of **2**. In the anion mode, the $[TEF]^-$ anion is detected. Vapor pressure osmometry yields a relatively high value for the mean molecular mass ($1700\ g{\cdot}mol^{-1} \pm 100\ g{\cdot}mol^{-1}$), suggesting that the compound is mainly associated in solution. The CO stretching frequencies in the IR spectrum in the solid state are broadened compared to the one of the free ligand and are shifted to higher wave numbers, probably due to the positive charge of the complex (Table 1). In the $^{13}C\{^{1}H\}$ MAS-NMR spectrum, the signals for the C atoms of the carbonyl groups are broadened as well and, interestingly, a differentiation between semi-bridging and non-bridging CO ligands is not possible.

2.4. Synthesis and X-Ray Structure of the Complex [Cu{Cp*Fe(μ,η⁵:η⁵-P₅)Mo(CO)₃}₂][TEF] (5)

The successful coordination of triple-decker complex **2** to Ag^I ions inspired us to study its coordination behavior towards Cu^I ions, specifically towards $[Cu(1,2\text{-}F_2C_6H_4)][TEF]$. This reaction was performed under similar reaction conditions as previously used and selectively yielded the complex $[Cu\{Cp^*Fe(\mu,\eta^5{:}\eta^5\text{-}P_5)Mo(CO)_3\}_2][TEF]$ (**5**) in good isolated yields (55%, Scheme 4).

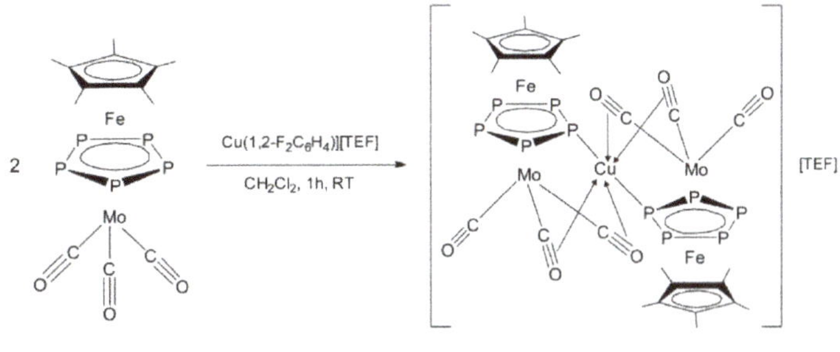

Scheme 4. The reaction of triple-decker complex **2** with $Cu(1,2\text{-}F_2C_6H_4)[TEF]$. Synthesis of the coordination compound **5**.

Compound **5** crystallizes in the monoclinic space group $P2_1/c$. The asymmetric unit contains two independent complexes $[Cu(\mathbf{2})_2]^+$ and two $[TEF]^-$ anions, which show slightly different bonding lengths and angles (entitled in the following as **5′** and **5″**). The monocationic units in **5** (**5′**, **5″**) exhibit the same geometry and coordination modes as the related Ag^I compound **4** (Figure 4). The Cu–P bond distances in **5** lie between 2.647(2) Å and 2.691(2) Å and the short contacts between Cu and the semi-bridging CO ligands range from 2.534 Å to 2.645 Å. Table 1 shows a comparison of selected structural and spectroscopic data of the independent molecules of **5′** and **5″** as well as of the silver-containing compounds **3** und **4** with the same ligand. The average P–P distances in the silver compounds are slightly larger than in the copper compound, but they are all elongated compared to the free triple-decker complex **2′**. The Cu–P distances are relatively large, but smaller than the Ag–P distances, as expected. The P–M–P angle amounts to nearly 180° in all cases. The largest deviation from linearity is observed in the $[BF_4]^-$ compound **3** with about 5°. The intermetallic distances within the copper compound are significantly smaller than those in the silver compounds. The complexes with $[TEF]^-$ as counterion show Mo–M–Mo angles between 173.56(3)° and 174.92(4)°, while compound **3** clearly differs from these (155.93(3)°).

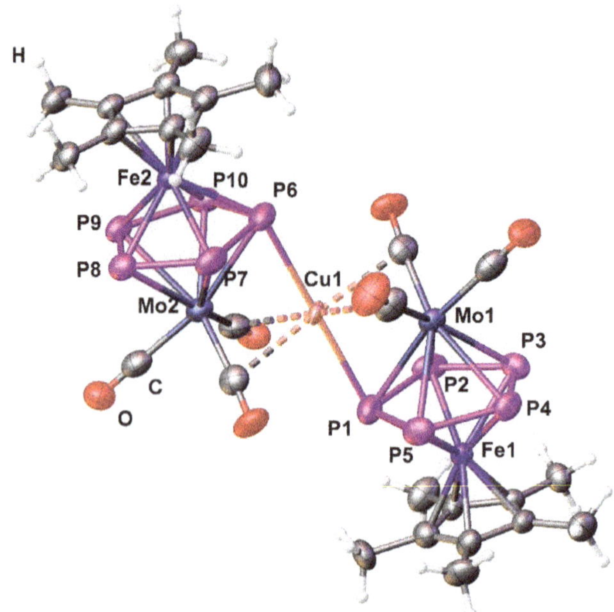

Figure 4. Molecular structure of the cation of **5** in the solid state.

In the $^{31}P\{^1H\}$ NMR spectra, all compounds show a low field shift compared to the signal of free complex **2**. The difference in the chemical shift is larger for $[TEF]^-$ than for $[BF_4]^-$ and larger for copper than for silver. The CO bands in the IR spectra of **3** are almost similar to those of the starting material. However, in regard to compounds **4** and **5**, they are broadened and shifted to higher wave numbers. In the ESI mass spectra of **5**, the molecular ion peak can be detected in the cation mode, as in the case of **4**, as well as peaks for the intact and partially decarbonylated ligands. In the anion mode, only a single peak can be observed for the $[TEF]^-$ anion.

3. Materials and Methods

3.1. General Information

All experiments were performed under an atmosphere of dry argon using standard Schlenk techniques. $[(C_7H_8)Mo(CO)_3]$ and $Ag[BF_4]$ were purchased from Sigma-Aldrich (Darmstadt, Germany) and used as received without further purification. The compound $[Cp^*Fe(\eta^5\text{-}P_5)]$ (**1**) [3] and the salts $Ag[Al\{OC(CF_3)_3\}_4]$ [33] and $Cu(1,2\text{-}F_2C_6H_4)[Al\{OC(CF_3)_3\}_4]$ [34] were synthesized according to literature procedures. Solvents were freshly distilled under argon from Na/benzophenone (THF), from Na (toluene), from CaH_2 (CH_2Cl_2) and from a Na/K alloy (*n*-pentane, *n*-hexane). IR spectra were recorded on a Varian FTS-800 spectrometer (Varian Inc., MA, USA). ^1H, ^{13}C, ^{31}P and ^{19}F NMR spectra were recorded on Bruker Avance 300 and 400 spectrometers (Bruker, Karlsruhe, Germany). MAS-NMR spectra were performed at 300 K on a Bruker Avance 300 spectrometer (Bruker, Karlsruhe, Germany). ^1H and ^{13}C NMR chemical shifts were reported in parts per million (ppm) relative to

Me_4Si as an external standard. ^{31}P NMR chemical shifts were expressed in ppm relative to external 85% H_3PO_4 and were decoupled from the proton. ^{19}F NMR chemical shifts were reported relative to $CFCl_3$. For the ESI-MS, a Finnigan Thermoquest TSQ 7000 mass spectrometer (Thermo Fisher Scientific, Frankfurt, Germany) was used. Elemental analyses were performed by the microanalytical laboratory of the University of Regensburg. Molecular mass determination performed by means of vapor pressure osmometry was carried out using a Knauer K-7000 vapor pressure osmometer (Knauer, Berlin, Germany) with CH_2Cl_2 as the solvent.

3.2. *Synthesis and Characterization of the Compounds* **2** *and* **3–5**

3.2.1. Synthesis and Characterization of [Cp*Fe(μ,η^5:η^5-P_5)Mo(CO)$_3$] (**2**)

To a stirred solution of [Cp*Fe(η^5-P_5)] (**1**, 1.28 g, 3.69 mmol) in 75 mL THF, [(C_7H_8)Mo(CO)$_3$] (1 g, 3.69 mmol) in 75 mL THF was slowly added. The crude reaction mixture was refluxed for 30 min. The solvent was removed under reduced pressure and the residue was extracted with 50 mL CH_2Cl_2. In order to obtain crystals of the polymorph **2′**, the crude mixture was stored for three days at −30 °C. Otherwise, the crude residue mixture was purified by column chromatography using silica gel (10 × 2 cm) and an eluent solvent mixture of hexane:toluene (9:1). The triple-decker complex **2** was obtained as a dark green solution, the solvent was removed under reduced pressure and **2** was isolated as an olive-green solid. Yield: (1.11 g, 57%). Crystal Data for $C_{13}H_{15}O_3P_5FeMo$, M_r = 525.89 g/mol, monoclinic, *C*2/*c* (No. 15), a = 15.8694(11) Å, b = 14.6679(10) Å, c = 15.9594(12) Å, β = 102.587(7)°, α = γ = 90°, V = 3625.6(5) Å^3, T = 101(2) K, Z = 8, Z' = 1, $\mu(CuK_\alpha)$ = 16.348, 10281 reflections measured, 3185 unique (R_{int} = 0.0235), which were used in all calculations. The final wR_2 was 0.0699 (all data) and the R_1 was 0.0237 (I > 2(I)).

3.2.2. Synthesis and Characterization of [Ag{Cp*Fe(μ,η^5:η^5-P_5)Mo(CO)$_3$}$_2$][BF_4] (**3**)

A solution of Ag [BF_4] (18 mg, 0.09 mmol) and [Cp*Fe(μ,η^5:η^5-P_5)Mo(CO)$_3$] (**2**; 95 mg, 0.18 mmol) in CH_2Cl_2 (15 mL) was stirred at r.t. for 1 h in the dark. Subsequently, the deep red solution was filtered through a G4 filter frit and layered with *n*-pentane (25 mL) using a Teflon capillary. Within two weeks, red crystals of **3** had formed after storage at 2 °C. These were filtered off, washed with *n*-pentane (3 × 2 mL), and dried in a vacuum. Yield (69 mg, 62%)

^{1}H NMR (300 MHz, CD_2Cl_2): δ = 1.36 ppm (s, $C_5(CH_3)_5$), $^{13}C\{^1H\}$ NMR (75.47 MHz, CD_2Cl_2): δ = 9.8 (br sext, $^3J_{P,C}$ = 1.8 Hz; $C_5(\underline{C}H_3)_5$), 89.0 (br s, $\underline{C}_5(CH_3)_5$), 214.3 ppm (br s, CO). $^{31}P\{^1H\}$ NMR (121.49 MHz, CD_2Cl_2): δ = 25.2 ppm (s, $\omega_{1/2}$ = 28 Hz). $^{19}F\{^1H\}$ NMR (282.40 MHz, CD_2Cl_2): δ = −151.7 ppm (br s, $\omega_{1/2}$ = 84 Hz; CF_3). IR (KBr): $\tilde{\nu}/cm^{-1}$ = 2957 (vw), 2906 (w), 2852 (vw), 1963 (vs), 1955 (vs), 1892 (vs), 1880 (vs), 1477 (w), 1447 (vw), 1427 (w), 1376 (w), 1309 (w) 1242 (m), 1209 (w), 1155 (w), 1124 (w), 1071 (w), 1020 (m), 986 (w), 600 (w), 584 (w), 557 (w), 544 (w) 493 (m), 477 (m), 440 (w). IR (CH_2Cl_2): $\tilde{\nu}/cm^{-1}$ = 3054 (vs), 2987 (w), 2349 (vw), 2305 (vw), 1970 (vs), 1892 (vs), 1897 (vs), 1422 (m), 1267 (vs). Elemental analysis, calcd. (%) for $C_{26}H_{30}AgBF_4Fe_2Mo_2O_6P_{10}$ (1246.46 g·mol^{-1}): C 25.05, H 2.43; found: C 25.81, H 2.54. Crystal Data for $C_{104}H_{120}Ag_4B_4F_{16}Fe_8Mo_8O_{24}P_{40}$, M_r = 4985.83 g·mol^{-1}, monoclinic, *P*2$_1$/*c* (No. 14), a = 8.77020(10) Å, b = 26.5995(3) Å, c = 19.5513(3) Å, β = 116.2050(10)°, α = γ = 90°, V = 4092.20(10) Å^3, T = 123(2) K, Z = 1, Z' = 0.25, $\mu(CuK_\alpha)$ = 18.417, 16649 reflections measured, 7769 unique (R_{int} = 0.0232), which were used in all calculations. The final wR_2 was 0.0955 (all data) and R_1 was 0.0354 (I > 2(I)).

3.2.3. Synthesis and Characterization of [Ag{Cp*Fe(μ,η^5:η^5-P_5)Mo(CO)$_3$}$_2$][TEF] (**4**)

A solution of [Ag(CH_2Cl_2)][TEF] (104 mg, 0.09 mmol) and two equivalents of [Cp*Fe(μ,η^5:η^5-P_5)Mo(CO)$_3$] (**2**; 95 mg, 0.18 mmol) in CH_2Cl_2 (15 mL) was stirred at r.t. for 1 h in the dark. The deep red solution was then filtered through a G4 filter frit and layered with *n*-pentane (25 mL) using a Teflon capillary. Within two weeks, red crystals of **4** had formed at 2 °C. These were filtered off, washed with *n*-pentane (3 × 2 mL), and dried in a vacuum. The mother liquor was further concentrated to 5 mL under vacuum, under which more product precipitated as a reddish brown precipitate upon addition of *n*-pentane (45 mL), likewise filtered off, washed with *n*-pentane (3 × 2 mL), and dried under vacuum. Yield (154 mg, 80%).

^{1}H NMR (400 MHz, CD_2Cl_2): δ = 1.34 ppm (s, $C_5(CH_3)_5$), $^{13}C\{^1H\}$ NMR (100.61 MHz, CD_2Cl_2): δ = 9.7 (br sext, $^3J_{P,C}$ = 1.8 Hz; $C_5(\underline{C}H_3)_5$), 90.3 (s, $\underline{C_5}(CH_3)_5$), 121.6 (q, $^1J_{F,C}$ = 292.1 Hz; CF_3), 213.9 ppm (br sext, $^2J_{P,C}$ = 1.6 Hz; CO). $^{13}C\{^1H\}$-MAS-NMR (75.47 MHz): δ = 9.0 (br s, $C_5(\underline{C}H_3)_5$), 92.5 (br s, $\underline{C_5}(CH_3)_5$), 122.3 (q, $^1J_{F,C}$ = 337.4 Hz; CF_3), 215.1 ppm (br s, CO). $^{31}P\{^1H\}$ NMR (161.98 MHz, CD_2Cl_2): δ = 30.1 ppm (s, $\omega_{1/2}$ = 4 Hz). $^{31}P\{^1H\}$ NMR (161.98 MHz, CD_2Cl_2, 183 K): δ = 30.3 ppm (s, $\omega_{1/2}$ = 6 Hz). $^{31}P\{^1H\}$-MAS-NMR (121.49 MHz): δ = 28.3 ppm (br s, $\omega_{1/2}$ = 93 Hz). $^{19}F\{^1H\}$ NMR (282.40 MHz, CD_2Cl_2): δ = −75.6 ppm (s, CF_3). IR (KBr): $\tilde{\nu}/cm^{-1}$ = 2962 (vw), 2918 (w), 2851 (vw), 1980(vs), 1928 (vs), 1914 (vs), 1543 (vw), 1480 (w), 1450 (w), 1428 (w), 1381 (m), 1353 (m), 1301 (vs), 1277 (vs), 1242 (vs), 1219 (vs), 1165 (m), 1136 (vw), 1075 (vw), 1021 (m), 974 (vs), 833 (w), 799 (vw), 755 (vw), 727 (s), 587 (m), 559 (w), 538 (m), 492 (w), 476 (w), 444 (m). IR (CH_2Cl_2): $\tilde{\nu}/cm^{-1}$ = 2984 (vw), 2959 (vw), 2915 (vw), 2849 (vw), 1993(vs), 1970 (vs), 1934 (s), 1902 (vs), 1478 (w), 1448 (vw), 1426 (m), 1379 (m), 1352 (m), 1300 (vs), 1277 (vs), 1242 (vs), 1225 (vs), 1167 (m), 1136 (vw), 1072 (vw), 1022 (m), 976 (vs), 833 (vw), 600 (w), 587 (w), 559 (w), 538 (w), 494 (w), 476 (vw), 444 (m). Osmometry (CH_2Cl_2): Average molar mass: 1700 g·mol^{-1} ± 100 g·mol^{-1}. Positive ion ESI-MS (CH_2Cl_2), m/z (%) = 1159.0 (7) $[Ag\{Cp^*FeP_5Mo(CO)_3\}_2]^+$, 527.9 (27) $[Cp^*FeP_5Mo(CO)_3]^+$, 499.9 (100) $[Cp^*FeP_5Mo(CO)_2]^+$. negative ion ESI-MS (CH_2Cl_2), m/z (%) = 967.2 (100) $[Al\{OC(CF_3)_3\}_4]^-$. Elemental analysis, calcd. (%) for $C_{42}H_{30}AgAlF_{36}Fe_2Mo_2O_{10}P_{10}$ (2126.79 g·mol^{-1}): C 23.72, H 1.42; found: C 23.73, H 1.44. Melting point > 200 °C. Crystal Data for $C_{42}H_{30}AgAlF_{36}Fe_2Mo_2O_{10}P_{10}$, M_r = 2126.79 g·mol^{-1}, monoclinic, $P2_1/c$ (No. 14), a = 19.6554(16) Å, b = 9.6564(9) Å, c = 36.745(3) Å, β = 98.912(9)°, $\alpha = \gamma$ = 90°, V = 6889.9(10) Å^3, T = 100(2) K, Z = 4, Z' = 1, $\mu(CuK_\alpha)$ = 12.144, 26789 reflections measured, 12207 unique (R_{int} = 0.0444), which were used in all calculations. The final wR_2 was 0.1467 (all data) and R_1 was 0.0498 (I > 2(I)).

3.2.4. Synthesis and Characterization of $[Cu\{Cp^*Fe(\mu,\eta^5:\eta^5\text{-}P_5)Mo(CO)_3\}_2][TEF]$ (**5**)

A solution of $[Cu(1,2\text{-}F_2C_6H_4)][TEF]$ (92 mg, 0.08 mmol) and $[Cp^*Fe(\mu,\eta^5:\eta^5\text{-}P_5)Mo(CO)_3]$ (**2**; 84 mg, 0.16 mmol) in CH_2Cl_2 (15 mL) was stirred at r.t. for 1 h. Subsequently, the deep red solution was then filtered through a G4 filter frit and layered with *n*-pentane (25 mL) using a Teflon capillary. Within two weeks, red crystals of **5** had formed at 2 °C, which were filtered off, washed with *n*-pentane (3 × 2 mL), and dried in a vacuum (5 h, 10^{-3} mbar). According to the elemental analysis, half an equivalent of *n*-pentane per unit formula could not be removed (The asymmetric unit of **5** also contains half an *n*-pentane molecule). Yield (93 mg, 55%).

^{1}H NMR (300 MHz, CD_2Cl_2): δ = 1.38 ppm (s, $C_5(CH_3)_5$), $^{13}C\{^1H\}$ NMR (100.61 MHz, CD_2Cl_2): δ = 9.7 (br s; $C_5(\underline{C}H_3)_5$), 90.5 (s, $\underline{C_5}(CH_3)_5$), 121.6 (q, $^1J_{F,C}$ = 291.4 Hz; CF_3), 212.9 ppm (br s; CO). $^{31}P\{^1H\}$ NMR (121.49 MHz, CD_2Cl_2): δ = 31.7 ppm (s, $\omega_{1/2}$ = 41 Hz). $^{19}F\{^1H\}$ NMR (282.40 MHz, CD_2Cl_2): δ = −75.6 ppm (s, CF_3). IR (KBr): $\tilde{\nu}/cm^{-1}$= 2962 (vw), 2919 (w), 2854 (vw), 1993(vs), 1961(vs), 1923 (vs), 1910 (vs), 1479 (w), 1451 (w), 1428 (w), 1381 (m), 1353 (m), 1302 (vs), 1277 (vs), 1241 (vs), 1220 (vs), 1166 (m), 1139 (vw), 1072 (vw), 1021 (m), 974 (vs), 833 (w), 756 (vw), 728 (vs), 586 (w), 561 (w), 537 (m), 492 (w), 475 (w), 443 (m). IR (CH_2Cl_2): $\tilde{\nu}/cm^{-1}$ = 3054 (vs), 2987 (w), 2348 (w), 2305 (vw), 1995 (s), 1969 (vs), 1925 (m), 1901 (s), 1422 (m), 1269 (vs). Positive ion ESI-MS (CH_2Cl_2), m/z (%) = 1114.6 (1) $[Cu\{Cp^*FeP_5Mo(CO)_3\}_2]^+$, 527.9 (100) $[Cp^*FeP_5Mo(CO)_3]^+$, 499.9 (94) $[Cp^*FeP_5Mo(CO)_2]^+$. Negative ion ESI-MS (CH_2Cl_2), m/z (%) = 967.1 (100) $[Al\{OC(CF_3)_3\}_4]^-$. Elemental analysis, calcd. (%) for $C_{42}H_{30}CuAlF_{36}Fe_2Mo_2O_{10}P_{10} \times 0.5C_5H_{12}$ (2118.55 g·mol^{-1}): C 25.23, H 1.71; found: C 25.36, H 1.73. Crystal Data for $C_{86.5}H_{66}Al_2Cu_2F_{72}Fe_4Mo_4O_{20}P_{20}$, M_r = 4200.99, monoclinic, $P2_1/c$ (No. 14), a = 32.1751(3) Å, b = 12.2014(2) Å, c = 35.8672(5) Å, b = 99.6750(10)°, $a = g$ = 90°, V = 13880.5(3) Å^3, T = 123(2) K, Z = 4, Z' = 1, $m(CuK_a)$ = 10.176, 50354 reflections measured, 24416 unique (R_{int} = 0.0606) which were used in all calculations. The final wR_2 was 0.1722 (all data) and R_1 was 0.0609 (I > 2(I)).

3.3. Crystallography

The crystals were selected and mounted on an Oxford Diffraction Gemini R Ultra diffractometer equipped with a Ruby CCD detector (**2**, **4**) and a SuperNova diffractometer equipped with an Atlas CCD detector (**3**, **5**), respectively. The crystals of compounds **2** and **4** were kept at T = 100(1) K and the crystals of the compounds **3** and **5** at T = 123(1) K during data collection, respectively. Data collection and reduction were performed with **CrysAlispro** (Version 171.33.41 (**2**), 171.34.9 (**3**), 171.33.42 (**4**), 171.33.61 (**5**)) [35]. For compounds **2** and **4**, a semi-empirical multi-scan absorption correction from equivalents [35] was applied. For compounds (**3**, **5**), an analytical numeric absorption correction using a multifaceted crystal model based on expressions derived by R. C. Clark & J. S. Reid was applied [36]. Using **Olex2** [37], the structures were solved with **SIR97** [38] and a least-square refinement on F2 was carried out with **ShelXL** [39]. All non-hydrogen atoms were refined anisotropically. Hydrogen atoms at the carbon atoms were located in idealized positions and refined isotropically according to the riding model. CIF files with comprehensive information on the details of the diffraction experiments and full tables of bond lengths and angles for **2–5** were deposited in Cambridge Crystallographic Data Centre under the deposition codes CCDC-1884710-1884713.

4. Conclusions

In conclusion, we found a new high-yielding synthesis of the triple-decker complex $[Cp^*Fe(\mu,\eta^5:\eta^5-P_5)Mo(CO)_3]$ (**2**) and used it as a ligand in coordination chemistry for the first time. Its reaction with $Ag[BF_4]$, Ag[TEF] or Cu[TEF] leads, in each case, to the selective formation of complexes with the general formula $[M\{Cp^*Fe(\mu,\eta^5:\eta^5-P_5)Mo(CO)_3\}_2][X]$ (M = Ag, Cu; X = $[BF_4]^-$, $[TEF]^-$). These results open a new chapter in the reactivity of triple-decker complexes with *cyclo*-P_5 middle-decks as promising candidates in supramolecular chemistry. Current studies are focused on extending the new synthesis approach to obtain other triple-decker compounds with *cyclo*-P_5 and *cyclo*-As_5 middle-decks and studying their coordination chemistry towards transition metal complexes.

Author Contributions: Stefan Welsch, synthesis and characterization of the coordination compounds **2'**, **3–5**. Luis Dütsch, Martin Piesch and Stephan Reichl, developing the new synthesis of the compound **2**. Michael Seidl, recalculating the X-ray structures of all compounds. Mehdi Elsayed Moussa, directing the new synthesis of compound **2**, writing the paper. Manfred Scheer, supervising the whole research work, writing the paper. All authors have read and approved the final manuscript.

Funding: This research was funded by the EUROPEAN RESEARCH COUNCIL, grant number ERC-2013-AdG-339072.

Acknowledgments: The authors thank Christian Gröger for the MAS-NMR spectra of **4**. Roland Neueder is gratefully acknowledged for the VPO measurement of **4**.

Conflicts of Interest: The authors declare no conflict of interest.

References

1. Whitmire, K.H. Transition metal complexes of the naked pnictide elements. *Coord. Chem. Rev.* **2018**, *376*, 114–195. [CrossRef]
2. Scheer, M. The coordination chemistry of group 15 element ligand complexes-a developing area. *Dalton Trans.* **2008**, 4372–4386. [CrossRef] [PubMed]
3. Scherer, O.J.; Brück, T. $[(\eta^5-P_5)Fe(\eta^5-C_5Me_5)]$, a pentaphosphaferrocene derivative. *Angew. Chem. Int. Ed.* **1987**, *99*, 59. [CrossRef]
4. Fleischmann, M.; Welsch, S.; Peresypkina, E.V.; Virovets, A.V.; Scheer, M. Highly Dynamic Coordination Behavior of P_n Ligand Complexes towards "Naked" Cu^+ Cations. *Chem. Eur. J.* **2015**, *21*, 14332–14336. [CrossRef] [PubMed]
5. Fleischmann, M.; Welsch, S.; Krauss, H.; Schmidt, M.; Bodensteiner, M.; Peresypkina, E.V.; Sierka, M.; Gröger, C.; Scheer, M. Complexes of Monocationic Group 13 Elements with Pentaphospha- and Pentaarsaferrocene. *Chem. Eur. J.* **2014**, *20*, 3759–3768. [CrossRef] [PubMed]

6. Scheer, M.; Gregoriades, L.J.; Merkle, R.; Johnson, B.P.; Dielmann, F. Formation of Spherical Giant Molecules and Dynamic Behaviour of Supramolecular Assemblies Based on P_n-Ligand Complexes. *Phosphorus Sulfur Silicon Relat. Elem.* **2008**, *182*, 504–508. [CrossRef]
7. Welsch, S.; Gregoriades, L.J.; Sierka, M.; Zabel, M.; Virovets, A.V.; Scheer, M. Unusual Coordination Behavior of P_n-Ligand Complexes with Tl^+. *Angew. Chem. Int. Ed.* **2007**, *46*, 9323–9326. [CrossRef] [PubMed]
8. Scheer, M.; Gregoriades, L.J.; Virovets, A.V.; Kunz, W.; Neueder, R.; Krossing, I. Reversible Formation of Polymeric Chains by Coordination of Pentaphosphaferrocene with Silver(I) Cations. *Angew. Chem. Int. Ed.* **2006**, *45*, 5689–5693. [CrossRef] [PubMed]
9. Dielmann, F.; Schindler, A.; Scheuermayer, S.; Bai, J.; Merkle, R.; Zabel, M.; Virovets, A.V.; Peresypkina, E.V.; Brunklaus, G.; Eckert, H.; et al. Coordination Polymers Based on [Cp*Fe(η^5-P_5)]: Solid-State Structure and MAS NMR Studies. *Chem. Eur. J.* **2012**, *18*, 1168–1179. [CrossRef] [PubMed]
10. Bai, J.; Virovets, A.V.; Scheer, M. Pentaphosphaferrocene as a Linking Unit for the Formation of one- and Two-Dimensional Polymers. *Angew. Chem. Int. Ed.* **2002**, *41*, 1737–1740. [CrossRef]
11. Schindler, A.; Heindl, C.; Balázs, G.; Gröger, C.; Virovets, A.V.; Peresypkina, E.V.; Scheer, M. Size-Determining Dependencies in Supramolecular Organometallic Host-Guest Chemistry. *Chem. Eur. J.* **2012**, *18*, 829–835. [CrossRef] [PubMed]
12. Scheer, M.; Schindler, A.; Bai, J.; Johnson, B.P.; Merkle, R.; Winter, R.; Virovets, A.V.; Peresypkina, E.V.; Blatov, V.A.; Sierka, M.; et al. Structures and Properties of Spherical 90-Vertex Fullerene-Like Nanoballs. *Chem. Eur. J.* **2010**, *16*, 2092–2107. [CrossRef] [PubMed]
13. Scheer, M.; Schindler, A.; Gröger, C.; Virovets, A.V.; Peresypkina, E.V. A Spherical Molecule with a Carbon-Free I_h-C_{80} Topological Frameworks. *Angew. Chem. Int. Ed.* **2009**, *48*, 5046–5049. [CrossRef] [PubMed]
14. Scheer, M.; Schindler, A.; Merkle, R.; Johnson, B.P.; Linseis, M.; Winter, R.; Anson, C.E.; Virovets, A.V. Fullerene C_{60} as an Endohedral Molecule within an Inorganic Supramolecule. *J. Am. Chem. Soc.* **2007**, *129*, 13386–13387. [CrossRef] [PubMed]
15. Scheer, M.; Bai, J.; Johnson, B.P.; Merkle, R.; Virovets, A.V.; Christopher, E.A. Fullerene-Like Nanoballs Formed by Pentaphosphaferrocene and CuBr. *Eur. J. Inorg. Chem.* **2005**, *2005*, 4023–4026. [CrossRef]
16. Bai, J.; Virovets, A.V.; Scheer, M. Synthesis of Inorganic Fullerene-Like Molecules. *Science* **2003**, *300*, 781–782. [CrossRef]
17. Welsch, S.; Gröger, C.; Sierka, M.; Scheer, M. An Organometallic Nanosized Capsule Consisting of *cyclo*-P_5 Units and Copper(I) Ions. *Angew. Chem. Int. Ed.* **2011**, *50*, 1435–1438. [CrossRef]
18. Elsayed Moussa, M.; Attenberger, B.; Peresypkina, E.V.; Scheer, M. Neutral two-dimensional organometallic-organic hybrid polymers based on pentaphosphaferrocene, bipyridyl linkers and CuCl. *Dalton Trans.* **2018**, *47*, 1014–1017. [CrossRef]
19. Elsayed Moussa, M.; Welsch, S.; Lochner, M.; Peresypkina, E.V.; Virovets, A.V.; Scheer, M. Organometallic-Organic Hybrid Polymers Assembled from Pentaphosphaferrocene, Bipyridyl Linkers, and Cu^I ions. *Eur. J. Inorg. Chem.* **2018**, *23*, 2689–2694. [CrossRef]
20. Scherer, O.J.; Brück, T.; Wolmershäuser, G. Pentaphosphaferrocene als Komplexliganden. *Chem. Ber.* **1989**, *122*, 2049–2054. [CrossRef]
21. Rink, B.; Scherer, O.J.; Heckmann, G.; Wolmershäuser, G. Neutrale 30-Valenzelektronen-Tripeldeckerkomplexe mit *cyclo*-E_5-Mitteldeck (E = P, As). *Chem. Ber.* **1992**, *125*, 1011–1016. [CrossRef]
22. Kudinov, A.R.; Rybinskaya, M.I. New triple-decker complexes prepared by the stacking reactions of cationic metallofragments with sandwich compounds. *Russ. Chem. Bull.* **1999**, *48*, 1636–1642. [CrossRef]
23. Kudinov, A.R.; Loginov, D.A.; Starikova, Z.A.; Petrovskii, P.V.; Corsini, M.; Zanello, P. Iron- and Ruthenium-Containing Triple-Decker Complexes with a Central Pentaphospholyl Ligand—X-ray Structures of [(η-C_5H_5)Fe(μ-η:η-P_5)Ru(η-C_5Me_5)]PF_6 and [(η-C_5H_5)Fe(μ-η:η-P_5)Ru(η-C_5Me_5)]PF_6. *Eur. J. Inorg. Chem.* **2002**, *2002*, 3018–3027. [CrossRef]
24. Kudinov, A.R.; Petrovskii, P.V.; Rybinskaya, M.I. Synthesis and the fluxional behavior of the 30-electron cationic iron-molybdenum triple-decker complex with a central pentaphospholyl ligand, [(η-C_7H_7)Mo(μ-η:η-P_5)Fe(η-C_5Me_5)]BF_4. *Russ. Chem. Bull.* **1999**, *48*, 1374–1376. [CrossRef]
25. Mädl, E.; Peresypkina, E.V.; Timoshkin, A.Y.; Scheer, M. Triple-decker sandwish complexes with a bent *cyclo*-P_5 middle-deck. *Chem. Commun.* **2016**, *52*, 12298–12301. [CrossRef] [PubMed]

26. Heinl, S.; Balázs, G.; Bodensteiner, M.; Scheer, M. Synthesis and characterization of manganese triple-decker complexes. *Dalton Trans.* **2016**, *45*, 1962–1966. [CrossRef] [PubMed]
27. Scherer, O.J.; Schwalb, J.; Wolmershäuser, G.; Kaim, W.; Gross, R. *cyclo*-P_5 as Complex Ligand—the Phosphorus Analogue of the Cyclopentadienyl Ligand. *Angew. Chem. Int. Ed.* **1986**, *25*, 363–364. [CrossRef]
28. Goh, L.Y.; Wong, R.C.S.; Chu, C.K.; Hambley, T.W. Reaction of $[\{Cr(cp)(CO)_3\}_2](cp = \eta^5\text{-}C_5H_5)$ with elemental phosphorus. Isolation of $[Cr_2(cp)_2(P_5)]$ as a thermolysis product and its X-ray crystal structure. *J. Chem. Soc. Dalton Trans.* **1990**, 977–982. [CrossRef]
29. Fleischmann, M.; Dielmann, F.; Gregoriades, L.J.; Peresypkina, E.V.; Virovets, A.V.; Huber, S.; Timoshkin, A.Y.; Balázs, G.; Scheer, M. Redox and Coordination Behavior of the Hexaphosphabenzene Ligand in $[(Cp^*Mo)_2(\mu,\eta^6{:}\eta^6\text{-}P_6)]$ Towards the "Naked" Cations Cu^+, Ag^+, and Tl^+. *Angew. Chem. Int. Ed.* **2015**, *54*, 13110–13115. [CrossRef]
30. Klingler, R.J.; Butler, W.M.; Curtis, M.D. Molecular Structure of Dicyclopentadienyltetracarbonyldimolybdenum (Mo≡Mo). Semibridging Carbonyls as Four-Electron Donors in Complexes with Metal-Metal Multiple Bonds. *J. Am. Chem. Soc.* **1978**, *100*, 5034–5038. [CrossRef]
31. Ghisolfi, A.; Fliedel, C.; de Frémont, P.; Braunstein, P. Mono- and polynuclear Ag(I) complexes of *N*-functionalized bis(diphenylphosphino)amine DPPA-type ligands: Synthesis, solid-state structures and reactivity. *Dalton Trans.* **2017**, *46*, 5571–5586. [CrossRef] [PubMed]
32. Croizat, P.; Sculfort, S.; Welter, R.; braunstein, P. Hexa- and Octanuclear Heterometallic Clusters with Copper-, Silver-, or Gold-Molybdenum Bonds and d^{10}-d^{10} Interactions. *Organometallics* **2016**, *35*, 3949–3958. [CrossRef]
33. Krossing, I. The Facile Preparation of Weakly Coordinating Anions: Structure and Characterisation of Silverpolyfluoroalkoxyaluminates $AgAl(OR_F)_4$, Calculation of the Alkoxide Ion Affinity. *Chem. Eur. J.* **2001**, *7*, 490–502. [CrossRef]
34. Santiso-Quiñones, G.; Higelin, A.; Schaefer, J.; Brückner, R.; Knapp, C.; Krossing, I. $Cu[Al(OR^F)_4]$ Starting Materials and their Application in the Preparation of $\{Cu(S_n)\}^+$ (n = 12, 8) Complexes. *Chem. Eur. J.* **2009**, *15*, 6663–6677. [CrossRef]
35. *CrysAlisPro Software System, Rigaku Oxford Diffraction*, version 1.171.38; Software for Data Reduction of the X-Ray Data; Rigaku: Tokyo, Japan, 2015.
36. Clark, R.C.; Reid, J.S. The analytical calculation of absorption in multifaceted crystals. *Acta Cryst.* **1995**, *A51*, 887–897. [CrossRef]
37. Dolomanov, O.V.; Bourhis, L.J.; Gildea, R.J.; Howard, J.A.K.; Puschmann, H. Olex2: A complete structure solution, refinement and analysis program. *J. Appl. Cryst.* **2009**, *42*, 339–341. [CrossRef]
38. Altomare, A.; Burla, M.C.; Camalli, M.; Cascarano, G.L.; Giacovazzo, C.; Guagliardi, A.; Moliterni, A.G.G.; Polidori, G.; Spagna, R. SIR97: A new tool for crystal structure determination and refinement. *J. Appl. Cryst.* **1999**, *32*, 115–119. [CrossRef]
39. Sheldrick, G.M. Crystal structure refinement with ShelXL. *Acta Cryst.* **2015**, *C27*, 3–8. [CrossRef]

Sample Availability: Samples of the compounds **2′–5** are available from the authors.

molecules

MDPI

Article

New Rigid Polycyclic Bis(phosphane) for Asymmetric Catalysis

K. Michał Pietrusiewicz [1], Katarzyna Szwaczko [1,*], Barbara Mirosław [1,*], Izabela Dybała [2], Radomir Jasiński [3] and Oleg M. Demchuk [1,4]

1 Faculty of Chemistry, Maria Curie-Sklodowska University, 33-Gliniana St., 20-031 Lublin, Poland; Kazimierz.Pietrusiewicz@poczta.umcs.lublin.pl (K.M.P.); O.Demchuk@IFarm.eu (O.M.D.)
2 Faculty of Pharmacy, Medical University of Lublin, 4[A]-Chodźki St., 20-093 Lublin, Poland; izalis1@gazeta.pl
3 Department of Organic Chemistry, Cracow University of Technology, 24-Warszawska St., 31-155 Cracow, Poland; radomir@chemia.pk.edu.pl
4 Pharmaceutical Research Institute, 8-Rydygiera St., 01-793 Warsaw, Poland
* Correspondence: katarzyna.szwaczko@poczta.umcs.lublin.pl (K.S.); barbara.miroslaw@poczta.umcs.lublin.pl (B.M.)

Academic Editor: Giovanni De Feo
Received: 11 January 2019; Accepted: 1 February 2019; Published: 5 February 2019

check for updates

Abstract: A simple, highly efficient synthesis of a series of novel chiral non-racemic rigid tetracyclic phosphorus ligands, applicable in important chemical asymmetric transformations, was performed. In a tandem cross-coupling/C-H bond activation reaction, a well-recognised and readily available ligand *(R,R)*-NORPHOS was used as the starting material. The palladium complexes of new ligands were obtained and characterised on the example of a crystalline dichloropalladium complex of [(1*R*,2*R*,9*S*,10*S*,11*R*,12*R*)-4-phenyltetracyclo[8.2.1.0^{2,9}.0^{3,8}]trideca-3,5,7-triene-11,12-diyl]bis(diphenylphosphane). A notably high activity and stereoselectivity of the palladium catalysts based on the new ligands were confirmed in a model asymmetric allylic substitution reaction. Herein, we discuss the geometry of the palladium complexes formed and its impact on the efficiency of the catalysts. A comparison of their geometric features with other bis(phosphane) ligand complexes found in the Cambridge Structural Database and built density functional theory (DFT) commutated models is also presented and rationalised.

Keywords: bis(phosphane) palladium complex; metallacycle; NORPHOS; allylic alkylation; asymmetric catalysis; chiral phosphines; C-H bond activation; polycyclic compounds; stereoselective synthesis; DFT calculations

---

1. Introduction

One of the most challenging approaches in contemporary synthetic organic chemistry is the selective activation of C-H bonds leading to the creation of new carbon-carbon bonds in the structure of fine organic compounds of commercial value. A palladium-catalysed Heck coupling reaction is one of the major tools used for the C-C bond formation. The reaction is widely used to prepare drugs, fragrances and agrochemicals [1–3]. An extension of the regular alkenes arylation according to the Heck coupling mechanism is the palladium-catalysed Heck–Catellani reaction, reported in 1985, which involves an intramolecular C-H bond activation and functionalisation [4]. This discovery provided the foundation for the majority of recent developments of direct C-H activation/C-C coupling reactions [5–9] that are conducted to improve the efficiency and atom economy of traditional couplings.

Chiral biphosphine ligands are a powerful synthetic tool for generating complex and diverse chemical compounds. Recently, we have reported a new methodology for a highly regio- and stereoselective reaction of the NORPHOS oxide (**1**) [10,11] with substituted aryl bromides (**2**), which furnished various pentacyclic NORPHOS oxide derivatives **3** (Scheme 1) [12]. This new type of the Heck–Catellani reaction proceeded through tandem cross-coupling/cross-dehydrogenative-coupling stages, mediated by highly active palladium complexes of bulky electron-rich bidentate mono phosphines (S-Phos, *Sym*-Phos and MeO*Sym*-Phos) [12,13]. Inspired by these achievements, we have been prompted to explore some new opportunities that a reaction of the Heck–Catellani type could provide. It had been mentioned in [12] that in the reaction leading to pentacyclic NORPHOS oxide derivatives (**3**), some by-products with molecular weights that formally satisfy chemical formulae of classical mono-arylated Heck products were formed in trace amounts. Since the formation of the Heck reaction product is rather non-typical for norbornene substrates, we supposed that the observed minor by-product could have the structure of a tetracyclic compound (**4**). Interestingly, the best chiral ligands used in the transition-metal-mediated asymmetric synthesis display chirality in the backbones of the ligand and a rigid structure weakly influenced by random conformational changes [14]. This phenomenon has its roots in the fact that the catalysts with restricted freedom of conformational movements more significantly differentiate the competitive transition states leading to enantiomeric reaction products. Thus, C-chiral chelating phosphanes whose stereometry is stabilised by polycyclic systems, especially those based on five- and four-membered rings, could be considered as promising candidates for efficient ligands to be used in catalytic transition-metal-mediated asymmetric transformations. A new, efficient, highly stereo- and regioselective approach to the synthesis of tetracyclic compound (**4**), using readily available substrates and inexpensive commercial palladium catalysts, should be a welcome addition to our previously reported methods for the synthesis of valuable polycyclic bis(phosphane) ligands.

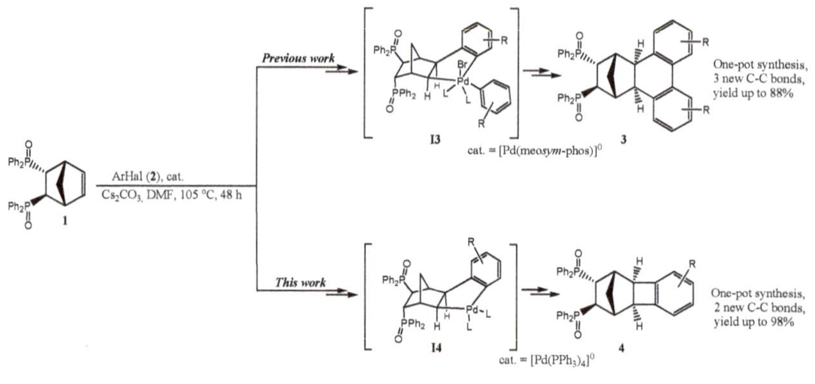

Scheme 1. Synthesis of polycyclic NORPHOS derivatives.

Herein, we disclose a transition-metal-catalysed, one-pot stereoselective synthesis of [tetracyclo[8.2.1.0^{2,9}.0^{3,8}]trideca-3,5,7-triene-11,12-diyl]bis(diphenylphosphane) dioxides: novel NORPHOS oxide derivatives bearing a benzocyclobutene motif. After deoxygenation, the obtained phosphanes could be used as ligands in transition-metal-mediated asymmetric transformations. We also present a short summary of the geometry in the pentametallacyclic complexes of bis(phosphones) with transition metals found in the Cambridge Structural Database, which were compared with the dichloropalladium complex of phosphane derived from **4**.

2. Results and Discussion

Utilisation of most active ligands, such as MeO*Sym*-Phos: Electron-rich and bulky, according to the basic concept [15,16], should accelerate the oxidative addition of aromatic halides to Pd(0) species and, therefore, a double addition of aromatic halides should predominate in the reaction [12]. Thus, in order to direct the reaction to an alternative path of mono-arylation, an electron-deficient ligand of a relatively small steric hindrance, which stabilises intermediate **I4** and facilitates the reductive elimination to product **4**, should be used (Scheme 1). Tetrakis(triphenylphosphine)palladium(0) was therefore selected to be used as the catalyst. The two-component reactions of racemic NORPHOS oxide (*rac*-**1**) and aryl halides (bromide or iodide) used in an equimolar amount or in a 3-fold excess were carried out in the presence of $Pd(PPh_3)_4$, bases and solvents at 105 °C for 12–48 h. We found that a proper choice of the base and solvent was crucial to achieving high yields. Among the tested bases, including K_3PO_4, KOAc, KO^{t}Bu, K_2CO_3 and others (see electronic supplementary information (ESI)), Cs_2CO_3 was selected as the most promising, which significantly increased the yield of desirable product **4**. Polar aprotic solvents are often used in the coupling reaction as they allow for the best solubility of the substrates and the catalysts. In the model reaction between **1** and *p*-bromotoluene (1.2 equiv.) run in dimethylformamide (DMF) in the presence of 5 mol% of $[Pd(PPh_3)_4]$ and caesium carbonate (3.0 equiv.) at 105 °C for 48 h, the isolated yield of product **4b** was 83%. A study of the solvent effects on the yield of the coupling reaction revealed that while the reaction proceeds well in toluene (yield 79%), in THF (yields 72%) and in acetonitrile (yield 65%), the catalytic process is greatly affected in DMF (see ESI). Compared to other combination of used bases and solvents, Cs_2CO_3 solubilised in DMF is less likely to coordinate to the palladium centre and, therefore, less likely to interfere with the catalyst activity. To obtain a satisfactory conversion in DMF, 5 mol% of $[Pd(PPh_3)_4]$ was required. The presence of 1 mol% and 0.1 mol% of the phosphine palladium complex resulted in the formation of 45% and 21% yields of the product, respectively (see ESI).

Having optimised the conditions, we prepared a series of phosphine oxides **4**. Table 1 presents the isolated yields of products **3** and **4** bearing diversified substituents in the introduced arenes.

The coupling reaction of ortho-substituted aryl halides with **1** gave cyclobutene derivatives **4** with a very good yield. A higher yield of **4** was observed when the proportion of the aryl halide was increased from 1.2 to 3.0 equiv. Thus, in the reaction of **1** with 1.2 equiv. or with 3.0 equiv. of *o*-bromotoluene (**2f**), the desired product was isolated in 78% and 85% of the yield, respectively (Table 1, entry 6). Under similar conditions, a tri-substituted aryl halide (Table 1, entry 8) allows us to achieve a similar 86% yield. Using $[Pd(PPh_3)_4]$, a generally low-activity catalyst, allows us to conduct the reaction between **1** and 2-bromochlorobenzene (**2g**) chemoselectively, and only product **4g** was isolated in a 54% yield. In the case of 2-bromobiphenyl (**2q**), the reaction furnished tetracyclic product [tetracyclo[8.2.1.0^{2,9}.0^{3,8}]trideca-3,5,7-triene-11,12-diyl]bis(diphenylphosphane) dioxides (**4q**) in an excellent 98% yield.

Surprisingly, we have found that the direction of the reaction depends not only on the catalyst used, but also on the nature of the aryl halide. The NORPHOS oxide underwent the reaction with highly active aromatic halides, e.g., possessing an electron-withdrawing groups (EWG) (**2j**, **2k**) or a trimethylsilyl (TMS) group (**2l**) and with unsubstituted phenyl bromide (**2a**) furnishing the mixture of products **3** and **4** (Table 1, entries 1, 10, and 11) with the ratio of **3** to **4** increasing in line with the increasing excess of substrates **2** in the reaction mixture. This observation supports the suggestion that the selection of the reaction path depends on the relation of the rates of conversion from intermediates **I4** to **4** and **I4** to **I3** (Scheme 2). Thus, using "fast", electron-rich and bulky catalysts and active substrates, due to rapid halo-arylation of the palladium atom in **I4**, facilitates the formation of **I4** and the next product **3**. In contrast to that, utilisation of less-active, electron-deficient and ligand-based catalysts and electron-donating group (EDG) substituted substrates resulted in the formation of product **4** in higher yields.

Our methodology can also be extended to non-benzenoid aromatic bromides (including naphthalenes and phenanthrenes). Both 1-bromonaphthalene (**2m**) and 2-bromo-6-methoxynaphthalene (**2n**) smoothly underwent the reaction with **1** to give the corresponding carbocyclisation products **4m** and **4n** in 53% and 68% yields, respectively (Table 1, entry 13, 14). A similar 54% yield of **4p** was obtained in the case of 9-bromophenanthrene (**2p**) (Table 1, entry 16). As could be expected, the halides that do not possess any hydrogen atoms in position ortho- to the aryl–halogen bond did not undergo the studied reaction (Table 1, entry 15). Moreover, our attempts to obtain heterocyclic products **3e** or **4e** failed, perhaps due to the strong coordination of the nitrogen atom of pyridine to palladium (Table 1, entry 5).

Table 1. Reaction of the NORPHOS oxide with aryl halides (**2**) in the presence of [Pd(PPh_3)$_4$].

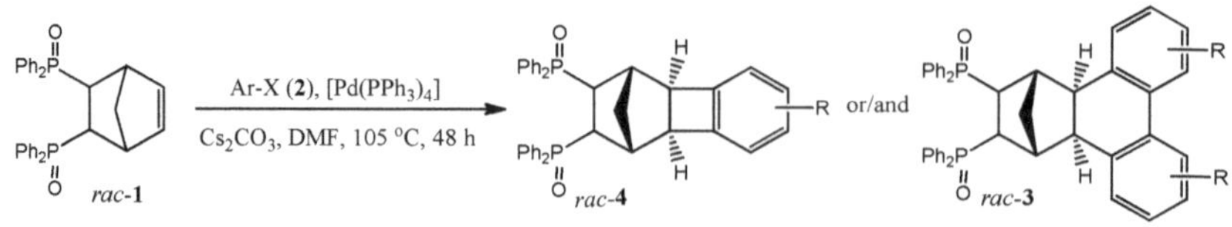

Entry	Aryl Halide	Equiv. of Aryl Halide	Product 3	Product 3 (Yield%)	Product 4	Product 4 (Yield%)
1	Br **2a**	3.0	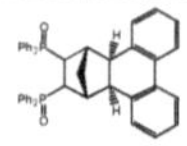	**3a** (69) [a]	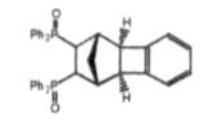	**4a** (no) [b]
2	H_3C–Br **2b**	1.2 3.0	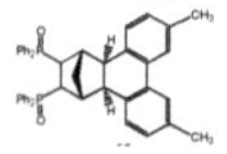	**3b** (no)	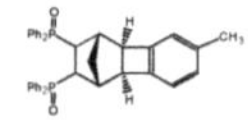	**4b** (59) **4b** (83)
3	OCH_3, Br **2c**	1.2	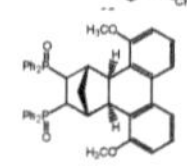	**3c** (no)	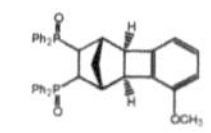	**4c** (76)
4	OCH_3, I **2d**	1.2 3.0	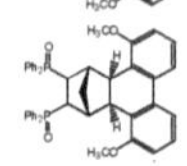	**3c** (no)	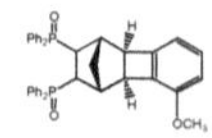	**4c** (35) **4c** (45)
5	N, Br **2e**	3.0	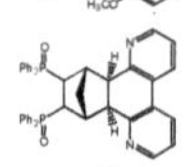	**3e** (no)	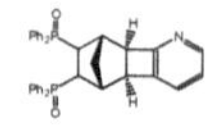	**4e** (no)
6	Br, CH_3 **2f**	1.2 1.2 3.0	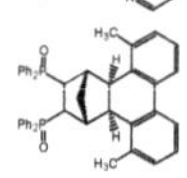	**3f** (no)	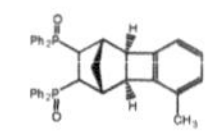	**4f** (25) [a] **4f** (78) **4f** (85)
7	Br, Cl **2g**	1.2 3.0	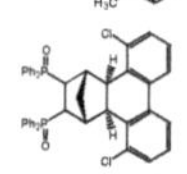	**3g** (no)	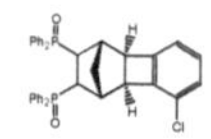	**4g** (48) **4g** (54)
8	H_3C, I, O_2N, CH_3 **2h**	1.2	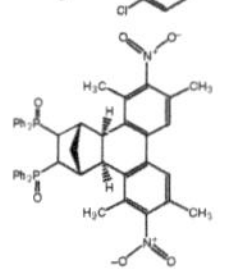	**3h** (no)	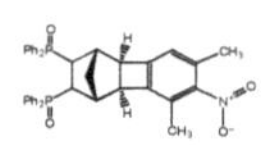	**4h** (86)
9	H_3CO–Br **2i**	3.0	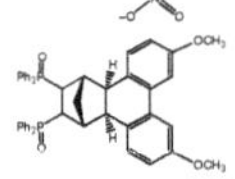	**3i** (no)	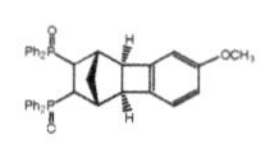	**4i** (18)
10	H_3CO_2C–Br **2j**	1.2 3.0	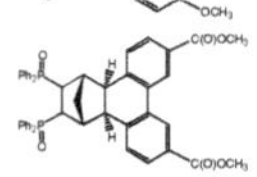	**3j** (<2) **3j** (51)	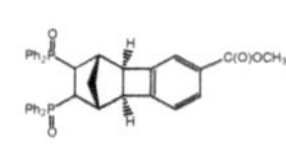	**4j** (84) **4j** (24)
11	F_3C–Br **2k**	1.2 3.0	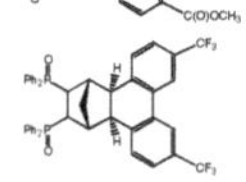	**3k** (32) **3k** (67)	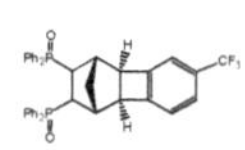	**4k** (no)
12	$(H_3C)_3Si$–Br **2l**	3.0	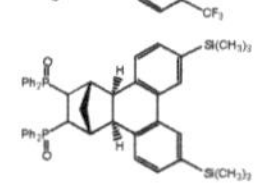	**3l** (63)	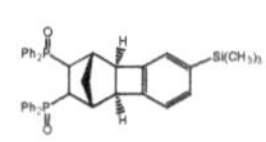	**4l** (no)
13	Br **2m**	1.2	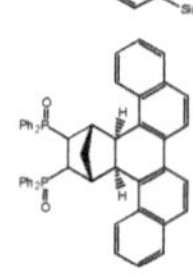	**3m** (no)	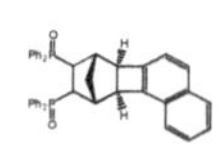	**4m** (68)
14	H_3CO, Br **2n**	1.2	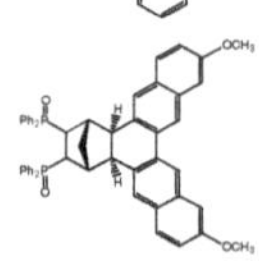	**3n** (no)	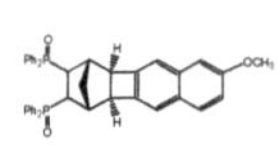	**4n** (53)

Table 1. *Cont.*

Entry	Aryl Halide	Equiv. of Aryl Halide	Product 3	Product 3 (Yield%)	Product 4	Product 4 (Yield%)
15	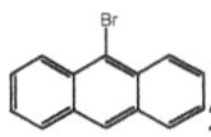 **2o**	1.2	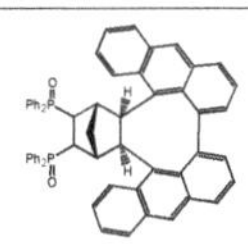	**3o** (no)	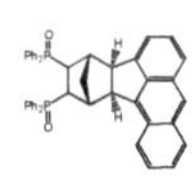	**4o** (no)
16	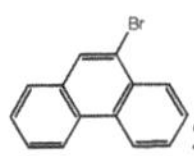 **2p**	1.2	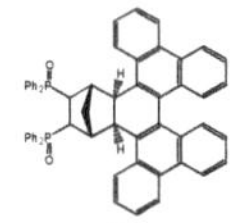	**3p** (no)	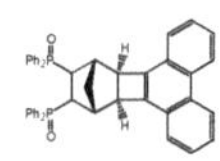	**4p** (54)
17	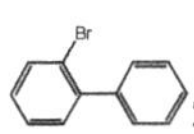 **2q**	1.2	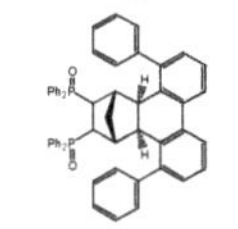	**3q** (no)	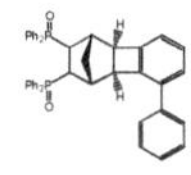	**4q** (98)

Conditions: NORPHOS oxide (0.2 mmol), aryl halide (3.0 equiv.), Cs_2CO_3 (3.0 equiv.), [$Pd(PPh_3)_4$] (5 mol%) and DMF (4 mL) at 105 °C for 48 h. [a] The reaction time was 18 h. [b] Product was not observed.

2.1. General Mechanism and Selectivity of the Reaction

The catalytic carbocyclisation reaction between the NOPHROS oxide and aryl halides that resulted in cyclobutene derivatives **4** is highly regioselective, giving the cis-exo isomer as the exclusive product. An unexpected side reaction product **4a** (2–10%) was detected in the reaction mixtures by the ^{1}H-NMR and HPLC-HR-MS spectral analyses in almost all cases when differently substituted aromatic halides were used (Figure 1). We could assume that, as an exclusive source of phenyls, the phenylidene fragment originated from triphenylphosphane. The migration of the phenyl substituent from the phosphorus atom to palladium had already been observed in the course of other studies [17]. The molecular mechanism of this rare transformation is not clear and its elucidation requires additional theoretical as well as experimental investigations.

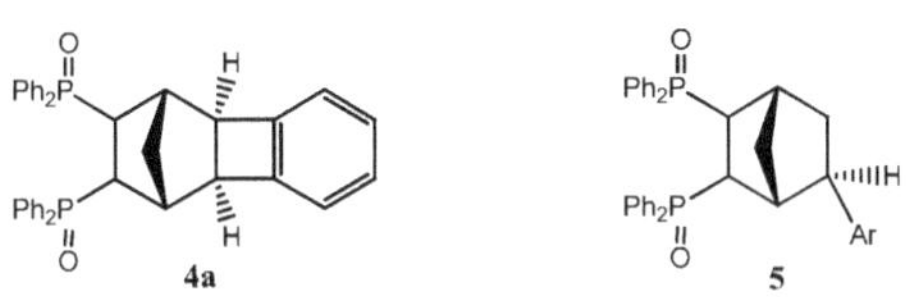

Figure 1. Side reaction products **4a** and **5**.

A tentative mechanism of the reaction leading to products **3**–**5** includes several stages (Scheme 2). The first obvious step of the reaction is an oxidative addition of **2** to the palladium catalysts; then, an insertion of the obtained arylpalladium intermediate into the double bond of the NORPHOS oxide furnished intermediate **I1**. Two possible isomeric intermediates **I1** and **I1′** could be formed. Nevertheless, only a single type of isomeric product was found in the reaction mixtures, and was identified as **3** and **4**. This could be rationalised by the steric hindrance created by P(O)Ph_2 groups, which forces large aryl substituents to occupy a less-hindered position. Obviously, both isomeric products of the alternative endo insertion of arylpalladium halides to **1** are significantly unfavorable and were not formed. The second reaction step is an intramolecular activation of the ortho C-H bond of the introduced arene by palladium, which resulted in the formation of intermediate **I2**. A base facilitated elimination of the hydrogen halide at the next step, which led to the reduction of the oxidation stage of palladium in intermediate **I4**. Next, when using electron-rich ligands of the *C,P*-type transition metal binding [12,13] or highly active aromatic halides, the oxidative addition of the second molecule of **2** occurs and the reaction runs via intermediate **I3** towards product **3**. Thus, in the case of less-active catalysts and non-activated substrates, intermediate **I4** undergoes reductive elimination to form product **4**.

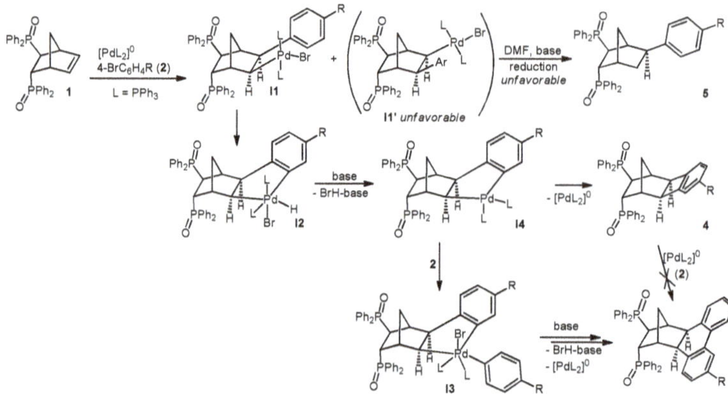

Scheme 2. A tentative suggestion for the molecular mechanism of the coupling reaction.

We have also examined the possibility of product **4**'s conversion to **3**, which theoretically may occur if an excess of the aromatic halide is present in the reaction mixture by the activation of a strained cyclobutane ring with reactive palladium species. For this purpose, in the presence of the palladium complex of the S-Phos ligand and caesium carbonate (3.0 equiv.), compound **4** was subjected to the reactions with 3.0 equiv. of aryl bromides (**2a** and **2b**). After 48 h in DMF at 105 °C, we did not observe the formation of compound **3** and did not find any evidence that such a reaction could be conducted under the studied conditions.

Since both products **3** and **4** are formed from the same intermediate **I4**, it could be anticipated that a lower rate of the oxidative addition reaction, inherent to the electron-poor, ligand-based catalyst, will facilitate the formation of product **4**.

Additionally, in the cases of Cs_2CO_3 (due to its notably strong basic character) and polycyclic aromatic substrates **2m** and **2p**, trace amounts of reductive Heck reaction products **5**, formed by reducing intermediate **I1**, presumably by a solvent [18,19], were detected in the postreaction mixtures (Figure 1).

2.2. *Preparation of Chiral Non-Racemic Ligand* **6q**

Having developed such a high-yielding procedure for the synthesis of desirable highly rigid bis(phosphane) dioxides, we aimed to conduct the synthesis of new chiral ligands to be used in asymmetric transition metal complex-catalysed reactions. We decided to use our new methodology to synthesise representative chiral non-racemic phosphane **6q**, since the corresponding bis(phosphane) dioxide **4q** was obtained in the highest yield. First of all, the scaleup of the synthesis of **4q** was performed. Notably, the synthesis conducted in a scale that increased up to one gram of *(R,R)*-**1** furnished the product in a 90% yield without the need of further optimisation of the reaction conditions. The obtained bis(phosphane) dioxide underwent a deoxygenation reaction in the presence of trichlorosilane and tributylamine, affording ligand *(R,R)*-**6q** in an 85% yield (Scheme 3). Ligand **6q** appeared to be sufficiently resistant to exposure to air and could be purified chromatographically. Using the enantiomerically pure, readily available substrate *(R,R)*-**1** in the synthesis allows us to obtain an enantiomerically pure product with no need of practical-scale enantiopurification.

2.3. *Synthesis of the Complex [Pd(6q)Cl$_2$]*

Since the validation of the proper geometry of the formed ligand by means of NMR spectroscopy is not a trivial matter, we have decided to perform it by analysing the single crystal X-ray diffraction data of the derived dichloropalladium complex [Pd(**6q**)Cl_2]. The complex was prepared from [Pd(^{t}-BuCN)$_2Cl_2$] and ligand **6q** in tetrahydrofuran (THF), and precipitated after a partial replacement of the solvent with C_6H_6. A single crystal has been grown from the hot ethanol/benzene mixture.

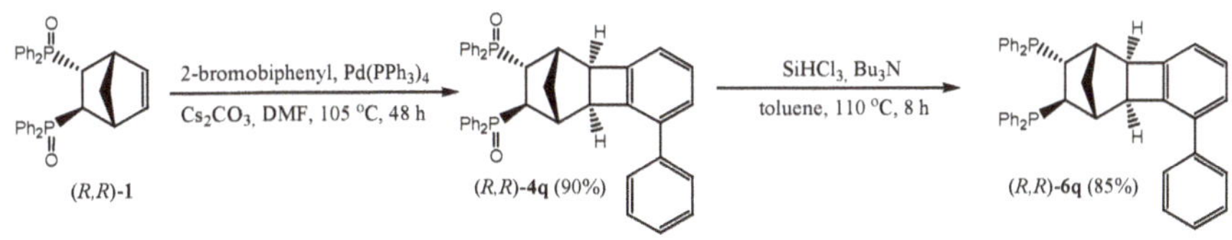

Scheme 3. Synthesis of the phosphine ligand.

2.4. Molecular and Crystal Structure of the Complex [Pd(6q)Cl_2]

The palladium coordination compound [Pd(**6q**)Cl_2] has been characterised by a single crystal X-ray diffraction analysis (CCDC No.: 1885894). [Pd(**6q**)Cl_2] crystallises in a tetragonal crystal system in the chiral $P4_32_12$ space group. The absolute structure assignment of the complex [Pd(**6q**)Cl_2] has been determined through anomalous scattering with the value of the Flack parameter refined to -0.007(3), using 2963 quotients (for more details, crystal data and structure refinements see (ESI).

The chiral centers are localised at the C1 (*R*), C2 (*R*), C9 (*S*), C10 (*S*), C11 (*R*) and C12 (*R*) atoms (Figure 2). The bond lengths and selected angles are given in Tables 2 and 3. An X-ray analysis confirmed the cis, exo configuration of the hydrocarbon skeleton at the C2–C9 joint atoms. The hydrogens attached to the C11 and C12 atoms are located in a trans configuration, which is in accordance with the synthesis route (Tables 2 and 3, Scheme 3, Figure 2). In the four-membered ring, the valence angles show strains resulting in a distortion of the sp^3- and sp^2- hybridised carbon atom centers from the ideal 109.5 and 120° to 86 and 94°, respectively, with higher values being observed at the side of the aromatic ring (Tables 2 and 3). The phenyl ring E is nearly coplanar with the ring F (torsion C2E–C1E–C4–C3 14.6(6)°). The complex [Pd(**6q**)Cl_2] is an example of metallacycles, which are known to be reactive intermediates in a catalysis. This five-membered ring is formed by one Pd, two P and two C atoms and has the shape of an envelope with the C12 atom located at the flap, which deviates by 0.835(4) Å from the mean plane of Pd1–P1–C11–P1′. The puckering amplitude (maximum out-of-plane deviation) q2 calculated for this metallacycle was 1.142(5) Å and the phase angle describing the puckering (phase shift) Φ2 was 153.8(1)° [20]. The Pd metal center has a planar square coordination with two cis-located Cl atoms. Ligand **6q** acts as a chelating bidentate agent forming two Pd–P coordination bonds of lengths equal to 2.2463(8) and 2.2722(9) Å. There are no classic hydrogen bonds in the crystal structure. The molecules interact through C–H ... Cl contacts and T-shaped π ... π interactions (Table 4, Figure 3).

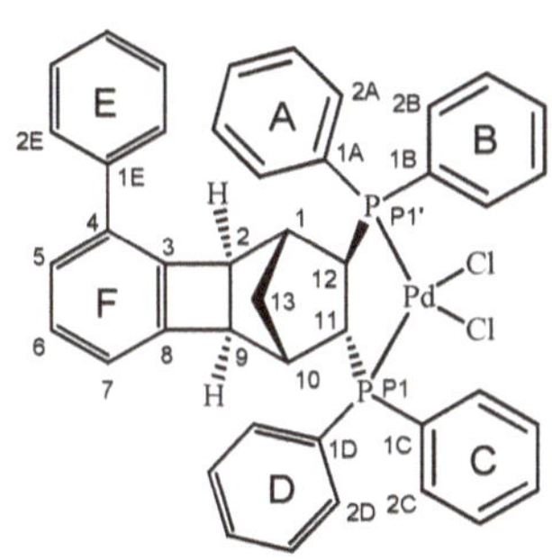

Figure 2. The molecular structure of the complex [Pd(**6q**)Cl_2].

Table 2. Bond lengths (Å) and valence angles (°) for [Pd(**6q**)Cl$_2$].

Bond	Length	Angle	Value
Pd1-Cl1	2.3452(12)	P1-Pd1-P1′	88.62(3)
Pd1-Cl2	2.3337(10)	P1-Pd1-Cl1	175.85(4)
Pd1-P1′	2.2463(8)	P1-Pd1-Cl2	89.30(4)
Pd1-P1	2.2722(9)	C1A-P1-Pd1′	120.58(13)
P1′-C1A	1.812(4)	C1A-P1′-C12	107.45(17)
P1′-C1B	1.802(4)	C1B-P1′-Pd1	113.58(12)
P1′-C12	1.824(3)	C1B-P1′-C1A	105.59(18)
P1-C1C	1.820(4)	C1B-P1′-C12	108.17(16)
P1-C1D	1.811(4)	C12-P1′-Pd1	100.75(11)
P1-C11	1.836(3)	C1C-P1-Pd1	118.19(13)
C1-C2	1.537(5)	C1C-P1-C11	105.81(16)
C1-C12	1.563(4)	C1D-P1-Pd1	114.56(14)
C1-C13	1.539(4)	C1D-P1-C1C	103.96(18)
C1E-C4	1.477(6)	C1D-P1-C11	110.03(16)
C2-C3	1.525(4)	C11-P1-Pd1	103.92(11)
C2-C9	1.602(5)		
C3-C4	1.388(5)	The valence angles of the four-membered ring	
C3-C8	1.386(5)	C2-C3-C8	93.9(3)
C4-C5	1.406(5)	C3-C8-C9	94.2(3)
C5-C6	1.389(6)	C8-C9-C2	85.8(3)
C6-C7	1.394(6)	C9-C2-C3	86.1(3)
C7-C8	1.374(5)		
C8-C9	1.525(5)		
C9-C10	1.540(5)		
C10-C11	1.545(4)		
C10-C13	1.550(5)		
C11-C12	1.555(5)		

Table 3. Torsion angles (°) for [Pd(**6q**)Cl$_2$].

Torsion	Value
Pd1-P1′-C12-C11	56.1(2)
P1′-C12-C11-P1	−61.2(2)
C12-C11-P1-Pd1	36.3(2)
C11-P1-Pd1-P1′-	−1.8(1)
P1-Pd1-P1′-C12	−26.5(1)
H11-C11-C12-H12	161
C3-C4-C1E-C2E	14.6(6)

Table 4. The geometry of intermolecular interactions in the [Pd(**6q**)Cl$_2$] (Å,°) crystal.

D–H . . . A	D–H	D...A	H...A	<D–H...A
C2C-H2C . . . Cl2	0.93	3.482(4)	2.73	139
C6-H6 . . . Cl2 ^{i}	0.93	3.557(4)	2.87	132
C6E-H6E . . . Cl1 ^{i}	0.93	3.623(4)	2.76	154
C7-H7 . . . Cl1 ^{ii}	0.93	3.756(4)	2.99	141
C3B-H3B . . . *Cg*1 ^{ii}	0.93	3.689(5)	2.77	169
C5B-H5B . . . *Cg*2 ^{iii}	0.93	3.521(4)	2.71	146

Symmetry codes: ^{i} −1 + x, y, z; ^{ii} $-\frac{1}{2}$ + y, $\frac{1}{2}$ − x, $\frac{1}{4}$ + z; ^{iii} −1 + y, 1 + x, 1 − z. *Cg*1 ring centroid C1E-C6E; *Cg*2 ring centroid C1A-C6A.

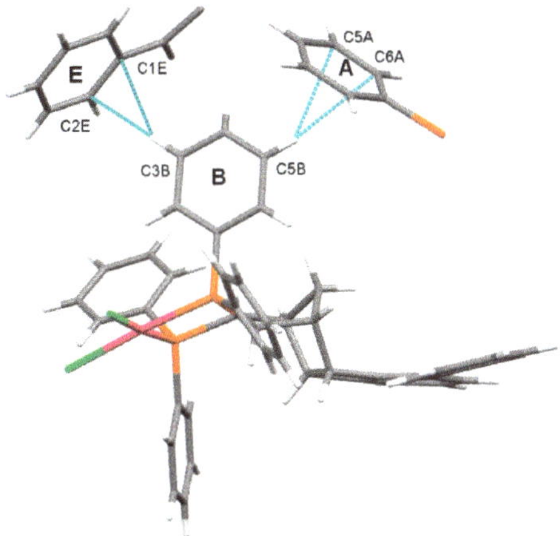

Figure 3. T-shaped $\pi \ldots \pi$ interactions in the complex [Pd(**6q**)Cl_2].

2.5. Geometry of Five-Membered Metallacycles of the Transition Metals and Bis(phosphine) Ligands Found in the Cambridge Structural Database (CSD)

In order to better understand the reactivity and stereochemistry of the bis(phosphane) metalorganic catalysts, a search of the CSD [21] for the structures of five-membered metallacycles with a transition metal and the P–C(HC)–C(HC)–P fragment of bis(phosphane) has been performed. It gave a result of 120 hits, 27 of which were the structures of complexes with Pd (CSD Refcodes: DEPXAX, GEPQUN, HOGBUA, MIDPIY, MIDPOE, NEXHOP, NIDYIK, NURKAM, OFAQET, OFAQIX, OFAQUJ, OKOHAX, PAQBIS, PAQBOY, POPZEA, QELSAB, QUBGID, QUBGOJ, RIYGUB, RONPOZ, RUDJAC, RUNVUR, TESBEZ, TIYPAS, VASXUI, WORDIQ and YOBWUH).

The endocyclic torsion angles characterising this system are summarised in Table 5, and the distribution of their absolute values is presented in Figure 4. There is a clear difference in the most frequently occurring values. The dihedral angles T4 and the corresponding one T5 (C–P–Tr–P) were most frequently synperiplanar, whereas torsions T1–T3 were usually synclinal. In most of the structures, the metallacycle had a conformation of an envelope, as in the presented complex [Pd(**6q**)Cl_2], with the exception of five entries where the ring was almost flat (CSD Refcodes: EQUDEA (Ru), HAYREE (Pt), HAYRII (Pt), VISRAR (Os) and ZOVCIY (Fe)) [22,23]. Usually, one of the carbon atoms was located at the envelope flap (the same as in [Pd(**6q**)Cl_2]), providing a favourable tetragonal flat coordination environment near the metal atom and the asymmetric structure of the metallacycle. Only in three palladium complexes was the metal atom positioned at the flap instead of the carbon atom (in the structures where the carbon atoms belonged to the four-membered hydrocarbon ring and the exocyclic H–C–C–H torsion indicated a cis configuration; CSD Refcodes: NIDYIK, QUBGID and QUBGOJ) [24,25]. However, in analogous Pd complexes with the four-membered hydrocarbon ring and the trans configuration at the same torsion, the metallacycle ring had, as usual, a carbon atom at the envelope flap (CSD Refcodes: OFAQET, OFAQIX and OFAQUJ) [26]. This asymmetry observed in the metallacycle structure is probably beneficial for transition-metal-promoted asymmetric transformations.

The analysis of the orientation of the H atoms within the metallacycle showed that cis stereoisomers were nearly half of the hits (54 of 120). In the presented complex [Pd(**6q**)Cl_2], the H–C–C–H torsion showed the trans configuration. To illustrate the differences within this group of crystal structures, they were divided into four subsets according to the value of the H–C–C–H torsion. There were 17 hits with synperiplanar torsions, 37 synclinal, 6 anticlinal and 60 antiperiplanar as in the presented complex [Pd(**6q**)Cl_2] (Tables 2–5). The results showed that the most frequent conformation for the five-membered metallacycles of transition metals and bis(phosphane) ligands was the envelope type conformation, regardless of the transition metal and phosphine substituent types.

Table 5. Absolute values of the endo- and exocyclic torsions within the five-membered metallacycles found in the Cambridge Structural Database (CSD).

Code	Torsion	Minimum	Maximum	Mean
abs(T1 or T3)	Tr–P–C–C	0.733	60.136	36.141
abs(T2)	P–C–C–P	0.353	69.677	44.108
abs(T4 or T5)	P–Tr–P–C	0.921	44.906	17.174
abs(T6)	H–C–C–H	0.164	179.96	113.52

T1–T5, endocyclic torsion angles defined in Figure 4; Tr, transition metal.

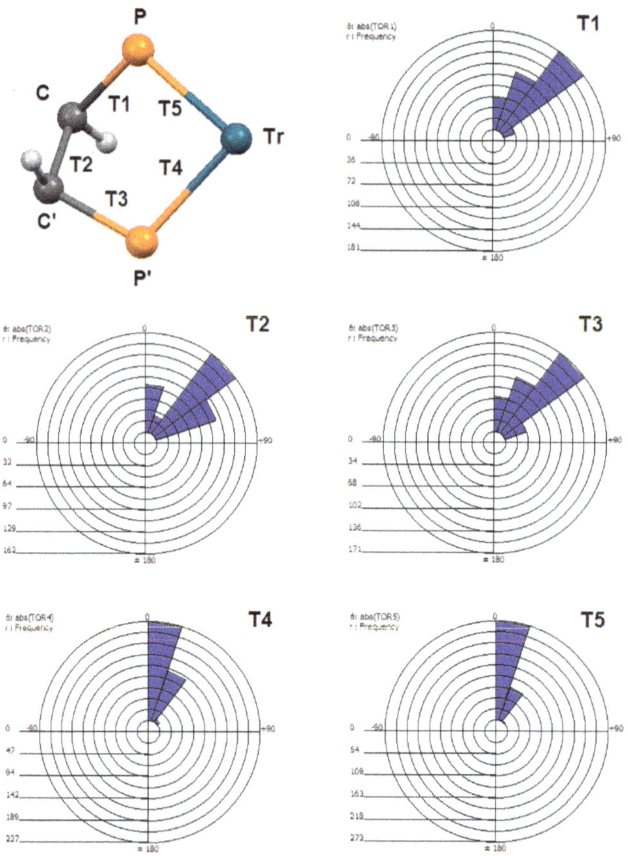

Figure 4. Distribution of the torsion values T1–T5 in the structures of five-membered metallacycles in the complexes of transition metals (Tr) with bis(phosphane) ligands found in the CSD.

2.6. *Quantum Chemical Considerations*

The quantum chemical calculation performed at the M062x/LANL2DZ theoretical level [27,28], which is specially dedicated to precise energetic considerations and has recently been applied in the simulation of energetic and structural analyses of molecules with similar structural moieties [29–33], allows us to compare the geometry of the molecules of phosphines **6**, their dioxides and complexes that were not crystallised in crystals, well-measurable by a single crystal X-ray diffraction. The geometry optimisation for the structure [Pd(**6q**)Cl_2] was therefore performed and the obtained geometry parameters of [Pd(**6q**)Cl_2]^{calc} were compared with those received from the X-ray diffraction experiments for [Pd(**6q**)Cl_2]^{xray}. It was found that all key interatomic distances (Table 6) are characterised by the values that are typical for the respective bonds in similar organic moieties [13,33–35]. The comparison of the obtained geometries with the respective X-ray structures confirms that the density functional theory (DFT) calculations at the M062x/LANL2DZ theoretical level illustrate well the key structural aspects of the considered complexes.

Table 6. Selected geometrical parameters of experimental and calculated [Pd(**6**)Cl_2] and **4**.

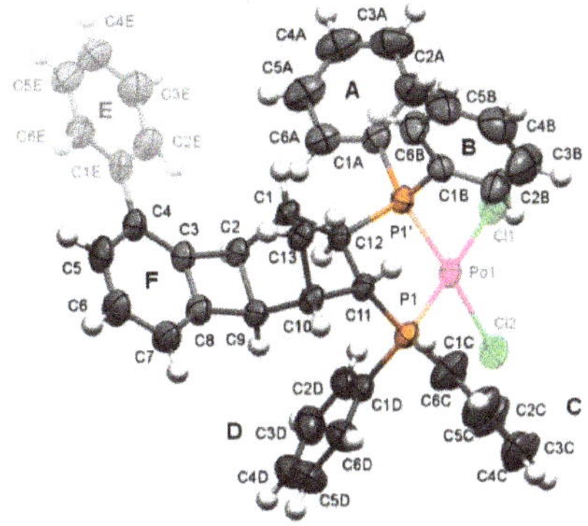

	[Pd(6q)Cl_2]^{xray}	[Pd(6q)Cl_2]^{calc}	[Pd(6a)Cl_2]^{calc}	4q^{calc}	4a^{calc}
Angle, (°)					
A/B	73.3	81.5	81.3	68.9	70.0
C/D	74.3	67.4	67.6	66.7	65.5
B/C	39.6	39.3	43.7	4.2	8.2
A/D	39.3	45.1	39.9	45.4	44.5
Torsion angle, (°)					
C3-C4-C1E-C2E	14.6	30.4	na	29.3	na
P1-C11-C12-P2	−61.2	−67.2	−66.9	−101.5	−103.8
Bond lengths, (Å)					
C1-C2	1.537(5)	1.551	1.552	1.550	1.550
C2-C3	1.525(4)	1.529	1.530	1.532	1.532
C3-C4	1.388(5)	1.397	1.390	1.397	1.391
C4-C5	1.406(5)	1.417	1.415	1.418	1.415
C5-C6	1.389(6)	1.406	1.408	1.406	1.407
C6-C7	1.394(6)	1.414	1.413	1.414	1.414
C7-C8	1.374(5)	1.390	1.392	1.391	1.391
C8-C3	1.386(5)	1.403	1.406	1.405	1.407
C8-C9	1.525(5)	1.533	1.532	1.530	1.531
C9-C2	1.602(5)	1.606	1.608	1.600	1.602
C9-C10	1.540(5)	1.549	1.550	1.547	1.546
C10-C11	1.545(4)	1.550	1.549	1.552	1.551
C11-C12	1.555(5)	1.561	1.561	1.564	1.566
C12-C1	1.563(4)	1.560	1.563	1.555	1.555
C1-C13	1.539(5)	1.552	1.552	1.549	1.550
C13-C10	1.550(5)	1.562	1.562	1.555	1.554
C12-P1	1.823(5)	1.881	1.881	1.872	1.878
C11-P2	1.836(5)	1.883	1.884	1.859	1.850
Pd-Cl1	2.345 (1)	2.423	2.425	na	na
Pd-Cl2	2.334(1)	2.420	2.419	na	na
Pd-P1	2.246 (1)	2.381	2.379	na	na
Pd-P2	2.272(1)	2.380	2.379	na	na

The fitting of molecular structures [Pd(**6q**)Cl_2]^{xray} and [Pd(**6q**)Cl_2]^{calc} clearly shows in which fragments the differences are observed (Figure 5). The molecular overlay indicates that, due to its rigidity, the polycyclic skeleton is identical both in the experimental and theoretical models. Thus, some minor differentiation is observed in the position of the phenyl substituted at the phosphorus atom. This could be explained by the fact that the calculation of a single molecule in vacuum, also with a consideration of a solvent, cannot correctly reflect all crystal net forces influencing the geometry of a molecule in a solid state. The structural comparison is summarised in Table 6. The comparison of the theoretical model and the experimental structure proved that there is a high degree of similarity of their geometries in the polycyclic core.

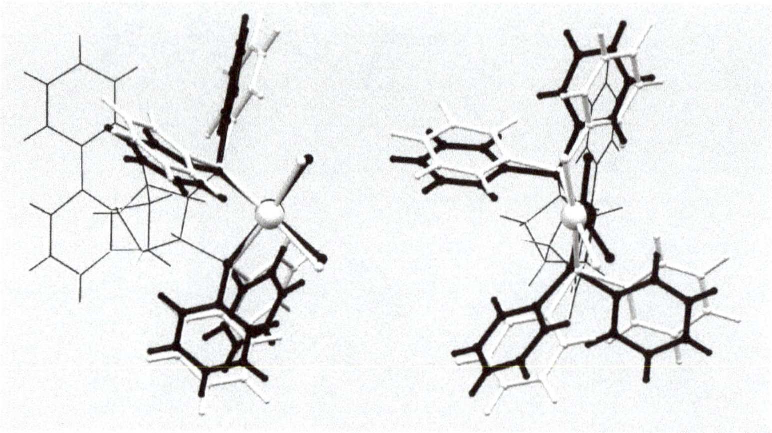

Figure 5. The structural overlay of the molecules of the coordination compounds [Pd(**6q**)Cl_2]^{xray} and [Pd(**6q**)Cl_2]^{calc}: front and side views. Black: measured X-ray structure [Pd(**6q**)Cl_2]^{xray}, white: calculated geometry of [Pd(**6q**)Cl_2]^{calc}. The fragments that match are marked with thin lines.

Moreover, analogous DFT calculations for the complex [Pd(**6a**)Cl_2]^{calc} and bis(phosphane) dioxides **4a** and **4q** were carried out to investigate the influence of the substitutes on the skeleton geometry. All geometries (experimentally determined and optimised using DFT calculations) were compared with each other. Selected bond lengths and angles are summarised in Table 6.

We found a regularity regarding the calculated bonds lengths of the molecules: they are slightly larger than the measured ones. This is a normal phenomenon, considering the fact that the calculation concerns an isolated molecule [36]. The key bond lengths and other geometrical parameters in [Pd(**6q**)Cl_2]^{calc} and [Pd(**6a**)Cl_2]^{calc} and the **4q** and **4a** molecules are the same (within 3σ). The presence of a phenyl substituent in [Pd(**6q**)Cl_2]^{calc} does not have any influence on the geometry of the hydrocarbon polycyclic fragment. In the crystal structure [Pd(**6q**)Cl_2]^{xray}, the phenyl ring E is nearly coplanar with the skeleton moiety (ring F, torsion 14.6°), while in the optimised model of [Pd(**6q**)Cl_2]^{calc}, the corresponding planes are twisted by 30°. The coplanarity of these fragments is stabilised in the crystal by the C-H···Cl bonds in which the carbon atoms of the phenyl ring are involved, and, therefore, it could not have been observed in the cases of isolated single molecules nor in the solution.

Some minor differences in the torsion angle P1-C11-C12-P2 were found in the geometries of molecules **4q** and **4a** in comparison to those of the complexes [Pd(**6q**)Cl_2]^{calc} and [Pd(**6a**)Cl_2]^{calc}. A smaller value of this parameter in the complex results from the necessity of flattening this molecular fragment in the palladium complexation process, but not from introducing a substituent in position C4 of the polycyclic core of the ligand. Differences in the molecular structures related to the position of the phenyl rings, which are caused by a free rotation around the P-1C_{Ph} σ-bonds, are not essential for the catalytic efficiency of the ligand-derived palladium complexes.

To summarise, different substitutions did not affect the stereochemistry of the polycyclic hydrocarbon moiety of the molecules. The main structural differences were observed in the orientation of the phenyl substituents attached to the phosphorus atoms that were involved in the thermal movements around single P-C bonds. Even the complexation of the ligand with palladium did not significantly change the rigid geometry of the ligands. Thus, such unique stereometrical properties of the ligands allow us to extrapolate, via computer modelling, the geometry of the selected measured structures to the stereometry of other ligands in this group.

2.7. *The Evaluation of the Efficiency of the Ligand* **6q** *Palladium Complexes in the Asymmetric Catalysis*

The potential of bis(phosphane) ligand **6q** was examined in a model Pd-catalysed allylic alkylation reaction between *rac*-(*E*)-1,3-diphenylallyl acetate and dimethyl malonate using BSA/KOAc and K_2CO_3/Cs_2CO_3 as bases in THF at ambient temperature (Table 7). The catalyst was generated in situ by pre-mixing allylpalladium(II) chloride dimer with **6q** in THF at ambient temperature for 30 min. The palladium-to-ligand ratios were 2/1 and 1/2. To our delight, bis(phosphane) **6q** proved to be suitable for the asymmetric allylic alkylation reaction, giving the alkylated product in excellent yields and with very good enantioselectivity (up to 90% ee). The results of the optimisation study show that the phosphine palladium complex obtained in the reaction with stoichiometry P/Pd = 1/1 was more beneficial in terms of the stereoselectivity of the reaction (Table 7, entry 1). Thus, it was proven that, in a benchmark asymmetric allylic alkylation reaction, the catalyst derived from the new *R,R*-**6q** phosphane was slightly more efficient than parent NORPHOS.

The enantiomeric composition and *S*- absolute configuration of the product of the model asymmetric reaction were determined by the peak integration and elution order from chiral HPLC using a Chiralcel OD-H column [37].

Table 7. The model asymmetric allylic substitution reaction.

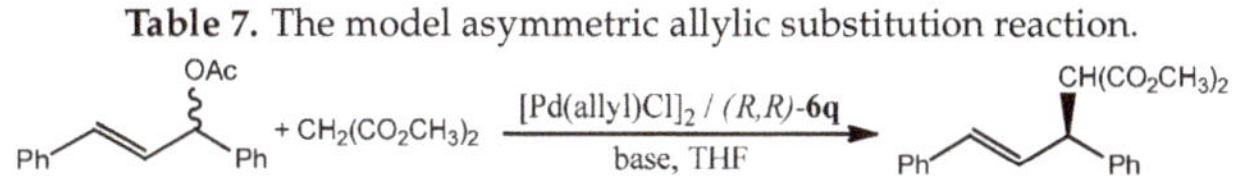

Entry	Ligand, mol%	$[Pd(allyl)Cl]_2$, mol%	Base	Yield, %	ee, %
1	2.0 (*R,R*-6q)	2.0	BSA */KOAc	97	90 **
2	4.0 (*R,R*-6q)	1.0	BSA/KOAc	98	67
3	2.0 (*R,R*-6q)	2.0	Cs_2CO_3/K_2CO_3	97	81
4	2.0 (*R,R*-6q)	2.0	Cs_2CO_3	96	83
5	2.0 (*R,R*-1)	2.0	BSA/KOAc	95	87
6	2.0 (*R,R*-1)	2.0	Cs_2CO_3	92	85
7	1.2 (*S,S*-1)	0.5	NaH	80	81 ***

6q (2.0 or 4.0 mol%), $[Pd(allyl)Cl]_2$ (1.0 or 2.0 mol%), THF (1.0 mL), 30 min; next 1,3-diphenylallyl acetate (0.5 mmol), THF (1.0 mL), dimethyl malonate (1.5 mmol), base (1.5 mmol), 12 h, ambient temperature. * BSA = bis(trimethylsilyl)acetamide. ** $[\alpha]_D^{20} = -14.8$ (c = 1.3, $CHCl_3$). *** Reported previously [38].

3. Materials and Methods

3.1. *General Information*

Unless specified otherwise, all starting materials and solvents were used as obtained from commercial suppliers without further purification. Organic solvents used in this study were dried over appropriate drying agents and distilled prior to use. Thin layer chromatography (TLC) was carried out using Merck silica gel 60 F_{254} plates (Merck, Kenilworth, NJ, USA). Visualisation of the TLC plates was performed by UV light, either $KMnO_4$ or I_2 stains. The NMR spectra were recorded on a Bruker Avance 500 MHz spectrometer, while chemical shifts are reported in ppm and calibrated to the residual solvent peaks at 7.27 ppm and 77.00 ppm for ^1H and ^{13}C, respectively, in $CDCl_3$ or internal reference compounds. A similar technique was applied recently for the structural analysis of many norbornene systems [39,40]. The following abbreviations are used in reporting NMR data: s (singlet), d (doublet), t (triplet), q (quartet), m (multiplet), br (broad). Coupling constants (J) are given in Hz. The spectra are reported as follows: chemical shift (δ, ppm), multiplicity, integration, coupling constants (Hz). The IR spectra were recorded on a Nicolet 8700A FTIR-ATR spectrometer; wave numbers are in cm^{-1}. Products were purified by flash chromatography on silica gel 60 (230–400 mesh) using a BUCHI chromatograph. The X-ray diffraction intensities were collected at room temperature on a SuperNova X-ray diffractometer equipped with an Atlas S2 CCD detector using mirror-monochromatised CuKα radiation (λ = 1.54184 Å). Low- and high-resolution mass spectra were obtained with a Shimadzu LC-MS (Kinetex® 2.6 μm Biphenyl 100 Å 50 × 2.1 mm LC-column, acetonitrile/water with the HCO_2H additive mobile phase) IT-TOF spectrometer. Commercially unavailable substrates were obtained by known literature procedures. The physical properties and spectra of the obtained products are available free of charge in the Supplementary Data.

3.2. Synthesis and Spectral Data

3.2.1. Typical Procedure for the Coupling Reaction Leading to the Formation of [(1*R*,2*R*,9*S*,10*S*,11*R*,12*R*)-4-phenyltetracyclo[8.2.1.$0^{2,9}$.$0^{3,8}$]trideca-3,5,7-triene-11,12-diyl]bis(diphenylphosphane) dioxide (*R*,*R*-**4q**)

(R,R)-**1** (1.0g, 1.54 mmol), **2q** (1.84 mmol), Cs_2CO_3 (4.63 mmol) and $Pd(PPh_3)_4$ (0.07 mmol) were placed in a Schlenk tube under argon. Dry DMF was added (15 mL) and the mixture was stirred at 105 °C for 18–24 h. Then, the mixture was cooled down to root temperature (RT). The solvent was evaporated under reduced pressure, and 50 mL of water was added. The products were extracted in DCM (3 × 30 mL). The combined organic phase was dried over $MgSO_4$, filtered and the solvent was evaporated. The product was purified by silica-gel column chromatography (Hexane/*i*PrOH 50/1) and finally crystallised from the mixture of hexane/DCM. The compound was isolated in 98% yield as a white solid, m.p.: 255–258 °C, $[\alpha]_D^{20}$ = +79.4 (*c* = 0.45, $CHCl_3$), ^{31}P-NMR (202 MHz, $CDCl_3$): δ = 32.5 (d, *J* = 10.1), 28.7 (d, *J* = 10.1), ^{1}H-NMR (500 MHz, $CDCl_3$): δ = 8.68–8.65 (m, 2H, Ar-H), 7.99–7.42 (m, 20H, Ar-H), 7.13–7.06 (m, 5H, Ar-H), 6.83 (d, 1H, *J* = 10.0, Ar-H), 4.29 (s, 1H, CH), 3.93–3.83 (m, 2H, CH), 3.52–3.47 (m, 1H, CH), 2.61 (d, 2H, *J* = 5.0, CH), 1.90 (d, 1H, *J* = 10.0, CH), 1.01–0.85 (m, 1H, CH), ^{13}C-NMR (125 MHz, $CDCl_3$): δ = 145.3, 141.8, 137.3, 134.1, 133.4, 132.1, 132.0, 131.9, 131.2, 130.9, 130.6, 130.5, 130.4, 130.1, 130.0, 128.8, 128.7, 128.5, 128.5, 128.5, 128.4, 128.3, 128.2, 127.1, 127.0, 126.8, 126.5, 126.1, 125.1, 120.7, 50.0, 49.9, 48.3 (d, *J* = 6.2), 41.8, 40.4, 38.4, 37.5, 32.8 (d, *J* = 12.5), LCMS (ESI) $[M + H]^+$ = 647 Da, HRMS calcd. for $C_{43}H_{36}O_2P_2$ 647.2259, found 647.2263 (diff. 1.18 ppm).

3.2.2. 1,2,3,4,4a,12b-Hexahydro-1,4-methanotriphenylene-2,3-diylbis(diphenylphosphane) dioxide (*rac*-**3a**)

The compound was isolated in 69% yield as a white solid. The analysis of the product spectra confirmed its identity with literature data [12].^{31}P-NMR (202 MHz, $CDCl_3$): δ = 31.60 (d, *J* = 10.1), 28.78 (d, *J* = 10.1), ^{1}H-NMR (500 MHz, $CDCl_3$): δ = 7.98–7.96 (m, 2H, Ar-H), 7.86–7.85 (m, 2H, Ar-H), 7.76–7.72 (m, 2H, Ar-H), 7.64–7.50 (m, 10H, Ar-H), 7.17–7.05 (m, 10H, Ar-H), 6.81–6.78 (m, 2H, Ar-H), 5.22 (d, 1H, *J* = 2.0, Ar-H), 4.00 (d, 1H, *J* = 4.0, CH), 3.95–3.87 (m, 1H, CH-P), 3.86–3.80 (m, 1H, CH-P), 3.71 (d, 1H, *J* = 4.0, CH), 2.64 (s, 1H, CH), 2.45 (d, 1H, *J* = 4.0, CH_2), 1.98 (d, 1H, *J* = 4.0, CH_2), 1.45–1.44 (m, 1H, CH).

3.2.3. (5-Methyltetracyclo[8.2.1.$0^{2,9}$.$0^{3,8}$]trideca-3,5,7-triene-11,12-diyl)bis(diphenylphosphane) dioxide (*rac*-**4b**)

The compound was isolated in 83% yield as a white solid, m.p.: 208–213 °C. ^{31}P-NMR (202 MHz, $CDCl_3$): δ = 32.4 (d, *J* = 8.08), 28.7 (d, *J* = 6.06), ^{1}H-NMR (500 MHz, $CDCl_3$): δ = 7.87–7.76 (m, 15H, Ar-H), 7.11–6.55 (m, 8H, Ar-H), 3.93 (s, 1H, CH), 3.87–3.83 (m, 1H, CH), 3.56 (s, 1H, CH), 3.31–3.27 (m, 1H, CH), 2.34 (d, 1H, *J* = 10.0, CH_2), 2.24 (s, 3H, CH_3), 2.03 (s, 1H, CH), 1.84 (d, 1H, *J* = 10.0, CH_2), 1.00–0.95 (m, 1H, CH), ^{13}C-NMR (125 MHz, $CDCl_3$): δ = 144.0 (d, *J* = 3.7), 142.0, 137.1, 134.3, 135.1, 134.4, 134.0, 133.6, 133.3, 132.6, 131.8, 131.2, 131.0, 130.8, 130.5, 130.4, 130.2, 130.1, 129.8, 128.7, 128.6, 128.5,

128.4, 128.3, 128.2, 128.2, 122.6, 121.4, 50.0 (d, *J* = 15.0), 45.2 (d, *J* = 6.2), 41.6 (d, *J* = 6.2), 38.3, 37.8, 32.5 (d, *J* = 12.5), 21.9 LCMS (ESI) $[M + H]^+$ = 585 Da, HRMS calcd. for $C_{38}H_{34}O_2P_2$ 585.2119, found 585.2108 (diff. 1.18 ppm).

3.2.4. (4-Methoxytetracyclo[8.2.1.$0^{2,9}$.$0^{3,8}$]trideca-3,5,7-triene-11,12-diyl)bis(diphenylphosphane) dioxide (*rac*-**4c**)

The compound was isolated in 76% yield as a white solid, m.p.: 276–278 °C. ^{31}P-NMR (202 MHz, $CDCl_3$): δ = 32,4 (s), 28.3 (s), ^{1}H-NMR (500 MHz, $CDCl_3$): δ = 7.87–7.79 (m, 4H, Ar-H), 7.55–7.42 (m, 11H, Ar-H), 7.07–7.00 (m, 7H, Ar-H), 6.56 (d, 1H, *J* = 5.0, Ar-H), 6.45 (d, 1H, *J* = 5.0, Ar-H), 4.08 (s, 1H, CH), 3.92–3.78 (m, 1H, CH), 3.59 (s, 1H, CH), 3.38–3.30 (m, 1H, CH), 2.93 (s, 3H, OCH_3), 2.58 (s, 1H, CH), 2.42 (s, 1H, CH), 1.99–1.97 (m, 1H, CH), 2.31 (s, 3H, CH_3), 2.27 (s, 3H, CH_3), 2.11–2.09 (m, 1H, CH), 1.15 (s, 1H, CH), ^{13}C-NMR (125 MHz, $CDCl_3$): δ = 153.2, 146.6, 131.4, 131.3, 131.0, 130.8, 130.5, 130.0, 1329., 128.8, 128.5, 128.4, 128.3, 126.6, 115.6, 114.5, 55.6, 53.4, 45.9, 43.4, 41.5 LCMS (ESI) $[M + H]^+$ = 601 Da, HRMS calcd. for $C_{38}H_{34}$ O_3P_2 601.2056, found 601.2067 (diff. 1.18 ppm).

3.2.5. (4-Methyltetracyclo[8.2.1.$0^{2,9}$.$0^{3,8}$]trideca-3,5,7-triene-11,12-diyl)bis(diphenylphosphane) dioxide (*rac*-**4f**)

The compound was isolated in 85% yield as a white solid, m.p.: 270–275 °C. ^{31}P-NMR (202 MHz, $CDCl_3$): δ 32.3 (d, *J* = 8.08), 28.5 (d, *J* = 8.08), ^{1}H-NMR (500 MHz, $CDCl_3$): δ = 7.87–7.77 (m, 4H, Ar-H), 7.65–7.45 (m, 11H, Ar-H), 7.17–7.00 (m, 7H, Ar-H), 6.86 (d, 1H, *J* = 5.0, Ar-H), 6.65 (d, 1H, *J* = 5.0, Ar-H), 3.93 (s, 1H, CH), 3.81–3.75 (m, 1H, CH), 3.54 (s, 1H, CH), 3.34–3.30 (m, 1H, CH), 2.43 (s, 2H, CH_2), 1.90 (d, 1H, *J* = 10.0, CH), 1.65 (s, 1H, CH_3), 1.01–0.97 (m, 1H, CH), ^{13}C-NMR (125 MHz, $CDCl_3$): δ = 144.1 (d, *J* = 3.7), 143.7, 135.4, 134.7, 134.1, 133.6, 133.4, 132.4, 131.7, 131.4, 131.3, 130.9, 130.7, 130.6, 130.5, 130.1, 130.0, 128.7, 128.6, 128.5, 128.3, 128.3, 128.3, 128.2, 128.2, 127.9, 49.9 (d, *J* = 3.7), 45.5, 41.3, 40.9, 38.4, 37.8, 32.6 (d, *J* = 13.7), 15.7 LCMS (ESI) $[M + H]^+$ = 585 Da, HRMS calcd. for $C_{38}H_{34}$ O_2P_2 585.2119, found 585.2107 (diff. 1.18 ppm).

3.2.6. (4-Chlorotetracyclo[8.2.1.$0^{2,9}$.$0^{3,8}$]trideca-3,5,7-triene-11,12-diyl)bis(diphenylphosphane) dioxide (*rac*-**4g**)

The compound was isolated in 54% yield as a white solid, m.p.: 250–254 °C. ^{31}P-NMR (202 MHz, $CDCl_3$): δ = 32.2 (d, *J* = 6.06), 28.7 (d, *J* = 8.08), ^{1}H-NMR (500 MHz, $CDCl_3$): δ = 7.87–7.73 (m, 4H, Ar-H), 7.67–7.46 (m, 11H, Ar-H), 7.17–7.03 (m, 7H, Ar-H), 6.77 (d, 1H, *J* = 10.0, Ar-H), 4.03 (s, 1H, CH), 3.83–3.77 (m, 1H, CH), 3.58 (s, 1H, CH), 3.33–3.29 (m, 1H, CH), 2.41 (s, 2H, CH_2), 1.93 (d, 1H, *J* = 10.0, CH), 1.65 (s, 1H, CH), 0.98–0.95 (m, 1H, CH), ^{13}C-NMR (125 MHz, $CDCl_3$): δ = 145.1 (d, *J* = 3.7), 142.6, 134.7, 134.3, 133.9, 133.7, 133.5, 133.0, 132.2, 131.4 (d, *J* = 2.5), 131.3 (d, *J* = 2.5), 131.2 (d, *J* = 2.5), 130.9 (d, *J* = 2.5), 130.6, 130.5, 130.4, 130.4, 129.4, 129.3, 128.7, 128.7, 128.6, 128.5, 128.3, 127.5, 127.4, 126.9, 126.8, 120.5, 50.0 (d, *J* = 2.5), 49.3(d, *J* = 15), 45.7 (d, *J* = 6.2), 41.3(d, *J* = 3.7), 40.4, 38.2, 37.6, 32.6 (d, *J* = 12.5) LCMS (ESI) $[M + H]^+$ = 605 Da, HRMS calcd. for $C_{37}H_{31}O_2P_2Cl$ 605.1561, found 605.1561 (diff. 1.18 ppm).

3.2.7. (4,6-Dimethyl-5-nitrotetracyclo[8.2.1.$0^{2,9}$.$0^{3,8}$]trideca-3,5,7-triene-11,12-diyl)bis(diphenylphosphane) dioxide (*rac*-**4h**)

The compound was isolated in 86% yield as a white solid, m.p.: 302–304 °C. ^{31}P-NMR (202 MHz, $CDCl_3$): δ = 32.3 (d, *J* = 8.08), 28.6 (d, *J* = 10.0), ^{1}H-NMR (500 MHz, $CDCl_3$): δ = 7.88–7.76 (m, 4H, Ar-H), 7.64–7.47 (m, 11H, Ar-H), 7.16–7.01 (m, 5H, Ar-H), 6.65 (s, 1H, Ar-H), 3.94 (d, 1H, *J* = 5.0, CH), 3.84–3.78 (m, 1H, CH), 3.54 (d, 1H, *J* = 5.0, CH), 3.34–3.30 (m, 1H, CH), 2.41 (d, 2H, *J* = 10.0, CH_2), 2.10 (s, 3H, CH_3), 1.97 (d, 1H, *J* = 10.0, CH), 1.60 (s, 1H, CH_3), 1.00–0.95 (m, 1H, CH), ^{13}C-NMR (125 MHz, $CDCl_3$): δ = 151.5, 145.0 (d, *J* = 3.7), 142.4, 135.1, 134.3, 134.1, 133.6, 133.5, 132.8, 132.0, 131.5 (d, *J* = 2.5), 131.4 (d, *J* = 2.5), 131.2 (d, *J* = 2.5), 130.9 (d, *J* = 2.5), 130.6, 130.5, 130.4, 130.0, 129.8, 128.7, 128.7, 128.5, 128.4, 128.3, 128.3, 128.2, 123.8, 122.7, 49.3 (d, *J* = 15.0), 45.2(d, *J* = 6.2), 45.7 (d, *J* = 6.2),

41.3, 40.4, 38.3, 37.7, 32.6(d, J = 12.5) LCMS (ESI) $[M + H]^+$ = 644 Da, HRMS calcd. for $C_{39}H_{35}NO_4P_2$ 644.2120, found 644.2114 (diff. 1.18 ppm).

3.2.8. (5-Methoxytetracyclo[8.2.1.0^{2,9}.0^{3,8}]trideca-3,5,7-triene-11,12-diyl)bis(diphenylphosphane) dioxide (*rac*-**4i**)

The compound was isolated in 18% yield as a white solid, m.p.: 279–285 °C. ^{31}P-NMR (202 MHz, $CDCl_3$): δ = 32.4(d, J = 8.08), 28.8 (d, J = 10.0), ^{1}H-NMR (500 MHz, $CDCl_3$): δ = 7.87–7.76 (m, 15H, Ar-H), 7.12–6.47 (m, 8H, Ar-H), 4.03 (d, 1H, J = 35.0, CH), 3.88–3.83 (m, 1H, CH), 3.70 (s, 1H, OCH_3), 3.59 (d, 1H, J = 35.0, CH), 3.31–3.27 (m, 1H, CH), 2.49–2.33 (m, 2H, CH_2), 1.00–0.95 (m, 1H, CH), ^{13}C-NMR (125 MHz, $CDCl_3$): δ = 159.7, 145.3, 134.3, 133.9, 133.5, 133.2, 132.9, 132.5, 132.1, 132.0, 131.9 (d, J = 6.2), 131.1, 131.6, 130.5, 130.4, 130.2, 130.1, 128.7, 128.6, 128.5, 128.4, 128.3 (d, J = 2.5), 128.2 (d, J = 2.5), 127.6, 127.4, 122.8, 122.01, 114.2, 55.3, 50.4 (d, J = 13.5), 49.8 (d, J = 13.5), 46.7, 45.7, 38.5, 37.8, 31.4, 29.7 LCMS (ESI) $[M + H]^+$ = 601 Da, HRMS calcd. for $C_{38}H_{34}O_3P_2$ 601.2067, found 601.2056 (diff. 1.18 ppm).

3.2.9. [Methyl 11,12-bis(diphenylphosphoryl)tetracyclo[8.2.1.0^{2,9}.0^{3,8}]trideca-3,5,7-triene-5-carboxylate (*rac*-**4j**)

The compound was isolated in 84% yield as a white solid, m.p.: 169–171 °C. ^{31}P-NMR (202 MHz, $CDCl_3$): δ = 32.7 (d, J = 8.08), 29.1 (d, J = 8.08), ^{1}H-NMR (500 MHz, $CDCl_3$): δ = 7.87–7.76 (m, 15H, Ar-H), 7.12–6.47 (m, 8H, Ar-H), 4.03 (d, 1H, J = 35.0, CH), 3.88–3.83 (m, 1H, CH), 3.70 (s, 1H, OCH_3), 3.59 (d, 1H, J = 35.0, CH), 3.31–3.27 (m, 1H, CH), 2.49–2.33 (m, 2H, CH_2), 1.00–0.95 (m, 1H, CH), ^{13}C-NMR (125 MHz, $CDCl_3$): δ = 167.3, 151.0, 144.0 (d, J = 3.7), 135.0, 134.2, 134.1, 133.6, 133.3, 133.3, 132.9, 132.4, 131.7, 130.5, 130.4, 130.4, 130.3, 130.2, 130.1, 129.7, 129.6, 129.9, 129.5, 128.7, 128.6, 128.5, 128.5, 128.3, 128.3, 128.3, 128.2, 123.1, 122.9, 122.0, 121.8, 51.9, 50.0 (d, J = 13.7), 47.9, 41.1, 38.2, 32.4 LCMS (ESI) $[M + H]^+$ = 629 Da, HRMS calcd. for $C_{39}H_{34}O_4P_2$ 629.2022, found 629.2005 (diff. 1.7 ppm).

Dimethyl 2,3-bis(diphenylphosphoryl)-1,2,3,4,4a,12b-hexahydro-1,4-methanotriphenylene-7,10-dicarboxylate (*rac*-**3j**)

The compound was isolated in 51% yield as a white solid. The analysis of the product spectra confirmed its identity with literature data [12]. ^{31}P-NMR (202 MHz, $CDCl_3$): δ = 31.10 (d, J = 10.1), 28.81 (d, J = 8.1), ^{1}H-NMR (500 MHz, $CDCl_3$): δ = 8.50 (d, 1H, J = 10.0, Ar-H), 7.97–7.93 (m, 2H, Ar-H), 7.88–7.85 (m, 2H, Ar-H), 7.74 (d, 1H, J = 5.0, Ar-H), 7.74–7.48 (m, 11H, Ar-H), 7.17–7.04 (m, 6H, Ar-H), 6.86 (d, 1H, J = 10.0, Ar-H), 5.26 (d, 1H, J = 10.0, Ar-H), 4.12 (d, 1H, J = 10.0, CH), 3.95–3.90 (m, 1H, CH-P), 3.87–3.81 (m, 1H, CH-P), 3.93 (s, 3H, CH_3), 3.92 (s, 3H, CH_3), 3.75 (d, 1H, J = 10.0, CH), 2.64 (s, 1H, CH), 2.42 (d, 1H, J = 10.0, CH_2), 2.02 (d, 1H, J = 10.0, CH_2), 1.94 (s, 1H, CH). The HNMR spectrum is in agreement with the literature [12].

3.2.10. [7,10-Bis(trifluoromethyl)-1,2,3,4,4a,12b-hexahydro-1,4-methanotriphenylene-2,3-diyl]bis(diphenylphosphane) dioxide (*rac*-**3k**)

The compound was isolated in a yield of up to 67% as a white solid. The analysis of the product spectra confirmed its identity with literature data [12]. ^{31}P-NMR (202 MHz, $CDCl_3$): δ = 31.04 (d, J = 8.1), 28.84 (d, J = 8.1), ^{1}H-NMR (500 MHz, $CDCl_3$): δ = 7.89–7.70 (m, 6H, Ar-H), 7.58–7.41 (m, 10H, Ar-H), 7.26 (d, 1H, J = 10.0, Ar-H), 7.10–6.97 (m, 7H, Ar-H), 6.81 (d, 1H, J = 10.0, Ar-H), 5.22 (d, 1H, J = 10.0, Ar-H), 4.02 (d, 1H, J = 10.0, CH), 3.83–3.76 (m, 1H, CH-P), 3.76–3.70 (m, 1H, CH-P), 3.67 (d, 1H, J = 10.0, CH), 2.54 (s, 1H, CH), 2.32 (d, 1H, J = 10.0, CH_2), 1.97 (d, 1H, J = 10.0, CH_2), 1.25–1.19 (m, 1H, CH). The HNMR spectrum is in agreement with the literature [12].

3.2.11. [7,10-Bis(trimethylsilyl)-1,2,3,4,4a,12b-hexahydro-1,4-methanotriphenylene-2,3-diyl]bis(diphenylphosphane) dioxide (*rac*-**3l**)

The compound was isolated in a yield of 63% as a white solid. The analysis of the product spectra confirmed its identity with literature data [12]. ^{31}P-NMR (202 MHz, $CDCl_3$): δ = 31.12 (d, *J* = 6.1), 28.60 (d, *J* = 10.1), ^{1}H-NMR (500 MHz, $CDCl_3$): δ = 7.98–7.93 (m, 4H, Ar-H), 7.67–7.51 (m, 10H, Ar-H), 7.21–7.05 (m, 6H, Ar-H), 6.93 (d, 1H, *J* = 5.0, Ar-H), 6.78 (d, 1H, *J* = 10.0, Ar-H), 5.22 (d, 1H, *J* = 10.0, Ar-H), 4.03 (d, 1H, *J* = 10.0, CH), 3.94–3.89 (m, 1H, CH-P), 3.85–3.81 (m, 1H, CH-P), 3.68 (d, 1H, *J* = 10.0, CH), 2.64 (s, 1H, CH), 2.48 (d, 1H, *J* = 10.0, CH_2), 1.96 (d, 1H, *J* = 10.0, CH_2), 1.66 (s, 1H, CH), 0.28 (s, 9H, CH_3), 0.26 (s, 9H, CH_3). The HNMR spectrum is in agreement with the literature [12].

3.2.12. 6b,7,8,9,10,10a-Hexahydro-7,10-methanobenzo[a]biphenylene-8,9-diylbis(diphenylphosphane) dioxide (*rac*-**4m**)

The compound was isolated in 68% yield as a white solid, m.p.: 225–227 °C. ^{31}P-NMR (202 MHz, $CDCl_3$): δ = 32.4 (d, *J* = 8.08), 28.7 (d, *J* = 6.06), ^{1}H-NMR (500 MHz, $CDCl_3$): δ = 7.96–7.48 (m, 16H, Ar-H), 7.36–6.28 (m, 4H, Ar-H), 7.19–7.03 (m, 6H, Ar-H), 6.83 (d, 1H, *J* = 10.0, Ar-H), 4.21 (d, 1H, *J* = 5.0, CH), 3.87–3.81 (m, 1H, CH), 3.59 (d, 1H, *J* = 5.0, CH), 3.42–3.39 (m, 1H, CH), 2.57 (s, 1H, CH), 2.46 (d, 1H, *J* = 15.0, CH_2), 1.88 (d, 1H, *J* = 15.0, CH_2), 0.90–0.85 (m, 1H, CH), ^{13}C-NMR (125 MHz, $CDCl_3$): δ = 145.9, 145.1, 142.6, 134.6, 134.2, 133.9, 133.7, 133.4, 133.3, 132.9, 132.2, 131.4, 131.3, 131.2, 130.9, 130.6, 130.5, 130.4, 130.4, 130.2, 130.1, 129.3, 128.8, 128.7, 128.6, 128.5, 128.3, 128.2, 127.5, 126.8, 120.5, 120.3, 49.8 (d, *J* = 15.0), 45.7 (d, *J* = 5.0), 41.3, 41.0, 40.4, 38.2, 37.7, 32.9 LCMS (ESI) $[M + H]^+$ = 621 Da, HRMS calcd. for $C_{41}H_{38}O_2P_2$ 621.2139, found 621.2166 (diff. 1.18 ppm).

[5-(naphthalen-1-yl)bicyclo[2.2.1]heptane-2,3-diyl]bis(diphenylphosphane) dioxide (*rac*-**5m**)

This minor reaction product was detected in trace amounts in the reaction mixtures by means of ESI-LC HRMS. HRMS calcd. for $C_{41}H_{36}O_2P_2$ 623.2312, found 623.2263 (diff. 7.8 ppm).

3.2.13. (7-Methoxy-1,2,3,4,4a,10b-hexahydro-1,4-methanobenzo[b]biphenylene-2,3-diyl)bis(diphenylphosphane) dioxide (*rac*-**4n**)

The compound was isolated in 53% yield as a white solid, m.p.: 209–214 °C. ^{31}P-NMR (202 MHz, $CDCl_3$): δ = 32.4 (d, *J* = 8.08), 28.8 (d, *J* = 8.08), ^{1}H-NMR (500 MHz, $CDCl_3$): δ = 7.91–7.42 (m, 16H, Ar-H), 7.28–7.02 (m, 10H, Ar-H), 4.21 (d, 1H, *J* = 5.0, CH), 3.92–3.85 (m, 1H, CH), 3.88 (s, 1H, OCH_3), 3.75 (d, 1H, *J* = 5.0, CH), 3.38–3.35 (m, 1H, CH), 2.60 (s, 1H, CH), 2.46 (d, 1H, *J* = 15.0, CH_2), 1.90 (d, 1H, *J* = 15.0, CH_2), 1.08–1.05 (m, 1H, CH), ^{13}C-NMR (125 MHz, $CDCl_3$): δ = 156.7, 144.1, 141.8, 135.2, 135.1, 134.3, 133.9, 133.5, 133.2, 132.5, 131.7, 131.3, 131.1, 130.9, 130.6, 130.5, 130.4, 130.2, 130.1, 129.5, 129.4, 129.3, 128.7, 128.6, 128.5, 128.3, 128.2, 128.2, 120.0, 119.9, 119.0, 117.1, 106.6, 55.2, 50.3, 46.6, 42.9, 42.2, 39.9, 37.8, 32.9 LCMS (ESI) $[M + H]^+$ = 651 Da, HRMS calcd. for $C_{42}H_{36}O_3P_2$ 651.2206, found 651.2212 (diff. 1.18 ppm).

3.2.14. 8c,9,10,11,12,12a-Hexahydro-9,12-methanobenzo[3,4]cyclobuta[1,2-*l*]phenanthrene-10,11-diylbis(diphenylphosphane) dioxide (*rac*-**4p**)

The compound was isolated in 54% yield as a white solid, m.p.: 295–298 °C. ^{31}P-NMR (202 MHz, $CDCl_3$): δ = 32.3 (d, *J* = 8.08), 28.8 (d, *J* = 8.08), ^{1}H-NMR (500 MHz, $CDCl_3$): δ = 8.68–8.65 (m, 3H, Ar-H), 7.99–7.42 (m, 20H, Ar-H), 7.13–7.06 (m, 6H, Ar-H), 6.83 (d, 1H, *J* = 10.0, Ar-H), 4.29 (s, 1H, CH), 3.93–3.83 (m, 2H, CH), 3.52–3.47 (m, 1H, CH), 2.61 (d, 2H, *J* = 5.0, CH), 1.90 (d, 1H, *J* = 10.0, CH), 1.01–0.85 (m, 1H, CH), ^{13}C-NMR (125 MHz, $CDCl_3$): δ = 139.3, 138.3 (d, *J* = 3.7), 135.5, 134.7, 134.4, 133.9, 133.6, 133.1, 132.3, 132.1, 132.0, 131.9 (d, *J* = 3.7), 131.4, 131.1, 130.9, 130.8, 130.7, 130.5, 130.2, 130.1, 128.7, 128.6, 128.5, 128.4, 128.3, 128.3, 127.8, 127.8, 126.6 (d, *J* = 3.7), 123.7 (d, *J* = 8.7), 122.6, 122.3, 49.2 (d, *J* = 15.0), 45.1 (d, *J* = 6.2), 40.4, 41.1, 38.7, 32.5 (d, *J* = 12.5), LCMS (ESI) $[M + H]^+$ = 671 Da, HRMS calcd. for $C_{45}H_{38}O_2P_2$ 671.2281, found 671.2263 (diff. 1.18 ppm).

[5-(phenanthren-9-yl)bicyclo[2.2.1]heptane-2,3-diyl]bis(diphenylphosphane) dioxide (*rac*-**5p**)

This minor reaction product was detected in trace amounts in the reaction mixtures by means of ESI-LC HRMS. HRMS calcd. for $C_{45}H_{38}O_2P_2$ 673.2422, found 673.2420 (diff. 0.3 ppm).

3.2.15. Synthesis of [(1*R*,2*R*,9*S*,10*S*,11*R*,12*R*)-4-Phenyltetracyclo[8.2.1.0^{2,9}.0^{3,8}]trideca-3,5,7-triene-11,12-diyl]bis(diphenylphosphane) ((*R*,*R*)-**6q**)

One gram (1.0 g) of (*R*,*R*)-**6q** (1.54 mmol) and 10 mL of dry toluene were added to a Schlenk flask and placed under a nitrogen atmosphere. Subsequently, 6 equiv. of $SiHCl_3$ and then 18 equiv. of Bu_3N were added and the mixture was heated at 110 °C. After the reaction had been completed, the mixture was cooled down and 30% NaOH solution was added drop-wise. The organic layer was dried over Na_2SO_4 and the crude was subjected to column chromatography with the ethyl acetate/hexane system. The phosphine was isolated as a white solid in 85% yield. $[\alpha]_D^{20} = -60$ (c = 0.4; $CHCl_3$); ^{31}P-NMR (202 MHz, $CDCl_3$): δ = 2.53, 1.10, −11.09, −13.67; ^{1}H-NMR (500 MHz, $CDCl_3$): δ = 7.94–7.42 (m, 16H, Ar-H), 7.23–7.02 (m, 11H, Ar-H), 6.83 (d, 1H, *J* = 5.0, Ar-H), 4.37 (s, 1H, CH), 3.99–3.93 (m, 1H, 2×CH), 3.69 (s, 1H, CH), 3.44–3.39 (m, 1H, CH), 2.70 (s, 1H, CH), 2.38 (d, 1H, *J* = 10.0, CH_2), 1.90 (d, 1H, *J* = 10.0, CH_2), 1.05–1.03 (m, 1H, CH), ^{13}C-NMR (125 MHz, $CDCl_3$): δ = 145.8, 141.8, 137.3, 134.3, 133.1, 131.3, 131.3, 130.9, 130.6, 130.5, 130.4, 130.1, 130.0, 128.6, 128.5, 128.3, 128.2, 127.2, 126.8, 126.5, 126.0, 125.1, 124.8, 120.8, 51.4 (d, *J* = 13.7), 50.0 (d, *J* = 13.7), 48.3 (d, *J* = 6.2), 45.3 (d, *J* = 6.2), 41.8, 41.6, 40.4, 38.4, 38.1, 37.8, 37.5.

3.2.16. Synthesis of [Pd(**6q**)Cl_2]

A solution of **6q** (0.1 g) in benzene (2 mL) was added to a stirred solution of [Pd($^{t-}$BuCN)$_2$$Cl_2$] (0.056 g) in 2 mL of THF. The reaction mixture was stirred overnight at ambient temperature, the volume of the solvents was reduced under reduced pressure down to about 2 mL and the precipitated complex was filtered off. The compound was isolated in 82% yield as a yellow solid. The samples for the X-ray measurements were grown from a warm ethanol/benzene mixture. $[\alpha]_D^{20} = -40.8$ (c = 0.25, CH_2Cl_2); ^{31}P-NMR (202 MHz, $CDCl_3$): δ = 31.3, 31.2, 28.0, 27.9, 27.9, 27.8, ^{1}H-NMR (500 MHz, $CDCl_3$): δ = 8.08–8.04 (m, 2H, Ar-H), 8.02–7.98 (m, 2H, Ar-H), 7.93–7.90 (m, 2H, Ar-H), 7.88–7.84 (m, 2H, Ar-H), 7.68–7.65 (m, 1H, Ar-H), 7.59–7.58 (m, 2H, Ar-H), 7.54–7.48 (m, 22H, Ar-H), 7.33–7.27 (m, 3H, Ar-H), 7.21–7.12 (m, 8H, Ar-H), 7.07–7.01 (m, 13H, Ar-H), 6.85 (d, 1H, *J* = 7.3, CH), 6.55 (d, 1H, *J* = 7.3, CH), 4.12–4.11 (m, 1H, CH), 4.00–3.99 (m, 1H, CH), 3.96–3.87 (m, 1H, CH), 3.48–3.43 (m, 2H, CH), 2.66–2.66 (m, 1H, CH), 2.55–2.55 (m, 1H, CH), 2.36–2.34 (m, 1H, CH), 2.27–2.25 (m, 1H, CH), 1.77–1.75 (m, 1H, CH), 1.68–1.66 (m, 1H, CH), 0.75–0.74 (m, 2H, CH). ^{13}C-NMR (dept 135, 125 MHz, $CDCl_3$): δ = 132.96, 131.81, 131.69, 131.63, 131.54, 131.53, 131.48, 131.46, 131.20, 131.14, 130.90, 130.84, 130.76, 130.70, 130.53, 130.46, 130.40, 130.25, 130.21, 130.18, 130.14, 129.49, 129/42, 129.34, 129.30, 129.28, 129.18, 129.09, 129.07, 128.98, 128.64, 128.62, 128.55, 128.53, 128.02, 127.63, 126.48, 125.96, 125.34, 124.75, 121.32, 121.10, 51.52, 51.41, 49.85, 49.73, 48.35, 48.31, 46.39, 46.35, 41.86, 41.61, 41.13, 41.10, 40.36, 40.18, 40.00, 39.84, 38.77, 38.74, 38.41, 38.36, 38.33, 38.19, 38.16, 37.87, 37.81, 37.77, 37.75, 32.26, 32.23, 32.17, 32.13. LCMS (ESI) $[M - Cl]^+$ = HRMS calcd. for C43H36P2ClPd 755.1010 Da, found 755.1034 Da (diff. 3.18 ppm).

3.2.17. General Procedure for Allylic Alkylation Reaction

A solution of ligand **6q** (2.0 mol % or 4.0 mol %) and [Pd(η^3-C_3H_5)Cl]$_2$ (1.0 mol % or 2.0 mol %) was stirred in 1.0 mL of dry THF for 30 min. Then, *rac*-(*E*)-1,3-diphenylallyl acetate (0.5 mmol) in 1.0 mL of THF was added, followed by dimethyl malonate (1.5 mmol) and base (1.5 mmol) of BSA with 2 mol % of KOAc or 1.5 mmol of K_2CO_3/Cs_2CO_3 (1:1). After stirring for 12 h at room temperature, the solution was concentrated and purified by chromatography on silica gel eluted with hexane/EtOAc (10:1). The product was isolated as a colorless oil. ^{1}H-NMR (500 MHz, $CDCl_3$): δ = 7.34–7.22 (m, 10H, Ar-H), 6.50 (d, 1H, *J* = 15.0, -CH=), 6.33 (dd, 1H, *J* = 10.0, 15.0, -CH=), 4.31–4.27

(m, 1H, CH), 3.98 (d, 1H, J = 10.0, CH), 3.77 (s, 3H, OCH_3), 3.54 (s, 3H, OCH_3), HPLC analysis (Chiralcel OD-H, hexane/*iso*-PrOH, 98:2, 1.0 mL/min, 254 nm): tr(minor) = 10.9 min, tr(major) = 14.5 min.

3.3. Computational Studies

All DFT calculations were performed using the "Prometheus" cluster in the "Cyfronet" computational centre in Cracow. A new generation M062x [27] functional, implemented in the Gaussian 09 package (Gaussian Inc.: Wallingford, UK) [28], was used. All stationary structures have been optimised using the LANL2DZ basis set with one f function for Pd and without pseudopotential. All structures were characterised by only positive eigenvalues in their diagonalised Hessian matrices. For the optimised structures, thermochemical data for the temperature T = 298 K and pressure p = 1 atm were computed using vibrational analysis data.

3.4. Crystallographic Studies

An X-ray diffraction experiment for the complex [Pd(**6q**)Cl_2] was conducted at room temperature on a SuperNova X-ray diffractometer equipped with an Atlas S2 CCD detector using mirror-monochromatized CuKα radiation (λ = 1.54184 Å). Multiscan absorption correction procedures were applied to the data [41]. The structure was solved by direct methods using the ShelXT [42] structure solution program with intrinsic phasing and refined with the Olex2.refine refinement package using Gauss–Newton minimisation [43]. Non-hydrogen atoms were refined anisotropically. The C-bound H atoms were positioned geometrically and refined with the 'riding' model. A summary of the experimental details and crystal structure refinement parameters are given in Tables 2–5 and ESI. The experimental details and final atomic parameters have been deposited with the Cambridge Crystallographic Data Centre as supplementary material (CCDC: 1885894). Copies of the data can be obtained free of charge on request via www.ccdc.cam.ac.uk/data_request/cif, or by emailing data_request@ccdc.cam.ac.uk.

4. Conclusions

Herein, we have demonstrated the design and straightforward synthesis of novel highly rigid polycyclic bis(phosphanes). The X-ray analysis of the coordination compound obtained with a new ligand and dichloropalladium complex confirmed the cis, exo configuration of the bis(phosphane) ligand molecule. The five-membered metallacycle has an envelope conformation with a puckering amplitude q2 of 1.142(5) Å and a phase angle Φ2 of 153.8(1)°. The CSD search showed that the most frequent conformation for the five-membered metallacycles of transition metals and bis(phosphane) ligands was an envelope with one of the carbon atoms positioned at the flap regardless of the transition metal and phosphine substituent type, just as in the presented complex [Pd(**6q**)Cl_2]. This stereochemical asymmetry observed in the analysed metallacycles is probably an important factor for efficient transition-metal-mediated asymmetric transformations. The comparison of the measured and calculated structures of the palladium complex and corresponding bis(phosphane) dioxides confirmed that, due to a structural rigidity, the conformation of the ligands depends only marginally on the substituents that those ligands bear in their polycyclic core. Notable efficiency, activity and selectivity of the new ligand in a model asymmetric allylic alkylation reaction have also been confirmed.

Supplementary Materials: The NMR spectra of the obtained compounds and some additional experimental details are available online.

Author Contributions: Conceptualisation: K.S. and O.M.D.; Methodology: K.S. and O.M.D.; Chemical Experiments: K.S. and O.M.D.; Crystallographic Studies and Structural Survey: B.M.; Software and Computation: R.J.; Writing (Original Draft Preparation): K.S., O.M.D., R.J., B.M. and I.D.; Writing (Review and Editing): O.M.D.; Visualization: I.D. and B.M.; Inspiration and Discussions: K.M.P.

Funding: This research received no external funding. In part the research was carried out with the PLGrid ('Prometheus' cluster) infrastructure (ACK 'Cyfronet' in Cracov), and equipment purchased thanks to the financial support of the European Regional Development Fund under the framework of the Operational Program Development of Eastern Poland 2007–2013 (Contract No. POPW.01.03.00-06-009/11-00, equipping the laboratories of the Faculties of Biology and Biotechnology, Mathematics, Physics and Informatics, and Chemistry for studies of biologically active substances and environmental samples) as well as Polish National Science Centre research grant (2012/05/B/ST5/00362).

Conflicts of Interest: The authors declare no conflict of interest. The funders had no role in the design of the study; in the collection, analyses or interpretation of data; in the writing of the manuscript; or in the decision to publish the results.

References

1. Oestreich, M. *The Mizoroki-Heck Reaction*; Wiley: Hoboken, NJ, USA, 2009; 587p, ISBN 9780470033944.
2. Beletskaya, I.P.; Cheprakov, A.V. The Heck Reaction as a Sharpening Stone of Palladium Catalysis. *Chem. Rev.* **2000**, *100*, 3009–3066. [CrossRef] [PubMed]
3. de Meijere, A.; Bräse, S.; Oestreich, M. *Metal-Catalyzed Cross-Coupling Reactions and More*; 3 Volume Set; Wiley: Hoboken, NJ, USA, 2014; 1576p, ISBN 9783527331543.
4. Catellani, M.; Chiusoli, G.P. Palladium-catalyzed synthesis of 1,2,3,4,4a,12b-hexahydro-1,4-methanotriphenylenes. *J. Organomet. Chem.* **1985**, *286*, c13–c16. [CrossRef]
5. Ye, J.; Lautens, M. Palladium-catalysed norbornene-mediated C−H functionalization of arenes. *Nat. Chem.* **2015**, *7*, 863–870. [CrossRef]
6. Li, R.; Dong, G. Direct Annulation between Aryl Iodides and Epoxides through Palladium/Norbornene Cooperative Catalysis. *Angew. Chem. Int. Ed. Engl.* **2017**, *57*, 1697–1701. [CrossRef] [PubMed]
7. Casnati, A.; Fontana, M.; Coruzzi, G.; Aresta, B.M.; Corriero, N.; Maggi, R.; Maestri, G.; Motti, E.; Della Ca', N. Enhancing Reactivity and Selectivity of Aryl Bromides: A Complementary Approach to Dibenzo[*b,f*]azepine Derivatives. *ChemCatChem Catal.* **2018**, *10*, 4346–4352. [CrossRef]
8. Liu, C.; Liang, Y.; Zheng, N.; Zhang, B.-S.; Feng, Y.; Bi, S.; Liang, Y.-M. Synthesis of indolines via a palladium/norbornene-catalyzed reaction of aziridines with aryl iodides. *Chem. Commun.* **2018**, *54*, 3407–3410. [CrossRef]
9. Demchuk, O.M.; Jasiński, R.; Formela, A. The Halogen-Less Catalytic Transition Metal-Mediated Cross-Coupling Reactions: A Sustainable Alternative for Utilisation of Organohalides. In *Chemistry Beyond Chlorine*; Tundo, P., He, L.-N., Lokteva, E., Mota, C., Eds.; Springer: Cham, Switzerland, 2016; pp. 17–94.
10. Brunner, H.; Pieronczyk, W. Asymmetric Hydrogenation of (Z)-(Acetylamino)-cinnamic Acid by a Rh/norphos Catalyst. *Angew. Chem. Int. Ed. Engl.* **1979**, *18*, 620–621. [CrossRef]
11. Marchand, A.P.; Romanski, J.; Kumar, T.P. (R,R)-(-)-NORPHOS and (S,S)-(+)-NORPHOS. In *e-EROS: Encyclopedia of Reagents for Organic Synthesis*; John Wiley & Sons, Ltd.: Hoboken, NJ, USA, 2003.
12. Szwaczko, K.; Demchuk, O.M.; Mirosław, B.; Strzelecka, D.; Pietrusiewicz, K.M. Straightforward approach to norbornene core based chiral ligands by tandem cross dehydrogenative coupling reactions. *Tetrahedron Lett.* **2016**, *57*, 3491–3495. [CrossRef]
13. Demchuk, O.M.; Kapłon, K.; Mazur, L.; Strzelecka, D.; Pietrusiewicz, K.M. Readily available catalysts for demanding Suzuki–Miyaura couplings under mild conditions. *Tetrahedron* **2016**, *72*, 6668–6677. [CrossRef]
14. Qiu, Z.; Sun, R.; Teng, D. Synthesis of highly rigid phosphine–oxazoline ligands for palladium-catalyzed asymmetric allylic alkylation. *Org. Biomol. Chem.* **2018**, *16*, 7717–7724. [CrossRef]
15. Snieckus, V.; Demchuk, O.; Yoruk, B.; Blackburn, T. A Mixed Naphthyl-Phenyl Phosphine Ligand Motif for Suzuki, Heck, and Hydrodehalogenation Reactions. *Synlett* **2006**, 2908–2913. [CrossRef]
16. Demchuk, O.M.; Kielar, K.; Pietrusiewicz, K.M. Rational design of novel ligands for environmentally benign cross-coupling reactions. *Pure Appl. Chem.* **2011**, *83*, 633–644. [CrossRef]
17. Chakravarty, A.R.; Cotton, F.A.; Tocher, D.A. Displacive transfer of a phenyl group from triphenylphosphine to a metal atom: Synthesis and molecular structure of the dibenzamidediphenyldiruthenium compound Ru2Ph2(PhCONH)2[Ph2POC(Ph)N]2. *J. Am. Chem. Soc.* **1984**, *106*, 6409–6413. [CrossRef]
18. Jin, L.; Qian, J.; Sun, N.; Hu, B.; Shen, Z.; Hu, X. Pd-Catalyzed reductive heck reaction of olefins with aryl bromides for Csp2–Csp3 bond formation. *Chem. Commun.* **2018**, *54*, 5752–5755. [CrossRef]

19. Wang, C.; Xiao, G.; Guo, T.; Ding, Y.; Wu, X.; Loh, T.-P. Palladium-Catalyzed Regiocontrollable Reductive Heck Reaction of Unactivated Aliphatic Alkenes. *J. Am. Chem. Soc.* **2018**, *140*, 9332–9336. [CrossRef] [PubMed]
20. Cremer, D. Calculation of puckered rings with analytical gradients. *J. Phys. Chem.* **1990**, *94*, 5502–5509. [CrossRef]
21. Groom, C.R.; Bruno, I.J.; Lightfoot, M.P.; Ward, S.C. The Cambridge Structural Database. *Acta Crystallogr. Sect. B Struct. Sci. Cryst. Eng. Mater.* **2016**, *72*, 171–179. [CrossRef]
22. Eller, S.; Trettenbrein, B.; Fessler, M.; Haringer, S.; Ruggenthaler, M.; Gutmann, R.; van der Veer, W.E.; Kopacka, H.; Müller, T.; Obendorf, D.; et al. Surprising photochemical reactivity and visible light-driven energy transfer in heterodimetallic complexes. *Dalton Trans.* **2011**, *40*, 3815–3829. [CrossRef]
23. Wilson, W.L.; Rahn, J.A.; Alcock, N.W.; Fischer, J.; Frederick, J.H.; Nelson, J.H. Thermal dimerization of 1-substituted-3,4-dimethylphospholes within the coordination sphere of platinum(II). *Inorg. Chem.* **1994**, *33*, 109–117. [CrossRef]
24. Oberhauser, W.; Ienco, A.; Vizza, F.; Trettenbrein, B.; Oberhuber, D.; Strabler, C.; Ortner, T.; Brüggeller, P. Regioselective Hydromethoxycarbonylation of Terminal Alkynes Catalyzed by Palladium(II)–Tetraphos Complexes. *Organometallics* **2012**, *31*, 4832–4837. [CrossRef]
25. Oberhauser, W.; Stampfl, T.; Bachmann, C.; Haid, R.; Langes, C.; Kopacka, H.; Ongania, K.-H.; Brüggeller, P. Palladium(II), platinum(II), and platinum(IV) complexes containing trans-1,2-bis(diphenylphosphino)ethene or cis,trans,cis-1,2,3,4-tetrakis(diphenylphosphino)cyclobutane: Complete X-ray structural characterization of binuclear compounds. *Polyhedron* **2000**, *19*, 913–923. [CrossRef]
26. Strabler, C.; Ortner, T.; Prock, J.; Granja, A.; Gutmann, R.; Kopacka, H.; Müller, T.; Brüggeller, P. Versatile Supramolecular Coordination Behaviour of a Bis(bidentate) Tetraphosphane. *Eur. J. Inorg. Chem.* **2013**, *2013*, 5121–5132. [CrossRef]
27. Zhao, Y.; Truhlar, D.G. The M06 suite of density functionals for main group thermochemistry, thermochemical kinetics, noncovalent interactions, excited states, and transition elements: Two new functionals and systematic testing of four M06-class functionals and 12 other functionals. *Theor. Chem. Acc.* **2007**, *120*, 215–241. [CrossRef]
28. Frisch, M.J.; Trucks, G.W.; Schlegel, H.B.; Scuseria, G.E.; Robb, M.A.; Cheeseman, J.R.; Scalmani, G.; Barone, V.; Mennucci, B.; Petersson, G.A.; et al. *Gaussian 09, Revision B.01*; Gaussian Inc.: Wallingford, UK, 2010.
29. Jasiński, R.; Demchuk, O.M.; Babyuk, D. A Quantum-Chemical DFT Approach to Elucidation of the Chirality Transfer Mechanism of the Enantioselective Suzuki–Miyaura Cross-Coupling Reaction. *J. Chem.* **2017**, *2017*, 1–12. [CrossRef]
30. Mirosław, B.; Babyuk, D.; Łapczuk-Krygier, A.; Kącka-Zych, A.; Demchuk, O.M.; Jasiński, R. Regiospecific formation of the nitromethyl-substituted 3-phenyl-4,5-dihydroisoxazole via [3 + 2] cycloaddition. *Monatshefte für Chemie (Chem. Mon.)* **2018**, *149*, 1877–1884. [CrossRef] [PubMed]
31. Demchuk, O.M.; Jasiński, R.; Pietrusiewicz, K.M. New Insights into the Mechanism of Reduction of Tertiary Phosphine Oxides by Means of Phenylsilane. *Heteroat. Chem.* **2015**, *26*, 441–448. [CrossRef]
32. Jasiński, R. Molecular mechanism of thermal decomposition of fluoronitroazoxy compounds: DFT computational study. *J. Fluor. Chem.* **2014**, *160*, 29–33. [CrossRef]
33. Demchuk, O.M.; Justyniak, I.; Miroslaw, B.; Jasiński, R. 2-Methoxynaphthylnaphthoquinone and its solvate: Synthesis and structure-properties relationship. *J. Phys. Org. Chem.* **2014**, *27*, 66–73. [CrossRef]
34. Demchuk, O.M.; Arlt, D.; Jasiński, R.; Pietrusiewicz, K.M. Relationship between structure and efficiency of atropisomeric phosphine ligands in homogeneous catalytic asymmetric hydrogenation. *J. Phys. Org. Chem.* **2012**, *25*, 1006–1011. [CrossRef]
35. Demchuk, O.M.; Jasiński, R. Organophosphorus ligands: Recent developments in design, synthesis, and application in environmentally benign catalysis. *Phosphorus Sulfur Silicon Relat. Elem.* **2015**, *191*, 245–253. [CrossRef]
36. Issa, T.B.; Sayari, F.; Ghalla, H.; Benhamada, L. Synthesis, crystal structure, DFT calculations and molecular docking of l-pyroglutamic acid. *J. Mol. Struct.* **2019**, *1178*, 436–449. [CrossRef]
37. Jin, M.-J.; Takale, V.B.; Sarkar, M.S.; Kim, Y.-M. Highly enantioselective Pd-catalyzed allylic alkylation using new chiral ferrocenylphosphinoimidazolidine ligands. *Chem. Commun.* **2006**, *663*. [CrossRef]
38. Yamaguchi, M.; Shima, T.; Yamagishi, T.; Hida, M. Palladium-catalyzed asymmetric allylic alkylation using dimethyl malonate and its derivatives as nucleophile. *Tetrahedron Asymmetry* **1991**, *2*, 663–666. [CrossRef]

39. Łapczuk-Krygier, A.; Ponikiewski, Ł.; Jasiński, R. The crystal structure of (1RS,4RS,5RS,6SR)-5-cyano-5-nitro-6-phenyl-bicyclo[2.2.1]hept-2-ene. *Crystallogr. Rep.* **2014**, *59*, 961–963. [CrossRef]
40. Jasiński, R.; Kwiatkowska, M.; Sharnin, V.; Barański, A. Experimental and theoretical studies of Diels–Alder reaction between methyl (Z)-2-nitro-3-(4-nitrophenyl)-2-propenoate and cyclopentadiene. *Monatshefte für Chemie (Chem. Mon.)* **2013**, *144*, 327–335. [CrossRef] [PubMed]
41. Rigaku Oxford Diffraction. *CrysAlis PRO ver. 1.171.38.46*; Rigaku Oxford Diffraction: Yarnton, UK, 2015.
42. Sheldrick, G.M. SHELXT—Integrated space-group and crystal-structure determination. *Acta Crystallogr. Sect. A Found. Adv.* **2015**, *71*, 3–8. [CrossRef] [PubMed]
43. Dolomanov, O.V.; Bourhis, L.J.; Gildea, R.J.; Howard, J.A.K.; Puschmann, H. OLEX2: A complete structure solution, refinement and analysis program. *J. Appl. Crystallogr.* **2009**, *42*, 339–341. [CrossRef]

Sample Availability: Samples of compounds **1–6** are available from the authors.

molecules

MDPI

Perspective

Protolysis and Complex Formation of Organophosphorus Compounds—Characterization by NMR-Controlled Titrations

Gerhard Hägele

Institute of Inorganic Chemistry and Structural Chemistry, Heinrich-Heine-University Düsseldorf, Universitätsstraße 1, D-40225 Düsseldorf, Germany; haegele@uni-duesseldorf.de

Academic Editors: György Keglevich and Kamyar Afarinkia
Received: 6 May 2019; Accepted: 2 September 2019; Published: 5 September 2019

Abstract: Phosphonic acids, aminophosphonic acids, and phosphonocarboxylic acids are characterized by an advanced hyphenated technique, combining potentiometric titration with NMR spectroscopy. Automated measurements involving ^{13}C, ^{19}F and ^{31}P nuclei lead to "pseudo 2D NMR" spectra, where chemical shifts or coupling constants are correlated with analytical parameters. Dissociation constants, stability constants, dynamic and specific chemical shifts are determined. Macroscopic and microscopic dissociation equilibria are discussed.

Keywords: phosphonic acids; aminophosphonic acids; phosphonocarboxylic acids; NMR-controlled titration; dissociation constants; stability constants; dynamic and specific NMR parameters

1. Introduction

NMR-controlled titration, also known as NMR titration, a useful tool combining NMR and analytical aspects, is based on fundamental observations in the dawn of NMR spectroscopy: "Early phosphorus NMR studies of condensed phosphates showed that raising the acidity of phosphate solutions increased the shielding of the phosphorus nucleus, causing a shift of the ^{31}P resonances to higher fields by several ppm" [1]. "Later studies on the short chain condensed phosphates exhibited that the pH variations of the chemical shifts and spin-coupling constants where, when measured to sufficient precision, sensitive functions of the molecular structure and the bonding". A first titration curve of H_3PO_4 shown as δ_P vs. pH was derived in this paper [2]. ^{13}C-NMR measurements on linear aliphatic acids revealed that COOH groups in $C_nH_{2n+1}COOH$ (n = 0 to 4) exhibit higher chemical shifts δ_C than COO^- groups of corresponding anions $C_nH_{2n+1}COO^-$. A characteristic downfield shift of δ_C ranging from 5.1 to 4.7 ppm was observed for deprotonation by addition of tetramethylammonium hydroxide to carboxylic acids [3].

In subsequent years, those phenomena attracted the attention of numerous studies dealing with inorganic and organic phosphorus chemistry. A higher level of sophistication was achieved by combining the analytical theory of protolysis and complex formation for acids and bases with advanced NMR technologies and expanding the range of sensor nuclei to ^{1}H, ^{13}C, ^{15}N, ^{19}F, ^{31}P and spin active metal nuclei.

The first NMR titration curves for phosphonoacetic acid $HOOC\text{-}CH_2\text{-}PO_3H_2$ using ^{13}C and ^{31}P NMR were reported as δ_C vs. pH and δ_P vs. pH functions. For the first time, a characteristic deprotonation sequence was established: $HOOC\text{-}CH_2\text{-}PO_3H_2 \rightarrow HOOC\text{-}CH_2\text{-}PO_3H^- \rightarrow {}^-OOC\text{-}CH_2\text{-}PO_3H^- \rightarrow {}^-OOC\text{-}CH_2\text{-}PO_3{}^{2-}$. In addition, $^1J_{PC}$ was related to the s-electron density around the central C-atom [4].

Several decades of creative work followed those early observations. Induced by synthetic, analytical, biological, or technical aspects, interests were concentrated on several classes of organophosphorus compounds. Particular attention was drawn towards analogues of amino acids, e.g., aminophosphonic acids and strong complexing agents like NTMP ($N(CH_2PO_3H_2)_3$) and EDTMP ($(H_2O_3PCH_2)_2NCH_2CH_2N(CH_2PO_3H_2)_2$), which are phospha analogues of the classical complexones NTA ($N(CH_2COOH)_3$) and EDTA ($(HOOCCH_2)_2NCH_2CH_2N(CH_2COOH)_2$). Dissociation constants, stability constants for protonation and metal complex formation were studied as quoted with a few selected key papers [5–16].

Further interests concentrated on phospha analogues of carboxylic acids, e.g., phosphonocarboxylic acids and geminal bisphosphonic acids. ^{31}P- and ^{13}C-NMR spectra of cyclohexyl- and phenylphosphonic acid showed that chemical shifts δ_P and δ_C including coupling constants $^nJ_{PC}$ (n = 1–4) of cyclohexanephosphonic acid and benzenephosphonic acid proved to be pH-dependent [17].

A key paper in understanding the NMR titration of geminal bisphosphonate structures described three asymmetric esters of chlodronic acid $(HO)_2(O)P\text{-}CCl_2\text{-}P(O)(OiPr)OH$, $(HO)_2(O)P\text{-}CCl_2\text{-}P(O)(OiPr)_2$, and $HO(iPrO)(O)P\text{-}CCl_2\text{-}P(O)(OiPr)_2$. Proton coupled ^{31}P-NMR titration spectra revealed the coupling constants $^2J_{PP}$ in a range between 15.6 and 17.9 Hz. This significant parameter is not accessible for the symmetric ester $HO(iPrO)(O)P\text{-}CCl_2\text{-}P(O)(OiPr)OH$ since this compound gives rise to a dynamic deceptively simple spectrum ranging from singlet to triplet as a result of the parent symmetric $[AM_6X]_2$ spin system [18]. ^{31}P-NMR measurements at 202.5 MHz showed that the chemical shift δ_P of $CH_3C(OH)[P(O)(OH)_2]_2$ (HEDP) is sensitive towards pH and the concentration of $[(CH_3)_4N]^+$ when $[(CH_3)_4N]Cl$ was used as an ion buffer [19].

The determination of high pK values (pK > 13) and low pK values (pK < 1) required specific, advanced techniques for NMR titration. Comments on measurements at high and low pH were reported [19,20]. $^1H/^{31}P$ NMR pH indicator series were used to eliminate the glass electrode in NMR spectroscopic pK determinations, leading to "electrodeless titrations" [21]. Comprehensive guidelines for NMR measurements for the determination of high and low pK values were given in a IUPAC Technical Report. Those sophisticated and detailed instructions should be followed for accurate analytical and NMR measurements, data evaluation and subsequent publications [22].

1.1. Developing Technical Setups for Automated NMR Titrations

In general, NMR titrations for various nuclei were performed in single sample techniques, which proved to be rather laborious and time consuming. For practical reasons, the number of data points were limited in those early titration curves. Hence, attempts were made to develop the technology of automated NMR titrations.

An innovative set up was constructed, which permitted the acquisition of spectra from spinning 20 mm NMR tubes, adding a solution of base under efficient mixing while monitoring the pH. This apparatus worked together with the wide-bore magnet of a Bruker CXP-300 spectrometer, yielding approximately 80 titration points within a couple of hours. This technology was successfully used to titrate H_3PO_4 vs. KOH and provided a smooth NMR titration curve [23].

Further progress for automated NMR titrations inside spinning 10 mm NMR tubes was described and the novel installation applied to monitor the complex formation between $Tl(I)^+$ and Cl^- in aqueous solutions. A Bruker CXP-100 spectrometer was used, operating at 51.9 MHz for ^{205}Tl [24].

Very recently, an elegant low-cost construction was developed for a gravity-driven pH adjustment inside a 5 mm NMR tube. No hardware modifications of the NMR spectrometer were requested. This technique was applied to site-specific protein pK measurements [25]. It might be useful for future studies of organophosphorus compounds.

A different route to automated NMR titrations was chosen by the Düsseldorf group. We were intrigued by the technology of 2D NMR spectra and the graphical representation of such spectra by standard spectrometer software. Hence, a hyphenated technique was envisaged, replacing the f2 axis of 2D NMR spectra by analytical parameters.

Bypass constructions were developed in several generations of increasing accuracy. A 10 mm NMR tube was attached to a special homemade insert and used with a Bruker AM 200 SY NMR spectrometer operating at 81 MHz for ^{31}P NMR. This insert acted as bypass to a precision titration equipment. A series of pH dependent 1D NMR spectra were recorded and processed (using standard spectrometer software) to yield instructive "pseudo 2D NMR" spectra (e.g., in analogy to COSY spectra). Chemical shift δ_P data were correlated with analytical data like pH or the degree of titration τ. The technical setup and two examples are shown in [26]. Phosphaalanine was used as an example where deprotonation and complex formation with Zn^{2+} cations were observed by titrations vs. tetramethylammonium hydroxide (TMAOH) [26].

This equipment was used to characterize a series of aminomethylphosphine oxides $(CH_3)_{3-n}(CH_2NH_2)_nPO$ (n = 1–3), adding n equivalents of HCl and back-titrating vs. NaOH. Ion-specific chemical shifts δ_P and pK data were obtained for those aminomethylphosphine oxide bases. In addition, technical details of NMR, analytics, software concepts and programs used were described [27].

A brief overview of "^{31}P NMR controlled titrations of Phosphorus-Containing Acids and Bases in Protolysis and Complex Formation" reported about the 81 MHz $^{31}P\{^1H\}$ NMR titration of phosphonoacetic acid [28]. The hardware and software concepts were shown.

Typical "pseudo 2D NMR" spectra correlating the chemical shift δ_P vs. the degree of titration τ were obtained for the pair of isomers 1- and 2-aminoethanephosphonic acids CH_3-$CH(NH_2)$-PO_3H_2 and NH_2-CH_2-CH_2-PO_3H_2 (α-Ala-P and β-Ala-P) and for diphosphaasparaginic acid H_2O_3P-$CH(NH_2)$-CH_2-PO_3H_2 (Asp-P_2). Macroscopic dissociation constants and ion-specific chemical shifts are reported. The *p*-aminophenylene-substituted phosphonic acid *p*-NH_2-C_6H_4-PO_3H_2 was compared [29]. Hardware and software concepts used in NMR titration were demonstrated. A subsequent UV-controlled titration revealed the microscopic dissociation scheme of *p*-NH_2-C_6H_4-PO_3H_2. Corresponding deprotonation patterns were discussed [30].

In practice, ^{13}C-NMR titrations in single sample techniques proved to be very time consuming. Hence, it seemed advisable to use the technology described above for automated 50.29 MHz $^{13}C\{^1H\}$ or ^{13}C-NMR measurements. As practical examples, the pair of isomers 1- and 2-aminoethanephosphonic acids were titrated vs. NaOH. Within this context, CH_3-$CH(NH_2)$-PO_3H_2 and the fluorinated analogue CF_3-$CH(NH_2)$-PO_3H_2 were compared using $^{31}P\{^1H\}$ and $^{19}F\{^1H\}$ NMR titrations. Replacing the CH_3 by a CF_3 group reduces the basicity of the NH_2 function, which is reflected in δ_P vs. τ and in the δ_F vs. τ correlations [31].

The experimental set up described above requested individual titrations for each nucleus wanted. Hence, multinuclear studies (e.g., ^{1}H and ^{13}C and ^{31}P) demanded high spectrometer times.

At this stage, special probe heads were developed by Bruker for another hyphenated technique combining liquid chromatography with HR NMR. In our laboratory, a Bruker LC probe head LC-TXO-NMR was successfully introduced to a DRX 500 NMR spectrometer and used for advanced NMR titrations. It became routine to run consecutively $^{31}P\{^1H\}$, ^{31}P, and ^{1}H-NMR spectra for each titration step, thus saving time, gaining higher sensitivity and reducing the necessary concentrations (and amounts) of titrands. Excellent spectra resulted with a high S/N ratio and high digital resolution in the chemical shift or frequency axis.

In addition, a special ^{19}F-LC probe head was available, combining ^{19}F and ^{1}H-NMR techniques. The high field stability of the supercon magnet allowed measurements in H_2O solutions (without D_2O), thus avoiding the problems with "mixed" stability and dissociation constants resulting from D_2O/H_2O mixed solvents.

A comprehensive report about the technical designs of NMR and analytical components, software, data evaluation, error calculations and applications was written in 2002 and incorporated into the Bruker NMR Guide collection, freely accessible for Bruker spectrometer users [32] only. This detailed review is now open for free downloads to all interested readers: (a) https://www.theresonance.com/nmr-controlled-titration-download-the-paper/, (b) https://www.bruker.com/fileadmin/user_upload/8-PDF-Docs/MagneticResonance/NMR/NMR_controlled_titration.pdf.

NMR titrations using 200 MHz and 500 MHz spectrometers were described using geminal bisphosphonic acids, e.g., HEDP and Pamidronic acid, as model systems. The TXO-HPLC probe head improved the signal-to-noise ratio of "pseudo 2D NMR" spectra and reduced the concentration of titrand required by this procedure: The following concentrations for sensor nuclei are recommended: ^{1}H: 0.25–0.01 mol/L, ^{13}C: 0.50–0.005 mol/L, ^{19}F: 0.01–0.005 mol/L, ^{31}P: 0.01–0.001 mol/L, and ^{113}Cd: 0.25–0.1 mol/L. ^{113}Cd NMR was used when studying protolytic and complex formation equilibria of $(H_2O_3P\text{-}CH_2)_2NCH_2CH_2N(CH_2PO_3H_2)_2$ (EDTMP).

Within this context, a special computer program MultipleNMRGraphics was developed which is able to generate four characteristic "pseudo 2D NMR" plots, e.g., δ_P vs. pH or δ_P vs. τ either as contour or as stacked plots, in black-and-white or color design [33]. Those graphics have a lower storage demand than the previously used "pseudo 2D NMR" spectra generated by the routine Bruker spectrometer software.

Some examples relevant to phosphorous chemistry and organic chemistry dealt with in [32] are listed in Table 1:

Table 1. Some examples for NMR-controlled titrations of phosphonic acids, phosphinic acids, and carboxylic acids as discussed in [32]. 1) Retro titration; 2) Micro dissociation; 3) Second-order ^{1}H-NMR spin systems.

Examples	NMR	Remarks
Phosphonic acids		
$CH_3P(O)(OH)_2$	$^{31}P\{^{1}H\}$	
$LiOOC\text{-}CH_2\text{-}P(O)(OLi)_2$	$^{31}P\{^{1}H\}$	1)
$(HO)_2(O)P\text{-}CH_2\text{-}CH_2\text{-}P(O)(OH)_2$	$^{13}C\{^{1}H\}$	
$CH_3\text{-}C(OH)[P(O)(OH)_2]_2$, HEDP, etidronic acid	$^{31}P\{^{1}H\}$	
$NH_2\text{-}CH_2\text{-}CH_2\text{-}C(OH)(P(O)(OH)_2)_2$, pamidronic acid	$^{31}P\{^{1}H\}$	
$HOOC\text{-}CH_2\text{-}CH(COOH)\text{-}CH(COOH)\text{-}P(O)(OH)_2$, PPTC	$^{31}P\{^{1}H\}$	
Phosphinic acids		
$(CH_3)_2P(O)OH$	$^{31}P\{^{1}H\}$	
$HOOC\text{-}CH_2\text{-}CH_2\text{-}P(CH_3)(O)OH$	$^{13}C\{^{1}H\}$, ^{1}H	2)
$HO(O)(CH_3)P\text{-}CH_2\text{-}CH_2\text{-}P(CH_3)(O)OH$	$^{13}C\{^{1}H\}$	
$HO(O)(CH_3)P\text{-}CH_2\text{-}CH_2\text{-}C(H)(NH_2)COOH$	^{1}H	2, 3)
Carboxylic acids		
CH_3COOH	$^{13}C\{^{1}H\}$	
$CH(CH_3)_2\text{-}CH_2\text{-}CH(NH_2)\text{-}C(O)\text{-}NH\text{-}CH(CH_3)\text{-}COOH$, peptide Leu-Ala	$^{13}C\{^{1}H\}$	
$CH_2{=}CF\text{-}CH_2\text{-}C(CH_3)(NH_2)\text{-}COOH$	^{19}F	

A modification of our design for automated NMR titrations shown above was adjusted to the local conditions of a Bruker 250 MHz spectrometer and applied to study the complexation of Zn^{2+}, Cd^{2+} and Pb^{2+} with diazacrown ethers substituted by phosphonate groups [34].

Particular attention was drawn towards microscopic dissociation constants going back to early studies on $NH_2\text{-}CH_2\text{-}CH_2\text{-}NH\text{-}CH_2\text{-}COOH$. 60 MHz and 100 MHz ^{1}H-NMR titrations evaluated the pH dependence of a singlet for the methylene group $NH\text{-}CH_2\text{-}COOH$, while the ethylene function $N\text{-}CH_2\text{-}CH_2\text{-}N$ appeared with the spectral character, changing from a deceptively simple singlet towards an AA′BB′ ($[AB]_2$) system. The analytical formalism and microscopic dissociation constants were derived [35]. For deeper reading, an up-to-date and comprehensive survey on the theory and practice of proton microspeciation based on NMR-pH titrations is recommended [36].

As an example, *S*-2-amino-4-(methylphosphinoyl)butyric acid (*S*-phosphinothricine, GLUFOSINATE) $HOOC\text{-}CH(NH_2)\text{-}CH_2\text{-}CH_2\text{-}P(CH_3)(O)OH$ was characterized by $^{31}P\{^{1}H\}$- and ^{1}H-NMR titrations. Microscopic dissociation and intramolecular rotational equilibria were discussed [32,37]. Within this context, a program LAOTIT was developed, which is able to simulate series of pH-dependent second-order NMR spectra. A practical example for $AFGMNQ_3X$ spin systems of GLUFOSINATE in a pH range from 1 to 6 was shown in [37].

The ring-chain tautomerism and protolytic equilibria of an effectively three-basic 3-hydroxy-3-phosphonoisobenzofuranone was studied by ^{1}H-, $^{13}C\{^{1}H\}$- and $^{31}P\{^{1}H\}$-NMR-controlled titrations. A complex pattern of macroscopic and microscopic deprotonation steps leading from the starting H_3L to the final L^{3-} (Scheme 1) was discussed.

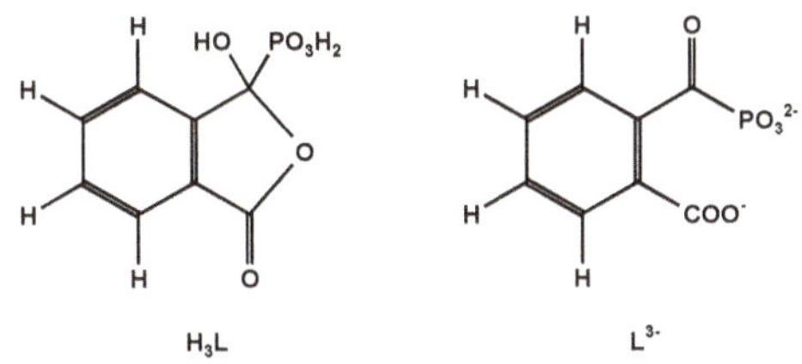

Scheme 1. Showing 3-hydroxy-3-phosphonoisobenzofuranone H_3L and its terminal anion L^{3-}.

The OPIUM program enabled the simultaneous evaluation of potentiometric, $^{31}P\{^{1}H\}$- and ^{1}H-NMR titrations using the four individual ^{1}H signals from the ABCD system [38]. Macroscopic dissociation constants, $pK_1 = 0.445 \pm 0.008$, $pK_2 = 5.792 \pm 0.003$, $pK_3 = 6.486 \pm 0.002$. δ_H, δ_C, and δ_P of H_3L (0.2510 mol/L in D_2O) and L^{3-} (0.1919 mol/L in 1 mol/L KOD), were determined. For details of the complex equilibrium system, see [39,40].

NMR-controlled titration was successfully used to analyze the mixture of diastereomers from 1-phosphonopropane-1,2,3-tricarboxylic acid, HOOC-CH_2-CH(COOH)-CH(COOH)-PO_3H_2 (PPTC). The genuine product from synthesis consisted of 64% of the RS/SR and 36% of the RR/SS forms. $^{31}P\{^{1}H\}$-NMR-controlled titration revealed two diastereospecific titration curves which were individually identified by additional 1D and 2D NMR studies using ^{1}H, ^{31}P and ^{13}C nuclei. Dissociation constants and ion-specific chemical shifts δ_P were calculated for the pair of diastereomers [41–43]. It seems evident to use automated NMR titration for production control in research and industrial chemistry.

1.2. Some Comments on Macroscopic Protolytic Equilibria—Dissociation and Stability Constants

Organophosphorus compounds studied by potentiometric or NMR-controlled titrations may be described by two numerical indices: a, the number of acidic functions (e.g., P(O)OH, C(O)OH, etc.) and b, the number of basic functions (e.g., NH_2, NHR, NR_2, etc.). The minimal protonated species corresponds to the n-valent base L^{-a} having a anionic centers and b neutral base centers in $(^{0}N)_b$-R-$(O^-)_a$. Total protonation leads to the n-valent acid H_nL^{b+} (n = a + b) having a neutral centers and b cationic acid centers in $(^{+}HN)_b$-R-$(OH^0)_a$.

Protonation equilibria of the n-valent base are described by Equation (1):

$$iH^+ + L^{-a} \rightleftharpoons H_iL^{i-a} \quad i = 1 \text{ to } n \tag{1}$$

and by brutto-stability constants following Equation (2):

$$\beta_i = \frac{[H_iL^{i-a}]}{[H^+]^i[L^{-a}]} \quad i = 1 \text{ to } n \quad \beta_0 = 1 \tag{2}$$

Stepwise dissociation equilibria of the n-valent acid are described by Equation (3):

$$H_{n+1-i}L^{b+1-i} \rightleftharpoons H^+ + H_{n-i}L^{b-i} \quad i = 1 \text{ to } n \tag{3}$$

while corresponding dissociation constants K_i are given by Equation (4):

$$K_i = \frac{[H^+][H_{n-i}L^{b-i}]}{[H_{n+1-i}L^{b+1-i}]} \quad i = 1 \text{ to } n \tag{4}$$

Stoichiometric stability constants and stoichiometric dissociation constants are connected by Equations (5) and (6):

$$pK_i = \lg\beta_{n+1-i} - \lg\beta_{n-i} \quad i = 1 \text{ to } n \tag{5}$$

$$\lg\beta_i = \sum_{j=1}^{i} pK_{n+1-j} \quad i = 1 \text{ to } n \tag{6}$$

This paper will use stoichiometric variables (containing concentrations instead of activities) in abbreviated forms: pK_i—macroscopic acid dissociation constant; pk_i—microscopic acid dissociation constant; pK_w—ion product of water. pH stands for the concentration-based pH = $-\lg(c_H)$. Glass electrodes were calibrated by blank titration. The more complex situation of activities and activity-based parameters exceeds the scope of this paper and hence will not be discussed at this stage.

The molar fractions x_i of protolytic species H_iL^{i-a} are derived from Equation (7):

$$x_i = \frac{10^{\lg\beta_i - i\cdot pH}}{\sum\limits_{j=0}^{n} 10^{\lg\beta_j - j\cdot pH}} \quad i = 0 \text{ to } n \tag{7}$$

Each protolytic species H_iL^{i-a} present in the equilibrium contributes specific NMR parameters $\delta(H_iL^{i-a})$ in an exchange reaction, which is rapid on the NMR timescale. Effectively, only one signal is observed when monitoring NMR during the course of titrations. A dynamically averaged chemical shift δ follows Equation (8):

$$\delta = \sum_{i=0}^{n} x_i \cdot \delta_{H_iL^{i-a}} \quad i = 0 \text{ to } n \tag{8}$$

A gradient called the deprotonation shift Δ_i [ppm] is given by Equation (9):

$$\Delta_i = \delta_{H_{n-i}L^{b-i}} - \delta_{H_{n+1-i}L^{b+1-i}} \quad i = 1 \text{ to } n \tag{9}$$

This gradient defines the change of chemical shift for each deprotonation step. Signs and magnitudes of gradients are used to elucidate the deprotonation and protonation pathways of multifunctional acids, bases and ligands as shown in examples below.

As deduced above, the dynamically averaged chemical shift δ is a function of pH. Experimentally, the pH of solutions may be varied by titration with a strong univalent base or a strong univalent acid. While the experiment directly provides the well-known titration curve pH = $f(V_{Titrator})$, it is more convenient to calculate the inverse function $V_{Titrator}$ = f(pH) with suitable computer programs. Within this paper, a reduced parameter τ, commonly called degree of titration, will be used to describe the status of a titration process. τ is a ratio defined by Equation (10):

$$\tau = \frac{n_{Titrator}}{n_{Titrand}} \tag{10}$$

The sign of τ is positive if $n_{Titrator}$ corresponds to the molar amount of a strong monovalent base (e.g., NaOH, KOH, TMAOH), but is negative for a strong monovalent acid (HCl, HNO_3, $HClO_4$). $n_{Titrand}$ corresponds to the molar amount of a n-basic acid H_nL.

Details of the basic principles and experimental equipment with hardware and software are described in [31,32] and in references given herein.

2. Results and Discussion

In the following sections, a few examples will be shown for automated NMR-controlled titrations using hardware and software concepts described above. Chemical shifts δ_C [ppm] quoted below were referenced vs. $(CH_3)_3Si$-CH_2-CH_2-SO_3Na, while δ_P [ppm] was virtually referenced towards external H_3PO_4. Coupling constants $^nJ_{XY}$ are given in [Hz].

2.1. Phosphonic Acids

Methanephosphonic acid **1** and phenylphosphonic acid **2** shown in Scheme 2 were chosen from [31,44], which will be presented below:

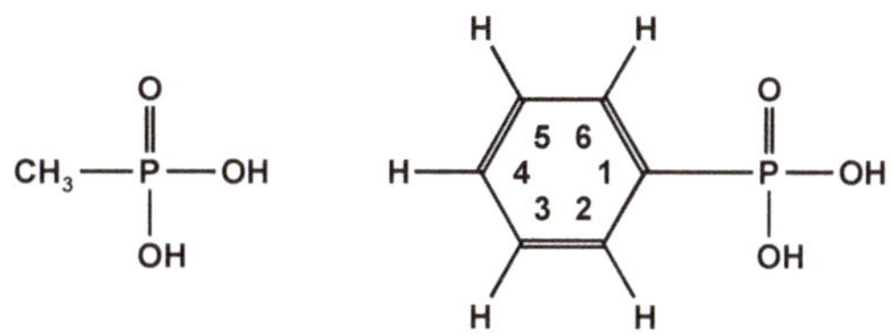

1 H2L **2** H2L

Scheme 2. Phosphonic acids: methanephosphonic **1** and phenylphosphonic acid **2**.

2.1.1. Methanephosphonic Acid 1

The results from a proton-coupled ^{31}P-NMR-controlled titration of methanephosphonic acid **2** vs. NaOH are shown as a contour plot in Figure 1. A quartet structure from the parent A_3X spin system of the P-CH_3 fragment is recognized. Numerical results are given in Table 2. The deprotonation of both P-OH functions induces a decrease in chemical shifts δ_P and a decrease in the absolute values of $^2J_{PH}$.

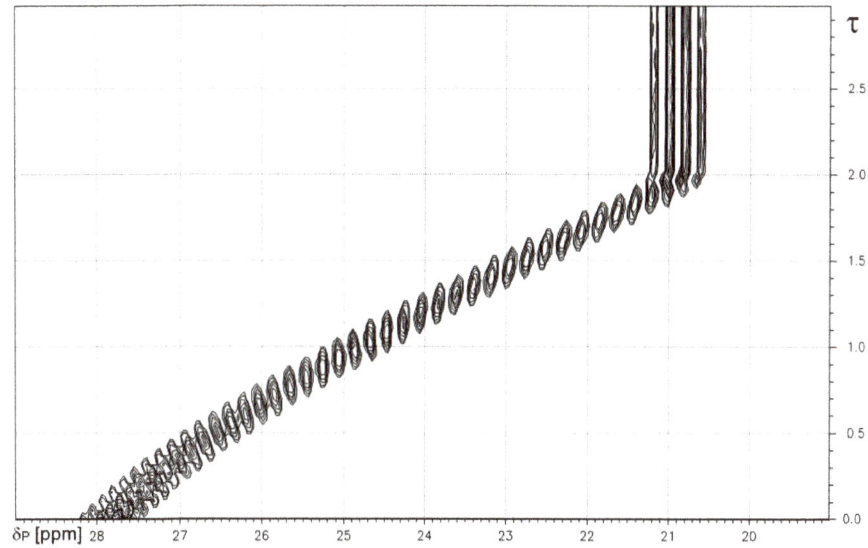

Figure 1. ^{1}H-coupled ^{31}P-NMR-controlled titration of methanephosphonic acid **1** vs. NaOH. Note: quartet fine structure from X-part of AX_3 system of P-CH_3 fragment. *X*-axis: δ_P [ppm]. *Y*-axis: degree of titration τ.

Table 2. Specific chemical shifts δ_C, δ_P and coupling constants $^1J_{PC}$ and $^2J_{PH}$ of methanephosphonic acid **1** were obtained by $^{13}C\{^1H\}$-, $^{31}P\{^1H\}$-, and ^{31}P-NMR-controlled titrations in H_2O. $\Delta_i = \delta(H_{n-i}L) - \delta(H_{n+1-i}L)$ or $\Delta_i = {^nJ_{PC}}(H_{n-i}L) - {^nJ_{PC}}(H_{n+1-i}L)$, respectively. i = 1 to n. n = 2. Experimental data: $C_{Titrand}$: a) 0.269 mol/L. b) 0.01220 mol. c) 0.0095 mol/L. $C_{Titrator}$: a) 4.82 mol/L KOH. b) 0.0971 mol/L NaOH. c) 0.0971 mol/L NaOH. d) Early data from results from titration vs. KOH [45].

	1 in H_2O					
Method	$^{13}C\{^1H\}$		$^{31}P\{^1H\}$	^{31}P		$^{31}P\{^1H\}$
Exp.	a)		b)	c)		d)
Species	δ_C	$^1J_{PC}$	δ_P	δ_P	$^2J_{PH}$	δ_P
	[ppm]	[Hz]	[ppm]	[ppm]	[Hz]	[ppm]
H_2L	14.27	135.92	33.03	33.03	−17.65	31.76
HL^-	15.53	133.82	24.79	24.79	−16.52	24.94
L^{2-}	16.61	129.95	21.08	21.08	−15.52	20.94
Gradients	δ_C	$^1J_{PC}$	δ_P	δ_P	$^2J_{PH}$	δ_P
Δ_1	1.26	−2.10	−8.24	−8.24	1.13	
Δ_2	1.08	−3.87	−3.71	−3.71	1.00	
pK_i						
pK_1	2.27		2.06	2.00		2.33
pK_2	7.85		7.66	7.68		7.78

pK_i values found are consistent with results from potentiometric titrations of $CH_3P(O)(OH)_2$ [46].

2.1.2. Phenylphosphonic Acid 2

Chemical shifts δ_P for protolytic species H_2L, HL^-, and L^{2-} of phenylphosphonic acid **2** together with dissociation constants pK_1 and pK_2 are listed in Table 3. The deprotonation of both P-OH groups leads to characteristic high field shifts for δ_P as indicated by negative gradients Δ_1 and Δ_2.

Table 3. Specific chemical shifts δ_P [ppm], gradients [ppm] and dissociation constants pK_i from $^{31}P\{^1H\}$-NMR-controlled titration of phenylphosphonic acid **2** vs. TMAOH and NaOH. Exp.: $C_{Titrator}$: a) 0.09894 mol/L TMAOH. b) 0.09925 mol/L NaOH. $C_{Titrand}$: a) 0.02 mol/L **3**. b) 0.008 mol/L **2**. Shifts and gradients given in ppm.

	Phenylphosphonic Acid 2	
	vs. TMAOH	vs. NaOH
	a)	b)
$\delta_P(H_2L)$	18.39	17.77
$\delta_P(HL^-)$	13.77	13.75
$\delta_P(L^{2-})$	11.69	11.72
Δ_1	−4.62	−4.02
Δ_2	−2.08	−2.03
pK_1	1.74	1.86
pK_2	7.28	7.16

Higher concentrations are required for $^{13}C\{^1H\}$-NMR-controlled titrations as shown for the titration of phenylphosphonic acid **2** vs. KOH in Figure 2:

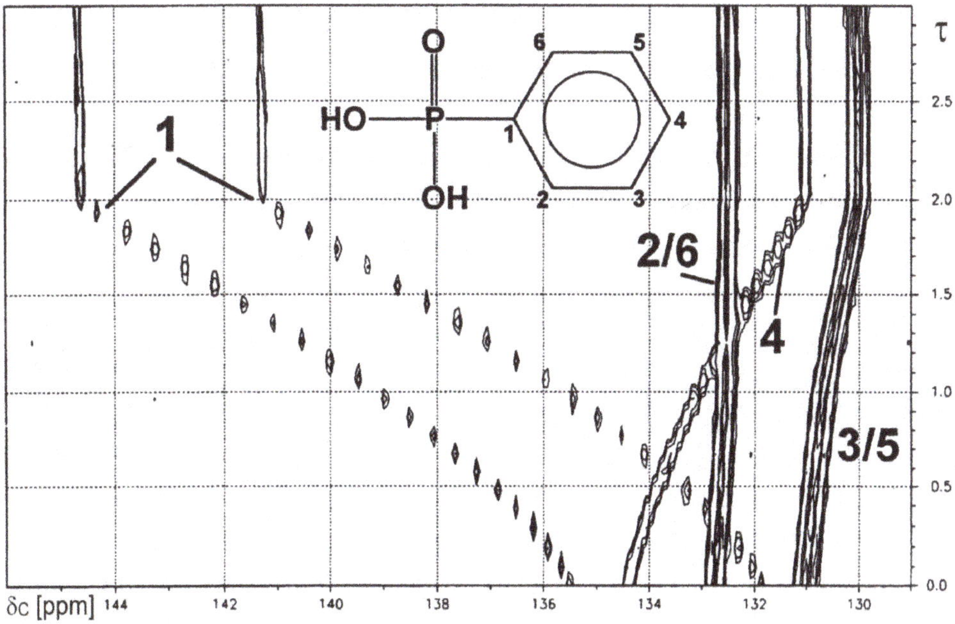

Figure 2. $^{13}C\{^1H\}$-NMR-controlled titration of phenylphosphonic acid **2** vs. KOH yielded a contour plot for chemical shift δ_C as a function of the degree of titration τ. *X*-axis: δ_C [ppm]. *Y*-axis: degree of titration τ.

The deprotonation of each of the two P-OH functions led to a strong low field shift for the *ipso*-C1 carbon. For the remaining carbons high field shifts are observed, an effect decreasing in the order *para*-C4 > *meta*-C3/5 > *ortho*-C2/5.

Semi-empirical calculations with VAMP 4.4 using parameter set AM1 showed that the electron density at C1 increases with deprotonation in the order $PhPO_3H_2$ < $PhPO_3H^-$ < $Ph\text{-}PO_3^{2-}$, while the electron density of C4 decreases in this order [44,47]. In addition, the deprotonation of P-OH led to a decrease for all $^nJ_{PC}$ (n = 1 to 4). Particularly indicative is $^1J_{PC}$ from the *ipso*-carbon C1, which reaches a minimum at total deprotonation. Numerical results for compound 2 are listed in Table 4:

Table 4. Specific chemical shifts δ_C [ppm] and coupling constants $^nJ_{PC}$ (n = 1 to 4) [Hz] for the $^{13}C\{^1H\}$-NMR-controlled titration of phenylphosphonic acid **2** vs. KOH. pK_1 = 1.75. pK_2 = 6.92. Experimental data: $C_{Titrand}$: a) 0.237 mol/L **2**. $C_{Titrator}$: 4.53 mol/L KOH.

	δ_C and $^nJ_{PC}$ for Species			Gradients	
Parameters	H_2L	HL^-	L^{2-}	Δ_1	Δ_2
δ_C **C1**	133.66	138.15	143.83	+4.48	+5.69
δ_C **C2/6**	133.68	133.45	133.43	−0.24	−0.02
δ_C **C3/5**	131.99	131.59	130.95	−0.40	−0.64
δ_C **C4**	135.66	134.03	131.98	−1.58	−2.06
$^1J_{PC}$	183.48	177.02	167.32	−6.46	−9.70
$^2J_{PC}$	10.51	9.72	8.79	−0.79	−0.93
$^3J_{PC}$	14.84	13.90	12.65	−0.94	−1.25
$^4J_{PC}$	3.08	2.91	2.74	−0.17	−0.17

Laborious and time-consuming single sample $^{31}P\{^{1}H\}$- and $^{13}C\{^{1}H\}$-NMR studies on phenylphosphonic acid **2** were performed, where δ_P, δ_C, and $^{n}J_{PC}$ data are consistent with findings derived from the automated titrations presented in this paper [17].

2.2. Comparison of Aliphatic and Aromatic Aminophosphonic Acids

α-Aminoethylphosphonic acid (α-Ala-P) **3** [29,44], β-aminoethylphosphonic acid (β-Ala-P, CILIATIN) **4** [29], and *p*-aminophenylphosphonic acid **5** [29,30,43] shown in Scheme 3 will be studied as examples in the following section.

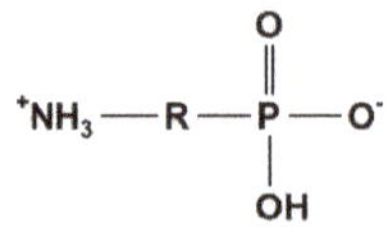

3: R = -CH(CH$_3$)- ; **4**: R = -CH$_2$-CH$_2$- ; **5**: R = -p-C$_6$H$_4$- .

Scheme 3. Aminophosphonic acids **3** to **5**.

2.2.1. Aliphatic Aminophosphonic Acids

Aminophosphonic acids NH_2-R-PO_3H_2, such as examples **3** to **5**, exist as betainic forms $^{+}NH_3$-R-PO_3H^{-} in solid and in solution state. Protolytic equilibria of aminophosphonic acids are described by macroscopic and microscopic formalisms as shown in Table 5 below:

Table 5. Macroscopic and microscopic dissociation species of aminophosphonic acids.

Dissociation Species		
Macroscopic	**Microscopic**	
H_3L^+	$^{+}NH_3$-R-PO_3H_2	
H_2L	$^{+}NH_3$-R-PO_3H^{-}	NH_2-R-PO_3H_2
HL^-	$^{+}NH_3$-R-PO_3^{2-}	NH_2-R-PO_3H^{-}
L^{2-}		NH_2-R-PO_3^{2-}

If R is aliphatic (e.g., in **3** and **4**), the deprotonation takes place following route a): $^{+}NH_3$-R-PO_3H_2 → $^{+}NH_3$-R-PO_3H^{-} → $^{+}NH_3$-R-PO_3^{2-} → NH_2-R-PO_3^{2-}. But if R is aromatic (e.g., in **5**. R = *p*-C_6H_4-), the deprotonation dominantly will follow b): $^{+}NH_3$-R-PO_3H_2 → $^{+}NH_3$-R-PO_3H^{-} → NH_2-R-PO_3H^{-} → NH_2-R-PO_3^{2-}. Consistent conclusions are supported by a combination of potentiometric titrations and $^{31}P\{^{1}H\}$-NMR-controlled titrations as shown for examples **3** to **5**, and in addition by $^{13}C\{^{1}H\}$-NMR-controlled titrations for examples **3** and **4**. Owing to its low solubility, **5** was not suitable for $^{13}C\{^{1}H\}$-NMR-controlled titrations.

Macroscopic dissociation constants pK_i of **3** and **4** are listed in Table 6. pK_i data of compounds **4** and **5** were discussed in [44,45,48–50].

Specific chemical shifts δ_P and gradients Δ for compounds **3** to **5** obtained by $^{31}P\{^{1}H\}$-NMR-controlled titrations are given in Table 7:

The deprotonation of the P-OH groups led to high field shifts for δ_P connected with negative gradients. The final deprotonation of the NH_3^{+} group gave rise to a low field shift for δ_P. This effect is stronger in α-aminophosphonic acid **4** than in β-aminophosphonic acid **5**. Earlier results for chemical shifts δ_P of H_2L, HL^- and L^{2-} species of **3** and **4** were mentioned in [5,45]. In addition, δ_P of H_3L^+ was accessible for **4** but not for **3**.

Table 6. Macroscopic dissociation constants pK_i of compounds α-Ala-P **3** and β-Ala-P **4** obtained by $^{13}C\{^1H\}$- and $^{31}P\{^1H\}$-NMR-controlled titrations and by potentiometric titrations [4a]. Note: $pK_3 - pK_2 > 3$ and $pK_2 - pK_1 > 3$ for compounds **3** and **4**. Exp.: **3:** $C_{Titrand}$: a) 0.0867 mol/L **3** + 0.0834 mol/L HNO_3. b) and c) 0.005 mol/L **3** + 0.00476 HNO_3 + 0.0917 mol/L $NaNO_3$. $C_{Titrator}$: a) 3.98 mol/L NaOH. b) and c): 0.100 mol/L NaOH. **4:** $C_{Titrand}$: d) 0.139 mol/L **4** + 0.139 mol/L HNO_3. e) 0.010 mol/L **4** + 0.010 mol/L HNO_3. f) 0.010 mol/L **4** + 9.747 mmol/L HNO_3. g) 0.005 mol/L **4** + 0.005 mol/L HNO_3 + 0.100 mol/L TMACl. $C_{Titrator}$: d) 0.98 mol/L NaOH. e): 0.0991 mol/L TMAOH. f): 0.0993 mol/L NaOH. g): 0.099 mol/L TMAOH.

		3			4		
	$^{13}C\{^1H\}$ a	$^{31}P\{^1H\}$ b	Pot. c	$^{13}C\{^1H\}$ d	$^{31}P\{^1H\}$ e	$^{31}P\{^1H\}$ f	Pot. g
	NaOH	NaOH	NaOH	NaOH	TMAOH	NaOH	TMAOH
pK_1	0.70	0.31	0.3	1.02	1.22	1.26	1.14
pK_2	5.72	5.63	5.58	6.38	6.23	6.24	6.34
pK_3	10.64	10.21	10.28	11.50	11.06	11.08	11.06

Table 7. Specific chemical shift δ_P [ppm] and gradients Δ [ppm] for compounds **3** to **5** [44]. *) Not iterated. Experimental details for **3** and **4** are given in Table 6. Titrator: a) NaOH; b) TMAOH; c) for **5** were used: $C_{Titrand}$ = 1.6953 mol/L **5** and 3.6935 mol/L TMAOH. Titrator = 0.09993 mol/L HCl.

	3	4		5
	$^{31}P\{^1H\}$ a	$^{31}P\{^1H\}$ b	$^{31}P\{^1H\}$ a	$^{31}P\{^1H\}$ c
Species	δ_P	δ_P	δ_P	δ_P
H_3L^+	15 *	22.9	23.4	13 *
H_2L	14.92	19.29	19.36	12.16
HL^-	13.08	16.80	16.81	15.29
L^{2-}	22.25	19.39	19.72	12.70
Gradients				
Δ_1	−0.08	−3.61	−4.04	−0.84
Δ_2	−1.84	−2.49	−2.55	+3.13
Δ_3	+9.17	+2.59	+3.91	−2.59

$^{13}C\{^1H\}$-NMR-controlled titrations of compounds **4** and **5** led to specific chemical shifts δ_C, coupling constants $^1J_{PC}$, and gradients Δ as listed in Table 8:

Table 8. Specific chemical shifts δ_C [ppm], coupling constants $^1J_{PC}$ [Hz], and corresponding gradients Δ for α-aminoethanephosphonic acid (α-Ala-P) **3** and β-aminoethanephosphonic acid (β-Ala-P, CILIATIN) **4**. Spin enumerations: **3**: C2-C1(N)-P; **4**: (N)C2-C1-P. $^1J_{PC}$ shows a minimum for species HL^- of **3** and **4**. $^2J_{PC}$ was not resolved for compounds **3** and **4**.

	3 in H_2O			4 in H_2O		
Species	δ_C(C1)	$^1J_{PC}$	δ_C(C2)	δ_C(C1)	$^1J_{PC}$	δ_C(C2)
H_3L^+	46.80	151.5	16.0	27.50	137.4	37.09
H_2L	47.70	143.8	16.43	28.73	131.4	38.22
HL^-	49.07	134.5	17.23	29.47	124.8	39.28
L^{2-}	48.15	138.0	19.79	35.45	126.5	39.76
Gradients						
Δ_1	+0.9	−8.7	+0.43	+1.23	−6.0	+1.13
Δ_2	+1.37	+9.3	+0.80	+0.74	−6.6	+1.04
Δ_3	−0.92	+3.5	+2.55	+5.98	+1.3	+0.47

NMR was used to monitor the complex formation of aminophosphonic acids with biorelevant cations in homogeneous solutions. An instructive example is the $^{31}P\{^1H\}$-NMR-controlled titration of CILIATIN/Mg^{2+} vs. NaOH where the formation of [MgL] and $[MgHL]^+$ was monitored [44].

2.2.2. Aromatic *p*-Aminophenylphosphonic Acid **5**

The deprotonation of PO_3H^- in aliphatic aminophosphonic acids **3** and **4** is affiliated with a high field shift (gradients Δ_2 are negative), while the deprotonation of the ammonium function $^+NH_3$ leads to a low field shift (gradients Δ_3 are positive).

The aromatic *p*-aminophenylphosphonic acid **5** exhibits a different pattern: while gradient Δ_2 is positive, Δ_3 is negative (Scheme 4).

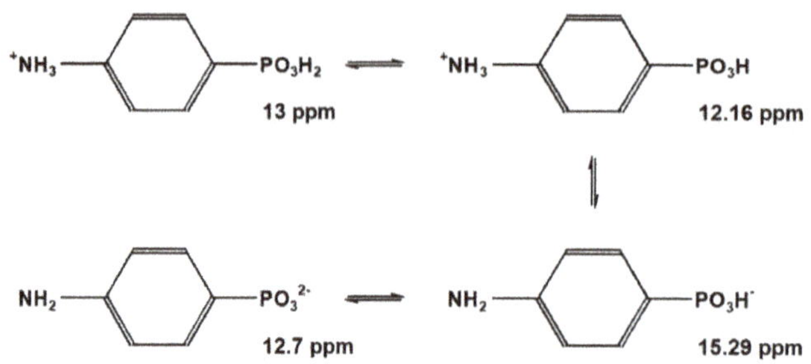

Scheme 4. Specific chemical shifts δ_P [ppm] derived from the $^{31}P\{^1H\}$-NMR-controlled retro titration of *p*-aminophenylphosphonic acid **5**. $\Delta_2 = +3.13$ ppm. $\Delta_3 = -2.59$ ppm.

But is it sufficient to assume a simple first-order macroscopic dissociation scheme for compound **5**? Deeper insight might be obtained from the microscopic dissociation scheme. In principle $^{13}C\{^1H\}$-NMR-controlled titration should lead to specific chemical shifts and coupling constants $^nJ_{PC}$ indicative for microscopic dissociations species of **5**. But *p*-aminophenylphosphonic acid **5** is less soluble in water than the aliphatic aminophosphonic acids **3** and **4**. The S/N-ratio of $^{13}C\{^1H\}$-NMR spectra of **5** is not sufficient to perform evaluable $^{13}C\{^1H\}$-NMR-controlled titrations. In this situation, UV/VIS-controlled titration, which allows for lower concentrations suitable for conclusive measurements, will help to study both the macroscopic and the microscopic dissociation equilibrium of **5** [30]. In addition, the parent compounds $C_6H_5PO_3H_2$ **2** and $C_6H_5NH_2{*}HCl$ **6** were compared. The following macroscopic pK_i data were found by potentiometric titration and listed in Table 9:

Table 9. Dissociation constants of compounds *p*-aminophenylphosphonic acid **5**, phenylphosphonic acid **2**, and anilinium hydrochloride **6**.

	5	2	6
pK_1	0.44	1.88	4.68
pK_2	3.95	7.15	
pK_3	7.56		

Those data point towards a dominating deprotonation sequence for **5** following $^+NH_3$-R-PO_3H_2 → $^+NH_3$-R-PO_3H^- → NH_2-R-PO_3H^- → NH_2-R-PO_3^{2-}. But is it justified to exclude the alternative route $^+NH_3$-R-PO_3H_2 → $^+NH_3$-R-PO_3H^- → $^+NH_3$-R-PO_3^{2-} → NH_2-R-PO_3^{2-}? Evaluating the macroscopic dissociation constants of **5** shows that between pH = 1.5 and pH = 10, only three macroscopic species exist: H_2L, HL^-, and L^{2-}. The UV/VIS-controlled titration of **5** [30] showed that the maximum concentration for macroscopic HL^- is reached at pH = 5.75, consisting of two microdissociation species NH_2-R-PO_3H^- and $^+NH_3$-R-PO_3^{2-} in a ratio of 9:1. Thus, the results from the UV/VIS-controlled titration of **5** [30] confirm the dominance of NH_2-R-PO_3H^- as previously assumed for the macroscopic deprotonation sequence derived from the $^{31}P\{^1H\}$-NMR-controlled titration of **5** [44].

2.3. *Phosphonocarboxylic Acids HOOC-(CH$_2$)$_n$-PO$_3$H$_2$* ***7a*** *to* ***7d*** *(n = 0 to 3)*

Phosphonocarboxylic acids HOOC-(CH_2)$_n$-PO_3H_2 gave rise to potentiometrically [4,44] and NMR-controlled titrations [4,28,31,44]. Those neutral acids of type H_3L deprotonate dominantly in a sequence: HOOC-(CH_2)$_n$-PO_3H_2 → $^-$OOC-(CH_2)$_n$-PO_3H^- → $^-$OOC-(CH_2)$_n$-$PO_3{}^{2-}$. Corresponding dissociations constants for compounds shown in Scheme 5 are listed in Table 10 below:

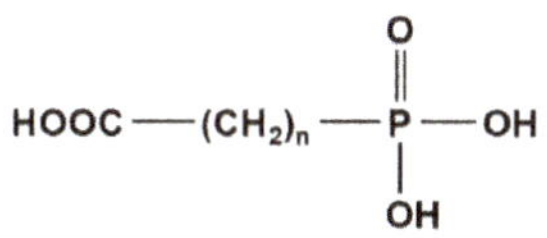

7a: n = 0 ; **7b**: n = 1 ; **7c**: n = 2 ; **7d**: n = 3 .

Scheme 5. Phosphonocarboxylic acids **7a** to **7d**.

Table 10. Dissociation constants of phosphonocarboxylic acids HOOC-(CH_2)$_n$-PO_3H_2 (n = 0 to 3) **7a** to **7d** [8,44]. Experimental data: [a] $C_{Titrand}$: 0.153 mol/L FOSCARNET (trisodium phosphonoformate hexahydrate), $C_{Titrator}$: 2.002 mol/L HNO_3, retro titration; [b] $C_{Titrand}$: 0.220 mol/L **7b**, $C_{Titrator}$: 0.980 mol/L NaOH. [c] $C_{Titrand}$: 0.200 mol/L **7c**, $C_{Titrator}$: 3.986 mol/L NaOH. [d] $C_{Titrand}$: 0160 mol/L **7d**, $C_{Titrator}$: 3.986 mol/L NaOH.

	HOOC-(CH$_2$)$_n$-PO$_3$H$_2$					
	7a	7a [4]	7b	7c	7c [4]	7d
	n = 0 [a]	n = 0	n = 1 [b]	n = 2 [b]	n = 2	n = 3 [b]
pK$_1$	0.78	1.7 ± 0.1	1.22 ± 0.166	2.58 ± 0.013	2.26 ± 0.04	2.276 ± 0.006
pK$_2$	3.60	3.59 ± 0.02	4.942 ± 0.004	4.633 ± 0.004	4.63 ± 0.02	4.776 ± 0.004
pK$_3$	7.57	7.56 ± 0.02	8.099 ± 0.003	7.738 ± 0.003	7.75 ± 0.02	7.969 ± 0.003

$^{13}C\{^1H\}$-NMR-controlled titrations yielded the specific chemical shift δ_C and coupling constants $^nJ_{PC}$ (n = 1 to 3) of phosphonocarboxylic acids HOOC-(CH_2)$_n$-PO_3H_2 (n = 0 to 3) **7a** to **7d** as listed in Table 11a. Gradients are given in Table 11b. Note: The deprotonation of P-OH groups and of C-OH led to a low field shift for all carbon atoms. Some chemical shifts and coupling constants $^1J_{PC}$ of **7a** and **7c** were obtained and discussed in an early key paper [4].

2.3.1. Compound **7c**: $^{13}C\{^1H\}$-NMR-Controlled Titration of 3-Phosphonopropionic Acid HOOC-CH_2-CH_2-PO_3H_2 7c.

3-Phosphonopropionic acid **7c** was chosen as an example to show practical results from $^{13}C\{^1H\}$-NMR-controlled titrations (see Figure 3a,b below). The deprotonation of C-OH and of both P-OH functions induces low field shifts for C1, C2, and C3. Hence, the corresponding gradients are negative. Lorentzian deconvolution of $^{13}C\{^1H\}$ signals yielded $^nJ_{PC}$, where absolute values follow the sequence: $^1J_{PC}$ >> $^3J_{PC}$ > $^2J_{PC}$. See Table 11a,b above.

Table 11. (**a**). Specific chemical shifts δ_C [ppm] and coupling constants $^nJ_{PC}$ [Hz] of phosphonocarboxylic acids HOOC-$(CH_2)_n$-PO_3H_2 (n = 0 to 3) **7a** to **7d**. For experimental data, see preceding Table 10. Remarks: n. r. = not resolved; (**b**). Gradients Δ_i of specific chemical shifts δ_C [ppm] and coupling constants $^nJ_{PC}$ [Hz] of phosphonocarboxylic acids HOOC-$(CH_2)_n$-PO_3H_2 (n = 0 to 3) **7a** to **7d**. For experimental data, see preceding Table 10.

(a)

7	n	Species	δ_C(C1)	δ_C(C2)	δ_C(C3)	δ_C(C4)	$^1J_{PC}$	$^2J_{PC}$	$^3J_{PC}$	$^4J_{PC}$
a	0	H_3L	176.8				246.6			
		H_2L^-	178.7				236.7			
		HL^{2-}	181.8				231.8			
		L^{3-}	187.3				220.0			
b	1	H_3L	172.91	37.68			128.6	n. r.		
		H_2L^-	175.44	39.30			117.8	n. r.		
		HL^{2-}	178.85	41.64			119.2	n. r.		
		L^{3-}	181.74	43.50			112.6	n. r.		
c	2	H_3L	179.25	30.09	24.51		138.5	3.6	17.3	
		H_2L^-	180.68	31.40	25.91		135.1	3.2	17.8	
		HL^{2-}	185.01	34.27	27.50		133.0	4.1	18.7	
		L^{3-}	186.71	35.61	28.99		130.3	3.6	19.8	
d	3	H_3L	180.38	36.67	20.44	28.19	135.2	4.0	17.4	n. r.
		H_2L^-	181.12	37.46	21.58	29.74	133.5	3.8	17.2	n. r.
		HL^{2-}	185.86	41.45	23.13	30.46	132.5	3.9	17.7	n. r.
		L^{3-}	186.41	42.01	24.08	31.92	130.1	3.4	17.9	n. r.

(b)

7	n	Gradients	δ_C(C1)	δ_C(C2)	δ_C(C3)	δ_C(C4)	$^1J_{PC}$	$^2J_{PC}$	$^3J_{PC}$
a	0	Δ_1	+1.9				−9.9		
		Δ_2	+3.1				−4.9		
		Δ_3	+5.5				−11.8		
b	1	Δ_1	+2.53	+1.62			−10.8		
		Δ_2	+3.41	+2.34			+1.4		
		Δ_3	+2.89	+1.86			−6.6		
c	2	Δ_1	+1.43	+1.31	+1.40		−3.4	−0.3	0.5
		Δ_2	+4.33	+2.87	+1.59		−2.1	0.9	0.9
		Δ_3	+1.70	+1.34	+1.49		−2.7	−0.5	1.1
d	3	Δ_1	+0.74	+0.79	+1.14	1.56	−1.7	−0.2	−0.2
		Δ_2	+4.74	+3.99	+1.55	0.72	−1.0	0.1	0.5
		Δ_3	+0.55	+0.56	+0.95	1.46	−2.4	−0.5	0.2

The 81 MHz $^{31}P\{^1H\}$-NMR-controlled titration of 3-phosphonopropionic acid **7c** vs. NaOH yielded Figure 3c. The deprotonation sequence reported in [4] corresponds to: HOOC-CH_2-CH_2-PO_3H_2 → HOOC-CH_2-CH_2-PO_3H^- → $^-$OOC-CH_2-CH_2-PO_3H^- → $^-$OOC-CH_2-CH_2-PO_3^{2-}. Deprotonation at PO_3H_2 or PO_3H^- is affiliated with high field shifts of δ_P, while deprotonation at HOOC induces a low field shift for δ_P.

Specific chemical shifts δ_P for **7c** and corresponding anions together with gradients are listed in Table 12.

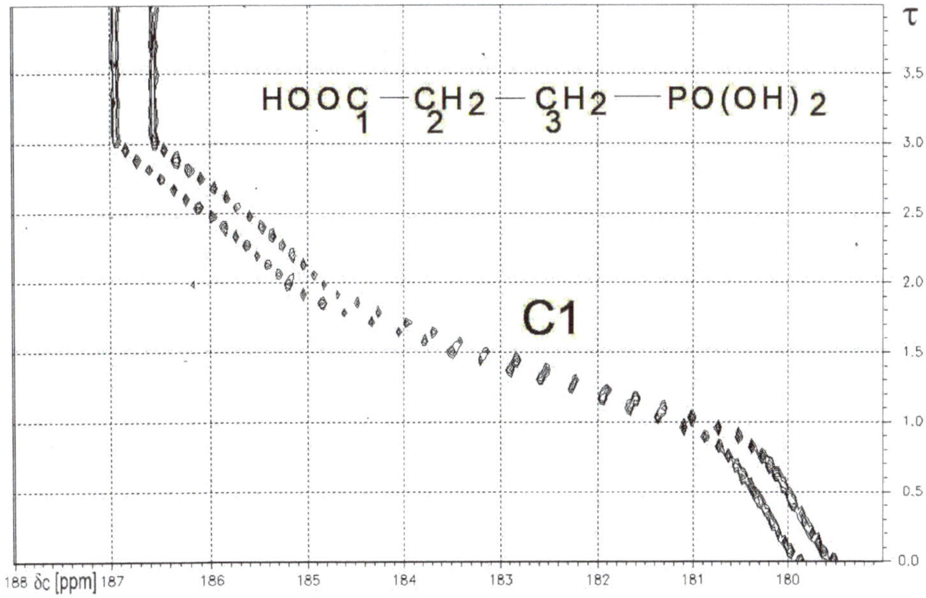

a

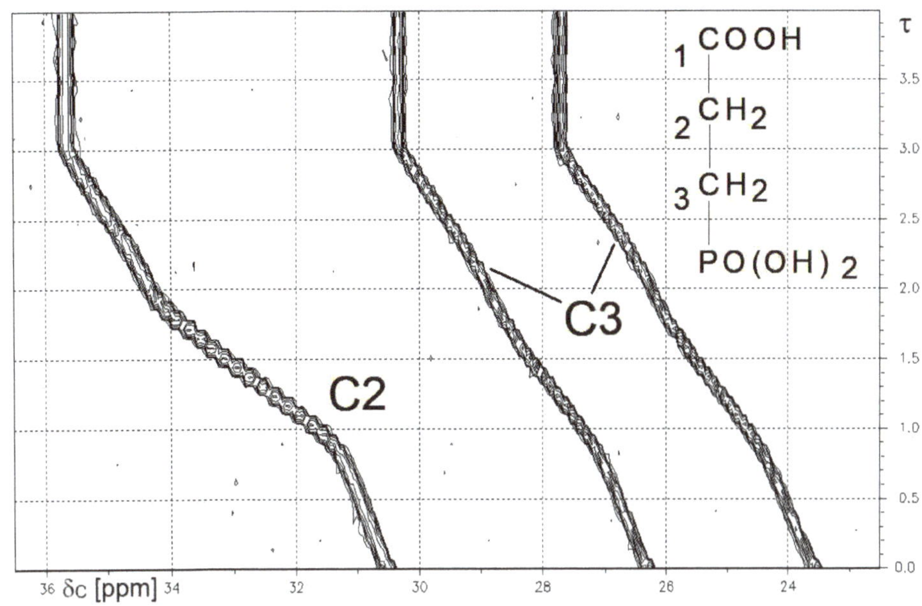

b

Figure 3. *Cont.*

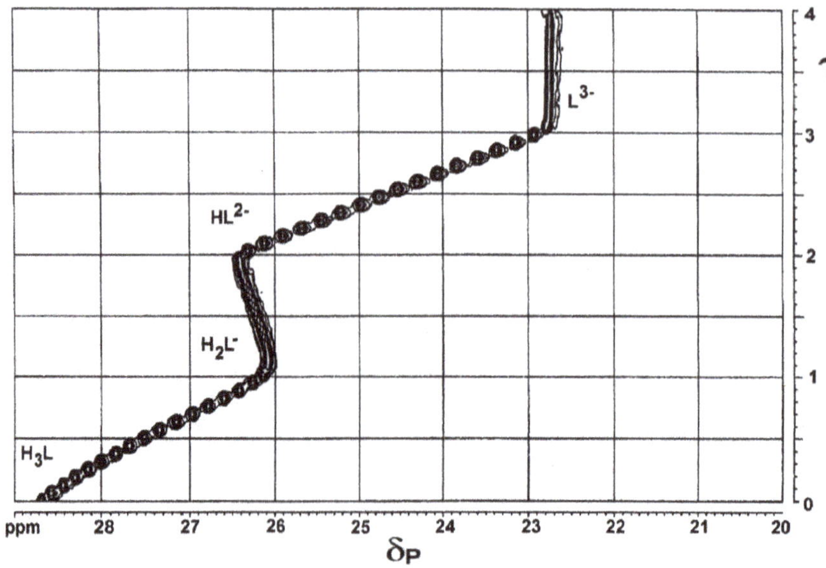

c

Figure 3. (**a**) $^{13}C\{^1H\}$-NMR-controlled titration of 3-phosphonopropionic acid **7c** vs. NaOH. Contour plot showing the range of the carboxylic carbon C1. For experimental data, see Table 10. *X*-axis: δ_C [ppm]. *Y*-axis: degree of titration τ; (**b**) $^{13}C\{^1H\}$-NMR-controlled titration of 3-phosphonopropionic acid **7c** vs. NaOH. Contour plot showing the range of the aliphatic carbons C2 and C3. For experimental data, see Table 10. *X*-axis: δ_C [ppm]. *Y*-axis: degree of titration τ; (**c**) 81 MHz $^{31}P\{^1H\}$-NMR-controlled titration of 3-phosphonopropionic acid **7c** vs. NaOH *X*-axis: δ_P [ppm]. *Y*-axis: degree of titration τ. $C_{Titrand}$: 0.010 mol/L of **7c**. $C_{Titrator}$: 0.10 mol/L TMAOH. Digital resolution: 0.6 Hz/point in *x*-axis. [51].

Table 12. Gradients Δ_i of specific chemical shifts δ_P [ppm] of **7c**. Experimental data: $C_{Titrand}$: 0.010 mol/L of **7c**. $C_{Titrator}$: 0.10 mol/L TMAOH.

Shifts	δ_P	Error
H_3L	29.93	±0.25
H_2L^-	24.56	±0.02
HL^{2-}	25.88	±0.01
L^{3-}	22.06	±0.01
Gradients		
Δ_1	−5.37	
Δ_2	+1.32	
Δ_3	−3.82	

2.3.2. ^{19}F-NMR-Controlled Retro Titrations of Lithium Salts $LiOOC\text{-}CH_{2\text{-}n}F_n\text{-}PO_3Li_2$ **8a** and **8b**

The trilithium salts $LiOOC\text{-}CH_{2\text{-}n}F_n\text{-}PO_3Li_2$ (**8a** and **8b**; n = 1 and 2, Scheme 6) were used for retro titrations vs. HNO_3, since the parent mono- and difluorophosphonoacetic acids **8c** and **8d** were not available for ^{19}F-NMR- and $^{31}P\{^1H\}$-NMR titrations. Corresponding dissociation constants pK_i of **8c** and **8d** were calculated as listed in Table 13, while chemical shifts δ_F and δ_P and coupling constants $^2J_{PF}$ are given in Table 14. As expected, the introduction of fluorine into the skeleton of the parent phosphonoacetic acid led to lower pK_1 and pK_2 values. The deprotonation of P-OH groups induces a low field shift for δ_F in fluorinated phosphonic acids **8c** and **8d**.

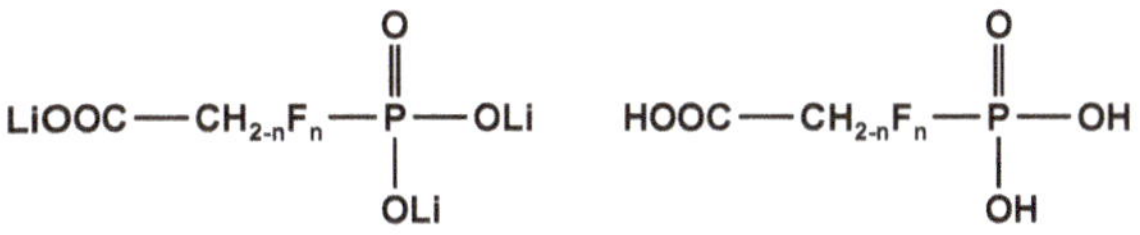

8a: n = 1 ; **8b**: n = 2 . **8c**: n = 1 ; **8d**: n = 2 .

Scheme 6. Trilithium salts LiOOC-$CH_{2-n}F_n$-PO_3Li_2 **8a** and **8b** and free acids HOOC-$CH_{2-n}F_n$-PO_3H_2 **8c** and **8d**.

Table 13. Dissociation constants of fluorinated phosphonocarboxylic acids HOOC-$CH_{2-n}F_n$-PO_3H_2 (**8c** and **8d**; n = 1 and 2) obtained by the retro titration of LiOOC-$CH_{2-n}F_n$-PO_3Li_2 (**8a** and **8b**; n = 1 and 2) vs. $HClO_4$. Experimental data: $C_{Titrand}$: 0.85 mol/L **8a** or **8b** resp. $C_{Titrator}$: 0.3928 mol/L $HClO_4$. No ion buffer.

	HOOC-$CH_{2-n}F_n$-PO_3H_2		
	8c n = 1	8d n = 2	
	[44]	[44]	[52]
pK_1	1.05	0.52	1.30
pK_2	3.43	2.22	1.95
pK_3	7.08	6.36	6.16

Table 14. Chemical shifts δ_F [ppm], coupling constants $^2J_{PF}$ [Hz] and corresponding gradients of fluorinated phosphonocarboxylic acids HOOC-$CH_{2-n}F_n$-PO_3H_2 (**8c** and **8d**, n = 1 and 2) obtained by ^{19}F-NMR-controlled titrations vs. HNO_3 of LiOOC-$CH_{2-n}F_n$-PO_3Li_2 (**8c** and **8d**, n = 1 and 2). For experimental data, see Table 13. δ_F is virtually referenced to $\delta_F(CF^{35}Cl_2{}^{37}Cl)$ = 0 ppm.

	HOOC-$CH_{2-n}F_n$-PO_3H_2			
	8c	8d	8c	8d
	n = 1	n = 2	n = 1	n = 2
	^{19}F		^{19}F	
Species	δ_F	δ_F	$^2J_{PF}$	$^2J_{PF}$
H_3L	−38.27	44.24	67.8	76.5
H_2L^-	−38.72	50.24	65.5	88.6
HL^{2-}	−29.31	53.05	70.3	92.8
L^{3-}	−27.15	55.24	63.4	82.0
Gradients				
Δ_1	−0.45	+6.00	−2.3	+12.1
Δ_2	−9.41	+2.81	+4.8	+4.2
Δ_3	−8.16	+2.19	−6.9	−10.8

2.3.3. 2,4-Diphosphonobutane-1,2-Dicarboxylic Acid (DPBDC) 9

Strong practical interests focused on polyfunctional phosphonocarboxylic acids, e.g., phosphonosuccinic acid (PBS), 1-phosphonopropane-1,2,3-tricarboxylic acid (PPTC), and 2-phosphonobutane-1,2,4-tricarboxylic acid (PBTC), which gave rise to analytical and NMR studies of protolytic and complex formation equilibria [42–44,53].

The following section will deal with 2,4-diphosphonobutane-1,2-dicarboxylic acid (DPBDC) **9** to demonstrate the potential of automated NMR titrations. Dissociation constants of this 6-valent acid DPBDC **9** were obtained from precision potentiometric [53] and by $^{13}C\{^1H\}$-NMR-controlled titrations vs. NaOH [44].

$^{13}C\{^1H\}$-technique yielded Figure 4a–d. The spin enumeration used in subsequent tables and figures is given in Scheme 7:

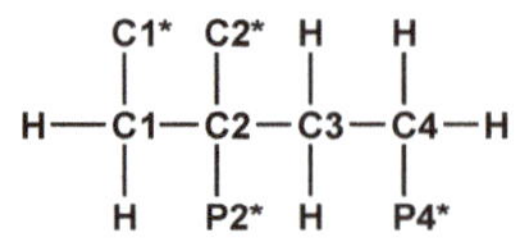

Scheme 7. Spin enumeration in DPBDC 9 used for $^{13}C\{^1H\}$-NMR C1* and C2* = COOH, COO^-. P2* and P4* = PO_3H_2, PO_3H^-, PO_3^{2-}.

Results for those six carbon atoms C1*, C2*, and C1 to C4 will be presented as (δ,τ))-contour plots in four separate spectral ranges shown in Figure 4a–d. Numerical results including specific chemical shifts δ_C and coupling constants $^nJ_{PC}$ of DPBDC are listed in Tables 15 and 16.

Table 15. Dissociation constants of 2,4-diphosphonobutane-1,2-dicarboxylic acid (DPBDC) as obtained from $^{13}C\{^1H\}$-NMR-controlled and potentiometric titrations vs. NaOH. * Concentrations given in mol/L.

	$^{13}C\{^1H\}$ NMR [44]	Potentiometric [44]	Potentiometric [53]
pK_1	1.07	0.6	1.806 ± 0.066
pK_2	2.73	2.42	2.250 ± 0.021
pK_3	4.82	4.32	4.078 ± 0.005
pK_4	7.05	6.46	6.562 ± 0.004
pK_5	8.95	8.18	8.664 ± 0.006
pK_6	11.62	10.75	12.839 ± 0.007
$C_{Titrand}$	0.262 (DPBDC) *	0.0050 (DPBDC) *	0.01119 (DPBDC) *
$C_{Titrator}$	3.986 (NaOH) *	0.0975 (NaOH) *	0.09863 (TMAOH) *
$C_{Ion\ buffer}$	0	0.1 (NaCl) *	0.09863 ($TMANO_3$) *

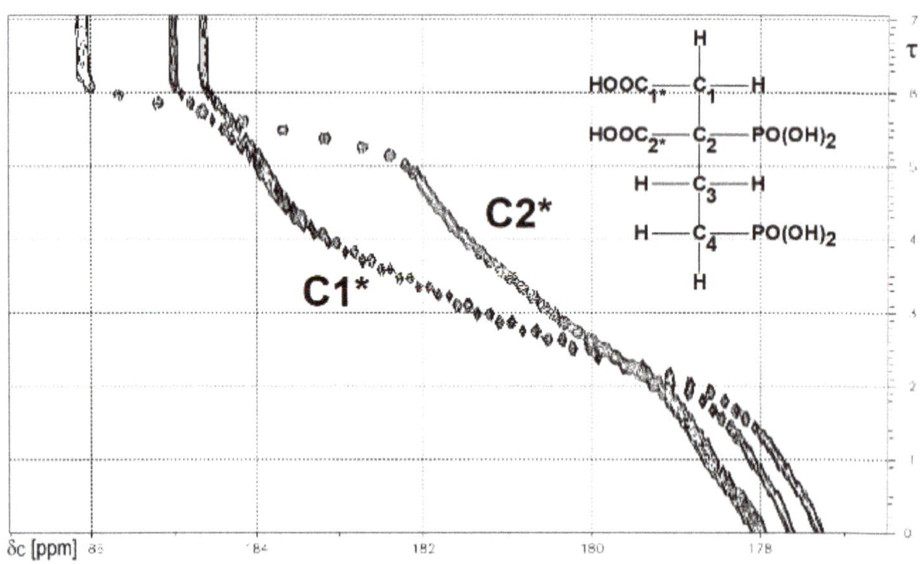

(**a**)

Figure 4. *Cont.*

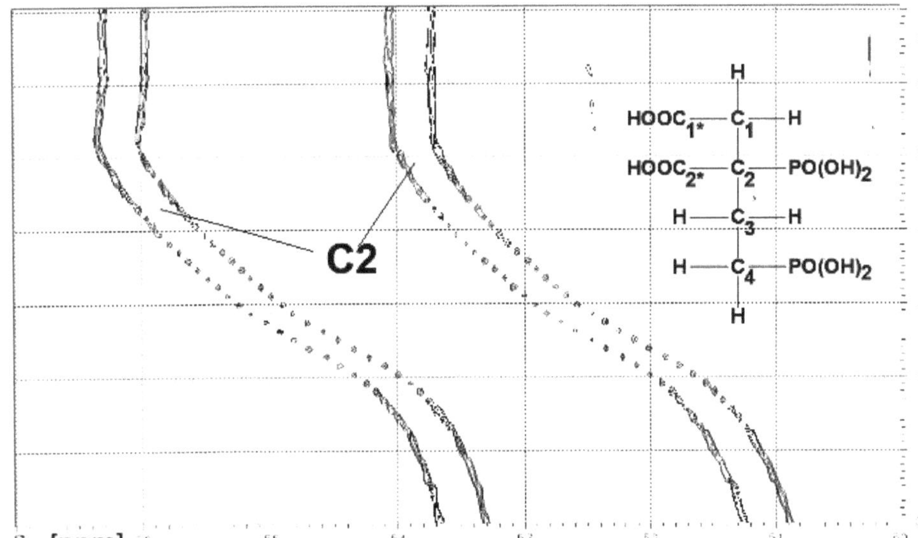

(**b**)

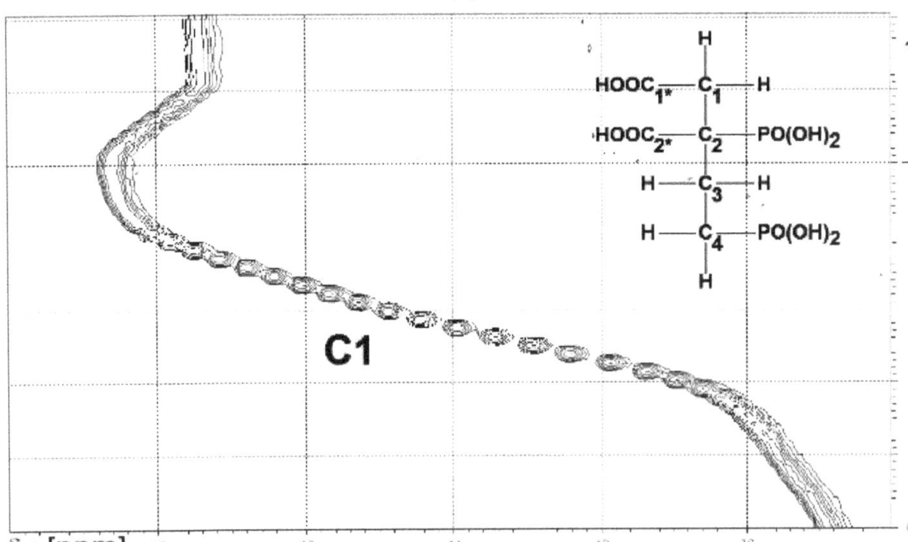

(**c**)

Figure 4. *Cont.*

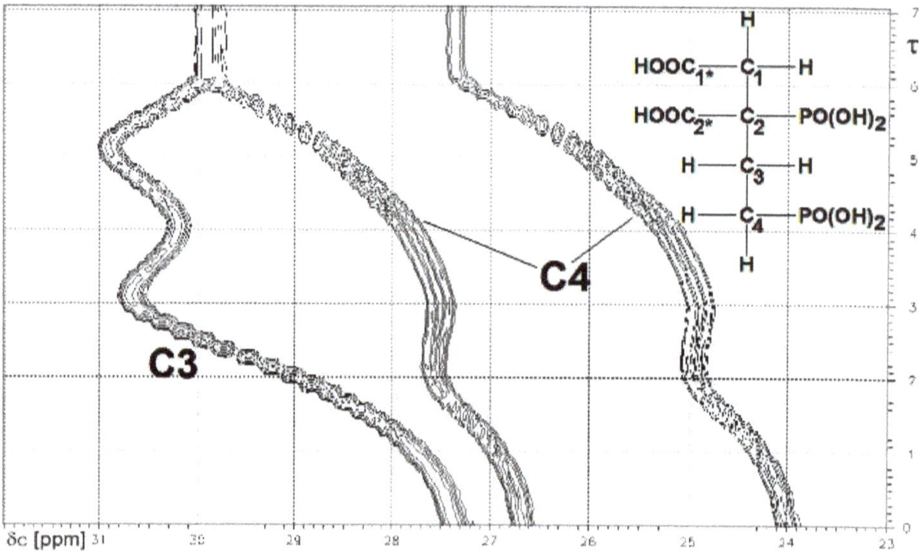

(d)

Figure 4. (**a**) The carboxylate functions C1* (note: $^{3}J_{PC}$) and C2*. δ_C = 186 to 177 ppm; (**b**) The skeleton carbon C2 (note: $^{1}J_{PC}$ and $^{3}J_{PC}$). δ_C = 56 to 50 ppm.; (**c**) The skeleton carbon C1. δ_C = 44 to 38 ppm; (**d**) The skeleton carbons C3 and C4 (note: $^{1}J_{PC}$). δ_C = 32 to 23 ppm.

Table 16. Specific chemical shifts δ_C [ppm], coupling constants $^{n}J_{PC}$ [Hz] and gradients Δ_i [ppm] of 2,4-diphosphonobutane-1,2-dicarboxylic acid DPBDC **9**. For experimental data, see Table 15.

	C1	C2			C3	C4			C1*		C2*
Species	δ_C	δ_C	$^{1}J_{PC}$	$^{3}J_{PC}$	δ_C	δ_C	$^{1}J_{PC}$	$^{3}J_{PC}$	δ_C	$^{3}J_{PC}$	δ_C
H_6L	38.29	52.19	124.2	18.4	27.26	25.11	134.0	4.7	177.23	17.5	177.53
H_5L^{-}	38.70	52.42	116.8	18.2	27.51	25.56	132.4	5.1	177.82	15.9	178.66
H_4L^{2-}	39.38	52.98	116.9	17.3	29.09	26.34	131.3	6.3	178.85	12.5	179.18
H_3L^{3-}	41.78	54.06	116.7	16.8	30.95	26.15	131.3	9.0	181.69	6.4	180.70
H_2L^{4-}	43.28	54.66	115.9	16.6	29.98	26.54	130.3	9.9	183.46	4.3	181.53
HL^{5-}	43.26	55.04	115.8	15.9	30.99	27.51	128.3	9.4	184.08	5.3	182.53
L^{6-}	42.70	55.01	111.7	16.3	29.87	28.65	127.7	-	184.82	18.9	186.15
Gradients											
Δ_1	+0.41	+0.23	−7.4	−0.2	+0.25	+0.45	−1.6	+0.4	+0.59	−1.6	+1.13
Δ_2	+0.68	+0.56	+0.1	−0.9	+1.58	+0.78	−1.1	+1.2	+1.03	−3.4	+0.52
Δ_3	+1.40	+1.08	−0.2	−0.5	+1.86	−0.19	0.0	+2.7	+2.84	−6.1	+1.52
Δ_4	+1.50	+0.60	−0.8	−0.2	−0.97	+0.39	−1.0	+0.9	+1.77	−2.1	+0.83
Δ_5	−0.02	+0.38	−0.1	−0.7	+1.01	+0.97	−2.0	−0.5	+0.62	+1.0	+1.00
Δ_6	−0.56	−0.03	−4.1	+0.4	−1.12	+1.14	−0.6	-	+0.74	+13.6	+3.62

Some Comments on DPBDC 9

A complicated example for NMR-controlled titration which needs some discussion is 2,4-diphosphonobutane-1,2-dicarboxylic acid DPBDC **9**. Measurements and data evaluation were performed according to the state of technique. But it is not possible to explain all the parameters found for compound **9** by comparison with data from analogous structural elements of HOOC-(CH_2)n-PO_3H_2 (n = 1 to3) **7b** to **7d**, H_2O_3P-$(CH_2)_3$-PO_3H_2, and phosphonopolycarboxylic acids **10** to **12** shown in Scheme 8:

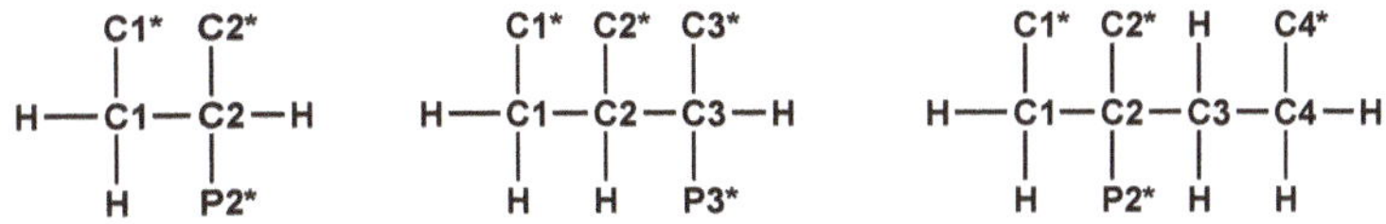

PBC **10** PPTC **11** PBTC **12**

Scheme 8. Phosphonopolycarboxylic acids **10** to **12** used for comparative $^{13}C\{^1H\}$-NMR-controlled titrations.

In a starting phase, 1D and 2D NMR techniques involving ^1H-, $^1H\{^{31}P\}$-, ^{13}C-, $^{13}C\{^1H\}$-, and C,H-COSY spectra were combined to assign the carbons C1*, C2*, C1 to C4 and phosphonate functions P2* and P4*.

For $^{13}C\{^1H\}$-NMR-controlled titration, the deprotonation steps may be divided into three sections (see Table 15 and Figure 4a,d). For $\tau = 0$ to 2 deprotonation $PO_3H_2 \rightarrow PO_3H^-$ takes place, first at P2* and then at P4*. In the second section for $\tau = 2$ to 4, the carboxylic units C1* and C2* are deprotonated. Finally for $\tau = 4$ to 6 the deprotonation $PO_3H^- \rightarrow PO_3{}^{2-}$ takes place at P2* and P4*.

(1) Comments on Chemical Shifts δ_C of Carbon Atoms in DPBDC **9**

The deprotonation of PO_3H_2, PO_3H^- and COOH functions in DPBDC **9** leads to a monotonous down field shift for δ_C C1* and C2* (see Figure 4a), while carbons C1 to C4 exhibit specific non-monotonous trends (see Figure 4b,d).

Since gradient Δ_6 for δ_C (C1*) > Δ_5 for δ_C (C1*), the final sixth deprotonation steps is affiliated with P2*. This conclusion is confirmed by Δ_6 for δ_C (C2*) >> Δ_5 for δ_C (C1*). Hence, the fifth deprotonation step of **9** is due to $PO_3H^- \rightarrow PO_3{}^{2-}$ of P4*. Dynamic chemicals shifts δ_C of C1* span a range from 177.5 to 184.61 ppm, while δ_C of C2* is found from 178 to 185.1, as shown in Figure 4a.

A tentative explanation may be found using Δ_3 for δ_C (C1*) > Δ_4 for δ_C (C1*) and Δ_3 for δ_C (C2*) < Δ_4 for δ_C (C1*). These findings imply that the carboxylic function C1* is more acidic than C*2.

Similar arguments for the relative acidity of C1* and C2* may be derived from the chemical shift δ_C of the skeleton carbon C2 (see Figure 4b). δ_C (C2) of H_6L corresponds to 52.2 ppm, while the totally deprotonated form L^{6-} is found at 55 ppm. Deprotonation at C1* and C2* is characterized again by Δ_3 of δ_C (C2) > Δ_4 of δ_C (C2).

Chemical shifts δ_C of C3 span a range of 38.4 to 43.3 ppm. Surprisingly, the final deprotonation $HL^{5-} \rightarrow L^{6-}$, due to $PO_3H^- \rightarrow PO_3{}^{2-}$ of P2* reduces δ_C (C3) from 43.26 to 42.70 ppm. This is the first observation (within this context) of a negative gradient ($\Delta_6 = -0.56$ Hz) connected to deprotonation at a PO_3H^- unit.

The situation is even more complex for the chemical shift δ_C of C2 covering a range from 27.4 to 30.9 ppm (Figure 4d). Two negative gradients are observed: $\Delta_4 = -0.97$ ppm for $H_3L^{3-} \rightarrow H_2L^{4-}$ due to deprotonation at C2* and $\Delta_6 = -1.12$ ppm for $HL^{5-} \rightarrow L^{6-}$ induced by deprotonation at P2*. For simpler compounds CH_3-$(CH_2)_n$-COOH (n = 0 to 3) and HOOC-$(CH_2)_n$-COOH (n = 1 to 3), solely positive gradients were described [54].

In addition, we did not observe negative gradients for compounds **7b** to **7d** and **10** (PBC), but in **11** (PPTC) and in **12** (PBTC) [43].

The major RR/SS diastereomer of PPTC **11** exhibited a negative gradient Δ_5 (C1) = −0.44 ppm for the final deprotonation step $PO_3H^- \rightarrow PO_3{}^{2-}$ at P3*. An upfield shift occurred, since δc (C1) of HL^{4-} = 42.17 ppm and δc (C1) of L^{5-} = 41.73 ppm. This effect might be due to opening of hydrogen bridges and conformational changes. In contrast here to is the minor RR/SS diastereomer of PPTC, it does not show a negative gradient Δ_5 (C1) [42,44].

Weaker negative gradients $\Delta_5 = -0.22$ ppm are observed for chemical shifts δ_C of both carbons C1 and C3 in PBTC **12.** The final deprotonation $PO_3H^- \rightarrow PO_3{}^{2-}$ at P2* is affiliated with following data:

δc (C1) of HL^{4-} = 43.54 ppm, δc (C1) of L^{5-} = 43.32 ppm, and δc (C3) of HL^{4-} = 32.60 ppm, δc (C3) of L^{5-} = 32.60 ppm. In contrast hereto carbons C2 and C4 in PBTC **12** exhibit positive gradients Δ_5.

Those unexpected observations for chemical shifts δc in **9** and conformational aspects will be mentioned in the following section on coupling constants $^nJ_{PC}$ as well.

(2) Comments on Coupling Constants nJPC (n = 1 to 3) of DPBDC **9**

The vicinal coupling $^3J_{PC}$ (P2*C1*) is remarkably sensitive towards the protonation state (see Figure 5):

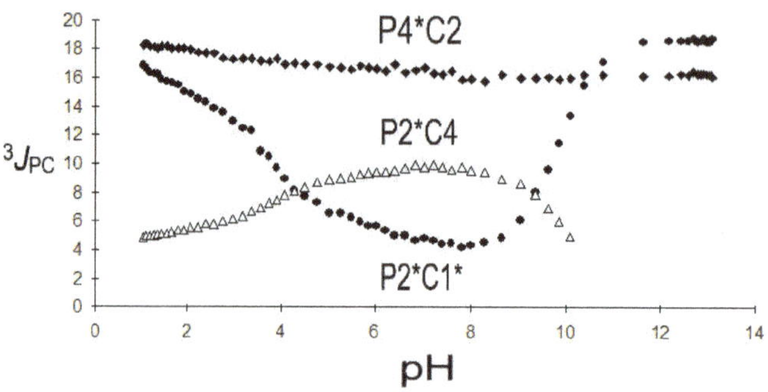

Figure 5. Three coupling constants $^3J_{PC}$ (P2*C1*), $^3J_{PC}$ (P2*C4), and $^3J_{PC}$ (P4*C12) in DPBDC **9**.

For the protolytic species H_6L to H_2L^{4-} of **9**, a decrease in $^3J_{PC}$ (P2*C1*) from 17.5 Hz down to a minimum of 4.3 Hz is observed, followed by an increase from 5.3 Hz to 18.6 Hz due to HL^{5-} and finally L^{6-}. Between pH = 7 and 8, a maximum of the protolytic species H_2L^{4-} is expected, while HL^{5-} dominates around pH = 9. Those observations indicate changes of the dihedral angle of P2*-C2-C1-C1* possibly involving hydrogen bridges as indicated by Scheme 9 below:

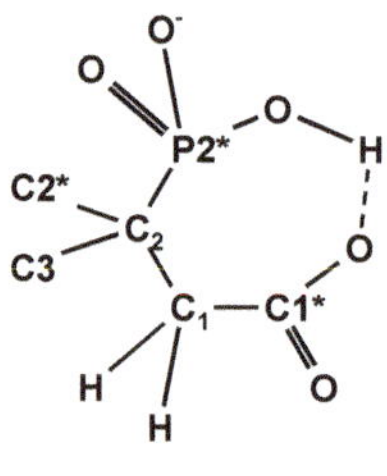

Scheme 9. Tentative hydrogen bridges for protolytic species H_3L^{3-} to HL^{5-}.

A corresponding bridge -C1-P-O—H—O-P-C2- was discussed for the HL^{3-} species of ethane-1,2-bisphosphonic acid [44].

$^2J_{PC}$ (P2*C1), $^2J_{PC}$ (P2*C3), $^2J_{PC}$ (P2*C2*), and $^2J_{PC}$ (P4*C3) were not resolved in $^{13}C\{^1H\}$-NMR spectra obtained by NMR-controlled titrations.

$^3J_{PC}$ (P2*C4) shows a monotonous increase from 4.7 Hz to a maximum of 9.9 Hz for the sequence H_6L to H_2L^{4-} followed by a decrease to 9.4 Hz for HL^{5-}. This observation points towards an increase in the dihedral angle in P2*-C2-C3-C4.

$^3J_{PC}$ (P4*C2) is less sensitive towards deprotonation but larger than $^3J_{PC}$ (P2*C4) and found in a range from 18.4 to 16.3 Hz possibly indicating a tendency towards trans-conformation of the fragment P4*-C4-C3-C2. For comparison, $^3J_{PC}$ in HOOC-$(CH_2)_3$-PO_3H_2 **7d** appeared in a corresponding range from 17.2 to 17.9 Hz.

$^1J_{PC}$ (P2*C2), ranging from 124.2 to 111.7 Hz, is markedly smaller than $^1J_{PC}$ (P4*C4), which is observed from 134.0 to 127.7 Hz. A $^1J_{PC}$, pH diagram is given in Figure 6 below:

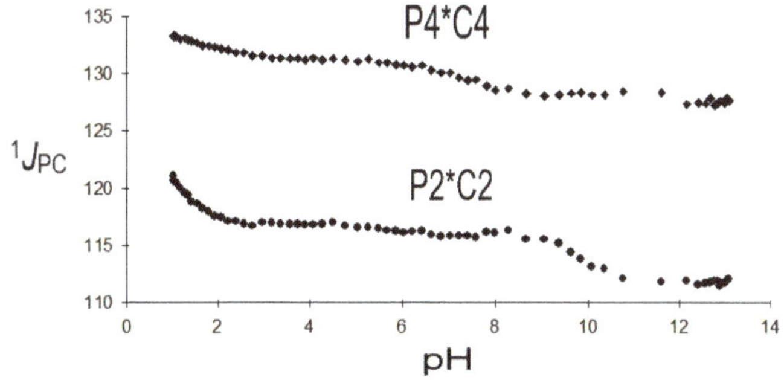

Figure 6. $^1J_{PC}$ (P2*C2) and $^1J_{PC}$ (P4*C4) of DPBDC **9**.

$^1J_{PC}$ (P2*C2) is very indicative and selective for the deprotonation of the phosphonic functions $PO_3H_2 \rightarrow PO_3H^-$ and $PO_3H^- \rightarrow PO_3^{2-}$. It indicates that the first deprotonation ($pK_1 = 1.07$) of DPBDC **9** takes place at P2* with a strong gradient Δ_1 $^1J_{PC}$ (P2*C2) = −7.4 Hz. The final deprotonation ($pK_6 = 11.62$) is affiliated with P2* as well as indicated by Δ_6 $^1J_{PC}$ (P2*C2) = −4.1 Hz. This assignment leaves $pK_2 = 2.73$ and $pK_5 = 8.95$ to the deprotonation of P4*. Deprotonation at the carboxylic groups ($pK_3 = 4.62$) and $pK_4 = 7.05$) does not significantly affect $^1J_{PC}$ (P2*C2) and $^1J_{PC}$ (P4*C4).

Results on $^1J_{PC}$ (P2*C2) of **9** are consistent with observations on phosphonosuccinic acid **10**, where $^1J_{PC}$ (P2*C2) is found in a range from 133.7 to 112.1 Hz. Strong gradients $\Delta_1 = -17.1$ Hz and $\Delta_4 = -4.5$ Hz are affiliated with the deprotonation of P2*, while deprotonations of C1* and C2* do not significantly influence $^1J_{PC}$ (P2*C2).

3. Conclusions

Automated NMR-controlled titrations efficiently combine ^{1}H, ^{13}C-, ^{19}F-, and ^{31}P-NMR spectroscopy with potentiometric titrations to determine dissociation constants, specific chemicals shifts and coupling constants. Results are presented in two-dimensional plots, where NMR parameters (chemical shifts, coupling constants) are correlated with analytical parameters (pH, degree of titration τ). High digital resolution and high S/N are achieved in time- and material-saving measurements. These hyphenated techniques are powerful instruments to identify the structure and purity of research and industrial compounds. Limitations of accuracy due to the nature of glass electrodes occur at very low and very high pH values, obscuring the lower and higher pK_i values. In those situations, single sample NMR methods are recommended. ^{13}C{^{1}H}-NMR-controlled titrations may be used for conformational analysis. For more complicated structures, additional studies using pH-dependent high-resolution ^{1}H- and ^{1}H{^{31}P}-NMR spectra, X-ray diffraction of selected salts and molecular modelling of acids and anions are needed to solve details of conformational problems. The latter topics are laborious and beyond the scope of this paper.

4. Experimental

Details of the hardware and software used in NMR-controlled titration together with references for this field are described in [31,32]. Most model compounds were obtained from external sources listed under the acknowledgements.

Funding: Our research was supported by Fonds der Chemischen Industrie e. V. (VCI), Frankfurt am Main, Germany, and by the German-Israeli Foundation (GIF), Jerusalem, Israel, Research Grant I-316-186-05/93. Material and technical help was obtained from BRUKER Spectrospin, Rheinstetten, Germany, and from SCHOTT Geräte GmbH, Hofheim, Germany.

Acknowledgments: α-Ala-P **1** and β-Ala-P **2** were obtained from R. Tyka and P. Mastalerz, Wroclaw, Poland. E. Breuer, Jerusalem, Israel, provided $LiOOC\text{-}CH_{2\text{-}n}F_n\text{-}PO_3Li_2$ (n = 0 to 2). Technical compounds were obtained from BAYER AG, Leverkusen, Germany, (PBS, PPTC, PBTC, DPBDC, $CH_3\text{-}CH_{2\text{-}n}F_n\text{-}PO(O)(OMe)_2$), HENKEL AG, Düsseldorf, Germany (PPTC), BOZZETTO, Bergamo, Italy (NTMP, EDTMP, DETPMP), and via K. Kellner, Halle, Germany (FOSCARNET). T. Clark, Erlangen, Germany, kindly provided VAMP 4.4 for CONVEX 220 (1989). Special thanks are due to members of the Düsseldorf team, M. Grzonka, M. Batz, J. Peters, A. Bier, H. J. Majer, J. Ollig, I. Reimann, C. Arendt, S. Hermens, and Z. Szakács, for dedicated cooperation and discussions concerning preparative, analytical and NMR studies including development of novel hardware and software.

Conflicts of Interest: The author declares no conflict of interests.

References and Notes

1. Van Wazer, J.R.; Callis, C.F.; Shoolery, J.N.; Jones, R.C. Principles of Phosphorus Chemistry. II. Nuclear Magnetic Resonance Measurements. *J. Am. Chem. Soc.* **1956**, *78*, 5715–5726. [CrossRef]
2. Crutchfield, M.M.; Callis, C.F.; Irani, R.R.; Roth, G.C. Phosphorus Nuclear Magnetic Resonance Studies of Ortho and Condensed Phosphate. *Inorg. Chem.* **1962**, *1*, 813–817. [CrossRef]
3. Hagen, R.; Roberts, J.D. Nuclear Magnetic Resonance Spectroscopy. ^{13}C Spectra of Aliphatic Carboxylic Acids and Carboxylate Anions. *J. Am. Chem. Soc.* **1969**, *91*, 4504–4514. [CrossRef]
4. Heubel, P.-H.; Popov, A.I.J. Acid Properties of Some Phosphonocarboxylic Acids. *Sol. Chem.* **1979**, *8*, 615–625. [CrossRef]
5. Appleton, T.G.; Hall, R.J.; Harris, A.D.; Kimlin, H.A.; McMahon, I.J.N.M.R. Study of Acid-Base Equilibria of Aminoalkylphosphonic Acids, $^{+}NH_3(CH_2)_nPO_3H^{-}$ (n = 1, 2, 3); Evidence for Cyclization in Solution. *Austr. J. Chem.* **1984**, *37*, 1833–1840. [CrossRef]
6. Appleton, T.G.; Hall, R.J.; McMahon, I.J. NMR Spectra of Iminobis(methylenebisphosphonic acid), $HN(CH_2PO_3H_2)_2$ and Related Ligands and of Their Complexes with Platinum(II). *Inorg. Chem.* **1985**, *25*, 726–734. [CrossRef]
7. Sawada, K.; Araki, T.; Suzuki, T. Complex Formation of Amino Polyphosphonates. 1. Potentiometric and Nuclear Magnetic Resonance Studies of Nitrilotris(methylenephosphonate) Complexes of the Alkaline Earth-Metal Ions. *Inorg. Chem.* **1987**, *26*, 1199–1204. [CrossRef]
8. Sawada, K.; Araki, T.; Suzuki, T.; Doi, K. Complex Formation of Amino Polyphosphonates. 1. Stability and Structure of Nitrilotris(methylenephosphonate) Complexes of the Divalent Transitions-Metal Ions in Aqueous Solution. *Inorg. Chem.* **1989**, *28*, 2687–2688. [CrossRef]
9. Sawada, K.; Kanda, T.; Naganuma, Y.; Suzuki, T. Formation and Protonation of Aminopolyphosphonate Complexes of Alkaline-earth and Divalent Transition-metal Ions in Aqueous Solution. *J. Chem. Soc. Dalton Trans.* **1993**, *17*, 2558–2562. [CrossRef]
10. Sawada, K.; Myagawa, T.; Sakaguchi, T.; Doi, K. Structure and Thermodynamic Properties of Aminopolyphosphonate Complexes of the Alkaline-earth Metal Ions. *J. Chem. Soc. Dalton Trans.* **1993**, *24*, 3777–3784. [CrossRef]
11. Sawada, K.; Ichikawa, T.; Uehara, K. Eight-membered chelate-ring complexes of cobalt(III)-polyamine complexes of aminopolyphosphonates in aqueous solution. *J. Chem. Soc. Dalton Trans.* **1996**, *14*, 3077–3085. [CrossRef]
12. Matczak-Jon, E.; Kurzak, B.; Kamecka, A.; Sawka-Dobrowolska, W.; Kafarski, P.; Lejjczak, B. Zink(II) complexes of phosphonic acid analogues of glutamic acid. *J. Chem. Soc. Dalton Trans.* **1996**, 3455–3464. [CrossRef]
13. Matczak-Jon, E.; Kurzak, B.; Kamecka, A.; Sawka-Dobrowolska, W.; Kafarski, P. Interactions of zinc(II), magnesium(II) and calcium(II) with iminodimethylenediphosphonic acids in aqueous solutions. *J. Chem. Soc. Dalton Trans.* **1999**, *20*, 3627–3637. [CrossRef]
14. Rohovec, J.; Kývala, M.; Vojtišek, P.; Hermann, P.; Lukeš, I. Synthesis, Crystal Structures, and Solution Properties of N-Methylene(phenyl)phosphinic Acid Derivatives of Cyclen and Cyclam. *J. Inorg. Chem.* **2000**, 195–203. [CrossRef]

15. Popov, K.; Popov, A.; Rönkkömäki, H.; Lajunen, L.H.J.; Hannu-Kuure, M.; Vendilo, A.; Tsirul'nikova, N. ^{31}P, ^{23}Na and ^{133}Cs NMR equilibrium study of iminobis(methylenephosphonic acid) complexes with alkali metals. *Inorg. Chimica Acta* **2002**, *344*, 1–6. [CrossRef]
16. Popov, A.; Rönkkömäki, H.; Popov, K.; Lajunen, L.H.J.; Vendilo, A. ^{31}P NMR protonation equilibria study of iminobis(methylenphosphonic) acid and its derivatives at high pH. *Inorg. Chimica Acta* **2003**, *353*, 1–7. [CrossRef]
17. Ohms, G.; Grossmann, G. Über die pH-Abhängigkeit der ^{31}P- und ^{13}C-NMR-Spektren von Cyclohexan-, Cyclohexen- und Benzenphosphonsäuren. *Z. Anorg. Allg. Chem.* **1987**, *544*, 232–240. [CrossRef]
18. Rönkömäki, H.; Jokisaari, H.; Lajunen, L.H. ^{31}P NMR and Potentiometric Studies on the Protonation of Isopropyl Esters of Chlodronic Acid. *Acta Chem. Scand.* **1993**, *47*, 331–337. [CrossRef]
19. Popov, K.; Niskanen, E.; Rönkkömäki, H.; Lajunen, L.H.J. ^{31}P NMR Study of organophosphonate protonation equilibrium at high pH. *New J. Chem.* **1999**, *23*, 1209–1213. [CrossRef]
20. Szakács, Z.; Hägele, G. Accurate determination of low pK values by ^{1}H NMR titration. *Talanta* **2004**, *62*, 819–825. [CrossRef]
21. Szakács, Z.; Hägele, G.; Tyka, R. $^{1}H/^{31}P$ NMR pH indicator series to eliminate the glass electrode in NMR spectroscopic pK_a determinations. *Anal. Chim. Acta* **2004**, *522*, 247–258. [CrossRef]
22. Popov, K.; Rönkkömäki, H.; Lajunen, L.H. Guidelines for The NMR Measurements for Determination of High And Low pK_a Values. *Pure Appl. Chem.* **2006**, *78*, 663–675. [CrossRef]
23. Yesinowski, J.P.; Sunberg, R.J.; Benedict, J.J. pH Control and Rapid Mixing in Spinning NMR Samples. *J. Magn. Res.* **1982**, *47*, 85–90. [CrossRef]
24. Glaser, J.; Henriksson, U.; Klason, T. A 205 TL NMR Titration Study of the Complex Formation between Tl(I) and Cl- in Aqueous Solution. *Acta Chem. Scand.* **1986**, *A40*, 344–349. [CrossRef]
25. Li, W. Gravity-driven pH adjustment for site-specific protein pK_a measurement by solution-state NMR. *Meas. Sci. Technol.* **2017**, *28*. [CrossRef]
26. Hägele, G.; Grzonka, M.; Kropp, H.-W.; Ollig, J.; Spiegl, H. Phosphonic and phosphinic acids: Monitoring protolytic and complex formation equilibria by titration dependent stopped-flow-NMR-techniques. *Phosphorus Sulfur Silicon* **1993**, *77*, 85–88. [CrossRef]
27. Hägele, G.; Varbanov, S.; Ollig, J.; Kropp, H.-W. Aminomethylphosphine Oxides: Synthesis, Dissociation and Stability Constants, $^{31}P\{^{1}H\}$-NMR-controlled Titration. *Z. Anorg. Allg. Chemie* **1994**, *620*, 914–920. [CrossRef]
28. Hägele, G. ^{31}P NMR controlled titrations of Phosphorus-Containing Acids and Bases in Protolysis and Complex Formation. In *Phosphorus-31 NMR Spectral Properties in Compound Characterization and Structural Analysis*; Quin, L.D., Verkade, J.G., Eds.; VCH Publishers: New York, NJ, USA, 1994; Chapter 30; pp. 395–409.
29. Hägele, G.; Ollig, J. NMR-controlled titrations of phosphorus containing acids and bases. *Comput. Chem.* **1995**, *19*, 287–294.
30. Ahrendt, C.; Hägele, G. The Photo-T-Concept: Hard- and software combination for the determination of macroscopic and microscopic dissociation constants. *Comput. Chem.* **1995**, *19*, 263–268. [CrossRef]
31. Hägele, G.; Szakács, Z.; Ollig, J.; Hermens, S.; Pfaff, C. NMR-Controlled Titrations: Characterizing Aminophosphonates and Related Structures. *Heteroat. Chem.* **2000**, *11*, 562–582. [CrossRef]
32. Hägele, G.; Grzonka, M.; Peters, J.; Spiegl, H.; Kropp, H.W.; Ollig, J.; Hermens, S.; Augner, S.; Uhlemann, C.; Pfaff, C.; et al. NMR-Controlled Titrations—Principles and Progress: Monitoring Protonation and Complex Formation Equilibria in Aqueous Solutions. Available online: https://www.theresonance.com/nmr-controlled-titration-download-the-paper/ (accessed on 2 September 2019).
33. Pfaff, C.G. Softwareentwicklung zur Auswertung und Visualisierung Kernresonanzspektroskopisch Kontrollierter Titrationen. Ph.D. Thesis, Heinrich-Heine-Universität, Düsseldorf, Germany, 2002.
34. Peters, M.; Siegfried, L.; Kaden, T.A. pH-metric and NMR studies of complexation of Zn^{2+}, Cd^{2+} and Pb^{2+} with diazacrown ethers having dangling phosphonate groups. *J. Chem. Soc. Dalton Trans.* **2000**, *24*, 4664–4668. [CrossRef]
35. Rabenstein, D.L.; Sayer, T.L. Determination of Microscopic Acid Dissociation Constants by Nuclear Magnetic Resonance Spectrometry. *Anal. Chem.* **1976**, *48*, 1141–1146. [CrossRef]
36. Szakacs, Z.; Kraszni, M.; Noszal, B. Determination of microscopic acid-base parameters from NMR-pH titrations. *Anal. Bioanal. Chem.* **2004**, *378*, 1428–1448. [CrossRef]

37. Szakacs, Z. NMR-Titrationen und neue Auswertekonzepte zur Aufklärung der Mikroskopischen Dissoziation und der pH-abhängigen Konformation von Biorelevanten Phophinsäuren und Carnosinderivaten. Ph.D. Thesis, Heinrich-Heine-Universität, Düsseldorf, Germany, 2002.
38. Kývala, M.; Lukeš, I. Chemometrics 95. In *Book of Abstracts, Proceedings of the 4th International Chemometrics Conference of the Czech Chemical Society, Pardubice, Czech Republic, 3–7 July 1995*; University of Pardubice: Pardubice, Czech Republic, 1995; p. 63.
39. Augner, S.; Kehler, J.; Szakács, Z.; Breuer, E.; Hägele, G. Ring-chain tautomerism and protolytic equilibria of 3-hydroxy-3-phosphonoisobenzo-furanone studied by ^{1}H-, ^{13}C-, and ^{31}P-NMR-controlled titrations. *New J. Chem.* **2008**, *32*, 1608–1616. [CrossRef]
40. Augner, S. Neue Entwicklungen zur Durchführung automatisierter NMR-Kontrollierter Titrationen von Phosphon- und Phosphinsäuren. Ph.D. Thesis, Heinrich-Heine-Universität, Düsseldorf, Germany, 2002.
41. Kropp, H.-W.; Hägele, G. Unpublished results from my student Kropp. Interested readers may contact G. Hägele. Unpublished work. 1994.
42. Lindner, A. 1-Phosphonopropan-1,2,3-tricarbonsäure—NMR- und Konformationsanalytische Untersuchungen. Ph.D. Thesis, Heinrich-Heine-Universität, Düsseldorf, Germany, 2000.
43. Hägele, G. NMR-controlled titrations characterizing organophosphorus compounds. *Phosphorus Sulfur Silicon Relat. Elem.* **2019**, *194*, 361–363.
44. Ollig, J. Untersuchungen zur Titrationsabhängigen Kernresonanzspektroskopie. Ph.D. Thesis, Heinrich-Heine-Universität, Düsseldorf, Germany, 1996.
45. Robitaille, P.-M.L.; Robitaille, P.A.; Brown, G.G., Jr.; Brown, G.G. An Analysis of the pH-Dependent Chemical-Shift Behavior of Phosphorus-Containing Metabolites. *J. Magn. Res.* **1991**, *91*, 73–84. [CrossRef]
46. Popov, K.; Rönkkömäki, H.; Lajunen, L.J.J. Critical evaluation of stability constants of phosphonic acids. *Pure Appl. Chem.* **2001**, *73*, 1641–1677. [CrossRef]
47. Clark, T.; Alex, A.; Beck, B.; Chandrasekhar, J.; Gedeck, P.; Horn, A.H.C.; Hutter, M.; Martin, B.; Rauhut, G.; Sauer, W.; et al. *VAMP 4.4*; Universität Erlangen-Nürnberg: Erlangen, Germany, 1990.
48. Wozniak, M.; Nowogrocky, G. Acidites et complexes des acides (alkyl- et aminoalkyl-) phosphoniques—I: Determination potentiometrique des constantes d'acidite par affinement multiparametrique: Prise en compte de l'impurete carbonate. *Talanta* **1979**, *26*, 1135–1141. [CrossRef]
49. Mohan, M.S.; Abbott, E.H. Metal complexes of biologically occurring aminophosphonic acids. *J. Coord. Chem.* **1978**, *8*, 175–182. [CrossRef]
50. Kiss, T.; Balla, J.; Nagy, G.; Kozzlowski, H.; Kowalik, J. Complexes of aminophosphonates. I. Transition metal complexes of aminophosphonic acid analogues of α-alanine, β-alanine, phenylalanine and tyrosine. *Inorg. Chim. Acta* **1987**, *138*, 25–30. [CrossRef]
51. Kropp, H.-W.; Hägele, G. Analytische und NMR-spektroskopische Untersuchungen an Organophosphorsäuren. Unpublished work.
52. Burton, D.J.; Sprague, L.G.; Pietrzyk, D.J.; Edelmuth, S.H. A Facile Synthesis of Difluorophosphonoacetic Acid. *J. Org. Chem.* **1984**, *49*, 3437–3438. [CrossRef]
53. Bier, A. Computereinsatz in der Analytischen Chemie zur Untersuchung von Protolyse- und Komplexbildungs-Gleichgewichten am Beispiel der Phosphonocarbonsäuren. Ph.D. Thesis, Heinrich-Heine-Universität, Düsseldorf, Germany, 1993.
54. Rabenstein, D.L.; Sayer, T.L. Carbon-13 Chemical Shift parameters for Amines, Carboxylic Acids, and Amino Acids. *J. Magn. Res.* **1978**, *24*, 27–39. [CrossRef]

molecules

MDPI

Review

Synthetic Strategies for Dinucleotides Synthesis

Lucie Appy, Crystalle Chardet, Suzanne Peyrottes and Béatrice Roy *

Nucleosides & Phosphorylated Effectors, Institut des Biomolécules Max Mousseron (IBMM), UMR 5247 CNRS, Université de Montpellier, ENSCM, Campus Triolet, cc 1705, Place Eugène Bataillon, CEDEX 5, 34095 Montpellier, France; lucie.appy@gmail.com (L.A.); crystalle.chardet@gmail.com (C.C.); suzanne.peyrottes@umontpellier.fr (S.P.)

* Correspondence: beatrice.roy@umontpellier.fr; Tel.: +33-4-6714-3879

Academic Editor: György Keglevich

Received: 7 November 2019; Accepted: 25 November 2019; Published: 27 November 2019

Abstract: Dinucleoside 5′,5′-polyphosphates (DNPs) are endogenous substances that play important intra- and extracellular roles in various biological processes, such as cell proliferation, regulation of enzymes, neurotransmission, platelet disaggregation and modulation of vascular tone. Various methodologies have been developed over the past fifty years to access these compounds, involving enzymatic processes or chemical procedures based either on P(III) or P(V) chemistry. Both solution-phase and solid-support strategies have been developed and are reported here. Recently, green chemistry approaches have emerged, offering attracting alternatives. This review outlines the main synthetic pathways for the preparation of dinucleoside 5′,5′-polyphosphates, focusing on pharmacologically relevant compounds, and highlighting recent advances.

Keywords: phosphorylation; dinucleotides; organophosphorus chemistry; mechanochemistry; dry eye syndrome; diquafosol; denufosol

1. Introduction

Dinucleoside 5′,5′-polyphosphates (DNPs), commonly abbreviated as Np$_n$Ns, are essential to human biological systems [1,2]. They contain two ribonucleosides, which are linked at the 5′-position of their sugar moiety through n phosphate groups (Figure 1).

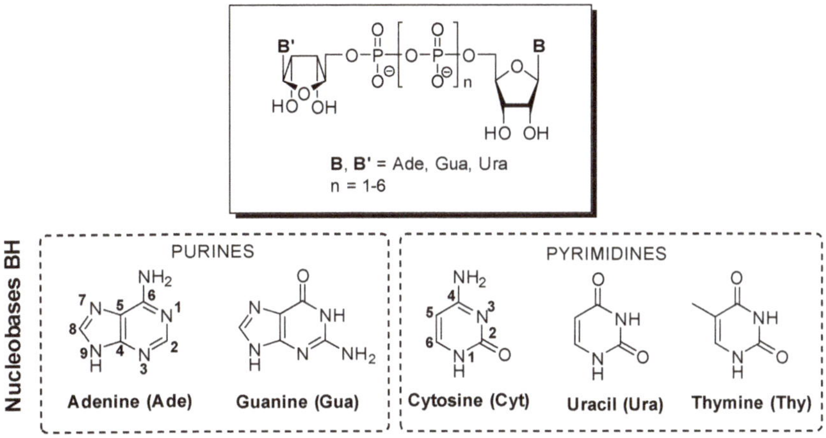

Figure 1. General structure of natural dinucleoside 5′,5-polyphosphates and canonical nucleobases.

Historically, dinucleoside 5′,5′-polyphosphates (i.e., Ap_2A and Up_2U) were first isolated by chemists [3]. In 1976, E. Rapaport and P.C. Zamecnik brought to light the presence of P^1,P^4-diadenosine-5′-tetraphosphate (Ap_4A) in mammals [4]. The authors assumed that this dinucleotide could have a role in cellular communication as a "signal nucleotide". Later, Up_4A was the first endogenous dinucleotide, possessing both purine and pyrimidine moieties, to be identified in living organisms [5,6]. Currently, 17 symmetrical or mixed Np_nNs have been isolated from human tissues and characterized [2]. DNPs are released into the extracellular space from different types of cells, such as platelets and endothelial cells, where they stimulate several cell-surface purinergic receptors. They have a strong physiological and pathophysiological impact on the cardiovascular system [2,7], and may also interact with enzymes, acting as inhibitors (e.g., kinases) or substrates [8–11]. Three classes of enzymes are known to catalyze the degradation of dinucleotides [10–12]: symmetrically cleaving dinucleoside polyphosphate hydrolases, asymmetrically cleaving dinucleoside polyphosphate hydrolases, and dinucleoside polyphosphate phosphorylases. Over the years, DNPs have gained increased attention and their therapeutic potential has been revealed. Likewise, structural analogues have been developed and some of them are drug candidates.

Herein, we first describe purinergic signalling, i.e., the role of nucleotides as extracellular signal molecules, with a focus on the P2Y receptor subtypes that interact with dinucleotides. Then, we present the synthetic strategies to obtain dinucleotides. Pyrophosphate bond formation has been addressed in some recent reviews and book chapters [1,13–15]. The current review focuses on methods reported for the synthesis of dinucleoside 5′,5′-polyphosphates, especially those of biological and pharmacological interest. We also highlight recent advances in the field, such as green chemistry approaches. It should be noted that methods for preparing cyclic dinucleotides [16] are beyond the scope of this review. Furthermore, the synthesis of nicotinamide adenine dinucleotides [17] and mRNA cap analogues such m^7Gp_nN (see reviews [18–20]) are not presented.

2. Interaction of DNPs with Purine and Pyrimidine Receptors

Purinergic receptors are extracellular receptors, currently consisting of four subtypes of P1 (or adenosine) receptors, seven subtypes of P2X ion channel receptors, and eight subtypes of P2Y receptors (Figure 2). Specifically, P2Y receptors (P2YRs) are G-protein-coupled receptors (GPCRs) activated by extracellular nucleotides, which are divided into two groups on the basis of sequence homology and the type of G protein they are primarily coupled to [21–23].

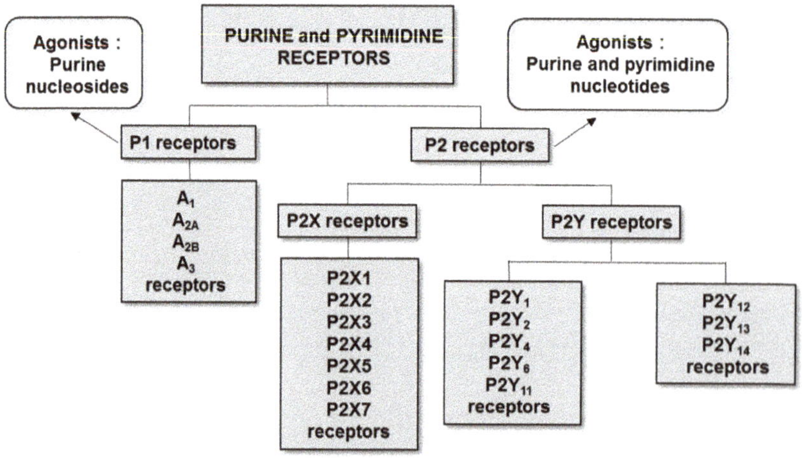

Figure 2. The classification of purine and pyrimidine receptors and their main physiological agonists.

Purinergic receptors are expressed in nearly all cell types, and have been found to play crucial roles in many biological processes, e.g., neurotransmission, neuroprotection in hypoxia and ischemia, regulation of cardiovascular function, platelet aggregation, smooth muscle contraction, secretion of hormones, modulation of immune response, control of cell proliferation, differentiation, and apoptosis. In this regard, different aspects of purinergic signaling such as the pathophysiology and the therapeutic potential have been extensively reviewed [21–28]. Several purinergic compounds are already on the market to treat thrombosis and stroke, and one dinucleotide is used to treat dry eye disease (see Section 2.1).

DNPs are strong endogenous agonists of the purinergic system [2]. In particular, they interact with the P2YRs, a family of high therapeutic relevance [2,23,29,30]. The $P2Y_1$, $P2Y_{11}$, $P2Y_{12}$, and $P2Y_{13}$ receptors are activated by adenosine 5′-di- or triphosphates (ADP and ATP), while uracil nucleotides (UDP and UTP) are the endogenous agonists for the $P2Y_4$, $P2Y_6$, and $P2Y_{14}$ receptors. On the other hand, the $P2Y_2$ receptor responds to both ATP and UTP. Natural dinucleoside tetraphosphates Ap_4A and Up_4U have agonist potencies comparable to those of ATP and UTP at $P2Y_2$ receptors [31–34]. It should be noted that dinucleotides are more resistant to hydrolysis than their parent mononucleotides, but are generally less potent, and lack selectivity in many cases [9,23]. Ligand development for the class of uracil nucleotide-activated P2Y receptors has been extensively reviewed by Rahefi and Müller in 2018 [23]. They have compiled existing data on, inter alia, dinucleotides such as Ap_nA (n = 2–6), Up_nU (n = 3–4) and structural analogues, as agonists for the uracil-activated P2YRs. Given the significant clinical potential of the P2YRs, substantial research efforts directed towards developing P2YR ligands for use as pharmacological tools and drugs have led to the discovery and development of a significant number of agonists but, so far, only a moderate number of antagonists. Two relevant dinucleotides, namely Up_4U and Up_4dC, have undergone clinical development to cure several diseases involving the $P2Y_2$ receptor (Figure 3).

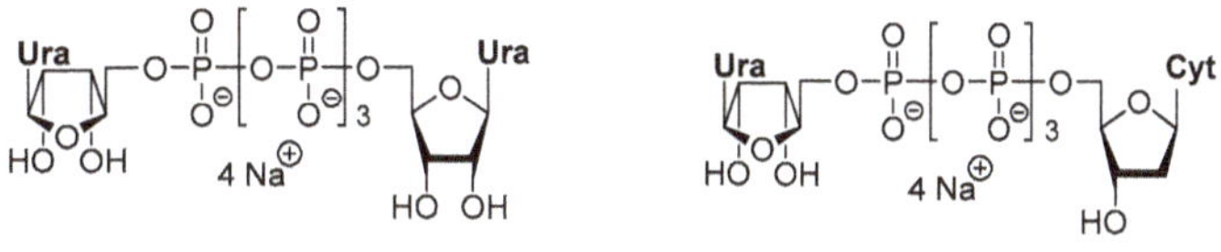

Diquafosol tetrasodium
INS365, Up_4U

Denufosol tetrasodium
INS37217, Up_4dC

Figure 3. Dinucleotides with clinical developments.

2.1. *Up_4U for the Treatment of Dry Eye Disease (DED)*

Diquafosol tetrasodium (*P^1*, *P^4*-diuridine 5′,5′-tetraphosphate, Up_4U, NS365, *Diquas®*) is the first agonist of the $P2Y_2$ receptor subtype that has been approved in Japan as 3% ophthalmic solution (Santen Pharmaceutical Co, Ltd., Osaka, Japan) for the management of dry eye disease (see reviews by Lau et al. [35], and Keating [36]). This pathology commonly causes symptoms including dryness, irritation, itching, and light sensitivity. It is associated with tear film instability, increased tear film osmolarity, and ocular surface inflammation. It impacts the daily life of patients and has a prevalence that varies from 5% to 35%. DED results from either decreased aqueous tear production (aqueous tear-deficient dry eye) or increased tear evaporation (evaporative dry eye), or both. Several studies demonstrated the presence of P2X and P2Y receptors in ocular tissues (retina, ciliary body, and lens), and indicated that $P2Y_2$ receptors may be the main subtype of purinergic receptor located at the ocular surface [33]. Clinical data show that diquafosol tetrasodium improves ocular surface staining and may improve tear film volume and stability.

2.2. Up_4dC for the Treatment of Cystic Fibrosis

Denufosol tetrasodium (P^1-uridine-P^4-2′-deoxycytidine 5′,5′-tetraphosphate, Up_4dC, dCp_4U, INS37217) is a mixed dinucleoside tetraphosphate. Along with Ap_4A and Up_4U, dCp_4U acts as a $P2Y_2$ agonist [29,37]. It has been developed by Inspire Pharmaceuticals (now part of Merck) as an inhaled drug for the treatment of cystic fibrosis [38]. This recessive genetic disease is characterized by pulmonary and reproductive tract dysfunctions, involving abnormal ion transport and defective mucociliary clearance. Denufosol acts on P2Y receptors expressed on the surface of airway epithelia to stimulate chloride secretion independent of the chloride channel, which is dysfunctional in cystic fibrosis. It was shown to significantly enhance tracheal mucus transport in an animal model [39,40]. The first clinical trials established a good safety profile. It was evaluated in two phase 3 clinical trials, TIGER-1 and TIGER-2, as a therapy for cystic fibrosis patients [41,42]. In the first trial (TIGER-1), it was found to significantly improve lung function in cystic fibrosis patients with normal to mildly impaired lung function. Unfortunately, less than 3 weeks after publication of the TIGER-1 data, Inspire Pharmaceuticals announced that the 466-patients, 48-week placebo-controlled phase 3 TIGER-2 clinical trial, had failed to demonstrate any benefit [41,42].

3. Synthesis Based on P(V) Chemistry

To date, the most widely used methods to access dinucleotides involve P(V) chemistry. They are based on the activation of a 5′-nucleotide (i.e., nucleoside 5′-mono, 5′-di and 5′-triphosphate) and the reaction of the corresponding intermediate with a second 5′-nucleotide or inorganic pyrophosphate to form dinucleotides containing up to six bridging phosphate groups. These steps are usually performed in dry, aprotic solvents (mostly *N*,*N*-dimethylformamide, DMF), and, therefore, require the use of nucleotides, as well as inorganic phosphate or pyrophosphate, in their tri- or tetra-*n*-butylammonium forms due to solubility issues. Divalent cations, especially Mg^{2+} and Zn^{2+}, are sometimes added as catalysts for anhydride bond formation. It is assumed that the metal ion could serve two roles, namely, the activation of the electrophilic P(V) center and templating the incoming phosphate nucleophile and the P(V) electrophilic center [13].

3.1. Synthesis via a Phosphoromorpholidate Intermediate

The first description of a chemical synthesis of dinucleotides was reported in the mid-60s by Moffatt and Khorana [43,44]. This strategy relies on the conversion of nucleoside 5′-monophosphates (NMPs) into their phosphoromorpholidate derivatives, followed by reaction with a phosphate salt, i.e., orthophosphate or pyrophosphate (Scheme 1). Briefly, the acidic form of the nucleotides were activated with N,N′-dicyclohexylcarbodiimide (DCC) in the presence of morpholine, leading to the nucleoside 5′-phosphoromorpholidates as 4-morpholine N,N′-dicyclohexylcarboxamide salts **1–3** (Scheme 1) [45]. In the second step, addition of pyrophosphate afforded the symmetrical dinucleoside tetraphosphates. Similarly, dinucleoside triphosphates were obtained when pyrophosphate was replaced by orthophosphate.

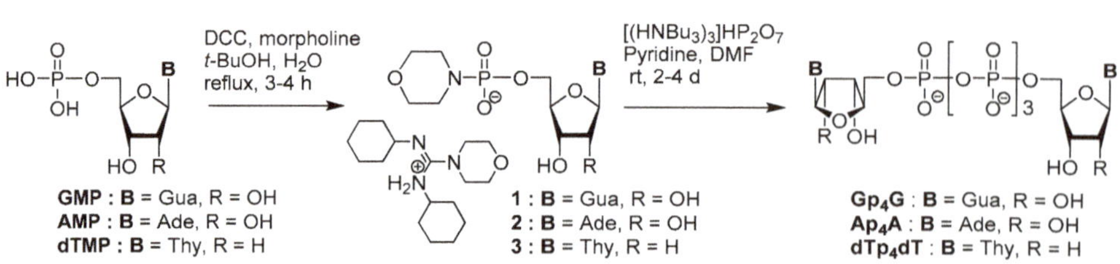

Scheme 1. Synthesis via phosphoromorpholidate intermediates [44].

Since then, this approach has been adapted by several research groups to access Np_2N, Np_3N, as well as some analogues. For example, Ap_2A was obtained in 90% yield, through the condensation of the bis-(tri-*n*-butylammonium) salt of AMP and the adenosine-5′-phosphoromorpholidate-4-morpholine-N,N′-dicyclohexylcarboxamide salt in anhydrous pyridine [46]. Mixed dinucleoside triphosphates were obtained in 10–30% yields, by reacting nucleoside 5′-phosphoromorpholidates with nucleoside 5′-diphosphates (NDPs), possibly in the presence of tetrazole [31,47,48]. However, due to long reaction times and generally low yields, this method has lost its relevance.

3.2. Synthesis via a Phosphorimidazolide Intermediate

Phosphorimidazolide derivatives exhibit high reactivity toward various nucleophiles. Therefore, they have been extensively used as intermediates for pyrophosphate bond formation. Typically, activation of a 5′-nucleotide (NMP, NDP or NTP) with *N,N′*-carbonyldiimidazole (CDI) in DMF, followed by the in situ condensation with a second 5′-nucleotide or pyrophosphate, affords dinucleotides containing two to four bridging phosphate groups in 10–60% yields [47–50]. Accordingly, the tri-*n*-butylammonium salt of UMP was activated with CDI to form a phosphorimidazolide intermediate **4**, which reacted with the remaining nucleotide to form Up_2U (Scheme 2) [47,49]. Slight modifications of this protocol allowed access to mixed dinucleotides [50]. First, activation of the bis(tri-*n*-butylammonium) salt of CMP was performed using ≈ 2 equiv of CDI in DMF for 2 h at rt, followed by the addition of dry methanol to quench the remaining CDI. Treatment with the bis(tri-*n*-butylammonium) salt of GMP for 2 h at rt afforded Gp_2C in 54% yield. Addition of GDP instead of GMP in the second step afforded Gp_3C, albeit in only 3% yield. Similarly, activation of NDPs with CDI followed by reaction with a NMP allowed the isolation of dinucleoside triphosphates in low yields [48].

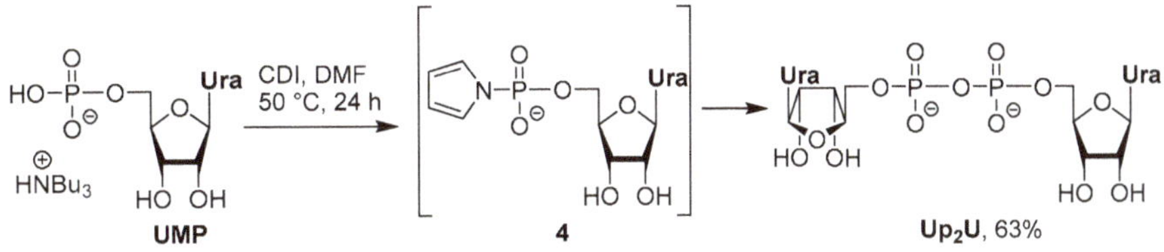

Scheme 2. Activation of UMP by CDI to access Up_2U [47,49].

Alternatively, symmetrical dinucleoside tetraphosphate Up_4U was obtained, either by activation of UMP with CDI followed by coupling with bis(tri-*n*-butylammmonium) pyrophosphate [47,49] or dimerization of UDP in the presence of CDI [51] (Scheme 3).

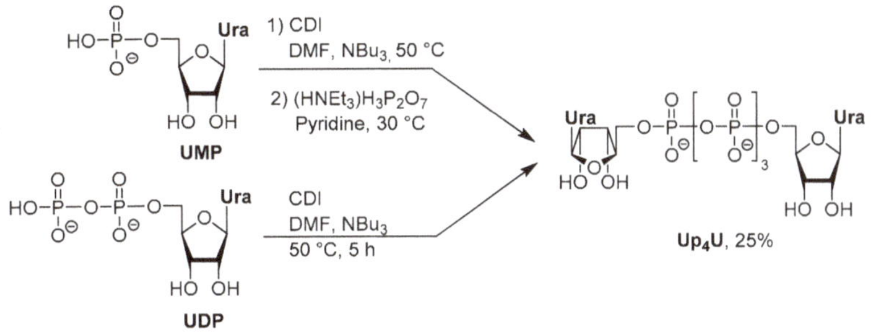

Scheme 3. Dimerization of nucleotides to access symmetrical Up_4U [49,51].

Activation of ribonucleotides in the presence of an excess of CDI usually results in the carbonation of the ribose moiety (Scheme 4). In ^{1}H-decoupled ^{31}P NMR spectroscopy, adenosine 5′-phosphorimidazolide is characterized by a singlet at −7.56 ppm, whereas phosphorimidazolide **5** exhibits a singlet at −7.76 ppm [52]. However, in some of the above-mentioned publications [47,49], carbonation was not observed, and may be due to the experimental conditions (number of equivalents of CDI, reaction time, temperature, adventitious presence of water). Nonetheless, the removal of the 2′,3′-carbonate protecting group can be easily performed under basic conditions [52,53].

Scheme 4. Reaction of AMP with 4 equiv CDI [52]. ^{31}P chemical shifts are indicated in ppm together with their multiplicity.

In 2011, a variant of the CDI method was developed by Yanachkov and co-workers [54], where the activation of pyrophosphate by CDI gave rise to *P^1*,*P^2*-di(1-imidazolyl)pyrophosphate (Scheme 5). Using ^{13}C-labeled CDI and monitoring the reaction by ^{31}P NMR (^{13}C-^{31}P and ^{1}H-^{31}P couplings), the authors proposed a reaction mechanism involving the fast formation of mixed anhydrides, which react slowly with imidazole to give diimidazolyl pyrophosphate. This activated pyrophosphate was isolated as its disodium salt **6** by precipitation with sodium perchlorate in acetone, and then reacted with AMP or UMP to afford the symmetrical 5′,5′-dinucleoside tetraphosphates Ap$_4$A and Up$_4$U in good yields (Scheme 5). Noteworthy, the use of 1*H*-tetrazole or zinc chloride during the second step allowed the reaction time to be shortened from 24–48 h to 3–16 h.

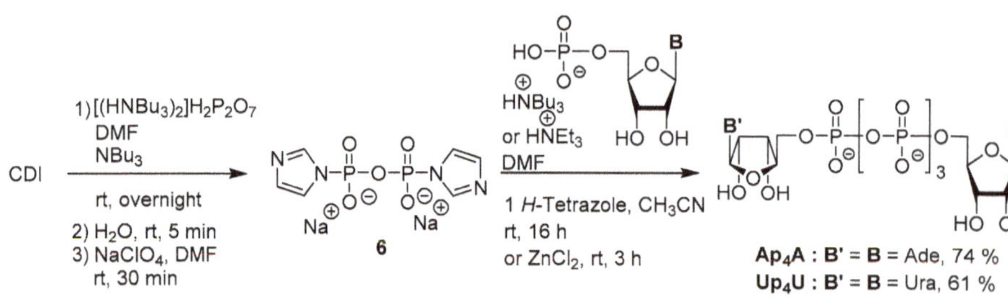

Scheme 5. Np$_4$Ns synthesis using *P^1*,*P^2*-di(1-imidazolyl)pyrophosphate **6** [54].

Interestingly, this approach allows the preparation of dinucleotide analogues, starting from modified pyrophosphate (the central oxygen being replaced with monofluoro, monochloro or dichloromethylene). These DNPs analogues were obtained in 70–80% yields.

While these methods involving phosphoromidazolides are straightforward and rapid, they all require dry conditions to prevent side-product formation and most often, use trialkylammonium salts. To avoid this drawback, our research group has recently developed a one-pot synthesis of dinucleotides, either in aqueous solution or mechanochemically, through the use of phosphorimidazolides intermediates [55,56]. These strategies will be detailed in Section 5 and are related to green chemistry approaches.

3.3. *Synthesis via a Cyclic Trimetaphosphate Intermediate*

The initial synthesis of trimetaphosphate esters from ATP was described by Khorana [57]. Later, Ng and Orgel reported that adenosine 5′-polyphosphates can react with 1-ethyl-3-(3-dim ethylaminopropyl)carbodidimide (EDC) in water to form diadenosine 5′,5′-polyphosphates as the main products [58]. Yields were improved by performing the reaction in organic solvents such as DMF or DMSO, in the presence of DCC and using the tris or tetrakis(tri-*n*-butylammonium) salts of nucleotides [31,47]. The activation of nucleoside 5′-triphosphates (NTPs) with DCC proceeds via a cyclic nucleoside 5′-trimetaphosphate intermediates (**7–8**), which can further react with another NMP to produce the corresponding dinucleoside tetraphosphates (Scheme 6). Accordingly, Up_4U and Up_4A were obtained in low yields starting from UTP and ATP, respectively. One should note that *N,N′*-dicyclohexylurea (DCU) may be removed as a precipitate either after the first or the second step of the reaction.

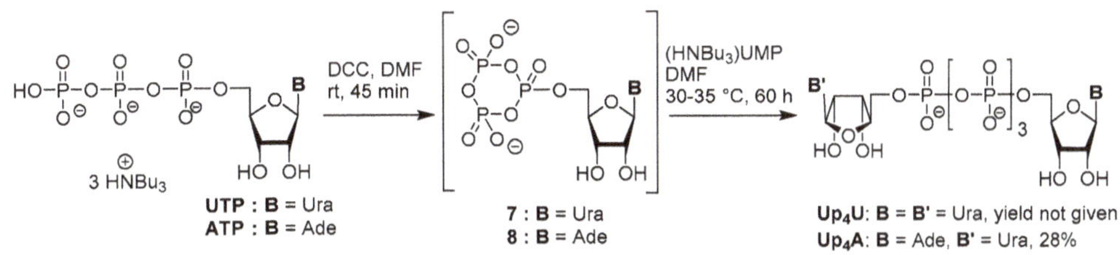

Scheme 6. Synthesis of Np_4Ns starting from nucleoside 5′-triphosphates (NTPs) [47].

Similarly, the activation of UTP or ATP with DCC in anhydrous DMF, followed by the addition of the tris(tri-*n*-butylammonium) salt of UDP or ADP in the second step of the reaction, afforded Up_5U and Ap_5A, in 10–60% yields [8,49,50]. Reactions must be performed under strictly anhydrous conditions and long reaction times are required.

In 2013, the Taylor research group reported a high-yielding synthesis of dinucleoside pentaphosphates (Np_5N) using the tri(tetra-*n*-butylammonium) salt of cyclic trimetaphosphate **9** as a phosphorylating agent [59]. The latter was prepared in almost quantitative yield by conversion of the trisodium salt of trimetaphosphate **10** into its pyridinium form, and then by titration of the solution to pH 7.0 with a dilute solution of tetrabutylammonium hydroxide and freeze-drying (Scheme 7).

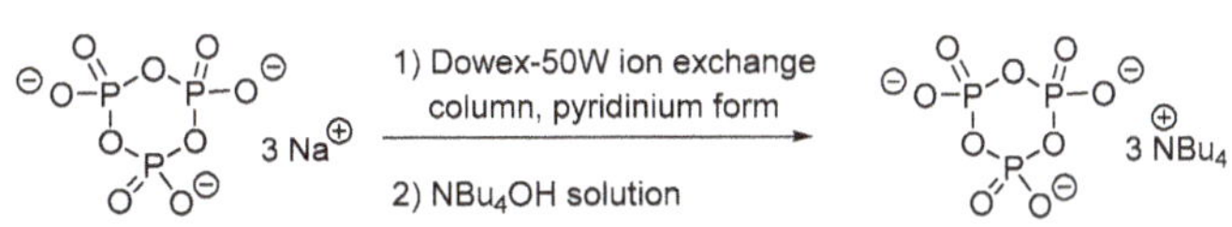

10 **9**

Scheme 7. Preparation of the cyclic trimetaphosphate **9** [59].

Then, the electrophilic reagent **9** was functionalized by a sulfonyl leaving group (LG) using *N*-methylimidazole (NMI) in the presence of either 2-mesitylenesulfonyl chloride **11** or 1-benzenesulfonyl-3-methyl-imidazolium triflate **12** in DMF (Scheme 8). Compound **11** was commercially available, whereas **12** was prepared in two steps starting from benzenesulfonyl chloride and imidazole [60]. Based on ^{31}P NMR monitoring of the reaction, the authors proposed that the activation proceeds via a mixed anhydride **13**, which reacts with NMI to give rise to intermediate **14**.

Addition of sub-stœchiometric amounts of NMPs to the activated cyclic trimetaphosphate **9** allowed their conversion to intermediates **15–18**, as suggested by ^{31}P NMR (Scheme 9). Addition of slight excess of the second NMP and anhydrous $MgCl_2$ gave rise to the corresponding dinucleoside pentaphosphates. In the absence of Mg^{2+} ions, the reaction was extremely slow.

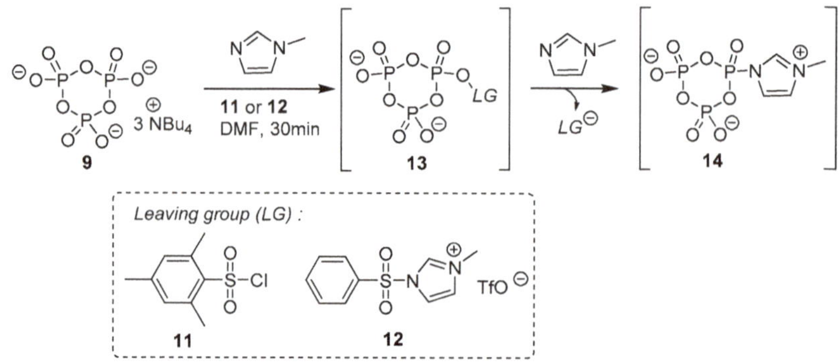

Scheme 8. Activation of the cyclic trimetaphosphate **9** [59].

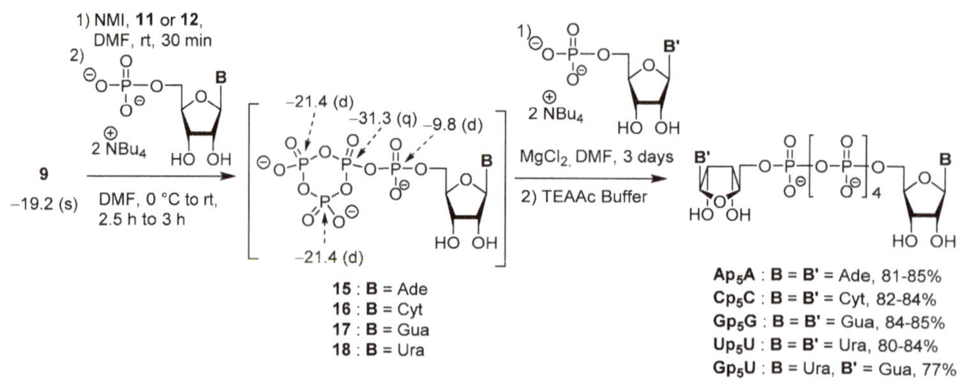

Scheme 9. Synthesis of dinucleoside pentaphosphates via an activated cyclic trimetaphosphate [59]. ^{31}P chemical shifts are indicated in ppm, together with their multiplicity.

This method constitutes an interesting approach as it allows formation of symmetric as well as mixed Np_5Ns in high yields, and uses inexpensive reagents (sodium cyclic trimetaphosphate, mesitylene chloride, NMI). It has recently been applied for the synthesis of dicaptides, used as substrates for polymerases [61]. However, it requires the preparation of the tetra-*n*-butylammonium salts of the cyclic trimetaphosphate and those of the nucleotides, reaction times are long (> 3 days), and must be performed under strictly anhydrous conditions.

3.4. Synthesis via a Phosphoromethylimidazolium Intermediate

Reacting the free acids of canonical nucleoside 5′-monophosphates with a large excess of NMI in the presence of triphenylphosphine/2,2′-dipyridyl sulfide as coupling agents and triethylamine in DMF or DMSO allowed dimerization to Np_2N in about 85% yields [62]. Unsymmetrical dinucleotides could also be obtained, albeit in much lower yields, by using a equimolar amount of the two NMPs to be coupled.

Another methodology to provide symmetrical and mixed dinucleoside polyphosphates Np_nN (n = 2–4) was inspired by the Bogachev procedure for the synthesis of deoxynucleoside 5′-triphosphates (dNTPs) [63]. It involves the reaction of nucleoside 5′-monophosphates-N-methylimidazolium as donors with an NMP, NDP or NTP to obtain dinucleoside di, tri and tetraphosphates (Scheme 10) [64]. These donors were prepared by treatment of the NMPs (disodium salts dihydrate or free acid hydrate) with a 16-fold excess of trifluoroacetic anhydride (TFAA) in acetonitrile (ACN) in the presence of excess triethylamine. This step resulted in the temporary protection of the hydroxyl and the amino (if present) groups and formation of a mixed anhydride, as shown by ^{31}P NMR (δ = 2 ppm for the mixed anhydride formed within a few minutes). It offered the additional benefit of drying the starting materials without deleterious effects on the reaction outcome. However, in the case of GMP, these conditions resulted in the complete decomposition of the starting material. This limitation could be circumvented by using GMP in its tri-*n*-butylammonium form, together with reducing the number of equivalent of TFAA.

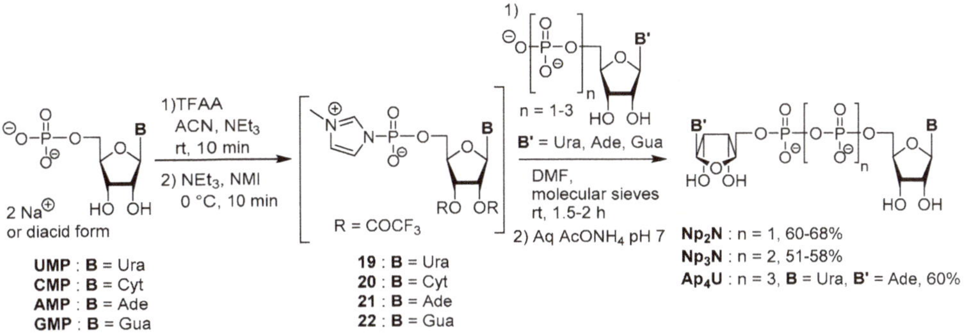

Scheme 10. Synthesis of dinucleotides via phosphoromethylimidazolium intermediates [64].

After removal of unreacted TFAA, the mixed anhydride was reacted with excess NMI in the presence of triethylamine to quantatively afford the highly reactive N-methylimidazolium salt donors **19–22** within 10 min. Formation of the N-methylimidazolium salts could also be easily followed by ^{31}P NMR spectroscopy as these species exhibit chemical shifts of approximately −9 to −10 ppm. The crude acetonitrile solutions were used directly in the next step. Thus, addition of the donors solutions to the bis(tri-*n*-butylammonium) salts of the nucleotide acceptors in DMF (in acetonitrile precipitation of the acceptors was observed) led to partially protected products. The latter were treated with aqueous ammonium acetate to remove the protecting groups, providing the desired compounds. For dinucleotides involving guanine nucleotides, N,N-dimethylaniline was found to be beneficial. Using this procedure, up to nine dinucleotides were obtained in yields of 51–68%. It should be noted that this method cannot be used to activate nucleoside 5′-di- or triphosphates. While the reactions were performed under argon, the authors pointed out that vigorous drying of the donor precursors was not always necessary when preparing the dinucleoside di- and triphosphates. Advantages of this one-pot method are short reaction time and good yields.

The same group also reported the synthesis of symmetrical di- and tetraphosphates by dimerization of nucleotides, using sulfonyl imidazolium salt **12** as a reagent (Scheme 11) [60,65]. The tetra-*n*-butylammonium salts of nucleoside 5′-mono or 5′-diphosphates were activated in DMF with 0.6–0.75 equiv of **12** in presence of 1–3 equiv of NMI and $MgCl_2$ (Scheme 11). Dimerization of the nucleotides led to Np_2N and Np_4N in high yields.

^{31}P NMR analyses showed that the reaction proceeds via the formation of the N-methylimidazolium salt intermediates **23–27** (Figure 4, ^{31}P δ = −9.2 ppm), which are able to react with the remaining nucleotide to form the corresponding dimer.

Mixed dinucleotides could also be obtained in high yields, by introducing minor modifications into the protocol. Accordingly, activation was performed in the presence of a small excess of sulfonyl imidazolium salt **12**, and NMI was replaced by diisopropylethylamine (DIPEA) to prevent dimerization (Scheme 12). Moreover, the time of reaction of the first step was reduced to 1 min. Cp_2A and Ap_3U were obtained in similar yields starting from AMP and UMP, respectively.

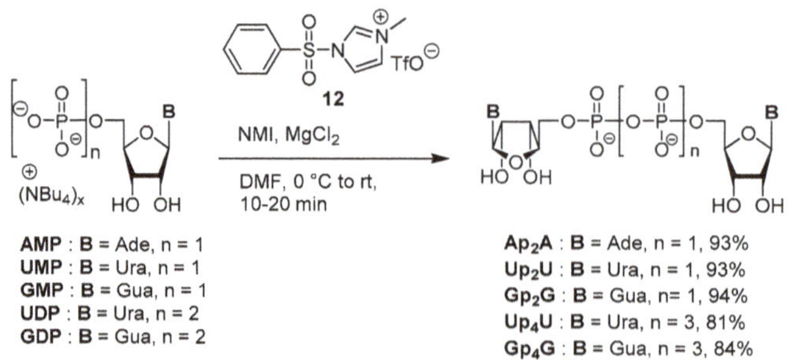

Scheme 11. Synthesis of symmetrical Np_2N and Np_4N [60].

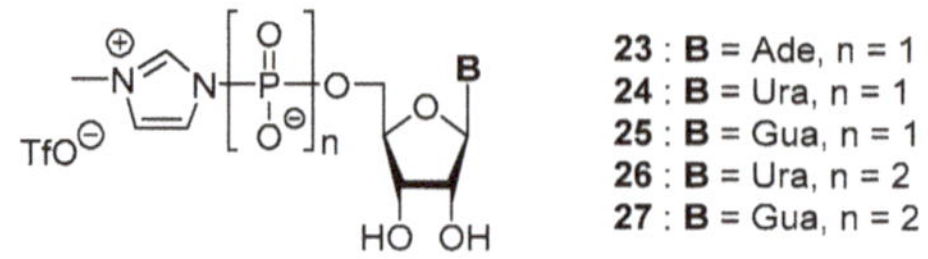

Figure 4. Phosphoromethylimidazolium intermediates [60].

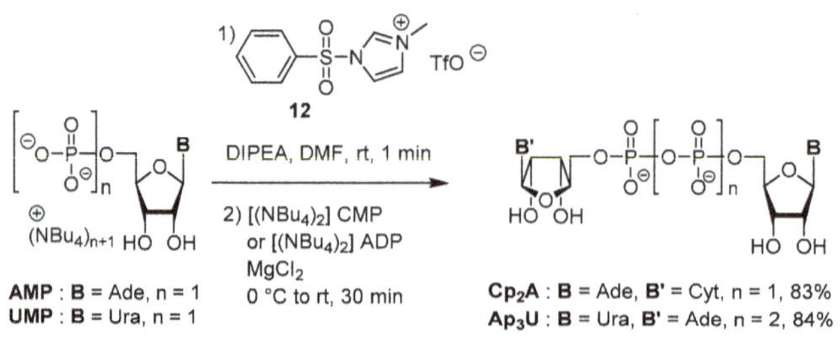

Scheme 12. Synthesis of mixed dinucleotides [60].

Finally, mixed adenine dinucleotides were obtained in high yields by activating the tetra-*n*-butylammonium salt of ATP with sulfonyl imidazolium salt **28**, followed by reaction with a second nucleotide (Scheme 13). Investigation of the first step by ^{31}P NMR revealed that the reaction proceeds via a cyclic adenosine trimetaphosphate intermediate **8**.

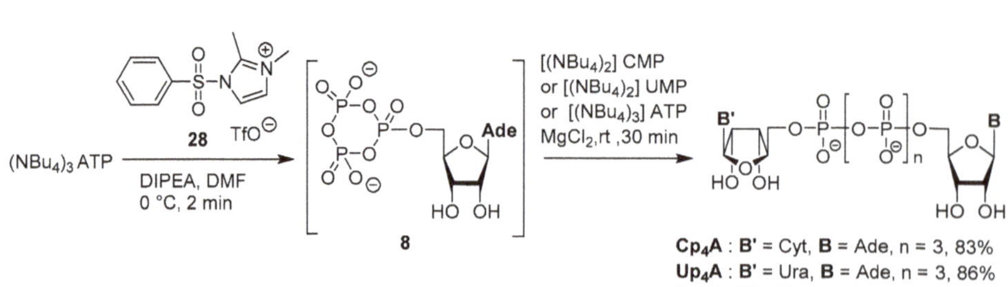

Scheme 13. Synthesis of mixed dinucleotides starting from ATP [60].

The advantages of this method using sulfonylimidazolium salts as key reagents, include short reaction times, access to a wide range of N_pNs (n = 2–6) and high yields. However, careful attention must be paid to maintain anhydrous conditions.

3.5. Synthesis via a Phosphoropiperidate Intermediate

In 2014, the Sun research group developed an approach based on the activation of nucleoside 5′-phosphoropiperidates with 4,5-dicyanoimidazole (DCI) as activator. The phosphoropiperidates **29–32** were obtained beforehand from unprotected NMPs via a redox condensation involving triphenylphosphine and 2,2′-dithiodianiline (Scheme 14) [66].

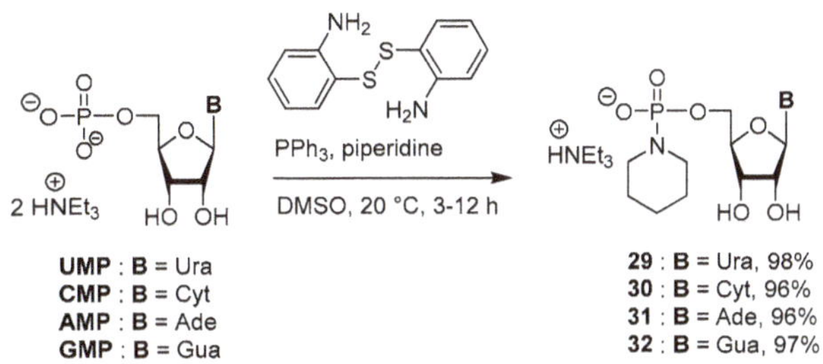

Scheme 14. Synthesis of nucleoside 5′-phosphoropiperidates [66].

As shown in Scheme 15, the reaction of nucleoside 5′-phosphoropiperidates with the tri-*n*-butylammonium salts of nucleotides (NMPs, NDPs, NTPs) in the presence of 2.5–3 equiv of DCI, led to the formation of more than 20 symmetrical or mixed dinucleoside di-, tri- and tetraphosphates [67,68]. Yields of Np$_n$N's dropped significantly along with the increase of the polyphosphate chain.

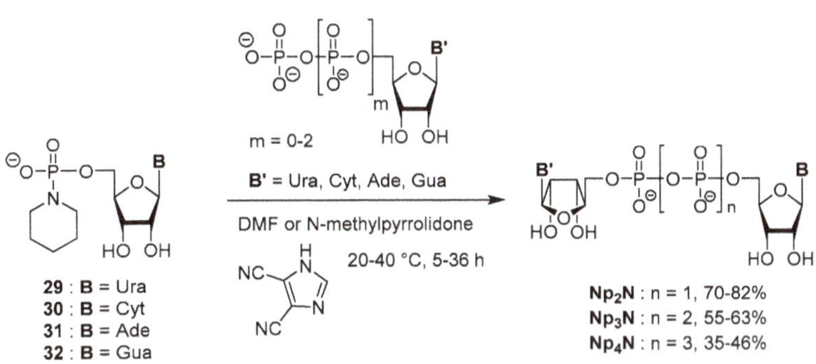

Scheme 15. Synthesis of dinucleotides from nucleoside 5′-phosphoropiperidates [67,68].

Among the polar aprotic solvents, *N*-methylpyrrolidone exhibited the best solvency for the tri-*n*-butylammonium salts of both NDPs and NTPs, while DMF was preferred for NMPs. Considering the cost of nucleoside 5′-polyphosphates and their similar chromatography properties with the products, only 0.5 equiv of NDPs or NTPs were used. This P(V) method is an efficient one, allowing access to a large range of Np$_n$Ns (n = 2–4) in short reaction times and good to high isolated yields.

Symmetrical Np$_n$N (n = 3–5) could also be obtained by reacting nucleoside 5′-phosphoropiperidates with sub-stœchiometric amounts (0.35–0.4 equiv) of inorganic phosphorylating agents in the presence of excess DCI [69]. In this alternative protocol, bis(tetra-*n*-butylammonium) hydrogen phosphate, tris(tetra-*n*-butylammonium) hydrogen pyrophosphate or tris(tetra-*n*-butyl-ammonium) dihydrogen triphosphate were used instead of a nucleotide (Scheme 16). ^{31}P NMR tracking experiments showed that pyrophosphate reacts first with the starting phosphoropiperidate to form an NTP, and then, with the remaining phosphoropiperidate to give rise to Np$_4$N. This methodology was also adapted to the synthesis of nucleotides with a higher number of phosphate groups (such as Np$_5$N), using inorganic triphosphate instead of pyrophosphate.

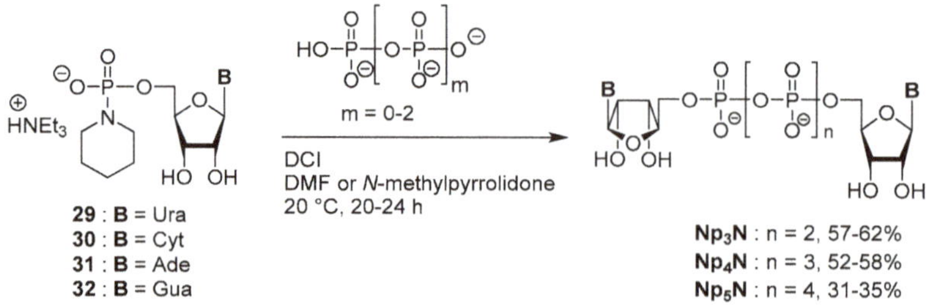

Scheme 16. Synthesis of symmetrical NpN by dimerization [67,68].

Interestingly, this strategy can be adapted to the synthesis of nucleotide analogues such as dinucleoside P^2,P^3-(dihalo)methylene tetraphosphates, by replacing pyrophosphate with bisphosphonate reagents.

4. Synthesis Based on P(III) Chemistry

Several methods involve P(III) reagents to provide phosphite or phosphoramidite intermediates, which are further oxidized to P(V). Initially developped for the synthesis of oligonucleotides, they have been adapted to dinucleotides by introducing minor modifications in the last steps of the reaction sequences. Because of the high reactivity of P(III) species, most of these approaches require the use of N-, and O-protecting groups on the starting nucleos(t)ides and strictly anhydrous conditions.

4.1. Synthesis via a Salicylphosphite Intermediate

This strategy is based on the well-known Ludwig–Eckstein procedure for NTP synthesis [70], and allows the one-pot preparation of symmetric dinucleoside tetra- and pentaphosphates starting from protected nucleosides [71]. Phosphitylation of triacetyl adenosine **33** with 2-chloro-4*H*-l,3,2-benzo-dioxaphosphorin-4-one (salicylchlorophosphite) led to intermediate **34**, then addition of bis(tri-*n*-butylammonium) pyrophosphate gave rise to the cyclic intermediate P(III)–P(V) **35** (Scheme 17). Oxidation to the cyclic trimetaphosphate, intermediate **36**, followed by reaction with adenosine 5′-monophosphate (AMP) or adenosine 5′-diphosphate (ADP) in dry DMF in the presence of $ZnCl_2$ as catalyst, afforded **37** and **38**, respectively. Removal of the acetyl groups under basic conditions, ion exchange to remove the Zn^{2+} cations and purification led to Ap_4A and Ap_5A as ammonium salts. It should be noted that symmetrical Ap_4A was obtained in much better yield than Ap_5A.

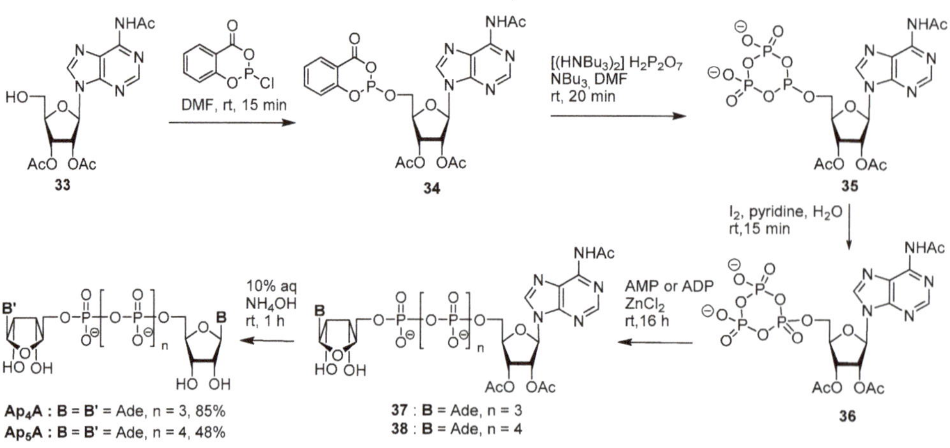

Scheme 17. Synthesis of the symmetrical dinucleotides Ap_4A and Ap_5A by phosphitylation [71].

The guanosine derivatives Gp_4G and Ap_4G were prepared in a similar manner, starting from 2′,3′-*O*-2-*N*-triacetylguanosine **39** (Scheme 18).

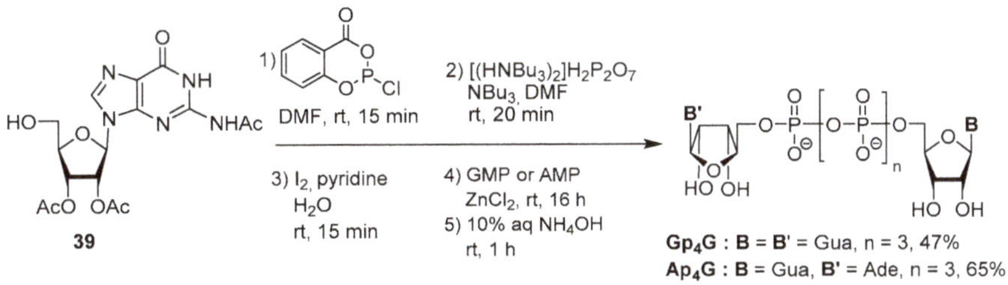

Scheme 18. Synthesis of dinucleotides Gp_4G and Ap_4G by phosphitylation [71].

4.2. *Synthesis via a Cyclosaligenyl Phosphite Intermediate*

This strategy initially developed by Meier and co-workers for the synthesis of prodrugs of nucleotides, called *cyclo*Sal (for *cyclo*saligenyl) pronucleotides [72,73], has been applied to the synthesis of nucleoside 5′-polyphosphates, sugar nucleotides and mixed dinucleotides [74,75]. Phosphochloridite **40** was prepared from the corresponding salicyl alcohol and phosphorus trichloride (Scheme 19). Reaction with 3′-*O*-acetylthymidine followed by oxidation with oxone (2 $KHSO_5.KHSO_4.K_2SO_4$) afforded 5-nitro-*cyclo*Sal-3′-*O*-acetylthymidine phosphotriester **41**.

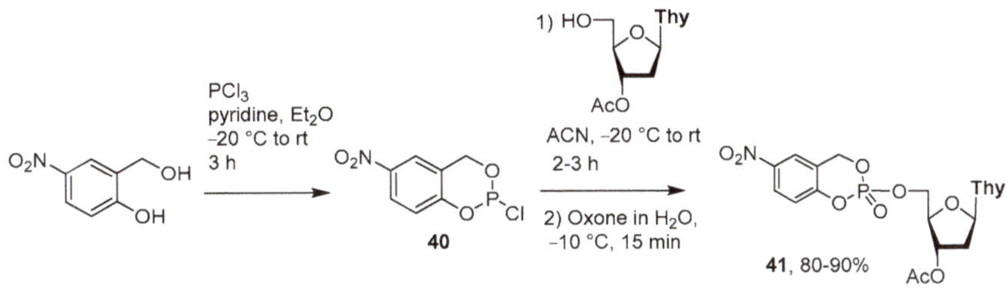

Scheme 19. Synthesis of a *cyclo*Sal phosphotriester derivative of thymidine [74].

This phosphotriester was too sensitive to moisture to be purified by chromatography. Therefore, the crude material (obtained after liquid-liquid extraction) was directly engaged in the next step, i.e., reaction with rigorously dried tetra-*n*-butylammonium UMP and tris(tetra-*n*-butylammonium) ATP in DMF (Scheme 20). After removal of the acetyl group and ion-exchange (n-$NBu_4{}^+$ to $NH_4{}^+$), Up_2T and Ap_4T were obtained in their ammonium form.

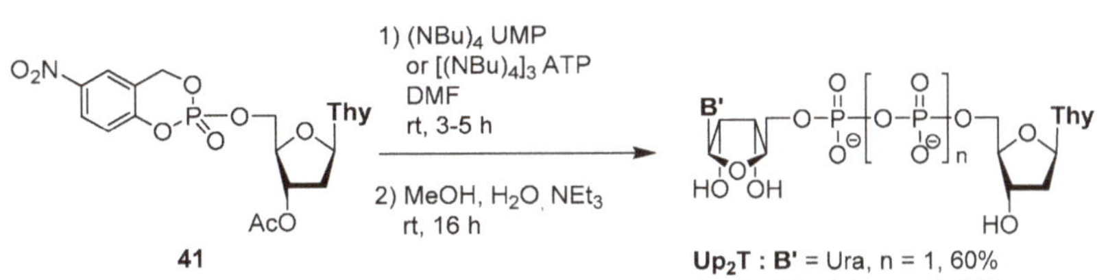

Scheme 20. Synthesis of the dinucleotides Up_2T and Ap_4T [74].

This approach has been adapted to solid phase [76]. The key intermediates of the synthesis are the 5′-substituted *cyclo*Sal nucleotides linked to a polystyrene support through the 3′-OH group and a succinyl linker (**42–44**, Figure 5).

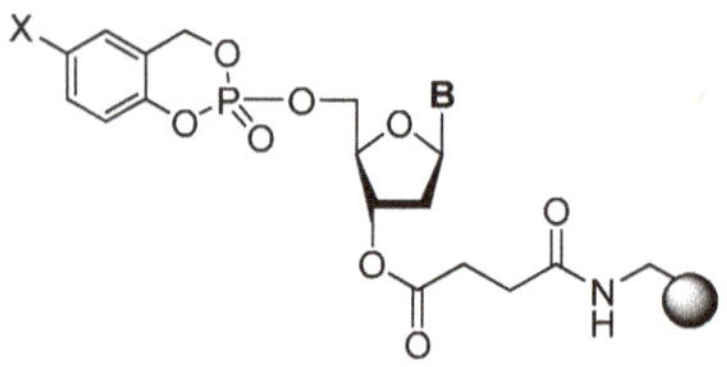

42 : X = Cl, **B** = Thy
43 : X = Cl, **B** = Ade
44 : X = $MeSO_2$, **B** = Thy

Figure 5. Structure of the supported nucleoside phosphotriesters [76].

Preparation of these intermediates started with the anchoring of the succinyl linker to the 5′-*O*-(4,4′-dimethoxytrityl)-protected nucleosides, and removal of the protecting group with trifluoroacetic acid (TFA). The 3′-*O*-succinylnucleosides were then treated with the appropriate chlorophosphites, followed by oxidation with oxone, to give rise to compounds **45–47** (Scheme 21). Coupling with the aminomethyl polystyrene support was achieved using 1-hydroxybenzotriazole (HOBt) and *N,N′*-diisopropylcarbodiimide (DIC) as coupling agents. Alternatively, 2-(1*H*-benzotriazole-1-yl)-1,1,3,3-tetramethyluronium tetrafluoroborate (TBTU) and 4-ethylmorpholine may be used. Conversion to Up_2T and Up_2dA was carried out by adding a large excess of tetra-*n*-butylammonium UMP, and then cleavage under basic conditions (70–78% purity, no overall yield was given). Compared to the solution phase synthesis, this supported approach is less straightforward and involves multiple steps.

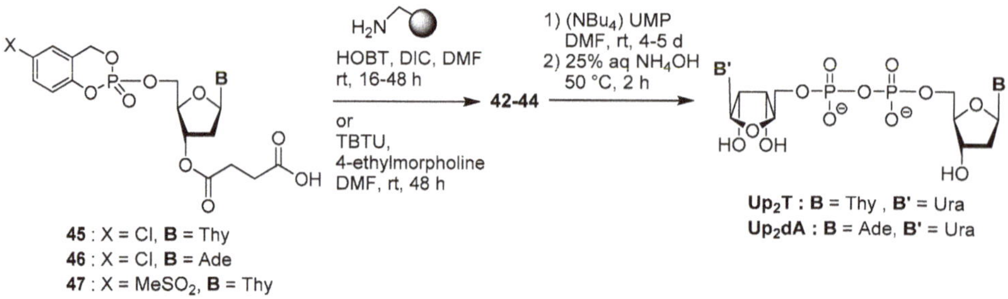

Scheme 21. Solid-phase synthesis of dinucleoside 5′,5′-diphosphates [76].

4.3. Synthesis via a Supported Phosphite Intermediate

Another P(III) strategy combined with a solid-phase approach was reported by Ahmadibeni and Parang [77]. Using these polymer-bound phosphitylating reagents, symmetrical 5′,5′-dinucleoside mono, di, tri, and tetraphosphates were prepared. Compounds **48–51** were obtained in 2 to 5 steps starting from phosphorus trichloride (Figure 6). Coupling with aminomethyl polystyrene resin-bound *p*-acetoxybenzyl alcohol yielded four classes of supported-phosphitylating reagents **52–55** (Figure 6).

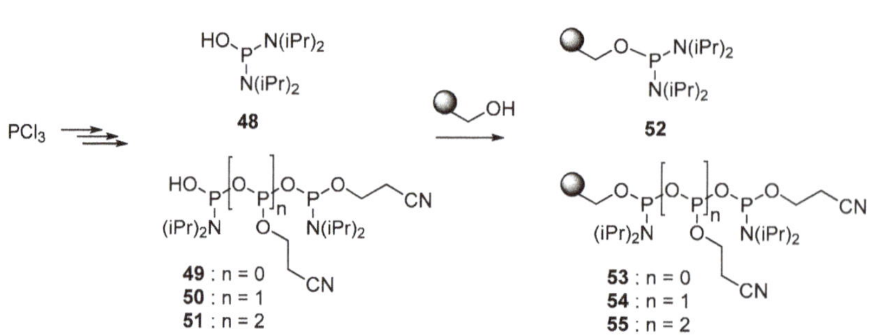

Figure 6. Synthesis of the immobilized phosphitylating reagents [77].

These supported reagents were reacted with unprotected nucleosides (e.g., thymidine, adenosine, cytidine, inosine or the analogue 3′-azido-3′-deoxythymidine) in the presence of 5-(ethylthio)-1*H*-tetrazole to afford compounds **56–59** (Scheme 22). Polymer-bound nucleotides underwent oxidation with *tert*-butyl hydroperoxide to **60–63**, then removal of the cyanoethoxy groups with 1,8-diazabicyclo[5.4.0]undec-7-en (DBU) was performed. Acidic treatment of **60** and **64–66** gave rise to the corresponding symmetrical dinucleotides in 59–78% yields starting from the supported reagents **52–55**.

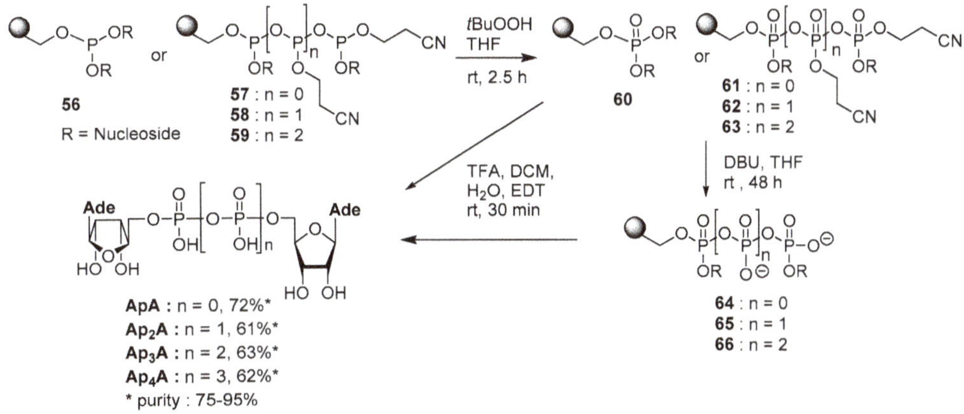

Scheme 22. Synthesis of symmetrical dinucleotides using polymer-bound phosphitylating reagents. Yields are given for the adenine derivatives [77].

This strategy allowed the synthesis of symmetrical dinucleotides starting from unprotected nucleosides without the need to purify the intermediates. It also offers the advantage of facile recovery of final products by filtration (trapped linkers on the resins). However, it requires the multistep synthesis of the supported polyphosphite reagents. It should be highlighted that part of this work could not be reproduced by others [78].

4.4. Synthesis via a 5′-H-Phosphonate Intermediate

This one-pot strategy developed by Sun et al. [79], and based on their previous work on the synthesis of nucleoside triphosphates [80], involves the formation of a pyridinium phosphoramidate intermediate starting from a nucleoside 5′-*H*-phosphonate monoester. Nucleoside 5′-*H*-phosphonate monoesters **67–70** were obtained from 2′,3′-*O*-isopropylidene ribonucleosides **71–74** using diphenylphosphite as a reagent (Scheme 23). Then, removal of the isopropylidene group with aqueous TFA afforded 5′-*H*-phosphonate monoesters **75–78** in yields ranging from 68% to 82% over two steps.

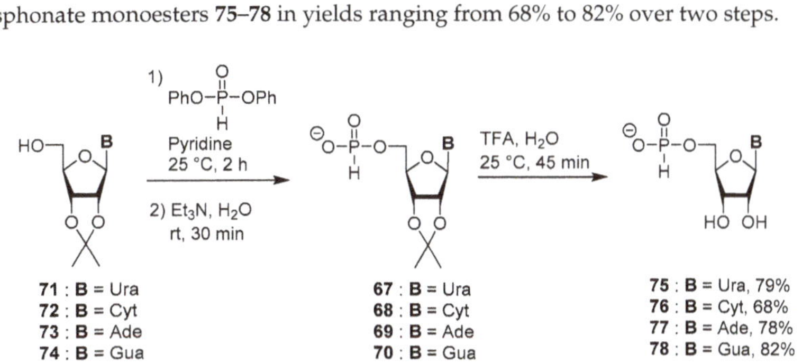

Scheme 23. Synthesis of nucleoside 5′-*H*-phosphonates from the 2′,3′-protected nucleosides [79].

H-phosphonates were silylated with trimethylsilyl chloride (TMSCl) in pyridine, and in situ oxidized with iodine to generate the highly reactive zwitterionic pyridinium phosphoramidate intermediates **79–82** (Scheme 24). Indeed, even trace amounts of water may result in the formation of polyphosphate by-products. The authors hypothesized that if 50% of the phosphoramidate intermediate could be hydrolyzed with precise control, NMP generated in situ would react with the remaining phosphoramidate to afford Np_2N as a major product. Accordingly, hydrolysis and coupling steps were optimized using ^{31}P NMR tracking experiments and accomplished by two sequential additions of H_2O in DMF. After purification and ion exchange, symmetrical dinucleoside diphosphates sodium salts were obtained in 68–91% yields starting from the nucleoside 5′-*H*-phosphonates. Advantages of this method are the short reaction time, simple purification procedure and high yields. However, it requires the preliminary synthesis of the *H*-phosphonate starting materials.

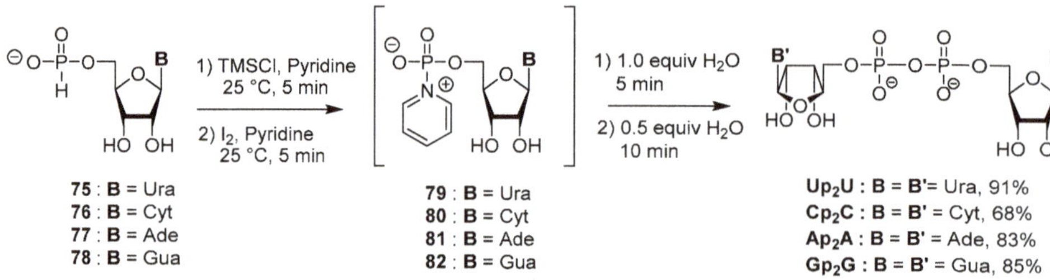

Scheme 24. Synthesis of symmetrical dinucleotides from nucleoside 5′-*H*-phosphonates [79].

4.5. Synthesis via a Phosphoramidite Intermediate

In 2015, the Jessen research group developed a very efficient strategy based on an iterative phosphoramidite approach to access nucleoside 5′-polyphosphates [81,82] and some derivatives [83]. The phosphoanhydride bond is formed by coupling a phosphoramidite (the donor) to a phosphate (the acceptor). In contrast to the other P(III) methods, all steps were performed under ambient conditions, i.e., in open flasks and using non-dried solvents. Regarding of the synthesis of dinucleotides, 2′,3′-*O*-diacetyl uridine **83** was treated with phosphorodiamidite **84** in the presence of 1*H*-tetrazole to afford phosphoramidite **85** (Scheme 25). Addition of the tetra-*n*-butylammonium salt of UMP and DCI, followed by oxidation with *m*CPBA, treatment with piperidine, and cleavage of the acetyl groups under basic conditions, gave rise to Up_2U as a mixed piperidinium/tetra-*n*-butylammonium salt. The mixed dinucleotide Ap_4U was also obtained by using the tetra-*n*-butylammonium salt of ATP and 5-ethylthio-1*H*-tetrazole (ETT) as an activator in the second step of the reaction (Scheme 25). It should be noted that these experiments were performed on a small scale (10 mg).

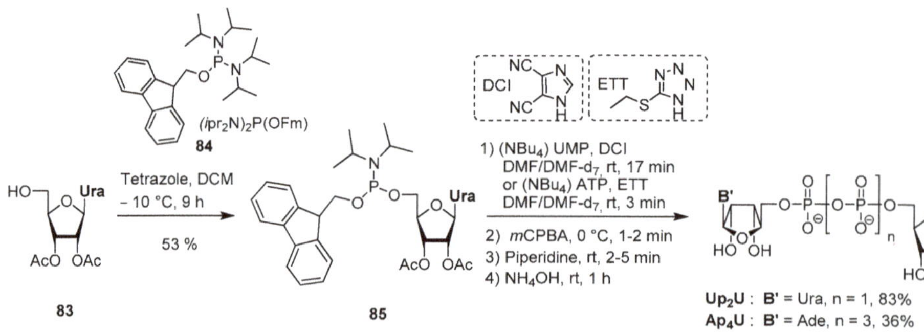

Scheme 25. Synthesis of dinucleotides via a phosphoramidite intermediate [83].

Symmetrical Np_nNs (n = 3, 5, 7) could also be accessed via chemoselective homologative dimerization of two phosphate monoesters with phosphorodiamidites [84]. The general strategy for the synthesis of dinucleoside triphosphates is shown in Figure 7. First, phosphordiamidites **86** and **87** (donors) were coupled with a NMP (the acceptor), affording terminal mixed phosphoric anhydride phosphoramidites **88–89** as unstable intermediates. Reaction with another phosphate monoester resulted in the formation of **90** containing a P^V–P^{III}–P^V bridge, which was then oxidized to yield **91**. The corresponding Np_3N was obtained after a final deprotection step.

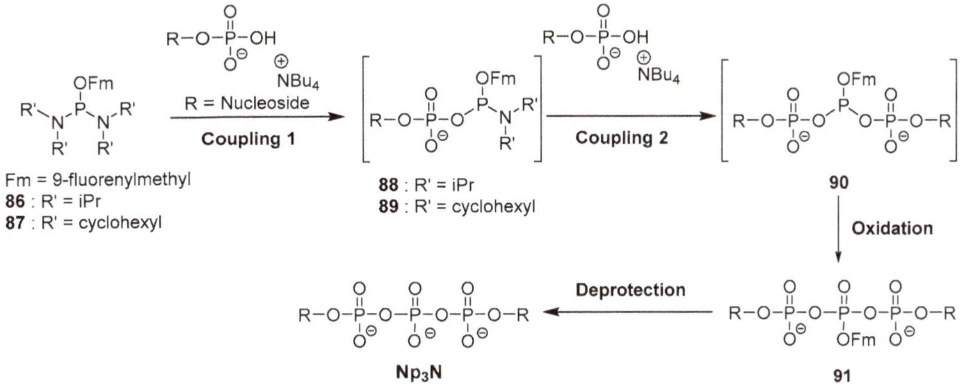

Figure 7. General approach for the preparation of symmetrical Np_3Ns based on dimerization [84].

The tetra-*n*-butylammonium salts of the canonical nucleoside 5′-monophosphates were treated in DMF with 9-fluorenylmethyl-*N*,*N*,*N*′,*N*′-tetraalkyl phosphorodiamidites **86–87** in the presence of ETT, followed by oxidation with *m*CPBA (Scheme 26). Progress of the reaction was monitored by ^{31}P NMR. The Fm-protected intermediates were isolated by precipitation with Et_2O/hexane, treated with piperidine, and purified by anion exchange chromatography to isolate the symmetrical dinucleoside triphosphates as ammonium salts in good yields.

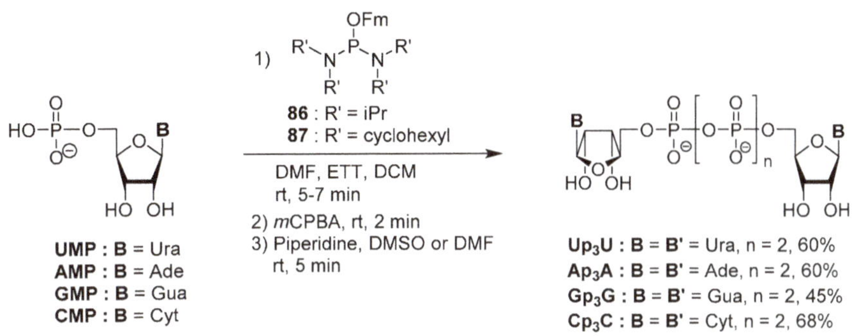

Scheme 26. Synthesis of dinucleotides via dimerization of phosphates [84].

This methodology was applied to the synthesis of symmetrical dinucleoside penta- and heptaphosphates by using as acceptor a NDP or a NTP, respectively. However, due to the water-sensivity of the intermediates, reactions had to be performed under anhydrous conditions, with a shorter reaction time and low temperature (coupling with ETT 30 s at 0 °C, oxidation 5 min at 0 °C). Accordingly, Ap_5A and Ap_7A were obtained in 50% and 18% yields, respectively. The chemoselective aspect of this method is a great advantage as it allows to prevent the use of nucleoside protecting groups. Compared to other procedures using P(III) chemistry, it offers attractive features, notably these rapid reactions can be run under ambient conditions.

5. Green Chemistry Approaches

Most of the reported methods present drawbacks, such as the use of non-volatile and toxic solvents (DMF, pyridine, *N*-methylpyrrolidone), high sensitivity to moisture, preparation of substrates or phosphorus reagents with lipophilic counterions due to solubility issues in organic media, anhydrous conditions (i.e., dry reagents and solvents), fastidious purification and possibly prior synthesis of the activated nucleotide substrates. An ideal procedure would meet the following requirements: Applicable to a wide range of substrates where protecting groups are not necessary, and giving rise to the products in short time and high yields.

In 2017, our research group reported a one-pot synthesis, in water medium, of nucleoside 5′-polyphosphates and dinucleotides starting from the corresponding NMPs [55,56]. This method uses 2-chloro-1,3-dimethylimidazolinium hexafluorophosphate (DMP) and imidazole as coupling reagents to activate the NMP into their phosphorimidazolides. Dinucleoside diphosphates were obtained by self-dimerization of the NMPs (Scheme 27).

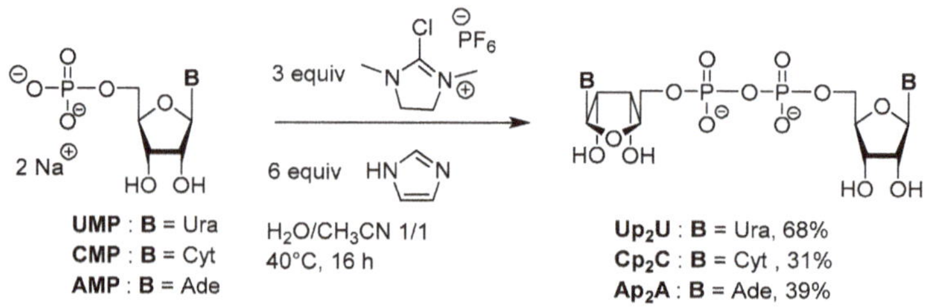

Scheme 27. Preparation of symmetrical dinucleoside diphosphates in aqueous media [56].

Our work originated from the small-scale synthesis of NDP-sugars described by Tanaka et al. [85]. The authors performed the coupling of a sugar-1-phosphate with a nucleotide in D_2O, using 2-chloro-1,3-dimethylimidazolinium chloride (DMC) and imidazole (Scheme 28). Based on ^1H and ^{31}P NMR spectroscopies, the reaction was shown to proceed via a 2-imidazolyl-1,3-dimethylimidazolinium chloride intermediate **92**, formed in situ by reaction of DMC and imidazole (Scheme 28). This key intermediate was able to activate an NMP into its corresponding phosphorimidazolide **93** (via intermediate **94**), which then reacted with a sugar-1-phosphate salt to form a sugar nucleotide.

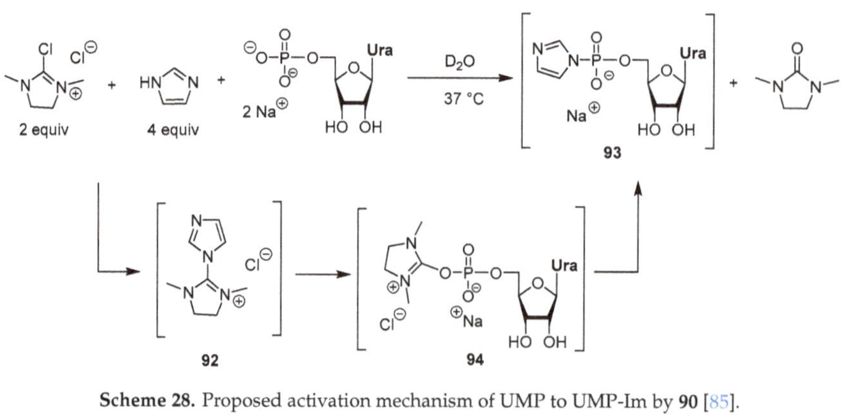

Scheme 28. Proposed activation mechanism of UMP to UMP-Im by **90** [85].

The main advantage of these reactions, which occur in D_2O [85] or in H_2O/CH_3CN [55,56] instead of the usual organic solvents, is that the NMPs are used in their commercially available and water-soluble forms, i.e., sodium and potassium salts. This greatly simplifies the handling of the reaction. Moreover, protecting groups on the nucleotide are not required and no carbonation of ribonucleotides are observed, in contrast to the use of CDI.

Alternatively, mechanochemistry has emerged in the field of nucleosides and, to a lesser extent, nucleotides [86]. This technique enables solid/solid reactions through mechanical grinding under solvent-free conditions [87–91]. In 2011, Ravalico et al. reported the mechanosynthesis of dinucleotides, namely Ap_nA (with n = 2–4), Ap_2dT and nicotinamide adenine dinucleotide (NAD^+), starting from adenosine 5′-phosphoromorpholidate **2**, in the presence of magnesium chloride, 1*H*-tetrazole as the acidic promoter and water (Scheme 29) [92]. Although advantageous, this methodology requires the use of costly reagents, namely tetrazole and the phosphoromorpholidate substrate, the latter may be prepared beforehand in solution using conventional solution synthesis [45].

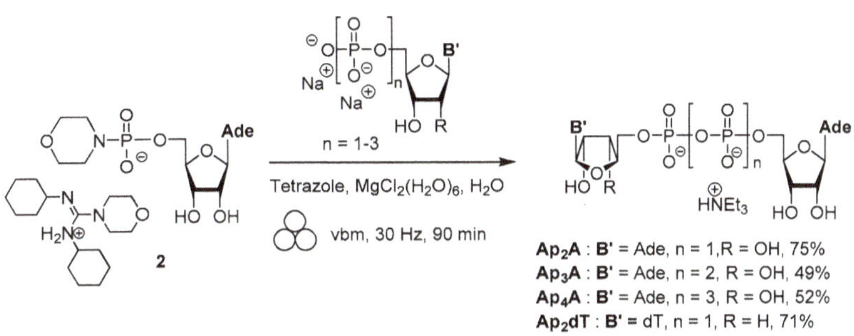

Scheme 29. Ball-milling route using commercially available nucleotide salts and morpholidate reaction partners [92].

In the course of our investigations on new strategies for phosphoanhydride bond formation, we also investigated the ball-milling technique [93]. To begin with, activation of UMP was performed according to the conditions previously reported in water medium (DMP/imidazole) [55,56], without adding any solvent in a vibratory ball-mill (vbm) at 30 Hz. A complete conversion to Up_2U was obtained within 30 min, suggesting that UMP was activated to its phosphorimidazolide and reacted promptly with the remaining UMP to form the dimer. Alternatively, we tested the use of CDI, which has recently gained interest in the field of mechanosynthesis due to the safety of its by-products (i.e., imidazole and carbon dioxide), its efficacy in *N*-acylation reactions and relatively low cost [94–97]. Activation of UMP (acidic form) to the corresponding 5′-phosphorimidazolide 2′,3-carbonate **95** was complete by grinding, for 1 h at 30 Hz, in the presence of 4 equiv. of CDI and acetonitrile (0.3 $\mu L.mg^{-1}$) as a liquid assistant (Scheme 30). The absence of reactivity observed using the disodium salt of UMP suggests that the acid form is required to protonate the imidazole of CDI and to facilitate its substitution by the phosphate monoester. This activation step was applied successfully to several ribonucleotides and a 2′-deoxyribonucleotide (dTMP).

In the second step, a nucleoside mono, di or triphosphate was added onto the activated NMP **95–96** in order to form the desired DNPs (Scheme 30). Thus, the mixed dinucleoside 5′,5′-diphosphates were successfully obtained by adding a slight excess of a NMP and a small amount of acetonitrile (0.6–0.95µL/mg) in the jar, and then ball-milling for 2 h at 30 Hz. Remarkably, the second grinding step also removed the carbonate protection. The reaction was less efficient with GMP, UDP and ATP, and, thus, required a larger excess of reagent to form the corresponding dinucleotides Gp_2U, Up_3U and Ap_4U, respectively. This user-friendly mechanochemical approach only requires a slight excess of reagents and limits the formation of side products. While isolated yields of symmetrical and mixed dinucleoside polyphosphates are similar to other approaches, the set-up of the experiments and the workup are greatly simplified, saving a considerable amount of time. In addition, compared to the previously reported mechanochemical synthesis [92], this one-pot two steps reaction starts from NMPs, which are readily available.

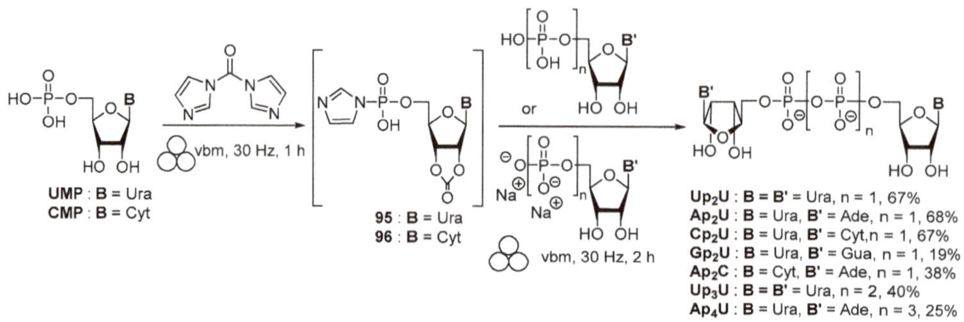

Scheme 30. Mechanochemical approach to obtain DNPs starting from NMPs [93].

In addition to their simplicity, the mechanochemical approaches alleviate issues associated with the low solubility of reagents in solution phase synthesis. Therefore, nucleotides and inorganic salts can be used in their commercially available sodium or potassium forms.

6. Purification and Characterization

6.1. Purification

In this section, we provide a brief overview of the purification procedures for dinucleotides. Due to their high polarity, they cannot be separated by normal-phase chromatography like most organic compounds. In addition, the byproducts or side products formed during synthesis share similar characteristics, resulting in a challenging separation from the desired compounds. The two main purification methods, reported in Table 1, use ion-exchange or reverse-phase (RP) chromatography.

Table 1. Main methods to isolate dinucleotides.

Types of Chromatography	Eluants	References
Anion-exchange		
DEAE-Sephadex®	0.02 M TEAB, pH 8; 10% ACN	[54]
TOYOPEARL® DEAE-650M	0.2 M TEAB, pH 8; 10% ACN	[54]
DEAE-Sephadex®	Gradient of aq. NH_4HCO_3	[67,69]
Q Sepharose® Fast Flow	Gradient of aq. NH_4HCO_3	[83,84]
DEAE Sepharose® Fast Flow	Gradient of aq. NH_4HCO_3	[83,84]
C-18 reverse-phase		
HPLC	0.01 M NH_4HCO_2, pH 4, 50% MeOH	[46]
Semi-preparative HPLC	TEAAc pH 7 or 9; 6–12% ACN	[59,60,64]
Preparative HPLC	0.1 M NH_4HCO_3; ACN	[71]
Preparative HPLC	TEAB pH 7.8; 65% ACN	[92]
RP-18 silica gel column	Water	[74]
RP-18 silica gel column	0.1 M TEAB, pH 7.5; 20% ACN	[55,56]
RP-18 silica gel column	0.1 M TEAAc, pH 7; 5–10% ACN	[93]
Gel filtration		
Sephadex® LH-20	Deionized water	[79]

The first type of ion-exchange support is positively charged, such as diethylaminoethyl (DEAE) Sephadex, and allows the chemical species to elute according to their number of charges. Elution is often performed using a gradient of triethylammonium hydrogen carbonate (TEAB) or ammonium hydrogen carbonate (NH_4HCO_3) buffers. Regarding C_{18} reverse-phase chromatography, either preparative or semi-preparative high pressure liquid chromatography (HPLC) or chromatography on RP-18 silica gel may be used. Compounds are often eluted using a gradient from an aqueous buffered solution, such as triethylammonium acetate (TEAAc) or TEAB, to acetonitrile or methanol. Alternatively, gel filtration was described for the purification of symmetrical Np_2N. Then, the samples are usually freeze-dried and converted to their sodium form by ion exchange using a Dowex-50W-Na^+ resin. The fact that purification is a time-consuming step is one of the reasons that pushed some research groups to develop solid phase strategies [76,77]. Indeed, those methods were designed in order to simplify purification of the final products, which can be collected by simple filtration. However, the recovered materials showed moderate purity and an additional purification step was required [76]. Consequently, solution-phase approaches are still predominant.

6.2. Physico-Chemical Properties

NMR, UV spectroscopy, and mass spectrometry (MS) are the main tools for the analysis of dinucleotides. Since the sugar and the phosphate moieties have no significant absorption above 230 nm, dinucleotides exhibit UV absorption profiles similar to those of their parent nucleosides with λ max values close to 260 nm [1]. MS analysis of nucleotides and derivatives such as dinucleotides has been extensively reviewed by Banoub et al. [98]. Alternatively, ^{31}P NMR is a very convenient tool to monitor reaction courses, characterize reaction intermediates, and study the conformation of dinucleotides and final compounds [1,50]. The chemical shift (δ) for most endogenous dinucleotides covers −8 to −24 ppm, and exhibits pH and counter ion dependence [1]. Table 2 summarizes the literature data for representative N_pN (n =1–7). In ^{1}H-decoupled ^{31}P NMR, the phosphorus atoms closest to the nucleosides are characterized by a singlet at approximately −10 ppm, while the phosphorus atoms inside the polyphosphate chain exhibit resonance signals around −22 ppm.

Table 2. ^{31}P Chemical shift ranges of dinucleoside polyphosphates in D_2O.

Np_nN	δ Ranges	References
Np_2N'	1 singlet (P^1 + P^2) from −9.6 to −11.5 ppm	[47,49,50,53,56,60,64,67,74,76,77,79,83,93]
Np_3N	1 doublet (P^1 + P^3) from −9.2 to −11.7 ppm 1 triplet (P^2) from −21.0 to −23.4 ppm	[47,49,50,54,60,64,67,69,84,93]
Np_4N	1 multiplet or 1 broad singlet or 1 doublet (P^1 + P^4) from −8.8 to −11.5 ppm 1 multiplet or 1 br singlet or 1 doublet (P^2 + P^3) from −20.6 to −23.1 ppm	[47,49,50,54,60,64,67,69,71,74,93]
Np_5N	1 multiplet or 1 br singlet or 1 doublet (P^1 + P^5) from −8.7 to −11.4 ppm 1 multiplet or 1 br singlet or 1 doublet (P^2 + P^3 + P^4) from −20.2 to −22.5 ppm	[49,50,59,69,71,84]
Np_6N	1 doublet at −10.1 ppm (P^1 + P^6) 1 multiplet at −21.7 ppm (P^2+ P^3 + P^4 + P^5)	[49]
Np_7N	1 broad singlet or 1 doublet from −10.1 to −11.6 ppm (P^1 + P^7) 1 multiplet from −21.7 to −23.5 ppm (P^2 + P^3 + P^4 + P^5 + P^6)	[49,84]

Interestingly, the study of dinucleotides using NMR techniques (^{1}H, ^{13}C and ^{31}P) can determine the conformation (*syn*/*anti*, sugar puckering), nucleobase properties (tautomerism, H-bonds and π-stacking interactions), as well as the anomeric α/β configuration. Such data are crucial for understanding the structure–activity relationship of either natural or synthetic dinucleotides as enzyme inhibitors and receptor ligands, and the design of potent therapeutic agents based on a dinucleotide scaffold. In this regard, Stern et al. have performed conformational studies on physiologically active Np_nN' sodium salts (N = A, G, U, C; N' = A, G, U, C; n = 2–5) in neutral aqueous solution, using NMR and computational techniques [50,98]. While no predominant conformation was observed for the ribose moiety in any of the studied dinucleotides, natural dinucleoside polyphosphates were more frequently found in a folded (stacked) rather than an extended conformation. Purine dinucleotides showed greater stacking interactions than pyrimidine dinucleotides. In addition, dinucleotides with longer phosphate chains were found to have weaker stacking interactions. Crystal structures of Ap_4A sodium salt [99], and some dinucleotides bound to kinases or hydrolases are available [100,101]. In protein-bound dinucleotides, the polyphosphate chain adopts various conformations: extended [100–103], S-shaped [101] or folded [100]. In some cases, the protein-bound dinucleotides coordinate with a Mg^{2+} ion.

Finally, Chen and Kohler investigated the dynamics of excited electronic states formed by UV excitation of the several diadenosine polyphosphates by femtosecond transient absorption (TA) spectroscopy [104]. They found that the excimer states seen in TA experiments on nucleobase dimers are only observed in π-stacked conformations, but the lifetimes of these states are insensitive to how the stacked bases are oriented.

7. Conclusions

Naturally occurring dinucleoside 5′,5′-polyphosphates are involved in a variety of cellular processes. Among them, symmetrical purine-containing dinucleotides are the most known representatives, whereas pyrimidine-containing and mixed derivatives are still intensively studied. Therefore, the development of synthetic methods required to obtain these compounds in few steps, good yields, high purity and comfortable amounts generated interest from the chemists and remains a field of intensive research. Herein, we have focused our attention on the synthesis of dinucleoside 5′,5′-polyphosphates, either symmetrical or mixed, and including non-modified pyrophosphate bonds. These methods may involve P(III) and/or P(V) intermediates and solution-phase or supported chemistry. P(V) synthetic approaches are usually based on the activation of a 5′-nucleotide using inexpensive reagents but require the conversion of the substrates, as well as the phosphate reactants, into the corresponding tri- or tetra-alkyl salts due to solubility issues. P(III) synthetic approaches are often effective processes. However, the presence of protecting groups on the nucleoside is often mandatory, thus increasing the overall number of steps. In most cases, strictly anhydrous conditions are essential and the length of the poly-phosphate chain affects the yields of the synthesis. Recently, environmental friendly alternatives were investigated and may constitute an area of substantial and growing interest for the preparation of these derivatives in the coming years.

Author Contributions: Conceptualization: S.P. and B.R.; Writing—original draft: B.R.; Writing—review & editing: L.A., C.C., S.P. and B.R.

Funding: This research received no external funding

Conflicts of Interest: The authors declare no conflict of interest.

References

1. Roy, B.; Depaix, A.; Périgaud, C.; Peyrottes, S. Recent trends in nucleotide synthesis. *Chem. Rev.* **2016**, *116*, 7854–7897. [CrossRef] [PubMed]
2. Jankowski, V.; Van Der Giet, M.; Mischak, H.; Morgan, M.; Zidek, W.; Jankowski, J. Dinucleoside polyphosphates: Strong endogenous agonists of the purinergic system. Br. *J. Pharmacol.* **2009**, *157*, 1142–1153. [CrossRef]
3. Christie, S.M.H.; Elmore, D.T.; Kenner, G.W.; Todd, A.R.; Weymouth, F.J. Syntheses of P^1P^2-diadenosine-5′ and P^1P^2-diuridine-5′ pyrophosphates. *J. Chem. Soc.* **1953**, 2947–2953. [CrossRef]
4. Rapaport, E.; Zamecnik, P.C. Presence of diadenosine 5′,5′′′ -P^1, P^4-tetraphosphate (Ap_4A) in mamalian cells in levels varying widely with proliferative activity of the tissue: A possible positive "pleiotypic activator". *Proc. Natl. Acad. Sci. USA* **1976**, *73*, 3984–3988. [CrossRef] [PubMed]

5. Jankowski, V.; Tölle, M.; Vanholder, R.; Schönfelder, G.; Van Der Giet, M.; Henning, L.; Schlüter, H.; Paul, M.; Zidek, W.; Jankowski, J. Uridine adenosine tetraphosphate: A novel endothelium- derived vasoconstrictive factor. *Nat. Med.* **2005**, *11*, 223–227. [CrossRef]
6. Zhou, Z.; Matsumoto, T.; Jankowski, V.; Pernow, J.; Mustafa, S.J.; Duncker, D.J.; Merkus, D. Uridine adenosine tetraphosphate and purinergic signaling in cardiovascular system: An update. *Pharmacol. Res.* **2019**, *141*, 32–45. [CrossRef]
7. Delicado, E.G.; Miras-Portugal, M.T.; Carrasquero, L.M.G.; Leon, D.; Pérez-Sen, R.; Gualix, J. Dinucleoside polyphosphates and their interaction with other nucleotide signaling pathways. *Pflügers Archiv. Eur. J. Physiol.* **2006**, *452*, 563–572. [CrossRef]
8. Hampton, A.; Kappler, F.; Picker, D. Species- or isozyme-specific enzyme inhibitors. 4. Design of a two-site inhibitor of adenylate kinase with isozyme selectivity. *J. Med. Chem.* **1982**, *25*, 638–644. [CrossRef]
9. Lee, S.-Y.; Müller, C.E. Nucleotide pyrophosphatase/phosphodiesterase 1 (NPP1) and its inhibitors. *MedChemComm* **2017**, *8*, 823–840. [CrossRef]
10. Götz, K.H.; Hacker, S.M.; Mayer, D.; Dürig, J.-N.; Stenger, S.; Marx, A. Inhibitors of the Diadenosine Tetraphosphate Phosphorylase Rv2613c of Mycobacterium tuberculosis. *ACS Chem. Biol.* **2017**, *12*, 2682–2689. [CrossRef]
11. Hacker, S.M.; Mortensen, F.; Scheffner, M.; Marx, A. Selective monitoring of the enzymatic activity of the tumor suppressor Fhit. *Angew. Chem. Int. Ed.* **2014**, *53*, 10247–10250. [CrossRef] [PubMed]
12. Guranowski, A. Specific and nonspecific enzymes involved in the catabolism of mononucleoside and dinucleoside polyphosphates. *Pharmacol. Ther.* **2000**, *87*, 117–139. [CrossRef]
13. Hodgson, D.R. Physicochemical aspects of aqueous and nonaqueous approaches to the preparation of nucleosides, nucleotides and phosphate ester mimics. *Adv. Phys. Org. Chem.* **2017**, *51*, 187–219.
14. Xu, Z. A review on the chemical synthesis of pyrophosphate bonds in bioactive nucleoside diphosphate analogs. *Bioorganic Med. Chem. Lett.* **2015**, *25*, 3777–3783. [CrossRef]
15. Bezold, D.; Dürr, T.; Singh, J.; Jessen, H.J. Cyclotriphosphate: A brief history, recent developments, and perspectives in synthesis. *Chem. Eur. J.* **2019**. [CrossRef]
16. Jenal, U.; Reinders, A.; Lori, C. Cyclic di-GMP: Second messenger extraordinaire. *Nat. Rev. Microbiol.* **2017**, *15*, 271–284. [CrossRef]
17. Mlynarska-Cieslak, A.; Depaix, A.; Grudzien-Nogalska, E.; Sikorski, P.J.; Warminski, M.; Kiledjian, M.; Jemielity, J.; Kowalska, J. Nicotinamide-containing di- and trinucleotides as chemical tools for studies of NAD-capped RNAs. *Org. Lett.* **2018**, *20*, 7650–7655. [CrossRef]
18. Shanmugasundaram, M.; Senthilvelan, A.; Kore, A.R. Recent advances in synthesis and biological activity of modified cap analogs. *Curr. Org. Chem.* **2017**, *21*, 2530–2560. [CrossRef]
19. Muttach, F.; Muthmann, N.; Rentmeister, A. Synthetic mRNA capping. *Beilstein J. Org. Chem.* **2017**, *13*, 2819–2832. [CrossRef]
20. Warminski, M.; Sikorski, P.J.; Kowalska, J.; Jemielity, J. Applications of phosphate modification and labeling to study (m)RNA caps. *Top. Curr. Chem.* **2017**, *375*, 16. [CrossRef]
21. Burnstock, G. Purinergic signalling: Therapeutic developments. *Front. Pharmacol.* **2017**, *8*, 661.
22. Burnstock, G. The therapeutic potential of purinergic signalling. *Biochem. Pharmacol.* **2018**, *151*, 157–165. [CrossRef] [PubMed]
23. Rafehi, M.; Muller, C.E. Tools and drugs for uracil nucleotide-activated P2Y receptors. *Pharmacol. Ther.* **2018**, *190*, 24–80. [CrossRef] [PubMed]
24. Jacobson, K.A. Structure-Based approaches to ligands for G-protein-coupled adenosine and P2Y Receptors, from small molecules to nanoconjugates. *J. Med. Chem.* **2013**, *56*, 3749–3767. [CrossRef] [PubMed]
25. Burnstock, G. Purinergic signalling. *Br. J. Pharmacol.* **2006**, *147*, S172–S181. [CrossRef] [PubMed]
26. Burnstock, G. Pathophysiology and therapeutic potential of purinergic signaling. *Pharmacol. Rev.* **2006**, *58*, 58–86. [CrossRef] [PubMed]
27. White, N.; Burnstock, G. P2 receptors and cancer. *Trends Pharmacol. Sci.* **2006**, *27*, 211–217. [CrossRef]
28. Burnstock, G. Purinergic signalling and disorders of the central nervous system. *Nat. Rev. Drug Discov.* **2008**, *7*, 575–590. [CrossRef]
29. Erlinge, D. P2Y Receptors in health and disease. *Adv. Pharmacol.* **2011**, *61*, 417–439.
30. Xu, P.; Feng, X.; Luan, H.; Wang, J.; Ge, R.; Li, Z.; Bian, J. Current knowledge on the nucleotide agonists for the P2Y2 receptor. *Bioorg. Med. Chem.* **2018**, *26*, 366–375. [CrossRef]

31. Douglass, J.G.; Patel, R.I.; Yerxa, B.R.; Shaver, S.R.; Watson, P.S.; Bednarski, K.; Plourde, R.; Redick, C.C.; Brubaker, K.; Jones, A.C.; et al. Lipophilic modifications to dinucleoside polyphosphates and nucleotides that confer antagonist properties at the platelet $P2Y_{12}$ receptor. *J. Med. Chem.* **2008**, *51*, 1007–1025. [CrossRef] [PubMed]
32. Hoyle, C.H.; Hilderman, R.H.; Pintor, J.J.; Schlüter, H.; King, B.F. Diadenosine polyphosphates as extracellular signal molecules. *Drug Dev. Res.* **2001**, *52*, 260–273. [CrossRef]
33. Carracedo, G.; Crooke, A.; Guzman-Aranguez, A.; De Lara, M.J.P.; Martin-Gil, A.; Pintor, J. The role of dinucleoside polyphosphates on the ocular surface and other eye structures. *Prog. Retin. Eye Res.* **2016**, *55*, 182–205. [CrossRef]
34. Pintor, J.; Díaz-Hernández, M.; Gualix, J.; Gómez-Villafuertes, R.; Hernando, F.; Miras-Portugal, M.T. Diadenosine polyphosphate receptors. from rat and guinea-pig brain to human nervous system. *Pharmacol. Ther.* **2000**, *87*, 103–115. [CrossRef]
35. Lau, O.C.F.; Samarawickrama, C.; Skalicky, S.E. $P2Y_2$ receptor agonists for the treatment of dry eye disease: A review. *Clin. Ophthalmol.* **2014**, *8*, 327–334.
36. Keating, G.M. Diquafosol ophthalmic solution 3%: A review of its use in dry eye. *Drugs* **2015**, *75*, 911–922. [CrossRef]
37. Kellerman, D.; Mospan, A.R.; Engels, J.; Schaberg, A.; Gorden, J.; Smiley, L. Denufosol: A review of studies with inhaled $P2Y_2$ agonists that led to phase 3. *Pulm. Pharmacol. Ther.* **2008**, *21*, 600–607. [CrossRef]
38. Accurso, F.J.; Moss, R.B.; Wilmott, R.W.; Anbar, R.D.; Schaberg, A.E.; Durham, T.A.; Ramsey, B.W. Denufosol tetrasodium in patients with cystic fibrosis and normal to mildly impaired lung function. *Am. J. Respir. Crit. Care Med.* **2011**, *183*, 627–634. [CrossRef]
39. Yerxa, B.R.; Sabater, J.R.; Davis, C.W.; Stutts, M.J.; Lang-Furr, M.; Picher, M.; Jones, A.C.; Cowlen, M.; Dougherty, R.; Boyer, J.; et al. Pharmacology of INS37217 [P^1-(Uridine 5′)-P^4- (2′-deoxycytidine 5′)tetraphosphate, tetrasodium salt], a next-generation $P2Y_2$ receptor agonist for the treatment of cystic fibrosis. *J. Pharmacol. Exp. Ther.* **2002**, *302*, 871–880. [CrossRef]
40. Deterding, R.R.; LaVange, L.M.; Engels, J.M.; Mathews, D.W.; Coquillette, S.J.; Brody, A.S.; Millard, S.P.; Ramsey, B.W. Phase 2 randomized safety and efficacy trial of nebulized denufosol tetrasodium in cystic fibrosis. *Am. J. Respir. Crit. Care Med.* **2007**, *176*, 362–369. [CrossRef]
41. Moss, R.B. Pitfalls of drug development: Lessons learned from trials of denufosol in cystic fibrosis. *J. Pediatr.* **2013**, *162*, 676–680. [CrossRef] [PubMed]
42. Ratjen, F.; Durham, T.; Navratil, T.; Schaberg, A.; Accurso, F.J.; Wainwright, C.; Barnes, M.; Moss, R.B. Long term effects of denufosol tetrasodium in patients with cystic fibrosis. *J. Cyst. Fibros.* **2012**, *11*, 539–549. [CrossRef] [PubMed]
43. Reiss, J.R.; Moffatt, J.G. Dismutation reactions of nucleoside polyphosphates. 3. The synthesis of α,ω-Dinucleoside 5′-polyphosphates. *J. Org. Chem.* **1965**, *30*, 3381–3387. [CrossRef] [PubMed]
44. Adam, A.; Moffatt, J.G. Dismutation reactions of nucleoside polyphosphates. V. Syntheses of P^1,P^4-Di(guanosine-5′) tetraphosphate and P^1,P^3-Di(guanosine-5′) triphosphate. *J. Am. Chem. Soc.* **1966**, *88*, 838–842. [CrossRef]
45. Moffatt, J.G.; Khorana, H.G. Nucleoside polyphosphates. X. synthesis and some reactions of nucleoside-5′phosphoromorpholidates and related compounds. Improved methods for the preparation of nucleoside-5′polyphosphates. *J. Am. Chem. Soc.* **1961**, *83*, 649–658. [CrossRef]
46. Millo, E.; Zocchi, E.; Galatini, A.; Benatti, U.; Damonte, G. Simple Synthesis of P^1P^2-Diadenosine 5′-Pyrophosphate. *Synth. Commun.* **2008**, *38*, 3260–3269. [CrossRef]
47. Shaver, S.R.; Rideout, J.L.; Pendergast, W.; Douglass, J.G.; Brown, E.G.; Boyer, J.L.; Patel, R.I.; Redick, C.C.; Jones, A.C.; Picher, M.; et al. Structure–Activity relationships of dinucleotides: Potent and selective agonists of P2Y receptors. *Purinergic Signal.* **2005**, *1*, 183–191. [CrossRef]
48. Ko, H.; Carter, R.L.; Cosyn, L.; Petrelli, R.; De Castro, S.; Besada, P.; Zhou, Y.; Cappellacci, L.; Franchetti, P.; Grifantini, M.; et al. Synthesis and potency of novel uracil nucleotides and derivatives as $P2Y_2$ and $P2Y_6$ receptor agonists. *Bioorganic Med. Chem.* **2008**, *16*, 6319–6332. [CrossRef]
49. Pendergast, W.; Yerxa, B.R.; Douglass, J.G.; Shaver, S.R.; Dougherty, R.W.; Redick, C.C.; Sims, I.F.; Rideout, J.L. Synthesis and P2Y receptor activity of a series of uridine dinucleoside 5′-polyphosphates. *Bioorganic Med. Chem. Lett.* **2001**, *11*, 157–160. [CrossRef]

50. Stern, N.; Major, D.T.; Gottlieb, H.E.; Weizman, D.; Fischer, B. What is the conformation of physiologically-active dinucleoside polyphosphates in solution? Conformational analysis of free dinucleoside polyphosphates by NMR and molecular dynamics simulations. *Org. Biomol. Chem.* **2010**, *8*, 4637–4652. [CrossRef]
51. Liu, K.K.-C.; Sakya, S.M.; O'Donnell, C.J.; Flick, A.C.; Ding, H.X. Synthetic approaches to the 2010 new drugs. *Bioorganic Med. Chem.* **2012**, *20*, 1155–1174. [CrossRef] [PubMed]
52. Zatorski, A.; Goldstein, B.M.; Colby, T.D.; Jones, J.P.; Pankiewicz, K.W. Potent inhibitors of human inosine monophosphate dehydrogenase type II. Fluorine-substituted analogues of thiazole-4-carboxamide adenine dinucleotide. *J. Med. Chem* **1995**, *38*, 1098–1105. [CrossRef] [PubMed]
53. Chen, L.; Rejman, D.; Bonnac, L.; Pankiewicz, K.W.; Patterson, S.E. Nucleoside-5′-phosphoimidazolides: Reagents for facile synthesis of dinucleoside pyrophosphates. *Curr. Protoc. Nucleic Acid Chem.* **2006**, *23*, 13.4.1–13.4.10. [CrossRef] [PubMed]
54. Yanachkov, I.B.; Dix, E.J.; Yanachkova, M.I.; Wright, G.E. P^1, P^2-diimidazolyl derivatives of pyrophosphate and bis-phosphonates - synthesis, properties, and use in preparation of dinucleoside tetraphosphates and analogs. *Org. Biomol. Chem.* **2011**, *9*, 730–738. [CrossRef] [PubMed]
55. Depaix, A.; Peyrottes, S.; Roy, B. Water-Medium synthesis of nucleoside 5′-polyphosphates. *Curr. Protoc. Nucleic Acid Chem.* **2017**, *69*, 13.16.1–13.16.11.
56. Depaix, A.; Peyrottes, S.; Roy, B. One-Pot synthesis of nucleotides and conjugates in aqueous medium. *Eur. J. Org. Chem.* **2017**, *2*, 241–245. [CrossRef]
57. Smith, M.; Khorana, H.G. Nucleoside Polyphosphates. VI. 1 An improved and general method for the synthesis of Ribo-and Deoxyribonucleoside 5′-Triphosphates. *J. Am. Chem. Soc.* **1958**, *80*, 1141–1145. [CrossRef]
58. Ng, K.-M.E.; Orgel, L.E. The action of a water-soluble carbodiimide on adenosine-5′-polyphosphates. *Nucleic Acids Res.* **1987**, *15*, 3573–3580. [CrossRef]
59. Mohamady, S.; Taylor, S.D. Synthesis of nucleoside tetraphosphates and dinucleoside pentaphosphates via activation of cyclic trimetaphosphate. *Org. Lett.* **2013**, *15*, 2612–2615. [CrossRef]
60. Mohamady, S.; Desoky, A.; Taylor, S.D. Sulfonyl imidazolium salts as reagents for the rapid and efficient synthesis of nucleoside polyphosphates and their conjugates. *Org. Lett.* **2012**, *14*, 402–405. [CrossRef]
61. Mohsen, M.G.; Ji, D.; Kool, E.T. Polymerase synthesis of four-base DNA from two stable dimeric nucleotides. *Nucleic Acids Res.* **2019**, *47*, 9495–9501. [CrossRef] [PubMed]
62. Kanavarioti, A.; Lu, J.; Rosenbach, M.T.; Hurley, T.B. Unexpectedly Facile Synthesis of Symmetrical P^1,P^2-Dinucleoside-5′pyrophosphates. *Tetrahedron Lett.* **1991**, *32*, 6065–6068. [CrossRef]
63. Bogachev, V.S. Synthesis of deoxynucleoside 5′-triphosphates using trifluoroacetic anhydride as an activating reagent. *Bioorg. Khim* **1996**, *22*, 699–705.
64. Mohamady, S.; Taylor, S.D. General Procedure for the Synthesis of Dinucleoside Polyphosphates. *J. Org. Chem.* **2011**, *76*, 6344–6349. [CrossRef]
65. Mohamady, S.; Taylor, S.D. Rapid and efficient synthesis of nucleoside polyphosphates and their conjugates using sulfonyl imidazolium salts. *Curr. Protoc. Nucleic Acid Chem.* **2012**, *51*, 13.11.1–13.11.24. [CrossRef]
66. Sun, Q.; Gong, S.S.; Sun, J.; Wang, C.J.; Liu, S.; Liu, G.D.; Ma, C. Efficient synthesis of nucleoside 5′-triphosphates and their β,γ-bridging oxygen-modified analogs from nucleoside 5′-phosphates. *Tetrahedron Lett.* **2014**, *55*, 2114–2118. [CrossRef]
67. Sun, Q.; Gong, S.-S.; Liu, S.; Sun, J.; Liu, G.-D.; Ma, C. 4,5-Dicyanoimidazole-promoted synthesis of dinucleoside polyphosphates and their analogs. *Tetrahedron* **2014**, *70*, 4500–4506. [CrossRef]
68. Sun, Q.; Sun, J.; Gong, S.-S.; Wang, C.-J.; Wang, X.-C. Synthesis of nucleoside tetraphosphates and dinucleoside pentaphosphates from nucleoside phosphoropiperidates via the activation of P(V)-N bond. *Chin. Chem. Lett.* **2015**, *26*, 89–92. [CrossRef]
69. Sun, Q.; Sun, J.; Gong, S.-S.; Wang, C.-J.; Wang, X.-C. One-pot synthesis of symmetrical dinucleoside polyphosphates and analogs via 4,5-dicyanoimidazole-promoted tandem P–O coupling reactions. *Tetrahedron Lett.* **2014**, *55*, 5785–5788. [CrossRef]
70. Ludwig, J.; Eckstein, F. Rapid and efficient synthesis of nucleoside 5′-*O*-(1-thiotriphosphates), 5′-triphosphates, and 2′,3′-cyclophosphorothioates using 2-chloro-4*H*-1,3,2-benzodioxaphosphorin-4-one. *J. Org. Chem.* **1989**, *54*, 631–635. [CrossRef]

71. Han, Q.; Gaffney, B.L.; Jones, R.A. One-Flask synthesis of dinucleoside tetra- and pentaphosphates. *Org. Lett.* **2006**, *8*, 2075–2077. [CrossRef] [PubMed]
72. Meier, C. cycloSal-pronucleotides design of chemical trojan horses. *Mini-Rev. Med. Chem.* **2002**, *2*, 219–234. [CrossRef] [PubMed]
73. Meier, C. cycloSal Phosphates as chemical trojan horses for intracellular nucleotide and glycosylmonophosphate delivery — Chemistry meets biology. *Eur. J. Org. Chem.* **2006**, *2006*, 1081–1102. [CrossRef]
74. Warnecke, S.; Meier, C. Synthesis of Nucleoside Di- and Triphosphates and Dinucleoside Polyphosphates with cycloSal-Nucleotides. *J. Org. Chem.* **2009**, *74*, 3024–3030. [CrossRef]
75. Wendicke, S.; Warnecke, S.; Meier, C. Efficient Synthesis of Nucleoside Diphosphate Glycopyranoses. *Angew. Chem. Int. Ed.* **2008**, *47*, 1500–1502. [CrossRef]
76. Tonn, V.C.; Meier, C. Solid-Phase synthesis of (poly)phosphorylated nucleosides and conjugates. *Chem. Eur. J.* **2011**, *17*, 9832–9842. [CrossRef]
77. Ahmadibeni, Y.; Parang, K. Solid-Phase Synthesis of Symmetrical 5′,5′-Dinucleoside Mono-, Di-, Tri-, and Tetraphosphodiesters. *Org. Lett.* **2007**, *9*, 4483–4486. [CrossRef]
78. Kistemaker, H.A.V.; Meeuwenoord, N.J.; Overkleeft, H.S.; Van Der Marel, G.A.; Filippov, D.V. On the synthesis of oligonucleotides interconnected through pyrophosphate linkages. *Eur. J. Org. Chem.* **2015**, *2015*, 6084–6091. [CrossRef]
79. Sun, Q.; Liu, S.; Sun, J.; Gong, S.; Xiao, Q.; Shen, L. One-pot synthesis of symmetrical P^1,P^2-dinucleoside-5′-diphosphates from nucleoside-5′-*H*-phosphonates: Mechanistic insights into reaction path. *Tetrahedron Lett.* **2013**, *54*, 3842–3845. [CrossRef]
80. Sun, Q.; Edathil, J.P.; Wu, R.; Smidansky, E.D.; Cameron, C.E.; Peterson, B.R. One-Pot Synthesis of Nucleoside 5′-Triphosphates from Nucleoside 5′-*H*-Phosphonates. *Org. Lett.* **2008**, *10*, 1703–1706. [CrossRef]
81. Jessen, H.J.; Ahmed, N.; Hofer, A. Phosphate esters and anhydride—recent strategies targeting nature's favoured modifications. *Org. Biomol. Chem.* **2014**, *12*, 3526–3530. [CrossRef] [PubMed]
82. Cremosnik, G.S.; Hofer, A.; Jessen, H.J. Iterative synthesis of nucleoside oligophosphates with phosphoramidites. *Angew. Chem. Int. Ed.* **2014**, *53*, 286–289. [CrossRef] [PubMed]
83. Hofer, A.; Cremosnik, G.S.; Müller, A.C.; Giambruno, R.; Trefzer, C.; Superti-Furga, G.; Jessen, H.J. Modular synthesis of modified phosphoanhydrides. *Chem. Eur. J.* **2015**, *21*, 10116–10122. [CrossRef] [PubMed]
84. Hofer, A.; Marques, E.; Kieliger, N.; Gatter, S.-K.N.; Jordi, S.; Ferrari, E.; Hofmann, M.; Fitzpatrick, T.B.; Hottiger, M.O.; Jessen, H.J. Chemoselective dimerization of phosphates. *Org. Lett.* **2016**, *18*, 3222–3225. [CrossRef]
85. Tanaka, H.; Yoshimura, Y.; Jørgensen, M.R.; Cuesta-Seijo, J.A.; Hindsgaul, O. A simple synthesis of sugar nucleoside diphosphates by chemical coupling in water. *Angew. Chem. Int. Ed.* **2012**, *51*, 11531–11534. [CrossRef]
86. Eguaogie, O.; Vyle, J.S.; Conlon, P.F.; Gîlea, M.A.; Liang, Y. Mechanochemistry of nucleosides, nucleotides and related materials. *Beilstein J. Org. Chem.* **2018**, *14*, 955–970. [CrossRef]
87. James, S.L.; Adams, C.J.; Bolm, C.; Braga, D.; Collier, P.; Friščić, T.; Grepioni, F.; Harris, K.D.M.; Hyett, G.; Jones, W.; et al. Mechanochemistry: Opportunities for new and cleaner synthesis. *Chem. Soc. Rev.* **2012**, *41*, 413–447. [CrossRef]
88. Bowmaker, G.A. Solvent-assisted mechanochemistry. *Chem. Commun.* **2013**, *49*, 334–348. [CrossRef]
89. Takacs, L. The historical development of mechanochemistry. *Chem. Soc. Rev.* **2013**, *42*, 7649–7659. [CrossRef]
90. Wang, G.W. Mechanochemical organic synthesis. *Chem. Soc. Rev.* **2013**, *42*, 7668–7700. [CrossRef]
91. Friščić, T.; Mottillo, C.; Titi, H.M. Mechanochemistry for synthesis. *Angew. Chem. Int. Ed. Engl.* **2019**, in press.
92. Ravalico, F.; Messina, I.; Berberian, M.V.; James, S.L.; Migaud, M.E.; Vyle, J.S. Rapid synthesis of nucleotide pyrophosphate linkages in a ball mill. *Org. Biomol. Chem.* **2011**, *9*, 6496–6497. [CrossRef] [PubMed]
93. Appy, L.; Depaix, A.; Bantreil, X.; Lamaty, F.; Peyrottes, S.; Roy, B. Straightforward ball-milling access to dinucleoside 5′,5′-polyphosphates via phosphorimidazolide intermediates. *Chem. Eur. J.* **2019**, *25*, 2477–2481. [PubMed]
94. Verma, S.K.; Ghorpade, R.; Pratap, A.; Kaushik, M. Solvent free, *N,N*′-carbonyldiimidazole (CDI) mediated amidation. *Tetrahedron Lett.* **2012**, *53*, 2373–2376. [CrossRef]

95. Metro, T.-X.; Bonnamour, J.; Reidon, T.; Sarpoulet, J.; Martinez, J.; Lamaty, F. Mechanosynthesis of amides in the total absence of organic solvent from reaction to product recovery. *Chem. Commun.* **2012**, *48*, 11781–11793. [CrossRef]
96. Métro, T.-X.; Bonnamour, J.; Reidon, T.; Duprez, A.; Sarpoulet, J.; Martinez, J.; Lamaty, F. Comprehensive study of the organic-solvent-free CDI-Mediated acylation of various nucleophiles by mechanochemistry. *Chem. Eur. J.* **2015**, *21*, 12787–12796. [CrossRef]
97. Métro, T.-X.; Martinez, J.; Lamaty, F. 1,1′-Carbonyldiimidazole and mechanochemistry: A shining green combination. *ACS Sustain. Chem. Eng.* **2017**, *5*, 9599–9602. [CrossRef]
98. Banoub, J.H.; Newton, R.P.; Esmans, E.; Ewing, D.F.; Mackenzie, G. Recent developments in mass spectrometry for the characterization of nucleosides, nucleotides, oligonucleotides, and nucleic acids. *Chem. Rev.* **2005**, *105*, 1869–1915. [CrossRef]
99. Watanabe, D.; Ishikawa, M.; Yamasaki, M.; Ozaki, M.; Katayama, T.; Nakajima, H. Tetrasodium P^1,P^4-Bis(5′-adenosyl)tetraphosphate Dodecahydrate. *Acta Crystallogr. Sect. C Cryst. Struct. Commun.* **1996**, *52*, 338–340. [CrossRef]
100. Simanshu, D.K.; Savithri, H.S.; Murthy, A.R.N. Crystal structures of Salmonella typhimurium propionate kinase and its complex with Ap4A: Evidence for a novel Ap_4A synthetic activity. *Proteins* **2008**, *70*, 1379–1388. [CrossRef]
101. Baker, M.D.; Holloway, D.E.; Swaminathan, G.J.; Acharya, K.R. Crystal structures of eosinophil-derived neurotoxin (EDN) in complex with the inhibitors 5′-ATP, Ap_3A, Ap_4A, and Ap_5A. *Biochemistry* **2006**, *45*, 416–426. [CrossRef] [PubMed]
102. Scheffzek, K.; Kliche, W.; Wiesmuller, L.; Reinstein, J. Crystal structure of the complex of UMP/CMP kinase from Dictyostelium discoideum and the bisubstrate inhibitor P^1-(5′-adenosyl) P^5-(5′-uridyl) pentaphosphate (UP_5A) and Mg^{2+} at 2.2 A: Implications for water-mediated specificity. *Biochemistry* **1996**, *35*, 9716–9727. [CrossRef] [PubMed]
103. Berry, M.B.; Phillips, G.N., Jr. Crystal structures of Bacillus stearothermophilus adenylate kinase with bound Ap_5A, Mg^{2+} Ap_5A, and Mn^{2+} Ap_5A reveal an intermediate lid position and six coordinate octahedral geometry for bound Mg^{2+} and Mn^{2+}. *Proteins* **1998**, *32*, 276–288. [CrossRef]
104. Chen, J.; Kohler, B. Base stacking in adenosine dimers revealed by femtosecond transient absorption spectroscopy. *J. Am. Chem. Soc.* **2014**, *136*, 6362–6372. [CrossRef] [PubMed]

MDPI
St. Alban-Anlage 66
4052 Basel
Switzerland
Tel. +41 61 683 77 34
Fax +41 61 302 89 18
www.mdpi.com

Molecules Editorial Office
E-mail: molecules@mdpi.com
www.mdpi.com/journal/molecules

www.ingramcontent.com/pod-product-compliance
Lightning Source LLC
Chambersburg PA
CBHW041749220926
51847CB00014B/153